Lecture Notes in Computer Science 3421

Commenced Publication in 1973
Founding and Former Series Editors:
Gerhard Goos, Juris Hartmanis, and Jan van Leeuwen

Pascal Lorenz   Petre Dini (Eds.)

Networking – ICN 2005

4th International Conference on Networking
Reunion Island, France, April 17-21, 2005
Proceedings, Part II

 Springer

Volume Editors

Pascal Lorenz
University of Haute Alsace
34 rue du Grillenbreit, 68008 Colmar, France
E-mail: lorenz@ieee.org

Petre Dini
Cisco Systems, Inc.
170 West Tasman Drive, San Jose, CA 95134, USA
E-mail: pdini@cisco.com

Library of Congress Control Number: 2005922556

CR Subject Classification (1998): C.2, K.4.4, H.4.3, H.5.1, H.3, K.6.4-5

ISSN 0302-9743
ISBN 3-540-25338-6 Springer Berlin Heidelberg New York

Springer is a part of Springer Science+Business Media

springeronline.com

© Springer-Verlag Berlin Heidelberg 2005
Printed in Germany

Typesetting: Camera-ready by author, data conversion by Scientific Publishing Services, Chennai, India
Printed on acid-free paper   SPIN: 11407072   06/3142   5 4 3 2 1 0

Preface

The International Conference on Networking (ICN 2005) was the fourth conference in its series aimed at stimulating technical exchange in the emerging and important field of networking. On behalf of the International Advisory Committee, it is our great pleasure to welcome you to the proceedings of the 2005 event.

Networking faces dramatic changes due to the customer-centric view, the venue of the next generation networks paradigm, the push from ubiquitous networking, and the new service models. Despite legacy problems, which researchers and industry are still discovering and improving the state of the art, the horizon has revealed new challenges that some of the authors tackled through their submissions.

In fact ICN 2005 was very well perceived by the international networking community. A total of 651 papers from more than 60 countries were submitted, from which 238 were accepted. Each paper was reviewed by several members of the Technical Program Committee. This year, the Advisory Committee revalidated various accepted papers after the reviews had been incorporated. We perceived a significant improvement in the number of submissions and the quality of the submissions.

The ICN 2005 program covered a variety of research topics that are of current interest, starting with Grid networks, multicasting, TCP optimizations, QoS and security, emergency services, and network resiliency. The Program Committee selected also three tutorials and invited speakers that addressed the latest research results from the international industries and academia, and reports on findings from mobile, satellite, and personal communications related to 3rd- and 4th-generation research projects and standardization.

This year we enriched ICN with a series of papers targeting emergency services and disaster recovery (the AICED section); this emerging topic hopefully will lead to more robust and fault-tolerant systems for preventing technical and human disasters.

We would like to thank the International Advisory Committee members and the referees. Without their support, the program organization of this conference would not have been possible. We are also indebted to many individuals and organizations that made this conference possible (Cisco Systems, Inc., France Telecom, IEEE, IARIA, Region Reunion, University of La Reunion, ARP). In particular, we thank the members of the Organizing Committee for their help in all aspects of the organization of this conference.

We hope that the attendees enjoyed this International Conference on Networking on Reunion Island, and found it a useful forum for the exchange of ideas and results and recent findings. We also hope that the attendees found time to enjoy the island's beautiful countryside and its major cultural attractions.

April 2005   Pascal Lorenz
                                                                     Petre Dini

International Scientific Committee

Advisory Committee Board

P. Dini (USA) — Cisco Systems, Inc.
P. Lorenz (France) — University of Haute Alsace
G. Parr (UK) — University of Ulster (for the AICED section on emergency
services and disaster recovery)

Tutorial Chairs

A. Jamalipour (Australia) — University of Sydney
M. Hur (USA) — Microsoft (for the AICED section on emergency services
and disaster recovery)

Program Chairs

M. Freire (Portugal) — University of Beira Interior
H. Debar (France) — France Telecom R&D (for the AICED section on
emergency services and disaster recovery)

International Advisory Committee

H. Adeli (USA) — Ohio State University
K. Al-Begain (UK) — University of Glamorgan
P. Anelli (France) — University of La Reunion
E. Barnhart (USA) — Georgia Institute of Technology
J. Ben Othman (France) — University of Versailles
B. Bing (USA) — Georgia Institute of Technology
D. Bonyuet (USA) — Delta Search Labs
J. Brito (Brazil) — INATEL
M. Carli (Italy) — University of Rome TRE
P. Chemouil (France) — France Telecom R&D
M. Devetsikiotis (USA) — North Carolina State University
P. Dini (USA) — Cisco Systems, Inc.
P. Dommel (USA) — Santa Clara University
Y. Donoso (Colombia) — University del Norte
I. Elhanany (USA) — University of Tennessee

A. Finger (Germany) — Dresden University of Technology
M. Freire (Portugal) — University of Beira Interior
E. Fulp (USA) — Wake Forest University
B. Gavish (USA) — Southern Methodist University
F. Granelli (Italy) — University of Trento
H. Guyennet (France) — University of Franche-Comté
Z. Hulicki (Poland) — AGH University of Science and Technology
A. Jamalipour (Australia) — University of Sydney
A.L. Jesus Teixeira (Portugal) — University of Aveiro
D. Khotimsky (USA) — Invento Networks
R. Komiya (Malaysia) — Faculty of Information Technology
S. Kumar (USA) — University of Texas
J.C. Lapeyre (France) — University of Franche-Comte
P. Lorenz (France) — University of Haute Alsace
M.S. Obaida (USA) — Monmouth University
A. Pescape (Italy) — University of Napoli "Federico II"
D. Magoni (France) — University of Strasbourg
Z. Mammeri (France) — University of Toulouse
A. Molinaro (Italy) — University of Calabria
H. Morikawa (Japan) — University of Tokyo
H. Mouftah (Canada) — University of Ottawa
A. Pandharipande (Korea) — Samsung Advanced Institute of Technology
S. Recker (Germany) — IMST GmbH
F. Ricci (USA) — The Catholic University of America
J. Rodrigues (Portugal) — University of Beira Interior
P. Rolin (France) — France Telecom R&D
B. Sarikaya (Canada) — UNBC
J. Soler-Lucas (Denmark) — Research Center COM
S. Soulhi (Canada) — Ericsson, Inc.
V. Uskov (USA) — Bradley University
R. Valadas (Portugal) — University of Aveiro
L. Vasiu (UK) - Wireless IT Research Centre
E. Vazquez (Spain) — Technical University of Madrid
D. Vergados (Greece) — University of the Aegean
S. Yazbeck (USA) — Barry University
V. Zaborovski (Russia) — Saint-Petersburg Politechnical University
A. Zaslavsky (Australia) — Monash University
H.J. Zepernick (Australia) — Western Australian Telecommunications Research
    Institute

International Committee for the topics related to the emergency services
and disaster recovery

M. Barbeau (Canada) — Carleton University
G. Candea (USA) — Stanford University

H. Debar (France) — France Telecom R&D
P. Dini (USA) — Cisco Systems, Inc.
S. Gjessing (Norway) — Simula Research Laboratory
P.-H. Ho (Canada) — University of Waterloo
M. Hur (USA) — Microsoft
N. Kapadia (USA) — Capital One
P. Lorenz (France) — University of Haute Alsace
D. Malagrino (USA) — Cisco Systems, Inc.
M. Moh (USA) — San Jose State University
G. Parr (UK) — University of Ulster
A. Pescapé (Italy) — Università degli Studi di Napoli "Frederico II"
H. Reiser (Germany) — Ludwig Maximilians University Munich
J. Salowey (USA) — Cisco Systems, Inc.
D. Schupke (Germany) — Siemens, AG
F. Serr (USA) — Cisco Systems, Inc.
C. Becker Westphall (Brazil) — Federal University of Santa Catarina
A. Zaslavsky (Australia) — Monash University

Table of Contents – Part II

MIMO

MPLS

Ad Hoc Networks (I)

TCP (I)

Routing (I)

Ad Hoc Networks (II)

TCP (II)

Routing (II)

Ad Hoc Networks (III)

Signal Processing

Routing (III)

Mobility

Performance (I)

Peer-to-Peer (I)

Security (I)

Performance (II)

Peer-to-Peer (II)

Security (II)

Multicast (I)

CDMA

Table of Contents – Part I

Wireless Networks (I)

QoS (I)

Optical Networks (II)

Wireless Networks (II)

QoS (II)

Optical Networks (III)

Sensor Networks (I)

Traffic Control (I)

Traffic Control (II)

Audio and Video Communications

Sensor Networks (III)

Traffic Control (III)

Streaming

Decoding Consideration for Space Time Coded MIMO Channel with Constant Amplitude Multi-code System

Jia Hou[1], Moon Ho Lee[1], Ju Yong Park[2], and Jeong Su Kim[3]

[1] Institute of Information & Communication, Chonbuk National University,
Chonju, 561-756, Korea
{jiahou, moonho}@chonbuk.ac.kr
[2] Department of Electronics Engineering, Seonam University, Seonam, 590-711, Korea
phyhsj@kornet.net
[3] Department of Computer, Information & Comm., Korea Cyber University,
Seoul, 120-749, Korea
kjs@mail.kcu.ac.kr

Abstract. The space time constant amplitude multi-code design applied to wireless communication is reported in this paper. The propose systems could lead to both simple implementation and the maximum diversity gain corresponding to an additional orthogonal encoding.

1 Introduction

The orthogonal multi-code system has been drawn much attention in high data rate transmission. It can achieve the code division multiplexing and support the variable rates to each user. To provide lower envelope variations, the constant amplitude code was proposed by [1,2,3]. In this paper, we proposed two schemes, which apply the special characters from constant amplitude code to the space time block codes [4,5]. One is used fro more than two transmit antennas with real modulation, but it leads to interferences of inter-codewords from the constant amplitude code, if the transmit antennas are two. And the other can provide both simple implementation and maximum diversity without interferences, if two transmit antennas and complex modulation are used.

Constant amplitude multi-code system is useful to transmit the high rate data for multimedia communication. Based on well known Hadamard matrices [6,7], the constant amplitude can be seen several groups of $2^2 \times 2^2$ Hadamard matrix. As illustrated in Fig.1, a simple system model of constant amplitude code is from the Hadamard matrix,

$$
[H]_2 = \begin{bmatrix} w_0^2 \\ w_1^2 \\ w_2^2 \\ w_4^2 \end{bmatrix} = \begin{bmatrix} 1 & 1 & 1 & 1 \\ 1 & -1 & 1 & -1 \\ 1 & 1 & -1 & -1 \\ 1 & -1 & -1 & 1 \end{bmatrix},
$$

(1)

and the input bit sequence $b = (b_0, b_1, b_2, b_3)$, which should be satisfied

$$
b_3 = 1 \oplus b_0 \oplus b_1 \oplus b_2,
$$

(2)

P. Lorenz and P. Dini (Eds.): ICN 2005, LNCS 3421, pp. 1–7, 2005.

Table 1. Amplitude pattern by using size four Hadamard code [1]

Sum of the information bits (module 2)	Input information bits $b_0 b_1 b_2 b_3$	Output codeword	Constant Amplitude
0	0000	-4 0 0 0	no
1	0001	-2 -2 -2 2	yes
1	0010	-2 2 -2 -2	yes
0	0011	0 0 -4 0	no
1	0100	-2 -2 2 -2	yes
0	0101	0 -4 0 0	no
0	0110	0 0 0 -4	no
1	0111	2 -2 -2 -2	yes
1	1000	-2 2 2 2	yes
0	1001	0 0 0 4	no
0	1010	0 4 0 0	no
1	1011	2 2 -2 2	yes
0	1100	0 0 4 0	no
1	1101	2 -2 2 2	yes
1	1110	2 2 2 -2	yes
0	1111	4 0 0 0	no

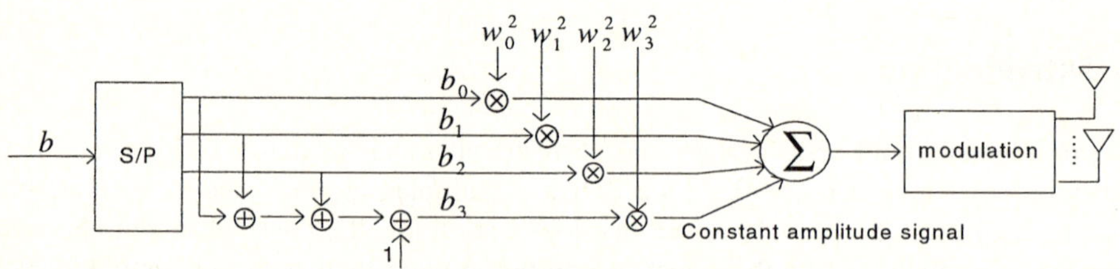

Fig. 1. The simple system model of four channels constant amplitude coding

where $\oplus$ is mod-2 addition. The four channels constant amplitude coding can be extended to the size of $2^m \times 2^m$ Hadamard matrices, where $m > 2$, by using the groups.

2   Constant Amplitude Coding Designed on Space Time Block Coded MIMO System

Recently, the antenna diversity from multiple input and multiple output (MIMO) system by using space time block codes (STBC) is reported as maximum capacity [9, 10]. In this paper, we investigate the space time coding combined with constant amplitude code and its simplified algorithms. In the proposed model, the output sequence from the constant amplitude code with matrix $[H]_2$, may be written as

$$T_n = b_0 h_0^n + b_1 h_1^n + b_2 h_2^n + b_3 h_3^n ,\tag{3}$$

where (b_0, b_1, b_2, b_3) are the information bits and they are satisfied as (2). The Hadamard matrix with size of $2^m \times 2^m$ could be expressed as

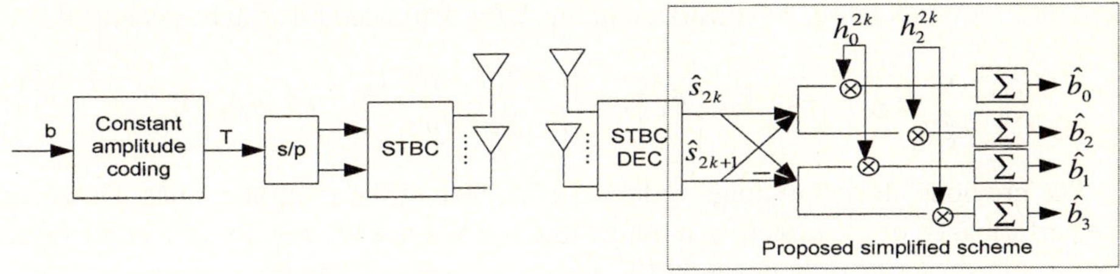

Fig. 2. The system model of the *Proposal 1*

$$[H]_m = \begin{bmatrix} h_0 \\ h_1 \\ \vdots \\ h_n \end{bmatrix} = \begin{bmatrix} h_0^0 & h_0^1 & \cdots & h_0^n \\ h_1^0 & h_1^1 & \cdots & h_1^n \\ \vdots & & \cdots & \vdots \\ h_n^0 & h_n^1 & \cdots & h_n^n \end{bmatrix}, \quad n \in \{0,1,\ldots,2^m-1\}. \tag{4}$$

Additionally, the Hadamard sequences have $h_{2k}^n h_{2k+1}^n = (-1)^n$, $k \in \{0,1,\ldots,\frac{n-1}{2}\}$. Thus we rewrite the (3) as

$$T_n = b_0 h_0^n + (-1)^n b_1 h_0^n + b_2 h_2^n + (-1)^n b_3 h_2^n . \tag{5}$$

Specially, in Hadamard matrices, we also have $h_n^{2k} = h_n^{2k+1}$, if $n = even$.

Proposal 1: By combined with Alamouti space time block code [5], the transmitted code words are

$$S = \begin{bmatrix} s_{2k} & s_{2k+1} \\ -s_{2k+1}^* & s_{2k}^* \end{bmatrix}, \qquad k = \{0,1\} , \tag{6}$$

where $X *$ denotes the conjugate of the X , and $s_n = T_n, n \in \{0,1,2,3\}$. Then the maximum ratio combing for estimated values at the receiver could be given as

$$\hat{s}_{2k} = (|\alpha_{2k}|^2 + |\alpha_{2k+1}|^2)(b_0 + b_1)h_0^{2k} + (|\alpha_{2k}|^2 + |\alpha_{2k+1}|^2)(b_2 + b_3)h_2^{2k} + n_{2k}, \tag{7}$$

and

$$\begin{aligned} \hat{s}_{2k+1} &= (|\alpha_{2k}|^2 + |\alpha_{2k+1}|^2)(b_0 - b_1)h_0^{2k+1} + (|\alpha_{2k}|^2 + |\alpha_{2k+1}|^2)(b_2 - b_3)h_2^{2k+1} + n_{2k+1} \\ &= (|\alpha_{2k}|^2 + |\alpha_{2k+1}|^2)(b_0 - b_1)h_0^{2k} + (|\alpha_{2k}|^2 + |\alpha_{2k+1}|^2)(b_2 - b_3)h_2^{2k} + n_{2k+1} \end{aligned} \tag{8}$$

Further, a simplified estimated algorithm could be written as

$$\hat{s}_{2k} + \hat{s}_{2k+1} = 2(|\alpha_{2k}|^2 + |\alpha_{2k+1}|^2)b_0 h_0^{2k} + 2(|\alpha_{2k}|^2 + |\alpha_{2k+1}|^2)b_2 h_2^{2k} + n_{2k} + n_{2k+1}, \tag{9}$$

$$\hat{s}_{2k} - \hat{s}_{2k+1} = 2(|\alpha_{2k}|^2 + |\alpha_{2k+1}|^2)b_1 h_0^{2k} + 2(|\alpha_{2k}|^2 + |\alpha_{2k+1}|^2)b_3 h_2^{2k} + n_{2k} - n_{2k+1}, \quad (10)$$

where $\alpha_{2k}, \alpha_{2k+1}$ denote the path gain from the transmit antennas and n_{2k}, n_{2k+1} are zero mean AWGN noise. As illustrated in Fig.2, the estimated bits can be generated by

$$\hat{b}_n = \sum_{k=0}^{1} (\hat{s}_{2k} + (-1)^n \hat{s}_{2k+1}) h_l^{2k}, \quad \begin{cases} l = 0, if & n \le 1 \\ l = 2, if & n \le 3 \end{cases}, \quad n \in \{0,1,2,3\}. \quad (11)$$

The previous algorithm could reduce the number of the multiplications. However, the performance of the system can not be improved, since we can not cancel the effect between the size two orthogonal STBC and size 4 constant amplitude code.

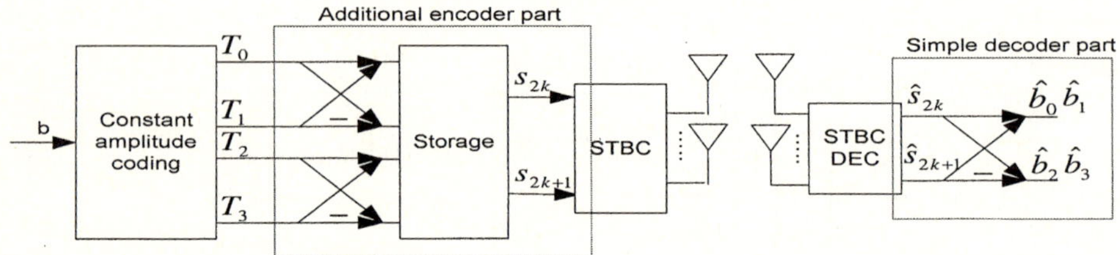

Fig. 3. The system model of the *Proposal 2*

Further, in this paper, we investigate a novel algorithm which could improve both the complexity of the computation and the BER performance. As shown in Fig.3, the proposed algorithm should make an additional encoder part.

Proposal 2: The transmitted matrix could be written as

$$S = \begin{bmatrix} s_{2k} & s_{2k+1} \\ -s_{2k+1}{}^* & s_{2k}{}^* \end{bmatrix} = \begin{bmatrix} T_0 + (-1)^k T_1 & T_2 + (-1)^k T_3 \\ -(T_2 + (-1)^k T_3)^* & (T_0 + (-1)^k T_1)^* \end{bmatrix}, \quad k = \{0,1\}, \quad (12)$$

and the estimation from the maximum combing ratio at the received can be given as

$$\hat{s}_{2k} = (|\alpha_{2k}|^2 + |\alpha_{2k+1}|^2)(T_0 + (-1)^k T_1) + n_{2k}, \quad (13)$$

$$\hat{s}_{2k+1} = (|\alpha_{2k}|^2 + |\alpha_{2k+1}|^2)(T_2 + (-1)^k T_3) + n_{2k+1}. \quad (14)$$

Similarly, we have

$$\hat{s}_{2k} + \hat{s}_{2k+1} = (|\alpha_{2k}|^2 + |\alpha_{2k+1}|^2)(\sum_{l=0}^{1} T_{2l} + (-1)^k T_{2l+1}) + n_{2k} + n_{2k+1}, \quad (15)$$

and

$$\hat{s}_{2k} - \hat{s}_{2k+1} = (|\alpha_{2k}|^2 + |\alpha_{2k+1}|^2)(\sum_{l=0}^{1} (-1)^l (T_{2l} + (-1)^k T_{2l+1})) + n_{2k} - n_{2k+1}. \quad (16)$$

The final estimated bits could be expressed as

$$\hat{b}_n = \hat{s}_l + (-1)^n \hat{s}_{l+1}, \quad \begin{cases} l = 0, if & n \leq 1 \\ l = 2, if & n \leq 3 \end{cases}, \quad n \in \{0,1,2,3\}. \tag{17}$$

By comparing with the *Proposal 1*, the *Proposal 2* could improve the BER performance without the complexity increasing. This algorithm added an orthogonal encoder part to cancel the effect between the size two orthogonal STBC and size four constant amplitude coding. Moreover, the additional encoder can provide the orthogonal signaling on bit-level, therefore, the BER performance of the proposed system could be obtained much enhancement.

Table 2. Computations and Performance of the Proposes Schemes

	Additional encoder part	Multicode decoder part	BER Performances (2Tx, 1Rx) (10^{-3})
Conventional	No	16 multiplications and 12 additions	10 dB
Proposal 1	No	8 multiplications and 8 additions	9.8dB
Proposal 2	4 additions	4 additions	2dB*

Furthermore, the union bound of space time constant amplitude coding by using orthogonal STBC could be derived as follows.
First, the symbol error rate is limited to

$$P_S = P(C \to \tilde{C}) \leq \frac{1}{2} \left\{ \left(\prod_{i=1}^{n} \lambda_i \right)^{\frac{1}{n}} \right\}^{-nm} \left(\frac{\rho}{4n} \right)^{-nm}$$

$$= \frac{1}{2} \left\{ \left| \det\left[\left(C - \tilde{C}\right)^H \left(C - \tilde{C}\right) \right] \right|^{\frac{1}{2}} \cdot \frac{\rho}{4n} \right\}^{-nm}, \tag{18}$$

where C and $\tilde{C}$ are the indistinct codewords, ρ is the SNR at the receive antenna,

and the diversity product $\lambda = \min_{\{C \neq \tilde{C}\}} \left| \det\left[\left(C - \tilde{C}\right)^H \left(C - \tilde{C}\right) \right] \right|^{\frac{1}{2n}}$ by using m transmit

antennas and n receive antennas, where C^H denotes the Hermitian of the codeword C. By exploiting the Hamming bound of the Hadamard codes, we could write the constant amplitude code with space time signaling as

$$P_b \approx \frac{1}{l} \sum_{j=t+1}^{l} l \binom{l-1}{j-1} p^j (1-p)^{l-j} = P_S (1 - P_S)^3 + 3P_S^2 (1 - P_S)^2 + 3P_S^3 (1 - P_S) + P_S^4$$

$$= P_S - 3P_S^3 + 3P_S^4$$

$$= \frac{1}{2} \left\{ \left| \det\left[\left(C - \tilde{C}\right)^H \left(C - \tilde{C}\right) \right] \right|^{\frac{1}{2n}} \cdot \frac{\rho}{4n} \right\}^{-nm} - \frac{3}{2} \left\{ \left| \det\left[\left(C - \tilde{C}\right)^H \left(C - \tilde{C}\right) \right] \right|^{\frac{1}{2n}} \cdot \frac{\rho}{4n} \right\}^{-3nm}$$

$$+\frac{3}{2}\left\{\left|\det\left[\left(C-\tilde{C}\right)^H\left(C-\tilde{C}\right)\right]\right|^{\frac{1}{2n}}\cdot\frac{\rho}{4n}\right\}^{-4nm}$$

$$=\frac{1}{2}\left\{\left|\det\left[\left(C-\tilde{C}\right)^H\left(C-\tilde{C}\right)\right]\right|^{\frac{1}{2n}}\cdot\frac{E[\det(H*H)]Eb}{3nNo}\right\}^{-nm}$$

$$+\frac{3}{2}\left(\left\{\left|\det\left[\left(C-\tilde{C}\right)^H\left(C-\tilde{C}\right)\right]\right|^{\frac{1}{2n}}\cdot\frac{E[\det(H*H)]Eb}{3nNo}\right\}^{-4nm}\right)$$

$$-\frac{3}{2}\left\{\left|\det\left[\left(C-\tilde{C}\right)^H\left(C-\tilde{C}\right)\right]\right|^{\frac{1}{2n}}\cdot\frac{E[\det(H*H)]Eb}{3nNo}\right\}^{-3nm},\qquad(19)$$

where $t=0$, $l=4$, $Rate=3/4$, $\rho=\left(\frac{4}{3}E[\det(H*H)]\right)\frac{Eb}{No}$, and $P_S=p$. The H is the transmitted channel matrix, and $C*$ denotes the conjugate of the matrix C. The

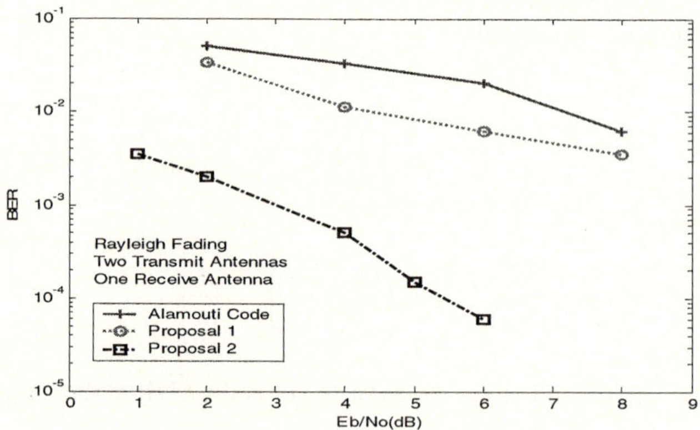

Fig. 4. The performance of the proposed systems

simulation results demonstrate that the proposed systems could lead to much enhancement from the conventional space time block codes on multi-code system, as illustrated in Fig.4.

3 Conclusion

This paper presents some interesting properties from the constant amplitude multi-code system. To combat the multi-path fading channel, we utilize the space time block codes and MIMO architecture to achieve the maximum transmit diversity for high data rate wireless communications. There are two proposed schemes. One can reduce the computations of the multiplications and the additions in the decoder without the BER performance loss. And the other could improve both the computations and BER performance, by using an additional orthogonal encoder and a butterfly decoding algorithm. In general, the second proposal is a very efficient scheme for realization the multi-code high data rate transmission over the multi-path fading channel, on low cost, high speed and high quality.

Acknowledgement

This work was supposed by university IT research center projects, Ministry of infomation & Communication, Ministry of Commerce industry & energy, Korea.

References

1. T. Wada, T. Yamazato, M. Katayama, and A. Ogawa: A Constant Amplitude Coding for Orthogonal Multi-code CDMA Systems. IEICE Trans. on Fundamentals. Vol.E80-A, no.12, (1997) 2477–2484
2. F. Adachi, M. Sawahashi, and H. Suda: Wideband DS-CDMA for Next Generation Mobile Communication System. IEEE Comm. Magazine. Vol.36, no.9, (1998) 56–69
3. I. C. Lin, and R. D. Gitlin: Multi-code CDMA Wireless Personal Communications Networks. Proc. IEEE ICC'95, Vol.2, Seattle, WA, (1995) 1060–1064
4. V. Tarokh, H. Jafarkhani, and A. R. Calderbank: Space Time Block Codes from Orthogonal Designs. IEEE Trans. on Info. Theory. Vol.45, no.5, (1999) 1456–1467
5. S. M. Alamouti: A Simple Transmit Diversity Technique for Wireless Communications. IEEE J. on Selected Areas in Comm. Vol.16, no.8, (1998) 1451–1458
6. Moon Ho Lee, and M. Kaveh: Fast Hadamard Transform Based on A Simple Matrix Factorization. IEEE Trans. on ASSP. Vol.34, no.6, (1986) 1666–1667
7. Moon Ho Lee, and Y. Yasuda: Simple Systolic Array Algorithm for Hadamard Transform. IEE Electronics Letters. Vol.26, no.10, (1990) 167–168
8. J. G. Proakis: Digital Communications. McGraw-Hill, Third edition (1995)
9. I. Telatar: Capacity of Multi-antenna Gaussian Channels. Technical Report, AT&T Bell Labs, (1995)
10. V. Tarokh, N. Seshadri, and A. R. Calderbank: Space Time Codes for High Data Rate Wireless Communication: Performance Criterion and Code Construction. IEEE Trans. on Information Theory, Vol.44, no.2, (1998) 744–765
11. Moon Ho Lee: The Center Weighted Hadamard Transform. IEEE Trans. on Circuits and Systems. Vol.36, no.9, (1989) 1247–1249
12. Moon Ho Lee: A New Reverse Jacket Transform and Its Fast Algorithm. IEEE Trans. on Circuits and Systems II, Vol.47, no.1, (2000) 39–47
13. Moon Ho Lee, and B.S. Rajan: Quasi-Cyclic Dyadic Codes in The Walsh-Hadamard Transform Domain. IEEE Trans. on Information Theory. Vol.48, no.8, (2002) 2406–2412

MIMO Frequency Hopping OFDM-CDMA: A Novel Uplink System for B3G Cellular Networks

Laurent Cariou and Jean-Francois Helard, *Member IEEE*⋆

Institute of Electronics and Telecommunications in Rennes,
INSA Rennes, 20, avenue des Buttes de Coesmes, 35000 Rennes, France
Phone: (+33)2 23 23 86 16, Fax: (+33)2 23 23 84 39
{laurent.cariou, jean-francois.helard}@insa-rennes.fr
http://www.ietr.org/

Abstract. In this paper, we propose and give the performance of a novel uplink system based on the combination of multi-carrier (MC), code division multiple access (CDMA) and multiple input multiple output (MIMO) techniques. We focus on the SS-MC-MA scheme, where each user spreads its data symbols on a specific subset of adjacent or multiplexed subcarriers, to facilitate the channel estimation and reduce complexity at the reception. In order to compensate for the lack of frequency diversity of SS-MC-MA with adjacent subcarriers, we combine it with an orthogonal space-time block code (STBC). Then, we propose to allocate the subsets to the different users by applying a frequency hopping pattern (FH). The performance of this scheme is evaluated over realistic MIMO channel with channel turbo coding for systems offering asymptotic spectrum efficiency of 1 and 3 bit/s/Hz. Thus, the efficiency of the novel proposed STBC FH SS-MC-MA system as a very promising multiple access and modulation scheme for the uplink of the future wideband wireless networks is successfully demonstrated.

1 Introduction

Nowadays, one of the most promising technologies for the air interface of the 4G terrestrial system downlink is multi-carrier code division multiple access (MC-CDMA). MC-CDMA combines the merits of orthogonal frequency division multiplex (OFDM) with those of CDMA [1]. In recent years, there has been a very strong emphasis on this technology including the realization of some fields trials, especially in Japan. At the European level, MC-CDMA has been studied for more than two years within the IST FP5 project MATRICE [2]. The European IST FP6 project 4MORE [3] aims at enhancing MATRICE by taking advantages of the potentialities offered by multiple input multiple output (MIMO) techniques, and advancing one step towards an optimized implementation.

⋆ This work was supported by the European FP6 IST project 4MORE (4G MC-CDMA Multiple Antenna System On Chip for Radio Enhancements).

P. Lorenz and P. Dini (Eds.): ICN 2005, LNCS 3421, pp. 8–17, 2005.

Based on the combination of OFDM and CDMA schemes, by spreading users' signals in the frequency domain, MC-CDMA offers for the downlink of the future cellular networks a strong robustness to multipath propagation and inter-cell interference as well as an interesting flexibility. In the uplink though, the MC-CDMA signal received at the base station (BS) is the result of a sum of N_u signals, that may be asynchronous, emitted by N_u mobile terminals (MT), each of which had undergone amplitude and phase distortions from N_u different channels. Thus, complex channel estimation and multi-user detection must be carried out to mitigate the multiple access interference (MAI).

To overcome those problems, spread spectrum multi-carrier multiple access (SS-MC-MA) has been proposed in [4] for the uplink context. In a few words, the basic difference between SS-MC-MA and MC-CDMA is that with SS-MC-MA the code division is used for the simultaneous transmission of the data symbols of a single user on a given subset of subcarriers, while in the case of MC-CDMA, it is used for the simultaneous transmission of different users' symbols on the same subset. SS-MC-MA assigns each user exclusively its own subset of subcarriers according to the FDMA scheme. Thanks to that, the BS only has to estimate, for each subcarrier, one channel compared to N_u for uplink MC-CDMA. Moreover, the symbols simultaneously transmitted on a subset of subcarriers undergo the same channel distortion and the MAI can now easily be cancelled by single user detection (SD).

MIMO communication systems, by using several antennas at the transmitter and at the receiver, inherit from space diversity to mitigate fading effects. Space-time block coding (STBC), based on orthogonal transmission matrix, as proposed by Alamouti in [5], provides full spatial diversity gains, no intersymbol interference and low complexity maximum likelihood receivers over frequency non-selective channels. Moreover, STBC can be easily combined with MC-CDMA [6].

In this paper, we describe a novel uplink system based on the combination of SS-MC-MA with STBC and frequency hopping. In order to compensate for the low frequency diversity of the SS-MC-MA scheme based on adjacent subcarriers, we combine it with STBC to take advantage of the spatial diversity. Then, we propose to allocate the subsets to the different users by applying a frequency hopping pattern. In that case, each user benefits from the frequency diversity linked to the whole bandwidth. The performance of this novel uplink scheme is evaluated over realistic MIMO channel with channel turbo coding. Particularly, the frequency hopping specific gain is stressed on.

The article is organized as follows. In Sect. 2, we describe the SS-MC-MA scheme. Section 3 presents the complete proposed uplink system. Section 4 introduces the 3GPP2-like MIMO channel model, developed within the IST-MATRICE project, and gives the channel and system parameters. Performance of different STBC SS-MC-MA systems with and without frequency hopping are given in Sect. 5 over this MIMO channel model. Different schemes offering with convolutional turbo coding an asymptotic spectrum efficiency of 1 and 3 bit/s/Hz are considered. Finally, we draw some conclusions in Sect. 6.

2 SS-MC-MA Uplink Scheme Description

In realistic systems concerned by flexibility and bandwidth efficiency, the MC-CDMA symbol duration is increased to reduce the loss in spectrum efficiency due to the guard interval insertion. N_c becomes larger than the length of the spreading sequences L. Consequently, several groups of flexible length spread symbols can be multiplexed on the different subcarriers. As for OFDM systems, the number N_c of subcarriers for a given bandwidth is chosen in order to obtain flat fading per subcarrier, while keeping time invariance during each MC-CDMA symbol. Thus, N_c can reach one thousand, or even more especially for outdoor applications, while L is generally less than or equal to 128.

Driven by the same realistic considerations, SS-MC-MA adds an FDMA scheme at subcarrier level in which each user $j = 1, \ldots, N_u$ exclusively transmits on a subset of L subcarriers out of a total of $N_c = N_u \cdot L$. After symbol mapping, a CDMA Walsh-Hadamard spreading process is applied to a defined number of complex-valued data symbols x_l, which is equal to the length L of the spreading code in the full-load case. SS-MC-MA and MC-CDMA systems share exactly the same process, the only difference lies on the selection of the data symbols x_l. In the former, those N_L symbols belong to the same user whereas in the latter, each user provides one data symbol to the selection. As we present here the SS-MC-MA system, the selected data symbols can now be denoted by $x_l^{(j)}$ as they all belong to user j.

Those selected data symbols are multiplied by their specific orthogonal Walsh-Hadamard spreading code $\mathbf{c}_l = [c_{l,1} \ldots c_{l,k} \ldots c_{l,L}]^T$ and superposed with each other. $c_{l,k}$ is the kth chip and $[.]^T$ denotes vector or matrix transposition. $\mathbf{c}_l$ is the lth column vector of the $L \times N_L$ spreading code matrix $\mathbf{C}$. The resulting CDMA spread symbol $\mathbf{s}^{(j)} = [s_1^{(j)} \ldots s_k^{(j)} \ldots s_L^{(j)}]$ can be expressed as follows:

$$s_k^{(j)} = \sum_{l=1}^{N_L} x_l^{(j)} c_{l,k} \qquad k = 1, \ldots, L$$

The chips $s_k^{(j)}$ of a CDMA spread symbol are then transmitted in parallel over a subset of L subcarriers among N_c proposed by the OFDM modulation. Hence, assuming that N_c is a multiple of L, the system can support $N_u = N_c / L$ spread symbols $s^{(j)}$, each user possessing its own spread symbol and its own subset of subcarriers. At the reception, for both systems, the OFDM demodulation is carried out by a simple and unique fast fourier transform (FFT) applied to the sum of the N_u different users' signals.

SS-MC-MA requires a very low complexity at the detection level to achieve good performance in terms of bit error rate (BER). Furthermore, compared to MC-CDMA, only one uplink channel has to be estimated for each subcarrier. When the mapping of the spread data symbols is made on adjacent subcarriers, this leads to a reduced complexity of the channel estimation. The major drawback of SS-MC-MA with adjacent subcarriers is its weakness in the exploitation of the available frequency diversity, essential in wireless systems to combat small scale fading caused by multi-path propagation.

One solution to this lack of diversity could be to introduce a frequency multiplexing on the spread signals as in downlink MC-CDMA, *i.e.* distribute the chips of a spread symbol on the whole bandwidth in order to maximize the frequency separation between chips and use all the frequency diversity available at the despreading process. However, this proposal is totally in contradiction with one of the criteria that led us to prefer SS-MC-MA to MC-CDMA in uplink: the need to reduce the number of pilots subcarriers required for the channel estimation.

The other solution, that allows to preserve the distribution of spread symbols on adjacent subcarriers and to respect the criteria expressed above, consists in introducing a frequency hopping component. In other words, the subset of adjacent subcarriers allocated to a user changes from one OFDM symbol to another by applying a frequency hopping pattern. This way, we tend to obtain perfectly uncorrelated channel distortions between successive OFDM symbols, thanks to frequency diversity. Unlike the previous solution, the use of channel coding is required to take advantage of it. This diversity can not be exploited during the de-spreading process but at the decoding step.

3   STBC FH SS-MC-MA System Description

3.1   MIMO Transmission Scheme

The use of multiple antennas at both ends of the communication link in conjunction with rich scattering in the propagation environment can provide multiplexing gain or diversity gain whether we prefer to increase the data rate or the reliability of the transmission. The uplink of the future cellular mobile systems will not need throughputs as large as the downlink and the system should be able to cope with spatially correlated channels. Furthermore, providing another form of diversity to SS-MC-MA system will fill in its imperfections. Then, giving more importance to spatial diversity gain and considering the case where the channel is not known at the transmitter, the use of space-time codes is straightforward. As OFDM transforms a frequency selective channel into N_c frequency non-selective parallel channels, the conditions required by the optimal use of those codes are fulfilled. Among them, orthogonal block codes have proved to be well suited to a combination with MC-CDMA [6]. We have focused on the well-known Alamouti's unitary-rate orthogonal STBC applied in space and time per subcarrier. Using two transmit and one or more receive antennas, Alamouti code provides full spatial diversity gain, no ISI and low complexity maximum likelihood (ML) receiver thanks to the orthogonality of its transmission matrix.

3.2   Frequency Hopping Pattern

The frequency hopping pattern used in our studies is given by the equation below:

$$s_{i,j} = (s_{i-1,j} + inc) \bmod N_t, \tag{1}$$

where $s_{i,j}$ is the subset number of subcarriers (between 1 and N_t) for OFDM symbol i and user j. N_t equal to N_c/L is the number of subsets of subcarriers in an OFDM symbol and corresponds also to the maximum number N_u of users. Here, N_c is equal to 736, L to 32 and thus N_t is equal to 23. The optimized value for the subset of subcarriers increment inc has been found to be inc=10. This simple law allows each user to experiment equiprobably each frequency subset and thus to take advantage of the frequency diversity related to the whole bandwidth, while avoiding collisions between users.

3.3   General Description of the Proposed System

Figure 1 shows a simplified MIMO SS-MC-MA system for user j based on Alamouti's STBC with $N_t = 2$ transmit antennas and $N_r = 2$ receive antennas. Channel coding and framing are not shown on the figure. User j simultaneously transmits N_L symbols $x_{j,l}^0$ and $x_{j,l}^1$ respectively from antenna 1 and 2 at time t, and the symbols $-x_{j,l}^{1*}$ and $x_{j,l}^{0*}$ at time $t+T_s$ where $l = 1, \ldots, N_L$. At the output of the space-time encoder, SS-MC-MA process, explained in Sect. 2, is applied on each antenna output. The spreading process can be carried out using fast hadamard transform (FHT). Note that we can swap the two linear processes STBC and spreading. The number N_c of subcarriers is set equal to N_u . L. The chip mapping component determines the subset of subcarriers on which the chips of the spread symbols of user j are distributed. One dimensionnal (1D) chip mapping on adjacent subcarriers is selected as explained before but a linear frequency interleaving or 2D-spreading can also be chosen. Each data symbol is then transmitted on L parallel and adjacent subcarriers. The vector obtained at the rth receive antenna at time t and $t+T_s$, after the OFDM demodulation and deinterleaving, on the subset of subcarriers associated with user j is given by:

$$\mathcal{R}_r = \mathcal{H}_r \mathcal{C} \mathcal{X} + \mathcal{N}_r \qquad with \qquad \mathcal{H}_r = \begin{bmatrix} H_{1r} & H_{2r} \\ H_{2r}^* & -H_{1r}^* \end{bmatrix} \qquad (2)$$

where $\mathcal{R}_r = [\mathbf{r}_r^T(t)\, \mathbf{r}_r^H(t + T_s)]^T$ with $\mathbf{r}_r(t) = [r_{r,1}(t) \ldots r_{r,k}(t) \ldots r_{r,L}(t)]^T$ the vector of the L received signals at time t and $[.]^H$ denotes the Hermitian (or complex conjugate transpose). $H_{tr} = diag\{h_{tr,1}, \ldots, h_{tr,L}\}(t, r \in \{1,2\})$ is a $L \times L$ diagonal matrix with $h_{tr,k}$ the complex channel frequency response, for the subcarrier k from the transmit antenna t to the receive antenna r. Time invariance during two SS-MC-MA symbols is assumed to permit the recombination of symbols when STBC is used. $\mathcal{C} = diag\{\mathbf{C}, \mathbf{C}\}$ where $\mathbf{C} = [c_1 \ldots c_l \ldots c_{N_L}]$ is the $L \times N_L$ matrix of user's spreading codes. $\mathcal{X} = [x^{0^T}\, x^{1^T}]^T$ where $x^0 = [x_1^0 \ldots x_l^0 \ldots x_{N_L}^0]^T$. $\mathcal{N}_r = [\mathbf{n}_r^T(t)\, \mathbf{n}_r^H(t+T_s)]^T$ where $\mathbf{n}_r(t) = [n_{r,1}(t) \ldots n_{r,k}(t) \ldots n_{r,L}(t)]^T$ is the Additive White Gaussian Noise (AWGN) vector with $n_{r,k}(t)$ representing the noise term at subcarrier k, for the rth receive antenna at time t with variance given by $\sigma_k^2 = \mathbf{E}\{|n_k|^2\} = N_0\ \forall k$.

In the receiver, in order to detect the $N_L \times 2$ transmitted symbols $x_{j,l}^0$ and $x_{j,l}^1$ for the desired user j, zero forcing (ZF) or MMSE SD schemes are applied to the received signals in conjunction with STBC decoding. In the SISO case, MMSE SD is the most efficient SD scheme [1]. In this paper, in order to evaluate the gain of MISO and MIMO systems, the performance of ZF and MMSE detection with SISO schemes are given as reference.

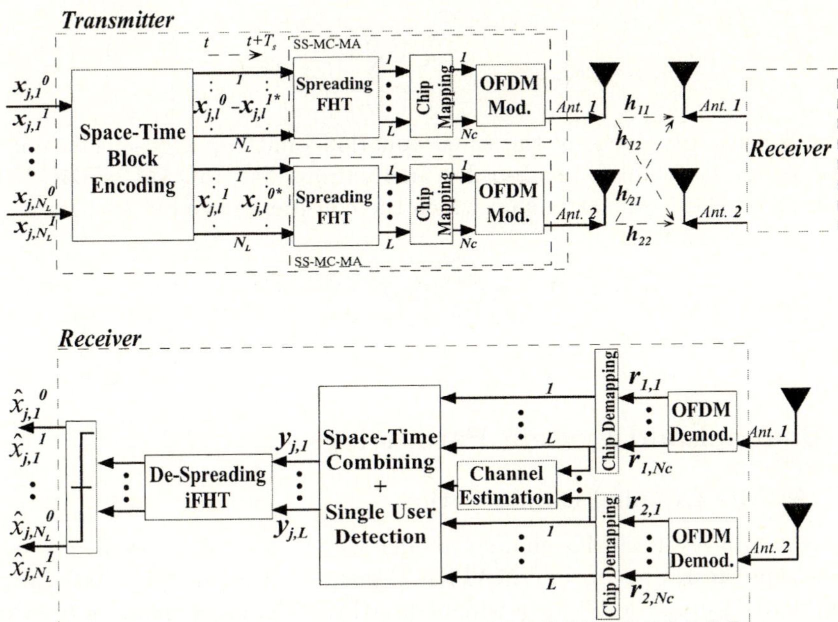

Fig. 1. SS-MC-MA transmitter and receiver for user j with two transmit and two receive antennas

3.4   Single User Detection in MIMO Case

After equalization, for each receive antenna r, the two successive received signals are combined. The resulting signals from the N_r receive antennas are then added to detect symbols $x_{j,l}^0$ and $x_{j,l}^1$. After de-spreading and threshold comparison, the detected data symbols $\hat{x}_{j,l}^0$ and $\hat{x}_{j,l}^1$ for user j are:

$$[\hat{x}_{j,l}^0 \ \hat{x}_{j,l}^0]^T = (\mathbf{I}_2 \otimes \mathbf{c}_l^T)\mathcal{Y} = (\mathbf{I}_2 \otimes \mathbf{c}_l^T)\sum_{r=1}^{N_r}\mathcal{G}_r\mathcal{R}_r \quad with \quad \mathcal{G}_r = \begin{bmatrix} G_{1r} & G_{2r}^* \\ G_{2r} & -G_{1r}^* \end{bmatrix} \quad (3)$$

where $\mathbf{I}_2$ is the 2×2 identity matrix, $\otimes$ is the Kronecker product, $\mathcal{Y} = [y_1^0 \ldots y_k^0 \ldots y_L^0 \ \ y_1^1 \ldots y_k^1 \ldots y_L^1]^T$ is the vector of the received signals equalized and combined from the N_r antennas, G_{tr} is a diagonal matrix containing the equalization coefficients for the channel between the transmit antenna t and the receive antenna r. To detect $x_{j,l}^0$, the MMSE SD coefficients $g_{tr,k}$ minimizes the mean square value of the error ϵ_k^0 between the signal transmitted on subcarrier k and the received signals combined from the N_r receive antennas by the Alamouti's decoding. In the same way, the ZF coefficients $g_{tr,k}$ restore the orthogonality between the different users. Knowledge of interfering spreading codes $\mathbf{c}_l$ is not required to derive the ZF and MMSE SD coefficients. Those for MMSE are given by the following equation:

$$g_{tr,k} = (h^{*}_{tr,k} \times \rho) / \left[\sum_{t=1}^{N_t=2} \sum_{r=1}^{N_r} |h_{tr,k}|^2 + \frac{1}{\gamma_{r,k}} \right] \tag{4}$$

ZF coefficients are given by the same equation omitting $1/\gamma_{r,k}$. The normalization factor ρ, useful for high-order modulations like 16QAM or 64QAM, is calculated on the L subcarriers on which the considered symbol is spread:

$$\rho = L / \sum_{k=1}^{L} \frac{\sum_{t=1}^{N_t=2} \sum_{r=1}^{N_r} |h_{tr,k}|^2}{\sum_{t=1}^{N_t=2} \sum_{r=1}^{N_r} |h_{tr,k}|^2 + \frac{1}{\gamma_{r,k}}} \tag{5}$$

4   Channel and System Parameters

4.1   MIMO Channel Configuration

We use a link level MIMO channel model which has been specifically developed within the European IST MATRICE project. This model is based on the 3GPP/3GPP2 proposal [7] for a wideband MIMO channel exploiting multipath angular characteristics. It consists in elaborating a MIMO spatial channel model following a hybrid approach between a geometrical concept depending on cluster positions and a tapped delay line model consisting in describing the average power delay profile (APDP) with a fixed number of taps. The spatial parameters are defined at the Base Station and the Mobile Terminal in a 2D-plane as the Clarke model. The angular distribution is modeled either as a Laplacian distribution or a Uniform distribution. The model parameters have been adapted to the 5 GHz bandwidth for an outdoor environment.

We use the BRAN E channel APDP, which refers to a typical outdoor urban multi-path propagation characterized by a large delay spread, possibly large Doppler spread and an asymmetrical antenna configuration. The measured coherence bandwidth mono-sided at 3 dB is roughly 1.5 MHz, close to the theoretical BRAN E one. Spatial correlation is inferior to 0.1 for an antenna spacing of 10 λ at the BS where λ stands for wavelength and close to 0.3 for 1 λ at the MT. Note that, at 5 GHz, 1 λ corresponds to a distance of 6 cm, which is realistic in a MT. Finally, although the MT velocity is 72 km/h, the time correlation remains close to the frame duration. Concerning the white additive gaussian noise, we assume the same noise level for each subcarrier or receive antenna.

4.2   System Configuration

The system parameters are chosen according to the time and frequency coherence of the channel in order to reduce ICI and ISI and to match the requirements of a 4G mobile cellular system. Channel information is available at the reception and is assumed to be perfect. The studied configuration proposed for outdoor scenarios is based on a sampling frequency equal to 57.6 MHz which is a multiple of the 3.84 MHz UMTS frequency and the carrier frequency is equal to 5 GHz.

The FFT size is 1024 with $N_c = 736$ modulated subcarriers leading to an occupied bandwidth equal to 41.46 MHz. The guard interval and the total OFDM symbol durations are equal to 3.75 and 21.52 μs respectively. Walsh-Hadamard spreading sequences with a length of 32 have been chosen.

Channel coding is composed of a rate 1/3 UMTS turbo-coding followed by puncturing pattern defined to achieve a global coding rate R of 1/2 or 3/4. The time interleaving depth has been adjusted to the frame duration. To obtain asymptotic spectrum efficiency of 1 and 3 bit/s/Hz, QPSK and 16QAM modulations have been considered, combined with 1/2 or 3/4 rate UMTS turbo code respectively. Taking into account the losses due to the guard interval insertion, the net bit rates corresponding to the asymptotic spectrum efficiency of 1 and 3 bit/s/Hz are equal to 34.2 and 102.6 Mbit/s respectively.

5   Simulation Results

All the following curves are given in terms of bit energy to noise ratio (E_b/N_0), without taking into account the guard interval efficiency loss and pilot symbols insertion overhead. The total transmit power is normalized and equal to P either for SISO and MIMO schemes and the presented results do not take into account the power gain provided by the use of multiple receive antennas. Full load systems, $N_L = 32$, are considered.

As mentioned before, the lack of diversity can be partially filled in by inserting a frequency hopping pattern. To have a clear idea of its advantages in our system, Fig. 2 presents Alamouti STBC22 and STBC24 SS-MC-MA performance with turbo coding and with or without frequency hopping. As a general rule, STBC $N_t N_r$ stands for a system using Alamouti STBC with N_t transmit and N_r receive antennas. Performance over the gaussian channel are also given as reference. System offering 1 bit/s/Hz spectrum efficiency with channel coding are considered. The diversity gain offered by the frequency hopping pattern for both spectrum efficiencies is superior or equal to 3.7 and 2.4 dB for a BER of 10^{-4} with 2 and 4 receive antennas respectively. This gain is all the more important as the number of receive antennas and consequently the spatial diversity is low. Performance of the same systems without any channel coding are also presented to evaluate the coding gain. The efficiency of the turbo-code is optimized as the performance gets closer to the performance over the AWGN channel, especially for four antennas at the reception.

Figure 3 shows the performance of the full load turbo-coded STBC SS-MC-MA system with frequency hopping and adjacent subcarriers proposed for the uplink. Systems with asymptotic spectrum efficiencies of 1 and 3 bit/s/Hz are presented. As previously, the realistic MIMO channel model based on the 3GPP/3GPP2 proposal is used to modelize a typical outdoor urban multipath propagation. Regarding the equalization schemes, it's well known that in the SISO case, ZF leads to excessive noise amplification for low subcarrier complex channel frequency response h_k. However, in the MIMO case, due to spatial diversity, this occurrence is statistically reduced, especially in a spatial uncorrelated

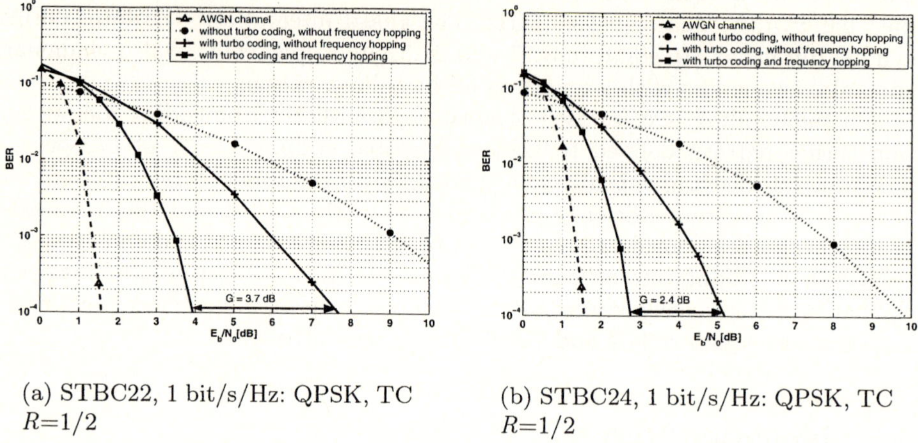

(a) STBC22, 1 bit/s/Hz: QPSK, TC $R=1/2$

(b) STBC24, 1 bit/s/Hz: QPSK, TC $R=1/2$

Fig. 2. Frequency hopping gain with full load uplink turbo-coded STBC SS-MC-MA systems, MMSE Single user Detection, adjacent subcarriers, 1D-spreading, 2 and 4 receive antennas

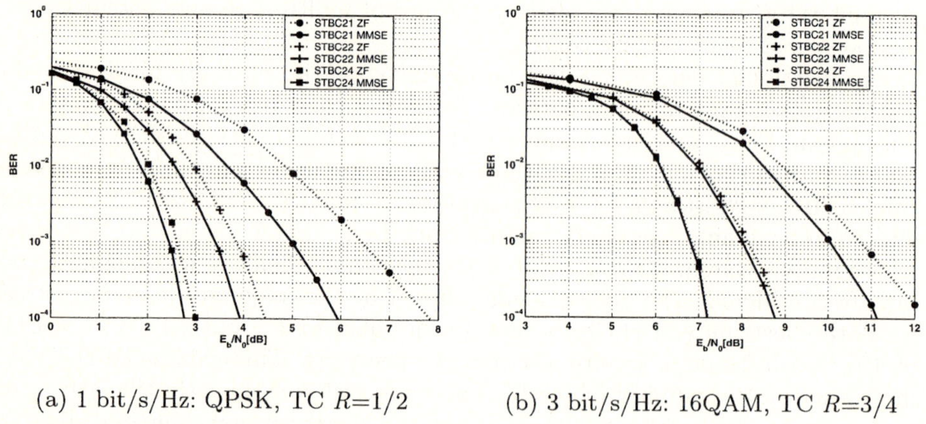

(a) 1 bit/s/Hz: QPSK, TC $R=1/2$

(b) 3 bit/s/Hz: 16QAM, TC $R=3/4$

Fig. 3. Performance of full load turbo-coded STBC SS-MC-MA systems with Frequency Hopping proposed for the uplink, ZF and MMSE Single user Detection, adjacent subcarriers, 1D-spreading, 1, 2 and 4 receive antennas

situation. Thus, as seen on the figure, with an increasing number of antennas, ZF tends to MMSE efficiency, while avoiding any SNR estimation.

Thanks to the spatial diversity gain and the frequency hopping gain, we can note that the performance obtained with very simple SD in this realistic case corresponding to correlated channels with 1 λ separation between the two antennas in the MT are very good. For example, with the uplink scheme offering 3 bit/s/Hz asymptotic spectrum efficiency, corresponding to a net bit rate of

102.6 Mbit/s, a BER of 10^{-4} is achieved with an E_b/N_0 equal to 7.2 dB. If we want to take into account the power gain provided by the use of multiple receive antennas, the curves have to be shifted left 3 dB for STBC22 and 6 dB for STBC24. These different results demonstrate that the proposed STBC SS-MC-MA system with frequency hopping takes full benefits from the frequency and spatial diversities.

6   Conclusion

In order to increase the spatial diversity exploited by the system for the uplink of the future wideband wireless networks, the combination of SS-MC-MA with Alamouti orthogonal STBC is considered. To improve the efficiency of the channel coding by increasing the frequency diversity, we have proposed to allocate the subsets to the different users by applying a frequency hopping pattern. In that case, each user benefits from the frequency diversity linked to the total bandwidth. It has been shown that the new STBC FH SS-MC-MA significantly outperforms the previous STBC SS-MC-MA. Using MMSE or ZF SD techniques, the global system exhibits a good trade-off between performance and complexity. Thus, the efficiency of this novel scheme as a promising system for the uplink of the future wideband wireless networks is successfully demonstrated.

References

1. Hara, S., Prasad, R.: Overview of multicarrier CDMA. *IEEE Communications Magazine*, vol. 35, no. 12, pp. 126-133, Dec. 1997.
2. IST MATRICE project, web site http://www.ist-matrice.org
3. IST 4MORE project, web site http://www.ist-4more.org
4. Kaiser, S., Krzymien, W.A.: Performance effects of the uplink asynchronism in a spread spectrum multi-carrier multiple access system. *European Transactions on Telecommunications*, vol. 10, no. 4, July/August 1999.
5. Alamouti, S.M.: A simple transmit diversity technique for wireless communication. *IEEE Journal on Selected Areas in Communications*, vol. 16, pp. 1451-1458, Oct. 1998.
6. Auffray, J-M., Baudais, J-Y., Helard, J-F.: STBC MC-CDMA systems : comparison of MMSE single-user and multi-user detection schemes over Rayleigh and MIMO METRA channels. *European Transactions on Telecommunications*, vol. 15, pp. 275-281, 2004.
7. 3GPP2 TR25.996 2003: Network Spatial Model for Multiple Input Multiple Output Simulations (Release 6). *3GPP, Spatial Channel AHG*, SCM-134, April 2003.

Transient Capacity Evaluation of UWB Ad Hoc Network with MIMO

Cheol Y. Jeon and Yeong M. Jang

School of Electrical Engineering,
Kookmin University,
861-1, Jeongneung-dong, Songbuk-gu, Seoul 136-702, Korea
{feon77, yjang}@kookmin.ac.kr

Abstract. In this paper we evaluate capacity of UWB ad hoc network which is based on direct-sequence code division multiple access(DS-CDMA) with multiple-input, multiple-output(MIMO). We propose an efficient and simple connection admission control(CAC) algorithm using transient quality of service(QoS) measure. The transient outage probability, one of the QoS measures, will be estimated using the Chernoff bound approach and the central limit approximation for real-time applications. Numerical results show that there is a substantial increment in the system capacity by adopting MIMO system and the central limit approximation is a good approach for transient QoS support.

1   Introduction

The gigabit Wireless Personal Area Network(WPAN) is one of the hottest area of wireless communications, which is standardized by the IEEE 802.15 alt PHY(IEEE 802.15.3a). They are mainly focused on the higher data rates and better supports for QoS such as multimedia applications in a home environment. The 802.15.3a has exposed several viable techniques to access the UWB spectrum as like OFDM, DS-CDMA and TD/FDMA pulses[1], [2]. In this paper we focus on DS-CDMA based UWB technique. To enhance the capacity of the network, some researches have been studied[3], [4]. Lately the multiple-antenna technology has garnered much attention in wireless communications. It is well known that the MIMO channel can offer spatial multiplexing gain, diversity gain, interference canceling gain, and antenna array gain[5], [6]. It has been shown that using the MIMO techniques can boost up the channel capacity significantly higher than that attainable by any known method based on a single-input, single-output(SISO) channel. That is to say, we have a spectacular increment in spectral efficiency, with the channel capacity being roughly doubled by doubling the number of antennas at both ends of the link[7], [8]. Although the system capacity increases, we need QoS guarantees for the different types of offered traffic[9]. Several CAC schemes for wireless ad hoc networks have been proposed and analyzed[10]. To fulfill such extensive demands for QoS support, we investigate an efficient transient algorithm for admission control in the UWB network with QoS support for different traffic classes.

P. Lorenz and P. Dini (Eds.): ICN 2005, LNCS 3421, pp. 18–25, 2005.

We focus on the MIMO technology to increase the system capacity for UWB networks, and then calculate the number of simultaneous DEVs in the DS-UWB network with MIMO. We propose an efficient and simple CAC algorithm for UWB ad hoc network with MIMO.

This paper is organized as follows: Section 2 discusses the UWB network based on DS-CDMA with MIMO system. We discuss CAC algorithm and derive the outage probability in section 3. Section 4 presents numerical results of this discussion and finally come to conclusion in section 5.

2   UWB Ad Hoc Network with MIMO System

2.1   UWB Ad Hoc Network Based on DS-CDMA

The UWB ad hoc network is piconet topology which is composed of a PNC and multiple DEVs. The piconet's basic component is the DEV. One DEV is required to assume the role of the PNC of the piconet. The PNC provides the basic timing for the piconet with the beacon. Furthermore, PNC must perform many functionalities to handle the correct control of the cluster and an efficient sharing of the radio resource, as like, the PNC manages the QoS requirements[11].

The 802.15.3a has exposed several viable techniques to access the UWB spectrum such as OFDM, DS-CDMA and TD/FDMA pulses[1]. In this paper we focus on DS-CDMA. DS-UWB ad hoc network with MIMO schemes allows us to achieve very high bit rate and low error probability matched to QoS requirement[4]. So we introduce the capacity of MIMO system in next subsection.

2.2   Capacity of MIMO System

When we use n_T transmit antennas and n_R receive antennas, $m=\min(n_T, n_R)$ and $n=\max(n_T, n_R)$. The MIMO capacity increases linearly with m. The ergodic capacity in flat fading channel is given by [7]

$$C = E[log_2\ det(I_{n_R} + \frac{SNR}{n_T}HH^{\dagger})][bits/s/Hz] \tag{1}$$

where $\dagger$ is transpose conjugate and I_{n_R} is $n_R \times n_R$ identity matrix. The ergodic capacity of a MIMO link involves the matrix product $HH^{\dagger}$. We assume that transmission at each antenna performs with constant power over the burst of n_T vector symbols and that each antenna transmits with the same power. Then we make the statistical assumption that the symbols transmitted are identically and independently distributed($i.i.d.$). According to [12], we may rewrite Eq. (1) in the equivalent form

$$det(I_{n_R} + \frac{SNR}{n_T}HH^{\dagger}) = \prod_{i=1}^{n_R}(1 + \frac{SNR}{n_T}\lambda_i) \tag{2}$$

where λ_i is i-th eigenvalue of $HH^{\dagger}$ finally, Eq. (2) into Eq. (1)

$$C = E[\sum_{i=1}^{n_R} log_2(1 + \frac{SNR}{n_T}\lambda_i)] = n_R log_2(1 + \frac{SNR}{n_T}\lambda_i)[bits/s/Hz]. \tag{3}$$

For simplicity, in the ideal case in which the eigenvalue (which means an ideal propagation environment)[7], [8], we can achieve maximum data rate

$$C = n_R log_2(1 + \frac{SNR}{n_T})[bits/s/Hz].$$ (4)

and we can derive capacity gain

$$G_M = \frac{C_{MIMO}}{C_{SISO}} = \frac{n_R log_2(1 + \frac{SNR}{n_T})}{log_2(1 + SNR)}.$$ (5)

Eq. (5) shows that using the MIMO techniques can boost up the channel capacity significantly higher than the SISO channel. If the number of antenna increases, the spectral efficiency increases.

3   Connection Admission Control

3.1   Model Assumptions

We consider a single piconet that consists of a PNC and multiple DEVs which adopted MIMO system. Actually piconet supports direct mode but DEV should receive the time slot from PNC and only one DEV can transmits at one time slot. CAC is performed by the PNC every frame one of the associated DEVs generates a request to open a new connection towards another DEV.

Currently, timing in the 802.15.3 piconet is based on the superframe, which is composed of the beacon, the contention access period(CAP) and the contention free period(CFP). The beacon is only used to set the timing allocations and to communicate management information for the piconet. To transfer data, DEV uses the CAP and the CFP. The total link capacity with MIMO technique of one superframe is obtained by Eq. (5)

$$C_{MIMO} = G_M \times C_{SISO} \times superframe\_duration\ [bits/frame].$$ (6)

For simple model, we assume that the duration of superframe is constant and the frame is consist of almost CFP. We neglect beacon duration and CAP duration. The system capacity using CFP is C_{MIMO} which is defined in Eq. (6).

Suppose that $N(= N_1 + \cdots + N_m + \cdots + N_M)$ independent heterogeneous On-Off sources are connected to piconet, where N_m denotes the number of connections of class-m. Therefore it can be modeled as a single server system with system capacity of C_{MIMO}bps. A QoS predictor predicts outage probability for a time $t(=$beacon interval which is the same as frame size) ahead.

3.2   Calculating QoS Measures

Transient Fluid Model Approach: We assume that a series of packets arrive in the form of a continuous stream of bits or a fluid. We also assume that 'ON' and 'OFF' periods of sources of class-m are both exponentially distributed with parameters λ_m and μ_m, respectively. The transitional flow rate from the 'ON' state to the 'OFF' state is λ_m and from 'OFF' to 'ON' is μ_m. In this traffic model, when a source is in the 'ON' state, it generates packets with a constant inter-arrival time, $\frac{1}{R_m}$ seconds/bit. When the source is in the 'OFF' state, it does not generate any packets. We assume that class-m sources of N, N_m, the connections sharing a link have the same traffic parameters(λ_m, μ_m, R_m).

We use a statistical bufferless fluid model to predict the $P_{out}(t)$ at a future point in time t. Let $\Lambda_m(t)(= R_m Y_m(t))$ be aggregate arrival rate from $Y_m(t)$ active sources. In a bufferless system, outage occurs when $\Lambda(t)(= \sum_{m=1}^{M} \Lambda_m(t))$ exceeds the link capacity, C_{MIMO}. Taking into consideration the fact that each of $N(= N_1 + \cdots + N_m + \cdots + N_M)$ existing connections belongs to one of the M connection classes, given by an arbitrary initial condition $Y(0) = I = [Y_1(0) = i_1, Y_2(0) = i_2, \ldots, Y_M(0) = i_M]$, we obtain the conditional moment generating function of $\Lambda_m(t)$, $s \geq 0$:

$$G_{\Lambda_m(t)|Y_m(0)}(s) = E[e^{sR_m Y_m(t)}|Y_m(0) = i_m] \tag{7}$$
$$= [p_m(t)(e^{sR_m} - 1) + 1]^{N_m - i_m}[q_m(t)(e^{sR_m} - 1) + 1]^{i_m}$$

where $p_m(t)$ and $q_m(t)$ are defined at [13]. $p_m(t)$ is the transition probability that a class-m source is active at a future point in time t, given that the source is active at time 0. $q_m(t)$ is the transition probability that a class-m source is active at a future point in time t, given that the source is idle at time 0.

Chernoff Bound Approach: To predict small $P_{out}(t)$, which is several standard deviations away from the mean, other approximations such as the Chernoff bound(CB) and the large deviations theory can be used. In the subsection, we provide an upper bound for $P_{out}(t)$. To this end, for any random process $\Lambda(t)$ and constant C_{MIMO}, let

$$\mu_{\Lambda(t)|Y(0)}(s) = lnG_{\Lambda(t)|Y(0)}(s) \tag{8}$$

$$= ln\, E[e^{s\Lambda(t)}|Y(0) = I] = \sum_{m=1}^{M} \mu_{\Lambda(t,m)|Y_m(0)=i_m}(s).$$

Note that $\mu_{\Lambda(t)|Y(0)}(s)$ is the logarithm of the conditional moment generating function for $\Lambda(t)$. $\mu_{\Lambda(t,N+1)|Y_{N+1}(0)=i}(s)$ is the logarithm of the conditional moment generating function for $\Lambda(t, N + 1)$, associated with a new connection request. The CB gives us an upper bound for $P(\Lambda(t) > C_{MIMO}|Y(0) = I)$. Assume that exists $E[\Lambda(t)]$ and $C_{MIMO} \geq E[\Lambda(t)] \geq -\infty$. Then the supreme in Eq. (9) is obtained within values $s^* \geq 0$, and minimizing the right hand side of Eq. (9) with respect to s yields the CB

$$P_{out}(t) = P(\Lambda(t) > C_{MIMO}|Y(0) = I) \leq e^{-s^* C_{MIMO}} G_{\Lambda(t)|Y(0)}(s^*) \tag{9}$$
$$= e^{-s^* C_{MIMO} + \mu_{\Lambda(t)|Y(0)}(s^*)}, \quad for\ s^* \geq 0\ and\ C_{MIMO} \geq E[\Lambda(t)]$$

where s^* is the unique solution to

$$\mu'_{\Lambda(t)|Y(0)}(s^*) = \sum_{m=1}^{M} \frac{(N_m - i_m)p_m(t)R_m e^{s^* R_m}}{p_m(t)(e^{s^* R_m} - 1) + 1} + \sum_{m=1}^{M} \frac{i_m q_m(t)R_m e^{s^* R_m}}{q_m(t)(e^{s^* R_m} - 1) + 1} \tag{10}$$

The CB can be used to predict the number of connections that is needed to satisfy a given the conditional outage probability bound, QoS_{out} at the link, i.e., $P(\Lambda(t) > C_{MIMO}|Y(0) = I) \leq QoS_{out}$. The CB gives the admission control condition

$$P_{out}(t) = e^{-s^* C_{MIMO}} G_{\Lambda(t)|Y(0)}(s^*) \leq QoS_{out}. \tag{11}$$

Using Eq. (11), the admission control condition scheme operates as follows. The logarithms of conditional moment generating functions, $\mu_{\Lambda(t)|Y(0)}(s^*)$, are first calculated for all connections. Now, suppose a new connection $(N+1)$-st is requested. We update $\mu_{\Lambda(t)|Y(0)}(s^*) \leftarrow \mu_{\Lambda(t)|Y(0)}(s^*) + \mu_{\Lambda(t,N+1)|Y_{N+1}(0)=i}(s^*)$ and find the s^* that satisfies Eq. (10). Finally, we admit the connection if and only if Eq. (11) is satisfied.

Central Limit Approximation Approach: In order to provide a practical CAC mechanism, we consider bounds of QoS measures. A central limit approximation(CLA) for transient outage probability is considered. Let us assume that the conditional aggregate traffic rate has a Gaussian distribution. The CLA is a quite accurate for predicting the distribution of the aggregated traffic within a few standard deviations from the mean [13]. For the class-m On-Off sources, the conditional mean arrival rate is

$$A_m(t) = R_m[i_m q_m(t) + (N_m - i_m)p_m(t)], \tag{12}$$

and conditional variance of arrival rate is

$$\sigma_m^2(t) = R_m^2[(N_m - i_m)p_m(t)(1 - p_m(t)) + i_m q_m(t)(1 - q_m(t))]. \tag{13}$$

By the central limit theorem, $\Lambda(t)$ is approximated by a normal random process with conditional mean and variance

$$A(t) = \sum_{m=1}^{M} A_m(t), \qquad \sigma^2(t) = \sum_{m=1}^{M} \sigma^2(t). \tag{14}$$

With established traffic model, P_{out} may now be easily computed from the tail of the normal distribution[12], [13]. Given a specific QoS requirement $P_{out} \leq QoS_{out}$, where QoS_{out} is small number such as 10^{-3}, the QoS requirement $P_{out} \leq QoS_{out}$ is met and only if.

$$P_{out} = Q(\frac{C_{MIMO} - A(t)}{\sqrt{\sigma^2(t)}}) \leq QoS_{out} \tag{15}$$

where Q(.) denotes the Q-function. Using this admission rule Eq. (15), a new $(N+1)$-st connection is established. We update $A(t) \leftarrow A(t) + A_{N+1}(t)$, $\sigma^2(t) \leftarrow \sigma^2(t) + \sigma_{N+1}^2(t)$ and $\mu(t) \leftarrow \mu(t) + \mu_{N+1}(t)$, where $A_{N+1}(t)$, $\sigma_{N+1}^2(t)$ and $\mu_{N+1}(t)$ are computed from traffic descriptor specified by new DEV. Then admit the new connection if and only if the connection in Eq. (15) is met.

4   Numerical Result

The goal of the simulations is to study the effect of the MIMO system and compare more efficient CAC algorithm. The first class is asynchronous class which has low requirements for bit error rate around 0.03. The second class, synchronous class, is specified to carry real-time traffic. This class has high requirements for bit error rate around 10^{-3}[15].

First we simulated the effect of MIMO system at each traffic classes, for example, $C_{SISO} = 112$Mbps, $super frame_duration = 65.535$msec, synchronous data rates, R_1, is 13kbps, $QoS_{out} = 0.03$ and asynchronous data rates, R_2, is 384kbps, $QoS_{out} = 0.001$. We assumed that the system is in steady state and the value of the initial condition, $Y(0)$, is zero. As presented in Table 1, when data rate increases, the system acceptance decreases. It shows that $m \times m$ MIMO system provides system capacity increment with m. This result represents that the piconet capacity is improved with MIMO.

Table 1. Maximum number of connections(using CLA)

Traffic class	Data rate R_m	Number of Tx and Rx Antenna			
		1×1	2×2	3×3	4×4
synchronous	$R_1 = 13kbps$	1,414	2,012	2,348	2,644
asynchronous	$R_2 = 384kbps$	29	45	55	62

Now we will investigate the proposed transient CAC algorithm. We consider asynchronous traffic source, for example, $R_2 = 384$Kbps, $QoS_{out} = 0.001$, $\lambda = 0.5$, $\mu = 0.833$, various values of the initial conditions, $Y(0)$ are 15 and 35. After approximately 4 seconds the predicted outage probability, $P_{out}(t)$, will converge to the steady state value, $P_{out}(\infty)$. We show differences in the results obtained as a function of the different initial conditions and QoS measures and differences in $P_{out}(t)$ between SISO and MIMO. $P_{out}(t)$ in SISO system is higher than in 4×4 MIMO system. We also observe in 4×4 MIMO system that the $P_{out}(t)$ at $t=1$sec and $Y(0) = 15$ is approximately 8.5×10^{-4} while the steady state, $t>5$sec, $P_{out}(t)$ is approximately 9.1×10^{-3} when CB approach is adopted. The other hand, when CLA is adopted, at $t=1$sec and $Y(0) = 15$, we observe that the $P_{out}(t)$ is approximately 5.0×10^{-5} while the steady state, $t>5$sec, $P_{out}(t)$ is approximately 8.8×10^{-4}.

We investigated the number of connections by using different QoS measures when asynchronous and synchronous classes with assumed above parameters, are existed. We observe that the maximum number of connections in MIMO system is higher than SISO system in Fig. 1. As the number of asynchronous connections increases, the maximum number of synchronous connections will be decreased. The resulting admissible region using the CB is conservative. The CLA can admit more connections at each case.

The CB approach is always conservative connection control policy under the light load situation. For the large number of active sources, CLA may be used because we can avoid the solution for Eq. (10) which has been used for the CB

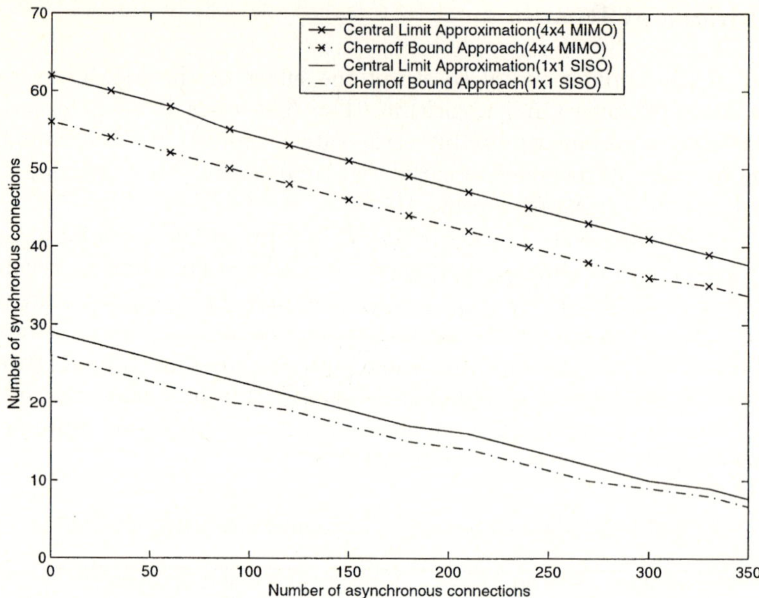

Fig. 1. Number of asynchronous connections v.s. number of synchronous connections

approach. The transient approaches are more complex than the steady state approaches. We, therefore, conclude that the connection control using the CLA is more simple for UWB ad hoc networks.

5   Conclusions

In this paper, we evaluated the capacity for UWB network based on DS-CDMA with MIMO system. We introduced the CAC algorithm for MIMO UWB network that guarantees the QoS. To do this, we derived the spectral efficiency of MIMO system and developed approximation using the transient QoS measure for admitting on-off sources in the piconet. Our analytical results show that there can be a substantial capacity increment in the UWB piconet capacity by adopting MIMO and the CLA is a good approach to guarantee the transient QoS.

Acknowledgement

This research was supported by University IT Research Center(INHA UWB-ITRC), Korea.

References

1. IEEE 802.15 WPAN High Rate Alternative PHY Task Group 3a(TG3a): http://grouper.ieee.org/groups/802/15/pub/TG3a.html
2. K. Mandke, et al., 'The Evolution of Ultra Wide Band Radio for Wireless Personal Area Networks,' *High Frequency Electronics*, 2003.
3. W. Siriwongpairat, et al., 'On the Performance Evaluation of TH and DS UWB MIMO Systems,' *IEEE WCNC*, 2004.
4. E. Baccarelli, et al., 'A Simple Multi-Antenna Transceiver for Ultra Wide Band based 4GWLANs,' *IEEE WCNC*, 2004.
5. H. Bölcskei, 'Fundamental performance tradeoffs in coherent MIMO signaling,' *Private Communication*, 2003.
6. L. Zheng and D. N. Tse, 'Diversity and multiplexing: a fundamental tradeoff in multiple-antenna channels,' *IEEE Trans. Inform. Theory*, vol. 49, pp. 1073-1096, May 2003.
7. Martone and Massimiliano, *Multiantenna Digital Radio Transmission*, Artech House, Inc., pp. 113-119, 2002.
8. P. J. Smith and M. Shafi, 'On a Gaussian Approximation to the Capacity of Wireless MIMO Systems,' *IEEE ICCC*, May 2002.
9. S. Shakkottai, et al., 'Cross-layer Design for Wireless Networks,' *IEEE Comm. Magazine*, Oct. 2003.
10. G. Razzano and A. Curcio ,'Performance comparison of three Call Admission Control algorithms in a Wireless Ad-Hoc Network,' *ICCT*, 2003.
11. Draft Standard for Telecommunications and Information Exchange Between Systems: Wireless Medium Access Control (MAC) and Physical Layer (PHY) Specifications for High Rate Wireless Personal Area Networks (WPAN), *Draft P802.15.3/D17*, Feb. 2003.
12. S. Haykin and M. Moher, *Modern Wireless Communications*, Pearson Prentice Hall, pp. 339-376, 2005.
13. Yeong M. Jang, 'Central limit approximation approach for connection admission control in broadband satellite systems,' *IEE Electronics Letters*, pp. 255-256, Feb. 2000.
14. Yeong M. Jang, 'Connection admission control using transient QoS measure for broadband satellite systems,' *HSNMC2004*, June 2004.
15. T. Janevski, *Traffic Analysis and Design of Wireless IP Networks*, Artech House, Inc., pp. 35-44, 2003.

Chip-by-Chip Iterative Multiuser Detection for VBLAST Coded Multiple-Input Multiple-Output Systems*

Ke Deng, Qinye Yin, Yiwen Zhang, and Ming Luo

Institute of Information Engineering, School of Electronics,and Information, Engineering Xi'an Jiaotong University, Xi'an 710049, China

Abstract. We propose a vertical Bell Lab's space-time (VBLAST) coded MIMO transceiver scheme and its iterative (turbo) multiuser detection algorithm. Unlike other multiuser schemes, users in this scheme are distinguished by different chip-level interleavers. Consequently, the maximum number of available users increases a lot. The proposed multiuser detection algorithm is based on the turbo multiuser one, which is not only computationally efficient compared with the conventional optimal maximum likelihood decoder, but also achieves a probability of error performance being smaller than the traditional VBLAST schemes under the same operation conditions.

1 Introduction

The space-time coding (STC) is an effective coding technique that uses transmit diversity to combat the detrimental effects in wireless fading channels by combining the signal processing at the receiver with coding techniques appropriate to multiple transmit antennas to achieve higher data rates [1]. Among various space-time codec schemes, the vertical Bell Labs layered space-time (VBLAST) [2] codec scheme is relatively simple and able to implement high data rate in the wideband wireless communication systems. And it was advised to some protocols in wireless local loop and wireless local area network (WLAN).

Recently, the iterative "turbo" receiver has received considerable attention followed by the discovery of the powerful turbo Codes [3]. Furthermore, because of the simple structure of the so-called turbo principle and its capability of achieving near-optimal performance, the turbo principle has been successfully applied to many decoding/detection problems such as channel decoding, channel equalization, coded modulation, multiuser detection, and joint source and channel decoding.

Till now, there are some researches on iterative decoding of the multiple-input multiple-output (MIMO) systems, [4] proposed an iterative receiver for so-called Turbo-VBLAST structure, achieving a probability of error performance that

* Partially supported by the National Natural Science Foundation (No. 60272071) and the Research Fund for Doctoral Program of Higher Education (No. 20020698024) of China.

P. Lorenz and P. Dini (Eds.): ICN 2005, LNCS 3421, pp. 26–33, 2005.

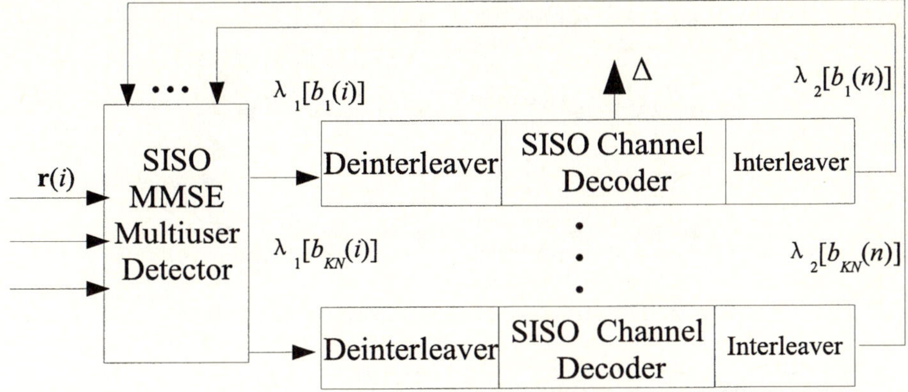

Fig. 1. Turbo receiver of VBLAST coded MIMO system

is orders of magnitude smaller that traditional BLAST schemes for the same operation conditions. But this paper only consider the single user scenario, and its performance decreased a lot when the users increased because there are much more transmit antennas than receive antennas.

In this paper, we propose a multiuser MIMO transceiver scheme coded by VBLAST, where active users are distinguished by different chip-lever interleavers. Consequently, the maximum available number of users increases a lot. At the receiver, an iterative chip-by-chip multiuser detection algorithm based on what proposed in [4][5] is employed. During iterations, extrinsic information is computed and exchanged between a soft MMSE receiver and a bank of maximum *a posteriori* probability (MAP) decoders to achieve successively refined estimates of users' signals. The proposed MIMO multiuser architectures can handle any configuration of transmit and receive antennas. Simulation results demonstrate that the performance is decided by the length of the spreading code, number of the transmit and receive antennas, and it can significantly outperform the turbo multiuser detection at the symbol level.

2   Transmitter Structure

The transmitter structure is shown in fig.2. There are K users equipped with N transmit antennas in the proposed scheme, while M receive antennas is equipped at the base station. the input data sequence d_k of user k is demultiplexed to N substream $d_{k,n}$, which is successively encoded by a recursive system convolution (RSC) coder g, and incorporated with a spreading operation where the length of the spreading code is G. The spreading operation will lead to an improved coding gain and some flexibility. By passing a chip-level interleaver $\pi_{k,n}$, the encoded data substream is sent to the transmitter antenna and each interleaved substream is transmitted using a separate antenna. The transmitted signals are received on

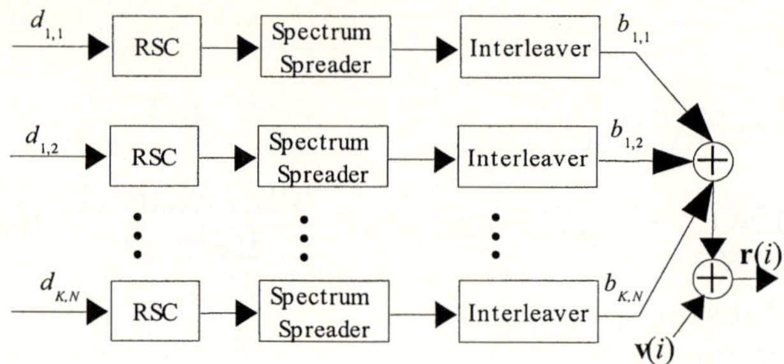

Fig. 2. The structure of the transmitter of one user

receiving antennas, whose output signals are fed to an iterative receiver. To every antenna of the total KN transmit antennas, either the spreading code or the interleaver is different.

Throughout this paper, we assume that the transmitters operate with synchronized chip timing at a rate of chips per second and that the sampling times of receivers are chip synchronous. The channel variation is assumed to be negligible over chip periods, comprising a packet of chips. Moreover, we only consider a narrowband frequency-flat communication environment, i.e, no delay spread. The extension of this scheme to a frequency-selective environment is straightforward.

After sampling at the chip rate, the received signal on the antenna m of the base station from K active users can be written as

$$r_m(i) = \sum_{k=1}^{K} \sum_{n=1}^{N} h_{m,k,n} b_{k,n}(i) + v_m(i) \tag{1}$$

where $b_{k,n}(i)$ is the i-th chip transmitted by the n-th antenna of user k, $h_{m,k,n}$ is the channel coefficient form the n-th transmit antenna of the user k to the m-th receive antenna, $v_m(i)$ is the sample of a zero-mean additive Gaussian noise (AWGN) with variance $\sigma^2 = N_0/2$.

The corresponding vector representation of (1) is

$$r_m(i) = \mathbf{h}_m \mathbf{b}(i) + v_m(i) \tag{2}$$

where $1 \times KN$ $\mathbf{h}_m$ represents the uplink channel vector from all users to antenna m, and is defined as $\mathbf{h}_m = (\,h_{m,1,1} \quad h_{m,1,2} \quad \ldots \quad h_{m,K,N}\,)$, $KN \times 1$ $\mathbf{x}(i)$ represents transmitted chips at interval i from all KN transmit antennas, and is defined as $\mathbf{b}(i) = (\,b_{1,1}(i) \quad \ldots \quad b_{K,N}(i)\,)^T$.

Stacking the received data from antenna 1 to antenna M at the base station, yields an extended uplink data vector with dimension $M \times 1$ as

$$\mathbf{r}(i) = \mathbf{H}\mathbf{b}(i) + \mathbf{v}(i) \tag{3}$$

where $\mathbf{H}$ represents the channel matrix between the receive antennas and the all transmit antennas with dimensions $M \times KN$, is defined as $\mathbf{H} = (\,\mathbf{h}_1^T \quad \mathbf{h}_2^T \quad \ldots \quad \mathbf{h}_N^T\,)^T$; $\mathbf{v}(i) = (\,v_1(i) \quad v_2(i) \quad \ldots \quad v_N(i)\,)^T$.

3 Multiuser Detection

3.1 Basic Concept

This section proposes a practical sub-optimum multiuser detection scheme based on iterative "turbo" detection principles. The intersubstream coding proposed as independent encoding and interleaving can be viewed as a serially concatenated code as illustrated in Fig. 2. In the iterative decoding scheme, we separate the optimal decoding problem into two stages and exchange the information learned from one stage to another iteratively until the receiver converges. The two decoding stages are 1) Inner decoder: soft-input/soft-output (SISO) minimum mean-square error (MMSE) multiuser detector; 2) Outer decoders: A set of KN parallel SISO RSC decoders. The detector and decoder stages are separated by interleavers and deinterleavers which are used to compensate for the interleaving operation used in the transmitter as well as to decorrelate the correlated outputs before feeding them to the next decoding stage. The iterative receiver produces new and better estimates at each iteration of the receiver and repeats the information-exchange process a number of times to improve the decisions and channel estimates. The iterative decoder is shown in Fig. 1.

The SISO MMSE multiuser detector delivers the *a posteriori* log-likelihood ratio (LLR) of a transmitted "+1" and a transmitted "−1" for every code bit of every antenna of every user,

$$\Lambda_1[b_k(i)] \stackrel{\text{def}}{=} \log \frac{P[b_k(i) = +1|\mathbf{r}(t)]}{P[b_k(i) = -1|\mathbf{r}(t)]} \tag{4}$$

Using Bayes' rule, (4) can be written as

$$\Lambda_1[b_k(i)] = \log \underbrace{\frac{P[\mathbf{r}(t)|b_k(i) = +1]}{P[\mathbf{r}(t)|b_k(i) = -1]}}_{\lambda_1[b_k(i)]} + \log \underbrace{\frac{P[b_k(i) = +1]}{P[b_k(i) = -1]}}_{\lambda_2^p[b_k(i)]} \tag{5}$$

where the second term in (5), denoted by $\lambda_2^p[b_k(i)]$, represents the *a priori* LLR of the code bit $b_k(i)$, which is computed by the k-th RSC decoder in the previous iteration, interleaved and then fed back to the SISO MMSE detector. (The superscript p indicates the quantity obtained from the previous iteration). For the first iteration, assuming equally likely code bits, i.e., no prior information available, we then have $\lambda_2^p[b_k(i)] = 0$.

Similarly, based on the prior information $\lambda_1^p[b_k(i)]$, and the trellis structure (i.e., code constraints) of the channel code, the k-th SISO RSC decoder computes the *a posteriori* LLR of each code bit

$$\Lambda_2[b_k(i)] = \lambda_2[b_k(i)] + \lambda_1^p[b_k(i)] \tag{6}$$

The SISO RSC decoder has the same structure and iterative algorithm as what proposed in [6].

3.2 Inner Decoder — MMSE Multiuser Detector

A minimum mean-square error (MMSE) receiver is used after receiving the signal. Let $b_k(i)$ be the desired signal, (3) can be rewritten as

$$\mathbf{r}(i) = \mathbf{h}_k(i)b_k(i) + \mathbf{H}_k(i)\mathbf{b}_k(i) + \mathbf{v}(i) \tag{7}$$

where $\mathbf{H}_k(i) = (\mathbf{h}_1(i), \ldots, \mathbf{h}_{k-1}(i), \mathbf{h}_{k+1}(i), \ldots, \mathbf{h}_{KN}(i))$, and $\mathbf{b}_k(i) = (b_1(i), \ldots, b_{k-1}(i), b_{k+1}(i), \ldots, b_{KN}(i))$.

The output of $\mathbf{r}(i)$ passing a linear filter $\mathbf{w}_k$ is

$$\mathbf{w}_k^H \mathbf{r} = \mathbf{w}_k^H \mathbf{h}_k(i)b_k(i) + \underbrace{\mathbf{w}_k^H \mathbf{H}_k(i)\mathbf{b}_k(i)}_{u_k} + \mathbf{w}_k^H \mathbf{v}(i) \tag{8}$$

where u_k represents the CAI. For brevity, we omit the time i.

We remove CAI from the linear filter output and write

$$\hat{b}_k = \mathbf{w}_k^H \mathbf{r} - u_k \tag{9}$$

According to the definition of MMSE, the weights $\mathbf{w}_k$ and the interference estimate u_k are optimized by minimizing the mean-square value of the error between each substream and its estimate.

$$(\hat{\mathbf{w}}_k, \hat{u}_k) = \arg \min_{(\mathbf{w}_k u_k)} \varepsilon[\|b_k - \hat{b}_k\|^2] \tag{10}$$

where the expectation ε is taken over the noise and the statistics of the data sequence.

The solution of the MMSE receiver is given by [6]

$$\hat{\mathbf{w}}_k = (\mathbf{P} + \mathbf{Q} + \sigma^2 \mathbf{I})^{-1} \mathbf{h}_k \tag{11}$$

$$\hat{u}_k = \mathbf{w}_k \mathbf{z} \tag{12}$$

where

$$\mathbf{P} = \mathbf{h}_k \mathbf{h}_k^H$$

$$\mathbf{Q} = \mathbf{H}_k [\mathbf{I} - \mathrm{diag}(\tilde{\mathbf{b}}_k \tilde{\mathbf{b}}_k^H)] \mathbf{H}_k^H$$

$$\mathbf{z} = \mathbf{H}_k \tilde{\mathbf{b}}_k$$

The $KN - 1 \times 1$ vector $\tilde{\mathbf{b}}_k = (\tilde{b}_k^1 \quad \ldots \quad \tilde{b}_k^{KN-1})$ is the expectations of the transmit bits, and according to [9] one of its elements is given by

$$\tilde{b}_k^i = \tanh(\lambda_2^p[b_k^i]/2) \tag{13}$$

The optimal linear MMSE filter is given at the last section, but the output is still "hard", not iterativable "soft". The distribution of the residual interference-plus-noise at the output of a linear MMSE multiuser detector can be approximated by a Gaussian distribution. In what follows, we assume that the output of the soft instantaneous MMSE filter in (9) represents the output of an equivalent additive white Gaussian noise channel having as its input symbol. This equivalent channel can be represented as

$$\tilde{b}_k = \mu_k b_k + v_k \tag{14}$$

where μ_k is the equivalent output amplitude of the k-th channel, and $v_k \sim \mathcal{N}(0, \sigma_k^2)$ is a Gaussian noise sample. The μ_k can be computed as

$$\mu_k = \mathrm{E}(\hat{b}_k)/b_k = \mathbf{w}_k^H \mathbf{h}_k$$
$$\mathrm{E}(\hat{b}_k^2) = (\mathbf{w}_k^H \mathbf{h}_k)^2 + \sigma_v^2 \mathbf{w}_k^H \mathbf{w}_k$$
$$\sigma_k^2 = \mathrm{E}(\hat{b}_k^2) - [\mathrm{E}(\hat{b}_k)]^2 = \sigma_v^2 \mathbf{w}_k^H \mathbf{w}_k \tag{15}$$

Then, the outer information delivered by the soft instantaneous MMSE filter is

$$\lambda_1[\hat{b}_k] = \log \frac{P[\hat{b}_k | b_k = +1]}{P[\hat{b}_k | b_k = -1]} = -\frac{(\hat{b}_k - \mu_k)^2}{2\sigma_k^2} + \frac{(\hat{b}_k + \mu_k)^2}{2\sigma_k^2} = \frac{2\mu_k \hat{b}_k}{\sigma_k^2} \tag{16}$$

4   Simulation Results

This section shows the performance of our scheme where the information bits are modulated with the binary phase shift keying (BPSK). At the transmitter, each substream at each transmit antenna of each user of 256 information bits is independently encoded by a rate-1/2 convolution code generator (7,5). After spread by a length 2 spreading code, each substream is interleaved by interleavers chosen randomly. We synthesize the received signal using the measured channel characteristics and evaluate the performance of the proposed multiuser detection algorithm over a wide range of signal-to-noise ratio (SNR), and the VBLAST system is configured with fewer receive antennas than total transmit antennas.

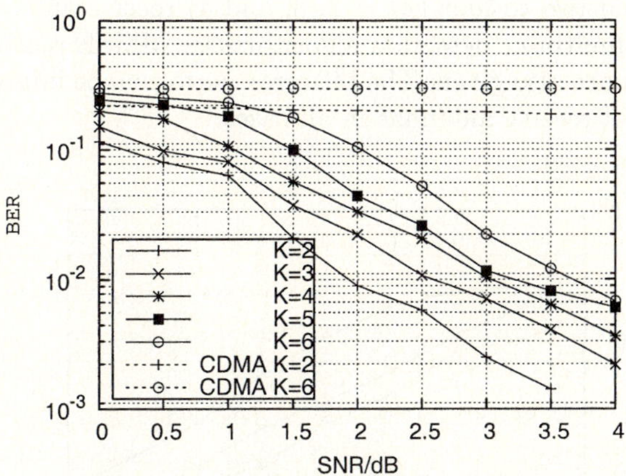

Fig. 3. BER versus SNR using SISO MMSE receiver

Simulation 1 — $M = 2$, $N = 2$, $K = 2,\ldots,6$: Fig.3 and 4 shows the bit-error rate performance of VBLAST coded MIMO multiuser detection for antenna configurations of two receive ($M = 2$) and from four to twelve ($N =$

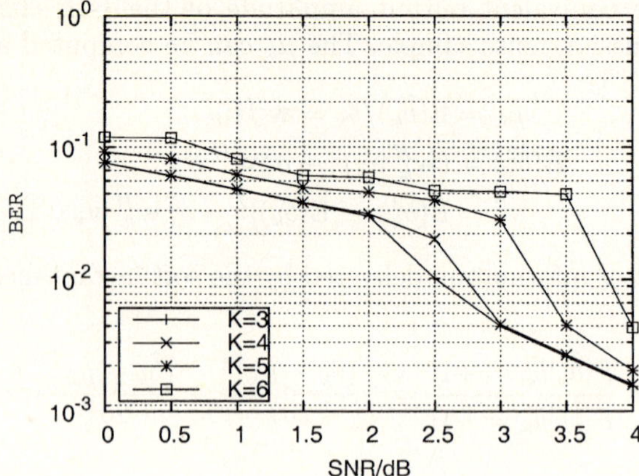

Fig. 4. BER versus SNR using RAKE Gaussian approach

$2, K = 2, \ldots, 6$) total transmit antennas. Two multiuser detection algorithms — MMSE receiver and Rake Gaussian Approach [5] are given. We use 10 iteratives for two multiuser detection algorithm. The bit-error performance improves with the decreasing number of transmit antennas in all cases. Although the total transmit antennas is six times of the receive ones, the performance is acceptable. In terms of VBLAST performance, a substantial gain in BER performance is realized with fewer transmit antennas.

Simulation 2 — $N = 2$, $K = 3$, $M = 2$, 3, and 4: Fig. 5 and Fig. 6 show the BER performance of proposed algorithm changes according to the different number of receive antennas. With antenna configurations of total six transmit ($N = 2$, $M = 3$) and two to four ($M = 2$, 3, and 4) receive antennas. Just like Simulation 1, Two multiuser detection algorithms — MMSE receiver and Rake Gaussian Approach are also given. The bit-error performance improves with the increasing number of receive antennas in all cases.

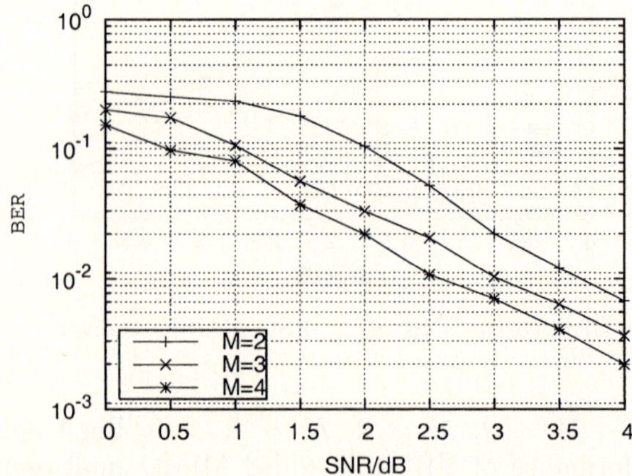

Fig. 5. BER versus the number of recieve antennas using SISO MMSE receiver

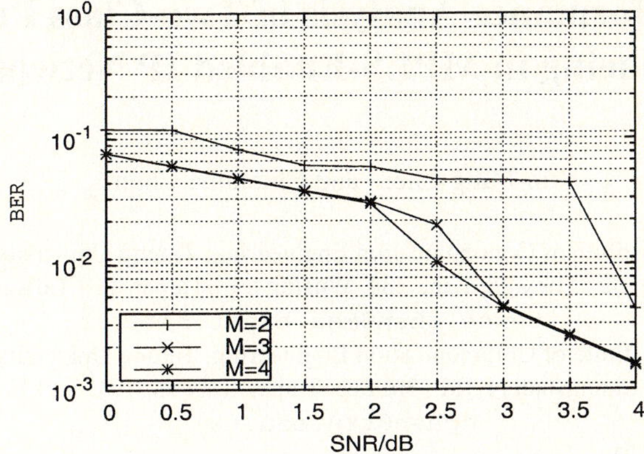

Fig. 6. BER versus the number of recieve antennas using RAKE Gaussian approach

Moreover, the performance of the conventional CDMA system is shown in Fig. 3. In these simulations, RSC encoder is omitted and we let $G = 4$ so that the code rate is the same. Random spectrum spread sequences are chosen for every antenna. At the receiver, the decorrelation multiuser detector and RAKE combine are exploited. From simulations, we observe that the performance of our scheme is much better than that of the conventional CDMA system.

5   Conclusions

In this paper, we studied the BLAST coded MIMO system and showed that the combination of BLAST and turbo principles provides a reliable and practical solution to high data-rate transmission for wireless communication. We demonstrated the performance of the BLAST coded MIMO system using turbo receiver with various antenna configurations, including the case of fewer receive antennas than transmit antennas in an flat-fading environment. The iterative detection decoding receiver improves the performance greatly.

References

1. Foschini, G. J.: Layered space-time architecture for wireless communication in a fading environment when using multi element antennas. Bell Labs Tech. J. **2**(1996) 41–59
2. Foschini, G. J., Golden, G. D., Valenzuela, R. A., Wolniansky, P. W.: Simplified processing for high spectral efficiency wireless communication employing multi-element arrays. IEEE J. Select. Areas Commun. **17**(1999) 1841–1852
3. Berrou, C., Glavieux, A.: Near optimum error- correcting coding and decoding: Turbo codes. IEEE Trans. Commun. 44(1996) 1261–1271
4. Sellathurai, M., Haykin, S.: TURBO-BLAST for Wireless Communications: Theory and Experiments. IEEE Trans. SP. **50**(2002) 2538–2543
5. Li, P., Liu, L., Leung W. K.: A Simple Approach to Near-Optimal Multiuer Detection: Interleave-Division Multiple-Access. IEEE Proc. WCNC 2003. 391-396
6. Wang, X. Poor, V.: Iterative (Turbo) soft interference cancellation and decoding for coded CDMA. IEEE Trans. Commun. **47**(1999) 1046–1061

The Performance Analysis of Two-Class Priority Queueing in MPLS-Enabled IP Network

Yun-Lung Chen[1] and Chienhua Chen[2]

[1] Graduate institute of Communication Engineering, Tatung University, Taiwan
2, Fl.-1, No. 52, Alley 28, Lane 284, Wusing ST., Taipei, 110 Taiwan, R.O.C.
`tense.chen@msa.hinet.net`
[2] Graduate institute of Communication Engineering, Tatung University, Taiwan
40, Chungshan N. Rd., 3rd Sec. Taipei, Taiwan, 104, R.O.C.
`cchen@ttu.edu.tw`

Abstract. This paper presents analysis of the performance modeling of the mean queue length and the mean queueing delay in MPLS (Multiprotocol Label Switching) LER (Label Edge Router) through the application of GE/G/1. We consider traffic with inter-arrival times of external arrivals at queue having generalised exponential (GE) distribution. The choice of the GE distribution is motivated that measurements of actual traffic or service times may be generally limited and only few parameters can be computed reliably and GE has a pseudo-momoryless property which makes the solution of queueing system and network analysis easily controlled or dealt with. From the analytical results we obtain the mean queue length and the mean queueing delay in a non-preemptive HOL (Head of Line) with two priority classes. We provide a framework including closed expressions and compare the mean queue lengths and the mean queueing delays at different traffic intensities. These numerical results can be useful for determining the amount of traffic entering from the external network for an MPLS LER system.

1 Introduction

Trend in business development in the broadband on-line service industry indicates that certain classes of customers, such as corporate subscriber may demand higher priority. From a report by Taiwan Network Information Center in January 2004, we can obtain the amount of time broadband users spend in on-line services (Fig. 1). It shows that those who spend more than 1 hour and less than 3 hours per year account for 42.1% of total subscribers. Therefore, the existing IP based ATM network at CT/ILEC (Taiwan's leading and largest ILEC) will not have enough bandwidth to satisfy its users' traffic demand and QoS needs. Therefore, there is urgent need to resolve effectively the inadequacy of edge node bandwidth. In addition, data published by Taiwan Network Information Center, inform us the ADSL on-line transmission rates selected by subscribers to be as follow: about 95% of subscribers (ADSL) select between 2Mbps and 8Mbps, about 4% (ADSL) 512 Kbps to 1Mbps, and the remaining 1% of the subscribers, including FTTB having selected other rates. We consider a two-class priority queue in an ingress MPLS LER, which handle two kinds of incom-

P. Lorenz and P. Dini (Eds.): ICN 2005, LNCS 3421, pp. 34–41, 2005.

ing traffic; one is ADSL traffic, and the other FTTB traffic. Because of the burstiness of the incoming traffic, we select a stable GE/G/1 queue with R ($\geq$2) priority classes under HOL scheduling disciplines. This paper we only discuss two-class priority queueing for FTTB corporate subscribers and ADSL general subscribers. Owing to current developments in Taiwan, we assign class 1 to incoming traffic of FTTB users and class2 to ADSL users.

The rest of this article is organized as follows: Section 2 describes the queueing model and derives a number of formulas. Section 3 presents the derived expression for the mean queue length and the mean queueing delay. Section 4 analyzes the numerical results for the two classes. The conclusion is given in Section 5.

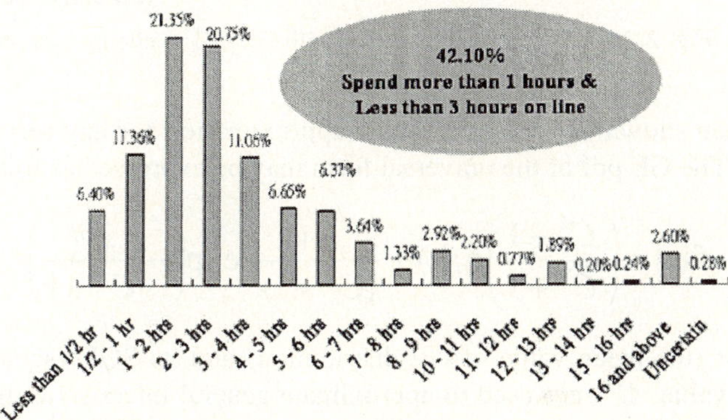

Fig. 1. The amount of time broadband users spent on-line services in Taiwan. Source: The report of Statistics Analysis Research and Consultancy Center, by Taiwan Network Information Center

2 System Model

Figure 2 shows a typical MPLS-Enabled IP network with queueing mechanism. In MPLS LER, various incoming traffic types arrive at the input port, e.g. 3G/WLAN, FTTB, ADSL, MOD, etc. As mentioned before, this paper is devoted to discussing a situation involving two types of traffic only. In order to identify the priority from the incoming packets, MPLS allows the precedence, or class of service (COS), to be inferred from the label, which represents the combination of a forward equivalent class (FEC) and a COS. In other words, FEC and COS can be used to identify the types of service [1].

In this paper, the performance analysis of MPLS LER system is completed with non-preemptive HOL GE/G/1 queues with two classes of jobs having been established, where arrivals are assumed to be of generalised distribution with mean rate λ_i, and the service rate μ_i which is slightly higher than, λ_i, i=1, 2. From Little's law, we have $\lambda_i = \mu_i \times \rho_i$ and $0 < \rho = \rho_1 + \rho_2 < 1$. The GE pdf (probability density function) is an extremal case of a family of distributions with a mixture of exponential models and can be interpreted as a CPP (Compound Poisson Process), where the squared coefficient of variation, $C^2 > 1$, which is a very important parameter for modeling process,

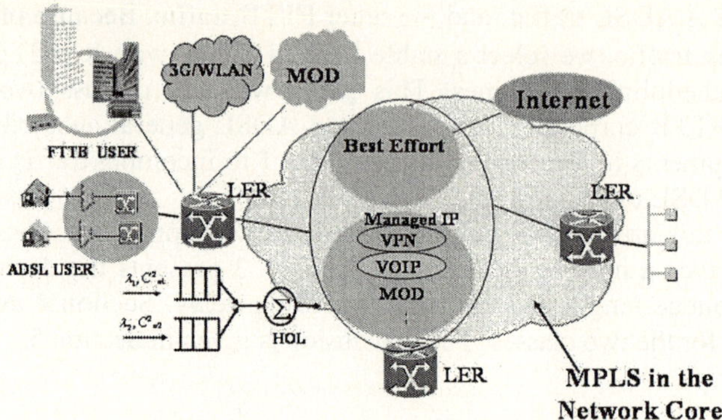

Fig. 2. MPLS-Enable IP network with GE/G/1 queueing system

it has also been shown that GE is a robust approximation for any service time distribution [2-3]. The GE pdf of the universal form may be expressed as follows :

$$f(t) = \left(\frac{C^2-1}{C^2+1}\right)\delta_0(t) + \frac{4v}{(C^2+1)^2}\exp\left(-\frac{2vt}{C^2+1}\right) \qquad t \geq 0$$

Where $1/v$ (the mean value of GE distribution) and its SQV (squared coefficient of variation) value, C^2, are used to approximate general interarrival time and service time with known first two moments, where C is the ratio of the standard deviation to the mean value. $\delta_0(t)$ is a unit impulse response. When the GE distribution model of the form is used to represent general service time distributions with μ and C_s^2 being the squared coefficient of variation, we find the mean value of service time $E(T)$ to be equal to $1/\mu$, by integrating $f(t)$ of equation (1)

$$E(T) = \int_0^\infty t\left(\frac{C_s^2-1}{C_s^2+1}\right)\delta_0(t)d(t) + \int_0^\infty \frac{4\mu t}{(C_s+1)^2}\exp\left(-\frac{2\mu t}{C_s^2+1}\right)dt = \frac{1}{\mu}$$

$$t \geq 0$$

3   Performance Analysis

In this paper, the mean queue length is defined as the mean number of FTTB and ADSL messages in the MPLS LER system, which could be as a real number other than an integer. Consider a stable GE/G/1 queue with R ($\geq$2) priority classes under HOL scheduling disciplines, i =1, 2, 3,....R. It is only assumed that class i jobs, i = 1, 2 (jobs in priority class 1 are given preference over jobs in class 2), arrive in the system according to a GE distribution with mean rate λ_i and squared coefficient of variation C_{ai}^2, and they are served by a single server with mean service rate μ_i, and squared coefficient of variation C_{si}^2, and we further assume that infinite queueing space is provided. We consider a queueing mechanism placed in an ingress MPLS

LER[4-5]. The ingress MPLS LER has an ATM STM-1 interface at its input link and a FE 100 Mbps interface at its output link. We use traffic data of FTTB and ADSL network from section 1, especially Figure 1. We can obtain the mean value of traffic from ADSL network and FTTB. The approximate value of the mean message length at link speed 100 Mbps is about 15143 bytes for FTTB and 15942 bytes for ADSL. These values are the ones used follow in a traffic model of guaranteed FTTB messages and non-guaranteed ADSL messages with a link speed of 100 Mbps. Thus we have defined the traffic parameters of proposed two-class traffic model for guaranteed FTTB message and non-guaranteed ADSL message for HOL.

3.1 Derivation Results of the Mean Queue Length

According to theorem 3 of Demetres Kouvatsos and Nasreddine Tabet-Aouvel [6]. For a stable GE/G/1 queue, with R ($\geq$2) priority classes under HOL scheduling disciplines, the exact marginal mean queue lengths $\{L_i\}$, i =1, 2,…R, are given by closed-form expressions. Thus, the derivation of the mean queue lengths per class i, L_i, i =1, 2 of a HOL GE/G/1 priority queue may be obtained, and the number of messages in the MPLS LER system can be derived from our two-class GE/G/1 queueing model.

The mean queue length L_{1HOL} and L_{2HOL} for HOL GE/G/1 priority queue can be obtained from the above-mentioned theorem 3. Then, the closed-form can be expressed as equation (2) and (3) respectively. Note that equations (2) and (3) for HOL GE/G/1 queues are strictly exact in a stochastic sense when C_{ai}^2 and $C_{si}^2 \geq 1$.

$$L_{1HOL} = \rho_1 + \frac{\rho_1}{2(1-\rho_1)}(C_{a1}^2 - 1) + \frac{1}{2(1-\rho_1)}\left[\rho_1^2(C_{s1}^2+1) + \frac{\lambda_1}{\lambda_2}\rho_2^2(C_{s2}^2+1)\right] \quad (2)$$

$$L_{2HOL} = \rho_2 + \frac{\rho_2}{2(1-\rho_1-\rho_2)}(C_{a2}^2 - 1) +$$

$$\frac{1}{2(1-\rho_1)(1-\rho_1-\rho_2)}\left[\rho_2^2(C_{s2}^2+1) + \frac{\lambda_2}{\lambda_1}\rho_1^2(C_{s1}^2+C_{a1}^2)\right] \quad (3)$$

3.2 Derived Results of Queueing Delay

The mean queueing delay from Little's formula. Let W_{1HOL} and W_{2HOL} be the mean queueing delay of guaranteed FTTB messages and non-guaranteed ADSL messages respectively for two-class GE/G/1 queueing model. We can then write $L_i = \lambda_i W_i$, where i = 1, 2. Thus we can derive the general result for the mean queueing delay of each i class.

4 Numerical Analysis

By MATLAB analysis, we observe L_i and W_i, i = 1, 2. In the numerical analysis, we get a set of performance curves and tables. For ease of illustration, we take four different combinations of C_{ai}^2 and C_{si}^2: 1 and 2, 2.5 and 5, 7 and 3, 3 and 7. These go together with $0 < \rho = \rho_1 + \rho_2 < 1$ and $\lambda_i = \mu_i \times \rho_i$, to form a set of curves and tables, such as the mean queue length L_i and the mean queueing delay W_i respectively, i =1,

2. Note that $\mu_1 = 10^8$ bits/s $\div$ (15143 bytes $\times$ 8 bits/bytes) = 825.464 s^{-1} for guaranteed FTTB messages and $\mu_2 = 10^8$ bits/s $\div$ (15942 bytes $\times$ 8 bits/byte) = 784.092 s^{-1} for non-guaranteed ADSL messages.

There are three significant numerical are discussed for the following categories $\rho_1 < \rho_2, \rho_1 = \rho_2, \rho_1 > \rho_2$. With a given difference between ρ_1 and ρ_2 selected at 0.2 (except for points near the origin) for the sake of illustration. Note that the suffixes "HOL1, HOL2, HOL3, and HOL4" in the curves and tables, correspond to four different combinations of C_{ai}^2 and C_{si}^2, i = 1, 2; i.e. 1 and 2, 2.5 and 5, 7 and 3, 3 and 7 respectively.

Example1: $\rho_1 < \rho_2$ (ρ_1 : 0.001~0.395, ρ_2 : 0.021~0.595)

We find that the curve of guaranteed FTTB messages increases slowly for L_{1HOL1} ~ L_{1HOL4} and W_{1HOL1} ~ W_{1HOL4}, approaches a piecewise straight line which as ρ_1 and ρ_2 increase. The curve of non-guaranteed ADSL messages for L_{2HOL1} ~ L_{2HOL4} and W_{2HOL1} ~ W_{2HOL4} rises gradually for lower values of ρ_1 and ρ_2 (i.e. combination values of ρ_1 and ρ_2). However, it increases steeply at higher values of ρ_1 and ρ_2 . For ease of comparison of guaranteed FTTB messages and non-guaranteed ADSL messages for HOL, we decide to use logarithmic scale on the vertical axis of L_i and W_i for HOL, i =1, 2 (Fig. 3 and Fig. 4). We find an interesting phenomenon: the mean queue lengths of L_{1HOL4} are smaller than L_{1HOL3} at all values of ρ_1 and ρ_2 except for higher values of ρ_1 and ρ_2 (0.395, 0.595) where L_{1HOL4} increases steeply starts to exceed L_{1HOL3}, their corresponding values being 19.678 and 19.399 at levels of C_{ai}^2 and C_{si}^2 at 3 and 7 and 7 and 3 respectively, i = 1, 2. Note that L_{2HOL4} is smaller than L_{2HOL3} at all values of ρ_1 and ρ_2 . Owing to space limitation, MATLAB-produced values are shown in the table only for L_{iHOL3} and L_{iHOL4}, i=1 ,2.

Same phenomenon occurs in W_{1HOL1} ~ W_{1HOL4}, i.e. the mean queueing delay of W_{1HOL4} starts to exceed W_{1HOL3}, their corresponding values being 0.060, 0.059 at levels of C_{ai}^2 and C_{si}^2 at 3 and 7, i = 1, 2. We also find that the value of L_{1HOL4} increases slowly in proportion to the values of C_{ai}^2 and C_{si}^2 as they increase. Once the difference between C_{si}^2 and C_{ai}^2 increases, the mean queue length of L_{1HOL4} slowly becomes larger than L_{1HOL3}. For example, when C_{ai}^2 and C_{si}^2 approaches 3 and 7.5 respectively, the mean queue length of L_{1HOL4} starts to exceed L_{1HOL3} for four additional ρ_1 and ρ_2 combination values; namely (0.38, 0.58), (0.36, 0.56), (0.34, 0.54) and (0.32, 0.52). If we further increase C_{si}^2 (e.g. from 7.5 to say 8), we expect the transition point of the mutual relationship L_{1HOL4} and L_{1HOL3} to shift toward the left of the table 1. From the MATLAB, we obtain the value of L_{1HOL4} as 13.877 at ρ_1 and ρ_2 values of 0.32 and 0.52, which starts to exceed L_{1HOL3} (13.655).

Similarly our calculation shows that W_{1HOL4} (0.052) is higher than W_{1HOL3} (0.051) at ρ_1 and ρ_2 values of 0.32 and 0.52. Note that L_{2HOL4} slowly starts to exceed L_{2HOL3} following C_{si}^2 increases to 7.5 from 7, where L_{2HOL4} starts to exceed L_{2HOL3}, their corresponding values being 195.77 and 190.81 at levels of C_{ai}^2 and C_{si}^2 at 3 and 7.5, i = 1, 2 and ρ_1 and ρ_2 values of 0.34 and 0.54.

Similarly, where W_{2HOL4} starts to exceed W_{2HOL3}, their corresponding values being 0.462 and 0.450 at levels of C_{ai}^2 and C_{si}^2 at 3 and 7.5, i = 1, 2, and ρ_1 and ρ_2 values of

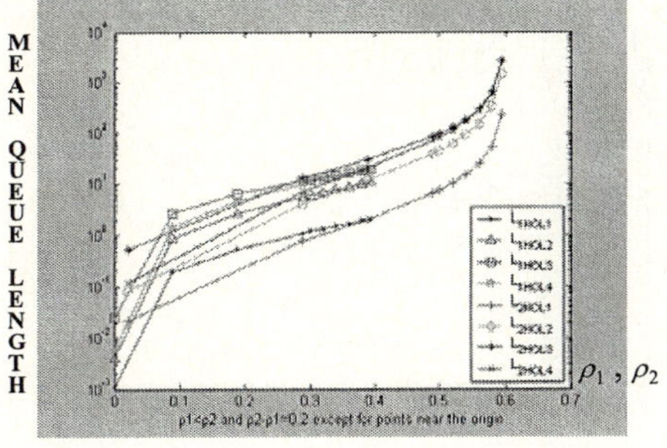

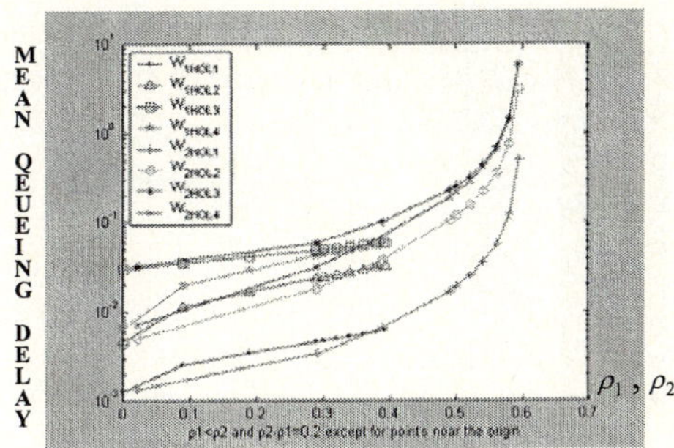

Fig. 3. Comparison of mean queue lengths $L_{1HOL1} \sim L_{1HOL4}$ and $L_{2HOL1} \sim L_{2HOL4}$ at $\rho_1 < \rho_2$

Fig. 4. Comparison of mean queuing delays $W_{1HOL1} \sim W_{1HOL4}$ and $W_{2HOL1} \sim W_{2HOL4}$ at $\rho_1 < \rho_2$

Table 1. At different traffic intensities $\rho_1 < \rho_2$ and $\rho_2 - \rho_1 = 0.2$

except for points near the origin   unit: messages/sec

ρ_1 :	0.001	0.09	0.19	0.29	0.3	0.32	0.34	0.36	0.38	0.395
ρ_2 :	0.021	0.29	0.39	0.49	0.5	0.52	0.54	0.56	0.58	0.595
L_{2HOL3}	0.539	13.535	30.911	86.690	98.943	133.08	190.81	307.65	661.21	2791.3
L_{1HOL3}	0.025	2.659	6.524	11.739	12.357	13.655	15.044	15.531	18.125	19.399
L_{2HOL4}	0.119	7.162	21.282	72.887	84.657	117.78	174.44	290.15	642.50	2771.6
L_{1HOL4}	0.006	1.463	4.650	10.153	10.868	12.407	14.101	15.963	18.010	19.678

0.34 and 0.54. With above mentioned, the choice of C_{ai}^2 and C_{si}^2 is very important for performance analysis and design.

Example 2: $\rho_1 = \rho_2$ (ρ_1 : 0.001~0.495, ρ_2 : 0.001~0.495)

From Fig.5 and Fig. 6, we find that there are the same phenomena as in Fig. 3 and Fig. 4. When the mean queue length of non-guaranteed ADSL messages rises gradually to reach a similar threshold level of ρ_1 and ρ_2 (i.e. $\rho_1 = \rho_2$) at about 0.48, the mean queue length increases steeply at values of ρ_1 and ρ_2 of 0.495 and beyond, where L_{1HOL4} slowly starts to exceed L_{1HOL3}, their corresponding values being 29.316 and 29 at levels of C_{ai}^2 and C_{si}^2 at 3 and 7 and 7 and 3 respectively. Same phenomenon occurs in $W_{1HOL1} \sim W_{1HOL4}$, e.g. the mean queueing delay of W_{1HOL4} starts to exceed W_{1HOL3}, their corresponding values being 0.072, 0.071. When C_{ai}^2 and C_{si}^2 approaches 3 and 7.5 respectively, from the MATLAB, we obtain the value of L_{1HOL4} as 21.188 at ρ_1 and ρ_2 values of 0.42, which starts to exceed L_{1HOL3} (20.921).

Example 3: $\rho_1 > \rho_2$ (ρ_1 : 0.021~0.595, ρ_2 : 0.001~0.395)

Figure 7 and Figure 8 show the same phenomena to occur as $\rho_1 = \rho_2$ or $\rho_1 < \rho_2$. Worthy of mention, from the point of view of traffic intensity, we find that the mean

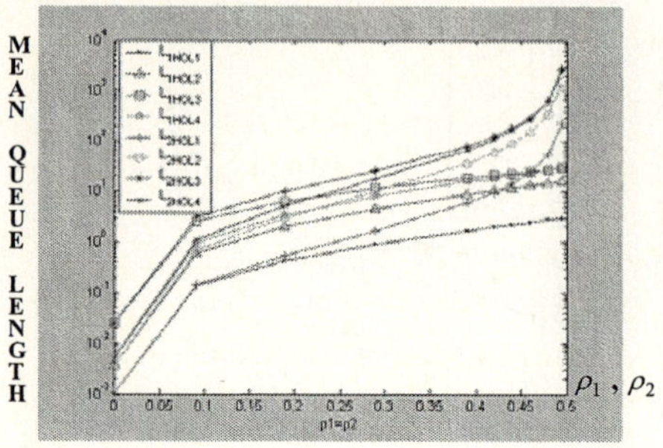

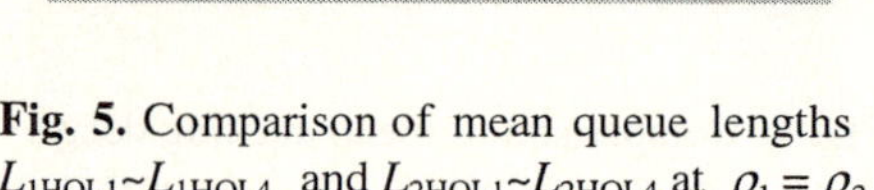

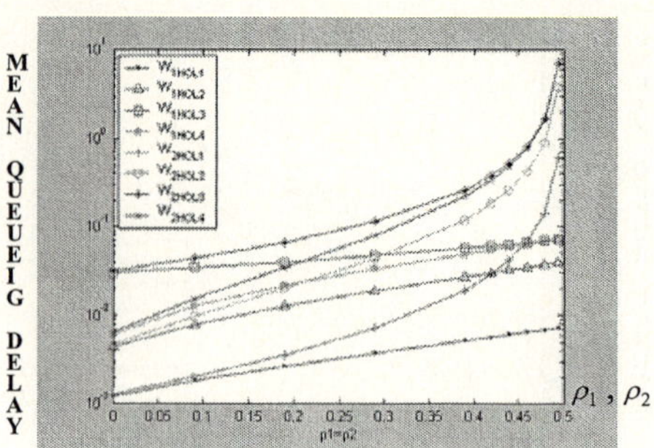

Fig. 5. Comparison of mean queue lengths L_{1HOL1}~L_{1HOL4} and L_{2HOL1}~L_{2HOL4} at $\rho_1 = \rho_2$

Fig. 6. Comparison of mean queuing delays W_{1HOL1}~W_{1HOL4} and W_{2HOL1}~W_{2HOL4} at $\rho_1 = \rho_2$

queue length of non-guaranteed ADSL messages is shortest among the three examples, both the mean queueing delay of non-guaranteed ADSL messages and guaranteed FTTB messages are the largest of the three. For example, we compare $\rho_1 > \rho_2$ (0.595, 0.395) with $\rho_1 < \rho_2$ (0.395, 0.595): The mean queue length of $\rho_1 > \rho_2$ at four different levels of C_{ai}^2 and C_{si}^2, we obtain $L_{1HOL1} \sim L_{1HOL4}$ and $L_{2HOL1} \sim L_{2HOL4}$ values of (4.307, 23.757, 43.280, 43.598) and (234.51, 1466.2, 2739.6, 2720.1) respectively. Similarly, we obtain corresponding $W_{1HOL1} \sim W_{1HOL4}$ and $W_{2HOL1} \sim W_{2HOL4}$ values of (0.0087, 0.048, 0.088, 0.089) and (0.758, 4.734, 8.845, 8.782) respectively. Returning to the example 1, we obtain $L_{1HOL1} \sim L_{1HOL4}$ and $L_{2HOL1} \sim L_{2HOL4}$ values of (2.062, 10.788, 19.399, 19.678) and (239.14, 1494.1, 2791.3, 2771.6) respectively, and $W_{1HOL1} \sim W_{1HOL4}$ and $W_{2HOL1} \sim W_{2HOL4}$, their values are (0.0063, 0.033, 0.059, 0.0060) and (0.513, 3.202, 5.983, 5.941) respectively. In other words, once this kind of incoming traffic occurs (i.e. $\rho_1 > \rho_2$), we must consider the impact on the mean queue lengths and the mean queueing delays.

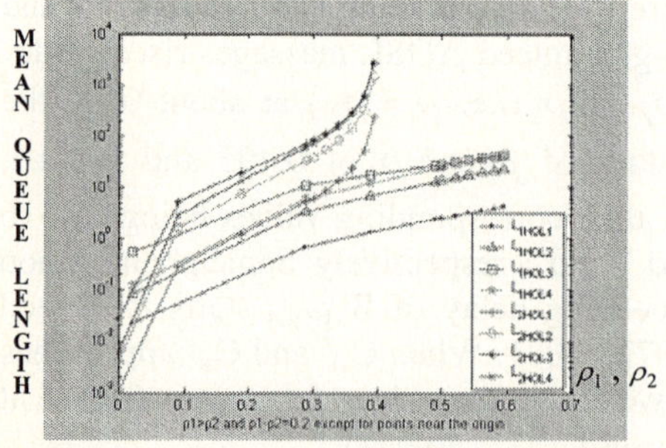

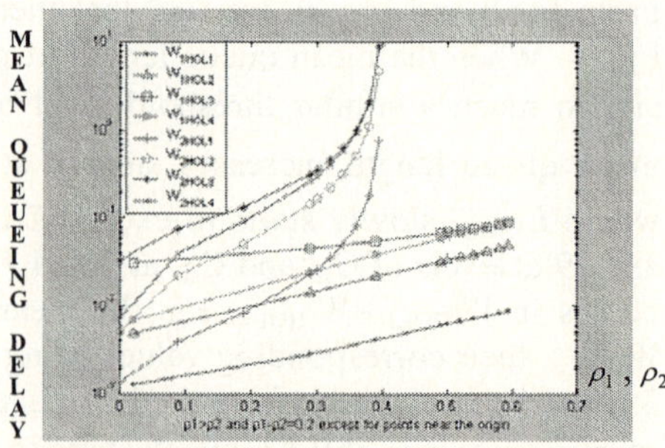

Fig. 7. Comparison of mean queue lengths L_{1HOL1}~L_{1HOL4} and L_{2HOL1}~L_{2HOL4} at $\rho_1 > \rho_2$

Fig. 8. Comparison of mean queuing delays W_{1HOL1}~W_{1HOL4} and W_{2HOL1}~W_{2HOL4} at $\rho_1 > \rho_2$

5 Conclusion

This paper is focused on providing an analysis of traffic intensity in a two-class non-preemptive priority queue for HOL. Using the results of performance analysis may lead to a guideline reference for designing a queueing mechanism in MPLS LER system. Further studies on the subject may include discussion of a scheduling discipline for a stable GE/G/1 queue with R ($\geq$2) priority classes under PR (Preemptive Resume). Besides, the performance analysis of multimedia service with multiclass priority based on the maximum entropy technique could be feasible topic.

References

[1] F. Le Faucheur, et al., "MPLS Support of Differentiated Services," RFC3270, May 2002.

[2] Demetres Kouvatsos, "Entropy maximization and queueing network models," Annals of Operation Research 48, pp. 63-126, 1994.

[3] Demetres Kouvatsos, "A maximum entropy analysis of the G/G/1 at equilibrium," J. Opl. Soc. Vol. 39, pp. 183-200, 1988.

[4] Gang UK Hwang and Khosrow Soharby, " Modelling and analysis of a buffer in an ATM-based MPLS LER system," CQRS-06-3, Taipei, Taiwan, Globecom. 2002.

[5] Hooi Miin Soo and Jong-Moon Chung, "Analysis of non-preemptive priority queueing of MPLS networks with bulk arrivals," IEEE, pp. 81-83, 2002.

[6] Demetres K. and Nasreddine T. "A maximum entropy priority approximation for a stable G/G/1 queue," Acta Information 27, pp. 247-286, 1989.

Constraint Based LSP Handover (CBLH)
in MPLS Networks

Praveen Kumar, Niranjan Dhanakoti, Srividya Gopalan, and V. Sridhar

Applied Research Group, Satyam Computers Services Limited,
#14, Langford Avenue, Lalbagh Road, Bangalore 560025 India
Phone: +91 80 2223 1696 Fax: +91 80 2227 1882
{Praveenkumar_GS, Niranjan_Dhanakoti,
Srividya_Gopalan, Sridhar}@satyam.com
http://www.satyam.com

Abstract. Multi Protocol Label Switching (MPLS) has become an important technology for Internet Service Providers (ISPs). MPLS features provide flexible traffic engineering solutions and Quality of Service (QoS) support for high priority traffic. This has resulted in an increasing interest in MPLS reliability for providing efficient restoration with guaranteed QoS. In this paper, we propose constraint based local path restoration technique in MPLS called *Constraint Based Label Switched Path (LSP) Handover (CBLH)* to dynamically reroute traffic around a failure or congestion in an LSP. The proposed scheme reduces the restoration time by QoS based Forward Equivalence Class (FEC), LSP classification and satisfies the traffic constraints like delay, packet loss and utilization for different Class of Service (CoS). Through simulation, the performance of the proposed scheme is measured and compared with existing schemes.

1 Introduction

Multi Protocol Label Switching (MPLS) is a link layer packet forwarding technology, which was proposed as a standard in [10]. The packet forwarding technology is based on a short, fixed length *label* in the packet header also called *shim header* and is now being adopted widely in IP backbones and other networks as well. MPLS enhances source routing and allows for certain techniques used in circuit switching through extensions to Label Distribution Protocol (LDP) [1] called Constraint based Routing - LDP (CR-LDP) [2] and RSVP-TE [3].

MPLS features like grouping all the packets/flows to a particular destination into an FEC and source routing provides an efficient traffic engineering solution and in effect enables QoS routing. In recognition of this, in our previous work [8], we proposed a four-stage mapping of the incoming IP flows on to the labels and *Constraint Based LSP Selection (CBLS)* algorithm to select LSP based on CoS at network edge, i.e., at the Label Edge Router (LER). As an extension to CBLS, we present *CBLH* to provide protection and recovery in case of link failure at the elements within the network, i.e., at the Label Switched Router (LSR) to facilitate reliable network operation.

P. Lorenz and P. Dini (Eds.): ICN 2005, LNCS 3421, pp. 42–49, 2005.

1.1 Motivation

In the Internet, multimedia applications are growing at a faster pace than the orthodox data applications. Current surveys estimate the media traffic to surpass the data traffic in the near future. As a result, a single link failure in a system may result in disruption of large number of real-time services. Since most of these services would involve time sensitive multimedia applications, it is most important for the Internet Service Providers (ISPs) to provide QoS, reliable transmission, cost and time efficient recovery mechanisms to these applications.

1.2 Objective

The main goals of LSP Handover mechanism are –

- Focus on facilitating LSP restoration that satisfies the end-to-end constraints of the traffic like packet loss, delay and bandwidth after handover.
- Should be subject to traffic engineering goals, of optimal allocation and use of network resources based on sub-FEC [8], which is classification of FEC based on CoS.
- Achieve restoration times comparable to the delay tolerance limits of time sensitive media applications.

This paper is organized as follows; Section 2 discusses some of the techniques outlined by IETF [12] for protection in MPLS networks and other related papers. Section 3 describes the proposed approach and Section 4 details the test environment and tools used in evaluating the proposed approach. Sections 5 and 6 discuss and summarize the measurements with final comments.

2 Related Work

RSVP-TE and CR-LDP are the two main signaling protocols to support QoS based LSP setup in MPLS networks. RSVP-TE is an extension of resource reservation protocol to support traffic engineering with fast reroute [9] and CR-LDP is an extension of LDP to set up constraint based LSPs.

Based on these configuration protocols, there are two basic path restoration models namely *protection switching* and *rerouting* with the scope of recovery being either *global* or *local repair* as described in [11]. Protection switching recovery mechanisms pre-establish a recovery path or path segment, based on network routing policies, the restoration requirements of the traffic on the working path, and administrative considerations. Rerouting involves establishing new paths or path segments on demand for restoring traffic after the occurrence of a fault based on fault information, network routing policies, pre-defined configurations and network topology information. In case of global repair as shown in figure 1, the idea is to provide an end-to-end backup path for each active path, which is link and node disjoint from the active path. However, the drawback with this approach is that when there is a failure, the information has to propagate back to the source, which in turn switches the traffic on to the backup path. Also, this scheme does not address any traffic constraints and is prone to delay and non-optimal resource utilization.

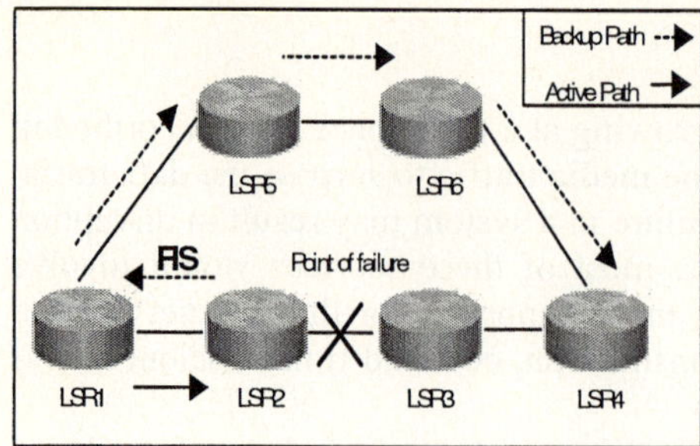

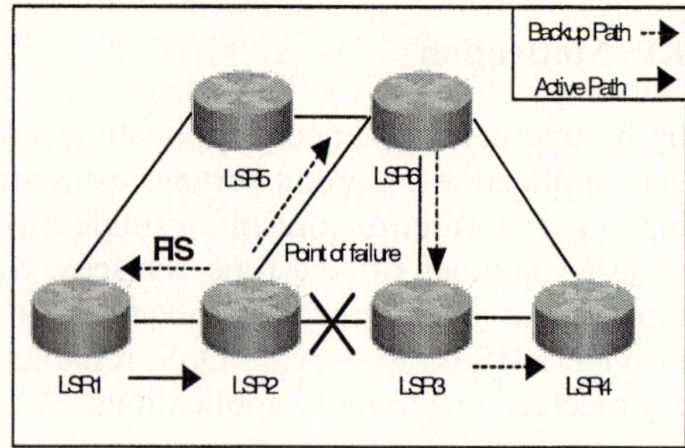

Fig. 1. Global Repair Model   **Fig. 2.** Local Repair Model

Local restoration schemes reroutes the traffic at the point of failure locally as shown in figure 2. [4] Is based on local restoration model and show that local repair has comparatively better utilization than the global repairs, but delay and loss from the point of restoration to the destination is not assured. CBLH addresses both restoration time and end-to-end QoS constraints by minimizing the delay, loss and maximizing bandwidth and utilization. The backup LSP search space is considerably reduced due to FEC and LSP sub-classification based on CoS as compared to the other schemes. The current network status is taken into consideration during selection of the backup LSP to dynamically facilitate reliable network operation.

3   CoS Based LSP Handover (CBLH)

The proposed CBLH scheme is based on local LSP handover with dynamic rerouting within the network. Following assumptions are made for CBLH:

- Alternate paths for the same destination (sub-FEC) exist at each hop (LSR).
- A source-destination pair has atleast two disjoint paths (LSPs) between them.
- Only single link failures are assumed, as protection against multiple link failures needs multiple backups, which is more complex and is too expensive in backup resource requirements in real network scenarios [6], [7].
- CBLH does not provide a backup path if the ingress or the egress of the traffic fails.

In our previous work [8], we had proposed a four-stage mapping of the incoming IP packet to the label in the control plane for creation of sub-FEC and sub-LSP and CoS Based LSP Selection (CBLS) algorithm for dynamic selection of sub-LSP. The mapping and the CBLS algorithm was executed at the network edge i.e., Label Edge Router (Ingress) to select LSP based on CoS. The four-stage mapping in the control plane was performed as described below:

1) IP-to-FEC Map   2) FEC-to-CoS Map
3) FEC-to-LSP Map   4) LSP-to-Label Map

The CBLS algorithm dynamically selects an optimally best LSP, based on the traffic type, its constraints like delay, loss, reliability requirements and network load conditions. In continuation to CBLS at the ingress, the CBLH provides CoS based recovery of the traffic within the network. The CBLH is based on CoS requirements, which are classified into three distinct classes as shown:

Class 1: Real - time (voice, interactive video)
Class 2: Semi-real-time (Streaming Video, network management, Business interactive, Oracle, People Soft, SAP, Telnet, etc)
Class 3: Best-effort data (E-mail, Web Browsing)

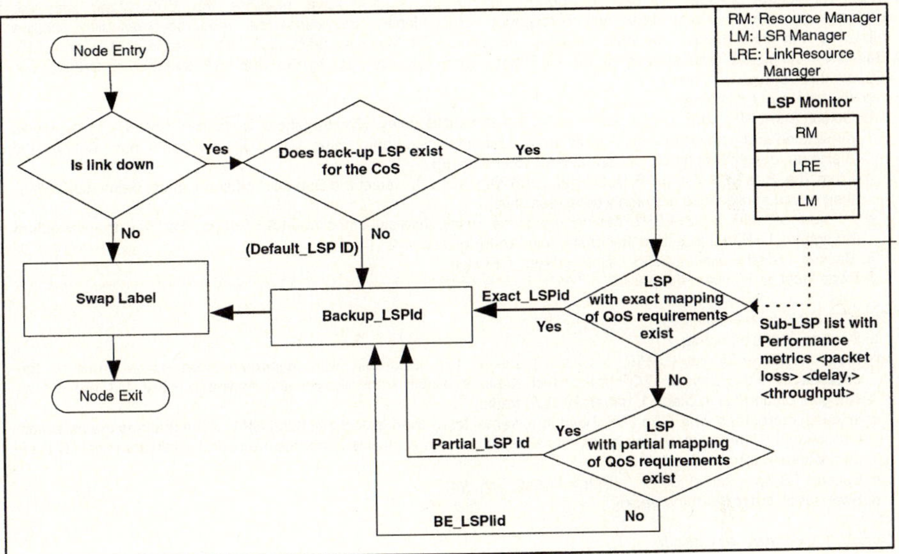

Fig. 3. Path Restoration Procedure Based on CBLH

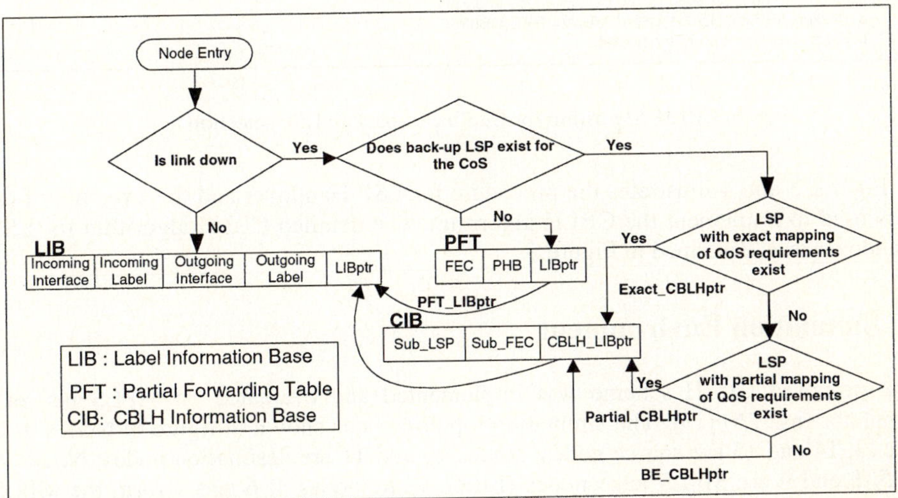

Fig. 4. Structure of Tables for CBLH

CBLH Algorithm to determine CoS based backup LSP for path restoration

Input: LSPid, CoS and FEC.
Output: Backup_LSPid.

Steps:
1. Upon link failure, Fault Indication Signal (FIS) is sent to all the neighboring nodes and further propagated to the edge nodes (LER) to indicate the failure and stop traffic towards the failed link.
2. For the failed LSP, check whether backup Sub-LSP exists for the particular CoS and FEC.
3. The LSP Monitor provides the backup Sub-LSP list {LSP[i]; $1 \le i \le n$ } that matches the CoS along with the performance metrics, Delay, Loss and Throughput. LSP Monitor maintains the resource availability across all LSRs and links through periodic updates from the LSR Manager (LM) and Link Resource Manger (LRE).
4. Based on constraint requirements of the CoS type, map the Sub-FEC to the Sub-LSP and then on to the Label.

    Case1: If (CoS_type == Class1)
    a. Perform exact mapping;
    b. Select Sub-LSPs with $Delay \le Delay\_Thresh$, the threshold delay. (The threshold is defined based on the media application and delay tolerance. For example, real-time voice traffic using ITU-T G711 vocoder has defined the maximum delay for optimum performace as 150 msec [6])
    c. From the Sub-LSP list {LSP[1], LSP[2].....LSP[k]; $k \le n$ }, select the Sub-LSP with minimum delay (since real-time media applications are highly delay-sensitive).
    d. In case of multiple Sub-LSP's having the same delay, then select the Sub-LSP LSP[m] that has the maximum throughput LSP[m].thput and minimum loss LSP[m].loss.
    e. Backup_LSPid = Exact_LSPid; LIBptr = Exact_CBLHptr;
    f. Swap label and forward the packet.

    Case2: If (CoS_type == Class2)
    a. Perform partial mapping;
    b. From the Sub-LSP list {LSP[i]; $1 \le i \le n$ }, select the Sub-LSP with minimum loss (since most of the applications in this class are TCP traffic which requires reliable transmission and minimum delay, but are not as stringent on time as in Class 1, generally UDP traffic).
    c. In case more than one Sub-LSP have the same loss, then select the Sub-LSP LSP[m] that has the maximum throughput LSP[m].thput. (Delay is not critical as long as the acknowledgements are bound within the initial RTT and do not lead to retransmissions).
    d. Backup_LSPid = Partial_LSPid; LIBptr = Partial_CBLHptr;
    e. Swap label and forward the packet.

    Case3: If (CoS_type == Class3)
    a. Perform Best effort mapping;
    b. From the Sub-LSP list {LSP[i]; $1 \le i \le n$ }, select the Sub-LSP LSP[m] based on the current network dynamics with no specific constraints that need to be satisfied.
    c. Backup_LSPid = BE_LSPid; LIBptr = BE_CBLHptr;
    d. Swap label and forward the packet.
5. If backup LSP does not exist for the particular CoS and FEC
    a. Backup_LSPid = Default_LSPid; LIBptr = PFT_LIBptr;
    b. Swap label and forward the packet.

Fig. 5. CBLH Algorithm for CoS based backup LSP selection

Figures 3 and 4 illustrates the procedure for LSP Handover and the structure of tables used to implement the CBLH algorithm. The detailed CBLH algorithm for LSP Handover is as described in Figure 5.

4   Simulation Environment

The proposed CBLH scheme was implemented and evaluated on MPLS network simulator (ns2.26) [13]. The simulation topology is as shown in figure 6. Nodes 0, 1, 12, 13, 14 and 15 are source nodes; Nodes 10 and 11 are destination nodes; Nodes 2, 3, 5, 8 and 9 are MPLS edge nodes (LER), while nodes 4, 6 and 7 form the MPLS core network (routers). Each node in the core is connected using a duplex link with 10millisecond delay and 100Mbps bandwidth. The access nodes are connected with a relatively higher delay (15 msec and 5Mbps bandwidth). All the nodes implement FIFO queue. To evaluate the proposed approach in a heterogeneous traffic environment, we have multiple sources, each belonging to different CoS and consequently having varied requirements. Nodes 0 and 1 generate voice traffic (CBR) over UDP at the rate of 64Kbps (PCM) and 32Kbps (ADPCM) respectively, Node 12 and 13 generate bursty traffic (VBR) to make the traffic arrival at the edge random and Node 14 and 15 are FTP sources over TCP connections, used to generate bulky background traffic for slowly loading the network.

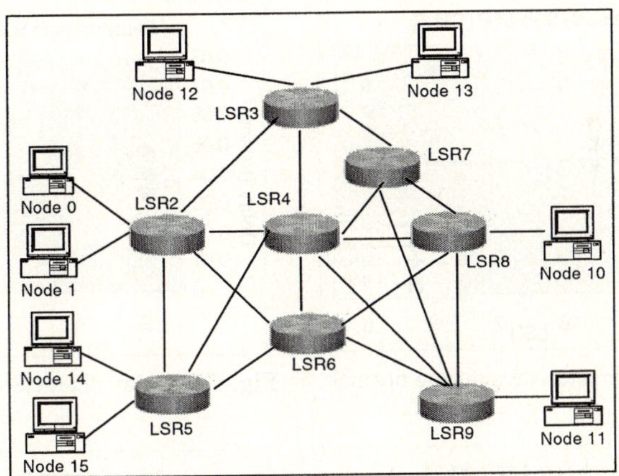

Fig. 6. Simulation Environment

5   Performance Evaluation

The objective of this simulation is basically to evaluate the performance of the proposed CBLH algorithm in case of failure with the normal operation (CBLS). This is basically to evaluate how close the recovery mechanism is to satisfy the constraints of different traffic classes based on CoS. To evaluate this scenario, we consider real time traffic like voice, to compare the performance between CBLS (normal operation) and CBLH (restoration) for various traffic constraints like delay, loss and utilization. In the case of voice termination at node 11 from node 1, there are six sub-LSPs available for voice termination. The six sub-LSPs are:

1) LSP91: 2_4_9   2) LSP92: 2_4_6_9   3) LSP93: 2_4_7_9
4) LSP94: 2_4_8_9   5) LSP95: 2_5_6_9   6) LSP96: 2_3_7_9

Since LSR4 is the common and critical node for all voice sub-LSPs, we inject fault (link failure) between LSR4 and LSR9 i.e. on LSP91. Figures 7a, 7c and 7e show CBLS operations without link failure. Figures 7b, 7d and 7f show CBLH operations upon *failure of LSP91*. The various performance results in this scenario are discussed in detail below.

A. Figures 7a and 7b compare the end-to-end delay characteristics for CBLS and CBLH respectively as described in scenario one in section 5.0. It can be observed that in case of failure of LSP91, the delay in the *CBLH restored path LSP93* is

48   P. Kumar et al.

132 msec, which is well within the permissible level defined for VoIP by ITU-T G711 standard [5].

B. Figures 7c and 7d compare the relative packet loss characteristics for CBLS and CBLH respectively. It can be observed that the net loss in CBLH restored path is around 2.8%, whereas in case of CBLS path LSP91 it is 2.13%. The higher loss percentage in case of CBLH can be attributed to the congestion at LSR4.

C. From figures 7e and 7f, it can be inferred that the utilization of the available bandwidth during normal operation is 84.5%, while in case of failure and conse-

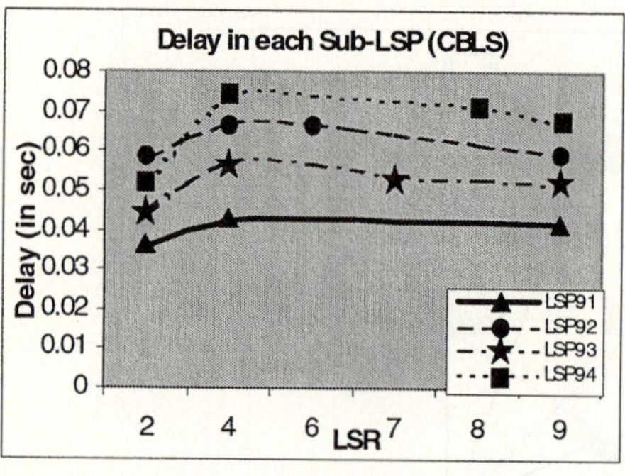

Fig. 7a. Delay characteristics during normal operation

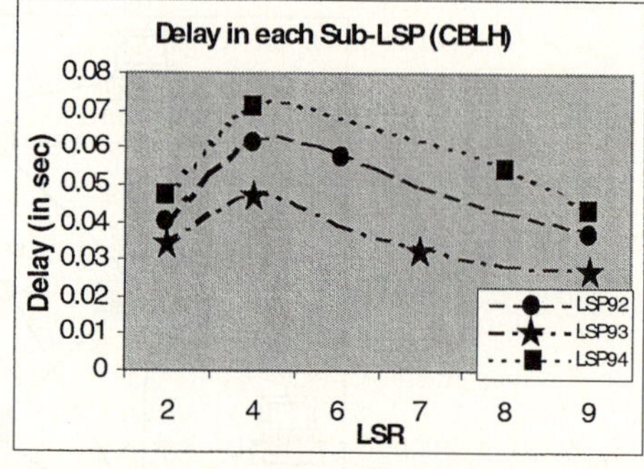

Fig. 7b. Delay characteristics during recovery

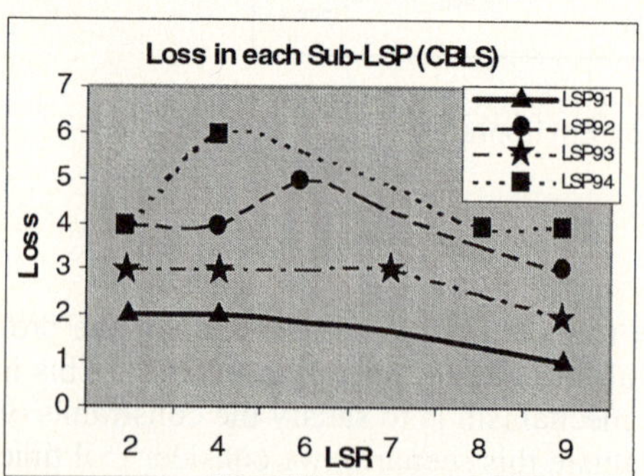

Fig. 7c. Loss characteristics during normal operation

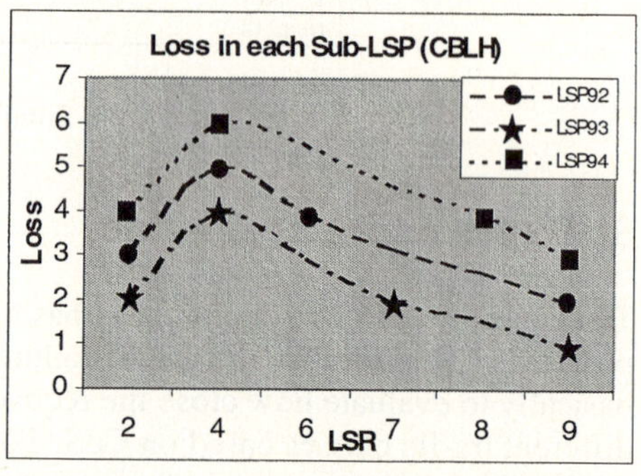

Fig. 7d. Loss characteristics during recovery

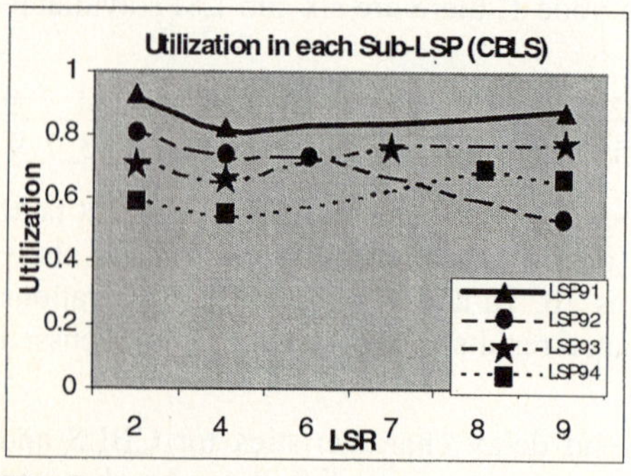

Fig. 7e. Utilization characteristics during normal operation

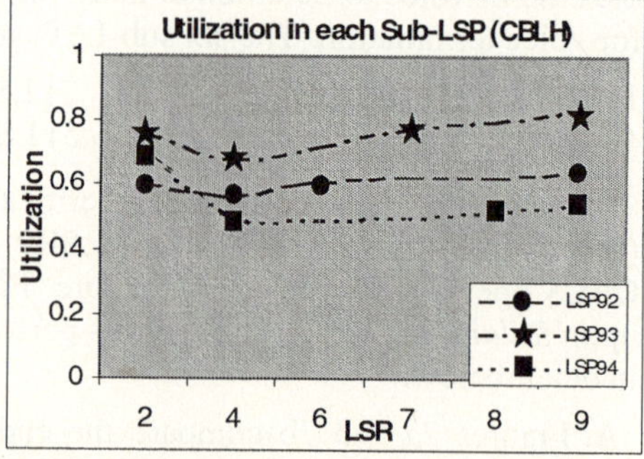

Fig. 7f. Utilization characteristics during recovery

quent restored path is 79.5%, which is comparably better than the other restoration schemes discussed in scenario 2.

6 Conclusion and Future Work

In this paper we presented the CBLH algorithm for dynamic and local restoration of LSP in case of failure in MPLS networks. The proposed scheme provides faster restoration time by CoS based FEC and LSP classification and satisfies the end-to-end traffic constraints like packet loss, delay and utilization in the restored path. In the simulation results, we showed that the CBLH algorithm effectively satisfies the requirements of the incoming traffic by selecting the most optimal path among the multiple backup sub-LSPs. In future, we plan to address multiple link failure scenarios by having multiple backup paths.

References

1. L. Andersson, P. Doolan, N. Feldman, A. Fredette and B. Thomas, "LDP Specification", Jan. 2001, Network Working Group, Request for Comments (RFC) 3036, http://www.ietf.org/rfc/rfc3036.txt.
2. J. Ash, M. Girish, E. Gray, B. Jamoussi and G. Wright, "Applicability Statement for CR-LDP", Jan. 2002, Network Working Group, Request for Comments, (RFC) 3213 http://www.ietf.org/rfc/rfc3213.txt.
3. Awduche, L. Berger, D. Gan, T. Li, V. Srinivasan and G. Swallow, "RSVP-TE: Extensions to RSVP for LSP Tunnels", Dec. 2001, Network Working Group, Request for Comments (RFC) 3209, http://www.ietf.og/rfc/rfc3209.txt.
4. Calle, J. L Marzo and A. Urra, "Protection Performance Components in MPLS Networks", in Computer Communications Journal, 2004.
5. S. Christensen, Voice over IP Solutions, White Paper, Juniper Networks, www.juniper.net.
6. M. Kodialam and T.V.Lakshman, "Minimum Interference Routing with Applications to MPLS Traffic Engineering", in the Proceedings of INFOCOM, 2000.
7. M. Kodialam and T. V. Lakshman, "Dynamic Routing of Locally Restorable Bandwidth Guaranteed Tunnels using Aggregated Link Usage Information", Appeared in Proceedings of INFOCOM 2001, April 22-26 2001, Anchorage, Alaska.
8. P. Kumar, N. Dhanakoti, S. Gopalan and V. Sridhar, "CoS Based LSP Selection in MPLS Networks", in Proceedings of 7th IEEE International Conference on High Speed Networks and Multimedia Communications (HSNMC), Vol. 3079/2004, pp. 314-323, June-July 2004.
9. P. Pan, G. Swallow, and A. Atlas, "Fast Reroute Extensions to RSVP-TE for LSP Tunnels", Internet Draft, draft-ietf-mpls-rsvp-lsp-fastreroute-07.txt.
10. Rosen, A. Viswanathan and R. Callon, "Multiprotocol Label Switching Architecture", Jan. 2001, Network Working Group, Request for Comments (RFC) 3031, http://www.ietf.org/rfc/rfc3031.txt.
11. V. Sharma and F. Hellstrand, "Framework for Multi-Protocol Label Switching (MPLS)-based Recovery", Network Working Group, Request for Comments (RFC) 3469, February 2003.
12. IETF, "MPLS working group", Feb. 2004, http://www.ietf.org/html.charters/mpls-charter.html
13. Network Simulator, Version2.26, http://www.isi.edu/nsnam /ns.

Optimizing Inter-domain Multicast Through DINloop with GMPLS

Huaqun Guo[1,2], Lek Heng Ngoh[2], and Wai Choong Wong[1,2]

[1] Department of Electrical & Computer Engineering, National University of Singapore,
4 Engineering Drive 3, Singapore 117576
[2] Institute for Infocomm Research, A*STAR, 21 Heng Mui Keng Terrace, Singapore 119613
`{guohq, lhn, lwong}@i2r.a-star.edu.sg`

Abstract. This paper proposes DINloop (Data-In-Network loop) based multicast with GMPLS (generalized multiprotocol label switching) to overcome the scalability problems existing in current inter-domain multicast protocols. In our approach, multiple multicast sessions share a single DINloop instead of constructing individual multicast trees. DINloop is a special path formed using GMPLS to establish LSPs (Label Switched Paths) between DIN Nodes, which are core routers that connect to each intra-domain. We adopt link bundling method to use fewer labels. Packet Processing Module analyzes the coming multicast message and GMPLS Manager assigns a label to it. Then the multicast packet is fast forwarded using label. Simulations show that DINloop-based multicast uses the smallest message number needed to form the multicast structure compared with conventional inter-domain multicast protocols. In addition, the routing table size in core routers does not increase as the number of multicast group increases, and therefore the routing scalability is achieved.

1 Introduction

IP Multicast is an extremely useful and efficient mechanism for multi-point communication and distribution of information. However, the current infrastructure for global, Internet-wide multicast routing faces some problems.

The existing multicast routing mechanisms broadcast some information and therefore do not scale well to groups that span the Internet. Multicast routing protocols like DVMRP (Distance Vector Multicast Routing Protocol) [1] and PIM-DM (Protocol Independent Multicast – Dense Mode) [2], periodically flood data packets throughout the network. MOPSF (Multicast Open Shortest Path First) [3] floods group membership information to all the routers so that they can build multicast distribution trees. Protocols like PIM-SM (Protocol Independent Multicast – Sparse Mode) [4] and CBT (Core Based Tree) [5] scale better by having the members explicitly join a multicast distribution tree rooted as a Rendezvous Point (RP). However, in PIM-SM, RPs in other domains have no way of knowing about sources located in other domains [6]. CBT builds a bidirectional tree rooted at a core router and the use of single RP can potentially be subjected to overloading and single-point of failure. Thus, those protocols are mainly used for intra-domain multicast.

P. Lorenz and P. Dini (Eds.): ICN 2005, LNCS 3421, pp. 50–57, 2005.

Multicast Source Discovery Protocol (MSDP) [7] and Border Gateway Multicast Protocol (BGMP) [8] were developed for inter-domain multicast. BGMP requires Multicast Address-Set Claim protocol to form the basis for a hierarchical address allocation architecture. However, this is not supported by the present structure of non-hierarchical address allocation architecture and it is not suitable for dynamic setup. MSDP requires a multicast router to keep forwarding state for every multicast tree passing through it and the number of forwarding states grows with the number of groups [9]. In addition, MSDP floods source information periodically to all other RPs on the Internet using TCP links between RPs [10]. If there are thousand of multicast sources, the number of Source Active (SA) message being flooded around the network would increase linearly. Thus, the current inter-domain multicast protocols suffer from scalability problems.

MPLS Multicast Tree (MMT) [9] utilizes multiprotocol label switching (MPLS) [11] LSPs between multicast tree branching node routers in order to reduce forwarding states and enhance scalability. However, its centralized network information manager system is a weak point and the number of forwarding states in inter-domain routers grows with the number of groups. Edge Router Multicasting (ERM) [12] is another work using MPLS reported in literature. ERM limits branching point of multicast delivery tree to only the edges of MPLS domains. However, it is related to MPLS traffic engineering and it is an intra-domain routing scheme.

To overcome the scalability problems in inter-domain multicast and centralized weakness posed by MMT, DINloop-based multicast with GMPLS [13] is proposed to optimize inter-domain multicast. The remainder of this paper is organized as follows. The solution overview is outlined in Section 2. Section 3 presents the details of DINloop-based multicast. To further investigate the proposed solution, we present experimental results in Section 4 followed by the conclusion in Section 5.

2   Solution Overview

In our approach, the core router referred to DIN Node, which connects to each domain, is chosen as the RP for that domain. Within a domain, the multicast tree is formed similarly to bidirectional PIM-SM, rooted at an associated DIN Node. In the core network, multiple DIN Nodes form a DINloop (see Section 3.2) using GMPLS for inter-domain multicast. The idea is that, in order to reduce multicast forwarding state and, correspondingly, tree maintenance overhead at core network, instead of constructing a tree for each individual multicast session in the core network, one can have multiple multicast sessions share a single DINloop. DINloop-based multicast achieves the scalability in the inter-domain multicast with below distinct advantages:

- In inter-domain, multiple multicast sessions sharing a single DINloop reduces multicast state and, correspondingly, tree maintenance overhead at core network.
- DINloop-based multicast consumes fewer labels than other multicast protocols.
- Link bundling mechanism (described in Section 3.1) further improves routing scalability, as well as reduces the routing look-up time for fast routing.
- DINloop-based multicast structure is setup fastest with least control messages for many-to-many multicast crossing multiple domains.

3   DINloop-Based Multicast

Our scheme uses GMPLS to set up the DINloop in the core network by the way that LSPs are established between DIN Nodes. One example of DINloop-based multicast is shown in Fig. 1. DIN Nodes A, B, C and D are the core routers and function as RPs for the associated intra-domain (e.g., dot-line square area) respectively.

The control modules in DIN Nodes are shown in Fig. 2. Intermediate System-to-Intermediate System (IS-IS) and Open Shortest Path First (OSPF) are routing protocols. Resource Reservation Protocol - Traffic Extension (RSVP-TE) and Constraint-based Label Distribution Protocol (CR-LDP) are signalling protocols for the establishment of LSPs. Link Management Protocol (LMP) is used to establish, release and manage connections between two adjacent DIN nodes.

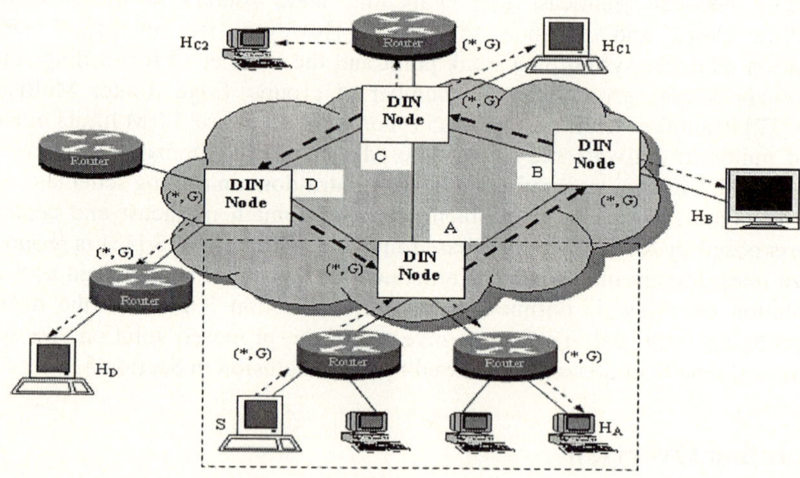

Fig. 1. DINloop-based multicast

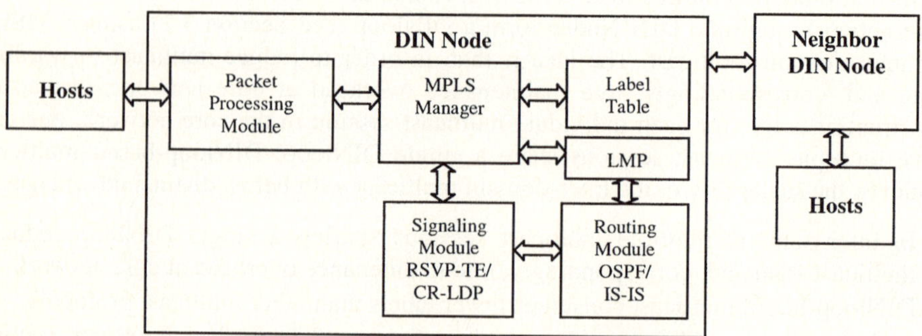

Fig. 2. Control Modules in DIN Node

3.1 Label Description

We use the Generalized Label that extends the traditional MPLS label by allowing the representation of not only labels that travel in-band with associated data packets, but also (virtual) labels that identify wavelengths.

To avoid a large size for the routing table and gain the speed in table look-ups, we adopt a method to store one multicast group into one wavelength and bind all multicast groups in a bundled link. When a pair of DIN Nodes is connected by multiple links, it is possible to advertise several (or all) of these links as a single link (a bundled link) into OSPF and/or IS-IS. Each bundled link corresponds to a label. The purpose of link bundling is to improve routing scalability by reducing the amount of information that has to be handled by OSPF and/or IS-IS.

The Generalized Label Request supports communication of characteristics required to support the LSP being requested. The information carried in a Generalized Label Request [13] is shown in Fig. 3. The length of LSP Encoding Type is 8 bits. It indicates the encoding of the LSP being requested. For our scheme, its value is 8 that it means the type is lambda. The length of Switching Type is 8 bits. It indicates the type of switching that should be performed on a particular link. In our scheme, the value is 150 that it means the switching type is Lambda-Switch Capable (LSC). The length of Generalized PID (G-PID) is 16 bits. It is an identifier of the payload carried by an LSP, i.e., an identifier of the client layer of that LSP. This is used by the nodes at the endpoints of the LSP, and in some cases by the penultimate hop.

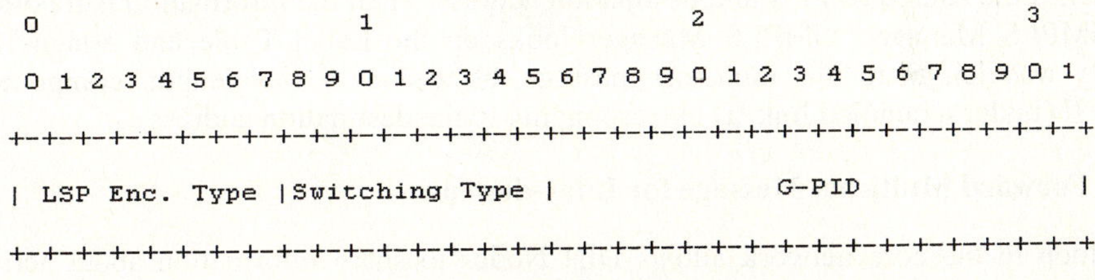

Fig. 3. Generalized Label Request

LSC uses multiple data channels/links controlled by a single control channel. The label with 32bits label space indicates the data channel/link to be used for the LSP.

3.2 Setting Up DINloop

One DIN Node can be elected as an initial node, said as DIN Node 1. The other DIN Nodes are called in loop sequence as DIN Node 2, DIN Node 3, …, DIN Node n. Starting from DIN Node 1, GMPLS Manager configures a control channel and data channels to connect to DIN Node 2. We assume that the control channel is set up in the Wavelength 0 and Wavelengths 1 to n are used for data channels. GMPLS Manager bundles Wavelengths 1 to n as a link with a bundle link label after looking up Label Table. Second, enable LMP on both DIN Node 1 and DIN Node 2. Third, configure an LSP on DIN Node 1 to DIN Node 2. DIN Node 1 sends an LSP request to the neighbour DIN Node 2 through signalling module.

For example in Fig. 4, Node 1 sends an initial PATH/Label Request message to DIN Node 2. The PATH/Label Request contains a Generalized Label Request. DIN Node 2 will send back a RESV/Label Mapping. When the generalized label is received by the initiator DIN Node 1, it then establishes an LSP with DIN Node 2 via RSVP/PATH message. Similarly, DIN Node 2 and DIN Node 3, ..., DIN Node n and DIN Node 1 establish an LSP.

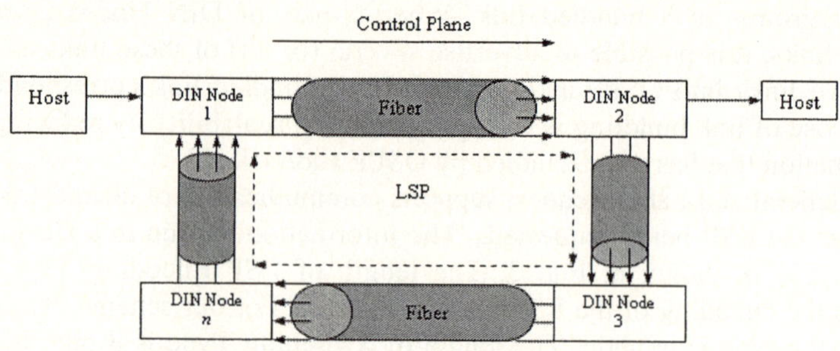

Fig. 4. LSP creation for DINloop

3.3 Assign Labels to Multicast Messages

When multicast packets arrive in the DIN Node, Packet Processing Module is activated. Packet Processing Module looks the IP header of multicast packets and identifies the source address and destination address. Then the information is reported to GMPLS Manager. GMPLS Manager looks up the Label Table and assigns an empty wavelength to this multicast message, as well as a wavelength (component link) ID under a bundled link ID corresponding to the destination address.

3.4 Forward Multicast Message for Inter-domain

DINloop in the core network allows DIN Nodes to share information about active sources. Each DIN Node knows about the receivers in their local domain respectively. For example in Fig. 1, when DIN Node B receives the join message for multicast group G from Receiver H_B, it posts the join message in the DINloop. When DIN Node A in the source domain receives this join message and knows that there is the receiver in the remote domain, DIN Node A will forward the multicast message into the DINloop. Then DIN Node B will use a combination of two identifiers (bundled link ID (label ID), component link ID) to retrieve the multicast message from the DINloop and forward it to its local receiver H_B. Multicast data can then be forwarded between the domains. Similarly, Receiver H_{C1}, H_{C2} and H_D in other domains can also join the multicast group G and receive the multicast information from Source S.

4 Experimental Results

4.1 Message Number

We use the message number to evaluate the performance of DINloop-based multicast versus MSDP and BGMP. The message load denotes the amount of signaling messages that is needed to form the multicast structure. The message load metric is incremented by one whenever a signaling message passes a link.

MSDP floods source information periodically to all other RPs using TCP links between RPs. MSDP also allows receivers to switch to shortest-path tree to receive packets from sources and we refer it as MSDP+source. BGMP builds bidirectional shared tree rooted at the root domain and is referred as BGMP+shared. In addition, BGMP also allows source-specific branches to be built by the domains. We refer to the tree built by BGMP consisting of the bidirectional tree and the source-specific branches as BGMP+source.

The simulations ran on the network topologies, which were generated using the Georgia Tech [14] random graph generator according to the transit-stub model [15]. We generate different network topologies. The number of nodes in transit domain was fixed, i.e., 25 that means 25 DIN Nodes for 25 domains. The number of nodes in stub domain was fixed to 1000 that spread in 25 domains. The number of source is varied from 2 to 25 and spread in 25 domains. The results are shown in Fig. 5.

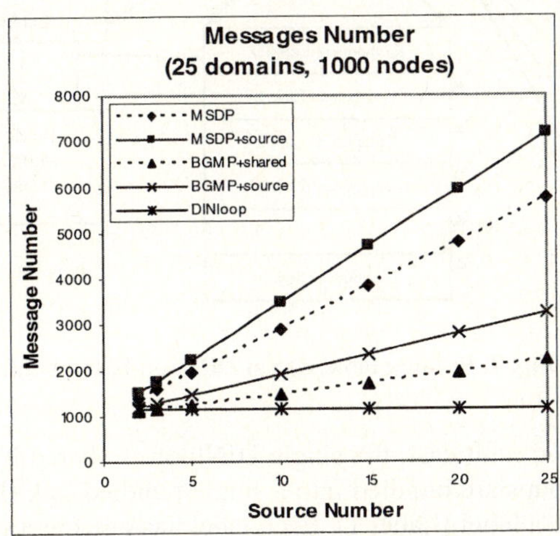

Fig. 5. Message number comparison on the effect of source number

From Fig. 5, the message numbers in MSDP and BGMP increase linearly as the source number increases. The slope in MSDP is higher than BGMP. However, the message number in DINloop-based multicast is the smallest and is not affected by the source number since all sources share the same DINloop.

4.2   Forwarding State Size

In this sub-section, we compare forwarding state size kept in core routers. We refer the current inter-domain multicast protocols as non-DINloop scheme.

In non-DINloop scheme, the core router looks up IP header of every packet and identifies the destination address. Then, the core router looks up the routing table and forwards the packet to the correct interface (Fig. 6). Therefore, the routing table size is increased linearly with the number of multicast groups.

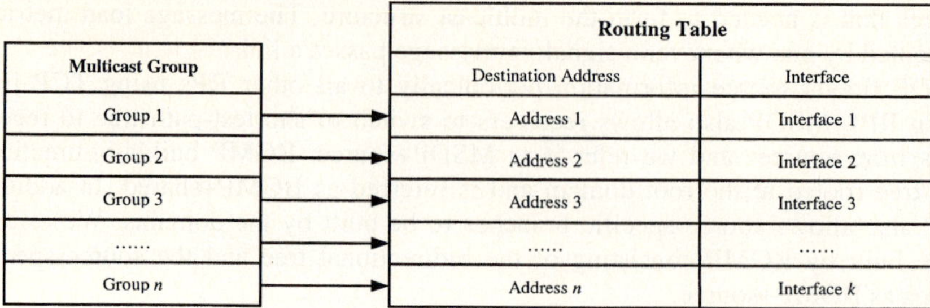

Fig. 6. Routing table size in non-DINloop scheme

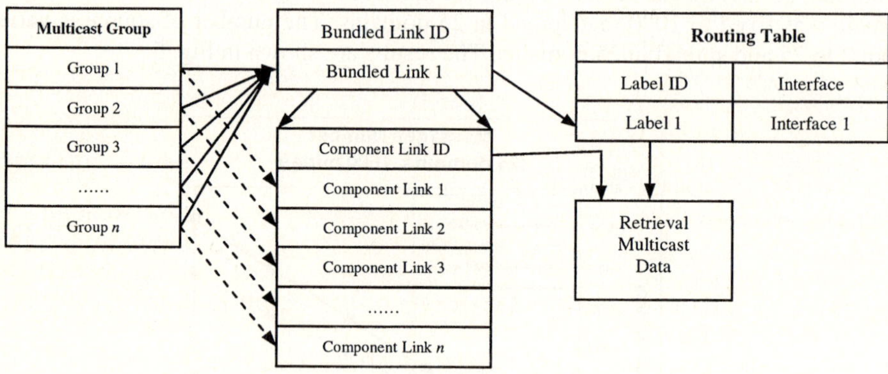

Fig. 7. Routing table size in DINloop-based multicast

In DINloop-based multicast, the single DINloop is shared by all multicast groups. So all multicast groups are bundled into a single bundled link (Bundled Link 1 in Fig. 7) corresponding to a label (Label 1). Each label has one egress interface, i.e., Label 1 corresponds to Interface 1 in the routing table.

From Fig. 7, we obtains the results that the routing table size does not increase as the number of multicast group increases, and therefore the DINloop-based multicast increases the routing scalability for inter-domain multicast, as well as reduces the routing look-up time for fast routing at the price of adding additional GMPLS header to the multicast messages.

5   Conclusion

DINloop-based multicast with GMPLS is presented to optimize inter-domain multicast. Instead of constructing a tree for each individual multicast session in the core network, multiple multicast sessions share a single DINloop formed using GMPLS LSPs between multiple DIN Nodes. We adopt link bundling method to use fewer labels. Packet Processing Module analyzes the coming multicast message and GMPLS Manager assigns an empty wavelength to it, as well as a component link ID under a bundled link ID corresponding to the destination address. Then the multicast packet is fast forwarded using label rather than address. Simulations show that DINloop-based multicast uses the smallest message number needed to form the multicast structure compared with conventional inter-domain multicast protocols. In addition, the routing table size in core routers does not increase as the number of multicast group increases, and therefore the routing scalability is achieved.

References

1. Waitzman, D., Partridge, C., Deering, S.: Distance Vector Multicast Routing Protocol. RFC-1075 (1988)
2. Adams, A., Nicholas, J., Siadak, W.: Protocol Independent Multicast - Dense Mode (PIM-DM): Protocol Specification (Revised). Internet Draft (2003)
3. Moy, J.: Multicast Extensions to OSPF. RFC-1584 (1994)
4. Estrin, D., Farinacci, D., Helmy, A., Thaler, D., Deering, S., Handley, M., Jacobson, V., Liu, C., Sharma, P., Wei, L.: Protocol Independent Multicast-Sparse Mode (PIM-SM): Protocol Specification. RFC-2362 (1998)
5. Ballardie, A.: Core Based Trees (CBT version 2) Multicast Routing. RFC-2189 (1997)
6. IP Multicast Technology Overview,
7. http://www.cisco.com/univercd/cc/td/doc/cisintwk/intsolns/mcst_sol/mcst_ovr.htm#53693
8. Fenner, B. (ed.), Meyer, D. (ed.): Multicast Source Discovery Protocol (MSDP). IETF RFC3618 (2003)
9. Thaler, D.: Border Gateway Multicast Protocol (BGMP): Protocol Specification. RFC3913 (2004)
10. Boudani, A., Cousin, B., Bonnin, J.M.: An Effective Solution for Multicast Scalability: The MPLS Multicast Tree (MMT). Internet Draft (2004)
11. Diot, C., Levine, B.N., Lyles, B., Kassem, H., Balensiefen, D.: Deployment Issues for the IP Multicast Service and Architecture. IEEE Network, January 2000 (2000) 78-88
12. Rosen, E., Viswanathan, A., Callon, R.: Multiprotocol Label Switching Architecture. RFC-3031 (2001)
13. Yang B., Mohapatra, P.: Edge Router Multicasting with MPLS Traffic Engineering. ICON (2002)
14. Berger, L.: Generalized Multi-Protocol Label Switching (GMPLS) Signaling Functional Description. RFC-3471 (2003)
15. Zegura, E., Calvert, K., Bhattacharjee, S.: How to model an internetwork. Proc. of IEEE Infocom (1996)
16. Modeling Topology of Large Internetworks. http://www.cc.gatech.edu/projects/gtitm/

A Fast Path Recovery Mechanism for MPLS Networks

Jenhui Chen[1], Chung-Ching Chiou[2], and Shih-Lin Wu[1]

[1] Department of Computer Science and Information Engineering,
Chang Gung University, Taoyuan, Taiwan 333, R.O.C.
{jhchen, slwu}@mail.cgu.edu.tw
[2] BroadWeb Inc., 24-1, Industry East Rd., IV,
Science Based Industrial Park, Hsin-Chu, Taiwan 300, R.O.C.
joe@broadweb.com.tw

Abstract. The major concept of the Multi-Protocol Label Switching (MPLS) network uses the Label Switch Path (LSP) technique that provides high performance in packet delivery without routing table lookup. Nevertheless, it needs more overhead to rebuild a new path when occurring link failure in the MPLS network. In this paper, we propose an efficient fast path recovery mechanism, which employs the Diffusing Update Algorithm (DUAL) to establish the working and backup paths concurrently and modify the Label Distribution Protocol (LDP) to establish the LSP by using Enhanced Interior Gateway Routing Protocol (EIGRP). Simulation results show that the proposed mechanism not only improves resource utilization but provides shorter path recovery time than the end-to-end recovery mechanism.

1 Introduction

In connectionless network protocols, an independent forwarding decision is made in each switch router when a packet is delivered from one router to next router. Traditionally, in the IP network, each router runs a network layer routing algorithm (e.g., Dijkstra algorithm) for supporting route setup procedure. Current routing algorithm, despite being robust and survivable, can take a substantial amount of time to recovery when a failure occurs, which can be on the order from several seconds to minutes and can cause serious disruption of service in the interim. This is unacceptable for many applications that require highly reliable service. Thus, the Internet service provider may need an efficient path protection mechanism to minimize the recovery time when link failure, and maximize the network reliability and survivability.

Path-oriented technologies such as Multi-Protocol Label Switching (MPLS), which is described in the RFC3031 of Internet Engineering Task Force (IETF) [4, 8], can be used to enhance the reliability of IP networks. A fundamental concept of MPLS networks, which consists of Label Edge Routers (LERs) around a core of meshed Label Switching Routers (LSRs), is to use small labels for

P. Lorenz and P. Dini (Eds.): ICN 2005, LNCS 3421, pp. 58–65, 2005.

routing. In order to carry the same labeled traffic in MPLS networks, a label assignment distribution scheme, Label Distribution Protocol (LDP) [1], is taken to establish the Label Switched Path (LSP) beforehand.

To protect an LSP in MPLS networks, a protection LSP is proposed to establish the working and backup paths at the same time in the initial setup process. When a network failure is detected, the MPLS will perform a path recovery mechanism by only switching working traffic onto the backup path. However, this recovery mechanism does not switch the working path to backup path efficiently and leads the degradation of network performance due to its long processing delay. Therefore, in this paper, we propose a fast path recovery mechanism, which employs the Enhanced Interior Gateway Routing Protocol (EIGRP) and the Diffusing Update Algorithm (DUAL) [5] together, to find the working and backup paths simultaneously and modify the LDP to establish the LSP by using the routing table of EIGRP. Moreover, the proposed path recovery mechanism would not occupy any available bandwidth in MPLS networks.

The remainder of this paper is organized as follows. Section 2 describes the proposed recovery mechanism in details. The simulation models and results are shown in Section 3. Finally, we give some conclusions in Section 4.

2   The Fast Path Recovery Mechanism

2.1   An Overview

The IETF has proposed two types of recovery models for the MPLS-based LSP: the protected switching model and the rerouting model [2, 6, 10]. However, there are two critical drawbacks that can be improved. One drawback is that the setup process needs at least two round-trip times to establish the working and backup paths. Another drawback is that the backup path cannot be found from the routing table due to its limited information. To overcome these drawbacks, we use the DUAL to converge after an arbitrary sequence of link cost or topological changes in a finite time and guarantee a loop-free routing. And employ the EIGRP, which is the Cisco proprietary protocol and a modified protocol stack of the MPLS, to establish working and backup paths by using the information of the successor and feasible successor indicated in EIGRP routing table.

2.2   The Recovery Mechanism

In the proposed LDP, we added additional parameters to the optional parameters field each of the label request message (LRM) and the label mapping message (LMM), which is shown in Fig. 1. The LRM contains a successor LRM (SLRM) and a feasible successor LRM (FSLRM), respectively. The SLRM is used to request the label of working path according to the successor of routing table and FSLRM is used to request label of backup path according to the feasible successor of routing table. The LMM contains the successor LMM and the feasible successor LMM, which are used to map label of working path and backup path, respectively. In addition, shown in Table 1, the Label Information Base (LIB)

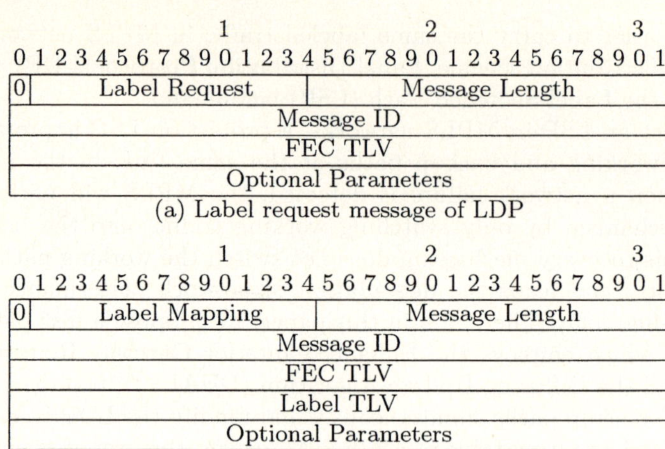

0	Label Request	Message Length
	Message ID	
	FEC TLV	
	Optional Parameters	

(a) Label request message of LDP

0	Label Mapping	Message Length
	Message ID	
	FEC TLV	
	Label TLV	
	Optional Parameters	

(b) Label mapping message of LDP

Fig. 1. The packet format of LRM and LMM

Table 1. An Example of modified LIB

LER S					
In/I	In/L	FEC	Out/I	Out/L	State
0	25	163.25.101	1	21	active
			2	20	backup

is modified to add an additional state field to record which label is assigned to working path or backup path.

Fig. 2 shows the proposed LDP algorithm in details. An example of using the proposed algorithm is given in Fig. 3. The Fig. 3(b) shows our recovery mechanism is belong to the protection switch model with partial backup. When some links of the working path experience failure, it can switch over the backup path rapidly since it does not need to send a notification to the source node.

3   Simulation Results

In this section, we evaluated the performance of proposed mechanism by carrying out simulation studies on regular network topologies in different mesh sizes: 5x10, 5x15, 5x20, 5x25, 5x30, 5x35 and 5x40, respectively. We also implemented the end-to-end backup scheme to compare its performance in terms of average path restoration time, throughput, average LSP establishing time, average resource consumption, and average hop counts of backup path.

The network topology is a partial mesh topology as shown in Fig. 4. The bandwidth of each link is set as 10 units and the link delay time is dependent on link cost. Thus, the link delay along any link is proportional to its link cost. The traffic arrival rate is constant and maximum transmitting unit is fixed with 1,500 bytes.

Input: $(G, V, \text{start\_node}, \text{end\_node}, RT)$
Output: Working path, backup path, and all of the LIBs
Begin
    For each vertexes V
        If neither a start_node nor a end_node
            If receive the successor label request message
                send SLRM to successor of its RT.
                send FSLRM to feasible successor of its RT.
            Else If receive the feasible successor label request message
                send FSLRM to successor of its RT.
            Else If receive the successor label mapping message
                send SLMM to previous node which sent the SLRM to it.
                put the assigned label to the outgoing label space and mark "active".
            Else If receive the feasible successor label mapping message
                send FSLMM to previous node that send FSLRM to this node.
                put the assigned label to the outgoing label space and mark "backup".
        Else If this node is a start_node
            If receive the successor label mapping message
                put the assigned label to the outgoing label space and mark "active".
            Else If receive the feasible successor label mapping message
                put the assigned label to the outgoing label space and mark "backup".
            Else
                send SLRM to successor of its RT.
                send FSLRM to feasible successor of its RT.
        Else If this node is a *end_node*
            If receive the successor label request message
                reply SLMM to previous node that send SLRM to destination node.
                put the assigned label to the incoming label space and mark "active".
            Else If receive the feasible successor label request message
                reply FSLMM to previous node that send FSLRM to destination node.
                put the assigned label to the incoming label space and mark "backup".
End

Fig. 2. The algorithm of modified LDP

In Fig. 5, we compare the packet delivery ratio of our scheme with the end-to-end backup scheme's by varying the mesh size. We can see that both packet delivery ratios of proposed scheme and end-to-end scheme degrades when the the size of mesh increases. This is because that the end-to-end scheme performs the path recovery mechanism after a link failure is detected. On the contrary, in our scheme, a backup path is established with the working path simultaneously and, therefore, the path recovery time will be minimized. As a result, many packets will be queued in buffer and wait for another available route to the destination. This drawback will degrade the performance of MPLS networks.

Fig. 6 shows the comparison of average restoration (path recovery) time of two schemes. The path restoration time of the end-to-end backup scheme is getting longer with the length of working path due to the end-to-end rerouting

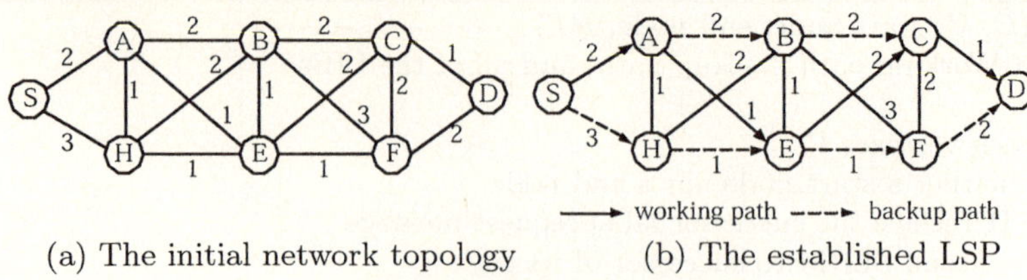

(a) The initial network topology   (b) The established LSP

S	Neighbor	FD	AD	State
(D)		6		FD
	Via A	6	4	Successor
	Via H	7	4	FS

H	Neighbor	FD	AD	State
(D)		4		FD
	Via A	5	4	
	Via B	5	3	FS
	Via E	4	3	Successor
	Via S	9	6	

A	Neighbor	FD	AD	State
(D)		4		FD
	Via B	5	3	FS
	Via E	4	3	Successor
	Via H	5	4	
	Via S	9	7	

E	Neighbor	FD	AD	State
(D)		3		FD
	Via A	6	5	
	Via B	4	3	
	Via C	3	1	Successor
	Via H	6	5	
	Via F	3	2	FS

B	Neighbor	FD	AD	State
(D)		3		FD
	Via A	6	4	
	Via C	3	1	Successor
	Via E	4	3	
	Via F	5	2	FS
	Via H	6	4	

FD: Feasible Distance   AD: Advertised Distance

(c) The convergent routing table of the EIGRP

Fig. 3. An example of the LSP establishment by proposed LDP

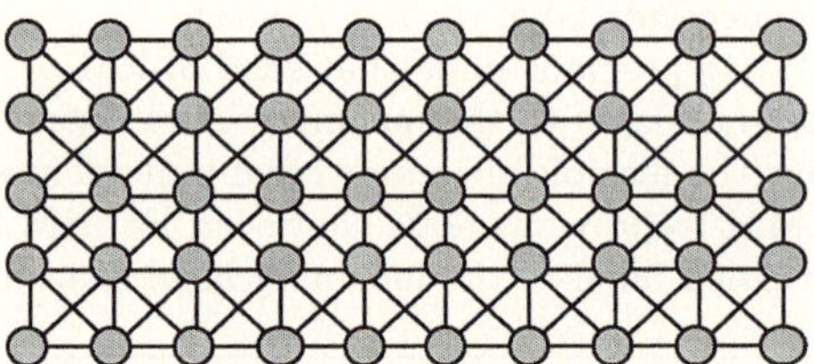

Fig. 4. An Example of mesh topology size with 5x10

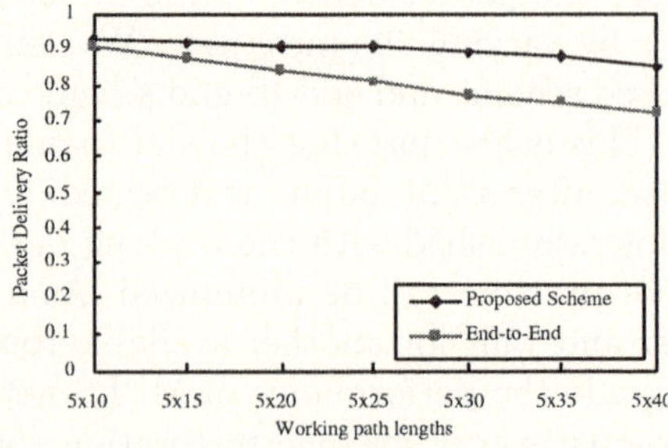

Fig. 5. The comparison of packet delivery ratio of proposed scheme and end-to-end backup scheme by varying the mesh size

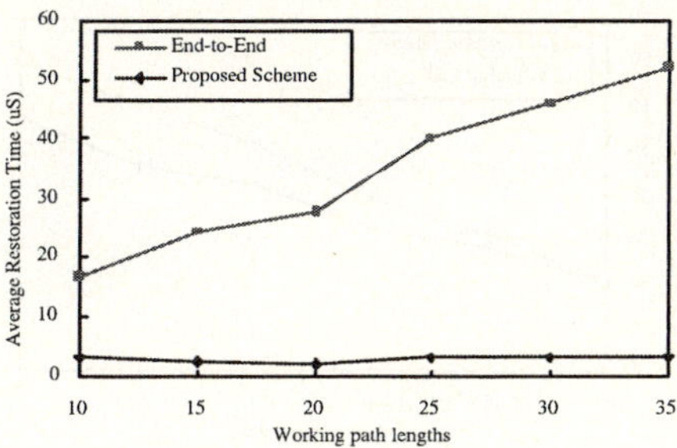

Fig. 6. The comparison of average restoration time of proposed scheme and end-to-end backup scheme by varying working path length

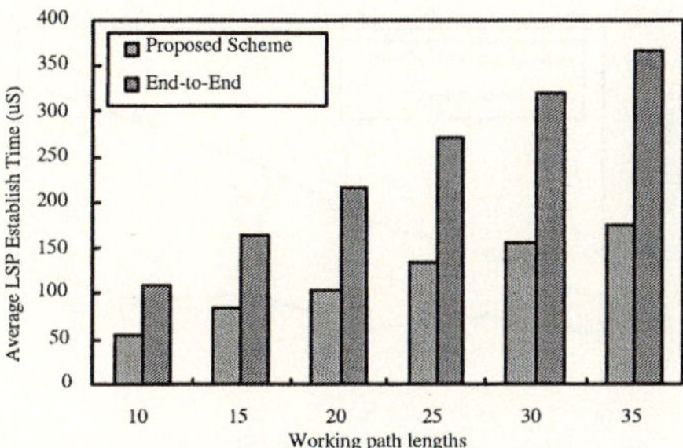

Fig. 7. The comparison of average LSP establish time of proposed scheme and end-to-end backup scheme by varying working path length

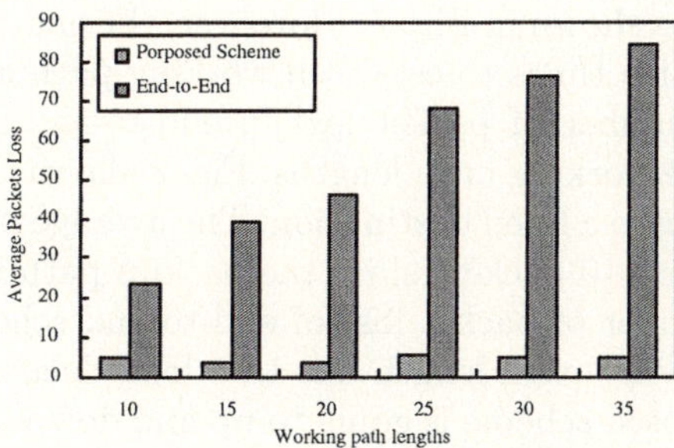

Fig. 8. The comparison of average packet loss of proposed scheme and end-to-end backup scheme by varying working path length

mechanism. However, the path restoration time of our scheme is a placid curve and keeps around $3\mu s$ since it has a pre-established backup path.

In addition, the establish time of LSP in our proposed scheme is half of the end-to-end backup scheme since our scheme establishes working path and backup

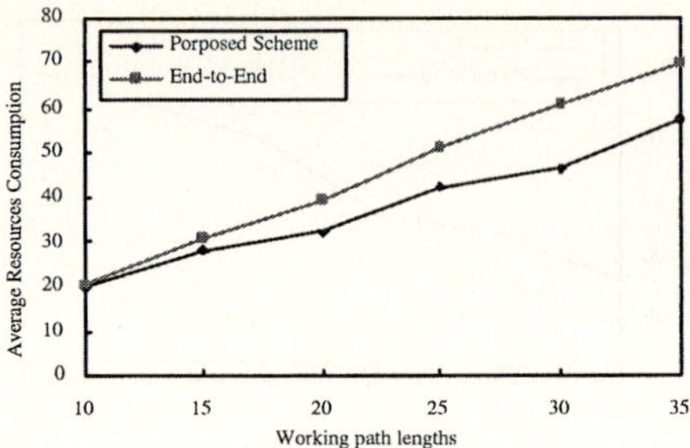

Fig. 9. The comparison of average resources consumption of proposed scheme and end-to-end backup scheme by varying working path length

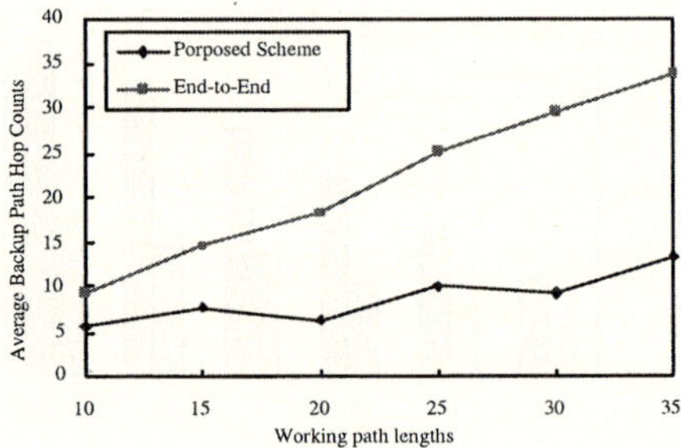

Fig. 10. The comparison of average hop counts of backup path of proposed scheme and end-to-end backup scheme by varying working path length

path simultaneously as shown in Fig. 7. Moreover, the end-to-end backup scheme must use two round-trip times to establish working path and backup path.

We observe the number of packet loss in end-to-end scheme and proposed scheme by varying the working path lengths. Fig. 8 shows the derived packet loss by sending 500 packets to a fixed destination. The average packet loss of proposed scheme does not exceed 10 packets since the backup path is pre-established. On the contrary, the number of packet loss of end-to-end scheme increases proportionally with the working path length due to a long delay of path recovery.

The curve of proposed scheme is going to up and down since current situation of EIGRP routing table might have the feasible successor or not. Thus, the resource consumption and backup path hop counts are depending on the number of feasible successors of the node in working path as shown in Fig. 9 and Fig. 10.

4   Conclusion

In this paper, a fast path recovery mechanism for LDP in MPLS networks is presented and investigated. The proposed mechanism using routing table of EIGRP base on Diffusing Update Algorithm (DUAL) can converge rapidly of finding the working and backup paths. This mechanism not only improves resource utilization but provides faster failure recovery time than the end-to-end recovery mechanism. Simulation results show that the proposed mechanism can fast migrate ongoing data streaming to pre-established backup path efficiently without wasting bandwidth. This mechanism enable MPLS networks to provide high quality-of-service (QoS) applications sufficiently.

Acknowledgment

This work was supported by National Science Council, Taiwan, R.O.C., under Contract NSC93-2213-E-182-022.

References

1. L. Andersson, P. Doolan, N. Feldman, A. Fredette, and B. Thomas, "LDP Specification," IETF RFC 3036, Jan. 2001.
2. D. Awduche, J. Malcolm, J. Agogbua, M. O'Dell, and J. McManus, "Requirements for Traffic Engineering over MPLS," IETF RFC 2702, Sept. 1999.
3. A. Farrel, "Fault Tolerance for LDP and CR-LDP," IETF draft draft-ietf-mpls-ldp-ft-01.txt, Feb. 2001.
4. Future Software, "Multi Protocol Label Switching," white paper of Future Software Limited, India, 2002. http://www.futsoft.com.
5. J.J. Garcia-Lunes-Aceves, "Loop-free Routing Using Diffusing Computations," *IEEE/ACM Trans. Networking*, vol. 1, no.1, pp. 130–141, Feb. 1993.
6. C.C. Huang, V. Sharma, K. Owens, and S. Makam, "Building Reliable MPLS Networks Using a Path Protection Mechanism," *IEEE Commun. Mag.*, vol. 40, no. 3, pp. 156–162, Mar. 2002.
7. J. Lawrence, "Designing Multiprotocol Label Switching Networks," *IEEE Commun. Mag.*, vol. 39, no. 7, pp. 134–142, July 2001.
8. E. Rosen, A. Viswanathan, and R. Callon, "Multiprotocol Label Switching Architecture," IETF RFC 3031, 2001.
9. J. Wu, D. Y. Montuno, H.T. Mouftah, G. Wang, and A.C. Dasylva, "Improving the Reliability of the Label Distribution Protocol," in *Proc. Local Comp. Networks*, pp. 236–242, 2001.
10. S. Yoon, H. Lee, D. Choi, Y. Kim, G. Lee, and M. Lee, "An Efficient Recovery Mechanism for MPLS-based Protection LSP," in *Proc. Int. Conf. ATM (ICATM'2001) and High Speed Intelligent Internet Symp.*, pp. 75–79, 2001.

A Study of Billing Schemes in an Experimental Next Generation Network

P.S. Barreto, G. Amvame-Nze, C.V. Silva, J.S.S. Oliveira, H.P. de Carvalho,
H. Abdalla Jr, A.M. Soares, and R. Puttini

Department of Electric Engineering, Faculty of Technology, University of Brasilia,
Campus Universitário Darcy Ribeiro - Asa Norte CEP 70910-900 - Brasília - DF – Brasil
{paulo, abdalla, martins, puttini}@ene.unb.br
{pris, georges, claudio, juliana}@labcom.unb.br

Abstract. In this paper, we present a discussion concerning the performance of four network scenarios for billing purposes. Using the results of packet losses in an experimental platform simulating a NGN environment, we evaluate on each scenario the impact in the billing process with different traffic flows comparing the total revenue calculus for two billing schemes: (1) charging per packet and (2) reducing the value corresponding to undelivered packets. Our results show that the environments that use Differentiated Services are both convenient for costumers and service providers.

1 Introduction

In the 80's, researches focused on the technological development that would allow the availability of a wide the variety of services in the telecommunications area. The efforts towards developing appropriate billing systems were little. However, in the recent years, the need to cover expansion costs and the possibility of using charging methods to influence the users' behavior in order to avoid congestion increased the interest in that topic [1].

Most of the billing systems are flat rate-based, which means that the costumers are charged indistinctly, without considering the real network resources utilization and the type of service which is provided. This method is abundantly used due to its simplicity, as it does not require any charging, accounting and metering policies. Despite of that, this method is not appropriate to implement congestion control through charging [2] [3].

The increasing demand for QoS (Quality of Service) induces the development of better adjusted billing mechanisms which discriminate each kind of traffic stimulating users to choose, according to their personal needs, the most appropriate type of service, avoiding excessive resources use and allocation. As the traffic is differentiated based on performance requirements, a QoS discriminative billing mechanism is expected to charge differently for different type of services [4].

In order to implement congestion control through charging, we must use a billing method that relates the bill to the real use of resources. Based on this concept we can propose one simple method: charge per sent packet. Nevertheless, due to the difficulty

P. Lorenz and P. Dini (Eds.): ICN 2005, LNCS 3421, pp. 66–74, 2005.

of tracing each packet, another possible mechanism is to count only the dropped ones and reduce their corresponding economic or monetary value from a certain amount.

In our discussion, we try to find which QoS environment would provide the best revenue for the service provider, but also regarding the users´ satisfaction. We will evaluate four different scenarios: IP network (best effort only), IP network with Diffserv, MPLS and MPLS with Diffserv.

This paper is organized as follows: the network topology and traffic patterns are presented in section 2 as well as the description of the different classes of traffic flows used. In section 3, we present a description of each scenario and the results regarding packet losses obtained on each experiment. In section 4, we present a discussion of the results and the revenue calculus in order to determine the best environment in which we could develop the most appropriate billing system for both the user and the provider. In section 5, we present our conclusions and future work.

2   The Experimental Testbed

The experimental environment is shown in figure 1. There are five different networks: a PSTN (Public Switched Telephony Network), an ADSL (Assymetric Digital Subscriber Line) access network, two local area networks (LANs), a wireless LAN and a MPLS/Diffserv core.

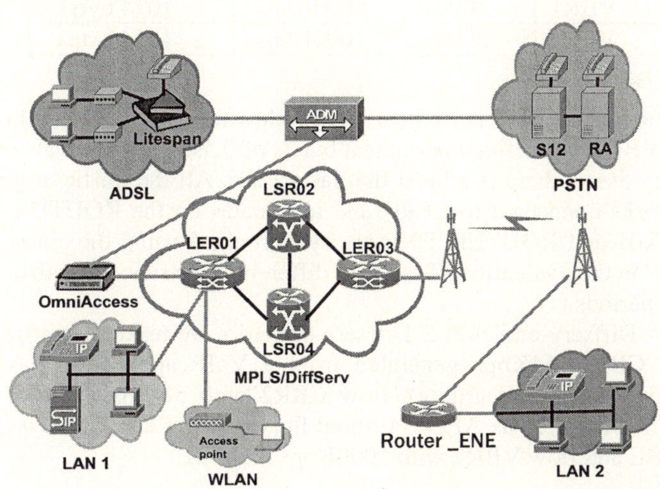

Fig. 1. Network Topology

The PSTN is formed by two local exchanges, Tropico RA and a S12, both from Alcatel. The MPLS/Diffserv core has four routers based on the Linux Operating System Kernel 2.4.21. For the scenarios using the MPLS protocol, we used an open source implementation available in [5]. The ADSL network, two local area networks and a wireless LAN are interconnected by the MPLS core, so in this way we concentrate the traffic treatment in the MPLS core.

The routers are four computers Pentium IV 2.1GHz which are interconnected by 10/100 Mbps links. The first router, LER01, connects three LANs to the core, and the LER03 via a radio link of 2 Mbps, connects the fourth LAN. The routers LSR02 and LSR04 are the forwarding elements of the core.

Regarding the links, some adjustments were made. The first adjustment concerns a bandwidth reduction. We decided to reduce these links to 1.2 Mbps in order to work with lower traffic volumes and still observe packet drop. The configuration files for this purpose were implemented using the CBQ (Class Based Queuing) discipline for traffic control in Linux systems.

On every experiment, we work with four traffic flows. The flows used to produce disturbances are VBR (Variable Bit Rate) traffic patterns, with several bursts that intend to overload the links within periodic time intervals.

The flows for evaluation are CBR (Constant Bit Rate) traffic patterns. In table 1, we show the traffic patterns used for the IP Best Effort and MPLS scenarios. The time interval for all traffic flows is 60 seconds.

Table 1. Traffic Patterns

Type	Port	Bandwidth	Packet size
CBR1	5000	64Kbps	256 bytes
CBR2	6000	384Kbps	512 bytes
VBR1	7000	1000Kbps	1024 bytes
VBR2	8000	1000Kbps	1024 bytes

Both of VBR traffics have periodical burst following and exponential distribution. In the case of VBR1, we defined periodical bursts of 0.5secs on intervals of 3secs. For VBR2, in every 5secs, there is a burst that lasts 1 sec. All the traffic originates in the different networks connected to LER01and terminates on the ROUTER_ENE. Both machines, LER01 and ROUTER_ENE are synchronized using the chrony utility for Linux systems. In the evaluation of the four different scenarios, the traffic is observed in 60 sec time periods.

For the IP - Diffserv and MPLS-Diffserv scenario, we used the traffic patterns as follows: flow CBR1, 64Kbps generated from a VoIP application has the higher priority EF (Expedited Forwarding), flow CBR2 with 384Kbps, which is a video streaming traffic classified as AF11 (Assured Forwarding Class 1), flow CBR12 with 1000Kbps as BE and flow VBR2 with 1000Kbps as AF21.

3   The Results

This first scenario simulates an IP network with Best Effort politics. Since our network is experimental, to reproduce a close to reality environment, we created routing tables with approximately 65500 entries on each router. Also, the routing cache was extremely reduced to less than 256 entries and a garbage loop traffic of 640 KB was simulated on each router to overload the routing cache update process. We decided to create an overloaded cache instead of disabling the cache, because we believe that this approach simulates in a better way a real network environment with nodes that have a considerable process load. The packet loss results obtained in this part of the experiment are shown in figure 2.

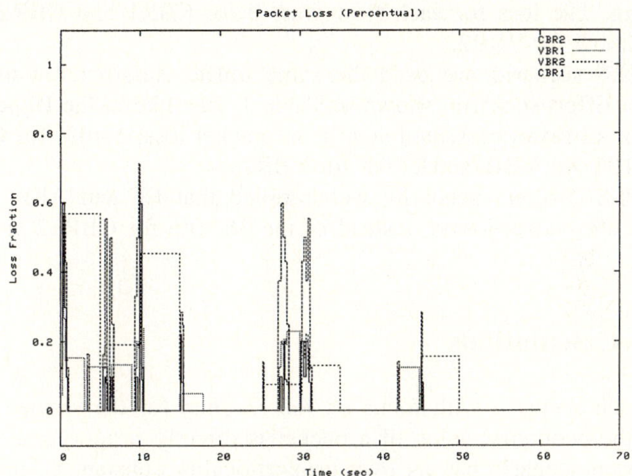

Fig. 2. Packet Loss (percentual) – IP Best Effort

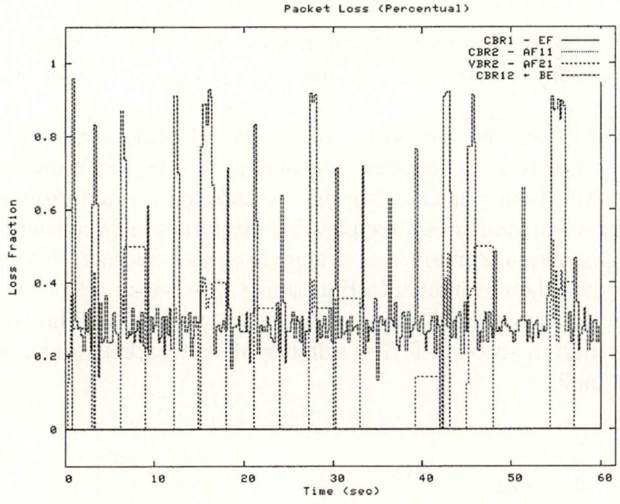

Fig. 3. Packet Loss (percentual) – IP DiffServ

As a result of the issued bandwidth limitation, packet loss occurred due to the peaks in the VBR traffic flows that promoted resources dispute and since all flows are treated the same, all of them were damaged with certain equality. The average loss for each flow was 3.90% for CBR1, 1.40% for CBR2, 7.30% for VBR1 and 12.40% for VBR2.

The results obtained for the IP-DiffServ scenario are shown in figure 3. Due to packet classification, we observed a greater percentage of packet losses in flows that have a lower priority. The packets from both EF and AF11 traffic were hardly ever dropped, which is a really important characteristic for applications such as VoIP and video streaming. The loss for each flow was: 0 for CBR1 and CBR2, 35.60% for CBR12 and 22.60% for VBR2.

For the MPLS scenario, we used the same traffic classification used in the IP network – Best Effort scenario, shown in Table 1. Just like in the IP network – Best Effort scenario, all flows presented significant packet loss: 5.40% for CBR1, 3,90% for CBR2, 14.10% for VBR1and 21.0% for VBR2.

For the MPLS-DiffServ scenario, we observed that EF and AF11 traffics were benefited with 0% packet losses, instead of the 36.70% for CBR12 and 26.30% for VBR2.

4 Revenue Calculations

To evaluate each scenario with focus on the revenue for the service provider, we assumed that the economic value of a packet is directly proportional to a standard value V and each scenario has its own proportionality constant C, in a way that a packet's economic value can be calculated as $C*V$. When using packet classification, the price of a packet is proportional to the priority. We assumed for each type of traffic the following proportionality constants: 1 for BE, 2 for AF21, 3 for AF11 and 4 for EF.

For example, a packet from a VoIP application should be classified as EF traffic, as it requires low latency and loss rate. So, the corresponding price for this packet would be 4V.

In the IP Best Effort and the MPLS scenarios, all packets are treated equally, as there is no classification, so appears reasonable to use the same constant for all packets indistinctly. Four our calculations, we selected a proportionality constant of 2.5, which is the mean value of all the possible proportionality constants.

To calculate the average number of dropped packets for each flow, represented by N, we will use the relations shown in Equation 1. The mean traffic rate in bps R, the average packet loss in percentage L, the mean packet size in bytes S and the experiment duration in seconds T. The number of sent packets P can be calculated as shown in equation 2.

$$N = \left[\frac{\frac{R}{8} \times T}{S} \right] \times L \qquad (1) \qquad\qquad P = \left[\frac{\frac{R}{8} \times T}{S} \right] \times (1 - L) \qquad (2)$$

For CBR traffic flows, R represents the transmission rate. In VBR traffics, R can be approximated using the burst transmission rate in bps Br, the burst duration in seconds Bt and the between bursts interval in seconds Bi, as shown in equation 3.

$$R = \frac{Br \times Bt}{Bt + Bi} \tag{3}$$

Using the first billing scheme (charge per packet), the revenue will be the total sum of all sent packets economic values. In the second scheme, we reduce the dropped packets corresponding economic values from a certain amount M.

4.1 IP Network-Best Effort Calculus

For the calculus of total revenue for the IP Network - Best Effort scenario we used the data summarized in table 2 and 3. The calculus of total revenue for each billing scheme is shown in table 4. The first and second rows indicate the variables used in the first and second billing schemes respectively.

Table 2. Main Variables – CBR Traffics

Traffic	R (Kbps)	T	S	L	N	P	C
CBR1	64	60	256	0,039	73	1801	2.5
CBR2	384	60	512	0,014	79	5546	2.5
VBR1	143	60	1024	0,073	76	971	2.5
VBR2	167	60	1024	0,124	152	1071	2.5

Table 3. Mean Rate - VBR Traffics

Traffic	Br(Kbps)	Bt	Bi	R(Kbps)
VBR1	1000	0.5	3	143
VBR2	1000	1	5	167

Table 4. Total Revenue Calculus for Best Effort

BS	Total Revenue Calculus
1	1801*(2.5V) + 5546*(2.5V) + 971*(2.5V) +1071*(2.5V)= 23472,5 V
2	M – 73*(2.5V) - 79*(2.5V) - 76*(2.5V) - 152*(2.5V)=M – 950 V

4.2 IP Network– Diffserv

The calculus for the IP Network – Diffserv scenario used the data shown in table 5 and 6. The calculus of total revenue for each billing scheme is shown in table 7.

Table 5. Main Variables – CBR Traffic

Traffic	R (Kbps)	T	S	L	N	P	C
CBR1	64	60	256	0,0	0	1875	4
CBR2	384	60	512	0,0	0	5625	3
CBR12	1000	60	1024	0,356	2607	4717	1
VBR2	167	60	1024	0,226	276	947	2

Table 6. Mean Rate-VBR Traffic

Traffic	Br (Kbps)	Bt	Bi	R(Kbps)
VBR2	1000	1	5	167

Table 7. Total Revenue Calculus for IP-Diffserv

BS	Total Revenue Calculus
1	1875*(4V) + 5625*(3V) + 4717*(1V) + 947*(2V)=30986 V
2	M − 0*(4V) - 0*(3V) - 2607*(1V) - 276*(2V)=M − 3159 V

4.3 MPLS Scenario

The calculus for the MPLS scenario used the data shown in table 8 and 9. The calculus of total revenue for each billing scheme is shown in table 10.

Table 8. Main Variables – CBR Traffics

Traffic	R (Kbps)	T	S	L	N	P	C
CBR1	64	60	256	0,054	101	1774	2.5
CBR2	384	60	512	0,039	219	5405	2.5
VBR1	143	60	1024	0,141	147	900	2.5
VBR2	167	60	1024	0,21	257	966	2.5

Table 9. Mean Rate - VBR Traffics

Traffic	Br (Kbps)	Bt	Bi	R(Kbps)
VBR1	1000	0.5	3	143
VBR2	1000	1	5	167

Table 10. Total Revenue Calculus For MPLS

BS	Total Revenue Calculus
1	1774*(2.5V) + 5405*(2.5V) + 900*(2.5V) + 966*(2.5V)=22612.5 V
2	M − 0*(4V) - 0*(3V) - 2607*(1V) - 276*(2V)=M − 3159V

4.4 MPLS and Diffserv

The calculus for the MPLS - Diffserv scenario used the data shown in table 11 and 12. The calculus of total revenue for each billing scheme is shown in table 13.

Table 11. Main Variables – CBR Traffics

Traffic	R(Kbps)	T	S	L	N	P	C
CBR1	64	60	256	0,0	0	1875	4
CBR2	384	60	512	0,0	0	5625	3
CBR12	1000	60	1024	0,367	2688	4636	1
VBR2	167	60	1024	0,263	321	901	2

Table 12. Mean Rate - VBR Traffic

Traffic	Br(Kbps)	Bt	Bi	R(Kbps)
VBR2	1000	1	5	167

Table 13. Total Revenue Calculus for MPLS-Diffserv

BS	Total Revenue Calculus
1	1875*(4V) + 5625*(3V) + 4636*(1V) + 901*(2V)=30813 V
2	M – 0*(4V) - 0*(3V) - 2688*(1V) - 321*(2V)=M – 3330 V

5   Conclusions and Future Work

For providing QoS benefits for both service provider and customers, the customer receives better performance in higher priority applications. On the other hand, packets with lower economic value are dropped more than those with higher value, probably increasing the provider's revenue.

For the billing schemes presented in this work, the results demonstrates that MPLS, despite of the benefits that it provides in terms of latency and jitter, does not affect the drop of packets that much, hardly affecting the final revenue. In fact, comparing the revenues of IP Best Effort and MPLS scenarios, the first one has a better result. The Diffserv scenario also presented a slightly better result comparing to the MPLS with Diffserv scenario, however in this case the difference has not great significance.

From the results, we can conclude that the scenarios with Diffserv have a better performance for both the provider and the customer.

To decide the best environment in which to develop a billing system, other aspects must be analyzed for that matter, as examples: in case of retransmission, should the customer be charged? Who should be charged in a call: who is calling, who is called or both?

Voice and data services require charging methods that differentiate the price depending on the application and the QoS which is provided for them. Diffserv appears to be a feasible architecture for developing an adequate billing system for these applications.

Acknowledgements. This work was supported by ANATEL (Agência Nacional de Telecomunicações), CAPES (Coordenação de Aperfeiçoamento de Pessoal em Nível Superior) and CNPQ (Conselho Nacional de Desenvolvimento Científico e Tecnológico).

References

1. M. Falkner, M. Devetsikiotis e I. Lambadaris: An Overview of Pricing Concepts for Broadband IP Networks, Carleton University, IEEE Commucations Surveys, 2000.
2. M. C. Caesar, S. Balaraman and D. Ghosal: A Comparative Study of Pricing Strategies for IP Telephony, Departament of Computer Science, University of California at Davis, Paper 0-7803-6451-1/00 IEEE, 2000, pp. 344 – 349.
3. [On-Line]: http://www.ushacomm.com/VOIPBilling.pdf, Usha.com: IP Billing, White Paper VoIP Billing.
4. [On-Line]: http://www.cs.arizona.edu/people/payalg/mobile/final3.pdf, P. Guha, S. Maitra, T. R Sahoo, J. Sharma: Billing and QoS Systems for Public Access 802.11 Networks, Dept. of Computer Science, University of Arizona.
5. [On-Line]: http://dsmpls.atlantis.rug.ac.be.

Overlay Logging: An IP Traceback Scheme in MPLS Network[1]

Luo Wen, Wu Jianping, and Xu Ke

Department of Computer Science and Technology, Tsinghua University,
Beijing, 100084, P.R.China
`{luowen, xuke}@csnet1.cs.tsinghua.edu.cn, jianping@cernet.edu.cn`

Abstract. IP traceback is an important task in Internet security area. Techniques have been developed to deploy in pure IP network, but, to date, no system has been presented to use the facility of MPLS in MPLS-enabled network. We present Overlay Logging, a technique combines the hash based logging (SPIE) and the convenience of setup overlay network in MPLS network. Our system can achieve a relatively lower false positive rate than SPIE, needs less hardware investment, and reduces the storage pressure. It is impervious to multi-path routing in the network. What's more, the network overhead and configuration cost of our system is low.

1 Introduction

Network attacks are having more and more significant influence on the Internet. DoS (denial-of-service) and DDoS (distributed DoS) attacks are widely reported[2]. They make a certain server inaccessible by exhausting its physical or logical resources. Another common attack is for the attacker to generate a few well-targeted packets to disable a system[3]. Therefore, identifying the sources of offending packets is an important task to make the attackers accountable.

IP traceback problem is defined in [1] as identifying the actual source of any packet sent across the Internet. The task of identifying the actual source of the packets is complicated by the fact that the IP address can be forged or spoofed. Unfortunately, IP traceback techniques neither prevent nor stop the attack; they are used only for identification of the source(s) of the offending packets during and after the attack. Furthermore, it may be impossible to precisely identify the source of the attack packets since it may be behind a firewall or have a private IP address. Consequently, IP traceback may be limited to identifying the point where the packets constituting the attack entered the Internet.

MPLS (Multi-Protocol Label Switch) [11] is new network architecture with an advanced forwarding scheme. In MPLS network, packet forwarding acts as label switch-

---

[1] Supported by: (1) the National Natural Science Foundation of China (No. 60473082 & No. 90104002 & No. 60373010); (2) the National Grand Fundamental Research 973 Program of China (No. 2003CB314801).

P. Lorenz and P. Dini (Eds.): ICN 2005, LNCS 3421, pp. 75–82, 2005.

ing. Two important and popular applications of MPLS are Traffic Engineering (TE) and Virtual Private Network (VPN). Now, more and more ISPs are deploying MPLS network, including AT&T, Bell Canada, and Deutsche Telekom etc.

Techniques have been developed to deploy in pure IP network, however, little work has been done on how to deploy IP traceback in MPLS network. Not all traceback solutions for IP network can be simply deployed in MPLS network. They do not take the difficulties in MPLS network into consideration, and say nothing of using the facility of MPLS.

We present Overlay Logging, a technique combines the hash based logging (SPIE [5]) and the convenience of setup overlay network in MPLS network. Compared with SPIE, Our system can achieve a relatively lower false positive rate, reduces the storage pressure, and needs less hardware investment. It is impervious to multi-path routing in the network. What's more, the network overhead and configuration cost is low.

The rest of this paper is organized as follows. In Section II, we describe related work concerning IP spoofing and solutions to the traceback problem. Section III outlines our basic approach. In Section IV, we discuss several problems. Finally, we summarize our findings in Section V.

2 Related Work

There have been several efforts to reduce the anonymity afforded by IP spoofing, as summarized in [4][5], including Ingress Filtering, Link Testing(Input Debugging, Controlled flooding), Logging, ICMP traceback, Packet Marking(PPM[9] & DPM[12]), and IPSec.

SPIE (Source Path Isolation Engine) is introduced in [5]. The key point of SPIE is exploiting hashing techniques to record individual packets passing through each router. In this scheme such routers are deployed with data generation agents (DGAs). Bloom filter is used to map one packet into a single array of bits. The false positive rate (FPR) is controlled by allowing an individual digest table to store limited number of digest sets. It was shown that the memory requirement is roughly 0.5% of the link capacity. The biggest problem of SPIE is that it is not scalable for higher speed routers, because its capability is limited by memory size and access rate.

An overlay network solution called CenterTrack is introduced in [8]. Logically, the solution introduces a tracking router (TR) in the network. By building a generic route encapsulation (GRE) tunnel from every edge router to this TR, this TR monitors all traffic that passes through the network. Then the TR can do input debugging to find the attack packets. This scheme adds about 20 bytes for each packet, which is a large bandwidth overhead. And it changes the packet's original route in the network.

PPM [9] and its improved version Advanced and Authenticated PPM [10] are both based on the idea that routers probabilistic mark packets that pass through them with their addresses or a part of their addresses. Packets are marked with a marking probability p, when the victim receives enough such packets, it can reconstruct the addresses of all the PPM-enabled routers along the attack path. This scheme also suffers from scalability problems. It cannot scale to large number of attackers [13].

Table 1. Traceback schemes comparison

	Management Overhead	Bandwidth overhead	Router overhead	Post-mortem capability	Packets required	Ability to handle major DoS
PPM[9]	Low	None	Low	Y	Thousands	Poor[13]
Center Track[8]	High	High	High	N	1	Good
SPIE[5]	High	None	High	Y	1	Good
Overlay Logging	High	Low	Moderated	Y	1	Good

Table 1 summarizes the comparison among these schemes.

Although both Overlay Logging and CenterTrack take advantage of tunnel and overlay network, they are actually quite different. Overlay Logging does not need to change the packet's route, and produces less bandwidth overhead. Because Overlay Logging is based on logging method, it can do traceback after attack.

PPM cannot be easily deployed in MPLS network. Although the authors of PPM[9] mention that router information can be stored in MPLS label stack for convenience, there are several factors which they have not taken into consideration. First, hosts are often not MPLS capable. Second, LSRs sometimes do penultimate hop popping[11]. Thus, router information in label stack can not be always received by victim host. What is worse, some nodes in MPLS network have no ability of handling IP header.

From table.1, we can see that SPIE is better than PPM in the following aspects: it can do single-packet traceback; it can handle DDoS attack; it does not need to modify IP header. But SPIE has its limitations. It cannot scale to higher speed routers. It needs to deploy special hardware in every router, which may need a lot of investment.

Several improved solutions are presented recently. [6] tries to reduce storage by flow based logging, but it is a problem to maintain per-flow state in high speed router. Sampling logging is introduced in [7]. It utilizes the relation of adjacent routers, combines mark and logging method, and reduces the storage required. However, it cannot do single-packet traceback. Our Overlay Logging solution does not make improvement by compromising SPIE's these advantages. It remains single-packet traceback, and does not introduce more overhead in router.

3 Overview

Overlay Logging extends SPIE in MPLS network. We aim to reduce the hardware investment and storage pressure while maintaining the false positive rate at a low level.

The new idea utilizes the following facts.

1. In a datagram packet network, the full-path traceback is as good as the address of an ingress point in terms of identifying the attacker [12]. It means that the full-path traceback costs a lot trying to get useless information. What's more, they may be puzzled by multi-path routing in the network and gain a worse result.

2. The cost of creating and maintaining a MPLS LSP tunnel is quite small, and the bandwidth overhead introduced by tunnel is few. LSP tunnels are very common in MPLS network, and some assistant tools have been developed to manage tunnels. LSP tunnel adds only 4 bytes (size of one label) to every packet. And LSP tunnels can be established inter-domain.
3. If we do logging in core network, the high bandwidth introduces high cost. If we only do logging in edge devices, the storage pressure can be reduced.

Based on the above facts, our idea is to simplify the attack graph by establishing MPLS tunnels. Consequently, the number of routers that need to deploy DGA is reduced and DGAs can be deployed in edge nodes as possible. We also modify SPIE's traceback mechanism by recording the previous hop information, thus we can achieve a lower false positive.

3.1 Simplify Attack Graph by LSP Tunnel

Two physically non-adjacent routers can become logical neighbors by establishing a LSP tunnel. Transit-only routers in the attack graph can be omitted by tunnel.

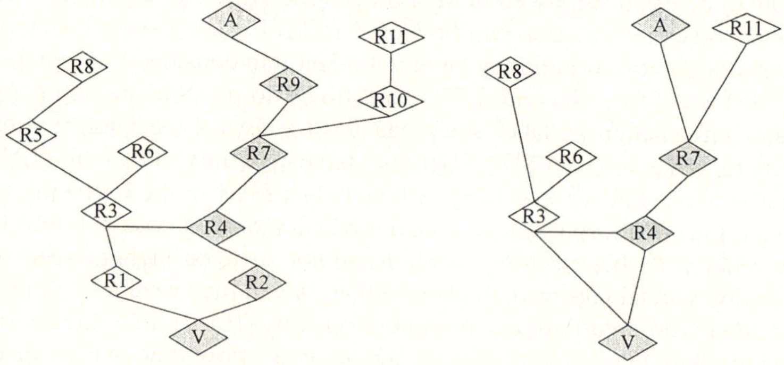

Fig. 1. Simplify attack graph by LSP tunnel

As shown in Fig.1, we simplify the left graph to the right one. Then, we do not need to deploy DGAs on the omitted routers. So the hop count of attack path is reduced, helping to decrease the false positive, as we will discuss in Section IV.

In some extreme situation, an overlay network that full connects all edge nodes can be established, as Fig.2. We just need to deploy DGAs on the edge nodes. And the attack graph becomes the simplest.

It is worth to mention that setting up tunnel and overlay network will not influence the network's origin route. It does not care about how packet route in the tunnel, so technologies such as traffic engineering can still work.

3.2 DGA Deployment and Packet Log

DGAs are only deployed on the routers in the simplified graph. We define other routers as internal routers, define internal interface as interfaces that only connect with internal routers, and define other interfaces as external interfaces.

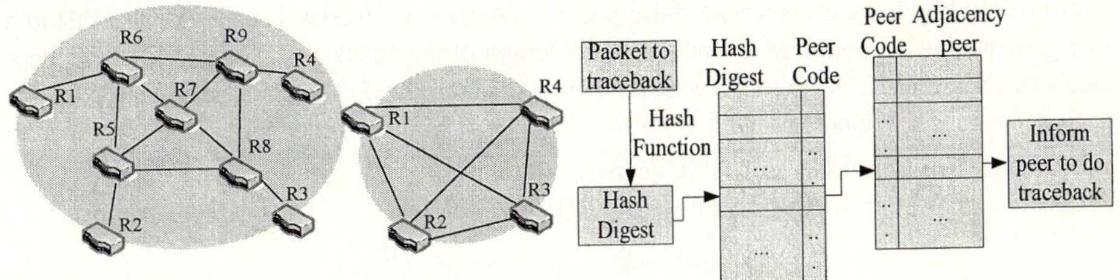

Fig. 2. Edge node's overlay network   **Fig. 3.** Traceback Process on DGA

Different from SPIE, our approach records all incoming packets from the external interfaces. Besides logging hash digest of packet, we also record which adjacency peer the packet came from. Hash digest is the same as SPIE, and the record of adjacency peer is done as following.

We maintain a table of adjacency peers, and encode with Huffman code. Just record Huffman code when record. Because of the limit number of adjacency peers, extra storage is quite limited, may be 8 bits are enough. When the adjacency peer changes, we update the table and Huffman code, and store the old one.

3.3 Traceback Processing

Our traceback processing is different from SPIE. No reverse-path flooding is needed because we have the incoming adjacency peer information. Then traceback in our approach can be done hop by hop. After receiving a request, we start from the router that is closest to the victim. It looks up in its storage, finds the match packet, and then gets the previous hop by the income adjacency peer record, as described in Fig.3. Then, it is turn for the previous router to do the same job. At last, we will find the attacker or the node where the packet entered the network.

4 Discussion

4.1 False Positive

According to [5], the maximum number of additional spurious nodes to the resulted attack graph is given as, $r = n * \dfrac{dP}{(1-dP)}$, where n is the number of nodes actually see a packet with the digest signature, d represents the maximum degree of neighbors among all nodes, and P denotes the FPR of a Bloom filter. If P is small enough, the maximum number of extra nodes approximates $n \cdot d \cdot p$.

In our scheme, $n' = k \cdot n, 0 < k \leq 1$, as we have simplified the attack graph by LSP tunnel. And because we need no reserve path flooding by recording incoming peer, $d' = 1$, then in our approach, $P' = p$. We get

$$r' = n' * \frac{d'P'}{(1 - d'P')} = k \cdot n \cdot \frac{P'}{(1 - P')}$$

Let k=1. In Fig.4, we compare the false positive number of Overlay Logging with SPIE in a SPIE-generated attack graph as a function of the length of the attack path, for p = 1/8. Our theoretical bound line with   is the same as SPIE' theoretical bound with P=1/(8*degree). Suppose d = 16, Our theoretical bound line with $P' = 1/(8 * degree)$ is even better than SPIE's simulation result with 100% utilization and P=1/(8*degree).

If we want to keep the value r equals SPIE, we get

$$n \cdot d \cdot P = k \cdot n \cdot P' \implies P' = \frac{d}{k} P$$

The memory scaling factor (MSF) of a digesting scheme is [5],

$$MSF = a(P) = 1.44 \cdot \log_2 \left(\frac{1}{P} \right)$$

Applying $P' = \frac{d}{k} P$ to the multiplier function in the MSF formula, we get

$$a\left(\frac{d}{k} P \right) = 1.44 \cdot \log_2 \left(\frac{k}{dP} \right) = a(P) - 1.44 \cdot \log_2 \left(\frac{d}{k} \right)$$

We come to the conclusion that if we keep P to a certain value, MSF value of our approach is quite smaller than SPIE.

When k=1, for instance, a false-positive rate of 0.00314, which corresponds to SPIE's degree-dependent simulation, P = p/d, with p = 1/8 for routers with as many as 40 neighbors, can be achieved using 8 digesting functions and memory factor of 12 by SPIE. With our approach, it can be achieved using 3 digesting functions and memory factor of 5.

When k<1, MSF can be more smaller, sometimes may be negative. It does not mean that no memory is needed. Actually it means that we reduce the number of DGAs by establishing tunnels and need much less memory than SPIE.

4.2 Deployment

Our solution reduces number of routers that need to deploy DGA by establishing overlay network. It can be gradually deployed ISP by ISP. Hash-based IP traceback with different digesting schemes can be deployed across the traceback-enabled network. Traceback in IP network and MPLS network can also cooperate together.

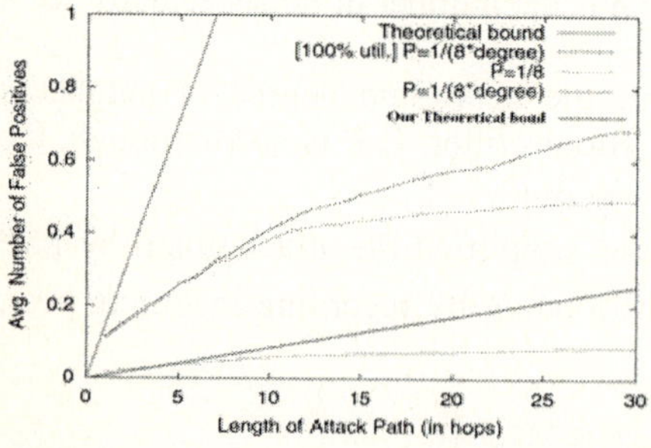

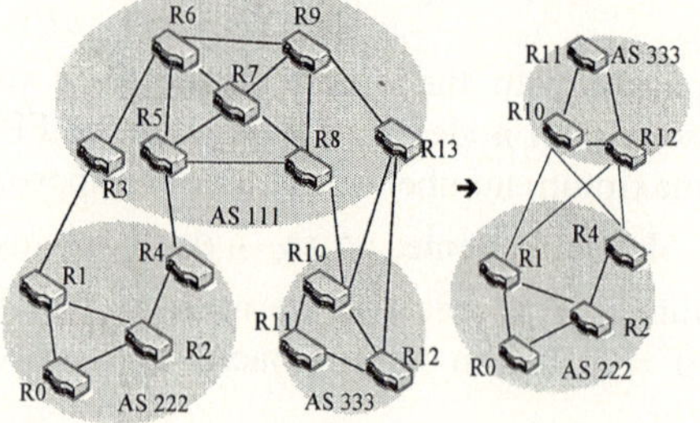

Fig. 4. False Positive compared with SPIE

Fig. 5. Deploy inter-domain LSP tunnel

Another advantage introduced by LSP tunnel is that we can steer clear of traceback-disable network by inter-domain LSP tunnel. In Fig.5, AS 111 does not support traceback, then we just connect AS 222 and AS 333 by inter-domain LSP tunnel to form an overlay network, thus AS 222 and AS 333 can cooperate in traceback.

4.3 Multi-path Routing

Multi-path routing occurs when packets are forwarded from a source s to a destination v through different sets of routers. It may be caused by load balancing or route changes. In MPLS network, Traffic Engineering is an important application, so multi-path routing happens more frequently than in traditional IP network. The existing of multi-path routing has a negative effect on several traceback schemes including PPM. But, it has no influence on our schemes. Our scheme does not care how packet travels in the ISP internal network.

4.4 Others

Overlay logging's vulnerabilities are almost the same as SPIE. Although in MPLS network there will be other attacks like label spoofing, we think that it is a basic MPLS network security problem and should be avoided by other approach.

IP packet transformations rarely happen in MPLS network. But if any transformation happens, methods in SPIE still work.

5 Conclusion and Future Work

In this paper, we have introduced an overlay logging method for IP traceback in MPLS network. The approach combines SPIE and the convenience of setup overlay network in MPLS network. It achieves a relatively lower false positive rate than SPIE, needs less hardware investment, and reduces the storage pressure. It is impervious to multi-path routing in the network. What's more, the network overhead and configuration cost is low. We also present some factors to be noticed when deploying an IP traceback solution in MPLS network.

How to reduce storage in logging still needs more research. Furthermore, the deployment of PPM in MPLS network is also an interesting topic.

References

[1] R. K. C. Chang, "Defending against Flooding-Based Distributed Denial-of-Service Attacks: A Tutorial," IEEE Commun. Mag., Oct. 2002, pp. 42-51.

[2] CERT Advisory CA-2000-01 Denial-of-service developments (2000,Jan.). [Online]. Available: http://www.cert.org/advisories/CA-2000-01.html

[3] R. Sekar, Y. Guang, S. Verma, and T.Shanbhag, "A High-Performance Network Intrusion Detection System," in In Proceedings of the 6th ACM conference on Computer and communications security, 1999, pp. 8-17.

[4] A. Belenky, N. Ansari, "On IP traceback", Communications Magazine, IEEE , Volume: 41 , Issue: 7 , July 2003, pp. 142 - 153

[5] A. C. Snoeren, C. Partridge, L. A. Sanchez, C. E. Jones, F. Tchakountio, B. Schwartz, S. T. Kent, and W. T. Strayers, "Single-Packet IP Traceback," IEEE/ACM Transactions on Networking, vol. 10, no. 6, December 2002, pp.721-734

[6] Tsern-Huei Lee, Wei-Kai Wu, Huang, T.-Y.W., "Scalable packet digesting schemes for IP traceback", Communications, 2004 IEEE International Conference on , Volume: 2 , 20-24 June 2004, pp. 1008 - 1013

[7] Jun Li, Minho Sung, Jun Xu, Li Li, "Large-scale IP traceback in high-speed internet: practical techniques and theoretical foundation", Security and Privacy, 2004. Proceedings. 2004 IEEE Symposium on , 9-12 May 2004, pp. 115 - 129

[8] R. Stone, "Centertrack: An IP Overlay Network for Tracking DoS Floods," Proc. 9th USENIX Sec. Symp., 2000, pp. 199-212.

[9] S. Savage et al., "Network Support for IP Traceback,"IEEE/ACM Trans. Net., vol. 9, no. 3, June 2001, pp. 226-237.

[10] D. X. Song and A. Perrig, "Advanced and Authenticated Marking Schemes for IP Traceback," Proc. INFOCOM,2001, vol. 2, pp. 878-886 vol.2.

[11] E. Rosen, A. Viswanathan, and R. Callon. "Multiprotocol Label Switching Architecture", RFC3031. January 2001.

[12] A. Belenky, N. Ansari, "IP traceback with deterministic packet marking", Communications Letters, IEEE , Volume: 7 , Issue: 4 , April 2003 pp. 162 - 164

[13] Kihong Park, Heejo Lee, "On the effectiveness of probabilistic packet marking for IP traceback under denial of service attack", INFOCOM 2001. Twentieth Annual Joint Conference of the IEEE Computer and Communications Societies. Proceedings. IEEE , Volume: 1 , 22-26 April 2001, pp. 338 - 347 vol.1

Monitoring End-to-End Connectivity
in Mobile Ad-Hoc Networks

Remi Badonnel, Radu State, and Olivier Festor

MADYNES Research Team,
LORIA-INRIA Lorraine,
Campus Scientifique - BP 239,
54600 Villers-les-Nancy Cedex, France
{remi.badonnel, radu.state, olivier.festor}@loria.fr

Abstract. The increasing interest in providing wireless networks, without fixed infrastructures, based on ad-hoc networking raises new challenges towards providing them the required quality of service (QoS). We propose in this paper a new QoS metric, called coverage degree, taking into account the possible end-to-end connectivity in a mobile ad-hoc network. Based on an initial model and extensive simulations, we show how different parameters affect this metric and we propose a monitoring approach capable to evaluate it.

1   Introduction

Wireless technologies show recently significant developments to meet the increasing need of mobility in networks and computing. Mobile Ad-hoc NETworks (MANETs) [1] are mobile networks which can be deployed spontaneously over a geographically limited area, without requiring a fixed infrastructure. They are formed from a set of mobile devices (laptops, personal digital assistants) and are self-organized. The routing task is performed by the mobile devices themselves: one device can interact as a router and forward packets of the others by a multi-hop mechanism. Several major service providers are interested in deploying new services in ad-hoc networks because these are adapted to dynamic environments and do not require costly infrastructures. For instance, the Internet service provider Ozone [2] uses ad-hoc technologies to provide commercial seamless Internet access in Paris. However, users expect a quality of service (QoS) from MANETs as close as possible as in fixed networks. In this paper, we propose a new QoS metric, called **coverage degree**, related to the end-to-end application level of connectivity for a mobile ad-hoc network (close to the notion of coverage used in usual mobile networks). We will study the parameters which have an impact on it and will analyze how we can consequently optimize it for MANETs, by configuring the network components.

This article is consequently structured as follow: we present in Section 2 a new end-to-end QoS measure for mobile ad-hoc networks and a monitoring approach to estimate it. We describe in Section 3 our experimental results, addressing

P. Lorenz and P. Dini (Eds.): ICN 2005, LNCS 3421, pp. 83–90, 2005.
© Springer-Verlag Berlin Heidelberg 2005

the impact of density and mobility on this measure, and proposing a simple provisioning method based on an analytical estimated model. A survey of related work is given in Section 4. Finally, Section 5 concludes the paper and presents future research efforts.

2   A New End-to-End QoS Metric for Ad-Hoc Networks

In this section, we define a QoS metric related to the end-to-end connectivity of an ad-hoc network and propose a monitoring approach for it.

2.1   Measure and Average Value of Coverage Degree

Ad-hoc network specificities make flow/class related type of metrics difficult to conceive and to justify for typical users. We consider that MANET users are interested in an additional metric corresponding to the percentage of all other users that they can interact with, thus a metric covering the communication potential available in the network. We call our proposed metric ad-hoc **coverage degree** and define it below. We consider an ad-hoc network as a set $V = \{v_1, v_2, v_3 \ldots v_n\}$ of mobile nodes moving in a square surface. We as-

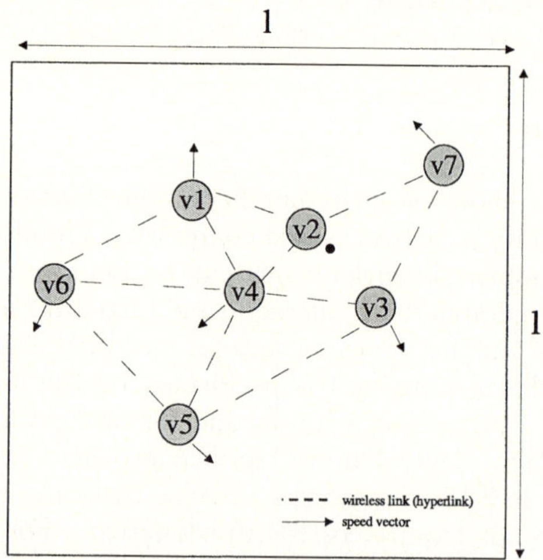

Fig. 1. Mobile ad-hoc networks model

sume nodes mobility references the random way-point (RWP) model [3]: each node moves at constant *speed s* to a destination point selected uniformly in the square surface and then waits during a *pause time p* before moving to a new destination. To assess the behavior of MANETs, we introduce the QoS measure τ which corresponds to an instant-time snapshot of the network coverage degree (see equation 1).

$$\tau : N \times L \times M \times R \times T \to [0,1]$$

$$(n,l,m,r,t) \mapsto \frac{1}{n} \cdot \sum_{i=1}^{n} reachability(v_i,t) \qquad (1)$$

The quintuplet (n,l,m,r,t) characterizes the ad-hoc network with:

- n, the number of nodes $n = |V|$ ($N = \aleph^+$),
- l, the side of the square surface ($L = \Re^+$),
- m, the mobility of nodes defined by the RWP model parameterized with the *speed s* and the *pause time p* ($M = P \times S = \Re^+ \times \Re^+$),
- r, the routing ad-hoc protocol chosen ; R = {AODV, DSR, OLSR, TBRPF},
- t, the time since the network deployment ($T = \Re^*$).

The reachability of a node v_i is the fraction of nodes that v_i can reach (as described in equation 2).

$$reachability(v_i,t) = \frac{|\{v_j \mid j \neq i \ \wedge \ v_i \ reachable \ v_j \ at \ t\}|}{n-1} \qquad (2)$$

v_i *reachable* v_j is true if and only if a possible path exists between the two nodes. Basically, high values of τ (close to 1) are associated to high coverage while low values mean low coverage. Assuming that an ad-hoc network evolves during a given time period $T = [0,t] \subset \Re^+$, the average coverage degree noted $\overline{\tau}$ is defined by equation 3.

$$For \ a \ given \ (n,l,m,r) \in N \times L \times M \times R, \ \overline{\tau}(T) = \frac{1}{T} \int_{0}^{T} \tau(n,l,m,r,t) \, dt \quad (3)$$

2.2 Monitoring Approach for the Average Coverage Degree $\overline{\tau}$

We propose a monitoring approach to evaluate the average coverage degree $\overline{\tau}$ in an operational network. This approach is active [4] i.e. control packets are sent over a time period T in the network to measure $\overline{\tau}$. T is divided in k equal time intervals and measures of coverage degree τ are periodically performed at time moments $T_{measures} = \{t_1, t_2, t_3 \ldots t_k\}$ with $t_p = (p-1)\frac{T}{k}$. Probing packets are used to assess the v_i *reachable* v_j condition i.e. to know if a route exists from v_i to v_j. For $t_p \in T_{measures}$ and $v_i \in V$, we measure $reachability(v_i, t_p)$ by sending one packet to each node $v_j|_{(j \neq i)} \in V$ from v_i and by checking the related packet loss. The coverage degree measures $\tau(t)$ at time moments $T_{measures}$ are averaged on the period T to estimate $\overline{\tau}(T)$ as follow in equation 4.

$$For \ a \ given \ (n,l,m,r) \in N \times L \times M \times R,$$

$$\overline{\tau}(T) \approx \frac{1}{k} \sum_{p=1}^{k} \tau(t_p) \approx \frac{1}{n \times k} \sum_{p=1}^{k} \sum_{i=1}^{n} reachability(v_i, t_p) \qquad (4)$$

3   Experimental Results

3.1   Methodology

Our coverage measures are performed using the discrete event network simulator ns-2 [5], with the OLSR routing protocol extension [6]. The physical layer is simulated with both the free space propagation and the two-ray ground reflection models while the data link layer is based on the IEEE 802.11 MAC model. At the routing layer, AODV[1] and OLSR[2] [7] $R_{simulations} = \{AODV, OLSR\} \subset R$ will be the routing protocols used in the simulation, because of their different fundamental underlying approaches [8]. AODV is a reactive protocol performing route discovery on demand, meaning that it builds routes between nodes only as desired by source nodes. OLSR operates as a table driven and proactive protocol, thus exchanges topology information with other nodes of the network regularly. Finally, the transport layer uses UDP and the application layer is simulated by a traffic source on each node.

The traffic source on a node $v_i \in V$ is configured to periodically send (at time moments t_p) probing packets of given size (32 bits) to each other nodes $v_j|_{(j \neq i)}$. $Reachability(v_i, t_p)$ is measured as the packet loss ratio and consequently, we estimate the average coverage $\overline{\tau}(T)$ using the previous equation. Note that packets sent from nodes v_i are spread over the period time between two coverage measures $t_{p+1} - t_p = \frac{T}{k}$ to optimize the network workload generated by the monitoring activity.

The random way-point mobility [3] is parameterized and simulated using the ns2 setdest tool. We assume the time period $T_{simulations}$ lasts 300 seconds for a simulation and the number of monitoring measures k is set to 30. Each experiment is performed 30 times to avoid random biases.

3.2   Impact of Network Density on Coverage Degree

The first experiment consisted in evaluating the impact of network density on the measured coverage degree using the routing protocol OLSR and respectively AODV. We used the ANOVA [9] method to check significant statistical differences between OLSR and AODV. We consider a simple mobility model scenario ($p = 10s$ and $s = 5m/s$) and we varied the number of nodes n from 10 to 80 (step 10) and the side of the square surface from 100 to 3000 m (step 100). The experimental set E_1 can be defined as follow:

$$E_1 = N_1 \times L_1 \times M_1 \times R_{simulations} \times T_{simulations}$$
$$= [10 - 80 \ nodes] \times [100 - 3000 \ m] \times \{10 \ s\}$$
$$\times \{5 \ m/s\} \times \{AODV, OLSR\} \times [300 \ s]$$

At a ($error \ probability < 0.005$) level, we concluded that the coverage degree is higher in OLSR (0.45) than in the AODV case (0.22). This result holds on the

---

[1] Ad-Hoc On Demand Distance Vector.
[2] Optimized Link State Routing.

overall simulations where every possible setting (node number, surface size) is considered. However, at a closer analysis, we observe that each routing protocol can be associated to a configuration settings where it outperforms the other protocol. For instance, AODV is better in cases, where the density (ratio $\frac{n}{l}$) is less than 0.015, while OLSR has better performance on cases where the density is above 0.015 in the experimental set E_1.

We performed a regression analysis in order to determine an analytical dependency among n, l and the coverage degree for each routing protocol (see table 1). Statistically significant dependencies (*error probability* < 0.005) are

Table 1. Standardized regression coef. with E_1 (Impact of density on coverage degree)

Regression coefficients	Routing Protocol	
	AODV	OLSR
Constant	0.72	0.99
Coefficient for n	-0.007	0.0007
Coefficient for l	-0.0001	-0.0003

detected among the coverage and both n and l. If we increase the number of nodes, the coverage degree tends to get lower in the AODV case, while the impact of node increase is positive (more nodes, better coverage) with OLSR. In AODV, the coefficient for n is a factor of magnitude higher than in OLSR: one added node affects 10 times more the degree. Comparing the coefficients for the surface, we note that a 1 unit variation in l (keeping the same n) affects 3 times more (decreasing) the coverage in the case of OLSR.

The practical applications of these experiences are the following:

1. Provisioning a given area for a desired QoS level: if a desired level of QoS must be provided on a given zone, and the number of nodes can not be controlled a priori then OLSR should be used.
2. Solving the regression inequality on 2 parameters can give us a bound on the third parameter. For instance, to assure a degree of 0.5 in a AODV routed zone of size l = 500 m, we must have $0.72 - 0.007 \times n - 0.0001 \times l > 0.5$ that is $n < 24$. This gives an upper bound on the number of mobile nodes that can be served at this coverage degree.
3. Self-configuration of the routing protocol can be performed. We observed that AODV outperforms OLSR for the lower density cases (low $\frac{n}{l}$) while OLSR was better for high density. So we could impose a routing protocol according to the measured density and reconfigures it if the observed density changes.

3.3   Impact of Network Mobility on Coverage Degree

In the second experiment, we checked the dependency of the mobility parameters (pause time and speed) on the coverage degree. We considered a constant number of nodes in the ad-hoc network (n = 30) and a constant surface size (l = 600 m). The mobility is characterized by pause time and speed so we varied the nodes mobility by choosing pause time from 300 to 10 s (step 10) and nodes speed from 2 to 50 m/s. Remark that a strong mobility means a high speed and a LOW pause time. Finally, the second experimental set E_2 considered is:

$$E_2 = N_2 \times L_2 \times M_2 \times R_{simulations} \times T_{simulations}$$
$$= \{30\ nodes\} \times \{600m\} \times [300 - 10\ s]$$
$$\times [2 - 50\ m/s] \times \{AODV, OLSR\} \times [300\ s]$$

We performed an ANOVA analysis [9] using pause time and speed as possible factors. At an ($error\ probability < 0.005$), we identified that the pause time and the speed affect the resulting coverage degree (see table 2). The impact

Table 2. Standardized regression coefficients with E_2 (Impact of mobility on coverage)

Regression coefficients	Routing Protocol	
	AODV	OLSR
Constant	0.86	0.92
Coefficient for p	-0.0007	0.0003
Coefficient for s	-0.001	-0.0006

of the pause time is antagonistic for the two routing procotols: the coverage is improved with higher pause time values in OLSR while the situation is reversed in the case of AODV. Moreover, the speed parameter impacts (decreasing) 2 times more with AODV than with OLSR. This regression based analysis gives us also simple formulas to use in order to provision an ad-hoc network. For instance, a QoS level of minimum 50% must be assured on 600 dimensional 30 node network and a given minimal pause time of 10 is known, so we have a limit of the maximal speed at which nodes can move.

4   Related Work

Other papers [10] [11] addressed the issue of the connectivity on mobile ad-hoc networks but their focus was on the impact of the transmitting range value r on the coverage (assuming r is the same for each node of the network). The coverage is estimated at the physical layer according to r. In [10], the author analyses the optimal transmitting range to ensure the global coverage for a percentage of time and the approach proposed in [11] evaluates the value of r to limit the number of isolated nodes (i.e. to ensure a number of 1-hop neighbors). In our approach, we consider that we do not have any control on the transmitting range (we assume a given value for r) and our work addresses connectivity from an end-to-end viewpoint, which is related to the way an application perceives the underlaying connectivity.

The MANET working group proposes in several other quantitative metrics such as end-to-end delay, route acquisition time, efficiency for routing procotols in MANETs [12]. The approach proposed in [13] presents also a performance comparison of ad-hoc network routing protocols concerning packet delivery ratio and routing overhead in packets. While we analyze the impact of the routing protocols AODV and OLSR on the coverage, we do not try to evaluate the performance of the routing protocols but their impacts on the coverage. In this way, [14] studies the impact of mobility on the delay and the throughput in MANETs: it exploits the patterns in the mobility of nodes to provide guarantees on these two metrics. The approach proposed in [15] focuses also on the throughput capacity but in an hybrid network (with some static nodes). These performance evaluations are based on probabilistic techniques and differ from our work, where we considered a close to real time estimation using a monitoring approach. A resource usage monitoring tool called WANMon is proposed in [16] to work over tcpdump and can be installed on a wireless node to monitor network, power, memory as well as CPU usage. The monitoring information must be next propagated periodically to the rest of nodes. In [17], an analysis is proposed to measure performance of ad-hoc networks using time-scales for information flow.

5   Conclusion and Outlook

Our paper proposes a new QoS metric, called **coverage degree**, which focuses on the potential end-to-end connectivity offered to a user or an application in ad-hoc networks. We are interested in providing an objective indicator of the potential of a node to communicate/interact with the others via multi-hop communications. After defining this metric, we propose a monitoring approach for it. We analyse the impact of several factors (routing, density, mobility) on it and describe an analytical formula for these dependencies based on statistical estimation. Our preliminary results could be used in a self-configuring ad-hoc infrastructure, where a monitoring phase could dynamically change/configure the network components. For instance, the routing protocol can be changed if some monitored parameters increase beyond predefined thresholds. Our monitoring process could be performed by a subset of elected nodes (light-weight approach) and be integrated in a more complete management approach for ad-hoc networks such as the ANMP architecture described in [18].

References

1. Internet Engineering Task Force: MANET (Mobile Ad-Hoc Networks) Working Group. (http://www.ietf.org/html.charters/manet-charter.html)
2. Haladjian, R.: OZONE Wireless Internet Provider. http://www.ozone.net (2003)
3. Bai, F., Sadagopan, N., Helmy, A.: Important: a Framework to Systematically Analyze the Impact of Mobility on Performance of Routing Protocols for Ad-Hoc Networks, San Francisco, CA, USA, International Conference on Computer Communications (IEEE INFOCOM) (2003)
4. Aida, M., Miyoshi, N., Ishibashi, K.: A Scalable and Lightweight QoS Monitoring Technique Combining Passive and Active Approaches, San Francisco, CA, USA, International Conference on Computer Communications (IEEE INFOCOM) (2003)

5. USC/ISI, PARC, X., LBNL, UCB: The VINT Project: Virtual InterNetwork Testbed. (http://www.isi.edu/nsnam/vint/)
6. Navy Research Laboratory OLSR Project: OLSR Extension for ns-2. http://pf.itd.nrl.navy.mil/projects/olsr/ (2003)
7. Toh, C.K.: Ad-Hoc Mobile Wireless Networks. Number ISBN 0-13-007817-4. Pearson Education, Prentice Hall (Eds.), New Jersey, USA (2002)
8. Perkins, C.E.: Ad-Hoc Networking. Number ISBN 0-201-30976-9. Pearson Education, Addison-Wesley (Eds.), New Jersey, USA (2000)
9. Muller, K.E., Fetterman, B.A.: Regression and ANOVA: An Integrated Approach Using SAS Software. Number ISBN 0-471-46943-2. (2003)
10. Santi, P., Blough, D.M.: An Evaluation of Connectivity in Mobile Wireless Ad-Hoc Networks, Bethesda, MD, USA, the International Conference on Dependable Systems and Networks (IEEE DSN), IEEE Computer Society (2002)
11. Bettstetter, C.: On the Minimum Node Degree and Connectivity of a Wireless Multihop Network, Lausanne, Switzerland, the 3rd ACM Interational Symposium on Mobile Ad-Hoc Networking and Computing (ACM MOBIHOC) (2002)
12. Corson, S., Macker, J.: Mobile Ad-Hoc Networking: Routing Protocol Performance Issues and Evaluation Considerations. http://www.ietf.org/rfc/rfc2501.txt (1999) IETF RFC 2501.
13. Broch, J., Maltz, D.A., Johnson, D.B., Hu, Y.C., Jetcheva, J.: A Performance Comparison of Multi-Hop Wireless Ad-Hoc Network Routing Protocols, Dallas, TX, USA, the 4th International Conference on Mobile Computing and Networking (ACM MOBICOM) (1998)
14. Bansal, N., Liu, Z.: Capacity, Delay and Mobility in Wireless Ad-Hoc Networks, San Francisco, CA, USA, International Conference on Computer Communications (IEEE INFOCOM) (2003)
15. Liu, B., Liu, Z., Towsley, D.: On the Capacity of Hybrid Wireless Networks, San Francisco, CA, USA, International Conference on Computer Communications (IEEE INFOCOM) (2003)
16. Ngo, D., Wu, J.: WANMON: a Resource Usage Monitoring Tool for Ad-Hoc Wireless Networks. Number ISBN 0-7695-2037-5, Bonn, Germany, the 28th Annual IEEE Conference on Local Computer Networks (IEEE LCN), IEEE Computer Society (2003) 738–745
17. D'Souza, R., Ramanathan, S., Land, D.: Measuring Performance of Ad-Hoc Networks using Timescales for Information Flow, San Francisco, CA, USA, International Conference on Computer Communications (IEEE INFOCOM) (2003)
18. Chen, W., Jain, N., Singh, S.: ANMP: Ad-Hoc Network Management Protocol. Journal on Selected Areas in Communications (IEEE JSAC) **17** (1999) 1506–1531

Multi-path Routing Using Local Virtual Infrastructure for Large-Scale Mobile Ad-Hoc Networks: Stochastic Optimization Approach[*]

Wonjong Noh and Sunshin An

Department of Electronics and Computer Engineering, Korea University,
Sungbuk-gu, Anam-dong 5ga 1, Seoul, Korea, Post Code: 136-701
Phone: +82-2-925-5377, FAX: +82-2-3290-3674
`{nwj, angus, sunshin}@dsys.korea.ac.kr`

Abstract. In this paper, we proposed new ad-hoc optimal redundant routing schemes for large-scale wireless mobile ad-hoc networks. First, we proposed locally and partially clustered network model and the concepts of synchronization. Second, we formulated a mathematical optimization model for the selection of an optimal redundant synchronous routing path. The performance evaluation results showed that the our optimal redundant routing scheme had better performance than SMR and TORA with respect to average throughput, average end-to-end delay and average control traffic overhead in large scale ad hoc networks.

1   Introduction

The diversification and integration of computing environment of wireless mobile terminals leads to large-scale ad-hoc networks [1]. Some typical problems, which get more important than ever, not yet solved but should be solved in large-scale ad-hoc networks, are as follows: First, frequent topology changes in large-scale ad-hoc networks caused by node mobility make more combinatorially unstable [2] routing than ever. In combinatorially unstable ad hoc network, global topology update message leads to incorrect network topology by its imprecise information, prevents find loop-free path, and reduces of available bandwidth of user data. Second, the real-time multimedia streaming traffics, the main traffic type in the future, require QoS-preserve [2] for its connection. There are so many hard researches to solve these problems with the hierarchical ad-hoc routings such as area-based hierarchy approach [3, 4] and landmark based hierarchy approach [5, 6].

The main concern of this research is to propose more stable and high performance routing through best-effort virtual infrastructure in large-scale ad-hoc network. Our routing scheme, Synchronization Degree based Optimal Redundant Routing (SDOR),

* This research was supported by the MIC (Ministry of Information and Communication), Korea, under the ITRC (Information Technology Research Center) support program supervised by the IITA (Institute of Information Technology Assessment).

P. Lorenz and P. Dini (Eds.): ICN 2005, LNCS 3421, pp. 91–98, 2005.

has the characteristics and differences such as: (i) Unlike [3], SDOR has no cluster-head selection algorithm. (ii) Unlike [3, 4], SDOR has no MCDS (Minimum Connected Dominating Sets), NP-hard problem. There is little overhead for state and spine management. (iii) Unlike [3, 4, 5, 6], SDOR is flooding based on-demand type redundant routing. (iv) Unlike [5], SDOR constructs the reduced topology graph of the entire network and compute the optimal routes. (v) SDOR uses the concepts of core nodes and synchronizations. The paper is organized as follows. Section II describes proposed ad-hoc network model and the concept of synchronization. Section III describes mathematical optimal problem formulation and its solution. The performance results are discussed in Section IV. The conclusions are presented in Section V.

2 Network Model

In this section, we describe our ad-hoc network model. We have some assumptions that (i) all nodes are homogeneous (ii) all nodes have sufficient computing power and battery power (iii) all nodes can know their position vector and mobility vector with the aids of Global Positioning System (GPS).

2.1 Network Model

The characteristics of our large-scale ad-hoc network model are as follows: There are special types of ad-hoc mobile nodes in our network model: hub and bridge. The clustering is performed partially, locally and in distributed way.

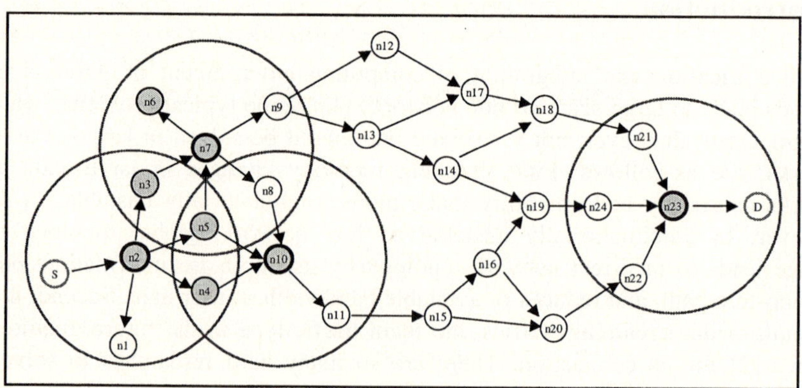

Fig. 1. Ad-Hoc Network Model

If a node moves with lower average speed than ε and lower current speed than δ, it plays a role of hub, e.g., n2, n7, n10 and n23. If a node directly connects two hubs, it plays a role of bridge, e.g., n3, n4, n5 and n8. There are only four clusters in the Fig. 1. The nodes of n12 ~ n20 have not been clustered because of there are not any hub nodes.

2.2 Synchronization

In this section, we introduce the time and mobility synchronization concepts. First, we introduce time synchronization concepts using (1). Where, $nbd(\bullet)$ and T_{ij}^e denote the neighbor nodes of $node(\bullet)$ and the expected link available time between node(i) and node(j) respectively. The T_{ij}^e is determined by current positions, mobility speeds and directions of node(i) and node(j) under the assumption of simple mobility model [7].

$$TimeSync(i,j) = \left\{ k \mid T_{ij}^e \leq T_{ik}^e, T_{ij}^e \leq T_{kj}^e, k \in \{ nbd(i) \cap nbd(i) \} \right\} \tag{1}$$

The (1) says that when the link(i, j) is broken, the link(i, j) can be locally repaired by still alive node(k), which is time-synchronized to link(i, j). Second, we introduce mobility synchronization concepts using (2). Where, v_i, mobility vector of node(i), denotes the mobility speed and direction of node(i).

$$MobilitySync(i) = \left\{ k \in nbd(i) \mid \left| \arccos\left(v_i \cdot v_k / |v_i| \cdot |v_k| \right) \right| \leq \frac{\pi}{2} \right\} \tag{2}$$

The (2) says that if the angle between node(i)'s mobility direction and node(k)'s mobility direction is less than $\pi/2$ degree, node(i) and node(k) is mobility-synchronized. When they are mobility synchronized, we can say that they move similar direction.

3 Optimal Redundant Route Decision

After route request procedures, the destination node determines optimal redundant route. In this section, we propose optimal route path selection scheme using mathematical formulation. The route request, route setup and route maintenance procedures of SDOR follows the procedures of [8]

3.1 Mathematical Formulation

We derive a mathematical formulation of Path-Constrained Path-Optimization type problem [9] to achieve the optimal redundant routing path. The problem data and decision variables are defined as follows.

$$p_{ij(H \to B)} \quad \sum_{j \in \, j \, and \, k} \min\left\{ |TimeSync(i,j)|, UpperBound \right\} \tag{3}$$

$$where, k \in \bigcup_{m \in \{nbdH(j)-i\}} \{ n \mid nbd(i) \, and \, m \in nbdH(n) \} \tag{4}$$

$$p_{ij(B \to H)} \quad \sum_{i \in \, i \, and \, k} \min\left\{ |TimeSync(i,j)|, UpperBound \right\} \tag{5}$$

$$where, k \in \bigcup_{m \in \{nbdH(j)\}} \{ n \mid nbd(i) \, and \, m \in nbdH(n) \} \tag{6}$$

$$p_{ij(H \to H)} \qquad \sum_{k} (p_{ik} + p_{kj}), \; where \; k \in TimeSync(i, j) \tag{7}$$

$$p_{ij(H \to N)} \qquad \min\{|TimeSync(i, j)|, UpperBound\} \tag{8}$$

$$p_{ij(N \to H)} \qquad \min\{|TimeSync(i, j)|, UpperBound\} \tag{9}$$

$$p_{ij(N \to N)} \qquad \min\left\{ \sum_{k \in TimeSync(i,j)} e^{-|E(v_{ij})|} \cdot I_{k,MobilitySync(i)}, UpperBound \right\} \tag{10}$$

$$E[q_{ij(H \to B)}] \qquad \sum_{k} [\pi_k \cdot \min\{|TimeSync(i,k)|, UpperBound\}] \tag{11}$$

$$E[q_{ij(B \to H)}] \qquad \sum_{k} [\pi_k \cdot \min\{|TimeSync(i,k)|, UpperBound\}] \tag{12}$$

$$E[q_{ij(H \to H)}] \qquad \sum_{k} [E[q_{ij(H \to B)}] + E[q_{ij(B \to H)}]], \; where \; k \in TimeSync(i, j) \tag{13}$$

$$E[q_{ij(H \to N)}] \qquad 0 \tag{14}$$

$$E[q_{ij(N \to H)}] \qquad 0 \tag{15}$$

$$E[q_{ij(N \to N)}] \qquad 0 \tag{16}$$

$$I_{k, MobilitySync(i)} \qquad \begin{cases} 1, k \in MobilitySync(i) \\ 0, k \notin MobilitySync(i) \end{cases} \tag{17}$$

$$x_{ij} \qquad \text{Link between node(i) and node(j);} \tag{18}$$

The p_{ij} means link profit, i.e., redundancy between node(i) and node(j), and $E[q_{ij}]$ means reward profit, i.e., expected link redundancy when it is impossible to connect node(i) and node(j). The p_{ij} and $E[q_{ij}]$ is measured as (3) ~ (10) and (11) ~ (16) according to the node(i)'s type and node(j)'s type. The π_k in (11) and (12) is a probability that node(k) is chosen. The (17) denotes the mobility synchronization indicator function. The (18) shows the decision variable. Using the definitions and problem data, we can formulate the optimization problem for expected redundancy maximization as follows:

Maximize

$$\sum_{(i,j) \in E} [p_{ij} + E_w[q_{ij}]] \cdot x_{ij} \tag{19}$$

Subject to

$$\sum_{(k,j) \in E} x_{kj} - \sum_{(i,k) \in E} x_{ik} = 1 \qquad \text{where, } k \text{ is a source node} \tag{20}$$

$$\sum_{(k,j) \in E} x_{kj} - \sum_{(i,k) \in E} x_{ik} = -1 \qquad \text{where, } k \text{ is a destination node} \tag{21}$$

$$\sum_{(k,j)\in E} x_{kj} - \sum_{(i,k)\in E} x_{ik} = 0 \qquad \text{where, } k \text{ is a relay node} \qquad (22)$$

$$\sum_{j\in nbdN(i)} x_{ij} \leq \sum_{j\in nbdB(i)} x_{ij} - 1 \qquad \text{for all hub} \qquad (23)$$

$$\sum_{j\in nbdN(i)} x_{ij} \leq \sum_{j\in nbdH(i)} x_{ij} \qquad \text{for all bridge} \qquad (24)$$

$$\sum_{j\in nbdN(i)} x_{ij} \leq x_{ihub(i)} \qquad \text{for all clustered normal node} \qquad (25)$$

$$\sum_{j\in nbdF(i)} x_{ij} \leq \sum_{j\in nbdC(i)} x_{ij} \qquad \text{for all non-clustered normal node} \qquad (26)$$

$$\sum_{(i,j)\in E} x_{ij} \leq 1 \qquad \text{for simple path} \qquad (27)$$

$$\sum_{(i,j)\in E} x_{ij} \leq m \qquad \text{for maximal path length constraint} \qquad (28)$$

$$x_{ij} = 0 \text{ or } 1 \qquad \text{for all node(i) and node(j)} \qquad (29)$$

We introduce new routing metric, 'expected path redundancy', which implies how much possible synchronized redundant paths on a route to be built up. A route's expected path redundancy is expressed by the sum of 'expected synchronization degree' of intermediate links and nodes involved in the route. In this formulation, our objective (19) is to select the optimal route, which maximizes the synchronization redundancy degree along the path. Constraints (20), (21) and (22) ensure that a continuous route starts from source and ends at destination. Constraint (23), (24), (25) and (26) denote the connection topology at hubs and bridges and normal nodes. Constraint (27) denotes that there should be no subtour. The optimal path should be simple path (that is, a path encountering no vertex more than once). Constraint (28) denotes that the length of optimal route path cannot be longer than m.

4 Performance Evaluations

In this section, we discuss experimental results for our optimal routing scheme. We evaluated the performance of SDOR with Split Multipath Routing with Maximally Disjoint Paths (SMR) [10] and the Temporally Ordered Routing Algorithm (TORA) [11] through simulations. For the simulation, the environmental factors we consider in our experimentals are as follows, Table 1.

Table 1. Evaluation Environment

MAC protocol	IEEE 802.11
Radio propagation range	200m
Channel capacity	2 Mbits/sec
Payload	512byte payload
Constant bit rate sources	2 packets/sec
Simulation time	300 sec
End-to-End Delay Constraint (EDCT)	2 sec
m	10
ε	2 m/sec
δ	4 m/sec
$\pi_k, k \in A$	$\dfrac{1}{\|A\|}$, uniform distribution

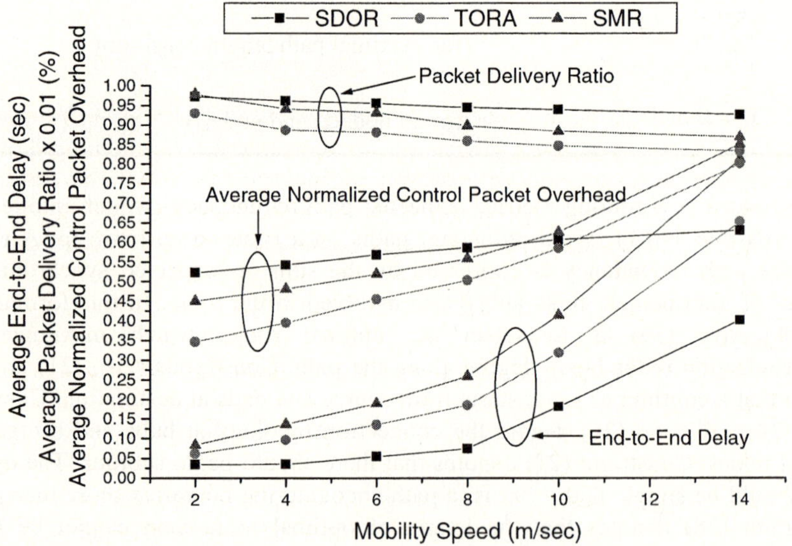

Fig. 2. Performance Evaluation: 1500 (m) x 1500 (m): 30 nodes: 10 sessions

Fig. 2 shows the performance of SDOR in comparison with SMR and TORA with respect to end-to-end delay, packet delivery ratio and control packet overhead.

First, SDOR has shorter end-to-end delay than SMR and TORA and this difference becomes more obvious as the mobility speed increases. There are three main reasons for this: longer route re-constructing delay and frequent route reconstructing. For SMR and TORA, "longer route re-constructing delay due to route recovers from source node instead of intermediate node [10, 11]", "frequent route re-discovery policy [10]" and "frequent link-reversal algorithm [11]" make SMR and TORA has longer end-to-end delay than SDOR. Second, SDOR and SMR have relatively robust packet delivery rather than TORA. The main reason is that SDOR and SMR has pre-setup loop free redundant routes, while TORA occurs packet drops due to looping occasionally after route reconstruction. And when there is a link failure on the route, fast route recovery or reconfiguration of SDOR and SMR makes the packet delivery more robust than TORA. Longer delays in reconstructing routes and the period of time the data packets are buffered at the source or intermediate node during route recovery results in larger packet loss. And then, over about 11 m/sec, SDOR has lower normalized control packet overhead, it is almost constant as the mobility speed increases, the than the others. For SMR, huge RREQ packets due to "permission of duplicate transmissions of route query messages" and "frequent route re-discovery" make SMR has higher control packet overhead than SDOR above about 9 m/sec. For TORA, "Update packets used for destination-oriented DAG at every link failure" and "hello packets" SMR has higher control packet overhead than SDOR above about 9 m/sec. Below about 9 m/sec, SDOR has a little larger control packet overhead than the others. This is because, unlike the others, SDOR uses hello packets. We, however, can control the hello packet intervals differently according to the type of hello packet to reduce the control packet overhead.

5 Conclusions

In this paper, we proposed new optimal redundant routing scheme using the concept of spatial, time and mobility synchronization, core nodes such as hub and bridge and local and partial clustering for large-scale wireless ad hoc networks. For optimal redundant routing path selection, we suggested mathematical optimization formulation model. The performance evaluation proved that SDOR outperformed SMR and TORA in terms of end-to-end delay, packet delivery ratio and control packet overhead. The main reasons are that proposed optimal redundant scheme could effectively localize the effects of route failures, reduce the route rediscovery procedures upon route failures and provides more practical and effective redundant paths. Our experiments confirm that our scheme is efficient to wireless mobile ad hoc networks over medium mobility speed of 9 m/sec of large-scale ad-hoc network. In the near future we intend to experiment with our routing scheme using more various network topologies and more extended synchronization concepts such as QoS synchronization and resource synchronization in real environments. We also intend to explore scheme for stable data dissemination of finite amount of sensing data in sensor networks consisting of mobile sensor nodes with low power energy and high mobility.

References

[1] M.Abolhasan, T.A.Wysocki, and E.Dutkiewicz: "A Review of Routing Protocols for Mobile Ad hoc Networks", Elsevier Journal of Ad hoc Networks, no. 2, 2004.
[2] S. Chakrabarti, et al, "QoS Issues in Ad Hoc Wireless Networks", IEEE Comm. Magazine, Feb., 2001.
[3] C. C. Chiang, "Routing in Clustered Multihop, Mobile Wireless Networks with Fading Channel", Proceedings of IEEE SICON, April, 1997.

[4] Raghupathy Sivakumar, Prasun Sinha, and Vaduvur Bharghavan "Spine Routing in Ad Hoc Networks", Journal of selected area in communication, August, 1999.

[5] Z. Hass and M. Pearlman, "The Performance of Query Control Schemed for the Zone Routing Protocol", Proceedings of ACM SIGCOMM'98, June, 1998.

[6] A. Iwata, C. C. Chiang, G. Pei, M. Gerla, and T. W. Chen, "Scalable Routing Strategies for Ad Hoc Wireless Networks", IEEE Journal on Selected Areas of Communications, August, 1999.

[7] Tracy Camp, Jeff Boleng and Vanessa Davies, "A Survey of Mobility Models for Ad Hoc Network Research", Wireless Communications & Mobile Computing (WCMC): Special issue on Mobile Ad Hoc Networking: Research, Trends and Applications, 2002.

[8] W. J. Noh, Y. K. Kim and S. S. An, "Node Synchronization Based Redundant Routing for Mobile Ad-Hoc Networks", LNCS, No. 2928. 2004.

[9] Shigang Chen, Klara Nahrstedt, An Overview of Quality-of-Service Routing for the Next Generation High-Speed Networks: Problems and Solutions, IEEE Network Magazine, Special Issue on Transmission and Distribution of Digital Video, vol. 12, num. 6, pp. 64-79, November-December, 1998.

[10] S.J. Lee and M. Gerla, "SMR: Split Multipath Routing with Maximally Disjoint Paths in Ad-Hoc Networks", Proceedings of ICC 2001, Helsinki, Finland, June 2001.

[11] V. Park and S. Corson, "Temporarally-Ordered Routing Algorithm (TORA) VERSION 1 Internet Draft", draft-ietf-manet-tora-spec-03.txt, June 2001.

Candidate Discovery for
Connected Mobile Ad Hoc Networks

Sebastian Speicher[*] and Clemens Cap

University of Rostock,
Institute of Computer Science,
Information and Communication Services Group
{speicher, cap}@informatik.uni-rostock.de

Abstract. Internet-connected multi-hop ad hoc networks enable Internet access for mobile nodes that are beyond direct communication range of an access point or base station. However, due to mobility or insufficient Quality of Service (QoS), mobile nodes need to perform handoffs between such networks. We propose a novel concept, which enables mobile nodes to query capabilities (e.g. available QoS or pricing) of adjacent Internet-connected multi-hop networks prior to the actual handoff. By this, mobile nodes can directly handoff to the most suitable network.

1   Introduction and Motivation

An emerging wireless networking concept to connect mobile clients to the Internet is the notion of Internet-connected multi-hop ad hoc networks [1]. In contrast to stand-alone mobile ad hoc networks, at least one Connected Mobile Ad Hoc Network (CMANET) node is also linked to the Internet and acts as an access router (AR) for the other ad hoc nodes. This concept is also known by the name of multi-hop hot spots [2], which extend the range of wireless access points by leveraging mobile nodes in range of the access point (AP) as packet forwarders for nodes further away. Implementations of this network model are already in place, providing Internet access to wireless community networks [3]. Another example for CMANETs are hybrid wireless/wired ad hoc sensor networks [4].

It is very likely that in many areas multiple CMANETs will coexist in parallel, based on different technologies, e.g. IEEE 802.11a/b/g or IEEE 802.16, or by using separate channels of the same technology. Hence, mobile nodes (MNs) could choose among multiple available networks and may handoff between them.

Up to now, MNs can select the CMANET to handoff to only based on signal strength to available points of attachment (i.e. packet forwarding nodes) of these networks, since this is usually the only information available when scanning for wireless nodes at the link layer. However, capabilities such as pricing, security or QoS are also important when selecting the handoff target. Nevertheless, querying

---

[*] Funded by a grant from the German Research Foundation (DFG), Graduate School 466, *Processing, Administration, Visualisation and Transfer of Multimedia Data.*

P. Lorenz and P. Dini (Eds.): ICN 2005, LNCS 3421, pp. 99–106, 2005.

such information from adjacent handoff candidate CMANETs already requires to perform full handoffs to them. This is a problem because each handoff adds additional delays. Furthermore, joining a network (even shortly) whose QoS is insufficient could already be disruptive for real-time applications.

Therefore, several papers [5, 6, 7] propose to avoid this by a) discovering geographically adjacent access routers, b) continuously exchanging capability information among these routers via the Internet and c) enabling each access router to provide this information to mobile nodes attached to it. Since [5, 6, 7] focus on one-hop wireless networks, capability information such as pricing or available QoS is considered equal for all nodes attached to an access router; thus, capabilities can only be defined per access router [7].

However, in the multi-hop CMANET case, capability information is not necessarily equal for the entire network. For instance, the QoS an individual CMANET node experiences when accessing the Internet depends on the QoS available at its point of attachment to the network, i.e. its next hop towards the access router. However, due to other traffic or differing channel conditions along the route, QoS available at different nodes can vary greatly. Thus, it is necessary to consider the capabilities of the likely point of attachment when mobile nodes request capabilities of candidate CMANETs. Furthermore, since each CMANET node is a potential point of attachment, capabilities must be available on a per node basis.

This paper contributes a candidate discovery architecture for connected mobile ad hoc networks, which allows to query capabilities of adjacent candidate CMANETs prior to handoff. In contrast to existing concepts, our approach also considers multi-hop specific aspects such as the likely point of attachment of MNs. As a result, mobile nodes can use these capabilities as the basis for the decision, which CMANET to handoff to; thus, MNs are enabled to *directly* handoff to the most suitable network.

The rest of this paper is structured as follows: In the next section we present the CMANET network model and related terminology in detail. While Sect. 3 discusses related work, we illustrate our concept in Sect. 4. Section 5 concludes this paper and lists tasks for future work.

2   Network Model and Terminology

Connected Mobile Ad Hoc Networks (CMANET), as depicted in Fig. 1, consist of mobile nodes (MN), which are equipped with wireless transceivers. All nodes a MN A can directly communicate with, without relying on the assistance of other forwarding nodes, form A's neighbourhood [8]. To communicate with other nodes, A's packets get routed to their destination via intermediate mobile nodes, which are determined by ad hoc routing protocols, such as AODV or DSR [9]. In contrast to stand-alone ad hoc networks, one CMANET node is also connected to a wide area network (WAN), e.g. the Internet, and offers gateway functionality, i.e. acts as an access router (AR) for other CMANET

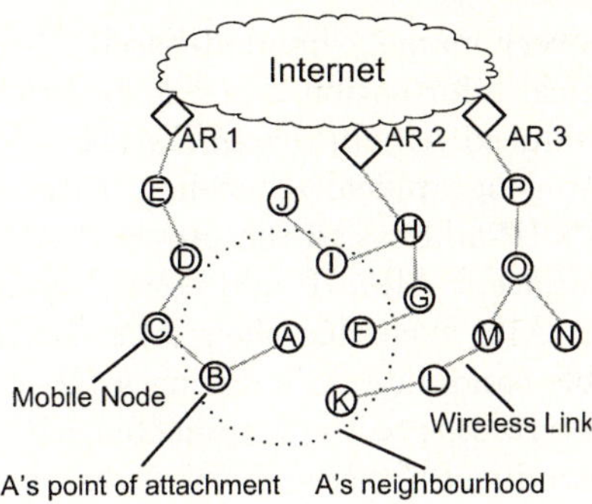

Fig. 1. Connected Mobile Ad Hoc Network Model

nodes. To define ARs' service range, i.e. to define CMANETs' size, a maximum hop-count - measured from the AR - is used. The same technique also determines CMANETs' broadcast coverage range [1].

Furthermore, we define a node's point of attachment (PoA) as its next hop on the path to the AR, e.g. in Fig. 1 A's PoA is node B. When nodes move, they change their PoA - within the same network managed by the ad hoc routing protocol. However, when they eventually reach the fringe of the network's coverage, they will have to select a PoA of another network. In A's case, candidate PoAs (CPoAs) are nodes F, K, I and B (note that these nodes can be on different channels or based on different wireless technologies than A is currently using).

3   Related Work

Previous research in handoff candidate discovery has primarily focused on one-hop wireless access networks. These networks consist of mobile nodes (MN), access routers (AR) and access points (AP). APs attached to one AR form a wireless access network, providing WAN access to MNs.

Candidate Access Router Discovery (CARD) emerged as a research topic due to seamless IP-handoff concepts such as [10] requiring to discover the IP addresses of candidate access routers (CAR) to communicate with them prior to the actual IP-handoff [11]. Furthermore, since mobile nodes are likely to have a set of CARs available for handoff, MNs need a way of querying each CARs capabilities to decide which AR to handoff to based on these capabilities [11].

Thus, the CARD problem is twofold [11]: (1) Mapping of layer-2 (L2) IDs (extracted from APs' beacon messages) to the IP addresses of their access routers and (2) identification of adjacent ARs' capabilities prior to handoff.

Mapping APs' L2-IDs to their ARs' IP addresses requires new discovery concepts; the Reverse Address Resolution Protocol (RARP) [12] is inapplicable, since different ARs are likely to be in separate subnets. The simplest approach, to statically configure each AR with the IP addresses of geographically adjacent ARs and their APs' L2-IDs is cumbersome, especially if adjacent ARs are under separate administrative control. Therefore, three approaches to dynamic candidate access router discovery, namely handoff-based discovery, L2 beacon-based discovery and geographical information-based discovery were developed.

Shim et. al [13] introduced *handoff-based discovery*, which uses handoffs between ARs to determine geographical adjacency. After a handoff to a new AR (nAR), MNs send nAR's IP-address to the previous AR (pAR); hence, all ARs eventually know their adjacent AR's IP addresses. Trossen et. al [5, 6, 7] extend this by having adjacent ARs exchange their APs' L2-IDs. This enables MNs, which discovered possible candidate APs during a L2-scan, to request their AR to reversely resolve these L2-IDs to the responsible AR's IP address.

Funato et. al [14] introduced *L2-beacon-based discovery*, which requires MNs to listen for L2 beacon messages of adjacent APs and to pass this information to the AR they are attached to. By this, each AR eventually knows all adjacent APs. To reversely resolve L2-IDs to ARs' IP addresses, [14] use the Service Location Protocol's (SLP) discovery mechanism. However, since SLP is based on multicasts, [14] does not cover inter-administrative domain discovery.

Geographical information-based discovery is based on distributing ARs' location and coverage information to enable ARs to determine whether their coverage areas overlap. However, coverage areas are difficult to define, especially in the 3D case (e.g. multi-storey buildings), and can change dynamically. Furthermore, flooding this information across domain boundaries does not scale well [15].

All these approaches enable MNs to reversely-resolve the L2-IDs of discovered adjacent APs to the IP addresses of the access routers these APs are attached to. The second aspect of the CARD problem, querying capabilities of these candidate access routers prior to IP-level handoff, is covered by [5, 6, 7], who suggest a two-step process: MNs request capabilities of CARs from their ARs, which in turn request it from the CARs via their WAN-connection (i.e. the Internet).

However, all discussed approaches exchange access router capabilities only. As pointed out in Sect. 1, candidate discovery for CMANETs requires per node capability information because each node is a candidate point of attachment.

Another problem is that all approaches require a centralised database per AR, storing the L2-IDs of all attached nodes. Since communication in mobile ad hoc networks can involve multiple hops, MANETs do not form a single ethernet broadcast domain. As a result, only adjacent nodes know each others L2-IDs; thus, a list of the L2-IDs of all CMANET nodes is not available by default.

4   Concept

Mobile nodes perform handoffs to other CMANETs for different reasons, e.g. when they are about to loose coverage, if QoS becomes insufficient or voluntarily, e.g. if handoffs to less congested networks lead to financial incentives. Thus, the handoff target should be selected based on capabilities such as QoS parameters, supported radio types or pricing. However, when performing a link layer scan to find candidate points of attachment (CPoAs) of adjacent CMANETs, MNs usually only know the L2-ID and the signal strength of these candidates.

Therefore, we propose - similar to [5, 6, 7] for one-hop wireless networks - MNs query capabilities of adjacent CPoAs indirectly via the Internet.

To achieve this, we introduce a *Capability Manager* into each CMANET, which provides capability information about all nodes of a CMANET. This does not only unburden resource-limited CMANET nodes from answering capability queries, but also reduces signalling overhead in the CMANETs' wireless part.

However, before MNs can query the capabilities of detected CPoAs, MNs need a means to determine the responsible capability managers. Thus, they need to determine the CMANETs that each detected CPoA is attached to. We suggest this to be done as a two step process: Firstly, MNs discover all adjacent CMANETs, which enables them in the second step to contact all these networks, to query to which of those a CPoA is currently attached to.

For step one, we introduce a *Neighbourhood Manager* into each CMANET, which keeps a list of adjacent CMANETs. Step two, determining whether MNs are members of a network, is trivial in ethernet broadcast domains. However, multi-hop MANETs do not form an ethernet broadcast domain.

Therefore, we propose a *Membership Manager* to pro-actively maintain an up-to-date list of all nodes forming a CMANET. This enables MNs to determine which CMANET a detected CPoA is attached to, upon which MNs can query that CMANET's capability manager to get information about the CPoA. This information can then be used to select the new target point of attachment. However, the selection algorithm itself is beyond the scope of this paper.

We suggest to deploy the Capability Manager, the Neighbourhood Manager and the Membership Manager to CMANETs' access routers, since this central entity is by definition part of any CMANET. In addition, this implicitly solves the discovery problem of the three managers, since MNs must discover the AR's IP address anyway. This can be achieved by gateway discovery protocols [16].

While Fig. 2 depicts an example interaction between a MN and the three managers, we discuss the latter in more detail in the following paragraphs.

4.1   Neighbourhood Management

To enable MNs to query the capabilities of CPoAs of adjacent CMANETs prior to handoff, each CMANET's *Neighbourhood Manager (NHM)* provides the IP addresses of the access routers of adjacent CMANETs. To achieve this, the NHM has to a) discover adjacent ARs and b) keep track of disappearing ARs. Finally, MNs must be able to query this neighbour list from the NHM.

To discover adjacent CMANETs, we propose to adapt *handoff-based discovery* [13] to CMANETs: Upon a handoff from one CMANET to another, MNs send the IP address of their new access router to the neighbourhood manager of their previous CMANET (via the Internet). By this, the NHM eventually knows the IP addresses of all adjacent ARs.

To detect the disappearance of ARs, neighbourhood information between two CMANETs are soft-states, which get refreshed in case of successful hand-offs. Otherwise, in case a state's lifetime has expired, the corresponding AR is

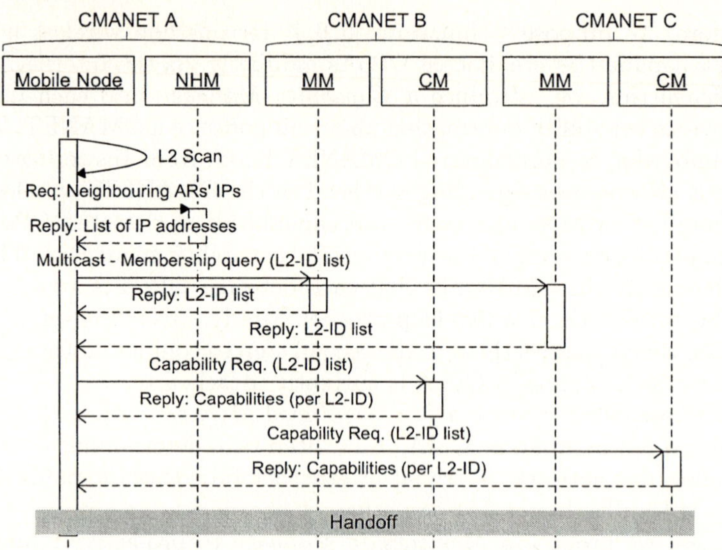

Fig. 2. Example interaction between a MN and the proposed managers: After performing a L2-scan, the MN queries the *Neighbourhood Manager (NHM)* of its own network for the IP addresses of adjacent ARs, upon which their *Membership Managers (MM)* are contacted to find the CMANETs, that the detected L2-IDs belong to. Then, the responsible *Capability Managers (CM)* are queried for the capabilities assigned with these L2-IDs. This information is now used to select the target point of attachment and to perform a handoff to it. Note that the membership queries and the capability queries are presented as separate messages for the sake of conceptual abstraction only; in an implementation they can be combined to reduce signalling overhead

deleted from the neighbour list after it was explicitly proven unreachable (e.g. by leveraging IPv6's Neighbour Unreachability Detection [17]).

Furthermore, we propose that MNs can request the neighbour list by a MN-initiated query/response protocol.

4.2   Membership Management

The *Membership Manager* maintains a list of the L2-IDs of all nodes in a CMANET. This information is necessary when deciding which CMANET a CPoA is attached to, in case MNs want to query that CPoA's capabilities.

Therefore, a CMANET's Membership Manager firstly has to acquire the L2-IDs of nodes joining the CMANET. Secondly, MNs' handoffs to adjacent CMANETs have to be tracked to ensure that MNs are only in the member list of one CMANET at a time. Finally, MNs that have disappeared from CMANETs for other reasons but active handoffs, e.g. because they were switched off or have lost connection, need to be detected to avoid the member list to grow indefinitely.

To achieve this, MNs that join a CMANET must send a message to the Membership Manager conveying their L2-ID. To address the second issue, MNs must send a *membership-change* message to the Membership Manager of their previous CMANET, which signals their successful handoff to another CMANET. Detecting MNs that have disappeared from CMANETs for other reasons, is realised by storing the membership information as soft-states. For each member node, the Membership Manager constantly decrements a lifetime counter, while on the other hand IP-traffic of a node that passes the AR refreshes that node's membership state again. In case a state expires, the corresponding node's reachability is explicitly checked by active probing (see above) before it is eventually deleted from the member list.

4.3   Capability Management

Each CMANET node is a potential candidate point of attachment for other MNs. To enable MNs to efficiently access CPoAs' capabilities when deciding which of them to handoff to, a CMANET's *Capability Manager (CM)* provides capability information about all CMANET nodes.

The conceptual data structure for node capabilities is a set of 3-tuples, each of which consists of the node's L2-ID, the capability type and the capability value. Table 1 shows an example capability table.

Table 1. Capability Table Example

Node ID	Capability Type	Value
00-02-3F-73-98-A9	if_802.11b_supported	1
00-02-3F-73-98-A9	if_802.11g_supported	0
00-02-3F-73-98-A9	average_MAC_delay	10ms

The CM has to acquire each node's capabilities, which we achieve by a simple request/response protocol. This protocol can be applied either *reactively*, i.e. upon the first request for a node's capabilities, or *pro-actively*, i.e. immediately after new nodes join the CMANET. The reactive approach's trade-off is a higher latency when a node's capabilities are requested for the first time by another MN. On the other hand, acquiring capabilities pro-actively causes additional overhead, even if the capabilities are never requested.

However, node capabilities are not only static, e.g. supported interface types, but also of dynamic nature, such as available bandwidth, forwarding willingness or mobility histories/predictions. Since all the latter are subject to change, capability information cached within the CM must be updated regularly. As MNs know when their capabilities such as forwarding willingness or bandwidth have changed, we suggest MNs send changed capabilities unsolicitedly to the CM.

5   Conclusion and Future Work

Candidate discovery for wireless networks is important since it enables handoff decisions based on parameters such as available QoS, pricing or security capabilities. We showed that existing candidate discovery concepts are not directly applicable to multi-hop CMANETs. Therefore, we have introduced our architecture, which is composed of the *Membership Manager*, the *Neighbourhood Manager* and the *Capability Manager*. In contrast to existing approaches, they jointly enable MNs to query candidate point of attachment-specific capabilities from adjacent CMANETs prior to handoff. As a result, MNs can directly handoff to the most suitable network.

As future work, we plan to integrate connected mobile ad hoc networks with multiple access routers into our candidate discovery concept. Another goal are efficient ways of acquiring QoS-relevant capability information in CMANETs.

References

1. Tseng, Y.C., et al.: Integrating Mobile IP with Ad Hoc Networks. In: Computer. Volume 36. IEEE (2003) 48–55
2. Balachandran, A., et al.: Wireless Hotspots: Current Challenges and Future Directions. In: Proceedings of the 1st ACM International Workshop on Wireless Mobile Applications and Services on WLAN Hotspots, ACM Press (2003) 1–9
3. Wikipedia: Wireless Community Network. `http://en.wikipedia.org/wiki/Wireless_community_network` (2004)
4. Dunkels, A., et al.: Connecting Wireless Sensor Networks with the Internet. Ercim News (2004) 61–62
5. Trossen, D., et al.: Protocol for Candidate Access Router Discovery for Seamless IP-level Handovers. INTERNET DRAFT (2001) Work in progress.
6. Trossen, D., et al.: A Dynamic Protocol for Candidate Access-Router Discovery. INTERNET DRAFT (2003) Work in progress.
7. Liebsch, M., et al.: Candidate Access Router Discovery. INTERNET DRAFT (2004) Work in progress.
8. Manner, J., Kojo, M.: Mobility Related Terminology. RFC 3753 (2004)
9. Perkins, C.E.: Ad Hoc Networking. Addison-Wesley, Boston, USA (2001)
10. Koodli, R.: Fast Handovers for Mobile IPv6. INTERNET DRAFT (2004) Work in progress.
11. Trossen, D., et al.: Issues in Candidate Access Router Discovery for Seamless IP-Level Handoffs. INTERNET DRAFT (2002) Work in progress.
12. Finlayson, R., et al.: A Reverse Address Resolution Protocol. RFC 903 (1984)
13. Shim, E., et al.: Fast Handoff Using Neighbor Information. INTERNET DRAFT (2000) Work in progress.
14. Funato, D., et al.: Geographically Adjacent Access Router Discovery Protocol. INTERNET DRAFT (2001)
15. Shim, E.: Mobility Management in the Wireless Internet. PhD thesis, Columbia University (2004)
16. Ghassemian, M., et al.: Performance Analysis of Internet Gateway Discovery Protocols in Ad Hoc Networks. In: Proceedings of the IEEE Wireless Communications and Networking Conference (WCNC) 2004. (2004)
17. Narten, T., et al.: Neighbor Discovery for IP Version 6 (IPv6). RFC 2461 (1998)

A Fault-Tolerant Permutation Routing Algorithm in Mobile Ad-Hoc Networks

Djibo Karimou and Jean Frédéric Myoupo

LaRIA, CNRS, FRE 2733, Université de Picardie-Jules Verne,
5, rue du Moulin Neuf, 80.000 Amiens, France
{myoupo, karimou}@laria.u-picardie.fr

Abstract. Mobile ad-hoc networks (MANET) is a distributed systems formed by wireless mobiles nodes, that can dynamically self organize into arbitrary and temporary networks with no pre-existing communication infrastructure. After initialization, one of a crucial issue in ad-hoc network, is how to transmit items held by the stations to their destinations called Permutation Routing. As mobiles devises are dependant on battery power, it is important that this permutation routing minimize their energy consumption. But, it can occur during the permutation routing that some stations develop faults and stop working. The existence of these faulty nodes can significantly affect the packets delivery rate. If a faulty node, participating in a permutation routing operations, drops packets, all these packets will be lost. Hence, the goal of a fault tolerant permutation routing is to provide certain packet delivery guarantee in spite of the presence of faulty stations.

The main contribution of this paper is to present a protocol that provided an adequate level of fault tolerant and minimized energy during the permutation routing. Unlike in [2], in this work, in presence of faulty nodes, items can be transmitted to their destinations without loss.

Keywords: Initialization Problem, Mobile Ad Hoc Network, Permutation Routing, Fault-Tolerant.

1   Introduction

Mobile ad-hoc networks are formed by a collection of mobile wireless nodes (stations) which can communicate with each other and dynamically self organize without any static network interaction. Each mobile host has the capability to communicate directly with another mobile host in its vicinity via a channel of transmission. It can also send packets destined for other nodes. Communication links are formed and disappear as nodes come into and go out of each other communication range. In this paper, we shall interest only to a single-hop network, in which each station can transmit or communicate directly with each other station. All the stations use the same channel to communicate, and the message broadcasted by one of station on the common channel is simultaneously received by all other stations. Nowadays, MANET is becoming more and more

P. Lorenz and P. Dini (Eds.): ICN 2005, LNCS 3421, pp. 107–115, 2005.

interesting research topic and the current development of wireless networks have made the research in mobile ad-hoc networks (MANET for short) attractive and desirable [1, 4, 6, 7, 9, 10, 11, 12].

Hence, the set of applications for MANET is diverse. Examples of such networks are Ad-hoc, packet radio, local area, and sensor networks, which are used in battlefields, disaster rescues, wireless conferences in the hall, monitoring objects in a possibly remote or dangerous environment, wireless internet etc .... In most applications, it is often assumed that the address of a station in wired networks and even in wireless networks [10, 13, 14] is either permanently stored in its local memory or is received at start-up time from a master station. In another word it is assumed that each station knows its address. This assumption is not always appropriate in some cases. The problem of assigning addresses to stations is somehow fundamental and is called the *Initialization problem*. Assigning unique IDS to the stations in ad-hoc network (initialization problem) has been solved in two different ways :

1. Probabilistic Form by Nakano et al [7]
They have introduced randomized protocols to initialize n stations of an n-MANET $((n, k)$ for short) with or without collision detection (CD) capability. In particular, they showed that with CD, a k-MANET can be initialized in $O(\frac{n}{k})$ broadcast rounds with probability $1-\frac{1}{n}$ wherenever $k \leq \frac{n}{3logn}$. Without CD, they showed that it can be initialized in $O(\frac{n}{k})$ broadcast rounds with the probability $1 - O(\frac{1}{n})$ wherenever $k \leq \frac{n}{4(logn)^2}$.

2. Deterministic Form by Myoupo [11]
He introduced a deterministic protocol to initialize n mobiles in ad-hoc networks. First, he derived a Single-channel protocol which requires $O(\frac{n}{log2})$ broadcast rounds in average case analysis. Next, he derived a k-channel initialization protocol running in $O(\frac{n-k+1}{log2}) + 4logk$ broadcast rounds in the worst case.

Another problem in Ad-hoc network is called permutation routing. Consider a MANET (n, p, k) of p stations with n items pretitled on it. Each item has a unique destination which is one of the p stations. Each station has a local memory of size $\frac{n}{p}$ items. But some of the $\frac{n}{p}$ items in the station say i, do not necessary have i as destination. And even, it can happen that one of the $\frac{n}{p}$ items belongs to it. The fault-tolerant permutation routing problem is to route, despite of the presence of faulty stations, the items in such a way that for all i, $1 \leq i \leq p$, station i contains all its own items.

The permutation routing problem in a single hop ad-hoc network has been explored in several papers. Nakano et al in [13] showed that the permutation routing of n items pretitled on mobile Ad-hoc network of p stations (p known) and k channels (MANET (n, p, k) for short) with $k \leq p$, can be carried out in $\frac{2n}{k} + k - 1$ broadcast rounds, if $k \leq \sqrt{p}$ and if each station has $O(\frac{n}{k})$-memory locations. If $k \leq \sqrt{\frac{p}{2}}$, and if each station has a $O(\frac{n}{p})$ memory locations, the permutations of these n pretitled items can be done also in $\frac{2n}{k} + k - 1$ broadcast rounds. Also, recently Myoupo et al in [8] presented another permutation routing protocol based on deterministic form initialization, which can be solved in $O(\frac{p}{ln2}) + (\frac{2}{k} + 1)n + k - 1$ broadcast rounds in the worst case. It can be solved in $O(\frac{p}{ln2}) + (\frac{2}{k})n + k - 1$ in the better case, without any restriction in k. Where ln is the logarithmic function in basis e.

In presence of faulty stations, in both of these protocols, the packets in faulty nodes can not be transmitted to their destinations. This constitutes a crucial problem because faulty nodes can significantly affect the performance of permutation routing. In order to solve this problem, we present in this paper a fault tolerant permutation routing which run regardless of faulty stations.

1.1   Previous Work

There are a few published talking about fault tolerant in permutation routing. Recently, Datta in [2] presented a fault tolerant permutation routing protocol of n items pretitled on mobile Ad-hoc network of p stations and k channels MANET(n, p, k) for short. He solved the fault-tolerant permutation routing problem in $\frac{2n}{k} + (\frac{p}{k})^2 + \frac{p}{k} + \frac{3p}{2} + 2k^2 - 1$ slots and each station needs to be awake for at most $\frac{4nf_i}{p} + \frac{2n}{p} + \frac{3p}{k} + k^2 + \frac{p}{2k} + \frac{p}{2} + 4k$ slots, where f_i is the number of faulty stations in a group of $\frac{p}{k}$ stations. But, in this protocol, he made assumptions : first, he considered that the senders stations knew the destinations of the items that they held. Secondary , he considered that in presence of faulty nodes, the packets in these faulty nodes will be lost. Also , he considered that the stations are partitioned in groups such that each group has $\frac{p}{k}$ stations, but it was not shown how this partition can be obtained in a MANET.

1.2   Our Contribution

This paper presents a fault-tolerant protocol which run regardless of the faulty stations in a MANET. In contrary of the work in [2], in our approach during this permutation routing, items in faulty nodes can be transmitted to their destinations without loss.

We consider a MANET (n, p, k) with n items, p stations and k channels with p unknown. We show that in presence of faulty stations, the fault tolerant permutation routing problem can be solved in $\frac{2n}{k} + k - 1$ broadcast rounds and each station should be awake for at most $\frac{2n}{k} + 1$ slots in the better case. In the worst case, it can be solved in $(\frac{2n}{k} + 1)n + k - 1$ broadcast rounds and each station should be awake for at most $\frac{2n}{k} + \frac{n}{p-f_i} + \frac{p}{k} + 1$ slots where f_i is the number of faulty stations.

The rest of this work is organized as follows : in section 2, a basic definitions and environment considered in this work is presented. Section 3, shows our fault tolerant permutation protocol, the case in presence of faulty stations. A conclusion ends the paper.

2   Basic Definitions

The system, we consider is composed of p mobile hosts, with n items, communicate via a k channels (MANET (n, p, k) for short). We supposed that the stations are initialized running an initialization protocol as in [11], then all stations are ID number named in $[1, p]$. Each station holds $\frac{n}{p}$ items but can stores $\frac{n}{k}$. Also, we consider that the p stations are partitioned in k groups G_i, $1 \leq i \leq k$, each containing $\frac{p}{k}$ stations. Partitioning stations into k groups each containing $\frac{p}{k}$ stations is an another problem and has been solved in [8].

An Ad-Hoc Network is a set S of n radio transceivers or stations which can transmit and/or receive messages through a set C of k channels. The time is assumed to be slotted and all stations have a local clock that keeps synchronous time. In any time slot, a station can tune into one channel and/or broadcast on at most one channel. A broadcast operation involves a data packet whose length is such that the broadcast operation can be completed within one time slot. Assuming that communication in the network proceeds in synchronous steps and all communications are performed at time slot boundaries, i.e., the duration of broadcast operations is assumed to be sufficiently short.

(i) Let us consider p stations which communicate through k distinct communication channels $C_1, C_2, \ldots, C_k$.

(ii) We assume that p is unknown, k is known, and the system is a *single hop* network.

(iii) *Status of resident in channel.* Each station may, in a given time unit, broadcast a message on one channel, and listen to the same channel. Thus it is then called a resident of that channel.

(iv) The status of channel is NULL if no station broadcasts on the channel in the current slot.

We make the following assumptions:

1. At the beginning, as the stations are equipped with new batteries, probabilities to develop faults are almost null, then we suppose that we have not faulty stations during the first broadcast round and at each time, it must have at least one station on a channel.

2. Let us consider p stations 1, 2,..., p which communicate through k distinct communication channels $C_1, C_2, \ldots, C_k$. We suppose that we have n items in the system. And each station of a MANET (n, k, p) is assumed to have a local memory of size at least $O(\frac{n}{k})$, but only $\frac{n}{p}$ items belong to it. Note that this assumption is reasonable, sine nowadays, capacity of storage is more and more meaningful.

Definition

We suppose that the n items denoted $a_1, a_2, ..., a_n$ are pretitled on a MANET (n, k, p) such that for every i, $1 \leq i \leq p$, station i stores the $\frac{n}{p}$ items $a_{(i-1)\cdot\frac{n}{p}+1}$, $a_{(i-1)\cdot\frac{n}{p}+2}, .., a_{(i)\cdot\frac{n}{p}}$. Each item has a *unique* destination station. For simplicity, we shall write $a_{i,j}$ for shorthand for $a_{(i-1)\cdot\frac{n}{p}+j}$. In other words, $a_{i,j}$ denotes the j-th item stored by i. Each item $a_{i,j}$ has a destination $d_{i,j}$ which specifies the identity of the station $d_{(i-1)\cdot\frac{n}{p}+j}$ abbreviated $d_{i,j}$, to which the item $a_{i,j}$ must be routed. For every v, $1 \leq v \leq p$, let h_v be the set of items whose destination is v, i.e. $h_v = \{a_{i,j} | d_{i,j} = v\}$.

The fault permutation routing problem is to route, despite of the presence of faulty stations, the items in such a way that for all v, $1 \leq v \leq p$, station v contains all the items in h_v. Consequently, each h_v must contain exactly $\frac{n}{p}$ items.

3   Fault Tolerant Permutation Routing

We assume that k channels are available in a MANET (n, p, k), say C_j, $1 \leq j \leq k$, channel C_i is assigned to group G_i, $1 \leq i \leq k$, p is unknown. We suppose that stations have identifiers and are partitioned in k groups each containing $\frac{p}{k}$ stations [8]. A solution of a problem of permutation routing with no faulty stations is presented in [8]. Here, we consider that, we are in presence of faulty stations.

Our approach involves two phases. The first phase is a local broadcast in which, each station in G_i, $1 \leq i \leq k$, broadcasts items that it holds. In the second phase, in each group, items which were recorded not to be in a group (outgoing items) are broadcasted to their destinations.

Phase 1 : Local Broadcastings
In this step broadcast is local in each group G_i. The channel C_i is assigned to group G_i, $1 \leq i \leq k$. We only focus on the routing that takes place in group G_1 using channel C_1, the broadcast on all other groups is similar. Each station, say v, $1 \leq v \leq \frac{p}{k}$, in G_1 maintains two types of counters :

- a counter, say $Count_{lacalv}$ (initialized to zero) which is used to record the number of items which do not belong to v.
- $\frac{p}{k} - 1$ counters named $Count_1$, $Count_2$,..., $Count_{\frac{p}{k}-1}$ initialized to zero. Each counter $Count_j$ is used to record the number of items broadcasted by station j and which do not belong to any station of G_1.

One after another, from the station named 1 to the station named $\frac{p}{k}$ in this order, stations in group G_1 broadcast items they hold in channel C_1 one by one. Each station in group G_1, monitors channel C_1 and copies into its local memory, in order in which they were broadcasted, every item that belongs to it and the items broadcasted by all stations in G_1. Assume that station v is carrying out the local broadcast. Consider an item broadcasted by v:

- if the item belongs to v, then, in the next slot, v broadcasts a positive acknowledgment in C_1 to inform the other residents of C_1 that it is its item.
- if the item belongs to another station w ($w \neq v$) in G_1, then, in the next slot, w broadcasts a positive acknowledgment in C_1 to inform the other residents of C_1 that it is its item.
- if none of these two situations above occurs, then v increments its local counter by 1 and all other stations increment their counters named by $Count_v$ by 1. Therefore the item is identified not belonging to a station of G_1.

So at the end of this step, all the items which belonged to a station in G_1 are stored by each station in G_1. All the items which are not in G_1 are identified, their number is known and are classified by each station in order in which they were broadcasted. Since each station in G_1 has $\frac{n}{p}$ items, this step needs $\frac{2n}{k}$ broadcast rounds, and each station should be awake for at most $\frac{2n}{k}$ slots.

Phase 2 : Multiplexing the items for outgoing broadcastings

This phase presents a procedure which helps to move the items which were recorded not to be in G_1 (outgoing items) to their final destinations in spite of the presence of faulty stations. In G_i, we denote f_i, the faulty stations which occur in G_i during the execution of the permutation routing. Again, we focus on the routing of items from G_1. The broadcasts from the other groups being similar. This phase is composed of two steps.

Step 1 : Observe that after computation of phase 1, all the items which are not in G_1 are identified, there number is known and are identified by each station in order in which they were broadcasted.

i.) We partition these items in $(\frac{p}{k} - f_i) \geq 1$ groups (working stations). Consider, if n_i is a set of outgoing items in G_i, then each station has $(\frac{n_i}{\frac{p}{k} - f_i})$ slice items to transmit to the other channels. We call this computation "partition items". ii). Now, we show how each station knows the faulty stations and when it can wake up to perform "partition items".

Considering the information in phase 1, then after performing firstly "partition items", each station can determine:

- When its turn of transmission has come, this is because the stations are numbered, there is a valuable ordering of the stations.
- The slice, exact number of transmissions, that will be executed by the stations preceding it and can determine also its slice of items to broadcast.

Hence, as each station knows exactly when each slice of transmission has finished, then at the end of each slice it will wake up and lesten to the channel, ($\frac{p}{k}$ slots are necessary). If the status of the channel is NULL, there is no broadcast in this slot s, $1 \leq s \leq \frac{p}{k}$. The stations consider that station which has to broadcast in slot s is faulty. These informations are sufficient to allow each station in G_i to know the number of faulty stations and the number of working stations. Working stations execute again "partition items".

Step 2 : The stations of G_1 broadcast first, followed by those in G_2, next G_3 and so on till G_k. In G_1, with the help of counters, each station knows the number of items which owners are not in G_1. Each item is multiplexed on channels $C_1, C_2, ..., C_k$, before broadcasting. One after another, stations in group G_1 broadcasts multiplexed items which final destinations are in group G_2, G_3,..., or G_k. Hence groups $G_2, G_3, ..., G_k$ receive the item simultaneously. The reader may wonder why the use of counters. The idea is to allow every station to know the exact moment at which it should start broadcasting the items which are in its local memory and which do not belong to any station in G_1 (say, outgoing items). The broadcasts are carried out, one after another, from the first station with smaller number to the last station with larger number, each station broadcasts its slice of outgoing items. for example in G_1, first station named 1 broadcasts the first slice of $\frac{n_i}{\frac{p-f_i}{k}}$ outgoing items, next, second station named 2 broadcasts the second slice, and so on.

But, if n_i is not a multiple of $\frac{p-f_i}{k}$, then the rest of items should be broadcasted by the station with larger address. After it has broadcasted all its outgoing items, the last station of G_1, broadcasts an information in C_2 to inform stations in G_2 that it is their turn to carry out multiplexed broadcast ($k-1$ broadcast are necessary). Now we continue by evaluating the number of broadcast rounds of this subsection.

(i) Suppose that in each group, no station contains an item belonging to the group. Clearly, all items are outgoing items. The reader should not have difficulty to confirm that, in this case, this step needs $n + k - 1$ broadcast rounds. It is the *worst case* and each station should be awake for at most $\frac{n}{p-f_i} + \frac{p}{k} + 1$ slots.

(ii) Suppose that all items of a group belong to the stations of the group. Therefore only $k - 1$ are necessary to inform the residents of each group that no broadcast is need. It is the *better case*.

Theorem
Let p mobile stations in a MANET (n, p, k) with p unknown, in the case where there can occur faulty stations, the fault tolerant permutation routing problem can be solved in $\frac{2n}{k} + k - 1$ broadcast rounds and each station should be awake for at most $\frac{2n}{k} + 1$ slots in the better case. In the worst case, it can be solved in $(\frac{2}{k} + 1)n + k - 1$ broadcast rounds and each station should be awake for at most $\frac{2n}{k} + \frac{n}{p-f_i} + \frac{p}{k} + 1$ slots where f_i is the number of faulty stations.

Remark : the above theorem supposes that faulty stations can not occur when a station develops faults during a broadcasting. Hence, when this problem occurs, our bellow approach allows us to resolve it.
We now go into the description of this approach, which is somewhat similar to an idea used in step 2 of our above procedure but with short modification. It is easy to see that, faulty stations can occur during the phase of transmission in two cases, when before transmitting or during a transmission of a station, it develops faults.

Case 1 : The main task on this approach is to solve the problem occurred when before transmitting one station develops faults. Recall that, as each station knows exactly when each slice of transmission has finished, then at the end of each slice, station will wake up and lesten to the channel. Then, if in a slot s, status of channel is NULL, there is no broadcast in this slot. All the stations consider that station which has broadcasted in slot s is faulty. Hence, as soon as station is faulty, each working stations computes "Partition item" to have its exact number of transmissions (its slice). Then, next station takes over the transmission.

Case 2 : The second case occurs when in current transmission a station develops faults and stops working. It is important that its next station takes over the transmission. This is a most difficult task in phase 2. For that we supposed that all the working stations remain awake for all this stage. In this case, even if a current station is faulty, the other stations have knowledge of this faulty station. Hence, they also compute "Partition items", and the next station take care over the transmission, and so on.

Lemma

As all the stations remain awake in task 2, this assumption modifies the awake time of our protocol. In this case, if task 2 occurs, each station should be awake for at most $n(\frac{2}{k} + 1) + \frac{p}{k} + 1$ slots.

4   Conclusion

In Ad-Hoc network, permutation routing is a crucial problem. In spite of development of many methods, faulty stations can significantly affect the packets delivery rate, and items in these faulty nodes will be lost.

We have suggested an approach in which items in faulty nodes can be transmitted to their destinations. We have presented two cases which can occur during the permutation routing. In this way, we show that the fault permutation routing can be solved in $\frac{2n}{k} + k - 1$ broadcast rounds and each station should be awake for at most $\frac{2n}{k} + 1$ slots in the better case. In the worst case, it can be solved in $(\frac{2}{k} + 1)n + k - 1$ broadcast rounds and each station should be awake for at most $\frac{2n}{k} + \frac{n}{p - f_i} + \frac{p}{k} + 1$ slots where f_i is the number of faulty stations.

References

1. B. S. Chlebus, L. Gasieniec, A. Gibbons, A. Pelc and W. Rytter. Deterministic broadcasting in Ad Hoc Networks. *Distributed Computing,* vol. 15, 2002, pp.27-38.
2. A. Datta. Fault-tolerant and Energy-efficient Permutation Routing Protocol for Wireless Networks. 17th *IEEE Intern. Parallel and Distributed Processing Symposium (IPDPS'03)*, Nice, France, 22-26, 2003.
3. A. Datta, A. Celik and V. Kumar. Broadcast protocols to support efficient retrieval from databases by mobile users. *ACM Transact on Database Systems*, vol.24, 1-79, 1999.
4. R. Dechter and L. Kleinrock. Broadcast communication and distributed algorithms, *IEEE Trans. Comput.,* C-35, 210-219, 1986.
5. S. Hameed and N. H. Vaidya. Log-time algorithm for scheduling single and multiple channel data broadcast. *In Proc. ACM MOBICOM,* 90-99, 1997.
6. T. Hayashi, K. Nakano and S. Olariu. Randomized Initialization Protocols for Ad-Hoc Networks. *IEEE Trans. Parallel Distr. Syst.* vol 11, pp.749-759, 2000.
7. T. Hayashi, K. Nakano and S. Olariu. Randomized Initialization Protocols for Packet Radio Networks. *IPPS/SPDP'99.*

8. D. Karimou, J. F. Myoupo. An Application of an Initialization Protocol to Permutation Routing in a Single-hop Mobile Ad-Hoc Networks. *Proc. International Conferences on Wireless Networks (ICWN'04)*. vol II, pp.975-981. Las Vegas, USA. June 21-24 , 2004.

9. N. Malpani, N. H. Vaidya and J. L. Welch. Distributed Token Circulation on Mobile ad-Hoc Networks. 9th *Intern. Conf. Networks Protocols* (ICNP), 2001.

10. J. F. Myoupo. Concurrent Broadcasts-Based Permutation Routing algorithms in Radio Networks. *IEEE Symposium on Computers and Communications,* (ISCC'03), Antalya, Turkey, 2003.

11. J. F. Myoupo. Dynamic Initialization Protocols for Mobile Ad-Hoc Networks. 11th *IEEE Intern. Conf. On Networks (ICON2003),* Sydney, Australia, 2003 pp.149-154.

12. J. F. Myoupo, V. Ravelomanana, L. Thimonier. Average Case Analysis Based-Protocols to Initialize Packet Radio Networks. *Wireless Communication and Mobile Computing,* vol.3, pp. 539-548, 2003.

13. K. Nakano, S. Olariu and J.L. Schwing. Broadcast-Efficient protocols for Mobile Radio Networks. *IEEE T.P.DS.,* vol.10, pp.12, 1276-1289, 1999.

14. K. Prabhakara, Kien A. Hua and J.H Oh. Multi-level multi-channel air cache designs for broadcasting in mobile environment. *Proc. IEEE Intern. Conf. On Data Engineering,* USA, 167-176, 2000.

Policy-Based Dynamic Reconfiguration of Mobile Ad Hoc Networks

Marcos A. de Siqueira, Fabricio L. Figueiredo, Flavia M. F. Rocha, Jose A. Martins, and Marcel C. de Castro**

CPqD Telecommunications Research and Development Center,
Rodovia Campinas Mogi-Mirim, km 118,
5 - 13086-902 Campinas - SP- Brazil
{siqueira, fabricio, flavia, martins, mcastro}@cpqd.com.br

Abstract. Ad hoc networks are intrinsically dynamic with respect to mobility, traffic patterns, node density, number of nodes, physical topology, and others. This scenario imposes several challenges to the ad hoc routing protocols, and there is no single solution for all scenarios. This paper proposes the application of Policy-Based Network Management (PBNM) for dynamic reconfiguration of ad hoc networks. PBNM uses the concept of policies formed by events, conditions and actions. The occurrence of an event triggers conditions evaluation, and if evaluated to true, a set of actions should be performed. The paper contribution comprises the proposal of an ad hoc policy information model based on DEN-ng policy model for the implementation of a policy manager prototype to be integrated in the NS-2 simulator allowing dynamic reconfiguration of routing protocol parameters in simulation time; and the proposal of policies for dynamically adjusting the ad hoc routing protocol behavior.

1   Introduction

Currently, telephony and Internet access services in the majority of the developing countries are mainly provided over the old cooper plant. In fact, fixed wireless (WLL - Wireless Local Loop) has failed in reaching a high number of subscribers around the world, due to poor service coverage, terminal cost and the lack of enough speed for Internet access (limited to 9.600bps).

Mobile Ad Hoc Networks (MANETs) are being widely studied by the academy and industry, and promises to be a quite suitable technology for the provision of network access services, including of voice services in uncovered areas. The main advantage of MANETs for this scenario is that there is no need of high cost infrastructure, which dramatically reduces the network deployment costs. One of the challenges of building MANETs is the design of a suitable ad hoc routing

** The research that resulted in this work was performed at CPqD and was funded by
FUNTTEL.

P. Lorenz and P. Dini (Eds.): ICN 2005, LNCS 3421, pp. 116–124, 2005.

protocol, each one suggested with a specific goal, such as low battery consumption, scalability, robustness and others. Besides a routing protocol, a MANET needs additional mechanisms, such as a suitable addressing scheme and a distributed algorithm for hierarchical address distribution; efficient physical and MAC (Media Access Control) layers and a distributed management system.

All these mechanisms may need to be dynamically adjusted, according to dynamics of topology, traffic and node state conditions. For instance, depending on the overall battery energy stored, the routing protocol could be changed from a "performance optimized protocol" to a "power consumption optimized" one.

In this context, an important goal of a network management system designed for ad hoc network is to set up network configuration according to several policies, thus allowing dynamic and optimized network configuration and adapting parameters, such as the routing protocol operation, addressing scheme and QoS mechanisms, to network conditions, which includes node density, signal propagation conditions, traffic patterns, and others. Moreover, the management system shall guarantee the uniformity of network configuration parameters throughout the network nodes, avoiding undesired scenarios, such as network loops and route instability, dynamic configuration of these parameters.

MANET management has not been a prioritized problem so far. The well-known proposed solutions for ad hoc network management, Ad Hoc Network Management Protocol (ANMP) [1] and Guerrilla Management Architecture [2] don't focus on QoS, neither on mechanisms for dynamic reconfiguration of routing protocol. The ANMP is the result of a secure message efficient protocol design focused on the development of a lightweight protocol that is compatible with SNMP. The Guerrilla Management Architecture envisions ad hoc networks to be self-organized and self-managed with the collaboration of autonomous nodes. It supports dynamic agent grouping and collaborative management in order to adapt to network dynamics and to minimize management overhead.

Recently, technologies such as the Policy-Based Network Management Architecture (PBNM) are emerging and allowing the operation of networks in a more automated way, adapting network devices configuration providing, for instance, suitable Quality of Service (QoS) for the different traffic flows or classes. Besides, PBNM proposes that high-level enterprize business policies could be mapped into network devices configuration by the management system, reducing network operation complexity, independently of the type, manufacturer model and operating system version of the network devices. The policy model DEN-ng [3] represents policies as a set of events, conditions and actions.

This paper proposes PBNM as a solution for ad hoc network management, which consists of extending DEN-ng abstract classes (PolicyEvent, PolicyCondition and PolicyEvent) for definition of ad hoc specific policies focusing dynamic routing protocol parameters configuration management.

This paper is organized as follows: section 2 presents a background in mobile ad hoc routing protocol requirements, section 3 proposes a policy information model for ad hoc networks, section 4 describes the design of a policy manager prototype for managing ad hoc nodes in the NS2 network simulator, section 5 presents the conclusion and future works.

2   Mobile Ad Hoc Networks Requirements

A real world ad hoc routing protocol must support continuous, efficient, secure connectivity, as well as assuring QoS parameters for the applications in MANETs. Nevertheless, this kind of network has characteristics that represent serious obstacles to the operation of routing protocols, such as dynamic topology, throughput limitation, route instability, battery limitations, broadcast difficulties mainly due to hidden node problem, lack of efficient dynamic and hierarchical addressing mechanisms and others.

CPqD has proposed a system based on ad hoc wireless network, aiming at providing voice and data services [4]. The main requirements of the system based on ad hoc wireless network being developed are; user data transmission rate greater that 64kbps, low cost terminal with connection to PSTN and IP networks, ability to identify system faults and to provide external visualization of them, operation at frequency bands according Brazilian spectrum regulation in multihop scenarios with quality of service.

In order to fulfill these requirements, ad hoc routing protocols shall support a large number of functionalities, including fully distributed operation; loop prevention; on demand, proactive or hybrid operation; security; sleep period support; unidirectional links support; link failure detection; ingress node buffering (while discovering routes), discovery and usage of multiple routes for load balancing; fast recovery; adoption of advanced metrics, such as path longevity, battery level, node processing power and others.

The success of a given routing protocol in a given network scenario can be measured in terms of a set of performance metrics achieved in the routes discovered by the protocol. Some of these metrics are: end-to-end throughput and latency, packet loss, rate of out of order packets delivered, routing efficiency - reflecting the rate of delivered payload versus routing protocol signaling overhead.

From Quality of Service point of view, the ad hoc routing protocol shall be able to measure and propagate available bandwidth as an additional metric, perform distributed admission control and resource reservation, fast route recovery and reliable route establishment.

We have performed an extensive research on ad hoc routing protocols and haven't found a single solution for satisfying all the requirements described above. Each routing protocol attends on a specific set of requirements and presents a limited set of functionalities. Indeed, some protocols are focused on node energy economy, others are focused on reduced overhead and others in reducing processing power requirements, and so on.

This is the main motivation for designing a policy-based auto-adaptation mechanism. It represents a feasible approach to dynamically modify ad hoc routing protocol behavior, aiming ate keeping the compliance to the network performance requirements.

3   Ad Hoc Policy Management

J. Strassner [3] proposed a policy architecture called DEN-ng that employs an UML (Unified Modelling Language) meta-model for granting that a set of building blocks (objects representing the network policies) are used in the model construction. The model is composed of three components: textual use cases, UML models and a data dictionary defining relationship semantics. Besides, DEN-ng defines a finite state machine for controlling policies life cycle.

DEN-ng defines policy rules as containers composed of four components: Meta data, event clause, condition clause and action clause. It is mandatory that the event, condition and action clause are present in a given policy rule. This approach provides a consistent rule structure. Each clause has OCL (Object Constraint Language) constraints, thus allowing the clarifications of the policy evaluation semantics, avoiding ambiguity and interoperability problems.

3.1   Ad Hoc Policy Events

As described in the previous sections, in the DEN-ng model a policy is represented as an association of a set of events, conditions and actions. This section proposes the modelling of events, conditions and actions for ad hoc networks.

We have modelled events as occurrences within the network domain. Table 1 illustrates some events that should lead to reconfiguration of network parameters.

These events represent an indication that network characteristics are changing and reconfiguration may be necessary. For instance, if the average dynamicity degree increases, it may lead to a rise of route breakages. A Policy Rule can be associated with one or more events and the execution of this policy (condition analysis) can be initiated by the occurrence of a particular event or by a set of events.

We propose to associate each event to a parameter measured in the network, as shown at table 1. The right column presents an illustrative default value for the thresholds, but these values could be used only initially and then they can be adapted according to dynamics of the system.

Table 1. Ad hoc events

Event variable	Operator and variable	Default value
Network availability (NA)	NA < Nai	NAi=97%
Avg packet loss (APL) for a sample of connections	APL > APLi	APLi=8%
"N" failures on connection establishment (CEF) in a time interval (Ti)	CEF > CEFi in Ti	CEFi = 3, Ti=10s
"N" route breaks (RB) in a time interval (Ti)	RB > RBi in Ti	RBi =3, Ti=10s

Table 2. Ad hoc conditions

Variable	Code and operator	Default value
Average node density	ANDE > ANDEi	5
Average Dynamicity Degree (ADD)	ADD > ADDi	0.2 events/s
Ratio between the number of Source /Destination pairs and the total number	DSD > DSDi	10%
Existence of real time applications (RTA)	RTA = RTAi	0 - no, 1- yes
Energy saving mode	ESM(ESM) = ESMi	0 - no, 1- yes
Number of nodes(NON)	NON > NONi	300
Physical topology clustering degree (PTC)	PTC < PTCi	40%

3.2   Ad Hoc Policy Conditions

The policy conditions are defined as parameters that allow the mapping of the adaptability of the routing protocol to current network state. In order to define network states that have influence on the routing protocol, seven network features are considered, according to Table 2.

The Average node density (ANDE) is calculated by counting the average number of directed connected neighbors for each node at the network. Some protocols such as DSDV don't operate adequately in these scenarios. Others such as DLAR operate optimally with many paths for load balancing.

The Average Dynamicity Degree (ADD) is defined as the average number of ticks each node can measure for the set of neighbors. It can be measured using the "number of ticks" defined by Toh [5]. Protocols such as ABR that use the path stability as a metric present better performance with high degree of mobility.

The Degree of source-destination pairs (DSD) is related to the application communication patterns. Client-server applications tend to concentrate all communications to the server. On the other hand, peer-to-peer applications tend to distribute communications uniformly between most of the nodes. Proactive protocols tend to obtain better results with many-to-many communication, while reactive protocols that use route caching are better with many-to-one communication patterns.

Networks that provide services with QoS requirements such as **real time applications** (RTA) need a QoS aware routing protocol such as AQOR [6], INSIGNIA[7] or SWAN (Stateless Wireless Ad hoc Networks)[8].

The condition **Energy Saving Mode** (ESM) should influence routing protocol parameters such as route update times and routing protocol choice for a "sleep period aware".

The **number of nodes** (NON) influences on the routing protocol choice. For instance, in a network with more than 300 nodes a scalable routing protocol such as ZRP should be configured.

The **Physical topology clustering degree** (PTC) is defined as the ratio between the average number of nodes per cluster and total number of nodes in a hierarchical ad hoc network. This parameter is difficult to measure and indicates

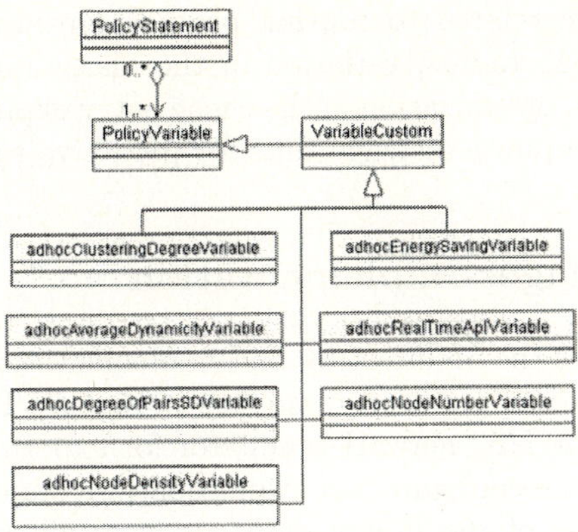

Fig. 1. Ad hoc condition policy variables

Table 3. Ad hoc actions

Action Variable	Type	Possible values
Routing protocol (RP)	Enum, int	DSDV (1), DSR (2), ABR (3), TDR (4), DLAR (5), OLSR (6), CBRP (7), ZRP (8).
Cluster or zone radius	int	< 10
Maximum number of paths per route	int	< 4
Cache entry expiration time	int	Positive
Route update messages time	int	> 5s

a tendency that a cluster-based routing protocol would provide better network performance.

Figure 1 presents the PIM (Policy Information Model) mapping for the proposed policy conditions. The conditions are created as VariableCustom classes. For policy evaluation, the PolicyVariables created should be matched against PolicyValues using PolicyOperators.

3.3 Ad Hoc Policy Actions

After the detection of an event and analysis of possible root causes, the Policy Manager performs a reactive action with the goal of re-adapting the system configuration. Some possible actions are shown at Table 3.

The main action proposed is the reconfiguration of the current routing protocol. This action should be executed in scenarios where overall network behavior has changed, such as mobility degree modification, communication pattern modification, power electricity company failure (nodes operating with battery), high modification in node density and others. The above actions implies in the reconfiguration of all network nodes, which is a very costly procedure.

Other actions are related to routing protocol optimization, such as variation of cluster or zone radius, variation of the maximum number of paths for multi-path protocols, optimization of the cache entry expiration time for reactive protocols and route update message time for proactive routing protocols.

4   Policy Manager Implementation

This section describes the architecture and implementation of a Policy Manager (PM) for dynamic reconfiguration of ad hoc networks. The main goal is to integrate the PM in the NS2 network simulator for experimentations of dynamic policy-based network reconfiguration evaluation in a simulation scenario.

The main purpose of the Policy Manager is to allow the creation, modification and removal of policies, as well as the evaluation of these policies and configurations at network device level.

The PDP architecture is shown in Figure 2, and the PM interfaces and components used are described below:

- I-PMT (Policy Management Tool Interface): provides access for network administrators, allowing them to insert, edit and delete policies;
- I-PR (Policy Repository Interface): provides access to the policies (events, conditions, actions and its relationships) stored at the policy repository;
- I-PSMR (Policy State Machine Repository Interface): provides access to the policies state repository;
- I-NELC (Network Element Layer Configuration Interface): provides access to the different network devices for configuration. This interface may translate policy actions into protocols and specific commands from the devices being managed;
- I-NELM (Network Element Layer Monitoring Interface): provides access to device for monitoring network and traffic state;
- PE (Policy Editor): implements the server side presentation logic of the policy editor. Policy edition encompasses a user friendly interface, policy translation to the adopted PIM (Policy Information Model), and policy conflict detection;

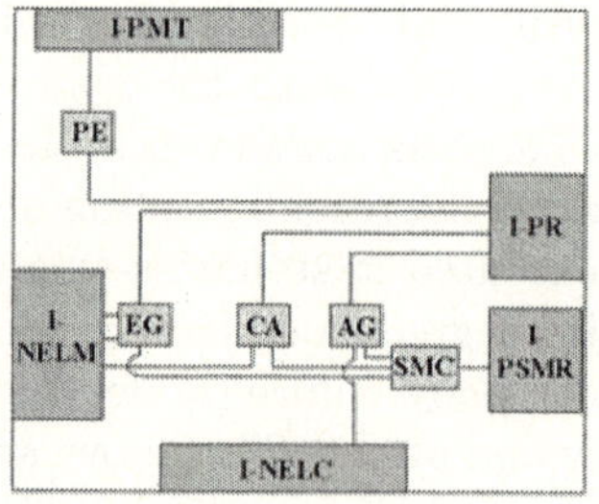

Fig. 2. Policy Manager internal architecture

- EG (Event Generator): provides generation of specific pre-configured events, mainly obtained through the I-NELM interface, subsequently triggering the analysis of the set of conditions from the policies associated with the given event;
- CA (Condition Analyzer): provides simple condition analysis based on Policy-Variables, PolicyOperators and PolicyValues;
- AG (Actions Generator): sends specific actions to the I-NELC interface after condition evaluation to TRUE or FALSE;
- SMC (State Machine Controller): communicate with EG, CA and AG modules for updating and providing up to date policy state, allowing the system to run free of policy evaluation errors and helping the network administrator to monitor the QoS levels applied through the policies.

5   Conclusion and Future Works

In this paper we have described the implementation architecture of a PBNM system designed specifically for dynamic configuration of wireless ad hoc routers. The paper contribution comprises the proposal of an ad hoc policy information model based on DEN-ng policy model for the implementation of a policy manager prototype to be integrated in the NS-2 simulator allowing dynamic reconfiguration of routing protocol parameters in simulation time; and the proposal of policies for dynamically adjusting the ad hoc routing protocol behavior.

As far as the policy-based management of ad hoc networks proves to be effective, not causing network instability (through simulation), the strategy shall be applied to the project proposed by CPqD of a system based on ad hoc wireless network, aiming at providing voice and data services.

Other future works includes the integration of the Policy Manager within NS2 network simulator, validation of the applicability of policy-based reconfiguration of ad hoc network parameters and the development of a policy conflict detector and a policy pluggable Policy Manager. This means that new kinds of events, conditions and actions could become available on the Policy Editor and Policy Manager without new generation of code. This task should be performed using mechanisms based on XML.

References

1. Chen, W., Jain, N., Singh, S.: ANMP: Ad hoc network network management protocol. IEEE Journal on Selected Areas in Communications, August (1999), 1506–1531
2. Shen, C., et al: The guerrilla management architecture for ad hoc networks. MILCOM 2002, vol. 21, no. 1, October (2002), 466–471
3. Strassner, J.: Policy-Based Network Management - Solutions for the Next Generation. Morgan Kauffmann Publishers (2004).
4. Figueiredo, F.L., Siqueira, M.A., Souza, H.J., Pacifico, A.L., Santos, L., Martins, J., Castro, M.C.: An Ad Hoc Wireless System for Small Towns and Rural Areas, to be published in the Journal of the Brazilian Telecommunications Society (2004).

5. Toh, C.K.: Ad Hoc Mobile Wireless Networks: Protocols and Systems. Prentice Hall; 1st edition December 3, (2001).
6. Xue, Q., Ganz, A.: Ad Hoc on-demand routing (AQOR) in mobile ad hoc networks. Journal of Parallel Distributed Computing (2003).
7. Ahn, G-S., et al.: INSIGNIA, IETF Internet Draft, draft-ietf-manet-insignia-01.txt, October (1999), expired.
8. Ahn, G.S., Campbell, A.T., Veres, A., Sun, L.H.: Supporting Service Differentiation for Real-Time and Best-Effort Traffic in Stateless Wireless Ad Hoc Networks (SWAN). IEEE Transactions on Mobile Computing, Vol. 1, No. 3, July (2002).

V-TCP: A Novel TCP Enhancement Technique

Dhinaharan Nagamalai[1], Beatrice Cynthia Dhinakaran[2],
Byoung-Sun Choi[1], and Jae-Kwang Lee[1]

[1] Department of Computer Engineering, Hannam University,
306-791, Daejeon, South Korea
dhinaharan2000@yahoo.com
{bschoi, jklee}@netwk.hannam.ac.kr
http://netwk.hannam.ac.kr
[2] Department of Computer Science & Engineering, Woosong University,
300-718, Daejeon, South Korea
dhinaharann@yahoo.com

Abstract. Transmission Control Protocol (TCP) is a reliable transport protocol tuned to perform well in habitual networks made up of links with low bit-error rates. TCP was originally designed for wired networks, where packet loss is assumed to be due to congestion. In wireless links, packet losses are due to high error rates and the disconnections induced are due to mobility. TCP responds to these packet losses in the same way as wired links. It reduces the window size before packet retransmission, initiates congestion control avoidance mechanism and resets its transmission timer. This adjustment results in unnecessary reduction of the bandwidth utilization causing significant degraded end-to-end performance. A number of approaches have been proposed to improve the efficiency of TCP in an unreliable wireless network. But researches only focus on scenarios where TCP sender is a fixed host. In this paper we propose a novel protocol called V-TCP (versatile TCP), an approach that mitigates the degrading effect of host mobility on TCP performance. In addition to scenarios where TCP sender is fixed host, we also analyze the scenario where TCP sender is a mobile host. V-TCP modifies the congestion control mechanism of TCP by simply using the network layer feedback in terms of disconnection and connection signals, thereby enhancing the throughput in wireless mobile environments. Several experiments were performed using NS-2 simulator and the results were compared to the performance of V-TCP with Freeze-TCP [1], TCP Reno and with 3-dupacks [2]. Performance results show an improvement of up to 50% over TCP Reno in WLAN environments and up to 150% in WWAN environments in both directions of data transfer.

1 Introduction

The research community has been working very hard to find the solutions to the poor performance of TCP over the wireless networks. Such work can be classified into 4 main approaches: a) Some researchers have focused on the problem

P. Lorenz and P. Dini (Eds.): ICN 2005, LNCS 3421, pp. 125–132, 2005.

at the data link layer level (LL) by hiding the deficiencies of the wireless channel from the TCP. b) Others believe in splitting the TCP: one for the wired domain and another for the wireless domain. c) A third group of researchers believe in modifying the TCP to improve its behavior in the wireless domain. d) A final group is those who believe in the creation of new transport protocols tuned for the wireless networks. While several approaches have been proposed for mitigating the effect of channel conditions,[3] [4] [5] [6] [7] of late approaches that tackle mobility induced disconnections have also been proposed [1] [2] [10]. Our approach V-TCP falls in this category. In this paper we propose a novel protocol V-TCP (versatile TCP) that mitigates the degrading affect of mobility on TCP performance. This paper is organized as follows. In section 2 we analyze the motivation and the related approaches. In section 3 V-TCP's mechanism is introduced. In section 4 experiments using the simulator are presented. Finally in section 5 we compare our approach with other approaches that were proposed earlier. We wrap up our contributions in section 6.

2   V-TCP Mechanism

The main idea of designing V-TCP is to improve the performance of TCP in wireless mobile environments in the presence of temporary disconnections caused by mobility. Unlike previous research approaches, V-TCP not only improves the performance when the TCP sender is a FH, but also when TCP sender is a MH. The only change required in the V-TCP mechanism is to modify the network stack at the MH and also it requires feedback regarding the status of the connectivity. V-TCP makes reasonable assumptions which are similar to the network layer in wireless mobile environments like mobile IP [11]. V-TCP assumes that a network layer sends connection-event signal to TCP, when MH gets connected to the network and a disconnection-event signal, when the MH gets disconnected from the network. V-TCP utilizes these signals for freeze/continue data transfer and changes the actions taking place at RTO (Retransmission Time Out-event), leading to enhanced TCP throughput. The mechanism of V-TCP is explained as follows.

2.1   Data Transfer from FH to MH

V-TCP will delay the Ack for the last two bytes by "x' milliseconds (nearly 800 milliseconds)[8].

Disconnection-Event Signal. Once disconnected the network connectivity status is updated.

Connection-Event Signal. TCP acknowledges the first bytes with ZWA (zero window advertisement) and second bytes with a FWA (full window advertisement). TCP at FH will process these acks as they have a higher sequence number than the previous Acks received [8]. The ZWA will cause the TCP sender at FH to freeze its retransmission timer, without reducing the cwnd. Thus TCP at FH is prevented from entering into congestion control mechanism when packets are lost and when the MH is disconnected.

2.2   Data Transfer from MH to FH

We consider 3 event signals here for our analysis.

Disconnection-Event Signal. V-TCP's behavior under different disconnection scenarios is explained s follows. *Case 1 Sending window open.* V-TCP cancels the retransmission timer and does not wait for Ack for the packets that were sent before disconnection. *Case 2 Sending window closed.* V-TCP waits for ack, and does not cancel the retransmission timer, but waits for the RTO to occur.

Connection-Event Signal. V-TCP assumes that a n/w layer sends a connection-event signal to TCP when MH gets connected to the network. *Case1 Sending window open.* V-TCP sets the retransmission timer after the data is sent. Because all the Acks are cumulative, any Ack for the data that had been newly sent also acknowledges the data sent before disconnection. *Case2 Sending window closed (RTO occurred).* V-TCP retransmits if the sending window is closed and RTO occurred. *Case3 Sending window closed (RTO not occurred).* V-TCP waits for an RTO event to occur.

Retransmission Time Out-Event Signal. V-TCP utilizes these signals for freeze/continue data transfer and changes the action taken place at RTO (Retransmission Time Out-event), leading to enhanced TCP throughput. V-TCP first checks, if there had been any disconnection in the network. If the disconnection had taken place then, V-TCP sets the ssthresh=cwnd at the time of disconnection, instead of reducing the ssthresh (behavior of TCP) and also sets cwnd=1. But in the case where a connection had occurred, V-TCP retransmits the lost packets without any modification to ssthresh and cwnd parameters. Thus V-TCP promptly salvages the cwnd value prior to disconnections, thus reducing under utilization of the available link capacity.

3   Implementation

We have performed several experiments using simulation. V-TCP, freeze-TCP and 3-dupack were implemented in the network simulator ns-2 [9]. In the ns-2 simulator TCP Reno is already implemented. The only modification required is mobile IP [11] for providing mobility information to the TCP agent. The mobility of the MH is maintained by a variable, Network status, in the TCP agent whose values changes from connection to disconnection or vice versa, determined by a disconnection timer handler.

3.1   Network Topology

The network topology is shown in Fig.1. An FTP application simulated a large data transfer, with a packet size 1000 of bytes and the throughput of TCP connections was measured. Values ranging from 50ms to 5s were chosen as disconnection duration. Larger values occur in WWAN and smaller values for WLAN. The disconnection frequency was chosen as 10seconds, indicating a high mobility. The RTT was chosen as 8ms for WLAN 800ms for WWAN. The link capacity

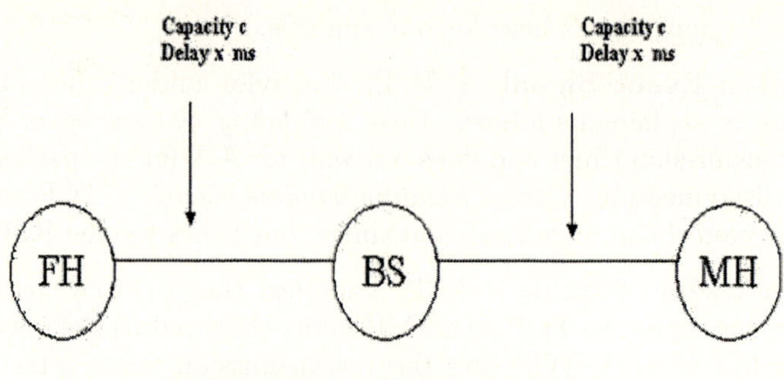

Fig. 1. Network Topology

(c) is 10Mbps for 8ms RTT (WLAN) and 100 kbps for 800ms RTT (WWAN). The capacity of both links, i.e. FH to BS and BS to MH are maintained equal to avoid any packet loss due to buffer overflow in the routers. The simulations were carried out for 100sec for WLAN environment and 1000sec for WWAN environment.

4   Performance Evaluation

The performance of V-TCP is explained for both directions and the results are compared with TCP Reno, Freeze-TCP and 3-dupack.

4.1   Data Transfer from MH to FH

As seen from Fig 2 and 3 V-TCP shows an increased throughput when compared to TCP Reno. Since there are no approaches in freeze-TCP and 3-dupacks for MH being TCP sender, therefore only TCP Reno is compared V-TCP in this case. We point the factors that lead to the increase throughput of V-TCP.

No Idle Period. When a RTO event occurs due to disconnection TCP Reno retreats exponentially. Upon MH reconnection, TCP Reno waits for the retransmission timer (RTX) to expire. Thereby the TCP Reno has to be idle until the RTX expires. As the disconnection period increases, the number of RTO events also increases. This result in exponentially increasing RTX values, thereby increasing the idle period for TCP Reno before it tries for retransmission. But in the case of V-TCP it does not have this idle period and thereby increases the performance.

RTO Event. At each event of RTO, TCP Reno reduces the ssthresh to half. If a RTO occurs when a MH is disconnected it is undesirable. But in the case of V-

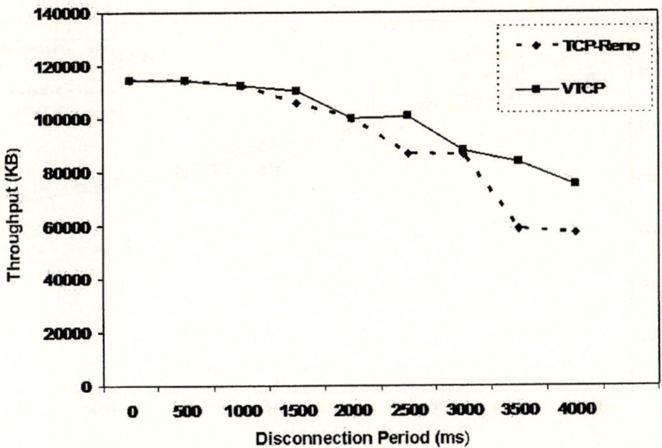

Fig. 2. Data transfer from MH to FH RTT (≈8ms)

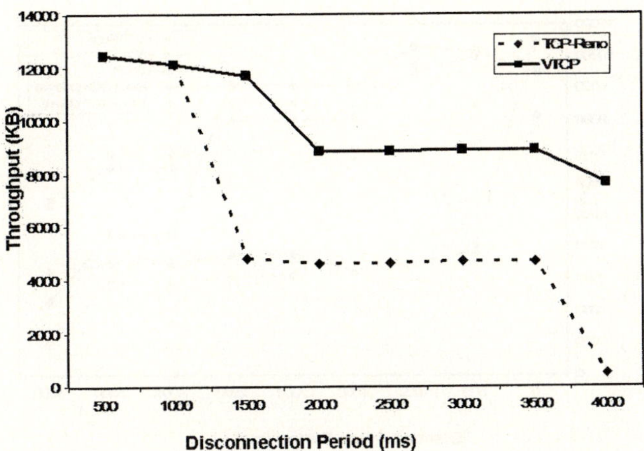

Fig. 3. Data Transfer from MH to FH RTT (≈800ms)

TCP, ssthresh is not reduced, instead it sets ssthresh equal to cwnd value reached at the time of disconnection. This results in V-TCP attaining a full window capacity faster than TCP Reno. There is a significant increase in throughput for V-TCP over TCP Reno for large RTT connections. This is because connections with large RTT have analogous large values of RTX,thereby increasing the idle period for TCP Reno. Performance results show an improvement of up to 50% improvement over TCP Reno for short RTT connections and up to 150% in the case of long RTT connections, with long periods of disconnection.

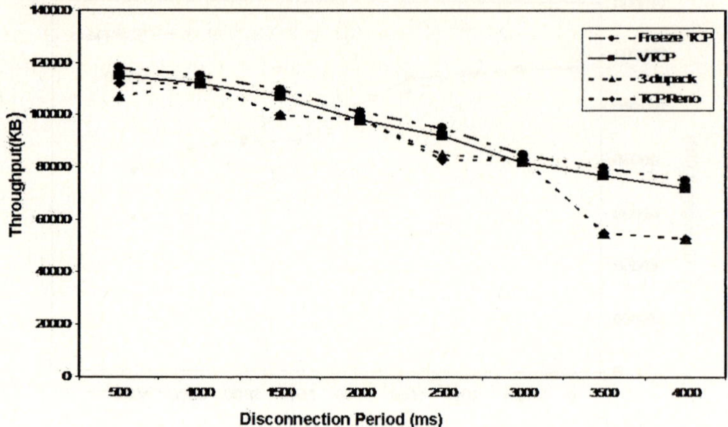

Fig. 4. Data transfer from FH to MH RTT (~8ms)

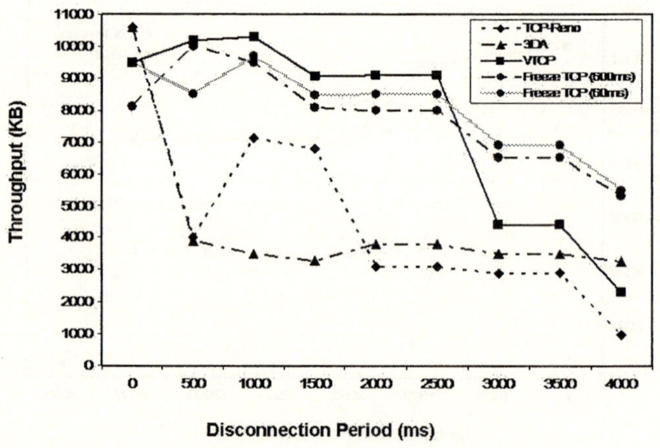

Fig. 5. Data transfer from FH to MH RTT (~800ms)

4.2   Data Transfer from FH to MH

We now compare the performance for FH to MH of V-TCP with TCP Reno, 3-dupacks and freeze-TCP. As seen in Fig 3 and 4, there is significant improvement in performance of V-TCP over TCP Reno and 3-dupacks.

WLAN Environment. For long disconnection intervals we see that V-TCP and freeze-TCP showed an improvement of 50 %. But in WLAN environments, the idle period after reconnection is the prime factor for degraded throughput rather than reduction of cwnd. Both V-TCP and freeze-TCP perform better than TCP Reno by reducing the idle period.

WWAN Environment. In the case of small disconnections periods the performance of V-TCP and freeze-TCP are almost very similar. For disconnections up to 1000ms, both V-TCP and freeze-TCP showed up to 150% improvement over TCP Reno.But for longer disconnection period V-TCP showed only 65% improvement whereas freeze-TCP showed 150% improvement over TCP Reno. However freeze-TCP depends on predicting impending disconnection and its throughput was observed to be sensitive to variations in the prediction period. In WWAN environments, the main features that degrade the performance are idle period and reduction of cwnd. For small disconnections periods where no RTO occurs, V-TCP doesn't change the cwnd value and hence achieves the same throughput as freeze-TCP. But in case of long disconnection period, V-TCP can prevent the reduction in cwnd value and hence the throughput is much better than Freeze TCP. V-TCP is also able to reduce idle period and perform much better than TCP Reno.

5   Conclusion

We have illustrated the V-TCP mechanism to alleviate the degrading effect of host mobility on TCP performance. It requires modifications only to TCP at MH and is optimized for data transfer from MH to FH as well as FH to MH. V-TCP uses feedback from the network layers at MH in terms of disconnection and connection signals, to swiftly regain the full window after the MH gets reconnected. Several simulated experiments were performed and the results of V-TCP were compared with 3-dupack, TCP Reno and freeze-TCP. V-TCP significantly performs better than TCP Reno in both directions of data transfer. Performance results show an improvement of up to 50% over TCP Reno in WLAN environments and up to 150% in WWAN environments in both directions of data transfer. As mentioned earlier 3-dupack and freeze-TCP approaches do not deal with data transfer from MH to FH and hence we compare them only for FH to MH. We wrap up by proposing a new approach called V-TCP, that performs better than TCP Reno and 3-dupack and generally analogous to that of freeze-TCP. Thus this new approach alleviates the degrading effect of host mobility in TCP.

Acknowledgement. This research was supported by the program for the Training of Graduate Students in Regional Innovation which was conducted by the Ministry of Commerce, Industry and Energy of the Korean Government.

References

1. Goff,T.,Moronski,J.,Phattak,D.S.,Gupta,V.: Freeze-TCP: A true end-to-end TCP enhancement mechanism for mobile environments. Infocom, Israel (2000)
2. Ramon Caceres, Liviu Iftode.: Improving the performance of reliable transport protocol in mobile computing environments.ACM Computer Communication review, vol 13. (1995)

3. Mascolo,S., Claudio Caseetti.: TCP Westwood: Bandwidth Estimation for enhanced Transport over wireless links. ACM SIGMOBILE 7/01 Rome Italy (2001)
4. Balakrishnan,H.,Padmanbham,V.N,Katz,R.: Improving Reliable Transport and Handoff Performance in Cellular Wireless Net. Wireless Networks, vol.1. (1995)
5. Sinha,P., Venkataraman,N.,Sivakumar,R.,Bharghavan,V.: WTCP: A reliable transport protocol for WWANs. ACM MOBICOM 99, Seattle, Washington (1999)
6. Ajay Bakre,Badrinath,B.R.: I-TCP:Indirect TCP for Mobile Hosts. Tech Rep, Reuters university (1995)
7. Balakrishnan,H.,Padmanabhan,V.N,Seshan,S.,Katz,R.H.: A Comparison of Mechanisms for Improving TCP Performance over Wireless Links.IEEE/ACM Transactions on Networking (1997)
8. Braden,R.:RFC 1122 for Internet Hosts-Communication Layers.(1989)
9. The network Simluator ns-2.1b8a, http://www.isi.edu/nsnam/ns
10. Brown,K.,Singh,S.: M-TCP: TCP for Mobile Cellular Networks. ACM Computer Communications Review, vol27, (1997)
11. Perkins,C.:RFC 2002. IP Mobility Support (1996)
12. Montenegro,G.,Dawkins,S.:Wireless Networking for the MNCRS. Internet drafts, (1998)

Optimizing TCP Retransmission Timeout

Alex Kesselman[1,*] and Yishay Mansour[2]

[1] Max Planck Institut für Informatik,
Saarbrücken, Germany
`akessel@mpi-sb.mpg.de`
[2] School of Computer Science, Tel-Aviv University, Israel
`mansour@cs.tau.ac.il`

Abstract. Delay spikes on Internet paths can cause spurious TCP timeouts leading to significant throughput degradation. However, if TCP is too slow to detect that a retransmission is necessary, it can stay idle for a long time instead of transmitting. The goal is to find a Retransmission Timeout (RTO) value that balances the throughput degradation between both of these cases. In the current TCP implementations, RTO is a function of the Round Trip Time (RTT) alone. We show that the optimal RTO that maximizes the TCP throughput need to depend also on the TCP window size. Intuitively, the larger the TCP window size, the longer the optimal RTO. We derive the optimal RTO for several RTT distributions. An important advantage of our algorithm is that it can be easily implemented based on the existing TCP timeout mechanism.

1   Introduction

In most cases the Internet does not provide any explicit information about the network conditions. Thus, it is up to the transport protocol to form its own estimates, and then to use them to adapt as efficiently as possible. For these reasons congestion avoidance and control have become critical to the use of the Internet. Jacobson [12] pioneered the concepts of TCP congestion avoidance and control based on additive increase/multiplicative decrease (AIMD). This scheme allows to avoid the congestion collapse as shown by Floyd and Fall [9]. TCP was later augmented with fast retransmission and fast recovery algorithms to avoid inefficiency caused by retransmission timeouts [13, 25].

Despite the conventional wisdom of relying less on timeout-based mechanisms, it has been indicated that a large number of lost packets in the Internet is recovered by retransmission timeouts [3, 21]. The problem is that delays on Internet paths can be highly variable resulting for instance from route flipping [4, 2]. On the one hand, underestimation of RTT leads to a premature retransmission timeout in case there is no loss or the retransmission could be handled by the fast retransmission mechanism. On the other hand, overestimation of RTT leads to a late retransmission timeout, in case there is a loss that cannot be captured by the fast retransmission mechanism. Therefore, it is crucial for the TCP performance to find a Retransmission Timeout (RTO) value that is an equilibrium point balancing between both of these cases.

---

* The work of the first author was supported by AvH-Stiftung.

P. Lorenz and P. Dini (Eds.): ICN 2005, LNCS 3421, pp. 133–140, 2005.
© Springer-Verlag Berlin Heidelberg 2005

Dolev et al. [6] study retransmission timeout setting for a simple transmission protocol by means of competitive analysis. Ludwig and Katz [17] propose the Eifel algorithm to eliminate the unnecessary retransmissions that can result from a spurious retransmission timeout. Gurtov and Ludwig [11] present an enhanced version of the Eifel algorithm and show its performance benefits on paths with a high bandwidth-delay product. Ekstr and Ludwig [7] propose a new algorithm for calculating the RTO, named the Peak-Hopper-RTO (PH-RTO), which improves upon the performance of TCP in high loss environments. Significant efforts have been also devoted to modeling such a complex protocol as TCP [16, 15, 18, 20].

In this paper we study how to find the optimal RTO maximizing the TCP throughput using the model of [20]. Our main contribution is to show that the optimal RTO need to depend on the TCP window size. We derive the optimal RTO as a function of RTT and the TCP window size for a general and some specific distributions of RTT. Intuitively, the larger the TCP window size, the longer the optimal RTO. We note that the heavy-tailed Pareto distribution has been shown to approximate most accurately the actual RTT distribution in the Internet [1, 2, 10, 5]. An important advantage of our algorithm is that it can be easily implemented on the top of the existing TCP timeout mechanism.

RFC 3649 [8] proposes a modification of TCP congestion control that adapts the increase strategy and makes it more aggressive for high bandwidth links (i.e. for large window sizes). In this work we demonstrate that for such scenarios TCP throughput could be further increased by selecting a larger RTO. Our results are strengthened by simulations in [11], which show that proper handling of spurious timeouts on paths with a high bandwidth-delay product can increase TCP throughput by up to 250%.

The rest of the paper is organized as follows. Summary of our results appears in Section 2. In Section 3 we describe the TCP model. Section 4 contains an analytic derivation of the optimal RTO. A general RTT distribution and some specific distributions are considered in Section 5 and Section 6, respectively.

2   Summary of Results

In this section we give an overview of our main results while the formal definitions and proofs are deferred to the following sections. We assume that RTT behaves like a random variable and derive the optimal retransmission timeout as a function of the mean and the variance of RTT and the TCP window size.

The input parameters to our algorithm are the RTT mean μ, the RTT variance σ^2 and the TCP window size W. (We assume that both μ and σ are finite.) Our goal is to find the optimal RTO maximizing the TCP throughput. We show that it is an increasing function on W.

First we obtain some upper bounds on the optimal RTO for a general RTT distribution. These bounds may be considered as worst-case bounds since they hold for any distribution. The results are presented in Table 1. We show that for any RTT distribution the optimal RTO is bounded from above by $W\sqrt{\log W}/3$ times the mean of RTT. Provided that higher moments of RTT exist, we establish bounds which are mostly driven by those moments while the effect of the window size becomes insignificant. Notice that when RTT is a fixed constant, we obtain an upper bound which tends to RTT.

Table 1. General distribution

Moment	RTO – Upper Bound
First moment	$\frac{1}{\sqrt{3}} W \sqrt{\log W}\, E[RTT]$
k'th moment	$\left(\frac{W^2 \log W}{3} \right)^{\frac{1}{k+1}} \left(E[RTT] E[RTT^k] \right)^{\frac{1}{k+1}}$

Table 2. Specific distributions

RTT Distribution	RTO – Optimal Value
Normal	$\mu + \sigma \cdot O\left(\sqrt{\ln W} + \ln \frac{\mu}{\sigma} \right)$
Exponential	$\mu \cdot O(\ln W)$
Pareto	$\left(\frac{W^2 \log W \mu}{3} \right)^{1-1/\mu}$

Next we derive the optimal RTO for some specific distributions. The corresponding results are presented in Table 2. Basically, we would like the probability of a premature retransmission timeout to be very small. The rational is that the throughput degradation due to a premature retransmission timeout is much higher than that due to a late retransmission timeout. Our model sets the probability of a premature retransmission timeout at about $1/W^2$, for optimizing the TCP throughput.

In case RTT is distributed according to the Normal distribution, one would expect the optimal RTO to be a sum of the mean plus the standard deviation times some factor, as our analysis indeed shows. The factor of μ/σ is due to the fact that when $RTO = \mu + \sigma \cdot A$, the expected number of rounds wasted as a result of a late retransmission timeout is $A \cdot \sigma/\mu$. This setting is similar to the RTO calculation of Jacobson [12] while the main difference is the dependence on the window size.

For the Exponential RTT distribution, we show that the optimal RTO is proportional to the mean of RTT and the logarithm of the window size. The logarithmic factor of the window size follows from the form of the density function.

Finally, we consider the heavy-tailed Pareto distribution of RTT and establish that the optimal RTO is the mean of RTT multiplied by a power of the window size. Such a dependence is due to the heavy-tail property of the Pareto distribution.

3   TCP Model

We adopt the model of [20] that is based on Reno version of TCP. The TCP's congestion avoidance behavior is modeled in terms of "rounds." The duration of a round is equal to the round trip time and is assumed to be independent of the window size. We define a round of TCP to be a time period starting from transmitting a window of W packets back-to-back and ending upon receiving the acknowledgments for these packets.

We make the following simplifying assumptions. There is always data pending at the sender, such that the sender can always transmit data as permitted by the congestion window while the receiver's advertised window is sufficiently large to never constrain the congestion window. Every packet is assumed to be individually acknowledged (the delayed acknowledgment algorithm is not in effect). A packet is lost in a round independently of any packets lost in other rounds. However, packet losses are correlated among the back-to-back transmissions within a round: if a packet is lost, all the consequent packets transmitted until the end of that round are also lost[1]. We define packet loss probability p to be the probability that a packet is lost, given that either it is the first packet in a round or the preceding packet in the round is not lost.

We call the congestion avoidance phase a *steady state*. We assume that timeout expiration does not occur during a slow start phase and concentrate on the timeout setting in a steady state. We also assume that the mean and the variance of RTT are available or could be estimated.

We approximate the packet loss probability as a function of the TCP window size in a steady state, which is a simplification of [20], as $p \approx \frac{1}{W^2}$. We note that the model of [20] captures the effect of TCP's timeout mechanism on throughput.

4   TCP Timeout Optimization

In this section we consider optimization of the retransmission timeout. The goal is to maximize the throughput of TCP. Notice that the optimal RTO is the actual RTT, which is unknown to our algorithm. Thus, the online decision must be based only on the available estimates of the mean and the variance of RTT. We try to find the value of RTO that balances throughput degradation between a premature retransmission timeout and a late retransmission timeout, which are so called "bad events" (that will be formally defined later). Recall that in our model bad events occur only in a steady state.

When a bad event happens, we consider the *convergence period* T during which TCP reaches a steady state. We compare the throughput of TCP during T with that of an optimal algorithm that uses the actual RTT as its RTO and sends in average W packets every round. We call to the number of extra packets sent by the optimal algorithm during T *throughput degradation*. The goal is to minimize the expected throughput degradation due to bad events.

First we will derive the expected duration of the convergence period. In the case of a premature retransmission timeout, it takes exactly $\log W$ rounds for TCP to reach a steady state since the TCP window grows exponentially during a slow start phase. In the case of a late retransmission timeout, TCP is idle instead of transmitting during $RTO - RTT$ time units. Thus, the expectation of the length of T in rounds is:

$$E[length(T) \mid RTO > RTT] = \frac{1}{P[RTO > RTT]} \int_0^{RTO} (RTO - RTT)dRTT.$$

We approximate the expected number of rounds using the Law of Large Numbers as $E[length(T)]/\mu$:

$$E[\# \text{ rounds in } T \mid RTO > RTT] \approx \frac{1}{P[RTO > RTT]} \int_0^{RTO} \frac{RTO - RTT}{\mu} dRTT.$$

---

[1] Such situation naturally occurs when the tail-drop policy is deployed by the bottleneck router.

Assuming that there is a sequence of one or more losses in a given round, the probability of retransmission timeout is $\min(1, \frac{3}{W})$ [20]. In the sequel, we assume that $W > 3$. Next we will define the bad events more formally.

Premature retransmission timeout. We say that a timeout occurred *prematurely* if no packet in the round is lost or the loss can be captured by the fast retransmission mechanism. Note that RTO must be smaller than RTT. The probability of this event is:

$$P_1 = P[RTO < RTT] \cdot \left((1-p)^W + \left(1 - (1-p)^W\right) \left(1 - \frac{3}{W}\right) \right)$$
$$\approx P[RTO < RTT].$$

The throughput degradation due to this event is: $L_1 = W \log W$. Observe that during the slow start phase, TCP sends at most W packets. We obtain that the expected throughput degradation as a result of a premature retransmission timeout is:

$$P_1 \cdot L_1 = P[RTO < RTT] \cdot W \log W.$$

Late retransmission timeout. We say that a timeout occurred *lately* if some packets in the round are lost and the loss cannot be captured by the fast retransmission mechanism. Note that RTO must be larger than RTT. The probability of this event is:

$$P_2 = P[RTO > RTT] \cdot \left(1 - (1-p)^W\right) \frac{3}{W} \approx P[RTO > RTT]\frac{3}{W^2}.$$

The throughput degradation due to this event is:

$$L_2 = W \frac{1}{P[RTO > RTT]} \cdot \int_0^{RTO} \frac{RTO - RTT}{\mu} dRTT.$$

We get that the expected throughput degradation as a result of a late retransmission timeout is:

$$P_2 \cdot L_2 = \frac{3}{W} \int_0^{RTO} \frac{RTO - RTT}{\mu} dRTT.$$

The optimal RTO, RTO^*, minimizes the expected throughput degradation, that is:

$$P_1(RTO) \cdot L_1(RTO) + P_2(RTO) \cdot L_2(RTO).$$

Thus, given the probability distribution of RTT, the optimal RTO minimizes:

$$P[RTO < RTT] \cdot W \log W + \frac{3}{W} \int_0^{RTO} \frac{RTO - RTT}{\mu} dRTT.$$

For simplicity, we will derive an approximation for the optimal RTO, the balanced RTO^{**}, for which the expected throughput degradation is the same for both of the bad events:

$$P[RTO < RTT] \cdot W \log W = \frac{3}{W} \int_0^{RTO} \frac{RTO - RTT}{\mu} dRTT). \qquad (1)$$

Note that in the worst case the expected throughput degradation for the balanced RTO is at most twice as large as that for the optimal RTO.

5   General Distribution

In this section we study what is the worst case effect of the TCP window size on the maximal value of the optimal RTO. We derive upper bounds on the optimal RTO that hold for any distribution of RTT. In our analysis we use a simplified form of (1): $P[RTO < RTT]W \log W = \frac{3}{W}\frac{RTO}{\mu}$.

First we show that for any RTT distribution with finite mean, the optimal RTO is bounded from above by $W\sqrt{\log W}/3$ times the mean of RTT. Applying Markov inequality to (1) we get: $\frac{\mu}{RTO}W \log W \geq \frac{3}{W}\frac{RTO}{\mu}$, and thus $RTO \leq \frac{1}{3}W\sqrt{\log W}\mu$.

In case higher moments of RTT exist, applying the general form of Chebyshev inequality and using (1) we obtain an upper bound that depends on both those moments and the window size: $\frac{E[RTT^k]}{RTO^k}W \log W \geq \frac{3}{W}\frac{RTO}{\mu}$, and we obtain

$$RTO \leq \left(\frac{1}{3}W^2 \log W\right)^{\frac{1}{k+1}} \left(E[RTT]E[RTT^k]\right)^{\frac{1}{k+1}} .$$

Notice that when RTT is almost constant, that is $E[RTT^k] \approx \mu^k$, for sufficiently large k the resulting upper bound tends to μ.

6   Specific Distributions

In this section we study the case in which RTT is distributed according to a given known distribution and derive the optimal value of RTO for some well-known distributions.

6.1   Normal Distribution

In this section we consider the Normal distribution of RTT with the mean μ and the variance σ^2, the density function $f(x) = \frac{1}{\sigma\sqrt{2\pi}}e^{-(x-\mu)^2/2\sigma^2}$ and distribution function $F(x) = \Phi(\frac{x-\mu}{\sigma})$. To avoid negative values, we can take RTT to be $\max(D, N(\mu,\sigma))$ for some $D < \mu$, which does not really affect the analysis that is concentrated on the tail of RTT values larger than μ. Since the Normal distribution is invariable under transforming the mean, one would expect the RTO bound to be a sum of the mean plus the standard deviation times some factor, which is indeed the case as we show.

Substituting to (1), $P[RTO < RTT] = 1 - \Phi(\frac{RTO-\mu}{\sigma})$, $E[RTT] = \mu$, $d(RTT) = \frac{1}{\sigma\sqrt{2\pi}}e^{-(x-\mu)^2/2\sigma^2}dx$ and $y = \frac{x-\mu}{\sigma}$ we obtain:

$$\left(1 - \int_0^r \frac{1}{\sqrt{2\pi}}e^{-\frac{y^2}{2}}dy\right)W \log W =$$

$$\frac{3}{W\mu}\left(RTO\int_0^r \frac{1}{\sqrt{2\pi}}e^{-\frac{y^2}{2}}dy - E[RTT\,|\,RTT < RTO]\right).$$

Provided that RTT is sufficiently large, we can assume that $E[RTT|RTT < RTO] \approx \mu$. Having done some calculations, we derive the following RTO: $RTO = \mu + \sigma \cdot O\left(\sqrt{\ln W + \ln \frac{\mu}{\sigma}}\right)$.

The interesting factor is $O(\sqrt{\ln W})$, which guarantees that the probability of a premature retransmission timeout is small.

6.2   Exponential Distribution

In this section we consider the Exponential distribution of RTT with the rate parameter λ, the mean $E[x] = 1/\lambda$, the density function $f(x) = \lambda e^{-\lambda x}$ and the distribution function $F(x) = 1 - e^{-\lambda x}$. We show that the optimal RTO is proportional to the mean of RTT and the logarithm of the TCP window size.

Substituting to (1), $P[RTO < RTT] = e^{-\lambda RTO}$, $E[RTT] = 1/\lambda$ and $d(RTT) = \lambda e^{-\lambda x} dx$ we get:

$$e^{-\lambda RTO} W \log W = \frac{3}{W} \lambda \left(\left(1 - e^{-\lambda RTO}\right) RTO - \int_0^{RTO} x \lambda e^{-\lambda x} dx \right).$$

This gives us the following RTO: $RTO \approx \frac{1}{\lambda} \ln\left(\frac{W^2 \log W}{3}\right) = \frac{1}{\lambda} O(\ln W)$. The logarithm of W achieves the effect of setting the premature retransmission timeout probability to be order of $1/W^2$.

6.3   Pareto Distribution

In this section we consider the heavy-tailed Pareto distribution of RTT with the shape parameter $a > 1$, the mean $E[x] = \frac{a}{a-1}$, the density function $f(x) = \frac{a}{x^{a+1}}$ and the distribution function $F(x) = 1 - (\frac{1}{x})^a$. We show that the optimal RTO is the mean of RTT multiplied by a power of the window size, which is due to the heavy-tail property of the Pareto.

Substituting to (1), $P[RTO < RTT] = \left(\frac{1}{RTO}\right)^a$, $E[RTT] = \frac{a}{a-1}$ and $d(RTT) = \frac{a}{x^{a+1}} dx$ gives us:

$$\left(\frac{1}{RTO}\right)^a W \log W = \frac{3}{W} \frac{a-1}{a} \cdot \left(\left(1 - \left(\frac{1}{RTO}\right)^a\right) RTO - \int_1^{RTO} RTT \frac{a}{x^{a+1}} dx \right).$$

Solving this equation derives the following RTO: $RTO \approx \left(\frac{W^2 \log W \mu}{3}\right)^{1-1/\mu}$. An interesting setting is $a = 2$ where $E[RTT] = 2$. In this case we get that $RTO \approx W\sqrt{\log W}$, which justifies the form of the bound we have for an arbitrary distribution.

References

1. A. Acharya and J. Saltz, "A Study of Internet Round-trip Delay," *Technical Report CS-TR-3736*, University of Maryland, December 1996.
2. M. Allman and V. Paxson, "On Estimating End-to-End Network Path Properties," *In Proceedings of SIGCOMM '99*, pp. 263-274.
3. H. Balakrishnan, S. Seshan, M. Stemm, and R. H. Katz, "Analyzing Stability in Wide-Area Network Per-formance," *In Proceedings of SIGMETRICS'97*.
4. J. C. Bolot, "Characterizing End-to-End Packet Delay and Loss in the Internet," *Journal of High Speed Networks*, 2(3), September 1993.
5. C. J. Bovy, H. T. Mertodimedjo, G. Hooghiemstra, H. Uijterwaal and P. Van Mieghem, "Analysis of End to end Delay Measurements in Internet," *In Proceedings of PAM 2002*, March 2002.

6. S. Dolev, M. Kate and J. L. Welch, "A Competitive Analysis for Retransmission Timeout," *15th International Conference on Distributed Computing Systems*, pp. 450-455, 1995.

7. H. Ekstr and R. Ludwig, "The Peak-Hopper: A New End-to- End Retransmission Timer for Reliable Unicast Transport," *In Proceedings of IEEE INFOCOM 04*.

8. S. Floyd, "HighSpeed TCP for Large Congestion Windows," *RFC 3649*, December 2003.

9. S. Floyd and K. Fall, "Promoting the Use of End-to-end Congestion Control in the Internet," *IEEE/ACM Transactions on Networking*, August 1999.

10. K. Fujimoto, S. Ata and M. Murata, "Statistical analysis of packet delays in the internet and its application to playout control for streaming applications," *IEICE Transactions on Communications*, E84-B, pp. 1504-1512, June 2001.

11. A. Gurtov, R. Ludwig, "Responding to Spurious Timeouts in TCP," *In Proceedings of IEEE INFOCOM'03*.

12. V. Jacobson, "Congestion Avoidance and Control," *In Proceedings of SIGCOMM'88*.

13. V. Jacobson, "Modified TCP congestion avoidance algorithm," *end2end-interest mailing list*, April 30, 1990.

14. P. Karn and C. Partridge, "Improving Round-TripTime Estimates in Reliable Transport Protocols," *In Proceedings of SIGCOMM '87*, pp. 2-7, August 1987.

15. A. Kumar, "Comparative Performance Analysis of Versions of TCP in a Local Network with a Lossy Link," *IEEE/ACM Transactions on Networking*, 6(4):485-498, August 1998.

16. T. V. Lakshman and U. Madhow, "The Performance of TCP/IP for Networks with High Bandwidth-Delay Products and Random Loss," *IEEE/ACM Transactions on Networking*, 3(3):336-350, June 1997.

17. R. Ludwig, and R. H. Katz, "The Eifel Algorithm: Making TCP Robust Against Spurious Retransmissions," *ACM Computer Communication Review*, 30(1), January 2000.

18. M. Mathis, J. Semske, J. Mahdavi, and T. Ott, "The macroscopic behavior of the TCP congestion avoidance algorithm," *Computer Communication Review*, 27(3), July 1997.

19. T. Ott, J. Kemperman, and M. Mathis, "The stationary behavior of ideal TCP congestion avoidance," November, 1996.

20. J. Padhye, V. Firoiu, D. Towsley and J. Kurose, "Modeling TCP Throughput: A Simple Model and its Empirical Validation," *In Proceedings of SIGCOMM'98*.

21. V. Paxson, "End-to-End Internet Packet Dynamics," *In Proceedings of SIGCOMM'97*.

22. V. Paxon and M. Allman, "Computing TCP's Retransmission Timer," *RFC 2988*, November 2000.

23. J. Postel, "Transmission Control Protocol," *RFC 793*, September 1981.

24. P. Sarolahti and A. Kuznetsov, "Congestion Control in Linux TCP," *In Proceedings of the USENIX Annual Technical Conference*, June 2002.

25. W. R. Stevens, "TCP Slow Start, Congestion Avoidance, Fast Retransmit, and Fast Recovery Algorithms," *RFC 2001*, January 1997.

26. L. Zhang, "Why TCP timers don' t work well," *In Proceedings of SIGCOMM'86*.

Stable Accurate Rapid Bandwidth Estimate for Improving TCP over Wireless Networks*

Le Tuan Anh and Choong Seon Hong

Computer Engineering Department, Kyung Hee Univerity,
1, Seocheon, Giheung, Yongin, Gyeonggi 449-701, Korea
letuanh@networking.khu.ac.kr
cshong@khu.ac.kr

Abstract. This paper presents a stable accurate rapid bandwidth estimate (SARBE) algorithm to improve TCP performance over wireless networks. The proposed algorithm estimates the bandwidth sample of the forward path of connection by monitoring the sending time intervals of ACKs. By using the stability-based filter, the current bandwidth samples are eliminated from transient changes, while keeping reasonably persistent changes in the absence of noise. TCP congestion control then uses the estimated bandwidth to properly set the slow start threshold ($ssthresh$) and congestion window size ($cwnd$) rather than halving the current value of the $cwnd$ as in TCP after fast retransmit. In the wireless environment, SARBE achieves robustness in aspect of stability, accuracy and rapidity of the estimate, and better performance compared with TCP Reno, Westwood and DynaPara. Furthermore, SARBE is fair in bottleneck sharing and friendly to existing TCP versions, and also tolerates ACK compression in the backward path.

1   Introduction

TCP was originally developed for the wired networks, where bit-error rate is trivial and packet losses are caused by network congestion. However, cellular/wireless networks and wired-wireless networks issue performance degradation challenges to TCP because TCP congestion control cannot distinguish between the packet losses caused by the random error of the wireless link, signal fading and mobile handoff processing, and those caused by network congestion.

At the TCP sender, there are two state variables, the congestion window size and the slow start threshold are maintained by congestion control to the control the transmission rate. In the slow start phase, at beginning of connection, $cwnd$ is increased by 1 for every arrived ACK at the sender. Until $cwnd$ reaches $ssthresh$, the sender TCP congestion control goes to the congestion avoidance phase, where $cwnd$ is increased by $1/cwnd$ for every arrived ACK at the sender.

---

* This work is supported by University ITRC of MIC. Dr. Hong is corresponding author.

P. Lorenz and P. Dini (Eds.): ICN 2005, LNCS 3421, pp. 141–148, 2005.

cwnd is additively increased to reach the network bandwidth and is abundantly decreased to one half of current value when the condition of network congestion occurs by indicating received Duplicate ACKs at the sender. TCP Reno then sets *ssthresh* to be the same as *cwnd*. If the packet losses occur by the random errors before *ssthresh* reaches the network capacity, *ssthresh* can be gotten a smaller value, thus blindly reducing the sending rate (i.e degrading TCP performance).

To improve TCP performance over wireless networks, several approaches have been proposed and [2] classified into three classes: the link-layer approach, which improves wireless link characteristics or hiding non-congestion caused packet losses from TCP; the split-connection approach, in which the sender have to be aware of the existence of wireless hop; the end-to-end approach, which retains TCP semantics, but requires to improve the protocol stack at either the sender side or the receiver side.

In this paper, we proposed SARBE algorithm for improving TCP over wireless networks. SARBE achieves robustness in aspect of stability, accuracy and rapidity of the estimate, and better performance compared with TCP Reno, Westwood and DynaPara. Furthermore, SARBE is fair in bottleneck sharing and friendly to existing TCP versions, and also tolerates ACK compression in the backward path.

The rest of this paper is organized as follows. Section 2 summarizes the previous works in improving the TCP performance for wired-wireless networks. Section 3 presents our proposed SARBE algorithm in detail. Various simulation results presented in section 4. Finally, section 5 concludes the paper.

2   Related Works

The end-to-end approach improves TCP performance at either the sender or receiver without violating end-to-end semantic. TCP SACK [9] (option in TCP header) improves retransmission of lost packets using the selective ACKs provided by the TCP receiver. In Freeze-TCP [5], before occurring handoff, the receiver sends Zero Window Advertisement (ZWA) to force the sender into the Zero Window Probe (ZWP) mode and prevent it from dropping its congestion window. Freeze-TCP can only improve TCP performance in handoff cases.

TCP Westwood scheme [3], [4], and TCP scheme with dynamic parameter adjustment (DynaPara) [7] are the end-to-end approach. In these schemes, the sender estimates available bandwidth dynamically by monitoring and averaging the rate of ACKs receiving. The sender then updates *cwnd* and *ssthresh* to the estimated bandwidth when fast retransmit or retransmission timeout occurs. DynaPara used a simple filter with dynamic adjusting of a filter parameter according to the sate of wireless link. Although, the filter of TCP Westwood is complex, it cannot reflect the rapid changes of the network condition, whereas in DynaPara algorithm, the estimated bandwidth fluctuates frequently. In addition, when the ACK packets encounter queuing with cross traffic along the backward path, their time spacing is no longer the transmission time upon leaving the queue. These time spacing may be shorter than original time spacing, called

ACK compression [10]. In this case, both schemes overestimate the available bandwidth.

3   SARBE Algorithm

3.1   Available Bandwidth Estimate

In comparison with TCP Westwood, we take the advantage of using ACKs sending time interval to achieve a more accurate available bandwidth estimate. SARBE employs the ACKs sending time intervals to compute the available bandwidth of the forward path via the timestamp in ACK. In SARBE approach, the estimate of the forward path is not be affected by ACK compression that results in overestimate.

When the kth ACK arrives, the sender simply uses information of the kth ACK to compute an available bandwidth sample, which can be written as

$$Bw_k = \frac{L_k}{ts_k - ts_{k-1}} \tag{1}$$

where L_k is the amount of data acknowledged by the kth ACK, ts_k is timestamp of the kth ACK; ts_{k-1} is the timestamp of the previous ACK arrived at the sender. It can be seen obviously, sample Bw_k represents the current network condition, which faces noises. So the bandwidth estimator has to eliminate transient noise but responds rapidly to persistent changes.

We used the stability-based filter [6] which is similar to the EWMA filter, except using a measure function of the samples' large variance to dynamically change the gain in the EWMA filter. After computing the bandwidth sample Bw_k from (1), the stability-based filter can be expressed in the recursive form

$$U_k = \beta U_{k-1} + (1 - \beta) \mid Bw_k - Bw_{k-1} \mid \tag{2}$$

$$U_{max} = max(U_{k-N}, ..., U_{k-1}, U_k) \tag{3}$$

$$\alpha = \frac{U_k}{U_{max}} \tag{4}$$

$$eBw_k = \alpha \cdot eBw_{k-1} + (1 - \alpha)Bw_k \tag{5}$$

where U_k is the network instability computed in (2) by EWMA filter with gain β, β was found to be 0.8 in our simulations; $U max$ is the largest network instability observed among the last N instabilities ($N = 8$ in our simulations); and eBw_k is the estimated smoothed bandwidth, eBw_{k-1} is the previous estimate and the gain α is computed as (4) when the bandwidth samples vary largely.

3.2   TCP Congestion Control Modification Algorithm

As mentioned in Section 1, *ssthresh* represents the probed network bandwidth; while the above estimated bandwidth value also represents the current available bandwidth of download link. Consequently, we have to transform the estimated value into equivalent size of the congestion window for updating *ssthresh*. [4] proposed the interrelation of estimated bandwidth with the optimal congestion window size (*oCwnd*) as

$$oCwnd = \frac{eBw \cdot RTT_{min}}{Seg\_size} \tag{6}$$

where RTT_{min} is the lowest Round Trip Time, Seg_size is the length of the TCP segment.

The pseudo code of the the congestion control algorithm for updating *ssthresh* and *cwnd* is the following:

```
if (Duplicate ACKs are detected)
    ssthresh = oCwnd;
    if (cwnd >= ssthresh)
        /* in the congestion avoidance phase*/
        cwnd = ssthresh;
    else /* in the slow start phase */
        cwnd = cwnd; /*keeping cwnd*/
end if

if ( retransmission timeout expires)
    ssthresh = oCwnd;
    cwnd = 1;
/* restarting from the slow start phase */
end if
```

Whenever the sender detects Duplicate ACKs representing that the sent packets were lost due to the light network congestion or the random error of the wireless link, the congestion control updates *ssthresh* to the optimal congestion window (*oCwnd*); and sets *cwnd* to *ssthresh* during the congestion avoidance phase, or keeps the current *cwnd* value during the slow start phase.

If the sender is triggered by the retransmission timeout event due to the heavy network congestion or the very high bit-error rate of wireless link, its congestion control sets *ssthresh* to the optimal congestion window, and sets *cwnd* to one for restarting from the slow start phase.

4   Simulation Results

Our simulations were run by the NS-2 simulation network tool. We used the recent Westwood module NS-2 [8] for comparison.

4.1   Effectiveness

We first evaluate the stability, accurateness and rapidity of SARBE. The simulation network scenario is depicted in Fig. 1(a). We used an FTP over TCP and an UDP-based CBR background load with the same packet size of 1000 bytes. The CBR rate varies according to time as the dotted line in Fig 1(b).

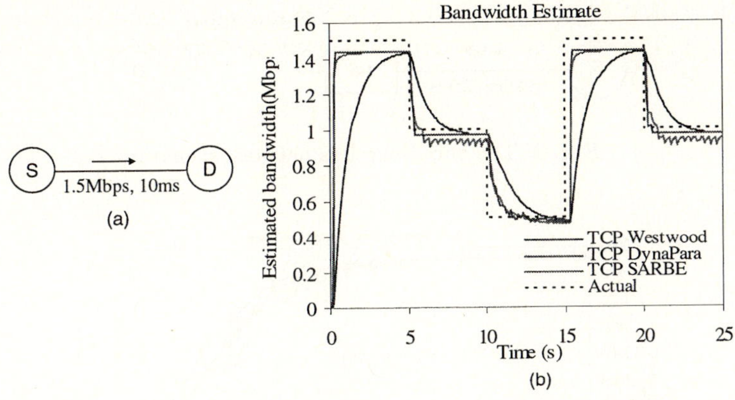

Fig. 1. (a) Single bottleneck link; (b) Comparison of Bandwidth estimate algorithms

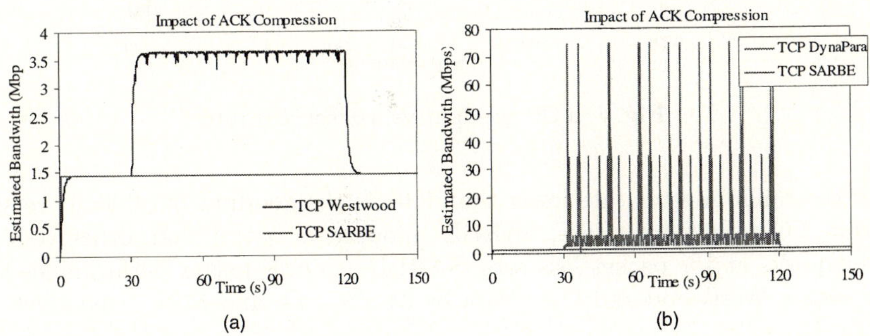

Fig. 2. The overestimated bandwidth of TCP Westwood (a) and TCP DynaPara (b) in the presence of the ACK compression

The result is shown in Fig. 1(b); TCP Westwood is very slow to obtain the available bandwidth changes, while DynaPara estimates fluctuantly and inaccurately when the bottleneck bandwidth is occupied by CBR. By contrast, SARBE reaches the persistent bandwidth changes rapidly, which closely follow the available bandwidth changes. This is due to adaptability of dynamic changes of gain α when the bandwidth samples vary largely.

To investigate the impact of ACK compression on estimate, we used the network scenario as Fig. 1(a) and supplemented a traffic load FTP in the reverse direction. The traffic load FTP was started at time 30 s and ended at 120 s for 150 s simulation time. In this interval, Westwood estimates over 2 Mbps more than SARBE, which is quite near the actual available bandwidth, as in Fig 2(a). In DynaPara, the estimated values fluctuate and are many times more than SARBE, shown in Fig 2(b).

Next, we evaluate TCP performance in the scenario as given in Fig. 3. The simulation was performed on one FTP in 100 s with the packet size of 1000 bytes, the wireless link random errors ranging from 0.001% to 10% packet loss. In Fig

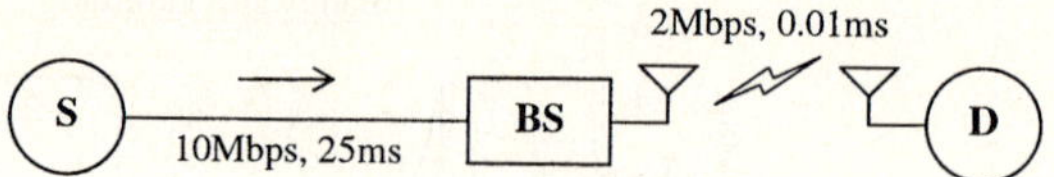

Fig. 3. The wired-wireless network scenario

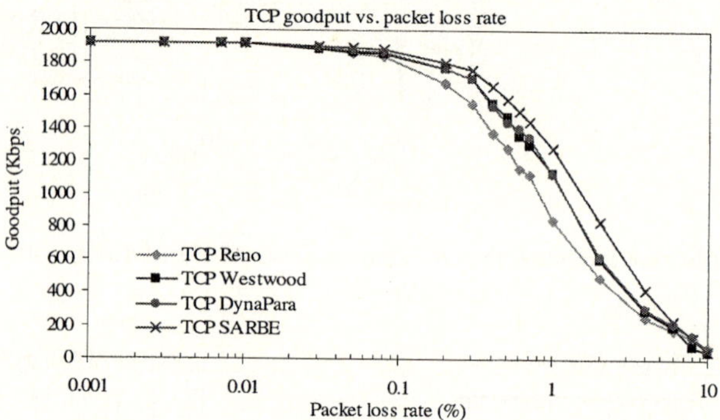

Fig. 4. TCP goodput vs. packet loss rate

4, for the random error rate lower than 0.01%, the goodput of all TCP versions is same. Over that error rate, SARBE' goodput is better than other versions. For example, at 1% packet loss rate, SARBE achieves better performance than TCP Reno, Westwood and DynaPara by 34.6%, 14% and 13%, respectively.

4.2 Fairness

Another evaluation for TCP is fairness that a set of connections of the same TCP, which can share fairly the bottleneck bandwidth. The index of fairness was defined in [1] as

$$fi = \frac{(\sum_{i=1}^{n} x_i)^2}{n(\sum_{i=1}^{n} x_i^2)}, \quad 1/n \le fi \le n \tag{7}$$

where x_i is the throughput of the ith TCP connection, n is the number TCP connections considered in simulation. The fairness index has a range from $1/n$ to 1.0, with 1.0 indicating fair bandwidth allocation. Using the same scenario as Fig. 3 with ten same TCP connections, we simulated the different TCP versions individually. The buffer capacity of bottleneck link is equal to the pipe size. The comparison result is shown in Fig. 5. TCP Reno is always manifested the best fair sharing, the second is SARBE, which achieves quite high fairness index.

4.3 Friendliness

The friendliness of TCP implies fair bandwidth sharing with the existing TCP versions. Our experiments were run on the scenario of Fig. 3. We considered a total of ten connections mixing SARBE with TCP Reno, Westwood and DynaPara

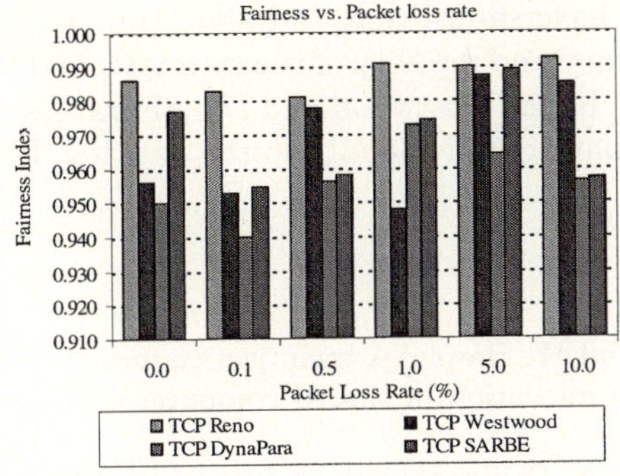

Fig. 5. Fairness vs. packet loss rate

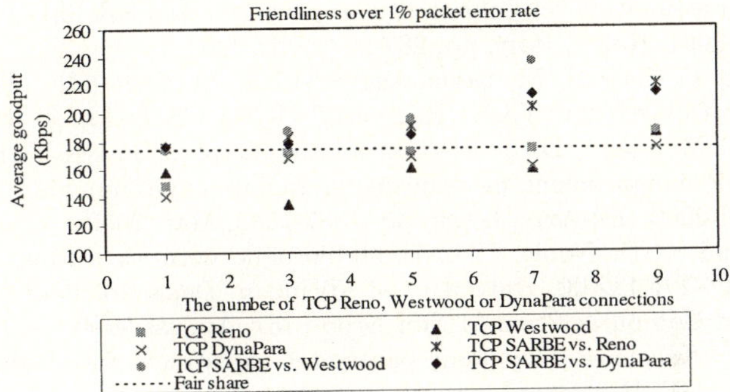

Fig. 6. Friendliness of TCP Reno, Westwood and DynaPara compared with SARBE, respectively, over 1% packet loss link

at 1% packet loss rate. The x-axis of Fig. 6 represents the number of TCP Reno, Westwood or DynaPara connections, the remaining connections use SARBE. The dotted line is the fair share. In Fig. 6, SARBE still preserves friendliness with the existing TCP versions, but outdoes in goodput. This result accords with the above performance evaluation result with the presence of 1% packet loss rate.

5   Conclusions

This paper has presented a novel approach in the available bandwidth estimate. Firstly, estimating the forward path of connection provides the true available bandwidth of download link and tolerates ACK compression. Secondly, applying the stability-based filter can resist transient changes, while keeping reasonably persistent changes of bandwidth samples. Thirdly, the modified congestion control algorithm sets *ssthresh* to the optimal congestion window size upon fast retransmit or retransmission timeout events.

Simulation results have shown that, our algorithm achieves robustness in aspect of stability, accuracy and rapidity of the estimate, and better performance in comparison with TCP Reno, Westwood and DynaPara. As a final point, SARBE is fair in bottleneck sharing and friendly to the existing TCP versions.

References

1. R. Jain, D. Chiu, and W. Hawe, "A quantitative measure of fairness and discrimination for resource allocation in shared computer systems," DEC, Rep.TR-301, 1984.
2. H. Balakrishnan, V. N. Padmanabhan, S. Seshan, and R. H. Katz, "A comparison of mechanisms for improving TCP performance over wireless links," IEEE/ACM Trans. Networking, vol. 5, no. 6, pp. 756769, 1997.
3. S. Mascolo, C. Casetti, M. Gerla, M. Y. Sanadidi,and R. Wang, "TCP Westwood: Bandwidth estimation for enhanced transport over wireless links," in Proc. ACM MobiCom 2001, Roma, Italy, pp. 287-297, July 2001.
4. S. Mascolo, C. Casetti, M. Gerla, and S.S. Lee, M. Sanadidi, "TCP Westwood: Congestion Control with Faster Recovery," UCLA CS Technical Report, 2000.
5. T. Goff, J. Moronski, D. S. Phatak, and V. Gupta, "Freeze-TCP: A true end-to-end TCP enhancement mechanism for mobile environments," in Proc. IEEE INFOCOM 2000, Tel-Aviv, Israel, pp. 1537-1545, Mar. 2000.
6. M. Kim and B. D. Noble, "SANE: stable agile network estimation," Technical Report CSE-TR-432-00, University of Michigan, Department of Electrical Engineering and Computer Science, Ann Arbor, MI, August 2000.
7. N. Itaya, S. Kasahara, "Dynamic parameter adjustment for available-bandwidth estimation of TCP in wired-wireless netwoks," Elsevier Com. Comm.27 976-988, 2004.
8. TCP Westwood - Modules for NS-2 [Online]. Available: http://www.cs.ucla.edu/NRL/hpi/tcpw/tcpw_ns2/tcp-westwood-ns2.html, 2004.
9. M. Mathis, J. Mahdavi, S. Floyd, and A. Romanow, "TCP selective acknowledgment options," RFC 2018, Oct. 1996.
10. L. Zhang, S. Shenker, and D. Clark, "Observations on the Dynamics of a Congestion Control Algorithm: The Effects of Two-Way Traffic," Proc. SIGCOMM Symp. Comm. Architectures and Protocols, pp. 133-147, Sept. 1991.

Performance Analysis of TCP Variants over Time-Space-Labeled Optical Burst Switched Networks

Ziyu Shao[1,*], Ting Tong[2], Jia Jia Liao[2], Zhengbin Li[2,*],
Ziyu Wang[2], and Anshi Xu[2]

[1] Duke University, Durham,
NC 27705, U.S.A
`zs@ee.duke.edu`
[2] Peking University, Electronics Department,
Beijing 100871, P.R.China
`{ttswift, jiajial, lizhb, wzy, xas}@ele.pku.edu`

Abstract. Future operators of core networks will use Optical Burst Switching (OBS). Therefore it is necessary to evaluate TCP congestion inside the optical core network. In this paper, we investigate the performance of TCP variants over TSL-OBS network. Simulation results show that TCP variants such as XCP and HS-TCP have a better performance than standard TCP in TSL-OBS network.

1 Introduction

The unprecedented growth of Internet traffic for the past few years have accelerated the research and the development in the optical network. To meet the increasing bandwidth demands and reduce costs, a new approach called Optical Burst Switching (OBS) that combines the best of optical circuit switching and optical packet switching was proposed by researchers [1,2], and has received increasing amount of attention for it achieves a good balance between OCS and OPS.

Time-Space-Labeled Optical Burst Switching (TSL-OBS) is a proposed variant of optical burst switching (OBS) paradigms, in which time-space label switching protocol (TSL-SP) is designed by using the idea that routing[1] function and signaling function should be integrated into one protocol. This eliminates poor scalability caused by the complicated interworking of multiple protocols. There have been some studies on TSL-OBS mechanisms [11, 12, 13], such as scheduling, contention resolution, synchronization and QoS. However, the impact of TCP variants over TSL-OBS network has not been studied earlier. This paper attempts to present such a study by considering the impact of congestion control mechanisms, deflection routing, data-burst drops, and burst-switching parameters, such as data-burst size and burst timeouts.

The remainder of the paper is organized as follows. Section II briefly introduces the main idea of TCP variants, then we analysis the existing research of TCP variants over OBS network. We present simulation results and analysis in Section III. Section IV concludes our work.

---

* Corresponding Authors are Zhengbin Li and Ziyu Shao.

P. Lorenz and P. Dini (Eds.): ICN 2005, LNCS 3421, pp. 149–155, 2005.

2 TCP Variants for High-Speed Networks

Standard TCP (TCP Reno) is a reliable transport protocol that is well tuned to perform well in traditional networks. However, several experiments and analysis have shown that this protocol is not suitable for bulk data transfer in high bandwidth, large round trip time networks because of its slow start and conservative congestion control mechanism. A lot of efforts are going on to improve performance for bulk data transfer in such networks. To solve the aforementioned problems, two main approaches are proposed [10]: One focuses on a modification of TCP and specifically the AIMD (Additive Increase Multiplicative Decrease) algorithm, the other proposes the introduction of totally new transport protocols. This is a very active research area in networks.

Different TCP variants such as XCP [3], High-speed TCP [4] and FAST [5] have been proposed to solve this problem in high speed networks. Of all these solutions, High-Speed TCP (HSTCP) [4] aims at improving the loss recovery time of standard TCP by changing standard TCP's AIMD algorithm. This modified algorithm would only take effect with higher congestion windows; XCP [3] introduces the concept of decoupling congestion control from fairness control. The congestion controller uses MIMD (Multiplicative Increase Multiplicative Decrease) principle to grab/release large bandwidth quickly and achieve fast response. The fairness controller uses the same principle TCP uses AIMD to converge to fair bandwidth allocation. So XCP is scalable and efficient for any bandwidth and any delay.

On the other hand, there are some researches [6, 7, 8, 9] on performance of standard TCP over JET or JIT based OBS, but the research of TCP variants over OBS have not been investigated yet. Since OBS network will become the core of next generation high-speed networks and TCP variants will become new standards for next generation high-speed networks, it is necessary to investigate the performance of TCP variants over OBS networks.

3 Performance Analysis

In this section we present an evaluation of the performance of TCP variants over a TSL-OBS network, obtained via simulation. Our simulations were conducted on ns-2 simulator [14]. The performance metrics measured are TCP Variants session throughput. The simulations were conducted for varying data-burst lengths, burst assembly times, data-burst dropping rates and deflection costs. Random data-burst drops are introduced at the edge nodes with a uniform probability. Initial simulations were conducted on a 3-node topology (that consisted of two edge nodes connected through a core node) with three TCP sources generating traffic at 1000 Mbps, 500 Mbps and 100 Mbps. The three sources fed traffic to a common edge node. We only consider the 1000Mbps Sessions since we mainly investigate high speed TCP variants.

3.1 Effects of Variation in Data-Burst Length

Figure 1, 2 presents the TCP variants throughput performance for the three-node topology. The burst time-out was set to 2 ms, data-burst dropping rate was set to

0.0001. From figure 1, figure 2 we can see for low burst drop probabilities, increasing the burst size significantly improves the throughput. With small burst sizes, more bursts are generated for the same input traffic rate and thus increased network burden. So more bursts are dropped during the simulation duration. Hence, more reductions in window size occur that leads to lower throughput, and a better throughput is achieved for larger burst sizes.

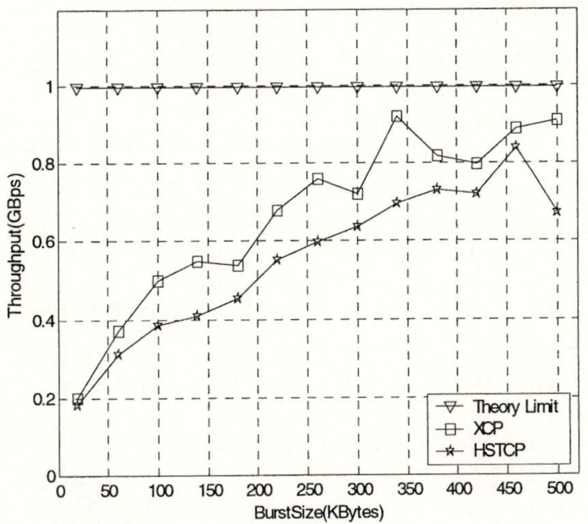

Fig. 1. One Throughput of XCP vs HSTCP

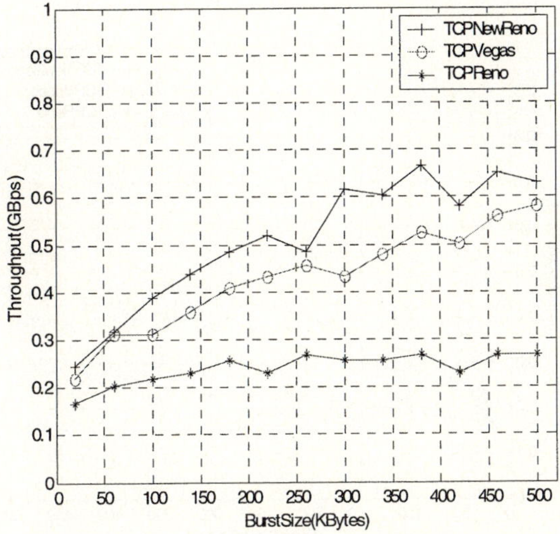

Fig. 2. Throughput of Three TCP versions

3.2   Effects of Variation in Data-Burst Dropping Rate

Figure 3, 4 presents the TCP variants throughput performance for the three-node topology. The burst time-out was not changed, while data-burst dropping rate was set to 0.001. From figure 3, figure 4 we can see , Lost packets trigger TCP's congestion control that reduces the sender window size; and it is a fact that the average (sender) window size of a session determines the achieved throughput. Hence, higher burst drop probabilities will result in lower throughput.

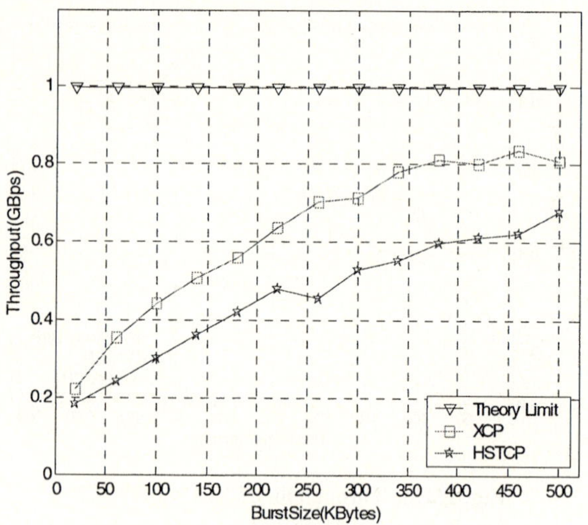

Fig. 3. Throughput of XCP vs. HSTCP

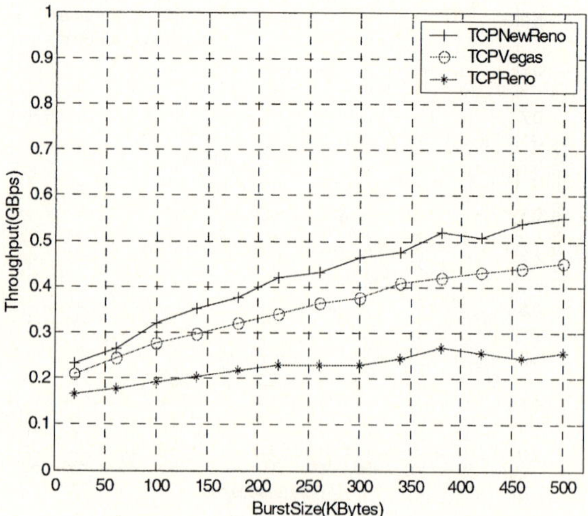

Fig. 4. Throughput of Three TCP versions

3.3   Effects of Variation in Burst-Assembly Times

Figure 5 presents the TCP variants throughput performance for the three-node topology. The burst time-out was varied. From figure 5 we can see XCP and HSTCP throughput changes more acute than TCP NewReno, which due to the different congestion control mechanisms

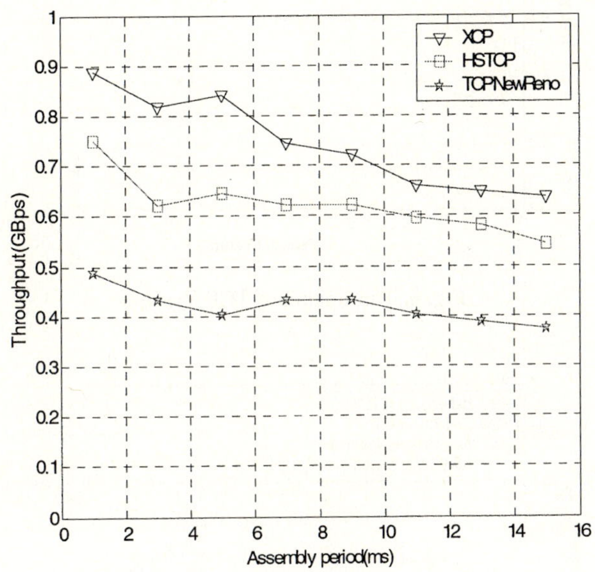

Fig. 5. Throughput of TCP Variants

3.4   Effects of Variation in Deflection Cost

Deflection Cost is defined delay of deflection route minus delay of shortest distance route. We compare in figure 6 and figure 7 TCP throughputs with and without routing versus the deflection routing versus the deflection cost. When deflection routing is not used, the TCP throughput does not change with the deflection cost, since the deflection path is not used at all. On the other hand, with deflection routing, there is a certain range of deflection cost that leads to substantially large TCP throughput (almost doubles in this case) compared with that without deflection routing, which is referred to as the best range henceforth. When the deflection cost is increased inside of the best range, since deflection routing reduces burst loss more than burst loss brought by small delay of deflection routing, higher throughput results. But as the deflection cost keep increasing and out of the best range, the throughput drops quickly to some value, which is close to (but still large than) that without using deflection routing . This due to long delay deflection routing brings more burst loss and hence lower throughput.

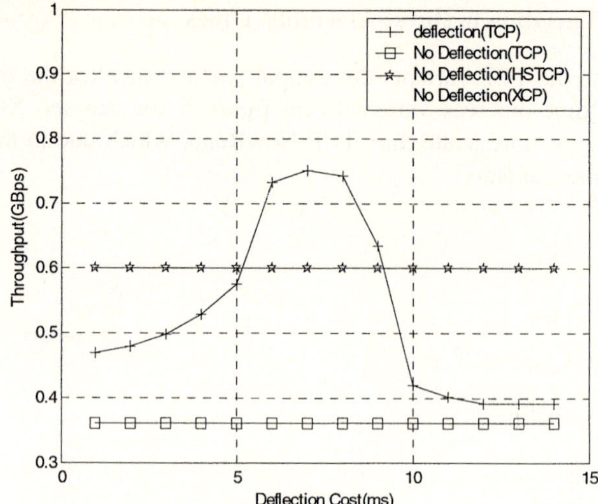

Fig. 6. Throughput of TCP Variants

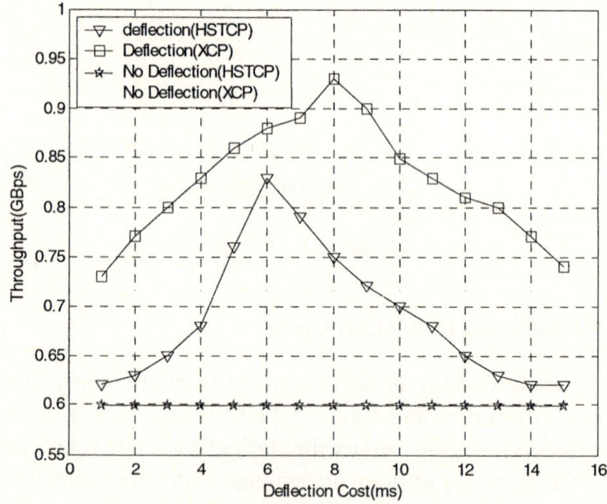

Fig. 7. Throughput of TCP Variants

4 Conclusions

This paper investigated TCP-level performance for time-space-label switched OBS networks. Performance of TCP variants including XCP, HSTCP, TCP New Reno, TCP Vegas and standard TCP (TCP Reno) have been compared with metrics of data-burst lengths, burst assembly times, data-burst dropping rates and deflection costs. As OBS is a strong candidate for next generation high-speed networks and TSL-OBS is a variant of OBS paradigm, it is meaningful to investigate performance of the TCP variants over TSL-OBS networks.

References

1. C. Qiao and M. Yoo, "Optical Burst Switching - A New Paradigm for an Optical Internet", Journal of High Speed Networks, Special Issue on Optical Networks, Vol. 8, No. 1, pp.69-84, 1999.
2. J. Ramamirtham and J. Turner, "Time Sliced Optical Burst Switching", Proceedings of IEEE Infocom 2003, volume 3, pages 2030-2038, 2003.
3. Dina Katabi and Charles Rohrs, "Internet Congestion Control for High Bandwidth-Delay Product Networks." ACM Sigcomm 2002, Pittsburgh, August, 2002.
4. Sally Floyd, "HighSpeed TCP for Large Congestion Windows", RFC 3649, Experimental, December 2003.
5. Cheng Jin, David X. Wei and Steven H. Low, "FAST TCP: motivation, architecture, algorithms, performance ", IEEE Infocom 2004, March 2004.
6. A. Detti and M. Listanti, "Impact of Segments Aggregation on TCP Reno Flows in Optical Burst Switching Networks", IEEE Infocom 2002, New York, NY, June 2002.
7. X. Cao, J. Li, and C. Qiao, "Assembling TCP/IP Packets in Optical Burst Switched Networks", Proceedings of IEEE Globe COM 2002, Taipei, Taiwan, November 2002.
8. Shie-Yuan Wang, "Using TCP Congestion Control to Improve the Performances of Optical Burst Switched Networks", Proceedings of IEEE ICC, volume 2, pages 1438-1442, 2003.
9. S. Gowda, R. Shenai, K. Sivalingam and H. C. Cankaya, "Performance Evaluation of TCP over Optical Burst-Switched (OBS) WDM Networks", Proceedings of IEEE ICC, volume 2, pages 1433-1437, 2003.
10. V. Jacobson, "Congestion Avoidance and Control", IEEE/ACM Transaction Networking, 1998, 6 (3): 314- 329
11. Anpeng Huang, Linzhen Xie, Zhengbin Li, Anshi Xu, "Time-space Label Switching Protocol (TSL-SP)- A New Paradigm of Network Resource Assignment," Photonic Network Communications, Vol. 6, No. 2 (September, 2003), pp.169-178.
12. Ziyu Shao, Anpeng Huang, Zhengbin Li, Ziyu Wang, Anshi Xu, "Service Differentiation in Time-Space-Labeled Optical Burst Switched Networks", Third IEEE/IEE International Conference on Networking (ICN 2004), France, March, 2004.
13. Ziyu Shao, Zhengbin Li, Ziyu Wang, Deming Wu, Anshi Xu, "Performance Analysis of TSL-SP For Optical Burst Switched Networks with Different Precisions of Global Network Synchronization", Third IEEE/IEE International Conference on Networking (ICN 2004), France, March, 2004.
14. http://www.isi.edu/nsnam/ns.

IPv4 Auto-Configuration of Multi-router Zeroconf Networks with Unique Subnets

Cuneyt Akinlar[1] and A. Udaya Shankar[2]

[1] Computer Eng. Dept., Anadolu University, Eskisehir, Turkey
`cakinlar@anadolu.edu.tr`
[2] Dept. of Comp. Science, Univ. of Maryland, College Park, 20742, USA
`shankar@cs.umd.edu`

Abstract. While administration may be necessary in today's large complex IP networks, it is unacceptable for emerging networks such as home networks, small office home office (SOHO) networks, ad-hoc networks, and many others. These networks require zero manual administration and configuration. Zero-configuration (Zeroconf) networks are a particular class of IP networks that do not require any user administration for correct operation. The first step in auto-configuration of zeroconf networks is IP address auto-configuration. While there are proposals for IPv4 host auto-configuration, a general solution for IPv4 auto-configuration of multi-router zeroconf networks does not exist. This paper reviews IPv4 host and single-router auto-configuration algorithms and extends them to more complex multi-router zeroconf network topologies. To the best of our knowledge, we are the first to propose a solution to IPv4 auto-configuration of multi-router zeroconf networks.

1 Introduction

IP hosts and network infrastructure have historically been difficult to configure, requiring network services such as DHCP and DNS servers, and relying on highly trained network administrators. This need for administration has prevented IP networks from being used in many environments such as in homes, in small businesses, in impromptu networks established at conferences etc.

As more intelligent, network-attached devices appear on the market, there is an increasing demand for internetworking of these devices for pervasive computing and communication. Given the usage and applications, the devices are interconnected to form ad-hoc, dynamic networks, with frequent connection/dis(re)-connection of devices from the network, and frequent changes in the network topology. Networking in such an environment, being so dynamic, should be done without any user intervention and administration. This demand recently initiated a new paradigm of IP networking called Zero-Configuration Networking.

Zero-configuration (zeroconf) networks [1] are a class of IP networks that do not need any user configuration and administration for correct operation. Typical applications of zeroconf networking are: Small office home office (SOHO)

P. Lorenz and P. Dini (Eds.): ICN 2005, LNCS 3421, pp. 156–163, 2005.

networks, home networks, i.e., networking of consumer electronic devices, impromptu networks among multiple devices, etc. To make zeroconf networking possible, two recent initiatives, Universal Plug and Play Forum (UPnP) [4] and IETF Zero-configuration working group [1], have emerged. Their goal is to design a set of auto-configuration protocols to enable plug-and-play, easy-to-use IP networking in small IP networks lacking administration.

The first step in auto-configuration of a zeroconf network is IP address auto-configuration. All devices in the network, i.e., hosts and routers, must have IP addresses before they can engage in any communication above the IP layer. Although there are proposals for IPv4 host auto-configuration [5, 3, 12] and single-router IPv4 auto-configuration [2, 4], a general solution for IPv4 auto-configuration of multi-router zeroconf networks does not exist. This paper reviews IPv4 host auto-configuration algorithms in section 2, single-router IPv4 auto-configuration in section 3, and proposes IPv4 auto-configuration algorithms in section 4.

2 IPv4 Host Auto-Configuration

In an IPv4 network, a host must have an IPv4 address before it can perform any communication above the IP-layer. An IPv4 host auto-configuration protocol provides the host with the ability to configure an IPv4 address.

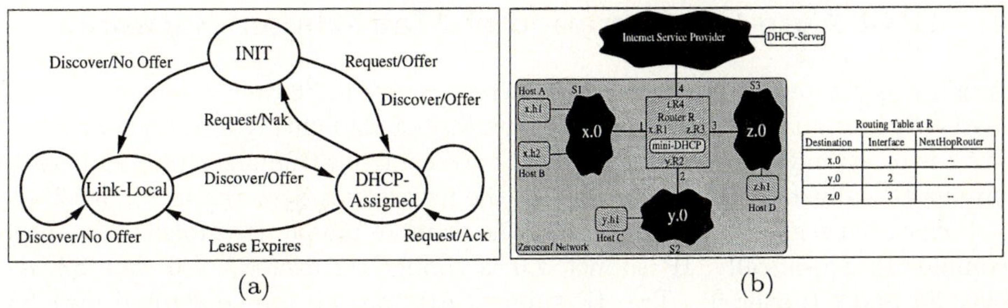

Fig. 1. (a) High-Level State Diagram of IPv4 Host Auto-Configuration Algorithm, (b) A single-router zeroconf network having 3 segments

The state diagram of a generic IPv4 host auto-configuration algorithm [5, 4, 3] is shown in figure 1(a). A host starts at the INIT state. In the presence of a DHCP server, it transits to the DHCP-Assigned state and configures from the DHCP server. In the absence of a DHCP server, the host transits to Link-Local state and randomly selects a link-local IP address, i.e., an IP address in the range 169.254/16 [12]. A host in Link-Local state periodically tests the uniqueness of its own IP address by broadcasting an ARP request and also looks for the presence of a DHCP server. When it finds a DHCP server, it transits to the DHCP-Assigned state and configures from the DHCP server. This protocol has been implemented by Microsoft 98, ME, XP and Apple MAC OS 8.x, 9.x.

The IETF zero-configuration working group [1] has standardized an IP host auto-configuration protocol similar to the one described above. In the rest of this paper, we assume that a host uses this protocol for IPv4 host auto-configuration.

3   IPv4 Auto-Configuration of Single-Router Networks

A single-router network is one where a router inter-connects several segments into a star-shaped topology. Figure 1(b) shows a single-router network where router R joins 3 segments S1, S2 and S3 together, and also connects the zeroconf network to an ISP over interface 4.

Aboba [2, 4] proposes auto-configuring a single-router network by having the router, e.g., a residential gateway, configure a unique IPv4 subnet over each segment attached to the router from the private IP address range 192.168.x/24. The router then starts a mini-DHCP server to configure hosts on directly attached segments. Figure 1(b) shows a single-router network configured using Aboba's method, where unique IPv4 subnets $x.0$, $y.0$ and $z.0$ have been created over segments S1, S2 and S3 respectively. All hosts have also finished configuring IPv4 addresses from the mini-DHCP server. Notice that the mini-DHCP server has configured interface 4 of the router from the DHCP server present on the ISP's network. The router can provide Internet connectivity by proxying, Network Address Translation (NAT) or a combination of both [6, 7].

4   IPv4 Auto-Configuration of Multi-router Networks

A multi-router zeroconf network is one where two or more routers connect several segments into an arbitrary topology. One way of configuring a multi-router network is to extend Aboba's single router auto-configuration algorithm and create unique IP subnets over all segments in the network as depicted in Figure 2(a).

Over each segment in Figure 2(a), one or more unique IP subnets have been configured. Specifically, IP subnet $a.0$ is configured over S1, $b.0$ over S2, $d.0$ over S4 and $e.0$ over S5. Two IP subnets $c.0$ and $h.0$ are configured over S3.

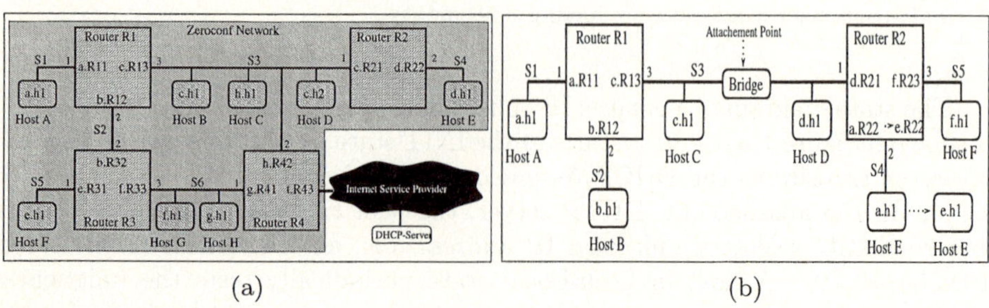

(a)   (b)

Fig. 2. (a) A multi-router network consisting of four routers connecting six segments, (b) A bridge joins two single-router networks to form a two-router network. R2 resolves the IP subnet conflict $a.0$ by changing its IP subnet assignment over S4 to $e.0$

Similarly, two IP subnets $f.0$ and $g.0$ are configured over S6. Notice that the *same* IP subnet can not be assigned to different segments in the network as this would make forwarding a packet impossible. So if the zeroconf network is in a state where the same subnet is temporarily assigned over different segments, which we call an **IP subnet conflict**, one of the IP subnets must be changed to resolve the conflict.

Figure 2(b) shows an example two-router network that has an IP subnet conflict at the time when two single-router networks are joined by a bridge installed on segment S3. The figure assumes that before the bridge is installed, router R1 configured IP subnets $a.0$, $b.0$ and $c.0$ over segments S1, S2 and S3 respectively, and that R2 configured IP subnets $d.0$, $a.0$ and $f.0$ over segments S3, S4 and S5 respectively using the algorithm described in section 3. After the bridge joins the single-router networks together, the new network has an IP subnet conflict: $a.0$ is assigned over two segments, S1 and S4. In the figure, R2 changes its IP subnet $a.0$ over S4 to $e.0$ to resolve the conflict. Host E then reconfigures from the mini-DHCP server at R2 and gets a new IP address $e.h1$.

4.1   IPv4 Router Auto-Configuration

The most difficult task in a multi-router zeroconf network is IPv4 router auto-configuration. Since there are several routers in the network, routers must (1) exchange routing information with other routers to establish loop-free routes to all IP destinations in the network, (2) detect and resolve any IP subnet conflicts and ensure that the same IP subnet is not configured over different segments.

Exchange of routing information and route computation is performed by routing algorithms. Two types of routing algorithms have been proposed in the literature [8]: (1) Shortest Path, a.k.a. Distance-Vector, Algorithms [11, 9], (2) Topology Broadcast, a.k.a., Link-State, Algorithms [10, 8]. In the next section, we describe how a router uses the basic link-state algorithm to establish loop-free routes to all IP destinations in an administered multi-router network where there are no IP subnet conflicts. In section 4.1, we augment the link-state routing algorithm to solve the problem of IP subnet conflict detection and resolution in addition to route computation in a multi-router zeroconf network.

Basic Link-State Algorithm in Administered IP Networks
Suppose that an IP network is represented as a graph $G = (N, L)$, called a routing graph, where N is the set of nodes in the network and L is the set of links between these nodes. Also assume that each node $i \in N$ is *uniquely identified* by an identifier n_i and each link $(i, j) \in L$ has a positive weight $w_{i,j}$, which may vary in time. The challenge is to construct the shortest loop-free paths (routes) from each node to every other node in the network and maintain these routes upon topological changes.

The idea in topology broadcast algorithms is to maintain full-topological information of the network at each node and use this information to calculate shortest loop-free paths to every destination [10, 8]. Let N denote the set of nodes in the network. Associated with each node $n \in N$, there is a sequence

---

Symbols used in the algorithm:
s_n: Sequence number at node n (kept in non-volatile memory).
N_n: Set of nodes neighboring to n.
$w_{n,m}$: Weight of link (n, m) for all $m \in N_n$.
$L_n = \{(m, w_{n,m}) : m \in N_n\}$.
$\mathcal{L}_n$: List of nodes known to n.
s_n^m: View of node n about s_m.
L_n^m: View of node n about L_m.

Basic Topology Broadcast Routing Algorithm for a Node n:
I. Node n comes up:
    I.1. If (comes up first time) then $s_n \leftarrow 0$.
    I.2. $N_n \leftarrow \emptyset$, $\mathcal{L}_n \leftarrow \{n\}$.
    I.3. Bring up all adjacent operating links.
II. Adjacent link (n, m) goes down:
    II.1. Delete m from N_n.
III. Adjacent link (n, m) comes up:
    III.1. $w_{n,m} \leftarrow$ measured weight for (n, m).
    III.2. Add m to N_n.
IV. Periodically:
    IV.1. $s_n \leftarrow s_n + 1$.
    IV.2. Send (n, s_n, L_n) to all neighbors in N_n.
V. Node n receives message (m, s, L):
    V.1. If $(m \notin \mathcal{L}_n$ or $s_n^m < s)$ then
        V.1.1. If $(m \notin \mathcal{L}_n)$ then Add m to $\mathcal{L}_n$.
        V.1.2. $(m, s_n^m, L_n^m) \leftarrow (m, s, L)$.
        V.1.3. Send (m, s, L) to all neighbors in N_n.

Fig. 3. Basic Topology Broadcast Routing Algorithm

number s_n and a list L_n of adjacent links l and their weights w_l. The topology data T for the network consists of the collection $T = \{(n, s_n, L_n) : n \in N\}$.

Figure 3 shows the basic topology broadcast routing algorithm [8]: Every node n periodically sends its local topology L_n to all neighbors while incrementing its sequence number (step IV). Whenever a node m receives a message (n, s_n, L_n) (step V), it consults its current view T_m of T. If n is not listed at all in T_m or if it is listed with a smaller sequence number, then T_m is modified by entering (n, s_n, L_n) into T_m. Then (n, s_n, L_n) is sent to all neighbors. If the sequence number is no larger, the message is discarded. Notice that a message originated from the destination node will eventually reach all other nodes in the network due to step V of the algorithm.

Augmenting the Link-State Algorithm for IPv4 Router Auto-Configuration in Multi-router Zeroconf Networks

For IP subnet conflict detection and resolution in a multi-router network, we use an augmented version of the topology broadcast algorithm (refer to Figure 4).

Although the basic topology broadcast algorithm allows nodes to establish loop-free paths to a destination node identified by its unique identifier, nodes in

Symbols used in the algorithm:
s_n: Sequence number at node n (kept in non-volatile memory).
subnetid$_n$: subnetid of node n (kept in non-volatile memory).
N_n: Set of nodes neighboring to n.
$w_{n,m}$: Weight of link (n,m) for all $m \in N_n$.
$L_n = \{(m, w_{n,m}) : m \in N_n\}$.
$\mathcal{L}_n$: List of nodes known to n.
s_n^m: View of node n about s_m.
L_n^m: View of node n about L_m.
subnetid$_n^m$: View of node n about **subnetid**$_m$.

Augmented Topology Broadcast Routing Algorithm for a Node n:
I. Node n comes up:
    I.1. If (comes up first time) then
        I.1.1. $s_n \leftarrow 0$, **subnetid**$_n \leftarrow$ A unique subnetid.
    I.2. $N_n \leftarrow \emptyset$, $\mathcal{L}_n \leftarrow \{n\}$.
    I.3. Bring up all adjacent operating links.
II. Adjacent link (n,m) goes down:
    II.1. Delete m from N_n.
III. Adjacent link (n,m) comes up:
    III.1. $w_{n,m} \leftarrow$ measured weight for (n,m).
    III.2. Add m to N_n.
IV. Periodically:
    IV.1. $s_n \leftarrow s_n + 1$.
    IV.2. Send $(n, s_n, \textbf{subnetid}_n, L_n)$ to all neighbors in N_n.
V. Node n receives message $(m, s, \textbf{subnetid}, L)$:
    V.1. If $(m \notin \mathcal{L}_n$ or $s_n^m < s)$ then
        V.1.1. If $(m \notin \mathcal{L}_n)$ then Add m to $\mathcal{L}_n$.
        V.1.2. $(m, s_n^m, \textbf{subnetid}_n^m, L_n^m) \leftarrow (m, s, \textbf{subnetid}, L)$.
        V.1.3. If $(\textbf{subnetid} = \textbf{subnetid}_n$ and $m \notin N_n$ and $m < n)$ then
            V.1.3.1. **subnetid**$_n \leftarrow$ A new subnetid. */ *Conflict* */
        V.1.4. Send $(m, s, \textbf{subnetid}, L)$ to all neighbors in N_n.

Fig. 4. Augmented Topology Broadcast Routing Algorithm

an IP network must establish loop-free paths to IP destinations, i.e., IP subnets. Since each router interface creates an IP subnet over the segment (or part of the segment) it is attached to, we augment the message exchanged during the basic topology broadcast algorithm with the subnetid of the destination router interface. So the message periodically sent by a node n becomes $(n, s_n, \textbf{subnetid}_n, L_n)$ (step IV.2).

Had router interfaces attached to different segments configured unique IP subnets, this simple addition to the routing algorithm would have solved the routing problem. But since a router randomly assigns a subnetid to each of its interfaces during initialization (step I.1) two or more routers might assign the same subnetid to their interfaces attached to different segments. This is called an **IP subnet conflict** and must be resolved. *Note that assigning the same subnetid to those interfaces attached to the same segment is in fact desirable and does not constitute an IP subnet conflict.*

To address the issue of IP subnet conflict detection and resolution, we augment the routing algorithm as follows: When a node n receives a message $(m, s, \text{subnetid}, L)$ about a node m (step V), it simply checks if its current subnetid subnetid_n is equal to the subnetid of this node (step V.1.3). If the subnetids are equal, there is an IP subnet conflict if the nodes are not attached to the same segment, i.e., they are not neighbors. To resolve an IP subnet conflict, one of the nodes involved in the conflict must change its subnetid. We propose that the node having the smaller identifier wins the IP subnet conflict battle (step V.1.3). That is, the node with the bigger identifier changes its subnetid while the one with the smaller identifier keeps its current subnetid. The loser node then simply selects a new subnetid in step V.1.3.1. Since the next message sent by the loser node will have a bigger sequence number due to step IV.1, all nodes in the network will start using the new subnetid of the node.

It can be shown that given a multi-router network having k router interfaces, if no further topological changes occur, then the network enters a state where no IP subnet conflicts exist and each router has established a loop-free path to all IP destinations, after $O(k \times e)$ message exchanges, where e is the total number of segments in the network. Informal proof is as follows: Observe that the router interface having the smallest id will beat every IP subnet conflict battle, if there is any. So after a message from this node reaches all nodes in the network, which takes e messages, all nodes in the network would have detected and resolved any IP subnet conflicts with this node and the IP subnet assigned to this smallest id node will not change after that. Next the node with the second smallest id would stabilize its IP subnet within the network in e messages and so on. Lastly the node with the maximum id will have its IP subnet fixed. So regardless of the number of IP subnet conflicts in the initial state of the network, they would all be resolved within $O(k \times e)$ message exchanges. Additionally, by the end of $O(k \times e)$ messages, all routers would have learned the complete topology of the network and can easily compute shortest routes to all IP destinations in the network. Also a router having Internet connectivity, e.g., router R4 in Figure 2(a), can advertise a "default router" address for outside communication.

4.2   IPv4 Host Auto-Configuration in a Multi-router Network

IPv4 host configuration is pretty simple after IPv4 router configuration completes. Recall that routers run a mini-DHCP server. So mini-DHCP servers would simply distribute IP addresses from within the IP subnet configured over their respective interfaces. Hosts attached to a segment would then configure from one of the active mini-DHCP servers on the segment. For example, in Figure 2(a), host G has configured $f.h1$ from the mini-DHCP server running at R3, while host H has configured $g.h1$ from the mini-DHCP server running at R4.

5   Concluding Remarks

We reviewed IPv4 auto-configuration algorithms for hosts and single-router zeroconf networks and proposed IPv4 auto-configuration algorithms for multi-router zeroconf networks. Among the proposed algorithms is a novel augmented basic link-state routing algorithm that solves both the problem of dynamic routing and consistent IP subnet assignment. To the best of our knowledge, we are the first to propose IPv4 auto-configuration algorithms for multi-router zeroconf networks.

References

1. Zero Configuration Networking (zeroconf). http://www.zeroconf.org/, http://www.ietf.org/html.charters/zeroconf-charter.html.
2. B. Aboba. The Mini-DHCP Server. draft-aboba-dhc-mini-04.txt, Sep. 29, 2001. A work in progress.
3. S. Cheshire, B. Aboba, and Erik Guttman. Dynamic Configuration of IPv4 Link-Local Addresses. RFC 3927, October 2004.
4. Microsoft Corp. A Universal Plug and Play Primer. http://www.microsoft.com/hwdev/homenet/upnp.htm, January 1999.
5. Microsoft Corp. APIPA. http://support.microsoft.com/support/kb/articles/Q220/8/74.ASP-Dec 2003.
6. E. Guttman. Zero Configuration Networking. *Proceedings of INET 2000, Internet Society, Reston, VA*, 2000. Available at http://www.isoc.org/inet2000/cdproceedings/3c/3c_3.htm.
7. E. Guttman. Autoconfiguration for IP Networking: Enabling Local Communication. *IEEE Internet Computing*, pages 81–86, June 2001.
8. J. M. Jaffe, A. E. Baratz, and A. Segall. Subtle Design Issues in the Implementation of Distributed, Dynamic Routing Algorithms. *Computer Networks and ISDN Systems*, 12:147–158, 1986.
9. J. M. Jaffe and F. H. Moss. A Responsive Distributed Routing Algorithm for Computer Networks. *IEEE Transactions on Communications*, COM-30(7):1758–1762, July 1982.
10. J. M. McQuillan, I. Richer, and E.C. Rosen. The new Routing Algorithm for the ARPANET. *IEEE Transactions on Communications*, COM-28, May 1980.
11. P. M. Merlin and A. Segall. A Failsafe Distributed Routing Protocol. *EE Publication No. 313, Department of Electrical Engineering, Technion, Haifa*, May 1978.
12. R. Troll. Automatically Choosing an IP Address in an Ad-Hoc IPv4 Network. draft-ietf-dhc-ipv4-autoconfig-05.txt, March 2, 2000, A work in progress.

K-Shortest Paths Q-Routing: A New QoS Routing Algorithm in Telecommunication Networks

S. Hoceini, A. Mellouk, and Y. Amirat

Computer Science and Robotics Lab – LIIA,
Université Paris XII, IUT de Créteil-Vitry
120-122, Rue Paul Armangot - 94400 Vitry / Seine - France
Tel.: 01 41 80 73 82 - fax. : 01 41 80 73 69
{hoceini, mellouk, amirat}@univ-paris12.fr

Abstract. Actually, various kinds of sources (such as voice, video, or data) with diverse traffic characteristics and Quality of Service Requirements (QoS), which are multiplexed at very high rates, leads to significant traffic problems such as packet losses, transmission delays, delay variations, etc, caused mainly by congestion in the networks. The prediction of these problems in real time is quite difficult, making the effectiveness of "traditional" methodologies based on analytical models questionable. This article proposed and evaluates a QoS routing policy in packets topology and irregular traffic of communications network called K-shortest paths Q-Routing. The technique used for the evaluation signals of reinforcement is Q-learning. Compared to standard Q-Routing, the exploration of paths is limited to K best non loop paths in term of hops number (number of routers in a path) leading to a substantial reduction of convergence time. Moreover, each router uses an on line learning module to optimize the path in terms of average packet delivery time. The performance of the proposed algorithm is evaluated experimentally with OPNET simulator for different levels of load and compared to Q-Routing algorithm.

1 Introduction

Internet has become the most important communication infrastructure of today's human society. It enables the world-wide users (individual, group and organizational) to access and exchange remote information scattered over the world. Currently, due to the growing needs in telecommunications (VoD, Video-Conference, VoIP, etc.) and the diversity of transported flows, Internet network does not meet the requirements of the future integrated-service networks that carry multimedia data traffic with a high QoS. First, it does not support resource reservation which is primordial to guarantee an end-to-end Qos (bounded delay, jitter, and/or bounded loss ratio). Second, data packets may be subjected to unpredictable delays and thus may arrive at their destination after the expiration time, which is undesirable for continuous real-time media.

Therefore, it is necessary to develop a high quality control mechanism to check the network traffic load and ensure QoS requirements.

P. Lorenz and P. Dini (Eds.): ICN 2005, LNCS 3421, pp. 164–172, 2005.
© Springer-Verlag Berlin Heidelberg 2005

Various techniques have been proposed to take into account QoS requirements. These techniques may be classified as follows: the congestion control (Slow Start [14], Weighted Random Early Detection [15]), the traffic shaping (Leacky Bucket [17], Token Bucket [16]), integrated services architecture, (RSVP [18]), the differentiated services (DiffServ [19], [20]) and QoS-routing. In this paper, we focus on QoS routing policies.

A routing algorithm consists of determining the next node to which a packet should be forwarded toward its destination by choosing the best optimal path according to given criteria. Among routing algorithms extensively employed in routers, one can note: distance vector algorithm RIP [10] and the link state algorithm OSPF [21]. These algorithms do take into account variations of load leading to limited performances.

For a network node to be able to make an optimal routing decision, according to relevant performance criteria, it requires not only up-to-date and complete knowledge of the state of the entire network but also an accurate prediction of the network dynamics during propagation of the message through the network. This, however, is impossible unless the routing algorithm is capable of adapting to network state changes in almost real time. So, it is necessary to develop a new intelligent and adaptive routing algorithm. This problem is naturally formulated as a dynamic programming problem, which, however, is too complex to be solved exactly. In our approach, we use the methodology of reinforcement learning (RL) introduced by Sutton [2] to approximate the value function of dynamic programming. One of pioneering works related to this kind of approaches concerns Q-Routing algorithm [5] based on Q-learning technique [13]. In this approach, each node makes its routing decision based on the local routing information, represented as a table of Q values which estimate the quality of the alternative routes. These values are updated each time the node sends a packet to one of its neighbors. However, when a Q value is not updated for a long time, it does not necessarily reflect the current state of the network and hence a routing decision based on such an unreliable Q value will not be accurate. The update rule in Q-Routing does not take into account the reliability of the estimated or updated Q value because it's depending on the traffic pattern, and load levels, only a few Q values are current while most of the Q values in the network are unreliable. For this purpose, other algorithms have been proposed like Confidence based Q-Routing (CQ-Routing) [22], Dual Reinforcement Q-Routing (DRQ-Routing) [6]

All these routing algorithms explore all the network environment and do not take into account loop problem in a way leading to large time of convergence algorithm.

In this paper, we propose a K-shortest paths Q-Routing algorithm. This algorithm improves standard Q-Routing algorithm in term of average packet delivery time. It reduces the search space to k best no loop paths in terms of hops number.

The K-shortest paths Q-Routing algorithm is presented in detail in section 2. The performances of this algorithm are evaluated experimentally in section 3 and compared to the standard Q-routing algorithm.

2 K-Shortest Paths Q-Routing Approach

This approach requires each router to maintain a link state database, which is essentially a map of the network topology. When a network link changes its state (i.e., goes up or down, or its utilization is increased or decreased), the network is flooded with a link state advertisement (LSA) message [11]. This message can be issued periodically or when the actual link state change exceeds a certain relative or absolute threshold [9]. Obviously, there is tradeoff between the frequency of state updates (the accuracy of the link state database) and the cost of performing those updates. In our model, the link state information is updated when the actual link state change. Once the link state database at each router updated, the router computes the K shortest paths and determines the best one from Q-Routing algorithm.

Before presenting our approach, we formulate the reinforcement learning process.

2.1 Reinforcement Learning

The RL algorithm, called reactive approach, consists of endowing an autonomous agent with a correctness behavior guaranteeing the fulfillment of the desired task in the dynamics environment [2]. The behavior must be specified in terms of Perception - Decision – Action loop. Each variation of the environment induces stimuli received by the agent, leading to the determination of the appropriate action. The reaction is then considered as a punishment or a performance function, also called, reinforcement signal. Thus, the agent must integrate this function to modify its future actions in order to reach an optimal performance. Reinforcement learning is different from supervised learning, the kind of learning studied in most current researches in machine learning, statistical pattern recognition, and artificial neural networks. Supervised learning learns from examples provided by some knowledgeable external supervisor. This is an important kind of learning, but alone it is not adequate for learning from interaction. In interactive problems it is not often practical to obtain examples of desired behavior that are both correct and representative of all the situations in which the agent has to act. Thus, RL seems to be well suited to solve QoS-routing problem.

2.2 Q-Learning Algorithm for Routing

In our routing algorithm, the objective is to minimize the average packet delivery time. Consequently, the reinforcement signal which is chosen corresponds to the estimated time to transfer a packet to its destination. Typically, the packet delivery time includes three variables: The packet transmission time, the packet treatment time in the router and the latency in the waiting queue. In our case, the packet transmission time is not taken into account. In fact, this parameter can be neglected in comparison to the other ones and has no effect on the routing process.

2.3 Evaluation of the Reinforcement Signal Using a Q-Learning Algorithm

Let's denote by $Q(s, y, d)$, the estimated time by the router s so that the packet p reaches its destination d through the router y. This parameter does not include the latency in the waiting queue of the router s. The packet is sent to the router y which determines the optimal path to send this packet.

The reinforcement signal T employed in the Q-learning algorithm can be defined as the minimum of the sum of the estimated Q (y, x, d) sent by the router x neighbor of router y and the latency in waiting queue q_y corresponding to router y.

$$T = \min_{x \in \text{neighbor of } y} \{ q_y + Q(y,x,d) \} \tag{1}$$

Once the choice of the next router made, the router y puts the packet in the waiting queue, and sends back the value T as a reinforcement signal to the router s. It can therefore update its reinforcement function as:

$$\Delta Q(s,y,d) = \eta(\alpha + T - Q(s,y,d)) \tag{2}$$

So, the new estimation $Q'(s,y,d)$ can be written as follows (fig.1):

$$Q'(s,y,d) = Q(s,y,d)(1-\eta) + \eta(T+\alpha) \tag{3}$$

α and η are respectively, the packet transmission time between s and y, and the learning rate.

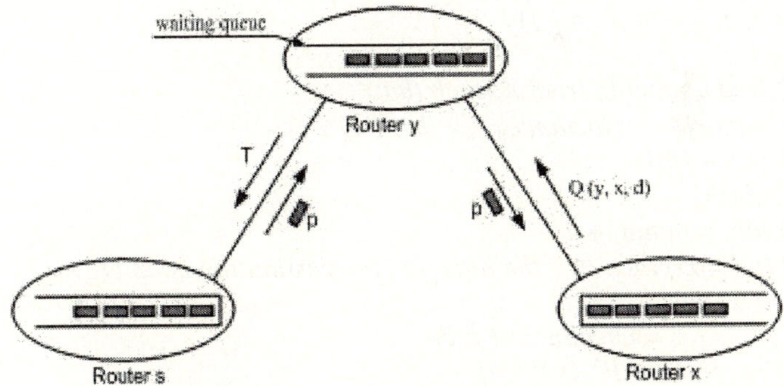

Fig. 1. Updating of the reinforcement signal

2.4 Constructing K-Shortest Paths

Several papers discuss the algorithms for finding K shortest paths [4, 3, 8, 1]. Our solution is based on a label setting algorithm (based on the Optimality Principle and being a generalization of Dijkstra's algorithm) [8]. The space complexity is O(Km), where K is the number of paths and m is the number of edges. By using a pertinent data structure, the time complexity can be kept at the same level O(Km) [8]. We modify the algorithm to find the K-shortest non-loop paths as follows:

Let a DAG (N; A) denote a network with n nodes and m edges, where N = {1,....., n}, and A = {a_{ij} /j,I ∈ N}. The problem is to find the top K paths from source s to all the other nodes.

Let's define a label set X and a one-to-many projection h: N → X, meaning that each node i ∈ N corresponds to a set of labels h(i), each element of which represents a path from s to i.

```
/* S the source node
 * N –set of nodes in network
 * X – the label set
 * Count_i – Number of paths determined from S to I
 * elm – Affected number to assigned label
 * P – Paths list from S to destination (D)
 * K – paths number to compute
 * h – corresponding between node and affected label number */
/* Initialisation */
count_i = 0   /* for all i ∈ N */
elem = 1
h(elem) = s
h^{-1}(s) = {elem}
distance_elem = 0
X = {elem}
P^K = 0
While (count_t < K and X != { })
    begin
        /* find a label lb from X, such that
          distance_lb <= distance_lb1 , ∀ lb1 ∈ X*/
        X = X – {lb}
        i = h(lb)
        count_i = count_i + 1
        if (i == D) then /* if the node I is the destination node D */
            begin
                p = chemin de 1 à lb
                P^K = P^K U {h(p)}
            end
        if (count_i <= K) then
          begin
            for each arc(i,j) ∈ A
                begin
                /* Verify if new label does not result in loop */
                v=lb
                While (h(v) != s)
                  begin
                    if (h(v) == j) then
                        begin
                           goto do_not_add
                        end
                    v = previous_v
                  end
```

/ Save information from new label */*
elem = elem + 1
$distance_{elem} = distance_k + c_{ij}$
$previous_{elem} = lb$
h(elem) = j
$h^{-1}(j) = h^{-1}(j)\ U\ \{elem\}$
$X = X\ U\ \{elem\}$
               do_not_add:
       end
     end
  end

---

3   Implementation and Simulation Results

To show the efficiency and evaluate the performances of our approach, an implementation has been performed on OPNET software of MIL3 Company. The proposed approach has been compared to that based on standard Q-routing [5].

OPNET constitutes for telecommunication networks an appropriate modeling, scheduling and simulation tool [7]. It allows the visualization of a physical topology of a local, metropolitan, distant or on board network. The protocol specification language is based on a formal description of a finite state automaton.

Exact comparison of different QoS Routing algorithms is difficult. Is is therefore important to evaluate methods on a standard topology.

The topology of the network employed here for simulations, which used in [5, 22], includes 33 interconnected nodes, as shown in Fig.2. Two kinds of traffic have been studied: low load and high load of the network.

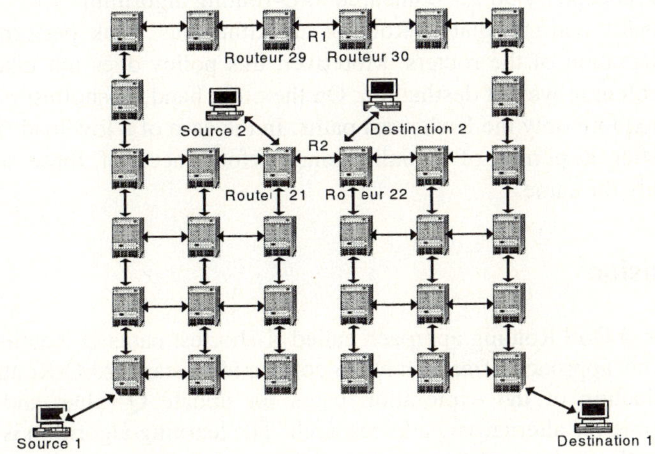

Fig. 2. Network topology for simulation

In the first case, a low rate flow is sent to node destination-1, from node source-1.

From the previous case, we have created conditions of congestion of the network. Thus, a high rate flow is generated by node source-2 to destination-2. Fig.3 shows the two possible ways R-1 (routers-29 and routers-30) and R-2 (routers-21 and routers-22) to route the packets between the left part and the right part of the network.

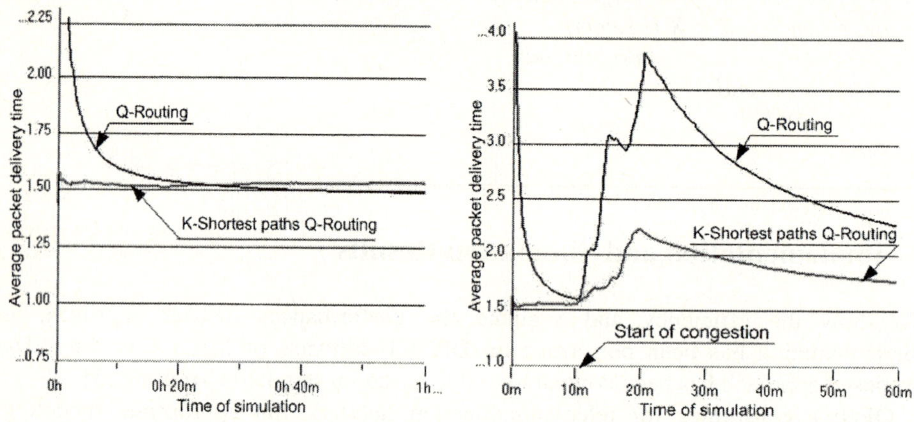

Fig. 3. Network with a low load   **Fig. 4.** Network with a high load

Performances of algorithms are evaluated in terms of average packet delivery time. Fig.3 and 4 illustrates the obtained results when source-2 send information packets during 10 minutes.

From fig.4, one can see, clearly, that after an initialization period, the K-shortest paths Q- Routing, exhibit better performances than standard Q-Routing algorithms. Thus, packet average packet delivery time obtained by K-shortest paths Q-routing algorithm is reduced by 30,2% compared to Q-routing algorithm.

These results confirm that Q-Routing algorithm has weak performances due to speed of adaptation of the routers. Moreover, this policy does not take into account the loop problem in way of destination. On the other hand, K-shortest paths Q-routing algorithms explore only the K-shortest paths. In the case of a low load (fig 3), one can note that after a period of initialization, performances of these algorithms are approximately the same.

4   Conclusion

In this paper, a QoS Routing approach called K-shortest paths Q-Routing algorithm is proposed. This approach offer advantage compared to standard Q-Routing algorithm, like the reduction of the exploration paths for update Q-value, and a reasonable computing time for alternative paths research. The learning algorithm is based on find K-shortest paths in term of hops router and the minimization of the average packet delivery time on these paths.

Simulation results show better performances of the proposed algorithm comparatively to standard Q-routing algorithms. In fact, at a high load level, the traffic is better distributed along the possible paths, avoiding the congestion of the network.

Finally, our work in progress concerns the metric use in finding K-shortest paths (residual bandwidth, loss ratio, waiting queue state ...) and take into account other parameters like the information type of each packet (voice, video, data) in paths selection

References

[1] G. Apostolopoulos, R. Guerin, S. Kamat, A. Orda, T. Przygienda, and D. Williams, "QoS Routing Mechanisms and OSPF Extensions", RFC, December, 1998.

[2] R.S. Sutton and A. G. Barto, "Reinforcement Learning" MIT Press, 1997.

[3] D. Eppstein, "Finding the K Shortest Paths", SLAM J. Computing 28:0(1999) pp. 652-673.

[4] C. Hsu and J. Y. Hui, "Load-Balanced K-Shortest Path Routing for Circuit-Switched Networks", In Proceedings of IEEE NY/NJ Regional Control Conference, August 1994.

[5] J. A. Boyan and M. L. Littman, "Packet Routing in Dynamically Changing Networks: A Reinforcement Learning Approach." In Cowan, Tesauro and Alspector (eds), Advances in Neural Information Processing Systems 6, 1994.

[6] S. Kumar and R. Miikkualainen, "Dual reinforcement Q-routing: an on-queue adaptive routing algorithm" In Proceedings of Neural Networks in Engineering, 1997.

[7] J.E. Nevers, M. J. Leitao, and L.B. Almeida, "Neural Networks in B-ISDN Flow Control: ATM Traffic Prediction or Network Modeling" In IEEE Com. Magazine Oct1995.

[8] E.Q.V. Martins, M.M.B. Pascoal and J.L.E. Santos, "The K Shortest Paths Problem", Research Report, CISUC, June 1998.

[9] G. Apostolopoulos, R. Guerin, and S. K. Tripathi, "Quality of Service Routing: A Performance Perspective", SIGCOMM'98, Vancouver, BC, September, 1998.

[10] G. Malkin, "RIP version2" : Carrying Additional Information, RFC 1388 RFC 1993.

[11] J. Yanxia, N. Ioanis, and G. Pawel, " Multiple path QoS Routing" Proc. Int. Conf. Communications (ICC2001), IEEE, Jun 2001, pp. 2583–2587

[12] Costa, L. H. M. K., S. Fdida, and O. C. M. B. Duarte: 2001, "A Scalable Algorithm for Link-state QoS-based Routing with Three Metrics". In: Proc. of IEEE ICC'2001. Helsink, Filand, 2001.

[13] C. J. Watkins, P. Dayan, "Q-Learning," Machine Learning, Vol.8, pp.279–292, 1989.

[14] V. Jacobson, "Congestion Avoidance of Network Traffic, Computer Communication" Review, vol. 18, no. 4, pp. 314-329, 1988.

[15] W. Feng, D. Kandlur, D. Saha, K. Shin, "Understanding TCP Dynamics in an Integrated Services Internet", NOSSDAV '97, 1997.

[16] S. Shenker, C. Partridge, R Guerin, "Specification of guaranteed quality of service", RFC2212, septembre, 1997.

[17] J. Turner, "New directions in communications (or which way to the information age)",IEEE Communications Magazine, 24(10), 1986.

[18] L. Zhang, S. Deering, D. Estrin et D. Zappala, "RSVP : A New Resource ReSerVation Protocol". IEEE Network, vol. 7, No 5, p. 8–18, Sep 1993.

[19] Y. Bernet, "Requirements of Diff-serv Boundary Routers", IETF Internet Draft, 1998.
[20] K. Nichols, S. Blake, "Differentiated Services Operational Model and Definitions", IETF Internet Draft, 1998
[21] J. Moy, "OSPF Version 2", RFC2328, IETF, 1998.
[22] S. Kumar and R. Miikkualainen, " Confidence-based Q-routing: an on-queue adaptive routing algorithm" In Proceedings of Neural Networks in Engineering, 1998.

Applicability of Resilient Routing Layers for k-Fault Network Recovery

Tarik Čičić[1], Audun Fosselie Hansen[1,2], Stein Gjessing[1], and Olav Lysne[1]

[1] Simula Research Laboratory, PB 134, 1325 Lysaker, Norway
[2] Telenor R&D, 1331 Forneby, Norway
{tarikc, audunh, steing, olavly}@simula.no

Abstract. Most networks experience several failures every day, and often multiple failures occur simultaneously. Still, most recovery mechanisms are not designed to handle multiple failures. We recently proposed a versatile recovery method called Resilient Routing Layers, and in this paper we analyze its suitability for handling multiple failures of network components. We propose a simple probabilistic algorithm for RRL layer creation, and evaluate its performance by comparing it with the Redundant Trees recovery mechanism. We show that not only does RRL provide better fault tolerance, but it also has qualitative advantages that make it very interesting in network systems design.

1 Introduction

The Internet is evolving into the main platform for business critical and real-time communications, and the quality and reliability of the communication have become of extreme importance. It is demonstrated that the Internet provides good service quality in the absence of failures, but also how services are disrupted during failures [1]. These disruptions may stem from fiber cuts, router outages or software and protocol related problems, as well as regular maintenance such as router software updates. Misconfigurations due to complex mechanisms and design may also contribute to those problems [2].

Methods for steering traffic around inoperative links and nodes are often referred to as network recovery. Network recovery is a common term for network protection and restoration. With protection the backup routes are calculated in advance, and stored as additional forwarding information in network routers or switches. With restoration the backup routes are calculated upon detection of failure. Consequently, protection has a shorter time scale of operation than restoration, on the cost of additional state. Restoration has the flexibility to optimize the backup route based on the type and localization of the failure. However, restoration may be inappropriate for recovering real-time and business critical communications due to the time-scale of operation.

Standard IP routing protocols such as OSPF perform network recovery by restoration through rerouting. Topology changes are signaled using routing messages, and new routes are calculated. This process often takes seconds to complete. Real-time applications require much faster recovery, which can only be

P. Lorenz and P. Dini (Eds.): ICN 2005, LNCS 3421, pp. 173–183, 2005.
© Springer-Verlag Berlin Heidelberg 2005

achieved through protection. Protection can be implemented by pre-calculating k disjoint paths between each source and destination to guarantee $(k$-$1)$-fault tolerance. However, this is a complex and resource-demanding task, not implementable in stateless pure IP networks without additional mechanisms such as MPLS. Protection schemes like Redundant Trees (RT, [3]), and p-Cycles [4] are proposed as scalable alternatives to constructing k disjoint paths, and their applicability to IP is also demonstrated. However, their main applicability is for circuit switched networks like WDM, and network resource utilization is a main parameter of optimization. Finally, many proposed recovery schemes can be classified as complex and hard to use in practice. Recent scientific discussion invites network management solutions that focus on easy, practical deployment and operation (e.g., [5]).

To answer the stated challenges of network recovery in IP and other packet-switched technologies, we developed a novel recovery method called Resilient Routing Layers (RRL), to isolate nodes and links in a simple and flexible manner that guarantees one-fault tolerance in biconnected networks [6]. The idea in our approach is that for each node in the network there should exist a *safe layer*, i.e., a spanning topology subset, that can handle any traffic affected by a fault in the node itself, or on any of its links. These layers should be calculated in advance and be used for protective packet forwarding upon the detection of a failure. We have designed RRL with a systems engineering mind-set. In other words, it is simple to understand and deploy, and it is made to be used by the network engineers in practice.

Most research on fault tolerance focuses on single failure cases. A failure however need not always appear alone. A fiber cut may cause multiple higher level links to fail. Several routers may undergo maintenance simultaneously, and a power outage or physical attack may disable whole network regions.

Protection schemes like RT, p-cycles and RRL guarantee recovery from a single failure. They can provide recovery in case of multiple failures as well, but it is not guaranteed. Rather, we speak about the likelihood of successful recovery under given conditions. RRL is not coupled with a specific algorithm for layer generation, and can make state/performance tradeoffs. Furthermore, RRL's main applicability domain is in packet networks, and it covers both link and node failures. Compared to the other recovery schemes, we can therefore say thay RRL has substantial advantages with respect to recovery of multiple failures. In this paper we evaluate RRL k-fault tolerance and compare it with the RT scheme.

The rest of this paper is organized as follows. In Sec. 2 we provide some background on the published methods for fault tolerance, and particularly redundant trees. Sec. 3 presents Resilient Routing Layers and the k-failure algorithm. In Sec. 4 the k-fault tolerance evaluation method is presented, while Sec. 5 presents the evaluation results. Finally, Sec. 6 provides a conclusion and some interesting directions for future work.

2   Background

We say that a graph G is k-connected if and only if for each pair (u, v) of distinct vertices there are at least k internally disjoint uv-paths in G (Menger's theorem, [7]). Thus, a 2-connected network can guarantee 1-fault tolerance, and generally a k-connected network can guarantee $(k\text{-}1)$-fault tolerance.

If restoration is used as the recovery strategy, and if the network survives the failure in connected state, the restoration procedure will find a backup route upon the detection of failures [8], [9]. On the contrary, protection methods must calculate backup routes for every possible failure in advance. The traditional way is to calculate k disjoint paths between every source-destination pair in the network to survive $k\text{-}1$ failures. If the topology does not allow totally disjoint paths one may use algorithms calculating maximum disjoint paths, allowing some of the paths sharing certain links and nodes. Otherwise algorithms mostly find the shortest disjoint paths [10], [11]. Establishing disjoint paths between every pair of nodes is a complicated task, providing the network with a great amount of state. For circuit-switched network where backup resources must be reserved, the disjoint path approach may cause inefficient resource utilization as well.

Several contributions have been proposed to accommodate such obstacles. Protection cycles by Grover and Stamatelakis [4] [12] provide circuit oriented mesh networks with the resource efficiency of mesh restorable networks and the recovery speed of rings. The idea is to pre-configure one or more cycles covering all nodes. When a link or node fails, the traffic is forwarded on its protection cycle instead of the original shortest path. Recent research has demonstrated that configuring one Hamiltonian cycle is most bandwidth-efficient [13].

Schupke and others have recently examined and improved the p-cycles applicability for dual link failures [14], [15]. In its original form, p-cycles offers about 60% double link fault tolerance, while 90% can be reached with an algorithm improvement about 90% dual link failure tolerance in a 3-connected network. However, this improvement comes at the cost of as much as 90% spare link capacity increase.

Although an elegant link-recovery method, p-cycles presents some major drawbacks like complicated node protection and increased backup path length. Furthermore, even though the concept could be extended to IP [16], the main field of application is in optical networks.

Several extensions of p-cycles have been proposed, such as using one or more trees instead of cycles to establish backup resources [17].

2.1   Redundant Trees

Medard et al. present Redundant Trees (RT, [18], [3]), as an elegant network recovery mechanism. RT does no optimization on resource efficiency, covers both link failures and node failures, and is applicable to packet networks [19]. Thus, RT is a suitable candidate for comparative evaluation of RRL performance.

In this approach two unidirectional trees with a single root, named red and blue, are constructed so that all nodes remain connected to the root of at least one of the trees in case of a node or link failure. In different implementations, the root can be either source or destination. The trees can be constructed as follows. First, a cycle that includes the root node is selected from the topology graph. The cycle is traversed in one direction to construct the first branch in the blue tree, and in the opposite direction to construct the first branch in the red tree. (The last link to the root is not included.) Then, the trees are extended by branches that traverse the nodes not yet in the trees. Direction of the blue and red trees on the new branches must always be opposite. RT does not specify the details on how the branches should be chosen, but for k-fault tolerance it is an advantage to use trees with high node degree and short distance between the root and the leaves.

Redundant Trees can be optimized with respect to QoS constraints [20].

3   Resilient Routing Layers

In [6] we presented RRL as a method that guarantees network recovery regardless of which node or link fails, unless the failure physically disconnects the network (i.e., the failed node is an articulation point). If articulation points exist, RRL provides recovery for all nodes except the articulation points.

RRL is based on spanning subsets of the network topology that we call *layers*—each layer includes all nodes but only a subset of the links in the network. We say that a *node is safe* in a layer if only one of its links is used in that layer. All nodes must be safe in at least one layer. We will use the term *safe layer* for a node to denote a layer for which the node is safe. We observe that a node will not experience transit traffic if the traffic is routed in its safe layer. In other words, its safe layer will provide an intact path between all pairs of sources and destinations, unless the node is self the source or destination.

A *safe layer of a link* is normally the safe layer of an adjacent node, alternatively in some rare cases RRL requires a controlled deflection. For details we refer to [6].

The layers are used as input to routing or path-finding algorithms, calculating a routing table or path table for each layer. RRL is suited for use in any packet-switching network technology, including IP and MPLS.

3.1   Layer Construction

RRL is associated with a construction method that generates the layers using the topology data as input. Different methods may be used. For smaller networks, manual construction may be feasible. In general, algorithmic construction is preferred. Different performance metrics of RRL may be optimized, such as the number of layers (state amount), protection path lengths, or the level of layer redundancy (k-fault tolerance).

We illustrate layer construction through a simple greedy algorithm and the sample network depicted in Fig. 1a). The layers are constructed one at a time as long as one or more nodes are not part of any layer. Each new layer is initially a copy of the original topology. For a node not already included in any layer, and which is not an articulation point in the current layer, the links are removed from the current layer except a random one to maintain connectivity (i.e., the current layer becomes safe for that node). At one point no more nodes can be removed without disconnecting the current layer. If all nodes are safe in at least one layer, the algorithm terminates, otherwise a new layer is constructed.

In our sample network, we assume that the nodes are analyzed in numeric sequence. The first constructed layer is safe for nodes 1, 2, 3 and 5. Note that node 4 is not a part of the first layer (Fig. 1b), because, when it was analyzed, links 1-2 and 2-3 were already removed, and node 4 could not be left with a single link without disconnecting the layer. Construction of the second layer starts with node 4. Nodes 4, 6, 7 and 8 are added to the second layer (Fig. 1c), and the algorithm terminates.

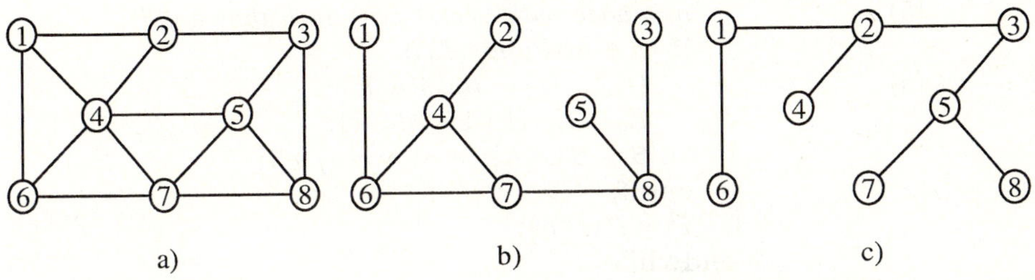

a)   b)   c)

Fig. 1. Sample network topology (a), and two layers that provide single-fault tolerance, (b) and (c)

Only two layers were needed to cover our sample network. We have previously tested the described simple algorithm on families of random and several real network topologies, and shown that very few layers suffice to cover even large networks [6]. In our tests, we never encountered a network that needed more than 6 layers, while 3-4 layers were most usual.

3.2 k-Failure Algorithm

For RRL to protect the network in a multi-failure situation, the failed components, i.e., nodes and links, must be safe within a common layer. The simple algorithm described above needs relatively few layers to protect any network topology. However, each node is safe in one layer only, which limits its suitability for handling multiple failures. To increase the probability of protection against multiple failures we instead propose a probabilistic algorithm with the following properties:

1. the minimum number of layers to generate is provided as a parameter, and can be used to increase the k-fault tolerance
2. the number of nodes that are safe within a layer is made as large as possible, constrained by the random selections made, to increase the likelihood of multiple nodes being contained within the same layer

3. nodes are assigned to the layers in a fair manner—nodes that are safe in fewest layers are prioritized for addition to the currently constructed layer.

We specify the k-failure algorithm using pseudo-code. The minimum number of layers l and the topology $G = (V, E)$, where V is the set of nodes and E is the set of links, are provided as the input:

```
(1)          S = artPoints(G);
             foreach n ∈ V
(2)             c(n) = 0;
             endfor
             i = 0;
(3)          while (i < l) or (|S| < |V|)
(4)             Li(Vi, Ei) = G;  P = {};
                while |P| < |V|
(5)                n = node with lowest c(n) such that n ∉ P;
                   if (n ∉ artPoints(Li))
(6)                   {l1, ..., lk} = links(n, Ei);
                      Ei = Ei\{lj| 1 ≤ j < k};
                      S = S ∪ {n};  c(n) = c(n) + 1;
                   endif
                   P = P ∪ {n};
                endwhile
                store layer Li;  i = i + 1;
             endwhile
```

Steps (1)-(6) deserve some comments. (1) Set S keeps track of the processed nodes, i.e., nodes that are either articulation points or safe nodes in an already computed layer. $artPoints(G)$ finds all articulation points in G, which are initial members of S. (2) $c(n)$ counts how many times node n has appeared in a safe layer. (3) We construct l layers or, for a small l, continue until all nodes are safe in at least one layer. (4) A new layer is generated as a clone of the full topology G. P keeps track of nodes already tested for this layer. (5) If more than a single node has the lowest safe node count, a random node is returned. (6) Function $links(n, L)$ returns all links adjacent to n in layer L in random order. In the next step, all but one are removed from the current layer.

4   Evaluation Method

Multiple network failures can be classified over the following parameter space:

- the number of simultaneous failures, $k = 2, 3, \ldots$
- link or node failures
- failure locality, i.e, are the failures occurring on independent locations in the network or on neighboring nodes.

To restrict the evaluation space while maintaining the practical relevance of this study we pay particular attention to the following failure scenarios:

- 2-5 independent node failures, covering cases such as network-wide router software update
- 2-5 localized node failures, mimicking power outages
- 2 independent link failures, such as physical link failure or multiple link card failures.

We study multiple node and link failures in families of random networks generated using BRITE network generation tool [21] and Waxman model [22]. There are at least 100 networks in each family, with n=32, 64, 128, 256, 512 or 1024 nodes and D=2 or 3 times as many links. We believe that this range provides a reasonable choice of practically relevant network parameters. All studied networks are biconnected. We implement the RRL algorithm described in Sec. 3.2 and the RT algorithm described in [3] as a routing-level simulation package in the Java programming language[1], and analyze the effect of failures in these schemes.

The software simulates the failure scenarios described above, registering the percentage of successful protection cases. We say that a protection case is successful if, after the multiple node or link failures, all remaining nodes retain connectivity using the applied protection scheme. For RRL that means that the failed components share a common safe layer. For RT it means that all nodes are still connected to the root by one of the trees.

The k-failure algorithm is set up to generate a minimal number of layers. In all cases, 3-5 layers were generated. Our RT implementation uses heuristics to generate trees with many leaf nodes for increased redundancy and thus objective evaluation.

We present the fault tolerance by the mean value among the 100 networks from each family, for all three scenarios. To improve the evaluation efficiency, we perform a number of tests with different combinations of the failed components for each network, rather than all combinations. The number of the tested failure combinations is chosen high enough to assure that the 99% confidence interval of the mean is within 1% of the values showed in Sec. 5. This was facilitated by the relatively low ($<$5%) standard deviation of tolerance within different network families, and the observation that the measured values seem to be normally distributed.

5 Evaluation Results

5.1 Node Failures

We test RRL and RT for 2-5 node failures, for independent and localized case.

For RRL, nodes are included into layers randomly and network-wide, while RT trees are built of neighbor nodes. Thus, it can be expected that RT is more tolerant for localized failures, while RRL is more tolerant for the independent failures.

---

[1] The software we implemented can be downloaded from *www.ifi.uio.no/~tarikc/software/RRL/protection.tar.gz*

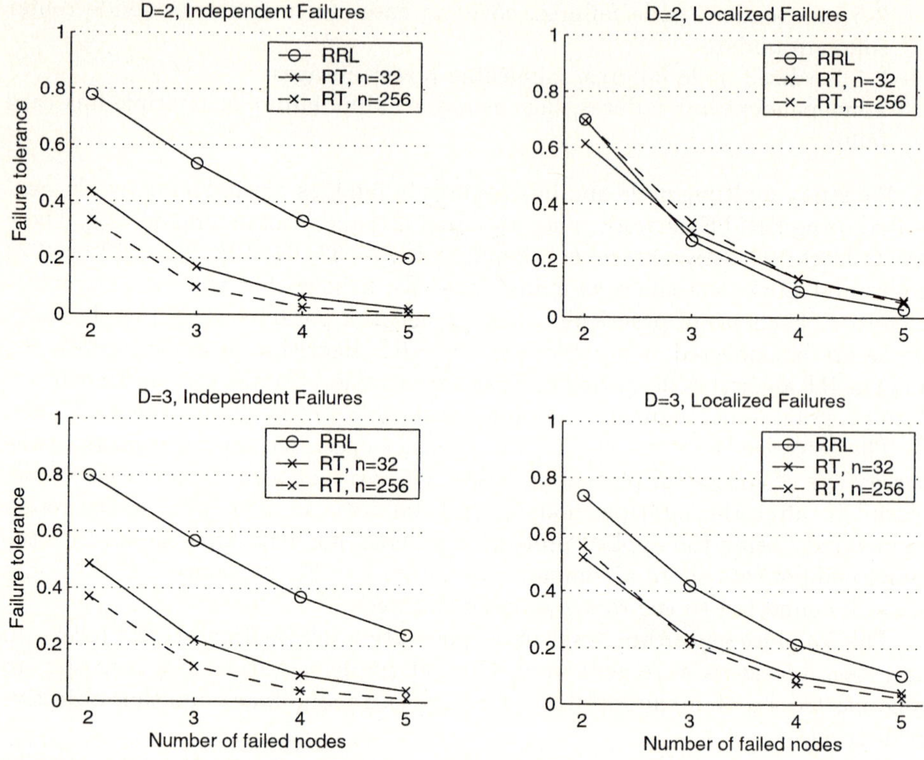

Fig. 2. Multiple node fault tolerance

Indeed, results shown in Fig. 2 confirm this hypothesis. RT shows better results for localized failures than for independent ones, while RRL performs better with independent failures than with the localized.

We observe that RRL k-fault tolerance is independent of the number of network nodes. For network sizes 32-1024 nodes, RRL tolerance ratios vary within 1%, possibly due to the method randomness. For RT, we show the difference for n=32 and 256.

Normally, adding more links to a network increases the fault tolerance. However, for RT neighbor failures this is not the case. Instead, fault tolerance is slightly lower for D=3 networks than for D=2. This is because the RT trees are shorter when there are more links to choose from during their construction (D=3). All non-leaf nodes then have more children. When several neighbor nodes fail, probability increases that a non-leaf node will fail as well, disconnecting its children, and yielding lower fault tolerance.

5.2   Link Failures

We test RRL and RT for two independent link failures. Denser networks show higher fault tolerance than sparser ones for both schemes (Fig. 3). RRL has

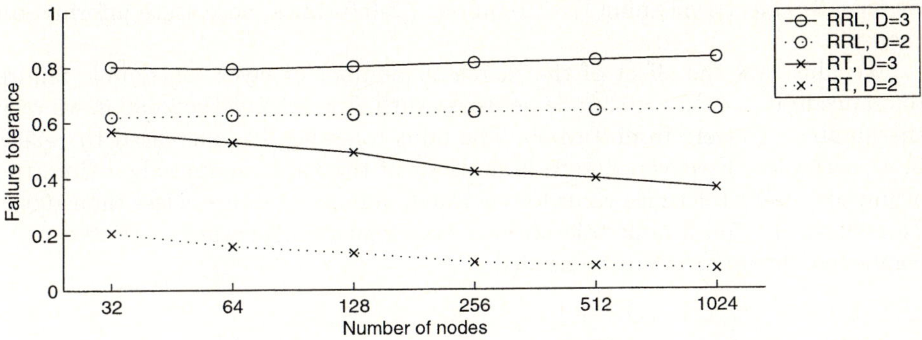

Fig. 3. Double link fault tolerance

higher double link fault tolerance than RT, and is practically independent of the network size.

5.3   Discussion

Our analysis so far clearly shows performance advantages of RRL in multi-failure context. In addition, flexibility is a qualitative advantage of RRL that RT cannot claim. Firstly, RRL can use an arbitrary number of layers and thus increase its protection ratio on the cost of increased state required by the routing mechanism. This is a useful trade-off for network designers. Secondly, the layer construction algorithm of RRL can be implemented to optimize any recovery performance metric, including the k-fault tolerance.

The RRL state increase is proportional to the number of layers used—l layers means up to l new entries for each original entry, $l + 1$ in total. The RT state increase is constant; to accomodate the blue and red trees we need two additional entries for each original entry, 3 in total. In our experiments earlier in this section

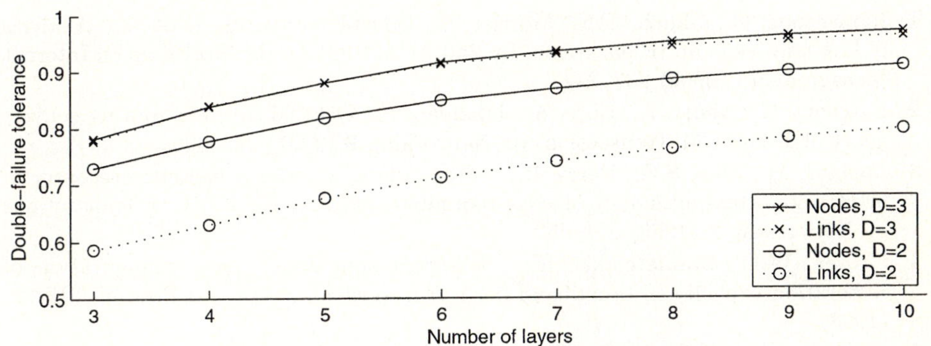

Fig. 4. RRL fault tolerance as a function of the number of layers

we used 3-5 layers, meaning RRL required 1.3-2.0 times more state information than RT.

We illustrate the effect of the increased number of layers on double failure tolerance in Fig. 4. In n=64 node networks with link density D=2 and 3, we vary the number of layers from 3 to 10. The fault tolerance has increased by $\sim$20% in all scenarios. However, due to limitations of the used random algorithm, the asymptotic fault tolerance value for very high number of layers is less than 100%. To achieve the full 2-fault tolerance, a deterministic algorithm restricted to 3-connected networks would be needed.

6   Conclusion and Future Work

In this paper we have evaluated the suitability of Resilient Routing Layers for multi-fault recovery in communication networks. We described the method, and illustrated its performance through a comparison with another popular protection method, Redundant Trees.

We provided a simple, probabilistic layer construction algorithm, which features the possibility to trade the number of layers and thus the network state amount for increased fault tolerance. We demonstrated that the fault tolerance is often more than doubled in RRL compared to RT. More importantly, we demonstrated RRL's flexibility with regard to the construction algorithm, which is a powerful network engineering tool.

This paper opens several directions for future work. We would like to understand the network state requirements for different protection schemes in more detail. Implementation strategies for RRL in multi-failure scenarios are not obvious. Also, an in-deep comparative study of k-failure recovery in the major protection schemes RT, p-cycles and RRL would provide a significant reference for network scientists and engineers.

References

1. Iannaccone, G., Chuah, C.N., Mortier, R., Bhattacharyya, S., Diot, C.: Analysis of link failures in an IP backbone. In: 2nd ACM SIGCOMM Workshop on Internet Measurement. (2002) 237–242
2. Labovitz, C., Ahuja, A., Bose, A., Jahanian, F.: Delayed Internet routing convergence. IEEE/ACM Transactions on Networking **9** (2001) 293–306
3. Medard, M., Finn, S.G., Barry, R.A.: Redundant trees for preplanned recovery in arbitrary vertex-redundant or edge-redundant graphs. IEEE/ACM Transactions on Networking **7** (1999) 641–652
4. Grover, W.D., Stamatelakis, D.: Self-organizing closed path configuration of restoration capacity in broadband mesh transport networks. In: Proc. CCBR'98. (1998)
5. Armitage, G.L.: Revisiting IP QoS: Why do we care, what have we learned?, ACM SIGCOMM 2003 RIPQOS Workshop Report. ACM/SIGCOMM Computer Communications Review **33** (2003) 81–88

6. Hansen, A.F., Cicic, T., Gjessing, S., Lysne, O.: Resilient routing layers: A simple and flexible approach for resilience in packet networks. Technical Report 13, Simula Research Laboratory (2004)

7. Menger, K.: Zur allgemeinen kurventheorie. Fund. Math. **10** (1927) 95–115

8. Bremler-Barr, A., Afek, Y., Kaplan, H., Cohen, E., Merritt, M.: Restoration by path concatenation: Fast recovery of MPLS paths. In: Proc. ACM Symposium on Principles of Distributed Computing. (2001) 43–52

9. Lau, W., Jha, S.: Failure-oriented path restoration algorithm for survivable networks. eTransactions on Network and Systems Management **1** (2004)

10. Suurballe, J.W., Tarjan, R.E.: A quick method for finding shortest pairs of disjoint paths. Networks **14** (1984) 325–336

11. Torrieri, D.: Algorithms for finding an optimal set of short disjoint paths in a communication network. IEEE Transactions on Communications **40** (1992) 1698–1702

12. Grover, W., Doucette, J., Clouqueur, M., Leung, D., Stamatelakis, D.: New options and insights for survivable transport networks. IEEE Communications Magazine **40** (2002) 34–41

13. Sack, A., Grover, W.D.: Hamiltonian p-cycles for fiber-level protection in homogeneous and semi-homogeneous optical networks. IEEE Networks **18** (2004) 49–56

14. Schupke, D.A., Grover, W., Clouqueur, M.: Strategies for enhanced dual failure restorability with static or reconfigurable p-cycle networks. In: Proc. ICC. Volume 3. (2004) 1628–1633

15. Schupke, D.A.: Multiple failure survivability in WDM networks with p-cycles. In: IEEE ISCAS. Volume 3. (2003) 866–869

16. Stamatelakis, D., Grover, W.D.: IP layer restoration and network planning based on virtual protection cycles. IEEE Journal on selected areas in communications **18** (2000)

17. Shah-Heydari, S., Yang, O.: Hierarchical protection tree scheme for failure recovery in mesh networks. Photonic Network Communications (Kluwer) **7** (2004) 145–159

18. Finn, S.G., Medard, M., Barry, R.A.: A novel approach to automatic protection switching using trees. In: Proc. ICC. Volume 1. (1997) 272–276

19. Bartos, R., Raman, M.: A heuristic approach to service restoration in MPLS networks. In: Proc. ICC. (2001) 117–121

20. Xue, G., Chen, L., Thulasiraman, K.: QoS issues in redundant trees for protection in vertex-redundant or edge-redundant graphs. In: Proc. ICC. Volume 5. (2002) 2766–2770

21. Medina, A., Lakhina, A., Matta, I., Byers, J.: BRITE: An approach to universal topology generation. In: IEEE MASCOTS. (2001) 346–353

22. Waxman, B.M.: Routing of multipoint connections. IEEE Journal on Selected Areas in Communications **6** (1988) 1617–1622

Network-Tree Routing Model for Large Scale Networks: Theories and Algorithms[*]

Guozhen Tan, Dong Li, Xiaohui Ping, Ningning Han, and Yi Liu

Department of Computer Science,
Dalian University of Technology,
Dalian, 116023, P.R.China
gztan@dlut.edu.cn

Abstract. For the first time, we propose Network-tree Model and its theorem of routing optimization, which greatly narrows the searching space of the routing procedure within much smaller sub-networks. We then show the routing scheme for Network-tree Model. Based on the communication capability of links, we design a network-tree clustering algorithm by employing the idea of multi-hierarchy partition and anomalistic regional partition and the network-tree routing algorithm (NTRA) which includes an aggregation scheme for network-tree (NTAS) that follows the network-tree Theorem of Routing Optimization. The work achieves a logarithmical reduction in communication complexity. Meanwhile, routing in network-tree reaches a high accuracy. While satisfying the two conditions which we finally addressed, NTRA can get the optimal routes. By simulations, we find that NTRA obtains high performance in convergence, routing accuracy and average throughput, as expected.

1   Introduction

Hierachical routing is introduced to solve the scalability for large scale networks[1], which brings the inaccuracy in routing information and degrade the performance of routing. To our knowledge, McQuillian[2] first proposed such routing scheme in 1974. Most proposals published for Hierachical routing were mainly based on distance vetor routing and link state routing(eg,[2, 3, 4, 5, 6]). For the investigation in the inaccuracy, Guerin and Oder[7] investigated the impact of inaccuracies in the available network state and metrics information on the path selection process. Recently, researchers turn their attention to MANTS [8].

Most of the hierarchical routing proposals published are based on the notion of regions which are made up of nodes according to their characters.In this paper, we first propose Network-tree Model which is constructed according to the communication capability of links of topology and the theorem of routing optimization based on the model. The main motivation for this new scheme is to provide routing algorithms with high routing accuracy using properties of network-tree model which benefit routing accuracy.

---

[*] This work is supported by the National Science Foundation of China(60373094).

P. Lorenz and P. Dini (Eds.): ICN 2005, LNCS 3421, pp. 184–191, 2005.

2   Network-Tree Model

Network-Tree Model is the combination of network and tree. Flat topology can be divided into sub-networks, which we call Domains. Domain will be regarded as the node of the network-tree.

2.1   Definition of Network-Tree Model

Let the network-tree be $N\_Tree = (T, H)$, where T is the domain set. If only one domain exists in T, $H = \phi$, otherwise,H is a binary relation defined on T. There is only one domain d_r called the root of T. If $T - d_r \neq \phi$, there exists a division $T_1, T_2, \cdots, T_n (n > 0)$ of $T - \{d_r\}$, with $\forall j \neq k, 1 \leq j, k \leq n, T_j \cap T_k = \phi$. For $\forall i, 1 \leq i \leq n$, the only domain $d_i \in T_i$ satisfies $< d_r, d_i > \in H$. As the counterpart of the $T - \{d_r\}$ division, there exists the only division of $H_1, H_2, \cdots H_n$ of $H - \{< d_r, d_1 >, \cdots, < d_r, d_n >\}$, with $\forall j \neq k, 1 \leq j, k \leq n, H_j \cap H_k = \phi$. For $\forall i, 1 \leq i \leq n, H_i$ is a binary relation defined on T_i, and (T_i, H_i) is also a network-tree defined above, and we call it the sub-domain of the root domain d_r.

2.2   Construction of Network-Tree Model

Let the original topology be $N = (V, A, W)$, where V is a finite set of nodes,$A \subseteq N \times N$ is a finite set of links,$W = w(a) \mid a \in A$,w is the mapping function of $A \to R^+$. First, we assign levels to all links according to characters in the original topology, that is, we use an integer h to denote the level of a link,$h \in I, I = 1, 2, \cdots, k$.Let the link set of level h be A_h.

Based on the link level, we create the network-tree $N\_Tree = (T, H)$ from $N = (V, A, W)$. Let the domain which is composed of the links of A_1 to be the root domain d_1 of N_Tree, $d_1 \in T$. If $N - \{d_1\} \neq \phi$, the other part left $N - \{d_1\}$ can be naturely divided into n_1 non-null domains $T_{11}, T_{12}, \cdots, T_{1n_1}$ when we extract all the links of A_1 from $N = (V, A, W)$, and for $\forall p \neq q$, $1 \leq p, q \leq n_1, T_{1p} \cap T_{1q} = \phi$. Then $\forall i, 1 \leq i \leq n_1$, suppose A^{1i} is made up of the highest-level links in T_{1i}, We regard the domain composed of the links of A^{1i} and the node set $T_{1i} \cap d_1$ as the ith sub-domain of d_1, called d_{1i}, that is $d_{1i} \in T, < d_1, d_{1i} > \in H$. If $T_{1i} - \{d_{1i}\} \neq \phi$, we continue to divide $T_{1i} - \{d_{1i}\}$ into n_{1i} non-null domains $T_{li1}, T_{1i2}, \cdots, T_{1in}$ by extracting A^{1i} from T_{1i}. Follow the step above recursively until we get the network-tree $N\_Tree = (T, H)$.

Definition 1(Connecting Node): In $N\_Tree = (T, H)$, arbitrary domain $d_c \in T$, if there exists the super-domain $d_p \in T, < d_p, d_c > \in H$. The node in $d_c \cap d_p$ is the connecting node. Connecting node set is denoted by R_{d_c}.

2.3   Theorem of Route Optimization

Let $sp(u, v)$ represent the shortest path from u to v in the network.

Theorem 1 (Nearest Optimization): In $N\_Tree = (T, H)$, if the path:
$$\pi = sp_{m_1}(s = v_0, v_1) + sp_{m_2(v_1, v_2)} + \cdots + sp_{m_h}(v_{h-1}, v_h) + sp_{m_{h+1}}(v_h, v_{h+1}) +$$
$$sp_{m_{h+2}}(v_{h+1}, v_{h+2}) + \cdots + sp_{m_{n-1}}(v_{n-2}, v_{n-1}) + sp_{m_n}(v_{n-1}, v_n = t), \text{ where}$$
$$n \geq 1, 1 \leq h \leq n, < m_h, m_{h-1} > \in H, < m_{h-1}, m_{h-2} > \in H, \cdots, < m_2, m_1 > \in$$

H, and $< m_{h+1}, m_{h+2} > \in H, \cdots, < m_{n-1}, m_n > \in H$ satisfies $u^*_{i+1} = \arg\min\limits_{u' \in R_{m_{i+1}}} sp(u_i, u'), 0 \le i < h$ and $u^*_i = \arg\min\limits_{u' \in R_{m_{i+1}}} sp(u', u_{i+1}), h < i < n,$ then $d(\pi_N(s,t)) = d(sp(s,t)) + \xi_N$, where $\xi_N \ge 0$.

Theorem 1 make the sub-path in domain with the connecting node being the nearest to starting node, and such optimal method is called nearest optimization.

3   Routing in Network-Tree

3.1   Definitions and Data Structures

Definition 2(definition of domain-class address): for a n-level network-tree, we define the domain-class address to be $a_1.a_2 \cdots a_{n-1}.0$. The address of each domain is recursively defined as: The address of root domain is $0.0 \cdots 0$. Each sub-domain inherits the address prefixion of its super-domain. Supposing that a domain address is $a_1.a_2 \cdots a_i \cdots a_{n-1}.0$ ($0 \le i < n$, and for $i < k < n$, $a_k = 0$),then the m-th sub-domain address of this domain is $b_1.b_2 \cdots b_i \cdots b_{n-1}.0$ ($0 \le i < n$,and for $k \le i$, $b_k = a_k$; for $k > i+1, b_k = 0$; and $b_{i+1} = m$). In this paper, uppercase represents domain-class address.

Definition 3((definition of node-class address): The node-class address is generated from the address of the domain which the node belongs to. If the domain-class address is $a_1.a_2 \cdots a_i \cdots a_{n-1}.0$, the k-th node of this domain is $a_1.a_2 \cdots a_i \cdots a_{n-1}.k$. In this paper, lowercase represents node-class address.

Definition 4(definition of destination address): Destination address referred in this paper is composed of domain-class address and node-class address.

Data structures maintained by a node: Array $\mathcal{D}$, each element $\mathcal{D}_j$ is the cost of the route from the local node to the destination address $|$. Array $\mathcal{P}$, each element $\mathcal{P}_j$ is the next hop of the route from the local node to the destination address $|$. Array $\mathcal{D}$ and $\mathcal{P}$ constitute the routing table. Table $\mathcal{T}$, two-dimensional cache table, each element $\mathcal{T}_{kj}$ is the cost of the route from the local node to the destination address $|$ via neighbor $\|$. Set $\mathcal{V}$, the routing information, and each element $\mathcal{V} < j, \mathcal{D}_j >$ is the cost of the route from the local node to destination address $|$ in the last computation.

3.2   Network-Tree Clustering Algorithm (NTCA)

Each node knows its level value by the highest level value of the links connected to it. Suppose local node is n, its level value is $\mathcal{L}_n$. The higher the level, the smaller the level value. i is one of the neighbors of node n. Set $\mathcal{R}$, every element $\mathcal{R} < e, \mathcal{L}_e >$, e is the node which has been sent by n. Set $\mathcal{S}$, every element $\mathcal{S} < e, \mathcal{L}_e >$, e is the node which has not been sent by n. Set $\mathcal{C}_i$ $(0 \le i \le m$) is the partition set of local node n; m is the number of the neighbor nodes of node n, whose level value is lower than $\mathcal{L}_n$.

[1] Initialization: $\mathcal{R} \leftarrow \emptyset$, $\mathcal{S} \leftarrow < n, \mathcal{L}_n >$, $\mathcal{C}_i \leftarrow \emptyset$
[2] Send$< e, \mathcal{L}_e >$of $\mathcal{S}$ to every neighbor node i, and add $< e, \mathcal{L}_e >$ to $\mathcal{R}$

[3] Receiving:When receiving $< e, \mathcal{L}_e >$ from i, do
[4] If $\mathcal{L}_e > \mathcal{L}_n$ Then $\mathcal{C}_e \leftarrow e$
[5] Else If not $< e, \mathcal{L}_e >\in R$ Then $\mathcal{S} \leftarrow < e, \mathcal{L}_e >$
[6] If $\mathcal{C}_j \bigcap \mathcal{C}_k \neq \emptyset$ Then $\mathcal{C}_j = \mathcal{C}_j \bigcap \mathcal{C}_k$

If $\mathcal{L}_k$ in $\mathcal{C}_j$ differs from each other, then $\mathcal{C}_j$ only contains the nodes of the highest level. Running this algorithm, each node knows the node information of its sub-domains. The addresses of nodes are recursively assigned by their connecting nodes beginning from the root domain.

3.3   Network-Tree Routing Algorithm(NTRA)

Initialization. Table $\mathcal{T}$ is composed of the neighbor information, array $\mathcal{D}$ and $\mathcal{P}$ are computed from table $\mathcal{T}$, and then routing information $\mathcal{V}$ is generated. $\mathcal{A}$ is the set of neighbor nodes; a_{ni} is the cost of link $< n, i >$. $\mathcal{B}$ is the set of the local node's sub-domains.

[1] Foreach $i \in \mathcal{A}$
[2] Add an element $\mathcal{D}_i$ to $\mathcal{D}$, $\mathcal{D}_i \leftarrow a_{ni}$;
[3] Add an element $\mathcal{P}_i$ to $\mathcal{P}$, $\mathcal{P}_i \leftarrow i$
[4] Begin
[5] If $i \in \mathcal{B}$ then
[6]   Begin
[7]   Add an element $\mathcal{D}_I$ to $\mathcal{D}$, $\mathcal{D}_I$;
[8]   Add an element $\mathcal{P}_I$ to $\mathcal{P}$, $\mathcal{P}_I$;
[9]   Add an element $\mathcal{V}_{Ij}$ to $\mathcal{V}$, $\mathcal{V}_{Ij} \leftarrow < I, 0 >$;
[10]   If $\mathcal{D}_N$ not exit
[11]   Then add an element $\mathcal{D}_N$ to $\mathcal{D}$, $\mathcal{D}_N \leftarrow 0$;
[12]   End.

Routing Information Aggregation Scheme(RIAS). The routing information set $\mathcal{V}$ can be divided into two parts: (1) for the domain-class address: If the neighbor node is in local node's sub-domain, send all the information except the information of the domain where this neighbor node belongs to. Otherwise, send all the information. (2) For the node-class address: Send all the information to neighbors that belongs to the local-domain and its super-domains. If the neighbor node is in local node's sub-domain, send all the information except the information of the domains who are the local node's sub-domains. F represents the set of domain-class address; G represents the set of node-class address.

[1] For every element $< j, \mathcal{D}_j >$ of $\mathcal{V}$ do
[2] If $j \in \mathcal{F}$ Then
[3]   Begin
[4]   For each neighbor node i
[5]   If $i \in B$
[6]   Then do nothing
[7]   Else send $< j, \mathcal{D}_j >$ to i
[8]   End

[9] If $j \in \mathcal{G}$ Then
[10]   Begin
[11]   For each neighbor node i
[12]   If $i \in \mathcal{B}$ Then
[13]   If $j \in \mathcal{B}$ Then send $< j, \mathcal{D}_j >$ to i
[14]   Else send $< j, \mathcal{D}_j >$ to i
[15]   End

Finally, Once the route or its cost in the cache table changes, nodes update routing table information according to the Bellman-Ford algorithm.

3.4   The Analysis of the Network-Tree Routing Algorithm (NTRA)

Property 1: The NTRA converges in finite time.

Theorem 2: The routes computed by NTRA are approximate shortest paths.

Proof: During the convergence of NTRA, each node receives the routing information, and assigns $\mathcal{D}_j$ a value according to $\mathcal{D}_j \longleftarrow min_i T_{ij}$, in which i and j are both arbitrary. It means that each node enters the super-domain via the nearest connecting node to itself. This meets the approximate optimal conditions given by Theorem 1, which is approximate, so the path computed by NTRA is approximate.

Let the number of nodes in a flat topology be n. Let the number of nodes in the root domain composed of the links of A_1(defined in Section 2) be $O(n^{d_1}), 0 < d_1 < 1$. Then the remainder of the network is divided by A_1 into $O(n^{m_1})$ non-null domains, $0 < m_1 < 1$, so that each domain T_{1i}has an average number of nodes of $O(n^{1-m_1})$. Suppose T_{1i} has an average number of nodes of $O(n^{(1-m_1)d_2}), 0 < d_2 < 1$ .T_{1i} is Recursively divided into $O(n^{m_2})$ non-null domains,$0 < m_2 < 1 - m_1$, so that the average number of nodes in each new sub-doman is $O(n^{1-m_1-m_2})$. Finally we get:

Theorem 3: The communication complexity of NTRA is $O(n^{max(\xi,\eta)})$, where

$$\xi = 1 - \sum_{j=1}^{i-1} m_j, \eta = \sum_{j=1}^{k-1} m_j.$$

Proof: In a k-level network-tree($k > 2$), the average number of nodes in each domain and its sub-domains in the i-th level is $O(n^{1-\sum_{j=1}^{i-1} m_j})$,and the total number of domains is $O(n^{m_1}) + O(n^{m_1+m_2}) + \cdots + O(n^{m_1+m_2+\cdots+m_{k-1}}) = O(n^{\sum_{j=1}^{k-1} m_j})$.According to our algorithm, during the convergence of NTRA, the total sum of information that has to be communicated is $O(n^{1-\sum_{j=1}^{i-1} m_j}) + O(n^{\sum_{j=1}^{k-1} m_j})$.

Let $\xi = 1 - \sum_{j=1}^{i-1} m_j, \eta = \sum_{j=1}^{k-1} m_j$,then the sum becomes $O(n^\xi) + O(n^\eta) = O(n^{max(\xi,\eta)})$, that is the communication complexity of NTRA.

Finally, according to algorithm NTRA and the properties of tree, we can easily address two special optimal conditions:(1)If $v_i (0 < i < n)$ of $SP(s, v_1, v_2, \cdots, v_{n-1}, d)$ is the node of the local-domain or super-domains of node d, $SP(v_i, \cdots, v_{n-1}, d)$ is optimal.(2)If each domain only has (corresponds to)one connecting node, the routes computed by NTRA are optimal.

4   The Simulations

We use Network Simulator 2 (NS2) as the simulation platform and Georgia Tech Internetwork Topology Models (GT-ITM) to generate the topologies for simulation scenarios. We mainly focus on the performance of convergence, the shortest path accuracy and the average throughput of the networks. Our experiments based on a Pentium IV at 2.0GHz and 256M RAM.

4.1   Performance of Convergence

In this section,we generate 5 topologies in which number of nodes are 70, 150, 300, 600 and 1200. Note that the nodes include only internal nodes (i.e., routers), with end hosts excluded. We compare the performance of NTAR, Vector-Distance and link state routing algorithm. In convergence, we focus on the whole simulation running time on the test-bed and the traffic traced. Fig.1 and fig.2 show the results. Note that the running time is not the real convergence time. we use it to show the different performance of the routing algorithms. We regard the time beyond 12 hours as Infinity.

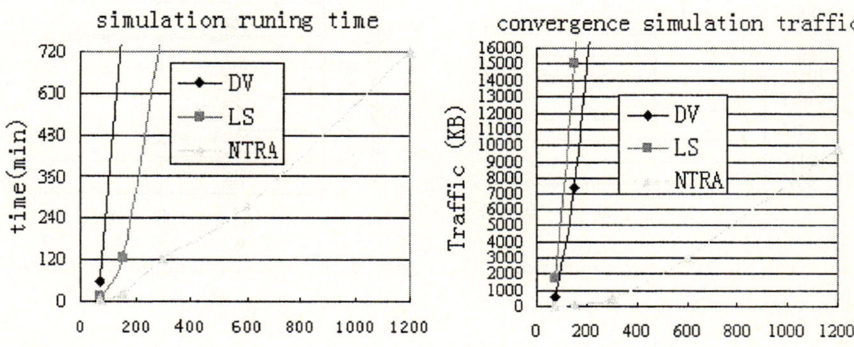

Fig. 1. Convergence running time   **Fig. 2.** Convergence simulation traffic

In fig.1, NTRA gets shortest period of convergence among the three algorithms. NTRA can keep the time grow linearly with the number of nodes, while the others can not even converge within available time. The similar behavior is repeated in fig.2

4.2   The Accuracy and Its Main Impact Factor

Definition 5 (definition of accuracy): Randomly select source-destination node pairs, trace their routing table, and then compute the accuracy according to the equation:$\overline{A} = 1 - \dfrac{\sum\limits_{i=1}^{n}((w_i'-w_i)/w_i)*h_i}{\sum\limits_{i=1}^{n}h_i}$. n denotes number of the node pairs randomly selected, w_i denotes the weight of the optimal path between one node pair and w_i' is the weight of the path got by NTRA. h_i is the hop counts of the optimal path. Table 1 shows the results of the 5 topologies generated before.

Table 1. The accuracy table for five topologies

node number	70	150	300	600	1200
Accuracy	99.37%	98.03%	96.54%	94.40%	93.01%

In Table 1, the accuracy does not decrease obviously. We investigate this problem, and find that only the total number of connect nodes is the main impact factor of accuracy, so we design two more detailed simulations to tradeoff the accuracy and the total number of the connecting nodes of all domains. Fig.3 shows the results.

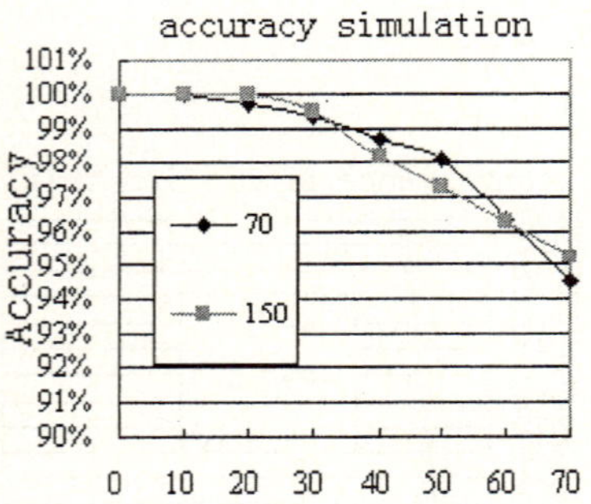

Fig. 3. Accuracy simulation

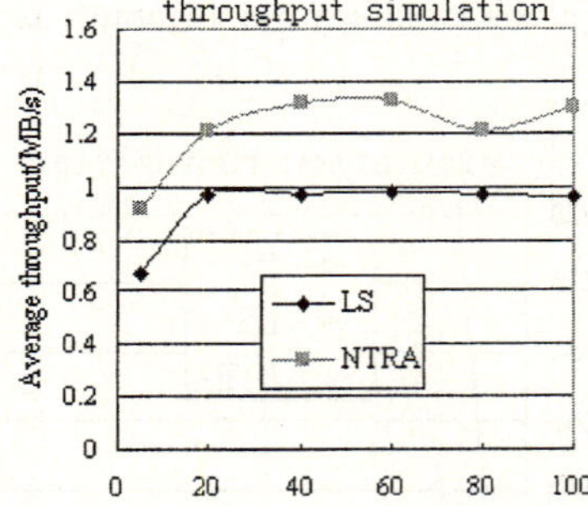

Fig. 4. Average throughput simulation

4.3   Average Throughput

In this subsection, we use the real router-level topology of China Education and Research Networks(CERNET) to compare the average throughput of Link State Algorithm and NTRA. Average throughtput is measured as the the average number of bytes that cross the network during a specific period of time. The advertising interval of NTRA is set to 10 seconds. We attach ten CBR traffic agents to the topology and configure the packet size at 800 bytes and a packet sending rate at 200 per seconds. This rate reaches the bandwidth of bottleneck links of the topology.

The results are shown in Fig.4. We obtain higher average throughput than link state algorithm as expected. It's mainly because network-tree model is constructed on the basis of the communication capability of links, and NTRA are designed to distribute more traffic first to the links which can burden more.

5   Conclusion

In this paper, we present the network-tree routing model for large-scale networks, which is built according to the links of topology. We design NTCA algorithm to implement the construction of network-tree. After we show the routing algorithm NTRA, we address two basic properties about the convergence and accuracy, then we analyze the communication complexity which indicates high performance of convergence and average throughput of NTRA. We also address two optimal conditions. Finally, we turn our attention to the simulations that focus on the performance of convergence, accuracy and average throughput, and the results show that NTRA can fast convergence, obtain high accuracy and average throughput.

References

1. Chen, S., Nahrstedt, K.: An Overview Of Quality Of Service Routing For Next-Generation High-Speed Networks: Problems And Solutions. Network. IEEE , Vol. 12 (1998) 64-79
2. McQuillan, J.: Adaptive routing algorithms for distributed computer networks. Bolt Beranek and Newman Inc., Cambridge MA, May 1974.
3. Tsuchiya P., The landmark hierarchy: A new hierarchy for routing invery large networks. Computer Communications Review, Vol. 18, No. 4 (1988) 43-54.
4. Garcia-Luna-Aceves, J., Murthy, S.: A path finding algorithm for loopfree routing. IEEE/ACM Transactions on Networking, Vol. 5, No. 1 (1997) 148-160.
5. Hao, F., Zegura, E.: On Scalable Qos Routing: Performance Evaluation of Topology Aggregation. INFOCOM 2000. Nineteenth Annual Joint Conference of the IEEE Computer and Communications Societies. Proceedings. IEEE. Vol 1, 26-30, 2000
6. Luo, Y., Bai, Y.: Topology aggregation with multiple QoS parameters for scalable routing problem. Communication Technology Proceedings, ICCT 2003. International Conference on. Vol 1, 458-461, 2003
7. Guerin, R., Orda, A.: Qos-based Routing in Networks with Inaccurate Information:Theory and Algorithms. INFOCOM '97. Sixteenth Annual Joint Conference of the IEEE Computer and Communications Societies. Proceedings IEEE. Vol 1, 75-83
8. Sucec, J., Marsic, I.: Hierarchical routing overhead in mobile ad hoc networks. Mobile Computing, IEEE Transactions on. Vol.3, Issue.1 (2004) 46-56

Failover for Mobile Routers: A Vision of Resilient Ambience

Eranga Perera[1], Aruna Seneviratne[2], Roksana Boreli[2], Michael Eyrich[3], Michael Wolf[4], and Tim Leinmüller[4]

[1] Dept. of Electrical Engineering, University of New South Wales,
Sydney, Australia
eranga@mobqos.ee.unsw.edu.au
[2] National Information and Communication Technologies Australia (NICTA),
Bay 15, Locomotive Workshop, Australian Technology Park, Eveleigh,
NSW 1430, Australia
{aruna, Roksana.Boreli}@nicta.com.au
[3] Technical University of Berlin, TKN, Germany
eyrich@tkn.tu-berlin.de
[4] DaimlerChrysler AG, Researchand Technology / Vehicle IT and services,
P.O. Box 2360, Ulm, Germany
{michael.m.wolf, tim.leinmueller}@daimlerchrysler.com

Abstract. The ambient networking approach includes the flexibility of every end system to be not just a node but also an entire network. The end user entities are, in the majority of use cases, mobile and they operate in a highly dynamic environment. The exponential growth of wireless devices and services contributes to the increasing number of these dynamic mobile networks, whose changeable characteristics in turn contribute to the high probability of failures. To maintain the high level of connectivity required for the overall ambient networks environment, it is imperative to maintain the same level of connectivity in various network parts. In this paper we consider the Mobile Router related failures that could occur in today's mobile networks and describe how the ambient networking environment with its in-built enhanced failover management functionality has a potential to create resilient networks.

1 Introduction

AMBIENT networking is geared towards increasing competition and cooperation in an environment populated by a multitude of user devices, wireless technologies, network operators and business entities. This architecture aims to extend all IP networks with three fundamental requirements of today's networking world. These requirements include dynamic network composition, mobility and heterogeneity. By encompassing these notions the Ambient Networks (AN) project [1] strives to achieve horizontally structured mobile systems that offer common control functions to a wide range of different applications and air interface technologies. In such an environment the user expectation for high availability is inevitable and providing this is vital in or-

P. Lorenz and P. Dini (Eds.): ICN 2005, LNCS 3421, pp. 192–201, 2005.

der to be viable in today's market. We therefore consider failover management to be an essential requirement for resilient ambient networking.

In this paper, we first describe the need for failover management of Mobile Routers. Section 3 will carry an introduction to the AN architecture from the perspective of failover management. Section 4 will investigate different types of mobile networking scenarios in order to present the problem. The requirements elicited from the scenarios will be stated in Section 5. In section 6 we will delve deeper into the Ambient Networks concepts to illustrate how this architecture can fulfill the elicited requirements. This will be followed by related work. Finally we will conclude the paper.

2   The Motivations

2.1   Why Failover Management as Opposed to Simple Fault Tolerance?

It is important to recognize that any system is prone to failures and this is especially applicable to wireless systems since wireless channels have adverse characteristics such as limited bandwidth and high jitter. Another contributing factor for unantici-pated failures in a mobile networking environment is the dynamicity. At any given time a failure could occur due to the sheer number of nodes attempting to connect to the Internet via a Mobile Router. Therefore in order to face such challenges transpar-ent to the users, there needs to be high-quality failure recovery mechanisms.

2.2   Why Investigating Mobile Routers as Opposed to any Other Entity in the AN Architecture?

AN can be defined as an All-IP based *mobile network* that adopts the much needed technological innovations of the beyond 3G networks. This relation of ANs to mo-bile networks creates the need for us to consider the existing IP based mobility archi-tectures. One such architecture is the NEtwork MObility (NEMO) [2] architecture. In the AN framework, Mobile Routers facilitate the mobility of entire networks and pro-vide a way to mount a complete network into the Internet's addressing hierarchy just as a mobile node can be mounted using Mobile IP [3]. Even though Mobile Routers bring about many benefits for the entire mobility environment, they create a single point of failure for an entire network in contrast to the failure of a single node. This motivated us to look into Mobile Routers when considering a resilient AN mobility management architecture.

3   Ambient Networking Approach

One of the main challenges in today's networking world is how to address mobility, heterogeneity and the integration of networks without having to plug and play various services in an **ad-hoc manner**. In order to answer this question the AN architecture introduces a novel concept of an Ambient Control Space (ACS). The Ambient Con-trol Space is a vision of an overlay network that would bring about the control func-tions for heterogeneous networks under one umbrella. By introducing a 'thick' control plane in Ambient Networks as opposed to the 'thin' control plane of the IP networks, the ACS would pave the way for seamless mobility. By making use of this well defined Ambient Control Space (ACS) via a well defined external interface named as the Ambient Network Interface (ANI) users can expose their communication resources to other users. The most primitive building block of an AN is a physical cluster, which can be defined as a group of nodes (or just one node) that are physically close to each other, are likely to stay near each other and are able to communicate. If the nodes of such a physical cluster are aware of each other then it can be considered as a routing group. This "awareness" involves, besides others, functionalities like address and mobility management, AAA service and gateway discovery, and failover functionality. Addressing the failover during the architectural design phase allows incorporating its requirements in the solution design rather than patching the system by adding e.g. hot-standby proxy-units. Failover management will utilize intra- and inter-ACS communication depending on whether the involved entities are part of the same or different AN domain. In the following figure we have depicted some of the core functional areas being researched in the AN project. The 'Failover for Mobile Routers' is a sub function of 'Mobility' functions.

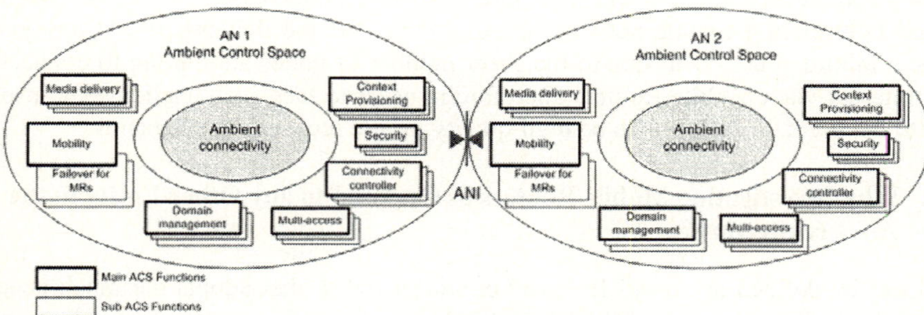

Fig. 1. Ambient Control Space

The functional area "Mobility" itself is organized in a toolbox like manner that allows multiple mechanisms to co-exist. It distinguishes between basic mobility management and advanced mobility management functions. The basic mobility management contains among others trigger processing, location management, session continuity and handover mechanisms. The latter type of mobility management in contrast provides add-on functionalities like multi-homing, context dependant handovers and inter-domain seamless handovers making use of inputs and outputs from other functional areas. Naturally the failover management belongs to the advanced mobility management functions.

In the next section we carry a problem description by making use of three different scenarios. This is followed by an analysis of how these problems can be tackled using the functionalities offered by the ACS.

4   Scenarios and Problem Description

We introduce three types of mobile networks in considering failover management, namely managed routing groups, unmanaged routing groups and hybrid routing groups. The former type of mobile network has a specific node as Mobile Router. The second type of mobile network has no such specific node and any device belonging to the network plays the role of a Mobile Router on the fly and the hybrid type has both these types of nodes which come into play in failover management.

Examples of managed routing grouped networks are public transportation systems with on-board Mobile Routers for Internet connectivity. The routers deployed on such networks are commercially available routers. We identify these routers as Specific Mobile Routers (SMR).The unmanaged routing groups bring about the notion of any capable device taking on the task of providing Internet connectivity for the rest of the nodes in the group. Personal Area Networks (PAN) fall within this category of routing groups. In an unmanaged routing group the devices that play the role of a router dynamically are identified as UnSpecific Mobile Routers (USMR) within the context of this paper. Private cars which have a combination of designated MRs (SMR) and external devices (such as passengers or owner's mobile devices) that could play the role of a MR on the fly (USMR) would constitute a hybrid type of routing group. We use 3 scenarios in order to describe the issues associated with each kind routing group with SMRs and USMRs.

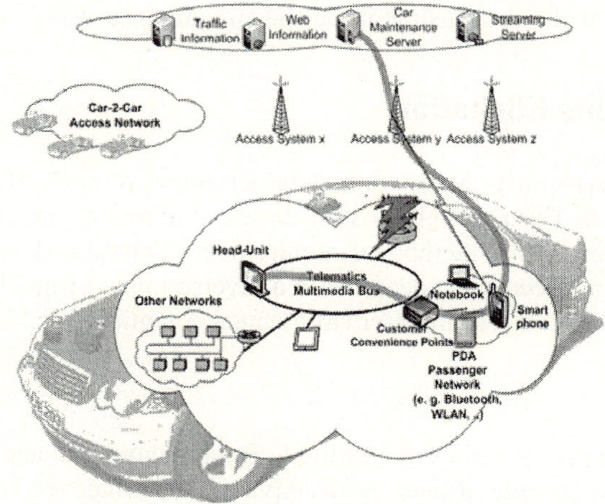

Fig. 2. Hybrid routing group with SMRs and USMRs

4.1   Scenario 1: Highlighting Failures of SMRs

Consider a scenario where a SMR experiences an outage which causes a connection failure with the base station due to the train traveling through a tunnel. In such a case a backup SMR deployed on the train would need to take over mobility of the network. It is important to note that these outages to the base stations could be quite common not only due to physical tunnels but also due to congestion, unavailability of channels etc. Another angle to this scenario is where these routers are connected to different providers for more robustness. Suppose in that case if one of the routers has a potential outage with its provider then the other router with the different provider would need to take over.

4.2 Scenario 2: Highlighting Failures of USMRs

Consider an unmanaged routing group (PAN) with many devices such as a PDA, mobile phone, laptop a digital camera and a MP3 player. Suppose the user is downloading an MP3 file to her MP3 player via the laptop. (In this case the laptop is playing the role of the MR). If the laptop fails for some reason for example by draining its battery then some other device of the PAN needs to take over routing of the mobile network. In this case an USMR would be providing failover for another USMR.

Furthermore consider the case where the home network of the mobile phone belongs to the user's office network. In this case the failover mechanisms would become more complex.

4.3 Scenario3: Failures in a Hybrid Combination of SMRs and USMRs

In this scenario we have depicted the case where there is a single SMR deployed in the car (mainly due cost reasons) and the failure of the SMR requiring another device in the car, an USMR to come into play. Moreover this scenario illustrates the case of the USMR providing failover for only a single connection out of the connections maintained by the SMR (assuming that the SMR had connections to the traffic information, web information, car maintenance and streaming servers).

5 Requirements Elicitation

In this section we identify the requirements pertaining to each of the scenarios described in section 4. Even though most of these requirements are applicable to all of the failover scenarios, this method of requirement elicitation demonstrates that the failover mechanisms need to be adaptive to a diverse range of mobility environments with a variety of devices, resources, environmental conditions etc.

5.1 Scenario 1

Mobility Predictions. A replacement Mobile Router should be activated best shortly before the current Mobile Router is deactivated for whatever reason in order to achieve seamless failovers. Route predictability is an exploitable characteristic of public transportation systems that we can use to avoid disruptions caused by unexpected outages. It is possible to gather information pertaining to link outages caused by interference, time of the day, hot spot areas, weather etc. This information together with location predictions of vehicles can be used to minimize the 'mean time to detect' contributing to seamless failover mechanisms.

Bandwidth Consumption in Failure Detection. Two main scarce resources in mobility management are low data rates and power. It is fair to assume that SMRs would have the necessary power (that is power supplied by the vehicle) in order to perform its routing functionalities. Therefore in the public transportation mobility environments the scarce resource pertaining to SMRs is the consumption of bandwidth.

Cost of Providing Unperturbed Internet Connectivity. In this scenario by deploying Mobile Routers redundantly would increase the probability of providing unperturbed Internet connectivity. But this would not be ideal because it would mean that this cost would contribute to increasing the charging for onboard connectivity from passengers. Therefore it is necessary to find the minimum number of Mobile Routers needed in order to achieve an optimal recovery from a failure. Furthermore consider the case of a 'bus'. Since it is possible to cover an area of a bus with one MR deploying another MR as a backup is not cost effective. In such cases it is necessary to utilize cooperating Mobile Routers in the vicinity for failover mechanisms. These cooperating MRs might be from a different administrative, naming, security, addressing or mobility domain.

5.2 Scenario 2

Power Consumption in Failure Detection. The typical nodes of a moving PAN would be laptops, PDAs, mobile phones etc. All these devices would typically rely on battery power. Therefore power consumption becomes a more imperative issue with USMRs than with SMRs that would be powered by the vehicle. For example if in order to minimize the 'mean time to detect a failure', if the devices continuously send liveliness messages this would not be worthwhile in terms of power consumption.

Discovering Candidate Backup Mobile Routers. In an USMR environment there are no nodes specifically assigned to be MRs, that is potential Mobile Routers are not known by definition. Therefore a discovery mechanism for devices that can take on the mobility management in case the primary device fails should be in situ.

State Synchronization. When the new device takes over it has to be synchronized with the device that was playing the role of the Mobile Router. In the USMRs case if the other potential back up devices are to be synchronized with the MR constantly this would dissipate power unnecessarily. On the other hand if state synchronization is not done in a timely manner this would cause delays and the failover will not be seamless.

5.3 Scenario 3

Connection Handling and Prioritization. The USMR in this case the PDA is not as capable as the SMR deployed in the automobile. Consequently it can support only a few connections out of the connections handled by the SMR. In such a case there needs to be priority based QoS interactions in the failover management procedures. Furthermore the QoS needs to be dynamically adaptive. Suppose at any given time the PDA is able to handle another connection and is prepared to do so, then the QoS mechanisms should adapt to the availability.

6 Failover Management as a Functionality of the Mobility Functional Area within the Ambient Control Space

The following section explains how the failover management becomes an easily achieved functionality within the ACS mobility Functionality Area (FA). The proposed AN architecture does not suggest any mechanisms to replace the networks that are in existence today, but rather a mechanism to coalesce today's networks in a coherent manner. By relating the elicited requirements to a failover functional model we demonstrate how the AN architecture facilitates the fulfillment of a seamless failover for Mobile Routers.

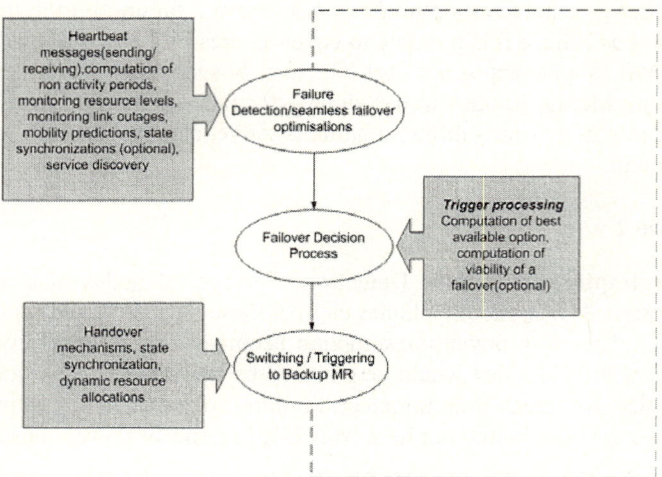

Fig. 3. Failover Functional Model

6.1   Failover Functional Model

The failover model (fig. 3) shows the functions and the functional flow that need to be in place for a complete seamless failover. It relies on the input from other mobility functions and also from the other functions supported by the generic Ambient Control Space functional areas. The aim is to react as early as possible to failover events and to start the appropriate actions. It is evident that the current IP architecture on its own cannot support a seamless recovery system for Mobile Routers. This is because there is no coherent control plane which encompasses today's diversified technologies and heterogeneous networks. The Ambient networking architecture with the concept of every networking entity as an 'Ambient Network' has introduced architectural principles that embrace heterogeneity in today's mobility environments [4].

6.1.1   Information – Oriented Functionalities

A common characteristic of all of the above mentioned functions is the need for a variety of information pertaining to each of the devices and to the networking environment as a whole. The liveliness of the devices, resources and capabilities of the devices and routing states for synchronization purposes are some of the information needed. Within ambient networking this information can be classified as context information i.e. information which represents the collection of any information that can be used to characterize the situation and/or operational state of an entity in different situations. The 'context provisioning' functionality of the ACS provides mechanisms to collect, store and disseminate context information. Since in the AN architecture any networking entity is also an AN, this information is available consistently regardless of the mobility environment. We do recognize that the granularity of information would change between different ANs, depending on the capability of the AN to produce this information (for example a mobile phone running on battery power (USMR) will not be able to contribute to the information base as much as a SMR). However, the important design principle in the AN architecture is that whatever information that is available can be used by other ANs in a similar manner. Therefore in failover management the USMRs, SMRs and any other networking entity in the mobility environment such as base stations, access points would be able to fulfill information oriented functionalities by making use of the "context provisioning" feature of the ACS. Moreover, QoS requirements pertaining to failovers which need to be dynamic and adaptive can be easily achieved by using the context provisioning functionality.

6.1.2 Triggering Functionalities

The 'trigger processing' sub FA of 'mobility' FA acts upon inputs and decides whether the triggering input should be forwarded to other 'mobility' functions or ignored. Link outages, failure of an interface of the MR, draining of the battery etc. are triggering inputs that would need to be processed as part of a failover mechanism. Trigger processing would also have a conflict resolution process which could be very useful in the network mobility environment. For example triggers from two potential devices that could step in as backup routers would create a conflict and this would require mechanisms for conflict resolutions.

It is evident that these triggers (failures) come from different levels such as service levels, context levels and also from different layers. For example the absence of router advertisement messages in the network layer would be a trigger that indicates the failure of the active router. Since the trigger processing functionality would take into account any type of input, this simplifies failover management as there is no additional requirement to provide separate failover management mechanisms.

6.1.3 Handover – Oriented Functionalities

When a MR fails a handover to another device takes place, provided there exists another device capable of taking over the routing functionalities on behalf of the mobile network. Handovers that would occur in failover management are extremely diversified. Inter-interface handovers (between interfaces of a multimodal MR), inter-device handovers, inter-point of attachment handovers (between base stations, access points) are types of handovers that we come across in network mobility. Seamless handovers is an essential requirement of all of the above handovers in the context of failover management of Mobile Routers. In order to provide seamless failovers it is necessary to give due consideration to handovers that are triggered by *"urgency parameters"*. Functionalities for advanced handovers are considered within the 'Mobility' FA. Furthermore other functionalities within the ACS such as context provisioning, cross domain mobility etc. would encompass necessary functions in order to provide a seamless handover between different domains.

6.1.4 Security – Oriented Functionalities

Security related functionalities are closely related to the aspects of trust relationships. Performing state synchronization, handover and other functions requires an existing trust between those entities. Existing trust relations can also be used and allow faster completion of failover functions. The security functional area within the ACS provides the necessary information whether a trust relationship is currently established or not. Accordingly, the failover function needs to establish a trust relation with the proxy/fallback MR before performing any other action. In the case of SMRs from different domains (in Scenario 1) the trust should be established in a proactive way to achieve apparently seamless operation. The security functionalities are also closely linked to the cross-domain functionalities introduced in 6.1.3.

7   Related Work

It is evident that the related work on failovers for Mobile Routers is fragmented and many researchers are working on specific solutions, for example, how to reroute via another interface of a MR if the active interface fails. None of the research work on this area pertained to a much needed all encompassing architectural solution for this problem such as the one we envision with the AN project. Nevertheless we present here some research related to Mobile Routers that could be used in failover mechanisms such as multi-homing solutions.

The IETF NEMO working group [2] has introduced a network mobility architecture where the Mobile Router handles the mobility of the entire mobile network. This is achieved by the NEMO Basic Support protocol [5] which is a logical extension to the MIPv6 protocol [6]. Multi-homing issues pertaining to IPv6 mobile networks are handled in [7], [8]. Paik et al in [9] describes a mechanism whereby a Home Agent is able to select the best MR.

The project Fleetnet [10] developed protocols for car-2-car based ad hoc networks. Accompanying to pure ad hoc communication the cars were able to communicate with the Internet using stationary roadside Internet Gateways. In the car a single router was deployed for handling all the communication with other cars and Internet Gateways. The field of failover management was only investigated with respect to the dynamicity of a managed routing group. The scenario introduced with the hybrid routing group described in section 4.3 was not covered in that project.

The BRAIN [11] and MIND [12] projects which developed protocols for extending edge networks have not considered the existence of several MRs.

8   Conclusions

It is evident that the IP architecture is not proficient enough to provide seamless failovers for Mobile Routers in different network mobility environments. As demonstrated by the variety of mobile network types and corresponding analysis of the failover scenarios, it is necessary to introduce an all encompassing simple networking architecture in order to facilitate mechanisms such as failovers. We have demonstrated that the Ambient Networking architecture which introduces a thick but simple control plane would be able to handle seamless failovers for Mobile Routers.

Acknowledgements

This document is a byproduct of the Ambient Networks Project, partially funded by the European Commission under its Sixth Framework Programme. It is provided "as is" and without any express or implied warranties, including, without limitation, the implied warranties of fitness for a particular purpose. The views and conclusions contained herein are those of the authors and should not be interpreted as necessarily representing the official policies or endorsements, either expressed or implied, of the Ambient Networks Project or the European Commission. The authors would like to thank Eisl Jochen and Vijay Sivaraman for reviewing this paper on numerous occasions.

References

1. http://www.ambient-networks.org/
2. http://www.ietf.org/html.charters/nemo-charter.html
3. Perkins C., "IP Mobility support for IPv4," RFC 3344, IETF, August 2002.
4. N. Niebert, A. Schieder, H. Abramowicz, G. Malmgren, J. Sachs, U. Horn, Ch. Prehofer and H. Karl, „Ambient Networks: An Architecture for Communication Networks beyond 3G", IEEE Wireless Communications, April 2004.
5. Devarapalli V., Wakikawa R., Petrescu A., Thubert P., "NEMO Basic Support Protocol" (draft-ietf-nemo-basic-support-03.txt), Internet Draft, IETF, June 2004, Work in Progress.
6. Johnson D., Perkins C., Arkko J., "Mobility Support in IPv6", RFC 3775, IETF, June 2004.
7. Kuntz R., Paik E., Tsukada M., Ernst T., Mitsuya K., "Evaluating Multiple Mobile Routers and Multiple NEMO- Prefixes in NEMO Basic Support" (draft-kuntz-nemo-multihoming-test-00.txt), Internet Draft, IETF, July 2004, Work in Progress.
8. Ng C., Paik E., Ernst T., "Analysis of Multihoming in Network Mobility Support" (draft-ietf-nemo-multihoming-issues-00.txt), Internet Draft, IETF, July 2004, Work in Progress.
9. Paik E., Cho H., Ernst t., Choi Y., "Load Sharing and Session Preservation with Multiple Mobile Routers for Large Scale Network Mobility," 18th International Conference on Advanced Information Networking and Applications (AINA 2004), IEEE Computer Society Press, Fukuoka, Japan, March 2004, pp. 393-398.
10. http://www.et2.tu-harburg.de/fleetnet/
11. IST 1999-10054 Project BRAIN, Deliverable D2.2, March 2001
12. IST 2000-28584 Project MIND, Deliverable D2.2, November 2002

Quality of Service Routing Network and Performance Evaluation*

Shen Lin, Cui Yong, Xu Ming-wei, and Xu Ke

Department of Computer Science, Tsinghua University, Beijing, P.R.China, 100084
{shenlin, cy, xmw, xuke}@csnet1.cs.tsinghua.edu.cn

Abstract. In order to provide QoS guarantees for data transmission, we developed the QoS router, which can seek feasible paths subject to QoS requirements for the IP packets. However, the QoS router cannot be deployed extensively in the Internet because the current routers need to be modified. Based on the overlay network, we propose QoS Routing Network (QOSRN), which consists of fewer QoS routers and virtual links. We go further by researching into the impact of the arrangement of QoS routers upon the performance of QOSRN. Extensive simulations show that by adopting reasonable arrangement scheme, e.g. deploying border routers first, QOSRN achieves high performance without increasing the load of the network excessively. The research result can give some guidance to the construction of the overlay network that provides QoS guarantees.

1 Introduction

In order to provide QoS guarantees for data transmission, we developed the QoS router to support QoS routing (QOSR)[1]. Compared with traditional routers, the QoS router has the following characteristics: (1) It gathers dynamical local link state information (e.g. available bandwidth, delay, loss ratio) by means of interface statistics, (2) QoS-aware routing protocols are used to exchange the gathered information between QoS routers, (3) Multi-constrained QoS routing algorithms[2][3] are employed to compute the routing table.

Although the QoS router can provide QoS guarantees for data transmission, it is not practical to transform all the routers into QoS routers at present. The only feasible scheme is to deploy a few QoS routers in the Internet, but it is also exposed to some problems. Firstly, QoS router needs to exchange the special measurement packets with neighbors to measure local links state (e.g. loss ratio). However, if the neighbor is a traditional router, the measurement cannot be carried out because the traditional router does not support the measurement protocol. Secondly, QoS-aware routing information cannot be exchanged by QoS routing protocol between QoS router and traditional router. Finally, for the above two reasons, QoS router cannot acquire the enough network state information, so it cannot compute the routing table by multi-constrained QoS routing algorithms.

---

* Supported by (1) the National Natural Science Foundation of China (No. 60403035, 60473082); (2) the National Major Basic Research Program of China (No. 2003CB314801).

P. Lorenz and P. Dini (Eds.): ICN 2005, LNCS 3421, pp. 202–209, 2005.

To solve these problems, we use overlay network for reference and propose QoS Routing Network (QOSRN), which consists of a few QoS routers and virtual links that connect all the QoS routers together. In QOSRN, each QoS router measures the QoS metrics of connected virtual links, exchanges with each other using QoS-aware routing protocol and computes paths consisting of virtual links by multi-constrained QoS routing algorithms.

In order to evaluate the performance of QOSRN, we conduct the simulation based on a 200-node hierarchical network topology generated by GT-ITM, from which a number of nodes are chosen to be QoS routers to construct the QOSRN. Two metrics (Improvement Ratio and Failure Ratio) are introduced to evaluate the performance of QOSRN. In the simulation, we research further into the arrangement of QoS routers in QOSRN and find that the rule of how QoS routers are chosen and their connection mode significantly influence the performance of QOSRN. According to the result of the simulation, we proposed some reasonable arrangement schemes, under which QOSRN achieves high performance while the load of the network does not increase excessively.

2   QoS Routing Network

2.1   Architecture

Definition 1. Virtual link
According to some rules (discussed in later section), each pair of QoS routers is configured connected or disconnected. The minimal-hop path between the connected QoS routers is called virtual link. It consists of the routers and the physical links on the path.

QoS Routing Network (QOSRN) is composed of interconnected QoS routers and the virtual links between them. Fig. 1 shows an example of QOSRN, which is com

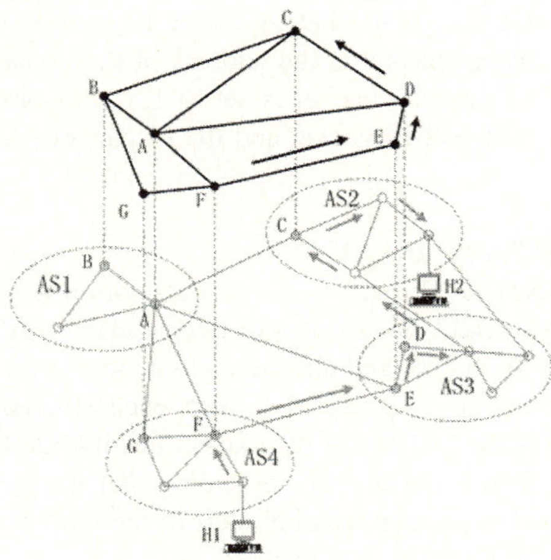

Fig. 1. QoS Routing Network

posed of 7 QoS routers A-G from 4 autonomous systems (ASes). The QoS routers are connected via virtual links. Each of the virtual links is usually composed of several physical links.

Several QoS-aware routing protocols are used in QOSRN. SQOSPF[4], whose advantages include easy implementation, multi-constrained QoS support, high-speed convergence and multiple QoS routing algorithms support, can be used in small-scaled QOSRN. In large-scaled QOSRN, hierarchical architecture must be adopted to enhance scalability, such as the combination of SQOSPF and the QoS extension of BGP[5].

2.2 QoS Routing

There are three phases in the IP packet transmission by QOSRN:

1. Sending IP packet from source to a chosen QoS router (source QoS router) in the same AS. The rule for choosing source QoS router will be discussed in the later section.

2. Transmitting IP packet within QOSRN using QoS routing and sending it to a chosen QoS router (destination QoS router) in the AS which contains the destination of IP packet. The rule of how destination QoS router is chosen will be discussed in the later section, too.

3. Sending IP packet from the destination QoS router to the destination.

Arrows in Fig. 1 show a complete packet transmission process using QOSRN. Host H1 in AS1 sends IP packets to Host H2 in AS2. Firstly, H1 sends the packet to QoS router F in AS1. F determines that the destination of the IP packet is in AS2 and chooses QoS router C as the destination QoS router from AS2. Then the routers in QOSRN forward the packet along a path that satisfies the QoS requirements of packet (in Fig 1, the chosen path is F-E-D-C). The black arrows show the forwarding path in QOSRN, and the grey arrows show the actual path in the underlying network.

2.2.1 Sending IP Packet from Source to Source QoS Router

We encapsulate the actual sent IP packet in a new IP packet called QoS IP packet. The destination of the new IP packet is the address of the chosen source QoS router. The protocol type field of the new packet is set to 126, indicating that the data after the new header is the actual sent IP packet and the IP packet is transmitted within the same AS.

2.2.2 Transmitting IP Packet in QOSRN

When QoS router receives an IP packet whose destination address is router 's local address and protocol type field is 126, it will determine the AS which contains the destination of the actual IP packet and choose a QoS router as destination QoS router from that AS. The actual sent IP packet is also encapsulated in a new IP packet whose destination field is the address of the chosen destination QoS router. The new packet 's protocol type field is set to 127, indicating that the packet is transmitted in QOSRN. Then, QoS router seeks a feasible path in the QoS routing table subject to the QoS requirements of the packet, and forwards it.

2.2.3   Sending IP Packet from the Destination QoS Router to the Destination

When QoS router receives an IP packet whose destination address is router 's local address and the protocol type field is 127, it is indicated that the destination of the actual IP packet is in the same AS. Then the router decapsulates the actual sent IP packet from the packet and forwards it to the destination.

2.2.4   Choosing Source and Destination QoS Routers

There may be several QoS routers in one AS. When end system wants to use QOSRN to transmit data packets, it must choose one as the source QoS router. Because routing from source to the source QoS router is best effort, we prescribe that the QoS router with minimal-hop path to source should be chosen as the source QoS router. The destination QoS router can be chosen in the same way. As mentioned above, QoS router distributes routes information acquired by exchanging OSPF protocol with traditional routers among QOSRN. According to these routes information, the source QoS router can choose destination QoS router with minimal-hop path to the destination.

3   Simulation Setup and Performance Metrics

Definition 2. QOSRN Routing

The routing mode in QOSRN, which combines Best Effort Routing and Complete QoS Routing, is called QOSRN Routing. The Best Effort Routing is widely used in current Internet, which selects the path with minimal hop count. The Complete QoS Routing is used in the network constructed entirely by QoS routers.

3.1   Simulation Setup

The goal of an effective QOSRN includes suitable proportion of QoS routers, high QoS guarantees to IP packets transmission and lower extra load. In order to construct such an effective QOSRN, we conduct simulation to evaluate the performance of QOSRN under different arrangement schemes. In the simulation, the rule of how QoS routers are chosen and their connection mode are taken into account.

The simulation is based on a 200-node hierarchical network topology generated by GT-ITM[6]. The 200 nodes are divided into 10 ASes, and each AS has 20 nodes. In an AS, the nodes connected with other ASes are named border nodes. Then, three types of QoS metrics which all obey uniform distribution between 1 and 100 are configured to each physical link. For convenience, we assume that these three types of QoS metrics are independent.

In the simulation, two nodes in different ASes are chosen as source and destination randomly, and 1000 QoS requirements with three metrics are used to test the performance of three routing mode (Best Effort Routing, Complete QoS Routing and QOSRN Routing). Each metric of QoS requirement obeys uniform distribution between 1 and 100D, where D is the diameter of the network topology. For each different arrangement scheme, we run the simulation 1,000 times and get the average result.

3.2 Performance Metrics

In order to evaluate the performance of QOSRN, we focus on two performance metrics: Improvement Ratio (IR) and Failure Ratio (FR).

Definition 3. Improvement Ratio (IR)

For a certain amount of QoS requirements, the Improvement Ratio (IR) is defined as follows: $IR = \dfrac{Num(\overline{BE} \wedge QOSRN)}{Num(\overline{BE} \wedge CompQoSR)}$, where $Num(\overline{BE} \wedge QOSRN)$

and $Num(\overline{BE} \wedge CompQoS)$ are the numbers of the QoS requirements which Best Effort Routing cannot satisfy but can be satisfied by QOSRN Routing and Complete QoS Routing respectively.

Definition 4. Failure Ratio (FR)

For a certain amount of QoS requirements, the Failure Ratio (FR) is defined as follows: $FR = \dfrac{Num(\overline{QOSRN} \wedge BE)}{Num(QoSReq.)}$, where $Num(\overline{QOSRN} \wedge BE)$ is the number of the

QoS requirements which QOSRN Routing cannot satisfy but can be satisfied by Best Effort Routing.

QOSRN consists of a few QoS routers. Compared with the network constructed entirely by QoS routers, it lacks sufficient global network state information. So QOSRN can only provide a certain degree of QoS guarantees. That is to say, QOSRN Routing can only satisfy a part of QoS requirements that Complete QoS Routing satisfies. Improvement Ratio is used to evaluate this relative degree, and reflects QOSRN 's ability of providing QoS guarantees. On the other hand, QOSRN Routing sometimes cannot satisfy some QoS requirements that Best Effort Routing satisfies for the lack of sufficient information. It's Failure Ratio that reflects the negative impact of QOSRN upon the IP packets transmission.

4 Simulation Result

During the simulation, we vary the rule of how QoS routers are chosen and their connection mode to investigate the factors that influence the performance of QOSRN. The simulation results are shown as follows.

4.1 Rule of Qos Router Choosing

In this simulation, full connection is also adopted to connect all the chosen nodes. Two rules of how QoS routers are chosen will be studied. One is to choose randomly; the other is to choose border node first.

The variations of IR and FR with different selection rules are shown in Fig. 2 and Fig. 3 respectively. The x-axis is the proportion of QoS routers. We observe that QOSRN performs better when the border nodes are chosen first. The reason is that the interdomain data transmission must pass border routers. Choosing border routers as QoS routers can provide better QoS guarantees according to dynamic network state.

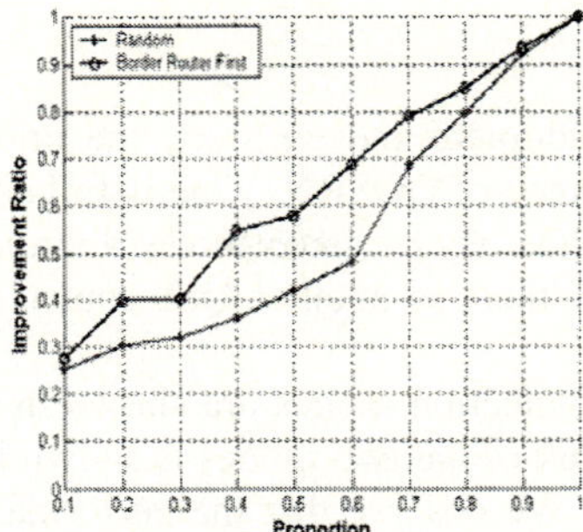

Fig. 2. Improvement Ratio vs. Proportion of QoS Routers (Random & Border Router First)

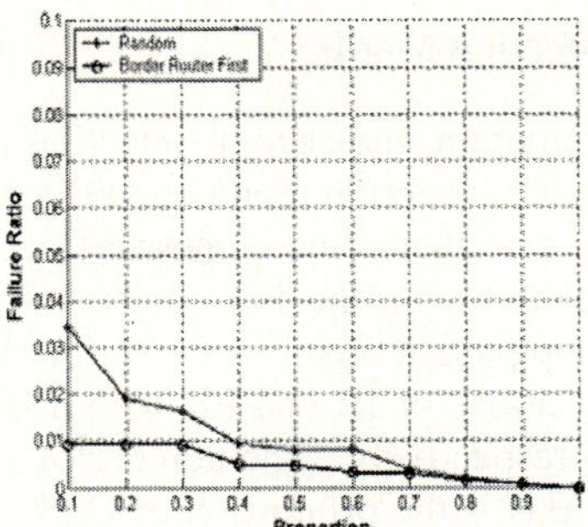

Fig. 3. Failure Ratio vs. Proportion of QoS Routers (Random & Border Router First)

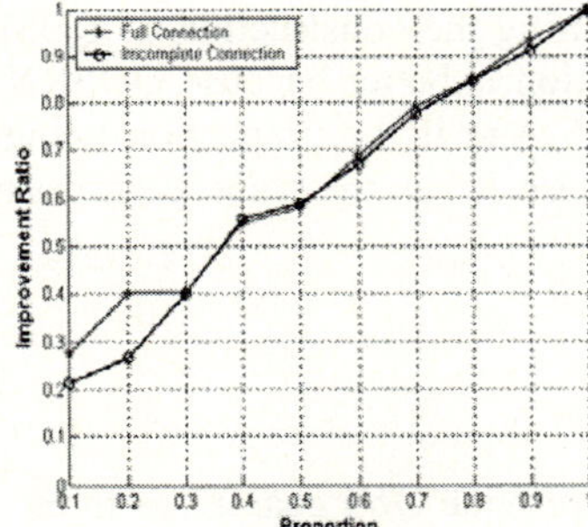

Fig. 4. Improvement Ratio vs. Proportion of QoS Routers(Full Connection & Incomplete Connection)

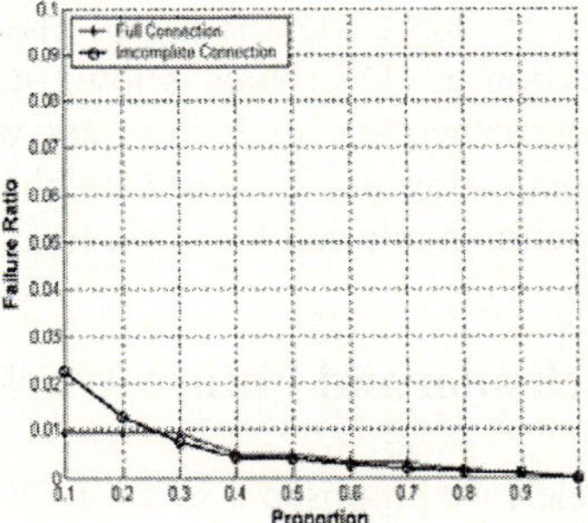

Fig. 5. Failure Ratio vs. Proportion of QoS Routers(Full Connection & Incomplete Connection)

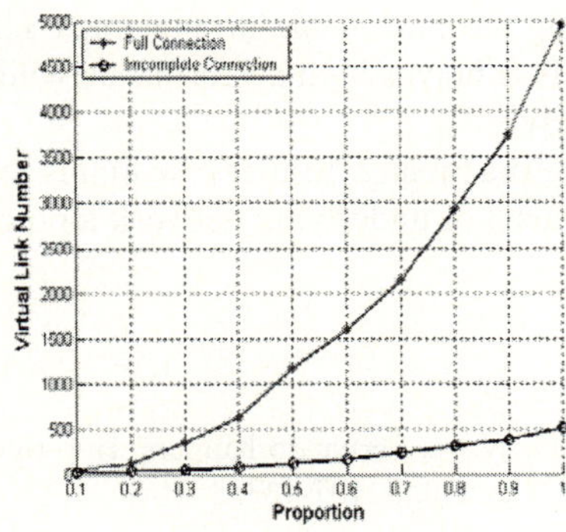

Fig. 6. Virtual Link Number vs. Proportion of QoS Routers (Full Connection & Incomplete Connection)

4.2 Connection Mode

In this simulation, the second selection rule is adopted, and we study the impact of two types of connection mode upon the performance of QOSRN. One is full connection mode; the other is incomplete connection mode. By the latter mode, a virtual link will be deleted from the full connected graph if there is another QoS router on the minimal-hop path of a pair of QoS routers.

The variations of IR and FR with different connection modes are shown in Fig. 4 and Fig. 5 respectively. The number of virtual link of the two modes is shown in Fig. 6. The x-axis is the proportion of QoS routers. We observe that the performance of the full connection mode is higher than that of the incomplete connection mode when proportion is below 20%, and the two modes have almost the same performance when proportion is above 20%. However, the number of virtual links of the second mode is far less than that of the full connection mode. Because the load of most QoS-aware routing protocols are proportional to the number of virtual links, the load of the incomplete mode is far less than that of the full connection mode, especially when the proportion of QoS routers is high. Therefore, during the construction of QOSRN, if the proportion of QoS router is low, full connection is better because QOSRN has a good performance and its load is not very high. And if the proportion is high, the second connection mode should be chosen because it has the same performance as full connection mode and far lower load.

5 Conclusion and Future Work

In this paper, we proposed a solution QOSRN to provide QoS guarantees in the Internet and evaluate its performance under different arrangement schemes. QOSRN is composed of a few interconnected QoS routers and the virtual links between them. Two metrics (Improvement Ratio and Failure Ratio) are introduced to evaluate the performance of QOSRN. In the simulation, the rule of how QoS routers are chosen and their connection mode are take into consideration to further analyze the impact of the arrangement of QoS routers upon the performance of QOSRN. Extensive simulations show that by adopting the reasonable arrangement scheme, e.g. deploying the border routers first, QOSRN achieves high performance while the load of the network does not increase excessively.

To deploy QOSRN more practically, further research is required on the QoS-aware routing protocol and the impact of underlying network topology upon QOSRN.

References

1. Piet Van Mieghem, Hans De Neve, Fernando Kuipers, Hop-by-hop quality of service routing, Computer Networks 37(2001), pp. 407-423.
2. Cui Yong, Xu Ke, Wu Jianping, Precomputation for multi-constrained QoS routing in high-speed networks, Proceedings - IEEE INFOCOM'03, 2003, vol. 2, pp. 1414-1424.
3. Cui Yong, Xu Ke, Wu Jianping, Yu Zhongchao, Multi-constrained routing based on simulated annealing, IEEE International Conference on Communications, 2003, vol. 3, pp. 1718-1722.

4. Shen Lin, Xu Mingwei, Xu Ke, Cui Yong and Zhao Youjian, Simple quality-of-service path first protocol and modeling analysis, IEEE International Conference on Communications, 2004, vol. 4, pp. 2122-2126.
5. Li Xiao, King-Shan Lui, Jun Wang, Klara Nahrstedt, QoS Extension to BGP, ICNP 2002, pp. 100-109.
6. GT-ITM: Georgia Tech Internetwork Topology Models, http://www.cc.gatech.edu/projects/gtitm/.

A Partition Prediction Algorithm
for Group Mobility in Ad-Hoc Networks

Namkoo Ha, Byeongjik Lee, Kyungjun Kim, and Kijun Han[*]

Department of Computer Engineering, Kyungpook National University, Korea
{adama2, leric, kjkim}@netopia.knu.ac.kr
kjhan@bh.knu.ac.kr

Abstract. One of important issues associated with group mobility in ad-hoc networks is predicting the partition time. The existing algorithms predict the partition time assuming that the partitioned groups move to opposite direction with the same speed and the same coverage. So, these algorithms could not accurately predict partition time in practical situation. In this paper, we propose a partition prediction algorithm considering network partition in any direction, at any speed, and with different coverage of group. To validate the proposed algorithm, we carried out a simulation study. We observe a sound agreement between numerical results obtained by our algorithm and computer simulation. Our algorithm can predict the partition time more accurately in real situations.

1   Introduction

Wireless ad-hoc networks are dynamically formed by mobile hosts without the support of pre-existing fixed infrastructures. As the mobile hosts are moving with diverse patterns, they cause frequent failures of the wireless links. Researchers have proposed mobility prediction schemes that attempt to predict the future availability of wireless links based on individual node mobility model [1-2]. The changes in link availability are caused by local topology changes, however, global scale topology change such as network partition cannot be predicted by these schemes [3][7].

The main cause of network partition is the group mobility behavior of the mobile nodes, in which some nodes belonging to a group exhibit similar movement characteristics while the others in another group show a different mobility pattern. When a network partitions, the partitioned parts are completely disconnected from other parts of the original network. Upper layer routing and other applications involving nodes in separate partitions are severely disrupted, and may terminate if the partitions do not merge in time. Such a situation is unacceptable in battlefield and rescue operations where every node must receive a certain level of Quality of Service (QoS) and have constant access to an important information repository. Therefore, to provision QoS guarantees for ad-hoc network applications, it is imperative to predict the occurrence of the network partitioning on a global scale [4-5].

---

[*] Correspondent author.

P. Lorenz and P. Dini (Eds.): ICN 2005, LNCS 3421, pp. 210–217, 2005.

If we predict the partition time in advance, we cannot avoid nodes receiving many redundant messages. If partition time is predicted behind time, nodes in partitioned group cannot receive some important messages. Some researchers proposed partition prediction algorithms by assuming that a group is partitioned into two clusters and they move to opposite angle with the same speed and the same coverage. Therefore, these algorithms could not accurately predict partition time in practical situation. In this paper, we propose a partition prediction algorithm considering any angle, any speed, and any coverage of cluster.

The organization of this paper is as follows. Section 2 reviews existing partition prediction algorithm in RVGM model. The proposed partition prediction algorithm is described in section 3. Section 4 presents validation of our algorithm through computer simulations. Section 5 concludes the paper.

2   RVGM Model and Partition Time

In realistic ad-hoc network application scenarios such as conference seminar sessions, conventional events, and disaster relief operations, the mobile users are often involved in team activities and exhibit collaborative mobility behavior. Such user mobility can be modeled by a group mobility model where the mobile users are organized into groups of different mobility pattern, mobility rate, and coverage area. One of group mobility models, RVGM(Reference vector Group Mobility) model [3] considers group partition. The node movement can be characterized by the velocity $W=(W_x, W_y)^T$, where W_x and W_y are the velocity components in the x and y directions. Each mobility group has a characteristic group velocity. The member nodes in the group have velocities close to the characteristic group velocity but deviate slightly from it. Hence, the characteristic group velocity is also the mean group velocity [3][6]. The membership of the ith node in the jth group is described by its velocity as follows

- Group velocity : $W_j(t) \sim P_{j,t}(w)$
- Local velocity deviation : $U_{j,i}(t) \sim Q_{j,t}(u)$
- Node velocity : $V_{j,i}(t)= W_j(t)+U_{j,i}(t)$

RVGM model is represented by modeling the group velocity, $W_j(t)$, and the local velocity deviation of the member nodes, $U_{j,i}(t)$, as random variables each drawn from the distribution $P_{j,t}(w)$ and $Q_{j,t}(u)$, respectively. Also, this model uses three factors, so that it is able to predict each group and node mobility as well as group partition [3][5].

As shown in Fig. 1, two groups, G_a and G_b move to the opposite angle at the $\overrightarrow{W_a}$ and $\overrightarrow{W_b}$, respectively, and then partition completely after some time. The RVGM model considers partition in the opposite angle, at the same speed and with the same coverage of cluster. Assuming that each group has the coverage of R and $\overrightarrow{W_a} = -\overrightarrow{W_b}$, according to [4], the partition time, T_p is given by

$$T_p = \frac{R}{\left\| \overrightarrow{W_a} \right\|} \tag{1}$$

where R means the radius of each group. Since this algorithm considers only partitioning in the opposite angle, it is very difficult to predict the actual partition time using this algorithm.

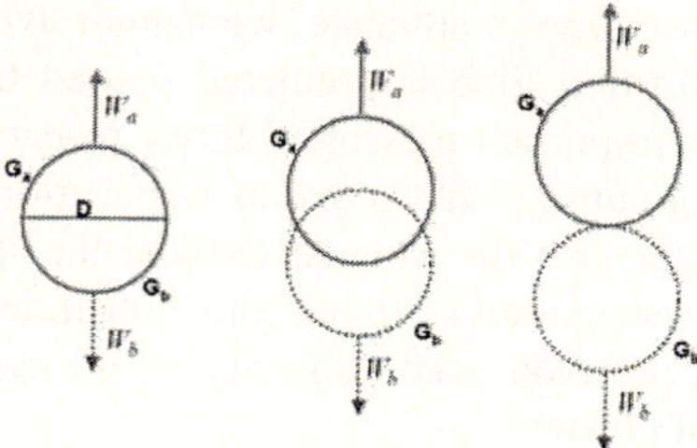

Fig. 1. Partition using RVGM model

3 Proposed Algorithm

In this paper, we propose a partition prediction algorithm considering network partition in any angle, at any speed, and with different coverage of group. Assume that two sub-groups, G_a and G_b, are partitioned to different angles with different speeds and with different coverage as depicted in Fig. 2 Initially, in Fig. 2(a), the coverage areas of the two groups overlap in Fig. 2(b), and the groups completely separate after some time in Fig. 2(c).

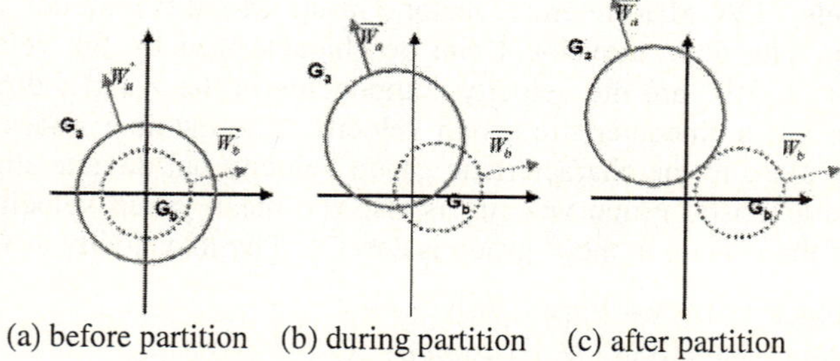

(a) before partition   (b) during partition   (c) after partition

Fig. 2. Partition in the practical situation

Fig. 3 shows a model for predicting the partition time. Let it be assumed that two groups, G_a and G_b, with the coverage of R_a, R_b, respectively, move at the velocity of $\overrightarrow{W_a}$ and $\overrightarrow{W_b}$, respectively.

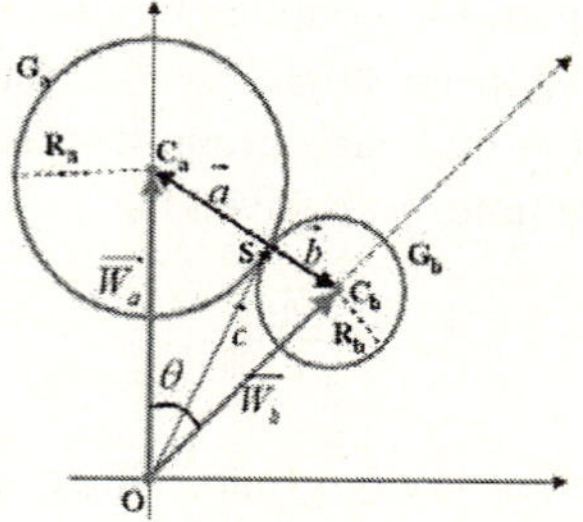

Fig. 3. Model for partition prediction

If two groups start to partition at point O and are completely separated at point S, we define vector $\vec{c}$ by $\vec{c} = \overrightarrow{OS}$. Then, we can easily derive

$$\vec{c} = \frac{R_b \overrightarrow{W_a} + R_a \overrightarrow{W_b}}{R_a + R_b} \tag{2}$$

If we define two vectors $\vec{a}$ and $\vec{b}$ by $\vec{a} = \overrightarrow{SC_a}$ and $\vec{b} = \overrightarrow{SC_b}$ where C_a and C_b denote the centers of two groups, respectively. Then, the two groups effectively partition to velocities, $\vec{a}$ and $\vec{b}$, respectively. As shown in Fig. 3, we have

$$\overrightarrow{W_a} = \frac{R_b \overrightarrow{W_a} + R_a \overrightarrow{W_b}}{R_a + R_b} + \vec{a} \tag{3a}$$

$$\overrightarrow{W_b} = \frac{R_b \overrightarrow{W_a} + R_a \overrightarrow{W_b}}{R_a + R_b} + \vec{b} \tag{3b}$$

Thus, $\vec{a}$ and $\vec{b}$ can be expressed by

$$\vec{a} = \frac{R_a (\overrightarrow{W_a} - \overrightarrow{W_b})}{R_a + R_b} \tag{4a}$$

$$\vec{b} = \frac{R_b (\overrightarrow{W_b} - \overrightarrow{W_a})}{R_a + R_b} \tag{4b}$$

Let $\|\vec{a}\|$ and $\|\vec{b}\|$ denote absolute values of $\vec{a}$ and $\vec{b}$ which mean the effective speeds of two groups, respectively. Then we can predict the partition time when two groups are partitioned at any angle, at different speed, and with different coverage as follows

$$T_p = \frac{R_a - R_b}{\|\vec{a}\| + \|\vec{b}\|} \tag{5a}$$

$$= \frac{R_a + R_b}{\sqrt{\|\overrightarrow{W_a}\|^2 + \|\overrightarrow{W_b}\|^2 - 2\|\overrightarrow{W_a}\|\|\overrightarrow{W_b}\|\cos\theta}} \tag{5b}$$

where we denote $\theta = \angle C_a O C_b$ as the partition angle which is formed when the two groups begin to be separated completely as shown in Fig. 3. Assuming that two groups, G_a and G_b have the same coverage (that is, $R_a = R_b = R$), then we have

$$T_p = \frac{2R}{\|\vec{a}\| + \|\vec{b}\|} \tag{6a}$$

$$= \frac{2R}{\sqrt{\|\overrightarrow{W_a}\|^2 + \|\overrightarrow{W_b}\|^2 - 2\|\overrightarrow{W_a}\|\|\overrightarrow{W_b}\|\cos\theta}} \tag{6b}$$

Furthermore, assuming that the speed and the coverage of each group are the same (that is, $R_a = R_b = R$ and $\|\overrightarrow{W_a}\| = \|\overrightarrow{W_b}\|$), the partition time can then be predicted by

$$T_p = \frac{R}{\|\vec{a}\|} \tag{7a}$$

$$= \frac{2R}{\sqrt{2\|\overrightarrow{W_a}\|^2 - 2\|\overrightarrow{W_a}\|^2 \cos\theta}} \tag{7b}$$

In particular, assuming that the speed and the coverage of each group is the same and that each group moves to the opposite direction (that is, $R_a=R_b=R$ and $\overrightarrow{W_a}=-\overrightarrow{W_b}=\vec{a}=-\vec{b}$), the partition time is identical to (1).

$$T_p = \frac{R}{\|\overrightarrow{W_a}\|} \tag{8}$$

4 Simulations

To validate our algorithm, we perform simulation study. The speed and angle of two groups follow Gaussian distributions, and also, those of each node in each group follow Gaussian distributions. Table 1 shows the parameters for simulation.

Table 1. Simulation parameters

Symbols	Remarks
θ (partition angle)	[0,360]
$\|\overrightarrow{W_a}\|$	2m/s
$\|\overrightarrow{W_b}\|$	2m/s~28m/s
R_a	20
R_b	10 ~20

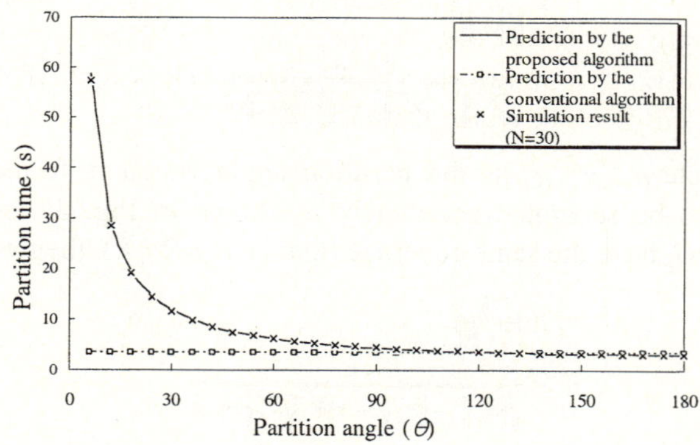

Fig. 4. Partition time as the partition angle is varied with when $\|\overrightarrow{W_a}\| = \|\overrightarrow{W_b}\| = 6$m/s, $R_a=R_b=20$

Fig. 4 shows the partition times predicted by the conventional algorithm and our algorithm, together the partition time observed by simulation when we assume that the speed of groups is $\|\overrightarrow{W_a}\| = \|\overrightarrow{W_b}\| = 6$ m/s and the radius of group coverage is $R_a=R_b=20$. In the simulation, the partition time is measured as the instant when all members of each group are located closer from its center than from the center of the other group. In this figure, we can see that our algorithm offers a good prediction whose value agrees with the simulation results over all partition angles from 0 to 180.

However, the conventional algorithm does not work well, especially for small values of the partition angle.

Fig. 5 shows the partition time as we vary the group speed. We assume that two groups move to the opposite direction ($\overrightarrow{W_a}=-\overrightarrow{W_b}$) and their coverage are the same ($R_a=R_b=20$). We can observe a good agreement between prediction by our algorithm and simulation results while the conventional algorithm does not offer a good prediction capability when the group speed is increased.

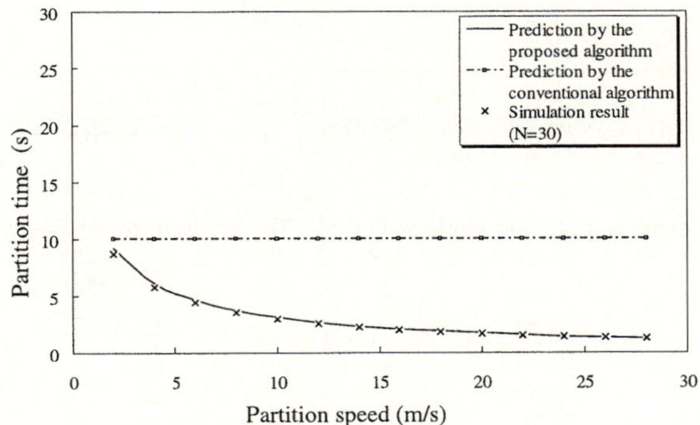

Fig. 5. Partition time as the group speed is varied with $\overrightarrow{W_a}=-\overrightarrow{W_b}$, $R_a=R_b=20$

Fig. 6 represents the partition time as the group coverage is varied. We assume that each group moves to the opposite direction and their speeds are the same ($\left\|\overrightarrow{W_a}\right\| = \left\|\overrightarrow{W_b}\right\| = 6$ m/s). This figure indicates that our algorithm offers a much better prediction capability than the conventional ones.

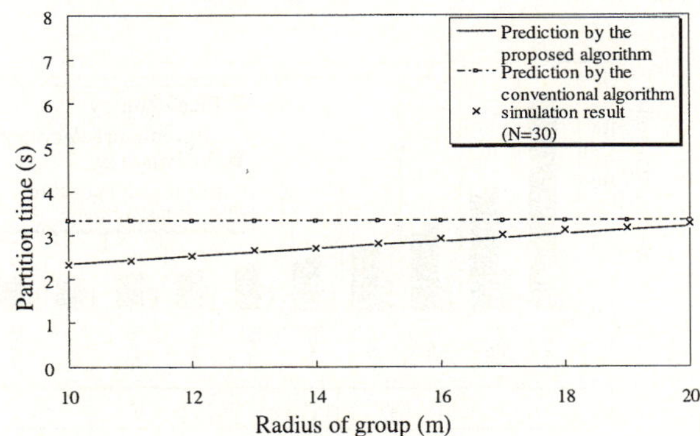

Fig. 6. Partition time as the group coverage is varied with $\overrightarrow{W_a}=-\overrightarrow{W_b}$, $\left\|\overrightarrow{W_a}\right\| = \left\|\overrightarrow{W_b}\right\| = 6$m/s

Fig. 7(a), 7(b), and 7(c) show the prediction error as we vary partition angle, group speed, and group size, respectively, for three different group sizes ($N = 10$, 30, and

50).   We can see that the prediction error is reduced when the group size becomes large. This is because there is a high probability that nodes are located on the edge of two groups when the group size is large.

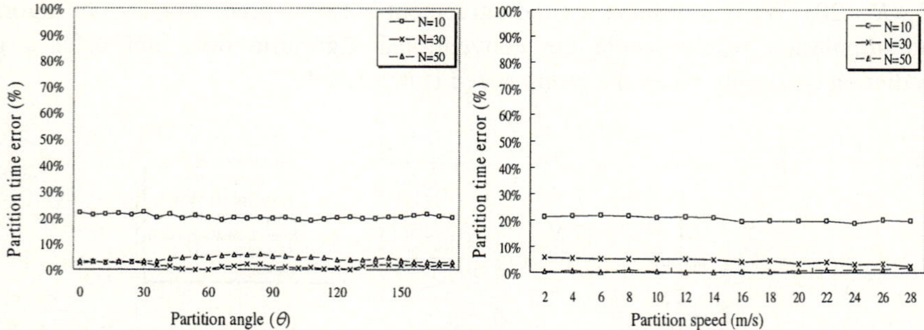

(a) Prediction error as partition angle is varied   (b) Prediction error as group speed is varied

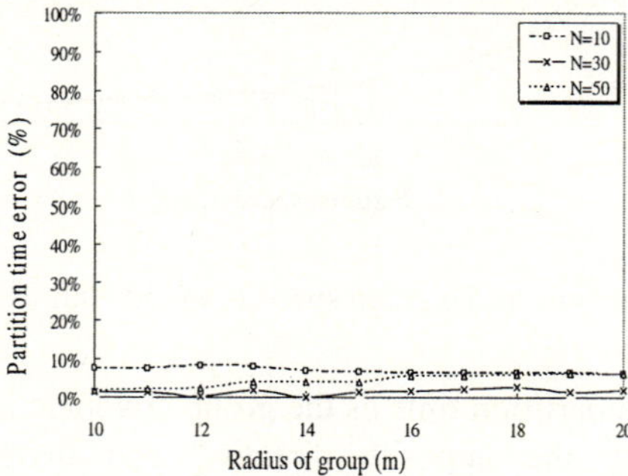

(c) Prediction error as the group coverage is varied

Fig. 7. Prediction error when the group size is 10, 30, and 50

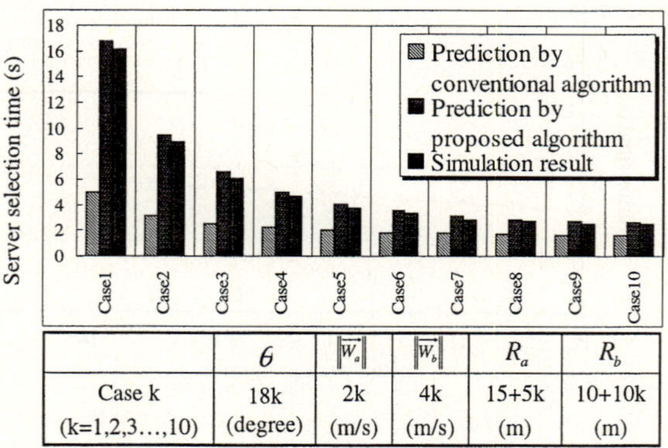

	θ	$\lVert \overline{W_a} \rVert$	$\lVert \overline{W_b} \rVert$	R_a	R_b
Case k (k=1,2,3…,10)	18k (degree)	2k (m/s)	4k (m/s)	15+5k (m)	10+10k (m)

Fig. 8. Partition time for various group velocities and coverage

Fig. 8 shows the partition time obtained in ten different cases by varying the partition angle, group speeds and coverage of two groups. This figure indicates that our algorithm gives a much better predictability in real situations than the conventional ones.

5   Conclusions and Future Work

The existing partition prediction algorithm predicts the partition time assuming that each group moves to the opposite direction, at the same speed, and with the same coverage. So, this algorithm could not predict the partition time accurately in the practical situation. In this paper, we proposed a partition prediction algorithm considering group movement to any direction, at any speed, and with different coverage. Simulation results showed our algorithm could predict the partition time much more accurately than the conventional ones. We are now studying to propose a server selection algorithm based on the partition prediction algorithm presented in this paper.

Acknowledgement

University Fundamental Research Program supported by Ministry of Information & Communication in Republic of Korea

References

[1] W. Su, S. J. Lee, and M. Gerla, "Mobility Prediction in Wireless Networks," *in Proceedings of IEEE MILCOM 2000*, Los Angeles, CA, October 2000.

[2] S. Jiang, D. He, and J. Rao, "A Prediction-based Link Availability Estimation for Mobile Ad Hoc Networks," in *Proceedings of IEEE INFOCOM*,Anchorage, Alaska, April 2001.

[3] K. H. Wang and B. Li, "Group mobility and partition prediction in wireless ad-hoc networks," *Communications, 2002. ICC 2002. IEEE International Conference*, Vol. 2 pp. 1017-1021, 28 April-2 May 2002

[4] M. Sanchez and P. Manzoni. Anejos: A java based simulator for ad-hoc networks. *Future Generation Computer Systems*, 17(5):573–583, 2001.

[5] T. Camp, J. Boleng, V. Davies, "A Survey of Mobility Models for Ad Hoc Network Research," *Wireless Communication & Mobile Computing (WCMC) Special issue on Mobile Ad Hoc Networking*, vol. 2, no. 5, pp. 483-502, 2002.

[6] K. H. Wang and B. Li "Efficient and Guaranteed Service Coverage in Partitionable Mobile Ad-hoc Networks," *INFOCOM 2002 Twenty-First Annual Joint Conference of the IEEE Computer and Communications Societies. Proceedings*. IEEE , Vol. 2, pp. 1089 – 1098, June 2002

[7] X. Hong, M. Gerla, G. Pei, and C. Chiang, " A Group Mobility Model for Ad hoc Wireless Networks," In Proceedings of the ACM International Workshop on Modeling and Simulation of Wireless and Mobile Systems (MSWiM), August 1999

Routing Cost Versus Network Stability in MANET

Md. Nurul Huda[1], Shigeki Yamada[2], and Eiji Kamioka[2]

[1] The Graduate University for Advanced Studies
[2] National Institute of Informatics,
2-1-2 Hitotsubashi, Chiyoda-ku, Tokyo 101-8430, Japan
huda@grad.nii.ac.jp
{shigeki, kamioka}@nii.ac.jp

Abstract. Recent MANET routing protocols minimize power cost because of the battery power limitation of participating mobile devices but may result in network disconnection. On the other hand lifetime prediction based algorithms help to maintain network stability with a cost penalty. In this paper we show how cost effective routing protocols affect network stability of MANET. We also find a tradeoff between routing cost and network stability to get overall best performance. The simulation results show that our proposed scheme needs less cost than in lifetime prediction based routing algorithms and results in a more stable network than cost-effective routing algorithms.

1 Introduction

Mobile Ad hoc Network (MANET) is an autonomous system with battery operated wireless devices connected by wireless links that works independent of any central control. Typically in MANET, it is assumed that all nodes are willing to act as intermediate nodes in a routing path by forwarding data for other network nodes. Historically cost efficiency has been a key objective in network routing protocols. The cost of forwarding messages can be defined and determined in various ways, taking into account factors such as cost of energy used to forward messages, hop count, delay etc. In MANET, nodes' battery power cost has been considered as the most important routing cost because of the limited power supply of the nodes. Hence recent MANET routing protocols focus on energy efficiency and power awareness [1],[2],[3],[4]. In these protocols the path with minimum power cost is used from a set of candidate paths. However, the selection of the least power cost route may possess a harmful impact on the network connectivity when the selected path contains some nodes with small remaining energy.

Roughly speaking the lifetime of a MANET is the time during which the network remains connected. The lifetime of an ad hoc network is reflected by the lifetime of its nodes. Authors of paper [5] targeted the maximum lifetime of the network by selecting the path that results in maximum lifetime of the

P. Lorenz and P. Dini (Eds.): ICN 2005, LNCS 3421, pp. 218–225, 2005.

network. They use the lifetime of nodes, which is a function of the remaining battery energy, as the metric for selecting a routing path. The lifetime of a node is predicted based on the residual battery capacity and the rate of energy discharge of that node. The authors calculated the lifetime of a route with the following equation.

$$Max(T_\pi(t)) = Min(T_i(t)).....\{i\epsilon\pi\} \tag{1}$$

Where, $T_\pi(t)$: lifetime of path π
$T_i(t)$: predicted lifetime of node i in path π
The lifetime of a path is predicted by the minimum lifetime of all nodes along the path. The minimum lifetimes of all the paths from the source to the destination are calculated and the path which has maximum value of calculated minimum lifetimes is selected for packet forwarding. The main objective of Lifetime Prediction based Routing (LPR) is to minimize the variance in the remaining energy of all the nodes and thereby prolong the network lifetime. Although LPR selects the path resulting in maximum lifetime of the network, this technique has not considered the cost of routing that results in higher routing cost. In most cases it may select a path with a higher cost than the minimum.

The routing protocol proposed in this paper is a reactive routing protocol like DSR [6]. There are two basic objectives in our scheme; one is to minimize the cost of routing and the other one is to maximize the network lifetime. But these two goals do not line up. Improvement of routing cost (i.e., less cost) degrades the stability (i.e., lifetime) of the network and improvement of network stability degrades the routing cost. So we find out a tradeoff between the two contradictory parameters. Our proposed route selection technique results in a more stable network than the Power Aware Routing (PAR) algorithms and also requires less routing cost than the LPR algorithms.

The remainder of this paper is organized as follows. In the next section we present a short overview of MANET routing protocols. Section 3 describes the rationale and details of the proposed Cost-effective Lifetime Prediction based Routing (CLPR) technique. Section 4 elaborates on the simulation environment and experimental results comparing CLPR with LPR and PAR.

2   MANET Routing Protocols

Among the two types of routing protocols, proactive (table driven) routing protocols are similar to and come as a natural extension of those for the wired networks. Each node contains the latest information of the routes to any node in the network. Any change in topology is updated and propagated through all nodes in the network. Several proactive routing protocols are addressed in [7], [8]. Reactive (on-demand) routing protocols do not maintain or constantly update their route tables with the latest route topology. Examples of reactive routing protocols are the dynamic source Routing (DSR) [6], ad hoc on-demand distance vector routing (AODV) [9] and temporally ordered routing algorithm (TORA)[10].

Due to the limited power supply of MANET nodes many energy efficient broadcast/multicast algorithms have emerged [1],[2],[11],[12]. One major approach for energy conservation is to route a communication session along the path which requires the lowest total energy consumption. This optimization problem is referred to as Minimum-Energy Routing [13]. Energy efficiency has also been considered from medium access control (MAC) and new MAC schemes were proposed in [14],[15]. Unfortunately energy efficient routing protocols may cause network disconnection if the minimum power loss route consists of nodes with small remaining energy. Keeping this in mind, paper [5] proposes a lifetime prediction based routing technique. The objective of this routing protocol is to extend the service life of MANET with a dynamic topology. This protocol favors the path whose lifetime is the maximum. But it suffers from poor cost effectiveness.

3   CLPR

3.1   Probability of Nodes Acting as Routers

Let us represent the ad hoc network with a graph where edges represent links between two devices and the vertexes represent network devices. We introduce and define the "path degree" of a node as the number of paths, between any two nodes of the network, for which the node is an intermediate node. The path degree of leaf nodes is zero while the path degree of non-leaf nodes would be non zero. For the ease of understanding and simplicity in figure 1 we show the path degree of the nodes of a tree. If node i has n number of subtrees with number of nodes $(N_1, N_2, ...N_n)$ then the path degree of that node (P_i) can be calculated with

$$P_i = \sum_{k=1}^{n-1} \sum_{j=k+1}^{n} N_k N_j \tag{2}$$

The higher the path degree of a node the higher it is likely to be used repeatedly for packet forwarding. Note that the degree of a node does not reflect the path degree of that node. A node with lower degree may have a larger path degree than a node with higher degree.

On one hand, in order to optimize cost, least cost routing path is desirable, while on the other hand, use of least cost route means that the energy of the nodes with a higher path degree are likely to be used up soon and they die. This may result in disconnection of the network. A disconnected network is useless in an ad hoc environment because of the infrastructure-less nature of the network. For network stability we should keep alive as many nodes as possible. Besides, loss of a node implies loss of all paths that pass through that node. The effect on the network of losing a node depends on the path degree of that node. A high path degree node is precisely the one that is very important to maintain the network connectivity. Absence of a high path degree node would compel to use longer routing paths (if exist) with probable higher cost and in the worst case disconnection of part of the network from the rest.

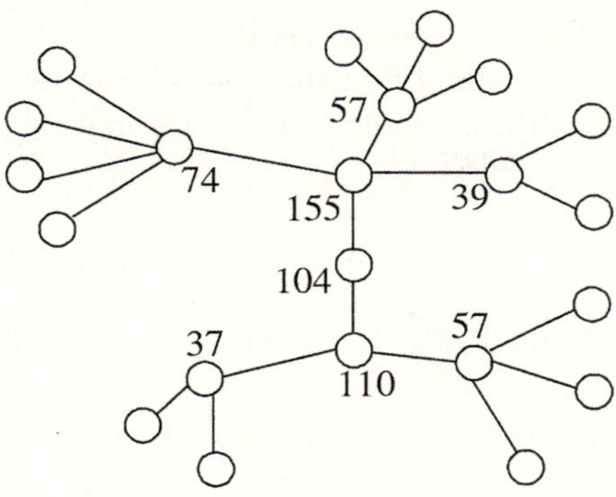

Fig. 1. Path degree of nodes

3.2   The Network Model

In our network model we consider a mobile ad hoc network N=(V,E,C) consisting
of a set of nodes $V = \{v_1, ..., v_n\}$ that represent mobile devices, a set $E \subseteq V \times V$
of edges $\{(v_i, v_j), 1 \leq i, j \leq n\}$ that connect all the nodes, and a weight function
ω for each edge (v_i, v_j) that indicates the transmission cost of a data packet
between node v_i and v_j. Each node has a unique identification number, but it
is not a priori known which nodes are currently in the network, nor is edge set
E or weight function ω known. A node can not control the direction in which
it sends data, and thus data are broadcast to all nodes inside its transmission
range. Nodes can move and the edge cost between any two nodes can change over
time. Also the lifetime of any node can change over time. However, for the ease
of presentation, we assume a static network during the route discovery phase.

3.3   Packet Forwarding Path

Our CLPR is a reactive routing protocol which only takes action and starts
computing routing paths when a network initiates a session. It uses a DSR-
like [6] route discovery protocol and channels all information regarding cost and
lifetime to the destination node. The destination node computes the cost and
lifetime of each path and sends this information back to the source.

Let us assume that the maximum possible lifetime of any node is L and the
maximum possible transfer cost between any two nodes is C. We define a scaling
factor ξ as the ratio of the two parameters.

$$\xi = \frac{L}{C} \tag{3}$$

Let there be n paths ($\pi_1, \pi_2, ...\pi_n$) from source to destination. The lifetime
of a path is bounded by the lifetime of all the nodes along the path. When a
node dies along a path we can say that the path does not exist any longer. So we
can consider the life-time of a path is the same as the minimum lifetime among
all the nodes along the path. The lifetime τ_i of a path π_i can be defined as:

$$\tau_i = Min(T_j(t)).......\{j\varepsilon i\} \tag{4}$$

$T_j(t)$: predicted lifetime of node j in path π_i

The cost of a path is the sum of all the costs calculated between two consecutive nodes along the path from source to the destination. Cost of a path i can be defined as:

$$\varsigma_i = \sum_{j=1}^{\pi_{i_m}-1} C_{\pi_{i_{j,j+1}}}(t) \tag{5}$$

where π_{i_m} is number of nodes in path π_i and $C_{\pi_{i_{j,j+1}}}$ is the cost between node j and j+1 of the path π_i.

Our path selecting parameter β is represented by

$$\beta_i = \frac{\tau_i}{\xi\varsigma_i} \tag{6}$$

CELP selects a path, which has the largest β i.e. $\max(\beta_i)$. If more than one path having highest β is found, any one of them can be selected. Thus, the proposed method is inclined to select a path having higher lifetime τ and lower cost ς.

Figure 2 shows an instance of an ad hoc network represented by a graph. Nodes are labeled with their lifetime values and the edges are labeled with the cost between its two adjacent nodes. In this instance there are six paths from source (S) to destination (D). They are S→A→B→D, S→A→B→C→F→G→D, S→E→F→C→B→D, S→E→F→G →D, S→C→F→G→D, and S→C→B→D.

If we calculate the total cost along each path, we get the cost 19 for the path S→A→B→D, 36 for the path S→A→B→C→F→G→D, 40 for the path S→E→F→C→B→D, 29 for the path S→E→F→G→D, 27 for the path S→C→F →G→D and 24 for the path S→C→B→D. Similarly we calculate lifetimes of each path and get the lifetime 100 for the path S→A→B→D, 100 for the path S→A→B→C→F→G→D, 400 for the path S→E→F→C→B→D, 450 for the path S→E→F→G→D, 400 for the path S→C→F→G→D and 400 for the path S→C→B→D.

If we select the path with minimum cost, as done in cost-effective routing, we get the path S→A→B→D having cost 19 and lifetime 100. While in LPR, the route S→E→F→G→D is chosen having lifetime 450 and cost 29. The minimum cost routing is greedy for cost minimization and LPR is greedy for highest

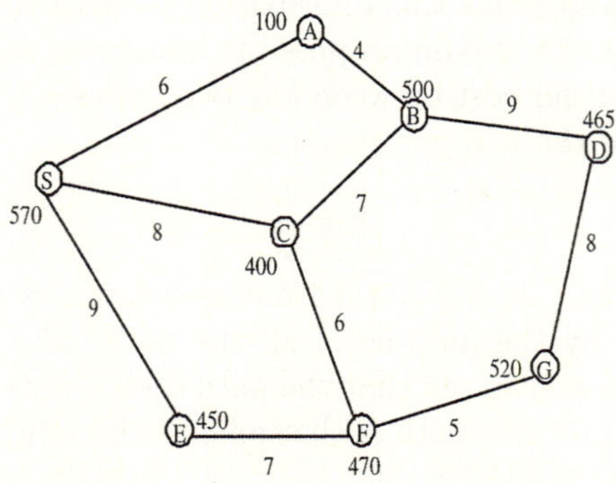

Fig. 2. An instance of a MANET

lifetime. Hence minimum cost routing suffers from poor lifetime of the path and LPR suffers from high routing cost.

For our CLPR algorithm let us assume that the maximum possible cost (C) between any two nodes is 15 and the maximum possible lifetime (L) of any node is 600. So the scaling factor ξ becomes 40. Hence, using CLPR algorithm the selecting parameter β for the paths S→A→B→D, S→A→B→C→F→G→D, S→E→F→C→B→D, S→E→F→G→D, S→C→F→G→D, and S→C→B→D are 0.1316, 0.069, 0.25, 0.3879, 0.3704 and 0.4166 respectively. The path S→C→B→D has the highest β value. So the selected path is S→C→B→D having cost 24 and lifetime 400.

We find that CLPR is better than LPR in cost perspective and also better than cost-effective routing in stability perspective. Although CLPR may select a path with cost little higher than a path with least cost and a path having little less lifetime than a path having highest lifetime, to achieve the balance between the two contradictory goals, this is acceptable considering both the stability and the cost-effectiveness of the route.

4 Simulation

We have had experiments on the performance of the proposed CLPR and have compared it with that of the LPR and PAR. In this section we describe the simulation environment, experimental results and comparison of the three related protocols.

4.1 Simulation Setup

In our discrete event driven simulation we used 20, 30, 40, 50, 60 and 70 nodes. The lifetime of nodes was between 1 and 600s while the transmission cost to neighboring nodes was varied between 1 and 15. Every node had fixed transmission power resulting in a 50m transmission range. The sources and sinks were spread uniformly over the simulation area; the size of the area 600m × 300m. Random connections were established between nodes within the transmission range. Data packet size was 512 bytes at a rate of 1 packets/sec. The simulation time was normalized to 200s. Nodes followed random waypoint mobility model. Each packet relayed or transmitted had a cost factor and that cost was considered as the cost at the transmitter node. The results of our simulation have been projected in figure 3 and figure 4. From figures 3 we see that as the number of nodes increase the data routing cost from the source to the destination increases. Power aware routing (PAR) requires minimum cost (i.e. power) among the three protocols, followed by cost-effective lifetime prediction based routing (CLPR). The difference of routing cost between PAR and LPR is increases with the increase of the number of nodes but this cost difference between CLPR and PAR decreases with the increase of the number of nodes. With the increase of the number of nodes in the network the cost increase rate of CLPR decreases and progresses towards PAR.

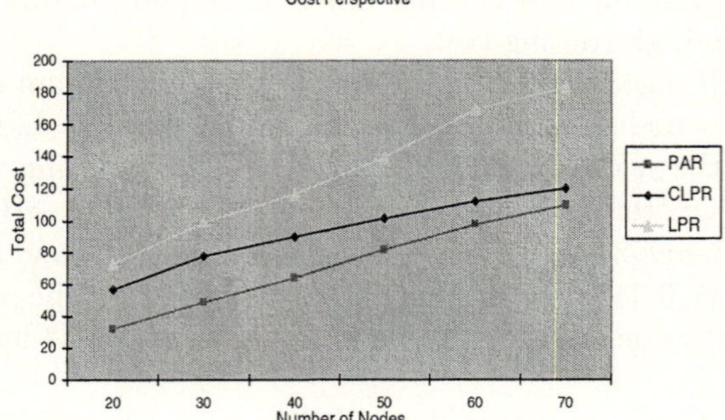

Fig. 3. Comparison of Cost among three related protocols

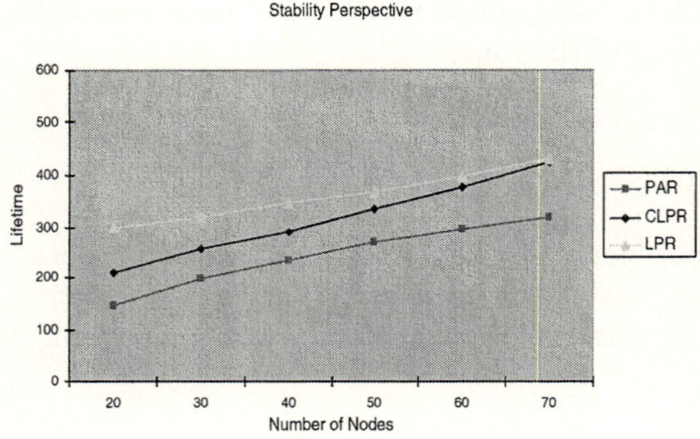

Fig. 4. Comparison of Lifetime/stability among three related protocols

Although PAR requires least cost its network stability is poor (figure 4). On the other hand, lifetime prediction based routing (LPR) has maximum network lifetime or stability but it require higher routing cost. The rate of increase in lifetime of PAR decreases with the increase on the number of nodes. The difference of lifetime between PAR and LPR increases with the increases of the number of nodes. However, this difference between LPR and CLPR decreases with the increase of the number of nodes in the network. and CLPR progresses towards LPR. The CLPR algorithm does not suffer extremely from either routing cost or network stability. It maintains a balance between the two and offers cost-effective routing maintaining maximum network stability.

5   Conclusion

A Cost-effective Lifetime Prediction based Routing protocol for mobile ad hoc networks that optimizes the network stability and routing cost has been proposed in this paper. Simulation results show that in average the proposed CLPR can increase the lifetime up to about 25 percent than that of power-aware routing and can cut routing cost about 33 percent than that of lifetime prediction based routing. Thus the proposed method cuts the cost short while it still tries to maintain maximum lifetime of the network. The lifetime and cost of such a network are two contradictory functions and improvement of one factor has a negative effect on the other. But if any one of these parameters is ignored totally, the network will suffer from poor efficiency. Our proposed method makes a tradeoff between the two and ensures a balanced use of both of them so that maximum utilization is achieved.

References

1. S. Doshi, S. Bhandare, T. X Brown, "An On-demand minimum energy routing protocol for a wireless ad hoc network," ACM SIGMOBILE Mobile Computing and Communications Review, vol. 6, issue. 2, (2002) 50–66.
2. J. E. Wieselthier, G. D. Nguyen and A. Ephremides: Energy-Efficient Broadcast and Multicast Trees in Wireless Networks. MONET vol. 7, (2002) 481–492
3. Suresh Singh, Mike Woo and C.S. Raghavendra: Power-Aware routing in Mobile Ad-hoc Networks. Proceedings of MOBICOM (1998) 181–190
4. Anderegg L., and Eidenbenz S.: Ad hoc-VCG: A Truthful and Cost-Efficient Routing Protocol for Mobile Ad hoc Networks with Selfish Agents, Proceedings of Mobi-Com (2003) 245–259
5. Maleki M., Dantu K., and Pedram M.: Lifetime Prediction Routing in Mobile Ad-Hoc Netowrks. Proceedings of IEEE WCNC vol.2 (2003) 1185–1190
6. David B. Johnson: The Dynamic Source Routing for Mobile Ad Hoc Wireless Networks. http://www.ietf.org/internet-drafts/draft-ietf-manet-dsr-09.txt, IETF Internet Draft, Apr. (2003)
7. C. Perkins and P. Bhagwat: Highly Dynamic Destination-Sequenced Distance-Vector Routing (DSDV) for Mobile Computers. Proc of ACM SIGCOMM (1994) 234–244
8. Murthy and J.J. Garcia-Luna-Aceves: An Efficient Routing Protocol for Wireless Networks. MONET vol. 1, (1996) 183–197
9. Charles E. Perkins, Elizabeth M. Belding-Royer, and Samir Das: Ad Hoc On Demand Distance Vector (AODV) Routing. IETF Internet draft November (2001)
10. V.Park and S.Corson: Temporally-Ordered Routing Algorithm (TORA). IETF Internet Draft, July (2001)
11. J. H. Chang and L. Tassiulas: Energy Conserving Routing in Wireless Ad Hoc Networks. Proc. of INFOCOM vol. 1(2000) 22–31
12. A. Michail and A. Ephremides: Energy Efficient Routing for Connection-Oriented Traffic in Ad Hoc Wireless Networks. Proc of the PIMRC vol.2 (2000) 762–766
13. J. WAN, G. C Alinescu LI and O. Frieder: Minimum-Energy Broadcasting in Static Ad Hoc Wireless Networks. Wireless Networks vol. 8, (2002) 607–617
14. Woesner. H, Evert. J, Schlager. M, and Wolisz A: Power-saving mechanisms in emerging standards for Wireless LANs: the MAC level perspective. IEEE Personal. Communication, vol. 5 (1998) 40–48
15. Jin, K. and Cho D: A MAC algorithm for energy-limited ad-hoc networks. IEEE Vehicular Technology Conference, vol. 1, Boston (2000) 219–222

Multipath Energy Efficient Routing in Mobile Ad Hoc Network

Shouyi Yin and Xiaokang Lin

Department of Electronic Engineering,
Tsinghua University
shouyi@gmail.com
linxk@atm.mdc.tsinghua.edu.cn

Abstract. In ad hoc networks, energy conservation is a very important design issue. Several energy-efficient routing protocols are proposed to maximize the network lifetime which is defined as the time until the first node in the network runs out of energy. In this paper, we propose a Multipath Energy Efficient Routing (MEER) mechanism. We consider a network that is shared by a set of sources, each one communicates with its corresponding destination using multiple paths. MEER operates at each source node in a distributed manner and adaptively adjusts the traffic allocation among multiple paths with the objective of maximizing the network lifetime. We evaluate the performance of MEER by comparing it with some present energy efficient routing protocols via simulation. Satisfyingly, our proposal achieves the goal of more residual energy and better balanced energy residual, as a result the network lifetime is elongated.

1   Introduction

Since years ago, a great portion of people's attention has been given to wireless ad hoc networks. Ad hoc network consists of several stationary or mobile nodes communicating with each other via wireless links without the need of existing infrastructure. Each node in ad hoc network operates not only as a host but also as a router to forward data.

Mobile devices are usually supplied by limited battery power. Depletion of batteries in some devices will lead to network partition and failure of application sessions. Therefore, energy conservation is one of the most important design issues in ad hoc networks. To address this issue, several energy efficient routing protocols were developed [1] [2] [3] [4] [5] [12]. However, most of existing energy efficient routing protocols are single-path and there is little research on multipath energy efficient routing protocols.

In this paper, we propose a Multipath Energy Efficient Routing (MEER) mechanism with the objective to extend the network lifetime, which is defined as the period from the time instant when the network starts functioning to the time instant when the first node runs out of energy. The motivation of using multiple paths is two-fold. One is that using multiple paths can provide an even distribution of traffic load among all nodes in all possible paths. Thus no host

P. Lorenz and P. Dini (Eds.): ICN 2005, LNCS 3421, pp. 226–233, 2005.

will be overused and the battery of each host will be used more fairly, in turn the lifetime of network will be extended. The other is for the need of reliability in wireless networks. The source-destination data transfer will not be obstructed if some paths fail because there are other paths still functioning.

We model an ad hoc network by a set of mobile nodes, and the network is shared by a set of traffic source-destination node pairs. Each source can establish a set of disjoint paths that can be used to reach its destination. And the traffic from source to destination is allocated among multiple disjoint paths. Several algorithms of finding disjoint source-destination path in a network is well studied in graph theory [6] [7]. Therefore, in this paper, we will not discuss how to find disjoint paths, and these algorithms are used as basic building blocks for MEER. The problem we focus on is how to allocate traffic among multiple paths to elongate the lifetime of network. We use an optimization approach to address this problem and a traffic allocation algorithm based on gradient projection method is proposed. This traffic allocation algorithm is the core function block of MEER. MEER can be easily implemented in a distributed manner. For each source-destination pair, MEER operates at source node and it can quickly converges to the optimal traffic allocation.

The rest of this paper is organized as follows. In section 2, we introduce our traffic allocation algorithm. We first describe the network model and the optimization problem formulation of traffic allocation, then the methodology for solving the optimization problem is presented. Performance evaluation based on simulation is conducted in Section 3. Section 4 concludes the paper.

2   Traffic Allocation Algorithm

2.1   Optimization Problem Statement

We model an ad hoc network by a set N of mobile nodes, with the nodes representing wireless devices. The network is shared by a set W of source-destination (SD) node pairs, indexed $1, 2, \ldots, W$. For each pair $w = (s, d)$, we introduce P_w, a set of disjoint paths from s to d.

For SD pair w, let r_w be total packet sending rate from s to d, and distribute x_{wp} amount of it on path p of P_w such that

$$\sum_{p \in P_w} x_{wp} = r_w, \quad x_{wp} \geq 0, \quad \text{for all} \quad w, p. \tag{1}$$

We define a vector x_w with components x_{wp}, $p \in P_w$, and let $x = (x_{wp}, p \in P_w, w \in W)$ be the vector of all rates in the network.

The packet transmission rates that are associated with node $n, n \in N$, includes three parts, x_s^n, x_f^n and x_d^n. x_s^n is the packet sending rate generated by node n,

$$x_s^n = \sum_{w \in W} \sum_{\substack{p \in P_w, \\ n \text{ is source of } p}} x_{wp}; \tag{2}$$

x_d^n is the packet arriving rate whose destination is n,

$$x_d^n = \sum_{w \in W} \sum_{\substack{p \in P_w, \\ n \text{ is destination of } p}} x_{wp}; \tag{3}$$

and x_f^n is the packet transmitting rate that is forwarded through n,

$$x_f^n = \sum_{w \in W} \sum_{p \in P_w, n \in p} x_{wp}. \tag{4}$$

($n \in p$ means that node n is in path p, but n is not the source or destination of p.)

For each node n, $n \in N$, the energy consumption due to traffic processing is neglected, and the energy consumed per packet while receiving and sending, denoted by e_r^n and e_s^n respectively, are assumed to be constant. Let γ^n be the energy consuming rate by node n. Then,

$$\gamma^n = x_s^n e_s^n + x_d^n e_r^n + x_f^n (e_r^n + e_s^n). \tag{5}$$

We assume that node $n, n \in N$, has a limited amount of remaining energy and denote by E^n, thus γ^n / E^n represents the normalized energy consumption rate. In [10], γ^n / E^n is also be defined as "remaining lifetime metric". It is a more accurate metric to describe the remaining lifetime of a node, since a node with relatively little remaining energy may have lager lifetime than one has large amount of remaining energy but is depleting energy rapidly [10].

We define an energy cost function of host n, denoted by D^n, as

$$D^n(\gamma^n) = \frac{1}{1 - \frac{\gamma^n}{E^n}}. \tag{6}$$

D^n will increase with growth of γ^n / E^n. Thus less D^n means more remaining lifetime. The reason of using equation (6) as energy cost function rather than using γ^n / E^n directly is that we always prefer a shorter path to transmit more traffic, which will decrease the total energy consumption of the whole network. The total energy cost of the network is defined as $D(x)$ (x is the rate vector of whole network),

$$D(x) = \sum_{n \in N} D^n(\gamma^n) \tag{7}$$

Therefore, to maximize the network lifetime, we should minimize the cost function of network, $D(x)$, by optimally allocating the traffic among all paths in $\cup_w P_w$:

$$min \quad D(x) = \sum_{n \in N} D^n(\gamma^n) \tag{8}$$

$$subject \quad to \quad \sum_{p \in P_w} x_{wp} = r_w, \quad for \ all \quad w \tag{9}$$

$$x_{wp} \geq 0, \quad for \ all \quad w, p \tag{10}$$

In other words, our objective is to balance the energy consumption rate γ^n/E^n, so that no host will be overused, thus the network lifetime will be extended.

Considering the optimization problem (8-10), since $D^n(\gamma^n)$ is continuous and convex in x_{wp}, this is a convex programming problem.

As observed in [8], we define the first derivative length of path p as the partial derivative of $D(x)$ with respect to x_{wp}, $\partial D(x)/\partial x_{wp}$, and

$$\frac{\partial D(x)}{\partial x_{wp}} = \frac{\partial D^s(\gamma^s)}{\partial x_{wp}} + \sum_{n \in p} \frac{\partial D^n(\gamma^n)}{\partial x_{wp}} + \frac{\partial D^d(\gamma^d)}{\partial x_{wp}}, \tag{11}$$

where s and d denotes source and destination node of p.

It has been proved that the rate vector x is optimal if and only if, for each pair w, all paths $p \in P_w$ with positive flows, $\partial D(x)/\partial x_{wp}$ are equal and minimum[9].

2.2   A Gradient Projection Approach

A standard technique to solve the convex programming problem is the gradient projection algorithm. In such an algorithm, the traffic on each path is iteratively adjusted in opposite direction of the gradient and projected onto the feasible space of the constrained optimization problem.

We define a set $T_w = \{t, 2t, 3t.., \}$ (t is a fixed time interval) of times at which SD pair w adjusts its rates, and $x_w(k)$ is used to denote the new rate vector at time kt. Each iterative adjustment takes the form as

$$x_w(k+1) = [x_w(k) - \lambda \frac{\partial D}{\partial x_w}(x(k))]^+, \tag{12}$$

where $x_w(k) = (x_{wp}(k), p \in P_w)$ and $\partial D/\partial x_w = (\partial D/\partial x_{wp}, p \in P_w)$. Here, λ is a positive step size, and $[z]^+$ is the projection of vector z onto the feasible space.

The adjustment will stop when all path $p, p \in P_w$ with positive flows have minimum and equal first derivative lengths. One important characteristic of this algorithm is that it can be distributively carried out by each pair w without the need to coordinate with other pairs.

In order to perform the iteration of (12), we calculate $\partial D(x(k))/\partial x_{wp}$ as follows:

$$\frac{\partial D}{\partial x_{wp}}(x(k)) = \frac{\partial D^s(\gamma^s(k))}{\partial x_{wp}} + \sum_{n \in p} \frac{\partial D^n(\gamma^n(k))}{\partial x_{wp}} + \frac{\partial D^d(\gamma^d(k))}{\partial x_{wp}} \tag{13}$$

According to (6),
$$\frac{\partial D^n(\gamma^n)}{\partial x_{wp}} = \frac{E^n}{(E^n - \gamma^n)^2} \frac{\partial \gamma^n}{\partial x_{wp}} \tag{14}$$

For source node s and destination node d of path p, we have

$$\frac{\partial D^s(\gamma^s(k))}{\partial x_{wp}} = \frac{E^s e_s^s}{(E^s - \gamma^s(k))^2} \tag{15}$$

$$\frac{\partial D^d(\gamma^d(k))}{\partial x_{wp}} = \frac{E^d e_r^d}{(E^d - \gamma^d(k))^2} \tag{16}$$

For node $n \in p$, we have

$$\frac{\partial D^n(\gamma^n(k))}{\partial x_{wp}} = \frac{E^n(e_r^n + e_s^n)}{(E^n - \gamma^n(k))^2} \tag{17}$$

Therefore

$$\frac{\partial D}{\partial x_{wp}}(x(k)) = \frac{E^s e_s^s}{(E^s - \gamma^s(k))^2} + \sum_{n \in p} \frac{E^n(e_r^n + e_s^n)}{(E^n - \gamma^n(k))^2} + \frac{E^d e_r^d}{(E^d - \gamma^d(k))^2} \tag{18}$$

The equation (18) implies that we only need to measure the energy consumption rate $\gamma^n(k)$ of all nodes in path p so that we can calculate traffic rate vector $x_w(k+1)$ for iteration.

In the implementation of this algorithm, it is required each node to record its energy consumption rate periodically. For each source-destination pair w, the source node is required to send probe packets periodically along each path $p, p \in P_w$. Once a node receives a probe packet, it send a reply packet to the source node to report its current energy consumption rate, γ^n. The source node collects the reply packets and records γ^n of all nodes n in p. Thus at each update time $t \in T_w$, source node could calculate the packets sending rate x_w according to equation (12) and (18) to adjust the traffic allocation.

Since we are using gradient projection method, the convergence is guaranteed [13]. And this method can be implemented in an asynchronous, distributed format, whereby the computation and information reception are not synchronized at each node [8].

3　Performance Evaluation

In this section some simulation results are presented to illustrate the performance of MEER. To evaluate the effectiveness and performance of MEER, we implement MEER in ns-2 [11]. In our simulations, IEEE 802.11 is used as the MAC and physical layer protocol. The initial energy of all nodes is set as 100 Joules, and transmitting and receiving powers are set as 2w and 1w respectively. We concentrate on two simulation scenarios. One is a stationary scenario of 8 nodes, the other is a mobile scenario of 30 nodes.

Figure 1 shows the topology of stationary scenario. Two disjoint paths connect node 0 and node 1, and one path connects node 6 and node 7.

The simulation of stationary scenario lasts for 80 seconds. At 10s, a 128kbps CBR connection starts from node 0 to node 1 and is distributed among path (0,2,4,1) and (0,3,5,1) equally. At 20s, a 64kbps CBR connection from node 6 to node 7 starts and transmits through path (0,2,4,7). Thus we have an unbalanced situation with one heavily-loaded path and one lightly-loaded path. Therefore the energy consumption of node 2 and 4 are increased, it should shift some traffic from path (0,2,4,1) to (0,3,5,1) in order to balance energy consumption and maximize the network lifetime. Figure 2 shows the traffic adjustment process of node 0. As shown in figure 2, MEER is able to successfully reduce the traffic

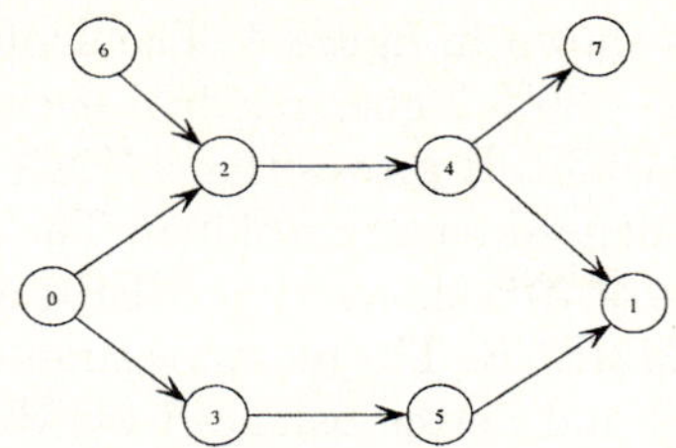

Fig. 1. Network topology of stationary scenario

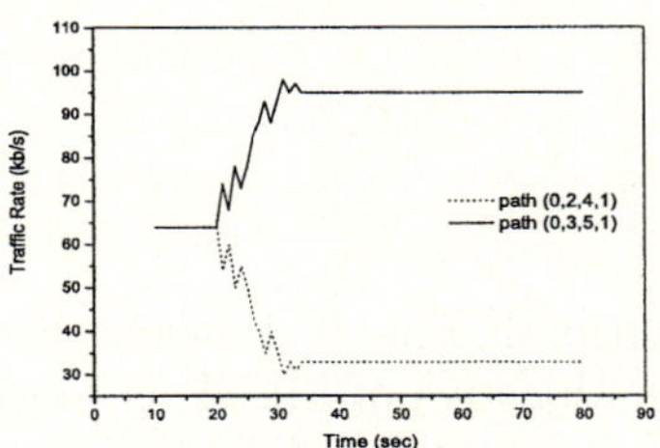

Fig. 2. Traffic adjustment process

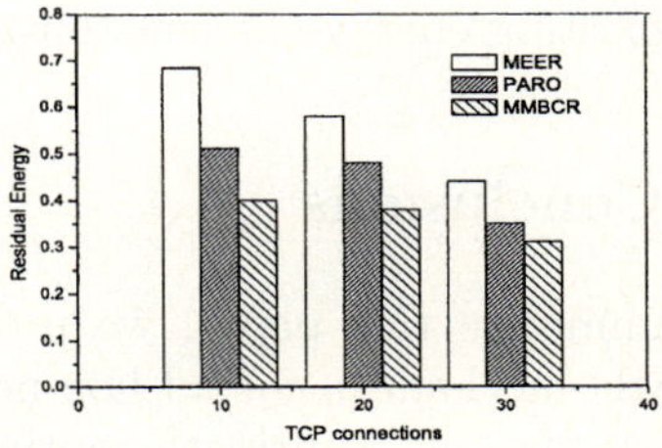

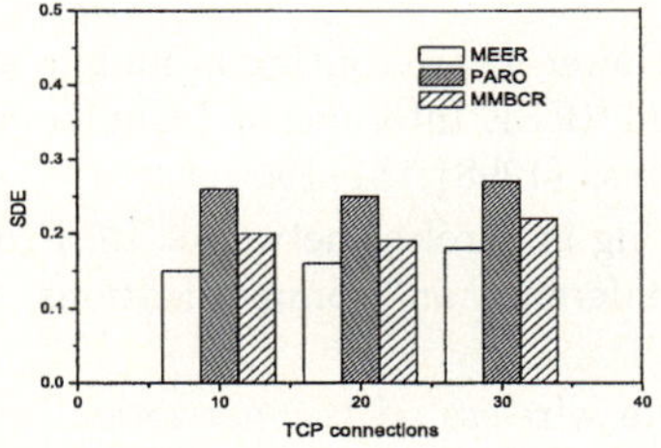

Fig. 3. The average residual energy of 30 nodes

from the path (0,2,4,1) and distribute them to the path (0,3,5,1). At 34s, the adjustment is completed and the final traffic distribution converges to a steady state. This result illustrates the convergence and stability of MEER algorithm.

The mobile scenario consists of 30 mobile nodes in a rectangular field, $1000m \times 1000m$, and the mobility model uses the random waypoint model. The max speed of nodes sets as 2m/s which is ordinary walking speed. Simulation results are all obtained from an average of 40 runs and each simulation run lasts for 500 seconds. In each run, FTP connections are generated and spread randomly over network. The number of connections is varied from 10 to 30 to change the traffic load in the network.

We compare MEER with PARO [12] and MMBCR [5] which are representatives of current energy-efficient routing protocols. We compare the performance in the case of 10, 20 and 30 TCP connections respectively. The comparison of average residual energy is shown in figure 3, while SDE (Standard Deviation

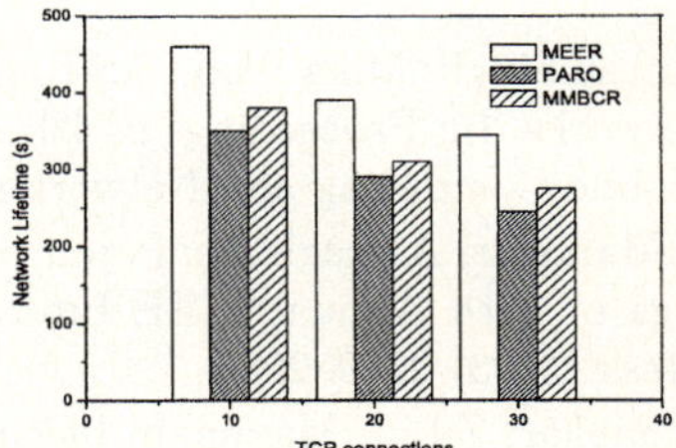

Fig. 4. SDE of 30 nodes

Fig. 5. The network lifetime

of Energy) comparison is shown in figure 4. The simulation results show that, MEER reach almost $20\% - 30\%$ higher residual energy and $10\% - 15\%$ lower SDE than PARO and MMBCR. It proves that MEER leads to less total energy consumption and more balanced energy residual. The network lifetime comparison is illustrated in figure 5. It is shown that MEER achieve $15\% - 30\%$ longer lifetime than PARO and MMBCR. The reason is simple: as the total energy consumption is more reduced and better balanced by MEER, the system lifetime increases consequently.

From all these simulation results, MEER shows better general performance in both extending network lifetime and minimizing total energy consumption than existing energy-efficient routing protocols.

4   Conclusions

To summarize this paper, we propose a Multipath Energy Efficient Routing (MEER) mechanism for ad hoc network. The objective of MEER is to extend network lifetime. Multipath routing can provide an even distribution of traffic load over the network and better reliability. In order to elongate the lifetime of ad hoc network, we address the problem of traffic allocation and energy efficient jointly. Between each source and destination node, there are multiple disjoint paths have been found by already known algorithms. A energy cost function is associated to each node, and we formulate the problem of traffic allocation as an optimization problem of minimizing the sum of node's energy cost in the network. A gradient projection method is used to resolve the optimization problem and it can be carried out in an asynchronous and distributed manner. MEER adjusts the traffic allocation based on the gradient projection algorithm to maximize the network lifetime. Some implementation issues of MEER are also discussed in this paper.

Based on simulations, we show that MEER works both for static networks and mobile ad hoc networks. We also compare MEER with some existing energy-efficient routing protocols. Simulation results show that, the total energy consumption in MEER is less than current energy-efficient routing protocols, also is more balanced. In general, MEER achieves much longer system lifetime.

References

1. Suresh Singh, Mike Woo, C.S.Raghavendra: Power-aware routing in mobile ad hoc networks. In: Proceedings of 4th Annual ACM/IEEE International Conference on Mobile Computing and Networking, ACM Press (1998) 181–190
2. R.Manohar, A.Scaglione: Power optimal routing in wireless networks. In: Proceedings of 38th Annual IEEE International Conference on Communications, IEEE Press (2003) 2979–2984
3. A.Sankar, Z.Liu: Maximum lifetime routing in wireless ad hoc networks. In: Proceedings of IEEE Conference on Computer Communications (INFOCOM) 2004, IEEE Press (2004) 1089-1097

4. V.Rodoplu, T.H.Meng: Minimum Energy Mobile Wireless Networks. IEEE Journal on Selected Areas in Communications **17** (1999) 1333–1344
5. C.K.Toh: Maximum Battery Life Routing to Support Ubiquitous Mobile Computing in Wireless Ad Hoc Networks. IEEE Communications Magazine **39** (2001) 138–147
6. J.W.Suurballe: Disjoint Paths in a Network. Networks **4** (1974) 125–145
7. J.W.Suurballe, R.E.Tarjan: A Quick Method for Finding Shortest Pairs of Disjoint Paths. Networks **14** (1984) 325–336
8. D.Bertsekas, R.Gallager: Data Networks. 2edn. Prentice-Hall, 1992
9. Anwar Elwalid, Cheng Jin, Steven Low, Indra Widjaja: MATE: MPLS Adaptive Traffic Engineering. In: Proceedings of IEEE Conference on Computer Communications (INFOCOM) 2001, IEEE Press (2001) 1300–1309
10. Frederick B., Carl B.: An energy-efficient routing protocol for wireless sensor networks with battery level uncertainty. In: Proceedings of IEEE Military Communications Conference (MILCOM) 2002, IEEE Press (2002) 489–494
11. The Network Simulator - ns-2, http://www.isi.edu/nsnam/ns/index.html
12. J.Gomez, A.T.Campbell, M.Naghshineh, C.Bisdikian: Conserving transmission power in wireless ad hoc network. In: Proceedings of 9th International Conference on Network Protocols, IEEE Press (2001) 24–34
13. D.Bertsekas, J.Tsitsiklis: Parallel and Distributed Computation. Prentice-Hall, 1989.

Performance of Service Location Protocols in MANET Based on Reactive Routing Protocols

Hyun-Gon Seo[1], Ki-Hyung Kim[2,*], Won-Do Jung[1], Jun-Sung Park[1],
Seung-Hwan Jo[3], Chang-Min Shin[3], Seung-Min Park[3], and Heung-Nam Kim[3]

[1] Department of Computer Engineering,
Yeungnam University, Gyungsan, Gyungbuk, Korea
`moses@yu.ac.kr`
[2] Division of Information and Computer Engineering,
Ajou University, Suwon, Korea
`kkim86@korea.com`
[3] Ubiquitous Computing Middleware Research Team,
Embedded S/W Technology Center, Basic Research Laboratory,
Electronics and Telecommunications Research Institute, Korea

Abstract. Automatic service discovery, the problem of discovering service providers by specifying desired properties of services, is an important and necessary component for collaboration in the ubiquitous computing environment such as MANET. This paper investigates the effects of the on-demand route discovery on the performance of service location protocols in MANET based on on-demand routing protocols. We first design a service discovery architecture in MANET based on AODV and evaluate both the distributed and centralized versions of the architecture by simulation. For evaluating the performance, we examine such performance metrics as service hit ratio, service discovery time, and control message overheads. The results show that the distributed scheme outperforms the centralized scheme in most simulation scenarios because of the on-demand routing overheads.

1   Introduction

Mobile ad hoc network(MANET) is an infrastructureless, self-configurable, multi-hop wireless network consisting of a set of mobile nodes. In contrast to traditional wired networks, it usually has several limitations such as low communication bandwidth, limited energy and battery, and node mobility. Typical MANET applications include situations in which a network infrastructure is not available but immediate deployment of a network is required, such as battlefields, outdoor assembly, or emergency rescue.

Discovery of services and other named resources, which allows devices to automatically discover network services with their attributes and advertise their

---

* Corresponding author: Ki-Hyung Kim (kkim86@korea.com).

P. Lorenz and P. Dini (Eds.): ICN 2005, LNCS 3421, pp. 234–241, 2005.

own capabilities to the rest of the network, is a major component for such self-configurable networks. Today, there exist several different industrial consortiums or organizations to standardize different service discovery protocols such as *Service Location Protocol (SLP)* [1] of IETF, Sun's *Jini*, Microsoft's *Universal Plug and Play (UPnP)*, IBM's *Salutation*, and Bluetooth's *Service Discovery Protocol (SDP)*.

Although some of the mentioned service location protocols claim to be usable over any type of networks, there are no enough performance results available for MANET, and the overhead incurred might be too great to be of practical use. The objective of this work is to address the topic of service location specifically from the performance perspective.

In this paper, we investigate the on-demand routing overheads on the performance of service discovery protocols in MANET based on on-demand routing protocols. We first design a service discovery architecture in MANET and evaluate both the distributed and centralized versions of the architecture by simulation. For evaluating the on-demand routing overheads on both the distributed and centralized service discovery protocols, we examine such performance metrics as service hit ratio, service search time, and control message overheads.

The novelty of this paper is two folds. First, the designed architecture considers both the centralized and distributed schemes and implements them in AODV[2] by adding minimal changes. There has been no detailed research on the implementation issues of the service discovery on specific MANETs. Second, this paper evaluates the on-demand routing overheads for service discovery by simulation. There has been no research on the routing overhead analysis with regard to the service discovery by the author's knowledge. For evaluating the routing overhead, we simulate the architecture by using ns-2[7].

The paper is organized as follows. Section 2 summaries related works on service discovery mechanisms in both wired and wireless networks. Section 3 designs a service discovery architecture for AODV-based MANET. Section 4 presents the performance evaluation results, and section 5 concludes the paper.

2   Related Works

In this section, we briefly describe service location protocols in general and in MANET.

2.1   Service Location Protocols

Service discovery protocols enable software components to find each other on a network and to determine if discovered components match their requirements. We briefly describe SLP as a representative service discovery protocol in the Internet and as a basis of the proposed service discovery architecture in MANET. SLP establishes a framework for service discovery using three types of agents that operate on behalf of network-based softwares; (*i*) a Service Agent(SA) advertises the location and attributes on behalf of services, (*ii*) a Directory Agent(DA) aggregates service information, and (*iii*) a User Agent(UA) performs service discovery by issuing a 'Service Request' (SrvRqst) on behalf of the client application, specifying the characteristics of the service which the client requires. The UA will receive a Service Reply (SrvRply) specifying the location of all services in the network which satisfy the request.

2.2   Service Discovery in MANET

There have been several research efforts on the service discovery architectures in MANET recently. A few service discovery protocols have been proposed for MANET, such as Lightweight service advertisement and discovery protocol [6] and cache mechanism based service discovery method [4]. W. Ma *et. al.* proposed a lightweight service advertisement and discovery protocol that piggybacks service advertisement and discovery information in ODMRP routing control packets [6]. S. Motegi *et. al.* exploited the effects of caching in service discovery, in which each node caches the configuration and service information during the lifetime [4].

The previous research works can broadly be classified into two ways: centralized and distributed. In the centralized service discovery approaches, there are a couple of DAs which advertise the existence of the DAs and collect service registrations/service requests from distributed UAs and SAs in MANET. Since it is a traditional way of the service discovery in fixed infrastructure networks like the Internet, there are many existing well-defined service discovery mechanisms such as SLP, Salutation, and Jini. L. Cheng proposed an approach of implementing the centralized service discovery architecture in MANET by employing ODMRP as the underlying multicast protocol[3].

As an alternative, the distributed service discovery approaches do not rely on particular nodes[4]. An example of this approach is the SLP without DAs, in which a UA multicasts (or broadcasts) a SrvRqst specifying the type of the desired service during a service discovery process. SAs that provide a service satisfying the specified service type respond with a SrvRply including the configuration information of the service. Performing such a service discovery can be an expensive process in MANET since it may cause a large number of generated SrvRqsts to be transmitted all over the network whenever UAs request service discoveries.

There are approaches to minimize spreading (or broadcasting) of SrvRqsts by introducing a cache mechanism[4][5]. Intermediate nodes which transfer a SrvRply cache the configuration information included in the SrvRply. When the intermediate nodes receive a SrvRqst, the nodes respond with a SrvRply including the cached configuration information and do not re-broadcast the received SrvRqst, thereby reducing the spreading of the messages.

3   A Service Discovery Architecture for MANET

In this section, we design a service discovery architecture for MANET based on on-demand routing protocols. The architecture integrates both the distributed and centralized schemes of SLP.

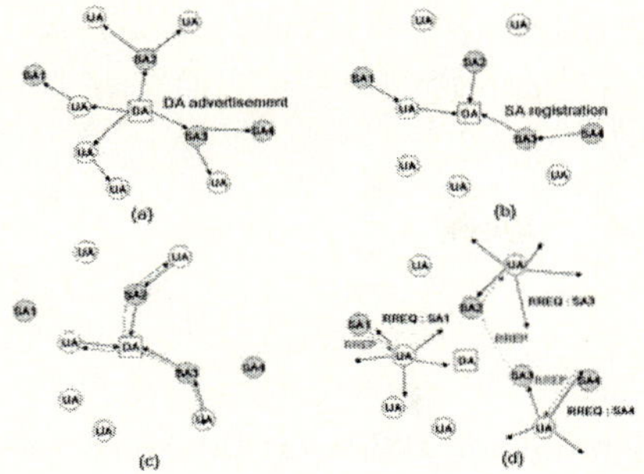

Fig. 1. Service discovery process in CLSP. (a) A DA advertises its existence periodically by multicasting/broadcasting. (b) SAs register services to the DA. (c) UAs query specific services to the DA (by using SrvRqst/SrvRply). (d) A UA discovers a route to the SA found in (c)

3.1   Centralized Service Location Scheme (CSLP)

The centralized version of the architecture ($CSLP$) consists of the following three entities as in SLP: User Agent(UA), Directory Agent(DA), and Service Agent(SA).

Fig. 1 shows the service discovery process of CSLP. A DA caches information about available services of registered SAs and advertises itself periodically by multicasting (or possibly broadcasting) an advertisement. In our architecture, we adopt broadcasting instead of multicasting to reduce the overhead of maintaining multicast information. Every mobile node (UA or SA) records the available DAs and maintains routing paths to the DAs in their routing tables. SAs listen to broadcasted advertisements of DAs.

Upon receiving an advertisement from a DA, a UA can choose its way of service discovery either in centralized or distributed schemes. If a UA chooses the centralized scheme for service discovery, it queries the DA about specific service by unicasting a SrvRqst instead of directly broadcasting it throughout a broadcast region; the DA then looks for the specified service in its cached information of registered services and responds to the UA by unicasting a SrvRply.

A SA registers periodically the location and attributes of the available services to the DA by unicasting a SrvReg and periodically renews its registration with the DA. The SrvReg message contains such information as the IP address and the service type.

When a UA needs a service, it assembles a SrvRqst and sends it to the DA to request the service's location. Then, the DA responds with a SrvRply that includes the address of the SAs of which services match against the UA's SrvRqst. Now, the UA can access one of the services by the returned addresses of the SAs. Notice that a UA knows only the address of the SAs, not the routing paths to the SAs. Therefore, a UA must establish routing paths to SAs. Fig. 1 (d) shows the route discovery process for accessing SAs. This process is inevitable in MANET based on on-demand routing protocols in which a route should be discovered before an actual data exchange occurs. Up to the author's knowledge,

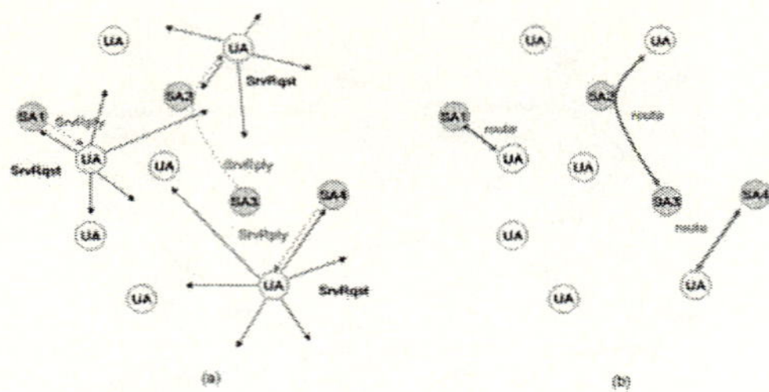

Fig. 2. Service discovery process in DSLP. (a) UA broadcasts a SrvRqst and receives SrvRplies from SAs. (b) Route establishment between UAs and SAs

its overhead has not been thoroughly investigated. Notice that wired networks and even MANETs with proactive routing protocols do not need this kind of route discovery steps. This extra route discovery step wastes precious bandwidth and battery resources and causes extra traffic in MANET. We will investigate this overhead in section 4.

3.2   Distributed Service Location Scheme (DSLP)

In DSLP, only two entities, the UA and the SA, are involved. When a UA attempts to use a service, and its configuration and existence is not yet known, it initiates a service discovery process. A UA broadcasts a SrvRqst specifying the type of the desired service which then floods a defined broadcast region. When a SrvRqst arrives at a SA within the broadcast region which provides service(s) satisfying the specified service type in the SrvRqst, it then responds to the UA with a SrvRply including the configuration information of the service.

A UA can receive multiple SrvRplies from several SAs, and it is up to the UA to select the best one. When an intermediate mobile node receives a SrvRply, it maintains the routing path toward the SA for later use of the service by the UA. Fig. 2 (a) shows the flooding of a SrvRqst and the reception of SrvRply(s) from SAs. Fig. 2 (b) shows a routing path between UAs and SAs.

4   Performance Evaluation

In this section, we use a packet-level simulation to explore the on-demand routing overheads in the service discovery process of the service discovery architecture in MANET.

4.1   Simulation Environment

In order to study the on-demand routing overheads on the performance of the two schemes of the architecture, we simulate them by using ns-2[7].

In the experiments, we generate a variety of MANET fields with different scenarios. The total number of nodes is 50, and they are randomly placed across a rectangular area of a 1000 x 800m square initially. Each node uses the IEEE

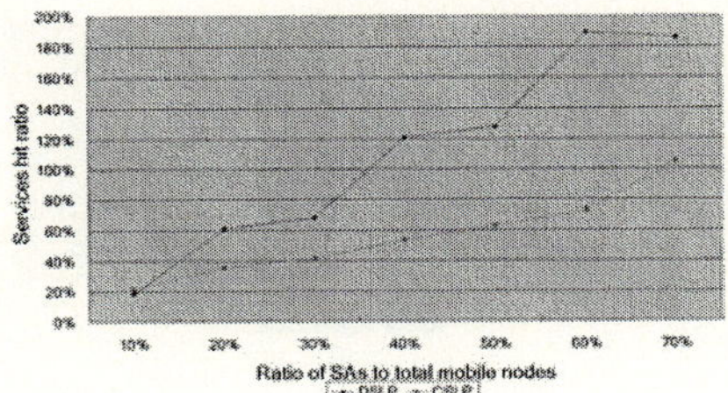

Fig. 3. Service hit ratio

802.11 standard MAC layer. The radio model is very similar to the first generation WaveLAN radios with nominal radio of 250m. All nodes move around in the area according to a mobility model (random waypoint). In this mobility model each node moves towards a random destination with a maximum speed of 10m/sec and pauses for certain time after reaching the destination before moving again.

Each node behaves as one of the three agents, *i.e.* a UA, a SA, or a DA. The number of SAs in the network is varied to see the impact of the service availability on the service discovery performance. For instance, if we select 5 nodes (out of 50 nodes) as SAs, the ratio of SAs becomes 10%. Total number of services specified during the simulation is 10, and each SA provides two distinct services. If the ratio of SAs is 20% (i.e. 10 SAs), total 20 services can be provided, and the same service can possibly be provided by different SAs. The choice of one service out of such multiple service instances is up to the favor of the UA. Every UA requests different service every 3 seconds during simulation. DAs advertise themselves every 5 seconds in CSLP.

We have conducted two kinds of simulations for both DSLP and CSLP. The first is the comparative performance of both schemes while varying the SA ratio from 10% (5 out of 50 nodes) to 70% (35 out of 50 nodes). The second is the performance while varying the pause time of nodes to study the impact of the node mobility.

4.2 Comparative Evaluation

We compare the performance of DSLP and CSLP while varying the SA ratio which is the ratio of the number of SAs to the total number of mobile nodes. The simulation result is the average of 5 identical traffic models, but different randomly generated mobility scenarios. Data points in each graph represent the mean of 5 scenarios.

Performance results are shown in Fig. 3 to 6. Fig. 3 shows the service hit ratio. The service hit ratio is directly proportional to the ratio of SAs. This is because the probability of whether the service requested by a UA exists in the services offered by SAs is proportional to the ratio of SAs. Remember that each SA offers two distinct services and the total number of service types is 10. Another interesting point is the hit ratio of DSLP becomes about 190% for the ratio of SAs 70%. This implies that a UA discovered two identical services offered by

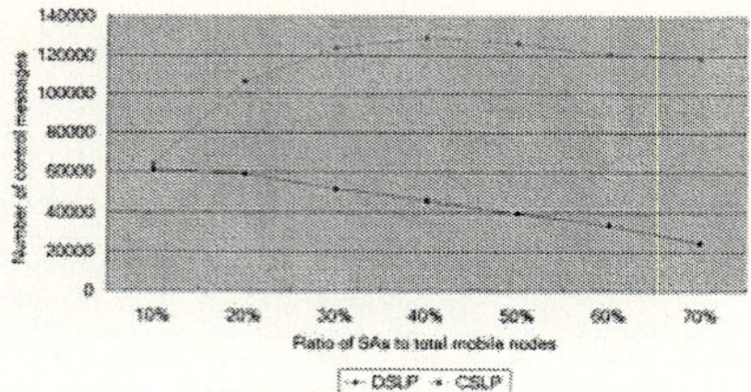

Fig. 4. Number of control messages

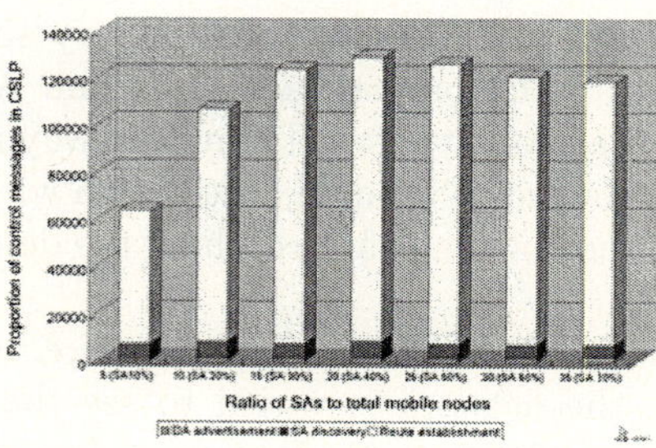

Fig. 5. Analysis of control messages in CSLP

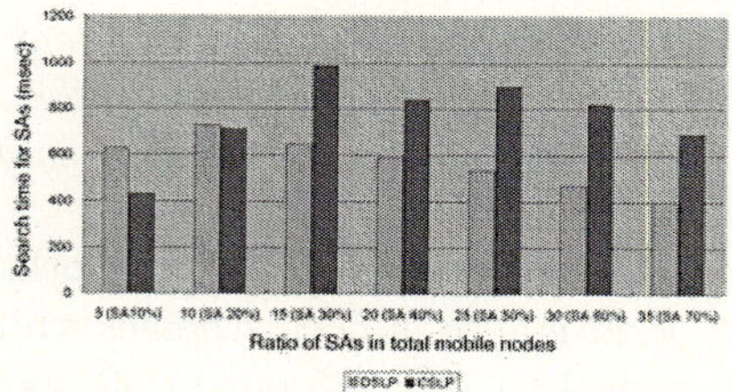

Fig. 6. Service discovery time for SAs

different SAs. DSLP shows almost twice performance than CSLP. The analysis of the performance difference can be possible by looking at Fig. 4 to 6. Fig. 4 shows the number of control messages generated for a service discovery. CSLP generates almost 5 times more control messages than DSLP for the ratio of SAs 70%. Fig. 5 shows the detailed proportional analysis of the control messages for CSLP. The control messages of CSLP consist of the following three types: DA's advertisement, SA's SrvReg and SrvRplies, and route establishment to the SA after finding the SA which provides the requested service. Most of the control messages are for the route establishment to SAs. That is, the DA just gives the address of the corresponding SA, not the routing path to the SA. A UA should find the route to the SA by itself after receiving a query reply from the DA.

Fig. 6 shows the service discovery time of both service schemes. DSLP shows better performance than CSLP as the number of SAs increases. This is because the two-phase transaction of CSLP takes more time than the one-phase transaction of DSLP.

5   Conclusion

This paper investigated the effects of the on-demand route discovery on the performance of service location protocols in MANETs based on on-demand routing protocols. We first design a service discovery architecture in MANETs based on AODV and evaluate both the distributed and centralized versions of the architecture by simulation. The simulation results show that the on-demand routing overhead occupies the largest portion in the overhead of CSLP. Since this overhead is too critical, the performance of CSLP has shown worse than DSLP for most of the simulation scenarios.

Acknowledgements. This research was supported by University IT Research Center Project.

References

1. E. Guttman, C. Perkins, J. Veizades, and M. Day, "Service Location Protocol(SLP), Version 2," *IETF RFC2608*, June 1999.
2. C. E. Perkins and E. Belding-Royer, "Ad Hoc On-Demand Distance Vector (AODV) Routing," IETF MANET Working Group, RFC3561, July 2003.
3. L. Cheng, "Service Advertisement and Discovery in Mobile Ad hoc Networks," in Conference on Computer Supported Cooperative Work (CSCW2002), 2002.
4. S. Motegi, K. Yoshihara, and H. Horiuchi, "Service discovery for wireless ad hoc networks," The 5th International Symposium on Wireless Personal Multimedia Communications, vol.1 , 27-30 Oct. 2002, pp. 232 - 236
5. P. Engelstad, Yan Zheng; T. Jonvik, and Do Van Thanh, "Service discovery and name resolution architectures for on-demand MANETs," 23rd International Conference on Distributed Computing Systems Workshops, 2003 pp. 736 - 742.
6. W. Ma, B. Wu, W. Zhang and L. Cheng, "Implementation of a lightweight service advertisement and discovery protocol for mobile ad hoc networks," Global Telecommunications Conference (GLOBECOM '03), IEEE, vol.2, 1-5 Dec. 2003. pp. 1023 - 1027.
7. K. Fall and K. Varadhan, Eds., "ns notes and documentation," 1999; available from http://www.isi.edu/nsnam/ns.

A New Scheme for Key Management in Ad Hoc Networks*

Guangsong Li and Wenbao Han

Department of Information Research, Information Engineering University,
P.O. Box 1001-745, Zhengzhou 450002, P.R. China
lggsss@sohu.com, wb.han@263.net

Abstract. Robust key management is one of the most crucial technologies for security of ad hoc networks. In this paper, a new scheme for key management is proposed using identity-based (ID-based) signcryption and threshold cryptography. It enables flexible and efficient key management while respecting the constraints of ad hoc networks. In our new scheme, public key certificates are not needed and every client uses its identity as its public key. It greatly decreases the need of computing and storage abilities of clients' terminals, as well as communication cost for system key management.

1 Introduction

An ad hoc network is a collection of autonomous nodes that communicate with each other by forming a multi-hop wireless network [1]. Ad hoc networks can be deployed very fast at a relatively low cost enabling communication when it is not possible or too expensive to deploy a support infrastructure. Significant applications of ad hoc networks include establishing survivable, efficient, dynamic communication for emergency operations, disaster relief efforts, and military networks that cannot rely on centralized and organized connectivity.

Providing security in ad hoc networks is challenging due to all the characteristics of these networks, such as the vulnerability of the wireless links, the limited physical protection of each node or the dynamically changing topology. Robust key management services are central to ensure privacy protection in ad hoc network settings. Traditionally, key management service is based on an on-line trusted entity called a certificate authority (CA) to issue public key certificate of every node. It is dangerous to set up a key management service using a single CA in an ad hoc network since the single CA will be the vulnerable point. How to set up a trusted key management service for an ad hoc network is a big issue.

One of the first notable schemes proposing a public key management service for ad hoc networks is by Zhou et al. [2], which is a partially distributed CA scheme. Threshold cryptography is used to provide robust and ubiquitous security support for ad hoc networks. At the initial time of network, a trusted third

---

* This work was supported by NSF of China with contract No.19971096 and No. 90104035.

P. Lorenz and P. Dini (Eds.): ICN 2005, LNCS 3421, pp. 242–249, 2005.

party is needed. Periodical share refreshing is also proposed to defend against powerful adversaries. Luo and others make an extension of [2] and they provide a fully distributed CA scheme. Under the assumption of no special nodes in the network, they propose that each node should hold a share of the private key of CA [3]. Hubaux et al. have proposed a self-organized public key infrastructure in [4] , which has similarity with PGP "web of trust" concept. Unlike the above publications, it does not require a trusted authority or any special nodes; instead, each node issues its own certificates to other nodes. Khalili et al. [5] provide a key distribution mechanism combining the use of ID-based and threshold cryptography. In their scheme, no trusted third party is needed at the initial time of network. However, the problem of clients' private keys update and share refreshing of system secret key is not addressed. Moreover, no mechanism handles a compromised node in [5].

In this paper, we propose a new approach that enables flexible and efficient key management while respecting the constraints of ad hoc networks. The rest of this paper is organized as follows. ID-based signcryption mechanism used in our scheme is briefly described in section 2. Section 3 details our new scheme for key management. A new communication protocol is provided in order to further reduce communication overhead of our scheme in section 4. Finally, section 5 concludes this work and presents future work.

2   ID-Based Signcryption Mechanism

In this section, we briefly describe the basic technology our scheme bases on. Let G_1 and G_2 be two cyclic groups of prime order p, writing the group action multiplicatively. Assume the discrete logarithm problem in these groups is hard.

Definition 1. An (efficiently computable,non-degenerate) map $e : G_1 \times G_1 \to G_2$ is called a bilinear pairing if, for all $x, y \in G_1$, and all $a, b \in Z$, we have $e[x^a, y^b] = e[x, y]^{ab}$.

Definition 2. The bilinear Diffie-Hellman problem for a bilinear pairing as above is described as follows: given $g, g^a, g^b, g^c \in G_1$, where g is a generator of G_1 and $a, b, c \in F_p^*$ are chosen at random, compute $e[g, g]^{abc}$.

It is believed that BDH problem is hard in some groups on certain supersingular elliptic curves or hyper-elliptic curves over finite fields. Boyen [7] gives an efficient ID-based signcryption scheme using the above hard problem, the most outstanding feature of which is that it accomplishes signature and encryption using the same group structures and parameters set. It is secure, compact, fast and practical, offers detachable signatures, and supports multi-recipient encryption with signature sharing for maximum scalability. In his scheme, no public key certificate is needed. Each user can select any string for its ID that acts as its public key.

A trusted private key generation authority, called PKG, generates system keys and clients' private keys. PKG picks two groups G_1 and G_2 of prime order

p, on which BDH problem is hard. Let $e : G_1 \times G_1 \to G_2$ be bilinear paring, and g be the generator of G_1. PKG picks $\sigma \in F_p^*$ randomly as system secret key and computes g^σ as system public key. σ is kept secret and g, e, g^σ, etc. are known to all clients. Let H be a one way function. Let $ID \in \{0,1\}^*$ be a client's identity, then the private key of the client is computed as follows: $i_{ID} = H(ID), d_{ID} = (i_{ID})^\sigma$, where d_{ID} is the private key of the client. For more details of the signcryption algorithm we refer the reader to [7].

3   A New Scheme for Key Management

In this section, we give a new scheme for key management in ad hoc networks using ID-based signcryption mechanism in [7] and threshold cryptography. This work considers an ad hoc network, where mobile nodes communicate with one another via the bandwidth-constrained, error-prone, and insecure wireless channel. We also make the following assumptions.

(1) There exists an off-line trusted authority.
(2) Each node has a unique identity.
(3) Nodes in the network have different abilities.
(4) Each node is equipped with some local detection mechanism to identify misbehaving nodes among its one-hop neighborhood, e.g., those proposed in [8].

3.1   Key Generation

At the initial time of the network, an off-line trusted authority (called PKG) generates system secret key σ and public key g^σ. Every client must contact PKG and register its identity before entering the network. The identity of a client can be any string selected according to its taste. After PKG authenticates identity of the client, PKG generates private key of the client using system secret key and the client's identity. Then the private key of the client and the system public key are given to the client securely. We rely on the off-line trusted PKG to form a trust anchor, which improves security level of the network. PKG chooses n nodes as on-line server nodes according to abilities of the nodes. Nodes with strong computation ability and large transmission range are selected. Then system secret key σ is shared by the n server nodes using an (n, k) threshold sharing scheme. Our design makes extensive use of the polynomial secret sharing due to Shamir. Any k nodes of the server nodes can act jointly like a virtual PKG which provides private key update service for clients on-line.

Signature algorithm can be used independently in the signcryption mechanism of [7]. In some cases, only authentication of a client's identity is needed, so a challenge-response process will do. When client A wants to send some messages secretly to client B, there are two methods for client A and B to establish a secret session key.

Key Agreement

Let δ be the generator of F_p^* . A selects $x \in F_p^*$ randomly and computes δ^x; then it signs δ^x using its private key and sends δ^x and the signature to B. When

B receives the messages from A, it verifies the correctness of the signature. If the signature is correct, B selects $y \in F_p^*$ randomly, computes δ^y, signs δ^y using its private key, and sends δ^y along with the signature to A. A also verifies the signature of B. If both the signatures are correct, A and B can ensure the identity of each other. Then A and B compute $(\delta^y)^x$ and $(\delta^x)^y$ as the shared session key respectively.

Session Key Generated by One Part

A generates a random session key, signcrypts the session key using its private key and the identity of B, then sends the ciphertext to B. B can decrypt the session key and verify the signature.

3.2   Localized Trust Model and Malicious Nodes List

Like [3], we present a localized trust model to deal with malicious nodes. The only difference is that we use malicious nodes list here, not certificate revocation list. We assume that k is set as a globally fixed parameter that is honored by each entity in the system. As long as k local trusted nodes think a node malicious, the node is not trusted in the whole network. In our assumption, each node is equipped with some local detection mechanism to identify misbehaving nodes among its one-hop neighborhood. Each node A maintains a table of malicious nodes, called malicious nodes list (MNL). Every entry of the MNL is composed of a node ID and a list of the node's accusers. If a node's accuser list contains less than k legitimate accusers, the node is marked as "suspect". Otherwise, the node is determined by A to be misbehaving or broken and marked as "convicted". We choose the threshold that convicts a node as k to ensure a legitimate node not be convicted by malicious accusations from an adversary. In two scenarios a node is marked "convicted". When by direct monitoring A determines one of its neighboring nodes to be misbehaving or broken, A puts the node into its MNL and directly marks the node "convicted". In this scenario A also floods a signed accusation against the node. The other scenario happens when A receives an accusation against some node. A firstly checks if the accuser is a convicted node in its MNL. If it is, the accusation is concluded to be malicious and dropped. If not, A updates its MNL entry of the accused node by adding the accuser into the node's accuser list. The accused node will be marked "convicted" if the number of accusers reaches k. When a node is convicted, A deletes the node from all accuser lists. A convicted node will be marked "suspect" if its number of accusers drops below k.

3.3   Update of a Client's Private Key

It is not believed to be secure that a private key is used for a long time. We present a secure and efficient mechanism for a client's private key update using signcryption scheme in [7].

It is suggested that the private key of a client should be updated after having been used for some time t. The lifetime of the network is divided into intervals of length t. We propose that clients use temporary IDs with time stamp as suffix.

For example, at the jth time interval, client A uses $ID_A\|j$ as its temporary ID, where $\|$ represents concatenation of strings. That is to say, its public key is $ID_A\|j$ at this time interval. Then, its private key is $d_{Aj}, d_{Aj} = (i_{Aj})^\sigma$, in which $i_{Aj} = H(ID_A\|j)$. Suppose that n server nodes selected by PKG are denoted respectively as $v_s, s = 1, 2, \cdots, n$. Each of them has a share σ_s of system secret key $\sigma, s = 1, 2, \cdots, n$. Any k nodes of these n sever nodes can act as a virtual PKG jointly to compute new private key for a client according to its new temporary ID. Before time interval j expires, client A must contact k server nodes to acquire new private key for time interval $j + 1$. A message of PREQ (Private key update REQuest) signed using its current private key is sent to k sever nodes by A. When a server node v_s receives PREQ, it verifies the correctness of the signature. If the signature matches the identity the node claims, v_s computes a partial private key of the client for time interval $j + 1$ using $\sigma_s, d^s_{A,j+1} = (i_{A,j+1})^{\sigma_s}$. Then v_s signcrypts $d^s_{A,j+1}$ and sends it in a PREP (Private key update REPly) message to A. After A decrypts and verifies the message, it needs to make a further check. It is possible that a malicious server node v_s may return a false partial private key generated without using its share. In order to check malicious server nodes, some measures must be taken. At the initial time of the network, PKG publishes a piece of verification information consisting of g^{σ_s} for each server node v_s. To check the validity of partial key it receives from v_s, A needs only to check whether the equation $e[i_{A,j+1}, g^{\sigma_s}] = e[d^s_{A,j+1}, g]$ holds. If it is true, $d^s_{A,j+1}$ is a valid partial value for A's $(j+1)$th time interval. Otherwise, A floods an accusation againgst v_s signed with its current private key. The accusation message should include the time that A makes the accusation. If A receives k valid partial values of private key for $(j + 1)$th time interval, it can compute its new private key using Lagrange interpolation.

3.4   Share Refreshing of Server Nodes

Proactive secret sharing is required to protect against attackers that might compromise k or more server nodes if there is enough time [2].

Like section 3.3, the lifetime of the network is divided up in time periods, where each time period consists of two phases, the operational phase and the share update phase. During the operational phase, all nodes including server nodes renew their private keys. During the share update phase, all server nodes compute new shares from old ones in collaboration. Thus, the adversary is challenged to compromise k server nodes between periodic refreshing. Share refreshing relies on the homomorphic property of shares of system secret key. Given n server nodes. Assuming all server nodes are correct, share refreshing proceeds as follows. Each server node v_s randomly generates $(\sigma_1^s, \sigma_2^s, \cdots, \sigma_n^s)$, an (n, k) sharing of 0. Each σ_l^s is called subshare. Then, every subshare σ_l^s is distributed to server node v_l through a secure link. When server node v_l gets the subshares $\sigma_l^1, \sigma_l^2, \cdots, \sigma_l^n$, it can compute a new share from these subshares and its old share $(\sigma_l' = \sigma_l + \sum_{s=1}^n \sigma_l^s)$. Share refreshing in [2] needs a secure channel for delivering subshares, Zhou et al. do not provide the implementation of this secure channel. In our scheme, signcryption exactly provides a way for secure transmission.

The above process may come across malicious server nodes' deceit. We can use verifiable sharing scheme in [9] to detect malicious server nodes. After isolating the malicious server nodes, share refreshing can be accomplished according to the above process.

4   A New Communication Protocol for Key Management

The communication pattern for a client updates its private key is one-to-many-to-one, which is called manycast in [6]. For manycast, Yi et al. propose β-uniscast to improve communication performance and successful ratio of update request. This method can also be used to improve our scheme. Moreover, we can give a further improvement of communication performance. In this section, we propose a new communication protocol using modified multicast protocol and key proxy technology. Our main idea is that all sever nodes form and maintain a few multicast groups [10] according to location. The PREQ packet is multicast to server nodes group in order to reduce traffic overhead in the network. When server nodes refresh their shares of the system secret key, they multicast their update information among the groups of server nodes they belong to. We first discuss the issues when sever nodes form one group as our basic scheme.

4.1   Basic Scheme

All server nodes in the network form a multicast group. The group routing structure is maintained through the lifetime of the network. Generally, a client that wishes private key update sends a multicast message of PREQ to the server nodes group on demand. Here, the client need not join the group and it only sends a multicast message to the group, which is different from most multicast protocols of ad hoc networks. So we introduce a key proxy for it. First, it floods its RREQ (Routing REQuset) to find a route to the server nodes group. When it receives RREPs (Routing REPly) from server nodes, it selects a server node, say u, which has the shortest path to itself as its key proxy. The routing information to the node u is stored. When it wants to update its private key later, it sends its PREQ to u and u multicasts the PREQ to all server nodes. In order to reduce communication cost and latency, when it searches the route to the group of server nodes, it floods to the network its PREQ piggybacked in the RREQ packet. Each server node having received the request computes a partial private key $d^s_{A,j+1}$ for the client using its share of system secret key as section 3.3, then it sends a PREP message containing the partial private key $d^s_{A,j+1}$ to the key proxy u. u waits for some time t and checks whether it has received k or more partial private keys from different server nodes. If it is not true, u will multicast the same PREQ again until the number of partial private keys received is larger than k. Otherwise, u will return to the client all these partial private keys in a single PREP packet. Then the client computes a valid private key for time interval $j + 1$. As long as routing information to its key proxy in its cache is still usable, the client unicasts its PREQ to the key proxy when it needs to update its private key. Otherwise, the client will flood its RREQ to find a new key proxy. Also, at this time, RREQ is integrated with PREQ in order to reduce traffic overhead and latency.

In this case, the success ratio is evidently very high since all server nodes will return a PREP theoretically. But partial private key generating is an operation costing much energy and time and the client needs only k PREPs of n PREPs from the server nodes. So the basic scheme results in unnecessary communication and energy cost.

4.2   Two Variations of Basic Scheme

A natural improvement of the basic scheme is the use of TTL (Time To Live) field in PREQ packets. When PREQ is unicast to it, the key proxy set TTL field of the PREQ packet according to the status of the network, such as maximum speed of nodes, and density of server nodes, or experiences in the past. When PREQ is forwarded one time, its TTL decreases 1. PREQ is only forwarded when its TTL is larger than 0. After a waiting time t, if the key proxy receives k or more PREPs from server nodes, it returns to the client all partial private keys in a single PREP packet. Then the client can complete private key update successfully. Otherwise, the key proxy sets larger TTL to the same PREQ packet and multicasts it again. If a server node has already answered this request, it only forwards this PREQ and decreases its TTL by 1. If it hasn't answered the request, it will return its PREP.

It needs more control packets and generates relatively heavier traffic overhead to maintain a large proactive multicast routing structure of all server nodes. Another reasonable improvement is that server nodes form multiple groups according to their current positions. For example, multiple areas are divided with unique identification. Then server nodes in the same area form and maintain a group. We should ensure that each group has $\beta = k + \alpha$ server nodes, where α larger than 0 is determined by local network status. And we do not strictly require that a server node belong to a certain group. It may belong to one or two multicast groups as needed. And it may leave current group and join a new group when it moves away. A client wanting private key update service only needs to find a routing to a key proxy on demand in its area.

4.3   Share Refreshing of Server Nodes Using Multicast

To fulfill the above share refreshing of server nodes in section 3.4, our scheme is as follows. We first discuss the case that all server nodes are in one multicast group. Each server node v_s randomly generates $(\sigma_1^s, \sigma_2^s, \cdots, \sigma_n^s)$, an (n, k) sharing of 0. Then it signcrypts $\sigma_l^s, l \neq s$, with its private key and server node v_l's public key. The ciphertext is denoted as c_l. Share refreshing information of server node v_s consists of a vector $(c_1, \cdots, c_{s-1}, 0, c_{s+1}, \cdots, c_n)$. Refreshing information is multicast to the server nodes group. Every server node v_l receiving refreshing information from node v_s can only decrypt ciphertext c_l to recover σ_l^s and learn nothing about other subshare $\sigma_r^s, r \neq l$. When server node v_l gets the subshares $\sigma_l^1, \sigma_l^2, \cdots, \sigma_l^n$, it can update its share of the system secret key. If share refreshing information is sent without using multicast, each server node v_s needs to find routing to other $n-1$ server nodes and then unicasts c_l to v_l respectively, where $l \neq s$. Altogether there will be $n(n-1)$ flooding for routing discovery and $n(n-1)$ unicast for share refreshing. Evidently, it is of heavy traffic.

As for the case server nodes maintaining multiple groups, share refreshing mimics the process above. The only difference is that each group refreshes its server node members' shares separately. Different versions of shares of the system secret key will appear in different server nodes group. As long as a client contacts enough server nodes that hold shares of the same version, it will update its private key successfully.

5   Conclusions and Future Work

Ad hoc networks are new paradigm in networking technologies. Key management is one of the most crucial technologies for security of ad hoc networks. This paper presents a new approach for key management using ID-based signcryption and threshold cryptography. Moreover, heuristic arguments of a new communication protocol have been provided for our key management scheme. We will further analyze the performance of the new scheme using NS-2 network simulator. Very interesting aspect would be to find more efficient communication protocols for threshold cryptography in ad hoc networks.

References

1. Perkins C.: Ad Hoc Networking. Addison-Wesley Newyork. (2001)
2. Zhou L., Haas Z.J.: Securing Ad hoc Networks. IEEE Networks, vol. 13(6). (1999) 24-30
3. Luo H., Zerfos P., Kong J., Lu S., Zhang L.: Self-securing Ad Hoc Wireless Networks. 7th IEEE Symp. on Computers and Communications. (2002) 567-574
4. Hubaux J. P., Buttyan L., Capkun S.: Self-organized Public-Key Management for Mobile Ad hoc Networks. IEEE Trans. on Mobile Computing, vol. 2(1). (2003) 52-64
5. Khalili A., Katz J., Arbaugh W. A.: Towards Secure Key Distribution in Truly Ad Hoc Networks. Proceedings of IEEE Workshop on Security and Assurance in Ad hoc Networks. (2003)
6. Yi S., Kravets R.: MOCA: Mobile Certificate Authority for Wireless Ad Hoc Networks. Available: http://mobius.cs.uiuc.edu/ seungyi.
7. Boyen X.: Multipurpose Identity-Based Signcryption: A Swiss Army Knife for Identity-Based Cryptography. Proceedings of Crypto'03, Lecture Notes in Computer Science, vol. 2729, Springer-Verlag, Berlin Heidelberg New York (2003) 383-399
8. Zhang Y., Lee W., Huang Y.: Intrusion Detection Techniques for Mobile Wireless Networks. ACM/Kluwer Wireless Networks Journal. Vol. 9(5). (2003) 545-556
9. Pedersen T.: Non-interactive and Information-theoretic Secure Verifiable Secret Sharing. Proceedings of Crypto'91, Lecture Notes in Computer Science, vol. 576, Springer-Verlag, Berlin Heidelberg New York (1992) 129-140
10. Royer E. M., Perkins C.: Multicast Operation of the Ad Hoc On-Demand Distance Vector Routing Protocol. ACM MOBICOM, (1999) 207-218

Robust TCP (TCP-R) with Explicit Packet Drop Notification (EPDN) for Satellite Networks

Arjuna Sathiaseelan and Tomasz Radzik

Department of Computer Science, King's College London,
Strand, London WC2R 2LS
Tel: +44 20 7848 2841
{arjuna, radzik}@dcs.kcl.ac.uk

Abstract. Numerous studies have shown that packet reordering is common, especially in satellite networks. Reordering of packets decreases the TCP performance of a network, mainly because it leads to overestimation of the congestion of the network. We consider satellite networks and analyze the performance of such networks when reordering of packets occurs. We propose a solution that could significantly improve the performance of the network when reordering of packets occurs in the satellite network. We report results of our simulation experiments, which support this claim. Our solution is based on enabling the senders to distinguish between dropped packets and reordered packets.

1   Introduction

A network path that suffers from persistent packet reordering will have severe performance degradation. Satellite links have high RTTs, typically in the order of several hundred milliseconds. Inorder to keep the pipe full, link-layer retransmission protocols send subsequent packets while awaiting an *ack* (acknowledgement)or *nak* (negative acknowledgement) for a previously sent packet. Here, a link-layer retransmission is reordered by the number of packets that were sent between the original transmission of that packet and the return of the *ack* or *nak* [3].

TCP uses cumulative acknowledgements it receives from the receiver to determine which packets have been successfully received at the receiver and retransmits the packets it believes to have been lost. For example, assume that four segments A, B, C and D are transmitted through the network from a sender to a receiver. When segments A and B reach the receiver, it transmits back to the sender an *ack* for B which summarizes that both segments A and B have been received. Suppose segments C and D have been reordered in the network. When segment D arrives at the receiver, it sends the *ack* for the last in-order segment received which in our case is B. Only when segment C arrives, the *ack* for the last in-order segment (segment D) is transmitted.

TCP has two basic methods of finding out that a segment has been lost.

P. Lorenz and P. Dini (Eds.): ICN 2005, LNCS 3421, pp. 250–257, 2005.

Retransmission timer
If an acknowledgement for a data segment does not arrive at the sender at a certain amount of time, the retransmission timer expires and the data segment is assumed to be lost and is immediately retransmitted [8].

Fast Retransmit
When a TCP sender receives three *dupacks* (duplicate acknowledgements) for a data segment X, it assumes that the data segment Y which was immediately following X has been lost, so it resends segment Y without waiting for the retransmission timer to expire [5]. Fast Retransmit uses a parameter called *dupthresh* which is fixed at three *dupacks* to conclude whether the network has dropped a packet.

Reordering of packets during transmission through the network has several implications on the TCP performance. The following implications are pointed out in [1]:

1. When a network path reorders data segments, it may cause the TCP receiver to send three successive *dupacks*, triggering the Fast Retransmit procedure at the TCP sender for data segments that may not necessarily be lost. Unnecessary retransmission of data segments means that some of the bandwidth is wasted.
2. The TCP transport protocol assumes congestion in the network when it assumes that a packet is dropped at the gateway. Thus when a TCP sender receives three successive *dupacks*, the TCP assumes that a packet has been lost and that this loss is an indication of network congestion and enters either slow start or the congestion avoidance phase and backs off its retransmission timer (Karn's algorithm) [5]. Satellite networks have high propagation delay and unnecessary reduction of congestion window leads to a poor throughput performance as it could take several round trip times to achieve the maximum window size.
3. TCP calculates the retransmission timeout (RTO) by sampling and averaging the round trip time (RTT) i.e. the time taken to send a data packet and receive a corresponding acknowledgement for the data packet. When a packet gets reordered in the network, the estimated round trip time is quite large which could falsely inflate the RTO estimation. This has a negative impact on the TCP performance, since if a packet was originally dropped then the TCP has to wait longer to retransmit the dropped packet.

We propose extending the TCP protocol to enable TCP senders to recognize whether a received *dupack* means that a packet has been dropped or reordered or corrupted. The extended protocol uses the Explicit Packet Drop Notification (EPDN) mechanism proposed by us in [10] to infer which packets have been dropped and uses this information to take an appropriate action. We call the resulting protocol Robust TCP (TCP-R).

In Section 2 presents the previous work related to our study. In Section 3 we present the details of our proposed solution. In Sections 4 and 5, we describe and discuss the evaluations of our solution via simulations. We conclude the paper with a summary of our work and a short discussion of the further research in Section 6.

2   Related Work

Several methods to detect the needless retransmission due to the reordering of packets have been proposed:

The DSACK option in TCP, allows the TCP receiver to report to the sender when duplicate segments arrive at the receiver's end. Using this information, the sender can determine when a retransmission is spurious [4]. Also in their proposal, they propose storing the current congestion window before reducing the congestion window upon detection of a packet loss. Upon an arrival of a DSACK, the TCP sender can find out whether the retransmission was spurious or not. If the retransmission was spurious, then the slow start threshold is set to the previous congestion window. Their proposal does not specify any mechanisms to proactively detect reordering of packets.

In [1], the authors use the DSACK information to detect whether the retransmission is spurious and propose various techniques to increase the value of *dupthresh* value. The main drawback in this proposal is that if the packets had in fact been dropped, having an increased value of *dupthresh* would not allow the dropped packets to be retransmitted quickly and the *dupthresh* value would be decreased to three *dupacks* upon a timeout.

In [11], the authors propose mechanisms to detect and recover from false retransmits using the DSACK information. They also extend TCP to proactively avoid false retransmits by adaptively varying *dupthresh*. The main drawback in their proposal is that, if a packet had originally been dropped, the packet loss could only be detected after a retransmission time out. Our proposal has a major advantage of being able to distinguish a packet drop event from a packet reorder event and perform actions based on the detection.

In [9], we proposed a method to enable the TCP senders to distinguish whether a packet has been dropped or reordered in the network by using the gateways to inform the 'sender' about the dropped packets. The gateway had to maintain information about all dropped packets for a flow, requiring considerable amount of dedicated memory at each gate. Moreover this method was proposed for networks that strictly follow symmetric routing and did not consider the case of asymmetric routing.

In [10], we proposed a novel method to enable the TCP senders to distinguish whether a packet has been dropped or reordered in the network by using the gateways to inform the 'receiver' about the dropped packets. This mechanism was called the Explicit Packet Drop Notification (EPDN). The gateway had to maintain minimal information about the dropped packets for a flow, requiring lesser amount of dedicated memory at each gate. The receiver then uses this information to inform the sender about which packets have been reordered by setting a 'reorder' bit. If the packets had been dropped in the network, the

TCP sender retransmits the lost packets after waiting for 3 *dupacks*. If the packets are assumed to be reordered in the network, the TCP sender waits for '3+k' *dupacks* ($k \geq 1$) before retransmitting the packets. RN-TCP supports asymmetric routing.

The methods mentioned above improve the performance of TCP in the case of packet reordering in wired networks. They do not test their proposed protocols on satellite networks. In satellite networks, the packet loss is mainly due to corruption of packets. There have been many proposals for extending TCP to improve the performance when the losses are mainly due to corruption.The proposals mentioned in the related work section and our previous proposals namely RD-TCP and RN-TCP do not consider error prone networks. The EPDN mechanisms inform the sender/receiver about dropped packets. In error prone networks, the packets could have been dropped by the data link layer due to corruption. If a packet had been actually dropped due to corruption and if the round trip delay is large, having an increased value of *dupthresh* requires a RTO to detect the packet loss. Thus increasing the *dupthresh* value to more than three when a packet has been assumed not to be dropped may have serious implications in the performance if the packet had actually been dropped due to corruption.

3   TCP-R: Robust TCP

We propose extending the TCP protocol to enable TCP senders to recognize whether a received *dupack* means that a packet has been dropped, reordered or corrupted. The extended protocol uses the Explicit Packet Drop Notification (EPDN) mechanism proposed by us in [10] to infer which packets have been dropped. The TCP receiver maintains two lists: the reorder/corrupted list and the drop list. The elements of these lists are packet sequence numbers. When a packet is received at the receiver, the receiver uses the sequence number of the current packet, the maximum and minimum dropped information in the packet and the sequence number of the last received packet in the buffer queue to detect which packets have been dropped or reordered/corrupted and inserts those sequence numbers into the drop list or the reorder/corrupt list accordingly. The receiver algorithm and implementation details are provided in [10]. The TCP receiver uses the information present in the lists to decide whether the gap between the out of order packets are caused by reordering or not and informs the TCP sender about its assumption. (Informing the sender is done by setting the 'drop-negative' bit in corresponding *dupacks*.) If the packets had been dropped in the network, the TCP sender retransmits the lost packets after waiting for 3 *dupacks* (fast retransmit) and reduces the congestion window by half (fast recovery). If the packets are assumed to be reordered in the network, the TCP sender retransmits the packet after receiving three *dupacks* with the 'drop-negative' bit set and enters our modified fast recovery mechanism where the procedure of reducing the slow start threshold (*ssthresh*) and the congestion window (*cwnd*) are bypassed i.e. we do not reduce the slow start threshold and the congestion window. We term our new version of TCP as TCP-R (Robust TCP).

3.1   Sender Side: Implementation Details

Sender Side Algorithm: Processing the Ack Packets. When an acknowledgement is received, the TCP sender does the following,

- If none of the three *dupacks* received have their 'drop-negative' bit set, then the TCP sender assumes that the packet has been dropped. So the sender retransmits the lost packet after receiving three dupacks and enters fast recovery.
- If all the three *dupacks* received have their 'drop-negative' bit set, then the TCP sender assumes that the packet has been reordered in the network and retransmits the packet immediately. The procedure of reducing the slow start threshold (*ssthresh*) and the congestion window (*cwnd*) in the fast recovery procedure are bypassed.

4   Simulation Environment

We use the network simulator ns-2 [6] to test our proposed solution. The nodes are connected to the routers via 10Mbps Ethernet having a delay of 1ms. The routers are connected to a satellite transmitter and receiver via a satellite repeater. Our simulations use 1000 byte segments. We used the drop-tail queuing strategy with a queue size of 100 segments. The experiments were conducted using a single long lived TCP flow traversing the network topology. The maximum window size of the TCP flow was 100 segments.The TCP flow lasts 1000 seconds. Reordering was introduced by delaying a percentage of packets traversing the link. To introduce delays, we used a mean of 200 ms and standard deviation of 80 ms.

5   Results

5.1   Throughput: Varying Packet Delay Rate

In this section, we vary the percentage of packet delays from 1% to 10% to introduce a wide range of packet reordering events and compare the throughput performance of the simulated network using SACK, DSACK-R, RR-TCP and TCP-R. As shown in the Figure 1, the throughput performance of TCP-R is much better compared to the throughput of other protocols. For e.g. when the link experiences 5% of packet delays, TCP-R performs almost ten times more than SACK and six times more than DSACK-R. TCP-R offers a two fold throughput improvement over RR-TCP. Similarly when the link experiences 10% of packet delays, TCP-R offers a three fold throughput improvement over RR-TCP and also performs almost ten times more than DSACK-R and twelve times more than SACK.

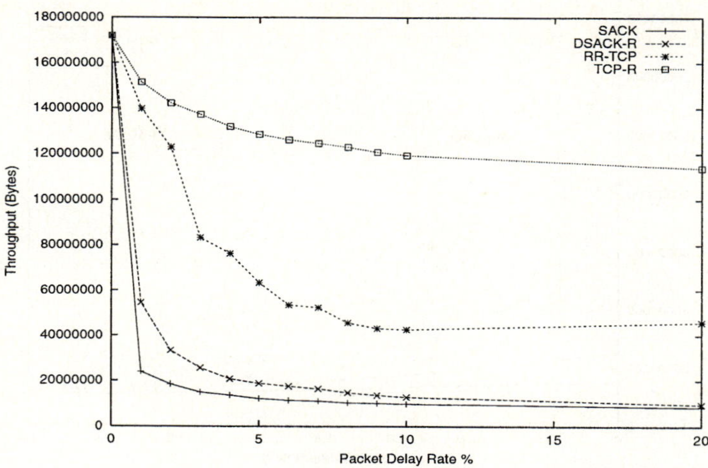

Fig. 1. Comparison of throughput performance of the network using SACK, DSACK-R, RR-TCP and TCP-R as a function of fraction of delayed packets

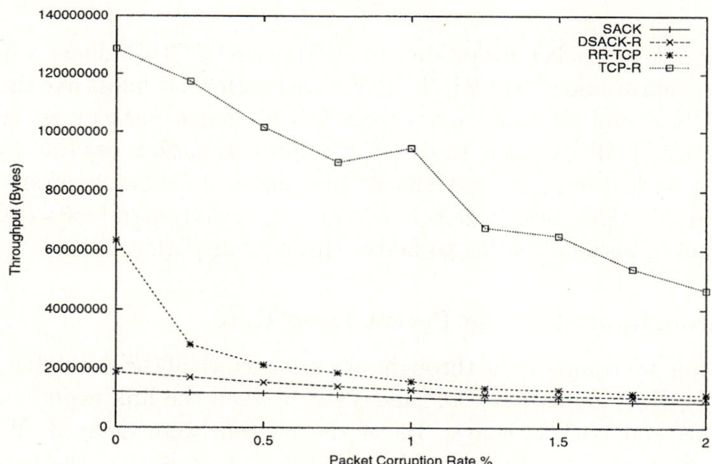

Fig. 2. Comparison of throughput performance of the network using SACK, DSACK-R, RR-TCP and TCP-R as a function of fraction of corrupted packets. 5% of packets delayed

5.2   Throughput: Varying Packet Corruption Rate

In this section, we compare the throughput performance of the simulated network using SACK, DSACK-R, RR-TCP and TCP-R when the link experiences both packet corruption and packet delays. 5% of the packets were delayed. We varied the packet corruption rate from 0% to 2%.

Figure 2, reveals that the throughput performance of TCP-R is much better compared to the throughput of SACK, DSACK-R and RR-TCP. When the link

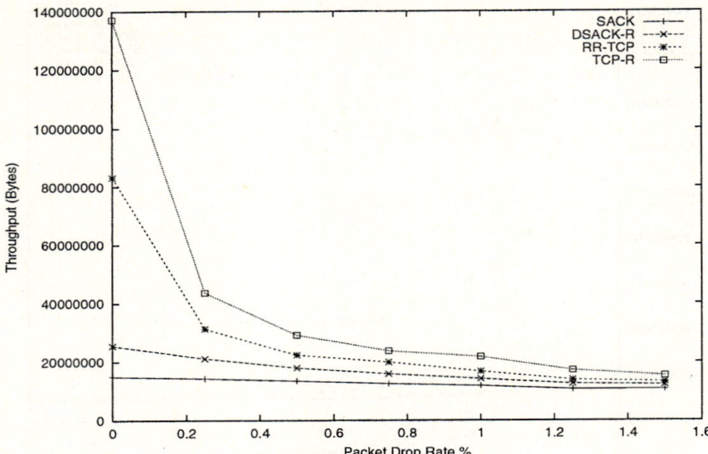

Fig. 3. Comparison of throughput performance of the network using SACK, DSACK-R, RR-TCP and TCP-R as a function of fraction of dropped packets. 3% of packets delayed

experiences 2% of packet drops due to corruption, TCP-R offers a four fold throughput improvement over RR-TCP and also performs almost five times more than DSACK-R and six times more than SACK. When packets get corrupted in the network, TCP-R is able to detect the cause of packet loss (i.e. the loss is not due to a packet drop) and retransmit the packet immediately after receiving three DUPACKs. Moreover it does not reduce the cwnd upon losses caused due to corruption of packets,leading to better throughput performance.

5.3   Throughput: Varying Packet Drop Rate

In this section, we compare the throughput performance of the simulated network using SACK, DSACK-R, RR-TCP and TCP-R when the link experiences both packet drops and packet delays. 3% of the packets were delayed. We varied the packet drop rate from 0% to 1.5%. Figure 3, reveals that the throughput performance of TCP-R is much better compared to the throughput of SACK, DSACK-R and RR-TCP. When packet drops occur, the throughput of any TCP variant would reduce drastically even when there is no reordering in the network. It is to be noted that the reduction in throughput of TCP-R is only due to packet drops and not due to false fast retransmissions caused by packet reordering.

6   Conclusions and Future Work

In this paper, we proposed a solution that allows the TCP sender to distinguish whether a packet has been lost or reordered in the satellite network and perform actions accordingly. This was done by maintaining information about dropped packets in the gateway and using this information to notify the sender, whether the packet has been dropped or reordered or corrupted in the gateway.

We believe the gateway could be modified to send the dropped information in an ICMP message to the sender. This requires further study and testing. Further simulations and testing needs to be carried out to find the efficiency of the protocol when there is an incremental deployment i.e. when there are some routers in a network which have not been upgraded to use our mechanism. Moreover, the simulated results presented in this paper needs verification in the real satellite network.

References

1. Blanton, E., Allman, A.: On Making TCP More Robust to Packet Reordering. ACM Computer Communication Review, 32(1), 2002.
2. Chinoy, B.: ACM SIGCOMM Computer Communication Review, Volume 23, Issue 4, New York, 45 - 52, 1993.
3. Ward, C., Choi, H., Hain, T.: A data link control protocol for LEO satellite networks providing a reliable datagram service, IEEE/ACM Transactions on Networking, 3(1):91103, Feb. 1995.
4. Floyd, S., Mahdavi, J., Mathis, M., Podolsky, M.: An Extension to the Selective Acknowledgement (SACK) Option for TCP, RFC 2883, 2000.
5. Jacobson, V.: Symposium proceedings on Communications architectures and protocols, California, 314 - 329, 1988.
6. McCanne, S, Floyd, S.: Network Simulator, http://www.isi.edu/nsnam/ns/
7. Pattan, B.: Satellite-Based Global Cellular Communications, McGraw-Hill, 1997.
8. Postel, J.: Transmission Control Protocol, RFC 793, 1981.
9. Sathiaseelan, A., Radzik, T.: RD-TCP: Reorder Detecting TCP, Proceedings of the 6th IEEE International Conference on High Speed Networks and Multimedia Communications HSNMC'03, Portugal, (LNCS 2720, pp.471-480), 2003.
10. Sathiaseelan, A., Radzik, T.: Improving the Performance of TCP in the Case of Packet Reordering, Proceedings of the 7th IEEE International Conference on High Speed Networks and Multimedia Communications HSNMC'04, Toulouse, France, 2004.
11. Zhang, M., Karp, B., Floyd, S., Peterson, L.: RR-TCP: A Reordering-Robust TCP with DSACK. 11th IEEE International Conference on Network Protocols (ICNP'03), Georgia, 2003.

Adapting TCP Segment Size in Cellular Networks

Jin-Hee Choi[1], Jin-Ghoo Choi[2], and Chuck Yoo[1]

[1] Department of Computer Science and Engineering,
Korea University
{jhchoi, hxy}@os.korea.ac.kr
[2] School of Electrical Engineering and Computer Science,
Seoul National University
cjk@netlab.snu.ac.kr

Abstract. In cellular networks, a frame size is generally made small to reduce the impact of errors. Thus, a segment of transport layer is splitted into multiple frames before transmission. A problem is that the whole segment is lost when a frame of a segment is lost. So, the segment error rate tends to be high even though the cellular network provides relatively the low frame error rate, which drops TCP performance. However, the relation between the frame size, the segment size and the error rate has not been closely investigated. In this paper, we analyze the impact of the segment size on TCP performance in cellular networks and propose a scheme alleviating the performance drop of TCP. Our result shows that our scheme reduces the drop by 82%.

1   Introduction

In cellular networks a frame, unit of transmission in physical layer, has very small size to reduce the frequent errors in hostile wireless channel conditions. Thus, a segment of TCP[1] is splitted into several frames and, then, transmitted one by one. A problem is that the whole segment is corrupted when even a frame is lost (See Fig.1). So, the segment error can occur very often though the frame error rate is low. It is very important to investigate the exact relationship among the frame size, the segment size and the frame errors closely since it makes severe influences on TCP and UDP performance.

We define some symbols to clarify the following discussions in this paper.

- s : payload size of a TCP segment.
- H : header size of IP and TCP layers.
- M : payload size of a frame in physical layer.
- n : number of frames consisting of a segment.
- p : segment corruption rate.
- e : frame error rate.

P. Lorenz and P. Dini (Eds.): ICN 2005, LNCS 3421, pp. 258–265, 2005.

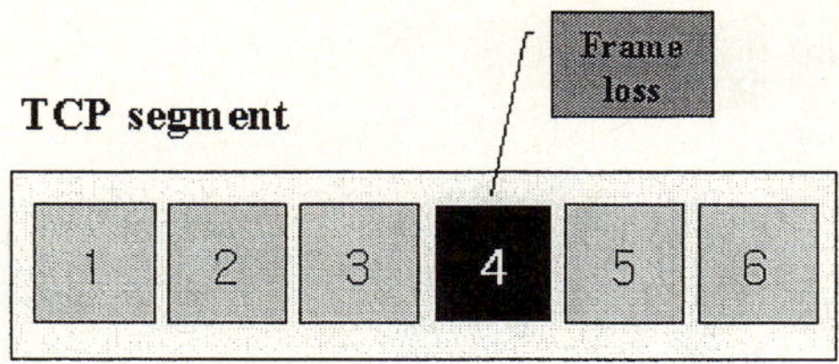

Fig. 1. Frame loss within a segment

The number of frames n is given as

$$n = \left\lceil \frac{H + s}{M} \right\rceil,\tag{1}$$

and the segment corruption rate is

$$p = 1 - (1 - e)^n.\tag{2}$$

For example, when $s + H$ is 1500, M is 128, and e is 1%, p becomes about 11.4%. It says that the segment corruption rate can be extremely high even with small frame errors, depending on n.

This paper is organized as follows. In the following Section, we show the simulation results to show how the UDP and TCP throughput is affected by the segment size and frame error rate and analyze the reason. Also, some valuable guidelines for determining the segment size is derived based on the analysis. Section 3 describes a simple heuristic scheme to reduce the performance drop in TCP over cellular networks. We discuss the limit of our scheme and the need of future work in Section 4. Finally, Section 5 concludes the paper.

2 Simulation Result and Analysis

2.1 Simulation Environment

The simulation study is performed by ns 2.27 version[2], and topology is a typical cellular network shown in Fig.2. There is a link that has 2Mbps bandwidth and 200ms latency between MH (Mobile Host) and BS (Base Station). Also, a wired link has 10Mbps bandwidth and 20ms latency, and it is placed between BS and CH (Corresponding Host). In all experiments, Random uniform and Gilbert-Elliot error models are used in channel modeling, and the frame size of the cellular link is set to 128 bytes. To focus on the cellular networks, we assume that there is no packet loss event except buffer overflow at the router.

2.2 UDP Experiments

UDP is a connectionless protocol that has only the function of port multiplexing and header checksum. Therefore, UDP's behavior totally depends on the characteristic of application because it does not have any flow control and congestion

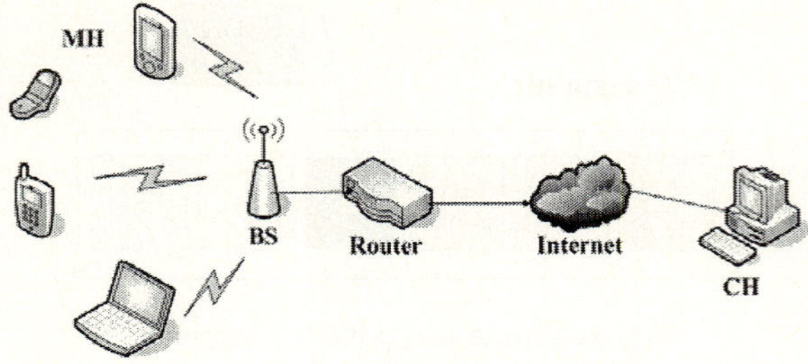

Fig. 2. Simulation topology

control mechanisms. Note that TCP's several complex mechanisms can make it difficult to observe the degree of the impact on throughput. Thus, with assumption that CBR (Constant Bit Rate) source is target application, our UDP experiment can be a basis model to understand an impact degree of the error rate and the difference of the frame and the segment size. To perform such experiments of UDP performance, we make a CBR source in the CH transmit data and a sink in the MH receive it. In all simulations, the frame size is set 128 bytes, and the throughput is measured varying the frame error rate from 0.2% to 20%, depending on the segment size. Also, same experiment is executed to observe the impact of the CBR's rate on the relation of the segment size and the error rate. Fig.3 shows how the segment size and the error rate affect the UDP throughput in case that CBR is 250Kbps. In Fig.3, we can see deeper decrease of the throughput when the error rate increases and the segment size gets large. Let us look in the case that the segment size is 1368 bytes. In this case, a segment split into 11 frames since the frame size is 128 bytes. If the link has 20% error rate in the average, only 7-8% throughput of non-error case can be achieved because about 92% segment error rate occurs. However, with 88 bytes segment, we can see almost 80% throughput of the non-error case, which means that the frame and the segment error rate are almost the same. As we can see in the simulation experiment, UDP performance is determined by the segment error rate and the header overhead. Note that the segment error rate increases as the frame number n increases by the equation 2. Fig.4 is the impact of 50Kbps CBR on the relation of the segment size and the error rate. It shows a similar decrease of the throughput although the throughput values are different.

2.3   TCP Experiments

It is generally believed that TCP throughput degrades as the frame error rate increases, regardless of the segment size. Based on our UDP experiment, our conjecture is that TCP throughput varies depending on the segment size. We will prove this conjecture by simulation.

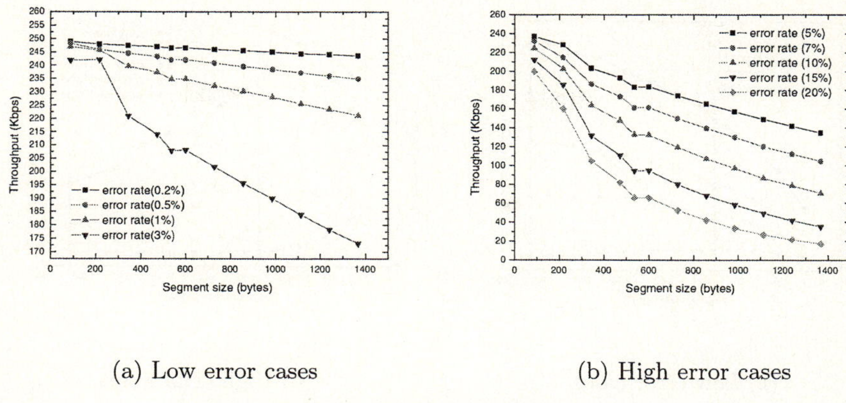

(a) Low error cases   (b) High error cases

Fig. 3. UDP throuhgput with 250Kbps CBR

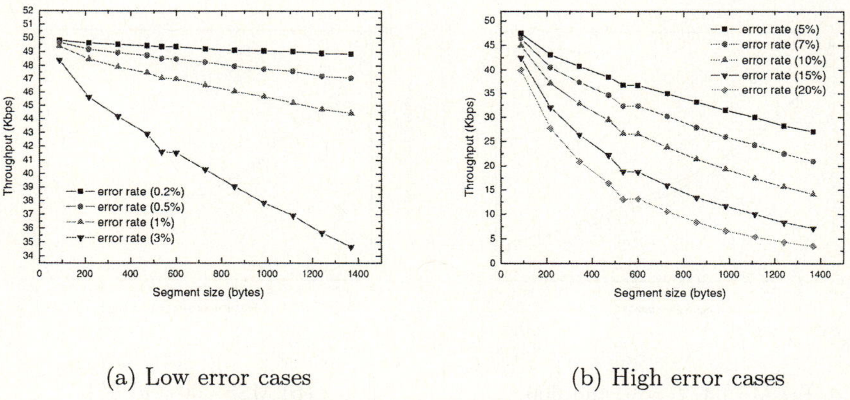

(a) Low error cases   (b) High error cases

Fig. 4. UDP throuhgput with 50Kbps CBR

2.4 Relationship of Frame Error Rate and Segment Size

We measured TCP Reno's throughput for several segment sizes varying the frame error rate from 0 to 20%. The frame size is fixed to 128 bytes. Fig.5 shows the results for the segment sizes 88, 216, 344, 472, 536, 728, 856, 984, 1112, 1368 and 1460, which includes packet sizes frequently observed in IP networks. The gaps of TCP throughput between the high and the low error rates are so large that we present two separate figures. Fig.5(a) is when the frame error rate is low - the throughput increases with the segment size. When the frame error rate is high, the throughput decreases with the segment size as in Fig.5(b). Note the dip in the throughput graphs at the segment sizes 536 and 1460, especially in Fig.5(b). Since the segment sizes 536 and 1460 are the most common packet sizes in the Internet (the two cases are marked with dash line), we did some detailed study that is explained in the following subsection.

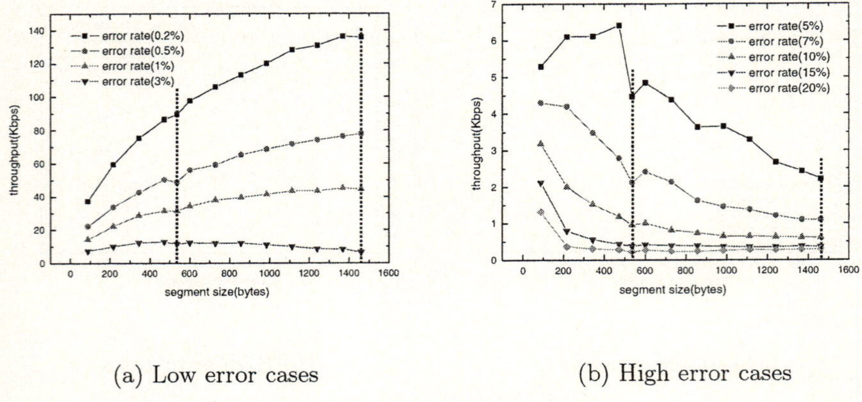

(a) Low error cases   (b) High error cases

Fig. 5. Throughput of TCP Reno

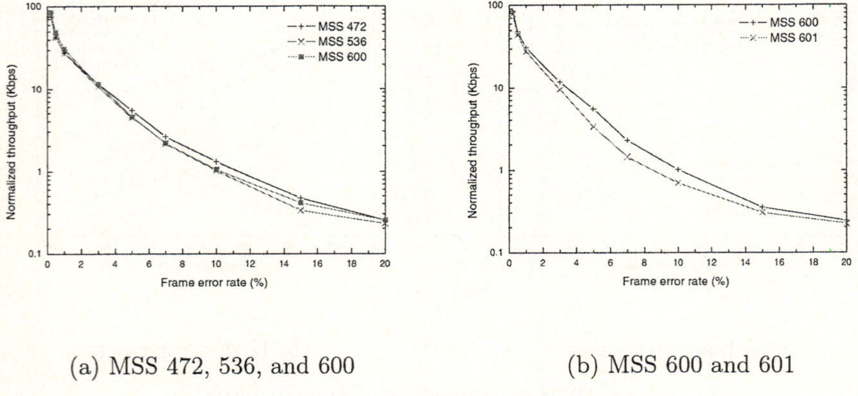

(a) MSS 472, 536, and 600   (b) MSS 600 and 601

Fig. 6. Normalized throughput of TCP Reno

2.5   Relationship Around the Segment Size 536

Fig.6 is the results of the detailed study around the segment size 536. Fig.6(a) says that the relationship of segment size and frame error rate is not always consistent. That is, 'MSS 600' shows better throughput than 'MSS 536' even when the error rate is high[1]. An outstanding difference of the two cases is that 'MSS 600' is aligned with the frame size[2] while 'MSS 536' is not, though they have the same number of frames. This shows that the protocol efficiency (defined as $\frac{s}{H+s}$) as well as the number of frames makes an effect on TCP performance though the degree of the influence is not high. Then, we observe the impact of

---

[1] We use MSS and the segment size by same meaning.
[2] Note that 600+40 (TCP/IP header size) is a multiple of 128, the frame size.

the number of frames on TCP performance by comparing 'MSS 600' with 'MSS 601'. Fig.6(b) clearly shows the difference between 5 frames ('MSS 600') and 6 frames ('MSS 601'). Even if the latter has better protocol efficiency, it shows fewer throughputs than the former.

Fig.5 and 6 give us the following guidelines to determine an "optimal" segment size.

– When the frame error rate is high, it is better to keep the segment size small.
– When the frame error rate is low, it is better to maintain the segment size large.
– In all cases, the segment size must be aligned to the frame size.

3   Heuristic Scheme

We propose a simple scheme adapting the MSS according to wireless channel conditions, whose algorithm is explained in Fig.7. Two different values of MSS are defined, small MSS and large MSS. The small MSS is used in bad channel condition and its size just fits in a frame. On the other hand, the large MSS

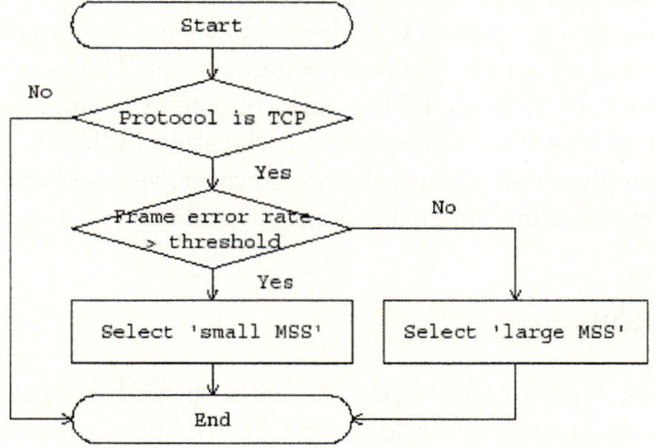

Fig. 7. Algorithm of proposed scheme

is used in good channel condition and its size is determined by the path MTU discovery mechanism of TCP[3]. The frame error rate is used as an indicator of channel states. Fig.8 shows the performance of the proposed scheme, which keeps the TCP throughput approximately optimal in almost every case of the experiments.

---

[3] WAP forum recommends that TCP should determine the maximum segment size (MSS) through the MTU discovery mechanism such as its optimization method for Wireless Profiled TCP[3] while RFC 793 recommends that the IP packet size is set to 576 bytes for external networks.

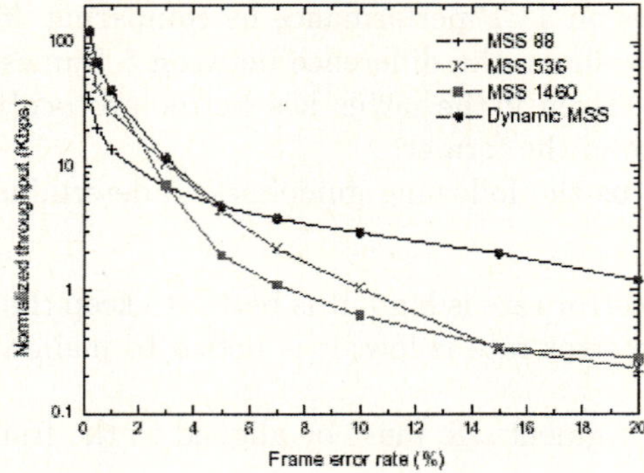

Fig. 8. Dynamic MSS adjustment

4   Discussion

The paper uses the frame error rate as an indicator of the channel state. That is
to say, the scheme judges that the channel condition is bad when the error rate
is over the threshold, and the state is good when the rate is under the threshold.
This threshold is selected by the simulation results (5%). It is not fitful to apply
our fixed threshold to all network environments since the good threshold may
be different depending on the network organization. The aim of this paper is
showing the impact of MSS on TCP's performance in cellular networks, so that
we do not focus on a specific mechanism for the state indicator. For the purpose
of the practical deployment of our scheme, however, we need additional research
about a novel scheme that infers the channel condition.

5   Conclusion

In this paper, we analyze the impact of MSS on TCP's throughput in cellular networks by extensive simulations, and derive some valuable guidelines for
determining an optimal segment size. Then, we propose a scheme that reduces
the performance drop of TCP by adapting the segment size dynamically by the
guidelines. We also find that the widely used MSS (536 and 1460 bytes) are not
optimal at all, and a better MSS can be found by considering alignment with the
frame size and protocol efficiency. A conclusion is that adjusting the segment size
dynamically is an effective way to handle the frame errors in cellular networks.

Acknowledgement

This work was supported by grant No.R01-2004-000-10588-0 from the Basic
Research Program of the Korea Science & Engineering Foundation.

References

1. J.Postel, *Transmission Control Protocol*, RFC 793, IETF, 1981.
2. UCB/LBNL/VINT Network Simulator, *ns2(version 2.27)*, http://www.isi.edu/nsnam/ns/.
3. Open Mobile Alliance, *Wireless Profiled TCP*, WAP-225-TCP-20010331-a, 2001.
4. H.Schulzrinne, *Internet Traffic Measurements*, http://www.cs.columbia.edu/ hgs/internet/traffic.html.
5. J.P.Ebert, A.Willig, *A Gilbert-Elliot Bit Error Model and the Efficient Use in Packet Level Simulation*, Technical Repoert, TKN-99-002, Technical University of Berlin, March 1999.
6. J.Mogul and S.Deering, *Path MTU Discovery*, RFC 1191, IETF, 1990.

AcTMs (Active ATM Switches) with TAP (Trusted and Active PDU Transfers) in a Multiagent Architecture to Better the Chaotic Nature of TCP Congestion Control[1]

José Luis González-Sánchez[1], Jordi Domingo-Pascual[2], and João Chambel Vieira[3]

[1] University of Extremadura. Escuela Politécnica de Cáceres Avda,
Universidad S/N. (10.071) Cáceres (Spain)
`jlgs@unex.es`
[2] Polytechnic University of Catalunya, Campus Nord, Modul D6,
Jordi Girona 1-3 (08034) Barcelona (Spain)
`jordi.domingo@ac.upc.es`
[3] University Moderna of Portugal, Polo de Beja Rua Marquês de Pombal, 1
(7800-067) Beja (Portugal)
`jccv@umoderna.pt`

Abstract. TAP (Trusted and Active PDU transfers) is a distributed architecture and a protocol for ATM networks that provides assured transfers to a set of privileged VPI/VCI. Our AcTMs (Active ATM switch) model supports the trusted protocol. This research also offers an attractive solution to the chaotic nature of TCP Congestion Control. Several simulations demonstrate the effectiveness of the mechanism that recovers the congested PDU locally at the congested switches with better end-to-end goodput in the network. Also, the senders are alleviated of NACK and end-to-end retransmissions.

1 Introduction

The ATM technology is characterized by its good performance with the different traffic classes and by its negotiation capacity of the QoS (Quality of Service) [1] parameters. Congestion causes the most common type of errors, and it is here that our work is intended to offer guaranteed transfers through our TAP (Trusted and Active Protocol) architecture. TAP adopts ARQ (Automatic Repeat Request) with NACK (Negative Acknowledgement) using RM (Resource Management) cells to alleviate the negative effect of implosion. The intermediate active nodes are responsible for local retransmissions to avoid end-to-end retransmissions. We have implemented a modification of EPD (Early Packet Discard) as a means of congestion control that we have called EPDR (Early Packet Discard and Relay) in order to alleviate the effect of congestion and PDU fragmentation. Currently, congestion control is delegated to

[1] This work is sponsored in part by the Regional Government of Extremadura under Grant No. 2PR03A090 and by the CICYT under Grant No. TIC2002-04531-C04 Advanced Mobile Services (SAM).

P. Lorenz and P. Dini (Eds.): ICN 2005, LNCS 3421, pp. 266–274, 2005.
© Springer-Verlag Berlin Heidelberg 2005

protocols that solve it with end-to-end retransmissions such as TCP. This is an easy technique to implement at high speeds by simplifying the switches and routers, but the whole network is overloaded with the retransmissions and this does not offer protection against egoist sources.

At present, the ATM networks are used as a technology to support all kinds of traffic, with predominance of TCP/IP protocols. Therefore we present the advantages that our mechanism of congestion retrievals can offer, not only for the native ATM traffic, but also for the traffic generated by TCP/IP sources. A considerable amount of research has intended to integrate two technologies as different as ATM and TCP/IP; however, their integration offers [2] poor results in the behavior of TCP throughput over ATM. While ATM is a connection-oriented technology of switched cells of 53 octets and uniform size, TCP and IP are based on routing mechanisms of segments and datagrams of variable size. These characteristics cause a very negative effect on the throughput when the TCP segments cross ATM switches with buffer size less than the window size of TCP. This causes loss of cells and retransmissions due to timeout. Moreover, the loss of only one cell causes the loss of a TCP segment at the receiver of the communication that will request the retransmission to the source, which must resend the whole segment and not only the lost cell.

Firstly, we shall comment on the general characteristics of TCP. Section 3 presents the TCP characteristics over ATM and the next section propose the use of TAP in an IPoverATM scenario and the evaluation of TAP performance. Our conclusions are presented at the end of this paper.

2 TCP Congestion Control Can Be Improved

The TCP protocol is a set of algorithms that sends packets to the network without any previous reservation, but can react if any event appears. Within this set of algorithms, the Congestion Control algorithm and the Loss Segment Retrieval algorithm are the most important.

A source TCP fixes the amount of data to be sent by using the CWND window, and transmits a window of segments for each RTT. The TCP adjusts the size of this window depending on the conditions of the network. Thus, the size of CWND increases to twice the segments for each ACK received in Slow Start algorithm, and increases by 1/CWND for each ACK received in the Congestion Avoidance algorithm. CWND increases exponentially while the size is less than SSTHRESH (using the Slow Start algorithm that progressively increases the number of segments (1, 2, 4...) when the ACKs are received). When the size of CWND is equal to the SSTHRESH, the congestion control of Congestion Avoidance works. Thus, the CWND window increases linearly by 1/CWND for each ACK.

The Slow Start algorithm is used by TCP to check the unknown capacity of the network and the amount of segments that it can support without congestion. When congestion is imminent, TCP passes the control to Congestion Avoidance which changes to a lineal increase of CWND until the congestion is detected.

We should point out that the TAP protocol solves the problems of loss that affect the decrease in size of the congestion window and also the subsequent retransmission of end-to-end losses. Thus, the source will not be obliged to reduce and to adjust its rate of transmission all the time, and also, when congestions appear, these are solved locally in the affected nodes.

Equation (1) calculates the transmitted bandwidth (BW) and expresses the throughput in the network and calculates the performance of TCP after all previous simplifications. *MSS* is the maximum segment size of TCP; *K* is a constant term and we can estimate the random packets lost with a *P* constant probability assuming that the link delivers *1/P* consecutive packets. Paper [3] presents other references with several approximations to the value of the *K* constant regardless of its value that is always less than 1.

$$BW = \frac{MSS}{RTT}\frac{K}{\sqrt{P}} \tag{1}$$

We can reorganize (1) and, considering *RTT* and *MSS* as constants, and *W* as the window size used by TCP, we will obtain the average loss rate *P* in equation (2),

$$P = \frac{0,75}{W^2} \tag{2}$$

Equation (2) can be understood as if the network discards a percentage of segments independently of actions that have been performed by the source. So, this describes how the source can react.

The loss probability *P* determinates the throughput of the TCP source, as the (2) previous expression intuitively indicates. When the loss probability increases, the throughput decreases logarithmically to achieve a lineal evolution. The negative logarithmic slope represents the fall of the throughput in the network when this is experiencing loss of packets. The problem is that TCP duplicates the intervals of retransmission times between successive loss of packets.

3 TCP over ATM

The research in the throughput evaluation of TCP over ATM is divided into three main groups [4]: 1) those research papers studying the dynamism of TCP; 2) those analyzing the throughput of ATM; and 3) those observing the interaction between TCP windows and the mechanisms of congestion control of the ATM layer. Although the throughput evaluation of TCP over ATM has been the objective of several research papers, the proposals only solve particular problems such as the fragmentation of TCP, the buffers required, the interaction between congestion schemes of TCP and ATM, and the degradation of TCP. There are a lack of proposals to solve all or even some of these problems. Our research has looked into these aspects and offers a MAS (MultiAgent System), optimizing the goodput with an improvement of entry queues. Moreover, an accurate policy of buffer management is used through the delegation of activities in agents that constitute the MAS.

Reference [2] presents the study of congestion in TCP networks over ATM and shows how the TCP throughput also falls when the discard of cells at ATM switches begins. The low throughput obtained is due to the waste of the bandwidth at congested links that transfer packets of corrupted cells; that is, packets with some dropped cells. Other research papers have demonstrated that TCP over UBR, with

EPD suffers a considerable degradation of fairness, and also need a big buffer size, even if there are few connections.

The literature [5, 6] also describes other ways to avoid the degradation of throughput at TCP sources over UBR. In order to do this, the discard of ATM cells is disconnected when there is congestion. So, the timeouts of TCP are avoided although they are the main cause of fall at TCP throughput, and also the periods of congestion are reduced, avoiding the big delay experienced with the fast retransmission algorithm of TCP before the source receives the duplicate ACK.

With all these differences, and as ATM is a protocol placed under the TCP transport layer, solutions are required to solve the throughput problems due to the integration of these different technologies. These solutions propose changes at ATM switches inside the network; or the implementation of new extensions for TCP; or perhaps, specialized protocols for nodes placed at the limits of the ATM network and the TCP network. Our TAP protocol solves these problems by working inside the network, with hardware (active ATM switches) and also software (multi-agent system with TAP protocol) mechanisms. All this configures the whole TAP architecture.

4 Advantages of TAP and Evaluation of TAP Performance

We propose EAAL-5 layer (Extended AAL-5), specifically designed for data communications over ATM. At TCP over ATM, the datagrams are transferred to the data field (payload) of EAAL-5, as we can see in Fig. 1 which shows the stack protocols of TCP over ATM. The TAP architecture is active, because it provides active nodes at strategic points that implement an active protocol to allow the user's code to be loaded dynamically into network nodes at run-time. TAP also provides support for code propagation in the network thanks to the RM cells.

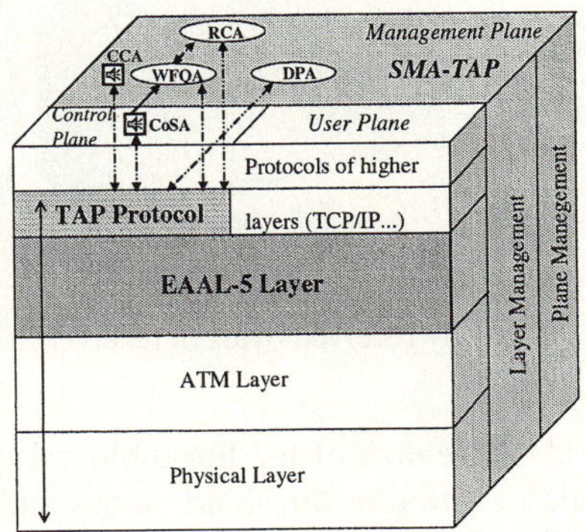

Fig. 1. TAP-MAS architecture

In Fig. 1 the TAP-MAS architecture is shown, focusing on the way to incorporate the modifications we propose for the ATM architecture in our proposals. Notice that the position of the MAS is part of the management plane because it provides new mechanisms to manage the ATM resources.

In [7] we have presented the architecture based on TAP-MAS, constituted by software agents and equipped with a DMTE dynamic memory. We have also implemented the PQWFQ (PDU Queues PDU based on Weighted Fair Queuing) algorithm to apply fairness at sources. Also, the EPDR algorithm manages the buffer congestion and avoids PDU fragmentation. The general motivation of this work is to find solutions to alleviate this negative problem of end-to-end retransmissions.

One of the most interesting aspects of the AcTMs is the implementation of the EPDR (Early PDU Discard and Relay) algorithm on the switch buffer. We propose this algorithm to retransmit the PDU between adjacent AcTMs and to avoid the existence of fragmented PDU in the network. In order to reduce the fragmentation, the algorithm controls the size of the input buffer threshold in each switch. Fig. 2 shows the input buffer of the AcTMs with the sizes controlled by the EPDR algorithm.

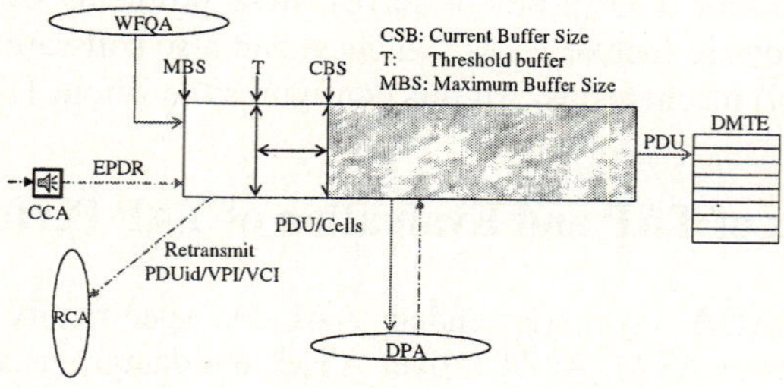

Fig. 2. Buffer of the AcTMs with EPDR

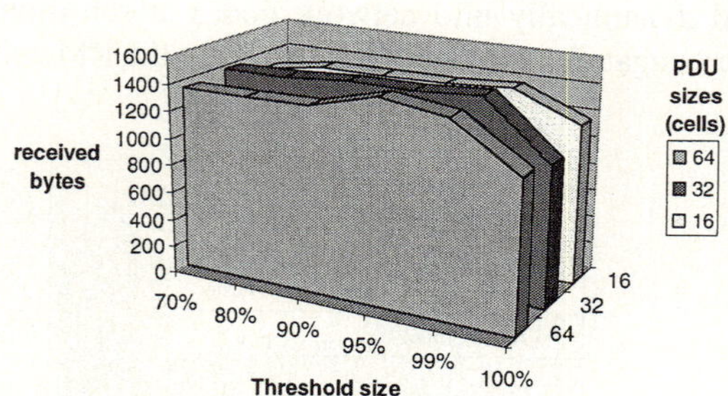

Fig. 3. Correctly received bytes in receiver R-EAAL-5

Fig. 3 shows the effect of the size of the threshold value of the EPDR algorithm. This study has been made of different threshold values and different PDU sizes. We have observed that the EAAL traffic generated by the ATM privileged sources is better than the AAL traffic of ATM sources without GoS. This is because less EAAL PDU are lost than AAL, and that avoids the retransmissions of higher layer protocols. The number of EAAL-5 PDU arriving rightly to the R-EAAL-5 sink, depending on the threshold size and the PDU size is shown. When the PDU size is minor, more PDU are received. The fluctuations are due to the in-flight cells at the links.

It is interesting to analyze the obtained results when the EPDR control does not act on the buffer. The Fig. 4 is obtained with the threshold at 100%. The EAAL traffic shown is significantly better than AAL traffic. The AcTMs-9 switch discards all the incorrect PDU coming from the S-EAAL-5 source. This is one of the tasks of the CoSA agent. While the corrupted AAL PDU go on transiting in the network making unnecessary use of resources, the correct PDU corresponding to privileged EAAL connections with GoS are discarded by the next active switch.

Moreover, the number of correct received PDU with EAAL traffic is higher than the received ones with AAL traffic. This is not due to the retransmissions because with the 100 % threshold an AcTMs switch cannot retransmit requests.

The previous figures have shown how the threshold value does not affect the number of PDU that arrive at a sink, except for 100% threshold. However, we know that congestions are produced and cells are lost in some PDU. For this reasons we now study the influence of the EPDR threshold in the appearance of the congestions and retransmit request of EAAL-5 PDU. Fig. 5 shows the discarded cells in i.e. AcTMs-9 switch due to discarded cells by EPDR or congestions. The cells are discarded depending on the PDU size and the buffer threshold.

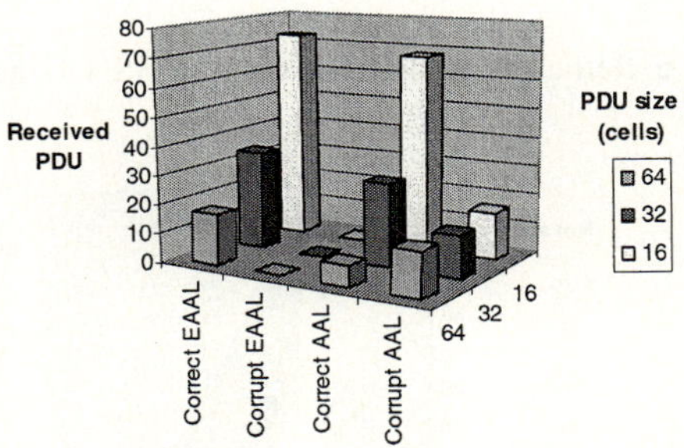

Fig. 4. Corrupt and corrects received PDU with 100% threshold

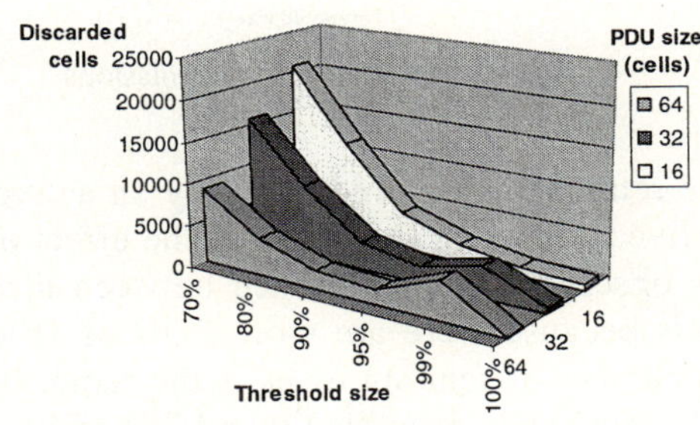

Fig. 5. Discarded cells depending on the PDU size

When the threshold per-cent increases, the number of discarded cells decreases, but only until a limited value. For PDU of 64 and 32 cells with a 99% threshold on a buffer of 1000 cells, the number of discarded cells increases. This is due to the EPDR algorithm behaviour. If the filling level of the buffer is higher than the threshold when the first cells of a PDU arrive, the whole PDU is discarded. For this reason when a threshold level is minor, the probability of discarding a PDU is higher.

When the difference of size between the threshold and the buffer size is very close to the PDU size, it increases the probability of discarding a cell because it does not fit in the buffer. This is corroborated because fewer cells are discarded (with a threshold close to 99%) as the PDU size decreases.

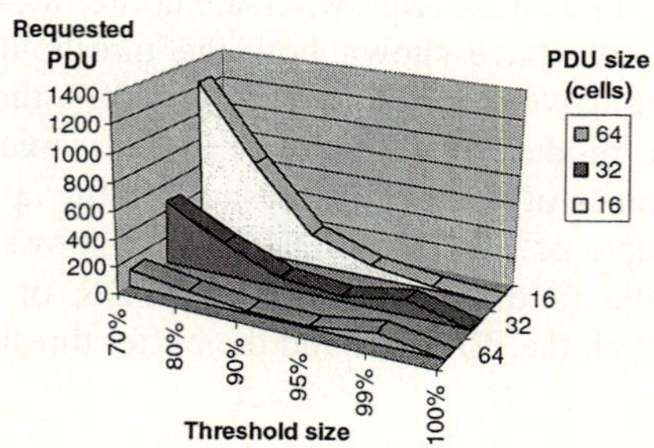

Fig. 6 Retransmitted PDU by the AcTMS 4 to the AcTMs 6

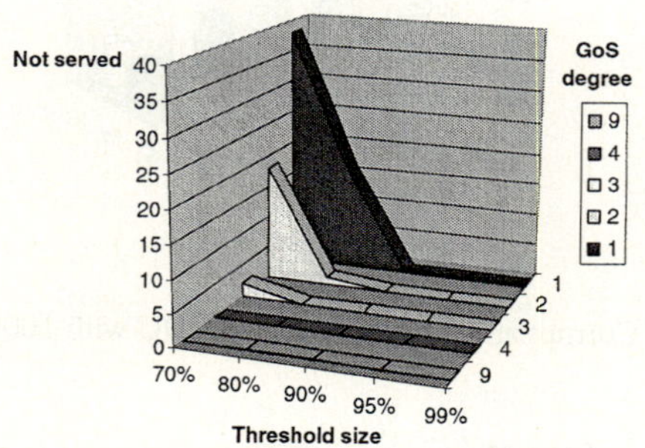

Fig. 7. Non-served retransmissions

Fig. 6 shows the retransmissions requested by an active switch (i.e. AcTMs-6) to a previous active switch (i.e. AcTMs-4). The effect of the threshold on the PDU discarding is observed. The difference between discarding with PDU of 16, 32 or 64 cells is because there are more PDU of 16 cells than PDU of 64 cells, although the number of sending bytes is the same. It is also observed that there are more PDU of 32 cells discarded than PDU of 64 with a 99% threshold, and this is because the retransmissions of the PDU of 32 are faster due to the lower limited number of cells.

Next, the influence of the GoS grade in the goodput of the EAAL connections is studied. The GoS grade is a parameter set to privileged EAAL connections that is negotiated with the active switches in the established connections process. This parameter represents the number of EAAL-5 PDU requested to the active switches to store in their DMTE memory in order to be able to serve the retransmissions. The results of these simulations are shown in Fig. 7 where PDU of 16 cells have been used in the AcTMs-6 switch. We studied the number of non-served retransmissions for different grades of GoS in EAAL connections depending on the threshold. Non-served retransmissions are those that cannot serve either the AcTMs-4 switch or the AcTMs-1 switch because there is not a copy of the PDU in the DMTE. This occurs when the required PDU has been substituted by another more recent one because there is not enough vacant space in the DMTE. Fig. 7 shows how the number of non-served PDU decreases quickly when the grade of GoS increases.

To manage the buffer and the input queues at each AcTMs switch we have implemented the PQWFQ (PDU Queues based on Weighted Fair Queuing) algorithm as part of the WFQ agent at TAP-MAS. This algorithm achieves a fair treatment of the PDU that arrive at AcTMs switches. We must treat the PDU of connections with GoS (Guarantee of Service) as privileged traffic.

These and other simulations demonstrate that the TAP architecture takes advantage of the AcTMs switches. We have verified that it is possible to retrieve an important number of PDU only with DMTE and a reasonable additional complexity.

5 Conclusions

In protocols of transport layer such as TCP over ATM, a packet is discarded by the network when one or several cells are lost, and the destination node requests the whole retransmission of the corrupted or lost packet. We have demonstrated through simulations the degradation experienced by the throughput of TCP. We have also studied how this falls logarithmically when the probability of loss of the ATM cells increases. The TAP protocol makes the retransmissions locally, and this decreases the loss probability and affects the throughput of TCP that alleviates the delays due to the end-to-end RTT. These simulations have demonstrated that the addition of active switches improves the throughput of a congested ATM network. Also the goodput of a privileged set of sources with GoS is increased. We conclude that the ATM network improves with the addition of the AcTMs switches.

References

1. Janusz Gozdecki, Andrzej Jajszczyk, and Rafal Stankiewicz, "Quality of Service Terminology in IP Networks," IEEE Communications Magazine, (March 2003).
2. Romanow, A. and Floyd, S., "Dynamics of TCP traffic over ATM networks," *IEEE JSAC*, pp. 633-641, (1995).
3. Matthew Mathis, Jeffrey Semke, Jamshid Mahdavi, and Teunis Ott, "The Macroscopic behavior of the TCP Congestion Avoidance Algorithm," *Computer Communications Review of ACM SIGCOMM*, vol 27, n. 3, (1997).
4. K. Djemame, and M. Kara, "Proposals for a Coherent Approach to Cooperation between TCP and ATM Congestion Control Algorithms," *Proceedings UKPEW'99*, (1999).
5. Hongqing Li, Kai-Yeung Siu, Hong-Yi Tzeng, Ikeda, C., and Suzuki, H., "A simulation study of TCP performance in ATM networks with ABR and UBR services," *Proceedings IEEE INFOCOM'96*, pp. 1269-1276, (1996).

6. Shunsaku Nagata, Naotaka Morita, Hiromi Noguchi, and Kou Miyake, "An analysis of the impact of suspending cell discarding in TCP-over-ATM," *Procceedings IEEE INFOCOM'2000*, pp. 1147-1156, (2000).
7. José Luis González-Sánchez, Jordi Domingo-Pascual, and Alfonso Gazo Cervero, "Robust Connections for TCP Transfers Over ATM Through an Active Protocol in a Multiagent Architecture," Proceedings 10th *IEEE IEE ICT'2003*, pp. 830-836 (2003).

AIMD Penalty Shaper to Enforce
Assured Service for TCP Flows

Emmanuel Lochin[1], Pascal Anelli[2], and Serge Fdida[1]

[1] LIP6 - Université Paris 6
[2] IREMIA - Université de la Réunion, France
{emmanuel.lochin, serge.fdida}@lip6.fr
pascal.anelli@univ-reunion.fr

Abstract. Many studies explored the guaranteed TCP throughput problem in DiffServ networks. Several new marking schemes have been proposed in order to solve this problem. Even if these marking schemes give good results in the case of per-flow conditioning, they need complex measurements. In this paper we propose a conditioning method to reduce these complex measurements and an AIMD[1] Penalty Shaper (APS) which is able to profile a set of TCP flows so as to improve its conformance to a desired target rate. The main novelty of this shaper is that the shaping applies an AIMD penalty delay which depends on the out-profile losses in a DiffServ network. This penalty shaping can be used with any classic conditionner such as a token bucket marker (TBM) or a time sliding window marker. We made an evaluation of the APS on a real testbed and showed that the proposed scheme is easily deployable and allows for a set of TCP flows to achieve its target rate.

Keywords: Bandwith allocation, Edge to Edge performance, Quality of service, Assured Service, TCP, Experimentation with real testbeds.

1   Introduction

The Differentiated Services architecture [1] proposes a scalable means to deliver IP Quality of Service (QoS) based on handling of traffic aggregates. This architecture advocates packet tagging at the edge and lightweight forwarding routers in the core. Core devices perform only differentiated aggregate treatment based on the marking set by the edge devices. Edge devices in this architecture are responsibles for ensuring that user traffic conforms to traffic profiles. The service called Assured Service (AS) built on top of the AF PHB is designed for elastic flows. The minimum assured throughput is given according to a negotiated profile with the user. Such traffic is generated by adaptive applications. The throughput increases as long as there are available resources and decreases when a congestion occurs. The throughput of these flows in the assured service breaks up into two parts. First, a fixed part that corresponds to a minimum

---

[1] Additive Increase Multiplicative Decrease.

P. Lorenz and P. Dini (Eds.): ICN 2005, LNCS 3421, pp. 275–283, 2005.

assured throughput. The packets of this part are marked like inadequate for loss (colored green or marked IN). Second, an elastic part which corresponds to an opportunist flow of packets (colored red or marked OUT). These packets are conveyed by the network on the principle of "best-effort" (BE). In the event of congestion, they will be dropped first. Thanks to an AIMD Penalty Shaper, we show that it is possible to provide service differentiation between two source domains, on a set of TCP flows, based on its marking profile. In this paper we evaluate the solution with long-lived TCP flows. The proposed solution provides the advantage of neither needing RTT[2] evaluation nor loss probability estimation. The solution takes care of the behavior of TCP flows only. Consequently, as it is easily deployable, it has been experimented on a real testbed.

2   Related Work

There have been a number of studies that focused on assured service for TCP flows but also on the aggregate TCP performance. In [2], five factors have been studied (RTT, number of flows, target rate, packet size, non responsive flows) and their impact has been evaluated in providing a predictable service for TCP flows. In an over-provisionned network, target rates are achieved regardless of these five factors. This result is corroborated by [3]. However the distribution of the excess bandwidth depends on these five factors. When responsive TCP flows and non-responsive UDP flows share the same class of service, there is unfair bandwidth distribution and TCP flow throughtputs are affected. The fair allocation of excess bandwidth can be achieved by giving different treatment to out-of-profile traffic of two types of flows [3]. Recently [4] demonstrates the unfair allocation of out-of-profile traffic and concludes that the aggregate that has the smaller/larger target rate occupies more/less bandwith than its fair-share regardless of the subscription level. In [5], a fair allocation of excess bandwidth has been proposed based on a traffic conditioner. The behavior of the traffic conditioner has a great impact on the service level, in terms of bandwidth, obtained by TCP flows. Several markers have been proposed to improve throughtput insurance, [6], [7], [8], [9]. These algorithms propose to mark aggressive TCP flows severely out-of-profile so that they are preferentially dropped. Even if these marking strategies work well in simulation, their main disavantage is their implementation complexity. Indeed, these algorithms need to measure a flow's RTT, its loss probability or have a per-state information of the flows.

3   The AIMD Penalty Shaper (APS)

Let $r(i)_{AS}$ be the assured rate of the flow i (*i.e.* in-profile packets throughput), n the number of AS TCP flows in the aggregate at the bottleneck level and C the link capacity. Precisely, this capacity corresponds to a bottleneck link in the network. If a number of i flows cross this link, the total capacity allocated for

---

[2] Round Trip Time.

assured service R_{AS} is : $\sum_{i=1}^{n} r(i)_{AS}$. Let C_{AS} be the ressource allocated to the assured service.

$$R_{AS} < C_{AS} \tag{1}$$

Equation (1) means an **under-subscription** network. In this case, there is excess bandwidth in the network. If $R_{AS} \geqslant C_{AS}$, this is an **over-subscription** network and there is no excess bandwidth. This configuration is the worst case for the AS. This service must provide an assurance until the over-subscription case is reached. Afterwards, not enough ressources are available and the service is downgraded.

$$TCP\ Throughput = \frac{C * \text{Maximum Segment Size}}{RTT * \sqrt{p}} \tag{2}$$

In a well-dimensioned network, inequity from (1) should be respected. When there are losses in the network, it corresponds to the losses of out-profile packets, and not in-profile packets. It means that a light congestion appears in the network and some out-profile packets must be dropped. In order to increase the loss probability of the opportunist flows, new conditionners presented in section 2 are based on increasing the out-profile part of the most aggressive traffic. Then, the loss probability raises and the TCP throughput of the opportunist traffic decreases. It's a logical behaviour because the latter has a reject probability higher than the non-opportunist traffic. [10] gives a model of TCP throughput represented by the equation (2). With W_{max} is the TCP maximum window size and p the loss probability. Changing the p value from the equation (2) thanks to a marking strategy is complex. Indeed, it is necessary to evaluate the loss probability of the network and estimate an RTT for each flow. As opposed to the marking strategy adopted by new conditionners, we propose a delay based shaper. This shaper applies a delay penalty to a flow if there are out-profile packets losses in the network and if it outperforms its target rate. The basic idea is that the penalty is a function of the out-profile packet losses. Instead of raising the p value, from equation (2), of the most opportunist flow, the AIMD Penalty Shaper raises a delay penalty to the flow. It results in a growth of the RTT. Mathematicaly, as shown in (2), increasing RTT value is similar to increasing p value in term of TCP throughput. [11] has shown that limiting out-profile packets is a good policy to achieve a target rate. Indeed, by avoiding packets dropping we avoid TCP retransmission. This is an efficient solution to optimize the bandwidth usage. Thus, our goal is to reduce out-profile losses by applying a delay penalty to the flows that are the most opportunist in the network. Therefore, when a RIO[3] [12] router in the core network is dropping out-profile packets, it marks the ECN flag [13] of the in-profile packets enqueued in the RIO queue. In a well-dimensioned network, there is no in-profile packet loss. Then, the edge device can be aware that there is a minimum of one flow or set of flows which are opportunists in the network. This opportunist traffic is crossing the same

---

[3] RED with IN and OUT.

```
K = 10ms
FOR each observation period T
 TSW gives an evaluation of the throughput : throughput_measured
 IF throughput_measured < target_rate OR there are no out-profile losses
 THEN reduce the penalty delay
      current_penalty = current_penalty - ((i/2) * K)
      i = 1
 ELSE
    raise the penalty delay
    current_penalty = current_penalty + (i * K)
    i = i * 2;
 ENDIF
ENDFOR
```

Fig. 1. APS algorithm

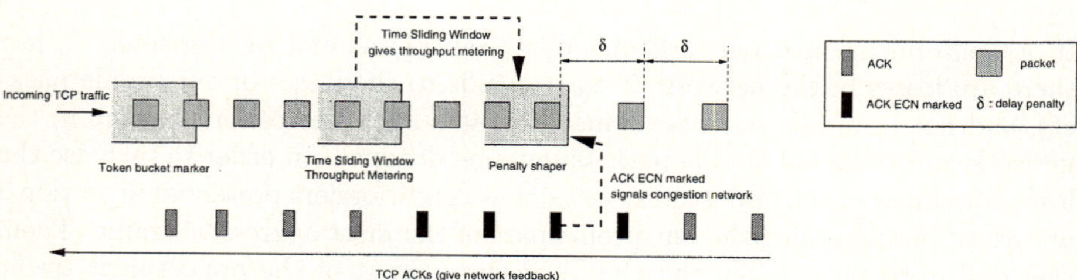

Fig. 2. conditioning with APS

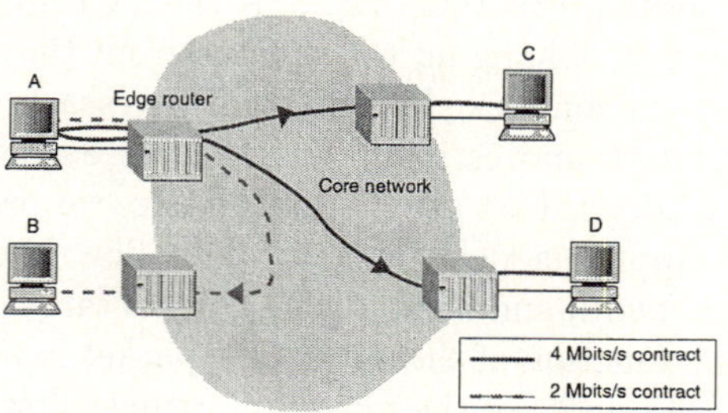

Fig. 3. Trafic conditioning sample

path. The edge device evaluates its sending rate thanks to a Time Sliding Window (TSW) algorithm [14]. If its sending rate is higher than its target rate, it considers that its traffic may be opportunist. Then, it applies a penalty to the incoming traffic until the network feedback that there are out-profile packets losses. This penalty allows a raise of the RTT and consequently, decrease the TCP throughput. We choose to use an AIMD penalty instead of a linear penalty in order to be in conformity with the TCP congestion control. If there is no loss anymore, the penalty decreases and the TCP throughput raises. The algorithm presented in figure 1 explains how the AIMD penalty is calculated and applied. As explained on figure 2, once incoming TCP traffic is shaped, it passes through a marker such as a TBM. This conditioner is setup on the edge device such an ingress edge router. Many are the conditioners presented in section 2 which will never leave the framework of simulation because of their conditioning constraints. We chose to make the traffic conditioning in the following way : each client emitting one or more flows towards one or more destinations will have one traffic profile per destination. As shown on figure 3, client A forces the edge router to setup three different traffic conditioners. Two conditioners with a contract rate of $4Mbits/s$ and one conditioner with a contract rate of $2Mbits/s$. The main advantage of this solution is that the conditioning can be made on flows with similar RTTs (i.e. in the same order of magnitude). This solution doesn't depend on the complex problem of RTT estimation necessary to the functionement of the conditioners presented in section 2. The solution of traffic shaping coupled to a conditioner/marker such as the TBM should be easily deployable and scalable.

4   Experimental Testbed

As shown in figure 4, we use the well-known dumbbell topology. The testbed is composed of computers running FreeBSD. On the edge routers, the token bucket marker from ALTQ[4] developement and the AIMD Penalty Shaper based on Dummynet[5]. On the core routers, a RIO queue with the ECN marking functionnality from ALTQ developement. Thus, the RIO queue is able to mark the ECN flag of the in-profile packets if it detects out-profile losses in its queue. We use the Iperf 1.7.0[6] traffic generator and two transmitting machines and two receivers for measurements. The main parameters and hypothesis are : traf-

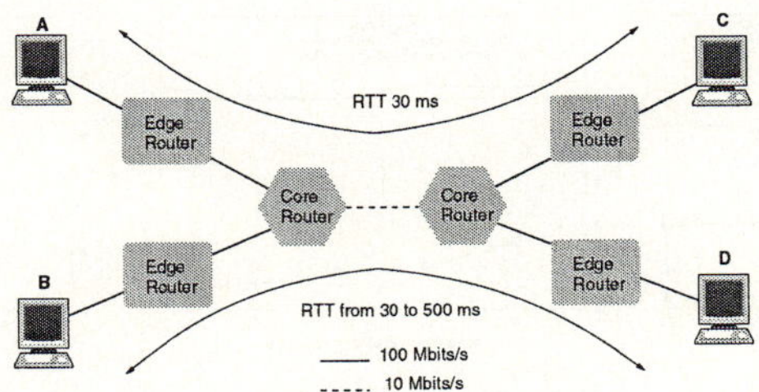

Fig. 4. Experimental testbed

fic generation is carried out in the following way: A to C (A, C) and B to D (B, D), after 120 seconds, Iperf gives an average throughput of the flow ; each AS flow is transmitted as TCP, packets have a size of 1024 bytes ; Iperf uses a TCP maximum window size $Wmax = 64packets$; each set of flows between two hosts is conditionned by one TBM with or without APS ; b parameter of the TBM is set to one packet ; r parameter is set to the desired target

---

[4] http://www.csl.sony.co.jp/person/kjc/
[5] http://info.iet.unipi.it/~luigi/ip_dummynet/
[6] http://dast.nlanr.net/Projects/Iperf/

rate ; the delay penalty is set to $10ms$ and the observation period to $1sec$. It means that each second the algorithm gives an estimation of the throughput and evaluates the penalty delay ; we use a non-overlapping RIO with parameters: $(min_{out}, max_{out}, p_{out}, min_{in}, max_{in}, p_{in}) = (1, 63, 0.1, 64, 128, 0.02)$, the queue size corresponds to $2 * Wmax$.

5   Performance Evaluation of the AIMD Penalty Shaper

This section presents the results obtained in a real testbed with the APS. We evaluate the performance of the APS when TCP traffic have the same or a different number of flows and identical or different RTTs. In these tests, the total capacity allocated for the assured service is $R_{AS} = 8Mbits/s$. The ressource allocated to the assured service is $C_{AS} = 10Mbits/s$ that corresponds to the bottleneck capacity. This is an under-subscription network because there are $2Mbits/s$ of excess bandwidth.

5.1   Impact of the Aggregates' Aggressiveness in an Under-Subscribed Network

Even if there is a different number of flows in the aggregates, the APS is able to reach the desired target rate. Results are presented in figure 5. When two aggregates with different number of flows are in a network, the higher outperforms the smaller [2]. In these tests, (A, C) has one flow and (B, D) has a variable number of flows ranging from 1 to 25. the RTT is set to $30ms$. After repeating each ex-

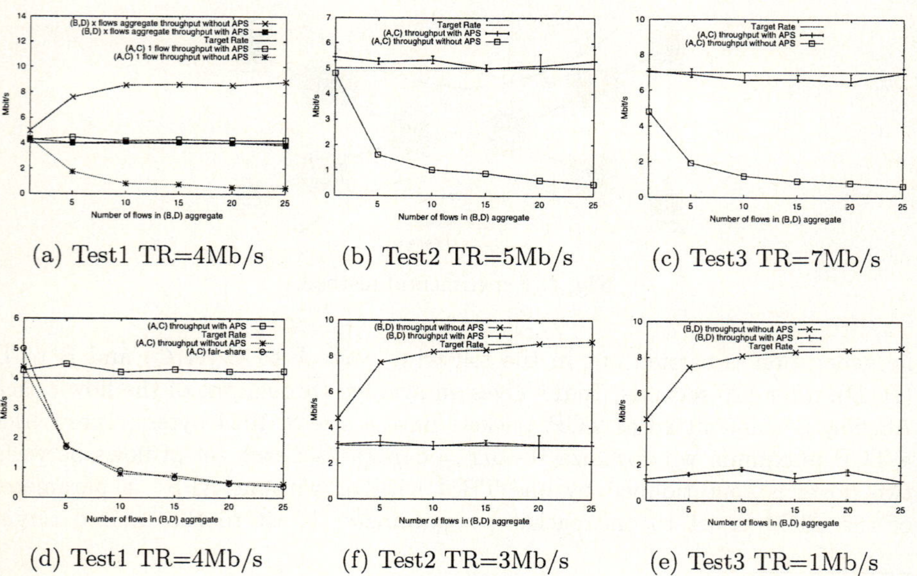

(a) Test1 TR=4Mb/s   (b) Test2 TR=5Mb/s   (c) Test3 TR=7Mb/s

(d) Test1 TR=4Mb/s   (f) Test2 TR=3Mb/s   (e) Test3 TR=1Mb/s

Fig. 5. TCP throughput versus aggregates' aggressiveness with various target rates

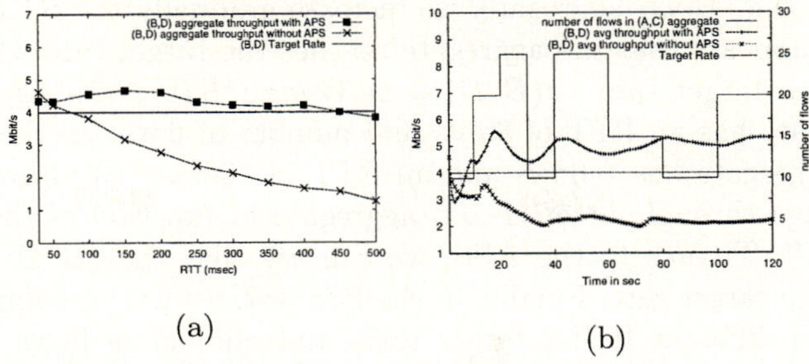

Fig. 6. TCP throughput versus RTT

periments five times, we calculate the average throughput value (for information purpose, min/max values are represented on the second and third test). In the first test, the target rate of both is set to $r(A,C)_{AS} = r(B,D)_{AS} = 4Mbits/s$. Figure 5 (a) shows the throughput obtained by both aggregates. For clarification, we draw on figure 5 (d) the throughput obtained by the (A,C) aggregate alongside the fair-share curve. Figure 5 (d) shows that the TBM stays close to the fair-share and that we obtain the desired target rate with the APS. Figures 5 presents the same scenario but the target rate on figures 5(b) and (e) is set to $r(A,C)_{AS} = 5Mbits/s$ and $r(B,D)_{AS} = 3Mbits/s$ and on figures 5(c) and (f) : $r(A,C)_{AS} = 7Mbits/s$ and $r(B,D)_{AS} = 1Mbits/s$. So, the second and the third tests illustrate both the case where the aggregates have near target rates and the case where they have distant target rates under under-subscription conditions. With APS, the target rate of TCP can be controlled and has a value over the target or near the target (in our worst case).

5.2   Impact of the RTT in an Under-Subscribed Network

Even if there is a high number of flows in the aggregate and a high RTT difference, the APS is able to reach the target rate requested by an aggregate. The target rate for (A,C) and (B,D) is $r(A,C)_{AS} = r(B,D)_{AS} = 4Mbits/s$ Figure 6(a) shows the throughput of a 10 flow aggregate (B,D) in competition with a 10 flow aggregate (A,C). For the (A,C) aggregate, the RTT is equal to

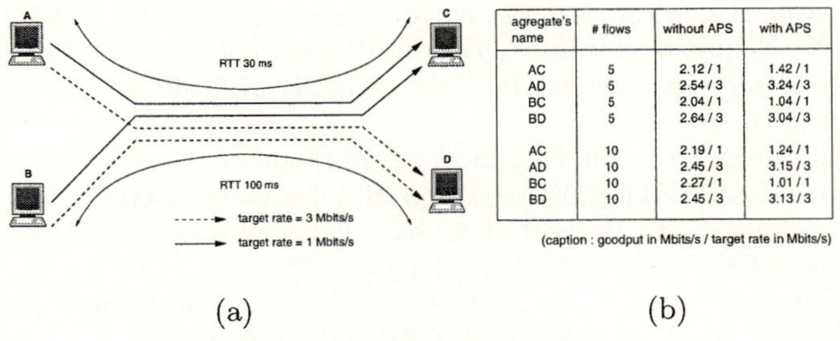

Fig. 7. Multi TCP agregates

$30ms$ and for the (B, D) aggregate, we increase gradually the RTT from $30ms$ to $500ms$. It appears that the aggregate reaches the target rate when it is feasible (i.e. when target rate : $r(B, D)_{AS} > Wmax/RTT$). On figure 6(b), the (A, C) aggregate has an RTT of $30ms$ and number of flows varying from 10 to 25. (B, D) aggregate has 5 flows and an RTT of $100ms$. We draw the instantaneous average throughput of (B, D) aggregate in function of the number of flows in (A, C). Thanks to the APS, we can see that the (B, D) throughput stays above the target rate. Finally, in the last test, we put in competition four agregates with different RTTs, target rates and number of flows. Figure 7(a) presents the scenario and figure 7(b) the results obtained. Thanks to the APS, all the aggregates reach their target rate.

6   Conclusion and Future Works

In this paper, we have studied on a real testbed an AIMD Penalty Shaper which provides throughput assurance between TCP flows. This is the first proposal that uses a delay penalty which depends on the out-profile losses in a DiffServ network. The TCP throughput is guaranteed because the conditioner works with the same dynamic than TCP (AIMD). The main consequence of these measurements is that we are able to obtain the guaranteed throughput if the profiled TCP aggregates in competition have the same or different number of flows. This is true whatever the differences between their RTTs and their target rates. The proposed solution has the advantage of being easily deployable because it doesn't require complex measurements. The solution is scalable and being likely to be used with the most frequently used conditionners such as token bucket marker or time sliding window marker. We are currently deploying this proposal on a wide area network with various traffic such as long-lived and short-lived TCP flows in order to improve this mechanism in general conditions. If the results are satisfying, then this proposal allows the effective deployment of a service adapted to the TCP traffic.

References

1. Blake, S., Black, D., Carlson, M., Davies, E., Wang, Z., Weiss, W.: An architecture for differentiated services. Request For Comments 2475, IETF (1998)
2. Seddigh, N., Nandy, B., Pieda, P.: Bandwidth assurance issues for TCP flows in a differentiated services network. In: Proc. of IEEE GLOBECOM, Rio De Janeiro, Brazil (1999) 6
3. Goyal, M., Durresi, A., Jain, R., Liu, C.: Performance analysis of assured forwarding. Internet draft, IETF (2000) draft-goyal-diffserv-afstdy-00.txt.
4. Park, E.C., Choi, C.H.: Proportionnal bandwidth allocation in diffserv networks. In: Proc. of IEEE INFOCOM, Hong Kong (2004)
5. Alves, I., de Rezende, J.F., de Moraes, L.F.: Evaluating fairness in aggregated traffic marking. In: Proc. of IEEE GLOBECOM. (2000)
6. El-Gendy, M., Shin, K.: Assured forwarding fairness using equation-based packet marking and packet separation. Computer Networks **41** (2002) 435–450

7. Kumar, K., Ananda, A., Jacob, L.: A memory based approach for a TCP-friendly traffic conditioner in diffserv networks. In: Proc. of the IEEE International Conference on Network Protocols - ICNP, Riverside, California, USA (2001)
8. Feroz, A., Rao, A., Kalyanaraman, S.: A TCP-friendly traffic marker for IP differentiated services. In: Proc. of IEEE/IFIP International Workshop on Quality of Service - IWQoS. (2000)
9. Habib, A., Bhargava, B., Fahmy, S.: A round trip time and time-out aware traffic conditioner for differentiated services networks. In: Proc. of the IEEE International Conference on Communications - ICC, New-York, USA (2002)
10. Floyd, S., Fall, K.: Promoting the use of end-to-end congestion control in the Internet. IEEE/ACM Transactions on Networking 7 (1999) 458–472
11. Yeom, I., Reddy, N.: Realizing throughput guarantees in a differentiated services network. In: Proc. of IEEE International Conference on Multimedia Computing and Systems- ICMCS. Volume 2., Florence, Italy (1999) 372–376
12. Clark, D., Fang, W.: Explicit allocation of best effort packet delivery service. IEEE/ACM Transactions on Networking 6 (1998) 362–373
13. Ramakrishnan, K., Floyd, S.: A proposal to add explicit congestion notification. Request for comments, IETF (1998)
14. Fang, W., Seddigh, N., AL.: A time sliding window three colour marker. Request For Comments 2859, IETF (2000)

Name-Level Approach for Egress Network Access Control

Shinichi Suzuki[1], Yasushi Shinjo[1,3], Toshio Hirotsu[2,3], Kazuhiko Kato[1,3], and Kozo Itano[1,3]

[1] Department of Computer Science, University of Tsukuba, Tsukuba, Ibaraki 305-8573, Japan
[2] Department of Information and Computer Sciences, Toyohashi University of Technology, Toyohashi, Aichi 441-8580, Japan
[3] Core Research for Evolutional Science and Technology (CREST), Japan Science and Technology Agency (JST)

Abstract. Conventional egress network access control (NAC) at the network layer has two problems. Firstly, wild card "*" is not allowed for a policy. Secondly, we have to run a Web browser for authentication even if we do not use the Web. To solve these problems, this paper proposes a name-level method for egress NAC. Since it evaluates the policy at the DNS server, this method enables a wild card to be used in the policy. Since each DNS query message carries user identification by using Transaction Signature (TSIG), the authentication for any service is performed without Web browsers. The DNS server configures a packet filter dynamically to pass authorized packets. This paper describes the implementation of the DNS server, the packet filter, and the resolver of the method. Experimental results show that the method scales up to 160 clients with a DNS server and a router.

1   Introduction

Network Access Control (NAC) is one of the most important research issues for improving security. NAC can be classified two types: ingress NAC and egress NAC. Ingress NAC protects internal resources from external attackers and intruders. Egress NAC prevents internal users to access unnecessary or dangerous external resources. Egress NAC is increasingly attracting the attention of educational organizations, government organizations, and companies.

Conventional egress NAC is performed by packet filters and proxies. Packet filters work at the network layer and can handle all protocols of the network layer. Furthermore, packet filters are faster than proxies. However, packet filters cannot use information at the application layer. On the other hand, proxies work at the application layer and can use information at the application layer. For example, an HTTP proxy can use application-level information such as URLs and can perform access control with URLs. However, a single proxy cannot handle all protocols of the application layer and proxies are slower than packet

P. Lorenz and P. Dini (Eds.): ICN 2005, LNCS 3421, pp. 284–296, 2005.

filters. Some systems combine packet filters and proxies for egress NAC. In this paper, we address problems of egress NAC at the network layer in packet filters.

In egress NAC, there are two ways to describe the policy: the IP-level and the name-level. In the IP-level description, an internal user is expressed as an IP address of the host the user is using, and an external server is expressed as an IP address of the host the server is running on. In the name-level description, an internal user is expressed as a string name, and an external server is expressed as a domain name. Although the policy can be described at the name-level, the policy is not evaluated at the name-level in a conventional packet filter. We have to translate these names into IP addresses in advance. Furthermore, the policy at the name-level cannot include a domain name with a wild card "*" because the domain name cannot be translated into IP addresses in advance. Furthermore, if we have to fix the binding between a user and an IP address, dynamic address assignment is not allowed. Some routers have a built-in Web server to bind a user with an IP address. In such an environment, we have to run a Web browser even though we do not use the Web.

To resolve these problems, we propose a method that performs egress NAC at the name-level. In our method, a network administrator describes a policy at the name-level by using user names and domain names, and the policy is evaluated at the name-level. Since we do not have to translate the name-level policy into IP addresses, we can include a wild card "*" in the policy. Furthermore, we can easily follow dynamically changing IP addresses.

To realize egress NAC at the name-level, we extend a DNS server. A DNS server is one of the best places for realizing egress NAC because virtually all users and applications use DNS servers and DNS query messages include domain names. If DNS query messages from clients include user names, DNS query messages include all pieces of information to evaluate a policy described at the name-level. To add user names to DNS query messages, we adopt Transaction SIGnature (TSIG)[16]. In TSIG, a signed DNS query message carries a user name and a domain name to the DNS server. The DNS server evaluates the policy at the name-level and configures a packet filter. The packet filter blocks all packets by default. The DNS server configures the packet filter dynamically to pass authorized packets from hosts of the internal user to external servers.

The rest of this paper is organized as follows. In Section 2, we present related work of egress NAC. In Section 3, we show the overview of our egress NAC at the name-level. In Section 4, we describe a DNS server that performs egress NAC at the name-level. In Section 5, we explain a packet filter that is configured by the DNS server. In Section 6, we describe a resolver that sends a user name to the DNS server. In Section 7, we present experimental results. Section 8 concludes this paper.

2   Related Work

Platform for Internet Content Selection (PICS)[12] enables labels (metadata) to be associated with Internet contents. PICS is designed to help parents and teachers control what children can access on the Internet. It also facilitates other uses for labels, including code signing and privacy. Based on the PICS platform, filtering software has been built. Some filtering products[15][1] can interpret HTTP, SMTP, NNTP, etc., with proxies. Since our method replaces packet filters, it can be used together with such proxies.

Authentication Gateway[19], Opengate[17], and Service Selection Gateway[3] are captive portals that perform egress NAC. A captive portal redirects all Web requests to a built-in Web server until a user is authenticated by the Web server. When the Web server authenticates the user, the router changes the filtering rules to pass the packets of the user. These routers need a Web browser for user authentication. Our method can perform user authentication without a Web browser.

SOCKS[10][9] is a proxy protocol at the network layer. SOCKS has capability of user authentication, a filter, and a DNS proxy. The current SOCKS implementations cannot configure filtering rules at the time of name resolution. If we modify the DNS proxy, we can realize egress NAC at the name-level. We did not adopt SOCKS because it works in the application-layer and it is slower than a kernel-based packet filter.

3   Egress NAC at the Name-Level in a DNS Server

In this section, first, we clarify the environment where our mechanism can be used. Next, we show the overview of our mechanism. Next, we describe a method of user identification by using TSIG. Finally, we discuss a security issue of DHCP.

3.1   The Environment

Figure 1 shows the environment where our mechanism can be used. Client hosts are connected with an internal network that is separated from external networks by a router. There are server hosts on external networks. A user runs application to access external servers at the client host .

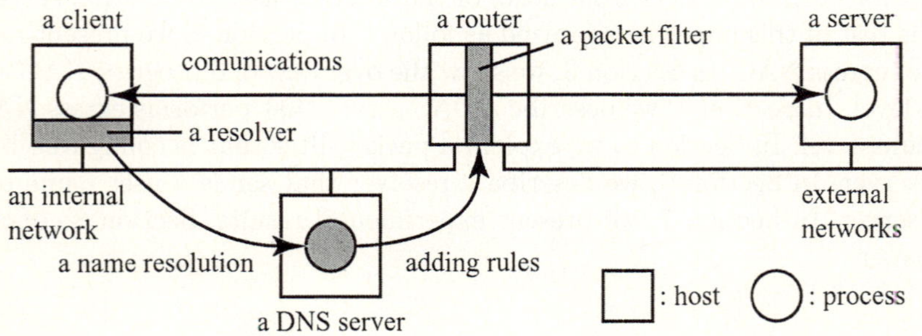

Fig. 1. The overview of egress NAC at name-level with a resolver, a DNS server, and a packet filter

An IP address must be allocate for a single user at a time either statically or dynamically. We can use DHCP. A single host is dedicated to a single user, and a user can work on multiple hosts at the same time.

Each server must have a unique IP address. In our method, we can use IP address-based virtual hosts. However, we cannot use name-based virtual hosts in HTTP and SMTP.

3.2   The Overview of Our Egress NAC

In our egress NAC, we run a DNS cache server for each network. A network administrator describes a policy for the DNS server. When a user runs an application to access an external server, the resolver linked with the application sends a signed DNS query message to the DNS server. The signed DNS query message includes the user name and a domain name.

When the DNS server receives the DNS query message from the resolver, the DNS server extracts the user name and the domain name from the DNS query message and evaluates the policy at the name-level. If the user is allowed to access the external server, the DNS server adds a permissive rule to the packet filter. After that, the DNS server returns IP addresses of the external server. By using one of these IP addresses, the application establishes a connection with the external server. The packet filter passes packets that matches with added rules and blocks all other packets.

3.3   User Identification with Transaction SIGnature (TSIG)

To perform egress NAC in a DNS server, we have to include a user name in a DNS query message. For this purpose, we adopt TSIG[16]. TSIG is a protocol that signs DNS messages based on shared private keys (*TSIG keys*) and one-way hash functions[1](Fig:2).

A TSIG key is distinguished by a string name (*a TSIG key name*). In TSIG, a resource record (RR) called *a TSIG RR* is included in the additional section of the DNS message. A TSIG RR includes a TSIG key name and a transaction signature (*a TSIG*). Based on the TSIG key name, a DNS server looks up a TSIG key and verifies the DNS message. Since DNS messages include a time stamp, TSIG can prevent replay attacks.

RFC2845 proposes two purposes of using TSIG: authentication in zone transfer and dynamic update. Currently, TSIG is not used for egress NAC. In our method, we allocate a different TSIG key name and TSIG key for each user, and we perform egress NAC with them.

The management cost of shared private keys can be expensive. We can lower the management cost of TSIG keys by using existing techniques. First, we can use TKEY[6] that is a protocol for exchanging TSIG keys. Second, we can also use Kerberos for managing TSIG keys.

---

[1] HMAC-MD5 is used by default.

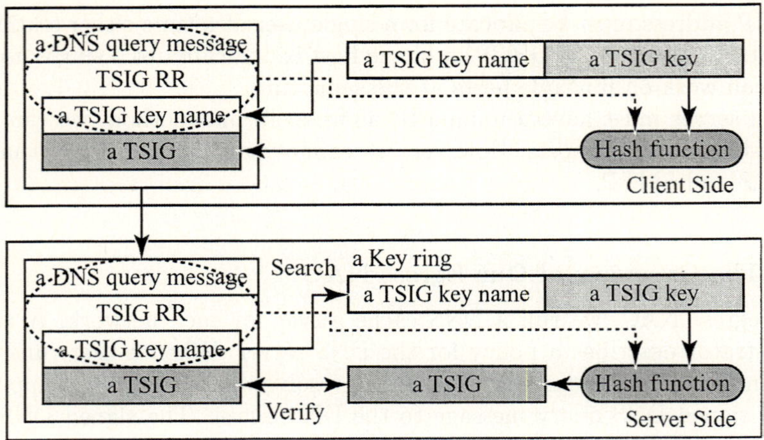

Fig. 2. The overview of Transaction SIGnature (TSIG)

SIG(0)[5] and DNSSEC[4] are other authentication protocols for DNS. SIG(0) and DNSSEC sign DNS messages by using public key cryptosystems. We did not adopt SIG(0) and DNSSEC because TSIG is lighter than SIG(0) and DNSSEC.

3.4   A Security Issue of DHCP

In our method, if an IP address is reused in DHCP, a user can access unauthorized servers the previous user accessed. To prevent this security problem, we give time-to-lives (TTLs) to packet filtering rules. If we configure TTLs that are shorter than reuse intervals of client IP addresses, the filtering rules expire before the IP addresses are reused.

4   Describing Access Control Policies at the Name-Level

We have implemented the DNS server by modifying the DNS server (named) of BIND9[7]. The size of the modified part is about 1800 lines.

In this section, we explain the method of describing policies for egress NAC. These policies are described by network administrators. We adopt Role Based Access Control (RBAC)[13] for describing access control policies because we would like to combine our method with RBAC in operating systems[14][11][18] in the future. We refined the RBAC configuration of the Trusted Solaris Operating System[14] for egress NAC.

The an access control policy is stored in four files: "user.conf", "role.conf", "profile.conf" and "server.conf". Each line of these files has the following syntax.

`name:attribute`

The name is a unique string and is used as an index of an attribute. The attribute is a list of key and value pairs separated by ";". The value is a list of names separated by ",".

The file "user.conf" defines roles and personal profiles of users (Fig:3). The file "role.conf" associates roles with profiles (Fig:4). The file "profile.conf" defines profiles as a list of allowed or denied servers (Fig:5). The file "server.conf" describes servers with domain names (Fig:6). We can use a wildcard "*" in the place of domain names. The file "server.conf" can also include protocols and port numbers which are used for configuring a packet filter.

```
Alice:role=MailUser
Bob:role=MailUser;profile=ProfExampleSSHOK
Carol:role=Admin
```

Fig. 3. An example of user.conf

```
MailUser:profile=ProfExampleMailOK
Admin:profile=ProfAllserversOK,ProfExampleMailNG
```

Fig. 4. An example of role.conf

```
ProfExampleMailOK:allow=ExampleMail
ProfExampleMailNG:deny=ExampleMail
ProfExampleSSHOK:allow=ExampleSSH
ProfAllserversOK:allow=Allservers
```

Fig. 5. An example of profile.conf

The DNS server extracts the user name and domain name from the DNS query message. By using these names, the server evaluates the policy. If the access is allowed by the policy, the evaluation returns the protocol and port numbers in the file "server.conf". The DNS server configures the packet filter to pass packets by using the IP address of the client host, the IP addresses of the external server host, and those port numbers. If the access is not allowed, the evaluation returns nothing, and the DNS server does not configure the packet filter.

For example, consider the following access control policy.

- The domain "example.com" has two mail servers: "mail1.example.com" and "mail2.example.com". These servers use port numbers 25 for SMTP and 110 for POP3.
- The domain "example.com" has three SSH servers: "host1.example.com", "host2.example.com" and "host3.example.com". These servers use the port number 22 for SSH.
- Alice and Bob are users who can read mail from servers in a domain "example.com".
- Bob can access SSH servers in the domain "example.com".
- Carol is one of the system administrators. The system administrators are allowed to access all servers except mail servers in the domain "example.com".

```
ExampleMail:domain="*.example.com";tcp=25,110
ExampleSSH:domain="host*.example.com";tcp=22
Allservers:domain="*";tcp=0-65535;udp=0-65535
```

Fig. 6. An example of server.conf

We can describe the above policy as follows.

1. In the file "server.conf" (Fig:6), we define the entry "ExampleMail" for mail servers "mail1.example.com" and "mail2.example.com", and we define the entry "ExampleSSH" for SSH servers "host1.example.com", "host2.example.com" and "host3.example.com". We define the entry "Allservers" for all external servers.
2. In the file "profile.conf" (Fig:5), we define the entry "ProfExampleMailOK" for allowing access to servers in "ExampleMail" and the entry "ProfExampleSSH" for allowing access to servers in "ExampleSSH". We define the entry "ProfAllserversOK" for allowing access to servers in "Allservers" and the entry "ProfExampleMailNG" for denying access to servers in "ExampleMail".
3. In the file "role.conf" (Fig:4), we define the entry "MailUser" to group users who can read mail with the profile "ProfExampleMailOK", and we define the entry "Admin" for the system administrators that have the profile "ProfAllserversOK" and the profile "ExampleMailNG".
4. In the file "user.conf" (Fig:3), we define entries "Alice", "Bob" and "Carol". The entry "Alice" and the entry "Bob" have the role "MailUser". The entry "Bob" also has the personal profile "ProfExampleSSHOK". The user "Carol" is one of the system administrators, so the entry "Carol" has the role "Admin".

5   The Packet Filter

In this section, we explain the implementation of a packet filter by using netfilter, a kernel-level packet filter in Linux.

As described in Section 3.4, the packet filter has to delete a rule when its TTL expires. Since the original netfilter of Linux cannot handle TTLs of rules, we have implemented a program called the netfilter manager that can handles TTLs of rules (Fig:7).

When the netfilter manager receives a rule with a TTL, the netfilter manager adds the rule to netfilter and records the rule, the time of adding, and the TTL to the internal list of rules. When a TTL of a rule expires, the netfilter manager deletes the rule from the netfilter and the internal list.

For example, suppose that the DNS server requests the netfilter manager to add the following rule.

- The client address: 192.168.0.10
- The server address: 216.239.51.99

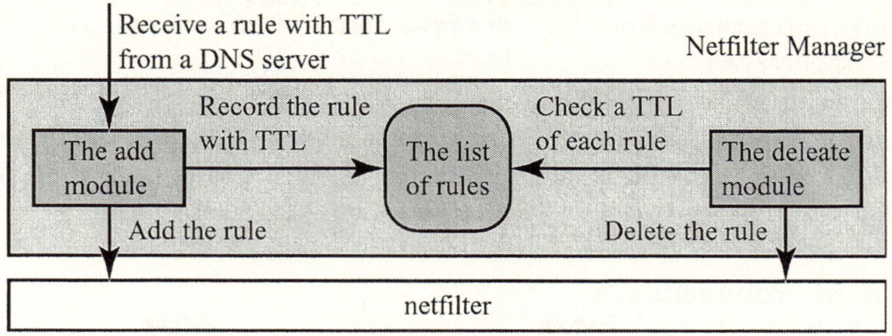

Fig. 7. Netfilter manager

- The protocol: TCP
- The port numbers: 25 and 110
- The TTL^2 : 3600

The netfilter manager adds a rule to netfilter as the following command line
does.

```
iptables -A FORWARD -i eth0 -s 192.168.0.10 -d 216.239.51.99\
-p tcp -m multiport --dport 25,110 -j ACCEPT
```

In this command line, `eth0` is the inside network interface. After 3600 seconds,
the netfilter manager deletes the rule from netfilter as the following command
line does.

```
iptables -D FORWARD -i eth0 -s 192.168.0.10 -d 216.239.51.99\
-p tcp -m multiport --dport 25,110 -j ACCEPT
```

6   The Resolver

We implemented a resolver that can handle TSIG in the shared library of
Linux. The resolver reads a TSIG key name and a TSIG key from the file
`$HOME/.tsig.key` by default. If the environment variable `TSIG_KEY` is set, the
resolver reads a TSIG key name and a TSIG key from the specified file.

When the resolver sends a DNS query message, it calculates the TSIG by
using the TSIG key. The resolver attaches a TSIG RR with the key name, the
calculated TSIG, and the time stamp to the DNS query message.

7   Experiments

In this section, we present experimental results of a microbenchmark that measured the performance of the DNS server and the resolver. Furthermore, we

---

^{2} in seconds.

measured router throughput and response times of an external HTTP server
and the DNS server.

All experiments were done on computers, each with an Intel Pentium4 3.0GHz
processor, 1GB of memory and one or two Intel PRO/1000 NICs. Their operating
system is Debian GNU/Linux 3.0 with Linux kernel 2.6, and they are connected
to a Gigabit Ethernet switch (DELL, PowerConnect 2616).

7.1   A Microbenchmark

We ran the microbenchmark for the original named and the modified named.
This microbenchmark calls the function `gethostbyname()` 1000 times in a single
process and measures the average and the standard deviation of execution times.
The result is shown in Table 1. In this table, "Original" means the original named
of BIND9. In this case, the resolver sends DNS query messages without TSIG,
and the original named does not handle TSIG. "Modified" means our modified
named. In this case, the resolver sends DNS queries with TSIG, and the modified
named handles TSIG. We gave the modified named a simple access control policy
that includes a user, a role, a profile and a server. Since we sent the same request
repeatedly, no new rule was added to netfilter.

The result in Table 1 means that the modified named is 0.28 milliseconds
slower than the original named. This time is used for compacting TSIG in both
the resolver and the DNS server. This difference is much shorter than the execution time in the cache miss case.

7.2   The Performance of the Router

First, we measured FTP throughputs according to the number of rules (1, 10000,
20000 and 30000). The result is shown in Figure 8. In this graph, the x-axis is
the number of rules, and the y-axis is FTP throughput in kilobytes per second.

The result means that the throughput of the router is independent of the
number of rules. This result was caused by the stateful nature of netfilter. Since
we had added a rule that allows passing established packets, all packets except
the first packet matched the first rule.

Next, we measured the response times of a HTTP server (Apache2.0) according to the number of rules in netfilter. In this experiment, since the client
established a new connection for each file transfer, the stateful nature of netfilter
is ineffective. The client once called the function gethostbyname() and repeatedly
called system calls connect(), write(), read() and close(). We averaged execution
times from connect() to clese(). The result is shown in Figure 9. The x-axis is

Table 1. The execution times of the function `gethostbyname()`

	Average time in milliseconds.	Standard deviation
Original	0.25	0.030
Modified	0.53	0.13

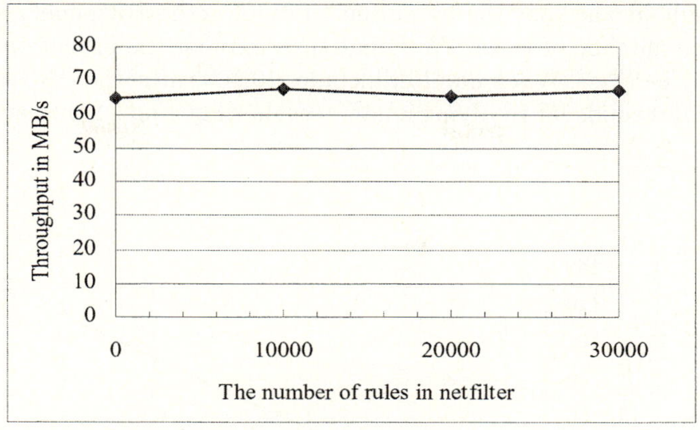

Fig. 8. FTP throughputs according to the number of rules

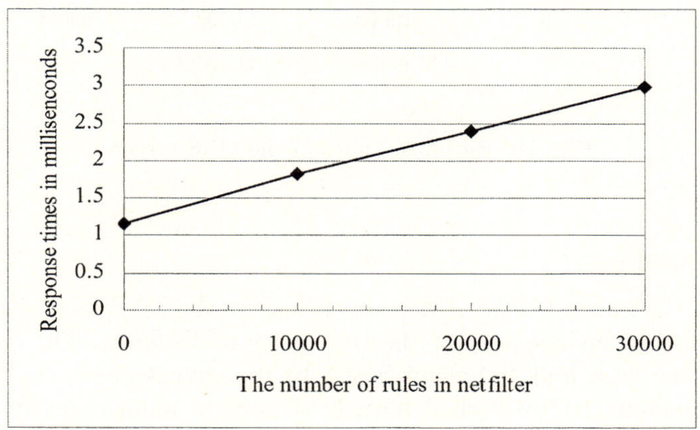

Fig. 9. Response times of the HTTP server (Apache2.0)

the number of rules in netfilter, and the y-axis is response times in milliseconds. The size of the file is 0 byte.

This result means that growing rules increases the response times of the HTTP server linearly. We found that most of the execution time was consumed by netfilter. When netfilter includes 10,000 rules, the HTTP server needs about 0.7 milliseconds more compared with the case of no rule. This difference is much shorter than typically response times of HTTP servers on the Internet, so internal users do not notice them.

Next, we measured response times of the DNS server according to the number of rules in netfilter. In this experiment, a single client called gethostbyname() repeatedly. The result is shown in Figure 10. The x-axis is the number of rules in netfilter, and the y-axis is the execution times of gethostbyname() in milliseconds.

This result means that the execution times of gethostbyname() are proportional to the number of rules. We found that most of the execution times was consumed by netfilter. When netfilter includes 10,000 rules, the DNS server needs about 35 milliseconds for resolving a name, evaluating a policy and adding a rule to netfilter.

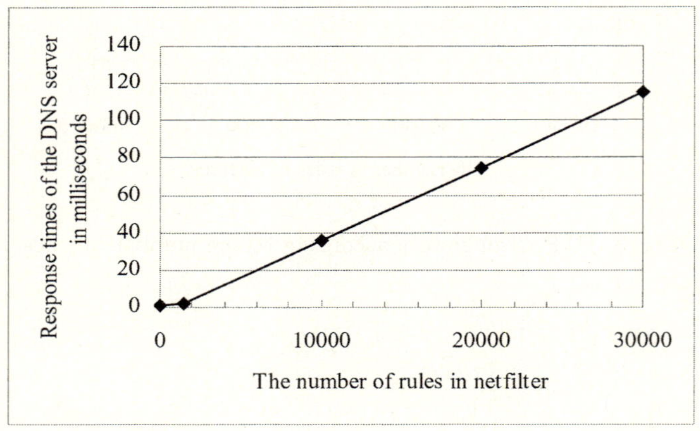

Fig. 10. Response times of the DNS server

7.3   Discussion

We counted the number of IP addresses cached by the DNS severs in an educational computing environment at the University of Tsukuba. The environment has 350 active users and 160 client hosts. In the environment, the DNS cache servers kept about 10,000 cached records of domain names and IP addresses. Normally, the TTL of a record is 86400 seconds (one day). If we set the TTL of rules to 86,400 seconds (one day), the packet filter of our mechanism keeps about 10,000 rules.

As described in Section 7.2, our DNS needs 35 milliseconds to resolve a name when netfilter includes 10,000 rules. This time is usually shorter than the time to resolve a name in the cache miss case. Therefore, we think that we can sustain such environments with our mechanism.

8   Conclusion

In this paper, we have proposed a method of egress network access control at the name-level by extending DNS servers. Our method uses a DNS server, a packet filter and a resolver. The DNS server extracts a user name and a domain name from a DNS query message signed by TSIG, performs access control with those names and adds a permissive rule to the packet filter. The packet filter passes packets in added rules. We have implemented the DNS server by modifying the DNS server (named) of BIND9 in Linux. The experimental results show that we can use TSIG with a small overhead (0.28 milliseconds). Our method needs 35 milliseconds to resolve a name when netfilter, a packet filter in Linux, includes 10,000 rules. This time is shorter than the time to resolve a name in the cache miss case of DNS.

Currently, our method uses HMAC-MD5. We have a plan to support GSS-API[2] by using GSS-TSIG[8]. We are also interested in combining our method with operating systems supporting Role Based Access Control.

Currently, we use netfilter of Linux as a packet filter. As shown in Section 7.2, the times for adding a rule in netfilter are proportional to the number of rules. We are trying other packet filters of FreeBSD.

References

1. Aladdin Knowledge Systems. eSafe 4 implementation guide. *http://www.eAladdin. com/*, July 2003.
2. E. Baize and D. Pinkas. The simple and protected GSS-API negotiation mechanism. *RFC2478*, December 1998.
3. Cisco Systems Inc. Service selection gateway. October 2003.
4. D. Eastlake. Domain name system security extensions. *RFC2535*, March 1999.
5. D. Eastlake. DNS request and transaction signatures (SIG(0)s). *RFC2931*, September 2000.
6. D. Eastlake. Secret key establishment for DNS (TKEY RR). *RFC 2930*, September 2000.
7. Internet System Consortium. BIND 9. *http://www.isc.org/*.
8. S. Kwan, P. Garg, J. Gilroy, L. Esibov, J. Westhead, and R. Hall. Generic security service algorithm for secret key transaction authentication for DNS (GSS-TSIG). *RFC3645*, October 2003.
9. M. Leech. Username/password authentication for SOCKS V5. *RFC1929*, March 1996.
10. M. Leech, M. Ganis, Y. Lee, R. Kuris, D. Koblas, and L. Jones. SOCKS protocol version 5. *RFC1928*, March 1996.
11. P. Loscocco and S. Smalley. Integrating flexible support for security policies into the Linux operating system. *2001 USENIX Annual Technical Conference (FREENIX '01)*, June 2001.
12. P. Resnick and J. Miller. PICS: Internet access controls without censorship. *Communications of the ACM*, 39(10):87–93, 1996.
13. R. S. Sandhu, E. J. Coyne, H. L. Feinstein, and C. E. Youman. Role-based access control models. *IEEE Computer*, 29(2):38–47, 1996.
14. Sun Microsystems Inc. System administration guide: Security services. *http://docs.sun.com/db/doc/816-4557/*, 2004.
15. Symantec Corporation. Symantec gateway security 5400 series refernece guide. *http://www.symantec.com/*, September 2003.
16. P. Vixie, O. Gudmundsson, D. Eastlake, and B. Wellington. Secret key transaction authentication for DNS (TSIG). *RFC2845*, May 2000.

17. Y. Watanabe, K. Watanabe, H. Eto, and S. Tadaki. An user authentication gateway system with simple user interface, low administrarion cost and wide applicability. *IPSJ Journal*, 42(12):2802–2809, December 2001.
18. R. Watson, W. Morrison, C. Vance, and B. Feldman. The TrustedBSD MAC framework: Extensible kernel access control for FreeBSD 5.0. *USENIX Annual Technical Conference, San Antonio, TX*, June 2003.
19. N. Zorn. Authentication gateway howto. *http://www.itlab.musc.edu/~nathan/ authentication_gateway/*, November 2002.

Efficient Prioritized Service Recovery Using Content-Aware Routing Mechanism in Web Server Cluster

Euisuk Kang, SookHeon Lee, and Myong-Soon Park

Department of Computer Science and Engineering,
Korea University, Seoul 136-701, Korea
prinskes@ilab.korea.ac.kr

Abstract. In the future, many more people will use web service in Ubiquitous environments. SMD-cluster, which is one of previous research, is a proposed scalable cluster system to meet the clients' requirement. SMD-cluster has a highly scalable architecture with multiple front-end nodes. However, SMD-cluster cannot provide continuous web service to clients, even though all real servers are alive when a front-end node fails. It causes considerable damage to the system, which provides mission-critical services such as E-business, Internet banking, and home shopping. To overcome this problem, we propose an efficient service recovery method in SMD-cluster. Service recovery information is maintained at each real server to minimize the overhead on front-end nodes. Waste of network usage is minimized by reducing communication messages among nodes in a cluster. We propose fast service recovery using a content-aware routing mechanism, which considers the specific content type of a request, to give priority to mission critical services.

1 Introduction

With the explosive growth of the World Wide Web, many of the popular web sites are heavily loaded with client requests. A single server hosting a web service is not sufficient to handle this aggressive growth. For handling many requests, a cluster-based web system is becoming an increasingly popular hardware platform for cost-effective, high performance network servers. Server clustering is a technology that has multiple computers providing web services, seen as a single server by the client, by connecting them through a network. Recently, the necessity of a scalable and highly available cluster-based web system has been more emphasized in Ubiquitous environments. Many projects for Ubiquitous environments are progressing based on the World Wide Web (e.g. Cooltown project [11], Aura project [12]). In the future, more people will use web services frequently, because people can access the Internet anytime and anywhere using diverse devices in ubiquitous environments. To satisfy client requirements for web services, we need a scalable and highly available cluster-based web system.

Research about cluster systems (e.g. Linux Virtual Server [5], MagicRouter [6], Resonate's Central Dispatch [7], ArrowPoint/Cisco's Content SmartTM Web Switch), shows that scalability and availability of cluster systems is limited because

P. Lorenz and P. Dini (Eds.): ICN 2005, LNCS 3421, pp. 297–306, 2005.
© Springer-Verlag Berlin Heidelberg 2005

the mechanism of efficient load balancing is operated by one node. Centralized clusters, in which client requests are distributed by one front-end node (load balancer) to an appropriate server in the cluster, has problems of bottlenecking and single-point-of-failure. [3][4] To solve these problems, SMD-cluster was proposed to use multiple front-end nodes effectively. [1] There is no communication among front-end nodes for determining which node processes the client request because each front-end node decides whether to process the client request. It reduces front-end node bottlenecking. Also, multiple front-end nodes solve the problem of single-point-of-failure. The SMD-cluster provides scalability of cluster systems with several active front-end nodes. However, there is a problem in that we lose connection information with client and real servers when a front-end node fails, so it can't provide web service continuously, even though all real servers are alive. It is a serious problem in mission-critical areas such as E-business, Internet banking, and home shopping, which circulate real money, goods, and valuable information.

In this work, we recognize this problem of SMD-clusters and propose a new, efficient method to solve the problem. The purpose of this paper is to propose a faster method of service recovery than the client requesting a new TCP connection which requires no additional hardware. In addition, communication overhead among front-end nodes is minimized without the limited scalability of the cluster system. [10] Also, we propose fast service recovery using a content-aware routing mechanism [2][8], which assigns the specific content type of request to give priority to mission critical services. The rest of the paper is organized as follows. In Section 2, we discuss SMD-cluster configurations with multiple front-end nodes for request distribution and their limitations. Section 3 describes our fast service recovery mechanism in SMD-cluster. The performance evaluation is illustrated in section 4. Finally, conclusions and future work are given in section 5.

2   Scalable Multiple Dispatchers

The centralized cluster has so many loads at the front-end node that the system is less scalable than the distributed cluster. [14][15] Even though the real server can afford to provide more services, the web server cluster system cannot expect the performance to increase. The restriction of the number of real servers results in the restriction of performance of the whole server cluster system. To overcome this limitation, an SMD-cluster that has more than one front-end node was proposed, so that the load of the front-end node is distributed. Thus, the bottleneck point to which a front-end node leads can be eliminated and packets can be forwarded into more real servers. Unlike previous research with multiple dispatchers, front-end nodes in SMD-cluster work together to distribute the load.

2.1   Architecture of SMD-Cluster

Figure 1 shows an architectural overview of an SMD-cluster in a LAN environment. An SMD-cluster consists of multiple nodes that are N front-end nodes and M real servers. All nodes in the cluster system are connected using TCP/IP protocol suite in a Local Area Network. Each front-end node has two NICs (Network Interface Card).

One NIC has a unique IP when the front-end node connects to the real server and the other is VIP that the front-end node connects to the client. The front-end should have VIP by IP aliasing. Each front-end node has a filtering program for processing the packet that is forwarded to the cluster with Virtual IP. Each of the N front-end nodes has Server-ID from 0 to (N-1), and Server-ID is applied for request distribution by packet scheduling.

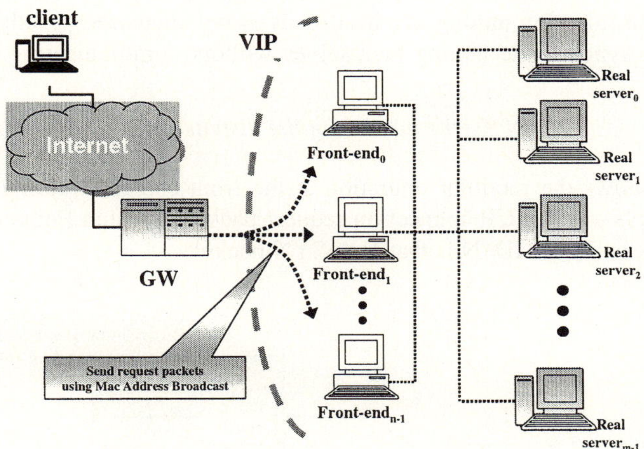

Fig. 1. SMD-cluster architecture

2.2 Operation of SMD-Cluster

We explain the process from a client to the front-end nodes via a gateway. Front-end nodes exist in the same subnet, connected to the gateway directly. When the client sends packets with VIP as a destination address to the SMD-cluster, the packets are routed by the ordinary manner used in the Internet until it arrives at the gateway. From there, the packet is routed from the gateway to all front-end nodes though broadcasting on the data link layer. In the ARP cache of a gateway, therefore, it is necessary for the entry for VIP to keep MAC_{br}. To achieve that, the kernel of a front - end node is modified in order to reply, not with an actual MAC address, but using MAC_{br} to the ARP request for VIP. This differs from ONE-IP, in that it permits the gateway to operate without modification. [13] Table 1 shows routing the packets from the gateway to the front-end nodes using the modified ARP for broadcasting the packets in the SMD-cluster. In this way, all the packets to the front-end nodes with VIP, after passing over the gateway, are broadcasted and then all front-end nodes receive the packets in the same order without any modification of the gateway.

Table 1. ARP Cache table

IP address	MAC address
VIP	MAC_{br}

2.3 Packet Scheduling by Round Robin

The problems of bottleneck and single-point-of-failure were solved by deploying several front-end nodes in an SMD-cluster. Each front-end node in the cluster using the Broadcast MAC address can receive the same packets from the Gateway simultaneously. A front-end node determines whether it is its turn to process the packet by means of a modular operation. According to the result, they process the packets. In this way, in spite of increasing the number of the front-ends, the overhead resulting from the synchronization among the front-ends is not increased greatly because the front-ends can synchronize among themselves without communicating with one another.

$$k = N \,\% \,(the\ number\ of\ the\ dispatcher) \tag{1}$$

Equation (1) shows the modular operation at the front-end for filtering. Each front-end node accepts a new TCP connection request packet in which Equation (1) condition was met. k is Server-ID, N is that N th SYN packet.

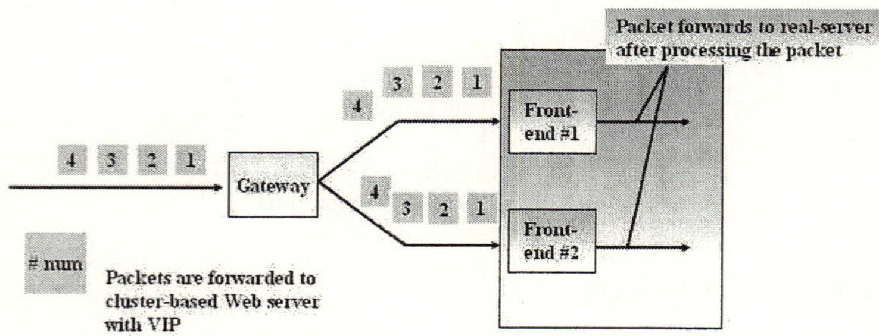

Fig. 2. Packet scheduling by Round Robin

2.4 Problem of SMD-Cluster

When a front-end node fails, another front-end node continues to provide web service for new client requests. However, the web cluster system does not provide web service when losing connection information at the failed front-end node. It is a serious problem in the mission critical areas, such as E-business, Internet banking, and home shopping. It makes users uncomfortable and it takes a long time to recover web service in order to preserve data consistency.

3 Proposed Mechanism

3.1 Service Recovery Method

When a front-end node fails, we lose all the connection information that is maintained by the front-end. To offer the availability of a cluster system, we propose a new service recovery mechanism. One of several approaches (e.g. Linux HA project[9],

Keepalived project [16]), which provides service recovery, is to use a backup node. The backup front-end node is ready to failover, and backs up the information of the primary front-end node during ordinary operation. If the primary front-end node goes down, the backup front-end node succeeds it and continues serving the request. The necessary information for the failover is connection information between client and front-end node as well as connection information between front-end node and real server. Table 2 shows maintaining connection table at each client for service recovery. #Num is difference of sequence number between two connections. (client and front-end node, front-end node and real server)

Table 2. Message Format for service recovery

Source IP Address	Source TCP Port number	Destination IP Address	Destination TCP Port number	Sequence Number
Client IP	Client Port number	Front-end node real IP	Front-end node Port number	#Num

Whenever a connection is established between a client and real server, the primary front-end sends the information to the backup front-end node. In addition, whenever the connection is terminated, the primary front-end node sends the information to the backup front-end node. This adds communication overhead on the front-end node, which causes front-end bottlenecking. Our proposed method is to maintain the information for service recovery at the real server instead of backing up the front-end node directly. The front-end node tries to establish a connection to the real server after the TCP connection is established between the client and front-end node. At this time, the front-end sends connection information to the real server. Each real server maintains its own connection information before a front-end node fails. Whenever a TCP connection is terminated between the client and real server, each real server deletes the connection information for service recovery by itself. To differ from previous works, it is necessary that only one message is sent so that the connection information of the front-end corresponds to the connection information of the real server. Accordingly, when the real server sends its own service recovery information to an active front-end node when another front-end node fails, the web server cluster can continue to provide uninterrupted web service.

We suppose that there are two front-end nodes and two real servers. The procedure for service recovery is as follows, as shown in figure 3:

1. Each front-end node watches for failure of other front-end nodes using Heartbeat demon
2. Front-end$_2$ node recognizes failure of front-end$_1$ node.
3. Front-end$_2$ node broadcasts a request packet to all real servers, offering service recovery. All real servers send the front-end$_2$ node the connection table information for front-end$_1$ that fails.
4. Front-end$_2$ node restores the connection table through the information that is transmitted from all real servers.
5. Front-end$_2$ node provides web service for client continuously.

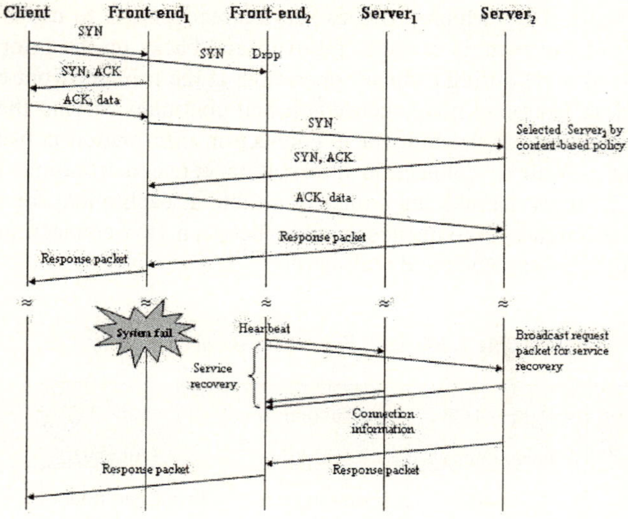

Fig. 3. Service recovery

Connection information for service recovery between the client and real server is maintained at each real server, so that it can greatly enhance the service recovery method efficiently without limiting the scalability of the cluster system.

3.2 Prioritized Service Recovery Using Content-Aware Routing

Each front-end node spends a lot of computing resources on maintaining connection information between the client and real server when many requests are forwarded to the web server cluster from the client. When all front-end nodes are overloaded, it is impossible for other front-end nodes to recover all connection information when a front-end node fails. To request a new TCP connection that the failed front-end node maintained is faster than trying to recover all connection information from the failed front-end. Therefore we propose an efficient service recovery method using a content-aware routing mechanism that considers priority of content.

Types of client requests are classified as static content (e.g. *.html) and dynamic content (e.g. *.asp, *.jsp, *.php). Dynamic content is classified into two groups; one that relates to security, and the other that does not. Content that relates to security takes an example data associated with SSL. When classifying into several types of content, high priority content recovers preferentially when a front-end node fails. By using a content-aware routing mechanism, service recovery is easier because the real server can be customized by type of content. Each real server customizes specific content when the cluster system is deployed. When a front-end node fails, another front-end node requests a specific real server that has important data to send service recovery information. In this way, we can achieve fast and efficient service recovery.

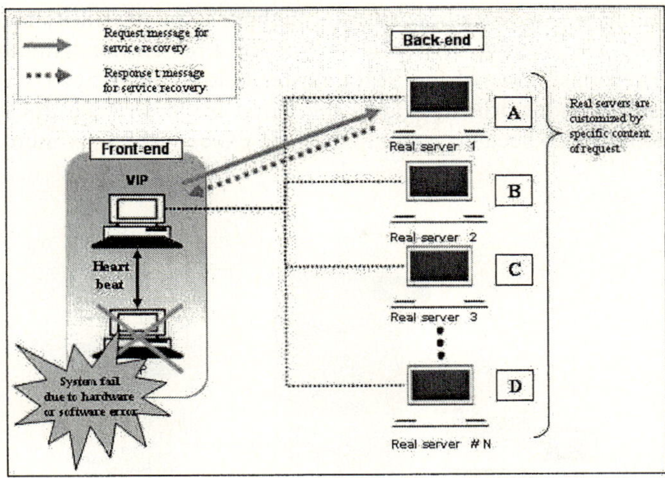

Fig. 4. Service recovery considering priority of content

Figure 4 shows a service recovery method according to priority of content type. Since only specific content is recovered, it can reduce the response time to recover service.

4   Performance Evaluation

4.1   Experimental Methodology

This section presents the simulation results of the performance analysis for the proposed method. As figure 5 shows, we organized the PSR-cluster that we proposed. The performance analysis is evaluated by comparing the previous research and the proposed scheme. We performed an experiment with 2 dispatchers, 6 real servers, and one router. All dispatchers had Pentium3-550MHz CPU, and 128MB RAM. We made a simple router that is adequate for a LAN environment. The real servers had Pentium MMX 200MHz CPU, 64MB RAM, and Apache 1.3 for web service. We limited the Apache web server to processing 2Mbps a second through MaxClient value included in the Apache configuration file. All computers used in the experiment are connected through Fast Ethernet with 100Mbps transfer rate. Only two real servers have priority over the others.

4.2   Simulation Results

The simulation result of the average response time is shown in figure 6, where the X-axis is the throughput and the Y-axis is the average response time. Throughput represents total throughput of the cluster with two front-end nodes before one front-end node fails. Response time represents the average value of response taken until the packet arrives at the client from the cluster after the front-end node fails. The result shows that the average response time of both the SMD-cluster and the proposed scheme are increased because throughput is increased by many requests from the client. The average response time of the PSR-cluster without priority is less than the response time of the SMD-cluster below 3000Kbytes/sec. However, the PSR-cluster without priority shows the worst response time after throughput passes 3000Kbytes/sec because a front-end node exceeds the maximum throughput limit.

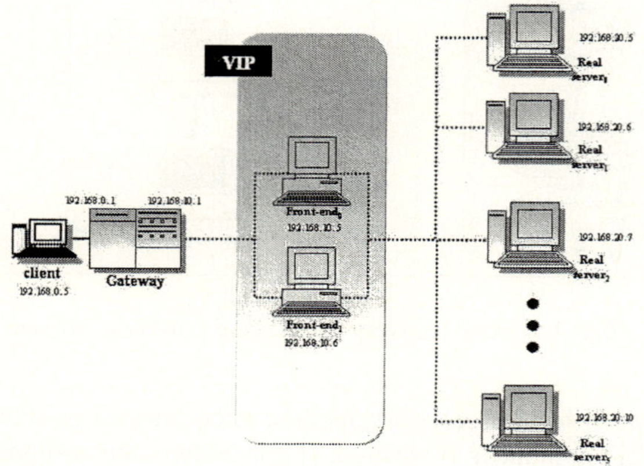

Fig. 5. Simulation model

The PSR-cluster with priority always shows less response time than the SMD-cluster. The response time of the PSR-cluster with priority reduces response time on average 14.7%, when compared to SMD-cluster. In addition, the PSR-cluster with priority is more efficient than the PSR-cluster without priority. As a result, our proposed method reduces response time in SMD-cluster.

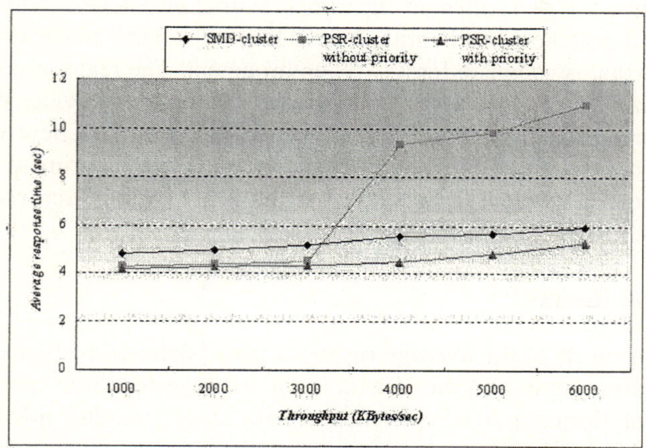

Fig. 6. Response time

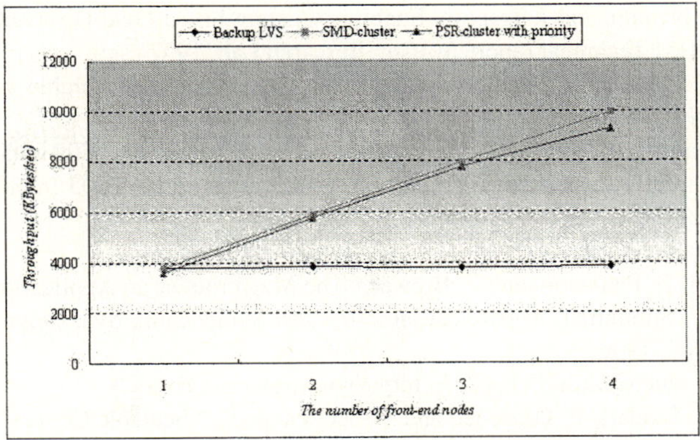

Fig. 7. Overhead of PSR-cluster

Figure 7 shows the result of the measured mean throughput (Kbytes/sec) that was processed in the PSR-cluster, SMD-cluster and Backup LVS per second as we increased the number of front-end nodes. Throughput of the PSR-cluster is increased accordingly as the number of front-ends is increased. It means that the PSR-cluster using prioritized service recovery has scalability similar to the SMD-cluster. However the PSR-cluster is less scalable than the SMD-cluster because of service recovery mechanism.

As a result, the proposed scheme provides scalable cluster system because communication overhead is not much.

5   Conclusions and Future Work

In this paper, we have presented a new scheme that can support service recovery using content-aware routing in SMD-clusters. It is less overhead for the front-end in that connection information between the client and real server for service recovery is stored at each real server. Therefore, we can recover connection information, which a failed front-end stored, without limiting scalability. To recover connection information between the client and real server immediately, we assign specific content type of requests to need mission critical service priority. In this way, it can reduce response time for service recovery. As we have shown, our recovery approach performs well in simulations with very little overhead and resource waste.

For future works, we will implement it inside the kernel. Since the overhead of content-aware routing in user-space is very high, it is best to implement it inside the kernel in order to avoid the overhead of content routing and memory copying between user-space and kernel-space.

References

1. Lee Jangho, SookHeon Lee, kyeongmo Kang, Myong-Soon Park, "Enhanced scalability with multiple dispatchers in cluster-based high web server system," The 7th World Multi conference on Systemics, Cybernetics, and Informatics (SCI 2003), Orlando, USA, July 27-30, 2003

2. G. Hunt, E. Nahum, and J. Tracey, "Enabling Content-based Load Distribution for Scalable Services," IBM Technical report. Watson Research Lab. 1997
3. D. Dias, W. Kish, R. Mukherjee and R. Tewari, "A scalable and highly available server," COMPCON 1996, pp. 85~92, 1996
4. D. Kerdlapanan, A. Khunkitti, "Content-based load balancing with multicast and TCP-handoff," Circuits and Systems, ISCAS 03. Proceedings of the 2003 International Symposium, May 25-28, 2003
5. Linux Virtual Server Project, http://www.linuxvirtualserver.org
6. E. Anderson, D. Patterson, and E. Brewer, "The Magicrouter, an Application of Fast Packet Interposing," Submitted for publication in the 2nd Symposium, Sys. Deign and Implementation, May 17, 1996
7. Resonate Products Central Dispatch, http://www.resonate.com
8. M. Aron, D. Sanders, P. Druschel, and W. Zwaenepoel, " Scalable Content-Aware Request Distribution in Cluster-Based Network Servers," Proceedings of 2000 USENIX Annual Technical Conference, San Diego, June 18-23, 2000
9. Linux High Availability project, Http://www.linux-ha.org
10. Clunix Co. Ltd, "EnclusterTM white paper," 1999, http://www.clunix.co.kr
11. T. Kindberg and J. Barton. "A Web-based nomadic computing system", Computer Networks (Amsterdam, Netherlands: 1999), 35(4): 443-456, 2001
12. D. Garlan, et al., "Project Aura: Toward Distraction-Free Pervasive Computing," IEEE Pervasive Computing, vol. 1, no. 2, Apr.-June 2002, pp. 22-31.
13. O. P. Damani et al., "ONE-IP: Techniques for Hosting a Service on a Cluster of Machines," J. Computer Networks and ISDN System, Vol.29, pp1019-1027, Sept. 1997.
14. A. Bestavros, M. Crovella, J. Liu, D. Martin, "Distributed Packet Rewriting and its Application to Scalable Server Architectures", In Proceedings of ICNP'98: The 6th IEEE International Conference on Network Protocols, Austin, USA, October 1998.
15. L. Aversa, A. Bestavros, "Load balancing a cluster of web servers: using distributed packet rewriting", Performance, Computing, and Communications Conference, IPCCC '00. Conference Proceeding of the IEEE international, pp.24-29, 2000.
16. Keepalived for Linux project, http://keepalived. Sourceforge.net

Queue Management Scheme Stabilizing Buffer Utilization in the IP Router

Yusuke Shinohara, Norio Yamagaki, Hideki Tode, and Koso Murakami

Department of Information Networking,
Graduate School of Information Science and Technology, Osaka University
{shinohara.yusuke, yamagaki, tode, murakami}@ist.osaka-u.ac.jp

Abstract. In the today's Internet, multimedia traffic have been increasing with the development of broadband networks. However, the Best-Effort service used in the IP network has trouble guaranteeing QoS for each user. Therefore, the realization of QoS guarantees becomes a very important issue. So far, we have proposed a new queue management scheme, called DMFQ (Dual Metrics Fair Queueing), to improve fairness and to guarantee QoS. DMFQ can improve fairness and throughput considering the amount of instantaneous and historical network resources consumed per flow. In addition, DMFQ has high-speed and high-scalability mechanism since it is hardware-oriented. However, DMFQ may be unable to adapt to fluctuations of networks due to its static set-up parameters. In this paper, we extend DMFQ to solve the above problem. This extended DMFQ stabilizes buffer utilization by controlling its parameter. Thus, it can dynamically adapt to various network conditions, which leads to the stability of queueing delay and the realization of QoS guarantee.

1 Introduction

Recently, various types of traffic have been increasing with the development of broadband networks. However, conventional Drop-Tail routers have trouble realizing Quality of Service (QoS) guarantee for each user, because they cause the throughput degradation and the unfairness, for example, in terms of bandwidth allocation per flow. In particular, Transmission Control Protocol (TCP)[1] connections are likely to decline the throughput significantly.

So far, various kinds of Active Queue Management (AQM) schemes have been studied to improve fairness and throughput. AQM discards arrival packets according to its policy in the output buffer of the router. Random Early Detection (RED)[2] is a typical AQM. Although RED is easy to implement, there is a risk of causing the unfairness per flow[3]. To provide fairer congestion control, Fair Random Early Drop (FRED)[3], Core Stateless Fair Queueing (CSFQ)[4] and Stochastic Fair BLUE (SFB)[5], etc. have been proposed. However, they have some problems such as processing speed, difficulties in implementation and poor adaptability to various network conditions due to its static parameters. In particular, to improve the adaptability issue, Adaptive RED (ARED)[6] that controls

P. Lorenz and P. Dini (Eds.): ICN 2005, LNCS 3421, pp. 307–317, 2005.

parameters according to buffer utilization has been proposed. Although ARED can remove the difficulty of setting some parameters, the unfairness problem may remain because of its RED-based mechanism.

In addition, some packet scheduling schemes such as Deficit Round Robin (DRR)[7] have been proposed. DRR can improve fairness and throughput effectively. However, routers with DRR are limited to a specific number of flows or classes because of its complexity.

Moreover, Differentiated Service (DiffServ)[8] has been proposed to support multi-class services. DiffServ classifies traffic and provides the classified service according to the class. However, the unfairness still remains within the same class because of its class-based mechanism.

Then, we have proposed a new queue management scheme, DMFQ (Dual Metrics Fair Queueing)[9], to improve fairness and to guarantee QoS. DMFQ discards arrival packets considering not only the arrival rate but also the flow succession time. Moreover, DMFQ has hardware-oriented mechanisms, and thus high-speed and high-scalability can be expected. However, DMFQ may be unable to adapt to fluctuations of networks due to its static set-up parameters, as well as other AQMs.

In this paper, to improve its adaptability to various network conditions, we extend DMFQ. The extended DMFQ has the same packet discard policy as DMFQ basically. However, in the extended DMFQ, we introduce the dynamic control mechanism of the buffer threshold which is the most influential parameter to calculate the packet discard probability of DMFQ. Concretely, the extended DMFQ can stabilize buffer utilization within a constant range by controlling the buffer threshold. Namely, the extended DMFQ can provide not only the fairness but also buffer stability.

The rest of the paper is organized as follows. Section 2 introduces DMFQ. Section 3 describes the extended DMFQ and Section 4 evaluates the performance of the extended DMFQ. Finally, Section 5 concludes our work so far.

2   Dual Metrics Fair Queueing (DMFQ)

2.1   Outline

DMFQ[9] is designed as the hardware-oriented mechanism for its high-speed and high-scalability, and improves the fairness in a single class.

DMFQ achieves the fairness from the viewpoint of not only the instantaneous amount of network resources consumed by each flow but also the historical ones. Concretely, DMFQ considers both the arrival rate and the flow succession time as instantaneous and historical conditions, respectively. In DMFQ, arrival packets are discarded with the packet discard probability calculated by using above information.

2.2   Flow Management

In DMFQ, the flow is defined as a stream of packets with the same reference data within the IP header, such as source and destination address and Type of Service

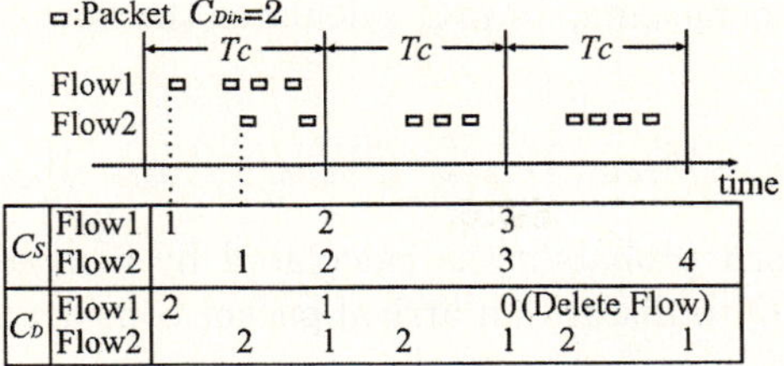

Fig. 1. An example of the succession and deletion counters

(ToS), consecutively arriving within a certain constant time interval. Moreover, DMFQ manages not only active flows which just have some backlogged packets in the buffer but also previous flows that have traversed the router so as to determine the succession time. Therefore, DMFQ defines "the managed flow" as a flow which is managed in the flow management table and distinguishes them from the active flows. That is, let F be the number of managed flow and $Fact$ be the number of active flows, where it is always $Fact \leq F$. To make the flow management simpler, DMFQ introduces the succession counter (C_S) and the deletion counter (C_D), as shown in Fig. 1. Separate counters are kept for each flow and all counters are operated every constant time interval T_C (we call this interval "the update cycle"). Concretely, C_S is used as the flow succession time. When the first packet of a flow arrives at the router, C_S is initialized to one. Then, C_S is incremented every update cycle. On the other hand, C_D is used to delete a flow entry. C_D is initialized to the initial value C_{Din} at every packet arrival. Then, C_D is decremented every update cycle. If C_D becomes zero, this means no packet has arrived within at least the time $(C_{Din} - 1) \times T_C$. Then, DMFQ deletes the flow entry from the management table.

2.3 Flow Coefficients

To calculate a packet discard probability, DMFQ uses coefficients of the arrival rate, W_{rate}, and the flow succession time, W_{con}. Simultaneously, W'_{rate} and W'_{con} are used as their average value within the class.

DMFQ calculates W_{rate} and W'_{rate} as Eq.(1).

$$W_{rate} \leftarrow \frac{W_{rate} \cdot R_c + pkt\_size}{now - T_f + R_c}, \quad W'_{rate} = \frac{Service\_rate}{Fact} \tag{1}$$

where R_c is the time window of Time Sliding Window (TSW)[10], pkt_size is the arrival packet size, now is the arrival time of the packet, T_f is the arrival time of the lastly arrived packet and $Service\_rate$ is the available rate in the class. Note that W'_{rate} means the fair share rate.

Besides, DMFQ just obtains W_{con} and W'_{con} according to C_S as follows.

$$W_{con} = C_S, \quad W'_{con} = \frac{\Sigma_F C_S}{F} \tag{2}$$

Applying these coefficients, DMFQ calculates the flow coefficient W and its average value W_{avr}.

$$W = W_{rate} \cdot W_{con}, \quad W_{avr} = W'_{rate} \cdot W'_{con} \tag{3}$$

The packet discard probability is calculated by comparing W with W_{avr}, which enables DMFQ to discard an arrival packet considering both the amount of instantaneous and historical network resources consumed by the flow.

2.4   Packet Discard Policy

Only comparing W with W_{avr} may still remain the unfairness. For example, W_{con} of a TCP connection with larger Round Trip Time (RTT) may become too large because it is difficult for such a connection from increasing the bandwidth. This leads to a larger packet discard probability P, and thus prevents to increase the available bandwidth. To solve this issue, DMFQ does not discard an arrival packet which belongs to the flow whose arrival rate is less than W'_{rate}. DMFQ calculates P by Eq.(4). Note that W_{base} is the standard value for comparison and calculated as Eq.(5).

$$P = \begin{cases} 0 & (W_{rate} < W'_{rate}) \\ \max\left[0, 1 - \dfrac{W_{base}}{W}\right] & (W_{rate} \geq W'_{rate}) \end{cases} \tag{4}$$

$$W_{base} = W_{avr} \cdot th/q_{avr} \tag{5}$$

Where, q_{avr} and th are the average queue length by Exponential Weighted Moving Average (EWMA)[2] and the buffer threshold. By calculating P with not W_{avr} directly but W_{base}, DMFQ discards arrival packets rarely at low buffer utilization and actively at high buffer utilization. Thus, DMFQ can use a buffer effectively.

3   The Extended DMFQ

3.1   Outline

DMFQ is the effective scheme that can improve fairness and throughput (See Section 4). However, DMFQ can not always attain the best performance in various network conditions due to its static set-up parameters. In particular, we focus on the buffer threshold (th) as the most important parameter. Because it influences W_{base} value (Eq.(5)), which is directly associated with the packet discard probability. However, there is no policy to set up th. Besides, it is important to stabilize buffer utilization though this property is out of concern in DMFQ. In this paper, to resolve the above problems and to attain better performance in spite of network condition, we extend DMFQ. In the extended DMFQ, we introduce the dynamic control mechanism of the buffer threshold th to stabilize buffer utilization. When th and q_{avr} are stabilized by our proposed mechanism, W_{base} is also stabilized. Therefore, the extended DMFQ can always compare W and stable W_{base} for all flows, which can improve fairness more effectively. The control of th is processed quickly because of its hardware-oriented mechanism. Thus, the extended DMFQ router may not be a bottleneck in broadband networks. Note that the extended DMFQ has the same algorithm as the packet discard policy of DMFQ.

3.2 Control Policy

The extended DMFQ controls th based on the following policy.

1. Introduction of the target queue length q_{tar} and the stability area.
We introduce the target queue length q_{tar} to control th and define the stability area which is between $\pm 10\%$ of q_{tar}. The extended DMFQ stabilizes q_{avr} around the stability area by controlling th flexibly according to network conditions. Concretely, when q_{avr} is smaller than the stability area, the extended DMFQ should insert an arrival packet into the buffer more actively. Therefore, the extended DMFQ increases th to calculate smaller P, and thus, q_{avr} increases. Conversely, if q_{avr} is larger than the stability area, th is decreased to calculate the larger P.

2. Control of th by Additive Increase Multiplicative Decrease (AIMD).
Because controlling th influences P value, th must be controlled carefully. That is, th should be increased slowly against a burst packet arrival, and decreased quickly to prevent a buffer overflow. From the above consideration, the extended DMFQ controls th by AIMD, i.e., it increases th additively and decreases th multiplicatively.

3. Introduction of sample point and update of th.
We introduce the update interval T_T. The extended DMFQ updates th every T_T (we call these points "the update points"). To prevent an excessive feedback of the change of network conditions, the extended DMFQ samples q_{avr} every time interval, T_T/n, and compares q_{avr} with the stability area at each point. We call these points the sample points. The extended DMFQ decides whether th is changed or not by these n comparison results.

With the above policy, the extended DMFQ can control th according to buffer utilization adequately.

3.3 Algorithm of Controlling th

The mechanism of controlling th is shown in Fig. 2. The stability area is between $0.9 \times q_{tar}$ and $1.1 \times q_{tar}$. The extended DMFQ samples q_{avr} and compares q_{avr} with the stability area at the sample point S_i $(i = 1, \cdots, n)$ whose interval is T_T/n. If q_{avr} exceeds the stability area at a sample point, the comparison result of the point indicates the recognition to the necessity of decreasing th. Conversely, if q_{avr} is below the stability area at a sample point, the comparison result indicates the recognition to the necessity of increasing th. After that, the extended DMFQ controls th every T_T (i.e., at the update point S_n), according to the comparison results of n times.

In the extended DMFQ, we introduce set-up parameters, α_{max} and β_{min}, which are the maximum increase and minimum decrease values. In addition, we

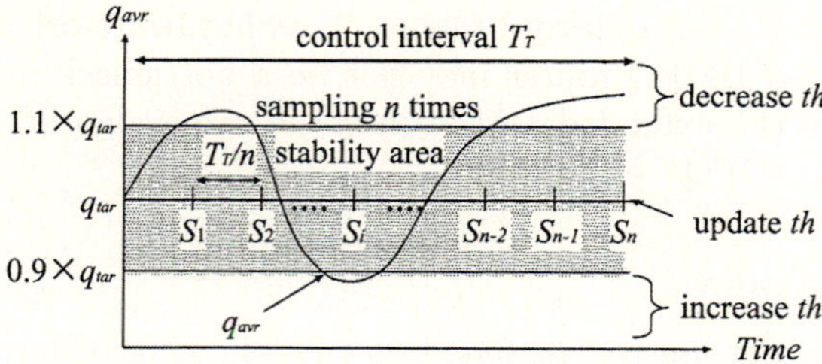

Fig. 2. The dynamic control mechanism of the extended DMFQ

use five variables, α , β, Bo, C_H and C_L. α and β are incremental and decremental values of th, respectively. Bo is a flag that indicates the occurrence of a buffer overflow within T_T. C_L and C_H are the increase and decrease counters. C_L is incremented when q_{avr} is smaller than the stability area and C_H is incremented when q_{avr} is larger than the stability area at each sample point. α, Bo, C_H and C_L are initialized to zero and β is initialized to one every T_T.

The concrete algorithm is shown as follows. When the extended DMFQ compares q_{avr} with the stability area at S_i, the following process are executed.

- In the case where $q_{avr} \leq 0.9 \times q_{tar}$, th must be increased to calculate smaller P. Therefore, α and C_L are updated as follows.

$$\begin{cases} \alpha \leftarrow \alpha + \alpha_{max}/n \\ C_L \leftarrow C_L + 1 \end{cases} \tag{6}$$

- In the case where $0.9 \times q_{tar} < q_{avr} < 1.1 \times q_{tar}$, th is considered to be proper value. Therefore, th have no need to be controlled.
- In the case where $1.1 \times q_{tar} \leq q_{avr}$, th must be decreased to calculate larger P. The extended DMFQ updates β and C_H as follows.

$$\begin{cases} \beta \leftarrow \beta - (1 - \beta_{min})/n \\ C_H \leftarrow C_H + 1 \end{cases} \tag{7}$$

Above operations at every sample point decide the variations of α and β. Note that if an arrival packet is discarded by a buffer overflow, Bo is set to one.

$$Bo = 1 \qquad \text{(when buffer overflow occurs)} \tag{8}$$

C_L, C_H, α and β determined by the above mechanism decide the degree of changing th at every update point as follows.

1. In the case where $C_L > C_H$, q_{avr} is relatively smaller than the stability area. This indicates the decline of buffer utilization. Therefore, the extended DMFQ increases th according to Eq.(9).

$$th \leftarrow \begin{cases} q_{tar} & (C_L = n \text{ and } q_{avr} < q_{tar}) \\ th + \alpha & (\text{Otherwise}) \end{cases} \qquad (9)$$

The extended DMFQ increases th additively by α. Note that if q_{avr} is smaller than its initial value ($q_{avr} < q_{tar}$) and all comparison results show that q_{avr} is always smaller than the stability area ($C_L = n$), the extended DMFQ initializes th to q_{tar} to stabilize q_{avr} into the stability area quickly.

2. In the case where $C_L = C_H$, q_{avr} is considered to remain stable. Thus, the extended DMFQ does not update th.

3. In the case where $C_L < C_H$, q_{avr} is relatively larger than the stability area. Therefore, extended DMFQ decreases th according to Eq.(10).

$$th \leftarrow \begin{cases} q_{tar} & (Bo = 1 \text{ and } q_{avr} > q_{tar}) \\ th \cdot \beta & (\text{Otherwise}) \end{cases} \qquad (10)$$

The extended DMFQ decreases th by multiplying β. If Bo is set to one ($Bo = 1$) and q_{avr} is larger than the initial value ($q_{avr} > q_{tar}$), th is forcibly initialized to q_{tar} to prevent a buffer overflow.

With the above algorithm, q_{avr} will stabilize in the stability area. Besides, q_{tar} of the extended DMFQ is easier to be set up than th of DMFQ because q_{avr} actually stabilizes around q_{tar}. Though we omitted the detail consideration, α_{max} and β_{min} should be set below $2/11 \times th$ and $9/11$ so that the control range of one update process does not exeed stability area range. Thus, parameters of the extended DMFQ can be determined more easily than the parameter of DMFQ, although the extended DMFQ needs more parameters, q_{tar}, α_{max}, β_{min} and n, than th of DMFQ.

4   Performance Evaluation

In this section, the performance of the extended DMFQ is evaluated through computer simulations. We use the OPNET10.0[11] package for this simulation.

4.1   Simulation Model

We evaluate the performance with the model shown in Fig. 3. In this model, there are 32 source-destination pairs ($N = 32$). Each source generates fixed size files as traffic, divides it into UDP or TCP segments and encapsulates it into IP packets. IP packet size is set to 1.5 Kbytes (constant). All traffic belong to the same class. We use TCP New Reno[12] and set to 100 ms as the maximum delay of the delayed ACK mechanism[1].

We compare the performance of the extended DMFQ with those of DRR, RED, ARED, FRED, SFB, CSFQ and DMFQ. Their parameters are set as follows: weighting factor for EWMA $w = 0.002$, Quantum Size = 1.5 [Kbytes] (DRR), $(min_{th}, max_{th}) = (5, 192)$ (RED, ARED), (64, 192) (FRED), max_p =0.1 (RED, FRED), $interval = 0.5$ [s], $\alpha = 0.01$, $\beta = 0.9$, $target = (79.8, 117.2)$ (ARED), $min_q = 2$ (FRED), Bin Size = 13, $\delta = 0.01$ (SFB), $th = 128$, $K = $

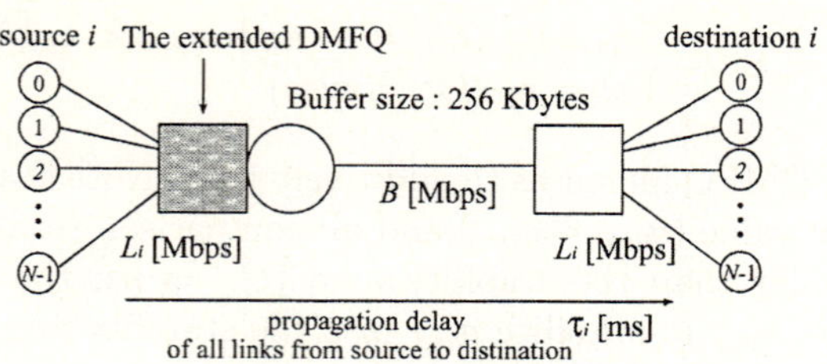

Fig. 3. Simulation Model

$K_\alpha = K_C = 400$ [ms] (CSFQ), $R_C = 100$ [ms], $T_C = 5.0$ [s], $C_{Din} = 2$ (DMFQ, the extended DMFQ), $\alpha_{max} = 0.8$ [Kbytes], $\beta_{min} = 0.95$, $n = 10$ (the extended DMFQ). In the simulations, the schemes which need flow management identify the flow by source and destination address. We evaluate the performance in the following three scenarios.

<**Scenario1**>. In this scenario, we evaluate the buffer stability as the basic performance of the extended DMFQ in changing network condition. Note that we compare the extended DMFQ with RED and ARED in this scenario. The bottleneck link speed B is set to 10 Mbps. The link speed L_i are set to 10 Mbps ($i = 0, \cdots, 29$) and 15 Mbps ($i = 30, 31$). The propagation delay $\tau_i (i = 0, \cdots, 31)$ are set to 10 ms. Source i ($i = 0, \cdots, 29$) generate TCP traffic with the Greedy Model, and each source starts to generate traffic randomly between 0 and 3 seconds. Source 30 and 31 send UDP traffic at a constant bit rate (CBR) of 15 Mbps from 40 to 60 seconds to change network conditions. The maximum window size of TCP is set to 32768 Kbytes. th of DMFQ and q_{tar} of the extended DMFQ are set to 128 Kbytes.

<**Scenario2**>. The propagation delay τ_i ($i = 0, \cdots, 31$) are set to 10 ms. Source 0 generates UDP traffic at CBR of 100 Mbps and source i ($i = 1, \cdots, 31$) generate TCP traffic. All sources generate traffic with the Greedy Model.

<**Scenario3**>. The propagation delay τ_i ($i = 0, \cdots, 31$) are set to 10, 20, 30, 40, 50, 60, 70 and 80 ms with four source-destination pairs, respectively. Source i ($i = 0, \cdots, 31$) transfer $F_i = 3.0$ Mbytes (constant) file with TCP. When a file transfer has been completed, each source waits for exponentially distributed time with a mean of 1.0 seconds until stating next one. Each source chooses a destination randomly from the four destination with the same propagation delay.

Moreover, in Scenarios 2 and 3, all sources start to generate traffic randomly between at 0 and 3 seconds. The bottleneck link speed B and the link speed L_i ($i = 0, \cdots, 31$) are set to 100 Mbps. The maximum window size of TCP is set to 65535 Kbytes. Both th and q_{tar} for DMFQ and the extended DMFQ are set to 64 Kbytes. The simulation results are the sample mean of 10 simulations whose duration is 30 (Scenario2) and 600 seconds (Scenario3).

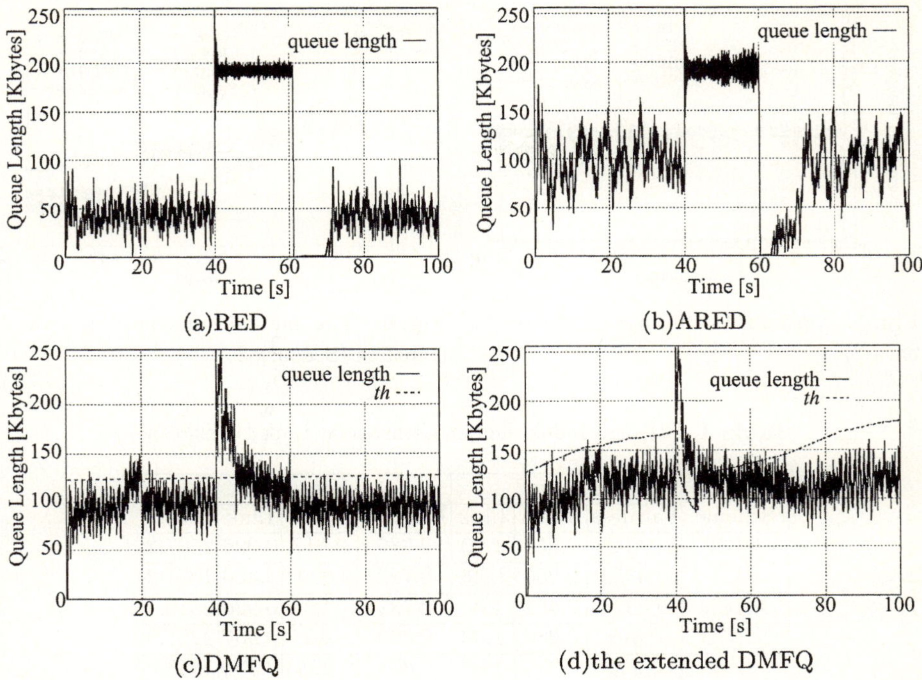

Fig. 4. Queue length (Scenario1)

4.2 Numerical Results

Figure 4(a), (b), (c) and (d) shows the queue length of RED, ARED, DMFQ and the extended DMFQ in Scenario1.

In the case of RED and ARED, queue length was stabilized around the 196 Kbytes ($= max_{th}$) by burst packet arrival of UDP traffic from 40 to 60 seconds. On the other hand, queue length of DMFQ and the extended DMFQ are hardly affected by burst packet arrival. Moreover, after 60 seconds, the extended DMFQ could stabilize queue length around q_{tar} quickly by controlling th.

Figure 5 reports normalized throughput and Tab. 1 shows the fairness index (fairness), the average buffer utilization (Buffer) and the link utilization (link) in Scenario2. The fairness index which can range between zero and one indicates the fairness, and the value closer to one means fairer[13]. Note that we sampled the normalized throughput of each connection and used them to calculate the fairness index in this paper. The average buffer utilization is defined as normalized value of the average queue length divided by the buffer size. The link utilization is defined as normalized value of the total throughput divided by the link capacity.

In Fig. 5 and Tab. 1, RED, ARED and FRED could not assign the fair share rate to all flows. In contrast, DRR, DMFQ and the extended DMFQ could assign the approximately fair share rate to all flows. Also, in particular, the extended DMFQ could stabilize the buffer utilization closely to q_{tar}.

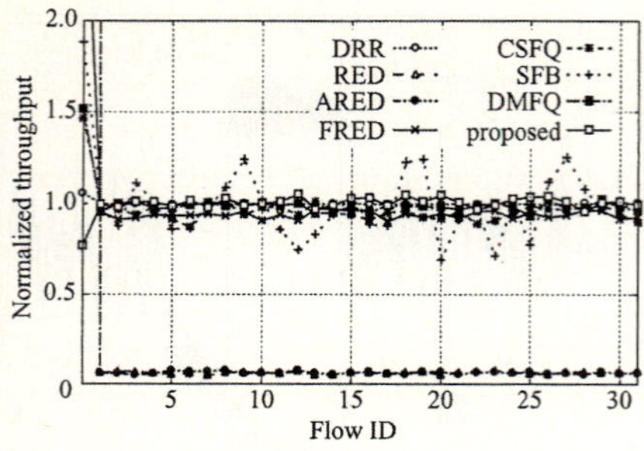

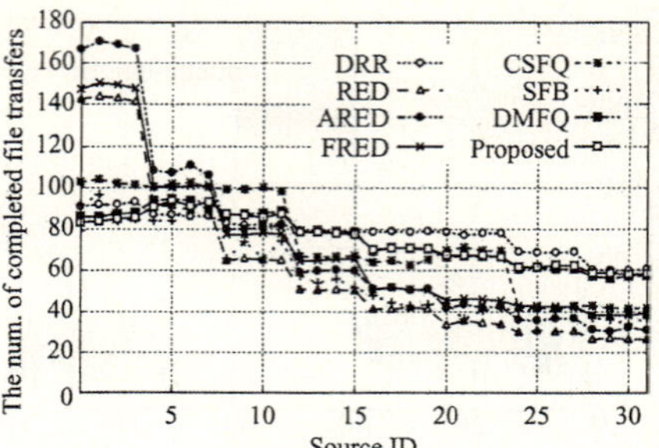

Fig. 5. Normalized throughput (Scenario2)

Fig. 6. The number of completed file transfers (Scenario3)

Table 1. Fairness Index, Buffer Utilization, Link Utilization

Scheme	Scenario2			Scenario3			
	Fairness	Buffer	Link	Fairness	Buffer	Link	Files
DRR	0.999	0.959	1.00	0.986	0.613	0.996	2507.2
RED	0.0352	0.250	0.999	0.722	0.0604	0.850	1927.2
ARED	0.0354	0.462	1.00	0.731	0.258	0.920	2312.1
FRED	0.895	0.280	1.00	0.800	0.126	0.907	2281.9
SFB	0.955	0.374	1.00	0.888	0.0253	0.734	1855.0
CSFQ	0.990	0.444	0.969	0.904	0.308	0.938	2352.9
DMFQ	0.991	0.373	1.00	0.971	0.273	0.960	2411.3
Proposed	0.998	0.307	0.999	0.975	0.252	0.957	2402.1

Figure 6 shows the number of completed file transfers per source in Scenario3, and Tab. 1 includes fairness index, buffer utilization, link utilization, and the total number of transfers of completed files in Scenario3.

In Fig. 6 and Tab. 1, RED, ARED and FRED could not achieve fair file transfers to all flows. Although the file transfers of SFB and CSFQ is fairer than RED, ARED and FRED, they transferred fewer files to the connections with long RTT (larger source ID). In contrast, DRR (not the packet discarding but the scheduling of multiple buffers), DMFQ and the extended DMFQ could achieve much fairer transfers to all flows and higher link utilization than the other schemes.

The above results show that the extended DMFQ can provide the fairness and the high throughput. In addition, the extended DMFQ can stabilize buffer utilization around the target queue length in the various network conditions.

5   Conclusion

In this paper, we extended DMFQ, to stabilize buffer utilization in various network conditions by controlling th dynamically. This leads to the stability of queuing delay and higher fairness. Numerical results as for fairness, throughput and stability of buffer utilization in various network condition indicate that the extended DMFQ is effective buffer management scheme. Future work includes the application of the extended DMFQ for the classified service environment and its hardware implementation.

References

1. W. Richard Stevens: "TCP/IP Illustrated, Volume 1," The Protocols, ADDISON WESLAY 1994.
2. S. Floyd and V. Jacobson: "Random Early Detection Gateways for Congestion Avoidance," IEEE/ACM Trans. on Networking, vol.1, no.4, pp.397-413, Aug. 1993.
3. D. lin and R. Morris, "Dynamics of Random Early Detection," Proc. of ACM SIGCOMM'97, vol.27, no.4, pp.127-137, Sept. 1997.
4. I. Stoica, S. Shenker, and H. Zhang: "Core-Stateless Fair Queueing: Achieving Approximately Fair Bandwidth Allocation in High Speed Networks," Proc. of ACM SIGCOMM'98, vol.28, no.4, pp.118-130, Sept. 1998.
5. W. Fang, D. D. Kandlur and D. Saha: "Stochastic Fair Blue: A Queue Management Algorithm for Enforcing Fairness," Proc. of IEEE INFOCOM 2001, vol.3 pp.1520-1529, Apri. 2001.
6. S. Floyd, R. Gummadi, and S. Shenker, "Adaptive RED: An Algorithm for Increasing the Robustness of RED's Active Queue Management," Aug. 2001 (available at http://www.icir. org/floyd/papers/adaptiveRed.pdf).
7. M. Shreedhar and G. Varghese: "Efficient Fair Queueing using Deficit Round Robin," IEEE/ACM Trans. on Networking, vol.4, pp.375-385, June. 1996.
8. X. Xiao and L. M. Ni: "Internet QoS: A Big Picture," IEEE Network, pp.8-18, Apr. 1999.
9. N. Yamagaki, H. Tode, and K. Murakami: "Dual Metrics Fair Queueing: Improving Fairness and File Transfer Time" Proc. of the 2005 International Symposium on Applications and the Internet (SAINT2005), Feb. 2005 (to appear).
10. D. D. Clark and W. Fang,: "Explicit Allocation of Best-Effort Packet Delivery Service," IEEE/ACM Trans. Networking, vol.6, no.4, pp.362-373, Aug. 1998.
11. OPNET Technologies Inc., http://www.mil3.com
12. S. Floyd and T. Henderson: "The NewReno Modification to TCP's Fast Recovery Algorithm," RFC2582, Oct. 1996.
13. R. Jain,D. Chiu, and W. Hawe:"A quantitative measure of fairness and discrimination for resource allocation in shared computer systems." Technical Report TR-301, DEC , Sep. 1984.

Two Mathematically Equivalent Models of the Unique-Path OSPF Weight Setting Problem

Changyong Zhang and Robert Rodošek

IC-Parc, Imperial College London,
London SW7 2AZ, United Kingdom
{changyong.zhang, r.rodosek}@icparc.ic.ac.uk

Abstract. Link weight is the essential parameter of OSPF, which is the most commonly used IP routing protocol. The problem of optimally setting link weights for unique-path OSPF routing is addressed. Existing methods solve the problem heuristically and generally result in local optima. In this paper, with the aim of finding a global optimal solution, two complete and explicit formulations with polynomial number of constraints are developed and mathematically proved equivalent. The two formulations are further compared on both model size and constraint structure for a proposed decomposition method to solve the problem.

1　Introduction

Open Shortest Path First (OSPF) [13] is the most widely deployed and commonly used protocol for IP networks. As with most other conventional IP routing protocols [6], OSPF is a shortest path routing protocol, where traffic flows between origin and destination nodes are routed along the shortest paths, based on a shortest path first (SPF) algorithm [5]. Given a network topology, the SPF algorithm uses link weights to compute shortest paths. The link weights are hence the main parameters of OSPF.

A simple way of setting link weights is to assign the weight of each link to one, so the length of a path is equal to the number of hops. Another default way recommended by Cisco is to set the weight of a link inversely proportional to its capacity, without considering any knowledge of traffic demands. More generally, the weight of a link may depend on its transmission capacity as well as its projected traffic load. As a result, an optimisation problem is to find an optimal weight set for OSPF routing, given a network topology, a projected traffic demand matrix [8], and a pre-specified objective function. It is known as the OSPF weight setting problem.

The OSPF weight setting problem has two instances, equal-cost multi-path (ECMP) and unique-path. For the first instance, traffic splitting is allowed and traffic flows of a demand can be routed over multiple paths. A number of heuristic methods have been developed, based on genetic algorithm [7] and local search method [9]. For the second instance, traffic flow of a demand is routed over a unique path. Lagrangian relaxation method [12], local search method [15], and

P. Lorenz and P. Dini (Eds.): ICN 2005, LNCS 3421, pp. 318–326, 2005.

sequential method [2] have been proposed to solve the problem. With these heuristic methods, the problem is not formulated completely and explicitly and so is generally not solved optimally.

From a management point of view, the unique-path instance requires much simpler routing functions to deploy and allows an easier monitoring of traffic flows [3]. As a result, the unique-path instance is taken into consideration in this paper. The problem is referred as the unique-path OSPF weight setting problem. It is a reduced problem of the NP-complete integer multicommodity flow problem [16].

The main goal of this paper is to formulate the problem mathematically such that an exact algorithm can be developed to solve the problem. The problem is specified in Section 2 and two complete formulations with polynomial number of constraints are developed in Section 3 and Section 4, respectively. The two formulations are mathematically proved equivalent in Section 5. The comparisons on model size and constraint structure are discussed in Section 6. Conclusions and future work are presented in Section 7.

2   Problem Specification

The unique-path OSPF weight setting problem, denoted by 1-WS, is defined as follows:

Given

- A network topology. This is a directed graph structure (V, E) where V is a finite set of nodes and E is a set of directed links. Every link (i, j) has a start node i, an end node j, and a capacity c_{ij} which is a non-negative real number.
- A traffic matrix. This is a set of demands D. Every demand d has an origin node $s_d \in V$, a destination node $t_d \in V$, and the required bandwidth b_d which is a non-negative real number.
- Weight bounds. The lower and upper bounds of the weight for each link are positive real numbers w_{min} and w_{max}, respectively.

find a non-zero weight w_{ij} for each link $(i, j) \in E$, subject to

- Flow conservation constraints. For each demand of D, at each node of V, the sum of all incoming flows (including demand bandwidth at the origin node s_d) is equal to the sum of all outgoing flows (including demand bandwidth at the destination node t_d).
- Link capacity constraints. For each link of E, the size of traffic flow over the link does not exceed the capacity of the link.
- Path uniqueness constraints. For each demand of D, there is only one assigned path.
- Path length constraints. For each demand d of D, the length of each path assigned to route d is less than the length of any other unassigned path from s_d to t_d.

- Link weight constraints. For each link $(i,j) \in E$, the weight w_{ij} is within the weight bounds, i.e., $w_{min} \leq w_{ij} \leq w_{max}$.

In networking area, the cost function is a quantitative measure of congestion. Without loss of generality, we propose to maximise the sum of residual capacity.

In the following, S denotes the set of origin nodes of all demands of D, and D_s denotes the set of all demands with origin node s.

3 A Link-Based Formulation

The 1-WS problem can be formulated as a link-based model LBM by defining a routing decision variable for each link with each demand.

Routing decision variables. Variable x_{ij}^d is equal to 1 if and only if the path assigned to demand d traverses link (i,j).

$$x_{ij}^d \in \{0,1\} \quad \forall d \in D, \forall(i,j) \in E \tag{1}$$

Link weight variables. Variable w_{ij} represents the routing cost of link (i,j).

$$w_{ij} \in [w_{min}, w_{max}] \quad \forall(i,j) \in E \tag{2}$$

Path length variables. Variable y_i^s represents the length of the shortest path from origin node s to node i.

$$y_i^s \in [0, \infty\} \quad \forall s \in S, \forall i \in V \tag{3}$$

$$y_s^s = 0 \quad \forall s \in S \tag{4}$$

Flow conservation constraints. Denote $a_i^d = -1$ if $i = s_d$, $a_i^d = 1$ if $i = t_d$, and $a_i^d = 0$ otherwise.

$$\sum_{k:(k,i)\in E} x_{ki}^d - \sum_{j:(i,j)\in E} x_{ij}^d = a_i^d \quad \forall d \in D, \forall i \in V \tag{5}$$

Link capacity constraints.

$$\sum_{d\in D} b_d x_{ij}^d \leq c_{ij} \quad \forall(i,j) \in E \tag{6}$$

Path length constraints.

$$x_{ij}^d = 0 \wedge \sum_{k:(k,j)\in E} x_{kj}^d = 0 \Rightarrow y_j^{sd} \leq y_i^{sd} + w_{ij} \quad \forall d \in D, \forall(i,j) \in E \tag{7}$$

$$x_{ij}^d = 0 \wedge \sum_{k:(k,j)\in E} x_{kj}^d = 1 \Rightarrow y_j^{sd} < y_i^{sd} + w_{ij} \quad \forall d \in D, \forall(i,j) \in E \tag{8}$$

$$x_{ij}^d = 1 \Rightarrow y_j^{sd} = y_i^{sd} + w_{ij} \quad \forall d \in D, \forall(i,j) \in E \tag{9}$$

Objective function. Maximise the sum of residual capacities, i.e. $\sum_{(i,j)\in E}(c_{ij} - \sum_{d\in D} b_d x_{ij}^d)$ which is equivalent to

$$\min \sum_{(i,j)\in E} \sum_{d\in D} b_d x_{ij}^d \tag{10}$$

Lemma 1. *LBM is a correct model of the 1-WS problem.*

4   A Tree-Based Formulation

In the previous section, the 1-WS problem is formulated as a link-based model, which defines one routing decision variable for each link with each demand. Based on the study of solution properties of the problem, it can be found that all routing paths of demands originating from the same node constitute a tree, rooted at the origin node. Accordingly, a more natural formulation of the problem is to define only one routing decision variable for each link with each origin node. In the following, the 1-WS problem is formulated as a tree-based model TBM.

Routing decision variables. Variable x_{ij}^s is equal to 1 if and only if the path of at least one of the demands originating from node s traverses link (i,j).

$$x_{ij}^s \in \{0,1\} \ \ \forall s \in S, \forall (i,j) \in E \tag{11}$$

Auxiliary flow variables. Variable z_{ij}^s represents the sum of traffic flows that transit link (i,j) from all demands originating from node s.

$$z_{ij}^s \in [0,\infty\} \ \ \forall s \in S, \forall (i,j) \in E \tag{12}$$

The *link weight variables* and the *path length variables* are the same as those in the link-based model LBM in Section 3.

Flow conservation constraints. Denote $a_i^s = -\sum_{d\in D_s} b_d$ if $i = s$, $a_i^s = b_d$ if $i = t_d, d \in D_s$, and $a_i^s = 0$ otherwise.

$$\sum_{k:(k,i)\in E} z_{ki}^s - \sum_{j:(i,j)\in E} z_{ij}^s = a_i^s \ \ \forall s \in S, \forall i \in V \tag{13}$$

Flow bound constraints. For each tree, the total flow over each link does not exceed the bandwidth of all demands originating from the root node and it is equal to zero if no demand originating from the root node traverses the link.

$$z_{ij}^s \le x_{ij}^s \sum_{d\in D_s} b_d \ \ \forall s \in S, \forall (i,j) \in E \tag{14}$$

Link capacity constraints.

$$\sum_{s\in S} z_{ij}^s \le c_{ij} \ \ \forall (i,j) \in E \tag{15}$$

Path length constraints.

$$x_{ij}^s = 0 \wedge \sum_{k:(k,j)\in E} x_{kj}^s = 0 \Rightarrow y_j^s \le y_i^s + w_{ij} \quad \forall s \in S, \forall (i,j) \in E \tag{16}$$

$$x_{ij}^s = 0 \wedge \sum_{k:(k,j)\in E} x_{kj}^s = 1 \Rightarrow y_j^s < y_i^s + w_{ij} \quad \forall s \in S, \forall (i,j) \in E \tag{17}$$

$$x_{ij}^s = 1 \Rightarrow y_j^s = y_i^s + w_{ij} \quad \forall s \in S, \forall (i,j) \in E \tag{18}$$

Path uniqueness constraints. For each tree, the number of incoming links with non-zero flows is equal to zero at the origin node, the number of incoming links with non-zero flows is equal to one at the destination node of each demand originating from the root node of the tree, and the number of incoming links with non-zero flows does not exceed one at any intermediate nodes.

$$\sum_{k:(k,s)\in E} x_{ks}^s = 0 \quad \forall s \in S \tag{19}$$

$$\sum_{k:(k,i)\in E} x_{ki}^s = 1 \quad \forall s \in S, \forall i \in \{t_d | d \in D_s\} \tag{20}$$

$$\sum_{k:(k,i)\in E} x_{ki}^s \le 1 \quad \forall s \in S, \forall i \in V \backslash \{s_d, t_d | d \in D_s\} \tag{21}$$

Objective function. Maximise the sum of residual capacities, i.e. $\sum_{(i,j)\in E}(c_{ij} - \sum_{s\in S} z_{ij}^s)$ which is equivalent to

$$\min \sum_{(i,j)\in E} \sum_{s\in S} z_{ij}^s \tag{22}$$

Lemma 2. *TBM is a correct model of the 1-WS problem.*

5   Proof of Equivalence Between the Two Formulations

Although the link-based model LBM and the tree-based model TBM represent the 1-WS problem in different perspectives, they are mathematically proved to be equivalent concerning both feasibility and optimality of the problem. The equivalence between the two models is presented based on the proof of equivalence between two corresponding models of a relaxed problem, the integer multicommodity flow problem [1] with sub-path optimality condition. The two models of the relaxed problem are denoted as RLBM and RTBM, respectively.

RLBM: Optimise(10) subject to (1), (5), (6)

RTBM: Optimise(22) subject to (11) − (15), (19) − (21)

Lemma 3. *RLBM has a feasible solution iff RTBM has a feasible solution.*

For every variable assignment satisfying the path length constraints (7)-(9) in model LBM there is a corresponding variable assignment satisfying the path length constraints (16)-(18) in model TBM. Furthermore, for every variable assignment satisfying the path length constraints (16)-(18) in model TBM there is a corresponding variable assignment satisfying the path length constraints (7)-(9) in model LBM. It follows:

Lemma 4. *The path length constraints in LBM are equivalent to those in TBM, concerning the feasibility of the 1-WS problem.*

Lemma 5. *The path length constraints in model LBM guarantee the resulting path for each demand $d \in D$ is a unique shortest path from node s_d to node t_d.*

Note, the path length constraints imply the sub-path optimality requirement on shortest paths, i.e. a sub-path of a shortest path is still a shortest path. More specifically, given two different nodes s and i, all demands originating from s and traversing node i use the same incoming link to i. Mathematically, it is formulated as follows:

$$\sum_{k:(k,i)\in E} \max\{x_{ki}^d | d \in D_s\} \leq 1 \ \ \forall s \in S, \forall i \in V\backslash\{s\} \tag{23}$$

Lemma 6. *LBM and TBM are equivalent concerning feasibility of the 1-WS problem.*

Proof. According to Lemma 5, path length constraints of the 1-WS problem imply sub-path optimality constraints. Therefore, LBM is equal to the following model:

$$\text{LBM': Optimise(10) subject to } (1) - (9), (23)$$

Clearly, TBM represents a reduced problem of RTBM and LBM' represents a reduced problem of RLBM. Besides the additional link weight variables and path length variables, the differences between TBM and RTBM represent path length constraints (16)-(18). Similarly, the differences between LBM' and RLBM represent path length constraints (7)-(9).

According to Lemma 3, RTBM and RLBM are equivalent concerning the feasibility of the relaxed problem. According to Lemma 4, path length constraints (16)-(18) in TBM are equivalent to the counterparts (7)-(9) in LBM and thereby those in LBM'. Thus, TBM and LBM' are equivalent concerning the feasibility of the 1-WS problem, and therefore TBM and LBM are equivalent concerning the feasibility of the 1-WS problem.

By constructing corresponding optimal solutions between the two models, using Lemma 6, the final result follows.

Theorem 1. *TBM and LBM are equivalent concerning optimality of the 1-WS problem.*

6   Comparisons of the Two Formulations

Model size. The path length constraints can be linearised without introducing auxiliary variables. The model sizes of the two formulations are shown in Table 1. Both sizes are polynomial in the number of nodes. The size of the tree-based model is smaller than that of the link-based model, if the number of demands with the same origin is larger than two.

Table 1. Sizes of link-based model and tree-based model of the 1-WS problem

Model	Link-based formulation	Tree-based formulation
Variables	$|D||E| + |S||V| + |E|$	$2|S||E| + |S||V| + |E|$
Linear constraints	$|D||V| + |E| + 2|D||E|$	$2|S||V| + 3|S||E| + |E|$

MIP solver. Forty-eight data sets with combinations of different parameter scenarios have been randomly generated within the following ranges: the numbers of nodes, edges and demands are 10-50, 22-648, 3-1000, respectively. Both models are coded in ECLiPSe [11] and the problem is solved using CPLEX 6.5 [10] on all data sets generated. The timeout is set to be 3600 seconds for each data instance. The objective value of the tree-based model and that of the link-based model are the same for each data instance solved optimally within timeout, which further demonstrates the equivalence between the two formulations. Unfortunately, the MIP solver does not find solutions to the problem on large data instances.

Constraint structures. Based on the study of the constraint structures of both the link-based model and the tree-based model, a proposed algorithm to solve larger instances of the 1-WS problem is Benders decomposition method [4].

Among the three types of constraints of the link-based model, flow conservation constraints and link capacity constraints contain routing decision variables, while path length constraints couple routing decision variables with link weight variables and path length variables. By using Benders decomposition method, the problem is decomposed into an integer multicommodity flow master problem and an LP subproblem. Compared with the initial MIP problem, the resulting master problem has a much smaller model size. Computational results show that path length constraints are the hardest constraints for the 1-WS problem, instead of link capacity constraints, which are recognised to be the hardest constraints for the integer multicommodity flow problem [14]. As a result, instead of solving a much larger and more complicated MIP problem in one step, Benders decomposition method solves the problem by dealing with a smaller and simpler master problem and an LP subproblem iteratively.

The tree-based model has a smaller model size and a more flexible constraint structure for decomposition algorithms such as Benders decomposition method to solve the problem. The problem can be globally decomposed into one master problem and two subproblems, instead of one master problem and one subproblem. The master problem contains only routing decision variables and path uniqueness constraints accordingly. The first subproblem deals with auxiliary flow variables and the second subproblem deals with link weight variables and path length variables. In addition, the master problem can be further decomposed, with one independent subproblem corresponding to each origin node.

7 Conclusions

Two complete and explicit mathematical formulations with polynomial number of constraints for the unique-path OSPF weight setting problem are developed and discussed. The problem is first formulated as a link-based model, based on the study of the relationship between the length of a shortest path and the weights of links within that path. The problem is further formulated as a tree-based model by analysing of solution properties of the problem. The two formulations are then proved to be equivalent concerning both feasibility and optimality of the problem. Based on the comparisons of the model sizes and constraint structures of the two formulations, the tree-based formulation has been identified to be the better one for decomposition algorithms such as Benders decomposition method to solve the problem.

Our future work includes developing the proposed algorithm completely and investigating possible improvements on both formulation aspect and algorithm aspect to accelerate the convergence rate of the algorithm.

References

1. Ahuja, R. K., Magnanti, T. L.,Orlin, J. B.: Network Flows: Theory, Algorithms, and Applications. Prentice Hall (1993)
2. Ameur, W. B., Bourquia, N., Gourdin, E., Tolla, P.: Optimal Routing for Efficient Internet Networks. In Proc. of ECUMN (2002) 10–17
3. Ameur W. B., Gourdin, E.: Internet Routing and Related Topology Issues. SIAM J. on Discrete Mathematics **17(1)** (2003) 18–49
4. Benders, J.: Partitioning Procedures for Solving Mixed-Variables Programming Problems. Numerische Mathematik **4** (1962) 238–252
5. Bertsekas, D., Gallager, R.: Data Networks, Second Edition. Prentice Hall (1992)
6. Black, U.: IP Routing Protocols: RIP, OSPF, BGP, PNNI and Cisco Routing Protocols. Prentice Hall (2000)
7. Ericsson, M., Resende, M. G. C., Pardalos, P. M.: A Genetic Algorithm for the Weight Setting Problem in OSPF Routing. J. of Combinatorial Optimization **6(3)** (2002) 299–333
8. Feldmann, A., Greenberg, A., Lund, C., Reingold, N., Rexford, J., True, F.: Deriving Traffic Demands for Operational IP Networks: Methodology and Experience. IEEE/ACM Transactions on Networking **9(3)** (2001) 265–279
9. Fortz, B., Thorup, M.: Internet Traffic Engineering by Optimizing OSPF Weights. In Proc. of INFOCOM (2000) 519–528
10. ILOG Inc.: ILOG CPLEX 6.5 User's Manual. (1999)
11. Imperial College London: ECLiPSe 5.7 Users Manual. (2003)
12. Lin, F. Y. S., Wang, J. L.: Minimax Open Shortest Path First Routing Algorithms in Networks Supporting the SMDS Services. In Proc. of ICC **2** (1993) 666–670

13. Moy, J.: OSPF Anatomy of an Internet Routing Protocol. Addison-Wesley (1998)
14. Ouaja, W., and Richards, B.: A Hybrid Multicommodity Routing Algorithm for Traffic Engineering. Networks **43(3)** (2004) 125–140
15. Ramakrishnan, K. G., Rodrigues, M. A.: Optimal Routing in Shortest-Path Data Network. Bell Labs Technical Journal (2001) 117–138
16. Wang, Y., Wang, Z.: Explicit Routing Algorithms for Internet Traffic Engineering. In Proc. of ICCCN (1999) 582–588

Fault Free Shortest Path Routing on the de Bruijn Networks*

Ngoc Chi Nguyen, Nhat Minh Dinh Vo, and Sungyoung Lee

Computer Engineering Department, Kyung Hee Univerity,
1, Seocheon, Giheung, Yongin, Gyeonggi 449-701 Korea
{ncngoc, vdmnhat, sylee}@oslab.khu.ac.kr

Abstract. It is shown that the de Bruijn graph (dBG) can be used as an architecture for interconnection networks and a suitable structure for parallel computation. Recent works have classified dBG based routing algorithms into shortest path routing and fault tolerant routing but investigation into shortest path in failure mode in dBG has been non-existent. In addition, as the size of the network increase, more faults are to be expected and therefore shortest path algorithms in fault free mode may not be suitable routing algorithms for real interconnection networks, which contain several failures. Furthermore, long fault free path may lead to high traffic, high delay time and low throughput.In this paper we investigate routing algorithms in the condition of existing failure, based on the Bidirectional de Bruijn graph (BdBG). Two Fault Free Shortest Path (FFSP) routing algorithms are proposed. Then, the performances of the two algorithms are analyzed in terms of mean path lengths. Our study shows that the proposed algorithms can be one of the candidates for routing in real interconnection networks based on dBG.

1 Introduction

For routing in dBG, Z. Liu and T.Y. Sung [1] proposed eight cases shortest paths in BdBG. Nevertheless, Z. Liu's algorithms do not support fault tolerance. J.W. Mao [4] has also proposed the general cases for shortest path in BdBG (case RLR or LRL). For fault tolerance issue, he provides another node-disjoint path of length at most $k + log_2 k + 4$ (in dBG(2,k)) beside shortest path. However, his algorithm can tolerate only one failure node in binary de Bruijn networks and it cannot achieve shortest path if there is failure node on the path.

Considering limitations of routing in dBG, we intend to investigate shortest path routing in the condition of failure existence. Two Fault Free Shortest Path (FFSP) routing algorithms are proposed. Time complexity of FFSP2 in the worst case is $0(2^{\frac{k}{2}+1}d)$ in comparison with $0((2d)^{\frac{k}{2}+1})$ of FFSP1 (in dBG(d,k)

* This research work has been partially supported by Korea Ministry of Information and Communications' ITRC Program joint with Information and Communication University.

P. Lorenz and P. Dini (Eds.): ICN 2005, LNCS 3421, pp. 327–334, 2005.
© Springer-Verlag Berlin Heidelberg 2005

and k=2h). Therefore, FFSP2 is our goal in designing routing algorithm for large network with high degree.

The rest of this paper is organized as follows. Background is discussed in section 2. In section 3, FFSP routing algorithms are presented. Performance analysis for FFSP routing algorithms is carried in section 4. Finally, some conclusions will be given in Section 5.

2   Background

The BdBG graph denoted as BdBG(d,k)[1] has $N=d^k$ nodes with diameter k and degree 2d. If we represent a node by $d_0d_1...d_{k-2}d_{k-1}$, where $d_j \in 0, 1, ..., (d-1)$, $0 \leq j \leq (k-1)$, then its neighbor are represented by $d_1...d_{k-2}d_{k-1}p$(L neighbors, by shifting left or L path) and $pd_0d_1...d_{k-2}$(R neighbors, by shifting right or R path), where $p = 0, 1, ..., (d-1)$. We write if the path $P = R_1L_1R_2L_2$ consists of an R-path called R_1, followed by an L-path called L_1, an R-path called R_2, an L-path called L_2, and so on, where subscripts are used to distinguish different sub-paths. Subscripts of these sub-paths can be omitted if no ambiguity will occur, e.g., $P = R_1LR_2$ or P=RL.

The following fig. 1a shows us an example for BdBG(2,4). Fig. 1b shows us eight cases of shortest path routing on BdBG. The gray areas are the maximum substring between source (s) and destination (d). The number inside each block represents the number of bits in the block.

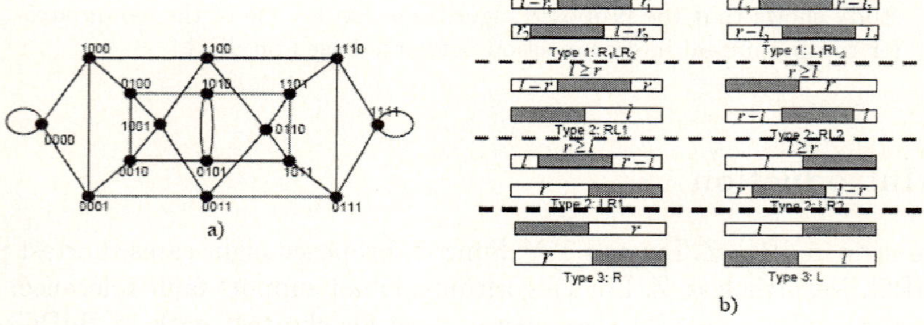

Fig. 1. a)The BdBG(2,4); b)Shortest path types[1]

3   Fault Free Shortest Path Routing Algorithms

By examining this example, finding shortest path between S 1110121100 and D 0012111001 (dBG(3,10)), we can easily see that methods provided by Liu and Mao [1][4] cannot provide FFSP. Their methods determine the maximum matched string, calculate path length corresponding with the matched string and then shifting are performed to finish routing process. In this case, the matched string is "00", path length is 8 and shortest path is 1110121100 → 1101211001 → 1012110012 → 0121100121 → 1211001211 → 2110012111 → 1100121110 →

$1001211100 \rightarrow 0012111001$ (A). If node 1012110012 is failed, then Liu's algorithm is failed in finding shortest path. Mao's algorithm can only work in binary dBG, hence it fails in dBG(3,10).

In order to find shortest path in the condition of failure existing, we cannot base on string matching concept. A shortest path found by string matching cannot be used if there is a fault in the path (as shown in the previous example). Therefore, a new concept of multi level discrete set is proposed (definition 1). By using multi level discrete set, several paths of a specific source destination pair are provided (redundancy is provided). In the above example, we can provide 3 shortest paths from S to D. Those are A; $1110121100 \rightarrow 1111012110 \rightarrow 1111101211 \rightarrow 1111012111 \rightarrow 1110121110 \rightarrow 1101211100 \rightarrow 1012111001 \rightarrow 0121110011 \rightarrow 0012111001$ (B); and $1110121100 \rightarrow 2111012110 \rightarrow 1211101211 \rightarrow 0121110121 \rightarrow 0012111012 \rightarrow 0001211101 \rightarrow 0000121110 \rightarrow 0001211100 \rightarrow 0012111001$ (C). In the case A is a failure path, we can use other 2 FFSP B and C. For building our algorithms, we assume that there is a separately protocol which detects failure nodes and then let other nodes know in periodically. Note that, level is defined simply in terms of distance from the root (level 1).

This section is organized as follows, from definition 1 to FFSP1, we state some concepts how to provide several shortest paths of a specific source and destination, and how to find FFSP among these shortest paths. Through definition 2 to FFSP2, we state how to improve the performance of our algorithm.

Definition 1: *the level m^{th} discrete set (DS_m) is a set which contains all neighbors of each element in discrete set level m-1; in the constraint that there is no existent element of discrete set level m coincides with another element of discrete set level q^{th} $(q \leq m)$ or failure node set.*

Lemma 1: *DS_m is fault free.*

Lemma 2: *all the neighbors of a node belong to DS_m are in DS_{m-1}, DS_m and DS_{m+1}, except failure nodes.*

Proof: obviously we see that DS_1 and DS_2 contain all the neighbors of DS_1 except failure nodes; DS_1, DS_2 and DS_3 contain all the neighbors of DS_2 except failure nodes. So Lemma 2 is right at m=1,2. Assuming that lemma 2 is right until p, now we prove it is right at p+1. Suppose it's wrong at p+1. That means there exist a neighbor A of an element B$\in DS_{p+1}$, and A $\in DS_i, i < p$. Because lemma 2 is right until p, hence all the neighbors of A are in DS_{i-1}, DS_i and DS_{i+1} except failure nodes. Therefore, there exists an element B'$\in DS_{i-1}, DS_i$ or DS_{i+1}, and B'=B. It contradicts with definition 1. So Lemma 2 is right at p+1. Following inductive method, lemma 2 is proved.

Lemma 3: *there exists no neighbor of any element of DS_m, which is a duplicate of any element of DS_h, $\forall h \leq m$-2.*

Proof: suppose there is a neighbor A of an element B $\in DS_m$ duplicates with an element A' of DS_h (h $\leq$ m-2). Following Lemma 2, all the neighbors of A' are in DS_{h-1}, DS_h and DS_{h+1}. Therefore, there must exist a neighbor B' of A' in level h-1 or h or h+1, and B'=B. It contradicts with definition 1.

Corollary 1: *for duplicate checking at the next level of DS_q, it is not necessary to check with any element of $DS_m, \forall m \leq q\text{-}2$.*

By assigning source node S to DS_1, then expanding to the higher level, we have the following theorem.

Theorem 1: *in BdBG(d,k), we can always find a FFSP from node $S \in DS_1$ to node $A_x \in DS_x (\forall x \leq k)$, if it exists.*

Proof: we use inductive method to prove this theorem. When x=1, 2, theorem 1 is right. Assuming that theorem 1 is right until m, m $\leq$ k. Now we prove it is right until m+1. Suppose that path from S to A_{m+1} is not the FFSP. Then we have the following cases,

- There exist $A_p \in DS_p$, $A_p = A_{m+1}$ and $p < m+1$. It contradicts definition 1.
- There exists a FFSP, $S \to B_1 \to B_2 \to ... \to B_k \to ... \to B_z \to ... \to A_{m+1}$, and B_k, $B_{k+1},..., B_z$ not belonging to any DS_i ($\forall i \leq m+1$). Because $B_{k-1} \in DS_j$ (j$\leq$m+1). Following Lemma 2, all the neighbors of B_{k-1} are in $DS_{j-1} or DS_j or DS_{j+1}$, except failure nodes. Therefore, B_k must be a failure node.
  $\to$ Theorem 1 is right at m+1. Theorem 1 is proved.

Corollary 2: *path length of a path from $S \in DS_1$ to $A_x \in DS_x$ is x-1.*

Fault free shortest path algorithm 1 (FFSP1) is proposed as a result of theorem 1 (shown in fig. 2a). It can always find FFSP in all cases (fault free mode, arbitrary failure mode) if the network still remain connected.

Proof of FFSP1: suppose path $s \to ... \to a_{ip} \to b_{jk} \to ... \to d$ is not FFSP, and then we have the following cases,

- There exist a FFSP $s \to ... \to a_{i'p'} \to b_{j'k'} \to ... \to d$ (i'$\leq$i, j'$\leq$j). It contradicts with the above assumption that a_{ip} and b_{jk} are the first neighbors between discrete sets A and B.
- There exist a FFSP $s \to ... \to a_{i'p'} \to c_1 \to ... \to c_m \to b_{j'k'} \to ... \to d$ ($i' < i, j' < j$), and $c_1, c_2, ..., c_m$ do not belong to any discrete set A_p or B_q (p$\leq$i, q$\leq$j). Due to $a_{i'p'} \in A_{i'}$ and following lemma 2, all the neighbors of $a_{i'p'}$ are in $A_{i'-1}, A_{i'}$ and $A_{i'+1}$ except failure nodes. Therefore c_1 must be a failure node.

Example 1: we want to find a FFSP from source 10000 to destination 01021, failure node 00102 (dBG(3,5)).

Applying FFSP1, we have, $A_1 = (10000)$ $B_1 = (01021)$ $A_2 = (00000, 00001, 00002, 01000, 11000, 21000)$ $B_2 = (10210, 10211, 10212, 00102, 10102, 20102)$.

However, 00102 is a failure node. So B_2=(10210, 10211, 10212, 10102, 20102).

$A_3 = (20000, 00010, 00011, 00012, 00020, 00021, 00022, 10001, 10002, 00100, 10100, 20100, 01100, 11100, 21100, 02100, 12100, 22100)$.

Then we find that 02100 and 10210 in A_3 and B_2 are the first neighbors. FFSP is found by tracking back from 02100 to 10000 and 10210 to 01021. We have FFSP $10000 \rightarrow 21000 \rightarrow 02100 \rightarrow 10210 \rightarrow 01021$. In this example, FFSP1 can provide 2 shortest paths (in the case of no failure node) $10000 \rightarrow 21000 \rightarrow 02100 \rightarrow 10210 \rightarrow 01021$ and $10000 \rightarrow 00001 \rightarrow 00010 \rightarrow 00102 \rightarrow 01021$. We pick up one FFSP $10000 \rightarrow 21000 \rightarrow 02100 \rightarrow 10210 \rightarrow 01021$ (node 00102 is fail).

Furthermore, we shall see that other elements like 00000, 00002, 01000, 11000 in A_2 are useless in constructing a FFSP. So, eliminating these elements can reduce the size of A_3 (reduce the cost at extending to next level) and improve the performance of our algorithm. It shows the motivation of FFSP2. Before investigating FFSP2, we give some definition and theorem.

Definition 2: *a dominant element is an element which makes a shorter path from source to a specific destination, if the path goes through it.*

Example 2: from the above example 1 we have 2 shortest paths (in the case 00102 is not a failure node) $10000 \rightarrow 21000 \rightarrow 02100 \rightarrow 10210 \rightarrow 01021$ and $10000 \rightarrow 00001 \rightarrow 00010 \rightarrow 00102 \rightarrow 01021$. Thus 00001 and 21000 are dominant elements of A_2, because they make shorter path than others of A_2.

Therefore, by eliminating some non-dominant elements in a level, we can reduce the size of each level in FFSP1 and hence, improve the performance of

```
1    DECLARE two array A[n] and B[n] of DS
         type

2    SET A[1] to s
3    SET B[1] to d
4    SET i and j to 1
5    WHILE (a_ip is not neighbor of b_jk)
6         CALL Expand function of A[i] and return
              value to A[i+1]
7         i = i+1
8         IF (a_ip is neighbor of b_jk) THEN
9               Exit while loop
10        ENDIF
11        CALL Expand function of B[j] and return
              value to B[j+1]
12        j=j+1
13   ENDWHILE

14   DEFINE Expand function of a DS M
15        DECLARE array N of DS type

16        SET N to DS of all neighbors of each
              element in M
17        CALL duplicate_check(N)
18        RETURN(N)
19   ENDDEFINE
```

a)

```
1    DECLARE two array A[n] and B[n] of DS type
2    SET A[1] to s
3    SET B[1]  to d
4    SET i and j to 1
5    WHILE (a_ip is not neighbor of b_jk)
6         CALL SPD function with A[i] and return value to A[i+1]
7         i = i+1
8         IF (a_ip is neighbor of b_jk) THEN
9            EXIT while loop
10        ENDIF
11        CALL SPD function with B[j] and return value to B[j+1]
12        j=j+1
13   ENDWHILE

14   DEFINE   SPD function of a DS M
15        DECLARE array N of DS type
16        SET N to DS of all neighbors of each element in M
17        WHILE (there exist set of elements T which differ in 1 bit
              at leftmost or rightmost in N)
18          IF (T is leftmost bit difference) THEN
19               CALL Pathlength function type RL2 and R
20               CALL Eliminate function to eliminate non-dominant
                    elements
21          ENDIF
22          IF (T is rightmost bit difference) THEN
23               CALL Pathlength function type LR2 and L
24               CALL Eliminate function to eliminate non-dominant
                    elements
25          ENDIF
26        ENDWHILE
27        CALL duplicate_check(N)
28        RETURN(N)
29   ENDDEFINE
```

b)

Fig. 2. a)Fault Free Shortest Path Algorithm 1 (FFSP1); b)Fault Free Shortest Path Algorithm 2 (FFSP2)

FFSP1. A question raised here is how we can determine some dominant elements in a DS_k and how many dominant elements, in a level, are enough to find FFSP. The following theorem 2 is for determining dominant elements and corollary 3 answer the question, how many dominant elements are enough.

Theorem 2: *If there are some elements different in 1 bit address at leftmost or rightmost, the dominant element among them is an element which has shorter path length toward destination for cases RL2, R (shown in fig. 1b) for leftmost bit difference and LR2, L for rightmost bit difference.*

Proof: as showing in fig. 1b, there are eight cases for shortest path. Only four cases RL2, R, LR2 and L make different paths when sources are different in leftmost bit or rightmost bit.

Example 3: following example 1, we check the dominant characteristic of three nodes A 01000, B 11000 and C 21000 (in A_2) to destination D 01021. Three nodes A, B and C are leftmost bit difference. So, type RL2, R are applied.

   • Apply type R: the maximum match string between A 01000 and D 01021 is 0, between B **1**1000 and D 01021 is 1, and between C **21**000 and D 010**21** is 2 → min path length is 3, in case of node C.
   • Apply type RL2: the maximum match string [5] between A **0**1000 and D 01021 is 1 (path length: 6), between B **1**1000 and D 01021 is 1 (path length: 7), and between C **21**000 and D 010**21** is 2 (same as case R) → min is 3, node C.

Therefore, minimum path length is 3 and dominant element is C.

Corollary 3: *when we apply theorem 2 to determine dominant elements, the maximum elements of DS_{m+1} are 2p(p is the total elements of DS_m).*

Proof: the maximum elements of DS_{m+1} by definition 1 are 2pd (dBG(d,k)). We see that in 2pd there are 2p series of d elements which are different in 1 bit at leftmost or rightmost. By applying theorem 2 to DS_{m+1}, we obtain 1 dominant element in d elements differed in 1 bit at leftmost or rightmost.

Fault Free Shortest Path Algorithm 2 (FFSP2) is proposed in fig. 2b.
The condition in line 5 and line 8 (fig. 2a, 2b) let us know whether there exists a neighbor of array A and B of discrete set, $\forall a_{ip} \in$A[i],$\forall b_{jk} \in$B[j] . The SPD(M) function, line 14 fig. 2b, finds the next level of DS M (DS N) and eliminates non-dominant elements in N followed theorem 2. Expand(M) function, line 14 fig. 2a, finds the next level of DS M. Pathlength type p function, line 19,23 fig. 2b, checks path length followed type p of each element in T toward destination. Eliminate function, line 20, 24, eliminates element in T, which has longer path length than the other. The duplicate_check(N) function, line 17 fig. 2a and line 27 fig. 2b, check if there is a duplicate of any element in N with other higher level DS of N. For duplication checking, we use the result from corollary 1. Then, we get FFSP by going back from a_{ip} to s and b_{jk} to d.

Example 4: we try to find FFSP as in example 1. By applying FFSP2, we have, $A_1 = (10000)$ $B_1 = (01021)$ $A_2 = (00001, 21000)$ $B_2 = (10210, 00102)$. However, 00102 is a failure node. So B_2 becomes (10210).

$A_3 = (00010, 10000, 10001, 02100)$. However, node 10000 coincides with 10000 of A_1. So A_3 becomes (00010, 10001, 02100). Then we find that 02100 and 10210 in A_3 and B_2 are the first neighbors. FFSP is found by tracking back from 02100 to 10000 and 10210 to 01021. We have FFSP 10000 $\rightarrow$ 21000 $\rightarrow$ 02100 $\rightarrow$ 10210 $\rightarrow$ 01021.

4   Performance Analysis for FFSP1 and FFSP2

Mean path length is the significant to analyze and compare our algorithm to others. Z. Feng and Yang [2] have calculated it based on the original formula, Mean path length $= \frac{Total internal traffic}{Total external traffic}$ for their routing performance. We can use the above equation to get the mean path length in the case of failure. We assume that failure is random, and our network is uniform. That means the probability to get failure is equal at every node in the network.

Table 1 shows the results in the simulation of mean path length using six algorithms, SCP[3], RFR, NSC, PMC[2], FFSP1 and FFSP2. Our two algorithms show to be outstanding in comparison with the four algorithms. They always achieve shorter mean path length than the other algorithms.

This section is completed with study in time complexity of our algorithms. As A. Sengupta [9] has shown that dBG(d,k) has connectivity of d-1. Hence, our

Table 1. Mean path length of FFSP1, FFSP2 in comparison with others

d,k	no. of nodes	SCP (fault free mode)	RFR (fault free mode)	NSC (fault free mode)	PMC (fault free mode)	FFSP1/2 (fault free mode)	FFSP1 (1 node fail)	FFSP2 (1 node fail)	FFSP1 (2 nodes fail)	FFSP2 (2 nodes fail)
2,2	4	1.167	1.167	1.167	1.167	1.167	1.333	1.333	1	1
2,3	8	1.643	1.643	1.643	1.643	1.643	1.762	1.762	1.733	1.733
2,4	16	2.258	2.188	2.146	2.142	2.142	2.1	2.21	2.286	2.286
2,5	32	2.984	2.796	2.794	2.766	2.754	2.787	2.794	2.285	2.832
2,6	64	3.801	3.653	3.551	3.495	3.453	3.467	3.483	3.487	3.506
3,2	9	1.417	1.471	1.417	1.417	1.417	1.429	1.429	1.476	1.476
3,3	27	2.128	2.105	2.007	2.077	2.077	2.095	2.095	2.117	2.117
3,4	81	2.978	2.911	2.865	2.849	2.833	2.84	2.842	2.846	2.848
3,5	243	3.907	3.800	3.755	3.736	3.674	3.676	3.696	3.678	3.698
3,6	729	4.875	4.775	4.704	4.722	4.572	4.573	4.626	4.573	4.627
4,2	16	1.550	1.550	1.550	1.550	1.550	1.552	1.552	1.56	1.56
4,3	64	2.369	2.343	2.321	2.321	2.321	2.327	2.327	2.335	2.335
4,4	256	3.298	3.251	3.214	3.218	3.178	3.18	3.185	3.184	3.189
4,5	1024	4.273	4.207	4.172	4.209	4.1	4.101	4.13	4.101	4.13
5,2	25	1.633	1.633	1.633	1.633	1.633	1.634	1.634	1.636	1.636
5,3	125	2.510	2.491	2.471	2.471	2.471	2.473	2.473	2.476	2.476
5,4	625	3.471	3.438	3.41	3.437	3.378	3.379	3.385	3.379	3.385
6,2	36	1.690	1.690	1.690	1.690	1.690	1.691	1.691	1.693	1.693
6,3	216	2.601	2.585	2.570	2.570	2.570	2.571	2.571	2.573	2.573

time complexity study is based on assumption that the number of failures is at most d-1 and our study is focused on large network with high degree (d>>1). Therefore, diameter of our network in this case is k. We have the following cases,

- For FFSP1, the second level DS lies in the complexity class $0(2d)$, the third level DS lies in the complexity class $0(2d(2d\text{-}1))\approx 0(4d^2)$, the fourth lies in $0(2d(2d-1)^2) \approx 0(8d^3)$, etc... Hence, time complexity of FFSP1 lies in the complexity class $0((2d)^n)$, the value of n equals to the maximum level DS provided by FFSP1. In the worst case, time complexity of FFSP1 lies in $0((2d)^{\frac{k}{2}+1})$ (k=2h), or $0((2d)^{\frac{k+1}{2}})$ (k=2h+1), k is maximum path length from source to destination (the diameter).

- The computation time of FFSP2 can be divided into 2 parts. One is performing computation on expanding to next level, checking for duplicate and neighboring checking between DS A[m] and B[q]. This part is like FFSP1, the difference is that each DS here grows following a geometric progression with common quotient 2 and initial term 1 (as shown in corollary 3). The other part is performing computation on finding dominant elements. Hence, the second level DS lies in the complexity class $0(2+2d)\approx0(2d)$, the third level DS lies in the complexity class $0(4+4d)\approx0(4d)$, the fourth lies in $0(8+8d)\approx0(8d)$, etc... Hence time complexity of FFSP2 lies in the complexity class $0(2^n d)$, the value of n equals to the maximum level DS provided by FFSP2. FFSP2 would cost us $0(2^{\frac{k}{2}+1}d)$ (k=2h), or $0(2^{\frac{k+1}{2}})$ (k=2h+1) time in the worst cases, k is maximum path length from source to destination (the diameter).

5   Conclusion

We have proposed new concepts, and routing algorithms in dBG(d,k). Our routing algorithms can provide shortest path in the case of failure existence. Our simulation result shows that FFSP2 is an appropriate candidate for the real networks with high degree and large number of nodes, while FFSP1 is a good choice for high fault tolerant network with low degree and small/medium number of nodes. Therefore, the algorithms can be considered feasible for routing in real interconnection networks.

References

1. Zhen Liu, Ting-Yi Sung, "Routing and Transmitting Problem in de Bruijn Networks" IEEE Trans. on Comp., Vol. 45, Issue 9, Sept. 1996, pp 1056 - 1062.
2. O.W.W. Yang, Z. Feng, "DBG MANs and their routing performance", Comm., IEEE Proc., Vol. 147, Issue 1, Feb. 2000 pp 32 - 40.
3. A.H. Esfahanian and S.L. Hakimi, "Fault-tolerant routing in de Bruijn communication networks", IEEE Trans. Comp. C-34 (1985), 777.788.
4. Jyh-Wen Mao and Chang-Biau Yang, "Shortest path routing and fault tolerant routing on de Bruijn networks", Networks, Vol. 35, Issue 3, Pages 207-215 2000.
5. Alfred V. Aho, Margaret J. Corasick, "Efficient String Matching: An Aid to Bibliographic Search", Comm. of the ACM, Vol. 18 Issue 6, June 1975.

Traffic Control in IP Networks with
Multiple Topology Routing

Ljiljana Adamovic, Karol Kowalik, and Martin Collier

Research Institute for Networks Communications Engineering (RINCE),
Dublin City University, Dublin 9, Ireland
{ljiljana, kowalikk, collierm}@eeng.dcu.ie

Abstract. The deployment of MPLS has been motivated by its potential for traffic engineering, and for supporting flows requiring QoS guarantees. However, the connectionless service available in MPLS incurs a considerable overhead due to the label management required. We advocate the use of a new protocol to support connectionless traffic in an MLPS-friendly way, called subIP. subIP uses the MPLS packet format with minimal changes, and thus interoperates easily with MPLS, but supports best-effort traffic with no signalling or label management overhead. It supports multipath routing through the use of multiple virtual network topologies. We present a Virtual Topology Algorithm to map incoming traffic onto multiple topologies for load balancing purposes. Simulations show that this algorithm can give rise to a significant improvement in network utilization.

1   Introduction

Over time Internet traffic has significantly increased hence there is a need to introduce control of traffic routes in IP networks in order to optimize network utilization and accommodate more traffic. Traditionally in IP traffic is routed along shortest paths in a network domain. While this approach conserves network resources, it may cause problems when several shortest paths overlap on a link, causing congestion on that link or when the capacity of the shortest path is insufficient for the traffic between a source and a destination, while a longer suitable path exists. A solution to this problem is to provide routers with multiple paths, not necessarily shortest, and enable them to balance the traffic among the available paths according to the current traffic load and path capacity.

Introducing multipath routing in IP networks can be simplified if we first simplify packet forwarding in a network domain by implementing label switching. In this case an additional header is added to each IP packet. This header contains a label which is in general a path identifier. Forwarding decisions through the domain are then based on labels, leading thus to a simpler and fast packet forwarding. One such solution is implemented in MPLS [1], a forwarding protocol which has received considerable attention in providing traffic engineering in IP networks. It uses a connection-oriented approach in providing multipath routing. A separate label distribution protocol acts as a signalling protocol for establishing label switched paths through an MPLS domain prior to packet forwarding. In [2] we proposed an alternative label switching solution for multipath

P. Lorenz and P. Dini (Eds.): ICN 2005, LNCS 3421, pp. 335–342, 2005.

intradomain routing with a connectionless forwarding protocol called subIP. subIP, although a separate protocol, may be considered as a connectionless service of MPLS since a minimal change is required in the MPLS header, as discussed in [2]. It can be used instead of currently proposed MPLS hop-by-hop routing [1]. subIP uses a distributed approach in providing multipath routing with a routing concept called Multiple Topology Routing (MTR) [2]. A control field in the packet header, set by the ingress router, determines which of the paths is used on forwarding. In this paper we present an algorithm called Virtual Topology Algorithm (VTA) to be used in MTR for determining multiple paths in a subIP area and framework for routing best-effort flows across these paths. We show that with subIP and a simple routing concept the network utilization in an IP network can be improved, while fairness in using network resources and flow throughput are increased.

The explicit routing concept of MPLS with a label distribution protocol may provide more paths between the area routers than subIP and its MTR concept and is thus more powerful in traffic engineering. The trade off is complexity of traffic control and protocol overhead. Another advantage of explicit routing is that it may provide bandwidth reservation per flow along a path, and thus quality-of service (QoS) as requested. However to provide end-to-end QoS to the customers MPLS needs to be globally implemented and traffic control protocols standardized. We expect that satisfying improvement in network performance may be achieved within subIP areas with our approach, where the simplicity of the connectionless packet forwarding is retained and that deploying subIP will assist migration towards a global MPLS network.

The paper is organized as follows. In section 2 we briefly explain subIP and Multiple Topology Routing concept. The Virtual Topology Algorithm is presented in section 3. Simulation results of MTR routing with VTA algorithm implemented are presented and discussed in section 4. The paper is summarized in section 5.

2   subIP and MTR

subIP [2] can be regarded as a best-effort service of MPLS [1], with one of the three experimental bits in the MPLS header, marked as bit T in Fig. 1, being reserved to differentiate between the two types of service: connectionless subIP, when set to 1, and connection-oriented MPLS, when set to 0.

The main difference between subIP and MPLS is that labels are global within a subIP area in subIP, with the format shown in Fig. 1, while labels in MPLS have local significance[1]. The size of a subIP area is bounded to 256 routers where each router is assigned a one byte long subIP address. Traffic is aggregated when it enters the area

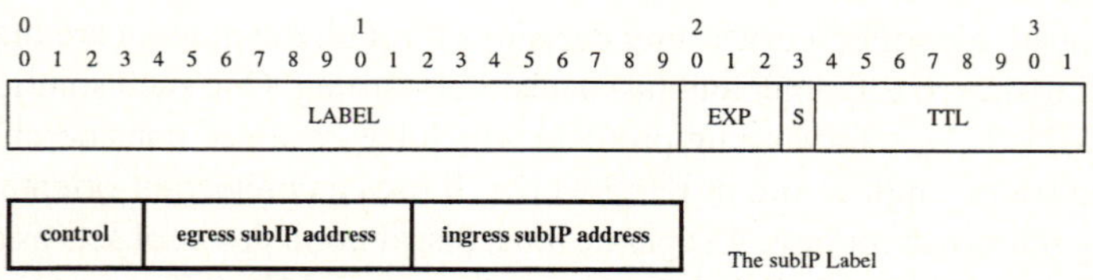

Fig. 1. The subIP Header and Label Format

based on the egress area router and the ingress and the egress router subIP addresses are stored in the label field. Within the area the IP routing concept is implemented. Forwarding decisions through the domain are based on the egress subIP address (the area destination). Packets are forwarded along the shortest paths, as calculated by each area router using the network topology information distributed by a routing protocol. Comparing to the MPLS concept, there is no label distribution when the topology changes (which introduces additional traffic and latency), no label swapping when forwarding and forwarding tables are smaller.

An additional novelty of subIP is Multiple Topology Routing (MTR), a multipath routing concept allowed with the *control* field in the subIP label and additional information distributed through the area using the subIP Control Message Protocol (sCMP) [2]. sCMP runs *on top* of subIP within a subIP area, its messages are sent in the data portion after the subIP header and identified with the *protocol field* in this header, which occupies the remaining two experimental bits. MTR is explained through an example in Fig. 2. The paths marked in Fig. 2 are the result of a shortest path calculation for the given networks as the path between routers R1 and R7, assuming that each link has unit cost. *Network 2* is obtained by removing the link R3-R5 from *Network 1*, which results in a different shortest path between routers R1 and R7. This feature, that various network topologies may provide different shortest paths through the network, is the basis of MTR. After a new *virtual* topology is derived from the physical network topology based on different criteria (an example of which is discussed in section 3), sCMP distributes the new topology information to all the routers within the area. The routers calculate shortest paths for the *virtual* topology also and store the results as a separate next hop routing table entry, providing thus alternative paths to the shortest paths calculated based on the physical network topology. In the example shown in Fig. 2, the *Network 2* topology is distributed to each router in *Network 1*. After the calculation of the shortest paths for both topologies, two paths are provided between routers R1 and R7, as shown in Fig. 2.

Which of the topologies is used for packet forwarding is determined by the value of the *control* field in the subIP packet header. It is set by the ingress router when a packet enters the area. Determining virtual topologies is the responsibility of one preselected

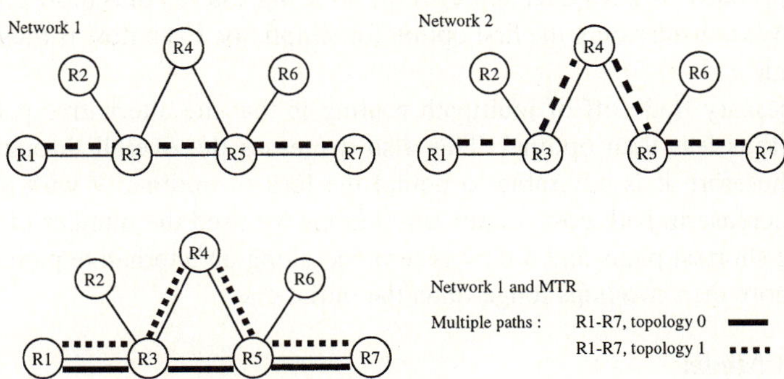

Fig. 2. Multiple R1-R7 Routes with MTR

area router, which we refer to as the traffic control router. It floods the virtual topology information throughout the area (which in general happens rarely) and controls the edge routers in using different topologies in packet forwarding. The proposed centralized approach in traffic control is expected to be based on the area-wide view of traffic, measured at the ingress points.

3   Virtual Topology Algorithm

The Virtual Topology Algorithm (VTA) is an algorithm designed for the MTR concept of subIP. The purpose of the algorithm is to improve IP network performance, where *network* refers to a subIP area. Given that subIP uses the IP routing concept, it faces the same problems as IP: multiple shortest paths may overlap on a link, causing congestion on that link or the shortest path bandwidth may be insufficient for the current traffic demands while there is a longer path with sufficient bandwidth. The goal of the algorithm is to reduce the probability of this happening. The probability of congestion on a link in the network will be reduced if the traffic is evenly spread among the network links, as much as possible. By increasing throughput per flow the probability that there is not enough bandwidth for the current traffic demands decreases. Briefly, VTA determines *critical* links in the physical topology of a subIP area and defines virtual topologies containing only non-critical links. If a flow follows a path which contains a critical link, and the alternative path exists in the new topology, the algorithm assigns the flow to the alternative path. Changes in traffic distribution and throughput before and after implementing MTR are monitored in evaluating the performance of the algorithm.

Each flow aggregation within the network between a pair of routers is identified with a unique subIP label and the individual flows within the aggregation which belongs to the different (source, destination) pairs cannot be distinguished. Therefore, in order to prevent splitting a flow belonging to the one (source, destination) pair, if there are multiple shortest paths between a pair of routers we do not allow load splitting. In this case our algorithm selects the path towards the neighbor with the lower subIP address.

When rerouting traffic along alternative routes all traffic from the shortest path between two routers may be shifted to an alternative path, or only a portion of traffic may be routed over an alternative path, so that a part of traffic still follows the initial route. The latter approach requires finer aggregation at the ingress and thus more complex traffic control. We consider only the first option for simplicity. Each flow is thus mapped to only one path.

The necessary trade off of multipath routing is that the alternative paths may be longer and thus less than optimal. This also decreases the overall throughput in the network. Therefore it is advisable to bound the loss of optimality with a maximum allowable increase in path cost. In our simulations we used the number of hops when determining shortest paths and a flow is rerouted along an alternative path only if this path is no more than two hops longer than the initial one.

3.1   Flow Model

Each flow in our network is assigned an amount of bandwidth which is represented in multiples of a constant unit of bandwidth B. Let L be the set of links in the observed network where each link l has capacity c_l. Let F be the set of flows in the network, while a set of flows through a link l is denoted as $F(l)$. Each flow $f_{i,j}$ between a pair of area routers i and j is assigned $b_{i,j}$ number of constant units of bandwidth B. The total bandwidth per flow $fbw_{i,j}$ can then be represented as $fbw_{i,j} = b_{i,j}B$. The $b_{i,j}$ values thus represent the relative size of the network flows. They can also be regarded as a measure of fairness and priority in using network resources. If each $b_{i,j}$ is set to 1, each flow is assigned an equal amount of bandwidth B, which corresponds to the maximum fairness regarding engagement of network resources for each flow. More realistic is however that this number varies between flows, given that the edge routers are connected to the networks of different size. The $b_{i,j}$ values can be determined by measuring the flows. This should be simple given that subIP flows are aggregated according to the egress router and identified with a label.

The traffic matrix with $b_{i,j}$ values for each flow f_i is the input of our algorithm. With the known $b_{i,j}$ values and link capacities c_l, maximal B_l bandwidth unit of each link can be expressed as:

$$B_l = \frac{c_l}{\sum_{\forall i,j \in F(l)} b_{i,j}}$$

The global bandwidth unit B is the minimal of the calculated B_l values: $B = \min_{l \in L}\{B_l\}$. In our simulations the traffic between each pair of routers i and j is constant and occupies maximal bandwidth allowed for this flow, which equals $fbw_{i,j} = b_{i,j}B$.

3.2 Critical Links

With the known bandwidth of each flow and known paths through the network of the flows, we may calculate utilization of each network link. The utilization u_l of a link l is calculated as:

$$u_l = \frac{\sum_{\forall i,j \in F(l)} b_{i,j}B}{c_l}$$

Let $\bar{u}$ be the average link utilization in the network. We consider a link l over-utilized and thus critical if its utilization is greater than the average link utilization in the network, $u_l \geq \bar{u}$. If $u_l < \bar{u}$ the link is considered under-utilized and non-critical.

3.3 The Algorithm

Given that each topology k is denoted as T_k, with the physical topology denoted as T_0, the algorithm is as follows:

1. Assign all flows to T_0.
2. Given the input $b_{i,j}$ values, determine B, if each flow is routed across shortest paths of the topology as assigned.
3. For the found B determine link utilization for each network link and determine mean link utilization $\bar{u}$ for the network.

4. Create a new virtual topology T_k containing only non-critical links and calculate shortest paths.
5. Assign flows for which a path exists in T_k to T_k if:

   - the path contains a critical link in the topology to which it is assigned
   - all links along the path in the new topology Ti will remain less utilized than the most utilized link of the path in the assigned topology
   - the increase in path cost is acceptable (not more than 2 hops).

6. repeat steps 2 to 6 until the improvement in traffic distribution becomes negligible (for example, no increase of B).

4   Simulation Results

The VTA performance was evaluated through simulations and the results are presented in this section. Simulations were performed with the *Network Simulator ns-2* [3] extended to support subIP and MTR. The network model used in simulations has a so-called ISP topology [4] shown in Fig. 3. In general the network consists of N nodes interconnected with L bidirectional links each of capacity C. In our example $N = 18, L = 30, C = 100$.

Performance evaluation of VTA is done based on the following metrics:

- mean link utilization in the subIP area, $\bar{u}$, as a measure of network utilization
- standard deviation of the link utilization in the subIP area, σ_u, as a measure of fairness in using network resources
- unit of bandwidth B, defined in section 3.3, as a measure of throughput.

We present simulation results when each $b_{i,j}$ is set to 1, assigning thus B bandwidth to each flow $f_{i,j}$. Simulations were also performed with different sets of $b_{i,j}$ values where randomly chosen routers were allowed to exchange more traffic and/or the capacity of some randomly chosen network links was changed. Introducing such asymmetries did not adversely effect the algorithm performance.

Figures 4(a) and 4(c) show changes of each of the monitored metrics with each new iteration of the algorithm, i.e. after a new topology is determined and the flows are assigned to this topology according to VTA. The mean $\bar{u}$ and the standard deviation σ_u of links utilization are presented in Fig. 4(a), while the bandwidth unit B is presented in Fig. 4(c). Numerical values are given in Table 4(d).

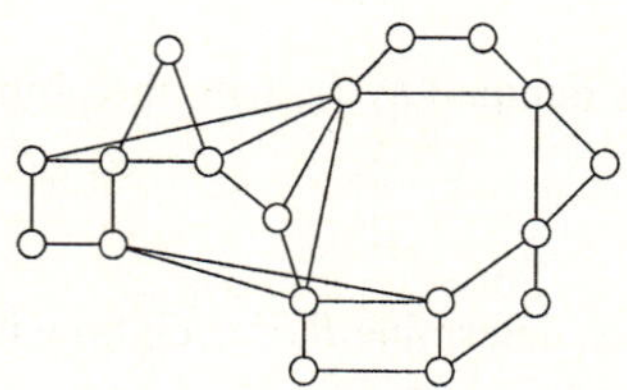

Fig. 3. The ISP Topology

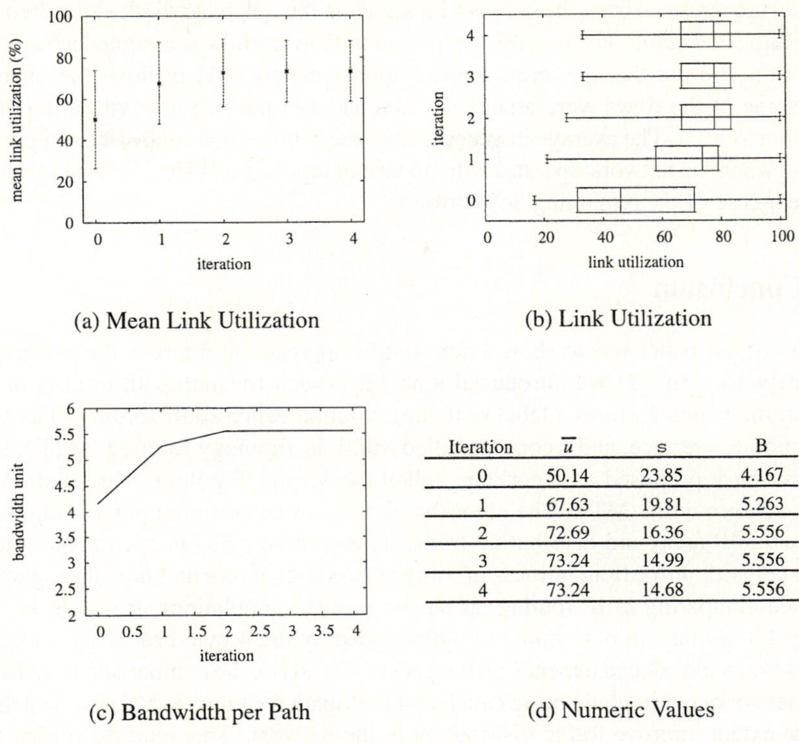

(a) Mean Link Utilization   (b) Link Utilization

(c) Bandwidth per Path   (d) Numeric Values

Fig. 4. Simulation Results

A significant improvement may be seen with only two virtual topologies introduced. Comparing to the initial state with no virtual topologies a 45% increase of $\bar{u}$, a 31.4% decrease of σ and a 33.3% increase of bandwidth unit B are obtained after the second iteration. This implies an increase in network utilization while more even traffic distribution is achieved across the network links and the throughput of the flows is higher. The shift of traffic across network links can be seen more clearly on the box plot presentation of link utilizations u_l for each iteration shown in Fig. 4(b). The plots show the minimum, median and maximum u_l values. The box contains 50% of u_l values, while 25% of these values are outside each side of the box. We may see that the overall range of u_l values narrows, showing us more uniform traffic distribution among network links. A shift of the plot to the higher values from one iteration to the next shows us the overall increase in link utilization.

With the next two virtual topologies introduced the change in traffic distribution is significantly smaller. The increase of mean link utilization is 0.8%, while decrease of standard deviation is 10.3%. There is no change of B.

We also compared bandwidth assigned to each flow using the *maximin* bandwidth allocation [5] after each iteration. The *maximin* technique allocates the entire capacity of a link to the flows traversing it in an equal share starting with the flows over the link which may allocate the minimum bandwidth per flow. Flows to which the bandwidth is assigned are removed from the network by reducing the link bandwidth along their paths for the allocated value. The process is repeated until every flow is assigned a bandwidth. Our results showed that the throughput of approximately 66% of flows had increased when some of the flows were assigned to the paths of the two new virtual topologies according to VTA. The average increase among these flows is around 40% comparing to the case when the network operates with no virtual topologies. (The increase is achieved at the expense of the remaining 34% flows.)

5   Conclusion

The aim of our paper was to show a new simple approach to improve the performance of IP networks. In [2] we introduced a new approach for multipath routing of best-effort traffic which requires a label switching solution called subIP, proposed as MPLS connectionless service, and a concept called Multiple Topology Routing (MTR). In this paper we have proposed an algorithm, called the Virtual Topology Algorithm (VTA), required in providing MTR. The algorithm is used in determining paths within a network (a subIP area) and distributing traffic among these paths in such a way that the overall network utilization, fairness in using network resources and flow throughput are increased comparing to IP routing, as shown through simulations. It should be noted that we did not take into account the traffic added by the 4 byte header, given that the overhead thus introduced depends on the packet size. Also, the comparison is made with the IP networks not implementing equal-cost multipath routing (ECMP) [6], which may to some extent improve traffic distribution in the network. This requires further more complex analysis. Additional tests with alternative algorithm parameters are required for more accurate performance analysis. Further improvement in performance may be achieved with congestion control at the edge of the domain, which is simple to implement with the subIP label switching concept. Given that some node and link failures are due to congestion, this would increase network stability and reduce rerouting of best-effort traffic. Overall network performance would be improved.

References

1. R. Callon E. Rosen, A. Viswanathan. RFC 3031: Multiprotocol Label Switching Architecture, January 2001.
2. Ljiljana Adamovic and Martin Collier. A New Traffic Engineering Approach for IP Networks. *in proceedings of CSNDSP 2004.*
3. VINT project. The Network Simulator - ns-2, 1989.
4. Karol Kowalik and Martin Collier.   ALCFRA - A Robust Routing Algorithm Which Can Tolerate Imprecise Network State Information. *Proceedings of ITC Specialist Seminar 2002.*
5. S. Chen and K. Nahrstedt. Maxmin fair routing in connection-oriented networks. *Tech. Rep., Dept. of Com. Sci., UIUC, USA,* 1998.
6. C. Hopps. RFC 2992: Analysis of an Equal-Cost Multi-Path Algorithm, November 2000.

Dynamic Path Control Scheme in Mobile Ad Hoc Networks Using On-demand Routing Protocol

Jihoon Lee[1] and Wonjong Noh[2]

[1] Comm.& Network Lab., Samsung Advanced Institute of Technology,
San 14-1, Nongseo-Ri, Kiheung-Eup, Yongin, Kyungki-Do, Korea, 449-712
vincent.lee@samsung.com
[2] Computer Network Lab., Dep. Of Electronics Engineering, Korea University,
Sungbuk-gu, Anam-dong 5-1ga, Seoul, Korea, 136-701
nwj@dsys.korea.ac.kr

Abstract. In existing on-demand routing protocols, the generated paths cannot dynamically adapt to the change of network topology even if another route with less hop is available. Routing efficiency affects network performance and lifetime. In this paper, we propose a dynamic path control (DPC) scheme that can actively adapt to network topology change due to node mobility. Each node monitors its active neighboring information and tries to meet the situation when a better local link is available. We evaluate the proposed DPC scheme using analytical modeling and prove that the path available time of existing routing schemes can be significantly improved by the application of the proposed scheme.

1   Introduction

Mobile ad hoc network (MANET) is composed of mobile nodes which communicate with each other using multi-hop wireless links [1]. These nodes can dynamically form a network without the aid of any fixed or pre-existing infrastructure. Each node in MANETs can operate as a host as well as a router. A node in the network can communicate directly with the nodes within its wireless range. On the other hand, nodes outside the wireless range would communicate indirectly, using a multi-hop route through other nodes in the network. The ease of deployment and the lack of the need of any infrastructure make MANETs an attractive choice for various applications. Examples of such applications can be embedded in the various forms (i.e., from personal area networks and sensor networks to intelligent transport systems).

A number of routing protocols have been proposed for ad hoc networks (e.g., AODV, DSR, OLSR, and TBRPF, etc.) [2, 3], and these protocols can be categorized as proactive and reactive. Proactive routing protocols maintain routing information for all network destinations independent of the traffic to such destination. On the other hand, reactive routing protocols query a route when there is a real demand for the route toward the destination. Many of research reports have shown that in terms of low control overhead, network scalability, and energy consumption, etc., reactive

P. Lorenz and P. Dini (Eds.): ICN 2005, LNCS 3421, pp. 343–352, 2005.

protocols perform better than proactive protocols because of the low routing overhead.

The existing reactive routing protocols accommodate route change only when route failure happens (i.e., an active path is disconnected). They cannot dynamically adapt to the change of network topology even if another route with less hop count becomes available by the movement of intermediate nodes or any neighboring nodes without any link disconnections. Given this factor, we propose Dynamic Path Control (DPC) scheme that regulates on-going paths by not any link disconnection events, but voluntary adaptation to node mobility based on the on-going path information. In order to change the on-going path, we introduce the new architecture of 'path set table' constructed in all nodes. We have evaluated the benefits of the proposed approaches in terms of path available lifetime. The results show significant improvements compared to the existing routing algorithms.

The rest of the paper is organized as follows. Section 2 briefly describes some related works in mobile ad hoc networks. Section 3 presents the design and detailed description of DPC scheme. Then, we present the evaluation results of analytical modeling and then finally we made a conclusion.

2   Related Works

In ad hoc networks, there are many concerns of routing performance. Our goal is to optimize routing issues without incurring any significance overheads. In this section, we describe the related previous on-demand routing protocols which do on-going path control in mobile ad hoc networks.

The Ad hoc On-Demand Distance Vector (AODV) [2] algorithm enables dynamic, self-starting, multihop routing between participating mobile nodes wishing to establish and maintain an ad hoc network. AODV allows mobile nodes to obtain routes quickly for new destinations, and does not require nodes to maintain routes to destinations that are not in active communication. AODV allows mobile nodes to respond to link breakages and changes in network topology in a timely manner. When links break, AODV causes the affected set of nodes to be notified so that they are able to invalidate the paths using the lost link. However, in AODV, route rediscovery happens only at source node when link failure exists. At that time, there is a high control message overhead by flooding-based route discovery. Moreover, it cannot dynamically adapt to the change of network topology even if another route with less hop count becomes available by the movement of intermediate nodes or any neighboring nodes if any link disconnections does not happen.

Roy et.al [4] presents the source-tree on-demand adaptive routing protocol (SOAR) based on link-state information. SOAR has the mechanism to shorten the active paths, but it achieves that by periodically exchanging link-state information in which a wireless router communicates to its neighbors the link states of only those links in its source tree that belong to the path it chooses to advertise for reaching destinations with which it has active flows. As this partial topology broadcast algorithms exchange relatively large control packets including the minimal source tree, total byte overhead due to SOAR control packets has been found to be 2-3 times more compared to the previous ad hoc routing protocols. High control packet overhead is undesirable in low bandwidth wireless environments. Additionally, periodical exchanging messages could collide with the data streams, thereby may degrade the performance.

Saito et.al [5] presents the dynamic path shortening (DPS) scheme based on smoothed signal-to-noise ratio (SSNR). DPS tunes up active paths by not any link disconnection event, but automatically adaptation to node mobility using SSNR as a link quality indicator. For this, [5] introduces the concept of proximity that represents the "nearness" of two communication nodes. Each node determines to shorten an active path by using proximity based on the local SSNR obtained from its own network interfaces. However, there is bad situation related to path failure. That is, by only considering intermediate nodes' mobility case in an active path, there is a probability that link failures in the path happen. Moreover, because of the use of link quality information, there is a probability that unreliable status estimation exists.

In contrast to the above works, the proposed scheme does not lead to much control overhead and unreliability using the promiscuous mode and neighboring path information, and does not require periodic information advertisements.

3   Dynamic Path Control (DPC) Scheme

This section describes the detailed design of DPC scheme. First, we show the situations in which an active path is efficiently controlled when the proposed DPC scheme is applied. After that, we explain DPC protocol performing the active path control in detail. It is assumed that the proposed scheme considers the situations under slow mobility or dense mobile ad hoc networks. In addition, we assume wireless interfaces that support promiscuous mode operation. Promiscuous mode means that if a node A is within the range of a node B, it can overhear communications to and from B even if those communications do no directly involve A. While promiscuous mode is not appropriate for all ad hoc network scenarios, it is useful in other scenarios for improving routing protocol performance [3].

3.1   Path Inefficiency Within On-going Path

In mobile ad hoc network, due to node mobility, there is a condition shown in Figure 1. In this case, we pay attention to node mobility without link disconnection in on-going path and node movement in the vicinity of the on-going path. For such node movement, we possibly find the less hop route (i.e., direct hop route shown in Fig. 1) than the current route in use. The primary goal of the solution approach is to discover short-cut routing paths as and when feasible.

Consider a routing path from a source node to a destination as shown in Figure 1(a). This initial path is determined through the route discovery process, in which the distance between the source and destination is the shortest in terms of the number of hops, or very close to it. A packet takes six hops while getting routed from node 1 to node 7. During the course of time, the mobility of the nodes may make the shape of the routing path similar to the one shown in Figure 1(b). In this new shape, node 5 is in the transmission range of node 2. However, because of the usage of route changes and the validity of the existing routing information, the routing table entries are not updated. Although the above functionality using the routing paths of Figure 1(b) is adequate, a packet still takes six hops to reach from node 1 to node 7. Ideally, the shortest path from node 1 to node 7 needs only four hops as shown in Figure 1(b). The goal of this paper is to recognize such situations and regulate the paths dynamically by modifying the entries of the routing tables.

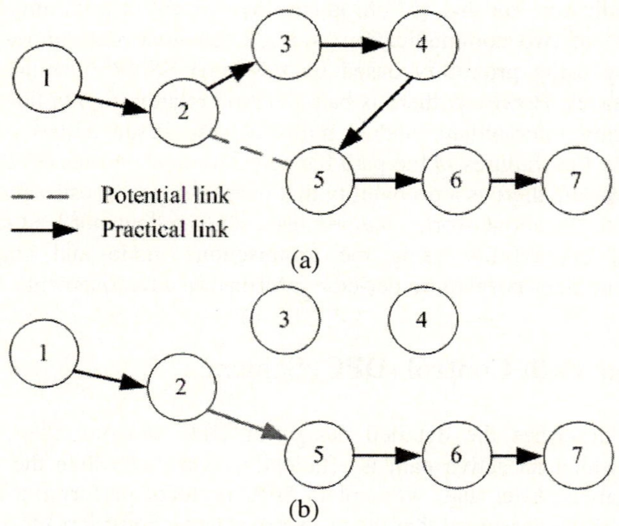

Fig. 1. Node 1 sends packets to node 7 via intermediate nodes. During active communication, there happens the node mobility in an intermediate node (e.g., node 5). Although node 2 can directly send packets to node 5, node 2 still sends packets to node 5 via node 3 and node 4

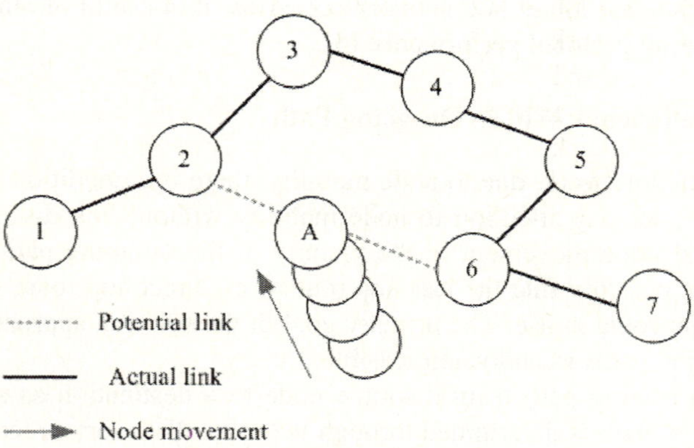

Fig. 2. Node 1 sends packets to node 7 via intermediate nodes. During active communication, there happens the node movement in the vicinity of intermediate nodes (e.g., node A). Although node 2 can directly send packets to node 6 via node A with shorter hops, node 2 still sends packets to node 6 via node 3, node 4, and node 5

Figure 2 shows another scenario. Some less hop routes may be available in the active path from source to destination. In Figure 2, it is shown that the routing path 2-

3-4-5-6 can be shortened to 2-A-6, since A is in the range of node 2, and node 6 is in the range of A.

Initially-organized paths that are much longer than the available shortcut paths are not desirable [6]. As data packets are delivered along these longer paths, unnecessary resource waste happens. Longer end-to-end delay and stronger probability of route change are expected as well. Since the number of hops in a longer path is high, shorter network lifetime is expected due to power consumption raised by control message delivery for route maintenance and data packet delivery. Moreover, longer path has higher possibility of performance degradation due to the interference among nodes, which the condition of interference among intermediate nodes in an on-going route is presented in [7]. A mechanism that can optimize or shorten the route length will result in significant performance gain over the underlying routing protocols.

DPC scheme is applicable in conjunction with any underlying ad hoc routing protocol that formulates the entries of the routing tables. The underlying routing protocol need not have to be very efficient or optimal. It could be very simple and adequate enough to ensure a reachable path from a source to the destination.

3.2   Operation Procedures of the Proposed Scheme

The point of the proposed scheme is to look for better routes during an ongoing packet flow. For long-lived flows, this may be beneficial, as the increased performance will be able to outweigh the additional overhead. Depending on the movement activity of each node, two nodes may become enough close to be within each other's radio range. At this moment, the current route from source node to destination node may have longer hops than the available path. The proposed scheme assumes that for efficient packet delivery through shortest path, all nodes in mobile ad hoc network collect the information for active path control using promiscuous mode. That is, it is assumed that to collect on-going neighboring flow information in nodes in mobile ad hoc network, the nodes can overhear the neighboring on-going packets.

In normal on-demand ad hoc routing protocol, the route discovery process goes first for data delivery. After that, when a node wants to forward a packet to another node, it unicasts the packet following the routing information and all the nodes within the transmission range can hear or receive the packet. From the overheard/received packet, the nodes configure the on-going flow information such as path information and hop count. Each of the nodes checks the packet header to see if the packet is destined for it (as the next hop) or not. The node that is the next hop destination, captures the packet and processes it for the next hop. A route can be determined when the route discovery packet reaches either the destination itself, or an intermediate node with a 'fresh enough' route to the destination. For forwarded packets in the discovered path, the destination address (DEST), source address (SRC), and the hop count (HC) information can be obtained from these packets. This information is maintained as an on-going flow list. The format of each entry in the list is <SRC, DEST, HC, Nb>, where Nb is the neighbor's address from which the packet was sent. Moreover, each entry has the lifetime field, which means the expiration time after which the entry is invalidated. The HC information can be obtained from the Hop limit field in the IPv6 packet header.

When node z receives or overhears a packet, which belongs to the <SRC x, DEST y> connection, the steps of the proposed DPC scheme can be enumerated as shown in Figure 3. If node z is the neighboring nodes (not the destination) for the received (or overheard) packet, <SRC, DEST, hop count, Neighbor> information is recorded in the on-going flow list if no entry exists corresponding to the <SRC x, DEST y> pair. If the same entry for the <SRC x, DEST y, Hop count, Neighbor> exists, then nothing is updated. If another entry for <SRC x, DEST y, Neighbor> exists, then path update is processed by the hop count difference. That is, in those nodes, if an entry corresponding to <SRC, DEST> does not exist then it is recorded. Otherwise the current hop count information is compared with the stored hop count. If the difference (highest hop count – lowest hop count) is three or more, then it implies that a shorter path exists. The node that detects this possibility modifies its routing table entry and that of another appropriate node to enable the shorter path formation. From Fig. 3, node J can overhear a packet from node A and E, so on-going flow list can be configured as shown in Fig. 3. Based on the difference of hop count information, node J can request the nodes to modify the current path to the available shorter path.

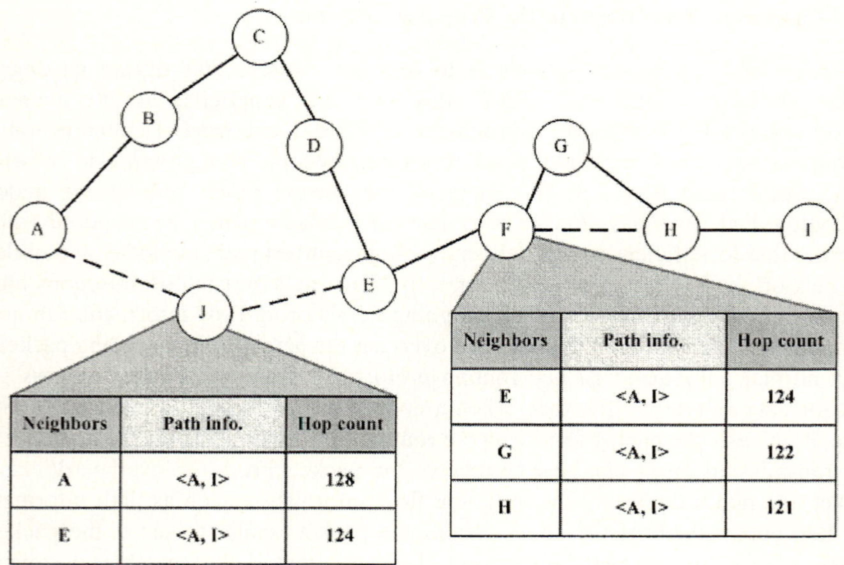

Fig. 3. An example of route changes in active path

The operation of DPC scheme is largely made of 4 steps. The detailed activity procedures are as follows.

Step 1. Route discovery: Similar to on-demand routing protocol, when a node wants to forward packets to another node, it floods the route discovery control packet to find the appropriate path towards the destination node. Each node in the ad hoc network to receive the flooded packet checks the packet header to see if the packet is destined for it (as the next hop) or not. A route can be determined when the route discovery packet reaches either the destination itself, or an intermediate node with a 'fresh enough' route to the destination.

Step 2. To receive or overhear packet and Path pair information: When a node receives or overhears a route control packet such as RREQ, RREP, RERR, the route control packet is processed without any dynamic path control events. If the received or overheard packet is unicasted, the destination address (DEST), source address (SRC), the neighbor address, and the hop count (HC) information can be obtained from these packets. This information is maintained as a neighbor lists. The format of each entry in the list is <SRC, DEST, HC, Nb>, where Nb is the address of the neighbor from which the packet was sent. Moreover, each entry has the lifetime field, which means the expiration time after which the entry is invalidated. The hop count (HC) information can be obtained from the Hop Limit field in the IPv6 packet header.

Step 3. Hop comparison: In those nodes, if an entry corresponding to <SRC, DEST> does not exist then it is recorded. Otherwise the current hop count information is compared with the stored hop count. If the HC difference is three or more, then it implies that a shorter path exists. The node that detects this possibility modifies or configures its routing table entry and that of another appropriate node to enable the shorter path formation.

Step 4. Route change request: To cope with the route change induced by node mobility, two kinds of messages are utilized such as RC_Req (Route change request), and RC_Conf (Route change confirm). Two messages have the similar type with Hello message of AODV. Let's consider that there happens the new link connectivity via node J between node A and node E in Fig. 3. When node J determines that there is a possibility for shorter path, it sends a RC_Req to upstream node A. The purpose is to notify node A of the possibility that there is a shorter path and to observe the bi-directional link between node A and itself. Upon receipt of RC_Req, node A sends RC_Conf to node J. As receiving RC_Conf, node J can confirm that node A can send packets directly to itself and vice versa.

In summary, the proposed scheme makes two main points. First, it shows that DPC scheme reduces the control message overhead induced by re-flooding route re-discovery. Our method is based on the promiscuous mode and on-going flow list. Second, it does not present the exceptional case appearing in the existing schemes. That is, even if there are simultaneous attempts for path control in an active path, the proposed scheme does not provide the additional processing overhead compared to the other schemes.

4   Analytic Evaluation

In this section, we analyze how much our scheme enhances the expected path available time. The terminologies used for this analysis and their definitions are listed in Table 1.

Table 1. Definitions

P_k	A route from S to D that consists of a sequence of k wireless links over k-1 intermediate nodes at route setup time.
L_i	The i-th link in the route
X_{L_i}	The lifetime of L_i is denoted by X_{Li}. Assume that X_{Li}, $i=1, 2, ..., k$, are independent and identically distributed (iid) exponential random variables, each with mean l.
X_{Pk}	The lifetime of a route P_k, $X_P = \min\left(X_{L_1}, X_{L_2}, \cdots, X_{L_k}\right)$
X_{O_i}	The inter-event time between i hop optimization event, $i=1, 2, ..., k-1$, independent and identically distributed (iid) exponential random variables, each with mean λ_o.
S_t	The hop count of a path, P_k, at time t, $1 \leq S_t \leq k$
T_n	The time of the hop count of a path, P_k, becomes n, $1 \leq n \leq k$

The distribution function of path available time of a path P_k can be derived as follows:

$$P\left(X_{P_k} \geq t\right) = \sum_{n=1}^{k} P\left(X_{P_k} \geq t, S_t = n\right)$$

$$= \int_{t`=0}^{t} \sum_{n=1}^{k} P\left(X_{P_k} \geq t, S_t = n \mid X_{P_k} \geq t`, T_n = t`\right) \cdot P\left(X_{P_k} \geq t`, T_n = t`\right) dt`$$

$$= \int_{t`=0}^{t} \sum_{n=1}^{k} P\left(X_{P_k} \geq t-t`, S_t = n, S_{t`} = n\right) \cdot P\left(X_{P_k} \geq t`\right) \cdot P\left(T_n = t`\right) dt`$$

where,

$$P\left(X_{P_k} \geq t-t`, S_t = n, S_{t`} = n\right) = e^{-n\lambda_l(t-t`)} \cdot e^{-(n-1)\lambda_o(t-t`)}$$

$$P\left(X_{P_k} \geq t`\right) = e^{-k\lambda_l t`}$$

$$P\left(T_n = t`\right) = \begin{cases} \left(\displaystyle\sum_{\{x_1+\cdots+x_{k-n-1}<k-n\}} \prod_{i=1}^{k-n-1} \frac{(\lambda_o t`)^{x_i}}{(x_i)!}\right) \cdot e^{-(k-1)\lambda_o t`} \cdot \lambda_o & (n \neq k) \\[20pt] \dfrac{1}{t} \cdot e^{-(k-1)\lambda_o t`} & (n = k) \end{cases}$$

$$\prod_{i=1}^{j} x_i = \begin{cases} x_1 \cdot x_2 \cdots x_j & (j \geq 1) \\[12pt] 1 & (j < 1) \end{cases}$$

The Expected Path Available Time of a path P_k is calculated as follows:

$$E[X_{P_k}] = \int\limits_{t=0}^{\infty} P[X_{P_k} \geq t] \cdot dt$$

$$= \int\limits_{t=0}^{\infty} \int\limits_{t`=0}^{t} \left( \sum_{n=1}^{k} \sum_{\{x_1+\cdots+x_{k-n-1}<k-n\}} \left[ \prod_{i=1}^{k-n-1} \frac{(\lambda_o t`)^{x_i}}{(x_i)!} \right] \cdot e^{-(k-1)\lambda_o t`} \cdot \lambda_o \cdot e^{-n\lambda_l (t-t`)} \cdot e^{-(n-1)\lambda_o (t-t`)} \cdot e^{-k\lambda_l t`} \right) \cdot dt` \cdot dt$$

$$+ \int\limits_{t=0}^{\infty} \int\limits_{t`=0}^{t} \left( \frac{1}{t} \cdot e^{-(k-1)\lambda_o t`} \cdot e^{-n\lambda_l (t-t`)} \cdot e^{-(n-1)\lambda_o (t-t`)} \cdot e^{-k\lambda_l t`} \right) \cdot dt` \cdot dt$$

In addition, we define the expected link available time improvement as the ratio of the expected link available time of the proposed scheme over the legacy scheme. Therefore, by applying dynamic link adaptation, we expect the improvement rate of expected path available time by:

$$Improvement\ rate = \left(\frac{E_{Adaptation}[X_{p_k}]}{E_{Legacy}[X_{p_k}]} - 1 \right) \times 100\% . \tag{1}$$

Let's take an example for $k=3$. That is, the path available time of a route that consists of a sequence of 3 wireless links over 2 intermediate nodes at route setup time is calculated by the above formulas.

$$E[X_{P_k}] = \int\limits_{t=0}^{\infty} P[X_{P_k} \geq t] \cdot dt$$

$$= \int\limits_{t=0}^{\infty} \int\limits_{t`=0}^{t} \left( \sum_{n=1}^{2} \sum_{\{x_1+\cdots+x_{k-n-1}<3-n\}} \left[ \prod_{i=1}^{2-n} \frac{(\lambda_o t`)^{x_i}}{(x_i)!} \right] \cdot e^{-2\lambda_o t`} \cdot \lambda_o \cdot e^{-n\lambda_l (t-t`)} \cdot e^{-(n-1)\lambda_o (t-t`)} \cdot e^{-3\lambda_l t`} \right) \cdot dt` dt$$

$$+ \int\limits_{t=0}^{\infty} \int\limits_{t`=0}^{t} \left( \frac{1}{t} \cdot e^{-2\lambda_o t`} \cdot e^{-n\lambda_l (t-t`)} \cdot e^{-(n-1)\lambda_o (t-t`)} \cdot e^{-3\lambda_l t`} \right) \cdot dt` \cdot dt$$

$$= \int\limits_{t=0}^{\infty} \int\limits_{t`=0}^{t} \left[ \begin{array}{l} e^{-\lambda_l (t-t`)} \cdot e^{-3\lambda_l t`} \cdot \left( \frac{(\lambda_o t`)^{1}}{1!} + 1 \right) \cdot e^{-2\lambda_o t`} \cdot \lambda_o + e^{-2\lambda_l (t-t`)} \cdot e^{-\lambda_l (t-t`)} \cdot e^{-3\lambda_l t`} \cdot e^{-2\lambda_o t`} \cdot \lambda_o \\ + e^{-3\lambda_l (t-t`)} \cdot e^{-2\lambda_o (t-t`)} \cdot e^{-3\lambda_l t`} \cdot \frac{1}{t} \cdot e^{-2\lambda_o t`} \end{array} \right] \cdot dt$$

$$= \int\limits_{t=0}^{\infty} \left[\begin{array}{l} \dfrac{\lambda_o^2}{(2\lambda_l + 2\lambda_o)^2} \cdot e^{-\lambda_l t} - \dfrac{\lambda_o^2}{(2\lambda_l + 2\lambda_o)^2} \cdot e^{-(3\lambda_l + 2\lambda_o)t} - \dfrac{\lambda_o^2}{(2\lambda_l + 2\lambda_o)} \cdot e^{-(3\lambda_l + 2\lambda_o)t} \\ - \dfrac{\lambda_o}{(\lambda_l + \lambda_o)} \cdot e^{-(3\lambda_l + 2\lambda_o)t} + \dfrac{\lambda_o}{\lambda_o + \lambda_l} \cdot e^{-(2\lambda_l + \lambda_o)t} + e^{-(3\lambda_l + 2\lambda_o)t} \end{array} \right] \cdot dt$$

$$= \frac{1}{3\lambda_l + 2\lambda_o} + \frac{\lambda_o}{2\lambda_l + 2\lambda_o} \cdot \left(\frac{1}{\lambda_l} - \frac{1}{3\lambda_l + 2\lambda_o} \right) - \frac{\lambda_o^2}{2\lambda_l + 2\lambda_o} \cdot \frac{1}{3\lambda_l + 2\lambda_o}$$

$$+ \frac{\lambda_o^2}{(2\lambda_l + 2\lambda_o)^2} \left(\frac{1}{\lambda_l} - \frac{1}{3\lambda_l + \lambda_o} \right) + \frac{\lambda_o}{\lambda_l + \lambda_o} \cdot \left(\frac{1}{2\lambda_l + \lambda_o} - \frac{1}{3\lambda_l + 2\lambda_o} \right)$$

According to the variable λ_l / λ_o values (i.e., x-axis, λ_l is variable & λ_o is fixed), the expected path available time is drawn for the length of wireless links (k= 3, 4, 5).

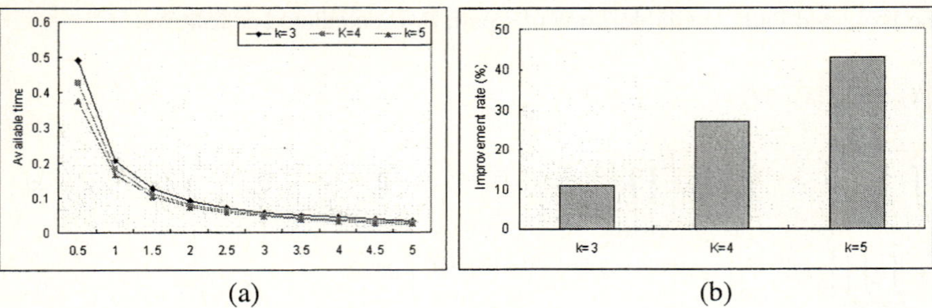

Fig. 4. (a) Expected path available time and (b) Improvement rate (%)

The larger x-axis value is, the higher the link disconnection probability is. From Fig. 4-(a), the longer the path length is, the shorter path available time is. From the Eq. (1), it has been shown in Fig. 4-(b) that the proposed scheme has longer network lifetime than the existing on-demand routing protocol.

5 Conclusion

This paper makes the following point. It shows that an active routing path control scheme using the information of neighboring active flows can enhance the path available time and incurs negligible overhead for route maintenance compared to the existing on-demand routing protocol. The basic idea of the method is to overhear the neighboring flows and to deliver the available changed route information. From this information, our approach is more adaptive to the route change status due to node mobility.

References

1. P. Trakadas, T. Zahariadis, S. Voliotis, and C. Manasis: Efficient Routing in PAN and Sensor Networks, ACM MC2R, Vol. 8 (1), Jan./Feb. (2004), 10-17
2. C. Perkins, E. Belding-Royer, and S. Das: Ad hoc On-Demand Distance Vector (AODV) Routing, IETF RFC 3561, July (2003)
3. D. Johnson, D. A. Maltz, and Y. Chu, The Dynamic Source Routing Protocol for Mobile Ad Hoc Networks (DSR) (Internet Draft), Mobile Ad-hoc Network (MANET) Working Group, IETF, July (2004)
4. S. Roy and J. J. Gracia Luna Aceves: Using Minimal Source Trees for On-Demand Routing in Ad Hoc Networks," IEEE Infocom, Vol. 2, April (2001) 1172-1181
5. Saito, M. Aida, H., Tobe, Y., and Tokuda, H.: A proximity-based dynamic path shortening scheme for ubiquitous ad hoc networks, Proc. of Int'l Conf. on Distributed Computing Systems, March (2004), 390-397
6. Gui, C. and Mohapatra, P.: SHORT: Self-Healing and Optimizing Routing Techniques for Mobile Ad Hoc Networks, ACM MobiHoc (2003) 279-290
7. Jiang, R., Gupta, V., and Ravishankar, C.V.: Interactions between TCP and the IEEE 802.11 MAC protocol, Proc. of DARPA Information Survivability Conference and Exposition (2003), Vol. 1, April (2003), 273-282

On the Capacity of Wireless Ad-Hoc Network Basing on Graph Theory[*]

Dai Qin-yun[1], Hu Xiu-lin[1], Yu Hong-yi[2], and Zhao Jun[2]

[1] Department of Electronic and Information, University of Hua Zhong Science & Technology WuHan, HuBei 430074, China
[2] Department of Communication Engineering in Institute of Information Engineering, University of Military Information Engineering ZhengZhou, HeNan 450002, China
daiqingyun874@sohu.com

Abstract. The network capacity is a very important parameter to evaluate the performance of ad hoc network. In this paper, the objective of our research is how to obtain the large capacity in ad hoc network. We establish the ad hoc network model from the novel point of view and consider the network capacity could achieve the maximum when the network resources (including area, bandwidth, power, load and so on) are used up minimally on the successful transmission condition. We set up the equation that makes the network resource consumed minimally and solve the problems of network capacity and the corresponding routing policy by the graph theory. Subsequently, we improve on these algorithms that are proved more efficient through the simulation experiment. Finally, the conclusion that the network capacity should combine with the routing policy is made.

Keywords: Ad Hoc Network, Network Capacity, Graph Theory, Multi-node Short Path Problem, Expending Matrix Algorithm.

1 Introduction

Ad hoc network consists of a number of nodes that communicate with each other over a wireless channel without any centralized control, whose main technology is the theory of self -organized network [8]. Nodes may cooperate in routing each other's data packets. Lack of any centralized control and possible node mobility give rise to many issues at the network, medium access, and physical layers, which have no counterparts in the wired networks like Internet, or in cellular network. At the network layer, the main problem is that of routing, which is exacerbated by the time-varying network topology, power constraints, and the characteristics of the wireless channel [9]. The choice of medium access scheme is also difficult in ad hoc network due to the time-varying network topology and the lack of centralized control. At the physical layer, an important issue is that of power control. The transmission power of nodes needs to be regulated so that it is high enough to reach the intended receiver while causing minimal interference at other nodes.

---

[*] This work was supported in part by the 863 National Science Foundation under Grant 2002AA123021.

P. Lorenz and P. Dini (Eds.): ICN 2005, LNCS 3421, pp. 353 – 360, 2005.

The common method on studying about the capacity of ad hoc networks is to establish the network model and analyze it under some assumptions. Gupta proposed a model [4] for studying the capacity of fixed ad hoc networks, where nodes are randomly located but are immobile. Each source node has a random destination in the network to which it wants to communicate. Every node in the network acts simultaneously as a source, a destination for some other node, as well as relays for others' packets. The main result shows that as the number of nodes per unit area n increases, the throughput per source-to-destination (S-D) pair decreases approximately like $1/\sqrt{n}$. Grossglauser set up the dynamic ad hoc model [3] where nodes move randomly in a circular disk such that their steady state distribution is uniform, which shows that it is possible for each sender-receiver pair to obtain a constant fraction of the total available bandwidth. This constant remains independent of the number of sender-receiver pairs. But in the scheme the delay to deliver the packet could be arbitrarily large. Yi showed that using directional antenna could improve the network capacity [10]. Bansal [1] considered how to make the network including static nodes and mobile nodes achieve the large network capacity on the low delay condition. Negi [6] restricted the transmission power that is not mentioned in the previous literatures and constructed the Ultra-Wide Band communication model (UWB [5]) where each link operates over a relatively large bandwidth W and with a constraint P_0 on the maximum power of transmission. Under the limiting case $W \to \infty$, the uniform throughput per node is $O\left((n\log n)^{\frac{\alpha-1}{2}}\right)$ (upper bound) and $O\left(\dfrac{n^{(\alpha-1)/2}}{(\log n)^{(\alpha+1)/2}}\right)$ (achievable lower bound). These bounds demonstrate that throughput increases with node density n, in contrast to previously published results.

In this paper, we attempt to set up the ad hoc network model basing on the graph theory from the new point of view and analyze network capacity according to this model. The model contains all factors of nodes transmission that are abstracted as a set and adapts to all kinds of environment such as the static network, the dynamic network, or the UWB network. All factors act as the subsets of the network resources set. The network capacity could attain the maximum when the resources consumed by all nodes transmissions in the ad hoc network are the minimal under successful transmissions. We give the linear equation of the network capacity resolved by the *Multi-node Short Path Problem* in the graph theory and obtain the corresponding routing policy.

The paper is organized as follows. In Section 2, we model the ad hoc network and give the linear equation of network capacity. Section 3, the equation is resolved by the graph theory and the algorithm is modified in order to improve the efficiency, at the same time, we also find the corresponding routing policy. Finally we end with discussions and conclusions in sections 4.

2 Ad Hoc Network Model

The basic idea of our model is that the network capacity can achieve the maximum when the network resources (including area, bandwidth, power, load and so on) are used up minimally on the successful transmission condition.

Model: The network is defined as N=(V, W), V denotes the set of nodes in the network, W is the wastage of nodes transmission in the network, where nodes are either in the plane or on the sphere and are classified as four types: source node v_s, destination node v_d, relay node v_r, and isolation node v_{is}, which satisfy the equation: $v = v_s + v_d + v_r + v_{is}$ or $V = V_S \cup V_R \cup V_D \cup V_{Is}$.

The minimal resources consumed between two nodes are defined as $c_{ij} = \min\limits_{P_l \in P_{ij}} \left\{ \sum\limits_{ij \in P_l} \omega_{ij} \right\}, i, j \in V$, where P_{ij} represents the set of all possible transmission paths from v_i and v_j and P_l is one of the paths belonging to P_{ij}. Let f_{ij} be data streams that need not relay between node i and j. s_{ij} is the transmission resource consumed at per unit data. $\omega_{ij} = s_{ij} f_{ij}$ denotes the wastage of direct transmission from v_i and v_j.

Equation: $c_{ij} = \min \sum\limits_{ij \in P_l} s_{ij} f_{ij}$

Restriction:
$$\sum_j f_{ij} \le a_i, i \in V_s$$
$$\sum_j f_{ij} - \sum_j f_{ji} = 0, i \in V_r$$
$$\sum_j f_{ji} \ge b_i, i \in V_d$$
$$0 \le f_{ij} \le u_{ij}$$

$\sum\limits_i a_i$ is the maximum capacity in the network. $\sum\limits_i b_i$ is the minimal capacity guaranteeing successful transmission in the network. u_{ij} is the maximum capacity between nodes not needing relay. If the number of nodes is large, the complexity of the arithmetic according to the linear equation is too great. We consider the characteristics of ad hoc network as additional conditions in order to reduce the complexity.

The traffic pattern of nodes transmission in ad hoc network could be classified as three types. The first type is no traffic. The second type is the traffic with which is communicated directly. The third type is the relaying traffic. Three conditions below are satisfied simultaneously when the network capacity attains the maximum:

1) The number of the first type node is as small as possible,

2) The second traffic satisfies: $c_{sd} = s_{sd} f_{sd}$,

3) The third traffic satisfies: $c_{sd} = \min \sum\limits_{sd \in P_l} s_{ij} f_{ij}$, $i \in (V_s \cup V_r), j \in (V_r \cup V_d)$.

Now we can utilize the graph theory to study network capacity.

3 Multi-node Short Path Problem in Ad Hoc Network Model

Multi-node Short Path Problem in the graph theory is equivalent to the shortest path problem between arbitrary two nodes in the network. In this paper, we study about the shortest path basing on the ad hoc network model which makes c_{sd} between source node and destination node attain the minimum and introduce the expending matrix algorithm to reduce the complexity.

3.1 The Application of the Relay Node

Let c^*_{sd} be the minimal resources consumed between v_s and v_d, ω_{ij} as wastage of the resources consumed between v_i and v_j with which communicates directly. We consider v_i has no choice to communicate directly with v_j if ω_{ij} goes to the infinity. v_i could reach v_j in theory when the condition $s_{ij}b_{ij} \leq \omega_{ij} \leq s_{ij}u_{ij}$ ($s_{ij}b_{ij}$ is the critical wastage under successful transmission) is satisfied. Now we calculate the resources consumed c_{ij} between v_i and v_j. The resources consumed are $\overline{c_{ij}} = c_{ik} + c_{kj}$ after the node gets across the arbitrary relay node v_r. The value of c_{ij} will not be changed if $c_{ij} < \overline{c_{ij}}$, otherwise c_{ij} is updated as $\overline{c_{ij}}$. The algorithm needs to check every relay node between v_s and v_d in the network and find some v_r through which makes c_{ij} decrease, and then we obtain the new value of c_{ij}. c^*_{sd} will be achieved after repeating the above process many times and then all relay nodes between v_s and v_d in the network will satisfy the condition $c_{ik} + c_{kj} \geq c_{ij}, i \neq j \neq k$.

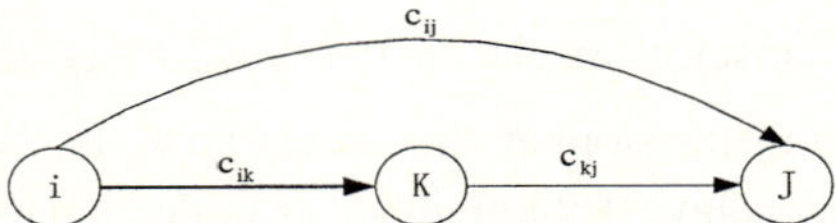

Fig. 1. The minimal resources consumed between v_i and v_j

3.2 The Basic Algorithm

As shown in Figure 1, $c_{ik} + c_{kj}$ substitutes for c_{ij} if $c_{ij} > c_{ik} + c_{kj}$. c_{ij} denotes the current value of the resources consumed between v_i and v_j. Let c^r_{ij} be the least value of resources consumed after the rth iterative algorithm and use calculation equation is:

$$c_{ij}^{r+1} = \min\{c_{ij}^r, c_{ik}^r + c_{kj}^r\}, i \neq j \neq k, i \in (V_s \cup V_k), k \in V_r, j \in (V_r \cup V_d). \tag{1}$$

3.3 The Approach and Process

We construct two matrixes during the iterative algorithm. The first matrix is the minimal resources consumed matrix. Every element in the matrix represents the current value of the least resources consumed. Let the first matrix be $C^r = \lfloor c^r_{ij} \rfloor$ after the rth iterative algorithm. The initialization of the matrix is $C^0 = \lfloor c^0_{ij} \rfloor$ and $c^0_{ij} = \omega_{ij}$. C^r is derived from C^{r-1}. The algorithm will be terminated until nodes in the network satisfy the relation of the equation (1).

The second matrix is the relay node matrix, which points out relay nodes when the resources consumed attain the minimum. Let the second matrix be $R^r = \left[R_{ij}^r\right]$ after the rth iterative algorithm. R_{ij}^r is the first relay node between v_i and v_j on the least resources consumed condition. The initialization of the matrix is $R^0 = \left[R_{ij}^0\right]$ and $R_{ij}^0 = j$. Elements in R^r are achieved according to the below equation:

$$R_{ij}^r = \begin{cases} r & c_{ij}^{r-1} > c_{ir}^{r-1} + c_{rj}^{r-1} \\ R_{ij}^{r-1} & \text{otherwise} \end{cases}$$

We calculate the above two matrixes after they are initiated and repeat the below steps:

Step 1: lay off all elements in the rth row and r^{th} line which are named after the axes of row and line;

Step 2: choose the elements not belonging to the axes of row and line and define them as $c_{ij}^{r-1}, j \neq r$.

We compare $c_{ir}^{r-1} + c_{rj}^{r-1}$ with c_{ij}^{r-1}. The values of i and j choose new values and continue the process to next c_{ij}^{r-1} when $c_{ir}^{r-1} + c_{rj}^{r-1} \geq c_{ij}^{r-1}$, and then we repeat step 2. $c_{ij}^r = c_{ir}^{r-1} + c_{rj}^{r-1}$ substitutes for c_{ij}^{r-1} and r substitutes for the corresponding element in relay node matrix if $c_{ir}^{r-1} + c_{rj}^{r-1} < c_{ij}^{r-1}$. The algorithm will return step 1 and r is added one after all elements are checked. The process of algorithm is terminated when r is equal to the value p, while we achieve the optimal minimal resources consumed matrix and the corresponding relay node matrix.

3.4   Expending Matrix Algorithm

The basic theory of the relay node to solve the problem of the multi-node short path is that the latter iterative algorithm remains the best result of the previous processes. The path will be changed only once the better scheme appears while the relay node is expended. This method increases the complexity of the algorithm due to the computation including nodes whose distance is far away from each other. Now we modify the algorithm in order to improve the algorithm efficiency.

Add elements of the rth row and line in C^r relative to C^{r-1}.

New increasing row:

$$c_{rj}^r = \min\left\{c_{rk}^0 + c_{kj}^{r-1}\right\} k = 1, \cdots, r-1 \tag{2}$$

New increasing line:

$$c_{ir}^r = \min\left\{c_{ik}^{r-1} + c_{kr}^0\right\} k = 1, \cdots, r-1 \tag{3}$$

Elements in C^r still satisfy Equation $c_{ij}^r = \min\left\{c_{ir}^r + c_{rj}^r, c_{ij}^{r-1}\right\} i, j = 1, \cdots, r-1$. The iterative process of the relay node matrix is: $R^0 = \left[R_{ij}\right], R^r = \left[R_{ij}^r\right]$. Elements of new increasing line are calculated by Equation (3). Let $R_{ir}^r = k_1$ if $c_{ir}^r = c_{ik_1}^{r-1} + c_{k_1r}^0$ and $R_{ir}^r = R_{ir}^0$ if $c_{ir}^r = c_{ir}^0$. Elements of new increasing row are calculated by Equation (2). Let $R_{rj}^r = k_h$ if

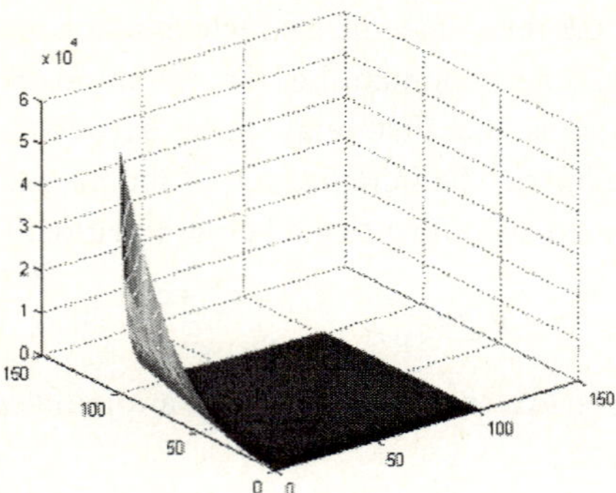

Fig. 2. Ratio of computation quantity of two algorithms on the minimal resources consumed matrix

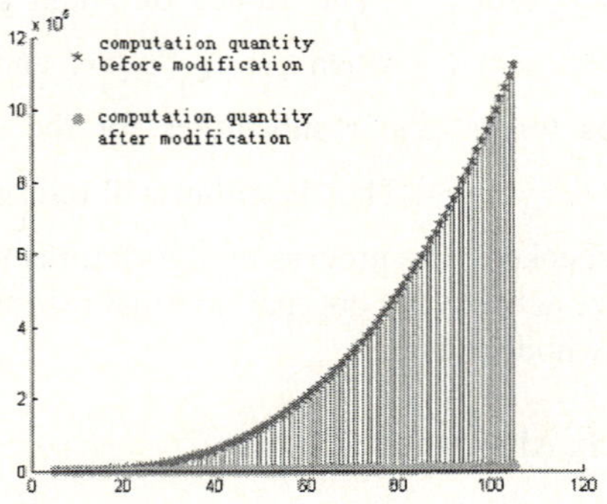

Fig. 3. Computation quantity of two algorithms on the relay node matrix

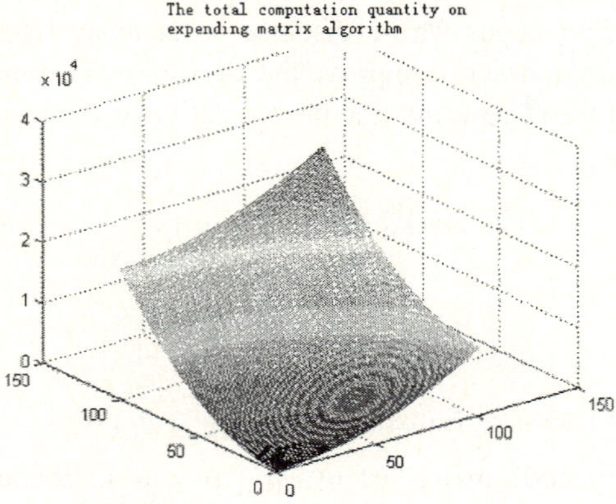

Fig. 4. Total computation quantity of the expending matrix algorithm

$c_{rj}^{r} = c_{rk_h}^{0} + c_{k_h j}^{r-1}$ and $R_{rj}^{r} = R_{rj}^{0}$ if $c_{rj}^{r} = c_{ij}^{0}$. Relations of the other nodes satisfy

$$R_{ij}^{r} = \begin{cases} r & c_{ij}^{r} = c_{ir}^{r} + c_{rj}^{r} \\ R_{ij}^{r-1} & \text{otherwise} \end{cases} \cdot c_{ii}^{r} = 0, R_{ii}^{r} = R_{ii}^{0}$$ for all i are satisfied.

3.5 Numerical Results

We change the number of nodes from 5 to 105 in the network of $1\text{km} \times 1\text{km}$ region and make a conclusion that the expending matrix algorithm is better than the previous algorithm from Figure 2 and 3. The two algorithms all include the computations of the minimal resources consumed matrix and the relay node matrix. The ratio of computation quantity between the algorithm unmodified and modified on the minimal resources consumed matrix increases approximately like the number of nodes v from Figure 2. There is the obvious difference between two algorithms on the relay node matrix from Figure 3. The expending matrix algorithm reduces the complexity of these problems, but the computation quantity still increases rapidly as the number of nodes from Figure 4, which shows the modified algorithm should be improved in the future work.

4 Discussions and Conclusions

The paper proposes the ad hoc network model where the resources consumed by nodes transmission (including area, bandwidth, power, load and so on) are acts as W. Gupta [4] provided the protocol model and the physical model concerning the expense of distance and power, but they ignored other factors such as bandwidth, interference. Section 2 of this paper demonstrates a standpoint that the network capacity could attain the maximum when resources are the minimum consumed under the successful transmission, but conditions of making all resources consumed minimally could not be satisfied synchronously.

The minimum resources consumed algorithm to analyze the ad hoc network capacity basing on the graph theory figures out the resources consumed minimally and obtains the corresponding the routing policy too, so we think the problem of network capacity should be considered combining with the routing policy. The computation quantity of the modified algorithm increasing rapidly as the number of nodes according to the simulations shows the method still needs to be improved. Finally, link of the relay transmission expending the additional resources such as time, we could adopt to the way of the additional coefficient in the graph theory for the ad hoc network model.

References

1. Bansal, M., Liu, Z.: Capacity, Delay and Mobility in Wireless Ad-Hoc Networks. In: Twenty-Second Annual Joint Conference of the IEEE Computer and Communications Societies 2 (2003) 1553-1563.
2. Bondy, J.A., Murthy, U.: Graph Theory with Applications. 2nd edn. Elsevier of New York (1976)

3. Grossglauser, M., Tse, D.: Mobility Increases the Capacity of Ad Hoc Wireless Networks. In: IEEE/ACM Transactions on Networking, Vol. 10. (2002) 477-486
4. Gupta, P., Kumar, P. R.: The Capacity of Wireless Networks. In: IEEE Transactions on Information Theory, Vol. 46. (2000) 388-404
5. Luo, H., Lu, S., Bharghavan, V.: A New Model for Packet Scheduling in Multi-Hop Wireless Networks. In: ACM MobiCom (2000) 76-86
6. Negi, R., Rajeswaran, A.: Capacity of Power Constrained Ad-Hoc Networks. In: IEEE INFOCOM (2004) 234-244
7. Proakis, J.G.: Digital Communication. 3rd edn. McGraw-Hill (1995)
8. Ramanathan, R., Redi, J.: A Brief Overview of Ad Hoc Networks: Challenges and Directions. In: IEEE communications Magazine 40 (2002) 28-31
9. Ramanathan, S., Steenstrup, M.: A Survey of Routing Techniques for Mobile Communication Networks. In: Mobile Networks and Applications, Vol. 1. (1996) 89-104
10. Yi, S., Pei, Y., Kalyanaraman, S.: On the Capacity Improvement of Ad Hoc Wireless Networks Using Directional Antennas. In: Proceeding of Third Annual Workshop on Mobile Ad Hoc Networking and Computing (2003) 108-117

Mobile Gateways for Mobile Ad-Hoc Networks with Network Mobility Support

Ryuji Wakikawa[1], Hiroki Matsutani[1], Rajeev Koodli[2],
Anders Nilsson[3], and Jun Murai[1]

[1] Keio University, 5322 Endo Fujisawa, Japan
[2] Nokia Research Center, 313 Fairchild Drive,
Mountain View, CA 94043 USA
[3] Lund University, Box 118 SE-221 00 Lund Sweden

Abstract. Providing "always-on" Internet connectivity is crucial for mobile networks in automobiles and personal area networks. The Network Mobility (NEMO) protocol provides such a support involving a bi-directional tunnel between a gateway for the mobile network and a Home Agent. However, the routing path between them is hardly optimal. In this paper, we investigate the use of ad-hoc routing protocols for route-optimized communication between mobile networks. We describe enhancements to a mobile gateway to choose the optimal route for a destination. The route can be provided by the ad-hoc routing protocol or by the basic nemo routing approach itself whenever it is desirable. Our results indicate that this approach provides better routes than the basic nemo routing approach. It also provides detour route(s) when a mobile network is disconnected from Internet access.

1   Introduction

Rapid advances in radio technologies, miniaturization and utilization of multiple network interfaces, and the ever increasing processing capacity are paving the way for even tiny devices to become IP end-points. This trend is naturally evolving to networks of such devices serving specific purposes. A mobile ad-hoc network (manet) is created dynamically when a set of nodes form a mesh routing state for their connectivity management, typically over a wireless network. Many routing protocols have been proposed in the Internet Engineering Task Force (IETF) MANET working group. These protocols aim to maintain localized routing at individual nodes despite movement of intermediate nodes that causes the routing path to change. Some recent work has focused on integrating manets into the Internet [4, 9]. However, the emphasis has been on nodes moving between ad-hoc networks and the Internet with Mobile IPv6 [3] to conceal their movements. We take a position that it is more efficient to apply network mobility support (i.e., the Network Mobility in IETF [2]) to an entire manet in moving vehicles or persons because only a representative mobile gateway needs to be mobility-aware; other individual nodes function without requiring

P. Lorenz and P. Dini (Eds.): ICN 2005, LNCS 3421, pp. 361–368, 2005.
© Springer-Verlag Berlin Heidelberg 2005

mobility-aware functionality. Nevertheless, all nodes need to be accessible from the Internet anywhere, anytime.

Efficient routing support between such mobile manets is necessary for "best-connected always-on" as well as for scenarios when manets are, perhaps temporarily, disconnected from the Internet when using the basic nemo protocol [2]. We introduce *mobile gateways* that provide enhanced routing in addition to the mobile router support specified in the basic nemo protocol. With this enhanced routing support, a manet can use direct route-optimized communication without incurring the overheads of bi-directional tunneling inherent in nemo. This enhanced routing also provides always-on connectivity when the manet is disconnected from the Internet for nemo purposes. Hence, this contribution provides crucial routing necessary in connecting mobile manets to the Internet.

The remainder of this paper is organized as follows. We first describe the reference model in Section 2 and provide a problem statement. We discuss the related work in Section 3. Then, we introduce the concept of a mobile gateway and the routing enhancements for multiple types of connectivity in Section 4. In Section 5, we report performance study. Finally, we provide concluding observations in Section 6.

2   Reference Model and Problem Statement

We primarily target scenarios involving networks in vehicles and Personal Area Networks. Figure 1 illustrates our reference model. A key feature of these networks is that the entire network moves along with the physical body (e.g., an automobile or a human body), while maintaining Internet connectivity. The intra-network mobility is managed by a manet routing protocol while the mobility with respect to external networks is managed by the basic nemo protocol.

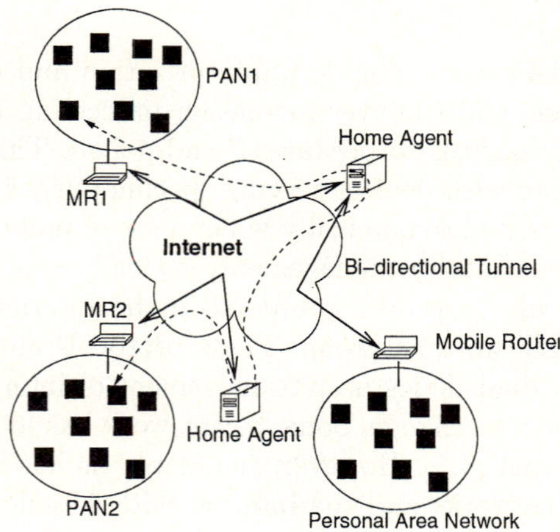

Fig. 1. Reference Model for Personal Area Network

When a node is moving on the Internet, Mobile IP provides the fundamental functionalities such as session continuity and application transparency. However, running Mobile IPv6 on every node inside a manet can be a very expensive. For instance, tiny sensor nodes will typically be incapable of supporting extended protocol stacks. In addition, when all the nodes within a network move simultaneously, it is feasible to aggregate the mobility support by introducing a single mobile router for each mobile network. The introduction of a mobile router that manages the mobility of the entire network can thus provide persistent connectivity as well as transparency to these sensors. Hence, we use the basic nemo protocol to provide a constant prefix for addresses as well as for movement transparency to an entire network.

Figure 1 also shows the overview of the basic nemo protocol. The basic Nemo protocol is an extension of Mobile IPv6 that provides a mobile router with a constant network prefix to enable the mobile ad-hoc network to be reachable from the Internet regardless of its movement. Typically, only the mobile router changes its point of attachment to the Internet. A mobile router has two network interfaces; ad-hoc routing is enabled on its *ingress interface* while basic nemo is managed on its *egress interface*. Typical egress interface is IEEE 802.11b and cell phone. The basic nemo protocol always uses bi-directional tunnel between a mobile router and a *Home Agent*. A home agent is a router on the mobile router's home link. The bi-directional tunnel is established by processing an extended *Binding Update* to the home agent. All data packets to and from the mobile network are then tunneled via the home agent. The basic nemo protocol does not support any route optimization. The existing nemo model only supplies redundant route due to bi-directional tunneling and dog-leg route. Packets are first routed to a home agent and tunneled to a mobile router. For example in Figure 1, when $manet_1$ and $manet_2$ communicate with each other, the routing traverses MR_1, HA_1, HA_2 and MR_2 with double bi-directional tunneling. The overhead caused by redundant route is especially significant for real time applications and streaming applications.

A mobile router can establish a better route without any signaling to a home agent and a correspondent node when physical distance between a source and a destination node is close enough to exchange routes by a manet routing protocol. The existing model assumes that the Internet connectivity is always available to communicate with other networks. Even if a destination is physically nearby, each manet needs to communicate via its home agent which could be many hops away on the Internet. Therefore, when the Internet connectivity is lost, for example, due to limited network coverage, the mobile network is isolated from other networks and hence cannot communicate.

Clearly, the integration of manet and nemo are consistent with the Internet architecture in terms of best effort routing. Given the above reference model, we address the following problems. First, how to enable a mobile manet to use the best available route to communicate with another manet. Second, how to allow a disconnected manet to retain its connectivity with the Internet by means of some other manet that does possess Internet connectivity.

3   Related Work

There are few existing solutions to provide Internet connectivity to the manet (e.g., [7, 8]) and to integrate Mobile IPv4 and Mobile IPv6 [3] and an ad-hoc network (e.g., [6, 5, 4, 9, 1]). A DSR-based manet is connected to the Internet with Mobile IPv6 in [6]. MIPMANET [4] integrates Mobile IPv4 and AODV utilizing *Foreign Agents* to support mobile nodes in a manet. Although Mobile IP can be integrated in our system, we focus on network mobility and not on host mobility. The approach in [9] proposes to use an Internet Gateway to connect a manet to the Internet, by extending manet messages for Gateway discovery and route establishment. Here, we integrate manet routing with a mobile router for best Internet connectivity.

4   Mobile Gateways System

As mentioned earlier, even when a mobile network meets different mobile networks nearby, it needs to communicate over the bi-directional tunnel via its home agent and vice versa. This overhead is not negligible. The existing hierarchical routing cannot take account of physical distance between source and destination nodes. Therefore, we propose to utilize manet routing protocols for the sake of direct route establishment to bypass both hierarchical routing and bi-directional tunneling.

4.1   Mobile Gateways

A mobile gateway is a nominated mobile router in a manet and is responsible for mobility and always-on Internet connectivity. As shown in Figure 2, mobile gateways have similar configuration as the mobile routers of the nemo basic protocol except for a third interface called an egress manet interface. Mobile gateways are attached to the visiting network with egress interfaces and belong to their manet via ingress interfaces. The egress manet interfaces are aimed to maintain inter-mobile gateways connectivity. Three interfaces of mobile gateways should be configured with different radio channels so that messages sent by each interface do not interfere with one another.

All packets sent from a local network to a remote network are always intercepted and routed by a mobile gateway, because the mobile gateway is a default router in its local manet. For efficient communication, mobile gateways maintain connectivity with adjacent mobile gateways and fixed Internet gateways by using egress manet interfaces in order to provide direct route, detour route, and backup route each of which are described below.

- A **direct route** is established when a destination and a source node can establish a route with egress manet interfaces and uses it as an optimized route without a bi-directional tunnel of the nemo basic protocol. Direct routes offer better communication than routes via the Internet in terms of round trip time and delay, etc. They are useful in scenarios such as when two PANs (using different access radios) come close to exchange business cards on PDAs.

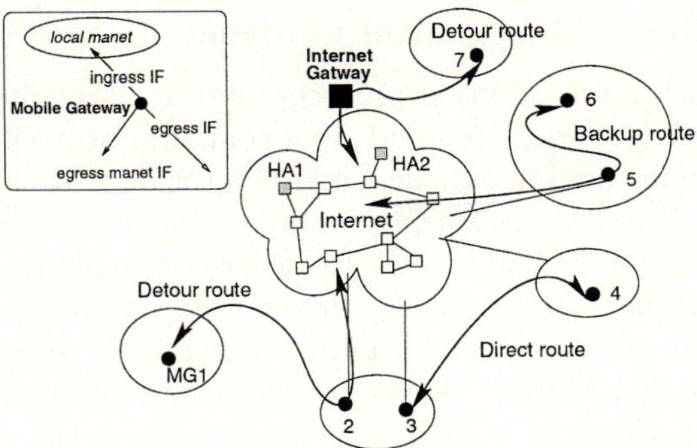

Fig. 2. Mobile Gateways

In addition, establishment of a direct route provides interoperability among different manet routing protocols. Mobile gateways are responsible for connecting manets running different manet routing protocol.

- A **detour route** is established when one of the manet loses Internet connectivity at its egress interface and obtains connectivity through either adjacent mobile gateway or another Internet gateway. If Internet connectivity is lost due to movement, a mobile gateway can use connectivity through the egress manet interface as a way out towards the Internet. In this case, a mobile gateway needs to obtain a care-of address at the egress manet interface and register this binding to its home agent. This configuration is called *nested mobility*.

- A **backup route** is used when multiple mobile gateways are configured in a manet. When a mobile gateway loses Internet connectivity, typically it notifies the manet nodes about invalid default route. Subsequently, the manet nodes re-select a default route to another mobile gateway to access a destination located the outside manet. However, such a switch-over to a new default route causes additional routing messages and delays, the mobile gateway that has lost Internet connectivity simply intercepts packets and routes them to another mobile gateway by using egress manet interface. This saves signaling to maintain a default route with low overhead when default routes are already diffused to a manet.

4.2 Adjacent Gateways Management

Mobile gateways are typically expected to be not power dependent in the same way as, for instance, the battery constraints of manet nodes are, since they are installed in vehicle or devices with richer resources. Thus, mobile gateways always maintain routes of adjacent manets by using proactive manet routing protocols such as OLSR and TBRPF. Proactive manet routing protocols periodically exchange topology information among mobile gateways and Internet gateways. Therefore, each mobile gateway always manages adjacent manets using the topology information. We use OLSR as a case study in this paper.

4.3   Default Route Management in Manet

Regardless of which manet routing protocol used, it is required that a default route is maintained on every manet node that communicates with external nodes. All packets with a destination outside a manet must be routed through a mobile gateway (i.e. default route). This default route can be discovered and established on-demand, if it is acquired through the use of a reactive manet routing protocol.

When there are multiple mobile gateways inside a manet, nodes could use more than one gateway for default routes by explicitly specifying a gateway address in the routing header. However, when a mobile gateway loses Internet connectivity, performing maintenance operations on the default route and switch-over to another default route causes significant overhead. Hence, the mobile gateway should not erase its route configured as a default route for the manet nodes. Instead, the mobile gateway keeps forwarding packets to another mobile gateway until each manet node expires the default route due to the lifetime expiration and stops using that default route. The mobile gateway can erase the default route by using route error notification messages. Thus, route discovery occurs randomly depending on the lifetime for reactive manet routing protocols. Manet nodes can obtain new mobile gateway address and switch to new one when the inactive default route is expired on proactive routing protocols.

4.4   Routing Examination

A manet node and a mobile gateway always examine routing table for packets sent from a manet node to the Internet, and vice versa.

A mobile gateway compares both a destination address and a source address with its mobile network prefix. If the destination address is matched to the prefix, the mobile gateway should not leak this packet outside its manet. It locally routes the packet to the destination. Otherwise, it checks availability of a direct route for the destination. If the direct route is active, it can directly forward packets to the destination. If all conditions are not met, it tunnels the packet to its home agent. The algorithm is fundamental and mobile gateway is required to control route selection depending on applications. The direct routes are not always better than a route through home agents, because of link stability, topology change frequency and the number of intermediate nodes. For example, round trip time of a direct route is faster than the route through home agents, but an application may want small delay and high bandwidth connectivity. It is often necessary to make more sophisticated decisions in real-life scenarios. An enhanced algorithm is discussed in Section 5.

A manet node needs a slightly different routing algorithm based on prefix comparison. The manet node searches a host route for the destination address, when the prefix part of the destination addresses is matched with the source address. If the host route is not found in the routing table, the manet node starts route discovery operations when a reactive routing protocol is used. If the host route is not available the manet node gives up sending packets. If the prefix part of both addresses is not equal, the manet node selects a default route. If the default route is not available in the routing table, the manet node needs to initiate default route discovery described in [9]. When the manet node occasionally has multiple default routes, it can transmit packet to one of mobile gateways explicitly with a routing header proposed in [9].

5   Performance Study

In our simulations, 30 mobile gateways were moving 10 km on a rectangular flat space [10000m x 50m] for 600 seconds of simulated time. The flat space model represents free ways and-or city roads. We tested two scenarios in which all the vehicles move either in the same direction or half of the vehicles move in opposite direction as shown in Figure 3. Essentially, each vehicle is moving just straight with varying speed (40km/h-120km/h). The distance between vehicles varies from 40m to 120m depending on the actual speed.

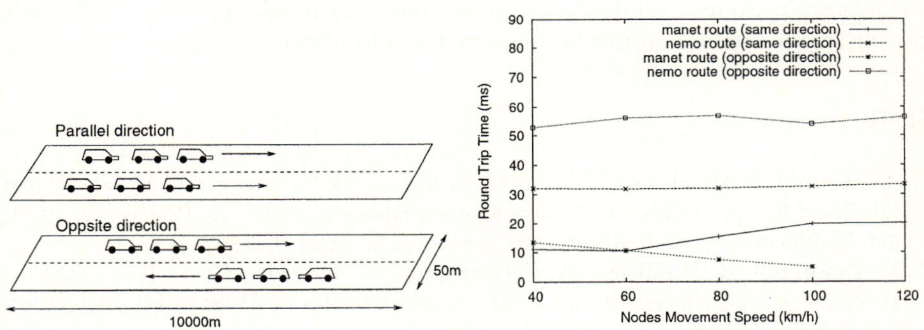

Fig. 3. Scenario and Result of Round Trip Time

A mobile gateway has two 802.11 compatible wireless interfaces for an egress manet interface and an ingress interface. The radio propagation range is set to 200 meters and the channel bandwidth 2 M-bits/sec. OLSR is used for inter-mobile gateway connectivity. Mobile gateways are capable of routing packets from local mobile network to the OLSR manet. The experiment compares the average round trip time when a mobile gateway selects either direct routes or a nemo route in the our simulation. A mobile gateway communicates directly with another mobile gateway by using OLSR routing when the routes are available.

Figure 3 shows the result. When a mobile gateway uses the nemo route, double bi-directional tunnels occur for both end mobile gateways. The link delay of the wired route through the two home agents is configured to be 6 ms. It is quite obvious that the round trip time of direct routes is faster than the nemo route because of absence of bi-directional tunnels. When each manet drives with 120km per hour, connectivity tends to break. In such a case, mobile gateway switches a route to the destination dynamically using the nemo route. Compared to the nemo routes, the manet routes always achieve the shorter round trip time. This indicates the mobile gateway achieves route optimization for the basic NEMO protocol by using manet routes.

6   Conclusion

In this paper, we investigated the integration of the nemo basic protocol and manet by using mobile gateways as enhancements to mobile routers of the nemo basic protocol. A mobile gateway establishes ad-hoc connectivity with other mobile gateways by a manet routing protocol and utilizes the additional route only when it is necessary for applications to exploit an optimized route and a route to the Internet if necessary. We studied the enhanced algorithm to select direct routes or the basic nemo route and provided parameters for the route selection. The mobile gateways were simulated for vehicles in motion, and the result of simulation shows that the model of mobile gateways is efficient. As for next step, we need to discuss security issues for mobile gateways and study parameters to switch a route between manet and nemo.

References

1. M. Benzaid, P. Minet, and K. Al Agha. A framework for integrating Mobile-IP and OLSR ad-hoc networking for future wireless mobile systems. In *Proceedings of the 1st Mediterranean Ad-Hoc Networks workshop*, September 2002.
2. V. Devaraplli, R. Wakikawa, A. Petrescu, and P. Thubert. Nemo Basic Support Protocol (work in progress, draft-ietf-nemo-basic-support-03.txt). Internet Draft, Internet Engineering Task Force, June 2004.
3. D. B. Johnson, C. Perkins, and J. Arkko. Mobility support in IPv6. Request for Comments (Prposed Standard) 3775, Internet Engineering Task Force, June 2004.
4. U. Johnsson, F. Alriksson, T. Larsson, P. Johannson, and G. Maquire Jr. MIP-MANET - Mobile Ip for Mobile Ad hoc Networks. In *Proceedings of First Annual Workshop on Mobile Ad Hoc Networking and Computing (Mobihoc)*, August 2000.
5. H. Lei and C. E. Perkins. Ad Hoc Networking with Mobile IP. In *Proceedings of 2nd European Personal Mobile Communication Conference*, September 1997.
6. D. A. Maltz, J. Broch, and D. B. Johnson. Quantitative Lessons From a Full-Scale Multi-Hop Wireless Ad Hoc Network Testbed. In *Proceedings of the IEEE Wireless Communications and Networking Conference*, September 2000.
7. G. Cirincione M.S. Corson, J.P. Macker. Internet-based mobile ad hoc networking. In *IEEE Internet Computing, Vol.3, No. 4*, pages 63–70, Jul 1999.
8. A. Striegel, R. Ramanujan, and J. Bonney. A protocol independent internet gateway for ad-hoc wireless networks. In *Proceedings of Local Computer Networks (LCN 2001)*, Tampa, Florida, Nov 2001.
9. R. Wakikawa, J. Malinen, C. Perkins, A. Nilsson, and A. Tuominen. Global Connectivity for IPv6 Mobile Ad Hoc Networks (work in progress, draft-wakikawa-manet-globalv6-02). Internet Draft, Internet Engineering Task Force, November 2002.

Energy Consumption in Multicast Protocols for Static Ad Hoc Networks[*]

Sangman Moh

Dept. of Internet Engineering, Chosun University,
375, Seoeok-dong, Dong-gu, Gwangju, 501-759 Korea
smmoh@chosun.ac.kr

Abstract. In *static ad hoc networks*, energy performance is as important as general performance since it directly affects the network operation time in the wireless environment. This paper evaluates the multicast protocols for static ad hoc networks in terms of energy consumption, proposes *analytical models* of energy performance in multicast protocols, and validates the models using computer simulation. In general, multicast protocols can be categorized into tree-based and mesh-based multicast. Tree-based protocols are more energy efficient than mesh-based ones, while the latter are more robust to link failure. According to the proposed analytical models, mesh-based multicast consumes around $(f + 1)/2$ times more energy than tree-based multicast, where f is average *node connectivity*. It is mainly due to broadcast-based flooding within the mesh. Our simulation study shows the models and simulation results are consistent with each other.

1 Introduction

Recently, *wireless ad hoc networks* [1, 2] attract a lot of attention with the advent of inexpensive wireless network solutions [3, 4, 5]. Such networks are emerging and have a great long-term economic potential but pose many challenging problems. Among them, *energy efficiency* may be the most important design criterion since a critical limiting factor for a wireless node is its operation time, restricted by battery capacity [6]. Since energy consumption due to wireless communication can represent more than half of total system power consumption [7], the key to energy efficiency is in the energy-aware network protocols.

This paper considers the energy performance of multicast on static ad hoc networks. *Multicasting* has been extensively studied for ad hoc networks [8, 9, 10, 11] because it is fundamental to many ad hoc network applications requiring close collaboration of participating nodes in a group. Since a static ad hoc network can be regarded as a special case of mobile ad hoc networks, where node mobility is zero, the multicast protocols designed for mobile ad hoc networks may be also used for static ad hoc networks even though they are not optimized for static environments.

It has been shown that *tree-based protocols* may consume less energy than *mesh-based protocols* while it is just the opposite with respect to general performance [10].

[*] This study was supported in part by research funds from Chosun University, 2004.

P. Lorenz and P. Dini (Eds.): ICN 2005, LNCS 3421, pp. 369–376, 2005.
© Springer-Verlag Berlin Heidelberg 2005

Tree-based protocols are much more energy efficient due to the following reasons: First, the power-saving mechanism, such as the one defined in IEEE 802.11 wireless LAN standard [3], puts a mobile node into sleep mode when it is neither sending nor receiving packets [4]. Second, a wireless *network interface card* (NIC) typically accepts only two kinds of packets; unicast and broadcast packets (all 1's). Since mesh-based protocols depend on broadcast flooding, every mobile node in the mesh must be ready to receive packets at all times during the multicast. In contrast, for unicast transmission along the multicast tree, only the designated receivers need to receive the transmitted data. Thus, a wireless node in tree-based protocols can safely put itself into a low-power sleep mode to conserve energy if it is not a designated receiver in each periodic beaconing message. Note here that all nodes listen and any the pending traffic is advertised during beaconing period.

In this paper, we evaluates the multicast protocols in terms of energy consumption, proposes *analytical models* of energy performance in multicast protocols, and validates the models using computer simulation. According to the proposed analytical models, the mesh-based multicast consumes around $(f + 1)/2$ times more energy than the tree-based multicast, where f is average *node connectivity*. Simulation study based on the QualNet simulator [12] shows that the models and simulation results are consistent with each other.

The rest of the paper is organized as follows: Earlier multicast protocols for ad hoc networks are briefly evaluated in the following section, focusing on energy performance. In Section 3, the analytical models of energy performance in multicast protocols for static ad hoc networks are formally presented. Section 4 presents our simulation study and shows that the models and simulation results are consistent with each other. Finally, concluding remarks are given in Section 5.

2 Multicast Protocols for Static Ad Hoc Networks

As in wired and infrastructured wireless networks (*i.e.*, wireless networks with base stations), *tree-based multicast* is used in ad hoc networks as well. Fig. 1 shows an example of a multicast tree. The tree consists of a root node (r), 3 nonmember intermediate nodes (p, s, and t), 7 member nodes of a multicast group (shaded nodes in the figure), and 10 tree links. A multicast packet is delivered from the root node r to 7 group members. For a member node u, for instance, the packet transmission is relayed through two tree links, *i.e.*, from r to q and then q to u. Now consider the last transmission from q to u. Even though all the nodes within node q's radio transmission range (*e.g.*, s, t and x) can receive the multicast packet, only node u will receive the packet and the rest of the nodes go into sleep-mode. Now, a single multicast packet requires 5 transmissions and 10 receives. However, since NIC typically accepts only unicast and broadcast addresses, nodes r, p, and s must use broadcast address because they have more than one receiver. This increases the number of receives to 17 including seven new receives ($r{\rightarrow}a$, $r{\rightarrow}b$, $p{\rightarrow}a$, $p{\rightarrow}r$, $s{\rightarrow}r$ and $s{\rightarrow}q$).

Fig. 2 shows an example of the mesh-based multicast with one-hop redundant links for the network of Fig. 1. Note that it includes 6 redundant links in addition to 10 tree links. A multicast packet is broadcast within a multicast mesh. Thus, sending a packet

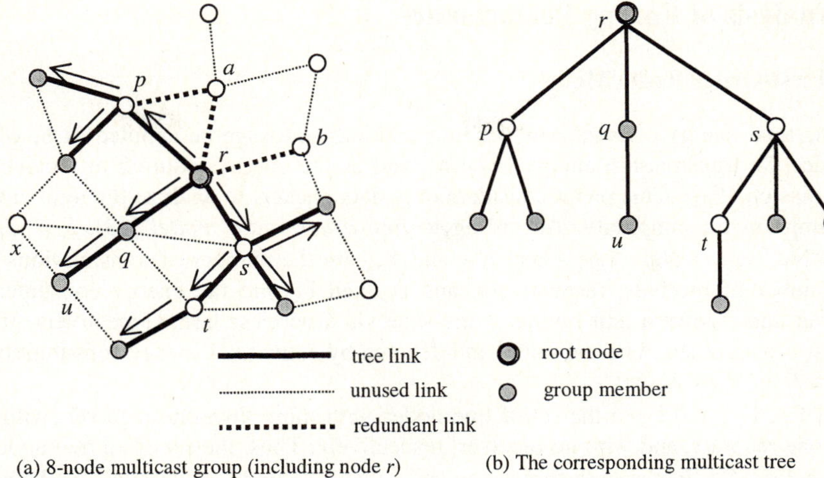

(a) 8-node multicast group (including node *r*)   (b) The corresponding multicast tree

Fig. 1. An example of tree-based multicast

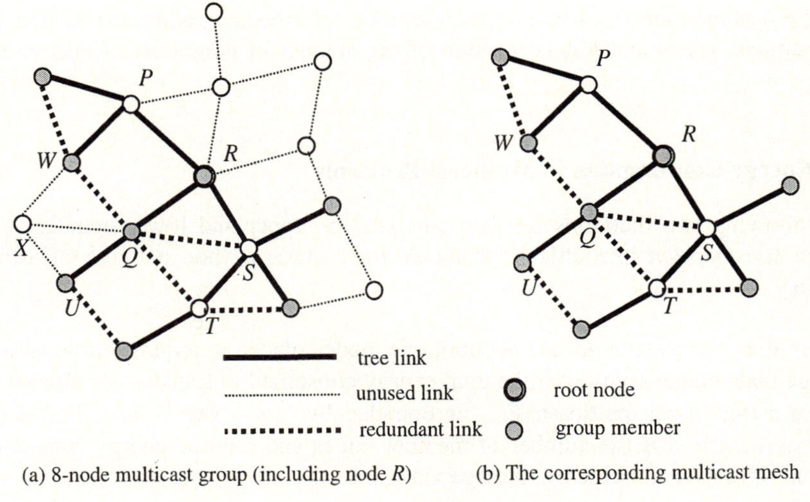

(a) 8-node multicast group (including node *R*)   (b) The corresponding multicast mesh

Fig. 2. An example of mesh-based multicast

from *R* to *U* involves three transmissions (*R*, *Q* and *U*) and fourteen receives (5 neighbors of *R*, 6 neighbors of *Q*, and 3 neighbors of *U*). For example, the transmission from node *Q* is received not only by *U* but also by neighbor nodes *R*, *S*, *T*, *W*, and *X*. The redundant link from *Q* to *W* may be useful when the path from *P* to *W* is broken. Although these redundant communications can be useful, they also waste more energy in battery-operated wireless nodes. Some redundant links are not used at all. For example, a transmission from *Q* to *X* is of no use because *X* is neither a member nor an intermediate node of the multicast group. Node *X* wastes energy receiving the packet but eventually discards it.

3 Analysis of Energy Performance

3.1 First-Order Radio Model

Let the total energy consumption per unit multicast message be denoted as E, which includes the transmission energy (E_{TX}) as well as the energy required to receive the transmission (E_{RX}). This paper considers only data packets to analyze the total energy consumption for simplicity. According to *first-order radio model* [13], $E = E_{TX} + E_{RX} = N_{TX} \cdot e_{TX} + N_{RX} \cdot e_{RX}$, where N_{TX} and N_{RX} are the number of transmissions and the number of receives, respectively, and e_{TX} and e_{RX} are the energy consumed to transmit and receive a unit multicast message via a wireless link, respectively. If e_{TX} and e_{RX} are assumed to be the same[1] and denoted by e, the total energy consumption is simply $E = (N_{TX} + N_{RX})e$.

Let Γ_+, Γ_1, and Γ_0 be the set of tree nodes with more than one receiver, with exactly one receiver, and with no receiver, respectively. Thus, the set of all tree nodes is $\Gamma = \Gamma_+ + \Gamma_1 + \Gamma_0$. It is straightforward to show that, in a multicast tree, N_{TX} is the number of tree nodes except the leaf receiver nodes (*i.e.*, root and intermediate nodes) and N_{RX} is $\sum_{i \in \Gamma_+} f_i + |\Gamma_1|$, where f_i is the number of neighbors of node i. In a multicast mesh, N_{TX} is the number of tree nodes (*i.e.*, root, intermediate, and receiver nodes) for the multicast group and N_{RX} is the sum of the number of neighbors of all tree nodes ($\sum_{i \in \Gamma} f_i$).

3.2 Energy Consumption in Multicast Protocols

The following two theorems formally analyze the upper and lower bounds of *total energy consumption* in multicast protocols for a static ad hoc network of arbitrary topology.

Theorem 1. For a static ad hoc network of n nodes placed in a quasi-circle area with average node connectivity of f, the total energy consumed to transfer a multicast message in a tree-based multicast, E_{tree}, is bounded by $(2m - O(m^{1/2}))e \le E_{tree} \le (2n - O(n^{1/2}))e$, where m is the number of member nodes and e is the energy consumed to transmit or receive a multicast message via a link.

Proof. Given a static ad hoc network of n nodes placed in a quasi-circle area with average node connectivity of f, the total energy consumption of a tree-based multicast for complete multicast can be regarded as the upper bound. In a complete multicast, $N_{TX} = n - O(n^{1/2})$, where $O(n^{1/2})$ is mainly due to the boundary nodes having less node connectivity than f, and $N_{RX} = n - 1$ since, given a tree of n nodes, the number of edges is $n - 1$. Note here that, in a quasi-circle area of n nodes as in Fig. 3, the number of boundary nodes is asymptotically $2\pi^{1/2}n^{1/2}$, resulting in $O(n^{1/2})$. Hence, $E_{tree} \le (N_{TX} + N_{RX})e \le (2n - O(n^{1/2}))e$. On the other hand, in the best case where a multicast

---

[1] In reality, e_{TX} and e_{RX} are silightly different. For example, e_{TX} is 300mA while e_{RX} is 250mA in an IEEE 802.11-compliant WaveLAN-II from Lucent [3].

tree consists of only the member nodes, $N_{TX} = m - O(m^{1/2})$ and $N_{RX} = m - 1$. Note here that, in the best case, all the member nodes form a quasi-circle area in nature because the multicast tree consists of only the adjacent member nodes and no intermediate nonmembers. Hence, $E_{tree} \geq (2m - O(m^{1/2}))e$. $\square$

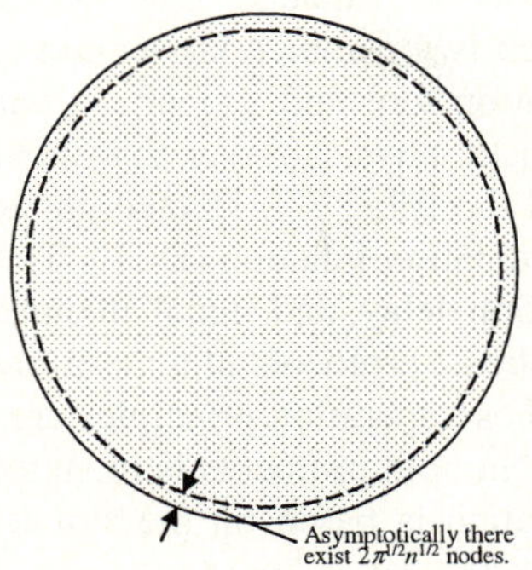

Fig. 3. A quasi-circle area of static n nodes

Theorem 2. For a static ad hoc network of n nodes placed in a quasi-circle area with average node connectivity of f, the total energy consumed to transfer a multicast message in a mesh-based multicast, E_{mesh}, is bounded by $((f + 1)m - O(m^{1/2}))e \leq E_{mesh} \leq ((f + 1)n - O(n^{1/2}))e$, where m is the number of member nodes and e is the energy consumed to transmit or receive a multicast message via a link.

Proof. Given a static ad hoc network of n nodes placed in a quasi-circle area with average node connectivity of f, the total energy consumption of a mesh-based multicast for complete multicast can be regarded as the upper bound. In a complete multicast, $N_{TX} = n$ and $N_{RX} = fn - O(n^{1/2})$ since the mesh-based multicast protocol uses broadcast-style communication, where $O(n^{1/2})$ is mainly due to the boundary nodes having less node connectivity than f. Note here that, in a quasi-circle area of n nodes as in Fig. 3, the number of boundary nodes is asymptotically $2\pi^{1/2}n^{1/2}$, resulting in $O(n^{1/2})$. Hence, $E_{mesh} \leq (N_{TX} + N_{RX})e \leq ((f + 1)n - O(n^{1/2}))e$. On the other hand, in the best case where a multicast tree consists of only the member nodes, $N_{TX} = m$ and $N_{RX} = fm - O(m^{1/2})$. Note here that, in the best case, all the member nodes form a quasi-circle area in nature because the multicast tree consists of only the adjacent member nodes and no intermediate nonmembers. Hence, $E_{mesh} \geq ((f + 1)m - O(m^{1/2}))e$. $\square$

According to Theorems 1 and 2, $E_{mesh}/E_{tree} \approx (f + 1)/2$ in the worst and the best cases. Since average node connectivity of f is usually much larger than 2 in order for any network partitioning not to occur, mesh-based multicast consumes around $(f + 1)/2$ times more energy compared to tree-based multicast. Conclusively, it is easily summarized that for a static ad hoc network, where wireless nodes are placed in a quasi-circle area with average node connectivity of f, *mesh-based multicast* consumes around $(f + 1)/2$ times more energy than *tree-based multicast*.

4 Energy Performance Evaluation

4.1 Simulation Environment

In order to validate the energy analysis, the energy performance of tree-based and mesh-based multicast protocols is evaluated. Our simulation study is based on the QualNet simulator [12], which is a commercial version of GloMoSim [14]. QualNet is a scalable network simulation tool and supports a wide range of ad hoc routing protocols. It simulates a realistic physical layer that includes a radio capture model, radio network interfaces, and the IEEE 802.11 medium access control (MAC) protocol using the distributed coordination function (DCF). The radio hardware model also simulates collisions, propagation delay, and signal attenuation.

We compared the two multicast protocols of a mesh-based protocol, ODMRP [15], and a shared tree protocol whose operation principles are described in [9]. The overhead due to control messages in both protocols is included in the simulation environment and results. Our simulation is based on the simulation of static nodes spread over a square area of 1000×1000 meter^2 for 15 minutes of simulation time, where the number of nodes is varied from 10 to 40 to see the effect of the node connectivity on the energy consumption. A free space propagation channel is assumed with a data rate of 2 Mbps. Note that omni-directional antennas and symmetric radio links are assumed in conjunction with the same transmission power.

In our simulation, a constant bit rate (CBR) source and its multiple destinations are randomly selected among the nodes, where group size (*i.e.*, the number of member nodes) is fixed as 10 for different node connectivity. In order to explore the connectivity effect, two different *radio transmission ranges* of 250 and 350 meters are taken into account for the simulation. Note here that, given network area A and radio transmission range R, the radio coverage C is $\pi R^2/A$ and the node connectivity f is given by $nC = \pi n R^2/A$, where n is the total number of nodes. For example, if $n = 20$ and $R = 250$ meters in the above network environment, $f = \pi \times 20 \times 250^2/10^6 = 3.93$. A CBR source sends a 512-byte multicast packet every 100 milliseconds during the simulation. For simplicity, we assume a multicast message consists of one data packet.

4.2 Simulation Results and Discussion

Fig. 4(a) shows *the total energy consumption* of two multicast protocols with respect to *node connectivity* of 2 to 8 for the radio transmission range (R) of 250 meters. The mesh-based multicast consumes more energy than the tree-based multicast by a factor of 2.1 ~ 3.9. Note here that, as the node connectivity increases, E_{mesh} is larger and larger compared to E_{tree} as expected. E_{mesh} rapidly increases along with node connectivity while E_{tree} slightly grows with the node connectivity.

Fig. 4(b) shows *the total energy consumption* of two multicast protocols with respect to *node connectivity* of 2 to 8 for the radio transmission range (R) of 350 meters. Notice that for a given number of nodes in an area, the longer radio transmission range means the higher node connectivity. The mesh-based multicast consumes more energy than the tree-based multicast by a factor of 2.2 ~ 4.0. Note here that with longer radio transmission range, the factors are a little larger than those in Fig. 4(a).

Even for longer radio transmission range, E_{mesh} rapidly increases along with node connectivity while E_{tree} slightly grows with the node connectivity.

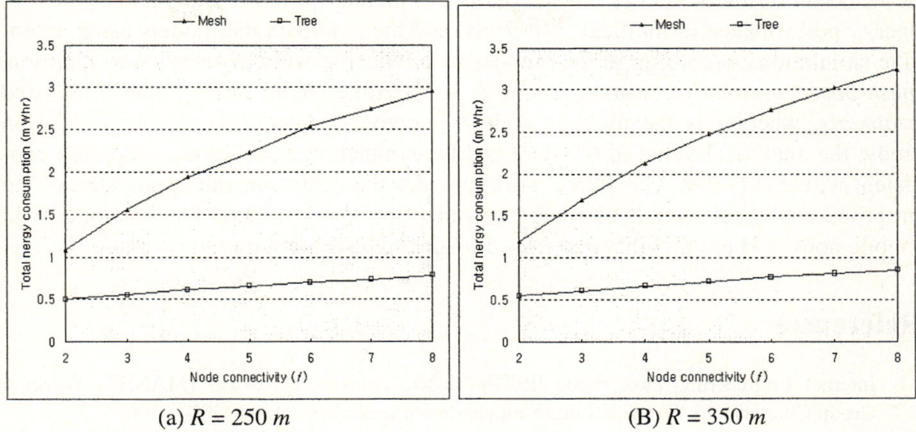

(a) $R = 250\ m$ (B) $R = 350\ m$

Fig. 4. Total energy consumption of multicast protocols

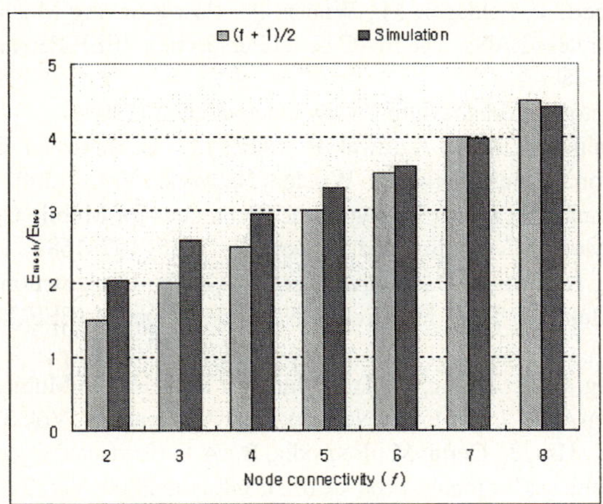

Fig. 5. Comparison of $(f + 1)/2$ and averaged simulation results

In order to validate the energy models developed in Section 4, the *analytical values* of $(f + 1)/2$ and the *simulation results* of E_{mesh}/E_{tree} are compared in Fig. 5, where the two sets of simulation results for the radio transmission range (R) of 250 and 350 meters are averaged. With low node connectivity, the simulation results of E_{mesh}/E_{tree} are larger than $(f + 1)/2$. This is mainly due to the fact that the *fixed overhead* of basic system energy is required no matter what the node connectivity is. As the node connectivity grows, however, they become closer to $(f + 1)/2$. Therefore, it is easily inferred that the analytical values of $(f + 1)/2$ and the simulation results of E_{mesh}/E_{tree} are consistent with each other if fixed overhead is excluded. Conclusively, it is clear that the tree-based multicast is the choice where energy is the primary concern.

5 Conclusion

The contribution of this paper is to evaluate the multicast protocols on static ad hoc networks in terms of energy consumption and thereby propose *analytical models* of energy performance in multicast protocols, and then validate the models using extensive simulation. According to the proposed analytical models of energy consumption, *mesh-based protocols* consume around $(f + 1)/2$ times more energy than *tree-based protocols*, where f is the average node connectivity. According to our simulation study, the analytical value of $(f + 1)/2$ and the simulation result of E_{mesh}/E_{tree} are consistent with each other. Our future work includes the extension and application of the proposed analytical models of energy consumption to wireless ad hoc networks with mobile nodes, where mobility and time domain analysis are required in nature.

References

1. Internet Engineering Task Force (IETF) Mobile Ad Hoc Networks (MANET) Working Group Charter. http://www.ietf.org/html.charters/manet-charter.html (2004)
2. Perkins, C.E.: Ad Hoc Networking. Addison-Wesley, Upper Saddle River, NJ (2001)
3. Kamerman A., Monteban, L.: WaveLAN-II: A High-Performance Wireless LAN for the Unlicensed Band. Bell Labs Technical Journal (1997) 118–133
4. Woesner, H., Ebert, J., Schlager, M., Wolisz, A.: Power-Saving Mechanisms in Emerging Standards for Wireless LANs: The MAC Level Perspective. IEEE Personal Communications, Vol. 5 (1998) 40–48
5. The Official Bluetooth Website. http://www.bluetooth.com/ (2004)
6. Jones, C.E., Sivalingam, K.M., Agrawal, P., Chen, J.C.: A Survey of Energy Efficient Network Protocols for Wireless Networks. Wireless Networks, Vol. 7 (2001) 343–358
7. Kravets R., Krishnan, P.: Power Management Techniques for Mobile Communication. Proc. of Int. Conf. on Mobile Computing and Networking (1998) 157–168
8. Wieselthier, J.E., Nguyen, G.D., Ephremides, A.: On the Construction of Energy-Efficient Broadcast and Multicast Trees in Wireless Networks. Proc. of IEEE Infocom 2000, Vol. 2 (2000) 585–594
9. Gerla, M., Chiang, C.-C., Zhang, L.: Tree Multicast Strategies in Mobile, Multihop Wireless Networks. Baltzer/ACM J. of Mobile Networks and Applications, Vol. 3 (1999) 193–207
10. Lee, S.-J., Su, W., Hsu, J., Gerla, M., Bagrodia, R.: A Performance Comparison Study of Ad Hoc Wireless Multicast Protocols. Proc. of IEEE Infocom 2000, Vol. 2 (2000) 565–574
11. Wan, P.-J., Calinescu, G., Li, X.-Y., Frieder, O.: Minimum-Energy Broadcast Routing in Static Ad Hoc Wireless Networks. Proc. of IEEE Infocom 2001, Vol. 2 (2001) 1162–1171
12. Scalable Network Technologies, Inc.: QualNet: Network Simulation and Parallel Performance. http://www.scalable-networks.com/products/qualnet.stm (2001)
13. Heinzelman, W.R., Chandrakasan, A., Balakrishnan, H.: Energy-Efficient Communication Protocols for Wireless Microsensor Networks. Proc. of the Hawaii Int. Conf. on System Sciences (2000) 3005–3014
14. Bajaj, L., Takai, M., Ahuja, R., Tang, K., Bagrodia, R., Gerla, M.: GloMoSim: A Scalable Network Simulation Environment. Technical Report, No. 990027, Computer Science Dept., UCLA (1999)
15. Lee, S., Gerla, M., Chiang, C.: On-Demand Multicast Routing Protocol. Proc. of IEEE Wireless Communications and Networking Conference (1999) 1298–1302

Weighted Flow Contention Graph and Its Applications in Wireless Ad Hoc Networks

Zeng Guo-kai, Xu Yin-long, Wu Ya-feng, and Wang Xi

Department of Computer Science & Technology,
University of Science & Technology of China,
National High Performance Computing Center at Hefei,
230027, Hefei, P.R.China
{Gkzeng, yfwu, rick_xi_wang}@ustc.edu
ylxu@ustc.edu.cn

Abstract. To fairly schedule flows in ad hoc networks, this paper defines a weighted flow contention graph, in which a maximal weighted independent set is always computed for scheduling. Two fair scheduling algorithms are then presented. One aims at maximizing network lifetime by setting the weight to the remaining energy, while the other tries to minimize the total delay by adopting the number of packets as the weight. Simulations are performed to compare these algorithms with the one based on unweighted flow contention graph [7]. Considering energy consumption, the network lifetime and the number of scheduled packets are improved evidently. While considering the number of packets, the transmission time and the total delay of the algorithm we propose are much less.

1 Introduction

Fair allocation of bandwidth and maximization of channel utilization have been identified as two important goals [1-7] for designing wireless networks, but they conflict with each other. In order to balance the trade-off between the fairness and the channel utilization, H. Luo, S. Lu, et al [6,7] proposed a flow contention graph to maximize the aggregate channel utilization subject to that the minimum channel allocation for each flow is guaranteed. L. Tassiulas and S. Sarkar [2] present a fair scheduling approach which schedules the flows which constitute a maximum token matching. H. Luo and S. Lu [3] propose a dynamic graph coloring approach which relies on the global information of the topology. A maximum clique based packet scheduling algorithm is presented by T. Ma et al [5], which promises to give the sub-maximum of the bandwidth utilization.

In order to improve performance with respect to fair scheduling, we present weighted flow contention graph, in which a maximal weighted independent set is always computed to schedule flows. The network performance is essentially sensitive to the choice of the weight. By adopting the energy value as the weight, one fair sched-

P. Lorenz and P. Dini (Eds.): ICN 2005, LNCS 3421, pp. 377–383, 2005.

uling algorithm can be applied in energy efficient issue. While setting the weight to the number of packets, a fair scheduling algorithm for multi-hop flows is presented.

The rest of the paper is organized as follows. We begin with a brief introduction of our network system model and weighted flow contention graph in Section 2. In Section 3 and 4, we apply the weighted flow contention graph in fair scheduling and empirically study its performance with respect to energy and the number of packets respectively. Finally, we conclude our work in Section 5.

2 Preliminary and Weighted Flow Contention Graph

This paper considers the multi-hop wireless networks in which the channel is shared among multiple contending hosts. Each host can transmit or receive packets, but cannot do both simultaneously. Transmissions are locally broadcast and only those within the transmission range of the sender can receive the packets. Two flows are contending with each other if the sender or the receiver of one flow is in the transmission range of the sender or the receiver of the other. Contending flows cannot be scheduled at the same time. Based on the generic ad hoc network environment, we make two assumptions: (1) time is divided into discrete slots, (2) packets are of the fixed size and each can be transmitted in a slot.

Definition 1. We define a *flow contention graph* as $G = (V, E)$, where V consists of all the flows and edge $(f_i, f_j) \in E$ if and only if flows f_i and f_j are contending with each other.

Definition 2. Each flow has a priority, which indicates the basic channel allocation the flow should deserve. A flow f_i with priority r_i should receive a lower bound on channel allocation of over an infinitesimal time period (t_1, t_2), where $B(t)$ is the set of

$$\frac{r_i}{\sum_{j \in B(t_1)} r_j} c(t_1, t_2) \tag{1}$$

backlogged flows in the entire network at time t, and c denotes the physical channel capacity. Formula (1) is called the *basic fairness* of flow f_i.

Definition 3. We define a *weighted flow contention graph (WFCG)* as $G = (V, E, w)$, where V consists of all flows in the network, $(f_i, f_j) \in E$ if and only if flows f_i and f_j are contending with each other, and $w: V \to R^+$ is a weight function.

Definition 4. A *maximum weighted independent set (MWIS)* of WFCG is a subset of vertices with the largest summation of weights and no two vertices in the subset are neighbors in the graph.

We present a greedy algorithm for computing MWIS in Fig.1. The algorithm incrementally grows an independent set by adding the vertex of the maximum weight, removing it along with its neighborhood from the graph, and iterating on the remaining graph until the graph becomes empty.

Algorithm

```
output set of nodes B←φ
set A ← {all the nodes in the graph}
While (A≠φ)
    choose node v such that its weight w(v)= max w(m)  m∈A
    B ← B ∪ {v}
    A ← A - {{v} ∪ v's neighborhood N(v)}
Return B
```

Fig. 1. MWIS Greedy Algorithm

Packet scheduling in ad hoc network is an inherent distributed computation problem. We adopt a CSMA/CA based MAC protocol. In each packet transmission, there is a RTS-CTS-DATA-ACK data handshake, and each transmitting host senses the carrier before sending out its RTS. Each backlogged flow will set an appropriate waiting time before it sends a RTS (the unit of waiting time is mini-slot). Each host maintains a local table which records the current weights of its neighbors. If the flow's weight is maximum in its table, then its waiting time is set to zero, otherwise, its waiting time is set to the difference between the maximum weight in its table and its weight. One flow transmits its RTS as soon as its waiting time is over, and its neighbors will sense the message and backoff themselves until the end of the slot.

3 Fair Scheduling with Energy Consumption

In ad hoc networks, the limited battery energy imposes a constraint on network lifetime. In this paper, we adopt a common definition of network lifetime, that is, the time until the energy of the first host uses up.

Although MIS is computed in Luo's algorithm to achieve channel reuse, unfortunately, a few flows with less energy may be much more often selected than some others with much more energy. Consequently, some hosts will use up their energy more quickly than others. At the end of the network lifetime, we may find that some hosts have much energy left, which shortens network lifetime and decreases total throughput. In the following, we present a fair scheduling algorithm on WFCG considering energy consumption.

Greatest Energy First (GEF) Algorithm

(1) At the start, the weight of a flow is set to the value of energy in the sender.

(2) Select the head of line packet of flow f^* according to STFQ algorithm [6-9], and increment its start and finish tag (used in STFQ) for flow f^*.

(3) Select MWIS S_{f^*} in $G - N[f^*]$, where G denotes the set of all flows in WFCG and $N[f^*]$ denotes the neighborhood of f^*.

(4) Schedule the flows in $\{f^*\} \cup S_{f^*}$, and decrease the weights of the flows by a certain number.

(5) If the weight of one flow equals zero, then the algorithm stops. Otherwise, return to Step (2).

Since STFQ is used in Step (2), we can provide the basic fairness of each flow. What is more, those flows with large weights (energy) are more possibly scheduled with MWIS. So the remaining energies of hosts become more balanced. Therefore, the network lifetime is prolonged, and the throughput is incremented as well.

We conduct a simulation to evaluate the performance of GEF. Random graphs are generated in 1000×1000 square units of a 2-D area, by randomly throwing hosts.. Transmission range R is 250 and the number of hosts n is 50. We use two scenarios to demonstrate the effectiveness of our model in terms of both network lifetime and average number of packets transmitted by each host. We prove it by running Luo's algorithm and GEF for each case, and the results are reported in Fig. 2 to Fig.5.

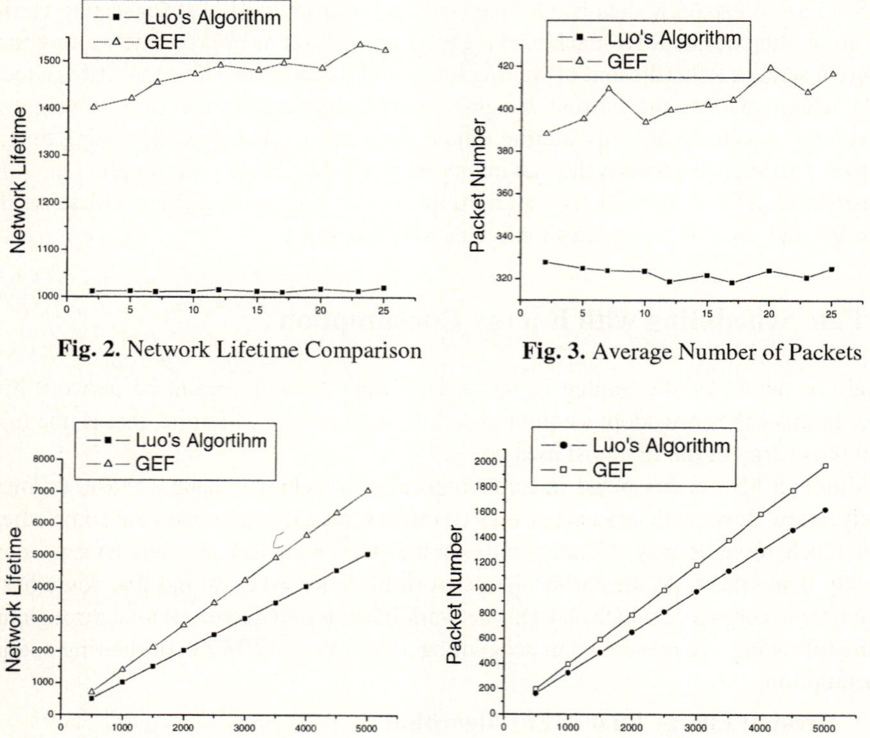

<table>
<tr><td>Fig. 2. Network Lifetime Comparison</td><td>Fig. 3. Average Number of Packets</td></tr>
<tr><td>Fig. 4. Network Lifetime Comparison</td><td>Fig. 5. Average Number of Packets</td></tr>
</table>

Scenario 1. At first, the value of energy of each host is 1000. We vary the number of multi-hop flows m from 2 to 25. For each m, generate a random connected graph 200 times. On the average, the network lifetime of GEF is about 45.8% longer than Luo's algorithm, and the hosts send more about 25.1% packets.

Scenario 2. The number of the multi-hop flows is 15. Generate a random connected graph 200 times, and we vary the value of each host e from 500 to 5000 on these graphs. On the average, the network lifetime of GEF is about 39.5% longer than Luo's algorithm, and the hosts send more about 21.5% packets.

4 Fair Scheduling of Multi-hop Flows

Packet scheduling in ad hoc networks is one-hop by one-hop. In [7] and in this paper, each multi-hop flow is broken into multiple one-hop flows, and its local host handles each one-hop flow. If multi-hop flow f is separated to a sequence of one-hop flows f_1, f_2, ..., f_h, f_{i-1} is the *prior flow* of f_i, and f_i is the *next flow* of f_{i-1}, $1 \le i \le h-1$. If f_{i-1} is the prior flow of f_i, the packet of f_{i-1} is transmitted from its sender to its receiver which is the sender of f_i. In other words, the packet is transmitted from f_{i-1} to f_i. The goals of fair scheduling of multi-hop flows are shown as follows: (1) provide the basic fairness for each flow, (2) minimize the total delay and the transmission time.

The packets of a communication request are transmitted one-hop by one-hop from the source to the destination. In other words, if a flow is scheduled, its packets are transmitted from its sender to its receiver which is its next flow's sender. So, there are different numbers of packets in different senders after a period of scheduling. What's more, it's possible that there are no packets in some senders, while there are many packets in others. And if there are more senders with no packets, fewer flows can be scheduled during the current slot.

Greatest Packets First (GPF) Algorithm

(1) At any slot, the weight of each flow is set to the number of packets in its sender currently.
(2) Select the head of line packet of flow f^* according to STFQ, and increment the start and finish tag for f^*.
(3) Select MWIS S_{f^*} in $G - N[f^*]$. Then, $S_{f^*} = S_{f^*} \cup \{f^*\}$.
(4) All flows in set S_{f^*} are scheduled at current slot. If flow f_i is just scheduled, its weight $W_{f_i} = W_{f_i} - 1$, and the weight of its next flow $W_{f_j} = W_{f_j} + 1$;
(5) If all the packets reach the destinations, stop the algorithm. Otherwise, go to Step (1).

At each slot, the flows with more packets are more likely to be selected with MWIS in step (3). Thus, we try to prevent the number of packets in one sender differing much from others. Therefore, the algorithm maximizes the channel reuse in a period and minimizes total delay of all packets as well as transmission time.

We evaluate our algorithms by simulations. Random graphs are generated in 1000×1000 square units of a 2-D area, by randomly throwing a certain number of hosts. The priority of each flow is generated randomly, but all the flows belonging to a multi-hop flow have the same priority. We use two scenarios to demonstrate the effectiveness of our model in terms of both transmission time and total delay of all packets. We prove it by running Luo's algorithm and GPF, and the results are reported in Fig.6 to Fig. 9.

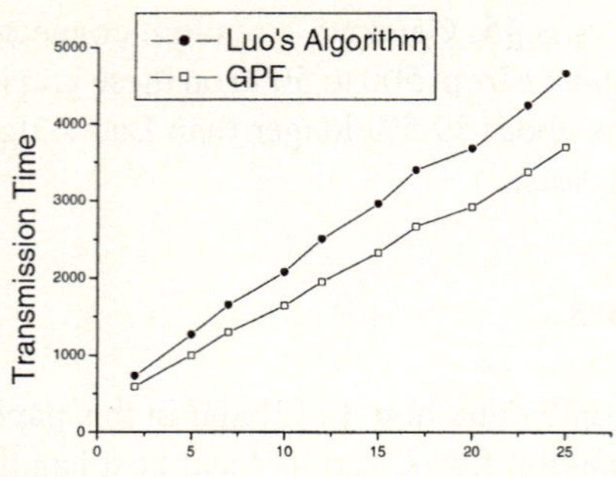

Fig. 6. Transmission Time Comparison

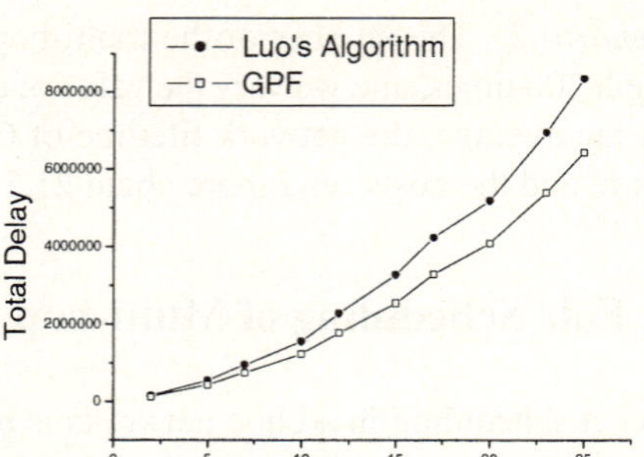

Fig. 7. Total Delxay Comparison

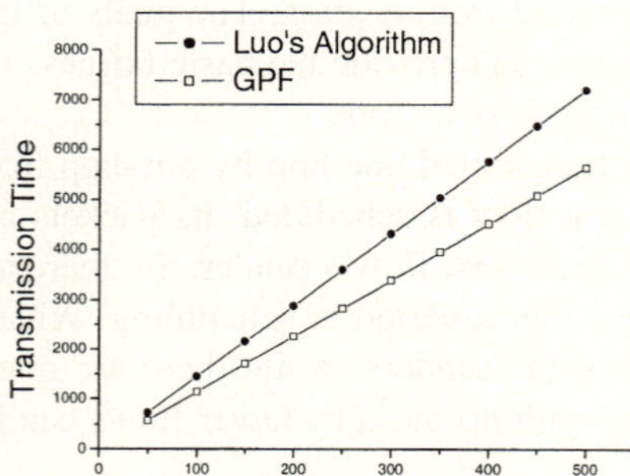

Fig. 8. Transmission Time Comparison

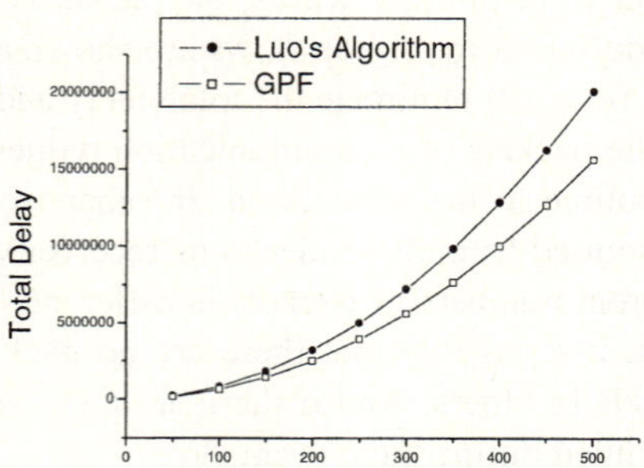

Fig. 9. Total Delay Comparison

Scenario 1. Transmission range R is 250 and the number of hosts n is 50. We vary the number of multi-hop flows m from 2 to 25, and the number of packets in the source of each multi-hop flow is 200. The results show that the transmission time of GPF is about 21.0% less than Luo's algorithm and the total delay is about 22.2% less.

Scenario 2. R is 250, n is 50 and m is 15. We vary the number of packets in the source of each multi-hop flow p from 50 to 500. For each p, generate a random connected graph 200 times. The results show that the transmission time of GPF is about 21.5% less than Luo's algorithm and the total delay is about 22.2% less.

5 Conclusion

This paper presents the weighted flow contention graph to balance the trade-off between fairness and channel utilization. The weighted flow contention graph can be used to achieve different network performances with different weights. As examples, a fair scheduling algorithm is applied for maximizing network lifetime by setting the weight to the remained energy. Another fair scheduling algorithm is proposed to minimize the total delay by setting the weight to the number of packets. We compare our algorithms with the one based on unweighted flow contention graph [7] by simulations. The results show that the network lifetime and number of scheduled packets are larger with GEF, and the transmission time and total delay are less with GPF.

References

[1] S. Lu, V. Bharghavan and R. Srikant, "Fair scheduling in wireless packet networks," *Networking, IEEE/ACM Transactions on, Aug. 1999, Volume: 7, Issue: 4, Pages: 473 - 489*

[2] L. Tassiulas and S. Sarkar, "Maxmin fair scheduling in wireless networks," *IEEE INFOCOM 2002, Vol. 2, pp. 763 – 772, June, 2002.*

[3] H. Luo and S. Lu, "A topology-independent fair queueing model in ad hoc wireless networks," *Network Protocols, 2000. Proceedings. 2000 International Conference on, 14-17 Nov. 2000, Pages: 325 – 335*

[4] X. Wu, C. Yuen, Y. Gao, H. Wu and B. Li, "Fair scheduling with bottleneck consideration in wireless ad-hoc networks," *Computer Communications and Networks, 2001. Proceedings. Tenth International Conference on, 15-17 Oct. 2001 Pages: 568 – 572*

[5] Tao Ma, Xinming Zhang, Guoliang Chen, "A Maximal Clique Based Packet Scheduling Algorithm in Mobile Ad Hoc Networks," *IEEE ICN'04, Feb, 2004, pp.690-695.*

[6] Haiyun Luo, Songwu Lu, Vaduvur Bharghavan, Jerry Cheng and Gary Zhong, "A packet scheduling approach to QoS Support in multihop wireless networks," *ACM Journal of Mobile Networks and Applications (MONET) Vol. 9, No. 3, June 2004*

[7] Haiyun Luo, Songwu Lu, Vaduvur Bharghavan, "A New Model for Packet Scheduling in Multihop Wireless Networks," *Proceedings of the 6th annual international conference on Mobile computing and networking, Oct, 2000 Page: 76 - 86*

[8] P. Goyal, H. M. Vin and H. Chen, "Start-Time Fair queueing: A Scheduling Algorithm for Integrated Service Access," *ACM SIGCOMM'96, August 1996.*

[9] J.C.R. Bennett and H. Zhang, "WF^{2}Q: Worst-case fair weighted fair queuing," *IEEE INFOCOM'96, 1996, pp. 120-127.*

Automatic Adjustment of Time-Variant Thresholds When Filtering Signals in MR Tomography

Eva Gescheidtova[1], Radek Kubasek[1], Zdenek Smekal[2], and Karel Bartusek[3]

[1] Faculty of Electrical Engineering and Communication,
Brno University of Technology,
Kolejni 4, 612 00 Brno, Czech Republic
[2] Purkynova 118, 612 00 Brno, Czech Republic
[3] Academy of Sciences of the Czech Republic,
Institute of Scientific Instruments
{gescha, smekal}@feec.vutbr.cz
bar@isibrno.cz

Abstract. Removing noise from an FID signal (a signal detected in MR measurement) is of fundamental significance in the analysis of results of NMR spectroscopy and tomography. Optimum solution can be seen in removing noise by means of a digital filter bank that uses half-band mirror frequency filters of the type of low-pass and high-pass filters. A filtering method using digital filters and the approach of automatic threshold adjustment is described in the paper.

1 Introduction

The MR imaging techniques of tomography and spectroscopy are exploited in many applications. For the MR instruments to function properly it is necessary to maintain high quality of homogeneity of the fundamental magnetic field.

Magnetic resonance (MR) is explained as the splitting of energy levels of atom nucleus due to the surrounding magnetic field. When the magnetic field changes, photons are emitted with an energy given exactly by the difference of these levels. The MR phenomenon can be detected as an electrical signal that will induce in a measuring coil a rotating magnetic moment of the nuclei being measured. This signal is referred to as free induction decay (FID). Since the frequency of an FID signal is identical with the natural frequency of nuclei, the kind of nuclei measured can be established on the basis of signal intensity and the number of nuclei. In the case of commonly used MR spectroscopes and tomographs the natural frequency of nuclei and thus also the fundamental frequency of an FID signal is of the order of tens to hundreds of MHz. But the signal spectral band is relatively small with respect to the fundamental frequency, of the order of tens to thousands of Hz. The signal is therefore commonly transformed into the fundamental band, i.e. zero natural frequency [1], [2]. Since the spectrum of FID signal is not symmetrical around the fundamental frequency, the FID signal after being transformed into the fundamental band is a complex signal. Because of technical considerations the MR phenomenon is in prac-

P. Lorenz and P. Dini (Eds.): ICN 2005, LNCS 3421, pp. 384–391, 2005.

tice called forth not by changing a (strong) fundamental magnetic field but by short-term superposition of another magnetic field. This further magnetic field is of rotational nature, which is achieved by excitation using a coil through which electric current is flowing and has a harmonic waveform of suitable amplitude modulated, so-called high-frequency excitation pulse. The frequency of the pulse (and thus also of field rotation) is chosen to be close to the natural frequency of the rotation of nuclei.

2   Noise in MR Signal

When defining the area being measured in localized spectroscopy and tomography, the gradient field is excited by very short pulses of sufficient magnitude. This gives rise to a fast changing magnetic field, which induces eddy currents in the conducting material near the gradient coils. These currents then cause retrospectively unfavourable deformation of the total magnetic field. The effect of eddy currents acts against the fast temporal changes in the magnetic field. The basic idea of a method that compensates this effect consists in replacing the missing leading edge of the field waveform by an overshoot of excitation current. To have the best possible compensation it is necessary to find an optimum shape of excitation pulse. Basically, this consists in obtaining the spectrometer response pulse, inverting it and using this inversion to filter the excitation (usually square) pulse. The term pre-emphasis compensation method is based on the fact that the compensation filter is in the nature of a differentiating element (high-pass filter).

Measuring the magnetic field gradient is converted to measuring the magnetic field induction in two symmetrically placed thin layers of the specimen being measured. A change in the magnetic field induction results in the phase (as well as amplitude) modulation of the FID signal. Instantaneous frequency of the FID signal is directly proportional to the magnetic field induction in the two regions defined. The dependence of instantaneous frequency on instantaneous gradient magnitude is linear while in the case of amplitude the dependence is more complicated. The frequency of an FID signal measured from the moment of the trailing edge of gradient excitation is in the nature of an exponentially descending function (it copies the drop in magnetic field induction) [3], [4].

In order to establish the optimum parameters of pre-emphasis filter it is necessary to know the exact time course of the induction of a gradient magnetic field induced by a certain exactly defined excitation pulse. Several measuring methods have been developed in the Institute of Scientific Instruments (Academy of Sciences of the Czech Republic) [5]. All of them establish the instantaneous magnitude of magnetic field induction from the instantaneous frequency of the FID signal being measured. A limiting factor of these methods is, above all, the signal-to-noise ratio (SNR), which deteriorates with time and the information sought gets usually lost in noise 10 ms or more after the arrival of the high-frequency excitation pulse. This situation is in Fig. 1.

Instantaneous frequency is the differentiation of instantaneous phase with respect to time. The method of numerical differentiation of the Newton interpolation polynomial is used. Fig. 2 is the flow chart of establishing the instantaneous frequency course of FID signal.

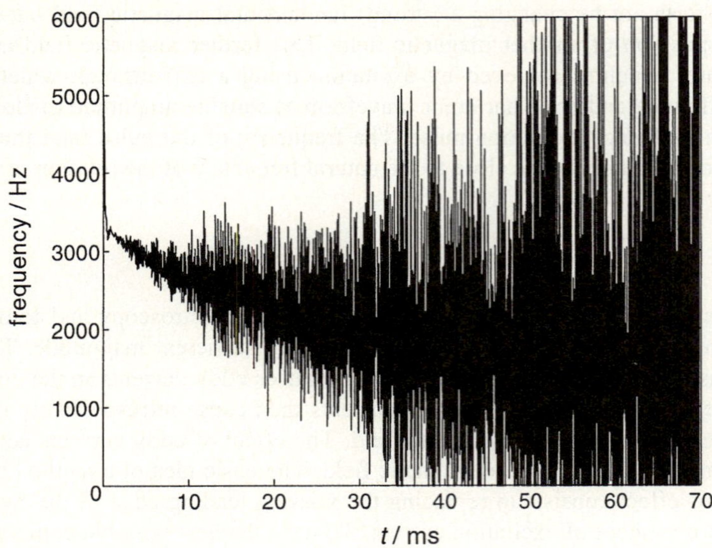

Fig. 1. Noise appearing in a signal that defines the course of instantaneous frequency

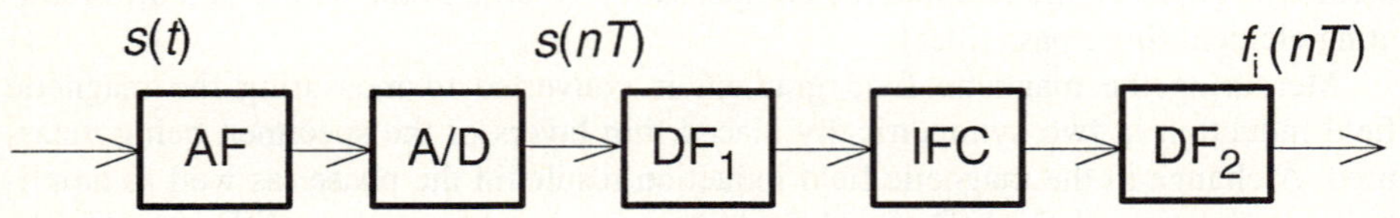

Fig. 2. Block diagram of establishing instantaneous frequency of FID signal

An analog signal $s(t)$ is first filtered by antialiasing low-pass filter AF. A/D conversion at the sampling frequency $f_s = 1/T$ is then performed using the A/D block. The digital signal is filtered again by low-pass filter DF_1.

The calculation of instantaneous frequency is realized by the IFC block. In the end, this signal is again filtered by low-pass filter DF_2 [6].

As mentioned above, noise greatly affects the calculation accuracy of numerical differentiation. While the error in phase calculation can be small even for a noisy signal, differentiating the phase will increase this error considerably since in the frequency domain it acts as a high-pass filter, which accentuates the high-frequency components. The basic requirement thus is to suppress noise in the FID signal as much as possible before calculating its instantaneous phase.

3 Noise Suppression Method

The basic idea of noise suppression consists in removing noise frequency components, which carry minimum information about the phase of FID signal but draw much energy. In practice, some distortion of useful signal is admissible if it is balanced by a sufficient increase in the SNR. If real-time processing is not a condition and post-acquisition processing (FID signal is first detected and stored in the memory and only then is it processed) is thus allowed, several noise suppression methods are available for application. In the beginning, methods of classical digital filtering were used but they did not bring the anticipated results [6]. Therefore an adaptive digital filtering method has been developed and tested [5]. [7]. Seen as most promising can be digital filtering by means of filter banks on the wavelet transform basis.

3.1 Filtering with the Aid of Digital Filter Banks

When making use of a digital filter bank, the principle of two-stage sub-band digital filtering is used [8], [9], [10]. The block diagram of this type of processing is given in Fig. 3. The principle of processing the signal in blocks WF_1 (Wavelet Filter 1) and WF_2 can be seen in greater detail in Fig. 4.

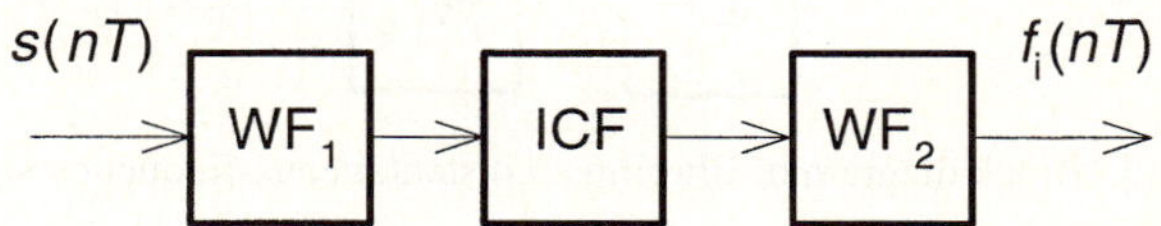

Fig. 3. Block diagram of two-stage digital filtering using digital filter banks

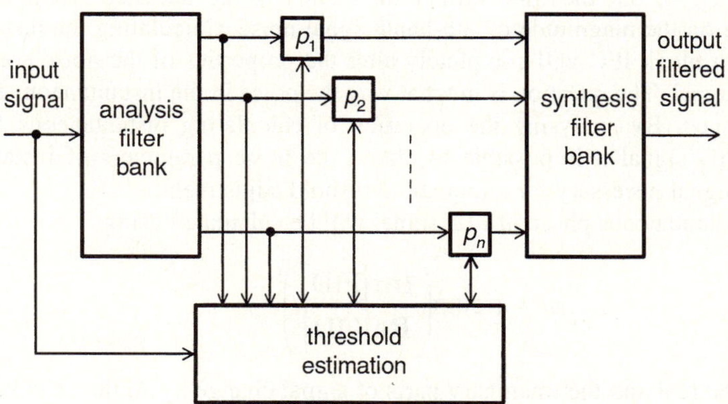

Fig. 4. Block diagram of digital filter banks in WF_1 and WF_2 blocks

The two processing stages contain a bank of analysing digital filters, AFB, which will divide the FID signal and the instantaneous frequency signal into frequency sub-bands, and a bank of synthesizing digital filters, SFB, which signal into subsequent to thresholding and noise removal will again synthesize the resultant signal. The banks are made up of several pairs of mirror frequency filters of the type of low-pass and high-pass filters. Since both the FID signal and the instantaneous frequency filter have a roughly exponential distribution of power spectrum density, both the analysing filter bank and the synthesizing filter bank are of octave spectrum division.

Noise thresholding in individual frequency sub-bands follows after the analysing filter bank. In both the WF_1 and WF_2 blocks the magnitudes of thresholds p_i are calculated by means of block TE (Threshold Estimation) on the basis of calculating the standard deviation at the end of the measured section of both the FID signal and the instantaneous frequency signal. In the former case, soft thresholding is utilized [11]. In the latter case the noise in block WF_2 is of non-stationary nature and the values of thresholds p_i are time-dependent. The magnitude of standard noise deviation increases with time and therefore it is necessary to use different types of thresholding.

3.2 Automating Adjustment of Time-Variant Thresholds

In automatic threshold adjustment we start from the block diagram in Fig. 5.

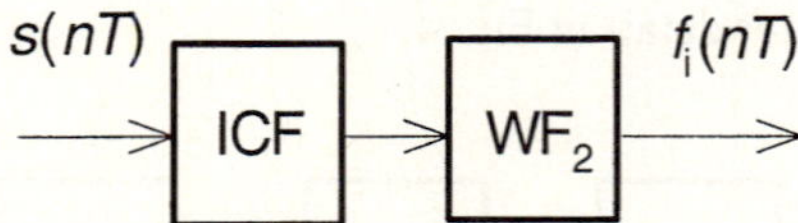

Fig. 5. Block diagram of filtering an instantaneous-frequency signal

Noise filtering is only applied in the instantaneous frequency region in the WF_2 block. The internal arrangement of the WF_2 block is the same as in the preceding case, see Fig. 4, but the time-variant thresholds p_i are adjusted automatically, in dependence on the magnitude of sub-band signal noise. Calculating the instantaneous frequency in block IFC will completely alter the properties of the noise contained in the useful signal. The question is in what way the noise in the instantaneous frequency signal changed. By analysing the operation of calculating instantaneous frequency from the FID signal it is possible to obtain the noise parameters of instantaneous-frequency signal necessary for automatic threshold adjustment.

The instantaneous phase of FID signal will be calculated using

$$\varphi = \arctan\left(\frac{\mathbf{Im}\left[\mathrm{FID}\right]}{\mathbf{Re}\left[\mathrm{FID}\right]}\right). \tag{1}$$

When the real and the imaginary parts of signal change by Δ, the calculated phase will change by

$$\Delta\varphi = \arctan\left(\frac{\mathbf{Im}\left[\mathrm{FID}\right]+\Delta}{\mathbf{Re}\left[\mathrm{FID}\right]-\Delta}\right) - \arctan\left(\frac{\mathbf{Im}\left[\mathrm{FID}\right]-\Delta}{\mathbf{Re}\left[\mathrm{FID}\right]+\Delta}\right) \tag{2}$$

It is exactly $\Delta\varphi$ that represents the noise contained in the instantaneous frequency signal of FID signal. The magnitude of the change Δ in the real and the imaginary parts of the signal is directly linked with the magnitude of standard noise deviation, $\delta \approx \Delta$. Since the standard noise deviation of FID signal is constant, $\Delta\varphi$ changes in dependence on the magnitude of the FID signal and is thus non-linearly dependent also on SNR. Fig. 6 gives the calculation procedure for the instantaneous phase of FID signal.

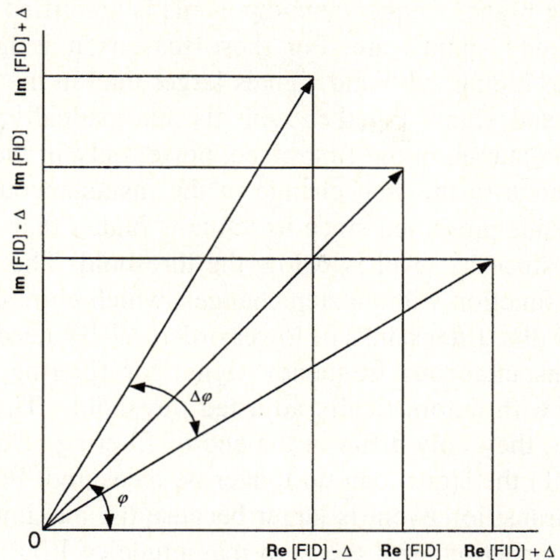

Fig. 6. Phasor diagram for phase calculation, noise $\Delta\varphi$

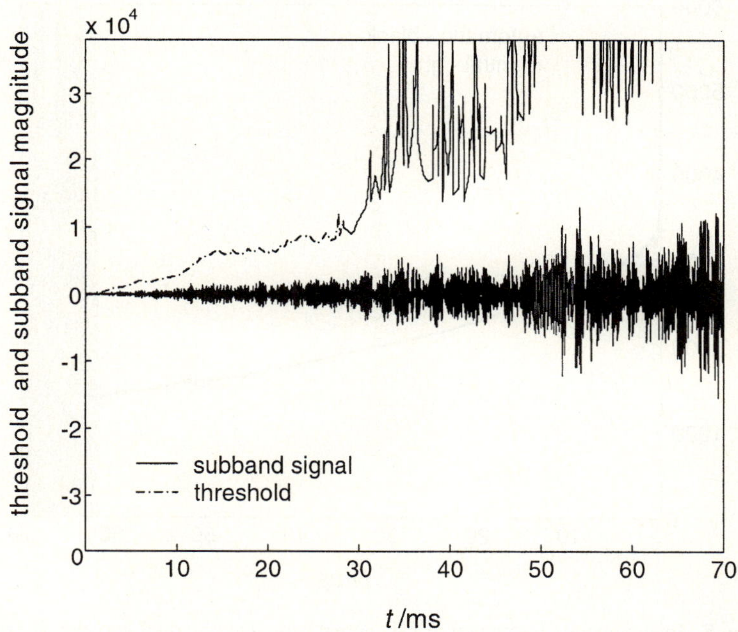

Fig. 7. Second sub-band signal (solid) and its threshold (dash-dotted)

The instantaneous frequency signal is obtained via differentiating the instantaneous phase. In our case, we calculate the derivative from two points only, in order to be able to follow the fast frequency changes in FID signal which occur at its beginning.

A signal containing higher frequency components is amplified together with noise more than low-frequency signals are. For these reasons it is necessary to set the threshold magnitude of higher sub-band signals larger than in the lower bands. Fig. 7 gives the other sub-band signal together with its automatically adjusted threshold. This sub-band signal contains, in the first place, noise; only in the beginning is there also useful signal, thanks to the step change in the instantaneous frequency signal. The useful signal remains preserved since its value is higher than the threshold while the noise is removed since its level is below the threshold. The threshold then proceeds as a continuous function without step changes, which eliminates possible transition events. In spite of this, filter banks of lower orders (5-10) need to be used.

Fig. 8 gives the instantaneous frequency signal for filtering with manually adjusted thresholds and with automatically adjusted thresholds. The two curves are of almost identical shape; they only differ at the end of filtering. This is because due to its very low SNR (<<1) the signal can no longer be measured. For automatically adjusted thresholds the transition event is larger because the maximum possible threshold is calculated in connection with the zero magnitude of FID signal, and thus the signal is step-changed.

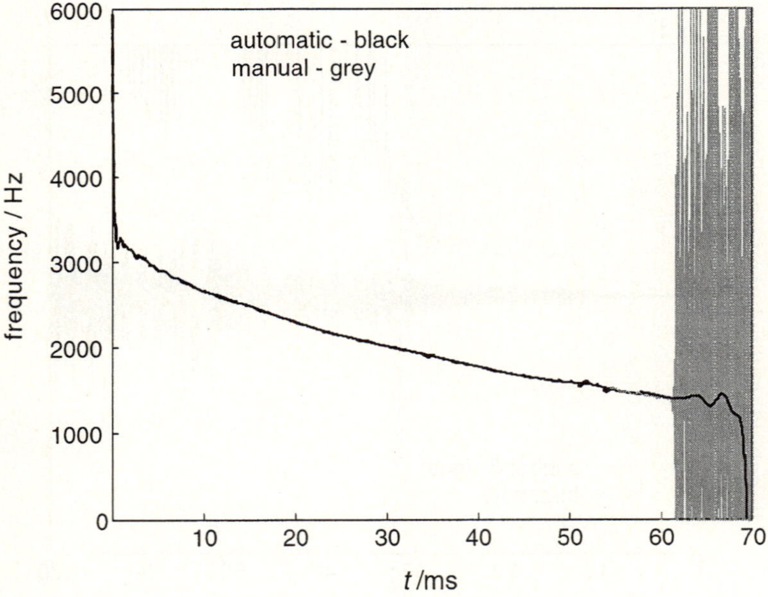

Fig. 8. Instantaneous frequency of FID signal processed by dual wavelet filtering

4 Conclusion

Removing noise from FID signals is of fundamental significance in analysing the results of MR spectroscopy and tomography. Optimum methods were sought for removing noise from a signal that describes the values of instantaneous frequency of FID signals during an MR experiment. The method of filtering MR by digital filter banks and using the wavelet transform yields very good results. Unlike filtering methods, which make use of current digital FIR filters and adaptive filtering as described in the introduction, it extends considerably the time for which MR signal with sufficiently high SNR can be detected. Automatic threshold adjustment does away with prolonged operations of adjusting the thresholds manually.

Acknowledgement

The paper was prepared within the framework of N°IAA2065201 project of the Grant Agency of the Academy of Sciences of the Czech Republic and with the support of the research plan MSM 262200022.

References

1. Vlaardingerbroek, M.: Magnetic Resonance Imaging. Springer-Verlag, (2000)
2. Mansfield, P., Chapman, B.: Active Magnetic Screening of Gradient Coils in NMR Imaging. Journal of Magnetic Resonance 66, (1986) 573-576
3. Bartusek, K., Puczok, V.: The MULTIFID Method for Measurement of Magnetic Field Gradients. Meas.Sci.Technol. 4, (1993) 357
4. Bartusek, K., Jílek, B.: Measurement of the Gradient Magnetic Field for NMR Tomography, in Proceedings of the 1st Nottingham Symposium on Magnetic Resonance in Medicine, Nottingham, (1994)
5. Svoboda, R.: Generation and Measurement of Gradient Magnetic Fields in NMR Spectrometer. PhD Thesis, Brno University of Technology (2003)
6. Bartusek, K., Gescheidtova, E.: Instantaneous Frequency of Spin Echo Method for Gradient Magnetic Field Measurement in MR Systems. Journal of Electrical Engineering, vol. 53, (2002) 49-52
7. Bartusek, K., Gescheidtova, E.: Adaptive Digital Filter for Gradient Magnetic Field Measurement in MR Tomography. In Proceedings of the IEEE International Conference APCCAS' (2002) 79-82
8. Vich, R., Smekal, Z.: Digital Filters. Academia, Praha (2000)
9. Fliege, N.: Multirate Digital Signal Processing. John Wiley & Sons, Chichester (1996)
10. Mitra, S. K.: Digital Signal Processing. McGraw-Hill, (1998)
11. Rajmic, P.: Statistical Properties of Wavelet Spectrum Thresholding Rules. In Proceedings of the 48th International Scientific Colloquium, Ilmenau, (2003) 87-88

Analytical Design of Maximally Flat Notch FIR Filters for Communication Purposes

Pavel Zahradnik[1], Miroslav Vlček[2], and Boris Šimák[1]

[1] Department of Telecommunications Engineering,
Czech Technical University Prague,
Technická 2, CZ-166 27 Praha, Czech Republic
Phone: +420-2-24352089, Fax: +420-2-33339810
{zahradni, simak}@fel.cvut.cz
[2] Department of Applied Mathematics,
Czech Technical University Prague,
Konviktská 20, CZ-110 00 Praha, Czech Republic
Phone: +420-2-24890720, Fax:+420-2-24890702
vlcek@fel.cvut.cz

Abstract. A novel fast analytical design procedure for the maximally flat notch FIR filters is introduced. The closed form solution provides recursive evaluation of the impulse response coefficients of the filter. The discrete nature of the notch frequency is emphasized. One design example is included in order to demonstrate the efficiency of the presented approach.

1   Introduction

The narrow band digital filters are frequently used in the digital processing of telecommunication signals. While narrow bandpass filters find their application in the detection of signals, the narrow bandstop filters are frequently used in order to remove a single frequency component from the spectrum of the signal. The narrow bandstop filters are usually called notch filters. The design of digital notch IIR filters is rather simple. These filters are frequently used despite of their infinite impulse and step responses which can produce spurious signal components that are unwanted in various applications. The notch IIR filters consist of an abridged all-pass second-order section that allows independent tuning of the notch frequency $\omega_m T$ and the 3-dB attenuation bandwidth [3]. The main drawback usually emphasized in connection with FIR filters is the higher number of coefficients compared to their IIR counterparts. However, this argument is weakened continuously due to the tremendous advance in DSP and FPGA technology. The decisive advantages of FIR filters is their constant group delay and superior time response [8]. Thus the implementation of FIR filters with one hundred coefficients has an practical impact in numerous applications. A few analytical procedures for the design of linear phase notch FIR filters have recently become available [5]. The methods which lead to feasible filters are

P. Lorenz and P. Dini (Eds.): ICN 2005, LNCS 3421, pp. 392–400, 2005.
© Springer-Verlag Berlin Heidelberg 2005

generally derived by iterative approximation techniques or by non-iterative, but still numerical procedures, e.g. the window technique. In our paper we are concerned with completely analytical design of maximally flat notch FIR filters. We introduce the degree formula which relates the degree of the generating polynomial, the length of the filter, the notch frequency, the width of the notchband and the attenuation in the passbands. We derive the differential equation for the generating polynomial of the filter. Based on the expansion of the generating polynomial into the Chebyshev polynomials, the recurrent formula for the direct computation of the impulse response coefficients is derived. Consequently, the FFT algorithm usually required in the analytical design of narrow band FIR filters is avoided. The proposed design procedure is recursive one. It does not require any FFT algorithm or any iterative technique.

2 Polynomial Approximation, Zero Phase Transfer Function

Here and in the following we use the independent transformed variable w [6] related to the digital domain by

$$w = \frac{1}{2}\left(z + z^{-1}\right)\Big|_{z=e^{j\omega T}} = \cos \omega T \ . \tag{1}$$

We denote $H(z)$ the transfer function of a notch FIR filter with the impulse response $h(m)$ of the length N as

$$H(z) = \sum_{m=0}^{N-1} h(m) z^{-m} \ . \tag{2}$$

Assuming an odd length $N = 2n+1$ and even symmetry of the impulse response

$$a(0) = h(n) \ , \quad a(m) = 2h(n \pm m) \ , \quad m = 1 \ldots n \tag{3}$$

we can write the transfer function of the notch FIR filter

$$H(z) = z^{-n}\left[a(0) + \sum_{m=1}^{n} a(m)\, T_m(w)\right] \tag{4}$$

where $T_m(w)$ is Chebyshev polynomial of the first kind. The frequency response of the filter $H(e^{j\omega T})$ can be expressed by the zero phase transfer function $Q(w)$

$$H(e^{j\omega T}) = e^{-jn\omega T} Q(\cos \omega T) = z^{-n}\, Q(w)\big|_{z=e^{j\omega T}} \ . \tag{5}$$

For $w = \frac{1}{2}(z + z^{-1})\big|_{z=e^{j\omega T}} = \cos \omega T$ the zero phase transfer function $Q(w)$ represents a polynomial of the real variable w. It reduces to a real valued frequency response of the zero-phase FIR filter. The zero phase transfer function $Q(w)$ of the narrow bandpass FIR filter is formed by the generating polynomial $A_{p,q}(w)$ while the zero phase transfer function $Q_A(w)$ of the notch FIR filter is

$$Q_A(w) = 1 - A_{p,q}(w) \ . \tag{6}$$

3   Maximally Flat Notch FIR Filter

For the design of maximally flat notch FIR filter we propose the generating polynomial $A_{p,q}(w)$ of the maximally flat narrow bandpass FIR filter introduced in [7]

$$A_{p,q}(w) = C(1-w)^p(1+w)^q \; . \tag{7}$$

The notation $A_{p,q}(w)$ emphasizes that p counts multiplicity of zeros at $w = 1$ and q corresponds to multiplicity of zeros at $w = -1$. Forming of the derivative of the polynomial

$$\frac{dA_{p,q}(w)}{dw} = -Cp(1-w)^{p-1}(1+w)^q + Cq(1-w)^p(1+w)^{q-1} \tag{8}$$

and by simple manipulation of (7)

$$(1-w)(1+w)\frac{dA_{p,q}(w)}{dw} = -p(1+w)A_{p,q}(w) + q(1-w)A_{p,q}(w) \tag{9}$$

we arrive at the differential equation for the generating polynomial $A_{p,q}(w)$

$$(1-w^2)\frac{dA_{p,q}(w)}{dw} + [p - q + (p+q)w]\,A_{p,q}(w) = 0 \; . \tag{10}$$

The differential equation (10) for the polynomial $A_{p,q}(w)$ forms a completely new concept in digital filter design as it provides the recursive evaluation of the impulse response coefficients of the filter described in Section 6. The normalization of the generating polynomial $A_{p,q}(w)$ constraints $A_{p,q}(w_m) = 1$ where w_m is the position of the maximum of the generating polynomial $A_{p,q}(w)$ as illustrated in Fig. 1. The normalization of the generating polynomial $A_{p,q}(w)$ results in

$$A_{p,q}(w) = \left[\frac{p+q}{2p}(1-w)\right]^p \left[\frac{p+q}{2q}(1+w)\right]^q . \tag{11}$$

The polynomial

$$Q_A(w) = 1 - A_{p,q}(w) = 1 - \left[\frac{p+q}{2p}(1-w)\right]^p \left[\frac{p+q}{2q}(1+w)\right]^q \tag{12}$$

represents the real-valued zero phase transfer function of the maximally flat notch FIR filter of the real variable $w = \cos \omega T$. The transfer function of the maximally flat notch FIR filter is

$$H(z) = \sum_{m=0}^{N-1} h(m)\, z^{-m} = z^{-n}\,(1 - A(p,q)(w)) \; . \tag{13}$$

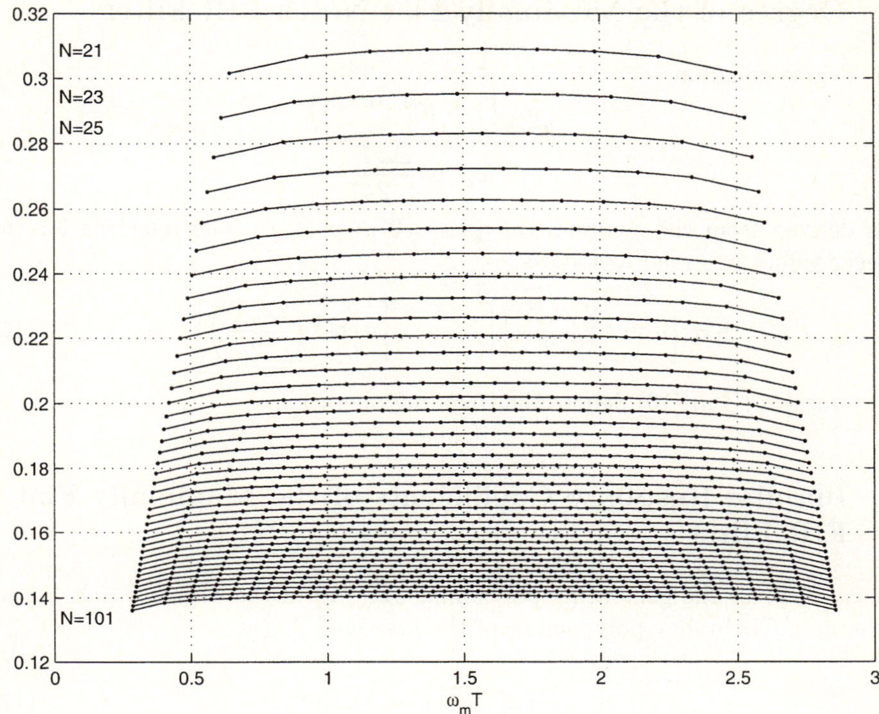

Fig. 1. Dependence of the width of the notchband $\Delta\omega T/\pi$ of the maximally flat flat notch FIR filter on the notch frequency $\omega_m T$ and the filter length N. The available notch frequencies are marked by the dots

4   Notch Frequency of the Maximally Flat Notch FIR Filter

The notch frequency $\omega_m T$ is derived from the minimum value of the zero phase transfer function $Q_A(w)$ (12) as

$$w_m = \cos\omega_m T = \frac{q-p}{q+p} \ . \tag{14}$$

The notch frequency $\omega_m T$ of the maximally flat notch FIR filter is given from (14) by the integer values p and q exclusively. It is obvious, that for the specified filter length $N = 2(p+q)+1$, exactly $p+q-1$ discrete notch frequencies $\omega_m T$ are available (Fig. 1).

5   Degree of the Maximally Flat Notch FIR Filter

The degree equation

$$n \geq \frac{\log\left(1 - 10^{0.05a[dB]}\right)}{\log \cos \dfrac{\Delta\omega T}{2}} \tag{15}$$

was derived from the symmetrical case $n/2 = p = q$. The relations for the integer values p, q read as follows

$$p = \left[n \sin^2\left(\frac{\omega_m T}{2}\right)\right] \ , \quad q = \left[n \cos^2\left(\frac{\omega_m T}{2}\right)\right] \ . \tag{16}$$

The brackets [ ] in (16) denote the rounding operation.

6   Impulse Response Coefficients of the Maximally Flat FIR Filter

We can express the generating polynomial $A_{p,q}(w)$ of the degree $n = p + q$ as the sum of Chebyshev polynomials of the first kind $T_m(w)$

$$A_{p,q}(w) = \sum_{m=0}^{n} a(m) \, T_m(w) \ . \tag{17}$$

The coefficients $a(m)$ define the impulse response $h(m)$ (3) of the length $N = 2(p+q)+1$. Assuming the generating polynomial $A_{p,q}(w)$ of the maximally flat narrow bandpass FIR filter in the sum (17) we can write

Table 1. Recursive Algorithm for the Evaluation of the Coefficients $a(m)$

given	$p, \ q$
initialization	$n = p + q$ $a(n + 1) = 0$

body
   (*for* $k = n + 1$ *to 3*)
 	
   (*end loop on* k)
	$a(k - 2) \ \stackrel{.}{=} \ -\dfrac{(n + k)a(k) + 2(2p - n)a(k - 1)}{n + 2 - k}$

	$a(0) = -\dfrac{(n + 2)a(2) + 2(2p - n)a(1)}{2n}$

$$(1-w^2)\frac{dA_{p,q}(w)}{dw} = \sum_{m=1}^{n} a(m)(1-w^2)\frac{dT_m(w)}{dw} = \sum_{m=1}^{n} a(m)\frac{m}{2}\left[T_{m-1}(w)-T_{m+1}(w)\right] .$$

(18)

By introducing (17) and (18) into the differential equation (10) and using the recursive formula for Chebyshev polynomials

$$T_{m+1}(w) = 2w\,T_m(w) - T_{m-1}(w)$$

(19)

we get the identity

$$\sum_{m=1}^{n} a(m)\frac{m}{2}\left[T_{m-1}(w) - T_{m+1}(w)\right] + (p-q)a(0)$$

$$+ \sum_{m=1}^{n} a(m)(p-q)T_m(w) + (p+q)a(0)w$$

(20)

$$+ \sum_{m=1}^{n} a(m)(p+q)\frac{1}{2}\left[T_{m-1}(w) + T_{m+1}(w)\right] = 0 .$$

By iterating eq. (20) we have deduced a simple recursive algorithm for the evaluation of the coefficients $a(m)$ of the generating polynomial $A_{p,q}(w)$ of the maximally flat narrow bandpass FIR filter. The recursive algorithm is presented in Table 1. The coefficients $h(m)$ of the impulse response of the maximally flat notch FIR filter are obtained from the coefficients $a(m)$ of the maximally flat narrow bandpass FIR filter as follows

$$h(n) = 1 - a(0) \,, \;\; h(n \pm m) = -\frac{a(m)}{2} \,, \;\; m = 1\ldots n \,.$$

(21)

7   Design of the Maximally Flat Notch FIR Filter

The goal of the maximally flat notch FIR filter design is to find the two integer values p and q in order to satisfy the filter specification as precisely as possible. The design procedure is as follows:

1. Specify the notch frequency $\omega_m T$, maximal width of the notchband $\Delta\omega T$ and the attenuation in the passbands a [dB] as demonstrated in Fig. 2.
2. Calculate the minimum degree n (15) required to satisfy the filter specification.
3. Calculate the integer values p and q (16).
4. Check the notch frequency (14) for the obtained integer values p, q.
5. Evaluate the coefficients $a(m)$ of the generating polynomial $A_{p,q}(w)$ recursively (Table 1).
6. Evaluate the coefficients of the impulse response $h(m)$ of the maximally flat notch FIR filter (21).

It is worth of noting that a substantial part of coefficients of the impulse response $h(m)$ of the maximally flat notch FIR filter has negligible values. From this fact follows the possible large abbreviation of the impulse response of the maximally flat notch FIR filter by the rectangular windowing without significant deterioration of the frequency properties of the filter as emphasized in [7].

8   Example of the Design

Design the maximally flat notch FIR filter specified by $\omega_m T = 0.35\,\pi$ and $\Delta\omega T = 0.15\,\pi$ for $a = -3.0103$ dB.

Using our design procedure we get $n = [43.8256] \rightarrow 44$ (15), $p = [11.9644] \rightarrow 12$ and $q = [31.8610] \rightarrow 32$ (16). The filter length is $N = 89$ coefficients. The actual filter parameters are $\omega_m T = 0.3498\,\pi$ and $\Delta\omega T = 0.1496\,\pi$. The attenuation at the frequency $0.3\,\pi$ amounts -168 dB. The coefficients $a(m)$ were evaluated recursively (Table 1). The coefficients of the impulse response $h(m)$ of the maximally flat notch FIR filter were evaluated by (21). Because $|h(m)| < 10^{-6}$ for $0 \le m \le 13$ and $75 \le m \le 88$, only the 71 central coefficients of the impulse response $h(m)$ for $14 \le m \le 74$ are summarized in Table 2. The amplitude frequency response $20\log|H(e^{j\omega T})|$ [dB] of the filter is shown in Fig. 2.

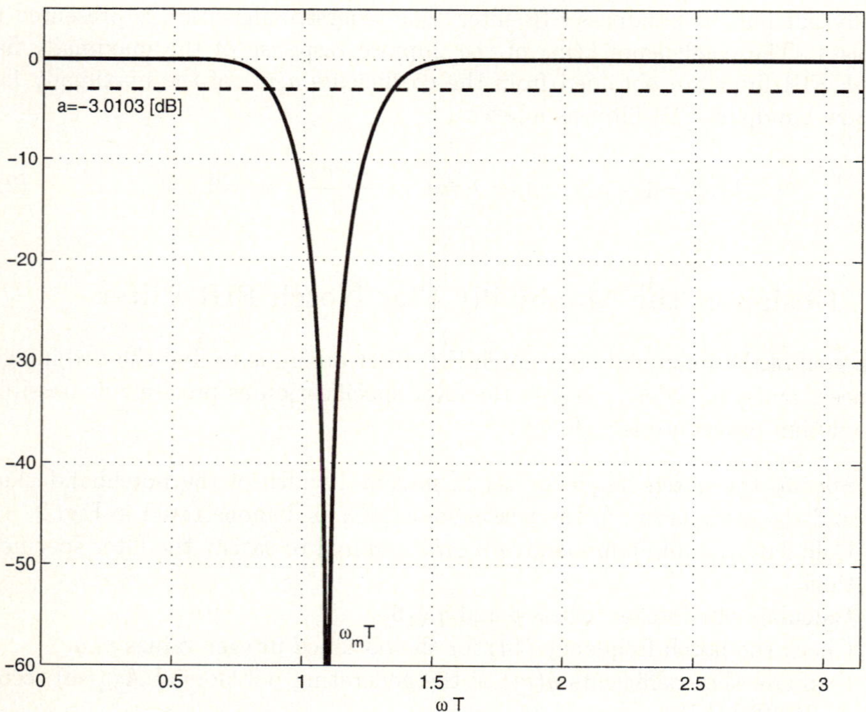

Fig. 2. Amplitude frequency response $20\log|H(e^{j\omega T})|$ [dB] based on the zero phase transfer function $Q(w) = 1 - A_{12,32}(w)$

Table 2. Impulse Response Coefficients

m		h(m)	m		h(m)
14	74	-0.000002	30	58	0.012289
15	73	-0.000003	31	57	0.002278
16	72	0.000000	32	56	-0.019427
17	71	0.000018	33	55	-0.027483
18	70	0.000037	34	54	-0.003357
19	69	0.000010	35	53	0.042804
20	68	-0.000111	36	52	0.048063
21	67	-0.000245	37	51	-0.009353
22	66	-0.000101	38	50	-0.075616
23	65	0.000537	39	49	-0.065324
24	64	0.001173	40	48	0.029196
25	63	0.000480	41	47	0.106554
26	62	-0.002149	42	46	0.068113
27	61	-0.004302	43	45	-0.053105
28	60	-0.001388	44		0.880514
29	59	0.007135			

9   Conclusions

A novel fast analytical design procedure for the design of maximally flat notch FIR filters for communication purposes was introduced. The closed form solution provides recursive evaluation of the impulse response of the filter. One example demonstrated the efficiency of the design procedure.

References

1. S. C. Dutta Roy, B. Kumar, S. B. Jain, FIR Notch Filter Design - A Review. *Facta Universitatis (Niš), Series Electronics and Energetics*. Vol. 14, No. 3, December 2001, pp. 295-327.
2. S. C. Pei, C. C. Tseng, IIR Multiple Notch Filter Design Based on Allpass Filter, *IEEE Transactions on Circuits and Systems*. Vol. 44, No. 2, February 1997, pp. 133-136.
3. P. A. Regalia, S. K. Mitra and P. P. Vaidyanathan, The Digital All-Pass Filter : A Versatile Signal Processing Building Block. *Proceedings of IEEE*. Vol. 76, No. 1, Jan. 1988, pp. 19 - 37.
4. I. W. Selesnick, C. S. Burrus, Exchange Algorithms for the Design of Linear Phase FIR Filters and Differentiators Having Flat Monotonic Passbands and Equiripple Stopbands. *IEEE Trans. Circuits, Syst.-II*. Vol. 43, Sept. 1996, pp. 671 - 675.
5. Tian-Hu Yu, S.K. Mitra and H. Babic, Design of Linear Phase FIR Notch Filters. *Sadhana*. Vol. 15, Nov. 1990, Iss. 3, pp. 133 - 55.
6. M. Vlček and R. Unbehauen, Analytical Solution for Design of IIR Equiripple Filters. *IEEE Trans. Acoust., Speech, Signal Processing*. Vol. ASSP - 37, Oct. 1989, pp. 1518 - 1531.

7. M. Vlček, L. Jireš, Fast Design Algorithms for FIR Notch Filters. *Proc. of IEEE International Symposium on Circuits and Systems ISCAS'94*. London, 1994, Vol. 2, pp. 297 - 300.
8. M. Vlček, P. Zahradnik. Digital Multiple Notch Filters Performance. *Proceedings of the 15th European Conference on Circuit Theory and Design ECCTD'01*. Helsinky, August 2001, pp. 49-52.
9. M. Vlček, P. Zahradnik, R. Unbehauen, Analytic Design of FIR Filters. *IEEE Transactions on Signal Processing*. Vol. 48, September 2000, pp. 2705-2709.

Iterative Decoding and Carrier Frequency Offset Estimation for a Space-Time Block Code System*

Ming Luo, Qinye Yin, Le Ding, and Yiwen Zhang

School of Electronics and Information Engineering,
Xi'an Jiaotong University, Xi'an, P.R.China
luomingmail@126.com

Abstract. In the paper, we address the problem of blind decoding and carrier frequency offset estimation for a full rate space-time block coded system. The system is subject to unknown frequency-selective fading channel and carrier frequency offsets between transmitter and receiver. An iterative method is proposed to estimate frequency offsets and decode the space time block code. The unknown frequency selective channel is first blindly equalized via subspace-based method, then decoding and frequency offsets estimation are achieved iteratively by exploiting structure of space time block code and finite alphabet property of encoded symbols. Simulation results are presented to demonstrate performance of the method.

1   Introduction

Space-Time Code is an effective technology to combat fading and increase data rates in wireless communication system by exploiting multiple transmit antennas [1]. Among all Space-Time Code, Space Time Block Code (STBC) is attractive because its maximum-likelihood decoding can be accomplished with linear complexity by exploiting its orthogonal structure.

In order to decode STBC, the Channel State Information (CSI) is indispensable on the receiver. The CSI can be estimated by transmitting training symbols. However, multiple transmit antennas mean more channel coefficients and more training symbols for channel estimation, therefore, training method consume more bandwidth and induce more spectral efficiency loss in space-time coded system. To save bandwidth or increase spectral efficiency, blind channel identification and blind decoding methods are applied to STBC system. Blind channel identification method is proposed in [2] for precoded Almouti STBC system over frequency selective channel. In [3] , blind equalization of "generalized" STBC is fulfilled by exploiting the structure of such STBC. In [4] a direct decoding method for a full rate STBC over frequency selective channel is proposed, the

* Partially supported by the National Natural Science Foundation (No. 60272071) and the Research Fund for Doctoral Program of Higher Education (No. 20020698024,20030698027) of China.

P. Lorenz and P. Dini (Eds.): ICN 2005, LNCS 3421, pp. 401–409, 2005.

method achieves the goal by applying subspace based blind equalization method on oversampled received signals and exploiting the orthogonal structure of space time block code.

Previous blind decoding or blind equalization methods[3, 4] for space time block code system always assume perfect carrier frequency synchronization between transmitter and receiver. In practice, due to mismatching between oscillators at transmitter and receiver or Doppler drift induced by relative motion of transmitter and receiver, there always exists CFO between transmitter and receiver. In STBC system, multiple transmit antennas should keep far enough each other in order that multiple channels among transmit antennas and receive antennas are uncorrelated, so, different transmit antennas possibly use separate oscillators. Without loss of generality, we consider the case that there are different CFOs between different transmit-receive antenna pairs.

In this paper, we consider blind decoding and carrier frequency offset estimation for 3 transmit antennas and 1 receive antenna STBC system over frequency selective channel in the presence of CFOs. An iterative method is proposed. We assume CFOs of different transmit-receive antenna pairs are different. By oversampling the receiver outputs, the proposed method blindly equalize unknown frequency selective fading channel by subspace-based method firstly, then decoding and CFO estimation are achieved by exploiting structure of STBC and finite alphabet property of encoded symbols.

The rest of the paper is organized as follows: In section 2, we describes the baseband model of the space time coded system over frequency selective channel with carrier frequency offsets, we emphasizes the effect of carrier frequency offsets and give an effective baseband model. In section 3, we exploit subspace-based blind equalization method to remove the inter-symbol interference induced by frequency selective channel. In section 4, based on the outputs of blind equalization, an iterative decoding and CFO estimation method is proposed. By exploiting the finite alphabet property and structure of the STBC, decoding and frequency offsets estimation are accomplished iteratively. In section 5 , we present simulation results.

2   System Model

We consider a system with $M = 3$ transmit antennas and one receive antenna. The extension of the result to the system with any number of receive antennas is straightforward. The information symbols $s(n)$ are encoded in Space-Time codec and the corresponding output encoded symbols $[c_1(n), c_2(n), \cdots, c_M(n)]$ are transmitted from M transmit antennas respectively. Let $h_j(t)$ be the channel impulse response between transmit antenna j and the receive antenna. Without loss of generality, an important assumption we make here is that the channel response is invariant within a data burst. Denotes CFO between transmit antenna j and receive antenna as Δf_j, and $\Delta w_j = 2\pi \Delta f_j$. Without consideration of white noise, the received signal due to signals transmitted from transmit antenna j can be written as

$$r_j(t) = e^{i\Delta w_j t} \sum_n c_j(n) h_j(t - nT) \tag{1}$$

where T is symbol period. Assuming the length of channel impulse $h_j(t)$ is $(L+1)T$. The received signal is oversampled by a factor of Q, that is, the sampling times are $t = kT + \frac{qT}{Q}$ and $k = 0, 1, 2, \cdots, q = 0, 1, \cdots Q - 1$. Let $m = k - n$, then, the over-sampled received signal is expressed as

$$r_j(kT + \frac{qT}{Q}) = \sum_{m=0}^{L} \tilde{c}_j(k - m) g_j(m, q) \tag{2}$$

where $\tilde{c}_j(n) = c_j(n)e^{i\theta_j n}$; $g_j(m, q) = h_j(mT + \frac{qT}{Q})e^{i\theta_j(m+q/Q)}$; $\theta_j = \Delta w_j T$ which is called normalized CFO. From above equation, we observed that the received signal can be expressed as convolution of effective symbols $\tilde{c}_j(n)$ and new time invariant effective channel $g_j(m, q)$, where the effective symbols are product of original encoded symbols $c_j(n)$ and time variant factor $e^{i\theta_j n}$ induced by CFO. With Defining

$$\mathbf{r}_j(k) = [r_j(kT), r_j(kT + \frac{T}{Q}), \cdots, r_j(kT + \frac{(Q-1)T}{Q})]^T$$

$$\mathbf{g}_{j,m} = [g_j(m, 0), g_j(m, 1), \cdots, g_j(m, Q - 1)]^T$$

the totally received signal from all transmit antennas can be equivalently written in vector form as

$$\mathbf{r}(k) = \sum_{j=1}^{M} \mathbf{r}_j(k) = \sum_{j=1}^{M} \sum_{m=0}^{L} \mathbf{g}_{j,m}\tilde{c}_j(k - m) + \mathbf{v}(k) \tag{3}$$

where $\mathbf{v}(k)$ is $Q \times 1$ white noise vector, whose elements are complex Gussian i.i.d with mean 0 and variance σ_v^2.

3   Blind Equalization

In order to eliminate Inter Symbol Interference, we exploit blind equalization method based on subspace method to equalize the effective frequency selective channel. For simplicity, we ignore the noise term in (3) temporarily. Collecting consecutive $N - L$ sampled vectors $\mathbf{r}(k)$ from $k = L + 1$ to $k = N$, we construct Hankel Matrix as below

$$\mathbf{X}(K) = \begin{bmatrix} \mathbf{r}(L+1) & \mathbf{r}(L+2) & \cdots \mathbf{r}(N-K+1) \\ \mathbf{r}(L+2) & \mathbf{r}(L+3) & \cdots \mathbf{r}(N-K+2) \\ \vdots & \vdots & \ddots & \vdots \\ \mathbf{r}(L+K) & \mathbf{r}(L+K+1) \cdots & \mathbf{r}(N) \end{bmatrix}$$

$$= \underbrace{[\mathbf{G}_1, \mathbf{G}_2, \cdots, \mathbf{G}_M]}_{\mathbf{G}} \underbrace{\begin{bmatrix} \tilde{\mathbf{C}}_1 \\ \tilde{\mathbf{C}}_2 \\ \vdots \\ \tilde{\mathbf{C}}_M \end{bmatrix}}_{\tilde{\mathbf{C}}(r)} \tag{4}$$

where $K = 1, 2, \cdots$, is *smoothing factor* and $r = K + L$. $\mathbf{G}_j$ is $KQ \times r$ block Toeplitz matrix defined as

$$\mathbf{G}_j = \begin{bmatrix} \mathbf{g}_{j,L} & \mathbf{g}_{j,L-1} & \cdots & \mathbf{g}_{j,0} & \mathbf{0} & \cdots & \cdots & \mathbf{0} \\ \mathbf{0} & \mathbf{g}_{j,L} & \mathbf{g}_{j,L-1} & \cdots & \mathbf{g}_{j,0} & & & \mathbf{0} \\ \vdots & \vdots & & & \vdots & & & \vdots \\ \mathbf{0} & \cdots & & \cdots & \mathbf{0} & \mathbf{g}_{j,L} & \mathbf{g}_{j,L-1} & \cdots & \mathbf{g}_{j,0} \end{bmatrix};$$

and $\tilde{\mathbf{C}}_j$ is $r \times (N - r + 1)$ Hankel matrix defined as

$$\tilde{\mathbf{C}}_j(r) = \begin{bmatrix} \tilde{c}_j(1) & \tilde{c}_j(2) & \cdots \tilde{c}_j(N - r + 1) \\ \tilde{c}_j(2) & \tilde{c}_j(3) & \cdots \tilde{c}_j(N - r + 2) \\ \vdots & \vdots & \ddots & \vdots \\ \tilde{c}_j(r) & \tilde{c}_j(r + 1) \cdots & & \tilde{c}_j(N) \end{bmatrix}.$$

We select appropriate K in order that $KQ > Mr$. Here, we assume the matrix $\mathbf{G}$ is full column rank when it is tall matrix. When $\mathbf{G}$ is full column rank, the rows of $\mathbf{X}(K)$ and $\tilde{\mathbf{C}}(r)$ span same space. The space spanned by rows of $\tilde{\mathbf{C}}(r)$ is "signal subspace" and its orthogonal complement space is "noise subspace", we denote it as $\mathbf{V}_o(r)$. The noise SUBSPACE can be obtained by performing singular value decomposition on $\mathbf{X}(K)$ as below

$$\mathbf{X}(K) = [\mathbf{U}_s(r), \mathbf{U}_o(r)] \begin{bmatrix} \mathbf{\Sigma} & \mathbf{0} \\ \mathbf{0} & \mathbf{0} \end{bmatrix} \begin{bmatrix} \mathbf{V}_s(r) \\ \mathbf{V}_o(r) \end{bmatrix} \tag{5}$$

where $\mathbf{V}_o(r)$ is $(N - (M + 1)r + 1) \times (N - r + 1)$ matrix. The signal space and noise space is orthogonal, so we have

$$\tilde{\mathbf{C}}(r)\mathbf{V}_o^H(r) = \mathbf{0} \Leftrightarrow \mathbf{V}_o(r)\tilde{\mathbf{C}}_j(r)^H = \mathbf{0}, j = 1, 2 \cdots, M \tag{6}$$

Due to Hankel structure of the matrix $\tilde{\mathbf{C}}_j(r)$ and Eq.(6), it is easy to verify that

$$\mathbf{V}\tilde{\mathbf{c}}_j^* = \mathbf{0} \tag{7}$$

where

$$\tilde{\mathbf{c}}_j = \underbrace{diag(1, e^{i\theta_j}, \cdots e^{i(N-1)\theta_j})}_{\mathbf{Z}_j} \underbrace{[c_j(1) \cdots c_j(N)]^T}_{\mathbf{c}_j}. \tag{8}$$

and $\mathbf{V}$ is matrix of $r(N - (M + 1)r + 1) \times N$ as below

$$
\mathbf{V} = \begin{bmatrix} \mathbf{V}_o(r) & \mathbf{0} & \cdots & \mathbf{0} \\ \mathbf{0} & \mathbf{V}_o(r) & \cdots & \mathbf{0} \\ \vdots & \vdots & \ddots & \mathbf{0} \\ \mathbf{0} & \mathbf{0} & & \mathbf{V}_o(r) \end{bmatrix}
$$

4   Iterative Decoding and CFO Estimation

In this section, by exploiting structure of the STBC and finite-alphabet property of encoded signals, we iteratively estimate the CFOs and decode the STBC code. The STBC adopted in this paper is full rate space time code with three transmit antennas. The input symbol stream are parsed into blocks of 4 symbols, and each block composed of 4 consecutive symbols are encoded in Space-time encoder as below

$$
\begin{bmatrix} s_1 \\ s_2 \\ s_3 \\ s_4 \end{bmatrix} \rightarrow \begin{bmatrix} s_1 & s_2 & s_3 \\ -s_2 & s_1 & -s_4 \\ -s_3 & s_4 & s_1 \\ -s_4 & -s_3 & s_2 \end{bmatrix} . \tag{9}
$$

Each column of the output encoded matrix will be transmitted by one of the three transmit antennas. For simplicity, we use binary phase-shift keying (BPSK) modulation. The encoded symbol vector $\mathbf{c}_j (j = 1, 2, 3)$ can be expressed by the information symbols vector $\mathbf{s} = [s(1), s(2), \cdots, s(N)]^T$ as

$$
\mathbf{c}_j = (\mathbf{I}_B \otimes \mathbf{A}_j)\mathbf{s}, \quad j = 1, 2, 3 \tag{10}
$$

where $\otimes$ denotes Kronecker product, $N = 4B$, B is the number of STBC blocks, $\mathbf{I_B}$ is $B \times B$ unit matrix and matrix $\mathbf{A}_1, \mathbf{A}_2, \mathbf{A}_3$ are defined as below

$$
\mathbf{A}_1 = \begin{bmatrix} 1 & 0 & 0 & 0 \\ 0 & -1 & 0 & 0 \\ 0 & 0 & -1 & 0 \\ 0 & 0 & 0 & -1 \end{bmatrix} \quad \mathbf{A}_2 = \begin{bmatrix} 0 & 1 & 0 & 0 \\ 1 & 0 & 0 & 0 \\ 0 & 0 & 0 & 1 \\ 0 & 0 & -1 & 0 \end{bmatrix} \quad \mathbf{A}_3 = \begin{bmatrix} 0 & 0 & 1 & 0 \\ 0 & 0 & 0 & -1 \\ 1 & 0 & 0 & 0 \\ 0 & 1 & 0 & 0 \end{bmatrix} \tag{11}
$$

Substitute Eq.(8),(11) into Eq.(7), we can obtain

$$
\mathbf{V}\mathbf{Z}_j^*(\mathbf{I}_B \otimes \mathbf{A}_j)\mathbf{s} = \mathbf{0}, j = 1, 2, 3 \tag{12}
$$

With consideration of noise, above equation will not hold, we replace it by minimization the Frobenius norm of left hand side of the equation.

Defining

$$
\mathbf{Q} = [\mathbf{Q}_1^T, \mathbf{Q}_2^T, \mathbf{Q}_3^T]^T
$$

$$
\mathbf{Q}_j = \mathbf{V}\mathbf{Z}_j^*(\mathbf{I}_B \otimes \mathbf{A}_j); \quad j = 1, 2, 3 \tag{13}
$$

then Eq.(12) can be replace by cost function

$$
J(\theta_1, \theta_2, \theta_3, \mathbf{s}) = \|\mathbf{Q}\mathbf{s}\|_F^2 = \sum_{j=1}^{3} J_j(\theta_j, \mathbf{s}) \tag{14}
$$

where $J_j(\theta_j, \mathbf{s}) = \|\mathbf{Q}_j\mathbf{s}\|_F^2$.

The information symbols and normalized CFOs can be estimate by minimize the above cost function as

$$\hat{\theta}_1, \hat{\theta}_2, \hat{\theta}_3, \hat{\mathbf{s}} = \underset{\theta_1,\theta_2,\theta_3,\mathbf{s}\in[+1,-1]}{\arg\min} J(\theta_1, \theta_2, \theta_3, \mathbf{s}) \tag{15}$$

In most mobile communication environment, the normalized frequency offset θ_j are small value, thus we proposed iterative method to solve Eq.(15).

Initialization: Iteration counter$k = 1$ set $\hat{\theta}_j^{(0)} = 0, j = 1, 2, 3$.

Iteration: Assuming $\hat{\theta}_j^{(k-1)}$ is known from $(k-1)th$ iteration, then, in kth iteration: *First*, substitute $\hat{\theta}_j^{(k-1)}$ in Eq.(15) to estimate the information symbol vector $\hat{\mathbf{s}}^{(k)}$ as

$$\hat{\mathbf{s}}^{(k)} = \underset{\mathbf{s}\in[+1,-1]}{\arg\min} J(\hat{\theta}_1^{(k-1)}, \hat{\theta}_2^{(k-1)}, \hat{\theta}_3^{(k-1)}, \mathbf{s}). \tag{16}$$

Second, substitute the estimated symbol vector $\hat{\mathbf{s}}^{(k)}$ into Eq.(15) to estimate normalized CFOs $\hat{\theta}_j^{(k)}(j = 1, 2, 3)$ respectively.

$$\hat{\theta}_j^{(k)} = \underset{\theta_j}{\arg\min} J_j(\theta_j^{(k)}, \hat{\mathbf{s}}^{(k)}), \quad j = 1, 2, 3 \tag{17}$$

Termination: Repeat until the counter k exceeds predefined values.

The minimization problem in Eq.(16) can be solved by performing singular value decomposition on $\mathbf{Q}^{(k)}$, which is obtained by substitute $\hat{\theta}_j^{(k-1)}$ into Eq.(14). Then, the right singular vector $\tilde{\mathbf{s}}^{(k)}$which associated with the smallest singular valued is projected into finite alphabet set to estimate information symbols. In case of BPSK modulation

$$\hat{\mathbf{s}}^{(k)} = sign(real(\tilde{\mathbf{s}}/\|\tilde{\mathbf{s}}\|)).$$

To estimate the CFOs from Eq.(17), we reformulate the cost function $J_j(\theta_j^{(k)}, \hat{\mathbf{s}}^{(k)})$ as

$$J_j(\theta_j, \mathbf{s}) = \left\| \mathbf{V} diag((\mathbf{I}_B \otimes \mathbf{A}_j)\hat{\mathbf{s}}^{(k)})\bar{\mathbf{Z}}_j \right\|_F^2 = \bar{\mathbf{Z}}_j^H \mathbf{P}_j^{(k)} \bar{\mathbf{Z}}_j; \quad j = 1, 2, 3 \tag{18}$$

where

$$\mathbf{P}_j^{(k)} = (\mathbf{V} diag((\mathbf{I}_B \otimes \mathbf{A}_j)\hat{\mathbf{s}}^{(k)}))^H (\mathbf{V} diag((\mathbf{I}_B \otimes \mathbf{A}_j)\hat{\mathbf{s}}^{(k)})); \bar{\mathbf{Z}}_j \left[1 \cdots e^{-i(N-1)\theta_j} \right]^T.$$

Eq.(18) is similar to MUSIC spectrum in DOA estimation, therefore, the normalized CFOs θ_j can be estimated by Root-MUSIC algorithm[5].

5   Simulation Results

In this section, numerical simulations are presented to demonstrate the performance of proposed method. The parameters are detailed as below: The frequency-selective channel is modeled as FIR filter whose coefficients are Raleigh distribution with mean 0 and variance 1, its length is $L = 2$. The oversampling factor is $Q = 5$ and the smoothing factor is $K = 2$. The block length is $N = 4B = 32$, where $B = 8$. The normalized CFOs among transmit antennas and receive antenna are set as: $\theta_1 = \pi/60, \theta_2 = \pi/70, \theta_3 = \pi/80$. In simulations, DBPSK modulation is used to solve the inherent phase ambiguity problem of subspace based method.

To evaluate the performance of CFO estimation, we define the Normalized Root Mean Square Error (NRMSE) of estimated CFO $\hat{\theta}_j$ as

$$NRMSE = \frac{1}{\|\theta_j\|} \left(\frac{1}{R} \sum_{i=1}^{R} \left\| \hat{\theta}_j - \theta_j \right\|^2 \right)^{\frac{1}{2}}$$

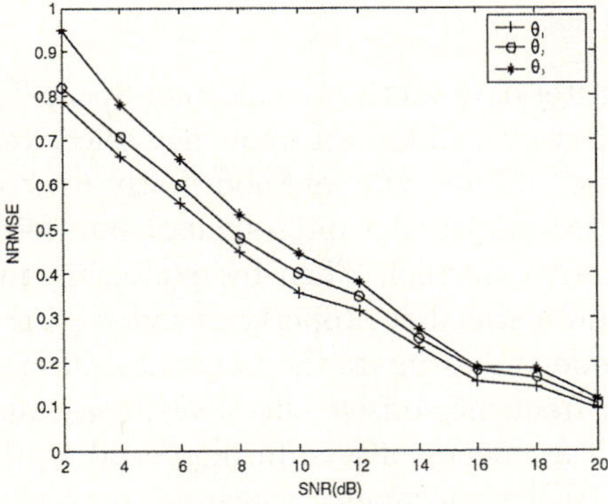

Fig. 1. NRMSE of estimated CFOs versus SNR

where θ_j is true value of normalized CFO, R is the number of Monte-Carlo trials. Figure 1 shows the NRMSE of CFOs estimated by the proposed method. The results are obtained by Monte Carlo simulation of 5000 runs and two iterations. As shown in the figure, the NRMSE decreases as SNR increases. In Figure 2, the BER versus SNR curves of the proposed decoding method are presents for different number of iteration ($N_i=1$, 2,3). We also draw the BER curves of direct decoding method proposed in [4] and No CFO case in Figure 2 for comparisons. The gap between curve of proposed method and direct decoding method enlarge as the number of iteration increases, this is because our method iteratively compensates the CFOs and thus the BER performance improved. The BER curves of the proposed method approximate that of No CFO case as the iterative procedure repeats.

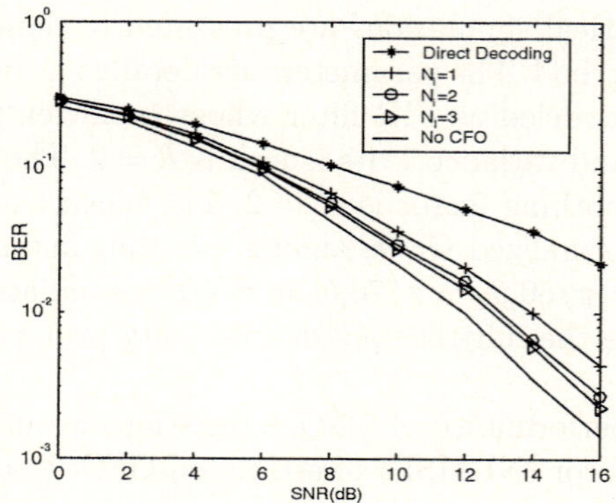

Fig. 2. BER versus SNR for different number of iteration

6   Conclusion

We have proposed an iterative method to decode Space-Time Block Code and estimate frequency offset over unknown frequency selective channel in the presence of carrier frequency offsets. Our method firstly uses subspace based blind equalization method to remove the inter-symbol interference induced by unknown frequency selective channel; Then by exploiting the structure of space time block code and finite alphabet property of coded symbol, we iteratively decode the space time code and estimate the frequency offsets. Our method is able to estimate the carrier frequency offsets effectively, consequently, it can compensate the impairment of frequency offsets during decoding the Space-Time Block Code, and the result BER performance converge to no carrier frequency offset case as the number of iteration increased.

References

1. A.F. Naguib, N. Seshadri, A.R. Calderbank, "Increasing data rate over wireless channels," IEEE Signal Processing Mag., vol. 17, pp. 76 -92 , May 2000
2. S. Zhou, B. Muquet, and G. B. Giannakis, "Subspace-based (semi-) blind channel estimation for block precoded space-time OFDM", IEEE Trans. on Commun., vol. 50, pp.1215-1228, May 2002.
3. Swindlehurst A L, Leus G. "Blind and semi-blind equalization for generalized space-time block codes". IEEE Transactions on Signal Processing, Vol.50:pp. 2489 -2498, 2002.

4. Zhao Z, Yin Q Y, and Zhang H, "Decoding of full rate space-time block code without channel state information in frequency selective fading channels," Proc.IEEE ICASSP 2003, Hong Kong, May 2003,vol.5,pp.121-124.
5. Liberti J C, Rappaport T S, "Smart Antenna for Wireless Communication IS-95 and Third Generation CDMA Application", Prentice Hall, Third Edition,1999, NJ,pp.263-264

Signal Processing for High-Speed Data Communication Using Pure Current Mode Filters

Ivo Lattenberg, Kamil Vrba, and David Kubánek

Dept. of Telecommunications, Faculty of Electrical Engineering and Communication,
Brno University of Technology,
Purkynova 118, 612 00 Brno, Czech Republic
`{latt, vrbak, kubanek}@feec.vutbr.cz`

Abstract. The paper deals with the novel filtration techniques for analog high-speed data signal preprocessing working on the base of a pure current mode filters. The paper describes a novel structure of the universal multifunction filter working in the pure current mode. Due to better frequency features of the current mode and a tendency to reduce the supply voltage because of the technology used (where in the voltage mode the dynamic range goes down in the voltage domain), we decided to design a universal filter working in exactly the pure current mode. It is a circuit where the active elements are only the current-controlled current sources. The current buffer with one input and one positive and one negative output appears to be the optimum element. The bipolar structure of such a current buffer is designed and the basic simulations are carried out.

1   Introduction

With increasing demands on a transfer rate, greater demands are laid on the preprocessing circuit bandwidth. Currently, we can ever more often meet with circuits working in the current mode. This is due to the usually wider bandwidth of these circuits. The next argument for using circuits in the current mode is their greater dynamic range. With decreasing supply voltage the dynamic range of circuits working in the voltage mode falls. The value of the supply voltage has no influence on the dynamic range of circuits in the current mode, which is another indisputable advantage.

The filter working in the pure current mode [1] has been designed – so it contains, except passive elements, only such active elements that have only current terminals. Present filter has been designed by using current buffers with unit current transfer and the passband current transfer of the universal filter designed has been only 0.5. The structure of current amplifiers used will be optimized with respect to achievement of unit current transfer of a filter in a passband. The further aspect of the optimization is a viability of such current amplifier by using the universal current conveyor UCCX 0349 [2], designed in our workplace and prototypal produced by AMI Semiconductor. It is exactly not using the voltage terminal that results in the total dynamic range of a filter not being affected by a particular dynamic range of the voltage terminal. This dynamic range is limited in consequence of the lower supply voltages and the given level of noise voltage

P. Lorenz and P. Dini (Eds.): ICN 2005, LNCS 3421, pp. 410–416, 2005.

2   Current Buffer with Symmetrical Output

The current buffer with one non-inverting and one inverting output (SOCB - Symmetrical Output Current Buffer) can be seen as the ideal active building element of circuits working in pure current mode. Its schematic symbol is shown in Fig. 1.

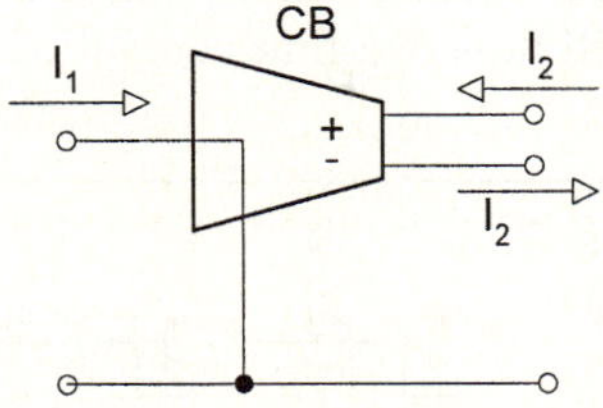

Fig. 1. Schematic symbol of the current buffer with symmetrical output (SOCB)

Bipolar structure of this active current element has also been designed (Fig.2).

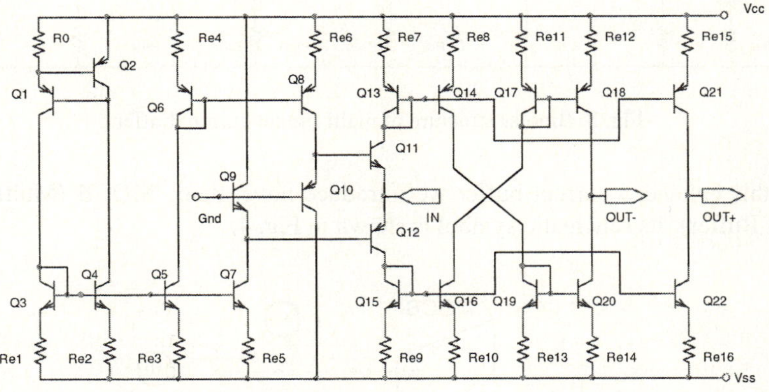

Fig. 2. Bipolar structure of current buffer with symmetrical output

This structure starts from the structure of current conveyor [3]. The current buffer can be thus substituted by a dual-output current conveyor CCX+/- [4,5] of 1st, 2nd or 3rd generation, with the Y terminal grounded. The input is then terminal X and the outputs are terminals Z+ and Z-. Using the second-generation current conveyor is, however, the most suitable. The 1st- and 3rd-generation current conveyors contain a current feedback, which is superfluous for the realization of the current buffer and, moreover, is returned to the Y-terminal, which is grounded anyway.

3   The Realization of the Current Amplifier with Finite Gain by Using Multi-output Current Buffer

It came out, that the use of unit transfer current buffers in the universal filter circuit eventuate the passband current transfer only 0.5 in some configurations. The higher current transfer of current amplifiers can be achieved either by using current amplifier with very high current gain and establishing the feedback or exploiting the fact that the output currents can be simply summed. It emerged that the circuits with high gain active elements in either voltage or current mode have worse frequency features than those with unit or low gain.

By enlarging the current buffer end stage by further four transistors (Fig. 3) we obtain further two outputs. By connection corresponding outputs we can then obtain the output through which then will flow two-times amplified input current.

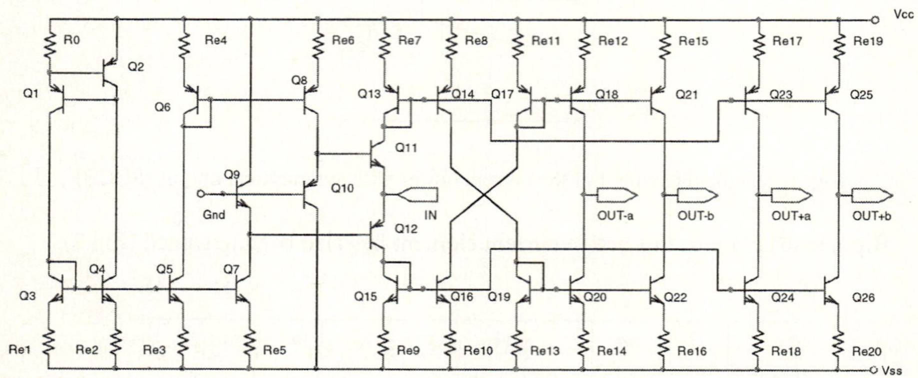

Fig. 3. Bipolar structure of multi-output current buffer

For this enhanced current buffer we introduce novel name MOCB (Multi Output Current Buffer). Its schematic symbol is shown in Fig. 4.

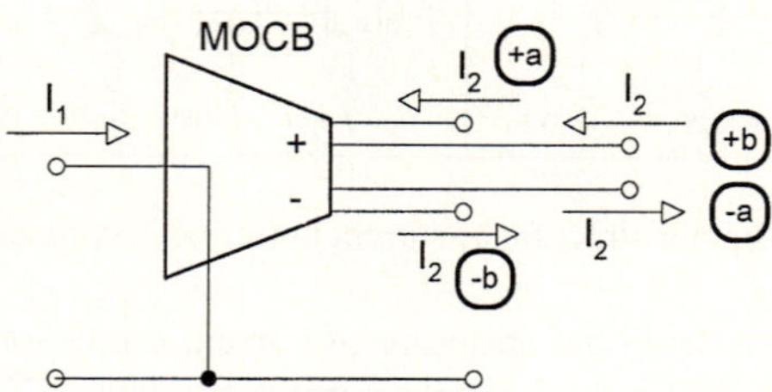

Fig. 4. Schematic symbol of multi-output current buffer MOCB

By using such designed multi-output current buffer we can realize current amplifier with symmetrical output and current gain of 2, so that the positive output we obtain by connecting terminals **+a** and **+b** and the negative output by connecting terminals **−a** and **−b**.

The four-output element MOCB shown in Fig. 4 can be easily realized by using the universal current conveyor UCCX 0349 [2], designed in our workplace. A circuit example of the universal current conveyor as a MOCB element is given in Fig. 5.

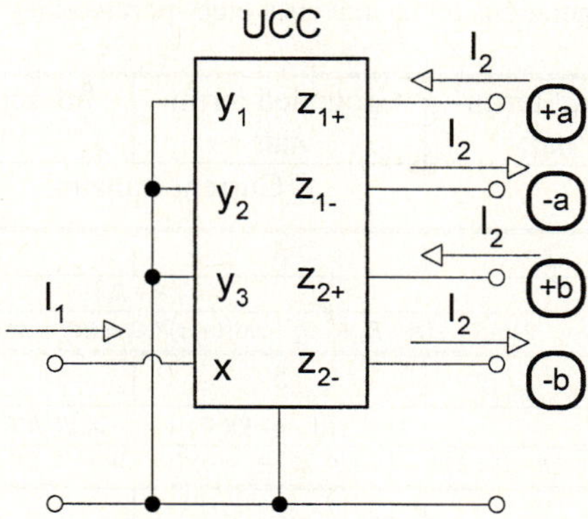

Fig. 5. Realization of MOCB by using universal current conveyor

4 The Design of Universal Filter with MOCB Elements

A second-order analog filter (biquad) working in the pure current mode and using MOCB active elements has been designed. The filter is conceived as a universal eight-port. Choosing an appropriate output terminal and interconnecting suitably the other terminals we obtain a lowpass, highpass or bandpass filter. The schematic diagram of such a circuit is shown in Fig. 6.

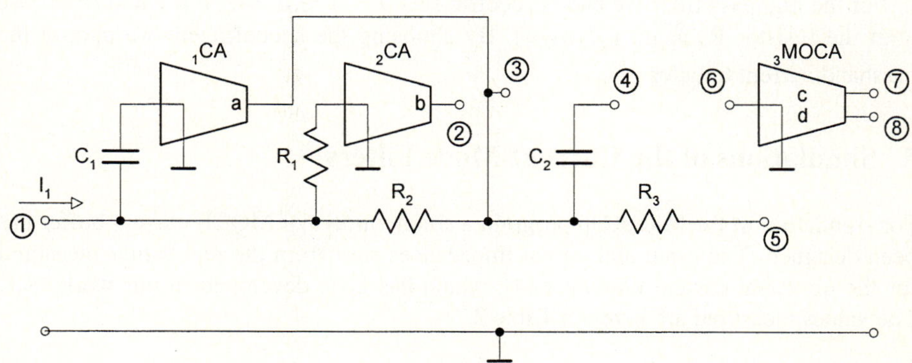

Fig. 6. Universal eight-port in pure current mode

Table 1 gives the configurations for the lowpass, highpass or bandpass filter. It also contains information about which terminals should be interconnected and which terminal is the output terminal for the given type of filter. The input is always terminal 1. The Table also shows the current transfer for each configuration. For simplistic reason we consider $R_1 = R_2 = R$ and current transfer $d = -1$.

Table 1. Possible configurations of the universal eight-port working in the pure current mode

Filter type	Output terminal	Grounded terminals	Interconnected terminals
			Current transfer
Lowpass	2	5	1-7, 3-8, 4-6
	$H(s) = \dfrac{b \cdot (R + R_3)}{2R + R_3 + (R + R_3)R / R_4 + sR((1+a)R_3C_1 + RC_1 + cR_3C_2) + s^2(-ac)R^2R_3C_1C_2}$		
Highpass	7	5	2-3, 4-6
	$H(s) = \dfrac{s(1-b)cRR_3C_2 + s^2(-ac)R^2R_3C_1C_2}{2R + (1+b)R_3 + (R + R_3)R / R_4 + sR((1+a)R_3C_1 + RC_1 + 2R_3C_2) + s^2R^2R_3C_1C_2}$		
Bandpass	7	4	2-3, 5-6
	$H(s) = \dfrac{(1-b)cR + s(-ac)R^2C_1}{2R + (1+b)R_3 + (R + R_3)R / R_4 + sR((1+a)R_3C_1 + RC_1 + 2R_3C_2) + s^2R^2R_3C_1C_2}$		

By detailed analysis of current transfer relationships we obtain recommended coefficient values or intervals of coefficient values a, b and c, which should be hold for proper filter functionality. For lowpass we choose $a \cdot c < 0$. For highpass and bandpass filters we choose coefficient $b = 1$.

For the lowpass filter we choose coefficients $a = -2$, $b = 2$ and $c = 1$. The passband current transfer can then be controlled by resistor R_4 in range of 0 and 4/3.

For the bandpass filter we choose coefficients $a = -2$, $b = 1$ and $c = 1$. We omit the resistor R_4 ($R_4 \to \infty$).

For the highpass filter we choose coefficients $a = -1$ or $a = -2$, $b = 1$ and $c = 1$. We omit the resistor R_4 again ($R_4 \to \infty$). By choosing the a coefficient we choose the passband current transfer.

5 Simulations of the Current-Mode Filters

For simulation in the MicroCap program a simple model of MOCB current buffer has been designed. The input and output impedances start from the real values measured for the universal current conveyor [4], which has been developed in our workplace. The values measured are given in Table 2.

Table 2. The measured impedance values of universal current conveyor input and output terminals

Frequency [MHz]	Z_x [Ω]	Z_z [kΩ]
1	4	400
10	20	100
30	90	-
50	150	-
100	450	10

The model designed is shown in Fig. 7.

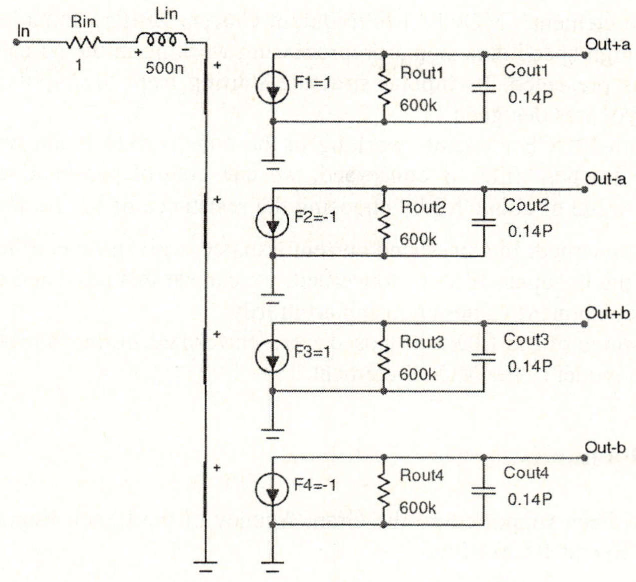

Fig. 7. Simple model of the MOCB current buffer

Filters with a cut-off frequency of 10 MHz have been designed. Fig. 8 gives the frequency response of biquads in the current mode. The passband current transfer chosen was $|K_0| = 1$ and resistors $R_1 = R_2 = 1\text{k}\Omega$.

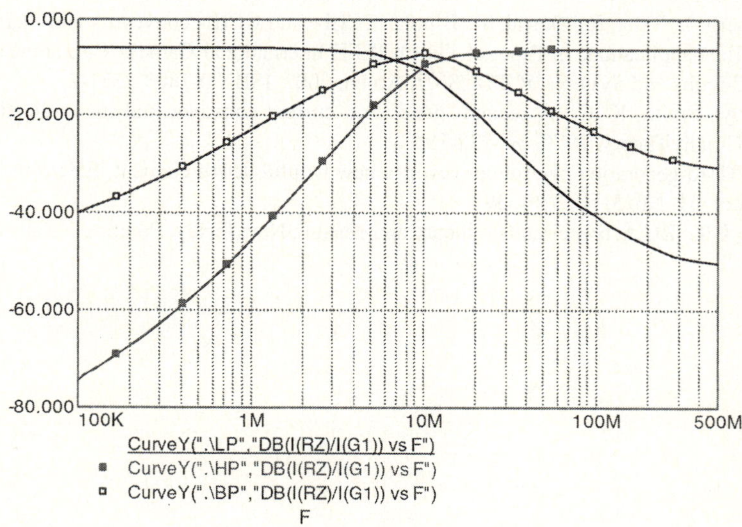

Fig. 8. Frequency response of LP, HP and BP filters

6 Conclusion

A novel active element - MOCB (Multi-Output Current Buffer) suitable for filter realizations for high-speed data signal preprocessing working in the so-called pure current mode was presented. Its bipolar structure starting from the bipolar structure of current conveyor was designed.

Biquads with MOCB elements working in the pure current mode were designed. As far as the lowpass filter is concerned, we can control passband current transfer $|K_0|$ till the value of about 1.3 by choosing the resistance of R_4. As far as the highpass filter is concerned, this passband current transfer is given by coefficient a value. And as far as the bandpass filter is concerned, we can set this passband current transfer by proper component values choosing arbitrarily.

The simulations of the filters proposed were performed in the MicroCap program using a simple model of the MOCB element.

Acknowledgements

This paper has been supported by the Grant Agency of the Czech Republic, projects no. 102/02/P130 and 102/03/1465.

References

1. Lattenberg, I., Vrba, K.: Universal filter with current buffers, In Proc. of International Conference, TSP 2004, TSP'04, Czech republic, Brno, 2004, pp. 75-78
2. Vrba, K., Čajka, J., Kubánek, D.: Biquad Filters Employing Universal Current Conveyors, In Proc. of ICN'04, ICN'04, 2004, Guadeloupe, ISBN 0-86341-326-9
3. Lattenberg, I., Vrba, K., Dostál, T.: Bipolar CCIII+ and CCIII- conveyors and their current mode-filter application, In Proc. of Fourth IEEE International Caracas Conference on Devices, Circuits and Systems, ICCDCS'02, Aruba, 2002, ISBN 0-7803-7381-2
4. Sedra, A., Smith, K. C.: A second- generation current conveyor and its application, IEE Trans. Circuit Theory, 1970, pp. 132-134.
5. Fabre: Third-generation current conveyor: a new helpful active element, Electronic Letters, 1995, Vol. 31, No. 5, pp. 338-339.
6. Bruton, L.T.: RC Active circuits – theory and design., New Jersey, Prentice – Hall, 1980.

Current-Mode VHF High-Quality Analog Filters Suitable for Spectral Network Analysis

Kamil Vrba, Radek Sponar, and David Kubánek

Department of Telecommunications, Faculty of Electrical Engineering and Communication,
Brno University of Technology (BUT), Purkynova 118, CZ 61200 Brno, Czech Republic
`{vrbak, kubanek}@feec.vutbr.cz`

Abstract. In modern spectral network analysis a particular spectral component needs to be extracted precisely in order to measure electrical quantities on the physical level of data networks, e.g. IEEE 802.11, 802.16 etc. Due to these facts, high-speed analog filters with satisfactory features are required. A new active device, UCCX, is presented as suitable for VHF current-, voltage-, and mixed-mode applications. This device contains realization structures of one UCC element and one CCII+/− element. Using UCCX device special filters operating in the current mode are derived.

1 Introduction to Current Conveyors

High-speed analog filters are irreplaceable in data network technology, in particular wide-bandwidth ARC filters equipped with modern devices operating in both the current and the voltage mode [1] to [3]. They are designed for harmonic signal production [4], modem construction [5] and modern spectral analyzer measurements [6]. A new network device, UCCX, was designed in BUT and developed in the CMOS 0.35 μm technology in the AMI Semiconductor Design Center [7] under the designation UCCX 0349. This device contains two network elements: an eight-port universal current conveyor UCC (Fig. 1a) and a four-port zero-class current conveyor CCII+/− (Fig. 1b) [8], [10].

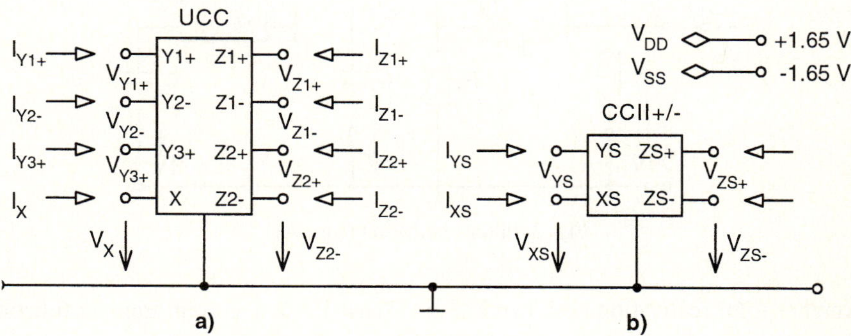

Fig. 1. UCC and CCII+/- elements in the UCCX 0349 device

P. Lorenz and P. Dini (Eds.): ICN 2005, LNCS 3421, pp. 417–424, 2005.

The UCC element is defined by the following equation set:

$$V_x = V_{y+} - V_{y-} + V_{y+}, \quad I_{y+} = I_{y-} = 0, \quad I_{z+} = +I_x, \quad I_{z-} = -I_x, \tag{1}$$

whereas the CCII+/− element is described as follows:

$$V_x = V_{y+}, \quad I_{y+} = 0, \quad I_{z+} = +I_x, \quad I_{z-} = -I_x. \tag{2}$$

2 Filter Realizations

Increasing a density of integration in new CMOS technology fabrication leads towards decreasing a supply voltage, which causes a output voltage limitation and thus signal-to-noise ratio decreasing. This drawback can be suppressed by changing the operation mode in the current type, therefore we will further deal with analog filters working in current mode.

We have proposed a few second-order filters using one UCCX device. The current transfer functions of all these filters can be described by the following general equations:

$$\frac{I_{o1}}{I_{in}} = a \, \frac{s^2 C_1 C_2}{D(s)}, \quad \frac{I_{o2}}{I_{in}} = b \, \frac{s C_2 G_2}{D(s)}, \quad \frac{I_{o3}}{I_{in}} = c \, \frac{G_1 G_2}{D(s)}, \tag{3}$$

where the coefficients a, b, and c equal to +1 or −1 according to a particular variant.

The characteristic equation is

$$D(s) = s^2 C_1 C_2 + s C_2 G_2 + G_1 G_2. \tag{4}$$

The first proposed filter is shown in Fig. 2. In this case it is valid: $a = -1$, $b = c = +1$.

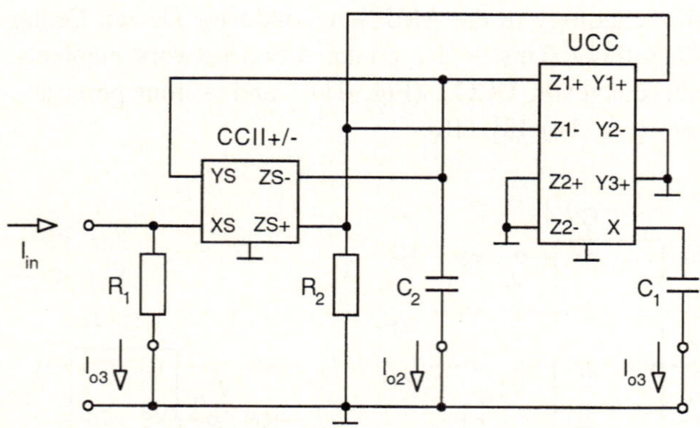

Fig. 2. Filter realization (case A)

Another filter realization is shown in Fig. 3 (case B). The current transfer functions have the following coefficients: $a = b = +1$, $c = -1$.

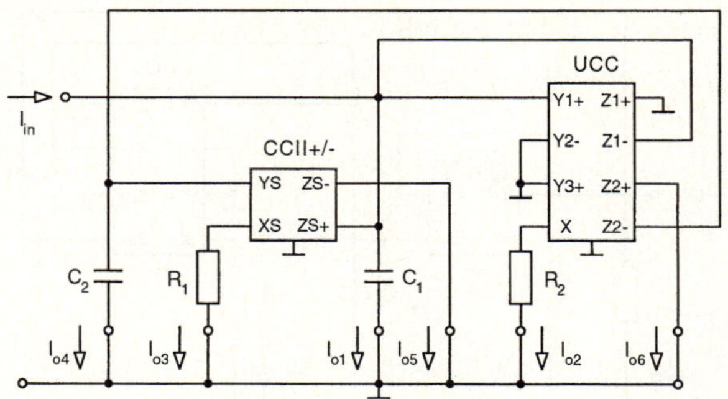

Fig. 3. Filter realization (case B)

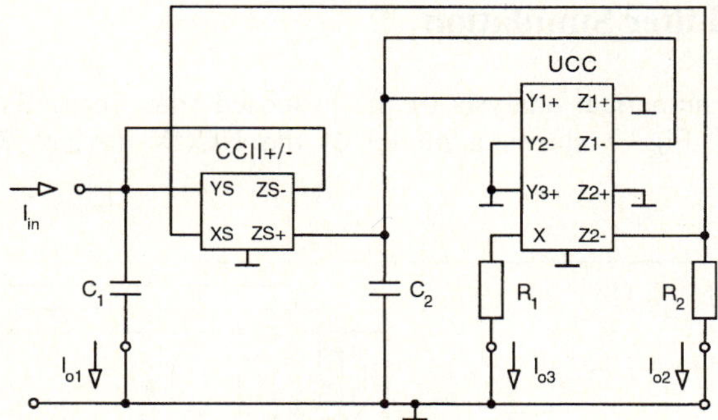

Fig. 4. Filter realization (case C)

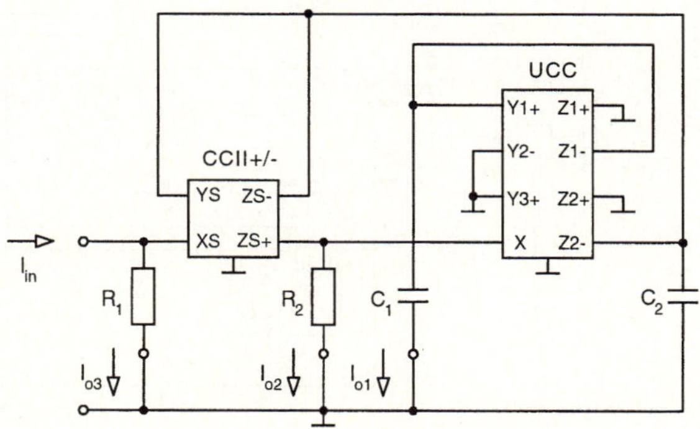

Fig. 5. Filter realization (case D)

Case C is shown in Fig. 4. For this connection we can write: $a = b = c = +1$.
Considering case D (Fig. 5), it is valid: $a = b = -1$, $c = +1$.
In Fig. 6 we find another filter realization (case E). Their coefficients are the same as in Fig. 4, i.e. $a = b = c = +1$.

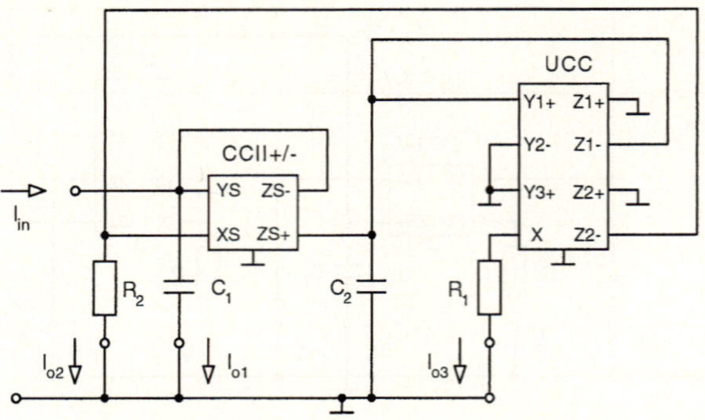

Fig. 6. Filter realization (case E)

3 Numerical Filter Simulation

We carried out a numerical analysis of the proposed filter (case B) in the and the frequency-domain. Fig. 7 shows a model of the UCCX device [7] improved by

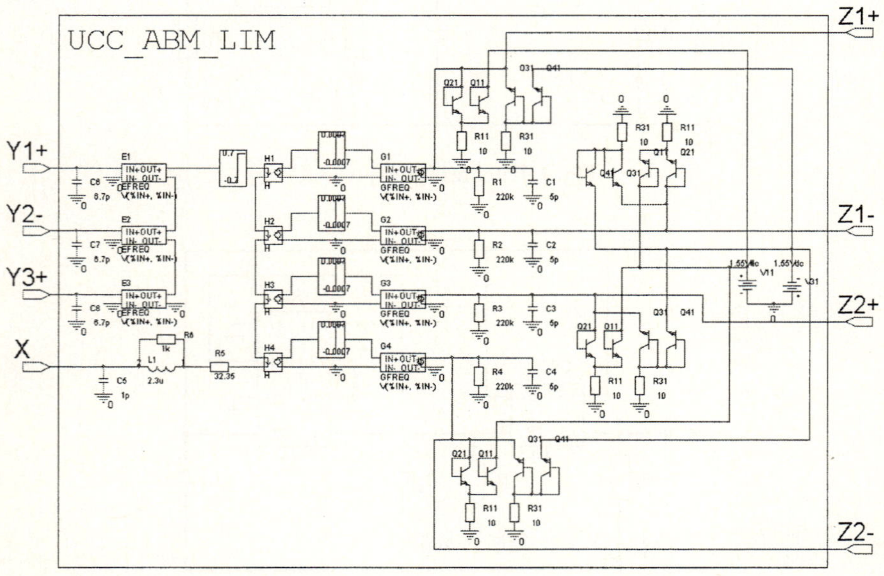

Fig. 7. UCC model (3rd level)

transistor-level voltage limiters at current outputs. It utilizes „Analog Behavioral Modeling" blocks and thus the frequency responses are modeled. The model of the CCII+/- device is similar.

The case B of the filter realization shown in Fig. 3 was chosen for analysis and experiments. This universal filter proves these features:

- inverting low pass: $I_{o3}/I_{in} - \mathrm{mag(o3)}$,
- non-inverting low pass: $I_{o5}/I_{in} - \mathrm{mag(o5)}$,
- non-inverting low pass: $I_{o5}/I_{in} - \mathrm{mag(o5)}$,
- inverting band pass: $I_{o4}/I_{in} - \mathrm{mag(o4)}$,
- non-inverting band pass: $I_{o2}/I_{in} - \mathrm{mag(o2)}$,
- inverting band pass: $I_{o6}/I_{in} - \mathrm{mag(o6)}$.

Simulation results of the magnitude frequency responses are depicted in Fig. 8. The numerical design originated from a low-pass filter having a characteristic frequency f_0 equal to approx. 5 MHz and a quality factor $Q_0 = 4$. It is used for a higher-order cascade of current-mode biquad filters.

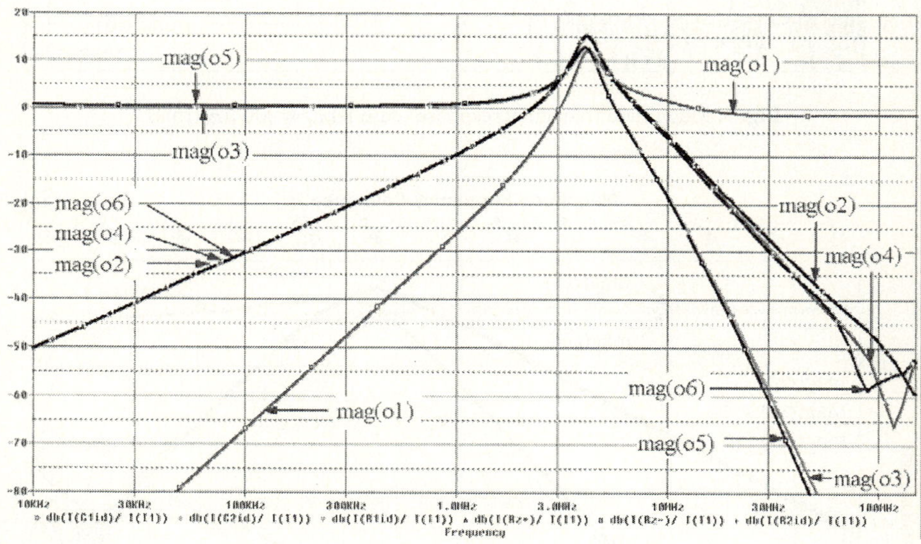

Fig. 8. Numerical analysis of current-mode filter using UCCX device, magnitude response

4 Filter Realization and Measurements

We constructed a filter (case B) with UCCX0349 for $Q_0 = 4$. The magnitude frequency responses are shown in Fig. 9. The input current was limited to $I_{in} = 200\ \mu\mathrm{A}$ (rms) in order not to damage the filter. The limiting effect arises at ± 1.35 V level as the UCCX device is supplied with voltage ± 1.65 V.

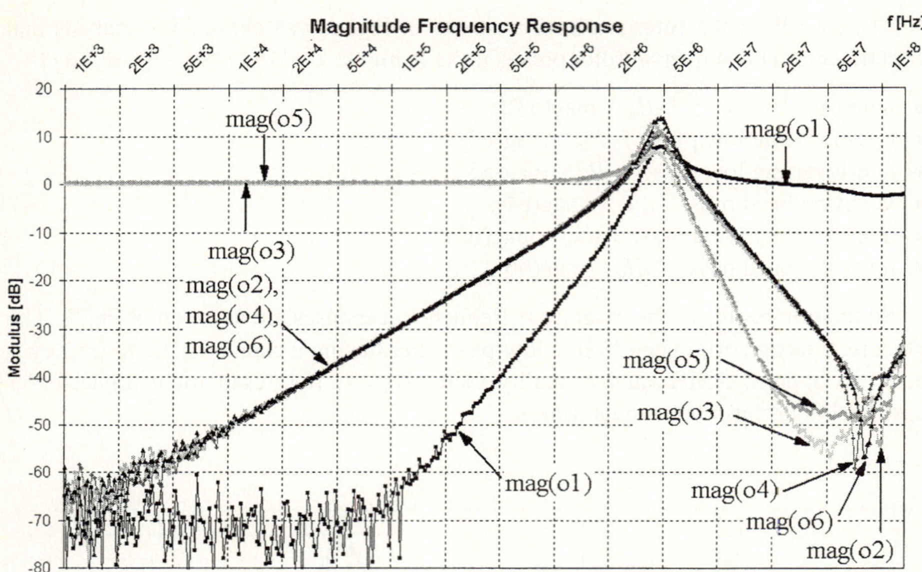

Fig. 9. Magnitude frequency response (case B) I_{in} = 200 μA (rms)

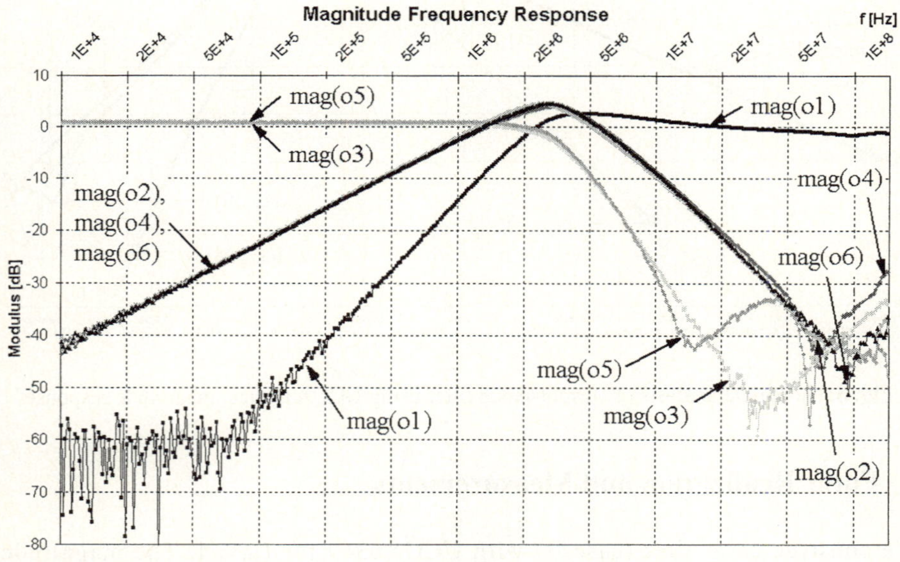

Fig. 10. Magnitude frequency response (case B) I_{in} = 400 μA (rms)

The measurements were also done for characteristic frequency f_0 equal to 3.5 MHz and a quality factor $Q_0 = 0.8$. The results are shown in Fig. 10. The input current was set to 400 μA (rms) for this case.

The UCCX device is also able to construct current-mode filters with a very high Q-factor (20 to 30). Examples of the band pass, low pass, and the high pass with highly stable central frequency are given in Fig. 11.

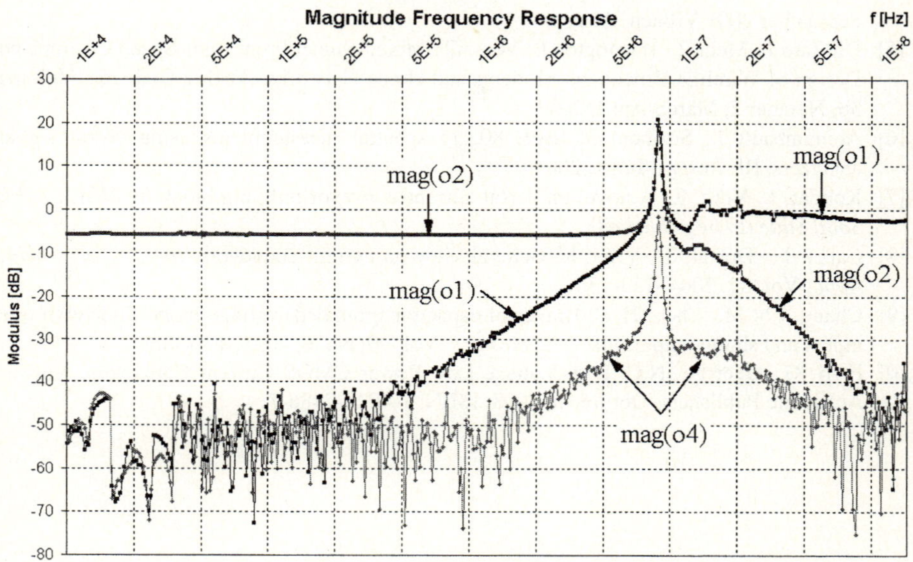

Fig. 11. Magnitude frequency response (case B) $I_{in} = 10$ µA (rms)

5 Conclusions

The current-mode biquad proposed in this paper is expected to be implemented in certain data network measuring systems [6], for high-speed modems [5] and for a highly challenging signal generator [4]. It is able to amplify great dynamic range signals of wide frequency range.

Acknowledgements

This work has been supported by the Grant Agency of the Czech Republic (Grant No. 102/03/1465) and by the Czech Republic Ministry of Education (Grant No. 1641/2003/G1).

References

[1] Carusone A. Ch., Johns, D. A. A 5th Order Gm-C Filer in 0.25 um CMOS with Digitally Programmable Poles & Zeroes, *IEEE Int. Symp. Circuits and Syst.*, May 2002

[2] Schmid, H. An 8.25-MHz 7th-Order Bessel Filter Built with Single-Amplifier Biquadratic MOSFET–C Filters, *Analog Integrated Circuits and Signal Processing*, January 2002, vol. 30, is. 1, pp. 69-81(13) Kluwer Academic Publishers

[3] Korotkov A.S., Morozov D.V., Unbehauen R. Low-Voltage Continuous-Time Filter Based on a CMOS Transconductor with Enhanced Linearity, *AEÜ – International Journal of Electronics and Communications, October 2002, vol. 56, no. 6*, pp. 416-420(5)

[4] Mäntyniemi, A., Rahkonen, T. A 30 MHz DDS Clock Generator with 8-bit, 130 ps Delay Generator and -50 dBc Spurious Level. *In Proc. ESSCIRC'01 conference*, 18-21 September 2001 Villach, Austria

[5] DiPilato J., Mehr, I., Harrington B. Versatile Mixed-Signal Front Ends Speed Customized Design of Wireline Broadband Modems and Home Networks, Analog Dialogue, Volume 36, Number 2, March-April, 2002

[6] Archambault J., Surineni S. IEEE 802.11 spectral measurements using vector signal analyzers, RF Design, Jun 1, 2004

[7] Koudar, I., Vrba, K. A novel universal current conveyor building block for VSLI. *IEEE Solid State (to be published)*

[8] Cajka, J., Dostal, T., Vrba, K. General view on current conveyors. *Int. J. Circ. Theor. Appl.,* Vol. 32, 2004, 133-138

[9] Chang, Ch. M., Chen, H. P. Universal capacitor-grounded voltage-mode filter with tree inputs and single output. *Int. J. Electronics*, Vol. 90, No. 6, 2003, 401- 406

[10] Ferri, G., Guerrini, N.C. Low-Voltage Low-Power CMOS Current Conveyors. Kluwer Academic Publishers, Dordrecht, 2003, ISBN: 1-4020-7486-7

Control of Digital Audio Signal Processing over Communication Networks

Jiri Schimmel and Petr Sysel

Faculty of Electrical Engineering and Communications,
Brno University of Technology,
Purkynova 118, 612 00 Brno, Czech Republic
Phone: + 420 541 149 167, Fax: + 420 541 149 192
`{schimmel, sysel}@feec.vutbr.cz`

Abstract. This paper deals with the design of a protocol for the control of digital audio signal processing over communication networks, primarily for TCP/IP networks, and its multi-platform parser. The application of this protocol in embedded audio processing units represents a modern approach to remote-controlled real-time processing of digital audio signals. The aim is to have the possibility to control audio signal processing in distributed processing networks over very long distances, e.g. via the Internet.

1 Introduction

Our objective was to design a universal communication protocol for the control of all types of systems for real-time digital audio signal processing. This protocol must be able to describe any partial process performed in digital audio signal processing.

The basic idea is that all system parameters related to digital audio processing must be unambiguously described in order to describe any partial process performed in digital audio signal processing. So every parameter transferred using the proposed protocol is unambiguously specified via the following identifiers: *Channel type*, *Channel group*, *Channel number*, *Parameter block*, *Parameter block index*, *Parameter type*, and *Parameter type index*.

Initially, the protocol was designed for the Dmatrixx system [3], so it is labelled DMP (DMatrixx Protocol) and it is available in version 1.3 at present.

2 Protocol Structure

Eight groups of messages, which show the scope of the protocol, are defined in the current version:

1. Communication messages,
2. Configuration messages,
3. Parameter messages,
4. Meter Map messages,

P. Lorenz and P. Dini (Eds.): ICN 2005, LNCS 3421, pp. 425–432, 2005.

5. Snapshot messages,
6. Parameter Group messages,
7. Parameter Mask & Protection messages,
8. Description messages.

The first group encapsulates all messages for establishing communication between a server and a client application, error handling messages, etc. The second group, configuration messages, contains messages for the transfer of information about system configuration, i.e. about channels, groups, blocks, and their interconnection. The controlled application provides these data for the controlling application so that it can build the system diagram, which is the basis of parameter identifiers. The third group contains messages for parameter requests and dumps. Messages of the fourth group allow the transfer of so-called meter map information. Meter map is a data container with meter point identifiers, i.e. points in a virtual audio signal path where the signal level is measured and transmitted to the control application.

Snapshot messages are used to store, recall, and manage the instantaneous parameter settings stored in the system memory. Parameter group messages are designed to transfer information about user-defined groups of parameters whose values are linked together in the relative or the absolute mode. The Parameter Mask & Protection group of messages contains messages for setting the parameter protection against user changes, and for setting the parameter masks, which are used to filter parameter changes. The last group, description messages, is designed for transferring text descriptions of channels, blocks, parameters, etc.

Every data block is supplemented with a header as shown in Table 1. It contains information about the protocol level, length and priority of the message, and timestamp. These data allow recognizing and handling the following situations:

1. Division of a message into several packets,
2. Linking of several messages into one packet,
3. Message delivery in a wrong sequence,
4. Decision on message priority when the same type of message is received from several clients at the same time.

Table 1. DMP message header

Data	Bits	Description
Timestamp	32	Time from the beginning of communication
Message length	16	Length of message in bytes
Protocol level	2	01 = version 1.x
Message priority	6	64 levels from "nobody" to "administrator" access level

3 Parameter Encoding

The protocol is intended for the control of large digital audio signal processing systems, such as mixing consoles for theatres, music and congress halls, etc., which need to send and receive thousands of parameters at the same time. Such systems need a high bit-rate data transfer to manage the data flow with low latency, so the parameter encoding must use parameter data blocks as small as possible. That is why the proposed protocol does not use float data types but supports the transfer of the following three data types:

- Continuous values within the range of <0, 1> quantized to 256 or 65536 levels,
- Integer values within the range of 0 to 65535, and
- Boolean values.

In addition to the parameter value, its type and resolution (8- or 16-bit) are transferred via the protocol. Boolean values are encoded directly in the data type and no data bytes are used. Continuous values can be used in the absolute and in the relative mode. Values within the range of <-1, 1> are transferred in the relative mode.

Table 2. Encoding of parameter block in DMP

Data	Bits	Description
Channel Group Range	3	
Channel Type	5	0 to 32
Channel Group and Channel Number	16	Variable bit span
Parameter Block	8	0 to 255
Parameter Block Index	8	0 to 255
Parameter Type	8	0 to 255
Parameter Type Index	5	0 to 32
Data format	2	Continuous/integer/bool
Data range	1	8- or 16-bit
Value	8	Low-resolution values
	16	High-resolution values
	0	Boolean values

Parameter block encoding is shown in Table 2. The first byte contains the *channel type* identifier and the first three bits determine the number of bits in two following bytes, in which the number of channel group is stored. User-defined channel groups are used in addition to the division of channels according to their types, i.e. the dynamic range coding of channel group numbers is used so that the maximum channel number is not limited by the number of channels in the group. The overall number of channels remains constant.

Two bits of *data format* determine one of the following data types: continuous value in the absolute mode, continuous value in the relative mode, integer value, and Boolean value. The *data range* bit determines the number of following bytes carrying the parameter value. This bit is used to transfer the Boolean-type value, so no additional bytes are transferred in the case of the Boolean value.

4 Audio Signal Level Measurement

The protocol supports the transfer of audio signal level information at any user-defined point of virtual audio signal processing path in the system. The identification of these points is similar to the parameter identification; it uses the *Channel type*, *Channel group*, *Channel number*, *Parameter block*, *Parameter block index*, and *Pre/post* identifiers.

Encoding of the meter data block in the protocol is shown in Table 3. The first six identifiers are the same as the first six identifiers of the parameter data block. The *range flag* determines whether the meter range definition follows the meter identification. Next identifiers determine the pre/post position and type of the meter (rms, peak, etc.). The last two optional bytes determine the maximum and minimum level of the meter in dB.

Table 3. Encoding of meter block in DMP

Data	Bits	Description
Channel Group Range	3	
Channel Type	5	0 to 32
Channel Group and Channel Number	16	Variable bit span
Parameter Block	8	0 to 255
Parameter Block Index	8	0 to 255
Reserved	4	
Range	1	Set if range follows
Position	1	Pre/post position
Type	2	Type of meter
Maximum range	8	−128 to +127 dB
Minimum range	8	−128 to +127 dB

The transfer of information about audio signal levels in the system has critical demands on the transfer rate because thousands of values must be transmitted up to fifty or more times per second to get a smooth visual indication of the signal level. In order to minimize the size of the meter value data block the protocol uses data dump of 8-bit meter values in series.

The values in the meter dump are transmitted as 8-bit values of linear scale in dB in the range of $<L_{min}, L_{max}>$. The real dB value of signal level L is encoded into integer value k according to the following rules:

$$
\begin{aligned}
k &= 0 && \text{for } L < L_{min} \text{ (no signal)} \\
k &= 1 && \text{for } L = L_{min} \\
k &= 253 \cdot (L - L_{min})/(L_{max} - L_{min}) + 1 && \text{for } L_{min} < L < L_{max} \\
k &= 254 && \text{for } L = L_{max} \\
k &= 255 && \text{for } L > L_{max} \text{ (overload)}
\end{aligned}
\tag{1}
$$

The meter levels are thus quantized to 254+2 levels. The data container called *Meter Map* determines the relation between meter points in the system and this dump. Any time a meter dump is received, the application goes through the container, decodes the values, and assigns them to the corresponding meter point. Meter map messages ensure that the meter map container is the same at both sides of the connection. This system reduces the data flow to 15%.

5 System-Specific Data

The protocol supports a transparent transfer of the system-specific data. Encoding of system-specific data is shown in Table 4. Every system-specific data block transferred in the system via DMP has its own identifier. There is a union defined in the protocol parser that stores the decoded data block and performs its typecast according to its identifier.

Table 4. Encoding of system specific data in DMP

Data	Bits	Description
Type	8	Data block identifier
Length	16	Length of data block in bytes
Data		Data block

6 Protocol Parser

The protocol parser is a session layer of the DMP. In other words, it is a process that enables applications to communicate with one another using the DMP. The basic functions the DMP parser are:

- Encoding and decoding of protocol messages,
- Linking of segmented data blocks,
- Separation of linked data blocks,
- Sequencing the incoming messages according to their timestamp and priority,
- Error correction and irrecoverable error reporting.

Only processing of received data and preparing data for transmitting are parser functions while transmitting and receiving as such are not. To transfer DMP data blocks, the application can use any communication interface, e.g. Ethernet, MIDI, etc.

Fig. 1 shows the class structure of the DMP parser. Round arrows represent the encoding direction and open arrows represent the decoding direction. The ANSI C++ language [1] and the STL (Standard Template Library) components [2] are used for the DMP parser implementation because of cross-platform application. The DMP parser is available in the form of source files, no statically or dynamically linked libraries are used. The following classes are defined in the parser.

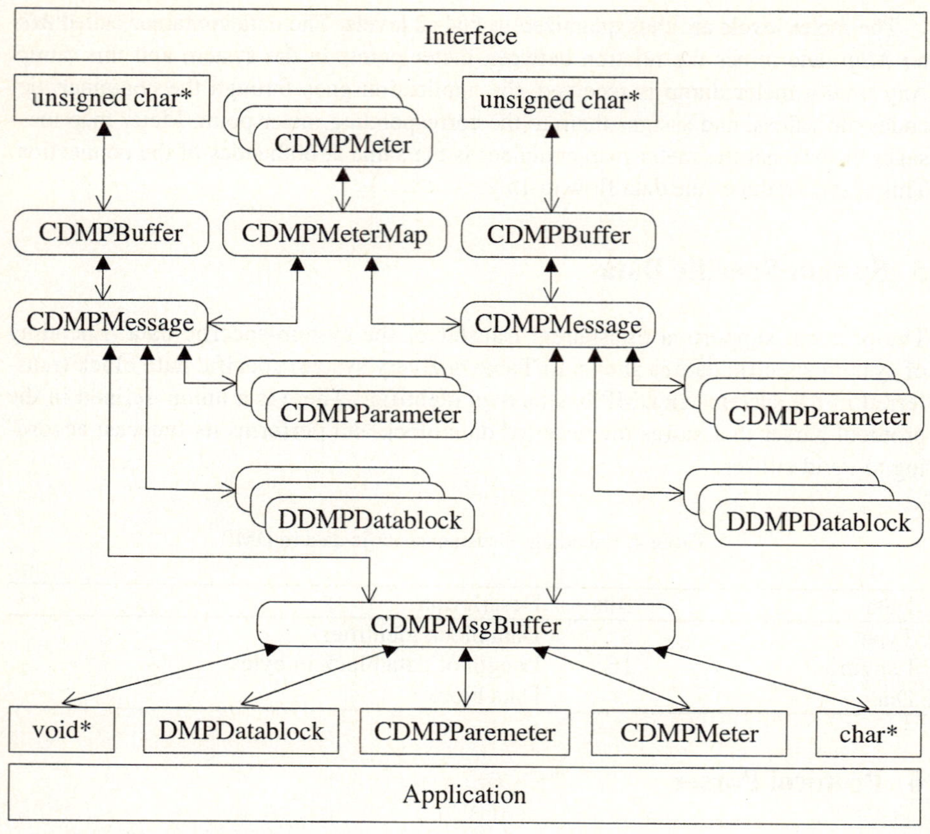

Fig. 1. Class structure of DMP parser

6.1 Error Handling

The *CDMPError* class is the standard C++ exception class. Other classes use this class to throw error exception if the use of exception is enabled.

6.2 Data Buffers

The *CDMPBuffer* class interconnects the communication interface with the parser. It implements the dynamic container of vector type over unsigned-char data type. It provides monitoring of data integrity. This class can be created using a pointer to the standard unsigned-char data type, and typecasting to this type can also be used.

6.3 Parameter Data Types

The *CDMPParameter* class contains the complete definition and value of parameter. Its basic function is conversion, = operator overloading and dynamic typecast of parameter value to float, integer, and Boolean data types. The class contains methods whose functions are doubled or identical. The reason for this is to enable the developer to use any programming techniques because *CDMPParameter* class objects form the basic data interface between the application and the DMP parser.

The *CDMPDatablock* class stores all the other data types used in the protocol such as simple values, text strings, system-specific data, etc.

6.4 Protocol Messages

The *CDMPMessage* class is the basic class of the DMP parser, which allows encoding and decoding protocol messages into and from objects of the *CDMPBuffer* class. The class contains two dynamic vector-type containers of *CDMPParameter* objects, which contain protocol parameters, and *CDMPDatablock* objects, which contain other data types. The class itself selects an optimum message type and the type of data encoding. The application requests data by calling the *ProcessData()* function, which returns the next data-ready supplemented with the data type identifier.

6.5 Message Buffers

The *CDMPMessageBuffer* class contains a dynamic container of pointers to the CDMPMessage class objects. It manages decoded messages, i.e. it organises them in the queue, changes their order according to the timestamp and priority, etc. It operates as a buffer for passing messages to the application. The application requests data from the parser by calling the *ProcessData()* function, which returns the next data-ready supplemented with the data type identifier.

6.6 Meters

The *CDMPMeter* class encapsulates meter properties. Its identification is similar to the parameter identification, with only the parameter type and index being replaced with pre/post fader identification.

6.7 Meter Map

The *CDMPMeterMap* class contains a definition of the meter map. Only the values of meters and not identifiers are transmitted in the meter data dump. The latter are arranged in the object of this class in a sequence corresponding to the sequence of meter values in the data dump. Objects of the *CDMPMessage* class assign the meter value to its identifier on the basis of data stored in the global object of the *CDMPMeterMap* class.

7 Conclusion

This paper gives a short overview of a protocol designed for the control of digital audio signal processing over communication networks and its multi-platform parser. Detailed information can be found in [4]. The protocol and its parser are still under development and they are currently experimentally used in the Dmatrixx mixing matrix system [3] for multi-client remote control of the mixing matrix, channel parameters, and level metering in 256 channels.

The following improvements are under development for the next version of the protocol:

- Binary data transfer for plug-in modules,
- Configuration data transfer for in-system configuration,
- Multimedia signal transmission,
- Support for distributed audio signal processing,
- Support for fault-tolerant systems.

Acknowledgement

The paper was prepared within the framework of N°102/04/1097 project of the Grant Agency of the Czech Republic and with the support of the research plan CEZ: J22/98:262200011.

References

1. STROUSTRUP, B. The C++ Programming Language, 3rd edition. Addison-Wesley Professional, 1997. ISBN 0201889544
2. JOSUTTIS, N., M. The C++ Standard Library: A Tutorial and Reference, 1st edition. Addison-Wesley Pub Co, 1999. ISBN 0201379260
3. Dmatrixx mixing matrix system [online]. D-Mexx, 2004, [2004-10-14]. http://www.d-mexx.com/english/prods_dmatrixx.htm
4. KRKAVEC, P. Universal Communication Interface for Control of Audio Signal Processing Systems. Research report on the solution of project No FD-K3/036

Fully-Distributed and Highly-Parallelized
Implementation Model of BGP4
Based on Clustered Routers[*]

Xiao-Zhe Zhang, Pei-dong Zhu, and Xi-cheng Lu

School of Computer, National University of Defense Technology,
410073 Changsha, China
`nudtzhangxz@263.net`

Abstract. With the explosive growth of the service providers topologies, the size of routing table and number of routing sessions have experienced a sharp increase, which makes the scaling of BGP implementation in backbone routers a rather stressing problem. Clustered router is a promising architecture in terms of forwarding capability. But traditional BGP implementation, based on single-processor architecture cannot utilize the distributed computing and memory resources to enhance the protocol's performance. Borrowing the idea of team working from Multi-Agent technology, this paper proposes a new fully-distributed and highly-parallelized implementation model of BGP, which gives equal chance to each routing node of the cluster to participate in routing selection and stores route entries among routing nodes. Chief algorithms are presented related with session dispatch and partition of prefixes computing and storage. Theory analysis and experiment show that the algorithm can achieve linear speedup over central control model and increase system reliability greatly.

1 Introduction

The Internet continues along a path of seemingly inexorable growth, at a rate that has almost doubled in size each year. In order to match the Internet expansion speed, network devices must provide more powerful computing ability and packet forwarding ability, and support huge density of physical interfaces. It makes traditional centralized control plane of network devices unable to meet the requirements of next generation Internet in terms of reliability, performance scalability and service scalability in future. *Clustered router* is a promising and attractive architecture to achieve forwarding scalability. Several multi-gigabit routers and future terabit routers support multiple-chassis interconnection and backplane extension technology, e.g. Juniper T640[1]. A number of PCs or low-cost routers can also be interconnected and work as a massive parallel router [9] using high-speed optical switching network. Such a clustered router can provide unprecedented capacity and two or three times the lifespan of typical

---

[*] This work is supported by National Sciences Foundations of China (NSFC), under agreement no. 90412011 and 90204005 respectively, National Basic Research Priorities Programme of China, under agreement no.2003CB314802.

P. Lorenz and P. Dini (Eds.): ICN 2005, LNCS 3421, pp. 433–441, 2005.

core routing equipment. But the evolution of software architecture of routers does not keep up with the development of hardware architecture. Typical router software supporting clustered architecture usually works in master-slave mode. The software scalability is not accomplished for clustered routers.

With the rapid development of Internet, Border Gateway Protocol (BGP) is facing with great challenges: the trend of exponential growth [2] of BGP table capacity consumes a lot of memory, the fast increase of prefix updates [3,4] and the number of BGP neighbors [5,12] also burden the control planes of backbone routers. However, the distributed computing resources and storage capability of clustered routers are not utilized efficiently to solve BGP's problems. There are a few research work and projects being helpful, such as CORErouter-1 [8], Zebra project [7] and parallel routing table computation algorithm for the OSPF [10]. But [7][8] are for the hybrid distributed routing platform and parallelism is between different protocol modules. [10] is aimed at OSPF and not feasible for BGP.

To utilize distributed resources of clustered routers efficiently and improve BGP to meet future demands of next generation Internet, this paper borrows the concept of agent to convert traditional BGP process into BGP Agents and have it resided on each routing node, extends the basic implementation model of IETF standard [6] to support team cooperation working, proposes a new implementation model of BGP based on distributed architecture and the partition algorithm of BGP route computation and storage among BGP Agents.

2   Distributed BGP Implementation Model

2.1   Basic Implementation Model

Networks under a single technical and administrative control can be defined as an Autonomous System (AS). BGP is a path vector protocol that allows import and export policies to modify the routing decision from the shortest-path default. Unit of routability provided by BGP is the network prefix, which is an aggregation of IP addresses in a contiguous block. A router advertisement is received from a neighbor AS over a BGP peering session. The basic implementation model of BGP, which is proposed by IETF, includes the following components:

- BGP neighbors: There are two types of BGP neighbors: Internal neighbor (IBGP) and External neighbor (EBGP). The behavior of BGP neighbor includes the execution of Finite State Machine (FSM), maintaining the TCP connection, processing prefix updates and routing policy computing.
- Routing Information Base (RIB): It includes all prefixes, announced by others routers, which are organized as TRIE structure.
- Routing selection procedure: When the prefixes in RIB are changed, routing selection procedure is ran to select the best prefixes, update IP forwarding table and generate BGP update packet.

There are two prefix sets, which influence the behavior of packet forwarding and BGP process. One is the set of all prefixes in RIB, the other is the set of best prefixes that will be filled in IP forwarding table. We define the former as *prefix view* and the latter as *best prefix view*. The precondition of multi-routing nodes cooperating as a single router is that the best prefix view of every routing node must be the same.

2.2 Distributed Implementation Model

Basic model of BGP has not the ability of supporting distributed computing environment. Traditional BGP process resides on one routing node, which is called master routing node, and other nodes do not take part in prefix storage and routing selection procedure. Based on the feature of Clustered router architecture, we import the idea of team working from Multi-Agents technology and propose a new distributed implementation model of BGP, which gives equal chance to each node to participate in routing selection procedure and prefixes are stored distributive among routing nodes.

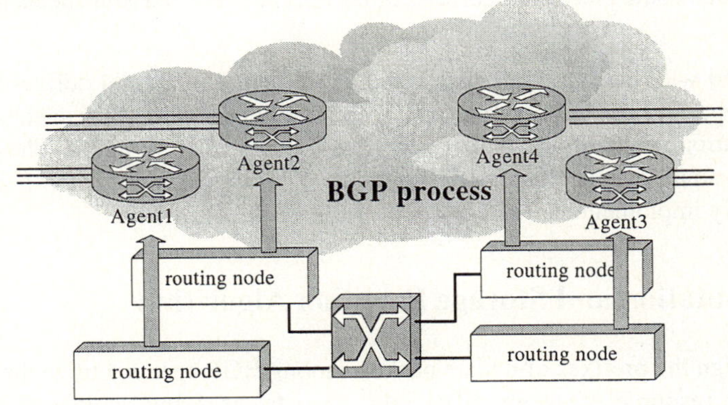

Fig.1. Distributed implementation model of BGP

Its core idea: extend the function of BGP process to be BGP agent and reside it on each routing node independently. Besides the behaviors of BGP neighbor establishing, prefix selecting and prefix storage, which are predefined in BGP standard [6], BGP agents synchronize local status each other and execute routing selection procedure cooperatively by internal communication protocol. BGP functions of clustered routers is accomplished by the set of BGP Agents, which are consistent with the external behavior of BGP process and provide uniform best prefix view to each routing node. Internal communication protocol also provides uniform user interface (UI) view in any BGP agent. The configuration of user in any agent is broadcasted to others.

Distributed implementation model includes BGP agent and communication protocol among agents.

1. *BGP agent*: a BGP agent resides on single routing node, which contains the basic functions of BGP. It can establish neighbor connections with peers independently, sends or receives update packets to or from BGP peers under user configuration. Single BGP agent can accomplish the external function of traditional BGP process. With the importing of Agent technology, there are a few new features in the internal behavior of BGP agent.

   a) *Distributed synchronization mechanism*: BGP agents execute user commands together. User command in any agent is broadcasted to others and there must be a synchronization mechanism to keep the consistency of user interface view among BGP agents.

b) *Distributed computing and storage ability*: Update packets are dispatched to BGP agents by the partition algorithm in section III. Each agent only processes local update packets, and the prefixes of local update packets are stored in agent's local RIB. Route election algorithm among agents can guarantee the consistency of *best prefix view* in every routing node and generates the *best prefix view* of BGP.

c) *Transfer ability*: when a routing node fails or overloads, prefixes and neighbors must be reallocated among agents or transferred to another lower load agent.

2. *Communication protocol*: it includes definition of internal communication packets, distributed route election algorithm of *best prefix view*, transfer mechanism among agents.

Compared with basic model, distributed implementation model defines a new BGP neighbor type — *I-IBGP* neighbor, which communicates with other agents by internal communication protocol and deals with new features of BGP agent. The new model only makes small extension onto BGP standard by defining new *I-IBGP* neighbor, so it can be easily implemented.

3 Computation and Storage Partition Algorithm

How to assign the prefixes of update packets among BGP agents, utilize the computing and storage resource of agents efficiently, keep fairness and achieve the maximum performance. It is the core problem of distributed implementation model of BGP. TCP/IP connection is used by BGP to exchange prefixes between peers. Only one of the prefixes to identical destination network address, which come from same BGP neighbor connection, can exist in RIB at most. Based on the analyzing result of section IV, there are so many redundant prefixes in RIB with huge BGP table capacity. It suggests that the redundancy of prefixes does come from multi-neighbors. We propose a computation and storage partition algorithm based on it.

3.1 Function Definition

Functions and macros to be used are defined here.

1) Physical interface of routing node is numbered in hierarchy order, such as: node / card / interface. Macro *NODE_ID* is used to get node number from interface number.

2) IP forwarding module provides routing lookup function, which is defined as: *Interface_ID Lookup* (dest_address).

3) The source neighbor of each prefix can be gotten by function: *Neighbor Get_Peer* (prefix).

4) Macro *ADDRESS* is used to get IP address of BGP peer from Neighbor structure. Macro *STATUS* is used to get current status of BGP neighbor connection.

3.2 Partition Algorithm

DEF .1: Let N = { n_1, n_2, ... , n_τ | $\tau \geq 1$} denotes the set of routing nodes in a clustered router. It is also the set of BGP agents for the relationship between routing node and BGP agent.

DEF .2: Let P = { p_1, p_2, ... , p_φ | $\varphi \geq 1$} denotes the set of internal neighbors and external neighbors of BGP.

DEF .3: Let R = { r_1, r_2, ... , r_υ | $\upsilon \geq 1$} denotes the set of prefixes in update packets which come from BGP peers.

DEF .4: Let A(n_i) = { p_x | $p_x \in$ P} denotes the set of BGP neighbors which can reside on routing node n_i.

DEF .5: Let C(n_i) = { r_x | $r_x \in$ R } denotes the set of prefixes which are processed by routing node n_i.

DEF .6: Function f : P × N → T , T = {true, false }

$$f(p_x, n_i) = \begin{cases} \text{true, NODE\_ID(Lookup(ADDRESS(p_x)))} = n_i \\[2ex] \text{false, otherwise;} \end{cases} \tag{1}$$

DEF .7: Function g : R × N → T , T = {true, false }

$$g(r_x, n_i) = \begin{cases} \text{true,} Get\_Peer(r_x) \in A(n_i) \cap STATUS(Get\_Peer(r_x)) = ESTAB \\[2ex] \text{false,otherwise;} \end{cases} \tag{2}$$

Every prefix in RIB is associated with one BGP neighbor. There exists associated operation of *replace* or *deletion* between the prefixes r_1, r_2 coming from same neighbor sometimes and it is very difficult to assign them into different *A(n_i)* for parallel computation. We propose a partition algorithm based on the assignment of BGP neighbors among BGP agents. Firstly each element p_φ in *P* is assigned into set *A(n_i)* by checking function f, then function *g* is used to assign r_υ into set C(n_i). Prefixes computing and storage are distributed among BGP agents naturally after the assignment of BGP neighbors. Each BGP agent just needs execute local prefix selection procedure, maintain local RIB and take part in distributed route election algorithm.

The procedure of partition algorithm:

```
INPUT: P, N, AND R
OUTPUT: O₁ = {  A(nᵢ) | nᵢ ∈ N }
O₂ = {  C(nᵢ) | nᵢ ∈ N }
PROCEDURE:
for each n ∈ N do
            A(n)   ← NULL
            C(n)   ← NULL
for each n ∈ N do
            for each p ∈ P do
               if f( p ,n ) = true then
                  A(n)   ←  A(n) ∪ p
```

```
for each n ∈ N do
for each r ∈ R do
            if g ( r ,n ) = true then
            C(n)  ←  C(n) ∪ r
            R ← R - r
```

4 Model Evaluation

4.1 Evaluation Attributes

BGP table capacity and system reliability are among the most important performance metrics of backbone routers. We use these two metrics to evaluate distributed implementation model. To reduce the capacity of BGP agents' RIB and the cost of internal protocol communication to a maximal extent, it is necessary that there are a lot of redundant prefixes whose destination network address are the same. The route redundancy of BGP table is represented as the proportion between the number of possible prefixes in BGP table and the number of different destination address of BGP table. The bigger route redundancy is, the more evenly prefixes are distributed among BGP agents.

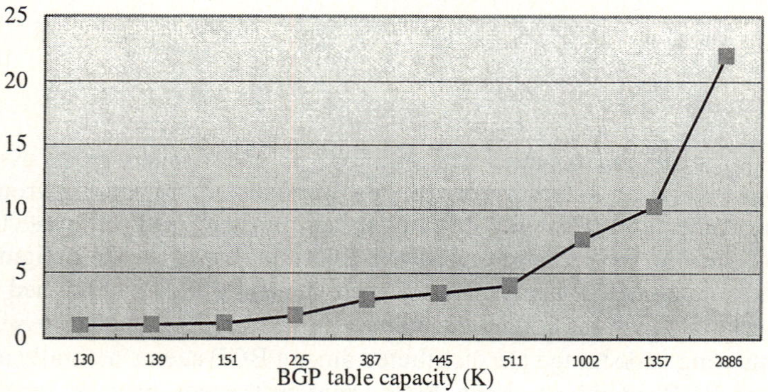

Fig. 2. Redundancy trend line of Internet BGP tables

The probability of two or more routing nodes failed simultaneously is almost zero. So the influence of single routing node failed reflects the reliability of clustered routers. In the following two subsections, we analyze Internet BGP table, which is collected from *RIPE* project [11]; exhibit the benefit of prefix storage and the influence to the *best prefix view* with single routing node failed.

4.2 Prefix Storage Evaluation

Routing Information Service (RIS) project [11] of *RIPE* collects default-free routing information of ten collectors using BGP4 from backbone routers in the Internet every eight hours. We have analyzed these collectors' BGP table of every fifth day from

October 1, 2003 to October 30, 2003 and found that the BGP table capacity does not flap dramatically in the condition of BGP neighbor connection being stable[1]. The flapping range does not exceed *2.1%* of average. We skip the influence of unstable BGP connections and use average capacity as BGP table capacity of these Internet collectors. Fig 2 indicates the relationship between these collectors' table capacity and the proportion of BGP table capacity to the number of destination address. X-coordinate denotes the average table capacity of Internet collectors in October and y-coordinate denotes the proportion. The proportion increases dramatically with the increase of BGP table capacity in Fig 2. We can know that the redundancy must come from different BGP neighbors from the feature of BGP in section □. It is an encouraging result to the evaluation of our partition algorithm.

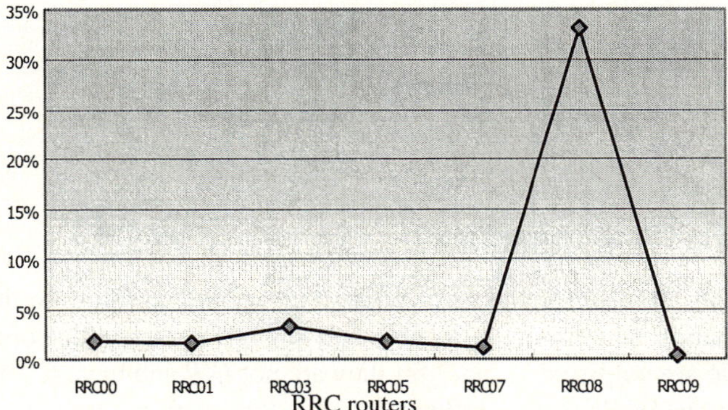

Fig. 3. The influence of single node failure

The worst results come from *RRC06* at Otemach of Japan, *RRC02* at Paris and *RRC04* at Geneva. Analyzing the connectivity of those routers shows that they are not *IXP*(Internet eXchange Point) but points of provider in single home mode. They establish a BGP neighbor connection with *IXP* and have few redundant routes and small table capacity, which does not burden control planes of core routers.

Let L_i denotes the local prefixes that are received from agent$_i$'s local BGP neighbor connection and not in *best prefix view*. Let BL_i denotes the local prefixes that are received from agent$_i$'s local BGP neighbor connection and in *best prefix view*. Let BM_i denotes the prefixes coming from other agents. Let S_i denotes the prefixes set in agent's local RIB, S denotes BGP table capacity and P_i denotes the percentage of S_i to S. Here we have:

$$S_i = L_i + BL_i + BM_i, i \in [1, n] \tag{3}$$

---

[1] The connections of *RRC09* to Internet Online AG and Lie-Comtel are lost from Oct-01 to Oct-05, Oct-10 to Oct-15, and Oct-15 to Oct-20. One of *RRC07*'s connections is also lost from Oct-01 to Oct-05.

$$S = \sum_{j=1}^{n} (L_j + BL_j) \tag{4}$$

$$P_i = \frac{S_i}{S} = \frac{(L_i + BL_i + BM_i)}{\sum_{j=1}^{n} (L_j + BL_j)} \tag{5}$$

For the clustered router containing n routing nodes, each agent exists only one established BGP neighbor connection whose peer is *IXP* and one of n parts of agent's table capacity is broadcasted to other agents as parts of *best prefix view* in the ideal condition. Then the equation (4) can be simplified into $2/n$. So the reduction of prefixes storage of distributed implementation model is linear and about $n/2$ times over the basic BGP model.

$$P_i = \frac{(L_i + BL_i + BM_i)}{\sum_{j=1}^{n}(L_j + BL_j)} = \frac{(N-1)BL_i + BL_i + (N-1)BL_i}{\sum_{j=1}^{n}((N-1)BL_i + BL_j)} = \frac{2N-1}{N^2} \approx \frac{2}{N} \tag{6}$$

4.3   The Influence of Single Node Failed

Fig.3 shows relationship between the percentage of un-redundant routes to the number of different destination address of BGP table and seven Internet collectors. *RRC02*, *RRC04* and *RRC06* are not listed in Fig.3 for they are not *IXP* and have few redundant routes. Single routing node failed influences only the prefixes coming from one neighbor connection in the ideal condition. The worst condition of single node failed is that the quantity of lost destination address equals to the quantity of un-redundant routes in BGP table. We can conclude that the influence is less than 5% for the most of collectors and the best one is about 0.4%. Reliability of clustered routers can be improved greatly by carefully arrangement between network topology and user configuration.

5   Conclusion and Future Work

This paper proposes a new protocol software architecture — *distributed BGP implementation model* and the *partition algorithm* of prefix storage and computation used for the model. The analysis of Internet BGP table and mathematic formula shows that our model can achieve linear reduction of storage requirement of single routing node and improve the reliability of clustered routers greatly. Compared with basic implementation model, distributed implementation model of BGP has many advantages.

1.   Avoid the problem of single node failed;
2.   Decrease table capacity of each agent and improve the storage ability of clustered router;
3.   Improve protocol-computing ability of clustered router;

4.   Fewer administrative cost compared with several independent routers at an *IXP* and more efficient;

Since clustered routers are a promising architecture for future router design, we believe parallelized protocol computation and protocol software architecture will become an interesting new research field. With popularity and maturity, Zebra project can be used as a good experimental platform for our *distributed BGP implementation model*. Our future work includes modifying the source code of Zebra BGPD and implementing a distributed BGP protocol on the hardware platform that is composed of PCs and high-speed Ethernet switch.

References

[1] Juniper Networks. http://www.juniper.net

[2] Geoff Huston. Internet BGP Table. http:// www.potaroo.net/

[3] Nina Taft . The Basics of BGP Routing and its Performance in Today's Internet. Sprint,Advanced Technology Labs California, USA May 2001

[4] Craig Labovitz ,G. Robert Malan, and Farnam Jahanian. Internet Routing Instability. ACM SIGCOMM 1997

[5] Zihui Ge, Daniel Ratton Figueiredo, Sharad Jaiswal, and Lixin Gao. On the hierarchical structure of the logical internet graph. In Proc. SPIE ITCom 2001.

[6] Y.Rekhter., A Border Gateway Protocol 4 (BGP-4),1995

[7] GNU Zebra (Routing Software). Http://www.Zebra.Org

[8] Mitsuru MARUYAMA, Naohisa TAKAHASHI, Members and Takeshi MIEI. CORErouter-1:An Experiental Parallel IP Router Using a Cluster of Workstations, IEICE TRANS.COMMUN., 1997, E80-B

[9] Sam Halabi. Pluris Massively Parallel Routing(White Paper). Pluris Inc., 1999

[10] Xipeng Xiao and Lionel M.Ni. Parallel Routing Table Computation for Scalable IP Routers, Proceedings of the IEEE International Workshop on Communication, Architecture, and Applications for Network-based Parallel Computing: 1998.

[11] Routing Information Service. http://www.ripe.net/ris/ris-index.html

[12] CIDR REPORT. http://www.cidr-report.org

A Routing Protocol for Wireless Ad Hoc Sensor Networks: Multi-Path Source Routing Protocol (MPSR)

Mounir Achir and Laurent Ouvry

Electronics and Information Technology Laboratory,
Atomic Energy Commission,
17, rue des Martyrs 38054 Grenoble Cedex 09, France
{mounir.achir, laurent.ouvry}@cea.fr

Abstract. One of the most compelling challenges of the next decade is the "last-meter" problem, how to extend the expanding data network into end-user data-collection and monitoring devices. This part of the network is called Wireless Personal Area Network (WPAN). A Wireless Sensor Network is a particular case of WPAN which has the following particular constraints : low bitrate, low cost and very low energy consumption. In this article, we present a Multi-Path Source Routing protocol (MPSR) which is adapted to Wireless Sensor Networks. This routing protocol exploit the different discovery paths to transmit data, this guaranteed to us a repartition of the energy consumption in all possible nodes that can route data to the destination.

Keywords: Wireless Sensor Network, Routing protocol, IEEE 802.15.4, Energy consumption minimization.

1   Introduction

Wireless sensor networks promise great advantages in terms of flexibility, cost, autonomy and robustness with respect to wired ones. These networks find a usage in a wide variety of applications, and particularly in remote data acquisition, such as climate monitoring, seismic activity studying, or in acoustic and medical fields. Unfortunately, the nodes are subject to strong constraints on power consumption due to their very reduced dimensions as well as their environment. As an example, frequent battery replacement is to be avoided in places where access is difficult or even impossible.

The considerable interest in wireless sensor networks led to the creation of a working group within the IEEE standardization committee, which specified a new standard IEEE 802.15.4 [1]. By specifying very low power consumption MAC (Medium Access Control) and Physical layers (less than few milliwatts), and thanks to the low complexity protocol and low data rate (no more than 250 Kbits/s), this standard makes wireless sensor networks possible. 802.15.4 is an enabling standard in the sense that it complements wireless standards

P. Lorenz and P. Dini (Eds.): ICN 2005, LNCS 3421, pp. 442–453, 2005.

such as WiFi and Bluetooth. It distinguishes itself from other wireless standards by various features, like low data rate, low power consumption, low cost, self-organization, and flexible topologies. It supports applications for which other standards are inappropriate. It not only opens the door to an enormous number of new applications, but also adds value to many existing applications. With various simple devices able to connect to networks, ubiquitous networking is closer than ever to us.

A host of applications can benefit from the new standard, including [2]:

- Automation and control: home, factory, warehouse;
- Monitoring: safety, health, environment;
- Situational awareness and precision asset location (PAL): military actions, firefighter operations, autonomous manifestering, and real-time tracking of inventory;
- Entertainment: learning games, interactive toys.

In this article, we present a Multi-Path Source Routing protocol which is adapted to Wireless Sensor Networks. This routing protocol exploit the different discovery routes to transmit data, this guaranteed to us a repartition of the energy consumption in all possible nodes that can route data to the base station.

The remainder of the paper is organized as follow : in section 2, we describe the IEEE 802.15.4 standard, in section 3, various approach and solutions which can be found in the literature are discussed. In section 4, we give assumptions and principle of our proposed routing protocol, in section 5 and 6 we see the simulation scenario and comment the different results.

2   An Overview of the IEEE Std 802.15.4

The new IEEE standard, 802.15.4, defines the physical layer (PHY) and medium access control (MAC) sublayer for low-rate wireless personnal area networks (LR-WPANs), which support simple devices that consume minimal power and typically operate in the personal operating space (POS) of 10 m or less. Two types of topologies are supported in 802.15.4: a one-hop star or, when lines of communication exceed 10 m, a multihop peer-to-peer topology. However, the logical structure of the peer-to-peer topology is defined by the network layer. Currently the ZigBee Alliance is working on the network and upper layers. In the following subsections, we give a brief description of some important design features in 802.15.4 [3].

IEEE Std 802.15.4 was designed to support multiple network topologies- from a star network to many network types based on peer-to-peer communication, including star, mesh, cluster and cluster-tree networks. We describe here different topologies that support this standard.

2.1   The Star Network

In the star topology, communication is controlled by a unique coordinator that operates as a network master, sending beacons for device synchronization and

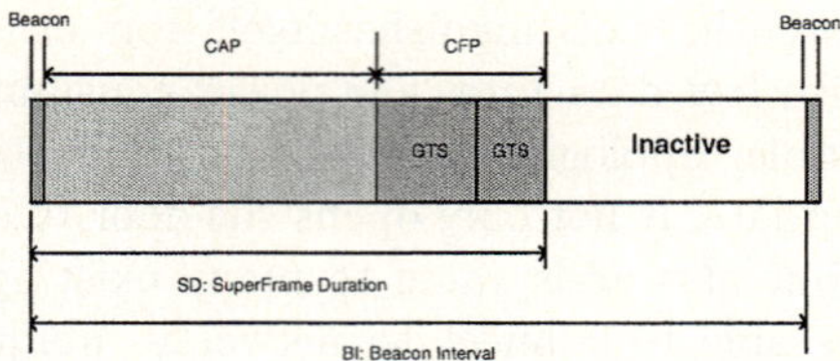

Fig. 1. Superframe structure of the IEEE Std 802.15.4

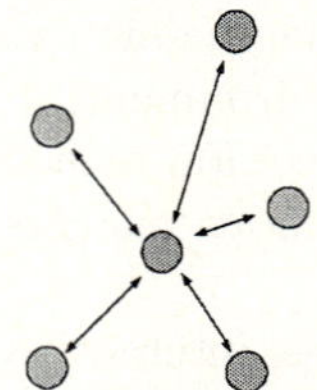

Fig. 2. Star network

maintaining association management, see figure 1. In this topology, the network devices communicate only with the coordinator, as shown in figure 2. An IEEE 802.15.4 operating in star network, can be a good alternative for a simple point-to-point applications that require an extremely low-cost.

2.2   The Peer to Peer Network

The peer-to-peer communication capability of IEEE 802.15.4 devices allow the creation of many types of peer-to-peer networks, each one with its own advantages and disadvantages. One can find Mesh network, Cluster network and Cluster tree network topologies, see respectively figures 3, 4 and 5

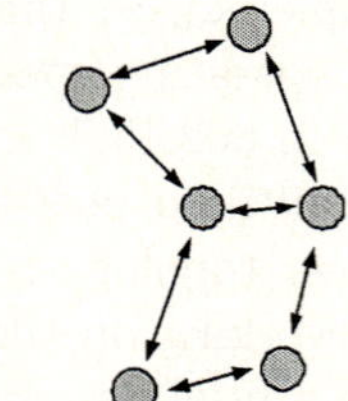

Fig. 3. Mesh network

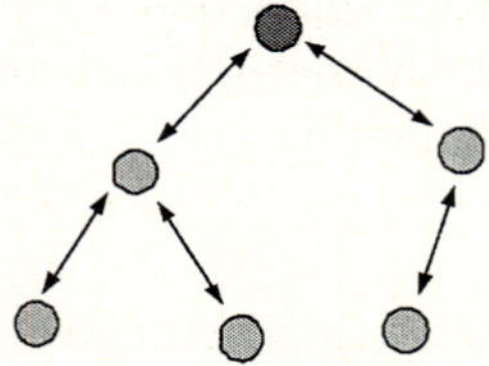

Fig. 4. Cluster network

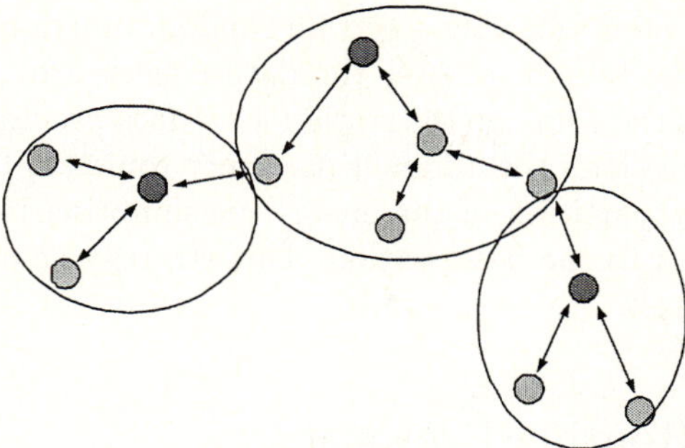

Fig. 5. Cluster Tree network

3   Related Works

Since research on ad hoc routing began with packet radio networks, numerous protocols have been proposed. These protocols have to contend with the wireless medium, i.e. low bandwidth, high error rates and burst losses, as well as the limitations imposed by ad hoc networks such as frequently changing topology and low power devices. Finally, the protocols have to scale well to a large number of nodes in the network. Considering the nature of the protocols, they can be categorized into proactive, like DSDV [4] or LSR [5], and reactive protocols, like AODV [6] or DSR [7].

The potential problem in current routing protocols is that they try to find the least expensive path, in term of latency or number of hop, and unfortunately the transmission is carried out on this path for each communication. However, this path is, perhaps, not the best one to make increase the life of the network. This was the principal motivation for [8] to propose an algorithm that thwart this problem and which, rather than choose the same path for each communication, use various paths at various moments by exploiting sub-optimal paths.

In the method suggested by [9], the author affirms that to choose the low cost path does not imply an increasing in the lifespan of the network!. Indeed, all will depend on the quality of the link, do multihop allows to make considerable profit on the level of consumption due to the fact that the transmissions are done in short range. Now, if the quality of the link is not very reliable and the communication is prone to non-neglected interferences, then making multihop lead likely to increase the number of retransmissions, from where more significant consumption. This justify [9] to introduce, into its cost function, a parameter which take into account the quality of the link which is the error rate in the link. Another solution is proposed by [10] in the μAMPS project (micro Adaptive Multi-Domain Power-Aware Sensors) which is developed at the Massachusetts Institute of Technology (MIT). It relates to miniature "power-aware" wireless sensors system. In this solution, the authors try to find an optimal method, in term of consumption, which allows the data acquisition resulting from a hundred of sensors. The data acquisition is made in the following way: the network is cut out in autonomous sub-networks in a dynamic way, in each sub-network a head of sub-network is elected following a certain random function. This head of sub-network will have the task to recover the data collected by nodes being in his group and transmits the whole to the station for a data processing. Our approach is different; MPSR exploits the different discovery routes to transmit data. This guarantees to us a repartition of the energy consumption in all possible nodes which can route data to the base station. This strategy enables to increase the lifespan of the network.

4   Proposed Routing Protocol

4.1   Assumptions and Principle

MPSR is a reactive routing protocol for Mesh network topology, and is adapted to wireless Ad-Hoc (Sensor) networks with a base station. This last has the responsibility of collecting data from all network nodes. MPSR must be robust with mobility and topology variations, because nodes can move, can fall out for any technical problem or can be in a moving environment. Consequently, the network will be limited in its number of nodes and thus will not be very wide, comparing to Tree and Cluster Tree topologies.

The protocol is made of three steps to establish a route from one source (which can be a sensor or any other device) to a destination which eventually process the received data (the destination is typically a base station). In the first step, the source broadcasts a signalization packet called RREQ (Route REQuest). These RREQ packets are broadcasted again by all nodes of the network until the packet is received by the destination. This last can receive several RREQ of the same source because the packet can use several routes. In the second step, the destination computes a cost for each discovered route, i.e. for each RREQ received, by an algorithm which which described in section 4.2. In the third step, the destination replies to the source by a RREP (Route REPly), this RREP is forwarded (not broadcasted) to the source. This last will have several routes with their respective costs, and can then transmit data by random choice of a route depending on its cost.

4.2   MPSR Protocol

As we see in the precedent paragraph, the protocol has three types of packets. Route request (RREQ) and route reply (RREP) for route discovering, and DATA packets for data transmitting.

Routes Setup Phase. RREQ packets are broadcasted at all neighbors which rebroadcast these packets to theirs neighbors until the destination is reached. In each forward, the node puts its address in the header of the packet. The destination will have so the address of all nodes between it and the source. After a RREQ packet reception, the destination executes its algorithm for cost computation and then replies to the source by a RREP. A DATA packet is send in the reverse route of a RREP packet route.

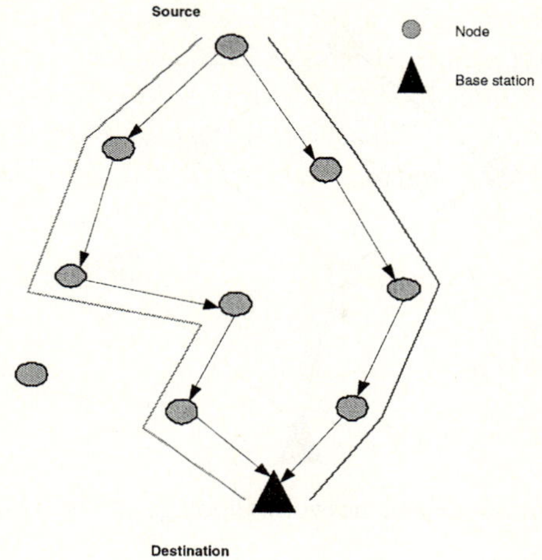

Fig. 6. Route setup phase in MPSR

Path Cost Computation Algorithm. We propose to look at the following example to illustrate the path cost computation algorithm. To do this we use the following notation:

- $\alpha_{i,j}$: cost of the jth route between node i and the destination, and represent the proportion of data (assuming that we use fixed packet length) that will be sent by node i in this route. These costs must satisfy the following condition: $\sum_j \alpha_{i,j} = 1$;
- E_i: energy consumed by node i to transmit current data, see its expression in the following example.

This algorithm computes the optimal configuration of $\alpha_{i,j}$ to have an equivalent repartition of traffic or data, i.e. energy consumption, between all network nodes. So to do this, we minimize the variance of E_i : $min(Variance(E_i(\alpha)))$ with $\alpha = \{\alpha_{i,j}\}$.

Example
Let's suppose the network topology represented below. This network is composed of five nodes plus one base station. All nodes transmit data to the base station by using multihop for some of them.

The different routes until the base station are:

node 1 : route 1: 1-3-5-BS, route 2: 1-2-4-BS

node 2 : 2-4-BS

node 3 : route 1: 3-5-BS, route 2: 3-4-BS

node 4 : 4-BS

node 5 : 5-BS

E_i are computed, in our example, by the following equation:

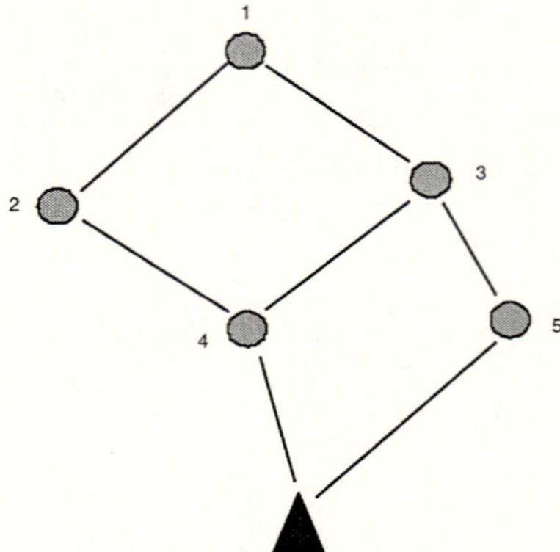

Fig. 7. Wireless sensor network topology of the example

$$\begin{cases} E_1 = \frac{\alpha_{11}+\alpha_{12}}{2} \\ E_2 = \frac{\alpha_2}{2} + \alpha_{12} \\ E_3 = \frac{\alpha_{31}+\alpha_{32}}{2} + \alpha_{11} \\ E_4 = \frac{\alpha_4}{2} + \alpha_{12} + \alpha_{32} + \alpha_2 \\ E_5 = \frac{\alpha_5}{2} + \alpha_{11} + \alpha_{31} \end{cases}$$

The node 1 send its data in two possible routes, route 1 and route 2. E_1, which is the energy consumed by node 1, is computed by summing α_{1j} for all j and divided the all by 2, because here the node 1 is the source and transmit data only. For the node 2, we sum $\alpha_2/2$ and α_{12}. Note that α_{12} is not divided by two because node 2 receives data from node 1 and then transmit it to node 4, so this node will consume twice more. We do the same principle for E_3 which is computed by summing $\frac{\alpha_{31}+\alpha_{32}}{2}$ and α_{11}, like for the computation of E_4 and E_5.

After the minimization of E_i's variance, by using Lagrange method, we obtain the following results: $\alpha_{11} = \alpha_{12} = 0.5$, $\alpha_2 = \alpha_4 = \alpha_5 = 1$, $\alpha_{31} = 1$ and $\alpha_{32} = 0$. So here, we do not use the second route of node 3 and we use once on two the first route of node 1 and once on two the second. This result can be also deduced by looking figure 7.

Sending and Routing Data. After the computation of the different weights by the destination, replies to the source i by a RREP which contain routes costs α_{ij}. The destination sends as much RREP as RREQ, i.e. discovered routes. Then, the source will have all nodes addresses which are between it and the destination, plus their respective costs. Each cost corresponds to the duty cycle of the route utilization.

5   Implementation and Simulation Scenario

To evaluate MPSR performances, we use SeNSE simulator [11](Sensor Network Simulator and Emulator). We search to compute the most important parameters which permit to appreciate the performance of a routing protocol in term of energy consumption. These parameters are the *Success Rate* and the *minimum remaining energy* in nodes batteries. The success rate represents the data percentage effectively received by the destination, and the minimum of nodes energies after a fixed simulation time, represent the network lifespan. These two parameters are computed as function of : nodes density (number of node), number of source nodes (we assume that not all nodes send data), bitrate and nodes speed.

During this simulation, and to have conditions which approach as much as possible the real case of an IEEE 802.15.4 network, we take into account a maximum possible specifications in the physical and MAC layers of the IEEE 802.15.4. We resume all hypothesis in table 1.

WPAN radio propagation model is used in several working group IEEE 802.11, IEEE 802.15.2, IEEE 802.15.3 and IEEE 802.15.4 for coexistence measurement:

$$d = \begin{cases} 10^{\frac{P_e - I - 40.2}{20}} & for \qquad d \leq 8m \\ 8 \times 10^{\frac{P_e - I - 58.5}{33}} & for \qquad d > 8m \end{cases}$$

With P_e and I in dBm, d in meter.

We have several parameters to set in the routing protocol to have an optimal routing, these parameters are resumed in table 2:

RreqTimeOut: the maximum attempt delay of the destination after a first RREQ reception. After this delay, all RREQ of the same source are ignored.

Table 1. Simulation parameters values

Parameter	Value
TX Consumption	10 mW
RX Consumption	10 mW
Idle Consumption	1 mW
Initial Energy in the batteries	3 Joule
Emission power	1 mW
Reception threshold	-85 dbm
Radio propagation model	WPAN Radio Model*
Dimension of the deployment area	200x200 m^2
Number of node	40
Nombre of source	8
Packet length	100 Bytes
Inter-packet interval	1 s
Speed	1 m/s
Simulation duration	100 s
Medium Access Control	CSMA/CA
traffic type	CBR (Constant Bit Rate)

WPAN Radio Propagation Model.

Table 2. MPSR parameters values

Parameter	Value
RreqTimeOut	0.05 s
RrepTimeOut	0.05 s
TTL (Time To Live)	10
PathDiscoveryTime	30 s
UpdateroutesTime	30 s

RrepTimeOut: the maximum attempt delay of the source after a first RREP reception. After this delay, all RREP of the same destination are ignored.

TTL: Time To Live.

PathDiscoveryTime: delay before deleting all RREQ entries. These entries permit to detect loops.

UpDateroutesTime: routing table up-date period.

6   Simulation Results

In this section, we present the simulation results. These results are averaged on 10 simulations of 100 seconds.

To have a good interpretation of these results, we compute the following parameter for each protocol:

$$S = \frac{SR}{1 - \frac{ME}{E_0}}$$

with :

SR: the Success Rate (in Joule);

ME: the Minimum Energy (in Joule);

E_0: the initial nodes energy (in Joule).

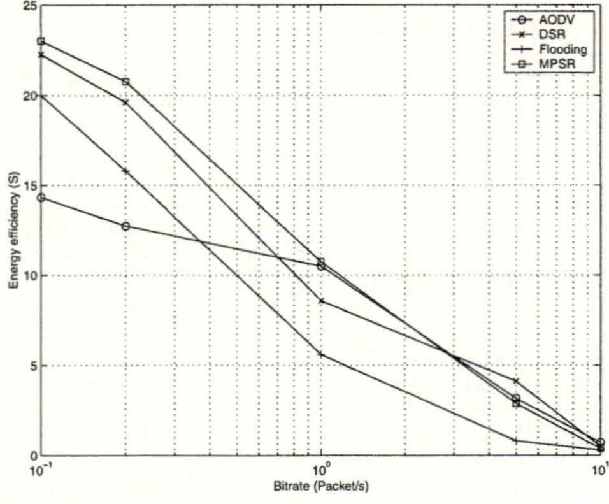

Fig. 8. Energy efficiency (S) versus Bitrate

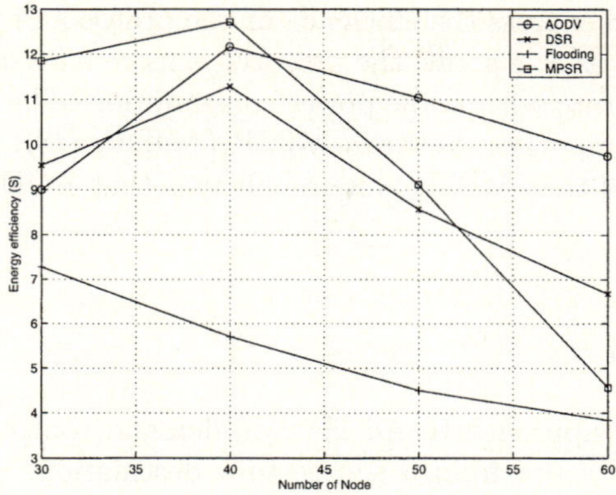

Fig. 9. Energy efficiency (S) versus Number of node

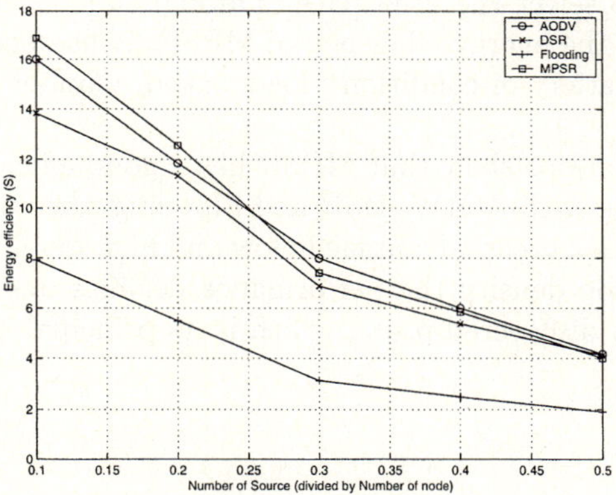

Fig. 10. Energy efficiency (S) versus Number of source node

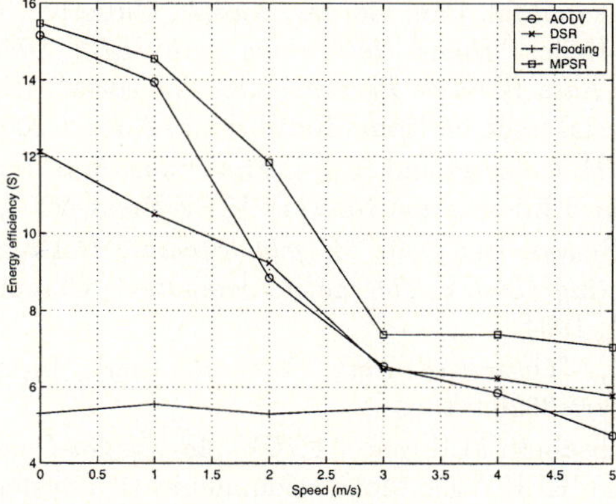

Fig. 11. Energy efficiency (S) versus Speed

The parameter S express the efficiency of one protocol in term of power consumption. As much as S is great, the protocol is more efficient, i.e, the protocol has a good success rate with a low power consumption. The figure 8, 9, 10 and 11 give a global comparison between MPSR, AODV, DSR and the Flooding. These figures proves that MPSR is more efficient that AODV, DSR and Flodding protocols.

7   Conclusion

MPSR offers a new approach to Ad Hoc, on-demand routing protocol. Rather than exploiting one route from a source to a destination, it exploits multiple routes with a certain cost for each route. Each source broadcasts RREQ in the network to localize its destination, this destination receives some of all broadcasted RREQ, computes a cost for each one and replies by RREP. The source exploits the cost to select the used path. Through simulation, we demonstrate the robustness and the energy efficiency of MPSR. It has been tested and characterized under a variety of condition : load, speed, number of node, number of source.

Results of simulations show that MPSR has a good efficiency in energy consumption comparing with AODV, DSR and Flooding; the curves representing S prove it. MPSR gives a great advantage in low and high mobility, also in low node density. In high node density the performance declines due to RREP packets, transmitted in each discovered route, which is its principal inconvenient.

References

1. Draft P802.15.4, Octobre-2003 : *"Wireless Medium Access Control (MAC) and Physical Layer (PHY) specifications for Low Rate Wireless Personel Area Networks (LR-WPANs)"*.
2. Ed Callaway, Venkat Bahl, Paul Gorday, Jose A. Gutierrez, Lance Hester, Marco Naeve and Bob Heile, *"Home Networking with IEEE 802.15.4: A Developing Standard for Low-Rate Wireless Personal Area Networks"*, IEEE Communications MAGAZINE, Special issue on Home Networking, August 2002, pp. 70-77.
3. J A. Gutierrez, E H. Calaway and R L. Barret, *"Low-Rate Wireless Personal Area Networks"*, Standard Information Network IEEE Press, 2004.
4. Charle Perkins, Pravan Bhagwat. *"Highly Dynamic Destination Sequenced Distance Vector Routing (DSDV) for mobile computer"*, Computer Communication revue pp. 234-244, 1994.
5. P. Jaquet et al, *"Optimized Link State Routing"*, Internet Draft, http://www1.ietf.org/rfc/rfc3626.txt.
6. Charle Perkins, Elisabeth M.Royer. *"AODV- Ad-hoc On-Demand Distance Vector Routing"*, Proc 2nd IEEE Wksp Mobile Communication Systems and Applications, 1999.
7. D. Johnson and D. Maltz, *"Dynamic Source Routing in Ad hoc wireless network"*, Mobile Computing, Kluwer pp. 153-181, 1996.

8. Rahul C.Shah, Jan M.Rabaey, *"Energy Aware Routing for Low Energy Ad hoc Sensor Networks"* IEEE Wireless Communications and Networking Conference (WCNC), March 17-21, 2002, Orlando, FL.
9. Archan Misra, Suman Banerjee, *"MRPC: Maximizing Network Lifetime for Reliable Routing in Wireless Environments"*. IEEE Wireless Communications and Networking Conference (WCNC), Orlando, Florida, March 2002.
10. Wendi Rabiner Heinzelman, Anantha Chandrakasan, Hari Balakrishnam, *Energy Efficient Communication Protocol for Wireless Microsensor Networks (LEACH)"*, in the Proceeding of the Hawaii International Conference System Sciences, Hawaii, January 2000.
11. SeNSE: Sensor Network Simulator and Emulator, http://www.cs.rpi.edu/ cheng3/sense/

Generalized Secure Routerless Routing
Extended Abstract

Vince Grolmusz[1] and Zoltán Király[1,2]

[1] Department of Computer Science, Eötvös University,
H-1117 Budapest, Hungary
[2] Communication Networks Laboratory, Eötvös University
{grolmusz, kiraly}@cs.elte.hu

Abstract. Suppose that we would like to design a network for n users such that all users would be able to send and receive messages to and from all other users, and the communication should be secure, that is, only the addressee of a message could decode the message. The task can easily be completed using some classical interconnection network and secure routers in the network. Fast optical networks are slowed down considerably if routers are inserted in their nodes. Moreover, handling queues or buffers at the routers is extremely hard in all-optical setting. The main result of the present work is the generalization of our earlier work [6]. That work gave two mathematical models for secure routerless networks when the Senders and the Receivers formed disjoint sets. Here we present mathematical models for routerless secure networks in the general case: now all users can send and receive.

Keywords: High speed optical networks, routerless routing, secure network protocols.

1   Introduction

The extreme bandwidth of a single optical fiber (25 000 GHz) is 1000 times larger than the total radio bandwidth of planet Earth (25 Ghz) [2]. Using this bandwidth effectively requires novel network designs.

We proposed two network designs for high-speed optical communication in [6]. It was supposed that there are given n Senders $S_1, S_2, \ldots, S_n$ and r Receivers $R_1, R_2, \ldots, R_r$. Let π be a function from $\{1, 2, \ldots, n\}$ to $\{1, 2, \ldots, r\}$. Our goal was to send long messages from S_i to $R_{\pi(i)}$, for $i = 1, 2, \ldots, n$ such that

(a) $R_{\pi(i)}$ can easily retrieve the message of S_i, for $i = 1, 2, \ldots, n$
(b) $R_{\pi(i)}$ cannot retrieve the message of S_j if $\pi(i) \neq \pi(j)$.

An obvious method for doing this is connecting S_i with $R_{\pi(i)}$ with private channels, that is, we use n channels for the n Senders and the r Receivers. The advantage of this solution is that n bits can be sent in parallel, and the transmission is private, in the sense that $R_{\pi(i)}$ receives only the transmission of S_i, for $i = 1, 2, \ldots, n$. The disadvantage of this solution is that the number of

P. Lorenz and P. Dini (Eds.): ICN 2005, LNCS 3421, pp. 454–462, 2005.

channels is equal to the number of communicating pairs, and this is infeasible in most cases. A much bigger problem with this solution is that if next time S_i wants to send messages to $R_{\sigma(i)}$, for $i = 1, 2, \ldots, n$ for some other function σ, then the whole network has to be reconfigured. If every Sender is directly connected to all Receivers, this solves the reconfiguration problem, but then the number of channels becomes nr.

Another obvious solution is that all the Senders and Receivers use the same channel, and they transmit their messages one after the other. Transmitting n bits this way needs n steps. In this case either a router has to be used just before the messages get to the Receivers, or some sort of encryption is needed for maintaining the privacy of the transmission. Using encryption has several drawbacks. Streamciphers, the most evident cryptographic tool which are fast and do not cause overhead in the communication have lots of recently proposed and successful attacks [8]. Block-ciphers are much slower, and may be infeasible in, say, in the 1000 Gbit/s range, and also, they causes non-negligible overhead in the communication. Using routers and addressing in the messages will also slow down the communication, especially in all-optical environments: with, say, 1000 Gbit/s throughput, by the best of our knowledge, no routers exist.

The main results of the work [6] were mathematical models of two networks, together with the associated network-protocols, in which the n Senders and the r Receivers are connected with only $r^{o(1)}$ channels in the first and $\log r$ channels in the second network.[1] Note, that in practice at most 32 channels are enough in both networks. The parallel channels do not speed up the transmission relative to the 1-channel network: the goal of using them is to facilitate the privacy of the communication and the distribution of the messages between the recipients, without any encryption or routers. Our first network in [6] used MOD m addition gates for some small modulus m (say $m = 6$ or $m = 10$) (see [11] for all-optical realizations), the second network used optically feasible MOD 2 addition (i.e., XOR) gates, (see [9] or [7]). The decoding at the user's side is done with very simple and optically feasible computations (counters or modular (XOR) gates).

The Generalized Design: Every Node Can Send and Receive. Lots of applications are excluded when the Receivers and the Senders form disjoint sets in the network. The main goal of the present work is to generalize the designs, given in [6], for the case when every node of the network can send and also receive messages.

We will present two different designs here: Network 1g and Network 2g, where letter g stands for the word "generalized".

From this point on, the Senders/Receivers will be called users, we may consider them as either computers or border routers, and we are going to plan a network connecting them.

---

[1] Here $o(1)$ denotes a quantity which goes to 0 as r goes to the infinity.

Our all-optical network consists of two parts, the *backbone* and the *connectors*. The backbone is an arbitrary, synchronized optical broadcast network with t parallel fibers. We need the synchronization by two distinct meaning, namely

- bits that are sent to different fibers at the same time must reach each individual user at the same time, and
- there is no collision, that is for example each user has time slots when only she can send bits.

Broadcast network means that each bit sent to the backbone network will reach all users' connector. (We use the fibers transmitting bits in both directions, but, due to the time slots, only in one direction at any given time.) Our goal is to minimize t, the number of fibers: in Network 1g it is $n^{o(1)}$, and in Network 2g it is $O(\log n)$. The sketch of one possible realization of the backbone network together with the connectors is shown on Figure 1.

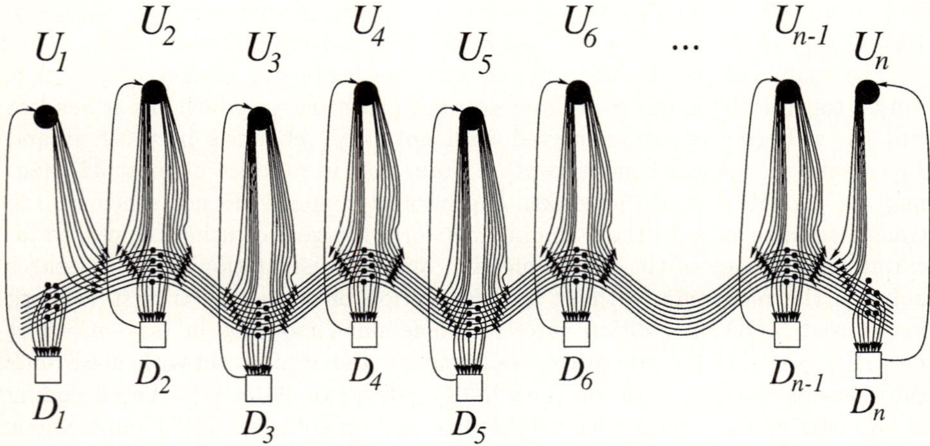

Fig. 1. One realization of our network. $U_1, \ldots, U_n$ are the users, $D_1, \ldots, D_n$ are the corresponding decoders

The second part, the connectors, consist of one transmitter and one decoder for each user. These connectors are uni-directional, a transmitter, consisting of simple fibers, plainly transmits the bits got from the user to the fibers of the backbone, while a decoder reads bits from the fibers and transmits bits to the user. Each decoder contains a MOD gate. In Network 2g a decoder gets bits from a subset of fibers, calculate the MOD 2 sum, ie., the XOR of them, and transmits the resulting bit to the user. Network 1g uses channels that are sets of 5 fibers. We are considering channels capable to transmit a MOD 6 number at a time, i.e. the sum of the 5 bits transmitted on the 5 fibers belonging to it. The MOD gate of the decoders of Network 1g calculate the MOD 6 sum of the numbers arriving on its channels, and transmit this sum to the user using one channel. For these decoders it is also allowed that from an input channel they get the input more then one time (maximum 5) using parallel channels.

In both of our networks the participants use random numbers for security. These random numbers are generated privately in both networks.

2   Network 1g

In packet-switched networks, the users should know their own identity (say, an IP or MAC address) in order to pick up only those packets from the transmission channels, which are addressed to them. Note, that in Network 1g, described in this section, the users need *not* know even their own identity: the bits, intended to be sent to them, will find them securely and automatically.

We assume that each user is going to communicate to every other user. In her time slot user U_i will send one bit to the other users; different users may get different bits.

We will use k channels, for we need $t = 5k$ fibers in the backbone.

Preliminaries. Let m be a non-prime power composite number; for simplicity, let us fix m to be 6. Note, that in this section all the additions and subtractions are meant modulo 6.

By the results of [4], for any n, there exists (effectively computable) $n \times k$ matrices $B = \{b_{ij}\}$ and $C = \{c_{ij}\}$, with entries from the set $\{0, 1, 2, 3, 4, 5\}$, where $k = n^{o(1)} << n$, such that for a vector of n variables $x = (x_1, x_2, \ldots, x_n)$, with the notion $z = xB$:

$$xBC^T = zC^T = x + 4xU + 3xV = x', \tag{1}$$

is satisfied, where $U = \{u_{ij}\}$ and $V = \{v_{ij}\}$ are some $n \times n$ matrices with 0's in the diagonal, and with the following property: for all i and j, either u_{ij} or $v_{ij} \equiv 0 \pmod 6$.

In other words, with simple linear transformations, length-n x is first compressed to length-k $z = xB$, then z is expanded to length-n $x' = xBC^T$. The ith coordinate of x' contains x_i plus the sum of some "surplus" variables (other than x_i), multiplied either by 3 or 4. Note, that matrices B and C are independent of vector x, that is, they are constant matrices.

We describe our Network 1g now. Assume that user U_i wants to send bit x_j to user U_j, for all $j \neq i$. In the first step U_i simply computes $z = xB$, and sends its k coordinates in parallel on the k backbone channels. At user U_j, the jth coordinate, x'_j, of $x' = xBC^T$ is computed by a linear transformation from the signals received from the k parallel channels of the backbone. This computation is performed by the decoder D_j.

Getting back the actual vector x needs some further steps, what we call *filtering*, and we detail it in the next section.

The Protocol. We describe the transmission-protocol. User U_i will transmit securely n bits, that is, bit x_j to the corresponding receiver, U_j, for $j = 1, 2, \ldots, n$.

A round is performed as follows:

Step 1 - Encoding - From the bits of $x = (x_1, x_2, \ldots, x_n)$ the mod 6 integers $z = (z_1, z_2, \ldots, z_k)$ are computed by linear combinations taken modulo 6: $z = xB \bmod 6$ at U_i.

Step 2 - Transmission - The mod 6 numbers $z_1, z_2, \ldots, z_k$ are sent on the k backbone channels.

Step 3 - Decoding - The linear transformation $x' = (x'_1, x'_2, \ldots, x'_n) = zC^T$ is computed modulo 6, x'_j at decoder D_j, and number x'_j is given to user U_j, who stores value x'_j in its memory. (Note, that because of the obvious information-theoretical reasons, generally it is not possible to retrieve bit x_j from integer x'_j).

Step 4 - Pre-Filtering - A random $\mu : \{1, 2, \ldots, n\} \rightarrow \{1, 2, \ldots, n\}$ permutation is generated privately at the user U_i. Then for $j = 1, 2, \ldots, n$ Steps 1, 2 and 3 are repeated for $x^{\mu(j)} \in \{0, 1\}^n$ instead of x, where $x^{\mu(j)}$ coincides with x, except on position $\mu(j)$, whereas $x^{\mu(j)}$ is 0 if it was 1 in x or 1 if it was 0 in x. Let x''_i denote the coordinate i of $x^{\mu(j)}BC^T$.

Step 5 - Post-Filtering - Now, user U_j follows the next program while receiving any new x''_j, originating in Step 4:

if $x''_j = x'_j$ or $x''_j \equiv x'_j \pm 3$ or $x''_j \equiv x'_j \pm 4$ it does nothing;
if $x''_j = x'_j - 1$ then U_j concludes that $x_j = 1$;
if $x''_j = x'_j + 1$ then U_j concludes that $x_j = 0$.

If at the end of the protocol user U_j did not learn bit x_j, she concludes that no bit was sent for her.

Remark 1. The security of the protocol depends on the fact that the user U_j does not know the random permutation μ.

Theorem 1. *After performing one round, user U_j retrieves the bit x_j, for $j = 1, 2, \ldots, n$.*

Proof. Clearly, x'_j is equal to the quantity, given in formula

$$x_j + 4(x_{i_1} + x_{i_2} + \cdots + x_{i_\ell}) + 3(x_{j_1} + x_{j_2} + \cdots + x_{j_v}) \qquad (2)$$

so decreasing any non-zero x_h in the sum of formula (2) by 1, leads to either a decrease of 1 of the sum (in the case when exactly x_j was decreased, or by 0 (when an x_g was decreased with a 0 coefficient in formula (2)), or by 4 or 3 modulo 6, (when the coefficient of the decreased variable was 4 or 3, respectively). Similarly, an increase by 1 will result an increase by 1 or 0 or 3 or 4 modulo 6, where 0, 3 and 4 means that not the variable with coefficient 1 was increased. Consequently, a change by value 1 means that $x_j = 0$, by value -1 means that $x_j = 1$.

The Security of Network 1g. Let us review, what U_j can learn from the bits, addressed to others. Clearly, U_j will know the number of the 1-bits with coefficient 4 (but will not know the number of the 1-bits with coefficient 3 in formula (2), since $+3 = -3 \bmod 6$). However, U_j will not know the identity of the 1-bits. So we have proved the following.

Theorem 2. *After each round of the protocol, user U_i learns its own bit, and also the number of the 1-bits with coefficient 4 in formula (2), and nothing else, for $i = 1, 2, \ldots, n$.*

3   Network 2g

Unlike in Network 1g, now we allow each user U_i to communicate to any number of other users. The communication goes by packets, in one time slot user U_i will send one packet to one user U_j. Consequently, the arrival of the information to a user will not be so equable.

Suppose, that the number of users is n, and let k denote the smallest integer such that $\binom{2k}{k} \geq n$. Clearly $k < 2 \log n$. We will use packets of length $n + 2k$ for communication, as an example, we will discuss the altering of the packet length later. Every user has a $2k$ bit long unique address, to be defined in the next paragraph. Note that, as we have a routerless all-optical network, the address cannot be used for directing the packets: when a user sends a packet, all other users will get a packet (but not the same one).

Suppose further that every user is capable to send one-one packet to the $t = 2k$ fibers simultaneously and we also assume that there are no collisions. The $t = 2k$ element set of fibers will be denoted by $F = \{f_1, f_2, \ldots, f_t\}$.

We assign a k element subset of F for each user uniquely, and the address of a user will be the characteristic vector of this subset. The decoders contain one optical XOR gate only (see [9] or [7] on constructions of optical XOR gates). For each user her decoder XORs the data of the k fibers assigned to her and forward the result to her.

The algorithm of a sender U_i is the following. Suppose we are going to send a message to user U_j having address x. Taking x we can calculate that U_j is connected to subset $A \subseteq F$, where $|A| = k$. Now we compose a packet that will start with the $2k$ bits of address x then followed by n "useful" bits (these will be the next n data bits we are going to send to U_j). Let y denote the $n + 2k$ bit long packet we made.

Now we choose a fiber $f \in A$ and for each $e \in F \setminus \{f\}$ we construct a random packet p_e of length $n + 2k$. Then we construct p_f as

$$p_f = y \oplus \bigoplus_{e \in A \setminus \{f\}} p_e.$$

After this procedure we may send the $2k$ p_e packets at the same time to the corresponding fibers.

The algorithm of the receiving users is quite simple. They know their own address of length $2k$. When (an XOR'd) packet arrives they check whether the first $2k$ bits are equal to their address, and if this is the case then they read the following n bits and consider them as data. So the receiver equipment of a user do not have to be more clever than an Ethernet card.

As the consequence of the algorithm of the senders, it is easy to see that the intended recipient of a packet will recognize the packet as hers, and also receives it.

What happens with the other users? We claim that

- they all will hear absolutely random noise, and
- they will recognize that the packet is not for them except with a negligible probability.

To see the first statement we prove slightly more.

Lemma 1. *For a subset B of fiber set F, vector-sum*

$$\bigoplus_{e \in B} p_e$$

is exactly y if $B = A$ and consists of $n+2k$ independent random bits if $\emptyset \neq B \neq A$.

Proof. The first part of this statement is an obvious consequence of the procedure of the senders. For the second part first observe, that if the selected fiber f is not an element of B then the sum above is a MOD 2 sum of $|B|$ independent random sequences. The next thing to understand is that p_f itself consists of independent random bits because it is a MOD 2 sum of a fixed sequence y and a random sequence $\bigoplus_{e \in A \setminus \{f\}} p_e$. Now suppose $f \in B$ but $B \neq A$. If $B \subset A$ then for $C = A \setminus B$ we have $C \neq \emptyset$ and $f \notin C$, so $\bigoplus_{e \in C} p_e$ is a random sequence, and so is $\bigoplus_{e \in B} p_e = y \oplus \bigoplus_{e \in C} p_e$. If there is an $e' \in B \setminus A$ then independently of what $q = \bigoplus_{e \in B \setminus \{e'\}} p_e$ is, $q \oplus p_{e'}$ consists of independent random bits.

For the defective recognition, the probability that a particular user other than U_j thinks that the packet is for her is at most

$$\frac{1}{2^{2k}} < \frac{1}{n}$$

Summarizing, the aimed user U_j gets the right data, all other users do not get any information neither about the data nor about the addressee. But with a very small probability a receiving user may think that the packet is for her, she also gets noise as data and none useful information. This probability is small enough if n is large, as we assumed here.

The overhead of this protocol is $\frac{O(\log n)}{n}$ ($2k$ address bits contrasted with the n data bits), of course, if we increase the length of the packets then this value can be made arbitrary small.

If n cannot be assumed so large, then the probability, that a user not intended by U_i falsely recognize the packet as her own, is not small enough. To fix this problem, we assign also an extended address for each user that is l copies of her original address, in succession. Now the packet length will be $2lk + n$ and starts with the extended address of the intended user and, of course, the receiving users look for their extended address at the beginning of each packet. Now, the above error probability can be decreased to $\frac{1}{2^{2lk}}$. On the other hand we increased the overhead.

It is worth to see some particular numbers that fit the practice. First of all, the packet length should not be too large, we may use packets of length 1024. Let $t = 2k = 32$. With this number we can choose the number of users $n = 500$ million. Further, we choose to repeat the address bits twice, so the length of an extended address is 64 and a packet contains 960 useful data bits. The probability, that a user not intended by U_i falsely recognize the packet as her own, is

$$\frac{1}{2^{64}} < \frac{1}{1000000000000000000}$$

This probability is so small that the expected time for a user to get a "bad frame" is more than 100 years.

The drawback of this method that some triple of users together can decipher the message for U_j, but not an arbitrary triple and not any two of them if k is odd. If we increase the number of fibers from $2k = O(\log n)$ to $k^2 = O(\log^2 n)$ then we can assign a set of fibers to each user in such a way, that no coalition of $k - 1$ users can decipher of a message sent to a user not in the coalition. For this assignment we use the momentum curve in the k-dimensional space over field $GF(2^k)$. It has 2^k points, any k of them are linearly independent over $GF(2^k)$. Each point can be represented as a binary vector of length k^2 in such a way that any k of them are linearly independent over $GF(2)$. These vectors can be used as the characteristic vectors of the "per user" fiber sets.

Acknowledgments

The authors acknowledge the partial support of the OTKA T 046234 grant.

References

1. Y. Azar, E. Cohen, A. Fiat, H. Kaplan, and H. Racke. Optimal oblivious routing in polynomial time. In *Proceedings of the thirty-fifth ACM symposium on Theory of computing*, pages 383–388. ACM Press, 2003.
2. S. Chatterjee and S. Pawlowski. All optical networks. *Communications of the ACM*, 42(6):74–83, 1999.
3. C. Dovrolis, D. Stiliadis, and P. Ramanathan. Proportional differentiated services: Delay differentiation and packet scheduling. In *SIGCOMM*, pages 109–120, 1999.
4. V. Grolmusz. Defying dimensions modulo 6. *ECCC Report TR03-058* http://eccc.uni-trier.de/eccc-reports/2003/TR03-058.
5. V. Grolmusz. Computing elementary symmetric polynomials with a subpolynomial number of multiplications. *SIAM Journal on Computing*, 32(6):1475–1487, 2003.
6. V. Grolmusz and Z. Király. Secure routerless routing *Proc. ACM SIGCOMM Workshop FDNA 2004, Portland, Oregon*, pages 21-27,
7. K. Hall and K. A. Rauschenbach. All-optical bit pattern generation and matching. *Electron. Lett.*, 32:1214, 1996.
8. P. Hawkes and G. Rose. Rewriting variables: the complexity of fast algebraic attacks on stream ciphers. Technical report, eprint.iacr.org/2004/081/, 2004.

9. M. Jinno and T. Matsumoto. Nonlinear Sagnac interferometer switch and its applications. *IEEE J. Quantum Electron.*, 28:875, 1992.
10. S. A. Plotkin. Competitive routing of virtual circuits in ATM networks. *IEEE Journal of Selected Areas in Communications*, 13(6):1128–1136, 1995.
11. A. Poustie, R. J. Manning, A. E. Kelly, and K. J. Blow. All-optical binary counter. *Optics Express*, 6:69–74, 2000.

A Verified Distance Vector Routing Protocol for Protection of Internet Infrastructure

Liwen He

Security Research Centre, BT Group CTO Office,
PP4, ADMIN 2, Antares Building,
BT Adastral Park, Ipswich, IP5 3RE, United Kingdom
Liwen.he@bt.com

Abstract. The routing protocols, one of the fundamental components in the operation of the Internet, lack basic efficient and effective security schemes to prevent internal and external attacks. Existing cryptographic techniques can protect IP routing infrastructure from external attack at the expense of performance but is difficult to protect a network from internal attacks. This paper describes a novel computational method for verifying routing messages in distance vector routing protocols that can effective and efficient to protect routing protocols from internal attacks such as mis-configuration or compromise.

1 Introduction

The modern Internet is gradually becoming a single, converged communication infrastructure, which delivers mission-critical applications and services for business. Routing protocols, the basis of today's Internet infrastructure, lack even the most basic mechanisms of authentication and data integrity. As dependence on the Internet infrastructure grows, its security becomes a major concern, especially in the light of the recent development of attack techniques and the increased attempts to compromise the Internet infrastructure. The injection of false routing information can easily degrade network performance, or even cause denial of service for a large number of hosts and networks over a long period of time. Such failures can affect wide area of the Internet and thousands or millions of users. Attacks on routing protocols are more damaging.

Currently, in distance vector routing protocols, the security of one router depends on the saintliness of other routers. Thousands of routers running distance vector routing protocols are operating in Internet. Mis-configuration or malicious attacks to these routers are inevitable. There are serious consequences associated with this attack such as *deception* (a legitimate router receiving false data and being convinced that the data is true.), *disclosure* (an unauthorised party gains access to routing information.), *disruption* (the function of legitimate routers is interrupted or prevented resulting in *congestion*, *delay*, or *blackhole*.). This paper focuses on describing a novel method to verify routing information and detect anomaly for distance vector routing protocols. This rest of this paper is organised as follows. Section 2 reviews the related work on

P. Lorenz and P. Dini (Eds.): ICN 2005, LNCS 3421, pp. 463–470, 2005.
© Springer-Verlag Berlin Heidelberg 2005

security of routing infrastructure. Section 3 introduces a novel computational method of routing message verification for distance vector routing protocol using numerical examples. This paper is terminated by the conclusions and further works in Section 4.

2   Related Work on Routing Security

Routing protocols are highly distributed and extremely dynamic, which makes security issues more challenging. Currently, two distinct approaches on securing distance vector routing protocols are preventive cryptography based techniques including Public Key Infrastructure and digital signatures and reactive Intrusion Detection techniques.

Kumar considers the general approaches to securing routing protocols [1]. He creates neighbour-to-neighbour digital signatures of routing updates, and adds sequence numbers, timestamps, acknowledgements and retransmissions to the updates so that modifications or replay of routing updates for distance-vector algorithm is prevented. Kumar and Crowcroft then describe the implementation of secret and public key authentication to encrypt neighbour-to-neighbour updates [2]. Murphy provides a scheme that requires the validation of a number of nested signatures equal to the number of routers in the route path. However, as the network size grows, this method will suffer in both update size and verification computation time [3]. Smith, Murthy and Garcia-Luna-Aceves develop security schemes with predecessor information which is called path finding algorithm (PFA) to protect the second-to-last information included in the *AS_PATH* attributes by digital signatures. This method is more efficient than Murphy's approach [3] by signing only the component link information in the form of predecessors, and performing a path traversal to validate full paths [4, 5]. Kent, *et. al.* present a novel scalable Secure-BGP (S-BGP) architecture to verify the authenticity and authorisation of BGP control traffic. The key component in the S-BGP overhead is the certificates that are used to validate the binding between an organisation and an AS, between an AS and a router, and between an organisation and a set of IP address prefixes. They also describe the experimental results on a testbed of examining interoperability, the efficacy of the S-BGP countermeasures in securing BGP routing protocol, the impact on BGP performance, and then evaluate the feasibility of real implementation in the Internet [6]. Murphy, *et. al.* consider using cryptographic protection of both OSPF and BGP. The implementation of this scheme is explained to enhance the security of routing protocols [7].

Cheung and Levitt design new protocols for detecting and responding to misbehaving routers to prevent against Denial of Service (DoS) [8]. Ramakrishna and Maarof describe various intrusion detection tools to defend DoS attacks and active sniffing in routing protocols [9]. Mittal and Vigna [10] describe an intrusion detection technique that uses information about both the network topology and the positioning of sensors to determine what can be considered malicious in a particular place of the network. They initialise a novel algorithm that automatically generates the appropriate sensor signatures and reports the deployment of this approach to an intra-domain distance-vector protocol and its experimental results on a test network. Another interesting method of detecting anomalies in the network is interactive visualisation of BGP data. Teoh, *et. al.* demonstrate the use of visualisation of BGP data to characterise routing pattern, understand potential weaknesses in connectivity which are dangerous to Internet operation, and detect and explain actual network anomaly scenarios [11].

In general, cryptographic measures are very effective in preventing external attacks but are still vulnerable to internal attacks but are computationally expensive and so carry performance penalties. Also, routing intrusion detection requires complex algorithms for the placement of sensors, and for analysing and differentiating abnormal behaviours from normal behaviours. This is not an easy task and its effectiveness is yet to be justified. Currently, distance vector routing protocols enable peers to transmit route announcements over authenticated channels using public key cryptography techniques, so that adversaries cannot impersonate the legitimate sender of a route announcements. So-called "man-in-the-middle" and similar types of attack are prevented. This effectively verifies which router is speaking but not what it says, and leaves the routing infrastructure extremely vulnerable to both accidental misconfigurations and deliberate attacks from legitimate routers. More recently, Wan, *et. al.* develop a new Secure-RIP scheme in which a router can confirm the consistency of an advertised route with those nodes that have propagated that route [12]. A reputation-based framework is designed to determine how many nodes need to be consulted. The experimental simulation shows that an honest router can identify inconsistent routing information in a network with many badly-behaving routers. There are some problems associated with this approach. First, it is based on assumptions, which might be difficult to be realised in practical routing operations. Second, it requires too much CPU computation and consumes too much bandwidth for transporting information, and may thus have difficulty if deployed in the Internet routing environment which is of high-speed and extremely dynamic. Therefore, these current schemes are either simple to compute but vulnerable to attack or very reliable against attack but with unaffordable computational cost. These call for our research attentions for effective and efficient solutions to secure distance vector routing protocol.

3   A Novel Method for Securing Distance Vector Routing Protocol

In distance vector routing protocols, each router maintains a routing table indexed by, and containing one entry for, each router in the network. The entry contains two parts: the preferred outgoing link to use for that destination, and an estimate of the *cost* to that destination [13]. The cost is a routing metric that consists of a value computed by the distance vector routing algorithm to determine which route is superior. For example, cost can be time delay in milliseconds (msec), hop count, bandwidth, path cost, load, reliability, total number of packets queued along the path or something similar. The general distance vector routing protocol, with a routing metric of delay, used as examples to illustrate the novel mechanism, are called Verified-Distance Vector Routing Protocol (V-DVRP) described as follow with the help of figure 1, which is a schematic illustration of a IP network from [13].

1. When a router receives the update message from its neighbours, it does not update its routing table immediately. Instead, it analyses the new message by comparing the proposed route with the correspondence entry in its current routing table.

2. As an example, using delay as a metric, a router J, keeps on receiving routing information updates from its neighbouring router A, I, H and K. In router J, the two consecutive (previous and current) routing update messages from all neighbouring routers are stored.

3. Each time router J receives an update, say from router A, it runs the routing algorithm as normal, and checks the outcome of this update with its current routing table to see if it indicates that a change should be made to the routing table.

4. If the outcome of the update indicates that there is not likely to be any impact on the router's decisions (e.g. J's next hops to all destinations in the routing table will not be modified), router J takes no action (and the method ends).

5. If however the outcome of the update indicates that there is likely to be an impact on the router's decisions (for example, it is found that router D can now reach much more quickly via router A than via router H previously set), router J then timestamps and sends two diagnostic packets with different sequence numbers and random amounts of padding to the router D via two different routes. One route is via the "new" router (e.g. router A), another one is via the current next router for the affected destination node H in the current routing table of router J. Upon receipt of the diagnostic packets the destination router (e.g. router D) just sends two reply packets back as fast as it can (possibly after timestamping the packets in some way if appropriate depending on the exact nature of the diagnostic packets). By noting the time of receipt of these two reply packets corresponding to the two diagnostic packets and comparing these receipt times with the times of transmission of the original diagnostic packets, router J calculates which route is shortest according to the packet travelling time between the two routes.

6. If it is determined that the new route advertised by router A has a packet travelling time which is less than that of the route currently suggested by the routing table, it is reasonable to assume that the information contained in the recently received update (e.g. from router A) is correct, and router J therefore updates its routing table as usual and renews its set of routing messages sent by all neighbours (i.e. by deleting the current "previous update" and moving the "current update" to the "previous update", and moving the newly received update into the "current update").

7. If, however, it is determined that the new route advertised by router A has a packet travelling time which is greater than that of the route currently suggested by the routing table, it is reasonable to be suspicious of the new routing information from router A and therefore router J does not update the routing table. Furthermore the newly received update is simply discarded and no change is made to the value stored in the "previous update" and "current update" field. If it is determined that the anomaly keeps on happening (e.g. if more than a predetermined number of consecutive suspicious and therefore discarded updates are received), router J sends an alert to its system administrator before ending the method.

In the above manner, using this simple verification, bogus routing message from mis-configured or malicious routers can be filtered and terminated by its honest neighbouring routers. (Note: if A advertises bogus routing information with a greater delay, A will be automatically isolated from the network.) The above process is illustrated by the following numerical example referring to Figure 1 and Tables 1 and 2.

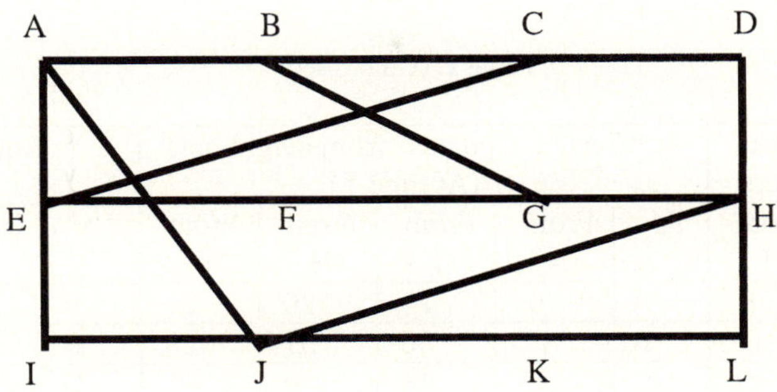

Fig. 1. A network topology (from [13])

Table 1 shows the measured times from router J to its 4 neighbours, the routing messages from A, I, H, K recorded in router J at time T and the routing table generated by router J from these two sets of information. Generally, if all of the routing information is normal, router J computes a new routing table. For illustration, consider how J computes its new route to router D. J knows that it can get to A in 8 msec, and A claims to be able to reach D in 40 msec, so J knows it can reach D via A on a delay of (8+40) 48 msec. Similarly, J knows it can reach D via I on a delay of (10+27) 37 msec, J knows it can reach D via H on a delay of (12+8) 20 msec, J knows it can reach D via K on a delay of (6+24) 30 msec. The minimal delay is 20. So in J's routing table the route to D specifies a delay of 20 and the next hop to D is router H.

Table 2 shows the routing messages from A, I, H, K recorded in J at time T+1 together with the new routing table that would result from the new updates either, in the conventional case or according to this novel method, once the relevant aspect or aspects of the newly received updates have been verified. It is obvious that A claims a decrease of delay to D from its previously advertised delay of 40 msec down to 10 msec. When J routinely calculates the delay to D via A, it determines that the new delay is only (8+10=) 18 msec via A. Clearly, the new delay of 18 msec is smaller than J's current minimal delay value 20 msec and the quickest route to D at time T+1 is via A with a delay of 18 msec (the previous quickest route is 20 msec via H). J then starts to verify the routing message from A. J timestamps and sends two similar diagnostic packets P and Q but with different sequence numbers to D via two different routes (P via A, Q via H). Upon receipt of P and Q, D may timestamp the packets and reply back with two response packets as fast as it can.

If the two response packets are both received back prior to a timeout occurring, J determines which diagnostic packet makes it to router D first. Provided the packet P has the least delay, J assumes that the recent update from neighbour A is correct and updates its table according to Table 2 as well as its "previous update" and "current update" fields. Otherwise, the new routing update message from router A (i.e. at time T+1) is deemed to be suspect and this event is recorded by router J in its system log file. J then takes no action to update its routing message from A or its routing table. If router A keeps sending suspect routing messages, J sends an alert to the system administrator.

Table 1. (Unit: msec)

Times to neighbours		"Current Updates" from neighbours (At time T)					Routing table (For J)		
To	From J	To	From A	From I	From H	From K	To	Delay	First Hop
A	8	A	0	24	20	21	A	8	A
I	10	B	12	36	31	28	B	20	A
H	12	C	25	18	19	36	C	28	I
K	6	**D**	**40**	27	8	24	**D**	**20**	H
		E	14	7	30	22	E	17	I
		F	23	20	19	40	F	30	I
		G	18	31	6	31	G	18	H
		H	17	20	0	19	H	12	H
		I	21	0	14	22	I	10	I
		J	9	11	7	10	J	0	-
		K	24	22	22	0	K	6	K
		L	29	33	9	9	L	15	K

Table 2. (Unit: msec)

Times to neighbours		"Current Updates" from neighbours (At Time T+1)					Routing table (For J)		
To	From J	To	From A	From I	From H	From K	To	Delay	First Hop
A	8	A	0	24	20	21	A	8	A
I	10	B	12	36	31	28	B	20	A
H	12	C	25	18	19	36	C	28	I
K	6	**D**	**10**	27	8	24	**D**	**18**	D
		E	14	7	30	22	E	17	I
		F	23	20	19	40	F	30	I
		G	18	31	6	31	G	18	H
		H	17	20	0	19	H	12	H
		I	21	0	14	22	I	10	I
		J	9	11	7	10	J	0	-
		K	24	22	22	0	K	6	K
		L	29	33	9	9	L	15	K

The successful transmission and return of the two diagnostic packets is critical to the success of this scheme. So the sequence number of one packet is randomly generated as Y with a significant bit length of 16; the sequence number for the other packet is set to (Y+1). This helps to prevent sequence number prediction attacks. Also, data padding is applied to the diagnostic packets, whereby a randomly selected amount of ostensibly useless bits of information is added to the packets, which makes it more difficult for network sniffers to analyse.

4   Conclusions and Further Works

The paper reports a novel distributed computational method for the verification of routing information and detection of routing anomaly in distance vector routing protocols. This scheme has a few unique features.

- It is simple and easily deployed into the present Internet with minor modifications of routing software and without any additional hardware, cryptographic key distribution or a central database or complicated IDS infrastructure.
- It is cost-effective with a very limited amount of management overhead being generated and small amount of bandwidth and CPU resource being consumed. The router starts to verify the advertised routing messages only when this advertisement will cause a change in its routing table.
- It is a distributed computing method that each router independently makes decisions on the information obtained locally, so it can scalable.
- It is straightforward: the faulty router that broadcasts bogus routing information can be accurately pinpointed and recorded by its normal neighbours.
- It is reliable and fault-tolerant. The router attempts to send the diagnostic packets up to a pre-defined number of times if one or both of the diagnostic packets are not received before the timeout occurs. Also, only when a router keeps on sending suspect routing updates, will its bad behaviour be notified to the system administrator. This reduces the number of false alarms. Also, it is unlikely to enable new attacks with the help of the large random sequence number and the data padding for diagnostic packet.
- It is very flexible, as the verification metric for diagnostic packets is changeable in accordance with routing metric such as delay, hop count, bandwidth, reliability, *etc*, used by the underlying routing protocols.

Further work is proposed as follows. First, this method is to be demonstrated in a Java-based IP network modelling and simulation platform like SSFnet toolkit [14]. Also, it will be tested in a real IP network testbed with mixed router types and several PCs running open source routing software. Second, the author is researching on how to implement this mechanism into IPv6, and actively looking for collaboration with members of the IPv6 research community who are considering routing security issues.

Currently, there is no industry-wide standard for detecting routing violation in the current Internet routing infrastructure. But it is important that network carriers, ISPs, network and security vendors work together to create a more secure IP network infrastructure. The new scheme discussed above does not claim to provide complete protection of Internet routing infrastructure. Instead, it raises the barrier for attacks, and is a promising complementary solution to cryptography and intrusion detection technologies for securing IP routing infrastructure.

References

1. B. Kumar: "Integration of Security in Network Routing Protocols", *ACM SIGSAC Review*, **11**(2): PP. 18-25, Spring, 1993.
2. B. Kumar and J. Crowcroft: "Integrating security in inter-domain routing protocols", *Computer Communications Review*, 23(5), October 1993.
3. B. Smith, J. Garcia-Luna-Aceves: "Securiing the border gateway routing protocol", *IEEE GLOBECOM 1996. Communications: The Key to Global Prosperity. GLOBAL INTERNET'96*. Conference Record, 18-22 Nov. 1996, IEEE pp 81-5.
4. B. Smith, S. Murthy, J. Garcia-Luna-Aceves: "Securing distance-vector routing protocols", *Proceedings of SNDSS '97: Internet Society 1997 Symposium on Network and Distributed System Security*, 10-11 Feb. 1997, IEEE Computer. Soc. Press, pp 85-92.
5. B. Smith, J. Garcia-Luna-Aceves: "Efficient security mechanisms for the border gateway routing protocol", *Computer Communications*, vol.21, no.3, 1998, pp 203-10.
6. S. Kent, C. Lynn, J. Mikkelson, and K. Seo: "Secure Border Gateway Protocol (S-BGP)", *Proceedings of ISOC Network and Distributed System Security Symposium, Internet Society*, Reston, VA, Feburary 2000.
7. S. Murphy, O. Gudmundsson, R. Mundy and B. Wellington: Retrofitting security into Internet infrastructure protocols, *Proceedings DARPA Information Survivability Conference and Exposition. DISCEX'00*, 25-27 Jan. 2000, IEEE Comput. Soc. pp 3-17 vol.1.
8. S. Cheung and K. Levitt: "Protecting Routing infrastructures from Denial of Service Using Cooperative: Intrusion Detection", *Proceedings of the New Security Paradigms Workshop (NSPW-97)*, (New York), pp. 94-106, ACM, Sept. 23-26 1997.
9. P. Ramakrishna, M. A. Maarof: "Detection and prevention of active sniffing on routing protocol", *2002 Student Conference on Research and Development. SCOReD2002. Proceedings. Global Research and Development in Electrical and Electronics Engineering*, 16-17 July 2002, IEEE pp 498-501.
10. V. Mittal and G. Vigna: "Sensor-based intrusion detection for intra-domain distance-vector routing", *Proceedings of the 9th ACM conference on Computer and communications security*, PP. 127 - 137, Washington, DC, USA. 2002.
11. S. Teoh, K. Ma, S. Wu, X. Zhao: "Case study: Interactive visualization for Internet security", *VIS2002. IEEE Visualization 2002. Proceedings, 27 Oct.-1 Nov. 2002*, IEEE pp 505-8.
12. T. Wan, E. Kranakis, P. C. Oorschot: "S-RIP: A Secure Distance Vector Routing Protocol", Applied Cryptography and Network Security, 2nd International Conference, China, 2004. *Lecture Notes in Computer Science*, pp103-119.
13. A. S. Tanenbaum: *Computer Networks*, third edition, Prentice-Hall Inc. 1996.
14. SSFnet IP network simulation tool: http://www.ssfnet.org/internetPage.html#ssfnet

Replay Attacks in Mobile Wireless Ad Hoc Networks: Protecting the OLSR Protocol

Eli Winjum[1], Anne Marie Hegland[1], Øivind Kure[2], and Pål Spilling[1]

[1] UniK - University Graduate Center at Kjeller,
P.O. Box 70, N-2027 Kjeller, Norway
[2] Q2S, NTNU, Trondheim, Norway
`ewi@unik.no`

Abstract. This paper discusses the possible effect of replay attacks on the Optimized Link State Routing protocol. It is investigated to which extent an adversary is able to achieve network control by replaying old protocol messages. Even without particular countermeasures the adversary has to overcome considerable difficulties in order to succeed. We propose a simple protection scheme based on the message sequence number and the procedures already embedded in the routing protocol. Our scheme is an alternative to replay protection based on clocks and timestamp exchange protocols.

1 Introduction

In a replay attack an adversary records a legitimate message in order to replay the message later on. The receivers of the replayed message are supposed to handle the message as if it were a fresh one. Replay attacks may be categorized in accordance with their purpose. If the intention is to burden the receivers with exhaustive computation, recorded messages may be poured into the network. Since any false messages may be utilized, we do not focus on jamming attacks. We focus on attacks, which intention is to manipulate information by inserting old messages into the network in a planned manner. For example, by replaying routing messages which announce routes that do not longer exist, an attacker may become able to control the traffic flow. The attacker may then perform remote jamming by rerouting large quantities of packets to a selected victim node. The attacker may also reroute traffic in order to ease eavesdropping. Attacks on routing information are described in [6].

Authentication and integrity services aim to detect insertion of false messages and unauthorized modification of data in transfer. A possible way of tricking an authentication service is to replay recorded messages, which are already signed by a legitimate message originator. In order to detect such attacks, specific replay protection is needed in addition to authentication and integrity services. Mechanisms which may enable a verification of the freshness of incoming messages are often based on *timestamps* or *sequence numbers*. The received information is processed only if the timestamp/sequence number is within a specified interval.

P. Lorenz and P. Dini (Eds.): ICN 2005, LNCS 3421, pp. 471–479, 2005.

Replay attacks are described and discussed in [2] and [7]. Reference [1] and [5] discuss replay attacks on the Optimized Link State Routing (OLSR) protocol and propose protection schemes based on timestamps. We propose a simpler protection scheme based on existing sequence numbers. The rest of the paper is organized as follows: Section 2 gives an overview of the OLSR protocol [3]. Section 3 is a brief description of the Timestamp Exchange Protocol [1]. We present and evaluate our proposals in section 4. In section 5 we compare the different schemes with regard to overhead. Conclusions are drawn in section 6.

2   The Optimized Link State Routing (OLSR) Protocol

The OLSR protocol is a proactive routing protocol proposed for mobile wireless ad hoc networks and is an optimization over classical link state protocols. Each node selects Multipoint Relays (MPRs) from its set of 1-hop neighbors such that all 2-hop neighbors can be reached through at least one of them. Since only MPRs retransmit protocol messages, flooding of protocol traffic is minimized. Four message types are specified: Nodes broadcast their links to 1-hop neighbors and their MPR selections through *Hello* messages. *Topology Change* (TC) messages disseminate topology information throughout the network. Only nodes, which are selected MPR by some other node, generate TC messages. Routing tables are computed from the link state information exchanged through TC messages. *Multiple Interface Declaration* (MID) messages declare a list of interface addresses in case a node has more than one. *Host and Network Association* (HNA) messages declare non-OLSR interfaces. TC, MID and HNA messages are retransmitted in contrast to Hello messages. All message types utilize the same message header format shown in Fig. 1. The header stems from the message originator. A packet header is attached to the message hop by hop.

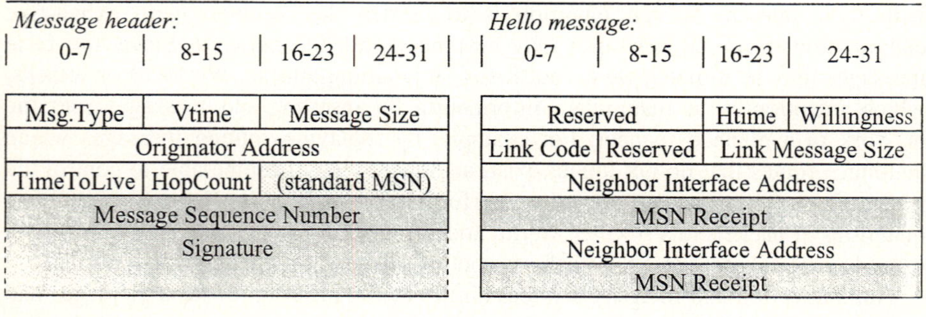

Fig. 1. OLSR formats. Proposed extensions are shaded

3   The Timestamp Exchange Protocol

In order to protect the OLSR protocol from replay attacks, the Timestamp Exchange Protocol (TEP) is proposed in [1]. TEP is based on global dissemination of local time information. Each node performs a timestamp exchange procedure with each of the other nodes by broadcasting the *Timestamp Exchange* (TE) message periodically. The message is flooded throughout the network. Each node keeps a timestamp table holding the information from the last TE message received from each of the other nodes. A TE message from node A contains its current clock and its current timestamp bounds. In this way node A advertises the intervals within which it will generate its timestamps. The message also contains at least one timestamp for each node, from which node A has received a timestamp. Upon receiving a TE message from node A, node B checks its timestamp table. If an entry of A does not exist, the new timestamp is recorded. So far node A has not delivered enough proof of freshness, but the timestamp will serve as a proof of freshness to node A when returned in the next message from node B. If an entry of node A exists, the handshake between node A and node B is completed. The entry is updated with the new timestamp.

Reference [1] also proposes a *separate* signature message type: A regular OLSR message and its timestamp are signed by the originator. The timestamp and the signature are sent in a *separate* message. This means that each regular OLSR message has a corresponding signature message containing its timestamp. In node B, a timestamp from node A is valid if it is within a time interval that node B has received from node A through TE messages. In the rest of this paper the term *TEP* refers to a scheme which comprises the utilization of separate signature messages.

4   Protection Based on Sequence Numbers

We suggest that the message sequence number (MSN) is used for replay protection. Together with the *originator address* field the MSN uniquely identifies a message. The message is then the unit to be checked for freshness, which corresponds to TEP.

According to [3] a *symmetric link* is a verified bi-directional link between two OLSR interfaces, while an *asymmetric link* is verified in only *one* direction. A *message originator* is the originator of the OLSR message. In case of a replayed message we call the node which records and replays the message an *attacker*, while the node which *initially* generated the message is still regarded the *message originator*. The *sender* may be the message originator or a forwarder.

We assume that all legitimate nodes perform the protocol operations correctly. Authentication and integrity services in OLSR are proposed in [1], [5] and [10]. We assume that message originators sign every OLSR message header. The receivers verify the signature and that the message was not tampered while in transit. Hence, any replayed OLSR message must have been correct at the time of recording.

4.1   Proposed Extensions and Modifications to the OLSR Protocol

When a limited amount of bits represents a sequence number which is incremented monotonically, the wrap-around from the maximum number to zero must occur. This means that two messages may have identical sequence numbers. Further, in case of wrap-around there has to be a rule to determine which message is the most recent. The rule recommended in [3] permits a replay when the MSN has increased with half the maximum number. Hence, if the MSN is incremented once a second, a 16 bit representation would allow a replay after 9 hours, while 32 bits would last for 68 years. In order to eliminate the wrap-around problem we propose to extend the MSN from 16 to 32 bits. We then propose that the MSN starts at zero when a particular OLSR application is associated with a particular message originator address. The MSN is then incremented monotonically during the lifetime of this association and irrespective of the device being switched off. We assume that authentication keys are renewed such that replayed messages, which reflect previous associations between an OLSR application and an IP-address, will be revealed in the authentication process.

Each node has to record one MSN pr message type pr node. We propose the following protocol modifications: Discard an incoming OLSR message if the MSN is lower than or equal to the most recent number recorded for this message type from this message originator. Else, the MSN is recorded, and the message is handled according to standard protocol rules. A MSN should not be deleted unless it is replaced by a new one. We call this scheme *MSN-Basic*.

To reveal a replayed message, the receiver must have stored a prior sequence number from the message originator. If the attacker has recorded a message which was broadcast in another part of the network or during network partition, the stored number may be too old. When the replay receiver compares the incoming MSN to the prior one, the replayed MSN may turn out to be the most recent. If the replay receiver has recently joined the network, it may not have received a previous MSN at all. Hence, a scheme based on solely a simple sequence number check has shortcomings in a network where nodes are mobile and join and leave the network dynamically.

In the next section we identify the conditions under which a previous MSN may either be missing or not applicable. To deal with this uncertainty we take advantage of the following: A two-way Hello message exchange has always taken place before a node utilizes the received information in protocol operations. We suggest that nodes have to sign for the reception of the previous Hello message by returning a receipt each time they advertise an *asymmetric* link. We propose the following protocol modifications: Attach the MSN of the most recent Hello message received from the node at the opposite end of the link to each advertised *asymmetric* link. Discard an incoming Hello message if the MSN receipt is deemed too old. A sliding window may be utilized. Consequently, we propose to extend the Hello message format with a MSN receipt field. We call this scheme *MSN-Receipt*. MSN-Receipt is meant to be implemented in addition to MSN-Basic.

All proposed extensions are shown in Fig. 1. The MSN extension implies a new version of the message header. Both the TEP scheme and the MSN-based schemes advise to discard a message, which is deemed invalid. This means that nodes, which do not take part in the protection process, are not supposed to operate within the network. Hence, backward compatibility is not an issue within a network.

4.2   Evaluation

An attacker has to repeat the replayed information constantly in order to avoid that the false information expires according to local time settings or is cancelled out by fresh messages from the message originator.

If the receiver of a replayed message also receives fresh messages from the message originator, MSN-Basic would enable the receiver to verify incoming MSNs and hinder the processing of replayed messages unless the replay was fresher than the most recent genuine message received. Then the replayed information would have been true information, and the node would eventually receive the original message as well. We now evaluate the case when the receiver of a replayed message *does not* receive fresh messages from the message originator, which means that a previous MSN may either be missing or not applicable:

Hello messages are not retransmitted. Therefore, receivers that might be tricked by a replay must be outside the transmission range of the message originator. If a receiver is not listed in the list of links, the message is the first one in the link establishing procedure between this particular pair of nodes, as shown in Fig. 2, step 1. In this case MSN-Basic would not enable node B to verify the incoming MSN. Also MSN-Receipt would be of no help: Since node A does not advertise a link to node B, node A has not recently received a Hello message from node B and consequently has nothing to sign for. If the message were a replay, MSN-Basic would permit node B to record an *asymmetric* link to the message originator node A. As long as the link stays asymmetric in node B's repositories, node B would not utilize the false information in protocol operation. Hence, the false information would have been harmless.

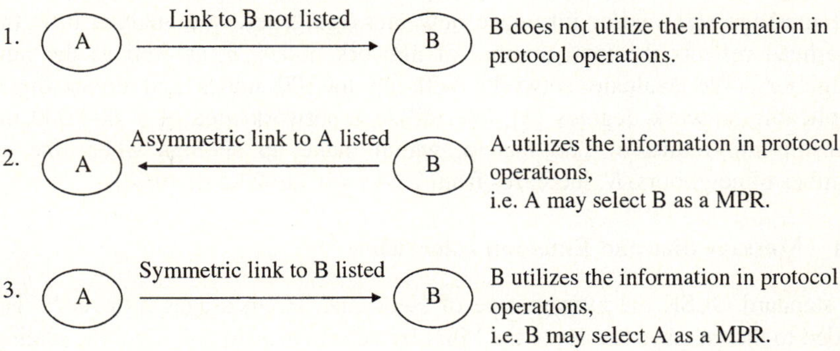

Fig. 2. Link establishing procedure by the exchange of Hello messages

If the receiver is listed as an *asymmetric* link, as shown in step 2, MSN-Basic might not be sufficient to verify incoming MSNs. However, in this case MSN-Receipt would help: Since node B advertises an asymmetric link to node A, node B must have received a recent Hello message from node A. Hence, node B would have been able to return a proper MSN receipt. Even though node A has no previous recorded MSN from node B, node A can easily check if the incoming message is fresh and thus reveal a replay. In this case the replayed information would have been destructive: Node A would record a *symmetric* link to node B and might select node B as a MPR.

If the receiver is listed as a *symmetric* link, as shown in step 3, the receiver must have received a recent Hello message from the message originator. If the message were a replay, the replay could not be fresher than the last message from the most recent neighbor relationship between the two nodes. Therefore, node B must have received the *original* message from node A. Thus MSN-Basic would enable node B to reveal the attack. A Hello message may also advertise links as lost. With regard to MSN-Basic, the case is equivalent to advertising symmetric links.

TC, MID and HNA message are retransmitted. Therefore, receivers that might be tricked by a replay must be situated in another partition than the message originator. If the replay is a TC message, the receiver will record the announced links. In order to compute a route to the advertised destinations, the receiver needs a path to the message originator. This path is missing since the message originator is outside the partition. Also, if the replay is a MID or HNA message, the receiver will record false information, but the information will never be utilized as long as the message originator stays in another partition. This means that a replay attack against receivers situated in another partition than the message originator is harmless with regard to the network information and serves solely as a jamming attack.

5   Overhead Analysis

We assume without further investigation that TEP and MSN-Receipt are equivalent regarding the ability to protect the routing information from replay attacks. MSN-Basic provides a lower level of protection.

OLSR overhead is analyzed in [4] and [9]. The replay protection schemes increase the message size and/or introduce new message types. We analyze how this extra overhead reflects the total number of network nodes, n, as well as the number of neighbors. We evaluate networks with up to 100 nodes and cover the span of applicable network degrees [8]. We utilize a network area of 1 000 000 m^2 and a transmission radius of 280 meters, which make up settings where the expected number of neighbors, N, increases from 2.5 ($n =10$) to 24.5 ($n =100$).

5.1   Message Size and Emission Intervals

In standard OLSR the average size of Hello and TC messages reflects N. The fields added to the message header in the MSN-based schemes do not affect the scaling factor. In the MSN-Receipt scheme the number of receipt fields reflects N, and the scaling factor is influenced accordingly[1]. In the TEP-scheme 32 bit timestamp fields are utilized. The size of the separate signature message shown in [1] is fixed, while the size of the TE message directly reflects n since all nodes are listed in each message. In the TE message we also calculate an 8 bit duration field and a signature field[2]. In all

---

[1] We apply a Hello message where half of the advertised links are asymmetric.

[2] The TE message content is listed in [1]. We assume that the TE messages are signed directly. We apply a TE message where only *one* timestamp interval is sent. Further, we assume that the timestamp exchange is completed with each of the other network nodes, so that the message contains only *one* timestamp pr node.

schemes a message header is added to each message body and a standard packet header is added to each message[3].

We utilize the standard Hello message interval of two seconds and the TC message interval of five seconds. In TEP these intervals also dictate the emission of the associated signature messages. Reference [1] does not discuss TE message emission intervals, but some important factors have to be taken into consideration. Since other OLSR messages are not processed unless the receiver already has received a valid timestamp from the message originator, the timestamp exchange between a pair of nodes has to be completed before the link establishing procedure between these nodes can start. Hence, when a new node joins the network, the node has to complete *two* procedures before a MPR can be selected and global communication enabled. With the standard intervals the link establishing procedure may last for six seconds. If TE messages are to be flooded by MPRs according to general OLSR rules, the forwarding can not start until the new node has selected a MPR. TEP has to deal with neighbor nodes selecting the new node as their MPR as soon as a symmetric link is established, while the new node has to complete the timestamp exchange with each network node before they accept its TC messages. Hence, a *global* handshake between the new node and each of the other network nodes has to be completed before the new node is fully integrated in the network. The timestamp exchange should not lead to a substantial lengthening of the time needed to join a network.

In order to cope with dynamic join and leave the timestamp exchange has to be performed rapidly. A TE message interval similar to the Hello message interval seems reasonable and is utilized in this analysis. The MSN-based schemes do not influence the number of OLSR messages generated pr second. In TEP every network node generates the new message types. Given the emission intervals set above, the number of messages generated pr second is increased by 183%. 45% of the increase is caused by the TE messages, while 55% stems from the separate signature messages. Overhead provided by the MAC layer should be especially considered when the number of messages generated pr second is increased.

5.2 Offered OLSR Traffic

Offered OLSR traffic as a function of n is calculated and shown in Fig. 3A. We assume that half of the nodes are MPRs that generate TC messages. Compared to the standard scheme, MSN-Basic leads to a minor increase in offered OLSR traffic. MSN-Receipt represents a larger increase. In TEP the growth reflects the rapid enlargement of the TE messages and the larger number of messages generated. Overhead reduction comes at the expense of increasing the emission interval. In order to reduce OLSR traffic to the same level as in MSN-Receipt, the interval has to be above one minute, which means that the timestamp exchange between a pair of neighbors may last for about 3 minutes. This is probably not tolerable, even in networks with low node mobility/speed and large transmission range.

---

[3] In all schemes we calculate a signature of 160 bits. TC messages advertise links to $N/2$ neighbors. MID and HNA messages represent a minor contribution to the overhead and are left out from the analysis.

The amount of offered OLSR traffic, which is supposed to be flooded, is shown in Fig. 3B. The MSN-based schemes lead to an increase caused by the new fields in the message header. In addition to the TC messages, both the signature messages and the TE messages are supposed to be flooded in TEP. This explains the rapid increase. If the TE message interval is set to about one minute as mentioned above, the amount of OLSR traffic meant for flooding would still be considerable. Another aspect is that a signature message is supposed to be flooded *independently* of its associated TC message. Since TC messages are not processed until the associated signature and timestamp are received and deemed valid, delay and jitter may affect the operation.

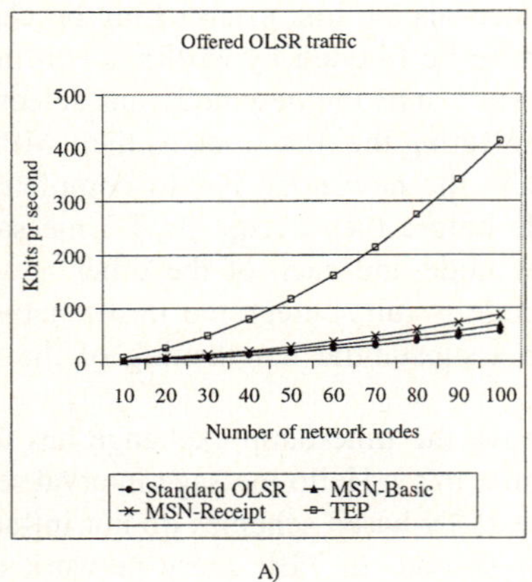

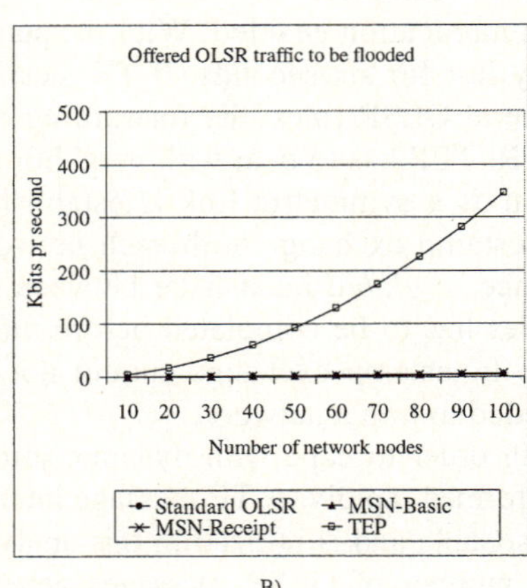

A)   B)

Fig. 3. OLSR traffic comparisons

6 Conclusions

Even without particular countermeasures, the OLSR protocol is rather robust to replay attacks, which intention is to manipulate the routing information. An attacker has to avoid that the replayed information is cancelled out by fresh messages from the message originator. Hence, in case of a Hello message, replay receivers have to be situated outside the range of the message originator, and in general the attack has to be targeted against at least one node, which corresponds to a link advertised in the replayed message. In case of other OLSR messages, the replay receivers have to be situated in another partition than the message originator. However, this attack is harmless with regard to routing information.

We have shown that a scheme based on a simple message sequence number check may be sufficient, even though nodes are mobile and join and leave the network dynamically. This scheme represents negligible overhead. We have identified a couple of scenarios where the scheme may fail. The shortcomings will be eliminated if nodes attach a receipt to each asymmetric link they announce in Hello messages. The receipt is the most recent Hello message sequence number received from the corresponding node. This scheme represents extra overhead but scales considerably better than the Timestamp Exchange Protocol, which is based on global dissemination of local time information and requires large messages to be emitted at short intervals in order to support dynamic join and leave.

References

[1] C. Adjih, T. Clausen, P. Jacquet, A. Laouiti, P. Mühlethaler, and D. Raffo. "Securing the OLSR protocol." Paper presented at the IFIP Med-Hoc-Net, Mahdia, Tunisia, 2003.

[2] T. Aura. "Strategies against Replay Attacks." Paper presented at the IEEE Computer Security Foundations Workshop, Rockport, Massachusetts, June, 1997

[3] T. Clausen, and P. Jacquet. (2003). RFC 3626: Optimized Link State Routing Protocol (OLSR). Mobile Ad Hoc Networking Working Group of the IETF.

[4] T. H. Clausen, P. Jacquet, and L. Viennot. "Investigating the Impact of Partial Topology in Proactive MANET Routing Protocols." Paper presented at the Wireless Personal Multimedia Communications, November, 2002.

[5] A. Hafslund, A. Tønnesen, R. B. Rotvik, J. Andersson, and Ø. Kure. "Secure Extension to the OLSR protocol." Paper presented at the OLSR Interop and Workshop, San Diego, California, 2004.

[6] M. Jakobsson, S. Wetzel, and B. Yener. "Stealth Attacks on Ad-Hoc Wireless Networks." Paper presented at the Vehicular Technology Conference, 2003.

[7] S. Malladi, J. Alves-Foss, and R. B. Heckendorn. "On Preventing Replay Attacks on Security Protocols." Paper presented at the International Conference on Security and Management, 2002.

[8] E. M. Royer, P. M. Melliar-Smith, and L. E. Moser. "An Analysis of the Optimum Node Density for Ad hoc Mobile Networks." Paper presented at the IEEE International Conference on Communications, Helsinki, Finland, June, 2001.

[9] L. Viennot, P. Jacquet, and T. Clausen. "Analyzing Control Traffic Overhead in Mobile Ad-hoc Network Protocols versus Mobility and Data Traffic Activity." Paper presented at the IFIP Med-Hoc-Net, Italy, 2002.

[10] E. Winjum, Ø. Kure, and P. Spilling. "Trust Metric Routing in Mobile Wireless Ad Hoc Networks." Paper presented at the World Wireless Congress, San Francisco, May, 2004.

S-Chord: Hybrid Topology
Makes Chord Efficient

Liu Hui-shan, Xu Ke, Xu Ming-wei, and Cui Yong

Department of Computer Science and Technology,
Tsinghua University, Beijing, 100084, China
{liuhs, xuke, xmw, cy}@csnet1.cs.tsinghua.edu.cn

Abstract. In the semi-structured Chord(S-Chord), every node maintains two static neighbors and $O(logN)$ dynamic neighbors, which can guarantee network connectivity and improve the success rate of fuzzy lookup. It can effectively reduce the average lookup length and improve transmission performance by adjusting dynamic neighbors and optimizing forward routes according to current traffic of network. The simulation results show that the semi-structured Chord can get higher lookup success rate with shorter average lookup length, and lower fuzzy lookup workload of network, so it can better support large-scale overlay service.

1   Introduction

Unstructured overlays, like Gnutella[1], Napster[2], organize nodes in a random graph. Each node is equal and maintains several neighbors selected at random. Unstructured overlays always use flooding or random walks[3] algorithms to find where the destination is. It's simple enough to implement but more query levels are needed to increase lookup success rate. In some cases, the query may fail even if the goal node has existed. Meanwhile, with the increase of the routing level, the number of query copies increases exponentially, which brings a large number of redundant packets in the network and reduces the utilization rate of the network. In addition, the unstructured topology structure can't support those services efficiently because the accurate keyword lookup is unavailable.

Structured overlays, like CAN[4], Chord[5] and Pastry[6], were developed to overcome the performance inefficiencies of unstructured overlays. They assign keywords to data items and organize the overlay nodes into a graph that maps each keyword to a responsible node. Every node maintains neighbors which key is closest to its. The route algorithms for structured topology can accurately lookup local routing table, and the forward path is one-way and repeatable. It is proved to adopt cache techniques to improve lookup efficiency. However, when the dynamic change takes place in nodes, the nodes must maintain the neighbors' information in order to keep specific position of the neighbors. It brings relatively heavy control workload for topology maintenance. With the increase of neighbors which each node has to maintain, the adoptability of the dynamic change of the nodes decrease.

P. Lorenz and P. Dini (Eds.): ICN 2005, LNCS 3421, pp. 480–487, 2005.

Three features that distinguish Chord from many other peer-to-peer lookup protocols are its simplicity, provable correctness, and provable performance. Chord is simple, routing a key through a sequence of $O(logN)$ other nodes toward the destination. A Chord node requires information about $O(logN)$ other nodes for efficient routing, but performance degrades gracefully when that information is out of date. This is important in practice because nodes will join and leave arbitrarily, and static $O(logN)$ state may be hard to maintain. Only one piece information per node need be correct in order for Chord to guarantee correct routing table; Chord has a simple algorithm for maintaining this information in a dynamic environment.

How to get an efficient Chord? In current researches, various techniques have been proposed to improve the performance of Chord by adopting new routing algorithms or balancing workload.

David refined the consistent hashing data structure that underlies the Chord P2P network and proposed two new load-balancing protocols to obtain provable performance guarantees[7].

Diminished Chord that allows for the creation of subgroups of nodes in the Chord was analyzed [8]. These subgroups are useful for efficiently carrying out computations or functions that do not require the involvement of all nodes.

NoN-greedy was investigated in [9], and it may reduce significantly the number of hops. NoN-greedy is implemented on top of the conventional greedy algorithms and maintains the good properties of greedy routing.

In semi-structured Chord(S-Chord), every node of S-Chord maintains two static neighbors and $\lfloor logN \rfloor$ dynamic neighbors, which can guarantee network connectivity and improve the success rate of fuzzy lookup. The S-Chord only maintains two static neighbors in a dynamic environment and chooses the routing with smaller hop-count as new dynamic neighbor according to the present traffic in the network. So, the average lookup length is reduced obviously among the whole network.

2   Topology Organization

2.1   Topology Design

Though each Chord node maintains a successor to guarantee the system scalability, the average lookup length is relatively long. Optimized Chord reduces the average lookup length by maintaining $\lfloor logN \rfloor$ neighbors. However, when the dynamic changes take place in nodes, every node must maintain all neighbors. Every node of S-Chord needs to maintain two static neighbors, a preview node and a next node, to reduce the average lookup length to $\frac{N}{2}$ in the subnet. In this way, S-Chord has kept the high scalability, connectivity and increased the stability of the network too. In order to reduce the average lookup length further, we have appended several dynamic neighbors for each node. To avoid the influence of dynamic change, we choose those neighbors at random and don't need to keep the special relation with the local node ID. We design S-Chord

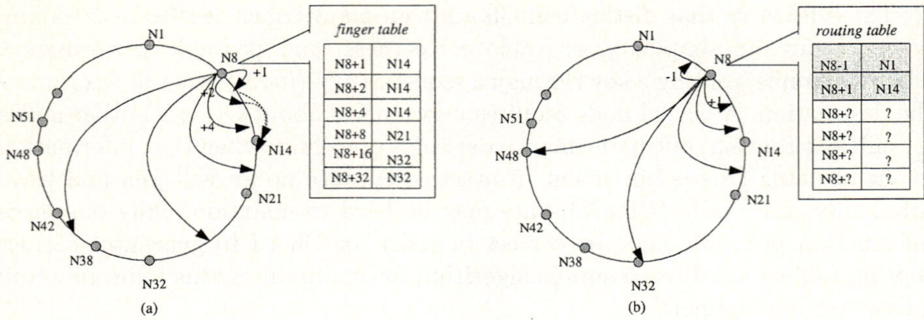

Fig. 1. (a) optimized Chord's structure and finger table; (b) S-Chord's structure and routing table

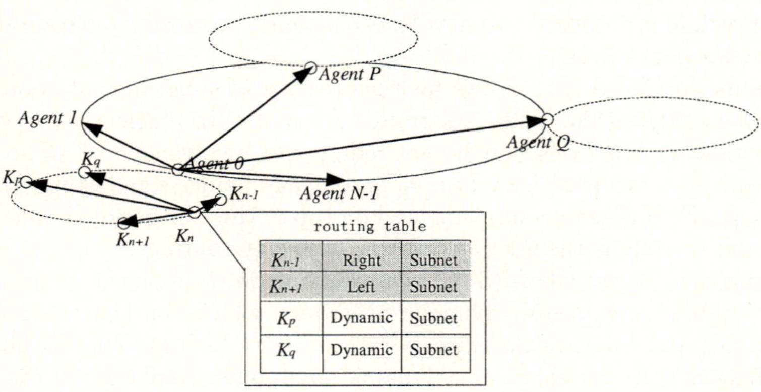

Fig. 2. hierarchical topology of S-Chord

with the structured and unstructured topological advantages in combination. (see figure 1)

To support large-scale overlay application, we recommend hierarchical topology to be adopted in S-Chord(see figure 2). The physically closed nodes construct subnet through distributed self-organizing method. Each subnet elects one node with good performance as agent, and all subnet agents form the upper network of the hierarchical structure. Both subnet and the upper sub network use semi-structured topology structure.

The routing table of ordinary node stores only the information of neighbors of local subnet, but the routing table of agent need to keep other neighbor agents' information in the upper subnet too. The identification of node is made up of two parts in S-Chord. The one is a subnet ID and the other is the local ID. The subnet ID and the node ID are all produced by applying IP address to SHA-1 [10] function, and they are unique in the subnet.

3   System Structure and Design of Components

3.1   Topology Maintenance

This module keeps topological integrality in distributed self-organizing method.

- **New Node Joining.** The applicant broadcasts the probe packet carrying node ID and IP address and use TTL to restraint the probe range. According to apperception, the applicant confirms its static neighbors and chooses $\lfloor logN \rfloor$ dynamic neighbors at random . If the applicant has not collected enough information to create connection with left and right neighbors, it will neglect this application and repeat again after some time. During this process, lots of the nodes in different subnets might send apperceptions to new applicant nodes which may choose one of the subnets randomly to achieve the joining operation.
- **Agent Bootstrap.** We adopt the procedure similar to Yallcast and YOID [11]. Suppose that 3Sons system has a related DNS domain name, and the domain name can be parsed into one or several bootstrap node IP address. New agent sends the joining application to its left and right agent neighbors, and creates the connection.
- **Node Quit.** The adjoint nodes use keep-alive packets to confirm forwarding path validity. The keep-alive packet carries information of static neighbors. So the node can keep the topological integrality when it breaks down or quit the system.

3.2   Substrate Support Routing

S-Chord adopts the distributed routing policy. It can support following three kinds of routing ways through combining the basic services in different subnet area:

- **Precise Routing:** Each node chooses the next hop node according to routing table of the local host, whose ID is the closest to destination ID. If the destination subnet ID differs from local subnet, the node will regard the local subnet agent as the middle node. if there is no closer node than local host, it is considered that the routing process fails. The pseudo-code of precise routing algorithm is shown in figure 3.Every S-Chord node needs to maintain $2 + \lfloor logN \rfloor$ routing table item, so, space and lookup complexity are all $O(2 + \lfloor logN \rfloor)$. When each node maintains only two static neighbors, the upper limit of average lookup length of S-Chord is $\frac{N}{2}$. Because we have also maintain $\lfloor logN \rfloor$ dynamic neighbors chosen at random, with the stability of the traffic of network, the average lookup length is close to optimized Chord.

- **Flooding Routing Algorithm:** In certain levels, S-Chord node duplicates and forwards the query to all neighbors of routing table. It is essential to transmit the query to more nodes of subnet in less flooding levels. We suppose that a node sends the query to x new node each time. After flooding k times, the query will be transmitted to $\sum_{i=1}^{k} x^i$ nodes. The flooding level should

---

Precise Routing Algorithm($dstSubnetID, dstInsideID$)

1.   **if**($dstSubnetID$ is not equal to local subnet ID){
2.   **while** (the node is not an agent) forward to the local agent;
3.   }
4.   **while** (agent's subnet ID is not equal to $dstSubnetID$){
5.   forward to the agent neighbor whose ID is closest to $dstSubnetID$);
6.   **if**(no closer agent neighbor) report(not existed);
7.   }
8.   **while** ($dstInsideID$ is not equal to local ID){
9.   forward to the neighbor whose ID is closest to $dstInsideID$;
10.   **if** (no closer neighbor) report(not existed);
11. }

Fig. 3. Precise routing algorithm

not be less than $log_x(N - \frac{N+1}{x})$ to ensure the query reach most nodes of the subnet. In practice, we fetch the upper limit $log_x(N)$. Every S-Chord node maintains two static neighbors and $\lfloor logN \rfloor$ dynamic neighbors. The number of query flooded through static neighbors is very limited. In order to simplify our design, we only consider the impact on flooding levels of the dynamic neighbor's count. It is L_F that we define the flooding routing levels to ensure the higher lookup success rate.

$$L_F = \lceil \frac{log_2 N}{log \lfloor log_2 N \rfloor} \rceil \tag{1}$$

– **Random Walks Routing Algorithm:** In certain levels, S-Chord node randomly duplicates and forwards the query to one or two neighbors of routing table. So, we can think approximately that a node sends the query to 1.5 new node each time. According to the similar analysis of the flooding routing algorithm, random walks routing is defined as L_R to ensure the higher lookup success rate.

$$L_R = \lceil \frac{log_2 N}{log 1.5} \rceil \tag{2}$$

The relevant experiment results is shown in figure 6.

3.3   Dynamic Neighbor Adjusting Module

Every 3Sons node checks the traffic statistics regularly, if it finds that the flow sent out from node A via itself to node B exceeds certain threshold, it will send the notice to node A and asks node A to create a direct route to node B. Node A selects the dynamic routing table item whose traffic statistic value is minimum, then replace next hop of the item with node C. A consultation is needed between node A and C to ensure the connect be correctly built. Through dynamic adjustment of the neighbors, we can reduce the average lookup length effectively.

4   Simulation and Experimental Results

We implemented a simple discrete event-based simulator which assigns each application level hop a unit delay. To reduce overhead and enable the simulation of large networks, the simulator does not model any queuing delays or packet loss on links. The simplified simulation environment was chosen for two reasons: first, it allows the simulations to scale to a large (up to 20K) number of nodes, and secondly, this evaluation is not focused on proximity routing depended on link status. Since our basic design is similar in spirit to Chord, we believe that heuristics for performing proximity-based routing can be adapted easily to S-Chord.

Experiment 1. In the subnet with 2000 nodes, we measure the influence of different dynamic neighbor's count to average lookup length on precise routing, and the result is shown in figure 4.

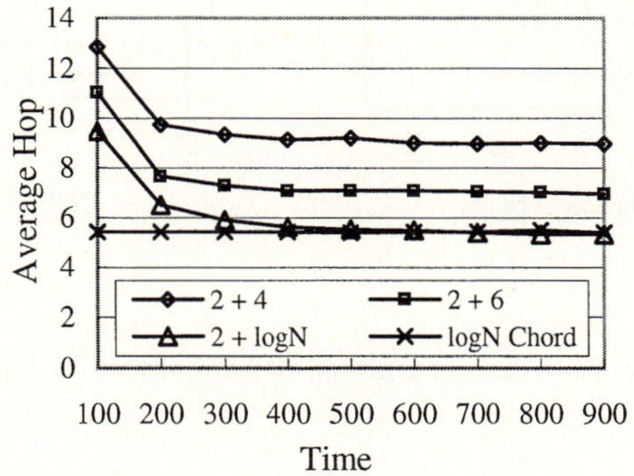

Fig. 4. Average lookup length in different dynamic neighbor's count

Because of the use of the dynamic neighbor adjusting algorithms, the forwarding path of larger traffic has been optimized as time goes. The average lookup length of the whole network decreases obviously. S-Chord adopts the policy of maintaining $\lfloor logN \rfloor$ dynamic neighbors, and its average lookup length is close to or slightly superior to optimized Chord which maintains $\lfloor logN \rfloor$ static neighbors.

Experiment 2. In the subnet with 20000 nodes, we measure the influence of different subnet scale to average lookup length on precise routing, and the result is shown in figure 5.

As the number of dynamic neighbors of S-Chord node is the logarithm of subnet scale, it increases slowly with the increase of the subnet scale. In addition, since we have used the dynamic neighbor adjusting algorithms, the average lookup length is not sensitive to subnet scale. Meanwhile, workload will not change greatly with the change of the subnet scale.

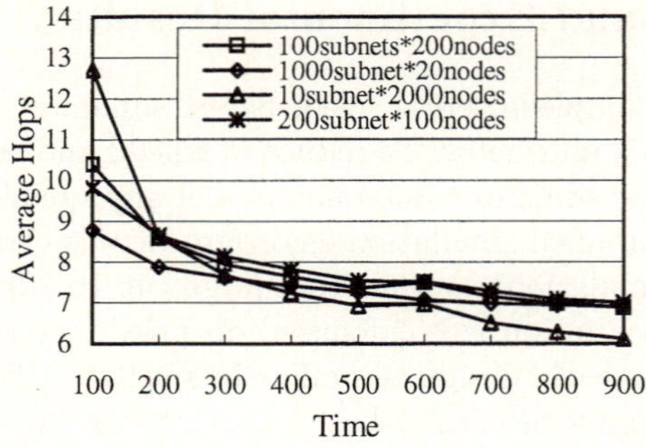

Fig. 5. average lookup length in different subnet scale

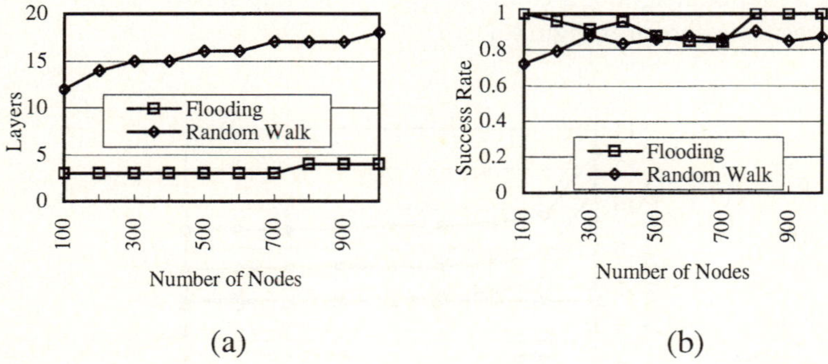

(a)   (b)

Fig. 6. (a)flooding levels; (b)random walks levels

Experiment 3. To compute the number of dynamic neighbor, flooding levels and random walks levels in different subnet scale. The result is shown in figure 6.

With the enlargement of the subnet scale, the number of dynamic neighbor will rise slowly, leading to slow growth of the flooding and Random walks algorithm level at the same time. A higher lookup success rate of single copies is ensured.

5   Conclusion and Future Work

The S-Chord node has maintained two static neighbors to construct foundational annular topology. Forwarding to static neighbor clockwise or counterclockwise, we can reach all nodes of subnet and guarantee the connectivity of the network. We replace the other neighbors with dynamic neighbors selected in random, and the dynamic neighbors can choose the route with smaller hop-count as new route according to the present traffic in the network. So, the average lookup length is reduced and the query success rate is increased. S-Chord can better support large-scale overlay services and improve flexibility in dynamic environment.

Acknowledgement

This work was supported by the National Nature Science Foundation of China (No 60473082, No 60403035) and National Key Fundamental Research Plan (973) of China (No 2003CB314801).

References

1. GNUTELLA: http://gnutella.wego.com
2. NAPSTER: http://www.napster.com
3. Lv, Q., Cao, P., Cohen, E., Li, K., Shenker, S.: Search and replication in unstructured peer-to-peer networks. Proceeding of the 16th international conference on Supercomputing. June. 2002
4. Sylvia, R., Paul, F., Mark, H., Richard, K., Scott, S.: A Scalable Content-Addressable Network. Proceeding of the ACM SIGCOMM. 2001
5. Ion, S., Robert, M., David, K., M.Frans, K., Hari, B.: Chord A Scalable Peer-to-peer Lookup Service for Internet Applications. Proceeding of the ACM SIGCOMM. 2001
6. Antony,R., Peter, D.: Pastry: Scalable, decentralized object location and routing for large-scale peer-to-peer systems. IFIP/ACM International Conference on Distributed Systems Platforms. 2001
7. David, R. Karger., Matthias, R.: Simple Efficient Load Balancing Algorithms for Peer-to-Peer Systems. In Proceedings IPTP. 2004.
8. David, R. Karger., Matthias, R.: Diminished Chord: A Protocol for Heterogrneous Subgroup Formation in Peer-to-Peer Networks. In Proceedings IPTP. 2004.
9. Moni,N., Udi, W.:Know thy Neighbor's Neighbor: Better Routing for Skip-Graphs and Small Worlds. In Proceedings IPTP. 2004.
10. FIPS 180-1. Secure Hash Standard. U.S. Department of Commerce/NIST, National Technical Information Service, SpringLeld, VA, Apr.1995
11. FRANCIS, P.:Yoid: Extending the internet multicast architecture. http://www.icir.org/yoid/docs/yoidArch.ps. 2000

Hierarchical Multi-hop Handoff Architecture for Wireless Network Mobility*

Yunkuk Kim, Sangwook Kang, Donghyun Chae, and Sunshin An

Computer Network Lab, Dep. of Electronics and Computer Engineering, Korea University,
5Ga 1, Anam-Dong, Sungbuk-Gu, Seoul, Korea, 136-701
{dbs1225, klogic, hsunhwa, sunshin}@dsys.korea.ac.kr

Abstract. Under the wireless mobile network environments, if a Mobile Router (MR) which has a wireless transceiver is more than one hop away from the Access Router (AR) connected to the internet, it cannot access directly to the AR and should use its neighbor MRs to access the Internet. Furthermore, a MR can receive the Router Advertisements (RAs) of all MRs in its coverage. In this case a MR cannot distinguish which Router is the upstream Router that destined to the internet. In these situations, it is necessary to support the network mobility by some mechanism. This paper propose a hierarchical multi-hop handoff architecture to provide internet connectivity and, in the same time, a route optimization scheme that can be applied to avoid the multi-angular routing in wireless mobile network (WNEMO) environments.

1 Introduction

Unlike host mobility support, Network Mobility (NEMO) [1] is concerned with the mobility management of an entire network which changes its point of attachment to the Internet. Such a kind of network is referred to as a *mobile network*, and includes one or more MRs which connected it to the internet. The typical examples of a mobile network are a Personal Area Network (PAN) and a network inside vehicles (train, aircraft, taxi, etc) connected to the Internet via multiple medium. The NEMO W/G [8] is developing a solution for a mobile network. The NEMO's basic approach [2] uses bidirectional tunneling between a MR and its Home Agent (HA) on Mobile IPv6 with minimal extensions to preserve session continuity while the MR moves. This approach is aimed at enabling network mobility with minimum change to existing nodes and protocols. However, when multiple MRs are deployed under the wireless mobile environments, the following two problems must be solved additionally to support network mobility.

One is to discover the default Router (MR or AR) that destined to the internet. If a MR is more than one hop away (i.e., outside of the propagation scope of RAs) from the AR, it cannot access directly to the AR and should use its neighbor MRs to access the internet. Furthermore, a MR can receive the RA messages of all MRs in its

* This research was supported by the MIC (Ministry of Information and Communication), Korea, under the ITRC (Information Technology Research Center) support program supervised by the IITA (Institute of Information Technology Assessment).

P. Lorenz and P. Dini (Eds.): ICN 2005, LNCS 3421, pp. 488–495, 2005.

coverage. Since this situation can cause a problem called *RA confliction* [3], a MR cannot distinguish which Router is the upstream Router that destined to the internet.

The other is to optimize routing in wireless mobile networks infrastructure. The mobile networks may be nested or hierarchy tree topology. Using NEMO basic solution on the nested (or tree-based) mobile networks, it brings a routing overhead to us which is well known as *dog-leg routing* problem [4]. In order to avoid this overhead, it is required to optimize the routing path from the MR in the nested (or tree-based) mobile network to the MR's HA, because it incurs very inefficient routing depending on the relative location of HAs.

To solve these problems, we propose a hierarchical multi-hop handoff architecture to provide internet connectivity and, in the same time, a route optimization scheme that can be applied to avoid the multi-angular routing in wireless mobile network environments. Our solution uses concepts similar to those defined in [5], such as Mobility Anchor Point (MAP), Regional care-of address (RCoA) and Local CoA (LCoA), to configure a hierarchical WNEMO handoff architecture. In our proposal a root-MR becomes a MAP, and a delegate CoA (DCoA) for WNEMO becomes a RCoA.

The remainder of this paper is organized as follows. We describe a hierarchical handoff architecture of WNEMO to support internet connectivity and optimal routing between WNEMO domain and the Internet in section 2 and section 3, respectively. Finally, section 4 concludes this paper with further work.

2 Hierarchical Multi-hop WNEMO Handoff Architecture

2.1 Extended MIPv6 Messages

To support the internet connectivity and network mobility under WNEMO environments, two MIPv6 message extensions are required as follows.

Extended RA Message. We propose to extend the *Prefix Information option* of the RA message with an extra flag "*D*" taken from the "*reserved 1*" field. When this flag is unset, it means that both the "*network prefix*" field indicates a network prefix as defined in the ICMPv6 Prefix Information Option specification and sending Router is a fixed AR directly connected to the internet. If this flag is set, it indicates that both the "*network prefix*" field is used for a DCoA of the nodes within WNEMO domain and sending Router is a MR. In this case, the network prefix of the received RA message computed by the leftmost "*Prefix Length*" bits of sending Router address.
We also employ another flag "*NLevel*" to establish parent-child relationships between MRs in WNEMO domain. This field defined as increasing value by the distance from root-MR and set to zero by root-MR.

Extended Binding Update (BU) Message. In this paper, the BU message is used when the current location of the MR registers to root-MR as well as updates its binding to its HA. The BU message is extended as follows. A new bit "*D*" is taken from the reserved set of bits in order to indicate whether or not this message is used for registering to root-MR. When this flag is unset, it indicates that the "*Care-of Address* "field contains the DCoA of the WNEMO domain for BU to its HA. If it is set, it means that the BU message is used for registering to root-MR, so that the "*Care-of Address* " field contains the LCoA of the MR.

2.2 Tree-Based Multi-hoped WNEMO Scheme

When a MR moves to a foreign link, it can receive RA that has been transmitted from AR or other MRs in its coverage. If a MR is one hop away from the AR, it received the normal RA as defined in [6]. In this case, a MR becomes a root-MR in WNEMO domain and can access directly to the internet via a fixed AR. The MR's CoA can be achieved using MIPv6 scheme. In this paper, we use a root-MR's CoA as a DCoA of all mobile network nodes (MNNs) in the WNEMO domain.

At the root-MR, the operation of registration to its HA is similar to that in base MIPv6 [7]. After this home registration, a root-MR adds its own CoA within "*Network Prefix*" field ("*D*" flag is set) and sets "*NLevel*" field to zero in the *Prefix Information option* of RA message. A root-MR then broadcasts an extended RA message to its subnet. An extended RA message that transmitted from a root-MR can be received by MRs in its coverage area. In this case, MRs recognize that it attached to another mobile network as the child mobile network by checking the "*D*" flag and the "*NLevel*" field within an extended RA message.

Generally, a MR that is more than one hop away from the AR cannot receive the AR's RA message directly. Accordingly, the child MRs should use its own parent MR to access the Internet as ad-hoc mobile networking. Receiving an extended RA message, the child MR stores path information to a root-MR in its Binding Cache to access the internet and increments network level by one, then broadcasts this message to its subnet. Also a parent MR can receive an extended RA message that transmitted from a child MR. In this case, a parent MR knows that sending MR is its child MR by checking the "*NLevel*" field, and stores the information about the child mobile network in its Binding Cache.

As described in the above, when a MR away from home link receives an extended RA message for the first time, a MR recognizes that sending Router is its parent MR. If MR received at least one RA again from its neighbors before the RA's lifetime of its parent MR expires, it performs the operation of handoff decision and relation establishment between MRs as shown in Fig.1.

Handoff Decision. In general, the principle movement detection relies upon either the Lifetime field within the main body of the RA message or network prefixes. A MR can receive the RAs message periodically sent by all neighbors in its coverage as described earlier in this section. In this case, because each MR having been assigned a unique network-prefix, the MR can assume that it could be changing links very rapidly, despite remaining completely still. This would generate a tremendous number of handoff operations. That is, as soon as MR alternately receives the RA messages that transmitted from neighbors, the handoff process is performed repeatedly.

To solve this problem, we use any policy to decide whether the MR actually moved to a new domain. In Fig.1, when a MR receives an extended RA again from other MRs before lifetime expiration, a Router Solicitation message is issued towards its own parent Router (MR or AR). If the solicited advertisement is received, it means that the MR is still in the coverage area of its own parent Router; otherwise, we can deduce that the MR has moved to another WNEMO domain and the handoff process is executed.

Relation Decision Between MRs. In our proposal, each MR knows both the root-MR's CoA and its network level in WNEMO domain. Accordingly, every time the MR receives the RA messages sent by a new MRs in its coverage area, it performs the operation of relation establishment between MRs as following.

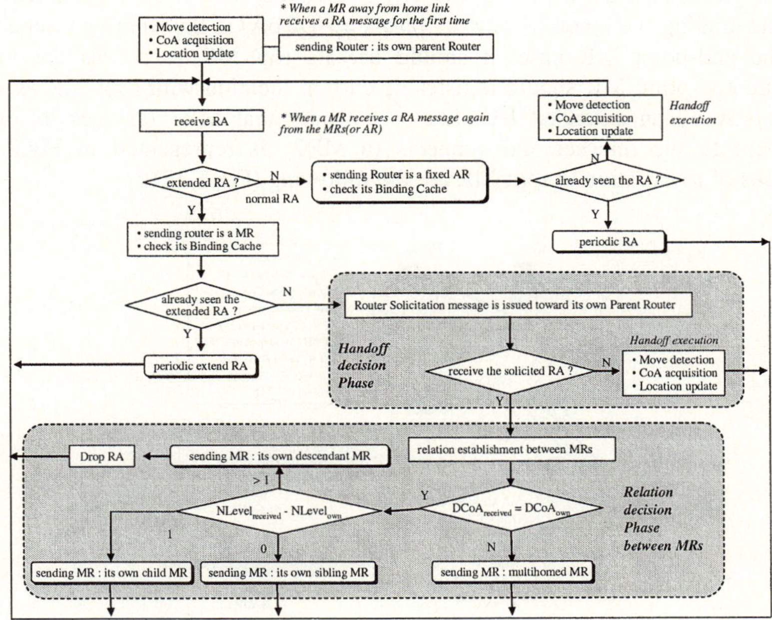

Fig. 1. Decision Tree of Hierarchical WNEMO Architecture

(1) If $DCoA_{received}$ = $DCoA_{own}$ and $NLevel_{received}$ = $NLevel_{own}$ + 1, a sending MR becomes its own child MR. In this case a MR stores the information of child mobile network in its binding Cache.

(2) If $DCoA_{received}$ = $DCoA_{own}$ and $NLevel_{received}$ > $NLevel_{own}$ + 1, a sending MR becomes its own descendant MR. In this case a MR ignores this message because a parent MR has the shortest hop distance to a root-MR than the descendant MR.

(3) If $DCoA_{received}$ = $DCoA_{own}$ and $NLevel_{received}$ = $NLevel_{own}$, a sending MR becomes its sibling MR. In this case a MR stores the information of sibling mobile network in binding Cache as an alternate upstream MR that destined to its root-MR.

(4) If $DCoA_{received}$ ≠ $DCoA_{own}$, this message was transmitted from a MR that has root-MR information of different WNEMO domain. In this case MR is multihomed. In this paper we make the assumption that the mobile network has only one root-MR and is not multihomed.

As described in this section, the wireless mobile networks can configure hierarchy tree which has parent-child relationships. In tree architecture, a parent MR has the location information of child mobile networks, and the child MRs maintain a path information to its parent MR as well as root-MR which attached to the internet.

2.3 Binding Updates

When a root-MR obtains a CoA on each subsequent point of attachment, it then informs its HA using a registration mechanism similar to that in [7]. That is, a root-MR sends an extended BU to its HA on the home link. This message contains MNN's common prefix and DCoA for WNEMO. If HA receives a valid extended BU, it stores the binding (i.e., mobile network prefix => DCoA) in the Binding Cache.

In the end-point MR case, it should use its root-MR to access the Internet. Therefore end-point MR should register its current location with root-MR as well as its own HA by an extended BU. Let us consider that MR3 changes its point of attachment to the internet and connects to MR2, as represented in Fig.2.   The operations of registration to root-MR are the following (Fig.2, (2)):

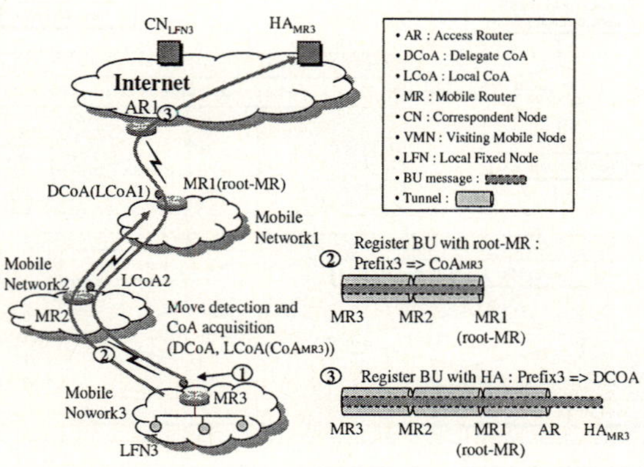

Fig. 2. The operation of registration at end-point MR

(1) Upon obtaining the information of new location, a MR3 checks its Binding Cache in order to request of registration with a root-MR using an extended BU message. An extended BU contains a MR3's LCoA (i.e., set to "D" flag) and the common prefix of mobile nodes. MR3 encapsulates this message to directly send to its parent MR2.

(2) MR3 transmits an extended BU message as a single local broadcast packet, which is received by all MRs currently within wireless transmission range of MR3. When MR3's neighbors receive this message they have two choices. If a MR is not targeted, it ignores the received message. But if a MR recognizes itself as the target MR of the received message (in Fig.2, MR2), it decapsulates a received message and searches the destination in its Binding Cache to determine the path to the destination of inner message. The MR2 encapsulates again inner message then forwards to its own parent MR1 as determined by examining its Binding Cache.

(3) Upon reception of a registration request message from the end-point MR3 via intermediate MRs, the root-MR stores it to its Binding Cache (i.e. Prefix3 => LCoA_MR3).

Then, the end-point MR3 should inform its HA of binding. This procedure is the same as the root-MR (Fig.2, (3)).

2.4 Handoff Management

Our proposal supports both an intra-domain and inter-domain handoff for the movement of a mobile network. In the former case, MR recognizes that it attached to another mobile network in same WNEMO domain by checking the DCoA within extended RAs transmitted from neighbors periodically. If a MR detects the movement of intra-domain, it can perform a handoff procedure with a binding update process only. This process is accomplished by transmitting an extended BU message, which consists of a mobile network prefix and new MR's LCoA, to a root-MR of the WNEMO domain. That is, there is no need to send extended BU to its HA, since DCoA does not change as long as the MR move within a WNEMO domain.

In the latter case, the MR receives an extended RA message contained the information of new WNEMO domain, and it obtains a new DCoA and a new LCoA. The MR then must update the bindings with the new root-MR and its HA by using extended BU message. Also, the MR may send an extended BU message to the previous root-MR requesting to forwards packets addressed to a new root-MR's CoA (new DCoA). This is similar to the smooth handover mechanism of MIPv6.

3 Route Optimization Using Proposed Architecture

When sending a packet to any node in WNEMO domain, a CN first checks if it has a binding for this destination. If a cache entry is found, a CN sends the packets directly to the CoA indicated in the binding. If no binding is found, the packet is sent to the MNN's home address, which tunnels it to the CoA stored in Binding Cache of its HA. Fig.3 (a) shows the flow of the routing when CN sending a packet to a MNN in the WNEMO domain.

(1) If a CN_{LFN3} has no valid binding for a LFN3, packets transmitted by CN_{LFN3} routed to the home link where the HA of its MR3.

(2) The HA_{MR3} intercepts these packets, and tunnels them to the DCoA of the LFN3 (i.e., root-MR's CoA). Upon reception of this tunneled packet from HA_{MR3}, the root-MR1 decapsulates the packets.

(3) MR1 finds its Binding Cache to determine the path to the LFN3. Because the prefix between LFN3 and Prefix3 are the same, MR1 first selects a MR3's CoA corresponding to the binding of Prefix 3 and $LCoA_{MR3}$ in Binding Cache. MR1 then searches Binding Cache recursively with a $LCoA_{MR3}$. At last, MR1 retrieves a $LCoA_{MR2}$ in Binding Cache, since MR3's CoA and Prefix 2 have a common prefix.

(4) After acquisition of path information to a LFN3, the root-MR1 first encapsulates these packets to send to a MR3. The root-MR then encapsulates again to send the tunneled packets to MR2 and forwards the nested tunneled packets to MR2.

(5) When a MR2 receives these packets it decapsulates the nested tunneled packets. A MR2 then finds the destination address of the inner packets and forwards these packets to the MR3. MR3 receives the packets.

(6) In the end-point MR3, the packets decapsulated and delivery to the LFN3. LFN3 receives the packet that is the packets originally sent from the CN_{LFN3}.

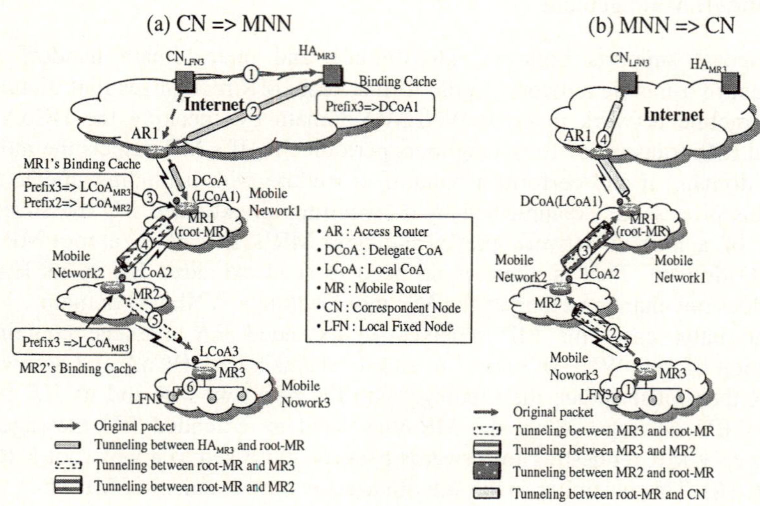

Fig. 3. The flow of the optimal routing

On the other hand, when a LFN3 sends a packet to its CN, this packet is encapsulated by a MR3 to be sent to root-MR which attached to the internet. A MR3 encapsulates the packet again to directly communicate with its own parent MR2. This packet goes to its parent MR, a MR2 does decapsulates this packet and encapsulation again for direct communication with its parent MR1. That is, this packet is performed encapsulation-decapsulation operation repeatedly by each MR located between root-MR and end-point MR. When this packet arrives on root-MR, a root-MR does decapsulates this packet and tunnel again to bypass the ingress filtering at AR that root-MR is attached. Tunneled packet transmitted by a root-MR routed to the CN via the AR. CN decapsulates the encapsulation headers made by the root-MR. This packet is the original packet that has been sent by the LFN3. Fig.3 (b) shows the flow of the routing when the MNN sending a packet to its CN in the WNEMO.

4 Conclusions and Future work

The WNEMO is an open research area in which we are conducting further studies. In this paper we propose a hierarchical WNEMO handoff architecture to provide internet connectivity and, in the same time, a delegate CoA based route optimization that can avoid the multi-angular routing under the WNEMO environments.

Although there is no any form of experiments or simulation results at this time, we can identify a couple of advantages and drawbacks as described follow.

Advantages. First, our solution can provide internet connectivity in WNEMO situation with minimal MIPv6 extensions. The proposed scheme enables any node on the WNEMO domain to communication with any node connected to the Internet via root-MR, because MR can distinguish the upstream MR that destined to the internet. Second, our route optimization method can avoid the dog-leg routing problem by using tree-based routing with a delegate CoA. In our proposal, packets destined for a node in the WNEMO domain are delivered to root-MR directly, which then relays the packets to the MNN via each MR located between root-MR and end-point MR. Also our solution can reduce transmission latency compare with the NEMO's basic approach, since the nested tunneling incurs in the WNEMO domain only.

Drawback. To provide internet connectivity and route optimization in WNEMO domain, root-MR must maintain the location information of all MRs in its administrative domain and the low level MRs must know a DCoA to access the Internet by using extended MIPv6 messages, such as extended BU and extended RA. If there are frequent movements of mobile networks and the administrative domain of the WNEMO is large scale, our approaches may have drawbacks due to the signaling overhead and bottleneck in the root-MR.

We believe that our solution is well adapted for supporting wireless NEMO environments. As a next step, we are going to evaluate our proposed scheme through simulation and leave the detailed security consideration into the future work.

References

[1] T. Ernst, "Network Mobility Support Goals and Requirements", <draft-ietf-nemo-requirements-02.txt>, Feb., 2004

[2] Vijay Devarapalli, Ryuji Wakikawa and Pascal Thubert, "Nemo Basic Support Protocol", < draft-ietf-nemo-basic-support-04.txt>, June 2004

[3] H. Cho, E. K. Paik, "Hierarchical Mobile Router Advertisement for nested mobile network s", < draft-cho-nemo-hmra-00.txt>, January 2004

[4] P. Thubert, M.Molteni, "Taxonomy of Route Optimization models in the Nemo Context", <draft-thubert-nemo-ro-taxonomy-02.txt>, Feb. 2004

[5] H. Soliman, C. Castelluccia, K. El-Malki, Ludovic Bellier, "Hierarchical Mobile IPv6 mob ility management (HMIPv6)", < draft-ietf-mobileip-hmipv6-08.txt>, Jun. 2003

[6] T. Narten, E. Nordmark and W. Simpson, "Neighbor Discovery for IP Version 6 (IPv6)", RFC 2461, December 1998

[7] D. Johnson, C. Perkins , J. Arkko , "Mobility Support in IPv6", RFC 3775, June 2004

[8] IETF NEMO W/G, http : //www.mobilenetworks.org/nemo

Mobility Adaptation Layer Framework for Heterogeneous Wireless Networks Based on Mobile IPv6

Norbert Jordan, Alexander Poropatich, and Joachim Fabini

Institute of Broadband Communications, Vienna University of Technology,
Favoritenstrasse 9/388, A-1040 Vienna, Austria
`norbert.jordan@ieee.org`

Abstract. Today's second generation's packet service GPRS and the emerging 3G mobile cellular networks are only some of the technologies moving towards a mobile IP future. Wireless access technologies such as IEEE 802.11, WiMAX (IEEE 802.16a) and the DVB-T standard will also change the mobile user behavior of today. Hence, the future 4G network will constitute the integration of heterogeneous networks, including a large number of different access technologies. But, regardless of whatever 4G networks may look like, it is foreseeable that heterogeneous IP networking will be a strong driver in future research and commercial deployment. Moreover, it looks as if the one common factor is that 4G networking requires to provide All-IP architectures and connetivity to anywhere and at anytime. Mobile IPv6 and its enhancements combined with the proposed Mobility Adaptation Framework, will "glue" together the different radio networks to provide pervasive access to the Internet in the near future.

1   Introduction

In the future mobile Internet, the mobile equipment will be considerably more diverse than nowadays, and the users will have a greater choice of access technologies. However, looking at technologies like Ethernet, ADSL, GPRS, UMTS, IEEE 802.11, WiMAX, or Bluetooth, this is not so far from what is possible today. These technologies offer different quality of service characteristics in terms of range (e.g., global or local coverage), bandwidth, delay, and error rate. Furthermore, the wide deployment of wireless technologies and the integration of various radio access interfaces into a single terminal, allows mobile end-users to be permanently connected to the IP network. However, this also means that a mobile end-user undergoes a dynamically changing environment. For this reason, adaptation mechanisms are strongly recommended to manage and optimize the IPv6 mobility management of the mobile terminal (e.g., between different access technologies and different IPv6 networks), and to make these transitions as transparent as possible for the user.

In order to allow the user to take advantage of the different available access technologies, this paper proposes a specific adaptation technique that will be implemented within the end-users mobile equipment supporting multiple access network interfaces. Particularly, this approach adopts Mobile IPv6 [1] as an integrative layer atop different "All-IP" access networks. Furthermore, an adaptation layer, referred to as the "Mobility Adaptation Layer" (MAL), is introduced in between the link-layer and the

P. Lorenz and P. Dini (Eds.): ICN 2005, LNCS 3421, pp. 496–503, 2005.

network-layer, hiding different access technologies behind a generic socket-type in-ter-face. This is a similar approach to the convergence layer presented in [2]. How-ever, in this case we focus on a unified means of link-layer triggering, per-flow movement [3], and multiple interface management [4] for various underlying radio technologies utilizing Mobile IPv6.

2 Mobility Adaptation Framework Definition

This section introduces the Mobility Adaptation Framework (MAF), representing an efficient architecture designed specifically for use in heterogeneous access networks. Mobile IPv6 together with reliable control management mechanisms for multiple interfaces is an enabling approach for continuous IPv6-bases sessions via different radio networks in an All-IP infrastructure. In fact, the Mobility Adaptation Frame-work enables a mobile node to manage multiple network interfaces and react to events such as card insertion or removal, cable connection and detection of available network infrastructures. This requires a smart client-based mobility management that is able to coordinate functions from the link-layer and the network-layer. Based on these re-quirements, the remainder of this section provides a brief specification of the MAF architecture along with the functional characteristics of its different parts, and to serve as a generic guide for possible reference design and implementation.

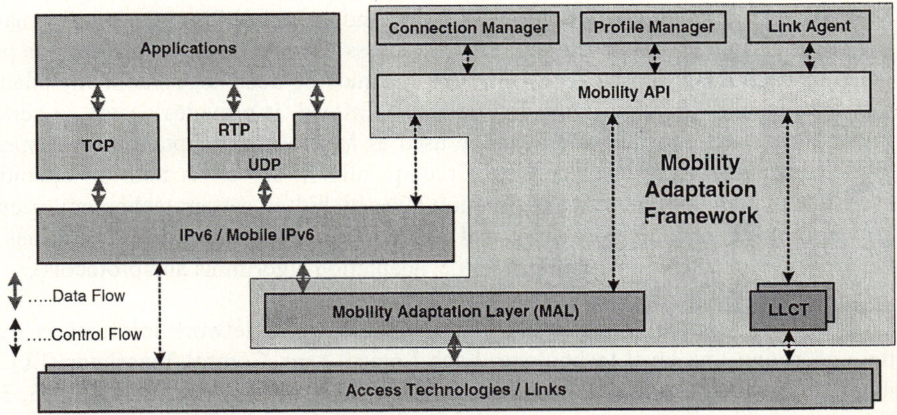

Fig. 1. Mobility Adaptation Framework High-Level Architecture

The major key feature for an adaptive framework in a wireless communication environment are listed below:

- Platform independent interface between the IP layer and a wide range of wire-less access technologies.
- Common link-layer trigger mechanisms for fast movement detection and low-latency horizontal and vertical handoffs.
- Smart client-based IP mobility management that is able to coordinate functions from the link-layer and the network-layer.

- Reliable control management mechanisms for multiple interfaces utilizing Mobile IPv6.
- User preferences and application requirements maintained and structured in profiles.
- Generic enhancement modules can be activated adaptively, aiming to improve performance via different wireless access network technologies.
- Open and extensible framework to future incorporation of adaptation mechanisms.

Figure 1 illustrates the MAF high-level architecture. As it can be seen, a MAF-enabled mobile node will have to incorporate six major parts, namely the Mobility Adaptation Layer (MAL), Mobility Application Programming Interface (API), Logical Link Control Translator, Link Agent, Connection Manager and the Profile Manager, all described in the following section.

3 Mobility Adaptation Framework Design

This section will briefly introduce the modules contained in the Mobility Adaptation Framework:

Mobility API: This is a core component of the MAF concept. Its role is to provide a uniform interface to applications and protocols, so as to drive their operation. The Mobility API provides configuration, statistics, and event handling services, by masking the underlying wireless drivers. That is, access to lower layer components is provided through the proposed API. However, this module does not contain any adaptation intelligence (i.e. algorithms and policies). Instead, it provides access to certain lower layer characteristics which can be used as feedback to adaptation mechanisms implemented in higher-level entities. Consequently, it serves as both an operating system and a network interface abstraction layer by hiding certain technology peculiarities, in order to achieve portable and technology independent implementations of the higher level adaptation intelligence (i.e. adaptation algorithms and protocols).

Logical Link Control Translator: This module extracts network information from the underlying link-level technology. Each Logical Link Control Translator (LLCT) module is aware of one access technology (e.g., IEEE 802.11, Bluetooth, GPRS, 3G, Ethernet,...) exporting one or more services and masking the OS and some wireless technology peculiarities of the specific driver interfaces. Such specific information related to that technology could comprise the interface availability (link up or down), the measured throughput, the bit-error rate, the actual data rate, the utilized coding scheme, the dropped frame rate, the actual power consumption, or available neighboring cells. The information is converted into a generic format and is then sent to other higher level MAF entities. This allows the Link Agent to continuously compare the capabilities of all interfaces and to periodically forward informative link statistics to the Connection Manager. In case some values cross a predefined threshold, the Link Agent triggers an unsolicited notification to the Connection Manager.

Link Agent: The coordination of all MAF functions and lower level mechanisms is achieved at the application layer, through a number of application entities classified into managers and agents. These components use the Mobility API to perform particular control functions and adaptations on other existing components or protocols. Such functions are the configuration, the statistics retrieval, the event handling, the neighbor capability discovery, the multiple interface management (horizontal and vertical handover), and the power management. The Link Agent is mainly based on network management agents, with main purpose to centralize and extend the accessibility of existing network management protocols (e.g., SNMP and local Mobility MIB) to the new parameters and statistics introduced by MAF. It continuously compares the capabilities of all available interfaces and periodically forwards informative link statistics to the Connection Manager. In case some values fall below a predefined threshold, the Link Agent directly triggers an unsolicited notification to the Connection Manager, similar to [5] and [6].

Connection Manager: The main function of the Connection Manager (CM) is to allow seamless and fast handover between various access networks when multiple interfaces are available. This ensures that a mobile client is always connected to the best available network. Several parameters can be taken into account to determine the best available network including signal-to-noise ratio, cost, bandwidth, and services available in a network.

- Communicates with the Link Agent to persistently evaluate the current charateristics of all available interfaces. This ensures that the MN always applies the best possible network connection.
- Predictive link monitoring to conduct horizontal handoffs with no change to another interface (if sufficient network coverage is available).
- Performing vertical handoffs when a mobile node is equipped with multiple interfaces and there might be the need to perform a vertical handoff.
- Management and fast signaling for Mobile IPv6 interaction during handoffs and making decisions for the best applied MIPv6 scheme (depending on the specific handoff scenario).

Profile Manager: In the future a mobile terminal will contain multiple network interfaces, and the "best" decision to select an interface and an access network from other possible combinations has to be taken. Besides link-specific parameters, this decision will also depend on user preferences and certain requirements from the applications. The Profile Manager assists the Connection Manager when it makes the choice of the best suited access option, by taking into consideration the application requirements and user preferences. All this information is maintained and structured in profiles, which are located at the mobile node. These profiles can be accessed by the CM in order to combine link-specific parameters (received from the Link Agent) with user preferences and application requirements.

Mobility Adaptation Layer: This is is an adaptation module that dynamically assigns the data flows to the network interfaces. As can be seen in Figure 1, the complete controlling part for the link-layer is located in the LLCT. However, the MAL Multiplexer is responsible to forward the whole data packet flow upwards to the Mobile IPv6 protocol stack and to downwards multiplex the data flow to the currently selected access network technology. As the IP layer is well established and also the wireless local area infrastructure is already widely deployed, there is little flexibility for change in either the IP or wireless network infrastructure layer. In order to overcome these problems, the MAL is to be added between the IP layer and the link-layer, maintaining transparency with respect to both the IP layer and the wireless network infrastructure layer. Using the MAL provides a platform independent interface between the IP layer and a range of wireless access technologies (i.e., the wireless drivers). Due to the fact that the MAL is a simple execution layer, the intelligence and controlling functionality is provided by the Connection Manager.

4 Performance Evaluation

In case a mobile node is geographically located inside several coverage areas of different access technologies, it may redirect its data flows between two of its interfaces to optimize the handoff procedure. We assume a global network covering wide area, like a 3G cellular network that is connected to the interface IF1. At the same time, the MN has internet access with much higher data rate on another interface IF2 (e.g, wireless LAN IEEE 802.11b). The coverage area of this second wireless technology is smaller but also geographically covered by the global network (umbrella coverage). As long as the mobile node is within the range of a WLAN base-station, it will prefer this technology as it provides much higher data rate in addition to less power consumption. If the signal quality at the interface IF2 decreases and no other WLAN base-station is in range to connect to, the MN may decide to seamlessly redirect the complete traffic or only the most important data flows (not exceeding the maximum bandwidth of the target interface) to the wide-area interface IF2. If the mobile node detects degradation of signal power in advance (predictive link-layer triggering), this procedure is called a seamless redirection, because the MN decides to redirect its flows on a new target interface IF1 while it still has the network connectivity through the source interface IF2. In order to demonstrate how a mobile node may benefit from the mobility management for multiple interfaces, testbed results for the seamless interworking of an IEEE 802.11b infrastructure and a 3G/UMTS cellular network in FDD mode are being presented in the followings. The mobile node in Figure 2 is equipped with two interfaces; IEEE 802.11b (Prism2 based chipset) at 11 Mbit/s and a 3G/UMTS interface (3G PCMCIA card) at 384 kbit/s maximum data throughput.

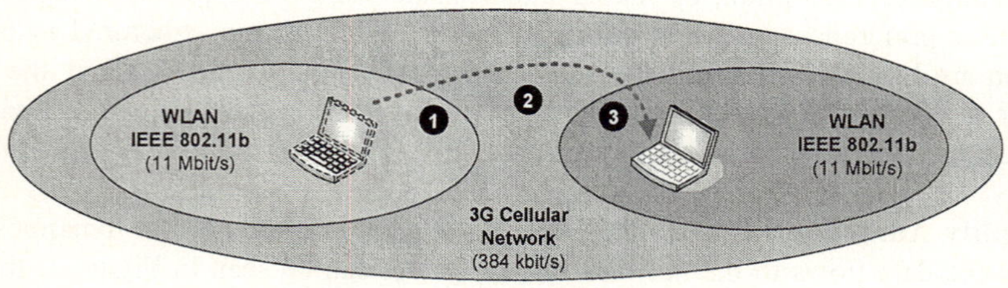

Fig. 2. Seamless Handoff Scenario for WLAN and 3G

The observed MN moves while receiving an audio-stream rated at 128 kbit/s on the IEEE 802.11b interface. In addition, another connection is established (3G attached) to the 3G cellular network which is not used while utilizing the WLAN. On its way to the alternative WLAN base-station (within the same IPv6 domain), the mobile node will loose the connectivity to WLAN for a short period of time (non-overlapping cells), if not taking advantage of multiple interfaces. Figure 3 presents the impact of a non-overlapping WLAN handover, resulting in a significant audio-stream interruption.

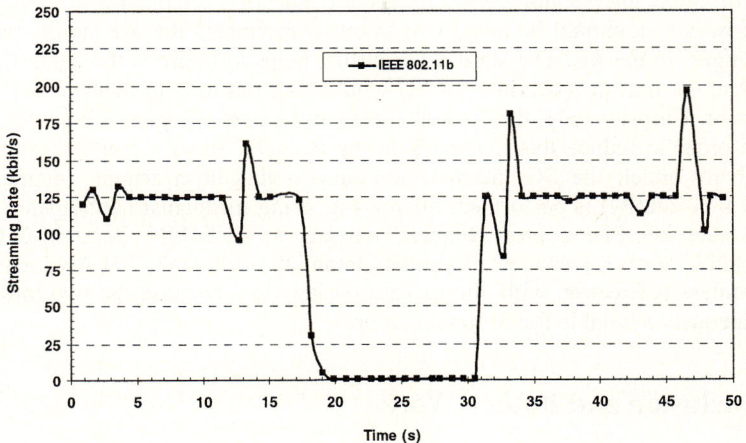

Fig. 3. Impact of a Non-Overlapping WLAN Handover

The Figure 3 illustrates the data rate received by the MN for the handoff period, without redirecting the audio-stream on the other interface (basic behavior). We can see that during a considerable period of time, no packets are received on the MN. This time includes the time to detect that the current connection between the MN and its AP is down, to discover if another AP is available in the MN range and to attaches to a

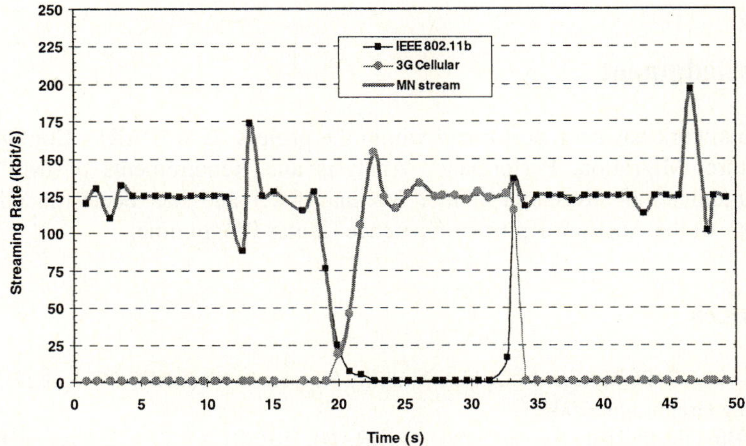

Fig. 4. Temporary Redirection of the Audio-Stream

new AP. In the second scenario, the MN redirects the audio-stream to the 3G network while the WLAN handover is going on. In case the MN detects that the WLAN radio link-quality decreases, it performs this redirection of the audio-stream on the other available interface. The Figure 4 illustrates the data rate for the seamless handoff procedure.

Due to the fact that the WLAN hotspots and the 3G service provider operate in different IPv6 domains, Mobile IPv6 has to be involved in order to keep the audio session up. Here, the data rate received by the MN decreases up to 0 for only a short periodof time. This short interruption comes from the connection establishment at the 3G interface. However, it should be noted that in our experiments the 3G mobile terminal already lingers in the READY state, what further helps to improve the handoff behavior. Otherwise the dial-in procedure for 3G would take too long to achieve any delay improvement. Unfortunately, this benefit comes at the cost of higher 3G power consumption. In order to reduce this drawback in the READY state, it could be suggestive to predictively attach the 3G interface not until a link-layer trigger (degrada-tion of WLAN link-quality) is performed. Such a triggering mechanism anticipates the build-up of a "Activate PDP context Request" message to the SGSN. Once the MN finishes its WLAN handover, it redirects its audio-streaming back on the WLAN interface. This is a seamless redirection with almost zero packets loss because the two interfaces are simultaneously available for communication.

5 Conclusion and Future Work

This paper presented our proposed Mobility Adaptation Framework for seamless interworking of multiple wireless access technologies based on Mobile IPv6. A first performanceevaluation confirms that our architecture may be suitable for future All-IP scenarios in heterogeneous 4G network environments. In order to refine this architecture, the MAL could also be responsible for running a process that enables and disables different enhancement modules. These generic modules can be selected and activated adaptively or through a setup process coordinated by the Link Agent and the Connection Manager. This modular approach may allow the Mobility Adaptation Framework to be flexible enough to enable the addition of new modules.

Acknowledgment

Part of this work has been performed within the project "CAMPARI - Configuration, Architecture, Migration, Performance Analysis and Requirements of IMS" at the Telecommunications Research Center Vienna (ftw.) and has been funded in the framework of the Austrian Kplus Competence Center Programme.

References

1. D. Johnson, C. Perkins, and J. Arkko. Mobility Support in IPv6. RFC 3775, IETF Network Working Group, June 2004.
2. ETSI. Broadband Radio Access Networks (BRAN); HIPERLAN Type 2; System Overview. Standard, ETSI Broadband Radio Access Networks, TR 101 683, August 2002.

3. H. Soliman, K.E. Malki, and C. Castelluccia. Flow movement in Mobile IPv6. Internet-draft, work in progress, IETF MIP6 Working Group, draft-soliman-mobileip-flow-move-03.txt, June 2003.

4. N. Montavont, T. Noel, and M. Kassi. Mobile IPv6 for multiple interfaces. Internet-draft, work in progress, IETF MIP6 Working Group, draft-montavont-mip6-mmi-01.txt, October 2003.

5. N. Jordan, R. Fleck, and C. Ploninger. Fast Handover Support in Wireless LAN based Networks. In The Proceedings of the Fifth IFIP-TC6 International Conference on Mobile and Wireless Communication Networks (MWCN 2003), Singapore, Singapore, pages 49–52, October 2003.

6. N. Jordan, A. Poropatich, and R. Fleck. Link Layer Support for Fast Mobile IPv6 Handover in Wireless LAN based Networks. In The Proceedings of the 13th IEEE Workshop on Local and Metropolitan Area Networks (LANMAN 2004), San Francisco, USA, pages 139–143, April 2004.

MiSC: A New Availability Remote Storage System for Mobile Appliance

Joo-Ho Kim, Bo-Seok Moon, and Myong-Soon Park

Department of Computer Science and Engineering,
Korea University, Seoul 136-701, Korea
{ralph, moony, myongsp}@ilab.korea.ac.kr

Abstract. The continued growth of both mobile appliances and wireless Internet technologies is bringing a new telecommunication revolution and has extended the demand of various services with mobile appliance. However, when working with wireless access devices, users have a limited amount of storage available to them due to their limited size and weight. In order to relieve this basic problem of storage capacity, we suggested iSCSI-based remote storage system for mobile appliance. However, iSCSI performance in the networks with high bit-error rates is degraded rapidly. Since iSCSI is designed to run on a TCP/IP network, it is influenced by an unnecessary reduction in link bandwidth utilization in the presence of high bit-error rates of wireless link. In this paper, we describe a new availability remote storage system for mobile appliance, called MiSC (as an abbreviation for multi-connectioned iSCSI), that avoids drastic reduction of transmission rate from TCP congestion control in wireless networks. Our experiments show that it is more powerful at handling wireless environments as compared to traditional iSCSI.

1 Introduction

Mobile appliances, including cell phones, PDAs, and smart phone, account for a large segment of the electronics and semiconductor industries. Due to their convenience and ubiquity, it is widely accepted that such mobile appliances will evolve into "Personal trusted devices" that pack our identity and purchasing power, benefiting various aspects of our daily lives. Moreover, the sheer rapidity of the spread of both mobile device and wireless Internet technologies are bringing a new telecommunication revolution and have extended the demand of various services using mobile appliance.

It means that the kind of data is used by mobile appliance becomes various and data size enlarges. However, When working with mobile appliances, users face many problem, such as: data which mobile appliances are storing are more vulnerable and lost than stationary data, because they can be easily damaged or stolen, and the limited storage capacity of mobile appliance will be obstruction for the adaptation of usable services of wired environment to mobile environment.

To alleviate these problems, we suggested a remote storage system for mobile appliance [1] which offers to its users the possibility of keeping large size of multimedia data and database in a secure space. Remote storage system provides allocated storage space to each client across networks. This means that a file system may actually be

P. Lorenz and P. Dini (Eds.): ICN 2005, LNCS 3421, pp. 504–520, 2005.

sitting on machine A, but machine B can mount that file system and it will look to the users on machine B like the file system resides on the local machine. Since remote storage system can make mobile appliance access storage on a remote host, mobile appliance could be free from the limitation of storage capacity. Through remote storage system, we could not only reduce additional costs to purchase high cost memory for mobile appliance but also access mass storage anytime and anywhere and prevent data loss from an unpredictable breakdown of mobile appliance.

Although remote storage system for mobile appliances offers several advantages, it has the problem with performance. Since iSCSI is used for a wide variety of environments and applications including remote storage access runs on a TCP/IP network, it utilizes TCP's features of guaranteed in-order delivery of data and congestion control. Although iSCSI benefits by using an existing and well-tested TCP, the performance in the networks with high bit-error rates is degraded rapidly. Since iSCSI is designed to run on a TCP/IP network as mentioned above, it is influenced by an unnecessary reduction in link bandwidth utilization in the presence of high bit-error rates of wireless link.

TCP performs well over wired networks by adapting to end-to-end delays and packet losses caused by congestion. The TCP sender uses the cumulative acknowledgments it receives to determine which packets have reached the receiver, and provides reliability by retransmitting lost packets. The sender identifies the loss of a packet either by the arrival of several duplicate cumulative acknowledgments or the absence of an acknowledgment for the packet within a timeout interval. TCP reacts to any packet losses by dropping its congestion window size before retransmitting packets.

Unfortunately, when packets are lost in wireless networks for reasons other than congestion, these measures result in an unnecessary reduction in end-to-end throughput. In a wireless environment, losses could also occur due to bad channel characteristics, interference or intermittent connectivity due to handoffs for example. Based on these losses, congestion control mechanism would lead to a reduction of the transmission rate and would only decrease the link utilization unnecessarily. iSCSI over TCP performance in such networks suffers from significant throughput degradation and very high interactive delays. Therefore, in this paper we propose Multi-connectioned iSCSI (MiSC) to avoid drastic reduction of transmission rate from TCP congestion control mechanism or guarantee fast retransmission of corruptive packet without TCP re-establishment. A MiSC session is defined to be a collection of one or more TCP connections connecting mobile client to remote storage server. If packet losses occur due to bad channel characteristics in a specific TCP connection between mobile client and remote storage server, MiSC will select another TCP connection opened for data transmission to prevent the congestion window from being divided by two.

The rest of this paper is organized as follows. In section 2, we describe two major technologies of file I/O-based remote storage system and block I/O-based iSCSI protocol that is the foundation of our proposed system. In Section 3, we describe briefly the problem of remote storage system over wireless links. We describe the details of our proposed solution which consists of parameter collector & Multi TCP connections controller, data distributor & gatherer and Q-Chained dispatcher in section 4. Section 5 presents the results and analysis of several experiments. We conclude with a summary and discuss our plan in section 6.

2 Background

2.1 Remote Storage System

Remote Storage System capacitates multiple clients to share and access data on allocated remote storage, which is recognized as its local storage. In addition, at present there are two major approaches to transfer data in storage network area. The one is 'file I/O' approach and the other is 'block I/O' approach.

The CIFS (Common Internet File System) and the NFS (Network File System) are widely used file I/O protocol in the area of remote storage system. The CIFS is the extended version of SMB protocol [2] that enables collaboration on the Internet by defining a remote file-access protocol that is compatible with the way applications already shared data on local disks and remote storage servers. CIFS incorporates multi-user read and write operations, locking, and file-sharing semantics that are the backbone of today's sophisticated enterprise computer networks. Moreover, CIFS runs over TCP/IP and utilizes the internet's global Domain Naming Service (DNS) for scalability, and is optimized to support slower speed dial-up connections common on the internet [2].

The NFS was originally developed by Sun Microsystems in the 1980's as a way to create a file system on diskless clients. NFS provides remote access to shared file systems across networks. The primary functions of NFS are to export or mount directories to other machines, either on or off a local network. These directories can then be accessed as though they were local. In additions, NFS uses a client/server architecture and consists of a client program, a server program, and a protocol used to communicate between the two. The server program makes file systems available for access by other machines via a process called exporting. NFS clients access shared file systems mounting them from an NFS server machine [3]. In addition, since multiple workstations in the NFS are operated as one single system through network, clients on the workstations can share heterogeneous file systems without regard to the kind of computer server.

The ability of NFS/CIFS to represent storage appliance as a local file system accessible by clients, allows file shared storage access. If we accommodate the remote storage system using the above protocol to mobile appliance, it can overcome the restriction of mobile appliance's storage capacity as the case of remote storage system in wired environment. However, there is number of applications that do not require file sharing. Moreover, there are applications, whose performance is significantly degraded when accessing data using "file I/O" approach, like databases and similar transaction-oriented applications. Therefore, we use block I/O protocol, iSCSI as a main technology to build remote storage system. While iSCSI cannot provide shared file access, it would be quite suitable for applications that do not need sharing [4].

2.2 iSCSI

iSCSI is Internet SCSI (Small Computer System Interface), an Internet Protocol IP-based storage networking standard for linking data storage facilities, developed by the Internet Engineering Task Force IETF. By carrying SCSI commands over IP networks, iSCSI is used to facilitate data transfers over intranets and to manage storage over long distances. Since the iSCSI protocol can make client access the SCSI I/O device of server host over an IP Network, client can use the storage of another host transparently without the need to pass server host's file system [5].

In iSCSI layer that is on top of TCP layer (Figure 1), Common SCSI commands and data are encapsulated in the form of iSCSI PDU (Protocol Data Unit). The iSCSI PDU is sent to TCP layer for the IP network transport.

Through the procedure, client who wants to use storage of the remote host can use it. Because the encapsulation and the decapsulation of SCSI I/O commands over TCP/IP enable storage user to access remote storage device of the remote host directly [6].

Likewise, if we build remote storage system for mobile appliance using iSCSI protocol, mobile clients can use the storage of server host directly like their own local storage. It enables mobile appliance to overcome the limitation of storage capacity as well as to adapt various application services of wired environment in need of mass scale data. Different from the traditional remote storage system that based on file unit I/O, iSCSI protocol provides block unit I/O. Therefore, it can make more efficient transmission throughput than the traditional remote storage system like CIFS and NFS does.

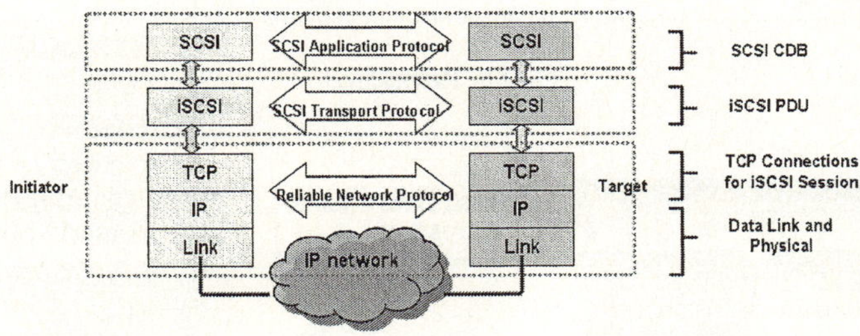

Fig. 1. iSCSI architecture and SCSI-3 layered model

Another characteristic of iSCSI protocol is that it operates on standard and commonly used network component like Ethernet. Since iSCSI protocol directly can be plugged in Ethernet environment, it is easier to manage than another transmission protocol of storage data as Fibre Channel. Moreover, iSCSI can reduce costs to build storage system due to the use of infra network without additional adjustment. The iSCSI protocol was defined as the standard SCSI transmission protocol lately by IETF and many related researches are being processed with the development of Gigabit Ethernet technology.

3   Motivation

In this section, we summarize some problem of iSCSI protocol used in wired environment to offer block I/O based remote storage service, adapting it to mobile environment.

3.1   Remote Storage System with iSCSI in Wired Environment

The iSCSI protocol is a mapping of the SCSI remote procedure invocation model over the TCP protocol. The following conceptual layering model (Figure 2) is used to specify client and remote storage server actions and the way in which they relate to transmitted and received iSCSI protocol data units.

The SCSI device driver layer builds/receives SCSI CDBs (Command Description Blocks) and passes them to iSCSI layer. iSCSI layer encapsulates SCSI CDB passed by upper layer protocol and sends it to TCP layer for the IP network transport. Additionally, communication between the initiator and target occurs over one or more TCP connections. The TCP connections carry control messages, SCSI commands, parameters, and data within iSCSI PDU. The iSCSI target that receives SCSI commands form client transfers or receives storage data. During the operation, client makes its own virtual storage which exists on remote storage server and can use it as if its own.

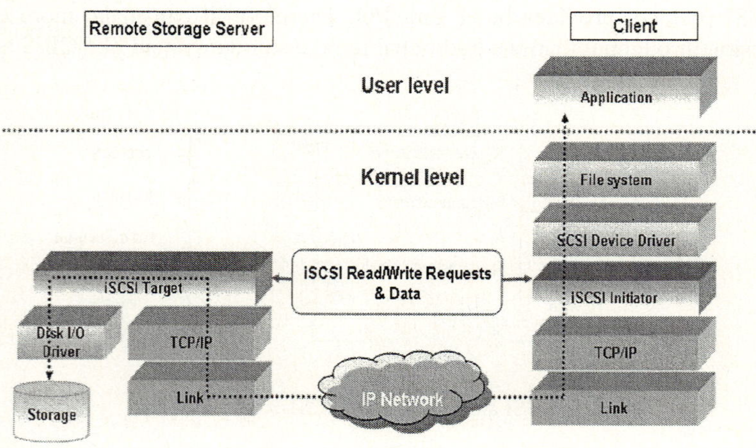

Fig. 2. iSCSI conceptual layering architecture

Table 1 shows that iSCSI significantly outperforms NFS which has been offering remote storage service conventionally.

Table 1. Performance comparison results of iSCSI and NFS

Reads, %	Service time, sec		Response time, sec		Throughput, KB/sec	
	iSCSI	NFS	iSCSI	NFS	iSCSI	NFS
0	3.12	3.77	31.18	37.65	321	266
25	2.98	3.64	29.82	36.43	335	275
50	2.64	3.45	26.44	34.48	378	290
75	2.95	1.94	19.43	29.49	515	339
100	0.64	2.78	6.41	27.81	1560	360

As it is seen from Table 1, iSCSI throughput rates are higher than NFS rates by ~20% for pure writes and by ~400% for pure reads. As for service and response times, these iSCSI parameters are ~20% lower for pure writes and almost 400% lower for pure reads than ones for NFS [4].

3.2   Remote Storage System Problem with iSCSI in Wireless Environment

Since performance of iSCSI is significantly affected by TCP protocol, there currently exist a number of efforts in the industry aiming to increase performance of TCP using hardware solutions. When we use iSCSI protocol in network with wireless and other lossy links in order to offer remote storage services with this situation, it suffer from significant non-congestion-related losses due to reasons such as a bit errors and hand-offs. So remote storage system in wireless networks needs several schemes designed to improve the performance of TCP as a try.

Reliable transport protocol such as TCP [8, 9] has been tuned for traditional networks comprising wired links and stationary host. This protocol assumes congestion in the network to be the primary cause for packet losses and unusual delays. TCP perform well over such networks by adapting to end-to-end delays and packet losses caused by congestion.

TCP congestion control algorithm has been designed by following the end-to-end principle and has been quite successful from keeping the Internet away from congestion collapse. The sender identifies the loss of a packet either by the arrival of several duplicate cumulative acknowledgments or the absence of an acknowledgment for the packet within a timeout interval equal to the sum of the smoothed round-trip delay and four times its mean deviation. Two variables, congestion window (*cwnd*) and slow start threshold (*ssthresh*), are used to throttle the TCP input rate in order to match the network available bandwidth. All these congestion control algorithms exploit the Additive-Increase/Multiplicative-Decrease paradigm, which additively increases the *cwnd* to grab the available bandwidth and suddenly decreases the *cwnd* when network capacity is hit and congestion is experienced via segment losses, i.e. timeout or duplicate acknowledgments [10, 11].

In a fixed network, a packet loss can in general be considered as an indication of overload and congestion situation. Communication over wireless link is often characterized by sporadic high bit-error rates, and intermittent connectivity due to handoffs. TCP reacts to packet losses as it would in the wired environment (Figure 3): it drops its transmission window size before retransmitting packets, initiates congestion control or avoidance mechanisms [12] and resets its retransmission timer. These measures result in an unnecessary reduction in the link's bandwidth utilization, thereby causing a significant degradation in performance in the form of poor throughput and very high interactive delays [13].

Since iSCSI is used for remote storage system in wireless environments runs on a TCP network, it requires new strategies to evade the forceful decrease of transmission rate from TCP congestion control mechanism without changing existing TCP implementations and without recompiling or relinking existing applications.

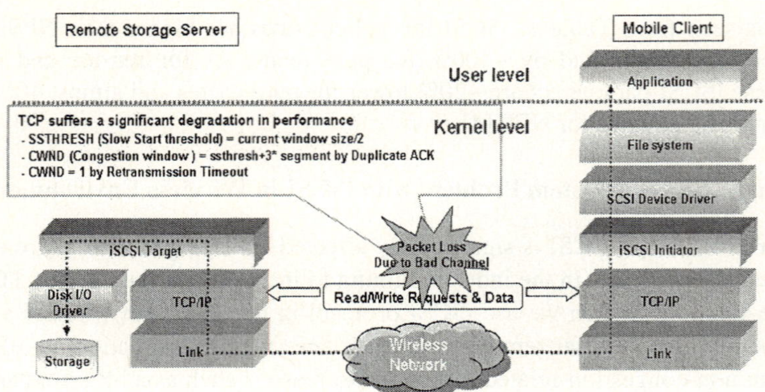

Fig. 3. Debasement in performance due to high bit-error rates

4   MiSC

In order to alleviate the degradation of remote storage services caused by TCP congestion control in wireless networks, we propose multi-connectioned iSCSI (MiSC) which uses the multiple existing TCP connections in the aggregate.

In the case of a single TCP connection, when congestion occurs indicated by a timeout or the reception of duplicate ACKs, one-half of the current window size is saved in ssthresh. Additionally, if the congestion is indicated by a timeout, cwnd is set to one segment. As mentioned in [13], it may cause a significant degradation in performance in wireless networks.

However, in the case of the multi connections method, if congestion occurs in some data connection, the takeover mechanism selects another TCP connection opened during an iSCSI service session and re-transmits all data in loss connection using the selected TCP connection. This can avoid the drastic reduction of transmission rate from TCP congestion control and speed up the re-transmission.

Figure 4 shows a general overview of the proposed multi-connectioned iSCSI (MiSC) scheme. An iSCSI session is defined to be a collection of multiple TCP connections connecting an initiator to a target. If packet losses occur due to bad channel characteristics in the data connection 3, Misc will randomly pick out the data connection opened during an iSCSI service session except loss connection between mobile client and remote storage server. This results in being able to transmit data without decreasing the *ssthresh* and the *cwnd*.

MiSC consists of three components for building or controlling multiple TCP connections. Three functional components of MiSC (Figure 5) are as follows: (1) parameter collector & multi TCP connections controller negotiate the number of TCP connections with round-trip time (RTT) determination algorithm (2) data distributor & gatherer define data assignment method to multi TCP connections (3) Q-Chained dispatcher balance the workload among the data connections in the event of packet losses due to bad channel characteristics.

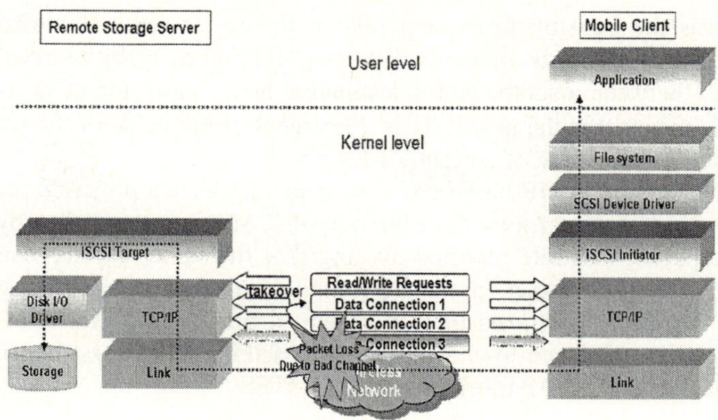

Fig. 4. Overview of MiSC scheme

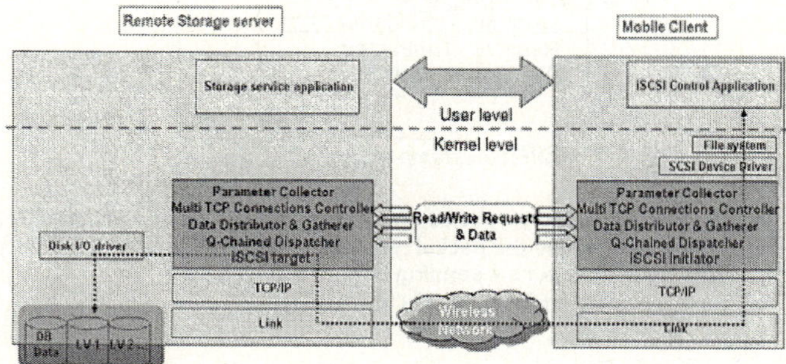

Fig. 5. Functional Components of MiSC

4.1 Parameter Collector and Multi TCP Connections Controller

Parameter collector gathers the parameters, which are necessary to determine the appropriate number of TCP connections for a specific destination. It measures the Round-trip time (RTT) in TCP three-way handshake.. RTT is used to indicate the incoming traffic size of TCP between mobile client and remote storage server. Shorter RTT usually signals smaller service time. Faster transfer rate can make multi TCP connections controller open a large number of TCP connections than longer RTT.

One of the common tools to measure RTT is ping. The tool works by sending an ICMP Timestamp request and reply, usually called a probe, which forces end-host to elicit a reply message. The RTT, then, is the elapsed time between the sending of the ICMP packet and the recipient of the reply. It provides a way to observe the dynamics of the RTTs.

The source creates a timestamp-request message. The source fills the original timestamp field with the Universal Time shown by its clock at departure time. The other two timestamp fields are filled with zeros. The destination creates the timestamp-reply message. The destination copies the original timestamp value from the request message into the same field in its reply message. It then fills the receive timestamp field with the Universal Time shown by its clock at the time the request was received.

Finally, it fills the transmit timestamp field with the Universal Time shown by its clock at the time the reply message departs [14]. Figure 6 shows the formulas to measure RTT between a source and a destination host. The value of α is usually 90 percent. This means that the new RTT is 90 percent of the value of the previous RTT plus 10 percent of the value of current RTT.

However, common ICMP-based tools, such as ping have a principle problem. Several host operating systems now limit the rate of ICMP responses, thereby artificially inflating the packet loss rate reported by ping. For the same reasons many networks filter ICMP packet altogether. Some firewalls and load balancers respond to ICMP requests on behalf of the hosts they represent, a practice we call ICMP spoofing, thereby precluding real end-to-end measurements. Finally, at least one network has started to rate limit all ICMP traffic traversing it [15].

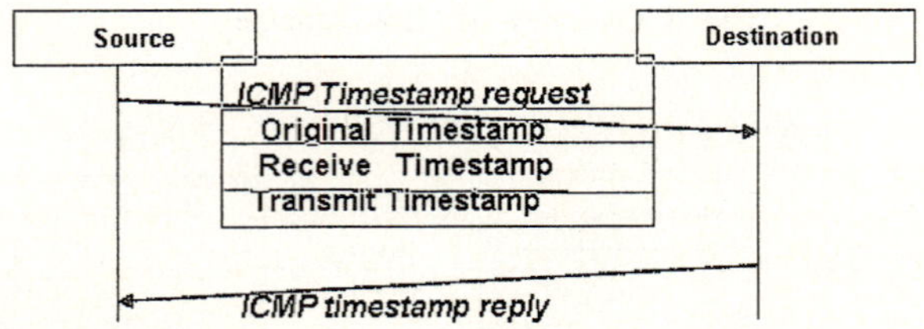

Fig. 6. Timestamp-request and timestamp-reply message

Due to the limitations and disadvantages of ICMP probes, parameter collector uses TCP probes to measure RTT, often called TCP ping. To measure RTT between the mobile client and the remote storage server using TCP ping, parameter collector utilizes the connection establishment, which is called three-way handshaking before any storage data transfer (figure 7).

Mobile client sends the first segment, a SYN segment. The segment includes the source and destination port number. The destination port number clearly defines the remote storage sever to which the mobile client wants to be connected. The segment also contains the client initialization sequence number used for numbering the bytes of data sent from the mobile client to the remote storage server. If the mobile client wants to define the Maximum segment size (MSS) that it can receive from the remote storage server, it can add the corresponding option here. In the second phase, the remote storage server sends the second segment, a SYN and ACK segment. This segment has a dual purpose. First, it acknowledges the receipt of the first segment using ACK flag. Second, the segment is used as the initialization segment for the remote storage server. In the final phase, the mobile client sends the third segment. This is just an ACK segment. It acknowledges the receipt of the second segment using the ACK flag.

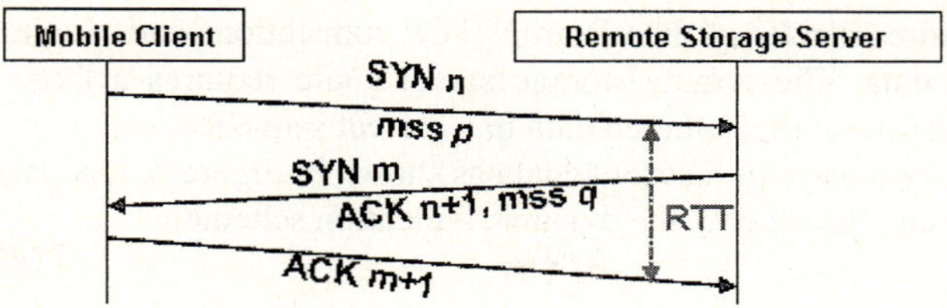

Fig. 7. Measurement of RTT in three-way handshaking

Mobile client sends the first SYN segment, starts a timer, and waits until it sends the final ACK segment. It measures the intervals between the sending time of the first SYN segment and the final ACK segment as RTT. As parameter collector utilizes TCP three-way handshaking to measure RTT between the mobile client and the remote storage server, it can be more efficient at avoiding filtering and inflation of packet than ICMP probes.

Multi TCP connections controller negotiates the number of connections between the mobile client and the remote storage server for storage data transmission according to equation (2) using parameter (RTT) which were collected by the parameter collector. Given a packet drop rate of p, the maximum sending rate for a TCP connection is T bps, for

$$T \leq \frac{1.5\sqrt{2/3} * B}{R * \sqrt{p}}, \tag{1}$$

for a TCP connection sending packets of B bytes, with a fairly constant RTT of R seconds. Given the packet drop rate p, the minimum Round-trip time R, and the maximum packet size B, the mobile client can use equation (1) to calculate the maximum arrival rate from a conformant TCP connection [16].

Equation (2) shows that the number of established TCP connections (N) used in MiSC depends on RTT (R_t) measured by parameter collector. The minimum RTT can determine the large number of connections to be opened between the mobile client and the remote storage server. However, while the use of concurrent connections increases throughput for remote storage service it also increases the packet drop rate. Therefore, it is important to obtain the optimal number of connections in order to set the expected throughput.

$$T \leq \frac{1.5\sqrt{2/3} * B}{R * \sqrt{p}} \leq \frac{N * W}{R_t}, \tag{2}$$

where W is window size of each TCP connection.

4.2 Data Distributor and Gatherer

At the mobile client's site, this module sequentially distributes the data to be transmitted into N TCP connections block by block. At the remote storage server's site, this module acquires blocks of data from N TCP connections and reconstructs them into the original data. The remote storage server's site requires a large size of receive buffer to reconstruct the gathered data in a correct sequence.

There are two ways to distribute data as shown in Figure 8. One is the static allocation scheme and the other is the dynamic allocation scheme.

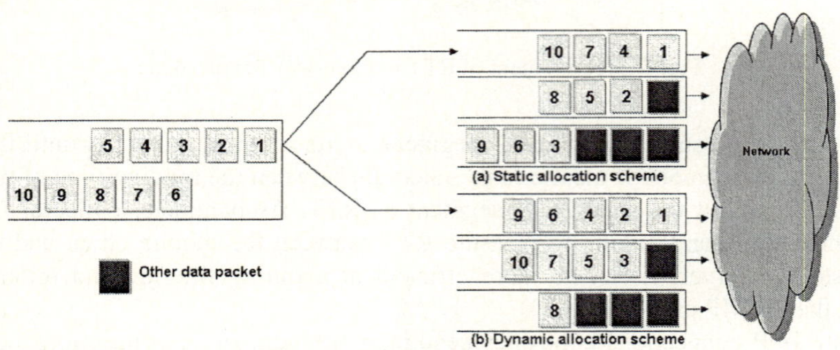

Fig. 8. Data allocation scheme to multiple TCP connections

In the static allocation scheme, a round robin algorithm is used to serve packets. Data distributor assigns the data block to a list of the connections on a rotating basis. The first data block is allocated to a connection picked randomly from the group, not all the first data blocks go to the same connection.. For the subsequent data blocks, data distributor follows the circular order. Once a connection is assigned a data block, the connection is moved to the end of the list. This keeps the connections equally assigned. This scheme is easy to implement and the load of the distributing process is low. However, after the data flows been restored in order to reorder the out-of-order packets by receiver, this scheme require a large size of receive buffer to deliver received data to application for remote storage service in sequence. The one with the lowest performance degrades the total performance of multi TCP connections.

In the dynamic allocation scheme, the fair queuing algorithm serves connections in order of their finish time. The connection with the minimum finish time is selected to send packet [17]. The finish time equation is

$$F_t(n,k,t) = F_t(n,k-1,t) + T(n,k,t)$$
$$T(n,k,t) = L(n,k,t) \div r$$
, (3)

Where $T(n,k,t)$ is the service time of k th packet on connection n, $L(n,k,t)$ is the length of k th packet that transmits on connection n at time t, r is the link service rate.

Therefore, each data block is allocated to the first non-blocked connection found at that time. This scheme requires a smaller buffer size than the former to reconstruct the data in sequence and the processing time becomes shorter.

MiSC uses the dynamic allocation scheme guaranteeing the minimum reordering buffer size.

4.3   Q-Chained Dispatcher

Q-chained dispatcher is able to balance the workload fully among data connections in the event of packet losses due to bad channel characteristics. When congestion occurs in a data connection, this module can do a better job of balancing the workload since the workload which is originated by congestion connection will be distributed among N-1 connections instead of a single data connection.

The performance of remote storage system can be measured in two different operational modes: the conventional mode, with no element congestion occurs, and the lossy mode, in which one data connection has packet losses. In the conventional mode of operation, remote storage system has turned out successful at transmitting storage data over a given time interval. However, when congestion occurs in a specific data connection, balancing the workload among the remaining connections can become difficult, as one connection must pick up the workload of the component where it takes place. In particular, unless the data placement scheme used allows the workload which is originated by congestion connection to be distributed among the remaining operational connections, the remote storage system will become unbalanced and the response time for a request may degrade significantly.

Figure 9 shows that a bottleneck organizes as a data connection 1 becomes target for takeover handles the heavy load of primary data and recovery data when congestion occurs in a data connection 4. Primary data is the allocated data evenly in each of the data connections by data distributor and recovery data is the remaining data in a congestion connection packet losses occur due to bad channel characteristics at that time. Since MiSC's takeover mechanism selects one data connection to take recovery data, the connection must manage all the time the heavy workload of primary data and recovery data. If the data connection , which is fully utilized, is given more burden for recovery when the congestion occurs, the response time for request that need to access remote storage server may double.

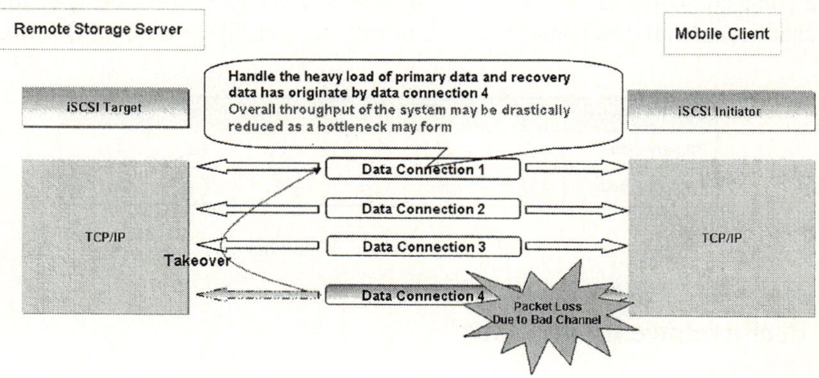

Fig. 9. Bottleneck in the lossy mode of operation

Figure 10 where M, the number of connections between mobile client and remote storage server, is equal to 6. The i-th primary data (designated D_i) is passed on the {i mod M}-th data connection. When congestion occurs in data connection 1, the recovery data (designated d_i) is passed on the $\{(i+1) \bmod M\}$-th data connection. The data connection 2 selected by MiSC's takeover mechanism has responsibility for transmitting the primary data and the recovery data at the same time.

Data Connection	0	1	2	3	4	5
Primary data	D_0	D_1	D_2	D_3	D_4	D_5
Recovery data	-	-	-	-	-	-

Data Connection	0	1	2	3	4	5
Primary data	D_0	F	D_2	D_3	D_4	D_5
Recovery data	-	F	d_1	-	-	-

Fig. 10. Takeover scheme without load balancing

Figure 11 illustrates how the workload is balanced in the event of congestion occurrence in a data connection (data connection 1 in this example) with Q-chained dispatcher. For example, with the congestion occurrence of data connection 1, primary data D_1 is no longer transmitted in congestion connection for the TCP input rate to be throttled and thus its recovery data d_1 of data connection 1 is passed to data connection 2 for conveying storage data. However, instead of requiring data connection 2 to process all data both D_2 and d_1, Q-chained dispatcher offloads 4/5ths of the transmission of D_2 by redirecting them to d_2 in data connection 3. In turn, 3/5ths of the transmission of D_3 in data connection 3 are sent to d_3. This dynamic reassignment of the workload results in an increase of 1/5th in the workload of each remaining data connection. We term this MiSC component to balance workload Q-chained dispatcher because all data connections are linked together, like a chain of Q form.

Data Connection	0	1	2	3	4	5
Primary data	D_0	F	$\frac{1}{5}D_2$	$\frac{2}{5}D_3$	$\frac{3}{5}D_4$	$\frac{4}{5}D_5$
Recovery data	$\frac{1}{5}d_5$	F	d_1	$\frac{4}{5}d_2$	$\frac{3}{5}d_3$	$\frac{2}{5}d_4$

Fig. 11. Q-chained dispatcher

5 Performance Evaluations

This section describes the experimental environment and then shows the results of performance evaluation for the proposed MiSC. The performance analysis is evaluated by comparing iSCSI and the MiSC in mobile environments. The NS-network simulator 2.27 was used to analyze the raw performance of the proposed scheme. We perform several experiments to determine the performance and efficiency of the proposed scheme on a point of view of the throughput for each number of connections in different RTTs and at different bit-error rates.

5.1 Experimental Methodology

The configurations of the experimental network are shown in Figure 12. The bandwidth of the link between node1 and node2 is limited by 2Mbps WaveLAN (lossy link). In order to measure the performance of MiSC under controlled conditions, we generate errors on the lossy link. The average error rate is 3.9×10^{-6} (this corresponds to a bit error every 256 kbits on average). In addition, to investigate the relation between the number of connections and the throughput by giving Round Trip Time (ms) change, the maximum possible number of connections from the initiator to the target across the wireless WaveLAN link is set by 9 and RTT range is from 50 to 500 (ms). The SCSI read is also used between initiator and target with data size (6 MByte) since the simulation results will emphasize not delay to wait R2T message as the case of SCSI write but only the relationship between number of connections and throughput.

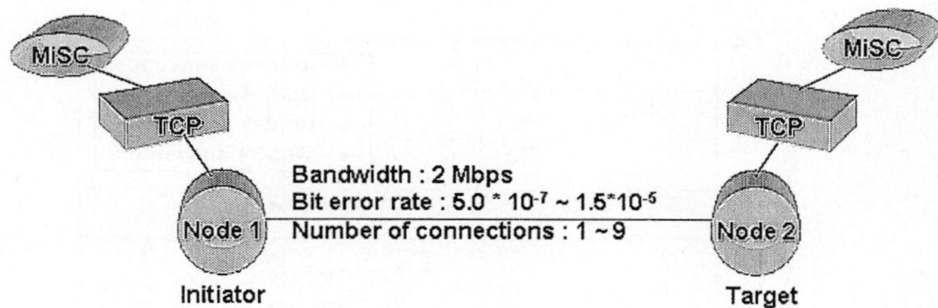

Fig. 12. Simulation scenario

5.2 Simulation Results

MiSC throughputs in different RTTs are measured for each number of connections in Figure 13. We see the slowness of the rising rate of throughput between 8 connections and 9 connections. This shows that throughputs are influenced by reconstructing the

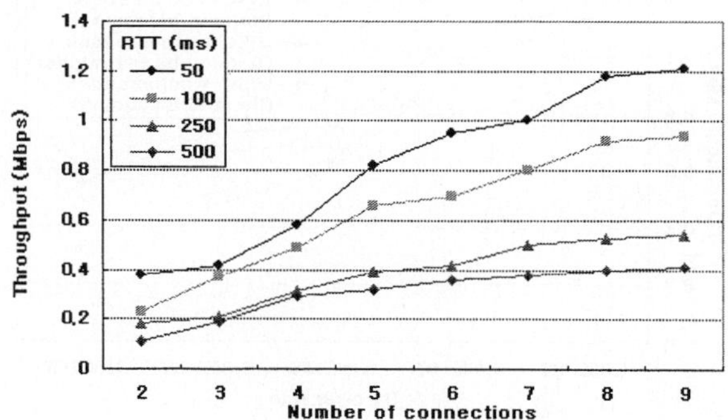

Fig. 13. Throughput of MiSC

data in turn and the packet drop rates are increased when the number of TCP connections is 9 as the maximum use of concurrent connections between initiator and target. Therefore, 8 is the maximum optimal number of connections from a performance point of view. MiSC mechanism also works effectively because the data transfer throughputs increase linearly when the round trip time is larger than 250ms.

In Figure 14, the performance comparison of MiSC and iSCSI at different bit-error rates is shown. We see that for bit-error rates of over 5.0×10^{-7} the MiSC (2 connections) performs significantly better than the iSCSI (1 connection), achieving a throughput improvement about 24 % in SCSI read. Moreover, as bit-error rates go up, the figure shows that the rising rate of throughput is getting higher at 33% in 1.0×10^{-6}, 39.3% in 3.9×10^{-6} and 44% in 1.5×10^{-5}. Actually, MiSC can avoid the forceful reduction of transmission rate efficiently from TCP congestion control using another TCP connection opened during a service session, while iSCSI does not make any progress. Under statuses of low bit error rates ($< 5.0 \times 10^{-7}$), we see little difference between MiSC and iSCSI. At such low bit errors iSCSI is quite robust at handling these.

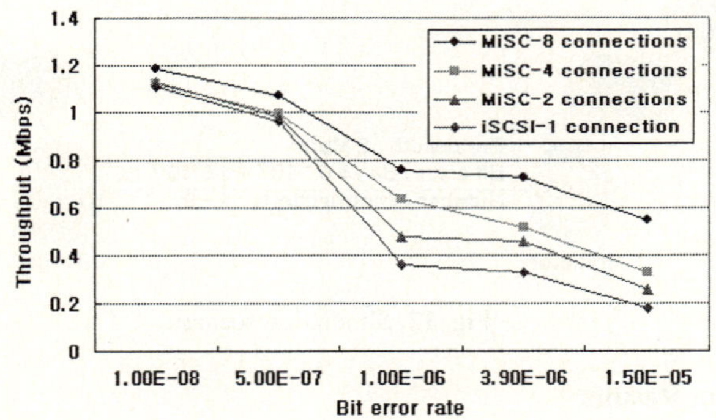

Fig. 14. Throughput of MiSC vs iSCSI at different bit-error rates

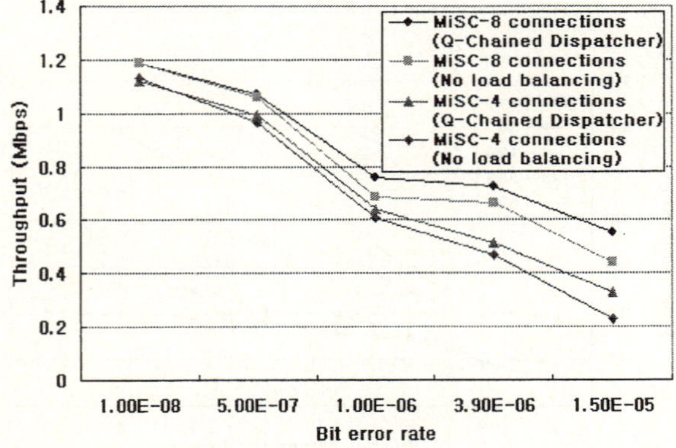

Fig. 15. Q-Chained Dispatcher vs No load balancer

As shown in Figure 15, MiSC (8 connections) with Q-Chained Dispatcher shows the better average performance about 11.5%. It can distribute the workload among all remaining connections when packet losses occur in any connection. To recall an example given earlier, with M = 6, when congestion occurs in a specific connection, the workload of each connection increases by only 1/5. However, if MiSC establishes a performance baseline without load balancing, any connection, which is randomly selected from takeover mechanism, is overwhelmed.

6 Conclusion

In this paper, we have introduced a new availability remote storage system for mobile appliance, termed MiSC, for avoiding drastic reduction of transmission rate from TCP congestion control mechanism in wireless environment. The proposed scheme provides improved performance in that it is able to transmit data between mobile client and remote storage server without decreasing the *ssthresh* and the *cwnd*. MiSC consists of three components for building or controlling multiple TCP connections opened during an iSCSI service session in order to prevent the sender from dropping congestion window when packet losses occur due to bad channel characteristics and interference in a specific data connection. These functional components can adapt the number of connections to mobile environment flexibly and balance the workload among all remaining data connections in the event of packet losses.

In the future we plan to allocate and reconfigure network bandwidth dynamically to guarantee storage QoS. It determines the appropriate number of connections considering not only diagnosing network but also the specification of the storage QoS requirement.

References

1. Sura Park, Bo-Suk Moon, Myong-Soon Park: Design, Implementation and Performance Analysis of the Remote Storage System in Mobile Environment, Proc. *ICITA 2004*, January (2004)
2. Microsoft: CIFS Protocol Operation, http://msdn.microsoft.com/library/default.asp?url=/library/en-us/cifs/protocol/cifs_protocol_operation.asp
3. Sun Microsystems, Inc.: NFS: Network File System Protocol Specification, RFC, Network Working Group, March (1989), RFC 1094
4. TechnoMages, Inc.: Performance Comparison of iSCSI and NFS IP Storage Protocol, White Paper, April (2003)
5. Shuang-Yi Tang, Ying-Ping Lu and David H.C. Du.: Performance Study of Software-Based iSCSI Security, Proc. *SISW'02*, December (2002)
6. Tom Clark: IP SANs : A Guide to iSCSI, iFCP and FCIP Protocols for Storage Area Neworks, Addison-Wesley, Reading, MA, October (2002)
7. Julian Satran: iSCSI Draft 20, http://www.ietf.org/internet-draft/draft-ietf-ips-iscsi-20.txt
8. J. B. Postel: Transmission Control Protocol, RFC, Information Sciences Institute, Marinadel Rey, CA, Septermber (1981), RFC 793
9. W. R. Stevens: TCP/IP Illustrated, Volume 1, Addison-Wesley, Reading, MA, November (1994)

10. Mascolo, S.: Congestion control in high-speed communication networks, *Automatica, Special Issue on Control Methods for Communication Networks*, December (1999) 1921-1935

11. Dah-Ming Chiu, Jain, R.: Analysis of the increase and decrease algorithms for congestion avoidance in computer networks, *Computer Networks and ISDN Systems*, June (1989)

12. V. Jacobson: Congestion avoidance and control, *In SIGCOMM 88*, August (1988)

13. R. Caceres and L. Iftode: Improving the Performance of Reliable Transport Protocols in Mobile Computing Environments, *IEEE JSAC*, June (1995)

14. B. A. Forouzan: TCP/IP Protocol Suite, McGraw-Hill, Reading, November (2000)

15. Stefan Savage: Sting: a TCP-based Network Measurement Tool, *USENIX symposium on Internet Technologies and Systems 99*, October (1999)

16. Sallly Floyd, Kevin Fall: Promoting the use of End-to-End Congestion Control in the Internet, *IEEE/ACM Transaction on Networking*, May (1999)

17. Bin Meng, Patrick B. T. Khoo, T. C. Chong: Design and Implementation of Multiple Addresses Parallel Transmission Architecture for Storage Area Network, *11th NASA Mass Storage Systems and Technologies Conference*, April (2003)

A Logical Network Topology Design
for Mobile Agent Systems

Kazuhiko Kinoshita[1], Nariyoshi Yamai[2], and Koso Murakami[1]

[1] Department of Information Networking, Graduate School of
Information Science and Technology, Osaka University,
Osaka 565-0871, Japan
[2] Information Technology Center, Okayama University,
Okayama 700-8530, Japan

Abstract. In a typical mobile agent system, the network used by the
agents is constructed logically. In such a logical network, agents migrate
and/or send messages to users or to other agents. Thus, the delay for
agent migration and communication and the total amount of traffic for
all agent activities depend on the form of the logical network. In addition,
plural paths between any pair of nodes on the network are necessary, to
minimize the effects of a node or link failure. In this paper, we propose
a method for designing a logical network dynamically. The proposed
method chooses nodes to which a new node is connected to satisfy the
demands for delay and total traffic on the physical network, which may
differ among various network applications. Moreover, our method works
well in practice, using only local node information. Finally, the performance of the proposed method is evaluated by simulation experiments.

1 Introduction

The recent rapid progress of information technologies has allowed computer networks such as the Internet to spread widely. Accordingly, the amount of information and the number of services over such networks increase very strongly.
To use this information and/or services easily, mobile agent technology has received much attention[1, 2]. A mobile agent is a kind of software application
that autonomously migrates from node to node on a network while performing
a given job using resources on the network. Mobile agent technology is developing actively, and many agent systems have been proposed, such as Telescript[3],
Aglets[4], Jade[5], Agent Gateway[6], and Anthill[7]. In a multi-agent system,
multiple agents execute their jobs on nodes in parallel, and are therefore more
efficient than execution by a single agent. Multi-agent systems are expected to
be a platform for distributed systems on large-scale networks[8].

In some agent systems, the network supporting the agents is constructed
logically, using node groupings for each application. On such a logical network,
agents migrate and/or send messages to a user or to other agents. Thus, the
transmission times for agent migration and communication with agents and the

P. Lorenz and P. Dini (Eds.): ICN 2005, LNCS 3421, pp. 521–530, 2005.

total amount of traffic for all agent activities depend on the form of the logical network. In addition, plural paths among arbitrary pairs of nodes on the network are necessary to avoid problems caused by node failure, such as agents being unable to migrate or communicate with others.

In this paper, we propose a design method for a logical network. The proposed method connects two logical links from a new node to the existing logical network, to make plural paths among any pair of nodes. On the other hand, demand for transmission time and the total traffic on a physical network vary for different network applications. To satisfy such demands, the proposed method chooses the nodes to which a new node is connected by the new logical link. The idea is to decrease the total number of hops between each pair of nodes to reduce the delay.

2   Problem Formulation

2.1   Definition of a Logical Network

We first show the relationship between a physical network and a logical network. In Fig. 1, there are three nodes A, B, C, and an agent platform (AP) is installed on nodes A and C. Broken lines and solid lines indicate physical links and logical links, respectively. In this case, the *physical hop count* between nodes A and C is two, that is, A $\rightarrow$ B $\rightarrow$ C. In contrast, the *logical hop count* between nodes A and C is one, that is, A $\rightarrow$ C.

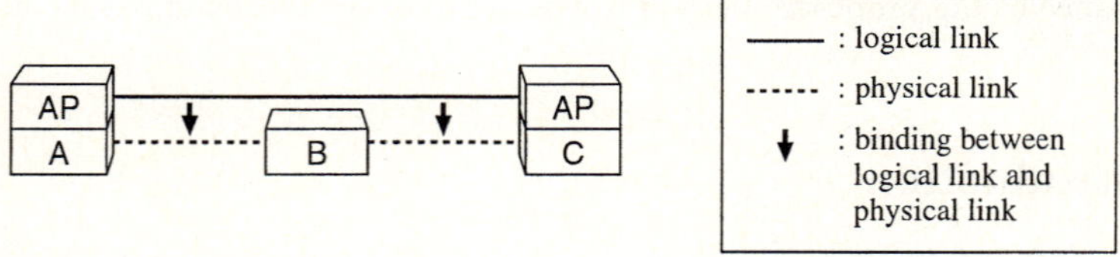

Fig. 1. Relationship between a physical network and a logical network

Next, in Fig. 2 (a), although the logical hop count between nodes A and C is one, the physical hop count depends on the mapping between the logical network and the physical network. For example, in case (b) in Fig. 2, the physical hop count is two, because the logical link A $\rightarrow$ C corresponds to the physical links A $\rightarrow$ B $\rightarrow$ C. On the other hand, in case (c) the physical hop count is three, in the same manner.

Obviously, a higher physical hop count causes more traffic load and increases the possibility of path disconnection by node or link failure. For this reason, we assume in this paper that a logical link corresponds to the shortest path on the physical network.

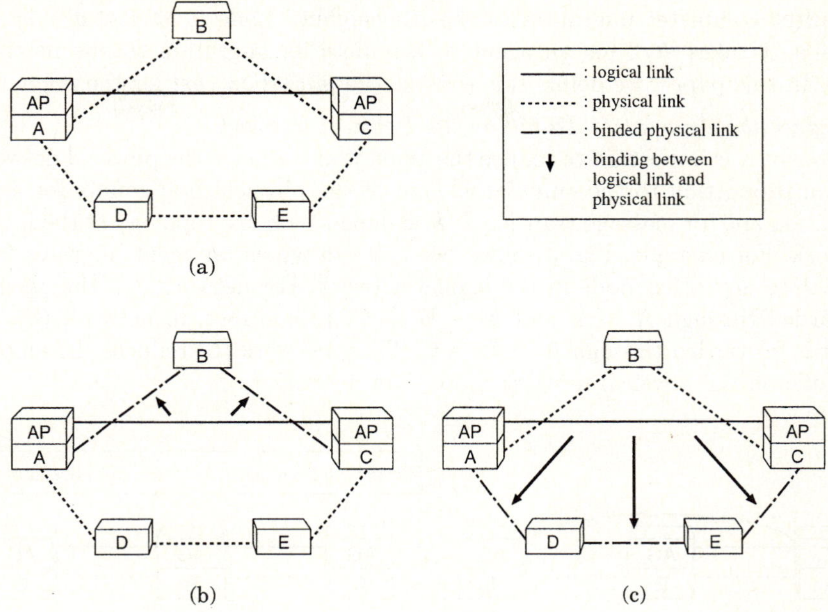

Fig. 2. Mapping between the logical network and the physical network

2.2   Ideal Topology of a Logical Network

In this subsection, we discuss the ideal topology of a logical network. We consider the following three purposes.

1. Reduce the Delay for Agent Migration and Message Transfer
For time-sensitive applications, it is preferable that an agent migrates to any node as quickly as possible and that messages for user–agent and interagent communications are also transferred as quickly as possible.

Note here that the delays for agent migration and message transfer depend on logical hop counts, rather than physical hop counts, between the originating node and the destination node because, compared with the time for agent migration or message transfer procedures in the application layer, the time for packet forwarding in the network layer is negligible. We conclude that the logical hop count between any two nodes should be short for faster agent migration and message transfer.

However, we should also take another factor into account. For agent migration in particular, the dominant procedure is duplication of the agent and this is required in each neighbor node for security reasons: all communication between each pair of nodes should be encrypted by an individual key. Therefore, to reduce the delay for agent migration, we should consider not only the logical hop count but also the degrees of intermediate nodes. For example, an agent migrates from a node to $d - 1$ neighbor nodes, where d means the degree of the node. Let t denote the time required to migrate from one node to another. It takes it until the agent completes migration to the ith neighbor node ($1 < i < d - 1$). On average, it takes $dt/2$ for an agent to complete its migration to one neighbor node. In this paper, we define this time as the *migration cost* for the node.

2. Reduce the Amount of Traffic on the Physical Network
Obviously, it is important to reduce the amount of traffic on the physical network. It is in proportion to the cumulative sum of the physical hop counts for agent migration and/or message transfer. It also depends on the topology of the logical network. For example, Fig. 3 shows two cases in which an agent migrates from node A to any other node in the logical network. For network (a), the agent is forwarded through A → B and A → B → C. In contrast, in network (b), the agent is forwarded through A → B → C. Thus, network (b) reduces the amount of traffic on the physical network more than network (a).

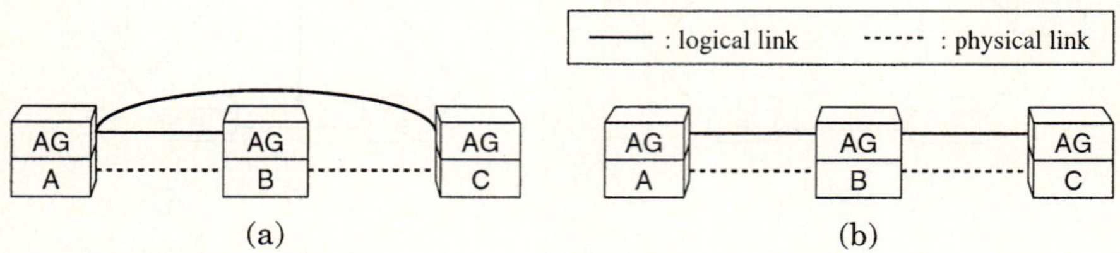

Fig. 3. Amount of traffic on the physical network

3. Enhance Robustness Against Node Failure
A node may fail. Moreover, in unstable networks such as ad-hoc networks, a node may move or leave. To avoid critical effects of node or link failures, a logical network should maintain a connected graph even if a node disappears. To achieve this, we ensure that any pair of nodes has at least two alternative paths into the logical network. Note here that we ignore node or link failures in the physical network, because hardware such as routers is generally much more stable than software such as an agent platform.

2.3   Design Policy

Among the three purposes shown in the previous subsection, there is obviously a tradeoff between the first and second issues. On the other hand, the third one can be treated as an essential requirement.

In addition, for practical use, scalability is an important factor. Thus, we assume that global information over the whole network cannot be obtained.

Therefore, we have developed a logical network topology design method with the following features.

1. It adaptively reduces the delay or the amount of traffic for agent migration and message transfer for a target application.
2. It designs a logical network where any pair of nodes has at least two alternative paths.
3. It works well without global topology information.

3   Proposed Method

The proposed method includes two processes: adding a node to the logical network, and deleting a node from the logical network. The following subsections explain these processes in detail.

Note here that to satisfy the second feature shown in subsection 2.3, a network must have at least three nodes, under the assumption that there is at most one direct link between any pair of nodes. Thus, we suppose that the number of nodes in a logical network is at least three and that when it is three, the nodes form a complete graph, e.g., a triangle.

3.1   Adding a Node

When a new node is added to the logical network, the proposed method chooses two nodes from the logical network and connects the new node to them directly. This process clearly satisfies the second feature shown in subsection 2.3.

The proposed method first chooses the nearest node in the physical network as the first node to be connected. This is reasonable because it minimizes the effect of a node failure in the physical network. On the other hand, it may be difficult for the new node to find a node in the logical network. We assume this process can be realized by using anycast technology[9].

To choose the second node to be connected, the new node obtains topology information via the first node connected. Here we introduce a parameter x. We assume that the new node can obtain such information from nodes that are x logical hops or less away from the first connected node. In other words, if x were unlimited, global topology information could be used.

To reduce the delay for agent migration and message transfer, it is preferable that the new link between the second node and the new node acts to shorten the logical hop count between them drastically. Recall, however, that migration cost is more important, as described in subsection 2.2. Therefore, for all candidate nodes, i.e., nodes that are x hops from the new node, the sum of migration costs on the least migration cost path to the new node is calculated as a measure. The proposed method chooses the node with the largest value as the second node.

Note here, however, that by adding a new link, the degree of the connected node is incremented. In other words, this increases the migration cost on the node by $t/2$. To express these effects quantitatively, we introduce an evaluation function $C(v)$. Let v_s denote the new node, v_{s1} the first node connected from the new node, $V_L = \{v_1, v_2, \cdots, v_n\}$ a set of nodes in the present logical network, and $V_L' = \{v_1, v_2, \cdots, v_m\}$ a set of candidates for the second node to be connected from the new node, where $(v_{s1} \notin V_L', V_L' \subset V_L)$. In addition, the migration cost $c(v_i)$ of the node v_i is expressed by $\frac{l(v_i)}{2}$, where $l(v_i)$ denotes the degree of the node v_i.

Using this notation, the evaluation function $C(v_i)$ is expressed by

$$C(v_i) = \{c(v_{s1}) + c(v_{s1}, v_i)\} - c(v_i), \tag{1}$$

where $c(v_i, v_j)$ denotes the sum of migration costs on the optimal path between the node v_i and the node v_j. The function's value is the difference in migration costs between the existing route $v_s \to v_{s1} \to v_i$ and the created route $v_s \to v_i$. Thus, to reduce the delay for agent migration, the node $v_i \in V_L'$ that has the largest value of $C(v_i)$ should be chosen as the second node connected to the new node.

On the other hand, to reduce the amount of traffic on the physical network, we introduce another evaluation function $H(v_i)$. Using the same notation as before, it is expressed by

$$H(v_i) = h(v_{s1}, v_i), \tag{2}$$

where $h(v_{s1}, v_i)$ means the physical hop count on the shortest path between the nodes v_{s1} and v_i. This function indicates the amount of traffic on the physical network when an agent migrates from node v_{s1} to node v_i.

Obviously, the node $v_i \in V_L'$ that has the minimum value of $H(v_i)$ should be chosen as the second node to be connected from the new node.

Finally, to find the tradeoff point between the delay and the amount of traffic, we combine these two functions. We first normalize them linearly between the largest and smallest values for each function. That is, the normalized functions $D_C(v_i)$ and $D_H(v_i)$ are expressed by

$$D_C(v_i) = \frac{C_{\max} - C(v_i)}{C_{\max} - C_{\min}} \quad (C_{\max} > C_{\min}), \tag{3}$$

$$D_H(v_i) = \frac{H(v_i) - 1}{H_{\max} - 1} \quad (H_{\max} > 1), \tag{4}$$

where

$$C_{\max} = \max\{C(v_i) \mid \forall v_i \in V_L'\}, \tag{5}$$

$$C_{\min} = \min\{C(v_i) \mid \forall v_i \in V_L'\}, \tag{6}$$

$$H_{\max} = \max\{H(v_i) \mid \forall v_i \in V_L'\}. \tag{7}$$

As an exception, $D_C(v_i) = 0$ for $C_{\max} = C_{\min}$, and $D_H(v_i) = 0$ for $H_{\max} = 1$.

Using these functions, the final evaluation function $D(v_i)$ is defined by

$$D(v_i) = (1 - k)^2 \cdot D_C(v_i)^2 + k^2 \cdot D_H(v_i)^2. \tag{8}$$

In this equation, k is a tunable parameter. It is set to a suitable value for the target application and the proposed method chooses the node v_i that has the minimum value of $D(v_i)$ as the second node to be connected to the new node.

3.2   Deleting a Node

When a node is removed from the logical network, the proposed method adds new links so that the logical network can maintain the third feature described in subsection 2.2. Moreover, the property for the delay and the amount of traffic should be kept unchanged as far as possible.

We assume that each node maintains the time when it joined the logical network, and which links were its first and second choices. Moreover, we propose that when a node has joined a logical network, the two nodes connected to it also record the connection time.

When a node leaves a logical network, it sends a message containing the list of neighbor nodes with their joining times to all neighbor nodes. Let v_r and $V_N(v_r) = \{v_r(1), v_r(2), \cdots, v_r(n)\}$ denote the leaving node and its set of neighbor nodes, respectively, where node $v_r(i)$ was the ith node to join the logical network in the set $V_n(v_r)$.

Just as when a node joins the network, nodes x hops away from the leaving node are candidate nodes for adding a new link. Let $V_L''(v_r(i))$ denote the set of nodes that joined before node $v_r(i)$ among the candidate nodes. In addition, let $V_N'(v_r)$ denote the set of nodes that joined before node $v_r(i)$ among the nodes $v_x \in V_N(v_r)$.

For each $v_r(i)$, when the link to the leaving node was the first choice, the proposed method chooses the node that has the minimum $H(v_x)$, where $v_x \in V_N'(v_r)$ or $v_x \in V_L''(v_r(i))$. The chosen node is connected to node $v_r(i)$.

On the other hand, when the link to the leaving node was the second choice, the proposed method chooses the node that satisfies $v_x \in V_N'(v_r)$ and $v_x \notin V_L''(v_r(i))$. If there is no such node, the node that has the minimum D among $V_L''(v_r(i))$ is chosen. For each $v_r(i)$, the chosen node is connected.

It has been proved elsewhere that this process can ensure that each pair of nodes in the logical network always has at least two alternative paths[10].

3.3   Execution Example

We demonstrate an example of the execution of the proposed method. In Fig. 4, there are 17 physical nodes and they are connected by the physical links indicated by the broken lines. Among these nodes, suppose that the three nodes A, B, and C form a logical network, and the nodes D and E join it, in that order. For simplicity, we set $x = 5$. Thus, all nodes are candidates for connection from a newly added node. In addition, we set $k = 0.5$.

In this figure, the node D first connects to the node C, because it is physically the closest node of A, B, and C. Next, the evaluation function $D(v_i)$ is calculated. For node A, $C(v_A) = -1$, because the degrees of node C and A are three and two, respectively, and $H(v_A) = 4$. Similarly for node B, $C(v_B) = -1$ and $H(v_B) = 3$. As a result, $D(v_A) = 1$ and $D(v_B) = 0.44$. Thus, node B is chosen as the second node connected to node D.

Next, Fig. 5 shows the situation after node E has been connected to node D, after each evaluation function is calculated as shown in the figure, in the same manner as for node D. As a result, node A is chosen as the second node connected to node E.

For space reasons, we omit an execution example of the node deletion process.

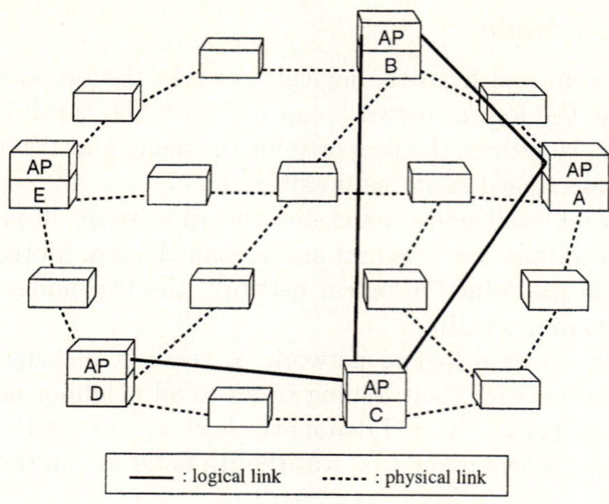

Fig. 4. Execution example of node adding process

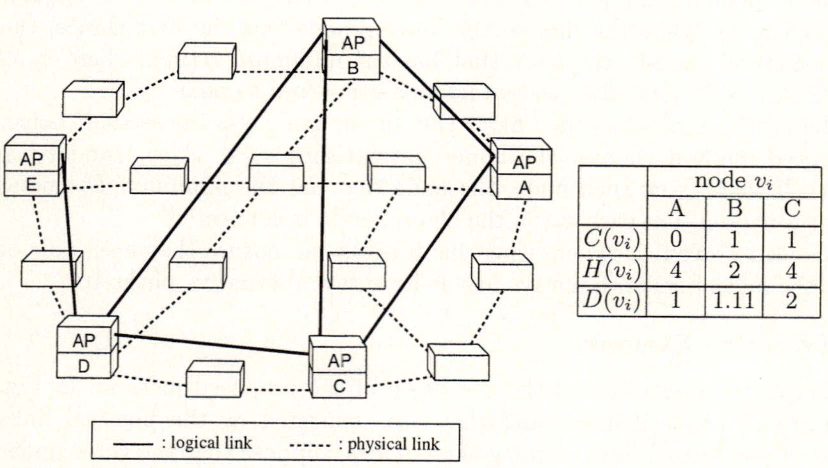

	node v_i		
	A	B	C
$C(v_i)$	0	1	1
$H(v_i)$	4	2	4
$D(v_i)$	1	1.11	2

Fig. 5. Logical network construction in process

4   Performance Evaluation

4.1   Simulation Model

We evaluated the performance of the proposed method by simulation experiments. The physical network topology was a lattice network with 400 nodes. From these nodes, 200 randomly selected nodes were added to the logical network sequentially.

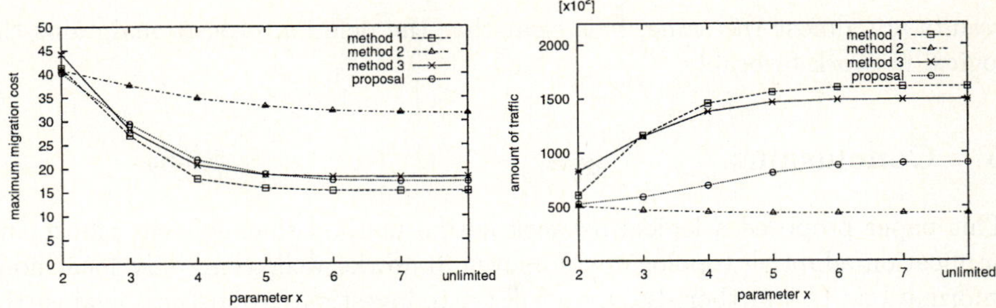

Fig. 6. Evaluation of the addition process

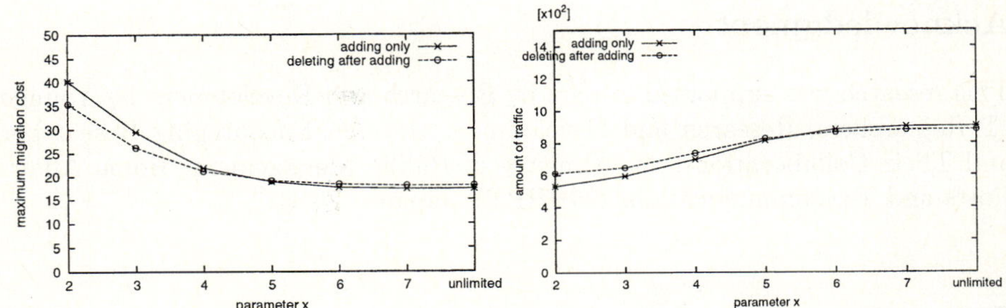

Fig. 7. Evaluation of the deletion process

We introduce the maximum sum of migration costs on paths between any pair of nodes and the amount of traffic on the physical network when 100 agents migrate from one node to any other node as performance measures.

We compared our method with three others. In method 1 a new node connects to the nearest and second nearest nodes. In method 2, a new node connects to the nodes that have the maximum and second maximum values of C, while in method 3, a new node connects to two randomly selected nodes.

4.2   Simulation Results

Figure 6 shows the maximum sum of migration costs and the amount of traffic as functions of the parameter x, averaged over 1000 experiments. These graphs show that the proposed method drastically reduces the amount of traffic while keeping the maximum sum of migration costs small. Values of the parameter $x \geq 4$ bring out the potential of the proposed method. It is clearly practical.

To evaluate the deletion process, we performed another simulation. With the same physical network, after 400 nodes were added to the logical network, 200 nodes were deleted sequentially.

Figure 7 shows the maximum sum of migration costs and the amount of traffic as a function of the parameter x, averaged over 1000 experiments. In these graphs, the solid lines indicate the results for 200 addition processes and the broken lines indicate the results for 400 additions and 200 deletions. These results are almost the same. It is clear that the deletion process maintains the logical network favorably.

5   Conclusions

This paper proposed a logical network design method that delivers robust and application-adaptive topologies. Moreover, it works well using only local node information. As a further study, we will try to investigate robustness against the topology of the physical network.

Acknowledgment

This research was supported in part by Research and Development by Younger IT Researchers, Research and Development through Encouraging Researchers and Their Collaborations, the Ministry of Public Management, Home Affairs, Posts and Telecommunications (MPHPT), Japan.

References

1. I. Satoh, "An Architecture for Next Generation Mobile Agent Infrastructure," Proceedings, International Symposium on Multi-Agent and Mobile Agents in Virtual Organizations and E-Commerce (MAMA'2000), pp. 281–287, ICSC Academic Press (Dec. 2000).
2. V. A. Pham and A. Karmouch, "Mobile Software Agents: An Overview," IEEE Communications Magazine, vol.36, no.7, pp. 26–37 (July 1998).
3. J. E. White: "Telescript Technology: in Mobile Agents," in J. Bradshaw (Ed.), Software Agents, MIT Press, Massachusetts (1996).
4. B. D. Lange and M. Oshima, http://www.trl.ibm.com/aglets/.
5. Jade; Java Agent Development Framework, http://jade.tilab.com/.
6. K. Watanabe, K. Kinoshita, N. Yamai, M. Kuwada, and K. Murakami, "Agent Gateway: A Location Transparent Agent Platform on Multi-Plane Logical Network," in Proceedings of the SoftCOM '99, pp.325–334 (Oct. 1999).
7. O. Babaoglu, H. Meling, and A. Monstresor, "Anthill: A Framework for the Development of Agent-Based Peer-to-Peer Systems," in Proceedings of the ICDCS22, pp. 15–22 (2002).
8. E. Steegmans, P. Rigole, T. Holvoet, and Y. Berbers, "Intelligent Buildings: A Multi-Agent System Approach," Artificial Intelligence and Applications Proceedings, pp. 771–776 (2003).
9. C. Partridge, T. Mendez, and W. Milliken, "Host Anycasting Service," RFC1546 (1993).
10. A. S. Tanenbaum, Computer Networks Third Edition, Prentice-Hall, New Jersey (1996).

Reduced-State SARSA Featuring Extended Channel Reassignment for Dynamic Channel Allocation in Mobile Cellular Networks

Nimrod Lilith and Kutluyıl Doğançay

School of Electrical and Information Engineering,
University of South Australia, Mawson Lakes, Australia

Abstract. This paper introduces a reinforcement learning solution to the problem of dynamic channel allocation for cellular telecommunication networks featuring either uniform or non-uniform offered traffic loads and call mobility. The performance of various dynamic channel allocation schemes are compared via extensive computer simulations, and it is shown that a reduced-state SARSA reinforcement learning algorithm can achieve superior new call and handoff blocking probabilities. A new reduced-state SARSA algorithm featuring an extended channel reassignment functionality and an initial table seeding is also demonstrated. The reduced-state SARSA incorporating the extended channel reassignment algorithm and table seeding is shown to produce superior new call and handoff blocking probabilities by way of computer simulations.

1   Introduction

Cellular systems organise a geographical area into a number of regularly sized cells, each with its own base station. By adopting this system of using a large number of low power transmitters rather than a singular high power transmitter, the carrying capacity of a given area for calls can be greatly increased. The available bandwidth is divided into a number of channels, which may be time slots or frequencies, each of which may be assigned to a call. Using a cellular system allows a given channel to be assigned simultaneously to multiple calls, as long as each assigning cell is at least a given distance apart, in order to avoid co-channel interference. This distance is termed the 'reuse distance'. Most modern mobile communication systems use a Fixed Channel Assignment (FCA) strategy, whereby channels are pre-allocated to given cells according to a regular pattern that minimises the distance between co-channel cells, i.e. cells that may assign the same channel to a call, whilst not violating the channel reuse distance constraint. Ongoing calls may move spatially about the cellular domain and this movement can lead to a call leaving one cellular area and entering an adjacent one. This then requires resources in the cell entered to be allocated to the new call and subsequently any resources in the cell left may be freed, a process known as 'handoff'. If resources cannot be allocated in the newly entered cell, then the handoff is blocked. In contrast to FCA, Dynamic Channel

P. Lorenz and P. Dini (Eds.): ICN 2005, LNCS 3421, pp. 531–542, 2005.
© Springer-Verlag Berlin Heidelberg 2005

Assignment (DCA) strategies do not permanently pre-allocate given channels to particular cells. Instead channels are assigned to cells as they are required, as long as these assignments do not violate the channel reuse constraint. This flexibility in channel assignment allows a cellular system to take advantage of possible stochastic variations in offered call traffic over a given area. A number of DCA schemes have been devised [3] which follow certain procedures in order to attempt to maximise the total traffic carried in a cellular area. One method of achieving this aim is to minimise the new call blocking probability, which is the probability that a new call will not be assigned a channel and thus be blocked. This paper proposes a new state reduction and limited channel reassignment mechanism to improve solution performance and reduce memory requirements, and examines the performance of the reduced-state reinforcement learning solutions to DCA in mobile cellular networks with handoffs. The proposed method of state reduction has been motivated by the weak link between one of the states and the blocking probability performance. Channel reassignment on call termination has been introduced in an attempt to partially mimic the optimal channel allocation method [3], which is computationally infeasible. The rest of the paper is organised as follows. Firstly, a brief introduction to reinforcement learning techniques is included in section 2. Section 3 details the specifics of the problem formulation employed, followed by simulation methods and results in section 4. Finally conclusions are drawn in section 5.

2   Reinforcement Learning

Assume $X = \{x_1, x_2, \ldots, x_N\}$ is the set of possible states an environment may be in, and $A = \{a_1, a_2, \ldots, a_N\}$ is the set of possible actions a learning agent may take. The learning agent attempts to find an optimal policy $\pi^*(x) \in A$ for all x which maximises the total expected discounted reward over time. The value of following policy π from state x at time t (denoted x_t) can then be expressed as:

$$V^\pi(x) = E\{ \sum_{k=0}^{\infty} \gamma^k r(x_{t+k}, \pi(x_{t+k})) \,|\, x_t = x\} \tag{1}$$

where $r_t = r(x_t, a)$ is the reward received at time t when taking action a in state x and γ is a discount factor, $0 \leq \gamma \leq 1$.

The action-value function for a given policy π, which is defined as the expected reward of taking action a in state x at time t and then following policy π thereafter, may then be expressed as:

$$Q^\pi(x, a) = E\{ r(x, a) + \sum_{k=1}^{\infty} \gamma^k r(x_{t+k}, \pi(x_{t+k})) \,|\, x_t = x, a_t = a\} \tag{2}$$

This formulation allows modified policies to be evaluated in the search for an optimal policy π^*, which will have an associated optimal state value function:

$$V^*(x) = V^{\pi^*}(x) = \max_\pi V^\pi(x), \forall\, x \in X, \tag{3}$$

and an optimal state-action value function:

$$Q^*(x, a) = \max_\pi Q^\pi(x, a), \forall\, x \in X \text{ and } a \in A, \tag{4}$$

which can be written in terms of (3) as:

$$Q^*(x, a) = E\{\, r(x, a) + \gamma V^*(x_{t+1}) \mid x_t = x, a_t = a\} \tag{5}$$

The state-action values of all admissible state-action pairs may be either represented in a table form or approximated via function approximation architecture, such as an artificial neural network, to save memory [1][5][10] . These state-action values are then updated as the learning agent interacts with the environment and obtains sample rewards from taking certain actions in certain states. The update rule for the state-action values is:

$$Q_{t+1} = \begin{cases} Q_t(x, a) + \alpha \Delta Q_t(x, a), & \text{if } x = x_t \text{ and } a = a_t \\ Q_t(x, a), & \text{otherwise} \end{cases} \tag{6}$$

where α is the learning rate, $0 < \alpha \le 1$, and

$$\Delta Q_t = \{r_t + \gamma \max_a Q_t(x_{t+1}, a_{t+1})\} - Q_t(x, a) \tag{7}$$

for Watkins' Q-Learning algorithm [11]. If α is reduced to zero in a suitable way and each admissible state-action pair is encountered infinitely often, then $Q_t(x, a)$ converges to $Q^*(x, a)$ as $t \to \infty$ with probability 1 [11].

The Q-Learning algorithm is an off-policy technique as it uses different policies to perform the functions of prediction and control. On-policy reinforcement learning methods, such as SARSA [10], differ from off-policy methods, such as Q-Learning, in that the update rule uses the same policy for its estimate of the value of the next state-action pair as for its choice of action to take at time t, that is for prediction and control. The update rule for the state-action values for SARSA takes the same form as that of Q-Learning (6) but with a different definition for $\Delta Q_t(x, a)$:

$$\Delta Q_t(x, a) = \{r_t + \gamma Q_t(x_{t+1}, a_{t+1})\} - Q_t(x, a) \tag{8}$$

3   Problem Formulation and Proposed Solution

3.1   Problem Structure

We consider a 7-by-7 cellular array with a total of 70 channels available for assignment to calls as the environment in which the learning agent attempts to find an optimal channel allocation policy. This is the same system simulated in [6][8][9].

Denoting the number of cells by N, and the number of channels by M, the state at time t, x_t, is defined as:

$$x_t = (i_t, L(i)_t), \tag{9}$$

where $i_t \in \{1, 2, \ldots, N\}$ is the cell index in which the call event at time t takes place, and $L(i)_t \in \{0, 1, \ldots, M\}$ is the number of channels allocated in cell i at time t, which can in turn be expressed as:

$$L(i)_t = \sum_{m=1}^{M} l(i, m)_t, \tag{10}$$

where

$$l(i, m)_t = \begin{cases} 1, & \text{if channel } m \text{ is allocated in cell } i \text{ at time } t \\ 0, & \text{otherwise} \end{cases} \tag{11}$$

This state formulation has been previously employed in reinforcement learning solutions to the DCA problem and is similar to that featured in [6] and [8], although our formulation defines $L(i)_t$ as the number of allocated channels, rather than the number of available channels as in those works. It was decided to define $L(i)_t$ as a direct measure of a given cell's currently allocated share of the total channels in the interference region centred on that cell which will enable extension of this work to incorporate call admission control.

In order to facilitate the computation of admissible actions and action rewards it was deemed necessary to be able to transform a cell's index into its constituent row and column values. Assuming i specifies the cell index in which a call event takes place in an $N \times N$ cellular system, j, the row of the cell can then be expressed as $j = \lceil i/N \rceil$, and k, the column of the cell can then be expressed as $k = i \bmod N$.

Admissible actions are restricted to assigning an available channel or blocking a call if no channels are available for assignment, i.e. a call will always be accepted and a channel assigned to it if there is at least one channel available in the originating cell. A given channel $m \in \{1, 2, \ldots, M\}$ is available for allocation in a given cell of index i, with corresponding row j and column k, of an $N \times N$ cellular system at time t if $v(i, m)_t = 0$, where:

$$v(i, m)_t = \sum_{p=j-2}^{j+2} \sum_{q=k-2}^{k+2} l(y, m)_t, \quad 1 \leq p \leq N \quad \text{and} \quad 1 \leq q \leq N, \tag{12}$$

where $y = (p - 1) \times N + q$.

An action can then be defined as $a = m, \quad m \in \{1, 2, \ldots, M\}$ and $v(i, m)_t = 0$.

Thus a state-action pair consists of three components, with a total of $N \times (M + 1) \times M$ ($49 \times 71 \times 70$) possible state-action pairs, which yields 243,530 distinct state-action pairs. In implementation this reduces to $N \times M \times M$ as a

Table 1. Simple Channel Reassignment Procedure

---

1. *Let c_{flag} = channel flagged for freeing*

2. *Get $c_{min} = \min_a Q_t(x, a)$, for x = current cell*

    $a \in$ *currently allocated channels in x*

3. *Free channel c_{flag}*

4. *If $c_{min} \neq c_{flag}$:*

    a) *Reallocate call on channel c_{min} to c_{flag}*

    b) *Free channel c_{min}*

---

cell with 70 channels already allocated, or 0 available, cannot accept a new call and thus action values for these states are not required for learning, resulting in 240,100 admissible state-action pairs.

The reward attributed towards each successful channel assignment event is the total number of ongoing calls in the system, which can be expressed in an $N \times N$ system as:

$$r = \sum_{p=1}^{N} \sum_{q=1}^{N} \sum_{m=1}^{M} l(y, m). \tag{13}$$

Rather than be concerned with the allocation patterns of specific channels over the system area, this reward structure is designed to minimise new call blocking probability by attempting to maximise the number of calls carried by the system at any given time, an approach similar to that in [9]. Handoff calls are treated as new calls by the cell being entered, and as such any channel allocated is rewarded and incorporated in the learning procedure.

The large number of possible state-action pairs signals that there may be some problems converging to an optimal or an acceptable near-optimal policy in a highly dynamic environment, especially if the learning is to be implemented online. Offered traffic patterns may vary not only spatially but also temporally [2] and thus an algorithm that can adapt to these variations and produce efficient policies in a minimal timeframe should be preferred.

3.2 Proposed Solution

Given the possibility of convergence issues in the implementation of our reinforcement learning solution we reduced the number of admissible state-action pairs using two different techniques. Firstly we reduced the number of possible state-action pairs by aggregating rarely encountered states. Specifically, the magnitude of the variable representing the number of channels currently allocated was limited to 30, rather than the original range of 0-69 (roughly half the number of channels available), i.e. $x_t = (i_t, L(i)_t),   0 \le L(i)_t \le 30$.

Simulations of this problem over a similar system have shown that over a 24 hour period a given cell will rarely, if ever, have more than a fraction of the total available channels allocated in it at any one time [4]. Consequently, the states associated with these configurations may rarely, if ever, be visited over this time period. By reducing the range for the number of channels allocated to 0-30, we are effectively trimming the size of the table required from 240,100 ($49 \times 70 \times 70$) to 106,330 ($49 \times 31 \times 70$), a reduction of over fifty percent in memory requirements. Secondly, we have deliberately excluded the state variable representing the number of channels currently allocated (m), i.e. $x_t = (i_t)$.

This resulted in reducing the number of state-action variables to two and bringing the number of possible state-action pairs from 240,100 ($N \times M \times M$) to 3,430 ($N \times M$), a reduction of over 98%. An important advantage of a reduced-dimension state space for reinforcement learning is that it may lead to a faster learning or convergence rate because of the increased rate of exploration, despite discarding potentially useful information.

Both of these memory saving techniques were simulated and compared to not only a full-table-based reinforcement learning solution, but also to a fixed channel allocation algorithm and a random channel allocation algorithm.

A simple channel reassignment mechanism was also included, whereby upon encountering a termination or handoff event the reinforcement learning agent considered a single channel reassignment (Table 1). The set of currently allocated channels in the cell, including the channel allocated to the call due for termination, was evaluated as if for a new call channel assignment, but rather than assigning the maximal valued channel the agent terminated the minimally valued channel, possibly requiring a reassignment if that minimally valued channel was not the one due for call termination. This required only a simple search over the currently learnt channel allocation values held in the agent's memory table, and as no learning was conducted on the reassignment actions this process was conducted in a strictly greedy manner.

This simple channel reassignment was then extended by invoking it on new call arrival events (Table 2). The computational overhead of this more aggressive reassignment strategy was still low, requiring a search over the same previously learnt state-action values, and assignments actions were still limited to at most one channel reassignment per call event.

Lastly, a scheme that initially placed an identical positive value in all entries of the agent's state-action value table corresponding to a fixed channel allocation scheme assuming a uniform offered traffic load was implemented prior to simulation, i.e.:

$$Q_0(x,a) = \begin{cases} 100.0, & \text{if } 0 < (a-c) \leq 10 \\ 0.0, & \text{otherwise} \end{cases} \tag{14}$$

where $c = ((x - 4 \times \lfloor (x-1)/7 \rfloor) \bmod 7) \times 10$ for the simulated 7×7 cellular system with 70 channels. This effectively 'seeded' the table so as to encourage the learning agent to favour actions consistent with an evenly distributed allocation of channels over the system. This method was simulated using both uniform and non-uniform traffic arrival patterns.

Table 2. Extended Channel Reassignment Procedure

---

1. Let $c_{max} = \max_a Q_t(x, a)$, for $x =$ current cell

    $\qquad\qquad a \in$ currently non-allocated channels in x

2. Let $v_{max} = Q_t(x, c_{max})$

3. Let $c_{min} = \min_a Q_t(x, a)$, for $x =$ current cell

    $\qquad\qquad a \in$ currently allocated channels in x

4. Let $v_{min} = Q_t(x, c_{min})$

5. If $v_{max} > v_{min}$:

    a) Reallocate call on channel c_{min} to c_{max}

    b) Free channel c_{min}

---

4 Simulation Methods and Results

4.1 Simulation Methods

New call arrivals were modeled as independent Poisson processes with a uniform distribution pattern, with mean call arrival rates λ between 100 to 200 calls/hour. The call duration obeyed an exponential distribution with a mean of $1/\mu$ equal to 3 minutes. New calls that were blocked were cleared. A proportion of calls ($p = 15\%$) were simulated as handoff traffic [7]. When a call was about to terminate a check was made to see if the termination was due to the call entering a new cell, i.e. if it fired a handoff event. The entered cell for handoff traffic was chosen randomly using a uniform distribution with all neighbouring cells as eligible. All handoff traffic was simulated according to an exponential distribution with a mean call duration $1/\mu$ of 1 minute in the new cell.

Both Q-Learning and SARSA were implemented and simulated. The Q-Learning algorithm simulated employed a different reward scheme, classifying channels according to their usage patterns with respect to an individual cell [6]. The simulated Q-Learning algorithm also used an alternate state representation, which was a combination of the cell index and the number of channels currently available in the cell. Channel assignment actions were selected using an ϵ-greedy algorithm, with ϵ being diminished over time. State-action values were stored using a table-based representation. An additional two variations of SARSA were also simulated, a trimmed-table SARSA (TT-SARSA) and reduced-state SARSA (RS-SARSA), both incorporating table-based state-action value storage and the same ϵ-greedy reduction mechanism. Versions of the SARSA algorithm which included channel reassignment functionality were also simulated.

TT-SARSA uses an identical state-action formulation as the SARSA algorithm but limits the value of the state variable representing the number of channels currently allocated in a cell to thirty, reducing the number of admissible state-action pairs. RS-SARSA further reduces the number of state-action pairs by eliminating the state variable representing the number of channels currently allocated in a cell. Both of these techniques drastically reduce the memory required for table-based storage of the state-action value function, as discussed in section 3.

4.2   Call Blocking Probabilities

Firstly the system was simulated with a uniform offered traffic load over all cells of 100 to 200 calls/per hour, corresponding to 5 to 10 Erlangs. All algorithms were simulated in an online manner, i.e. there was no initial learning period where Q values could be learnt via simulation prior to implementation. After 24 simulated hours the respective new call and handoff blocking probabilities were calculated (Fig. 1 & Fig. 2) . As can be expected, the fixed channel allocation algorithm performs the worst over the entire range of traffic loads, whilst all the reinforcement learning algorithms exhibit superior performance in terms of both blocking probabilities.

The channel reassignment functionality, denoted (RA), has a marked effect in decreasing the blocking probabilities of all of the SARSA algorithms, particularly RS-SARSA(RA) which has a much better new call and handoff blocking probability than all other assignment strategies simulated. RS-SARSA requires only a fraction of the memory of SARSA to represent its state-action values, and as was pointed out in section 3, this reduction in total state space can lead to an improvement in learning performance. Furthermore, TT-SARSA and TT-SARSA(RA) have identical performances to SARSA and SARSA(RA) respectively whilst requiring less than half the memory for state-action value representation, indicating that the aggregated states are rarely if ever visited and their aggregation has no effective bearing on the algorithm's performance.

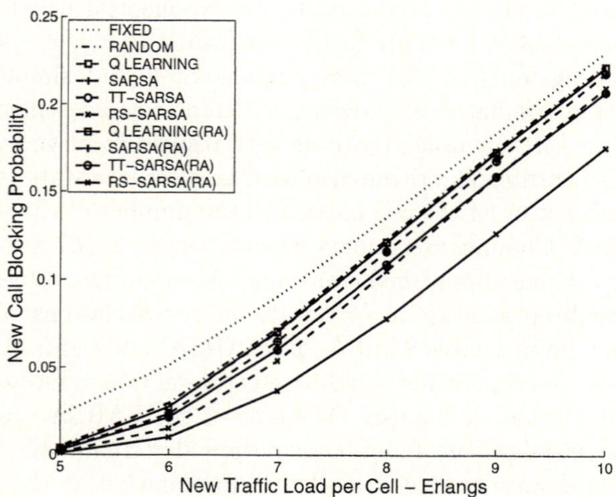

Fig. 1. New call blocking probability

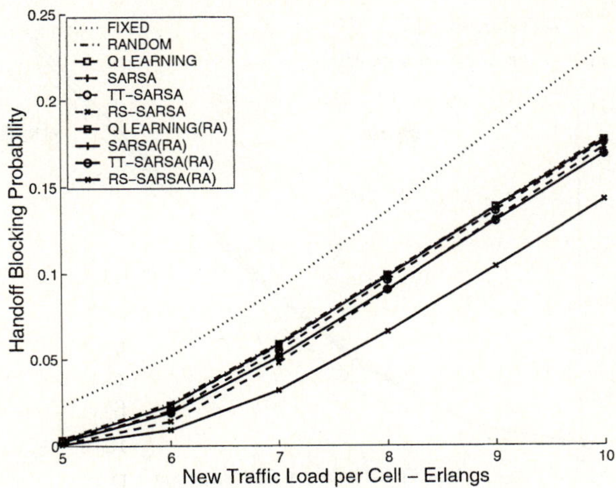

Fig. 2. Handoff blocking probability

4.3   Aggressive Channel Reassignment and Table Seeding

The simple channel reassignment scheme simulated in subsection B was extended
to be invoked on new call arrivals as well as call departures, as detailed earlier
in section 3. Furthermore, prior to simulation the state-action value table was
seeded with a value of 100.0 in elements corresponding to a uniform fixed channel
allocation scheme, i.e. a band of ten channel assignment actions per cell.

Figures 3 and 4 show the new call and handoff blocking probabilities respectively for RS-SARSA with the simple channel reassignment scheme, RS-SARSA
with channel reassignments also enacted on new call events, denoted RS-SARSA
- AGG.RA, and RS-SARSA incorporating both the aggressive channel reassignment scheme and a fixed seeding of 100.0 prior to simulation initialisation as
described above, RS-SARSA - AGG.RA(FS).

For low traffic levels RS-SARSA - RA and RS-SARSA - AGG.RA perform
similarly, whilst at higher traffic levels RS-SARSA - AGG.RA shows both a
superior new call and handoff blocking probability, as more channel reassignment
events are possible which can in turn lead to a more efficient use of channels when
they are in a higher demand, leading to greater channel availability and therefore
a lower blocking probability. Over all traffic loads simulated the seeding of the
agent's state-action value table produced an appreciable improvement of both
the new call and handoff blocking probability performance.

Seeding the agent's value table according to a fixed channel allocation scheme
assuming uniform traffic led to performance improvements when simulated with
a uniform traffic arrival. However making a priori assumptions about offered traffic patterns potentially weakens one of the strengths of a reinforcement learning
approach, namely that a reinforcement learning agent requires no model of the
environment beforehand and can adapt to environment dynamics. In view of this

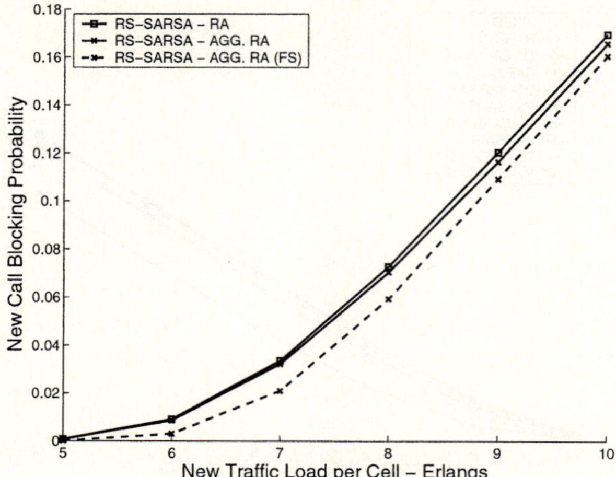

Fig. 3. Uniform new call blocking probability

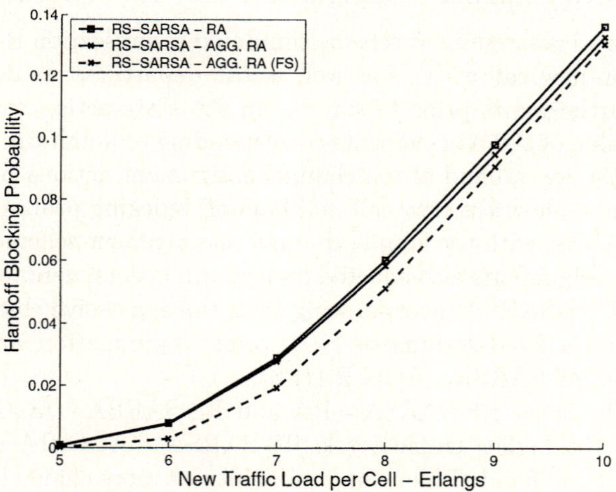

Fig. 4. Uniform handoff blocking probability

the efficacy of seeding the agent's value table assuming a uniform offered traffic load was simulated with an actual non-uniform new traffic pattern with a base traffic arrival rate of between 20 and 200 calls/hour which was then increased by up to 100 percent, the results of which are shown in Fig. 5.

Incorporating both the more aggressive channel reassignment functionality and an initial table seeding according to a uniform channel allocation scheme actually improved both the new call and handoff blocking probabilities for the non-uniform traffic arrival pattern simulated across all traffic loads.

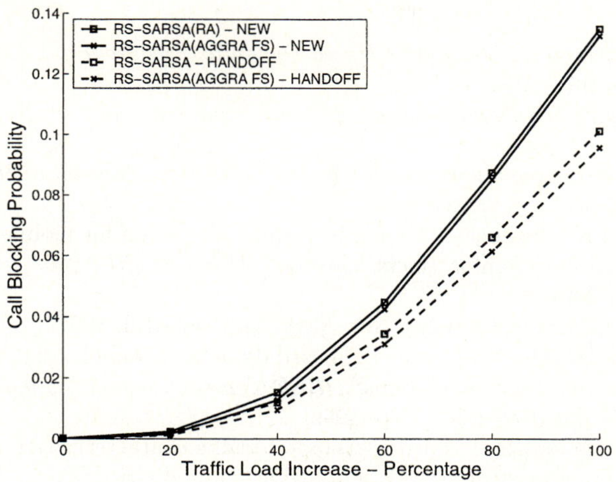

Fig. 5. Non-uniform new call and handoff blocking probability

5   Conclusions

We have proposed a novel reduced state SARSA algorithm featuring channel reassignment for dynamic channel allocation in mobile cellular networks with handoffs. By carefully reducing the state space the problem of dynamic channel allocation in cellular mobile networks with handoffs can be solved, gaining significant savings in terms of memory requirements compared to currently available table-based reinforcement learning schemes. Furthermore, this cost reduction can be achieved with effectively no performance penalty.

Additionally, by eliminating a state component of secondary importance we achieved a further reduction in memory requirements of over 98% in total, as noted in section 3.2. The reduced-state representation incorporating channel reassignment functionality also achieved improved performance in terms of both new call and handoff blocking probability over all other reinforcement learning solutions simulated.

Lastly, by extending the channel reassignment functionality and incorporating an initial state-action value table seeding according to a uniform channel allocation scheme, superior new call and handoff blocking probabilities were demonstrated over both uniformly and non-uniformly distributed offered traffic loads of varying traffic intensities in all simulations undertaken.

Acknowledgements

This work was supported by an Australian Research Council Linkage Project Grant and the Motorola Australia Software Centre.

References

1. D. Bertsekas and J. Tsitsiklis, *Neuro-Dynamic Programming*, Belmont, Mass.: Athena Scientific, 1996.
2. R. L. Freeman, *Telecommunication System Engineering*, 3rd ed. New York: Wiley, 1996.
3. S. Jordan, "Resource allocation in wireless networks", *Journal of High Speed Networks*, vol. 5 no.1, pp. 23-34, 1996.
4. N. Lilith and K. Doğançay, "Dynamic channel allocation for mobile cellular traffic using reduced-state reinforcement learning", *IEEE WCNC. 2004*, vol. 4, pp. 2195-2200, March 2004.
5. T. Mitchell, *Machine Learning*, New York: McGraw-Hill, 1997.
6. J. Nie and S. Haykin, "A Q-learning-based dynamic channel assignment technique for mobile communication systems", *IEEE Transactions on Vehicular Technology*, vol. 48 iss. 5, pp. 1676 -1687, Sep. 1999.
7. M. Rajaratnam and F. Takawira, "Handoff traffic characterization in cellular networks under nonclassical arrivals and service time distributions", *IEEE Transactions on Vehicular Technology*, vol. 50 iss. 4, pp. 954-970, Jul. 2001.
8. S. -M. Senouci and G. Pujolle, "Dynamic channel assignment in cellular networks: A reinforcement learning solution", *ICT 2003, 10th International Conference on Telecommunications*, vol.1, pp. 302-309.
9. S. Singh and D. Bertsekas, "Reinforcement learning for dynamic channel allocation in cellular telephone systems", *Advances in NIPS 9*, MIT Press, pp. 974-980, 1997.
10. R.Sutton and A. Barto, *Reinforcement Learning: An Introduction*, MIT Press, 1998.
11. C. J. C. H. Watkins and P. Dayan, "Q-learning", *Machine Learning*, vol. 8, pp. 279-292, 1992.

Call Admission Control for Next Generation Cellular Networks Using on Demand Round Robin Bandwidth Sharing

Kyungkoo Jun and Seokhoon Kang

Department of Multimedia System Engineering,[*]
University of Incheon, Korea
{kjun, hana}@incheon.ac.kr

Abstract. In this paper, we propose a novel call admission control scheme which improves call blocking probabilities as well as the packet–level QoS in the multiservice environment of next generation cellular networks. In our scheme, we suggest that bandwidth–inefficient service sessions such as WWW browsing of which traffic pattern contains frequent *think time* are grouped dynamically into a set, called *lock*, and forced to share the bandwidth with others on the basis of on demand round robin instead of being allocated separate bandwidth especially when the resource availability is unfavorable. Compared with the corresponding results of a conventional reconfiguration strategy by the use of simulation study, our scheme shows improved packet delay and loss rate in the case of WWW traffic, and superior performance in terms of blocking probabilities of both new call and handoff cases.

1   Introduction

Call admission control schemes ever developed with regard to the multiservice of next generation cellular networks have mostly focused on making admission decisions by considering the priorities among service classes or by exploiting the ability to adjust required bandwidth of services according to resource availability [1][2][5][6][8][7]. Among them, in [2], traffic with strict QoS requirements can "borrow" bandwidth from traffic which presents less strict QoS requirements. In [1], handoff calls are given priority over new calls by reserving a portion of bandwidth of the target cells. It also employs a *bandwidth reconfiguration scheme*; assuming that services have multiple levels of required bandwidth, services can downgrade their required bandwidth according to resource availability, or deprive bandwidth from low priority services by coercing them to drop their bandwidth level. In [7], besides bandwidth utilization and blocking/dropping probabilities, two new user-perceived QoS metrics, degradation ratio and upgrade/degrade frequency, are introduced in the process of call admission.

----

[*] This work was supported by the University of Incheon.

P. Lorenz and P. Dini (Eds.): ICN 2005, LNCS 3421, pp. 543–550, 2005.

One of the common problems observed in the aforementioned strategies is that, even with the bandwidth reconfiguration capability, it is inevitable to allocate low priority services certain portion of bandwidth even though most of which are not fully utilized because of the traffic characteristic of the services. For example, the bandwidth allocated for WWW service is mostly underutilized because of the presence of user think time, the time period until clicking to access other pages, which will be discussed shortly.

In this paper, we propose a novel call admission control scheme to remedy the above problem by adding a bandwidth sharing feature to the reconfiguration strategy proposed in [1], which will be discussed shortly. The key idea of our proposed strategy is that the services which do not fully utilize their assigned bandwidth are grouped into a set, called *lock*, and then forced to share the bandwidth in round robin with others in the same group rather than being allocated separate bandwidth, thus benefiting *bandwidth-efficient* services, which are mostly real-time and, at the same time, have higher priority than others.

We assume in this paper the services belong to one of four classes of services which were recommended in UMTS domain as follows : *conversational class* (class 1) for voice or video conference traffic, *streaming class* (class 2) for real-time video streaming, *interactive class* (class 3) for WWW or database access, and *background class* for email or downloading. Note that, when we refer to the classes, we will use their names and numbers interchangeably in the remaining sections.

This paper is organized as follows. Section 2 discusses our proposed call admission control strategy in detail. Section 3 shows the simulation results for the performance analysis of our proposed scheme in comparison with the reconfiguration strategy. Section 4 concludes this paper.

2   The Proposed Call Admission Control Scheme

The proposed call admission control scheme is motivated by the fact that the traffic patterns of some service sessions, e.g., the class 3 service such as WWW browsing or database access consist of several bursts of packets, interspaced by idle periods. It starts with authentication and registration to the cellular networks, and then this will follow a series of packet bursts, i.e., packet calls, with actual data traffic. Especially, the idle periods between the packet calls are "think time (read time)", which is the time duration elapsed between the completion of the transfer of the previous request and the beginning of the transfer of the current request [9].

Considering the presence of think times, when the resource in a cell is insufficient, it can be advantageous in terms of system utilization to group such services into a set, and then have them share bandwidth BW_s in round robin rather than to allocate each service the separate BW_s. We call this grouping mechanism *lock*. The round–robin bandwidth sharing works on a rotating basis in that the services of the lock are allowed to own the bandwidth at their turns for a predefined maximum time. When they finish their job earlier than allowed or do not need the bandwidth at their rotation slot, they just pass the turn to next neighbor promptly. Thus, the lock enables a set of the services to continue with less bandwidth than in conventional allocation schemes by exploiting the bandwidth which otherwise might be wasted.

In our proposed call admission control scheme, we use the lock to benefit the admission of the service requests of high priority services, which belong to class 1 and 2, especially when the available bandwidth of a cell is so insufficient that even the reconfiguration effort to lower the bandwidth level of ongoing low priority services belonging to class 3 and 4 services, fails to accept the requests. On receiving such requests, a lock is formed to deprive from the class 3 or 4 services the bandwidth of which amount is large enough to grant those requests. In other words, the bandwidth arranged by this lock-based sharing can be assigned to higher priority or more real-time constrained services, in this case class 1 and class 2, to meet their QoS requirements.

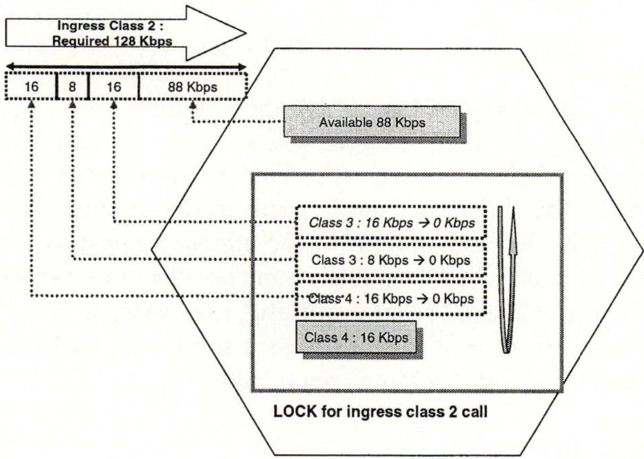

Fig. 1. The lock formation

2.1   The Lock Formation

When new–call or handoff requests of class 1 or 2 services are received at a cell which however has too little remaining bandwidth to admit the requests even by the reconfiguration, a lock is formed to arrange bandwidth by dynamically grouping class 3 or 4 services, called *locked services*. Multiple locks can be formed in a cell, and each lock is tightly coupled with the service which triggered its creation, called *lock initiator*, in the sense that the lock can be deleted only when the life time of its initiator expires.

The lock formation proceeds in three steps. At first, the amount of bandwidth to be collected, $BW_{collected}$, by the lock is calculated as

$$BW_{collected} = BW_{requested} - BW_{available} \tag{1}$$

where $BW_{requested}$ is the bandwidth amount required by ingress class 1 or 2 service, $BW_{available}$ is the available bandwidth at the moment in the cell. Obviously, a prerequisite condition $BW_{requested} > BW_{available}$ should be hold. The second step is to determine whether $BW_{collected}$ can be deprived from the lock creation by summing up the current bandwidth of class 3 or 4 services which will be locked but not already locked for others until the following two conditions are met;

1) The added amount satisfies

$$\sum_{i=1}^{c} BW_i \geq BW_{collected} \tag{2}$$

where c is the number of services whose bandwidth is included in this accumulation.

2) There exists another unlocked class 3 or 4 service whose bandwidth, $BW_{unlocked}$, satisfies the following.

$$BW_{unlocked} \geq max(BW_i), \quad i = 1, \ldots, c \tag{3}$$

The second condition is required to determine whether $BW_{unlocked}$, which will be shared by the lock members, is sufficient to accommodate even the largest required bandwidth of them. Finally, the lock is constructed by removing the bandwidth occupied by the locked services except one service ($BW_{unlocked}$) and then reallocating the collected amount to the ingress request as shown in Figure 1 where there are only 88 Kbps available when class 2 service requiring 128 Kbps enters a cell. Then the class 2 service initiates the creation of a lock to obtain 40 Kbps by locking two class 3 and one class 4 services. The 16 Kbps of class 4 in the lock is shared in round robin as explained.

2.2   The Lock Management

The Lock Update. The update on a lock occurs either when locked services withdraw from the lock or when incoming class 3 or 4 services try to join the lock. The withdrawal is the case of service termination or egress handoff.

If a service which withdraws from a lock occupies shared bandwidth at the moment, it should transfer the bandwidth to a next locked service before leaving the lock. If it is the last one residing in the lock, the bandwidth is released to be added to the available bandwidth of the cell. However, the lock itself continues to exist until the lock initiator service ends.

The other type of the lock update is the join of class 3 or class 4 services to existing locks. If a class 3 or 4 service enters a cell but there is not enough bandwidth, the service is still able to be admitted to the cell by joining one of the locks which have extra capacity to share the bandwidth. In other words, the join is allowed only if the following two conditions are met; 1) there is a space in the lock, i.e., the current number of locked services should be less than the number at the lock creation time, 2) the required bandwidth by the joining service is equal or less than the bandwidth shared in the lock. If a lock satisfying these two requirements is found, the requesting service is allowed to enter the cell by joining the lock. In this case, no bandwidth is needed to be allocated for this service.

The Lock Extermination. The lock can be deleted only when the lock initiator service ends at service time expiration or at handoff to other cell. The advantage of this limitation is two fold; simpler lock implementation and more consistent lock management with regard to creation and deletion.

When a lock is removed by one of the aforementioned reasons (expiration or handoff of an initiator), each bandwidth of the locked services is restored to the level before they are joined to the lock. If there is more bandwidth than needed for this restoration, the rest of bandwidth is reinstated to the cell for future allocation. It should be noted that, according to the Equation 1, the released bandwidth at the lock deletion must be larger than or equal to the sum of bandwidth required for the restoration.

2.3   Algorithm Details

For the new call connection, the algorithm works as follows. For class 1 and 2 new services, they first attempt to obtain their maximum bandwidth BW_{req}^{max} if this bandwidth is available in a cell. If it is not, they try to enter a cell at a degraded level by requesting BW_{req}^{min}. If there is sufficient bandwidth to do so, then the service is accepted with corresponding bandwidth allocation. If there is still insufficient bandwidth available, it is judged whether bandwidth enough to accept the services can be gathered by forming a lock: depriving bandwidth by locking class 3 and 4 services currently supported in the cell. If it is true, the services are accepted with BW_{req}^{min}. Otherwise, they are rejected. For class 3 and 4 new services, they first attempt to obtain their maximum bandwidth BW_{req}^{max} if this bandwidth is available in a cell. If it is not, they try to join any existing locks in the cell by searching for locks with capacity; less number of locked services than at the lock creation time remains in the lock. If such a lock is found, they can be admitted without bandwidth allocation. Otherwise they are rejected.

For the handoff case, the algorithm works in the same way as that of the new call connection above mentioned except the following two differences; firstly, for class 1 and 2 handoff services, the bandwidth reconfiguration of ongoing class 3 and 4 services is attempted before the lock formation, and secondly, for class 3 and 4 handoff services, they attempt to acquire BW_{req}^{max} and BW_{req}^{min} in order before deciding to join exiting locks.

3   Simulation and Performance Evaluation

In this section, we compare the performance of our proposed lock–based scheme with that of the conventional reconfiguration strategy by the use of the simulation. The comparison between them is made in both packet level and system level; in the packet level, the packet delay and loss rate measured during simulated WWW browsing sessions under each scheme are examined, whereas the

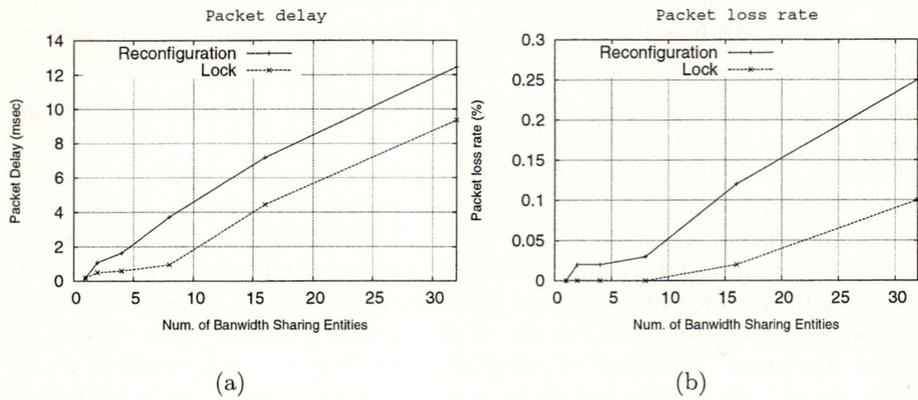

(a)　　　　　　　　　　　　　　(b)

Fig. 2. Comparison of the lock–based scheme with the reconfiguration strategy in terms of average packet delay and packet loss rate of WWW traffic

Table 1. Service Classes and Corresponding Required Bandwidth for Simulation

Service Class	Max. Bandwidth	Min. Bandwidth
Class 1 (Conversational)	32 Kbps	16 Kbps
Class 2 (Streaming)	128 Kbps	64 Kbps
Class 3 (Interactive)	16 Kbps	8 Kbps
Class 4 (Background)	16 Kbps	8 Kbps

system–level investigation focuses on the call blocking probabilities observed through the cellular network simulation.

3.1 Packet–Level QoS of WWW Traffic: Packet Delay and Loss Rate

Figure 2(a) and 2(b) show the packet delay and loss rates of WWW traffic under our proposed lock–based scheme and the reconfiguration strategy when using the WWW traffic model suggested in [10] [11]. With the X axes for the number of users sharing the allocated bandwidth (32 Kbps in our simulation) and the Y axes for the packet delay and the packet loss rate, respectively in Figure 2(a) and 2(b), our proposed lock–based scheme outperforms the reconfiguration strategy.

3.2 System–Level Evaluation: Call Blocking Probabilities

For the system–level evaluation, we use a simulation map consisting of 19 hexagon-shaped cells. Each cell has 1 Km radius and can support up to 2 Mbps bandwidth. The boundaries of the cells located at the borders of the map are wrapped around to those of diagonally opposite side cells, resulting in the effect of using a larger map than the actual size. As the radio propagation model in the simulation, we use both the pathloss model and the shadowing model, thus the effective radiuses of the cells are determined in a probabilistic manner.

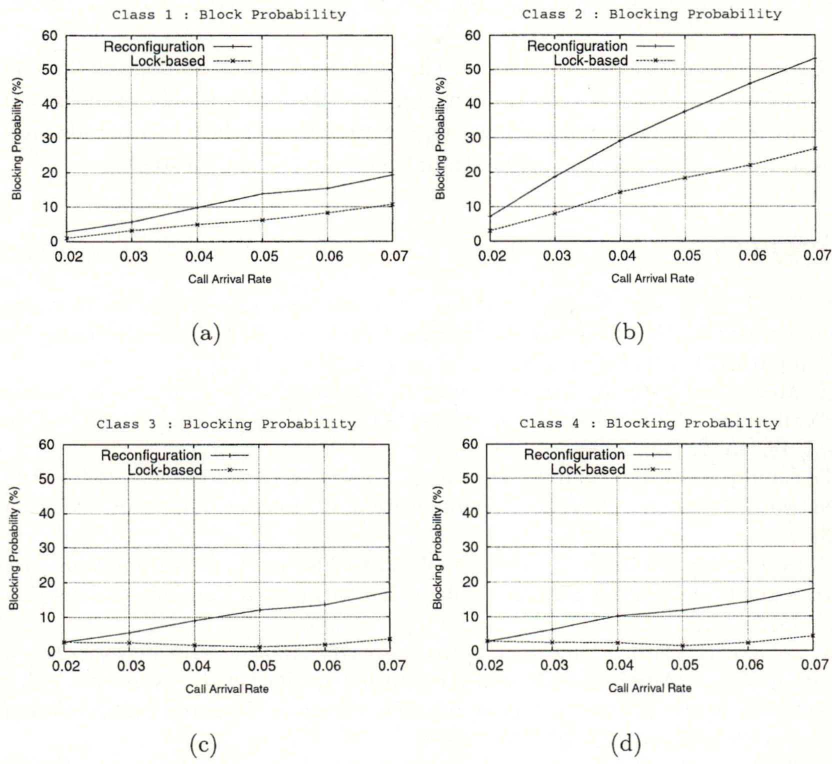

Fig. 3. Blocking probabilities of new and handoff calls : (a) class 1 (conversational) (b) class 2 (streaming) (c) class 3 (interactive) (d) class 4 (background)

Figure 3 shows the blocking probabilities of new and handoff services for each service class of the proposed scheme with the corresponding results of the reconfiguration strategy as arrival rates λ change. In the legends of the figures, "reconfiguration" indicates that the results are obtained under the reconfiguration scheme, while "lock-based" for our proposed system. In this simulation, we assume that the numbers of services per class are same.

From the figure, we observe that, for all classes, our proposed system can decrease the blocking probabilities, which demonstrates that our proposed system can admit more users than the reconfiguration system while still guaranteeing the same level of QoS for admitted services.

4   Conclusions

In this paper, we have proposed the lock-based call admission control scheme to prioritize class 1 and 2 services by having class 3 and 4 services be grouped into a bandwidth sharing set when the resource of a cell is insufficient. Our proposed strategy was motivated by the fact that, in the case of interactive services like class 3, a large portion of assigned bandwidth is wasted because of the user think time.

The performance evaluation study through the simulation demonstrated that our proposed lock-based scheme can significantly improve the system performance in terms of both packet level and system level when compared with the corresponding results of the conventional reconfiguration strategy.

References

1. J. Ye, J. Hou, and S. Papavassiliou, A Comprehensive Resource Management Framework for Next Generation Wireless Networks , IEEE Transactions on Mobile Computing, vol. 1, No. 4, 2002.
2. H. Meer, A. Corte, A. Puliafito, and O. Tomarchio, Programmable Agents for Flexible QoS Management in IP Networks , IEEE J. Selected Areas in Comm., vol. 18, no. 2, Feb. 2000.
3. T. Camp, J. Boleng, and V. Davies, A Survey of Mobility Models for Ad Hoc Network Research , Wireless Communication & Mobile Computing, vol. 2, no. 5, 2002.
4. C. Bettstetter, Mobility Modeling in Wireless Networks , Categorization, Smooth Movement, and Border Effects , Mobile Computing and Communications Review, vol. 5, no. 3.
5. A. Solana, A. Bardaji, and F. Palacio, Capacity Analysis and Performance Evaluation of Call Admission Control for Multimedia Packet Transmission in UMTS WCDMA System , Proceedings of 5th IEE European Personal Mobile Communications Conference (EPMCC'03), 2003.
6. J. Chak and W. Zhuang, Capacity Analysis for Connection Admission Control in Indoor Multimedia CDMA Wireless Communications , Wireless Personal Communications 12, 2000.
7. C. Chou and K. Shin, Analysis of Adaptive Bandwidth Allocation in Wireless Networks with Multilevel Degradable Quality of Service , IEEE Transactions on Mobile Computing, vol. 3, no. 1, 2004.
8. Y. Xiao and C. Chen, QoS for Adaptive Multimedia in Wireless/Mobile Networks , Proceedings of 9th International Symposium in Modeling, Analysis, and Simulation of Computer and Telecommunication Systems, 2001.
9. Z. Liu, N. Niclausse, and N. Jalpa-Villanueva, Traffic model and Performance Evaluation of Web servers , Performance Evaluation vol. 46, 2001.
10. J. Castro, The UMTS Network and Radio Access Technology: Air Interface Techniques for Future Mobile Systems, 2001.
11. E. Anderlind and J. Zander, A Traffic Model for Non–Real–Time Data Users in a Wireless Radio Network, IEEE Communications Letter, 1(2), 1997.
12. R. Bruno, R. Garroppo, and S. Giordano, Estimation of Token Bucket Parameters of VoIP Traffic, Proceedings of IEEE ATM workshop, 2000.

Performance Evaluation and Improvement of Non-stable Resilient Packet Ring Behavior

Fredrik Davik[1,2,3], Amund Kvalbein[1], and Stein Gjessing[1]

[1] Simula Research Laboratory
[2] University of Oslo
[3] Ericsson Research, Norway
{bjornfd, amundk, steing}@simula.no

Abstract. Resilient Packet Ring (RPR) is a new networking standard developed by the IEEE LAN/MAN working group. RPR is an insertion buffer, dual ring technology, utilizing a back pressure based fairness algorithm to distribute bandwidth when congestion occurs. In its attempt to distribute bandwidth fairly, the calculated fair rate in general oscillates and under some conditions the oscillations continue indefinitely even under stable load conditions. In this paper, we evaluate the performance of the RPR ring during oscillations. In particular, we analyze transient behavior and how the oscillations of the fairness algorithm influence the throughput, both on a per node basis and for the total throughput of the ring. For congestion-situations, we conclude that, in most cases, RPR allows for full link-utilization and fair bandwidth distribution of the congested link. A modification to the RPR fairness algorithm has previously been proposed by the authors. We compare the improved fairness algorithm to the original, and find that the modified algorithm, for all evaluated scenarios perform at least as well as the original. In some problem scenarios, we find that the modified algorithm performs significantly better than the original.

Keywords: Resilient Packet Ring, Fairness, Performance evaluation, Simulations, Next generation protocol design and evaluation, Communications modeling, Next Generation Networks Principles, High-speed Networks.

1   Introduction and Motivation

Resilient Packet Ring (RPR) is a new networking standard developed by the IEEE 802 LAN/MAN Standards Committee, assigned standard number IEEE 802.17-2004 [1, 2]. Although RPR was developed by the LAN/MAN committee, it is designed mainly to be a standard for metropolitan and wide area networks.

RPR is a ring topology network. By the use of two rings (also called ringlets), resilience is ensured; if one link fails, any two nodes connected to the ring, still have a viable communication path between them. When a node wants to send a packet to another node on the ring, it adds (sends) the packet onto one of the two ringlets. For bandwidth efficiency, the ringlet that gives the shortest path is

P. Lorenz and P. Dini (Eds.): ICN 2005, LNCS 3421, pp. 551–563, 2005.

used by default, but a sender can override this (on per packet basis) if it for some reason has a ringlet preference. When the packet travels on the ring, it *transits* all nodes between the sender and the receiver. When it reaches the destination, the packet is removed (stripped) from the ring. Hence the bandwidth that would otherwise be consumed by the packet on its way back to the sender (as is the case in a Token Ring), can be used by other communications. Such destination stripping of packets leads to what is commonly known as *spatial reuse*.

RPR uses *insertion buffer(s)* for collision avoidance [3, 4]. When a packet in transit arrives at a node that is currently adding a packet to the ring, the transiting packet is temporarily stored in an insertion buffer, called a *transit queue* in RPR. In order to get some flexibility in the scheduling of link bandwidth resources between add- and transit traffic, the transit queues may be in the order of hundreds of kilobytes large. In a buffer insertion ring like RPR, a *fairness algorithm* is needed in order to divide the bandwidth fairly between contending nodes, when congestion occurs[1] [5, 6]. The RPR fairness algorithm runs in one of two modes, called respectively the conservative and the aggressive mode. The aggressive mode of operation is simpler than the conservative one, and is used by e.g. Cisco Systems. The aggressive mode of the fairness algorithm is used in this paper.

The feedback control system nature of the RPR fairness algorithm makes the amount of add traffic from each sending node oscillate during the transient phase where the feedback control system tries to adjust to a new load [7]. Several papers have reported that in some cases the oscillations decreases and (under a stable traffic pattern) converges to a fair distribution of add rates, while under other conditions the algorithm diverges, and oscillations continues [7, 8, 9, 10].

A stable throughput per sender, decreases the jitter observed by the users of the network, and is hence obviously desirable. In this paper, we analyze and discuss RPR oscillations and, among other things, try to answer a crucial question; do these oscillations degrade the throughput from each node, and also, do they degrade the aggregate throughput performance of the ring?

The main contribution of this paper is the development of understanding of how bandwidth is divided among contending nodes when the system has not reached a stable state, or when a stable state is not possible to reach.

The RPR algorithm is complex, involving a number of different process-models interacting in parallel at both the intra- and inter-node levels as well as variable queueing delays. This makes it hard to analyze an RPR system using analytical methods, although some attempts using simplistic analytical models to study different algorithmic properties for specific load scenarios do exist [7, 8]. Hence, in this paper we use simulations to gather knowledge about the behavior we want to study. We will discuss which scenarios and benchmarks we consider important in trying to acquire knowledge that will be valid for a broad range of RPR traffic patterns.

---

[1] RPR nodes may have different weights, so a fair division might not be an equal one. In this paper, however, we assume all nodes have the same weight.

The rest of this paper is organized as follows. In the next section we give a short introduction to the RPR fairness algorithm, and describe some of its strengths and weaknesses. In section 3, we discuss how to find a good set of scenarios to base our evaluation on. In sections 4 and 5, we present and discuss results from the execution of the different scenarios. In sections 6 and 7, we discuss our proposed modification and assess how it improves the behavior of the fairness algorithm. In section 8, we discuss a notorious worst-case scenario. In the final sections we present related work, conclude and outline further work.

2   The RPR Fairness Algorithm

When several sending nodes try to send over a congested link concurrently, the objective of the RPR fairness algorithm is to divide the available bandwidth fairly between the contending nodes. RPR has three traffic classes, high, medium and low priority. Bandwidth for high and medium traffic is pre-allocated, so the fairness algorithm distributes bandwidth to low priority traffic only. In this paper all data traffic is low priority.

The fairness algorithm is a closed-loop control system. The goal of the fairness algorithm is to arrive at the "Ring Ingress Aggregated with Spatial Reuse" (RIAS) fair division of rates over the congested link [8]. The con-

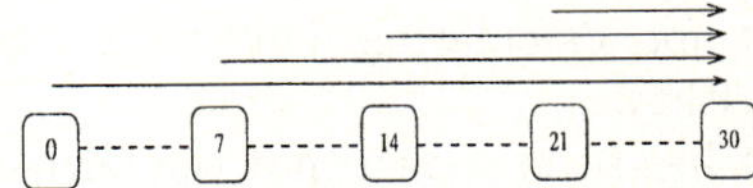

Fig. 1. Nodes 0, 7, 14 and 21 all send traffic to node 30

trol system encompasses all nodes that send over the same congested link, known in RPR as a *congestion domain*. The node directly upstream of the most congested link is called the *head* of the congestion domain. The node in the congestion domain that is furthest away from the head is called the *tail* of the congestion domain. Later in this paper we are going to use a scenario depicted in figure 1. Here nodes 0, 7, 14 and 21 all send traffic to node 30. When these nodes in total want to send more than the available bandwidth, the most congested link will be the link immediately downstream of node 21 (as well as all other links between node 21 and 30). The congestion domain will consist of all the 22 nodes from node 0 to node 21, i.e., 18 passive and 4 active nodes. Node 0 will be the tail of the domain, node 21 the head. When a node's total transit and add traffic amounts to more than the full bandwidth, the transit queue of the node with a congested out-link will fill up[2]. When the transit queue occupancy is above a threshold called *low*, the node enters a state called *congested* and if it is has not observed that there are downstream nodes that are more congested, it becomes head of a congestion domain. As head, it starts sending

---

[2] RPR nodes may have one or two transit queues. In the case of a node with two transit queues, the highest priority traffic will use one transit queue, while the two lower priority classes will use the other transit queue. In the RPR model used in this paper there are two transit queues, but since all data traffic will be of the lowest priority, the high priority transit queue will be empty (except for control traffic).

fairness messages, which is the feedback mechanism of the control system. These feedback messages instruct the upstream nodes to restrict their add rate to the head's own current send rate. The head node continues to add traffic until the transit queue occupancy reaches another, higher, threshold called *high*. At this point the head stops its add traffic until the upstream nodes have reduced their send rate so much that the head's transit queue occupancy decreases below the high threshold.

In the aggressive version of the fairness algorithm, the value used by the head as its approximation to a fair rate, is the head's own add rate run through a low-pass filter. When received by the upstream nodes, these nodes restrict their add rate to the received fair rate. The head estimates and advertises new fair rates with short intervals (by default every 100 microseconds).

The time it takes from the head advertises a new fair rate, until the head sees the effect of this action, is the time it takes for a fairness message to reach an upstream node, and then the time it takes for the packets from this node, generated in accordance with the newly received rate message, to reach the head. Hence, in general there is a considerable feedback latency in this system. This latency, combined with the configuration of the algorithm in the head for calculating fair rates, decides the stability of the RPR fairness algorithm.

It has been shown that the fairness algorithm does not always reach a stable state [7]. In general, the calculated fair rate always varies (oscillates) initially in response to transient load conditions. If the (new) traffic load is stable, these oscillations should decay as the fairness algorithm converges to the new fair division of sending rates. For some scenarios however, even under (new) stable load conditions, the fairness algorithm does not converge, and the rate at which each different node is allowed to send, continues to oscillate.

3   Scenario Discussion

The goal of this paper is to shed light on the behavior of the RPR ring when the fairness algorithm calculates and advertises oscillating fair rate values. In particular, we want to investigate how throughput performance is affected by transient load- and congestion conditions.

We have written two RPR simulators that run the fairness algorithm according to the final RPR standard text [1]. One simulator is based on OPNET Modeler [11], the other on J-Sim [12]. The results shown in the sequel are results from running the J-Sim simulator (but the OPNET simulator gives the same results).

As described above, congestion occurs when several senders at the same time want to transmit more data than the link capacity (bandwidth) over the same link can sustain (all links in RPR have the same capacity). Some senders may be greedy, i.e. they want to send as much as possible. Other senders send at a limited rate. For very modest senders, i.e. senders sending less than their fair share, RPR should not impose any rate restrictions. For nodes having more to send than the fair share, RPR should restrict their sending rate to their fair share.

In order to clearly illustrate this behavior of the fairness algorithm, and at the same time not use too much space, we have selected three sets of scenarios. The first uses four greedy senders, the second also uses four senders, but now each sender sends much less than the available link-bandwidth. In the final set there are only two senders, one very modest and one greedy sender. We have run several other scenarios, and to our experience, the main results shown in the sequel are the same as the ones obtained even if there are more senders, or the mix of senders are more complex than what is shown by the presented examples. For all scenarios, the link-length used is 40km (0.2 ms) and the packet size is 500B.

We will use stable traffic load from senders that all start at the same time. We have run other scenarios where all senders do not start at the same time, and, again, we found that this does not contribute more to the understanding. A long term stable traffic load is not very realistic, but we mainly observe the behavior of RPR during the time immediately after traffic startup. Hence we observe RPR during a transient phase. In fact, such transient phases will occur all the time when the traffic load changes in a real system, hence we indeed get realistic information from our scenarios.

One of our main objectives is to see how bandwidth is divided between nodes on a longer time scale compared to a shorter time scale. As described above, the fairness algorithm computes a new fair rate every 100 microsecond, a period that is called an *aging interval*. By experimenting with the sample period, it seems that 20 aging intervals, or 2 ms, i.e. when 2 Mbit has been transmitted over a fully utilized link in a one Gbit/sec system, is a reasonable short term sampling period. Using even shorter sampling periods gives visually the same results.

To see a more coarse grained behavior, the sampling period must be extended. By combining 25 of the short samples (into 50 ms sampling periods) we have found that the (relatively) long term behavior of the fairness algorithm is well illustrated.

Our greedy traffic senders send traffic as fast as they can while our non-greedy senders send packets with fixed intervals. These traffic models are very unlike the self-similar traffic patterns reported in [13, 14], believed by many to represent the best models of traffic in real Ethernet- and Internet environments. However, the simple traffic models we have chosen illustrate very clearly the transient behavior of an RPR system trying to adjust to a changing traffic load.

4   All Greedy Senders – Convergence

The scenario we run first is depicted in figure 1. In order to easily understand the results, we have only 4 senders, namely nodes 0, 7, 14 and 21 sending to node 30. (We wanted the distance between the senders to be long. We would have reached the same results with 4 senders with no other nodes in between, but with approximately 7 times as long links.) All senders start transmitting at full speed at time 0.10 sec. Node 21 becomes the head and node 0 the tail of the congestion domain. Figure 3a shows how the fairness algorithm converges to an approximate fair division of the bandwidth after about another 0.15 sec. From figure 3a we also see that the head (21) and the tail (0) are the two nodes whose add traffic are oscillating the most. We want to investigate how this oscillation influence the throughput of the individual nodes as well as the aggregate system throughput.

We measured the utilization of the link from node 21 to node 22, and found that this congested link is fully utilized all the time. This is good news, because it means that the most scarce resource does not become underutilized under heavy load. In order to see if the bandwidth has been divided equally among the sending nodes before the algorithm converges, we plot the throughput from each node on a more coarse grained scale, using 50 ms sampling periods. The result is depicted in figure 3b. Here we see that initially, the tail node is sending much more than the other nodes, and also that the head is sending more than the nodes inside the congestion domain. It is unfortunate that the tail node is able to send more than the other nodes. A remedy for this has been found by the authors [15]. We return to this problem and its solution in section 6.

The reason the head sends more than the two middle nodes is that whenever the upstream nodes have been slowed down below their real fair rate, the head takes all the rest of the available bandwidth over the congested link. This is also the reason why this link is 100% utilized all of the time. We measure that the transit queue in the head is almost all the time filled to the high threshold, while the transit queues in the other nodes are empty. (Except the transit queue in the active node downstream of the tail, i.e. node 7. Also this behavior will be discussed in section 6.)

Both from figures 3a and 3b, it can be observed that the throughput of the middle nodes are not oscillating much. In fact, from figure 3b, we see that both middle nodes in our scenario, nicely converges towards the fair rate.

5   All Greedy Senders – No Convergence

The RPR fairness algorithm does not converge if the communication latency between the head and the tail is long, and the low-pass filter is not smoothing its own add-rate measurements enough [7]. The degree of smoothing by the low-pass filter is determined by the filter's time-constant. Given a low time-constant, the filter's output-value is affected more (smoothed less) by short transients on the filter-input than is the case of a higher time-constant. The exact same scenario as above is run again, but this time we equip the fairness algorithm with a slightly less smoothing (quicker) low-pass filter. This is done by setting a configurable parameter called *lpCoef* to 128 instead of 256 (allowed values for *lpCoef* are 16, 32, 64, 128, 256 and 512). In figure 3c the outcome of a run is plotted with a 2 ms sampling period, to illustrate how the add rate of all the four senders oscillates. Figure 3d shows the same run with a sampling period of 50 ms. We see also here that node 0 initially gets far more of the bandwidth than its fair share. After 0.1 sec of packet sending (at time 0.2 sec), node 0 has transmitted 32.9 Mbytes, node 7 has transmitted 20.2 Mbytes, node 14 21.2 Mbytes and node 21 23.8 Mbytes of data. After time 0.2 sec, however, the fairness algorithm manages, on a coarse scale, to divide the capacity of the congested link relatively equally between the nodes. Node 0 (the tail) gets approximately 27.3% of the bandwidth of the congested link, while the others on average gets 24.2%. In the long run, it seems again that it is node 0 (the tail) that gets a little too much, and node 21 (the head) that gets too little.

6   Performance of an Improved Fairness Algorithm

We now return to the problem we identified above, i.e. that the tail node is sending significantly more than the other nodes. Going back to our scenario, and looking at figure 1, node 21 is the head, node 0 is the tail and node 7 is the node in the congestion domain closest to the tail that is active. If node 7 detects that its upstream active neighbor(s) (node 0) currently send less than the fair bandwidth advertised by the head, node 7 informs its upstream neighbor(s) (node 0) that it can transmit without any rate restrictions. The reason for this is that node 7 believes that node 0 is not (currently) contributing to the downstream congestion. Node 7 now assumes the role as the tail of the congestion domain (for a short while), until node 0 has sent so much that node 7 again detects that node 0 indeed contributes to the downstream congestion. During this period the transit queue of node 7 also fills up. Because of low-pass filtering, it takes some time for node 7 to find out that node 0 indeed contributes to the congestion.

In order to avoid this unfair extra sending from node 0, a modification to the RPR fairness algorithm has been proposed by the authors [15]. The proposal includes never sending full rate messages upstream when there is a downstream congestion. Instead the fair rate is propagated (by the tail) further upstream, either all the way around the ring, or until a new congestion domain is encountered. Such fairness messages do no harm, because before they reach the head of a new congestion domain they will mostly pass nodes that send very little (less than the fair rate). The only effect they have is the wanted one; to stop excessive sending from nodes that in reality (but not formally) are part of the congestion domain in which this fairness message originate from the head. Also, the forwarding of the message does not consume any extra resources (bandwidth, per node processing and per node state information) as the messages are sent anyhow, but with a different value in the rate field.

We have looked into the possibility of changing the formal definition of congestion domain tails, by changing the tests and some state variables in the tail. This however seems not to be an easy task, and such changes will also be harder to introduce in a future improved version of RPR.

The described improvement leads to faster convergence of the fairness algorithm and also that the algorithm converges in several scenarios when it used not to converge. Figure 3e shows how the traffic load that lead to persistent oscillations in figure 3c, now, with our simple modification, stabilize quite quickly (in approximately 0.1 s). In the present paper we, for the first time, show how our improvement also leads to more fair allocation of bandwidth; From figure 3f it can be observed that the tail (node 0) has no upstream advantage at all and that all nodes, except the head, smoothly converges to the fair rate. For reasons described above, the head still has a small initial advantage, but it is seen that also the head soon converges to the fair rate.

7   Non-greedy Senders

In this scenario, we use the same set of active senders as above (figure 1), but now the senders are more modest. When we run with a total load from all four active nodes less than the full bandwidth, RPR behaves very nice, and the fairness algorithm does not even kick in. Then we let each of the active senders transmit at 30% of the full bandwidth, resulting in an aggregate demand of 120% of the full bandwidth. The results are seen in figures 3g and 3h. This time the two plots are not that much different. Figure 3g shows some oscillations, but the most interesting period is clearly visible in both plots, i.e. from time 0.2 to about 0.45 sec. In this period the tail (node 0) is allowed to send much more than its fair share. The reason is the same as explained above in section 6. This time the unwanted behavior is even more noticeable. The reason is that when node 0 this time gets a signal from node 7 that it can send at full speed, it does not, but instead send at 30% of full rate, that is, just above its fair rate. Hence it will take much longer for node 7 to again understand that it has a serious contributor to congestion upstream. After time 0.5 sec, we have reached the steady-state (fair division of rates).

Our modified algorithm alleviates the problem of the tail sending too much. By continuing to send the fair rate upstream, node 0 continues to send at a rate much closer to the real fair rate. The results are shown in figures 3i and 3j. Notice how fast the algorithm now stabilizes, and that it is only node 21 (the head) that has an initial small downstream advantage. The reason for this advantage is, as described above, that the head will utilize any spare bandwidth on the congested link (limited upwards by its demand). Initially, there will be periods with spare bandwidth as the fairness algorithm converges towards the fair rate.

8   Mixed Greedy and Non-greedy Senders

We want to understand how RPR behaves when some senders are greedy and others are not. In order to get an example that illustrates this clearly and simply, we use a notorious worst-case scenario, first presented by Knightly et al. [16].

The scenario contains two senders only; one greedy and one that sends at only 5% of the total bandwidth. The ring used for this experiment is the same as the one used previously in this article, but this time long latencies between nodes do not matter, hence we now let nodes 0 and 1 send to node 2. First we let the 5% sender be the head of the congestion domain (node 1 in our experiment), and the greedy sender the tail (node 0) . Figure 2 shows the throughput of the two senders.

Initially both nodes send at their assigned rate. Node 0 will send, at full rate, traffic that transits node 1. Because node 1 sends at 5% only, the transit queue in node 1 will then fill up very slowly. When it is filled to the low threshold, a fairness message is sent to node 0. As we now know, the contents of this message is the fair rate, as seen by node 1, and this is the 5% rate, run through a low-pass filter, so it is even less. Consequently, node 0 must reduce its rate to less than 5% of full

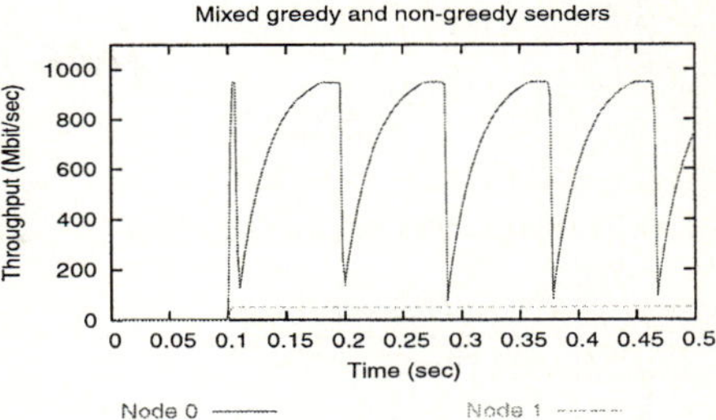

Fig. 2. Modest congestion head node. The figure plots throughput per sender node measured at node 2 using a quick low-pass filter (lpCoof=64) and 2 ms sampling intervals

rate. In fig. 2 this can be seen as the dip at about time 0.11 sec. When node 0 has sent at this low rate for a short time, node 1 is no longer congested, and notifies upstream nodes of this, by advertisement of full-rate messages. Having received full-rate messages, the fairness algorithm gradually allows node 0 to increase its send rate, as seen in fig 2. When the send rate has increased above 95% of full bandwidth usage, node 0 again becomes congested, and once more it will advertise its low-pass filtered add rate, that is below 5% of full bandwidth, and node 0 will again (at about time 0.2 sec) have to decrease its send rate, etc. Because of the large, long-lasting oscillations that can be observed, the total throughput is no longer 100%. In fact, after some time, the average total throughput arrives at approximately 70% of the link-bandwidth, i.e 700 Mbit/sec. Our findings in this scenario confirms what have been reported by others [8, 9, 10, 17].

If the sequence of the greedy and the humble senders is reversed, the greedy head will tell the tail to slow down, but the tail's rate is already below the fair rate calculated by the (greedy) head. The only effect of the fairness algorithm is then that the head has to stop sending in the periods when its transit queue occupancy exceeds the high threshold. The result is a stable division of bandwidth, where the head utilizes 5% of the congested link's bandwidth, while the head utilizes 95% of the congested link's bandwidth. Thus, in total, the congested link is fully utilized all the time. No plot is shown for this scenario.

9   Related Work

In an insertion-buffer ring, where the demand for link bandwidth is larger than the available capacity, a fairness algorithm is required to provide fair sharing of bandwidth resources between contending nodes. Many groups have studied the performance and implementation of different algorithms for various insertion-ring architectures [18, 8, 19, 20, 21]. Several papers have been published studying

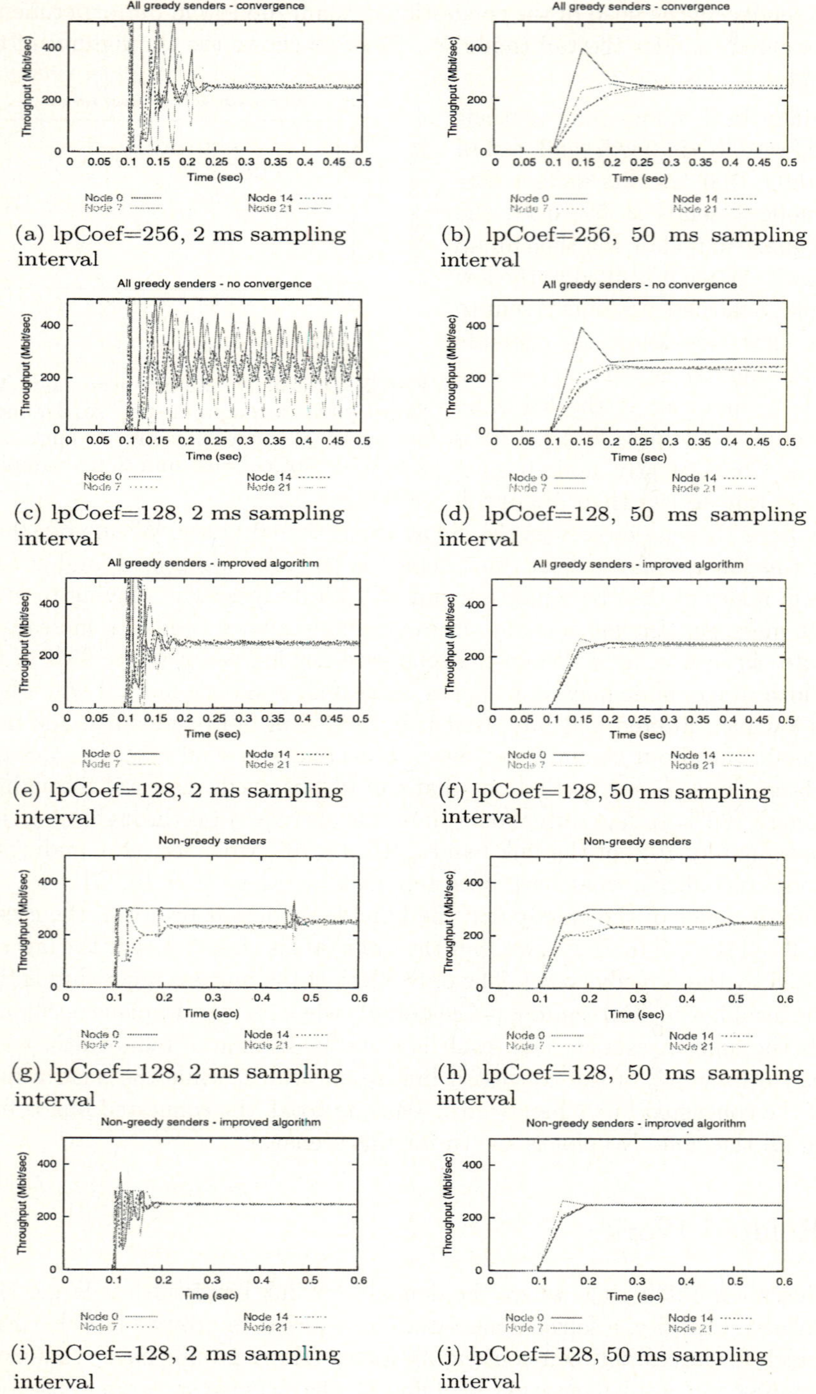

(a) lpCoef=256, 2 ms sampling interval

(b) lpCoef=256, 50 ms sampling interval

(c) lpCoef=128, 2 ms sampling interval

(d) lpCoef=128, 50 ms sampling interval

(e) lpCoef=128, 2 ms sampling interval

(f) lpCoef=128, 50 ms sampling interval

(g) lpCoef=128, 2 ms sampling interval

(h) lpCoef=128, 50 ms sampling interval

(i) lpCoef=128, 2 ms sampling interval

(j) lpCoef=128, 50 ms sampling interval

Fig. 3. Throughput per sender node measured at node 30 using various settings for the lpCoef parameter and two different sampling interval settings

different RPR performance aspects, both for hardware implementations [8, 22] and simulator models [2, 8, 23, 24]. Huang et al. presents a thorough analysis of ring access delays for nodes using only one transit queue [23]. Robichaud et al presents ring access delays for class B traffic for both one- and two transit queue designs [24]. Gambiroza et al. focus on the operation of the RPR fairness algorithm and their alternative proposal, DVSR, and their ability, for some given load scenarios to converge to the fair division of rates according to their RIAS fairness reference model [8].

10   Conclusion

In this paper we have investigated how the bandwidth of a congested RPR system is utilized when the operation of the aggressive mode fairness algorithm results in oscillatory sending behavior. We have executed performance evaluation experiments on our RPR simulation platforms in order to observe RPR behavior during non-stable executions. The set of traffic scenarios have been carefully selected in order to learn as much as possible about general RPR performance. We have also argued that even though we run with stable load, we mainly analyze the initial (transient) phase of the runs. This transient phase resembles the transient phase in a running network where the traffic pattern suddenly changes. Hence, we believe that the knowledge obtained is valid for many real traffic scenarios as well.

We have found that even when the estimated fair rate oscillates, the bandwidth allowances distributed to the contending nodes are relatively equal measured over a longer time scale. In several scenarios we have, however, observed that the most upstream node (the tail) in a congestion domain is initially assigned more bandwidth by the RPR fairness algorithm, than its fair share.

The authors have previously developed a modification to the fairness algorithm. This modification is designed to prevent oscillations caused by the tail getting more bandwidth than the other nodes. In the present paper it is shown, for the first time, that this modification also distributes bandwidth fairly to the tail, both initially, while the modified algorithm stabilizes the fair rate, and in the long run (in case of a stable traffic scenario).

In this paper we have also shown that the RPR fairness algorithm has the nice property that when a link is congested, it is usually utilized fully. The reason for this is that the node just upstream of the most congested link (the head of the congestion domain), may use all free capacity on this link. Hence the most congested link will be fully utilized as long as the head has enough data to send. Sometimes this causes the head to get a little more than its fair share of the bandwidth. In our experiments, the only time the most congested link is not fully utilized is when the head node is sending at a very low rate. In this paper we have also confirmed that such a head node, sending at a very low rate, may cause upstream nodes to send below their fair rate.

11   Further Work

In future work we want to see if there are other ways to modify the fairness algorithm to give it more wanted properties. Specifically, we want to investigate methods that addresses the problem, discussed in section 8, of throughput loss in congestion domains where the head sends traffic at a rate lower than the fair rate. We would also like to investigate how the oscillations observed in unstable configurations of an RPR ring, running the aggressive mode fairness algorithm, affects packet delays and jitter.

References

1. IEEE Computer Society: IEEE Std 802.17-2004 (2004)
2. Davik, F., Yilmaz, M., Gjessing, S., Uzun, N.: IEEE 802.17 Resilient Packet Ring Tutorial. IEEE Commun. Mag. **42** (2004) 112–118
3. Hafner, E., Nendal, Z., Tschanz, M.: A Digital Loop Communication System. IEEE Trans. Commun. **22** (1974) 877 – 881
4. Reames, C.C., Liu, M.T.: A Loop Network for Simultaneous Transmission of Variable-length Messages. In: Proceedings of the 2nd Annual Symposium on Computer Architecture. Volume 3. (1974)
5. Van-As, H., Lemppenau, W., Schindler, H., Zafiropulo, P.: CRMA-II a MAC protocol for ring-based Gb/s LANs and MANs. Computer-Networks-and-ISDN-Systems **26** (1994) 831–40
6. Cidon, I., Ofek, Y.: MetaRing - A Full Duplex Ring with Fairness and Spatial Reuse. IEEE Trans. Commun. **41** (1993) 110 – 120
7. Davik, F., Gjessing, S.: The Stability of the Resilient Packet Ring Aggressive Fairness Algorithm. In: Proceedings of The 13th IEEE Workshop on Local and Metropolitan Area Networks. (2004) 17–22
8. Gambiroza, V., Yuan, P., Balzano, L., Liu, Y., Sheafor, S., Knightly, E.: Design, analysis, and implementation of DVSR: a fair high-performance protocol for packet rings. IEEE/ACM Trans. Networking **12** (2004) 85–102
9. Alharbi, F., Ansari, N.: Low complexity distributed bandwidth allocation for resilient packet ring networks. In: Proceedings of 2004 Workshop on High Performance Switching and Routing, 2004. HPSR. (2004) 277 – 281
10. Zhou, X., Shi, G., Fang, H., Zeng, L.: Fairness algorithm analysis in resilient packet ring. In: In Proceedings of the 2003 International Conference on Communication Technology (ICCT 2003). Volume 1. (2003) 622 – 624
11. : (OPNET Modeler. http://www.opnet.com)
12. Tyan, H.: Design, Realization and Evaluation of a Component-Based Compositional Software Architecture for Network Simulation. PhD thesis, Ohio State University (2002)
13. Leland, W.E., Taqqu, M.S., Willinger, W., Wilson, D.V.: On the self-similar nature of Ethernet traffic (extended version). IEEE/ACM Trans. Networking **2** (1994) 1–15
14. Paxson, V., Floyd, S.: Wide area traffic: the failure of Poisson modeling. IEEE/ACM Trans. Networking **3** (1995) 226–244
15. Davik, F., Kvalbein, A., Gjessing, S.: Congestion Domain Boundaries in Resilient Packet Rings. Submitted to ICC05 (2004)

16. Knightly, E., Balzano, L., Gambiroza, V., Liu, Y., Yuan, P., Sheafor, S., Zhang, H.: Achieving High Performance with Darwin's Fairness Algorithm. http://grouper.ieee.org/groups/802/17/documents/presentations/mar2002 (2002) Presentation at IEEE 802.17 Meeting.
17. Yue, P., Liu, Z., Liu, J.: High performance fair bandwidth allocation algorithm for resilient packet ring. In: Proceedings of the 17th International Conference on Advanced Information Networking and Applications. (2003) 415 – 420
18. Cidon, I., Georgiadis, L., Guerin, R., Shavitt, Y.: Improved fairness algorithms for rings with spatial reuse. IEEE/ACM Trans. Networking 5 (1997) 190–204
19. Kessler, I., Krishna, A.: On the cost of fairness in ring networks. IEEE/ACM Trans. Networking 1 (1993) 306–313
20. Picker, D., Fellman, R.: Enhancing SCI's fairness protocol for increased throughput. In: IEEE Int. Conf. On Network Protocols. (1993)
21. Schuringa, J., Remsak, G., van As, H.R.: Cyclic Queuing Multiple Access (CQMA) for RPR Networks. In: Proceedings of the 7th European Conference on Networks & Optical Communications (NOC2002), Darmstadt, Germany (2002) 285 – 292
22. Kirstadter, A., Hof, A., Meyer, W., Wolf, E.: Bandwidth-efficient resilience in metro networks - a fast network-processor-based RPR implementation. In: Proceedings of the 2004 Workshop on High Performance Switching and Routing, 2004. HPSR. (2004) 355 – 359
23. Huang, C., Peng, H., Yuan, F., Hawkins, J.: A steady state bound for resilient packet rings. In: Global Telecommunications Conference, (GLOBECOM '03). Volume 7., IEEE (2003) 4054–4058
24. Robichaud, Y., Huang, C., Yang, J., Peng, H.: Access delay performance of resilient packet ring under bursty periodic class B traffic load. In: Proceedings of the 2004 IEEE International Conference on Communications. Volume 2. (2004) 1217 – 1221

Load Distribution Performance of the Reliable Server Pooling Framework

Thomas Dreibholz[1], Erwin P. Rathgeb[1], and Michael Tüxen[2]

[1] University of Duisburg-Essen,
Institute for Experimental Mathematics,
Ellernstraÿe 29, D-45326 Essen, Germany
Tel: +49 201 183-7637, Fax: +49 201 183-7673
`dreibh@exp-math.uni-essen.de`
[2] University of Applied Sciences, Münster,
Fachbereich Elektrotechnik und Informatik,
Stegerwaldstraÿe 39, D-48565 Steinfurt, Germany
Tel: +49 2551 962550
`tuexen@fh-muenster.de`

Abstract. The Reliable Server Pooling (RSerPool) protocol suite currently under standardization by the IETF is designed to build systems providing highly available services by providing mechanisms and protocols for establishing, configuring, accessing and monitoring pools of server resources.

While availability is one main aspect of RSerPool, load distribution is another. Since most of the time a server pool system runs without component failures, optimal performance is an extremely important issue for the productivity and cost-efficiency of the system. In this paper, we therefore focus especially on the load distribution performance of RSerPool in scenarios without failures, presenting a quantitative performance comparison of the different load distribution strategies (called pool policies) defined in the RSerPool specifications. Based on the results, we propose some new pool policies providing significant performance enhancements compared to those currently defined in the standards documents.

1 The Reliable Server Pooling Architecture

The convergence of classical circuit-switched networks (i.e. PSTN/ISDN) and data networks (i.e. IP-based) is rapidly progressing. This implies that SS7 PSTN signalling [1] has to also be transported over IP networks. Since SS7 signalling networks offer a very high degree of availability (e.g. at most 10 minutes downtime per year for any signalling relation between two signalling endpoints; for more information see [2]), all links and components of the network devices must be fault-tolerant, and this is achieved through having multiple links, and using the link redundancy concept of SCTP [3]. When transporting signalling over IP networks, such concepts also have to be applied to achieve the required availability. Link redundancy in IP networks is supported using the Stream Control Transmission Protocol (SCTP) providing multiple network paths and fast failover [4, 5]; redundancy of network device components is supported by the SGP/ASP (signalling gateway process/application server process) concept. However,

P. Lorenz and P. Dini (Eds.): ICN 2005, LNCS 3421, pp. 564–574, 2005.

this concept has some limitations: there is no support of dynamic addition and removal of components; it has only limited ways of server selection and no specific failover procedures and inconsistent application to different SS7 adaptation layers.

To cope with the challenge of creating a unified, lightweight, realtime, scalable and extendable redundancy solution (see [6] for details), the IETF Reliable Server Pooling Working Group was founded to specify and define the Reliable Server Pooling Concept. An overview of the architecture currently under standardization and described by several Internet Drafts is shown in figure 1.

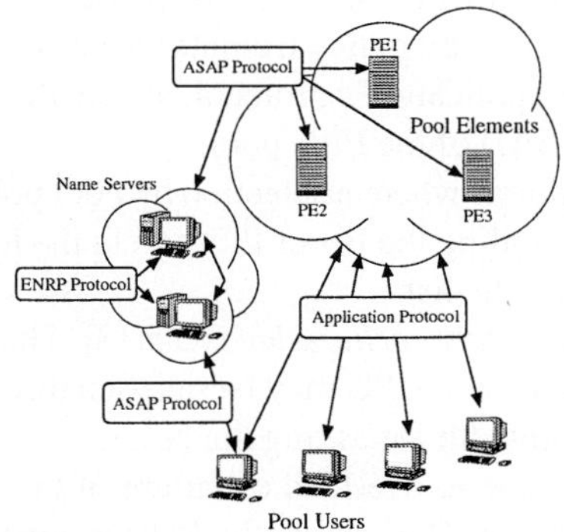

Fig. 1. The RSerPool Architecture   **Fig. 2.** PE Registration and Monitoring

Multiple server elements providing the same service belong to a *server pool* to provide redundancy on one hand and scalability on the other. Server pools are identified by a unique ID called *pool handle* (PH) within the set of all server pools, the *namespace*. A server in a pool is called a *pool element* (PE) of the respective pool. The namespace is managed by redundant *name servers* (NS). The name servers synchronize their view of the namespace using the Endpoint Name Resolution Protocol (ENRP [7]). NSs announce themselves using broadcast/multicast mechanisms, i.e. it is not necessary (but still possible) to pre-configure any NS address into the other components described in the following.

PEs providing a specific service can register for a corresponding pool at an arbitrary NS using the Aggregate Server Access Protocol (ASAP [8]) as shown in figure 2. The *home NS* is the NS which was chosen by the PE for initial registration. It monitors the PE using SCTP heartbeats (layer 4, not shown in figure) and ASAP Endpoint Keep Alives. The frequency of monitoring messages depends on the availability requirements of the provided service. When a PE becomes unavailable, it is immediately removed from the namespace by its home NS. A PE can also intentionally de-register from the namespace by an ASAP de-registration allowing for dynamic reconfiguration of the server pools. NS failures are handled by requiring PEs to re-register regularly (and therefore chosing a new NS when necessary). Re-registration also makes it possible for the PEs to update their registration information (e.g. transport addresses or policy states).

The home NS, which registers, re-registers or de-registers a PE, propagates this information to all other NS via ENRP. Therefore, it is not necessary for the PE to use any specific NS. In case of a failure of its home NS, a PE can simply use an arbitrarily chosen other one.

When a client requests a service from a pool, it first asks an *arbitrary* NS to translate the pool handle to a list of PE identities selected by the pool's selection policy (*pool policy*), e.g round robin or least used (to be explained in detail in section 2). The NS does not return the total number of identities in the pool, instead it has a constant value, *MaxNResItems*, which dictates how many PE identities should be returned. For example, if there were 5 PEs and *MaxNResItems* was set to 3, then the NS would select 3 of the 5; conversely, if *MaxNResItems* were set to 5, and there were only 3 PEs, then all 3 PE identities would be returned. The PU adds this list of PE identities to its local cache (denoted as PU-side cache) and again selects *one* entry by policy from its cache. To this selected PE, a connection is established, using the application's protocol, to actually use the service. The client then becomes a *pool user* (PU) of the PE's pool.

It has to be emphasized, that there are two locations where a selection by pool policy is applied during this process: at the NS when compiling the list of PEs and in the local PU-side cache where the target PE is selected from the list.

The default timeout of the PU-side cache, called *stale cache value* is 30s [8]. That is, within this time period, subsequent name resolutions of the PU may be satisfied directly from the PU-side cache, saving the effort and bandwidth for asking the NS.

If the connection to the selected PE fails, e.g. due to overload or failure of the PE, the PU selects another PE from its list and tries again. Optionally, the PU can report a PE failure to a NS, which may then decide to remove this PE from the namespace. If the PE failure occurs during an active connection, a new connection to another available PE is established and an application-specific failover procedure is invoked.

RSerPool supports optional client-based state synchronization [9] for failover. That is, a PE can store its current state with respect to a specific connection in a *state cookie* which is sent to the corresponding PU. When a failover to a new PE is necessary, the PU can send this state cookie to the new PE, which can then restore the state and resume service at this point. However, RSerPool is not restricted to client-based state synchronization, any other application-specific failover procedure can be used as well.

The lightweight, realtime, scalable and extendable architecture of RSerPool is not only applicable to transport of SS7-based telephony signalling. Other application scenarios include reliable SIP based telephony [10], mobility management [11] and the management of distributed computing pools [12, 13]. Furthermore, additional application scenarios in the area of load distribution and balancing are currently under discussion within the IETF RSerPool Working Group.

Currently, there are two existing implementations of RSerPool: the authors' own GPL-licensed Open Source prototype *rsplib* [12] and a closed source version, by Motorola [14]. The standards documents are currently Internet Drafts, having some open issues. These open questions are: evaluation of reliability aspects, state synchronization between PEs and load distribution among PEs.

In this paper, we focus our attention on the third point above – the open topic of load distribution among PEs, which is crucial for the efficient operation of server pools.

2   Load Distribution and Balancing

Whilst reliability is one of the obvious aspects of RSerPool, load distribution is another important one: the choice of the pool element selection policy (*pool policy*) controls the way in which PUs are mapped to PEs when they request a service. An appropriate strategy here is to balance the load among the PEs to avoid excessive response times due to overload in some servers, while others run idle.

For RSerPool, *load* only denotes a constant in the range from 0% (not loaded) to 100% (fully loaded) which describes a PE's actual normalized resource utilization. The definition of a mapping from resource utilization to a load value is application-dependent. Formally, such a mapping function is defined as $m(u) := \frac{u}{U_{max}-U_{min}}$, $u \in \{U_{min}, ..., U_{max}\} \subset \mathbb{R}$, where U_{min} denotes the application's minimum and U_{max} the maximum possible resource utilization.

A file transfer application could define the resource utilization as the number of users currently handled by a server. Under the assumption of a maximum amount of 20 simultaneous users: $U_{min} = 0$ and $U_{max} = 20$. Therefore, $m(u) := \frac{u}{20}$.

For an e-commerce database transaction system, response times are crucial; e.g. a customer should get a response in less than 5 seconds. In this case, utilization can be defined as a server's average response time. Then, $U_{min} = 0s$ and $U_{max} = 5s$ and $m(u) := \frac{u}{5s}$. Other arbitrary schemes can be defined as well, e.g. CPU usage, memory utilization etc.

Depending on the used pool element selection policy, RSerPool can try to achieve a balanced *load* of the PEs within a pool. That is, if the application defines its load as a function of the amount of users, RSerPool will balance the amount of users. And if load is defined as average response time, RSerPool will balance response times.

Currently, the drafts [8] and [15] define the following four pool policies: Round Robin (RR), Weighted Round Robin (WRR), Least Used (LU) and Leased Used with Degradation (LUD). The RR and WRR policies are called *static* policies because they do not require and incorporate any information on the actual load state of the active PEs when making the selection. However, they are "stateful" in a sense that the current selection depends on the selection made in the previous request. This can – if carelessly implemented - lead to a severe performance degradation in some situations as shown in section 5.1. The LU and LUD policies try to select the PEs which currently carry the least load. Therefore, the PEs are required to propagate their load information into the namespace (by doing a re-registration) regularly or upon changes. These required dynamic policy information changes lead to the term *dynamic policy*. It is obvious that the dynamic policies have the potential to provide a better load sharing resulting in a better overall performance. However, the tradeoff is that these policies require additional signalling overhead in order to keep the load information sufficiently current. These effects will be quantified in section 5.2.

3   Detailed Definition of Pool Policies

Even though the pool policies are mentioned in the current standards documents, their definitions are not sufficient for a consistent implementation. E.g. it was not defined how to perform the load degradation in the LUD policy. Therefore, to be able to do the implementation as well as the simulation study, we refined the definition of these policies in a first step as described below and introduced these definitions into the standardization process as Internet Draft [16]. Furthermore, based on our quantitative evaluation described in section 5 we propose modifications as well as additional, more efficient policies.

3.1   Policies Defined in the Standards Documents

Round Robin (RR) and Weighted Round Robin (WRR). Using this policy, the elements in a list of PEs should be used in a round robin fashion, starting with the first PE. If all elements of the list have been used, selection starts again from the beginning of the list.

The RR policy does not take into account the fact that servers may have different capacities. Therefore, WRR tries to improve the overall performance by selecting more powerful servers more often. The capacity of a PE is reflected by its integer *weight* constant. This constant specifies how many times per round robin round a PE should be selected. For example, this can be realized using a round robin list where each PE gets as many entries as its *weight* constant specifies. Obviously, RR can be viewed as a special case of WRR with all weight factors set to identical values.

Least Used (LU). The effort to serve a request may – in some application scenarios – vary significantly. Therefore, the LU policy tries to incorporate the actual load value of a server into the selection decision. When selecting a PE under the LU policy, the least loaded PE is chosen. That is, a NS executing the selection has to know the current load states of all PEs. For the case that there are multiple PEs of the same load, round robin selection between these equal-loaded PEs should be applied.

Least Used with Degradation (LUD). When using the LU policy, load information updates are propagated among the NSs using ENRP Peer Update messages [7]. However, they are not propagated immediately to the PU-side caches as these are only updated by ASAP Name Resolutions [8]. To keep the resulting namespace inconsistencies small, the LUD policy extends LU by a per-PE load degradation constant (this was not defined by the original ASAP draft, it had to be added by us to make this policy useful [16]). This load degradation constant specifies in units of the load how much a new request to the PE will increase its load. For the file transfer example above, a new request means a new user on this PE. Therefore, its load increases by $m(1) = \frac{1}{20} = 5\%$. That is, the load degradation constant should be set to 5%.

Each selecting component, i.e. NS or PU-side cache, has to keep a local per-PE *degradation counter* which is initialized with 0%. Whenever a PE is selected, this local counter is incremented by the load degradation constant. On update, i.e. the PE re-registers with its up-to-date load information and the information is spread via ENRP, the local degradatation counter is reset. For selection, the PE having the lowest sum of load value and degradation counter is selected. If there are PEs having equal sums, round robin selection is applied.

For example, there is a PE of load 50% and load degradation 5% in a PU-side cache. At first, its degradation counter is 0%. When it is selected for the first time, it is incremented to 5%, and then to 10% for the second time. Now, a new selection in the PE's pool is invoked. In this case, it is only selected when its sum of load and degradation counter (50% + 10% = 60%) is lowest within the pool (or if there are PEs of equal sum, by round robin selection among them).

First experiments have shown that the LUD performance is highly dependent on the scenario and generally rather unpredictable. Since no generally applicable results for LUD have been obtained so far, this policy is not recommended for use and not included in the quantitative comparison.

3.2   Modified and New Policies

Modified Round Robin (RRmod) and Modified Weighted Rd. Robin (WRRmod). In some situations, the RR selection degenerates due to its statefulness as shown in section 5.1. To avoid this, the cyclic pattern has to be broken up. Therefore, we propose to modify the RR policy by instead of incrementing the round robin pointer by the number of items actually selected, simply to increment it by one. The same modification should also be applied to WRR, which is a generalization of RR, resulting in the modified policy WRRmod.

Random Selection (RAND) and Weighted Random Selection (WRAND). Another solution to avoid the degeneration problem is to use a static and completely stateless selection mechanism. To achieve this, PEs are randomly selected from the pool with equal probability (RAND) or with a probability proportional to the weight constant of a PE (WRAND). RAND can be viewed as a special case of WRAND with all weight factors set to identical values (as for RR and WRR).

Priority Least Used (PLU). PLU is a dynamic policy based on LU with the difference that PEs can provide a load increment constant similar to LUD (see section 3.1). Then, the PE having the lowest value of load + load increment constant is selected. But unlike LUD, no local incrementation is applied to the load information by the selecting component (NS or PU-side cache) itself. This makes the policy simpler and avoids its sensitivity to variances of update timing and fraction of selected PEs actually used by the PU for service.

4   The Simulation Model

To quantitatively evaluate the RSerPool concept, we have developed a simulation model which is based on the discrete event simulation system OMNeT++ [17]. Currently, it includes implementations of the two RSerPool protocols – ASAP [8] and ENRP [7] – and a NS module. Furthermore, it also includes models for PE and PU components of the distributed fractal graphics computation application described in [13]. This application was originally created using our RSerPool prototype *rsplib* and tested in a lab testbed emulating a LAN/WAN scenario.

Figure 3 shows the simulation scenario. The modelled RSerPool network consists of 3 LANs, interconnected via WAN links. The LAN links introduce an average delay of 10ms, WAN links an average one of 100ms (both settings are based on the testbed

LAN/WAN scenario). Each LAN contains 1 NS, 6 PEs (the local NS is their home NS) and 12 PUs (using the local NS for name resolutions).

Unless otherwise specified, a PE has a default computation capacity of $C = 10^6$ calculations per second. A PE can process several computation jobs simultaneously in a processor sharing mode as commonly used in multitasking operating systems. At most, $MaxJobs = \left\lfloor \frac{C}{2.5*10^5} \right\rfloor$ simultaneous jobs are allowed on a server to avoid overloading and excessive response times. Therefore, for a server with the default capacity of 10^6 calculations per second, at most 4 jobs can be

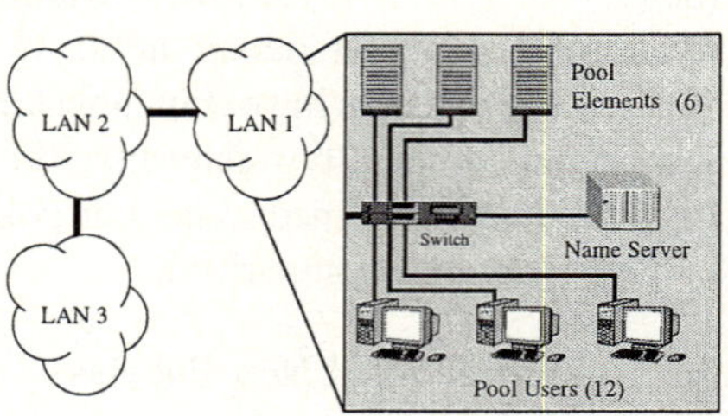

Fig. 3. RSerPool Simulation Scenario

processed simultaneously. The *load* of a server in our scenario has been defined as the number of currently running jobs, divided by its respective job limit *MaxJobs*. That is, if server B has twice the computation capacity of A, B may have a load of 50% with 4 jobs while A is already loaded 50% with 2 jobs. A PE rejects an additional job if it is fully loaded. In this case, the PU will try another PE (selected by pool policy, of course) after an average timeout of 100ms to avoid overloading the NS and network with unsuccessful ASAP Name Resolution requests (recommendation based on the results from [13]).

In our scenario, PUs sequentially request the processing of jobs by the pool, having an average job size of 10^7 calculations and a negative exponential distribution (approximation of the real system behavior, see [13]). After receiving the result of a job, a PU waits for an average of 10 seconds (again, negative exponentially distributed) to model the reaction time of a user. The stale cache value for the PU-side cache is set to the default of 30s (see [8] and [15]), i.e. a stale cache period contains about 2 to 3 service times.

The length of the simulation runs was set to 20 minutes simulated realtime. All simulation runs have been repeated 4 times using different seeds to be able to compute confidence intervals. For the statistical post-processing and plotting, GNU Octave and GNU Plot have been used. The plots show mean values and their 95% confidence intervals.

5   Simulation Results

5.1   RR Policy Degeneration

To show the effects of inappropriately implemented stateful policies (see section 2), we first examine a homogeneous scenario where all PEs have the same capacity of $C = 10^6$ calculations per second.

Figure 4 shows the total number of completed jobs for different values of *MaxNResItems*, i.e. for a different number of PE identities returned per name resolution request.

For the original RR policy, a significant periodic degeneration can be observed. If the number of PEs in the pool is an integer multiple of the number of entries in the list sent back by the NS, specific PEs will be systematically overloaded while the others will be hardly used. Assume, e.g. that the pool consists of the pool elements PE1 to PE6 and that the configured amount of PE identities delivered by the NS in a name resolution response is 3. Now, the first resolution query to the NS returns the set {PE1, PE2, PE3}, the following one will return {PE4, PE5, PE6}. Then, new resolutions again start with {PE1, PE2, PE3} and so on. In the worst case, the pool size is lower than the configured amount. Then, the reply is always the same (that is, the complete pool).

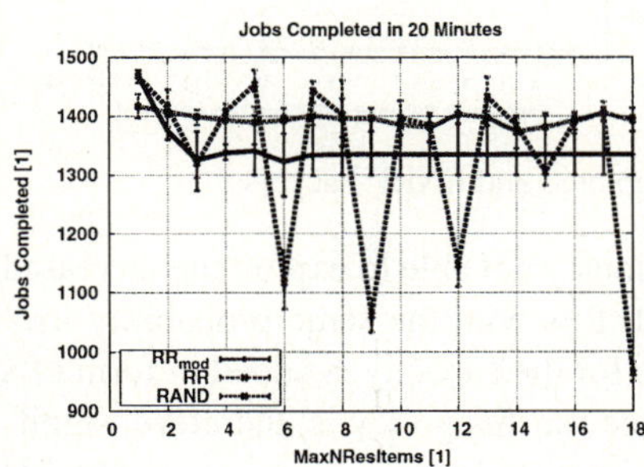

Fig. 4. Round Robin Policy Behaviour

From the list received from the NS, the PU selects again one PE to establish the application connection to (see section 2). Using round robin, this will always be the first PE of the list after a refresh of the PU-side cache. The result is of course that some PEs will be systematically overloaded. Subsequent service requests within the stale cache period also select subsequent elements of the list with decreasing probability.

The curve for the RRmod policy shows the behaviour if the RR policy is modified as described in section 3.2. Obviously, the problem of periodic variation has been solved.

The RAND policy shows a slightly higher performance as RRmod for this scenario. Only if a single PE identity is returned per request, the stateful policies RR and RRmod perform better, because PU-side caching always results in using the one PE selected by the NS first. For higher values, the PU-side cache contains a list of multiple elements, which are used for local round robin selection. For example, the NS replies with the list PE1, PE2, PE3 to PU1 and PE2, PE3, PE4 to PU2. Since the elements are ordered for round robin selection, the probability of simultaneous requests from PU1 and PU2 to PE2 is higher than for random selection.

5.2   Performance in Heterogeneous Scenarios

In real-world application scenarios, servers usually do not have equal capacities. Therefore, we examine two types of heterogeneous scenarios. In the first case, a specific server has a significantly higher capacity. The second case models an evolutionary scenario, where capacities vary linearly over a certain range due to several server generations in use.

Fast Server Scenario. In this scenario, one server ("fast server") in each LAN has a capacity which is scaled up from 1 to 15 million calculations per second while all the others have the standard capacity of 1 million calculations per second. *MaxNResItems* has been set to 5.

Figure 5, shows how well the increased total processing capacity can be used by various policies (left side) and how much control overhead (number of ENRP packets) is required (right side). It can be observed that for a homogeneous or nearly homogeneous scenario (fast server capacity up to 4×10^6), the dynamic policies generally perform better than the static ones. However, for more heterogeneous scenarios, the behaviour changes significantly.

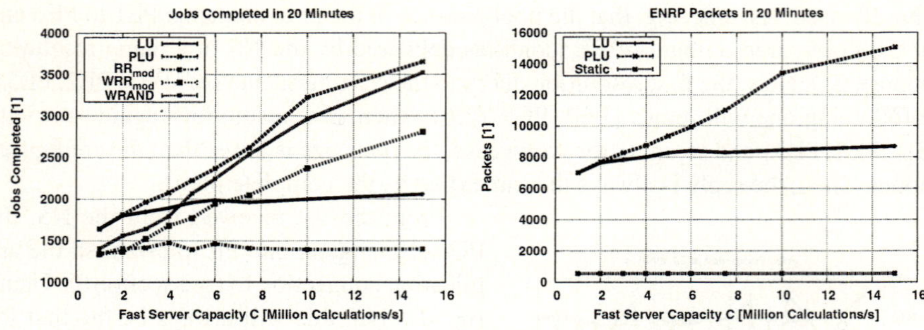

Fig. 5. Fast Server Scenario: Jobs Completed and ENRP Packets

As expected, the simple RRmod policy is obviously not able to exploit the increased capacity due to the fact that the policy selects all PEs with the same probability irrespective of their capacity. For WRRmod, a metric for the capacity is added in form of a weight constant per PE. The resulting performance for this policy is, therefore, significantly higher: up to 500 additional jobs can be completed within 20 minutes. The LU policy achieves a slightly better performance than WRRmod. However, it also requires about 7000 to 15000 ENRP control packets in 20 minutes, compared to 444 for the static policies.

The LU policy does not take into account the fact that a new job on a high-capacity PE increases the load less than a new job on a low-capacity one. The PLU policy proposed in section 3.2 does exactly that resulting in a significantly improved performance. However, since the number of re-registrations to update the load information in the NS is directly related to the number of accepted jobs, the overhead increases in a similar way[1].

Finally, the result for the WRAND policy proposed in section 3.2 is very promising. It achieves a performance close to that of the dynamic PLU policy with the minimum overhead of a static policy. Again, as explained in section 5.1 for the comparison of RRmod and RAND, the local selection in the PU-side cache leads to a higher probability of simultaneous requests to the same PE for WRRmod. WRAND therefore obviously performs much better.

Evolutionary Scenario. While the scenario examined in section 5.2 used selected PEs with high capacities, this examination uses an evolutionary scenario using PEs with linearly varying capacities. For each LAN, the PEs have capacities of $c(n) = n * \vartheta * 10^6$ calculations per second, where n denotes the PE number and ϑ is a constant scale factor. That is, for $\vartheta = 1$, the first PE in a LAN has a capacity of 10^6 calculations per second, the second one 2×10^6 and so on. All other parameters remain unchanged.

As expected, the number of completed jobs for RRmod is lowest, since this policy does not take into account the different server capacities. As in the fast server scenario, the LU and WRRmod policies again show similar performance with the static WRRmod policy requiring significantly less control overhead. While the dynamic policies again generally perform slightly better in relatively homogeneous scenarios (scale factors up

---

[1] The average size of an ENRP packet is usually less than 250 bytes, so even a slow 10 MBit/s Ethernet could handle this amount in less than 10 seconds.

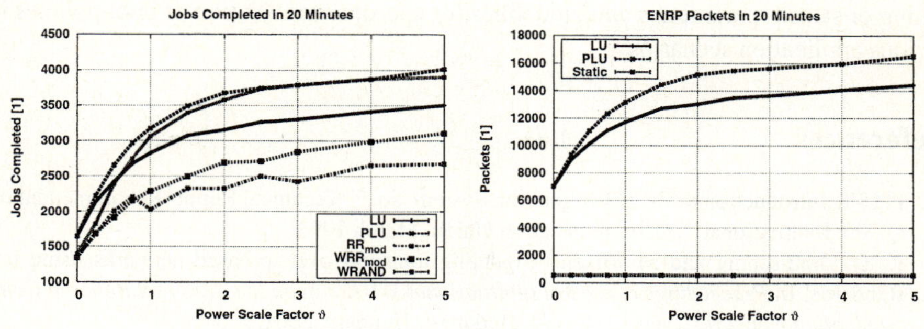

Fig. 6. Evolutionary Scenario: Jobs Completed and ENRP Packets

to 1), the picture changes for heterogeneous settings. The performance ranking among the policies is the same as for the fast server case although the differences are smaller. Again, the static WRAND policy performs remarkably well with minimum overhead.

6   Conclusion and Outlook

In this paper, we have presented the results of a study looking in detail at the issue of policy based load distribution in the Reliable Server Pooling concept currently under standardization in the IETF. Load distribution is one of the crucial aspects for RSerPool, as it significantly influences RSerPool's scalability and its capability to cope with realtime requirements while remaining "lightweight" in a sense that it keeps control overhead at an acceptable level.

As part of this work, we have detailed the incomplete specifications for the pool policies contained in the existing IETF RSerPool drafts [8] and [15] to a level which allows a consistent implementation.

Based on the results of our simulation studies, we proposed a modification of the Round Robin policy which avoids pathological patterns resulting in severe performance degradation in some cases. In addition, we proposed new static (RAND and WRAND) and dynamic (PLU) policies which perform significantly better than the original policies in realistic scenarios where the servers of a pool have different processing capacities.

As a consequence of this study, we have created the Internet Draft [16] containing the refinements and modifications for the already defined policies as well as the proposal to add our WRAND and PLU policies to the list of standard policies. This draft has been presented at the IETF RSerPool Working Group meeting at the 60th IETF meeting and has become a working group draft [18] of the IETF RSerPool Working Group.

After these first promising results, we are currently continuing the evaluation of the load distribution mechanisms by examining the sensitivity with respect to a broad range of system parameters including, e.g., the stale cache timer value, the length of the PE list returned per name resolution request, the maximum number of simultaneous jobs per PE and the PE capacity. In addition, we also investigate the scalability with respect to PEs, NSs and PUs as well as the influence of different traffic and job patterns. Our goal is to provide recommendations to implementers and users of RSerPool with respect to tuning of system parameters and guidelines for appropriate selection of pool policies in various application scenarios.

References

1. ITU-T. Introduction to CCITT Signalling System No. 7. Technical Report Recommendation Q.700, International Telecommunication Union, March 1993.
2. K. D. Gradischnig and M. Tüxen. Signaling transport over ip-based networks using ietf standards. In *Proceedings of the 3rd International Workshop on the design of Reliable Communication Networks*, pages 168–174, Budapest, Hungary, 2001.
3. R. Stewart, Q. Xie, K. Morneault, C. Sharp, H. Schwarzbauer, T. Taylor, I. Rytina, M. Kalla, L. Zhang, and V. Paxson. Stream Control Transmission Protocol. RFC 2960, IETF, Oct 2000.
4. A. Jungmaier, E.P Rathgeb, and M. Tüxen. On the Use of SCTP in Failover-Scenarios. In *Proceedings of the SCI 2002, Volume X, Mobile/Wireless Computing and Communication Systems II*, volume X, Orlando/U.S.A., Jul 2002.
5. A. Jungmaier, M. Schopp, and M. Tüxen. Performance Evaluation of the Stream Control Transmission Protocol. In *Proceedings of the IEEE Conference on High Performance Switching and Routing*, Heidelberg, Germany, June 2000.
6. M. Tüxen, Q. Xie, R. Stewart, M. Shore, L. Ong, J. Loughney, and M. Stillman. Requirements for Reliable Server Pooling. RFC 3227, IETF, Jan 2002.
7. Q. Xie, R. Stewart, and M. Stillman. Endpoint Name Resolution Protcol (ENRP). Internet-Draft Version 08, IETF, RSerPool WG, Jun 2004. draft-ietf-rserpool-enrp-07.txt, work in progress.
8. R. Stewart, Q. Xie, M. Stillman, and M. Tüxen. Aggregate Server Access Protcol (ASAP). Internet-Draft Version 09, IETF, RSerPool WG, Jun 2004. draft-ietf-rserpool-asap-08.txt, work in progress.
9. T. Dreibholz. An efficient approach for state sharing in server pools. In *Proceedings of the 27th Local Computer Networks Conference*, Tampa, Florida/U.S.A., Oct 2002.
10. P. Conrad, A. Jungmaier, C. Ross, W.-C. Sim, and M. Tüxen. Reliable IP Telephony Applications with SIP using RSerPool. In *Proceedings of the SCI 2002, Mobile/Wireless Computing and Communication Systems II*, volume X, Orlando/U.S.A., Jul 2002.
11. T. Dreibholz, A. Jungmaier, and M. Tüxen. A new Scheme for IP-based Internet Mobility. In *Proceedings of the 28th Local Computer Networks Conference*, Königswinter/Germany, Nov 2003.
12. T. Dreibolz and M. Tüxen. High availability using reliable server pooling. In *Proceedings of the Linux Conference Australia 2003*, Perth/Australia, Jan 2003.
13. Y. Zhang. Distributed Computing mit Reliable Server Pooling. Masters thesis, Universität Essen, Institut für Experimentelle Mathematik, Apr 2004.
14. Qiaobing Xie. Private communication at the 60th IETF meeting, San Diego/California, U.S.A., August 2004.
15. R. Stewart, Q. Xie, M. Stillman, and M. Tüxen. Aggregate Server Access Protocol (ASAP) and Endpoint Name Resolution Protocol (ENRP) Parameters. Internet-Draft Version 06, IETF, RSerPool WG, Jun 2004. draft-ietf-rserpool-common-param-05.txt, work in progress.
16. M. Tüxen and T. Dreibholz. Reliable Server Pooling Policies. Internet-Draft Version 00, IETF, RSerPool WG, Jul 2004. draft-tuexen-rserpool-policies-00.txt, work in progress.
17. OMNeT++ Discrete Event Simulation System. http://www.omnetpp.org.
18. M. Tüxen and T. Dreibholz. Reliable Server Pooling Policies. Internet-Draft Version 00, IETF, RSerPool WG, Oct 2004. draft-ietf-rserpool-policies-00.txt, work in progress.

Performance of a Hub-Based Network-Centric Application over the Iridium Satellite Network

Margaret M. McMahon[1] and Eric C. Firkin[2]

[1] United States Naval Academy, Computer Science Department,
572M Holloway Rd, 9F, Annapolis, MD, US 21402
`mmmcmahn@usna.edu`
[2] Raytheon Solipsys Corporation
6100 Chevy Chase Drive, Laurel, MD, US, 20707
`eric.firkin@solipsys.com`

Abstract. The hub-and-spoke architecture of a star network dates to the telecommunications central switching office. However, its benefits are only beginning to be explored in network-centric global applications. The hub facilitates interaction between dissimilar and distant systems. It serves multiple roles: translator between communication formats; subscription manager; multi-level data access controller; and data repository.

A network application has expectations of latency. Network-centric application developers need to understand the underlying performance of the spokes used to communicate with the hub. This understanding becomes more difficult when commercial spokes are used.

This paper describes the requirements for a hub that performs global networking, the implementation efforts to date, and the plans for future work. It addresses the methods used to gain insight into the performance of spokes when they are implemented by commercial network services. Results are presented for an example spoke technology, the Iridium satellite network.

1 Introduction

One goal of network-centric applications is the ability for multiple organizations to communicate seamlessly while pursuing a common goal. The different organizations may have dissimilar communication equipment and data formats. Communications capabilities and performance characteristics of the organizations' systems may vary widely; their data may differ in format and accuracy. Communication systems may vary in frequency, media used, effective transmission range, and data rate.

Currently, any user wanting to communicate with other groups of users must be equipped with the same types of communications equipment used by every group, or must establish a method to relay messages through groups that do have matching equipment. Relay of communications is difficult to arrange and maintain; it is also wasteful of what may be limited bandwidth of network-centric application participants. Possession of multiple communication systems is an obstacle to the *adhoc* networks that a network-centric application forms. Use of multiple systems

P. Lorenz and P. Dini (Eds.): ICN 2005, LNCS 3421, pp. 575–584, 2005.
© Springer-Verlag Berlin Heidelberg 2005

puts additional requirements on the user at the intersection of these data paths to integrate the data. This redundancy of effort and processing power is contrary to the benefits of network-centric applications. Use of a hub can overcome these limitations by providing flexible, reliable, and Beyond Line-Of-Sight (BLOS) communications. Ubiquitous communication can be enabled by a hub-and-spoke architecture. Users can connect to the hub using various technologies such as: Low Earth Orbit (LEO) satellite phones, land-lines, or radios. Each user connects to the hub via a separate spoke; the bandwidth of each spoke dictates the capacity of the communication link and user participation in the network-centric application. Using the hub also eliminates range limitations, and topographical obstacles.

Composing applications for the hub is challenging. One challenge is the use of public communications because commercial network providers give little insight into the internal performance of their networks. We need to use a black-box approach to characterize commercial network performance to assess its impact upon the applications. In this article, we focus on a method to characterize the performance of an application that uses a central hub with LEO satellites and land-lines as spokes.

2 Background

The hub-and-spoke architecture describes a star network topology, where each spoke in the network is a link connected to the central hub. Example systems that use this topology are: switched Ethernet, multi-network storage systems, and satellite networks. Switched Ethernet uses a hub to contain the collision domains of each computer. In a prototype storage system, local area networks (LANs) containing national information assets used the star topology and high-speed lines. A hub receives all satellite channels in Multichannel ALOHA networks [2], [3], [6], [10].

A star network is simple in structure; its complexity resides in the hub, not in the end stations. These networks are capable of high throughputs [1], [8]. The hub in a star network is the single, central potential point of failure. However, connecting multiple star networks together in a mesh can increase the reliability of the network.

There are several approaches to sharing data between different groups of users. When sharing within a multinational coalition, some systems require that networks be physically separated [12]; others rely on a common network with access controls in place [7]. When networks are physically separated, physical media can be hand carried between the networks, using a networking technique known as "sneakernet". These approaches may be of use during planning, but do not support real-time data sharing. When using a common shared medium, there is also the possibility of access to inappropriate data; security depends solely on user access controls. With the hub-and-spoke approach, data is not exposed on a shared network. Since there is an isolated link to each user there is no shared media, and the possibility of inadvertent disclosure during transmission is avoided. Separate and programmable filters from the hub to each spoke can provide access controls; limit the transmissions of specified data to and from an end user; adjust the fidelity and accuracy of data; match the send and receive capacity of a spoke transport media; and hide a specific user from all others to mask the source of data.

3   Related Work

The hub can be applied to business organizations of personnel as well as systems. The central hub serves as the center of decision making [4]. By organizing business-to-business transactions into an n-tier hub technology, data can be pushed to spokes as it is required; use of separate spokes prevents leakage of proprietary data. The hub can also provide a central repository of documents, databases, and applications [11].

Collaboration requires an ad hoc environment where spontaneous communication can occur and locating expertise in a distributed organization is problematic [4]. When a hub is used, expertise can be located through use of hub applications, facilitating collaboration.

In [5], the authors describe an application of the Iridium network to support communication needs of passenger, flight and cabin crews for voice, data and paging. Voice latency was estimated to be between 270-390 milliseconds (ms). Effective channel throughput delay for data (1024 bit message) is estimated between 427 ms and 1.7 seconds. Simulation of Iridium satellite networks performed by [9] estimated an upper limit of average end-to-end delay to be 210 ms when passing through 12 satellites. These studies were based on assumptions, rather than measurements.

4   The Hub

The hub is similar to the telephone central switching office. In addition to facilitating communications between dissimilar systems, it contains a real-time repository of historical and current data. Combining these features enables the hub to be the central point host for network-centric applications.

Although the hub is a centralized concept, its functions can be replicated to prevent having a single point of failure. Links to the hub, as well as the hub, can be redundant to maintain communication capability. A backup hub maintains the current hub configurations and can prevent disruption of the services and connections.

4.1   Hub Features

The hub hosts applications and facilitates effective information exchange. Since the spokes of the hub can provide communication links more reliably than Line of Sight (LOS) systems, participants can be aware of changing situations more quickly. Providing more current information translates to less chance of users inadvertently interfering with each other's objectives through ignorance, or use of obsolete data.

The persistent data that is maintained at the hub is provided to users via a validated request. Users may request specific data or subscribe to a data service. The hub maintains and controls publisher and subscriber associations, as well as data access controls based on users' needs and privileges. Experts and analysts can be stationed at the hub, and from there communicate decisions to application users.

To support ongoing operations, redundant hub sites are required to ensure continuous service. A back up hub will assume control when an error state exists, or there is loading beyond specification limits. A protocol will keep back-up hub(s) informed of the current state of connections and services. Upon failure, the spokes of the hub will be reconnected through the backup hub. The protocol will ensure a seamless service to the network-centric application users.

4.2 Hub Roles

The hub performs several distinct roles. It may be an information destination, an information filter, an intermediate point for the flow of information, or a gateway to other services. The resident network-centric applications are at the heart of the hub. Incoming data is processed at the hub, and the resulting information may be stored for future use, sent to users, transferred to human experts, or forwarded to other services.

The hub stores data and provides multi-level, asynchronous data access to subscribing users. Individual users may be more concerned with access to only certain types of data. For example, a ground-based user has a more urgent interest in road closures than an airborne user.

The hub can be an intermediate point, linking local LOS networks of communication and sensor systems. The users of separate physical networks can exchange data, regardless of format or data rate, and seamlessly become users in a global network.

4.3 Hub Components

The hub is composed of an operating system (OS), a connection manager, a database manager, and data archives. The OS starts the hub functions and monitoring the health of components. It also detects deadlock and overloading conditions. The hub OS must have a real-time kernel, guaranteeing quality of service for time-critical applications.

The Connection Manager is responsible for routing incoming data to the appropriate applications and outgoing links. Outgoing information is formatted for the intended receiver. Connections made into and out of the hub are implemented in an automated, flexible software solution. Information records are stored for each connection, including the access permissions associated with each connection. These records will be transferred to backup hubs, so that current connections may be maintained after a failover.

The Database Manager handles publishing and subscription requests. It works with the Connection Manager to fulfill subscribing users' needs. It is also responsible for saving and performing backups on archival media.

4.4 Hub Input/Output

The inputs into and outputs out of the hub may be of varying bandwidths, and various technologies. Examples of specific input and outputs technologies are Iridium phones, T-1 lines, Public Switched Telephone Network (PSTN), or a dedicated line to a Super Computer Center. By installing receiving and transmitting hardware/software on the hub, a connection of any type is possible.

4.5 Security

The security of transmissions can be provided by encryption. This requires encryption and decryption devices or software that can be unique to specific applications or communication services. Security processes and procedures may include various levels of physical security and configuration control and management of all external links.

4.6 Hub Applications

It is difficult to implement applications that require large amounts of data and processing power in fielded systems. A hub could combine data and provide appropriate to each user.

Rapidly evolving situations would benefit from these hub-based applications. A near-real-time radio frequency (RF) planning application that runs on the hub could assist in reducing RF interference in a critical area, without taxing the processing of fielded computing systems.

Developing and testing hub applications primarily affects only the hub. Any addition or update of applications should not impact hub users; the majority of the testing will be performed on the hub. Client/server interaction will use thin clients that require little or no maintenance on users' systems.

5 Implementing the Hub

5.1 Current Implementation

An initial layer of this global network architecture has been implemented. The operational hub is located in Lihue, Kauai, HI, with the backup hub located in Laurel, MD. Currently, users are linked to the hub using Iridium phones. Iridium phones are only one of many ways to connect to a hub; by installing other types of communicating equipment, the hub can be adapted to exchanging data between any sources.

The current hub-and-spoke architecture has already demonstrated its usefulness during U.S. Navy exercises. The hub permitted geographically distributed combatants to integrate sensor data to create a single integrated picture. At the end of each user/provider spoke into the hub was a Tactical Component Network (TCN[©]) segment, typically consisting of a LAN, multiple processors, and various peripherals. Some TCN segments provided sensor data to the hub; others functioned as subscribers to a fused sensor product. Each spoke into the hub used the Iridium satellite constellation.

5.2 Technical Challenges

The use of Iridium phones as the spokes has been successful, but challenging. The phones were not designed for the continuous communication required in a network. At times, they may require recycling power to accept data calls.

Currently, system operators manually perform most of the OS functions. For example, when a connection is dropped, it must be reestablished by redialing the Iridium phone. At this time, a hub power outage will not automatically switch connections to the secondary shadow hub.

Composing hub applications presents several technical challenges. The publish/subscribe method is a suitable paradigm for hub applications. More importantly, hub applications must be robust because of the variety of spoke technologies. Additionally, the applications must be developed to work consistently with hub backup and recovery protocols.

6   Black-Box Performance Characterization

To better understand the underlying network performance, we eliminated the global TCN application. We used a custom, instrumented, User Datagram Protocol (UDP) stateless echo application implemented in the C programming language. The programs used blocking socket functions. Both computers ran Windows XP, and were synchronized to Universal Coordinated Time (UCT) attained from GPS satellites.

In this initial phase of characterizing hub spoke performance we ran several experiments. In these experiments, the computers ran only the Windows XP operating system (OS) and the echo application. Communication functions were supported by the OS. Round trip time (RTT) was calculated by calculating the time elapsed by the client between sending and receiving a packet, subtracting the time between receiving and sending text at the server program.

We first ran the client and server on the same computer, shown in Table 1. These experiments showed values of either 0 or 10 milliseconds (msec), with the majority of values of were 0 msec. The smallest unit of measurement was msec, suggesting that the application itself had usually required less than 0 msec. Larger values were potentially experienced during a context switch. The second set of experiments was run on a 10BaseT local area network. Values greater than 10 msec could be explained by collisions on the Ethernet network.

Table 1. Same machine and same LAN performance

	Same workstation	10Base T LAN
Average RTT (msec)	0.86	36.18
Standard Dev	2.84	59.45
Max RTT (msec)	10	191
Min RTT (msec)	0	0
Mode (msec)	0	10
Average Packet Size (Bytes)	11.11	9.11

For the remaining experiments, clients were located in Annapolis, MD, and the hub was in Laurel, MD. The direct distance is 21 statute miles (42 miles round trip).

The next set of experiments was run over a dial-in connection using a local PSTN phone line to make an in-state call to the hub. A 56k modem was used with the point-to-point protocol (PPP) as the data link protocol for these experiments. This was seen as a baseline test with one of the most geographically direct type of spoke into the hub. The summary statistics for the dial-in tests are given in Table 2. The frequencyy of RTTs are shown graphically in Figure 1. The larger latencies were attributed to the reliability functions in the data link protocol over noisy analog voice lines. 75% of the RTTs were less than 200 msec; 81% were less than 400 msec.

Table 2. Summary statistics of dial-in tests

	Dial-in
Average RTT (msec)	1169.07
Standard Dev	3665.62
Max RTT (msec)	39246
Min RTT (msec)	134
Mode (msec)	160
Average Packet Size (Bytes)	15.86

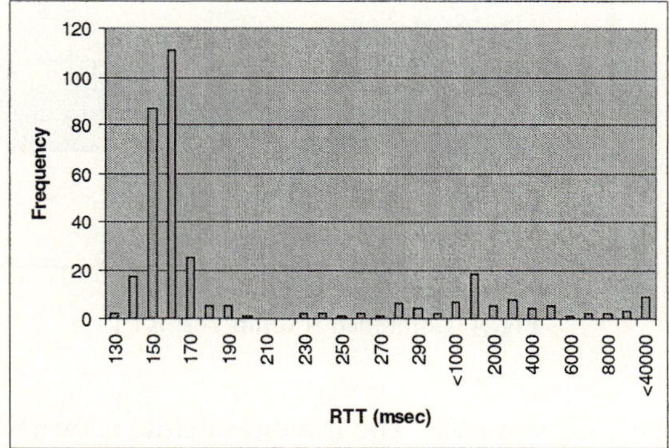

Fig. 1. Dial-in RTTs

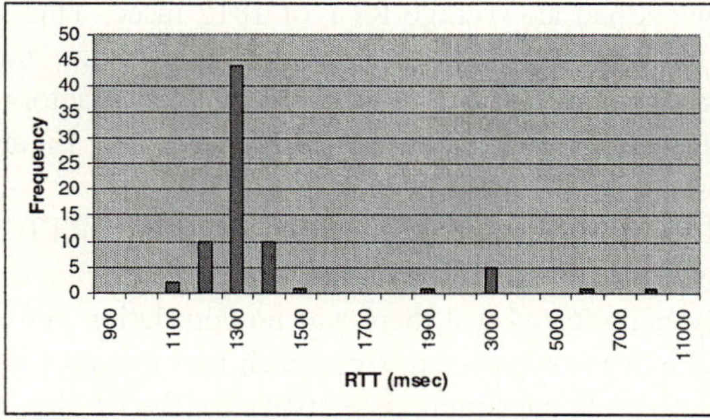

Fig. 2. Static Iridium Test RTTs

The final set of experiments used the Iridium network and land-line as the spoke into the hub. They were run on both static and on dynamic platforms. The Iridium

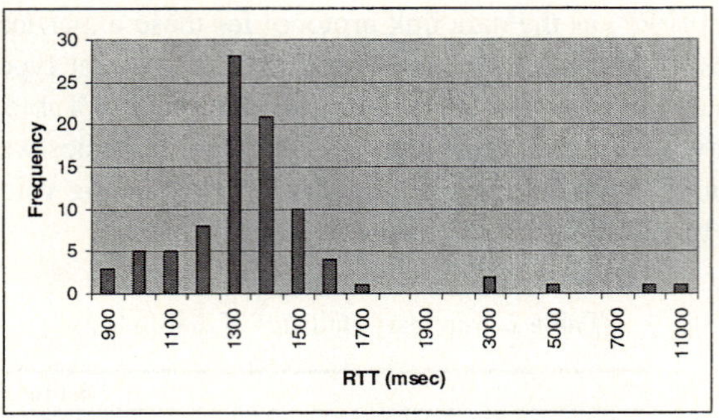

Fig. 3. Dynamic Iridium Test RTTs

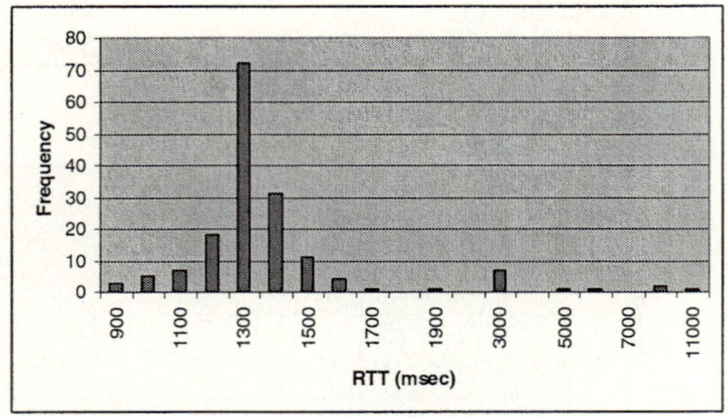

Fig. 4. Combined Iridium Test RTTs

gateway in Phoenix, AZ, was used. The Iridium satellite network carried traffic from the client laptop to the ground station in Arizona (2008 miles). The trip to the hub in Laurel was via land-line (1989 miles). The static experiments results, shown in Figure 2, had an average RTT of 1686 msec. Dynamic tests were conducted shipboard. Dynamic experiments had an average RTT of 1812 msec. This greater average was unlikely due to the moving platforms when using Iridium; the lowest RTT was recorded during a dynamic test (981 msec). More likely factors are traffic, routing delays, or the error correction Iridium uses for data service. Figure 3 contains results of the dynamic tests.

The combined data in the experiments had an average RTT of 1755 msec. This combined data is shown graphically in Figure 4.

Statistical analysis indicated that there was no correlation between packet size and round-trip time. This was as expected, since each text message was sent in one UDP datagram. Table 3 presents the summary statistics for the Iridium tests. The modes for both the static and dynamic tests are between 1300 and 1400 milliseconds, providing us a more realistic round-trip time to expect than using the average time.

The minimum Iridium RTT of 981 msec compared to the dial-in minimum 134 yields a ratio of 7.32. Collecting data to form ratios such as this supports the development of hub applications. A ratio gives insight into how the spokes compare in one geographic region, and can be used to tailor the application to be able to respond and react to missing or late communications.

Table 3. Summary statistics for the Iridium tests

	Static	Dynamic	Combined
Average RTT (msec)	1686.21	1811.89	1755.11
Standard Dev	1199.18	2059.99	1721.50
Max RTT (msec)	8832	16073	16073
Min RTT (msec)	1161	981	981
Mode (msec)	1362	1332	1091
Average Packet Size (Bytes)	11.48	16.14	13.82
Percent <= 1700 msec	89.33	93.41	91.57

7 Conclusions

Using a hub-and-spoke architecture is a viable approach to implementing a network-centric application. The hub can allow the exchange of data between dissimilar spoke users and serve as a host for applications that combine this data.

Developing hub applications requires insight into the performance of the spokes. Characterizing the performance of spokes such as Iridium needs to be done in the clients' geographic areas. Understanding potential performance, and parameterizing clients with this information will add to the robustness of hub applications.

8 Future Work

Additional data communication tests of the performance of client-server applications are planned. In this initial characterization phase, the spokes using Iridium had a significant land-line component. By using the primary hub in Kauai, the amount of land-line use will be minimized compared to the distance traveled via the LEO. Additionally, there is an analytic effort underway to develop a detailed analytic model for the whole Global TCN application, in an effort to reduce latency and increase bandwidth.

Acknowledgements

Raytheon Solipsys supported this research, and loaned the equipment used in the experiments. Many thanks go to LCDR Lori DeLooze, USN, for her invaluable participation in the LAN, static, and dynamic testing. Much gratitude goes to the students of the Spring 2004 SI455 Advanced Networks of the US Naval Academy course for their assistance in carrying out the shipboard portion of the experiments.

References

1. Abeysundara, B.W., and Kamal, A. E. (1991), "High-speed Local Area Networks and Their Performance: A Survey", ACM Computing Surveys (CSUR), vol. 23 no. 2, pp 221-264
2. Baron, D., and Birk Y. (2002), "Multiple Working Points in Multichannel ALOHA with Deadlines", Wireless Network, vol. 8, no. 1, pp 5–11
3. Coyne, R.A., Hulen, H., and Watson, R. (1992), "Storage Systems for National Information Assets", SIGARCH: ACM Special Interest Group on Computer Architecture, pp 626 – 633
4. Grinter, R.E. Herbsleb, J.D., and Perry, D.E. (1999), "The Geography of Coordination: Dealing with Distance in R&D Work", Proceedings of the international ACM SIGGROUP Conference on Supporting Group Work, Phoenix, Arizona, United States, pp 306 – 315
5. Lemme, P.W., Glenister, S.M., and Miller, A.W. (1999), "Iridium(R) Aeronautical Satellite Communications", Aerospace and Electronic Systems Magazine, IEEE , vol. 14 no. 11, pp 11 –16
6. Leo-Garcia, A., and Widjaja, I. (2000), Communication Networks. McGraw Hill.
7. Logsdon, R.C. (2000), "Coalition MLS Hexagon Prototype (CMHP) Demonstration", U.S. Joint Force Command (USJFCOM), available from http://www.mitre.org/support/multinat-conf02/BackgroundDocs/TMLSHEXAGONInitiative.ppt
8. Meddeb, A., Girard, A., and Rosenberg, C. (2002), The Impact of Point-to-multipoint Traffic Concentration on Multirate Networks Design", IEEE/ACM Transactions on Networking, vol. 10, no. 1, pp 115-124
9. Pratt, S.R., Richard A. Raines. R.A. (1999), Carl E. Fossa JR., and Michael A. Temple "An Operational and Performance Overview of the Iridium Low Earth Orbit Satellite System ", IEEE Communications Surveys, Second Quarter 1999 vol. 2, no. 2, available from http://www.comsoc.org/livepubs/surveys/index.html
10. Sharda, N. (1999), Multimedia Networks: Fundamentals and Future Directions", Communications of the Association for Information Systems, vol.1, article 10, pp 2-34.
11. Sommer, R.A., Gulledge, T.R., and Bailey, D. (2002), "The n-tier hub technology", ACM SIGMOD Record, vol. 31 , no. 1, pp 18 – 23
12. Treece, D. (1999), "Moving Sensitive U.S. Electrons Around in a Coalition Environment Without Spilling Any", IANewsletter, vol.2, no. 4

Performance Evaluation of Multichannel Slotted-ALOHA Networks with Buffering

Sebastià Galmés and Ramon Puigjaner

Dept. de Ciències Matemàtiques i Informàtica, Universitat de les Illes Balears,
Cra. de Valldemossa, km. 7.5, 07122 Palma, Illes Balears, Spain
{sebastia.galmes, putxi}@uib.es

Abstract. In the area of networking, as in computer systems, performance evaluation problems typically arise in scenarios where the availability of resources is limited. In many situations, buffering is provisioned at the resource site. This is the case of multiplexers and switches, where a buffer may precede a processing unit or a communication link. In other situations, buffering cannot be provisioned at the resource site, as it happens when several data sources directly transmit over a shared medium. In this case, some multiple access mechanism is required. In general, both problems have been separately addressed in the literature. In this paper, we perform an exact analysis of a combined problem, which includes a contention segment followed by a buffering segment. Particularly, we focus on a multichannel slotted ALOHA network with buffering. The analysis points out some interesting tradeoffs between the contention segment and the buffering segment.

1 Introduction

In the area of networking, as in computer systems, performance evaluation problems typically arise in scenarios where the availability of resources to requests is limited. There is a vast amount of literature on performance evaluation, and any attempt to provide a survey of references would be unfair. However, it is worth to mention some relevant works on the historical evolution and current state of the art of performance evaluation. This is the case of [1, 2], which offer comprehensive views regarding various aspects of the performance discipline, such as application areas, methodologies and tools.

An interesting point of view with regards to performance evaluation of communication networks is concerned about the provision of buffering at the resource site. When a buffer is present, it can be viewed as a way to regulate access of requests to resources. This is the case, for instance, of a switch, where a buffer may precede a processing unit (to decide how to reroute a packet), or a communication link (to forward the packet to the next switch). However, in other situations, buffering cannot be provisioned at the resource site, as it happens when several data sources directly transmit over a shared medium. In these cases, some multiple access mechanism is required to regulate access. In general, we observe that these two problems have been treated separately in the literature.

P. Lorenz and P. Dini (Eds.): ICN 2005, LNCS 3421, pp. 585–596, 2005.
© Springer-Verlag Berlin Heidelberg 2005

In this paper, we analyze a combined problem that includes a contention segment followed by a buffered segment. Particularly, we focus on the performance analysis of a multichannel slotted ALOHA network with buffering. This configuration may arise in different scenarios, such as packet radio, satellite or optical networks. Basically, the objective of the paper was to formalize mathematically the combined response time of the contention segment and the buffered segment, and analyze possible tradeoffs.

The paper is organized as follows. In Section 2, we provide a brief overview of ALOHA-based protocols. In Section 3, we fix the problem we are dealing with. Sections 4 and 5 analyze respectively the contention and buffered segments. Section 6 considers the global mean response time and the design tradeoffs between the two segments. It also provides some numerical results. Finally, Section 7 summarizes the main conclusions and proposes some future lines of research.

2 Overview of ALOHA-Based Protocols

As the reader may know, the ALOHA protocol was introduced by Norman Abramson in 1970 to support packet radio communications at the University of Hawaii [3]. In spite of its simplicity and low efficiency (about 18%), it constituted the basis of random access protocols working nowadays.

An immediate improvement was obtained by organizing transmissions according to regular time slots. The resulting protocol was named Slotted-ALOHA [4]. Basically, the Slotted-ALOHA reduced to the half the vulnerability period of the original pure ALOHA, thus increasing its efficiency up to 36%. However, this efficiency was still low, and then further improvements were proposed. These can be summarized as follows:

- Reservation ALOHA. Here, the most representative strategy is represented by the Packet Reservation Multiple Access (PRMA) protocol, introduced in [5].
- ALOHA with capture [6-7].
- Spread ALOHA [8].

In general, good overviews of multiple access schemes including the ALOHA protocols are provided in [9, 10]. A more detailed review of multiple access protocols, as well as their performance analysis, are better addressed in [11].

Another way to achieve higher throughputs, while keeping the simplicity of the traditional Slotted-ALOHA scheme, is by allowing for the coexistence of multiple channels. Multichannel Slotted-ALOHA schemes can be basically implemented in two ways: by using FDMA (Frequency Division Multiple Access), or by using SDMA (Space Division Multiple Access) [10, 12]. A good overview and detailed performance analysis of multichannel schemes for wireless communication networks can be found in [13].

Scenarios where ALOHA-based protocols can be currently found as part of the communications architecture are the following:

- Radio packet and satellite networks (both GEO and LEO systems) [14-18].
- Cellular wireless networks [5, 10].
- Optical networks [19-20].
- Indoor systems [21].

3   Problem Definition

We consider the system configuration shown in Figure 1, where users access a single queue through a multichannel Slotted-ALOHA network. This is constituted by a set of N synchronized Slotted-ALOHA channels ($N = 3$ in the figure). The queuing system is modeled as an infinite capacity queue followed by a single server. The service time is deterministic and equal to the slot duration of any single ALOHA channel. As it is illustrated in Figure 1, data packets only collide when they are transmitted over the same channel at the same time slot. Collided packets are lost, and must necessarily be retransmitted. Accordingly, the sequence of successfully transmitted packets defines a compound traffic entering the queue.

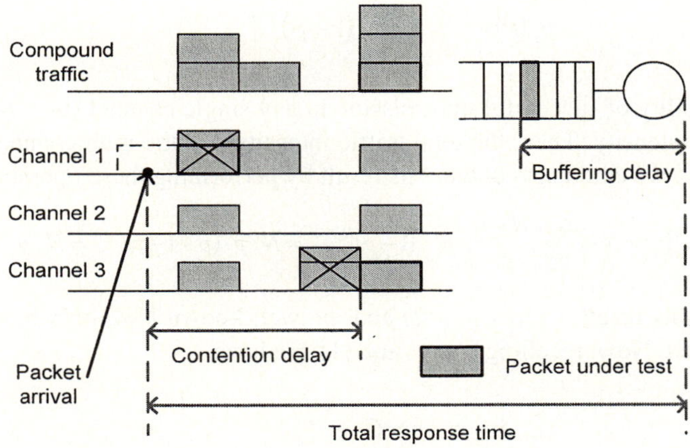

Fig. 1. Network configuration under study

In this paper, we analyze the total mean response time of the proposed configuration. This corresponds to the average time interval between the instant of arrival of a packet and the end of its service time (see Figure 1). In turn, this time interval can be decomposed into two contributions: the response time of the contention segment (contention delay), and the response time of the buffering segment (buffering delay).

4   The Contention Segment

In general, any packet in the contention segment experiences a set of collisions before being successfully transmitted. Each time the packet collides, it is retransmitted according to some retransmission probability. Initial transmission and retransmissions do not necessarily take place over the same channel; in fact, the sending station selects a channel randomly for each transmission attempt. In this paper, we will assume an infinite population of users (acceptable if the number of users is much larger than the number of channels) and the Poisson model for the overall traffic of new transmissions and retransmissions. Then, the analysis followed in this section is basically an extension to the multichannel case of that performed in [22] for single Slotted-ALOHA channels. Let λ and G be respectively the traffic intensities generated by new arrivals and retransmissions, with $G > \lambda$. If stations select channels randomly, then the traffic intensity offered to each channel is G/N. Consequently, the probability p of successful transmission in any particular channel, according to the Poisson distribution, is given by the following expression:

$$p = \frac{G}{N} \cdot e^{-\frac{G}{N}} \tag{1}$$

Then, the probability of successful transmission in exactly n out of N channels is:

$$p_s(n) = \binom{N}{n} \cdot p^n \cdot (1-p)^{N-n} \tag{2}$$

The probability of successful transmission in any single channel (p) is nothing else but its traffic intensity. Then, the total traffic intensity I of the multichannel system is $N \cdot p$. Of course, we could also obtain this result by performing these operations:

$$I = \sum_{n=1}^{N} n \cdot p_s(n) = N \cdot p \sum_{n=0}^{N-1} \binom{N-1}{n} \cdot p^n \cdot (1-p)^{N-1-n} = N \cdot p \cdot (p+1-p)^{N-1} = N \cdot p \tag{3}$$

To derive this result, expression (2) and the well-known Newton's binomial have been introduced. Now, recalling expression (1), we have:

$$I = N \cdot p = G \cdot e^{-\frac{G}{N}} \tag{4}$$

Note that for $N = 1$ this expression yields the well-known result for the single-channel Slotted-ALOHA system [22]. In equilibrium, the total traffic intensity in the multichannel system must be equal to the traffic intensity offered by new arrivals, that is, $I = \lambda$. As depicted in Figure 2, where a value of $N = 21$ has been adopted, this produces the usual bistable behavior of ALOHA-based systems.

Typically, some kind of stabilization algorithm must be introduced in order to keep the dynamics of the system around a single cross point. An example is the Pseudo-Bayesian algorithm, described in [22] for the single-channel system. Its extension to the multichannel system is quite simple: basically, if k is the number of backlogged stations (in the Pseudo-Bayesian algorithm, new arrivals are treated as retransmissions, and thus all stations having a pending packet are considered as backlogged) and q_r the retransmission probability, the average attempt rate A at any particular time slot is given by the following expression:

$$A = k \cdot \frac{1}{N} \cdot q_r \tag{5}$$

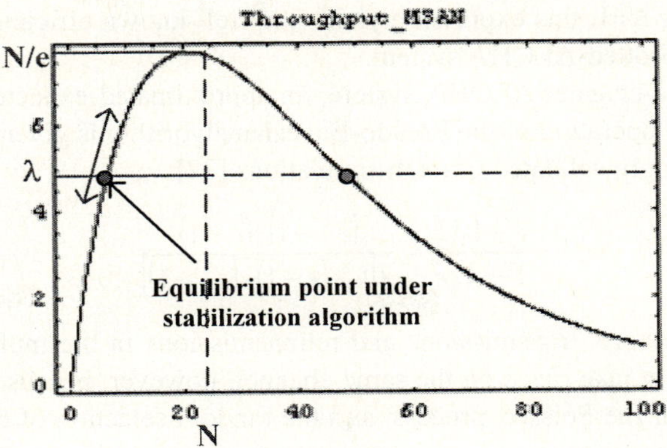

Fig. 2. Throughput of the Multichannel Slotted-ALOHA Network

Here, the fact that stations select the transmission channel randomly has been taken into consideration. In order to compensate the bistable behavior, the retransmission probability must be adjusted dynamically by stations, in such a way that the attempt rate is always equal or close to 1:

$$A = k \cdot \frac{1}{N} \cdot q_r = 1 \Rightarrow q_r = q_r(k) = \frac{N}{k} \tag{6}$$

In practice, stations may only have an estimate $\hat{k}$ of the number of backlogged stations, and, on the other hand, the retransmission probability must always be no greater than 1. Thus:

$$q_r = q_r(\hat{k}) = min\left(1, \frac{N}{\hat{k}}\right) \tag{7}$$

The introduction of the Pseudo-Bayesian algorithm fixes the equilibrium condition at the left-hand side point in Figure 2. For a given number of channels N, the condition for maximum throughput can be easily obtained by taking the single derivative of I with regards to G:

$$\frac{dI}{dG} = \frac{d\left(G \cdot e^{-\frac{G}{N}}\right)}{dG} = 0 \Rightarrow G = N \tag{8}$$

In this case, the maximum throughput of the system is given by:

$$I\big|_{G=N} = G \cdot e^{-\frac{G}{N}}\bigg|_{G=N} = \frac{N}{e} \tag{9}$$

Note that, for $N=1$, this expression yields the well-known efficiency of 36% of the single-channel Slotted-ALOHA system.

For the single-channel ALOHA system, an approximated expected contention delay w_c under the operation of the Pseudo-Bayesian algorithm is given by the following expression, strongly validated through simulations [22]:

$$w_c^{(1)} = \frac{e-1/2}{1-\lambda \cdot e} - \frac{\left(e^\lambda -1\right)\cdot(e-1)}{\lambda\left[1-(e-1)\cdot\left(e^\lambda -1\right)\right]} \tag{10}$$

As we said before, transmissions and retransmissions in the multichannel system do not necessarily take place on the same channel; however, because of the memoryless properties of the Poisson process, and the random selection of channels at transmission attempts, all time slots are subject to the same traffic conditions. Thus, we can think of the multichannel system as equivalent to a virtual single-channel Slotted-ALOHA, with an offered traffic rate of λ/N rather than λ. Accordingly, expression (10) can be transformed into the next one for the multichannel case:

$$w_c^{(N)} = w_c = \frac{e-1/2}{1-\dfrac{\lambda}{N}\cdot e} - \frac{\left(e^{\frac{\lambda}{N}} -1\right)\cdot(e-1)}{\dfrac{\lambda}{N}\left[1-(e-1)\cdot\left(e^{\frac{\lambda}{N}} -1\right)\right]} \tag{11}$$

5 The Buffering Segment

In order to analyze the buffering segment, we first need to characterize the traffic generated by successful transmissions. From Figure 1, we see that basically the compound traffic is of on/off type. Because of the memoryless property of the Poisson process, all time slots behave independently. Thus, we can expect that both the durations of the activity (on) and inactivity (off) periods are independent. Moreover, these durations will follow a geometric distribution, the only memoryless distribution in the discrete-time domain.

The compound or aggregate process is active whenever at least one successful transmission takes place in any of the channels. Consequently, the probability α that the aggregate process is active is given by the following expression:

$$\alpha = 1-(1-p)^N \tag{12}$$

Accordingly, the geometric distribution of the activity period $a(k)$ can be expressed as follows:

$$a(k) = (1-\alpha)\cdot\alpha^{k-1}, k \geq 1 \tag{13}$$

Similarly, the probability that the aggregate process is inactive (because of idle slots or collisions) is given by:

$$\beta = (1-p)^N = 1 - \alpha \tag{14}$$

Then, the geometric distribution of the inactivity periods is as follows:

$$b(k) = (1-\beta) \cdot \beta^{k-1} = \alpha \cdot (1-\alpha)^{k-1} , k \geq 1 \tag{15}$$

Finally, we need to characterize the batch sizes. Again, because of the properties of the Poisson process, successive batch sizes are independent, and consequently we only need to discover their distribution. To start with, let us express the probability $c^*(k)$ of k successful packets at a given instant of time:

$$c^*(k) = \binom{N}{k} \cdot p^k \cdot (1-p)^{N-k} , 0 \leq k \leq N \tag{16}$$

This is in fact the binomial distribution. However, typically batch sizes are only defined within active periods, what means that their value is assumed to be at least 1. Thus, we need to manage a modified version of the above distribution, which can be characterized as follows:

$$c(k) = prob(k \; successful \; packets / k \geq 1) = \frac{c^*(k)\big|_{k \geq 1}}{1 - c^*(0)} = \frac{\binom{N}{k} \cdot p^k \cdot (1-p)^{N-k}}{1 - (1-p)^N} , 1 \leq k \leq N \tag{17}$$

So, definitively the arrival traffic generated by successful packets can be modeled as a discrete batch-on/off process, where both on and off periods follow a geometric distribution, while the batch size corresponds to a modified binomial distribution. Independence assumptions hold for all samples of these distributions.

Accordingly, the analysis of the buffering segment relies on a (Batch-On/Off)/D/1 queue. The response time of a (Batch-On/Off)/D/1 queue was already analyzed in [23]. Particularly, an exact iterative algorithm producing an estimate of the expected response time was provided. This is the buffering delay w_b in our problem. As it is shown in [23], this algorithm makes use of generating functions evaluated at zero. For the present paper, we have used a modified version that works only in the time domain, thus avoiding the generation of indeterminate expressions and reducing the computational complexity. With regards to iterations, these are controlled by ε, for which we have adopted a value of $\varepsilon = 10^{-6}$. Finally, to make the execution of the algorithm feasible, we have considered truncated versions of the geometric distributions corresponding to the activity and inactivity periods. Particularly, we have used the same truncation value T for both distributions, as follows:

$$\tilde{a}(k) = \frac{1-\alpha}{1-\alpha^T} \cdot \alpha^{k-1} , 1 \leq k \leq T \tag{18}$$

$$\tilde{b}(k) = \frac{1-\beta}{1-\beta^T} \cdot \beta^{k-1} , 1 \leq k \leq T \tag{19}$$

We could previously test that beyond a relatively small truncation value of a few tens, final results hardly changed.

6 Evaluating the Total Mean Response Time

The contention and buffering delays analyzed in the previous sections can be combined to obtain the total mean response time of the system (w):

$$w = w_c + w_b \tag{20}$$

The response time w will depend on just two input parameters: the offered load λ and the number of channels N. However, they cannot be fixed in a completely independent way, because some considerations regarding the stability of the system must be considered. In particular, the contention segment is stable as long as the offered load does not exceed the maximum achievable throughput:

$$\lambda < \frac{N}{e} \tag{21}$$

This means that, with regards to the multichannel Slotted-ALOHA segment, for a given offered load there exists a minimum number of channels necessary to preserve stability: $N > \lambda \cdot e$.

On the other hand, since the buffering segment contains only one server (for instance, a single broadcast channel in a satellite system), the condition for stability is as follows:

$$\lambda < 1 \tag{22}$$

This means that in fact the buffering segment behaves like a bottleneck for the contention segment. Thus, the role of the multichannel network in our system configuration, rather than allowing for achieving a higher throughput, consists of reducing the contention delay – see expression (11), and, hopefully, the delay of the whole system. This is basically what we want to mathematically check in the present section.

As we said before, our input data are the offered load and the number of channels. If condition (21) is fulfilled, the contention delay can be calculated straightforward through expression (11). The highest computational complexity of the methodology proposed in this paper comes from the algorithm evaluating the buffered segment. This algorithm involves a procedure called Wiener-Hopf factorization, which introduces the major complexity. For more details on the computational complexity of this procedure, see [24]. Note also that the input data for the buffering segment are G and N, rather than λ and N: see expression (1) for the probability of success in any single channel, from which the distributions of the activity and inactivity periods and the batch size are defined. Thus, a simple algorithm is necessary to evaluate G from λ, that is, to determine the equilibrium point in Figure 2:

Algorithm

```
Step 1. Initialization phase:
  Set x_u = N  and  x_l = λ
```

$$x = \frac{x_u + x_l}{2} \; , \; y = x \cdot e^{-\frac{x}{N}}$$

$$\Delta = |y - \lambda|$$

```
Step 2. While Δ > ε do:
  if y > λ then x_u = x
  else x_l = x
```

$$x = \frac{x_u + x_l}{2}$$

$$y = x \cdot e^{-\frac{x}{N}}$$

$$\Delta = |y - \lambda|$$

```
Step 3. Output:
  G = x
```

When applying this algorithm, a value of $\varepsilon = 10^{-4}$ is enough.

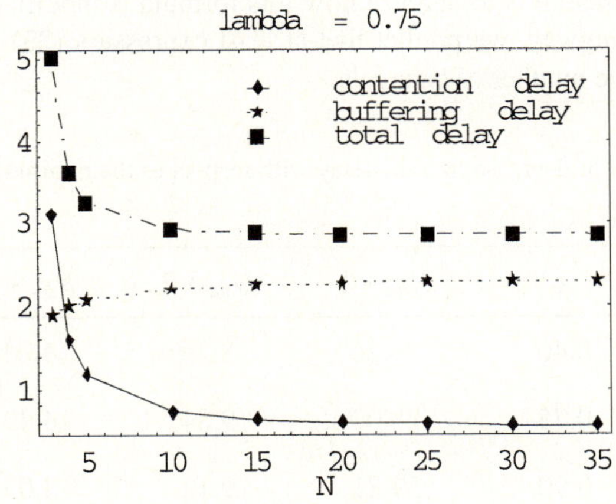

Fig. 3. Contention, buffering and total delays in terms of N for $\lambda = 0.75$

We performed several sets of experiments by combining different values of the offered traffic load (λ) and the number of channels (N). Intuitively, for a given traffic load, as the number of channels increase, the contention delay should decrease, but the buffering delay should also increase, since more successful packets tend to arrive in the buffer simultaneously. Thus, we can expect some kind of tradeoff. In general, we observed that for a wide range of traffic loads (from light to moderately high or even high) the decrease of the contention delay was stronger than the increase of the buffering delay, and consequently we only could appreciate a monotonic decrease of the total delay as a function of the number of channels. Only for a very high traffic load ($\lambda = 0.9$) we could appreciate an optimal number of channels for which the total delay reaches a minimum. However, even in this case, this optimal value was too high, and thus of low interest from a practical point of view. For illustration purposes, we plot in Figure 3 the evolution of the contention, buffering and total delays in terms of the number of channels, for a moderately high traffic load ($\lambda = 0.75$). Other traffic loads exhibited the same type of curves from a qualitative point of view.

On the other hand, by observing Figure 3 for small values of N, we see that the total delay initially experiences a considerable reduction, and then it evolves in a smoother way. In fact, we could observe this behavior for other traffic loads. Table 1 shows the percentage of decrease in response time produced by using a number of channels exceeding in one, two and three units the required number of channels for stability (N_{sta}). Note that the most important reduction takes place for only one channel in excess. Thus, from the engineering point of view (design, network planning), a cost-effective solution for the number of channels is represented by the following expression:

$$N_{opt} = \lceil \lambda \cdot e \rceil + 1 \tag{23}$$

Note that under the Poisson assumption formulated in this paper, the only input data to the network planning formula given by (23) is the offered traffic load. A question for further research is to analyze how this formula is modified for more complex traffic patterns. Anyway, we predict that at least expression (23) provides a practical lower bound on the number of channels.

Table 1. Percentage of decrease in total delay with respect to the minimum number of channels for stability

λ	$N_{sta} + 1$	$N_{sta} + 2$	$N_{sta} + 3$
0.40	24.26	8.35	4.04
0.75	28.00	9.84	4.49
0.90	39.81	9.12	3.03

7 Conclusions

In this paper, we have formulated and analyzed mathematically a combined problem of contention (multichannel Slotted-ALOHA network) and buffering (single-server infinite-capacity queue). We have assumed the Poisson model for the traffic of new arrivals and retransmissions on the contention segment, and we have developed an exact methodology to determine the expected total delay of the system. With regards to the contention and buffering delays, we have also provided a cost-effective tradeoff solution for the number of channels in the access network.

Questions for further consideration include the multiserver case and more complex traffic patterns, rather than the Poisson model.

References

1. LNCS 1769: Performance Evaluation: Origins and Directions (Günter Haring, Christoph Lindemann, Martin Reiser, eds.). Springer (2000)
2. LNCS 2965: Performance Tools and Applications to Networked Systems (Maria Carla Calzarossa, Erol Gelenbe, eds.). Springer (2004)
3. Abramson, N.: The ALOHA system. Another alternative for computer communications. Proceedings of the Fall Joint Computer Conference (1970)
4. Roberts, L. G.: ALOHA packet system with and without slots and capture. Computer Communications Review (April 1975)
5. Goodman, D. J., Valenzula, R. A., Gayliard, K. T. and Ramamurthi, B.: Packet reservation multiple access for local wireless communication. IEEE Transactions on Communications, Vol. 37, No. 8 (August 1989) pp. 885-890
6. Nelson, R. and Kleinrock, L.: The spatial capacity of a slotted ALOHA multihop packet radio network with capture. IEEE Transactions on Communications, Vol. COM-32, No. 6 (1984) pp. 684-694
7. Goodman, D. J. and Saleh, A. A. M.: Local ALOHA radio communications with capture and packet buffers. Proceedings of INFOCOM '85 (1985) pp. 984-990
8. Sato, I. T. *et al.*: Throughput analysis of DS/SSMA unslotted systems with fixed packet length. IEEE JSAC, Vol. 14, No. 4 (May 1996) pp. 750-756
9. Stallings, W.: Wireless Communications and Networks. Prentice-Hall (2001)
10. Rappaport, T. S.: Wireless Communications. Principles and Practice. Prentice-Hall (2002) (Second edition)
11. Abramson, N.: Multiple Access Communications. Foundations for Emerging Technologies. IEEE Press (1993)
12. Ward, J. and Compton, R. T.: High throughput slotted ALOHA packet radio networks with adaptive arrays. IEEE Transactions on Communications, Vol. 41, No. 3 (March 1993) pp. 460-470
13. Yue, W. and Matsumoto, Y.: Performance Analysis of Multi-Channel and Multi-Traffic on Wireless Communication Networks. Kluwer Academic Publishers (2002)
14. Gavish, B.: LEO/MEO systems – Global mobile communication systems. Telecommunication Systems, Vol. 8 (1997) pp. 99-141
15. Jamalipour, A., Katayama, M. and Ogawa, A.: Traffic characteristics of LEOS-based global personal communications networks. IEEE Communications Magazine (February 1997) pp. 118-122
16. Lutz, E.: Issues in satellite personal communication systems. Wireless Networks, Vol. 4 (1998) pp. 109-124
17. Le-Ngoc, T., Leung, V., Takats, P. and Garland, P.: Interactive multimedia satellite access communications. IEEE Communications Magazine (July 2003) pp. 78-85
18. Maral, G., Bousquet, M. and Nelson, J. C. C.: Satellite Communications Systems: Systems, Techniques and Technology. Wiley (1993) (Second edition)
19. Mukherjee, B.: Optical Communication Networks. McGraw-Hill (1997)
20. Sivalingam, K. M. and Subramaniam, S.: Optical WDM Networks. Principles and Practice. Kluwer Academic Publishers (2000)
21. Shad, F. and Todd, T. D.: Indoor slotted-ALOHA protocols using a smart antenna basestation. International Journal of Wireless Information Networks, Vol. 6, No. 3 (2003)
22. Bertsekas, D. and Gallager, R.: Data Networks. Prentice-Hall (1992) (Second edition)

23. Galmés, S. and Puigjaner, R.: An algorithm for computing the mean response time of a single server queue with generalized on/off traffic arrivals. Proceedings of the ACM Sigmetrics 2003 (2003)
24. Grassmann, W. K. and Jain, J. L.: Numerical solutions of the waiting time distribution and idle time distribution of the arithmetic GI/G/1 queue. Operations Research, Vol. 37, No. 1 (1989)ndoff calls

Towards a Scalable and Flexible Architecture for Virtual Private Networks

Shashank Khanvilkar and Ashfaq Khokhar

Department of Electrical and Computer Engineering,
University of Illinois at Chicago
`{skhanv1, ashfaq}@uic.edu`

Abstract. Virtual Private Networks (VPNs) are commonly used to provide secure connectivity over public networks. VPNs use tunnels to provide encryption, compression, and authentication functions, which are identically applied to every packet passing through them. However, this behavior may be overly rigid in many situations where the applications require different functions to be applied to different parts of their streams. Current VPNs are unable to offer such differential treatment, posing a heavy computational burden on edge routers, reducing their flexibility and degrading network performance. Additionally, the administrative cost of maintaining these tunnels is as high as $O(N^2)$ for an N node VPN. In this paper, we propose and evaluate a flexible VPN architecture (called Flexi-Tunes) where within a single VPN tunnel, different VPN functions are applied to different packet streams. Flexi-Tunes also replaces the traditional point-to-point tunnels with packet switched tunnels that improve scalability by reducing administrative costs to $O(N)$. Simulation of this enhanced model demonstrates 170% improvement in bandwidth and 40 times improvement in end-to-end delay.

1 Introduction

Virtual Private Networks are increasingly replacing traditional leased lines for secure global connectivity [1] [2]. All VPNs use point-to-point (p2p) tunnels that are characterized by the *encryption*, *compression* or *authentication* functions (collectively called as *VPN functions*) applied to its packets and the type of underlying transport (TCP/UDP) typically used. These features are pre-configured prior to tunnel establishment and remain static during its lifetime. Additionally, once the algorithms for performing such functions are selected, they cannot be changed without resetting the tunnel. Several practical scenarios, however, do not require such rigid constraints (*for e.g. an SSH session does not need any VPN functions to be applied to its packets*). Similarly not all information streams are equally important. While some may require strong cryptography, others may settle for weaker protocols that are not as compute intensive and hence faster (*for e.g. a confidential email sent by a manager as opposed to public memos*). Additionally, a fully connected N node VPN requires N(N-1)/2 tunnels and the cost of maintaining them affects the overall scalability of the VPN solution [3].

P. Lorenz and P. Dini (Eds.): ICN 2005, LNCS 3421, pp. 597–605, 2005.
© Springer-Verlag Berlin Heidelberg 2005

One simple approach to achieve application-level flexibility is to create separate tunnels tailored for every application using some form of centralized management. This approach, however, worsens the scalability problem, requiring the central entity to maintain $O(N^N)$ tunnels for N applications on an N node VPN.

In this paper we introduce Flexi-Tunes; a flexible and scalable tunnel architecture for VPNs that allows application-specific VPN treatment for different traffic streams. It also replaces the traditional p2p tunnels with packet switched tunnels, which improve scalability by reducing administrative costs to O(N). These objectives are achieved by simplifying the edge router architecture and introducing two new components: (a) a new IP Option header and (b) the VPN Subnet Server (VSS). The IP Option header adds additional fields to the IP header that can be used by applications to signal edge routers to apply specific VPN functionality to their streams (in *noncollaborative* mode) or apply all such functions at the end-host itself (in *collaborative* mode). The VSS, on the other hand, maintains a distributed database that aims to improve scalability, increase reliability, load balancing, site sharing and access control. Simulation of the proposed architecture, using voice traffic, shows 170% improvement in bandwidth utilization over conventional VPNs, which utilize only 35% of the available bandwidth. Similarly, end-to-end delay is improved by around 40 times. The rest of the paper is organized as follows. Section II starts with an introduction to the basic software architecture of a typical VPN edge router. Section III provides detailed description of the proposed Flexi-Tunes architecture and section IV discusses the test-bed and simulation results. Finally we conclude in Section V.

2 Software Architecture of Typical VPN Edge Router

Consider three VPN edge routers, R1, R2 & R3 (see Fig. 1(A)) serving private networks PN-1, PN-2 & PN-3 respectively, and connected in a full mesh using three tunnels. The routers could have been connected in a star topology requiring two tunnels instead of three. However, since every node in an overlay network (like VPN) can be considered to be a hop away from every other node, direct tunnels provide the shortest path. Hence we consider a full mesh to be the only viable topology for the rest of this paper. The bottom part of this figure shows the data-path for packets traveling between PN-1 and PN-2 (or to PN-3).

An analysis of several VPN solutions [3] suggests that every such router consists of two components: a Virtual Networking Interface (VNI) and a VPN Daemon. The VNI is an abstract interface that is used to exchange packets between the IP Routing Daemon (IRD) and the VPN Daemon and is separately initialized for every new tunnel. Like a real network interface, data-link protocols (like PPP) control packet delivery over the VNI. For e.g., in Fig. 1(A), VNI-0 and VNI-1 are created for tunnels between (R1, R2) and (R1, R3) respectively. The VPN daemon (see Fig. 1(B)), on the other hand, is a user- or kernel-level process consisting of a *Control-Plane* for connection maintenance and a *Data-Plane* for data processing. The *Control-plane* handles all issues related to tunnel security and maintenance, while the serial pipeline-like Data-plane uses multiple stages to provide encryption/authentication/compression functions.

Now consider a packet traveling between PN-1 and PN-2 (dark line in Fig. 1(A)). On arriving on *eth1*, the packet is handed to the IRD, which uses its destination address to determine the next hop router. Since the packet is destined for a private address, it needs to be tunneled towards R2 through the existing VPN. This is made possible by having the following entries in the routing table, where "X" and "Y" are the IP addresses assigned to VNI-0 and VNI-1 respectively.

Destination	Gateway	...	Interface
PN-2	X	...	VNI-0
PN-3	Y	...	VNI-1

Thus the packet is ultimately sent through VNI-0, which invariably passes it to the VPN Daemon. The VPN Daemon on its part, treats every packet as data and subjects it to different VPN functions that have been pre-configured for that tunnel, wraps the packet in a new IP header having the address of the destination router R2 (*which again is statically specified during configuration*) and sends it as any normal IP packet. Thus, in our example, every packet that is received by the VPN Daemon on VNI-0 is always tunneled towards R2. At the receiver the exact reverse steps take place as the packet is finally delivered to the correct destination host.

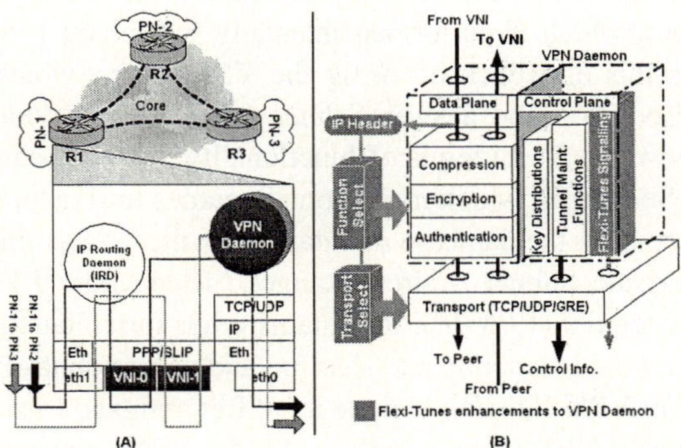

Fig. 1. (A) Basic VPN Edge router architecture, (B) Internal details of a VPN Daemon

In this way, the IRD actually makes all routing decisions (i.e. *which VNI to send the packet to?*). This allows arbitrary topologies to be created, but at the expense of requiring manual updates to the routing table. This is one of the primary sources for unscalability and one popular approach to eliminate this drawback is to use Virtual Routing [4]. However, in our opinion, virtual routing is overly redundant and often difficult. Flexi-tunes uses a simpler alternative to achieve the same objective as discussed next.

3 Flexi-Tunes VPN Architecture

In this section, we discuss Flexi-Tunes architecture. For simplicity, we first discuss the modifications that impact scalability followed by those that increase flexibility.

(A) Towards Higher Scalability: As said earlier, in conventional VPN edge routers, the IRD maps the destination private address to a VNI and the VPN daemon maps the VNI to the address of the peer VPN router. Flexi-Tunes eliminates the need to maintain this 2-level mapping, and enhances the VPN Daemon to maintain only a single association between the destination private address space and its corresponding VPN edge router. Thus in this new architecture ONLY one VNI and one VPN daemon is used irrespective of the number of nodes in the VPN and the IRD is demoted to the role of passively sending all VPN packets to it. This simple enhancement has several advantages. Since a single VNI is used, only one new routing entry, as show below, is required to be made, irrespective of the number of tunnels.

```
Destination      Gateway      …    Interface
   PN-*             X          …      VNI-0
```

Moreover, since no abstract p2p connection exists, state-full data-link protocols, like PPP, that were previously used for packet delivery over the VNI can now be replaced by null data-link protocols. Having fewer networking interfaces also bodes lower memory consumption for the operating system. Since the p2p virtual connection no longer exists and every VPN packet is tunneled independently, these tunnels are called packet-switched tunnels. The only apparent disadvantage of this enhancement is that the VPN daemon has to maintain an association for every private subnet and its peer router address, which if performed manually will result in un-scalability. Flexi-Tunes overcomes this drawback by using the VSS server, which uses a database to maintains this association and answer *Subnet Queries* like *"Which VPN edge router serves private-network-x?"*. In terms of functionality, the VSS can be compared to the Domain Naming Server [5], which maps domain names to IP addresses.

A sample schema for the database maintained by the VSS is illustrated in Fig. 2(A). Apart from the Private Subnet Address Ranges (listed under *PSAR*) and their corresponding edge router IPs (*VPN-IP*), it contains other miscellaneous fields to increase reliability, achieve load balancing, site sharing, and access control (as explained later). These fields are: (a) VPN-ID *(ID)*: Unique identifier assigned to an organization within a domain. (b) Priority *(Pr)*: Priority with which the VPN edge router information will be sent in reply to subnet queries; (c) *Cert*: Digital certificate of all registered VPN edge routers issued by a trusted authority (This also gives centralized powers to the VSS to periodically verify the authenticity of registered edge routers); and (d) Allow/Deny Access Control List (*ACL-A/-D*): Explicitly specifies nodes, which have (or don't have) access to router information present in the VSS. The first 3 lines in Fig. 2(A) is the configuration for our example network of Fig. 1(A).

Every packet-switched tunnel in Flexi-Tunes is defined by a *Security Association (SA)* that is dynamically created only when data flow occurs. It contains different tunnel specific parameters like lists of acceptable cipher/authentication/compression algorithms, as well as secret-keys to be used. A typical interaction between the VPN edge router and the VSS, leading to a successful SA negotiation is illustrated in Fig. 2(C). In the *Registration* phase, the site administrators register their routers with the VSS server (messages:1, 2 in Fig. 2(C)). Both R1 and R2 are initially unaware of each other's existence. Now if R1 receives a packet destined for the private subnet served by R2, it enters the *Query* phase and sends a digitally signed query to the VSS (msg: 3). The VSS server, on its part, authenticates the message, checks for the revocation of the R1's digital certificate and checks if R1 has access to R2's information. If all conditions are satisfied, a digitally signed reply containing R2's IP address and its certificate is sent back to R1 (msg: 4). At the same time, an information message containing R1's certificate may also be forwarded to R2 (msg: 5), which eliminates the need for R2 to re-authenticate R1. The VPN daemons at R1 and R2 can then create an SA using any standard negotiation protocol having a sequence of *offer/reject* messages (msg: 6, 7), leading to the *Data-Transfer Phase.* All packets between two sites can now be tunneled over this connection, in the same way as any conventional VPN. At the end of the data-transfer, the SA is broken and new streams restart from the "Query Phase".

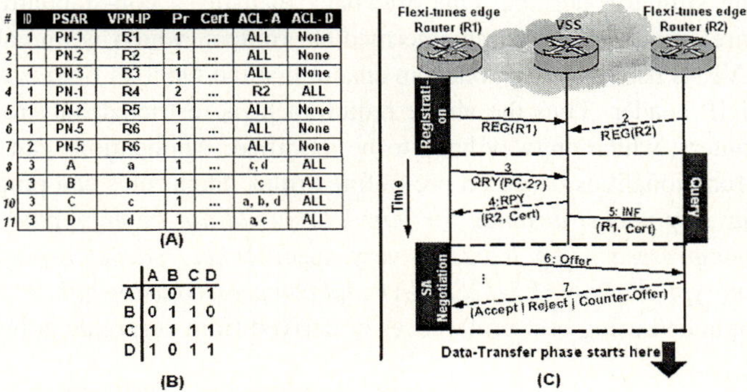

#	ID	PSAR	VPN-IP	Pr	Cert	ACL-A	ACL-D
1	1	PN-1	R1	1	...	ALL	None
2	1	PN-2	R2	1	...	ALL	None
3	1	PN-3	R3	1	...	ALL	None
4	1	PN-1	R4	2	...	R2	ALL
5	1	PN-2	R5	1	...	ALL	None
6	1	PN-5	R6	1	...	ALL	None
7	2	PN-5	R6	1	...	ALL	None
8	3	A	a	1	...	c, d	ALL
9	3	B	b	1	...	c	ALL
10	3	C	c	1	...	a, b, d	ALL
11	3	D	d	1	...	a, c	ALL

(A)

	A	B	C	D
A	1	0	1	1
B	0	1	1	0
C	1	1	1	1
D	1	0	1	1

(B)

(C)

Fig. 2. (A) VSS Data-base, (B) Access Control Matrix, (C) Interaction between edge routers & VSS

Let us now see how the VSS improves reliability, load balancing, site-sharing and access control. Reliability can be increased by using backup routers that automatically replace the main router during a breakdown. Rather than explicitly configuring separate backup tunnels as in conventional VPNs, backup routers are registered with a lower priority value in the VSS and are contacted for establishing tunnels, only when all high priority routers fail. For e.g., R4 (Line 4 of Fig. 2(A)) acts as a backup router for R1. Similarly, by registering edge routers at the same priority level, load balancing can be achieved (*for e.g., R5 is a load-sharing router for R2 and is contacted in a round robin fashion*). To share a subnet between two organizations (with disjoint address space), the edge router is registered twice under two different VPN-IDs (*for e.g., R6 allows PN-5 to be shared between two different organizations*). Lastly to realize the access control rules of Fig. 2(B), the last four lines of VSS are quite sufficient.

(B) Towards Higher Flexibility: To incorporate application-level flexibility during data-transfer, end-hosts directly interact with the edge routers to work in either *Collaborative (Co)* or *Non-Collaborative (NC)* mode. In either mode, the end-hosts specify the desired security functionality in a newly defined IP Option field. The VSS Daemons need to be modified with extra *function Select & Transport Select* blocks (see Fig. 1(A)) that parse through this field to determine the type & kind of VPN functions/transport protocol to be used. In *Non-Collaborative* mode end-hosts only select the desired security functionality for their packets and the ingress routers recognize these options to suitably apply them. In *Collaborative* mode, all essential VPN functionality is applied by the end-hosts, and the edge routers only tunnel/de-tunnel the packets. Packets with no option field are treated as conventional VPN packets.

Fig. 3(A) gives further details on the new *IP Option* field, which is added at the end of the IP header, increasing its size by 8 bytes. The *Option-Type*, uses a reserved value ensuring that there are no clashes with any existing standards. *Mode* is a 1-bit field that specifies the mode (*Co* or *NC*) under which the application is currently operating. Four single byte fields have been used for specifying different tunnel parameters. For example, the *Comp-Sel* field uses the first 4 bits to specify the compression algorithm and the last 4 bits to specify the compression level. *Encrypt-Sel*, *Auth-Sel* and *Transport-Sel* fields are similarly defined. One drawback of using a non-standard *IP Option* is that it requires modification to all intermediate routers, which is unrealizable. However, since VPNs use tunneling, one can ensure that the Options field is restricted only to the inner IP header. Thus the above requirement is restricted only to private hosts and edge routers, which anyway have to be modified. Applications can select the desired VPN functionalities based on pre-defined rules. Such rules can range from being very generic: *any packet destined for port 22, should not be encrypted nor authenticated but compressed using lzo:2;* to very specific: *any packet generated by PC-1 should be encrypted using AES-128/SHA1 and compressed using zlib:9.* Such rules can be easily populated using user preferences or derived from corporate policies.

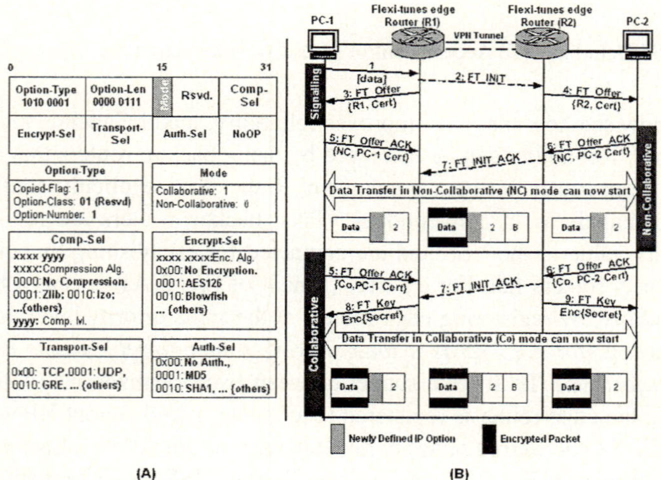

Fig. 3. (A) New IP Option field, (B) Interaction between end-hosts and Flexi-tunes enhanced edge router

Fig. 3(B) illustrates a typical *Data-Transfer* starting with initial signaling and leading to either *Co* or *NC* mode of operation. Flexi-Tunes operation is initiated by the edge routers on receiving the first packet (msg: 1 in Fig. 3(B)) or on receiving an *FT_INIT* message (msg: 2) from a peer. *FT_Offer* messages (msg: 3, 4) are then sent by the edge routers to their respective end-hosts offering them to use one of the two modes. For non-collaborative mode, the end-hosts acknowledge this offer (using *FT_Offer_ACK,* msg: 5, 6) and data-transfer can start after the initiating router (R1) receives an acknowledgement (*FT_INIT_ACK*, msg: 7) from its peer, signaling the end of Flexi-Tunes negotiations. PC-1 can now send packets with appropriately set *IP-Option* and the edge router parses through this option to apply the required functions. For collaborative mode, the edge router has to send back an extra *FT_KEY* message (msg: 8, 9) containing the shared secret. This message may be encrypted using the end-host's public key and digitally signed by the edge router. The *Flexi-Tunes Signaling* functional block (see Fig. 1(B)) is used to create such a trusted domain.

Collaborative mode is mainly targeted for desktop computers having sufficient processing power, while handheld devices benefit from non-collaborative mode. Note that, in Fig. 3(B), the inner IP headers were not encrypted, which may present a security risk for several organizations. One way to circumvent this problem is to have edge routers independently encrypt the inner IP headers. Since these headers are a fixed 28 bytes, encrypting them takes relatively short time.

4 Network Performance and Scalability Analysis

To evaluate network performance, we used OPNET to simulate Flexi-Tunes enhanced VPN tunnels and measured bandwidth utilization and end-to-end delay for four different scenarios: No_VPN, Non-Collaborative mode, Collaborative mode and Full_VPN. Voice streams (ON-OFF traffic) were used to measure network performance. We assume that the application can intelligently decide the percentage of packets that need VPN functions (termed as traffic Grade) to achieve a desired security level. For simplicity, we assumed a linear relationship between security and traffic grade. Bandwidth Utilization (BU) was measured by calculating the maximum number of voice users that can be accommodated with average packet loss limited to 2%, while end-to-end delay (ETE) was measured for a fixed number of voice users sending traffic at different grades. Other important simulation parameters are tabulated in Table I.

Table 1. Simulation Parameters

VPN Parameters	Value	Traffic Parameters	Value
Max. data rate	100 Kbps	Avg. length of ON period	Exp(1 sec)
time for all VPN funcs/bit	1.7 usec (=X)	Avg. pkt inter-arrival time.	Exp(9 msec)
Tunneling	exp(30% of X)	Avg. length of OFF period	Exp(1.35 sec)
Encryption	exp(33% of X)	Avg. packet size	Exp(208 bits)
Authentication	exp(23.5% of X)	Avg. packet data rate	~9.8Kbps
Compression	exp(11.5% of X)	Allowed packet drop rate	< 2%
Total Simulation Time =10 minutes.			

BU results for all four scenarios are illustrated in Fig. 4(A). The results have been normalized w.r.t. the No_VPN case, which also represents the maximum bandwidth that can be achieved. Since Full_VPN and No_VPN scenarios cannot provide differential packet treatment, the bandwidth utilization graphs are just straight lines, which present extreme limits for the other two scenarios. BU under the Non-collaborative mode decreases with increase in traffic grade, because the edge router has to spend more time per stream. The collaborative mode, on the other hand, shows optimal BU under all conditions, due to distribution of load to end-hosts. The cumulative distributive function (CDF) for the end-to-end delay is illustrated in Fig. 4(B). The abscissa shows the end-to-end delay faced by 90% of the packets, again normalized w.r.t No_VPN. Here collaborative mode achieves comparable performance at 10-Grade traffic and shows sustained degradation (up to 2.5 times) for increasing grade level. The performance degradation for Non-Collaborative mode is higher with 100-Grade streams achieving end-to-end delay that are ~40 times more (not shown) than the base case.

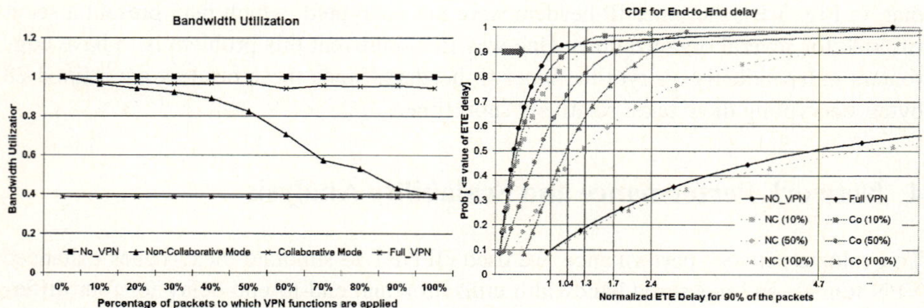

Fig. 4. (A) Bandwidth Utilization, (B) CDF for End-to-End delay

To analyze the scalability for Flexi-Tunes, we consider an N node VPN and calculate the total efforts required for updating routing tables, distributing security keys and editing configuration files, as was calculated for conventional VPNs in [3]. As discussed earlier, because of the single VNI only one general routing entry is needed at each of the N nodes, which can be done automatically. Thus ZERO efforts are spent on updating routing tables. Our enhanced architecture uses digital certificates, which allow edge routers to securely exchange shared secrets, which eliminates manual exchange of secret keys. The site administrator, however, has to obtain a signed certificate from a trusted authority and manually register with the VSS. Since every node has to do this, total efforts required is O(N). Finally the administrator at every site also has to update the configuration files with preferred cipher algorithms, key sizes etc. Adding all, the total cost can be calculated as varying in O(N).

5 Conclusion

In this paper, we have introduced Flexi-Tunes; a flexible and scalable tunnel architecture for VPNs that allows application-specific VPN treatment for different traffic streams. Simulation results with Flexi-tunes reveal 170% improvement in bandwidth utilization over conventional VPNs, which can achieve only 35% of the bandwidth. The total administrative cost for creating an N node Flexi-Tunes enhanced VPN was also shown to vary as O(N) as opposed to $O(N^2)$ for conventional VPNs.

References

1. R. Venkateswaran, "Virtual Private Networks", IEEE Potential, Mar 2001.
2. O. Kolesnikov and B. Hatch, "Building Linux VPNs", New Riders, 2001.
3. S. Khanvilkar, and A. Khokhar, "Virtual Private Networks: An Overview with performance evaluation", IEEE Comm. Mag., Oct 2004.
4. J. De Clercq and O. Paridaens, "Scalability implications of virtual private networks", IEEE Comm. Mag, May 2002.
5. P. Albitz, C. Liu,"DNS and BIND, 4ed.", O'Reilly, 2001.

A Simple, Efficient and Flexible Approach to Measure Multi-protocol Peer-to-Peer Traffic

Holger Bleul and Erwin P. Rathgeb

Computer Network Technology Group, Institute for Experimental Mathematics,
University of Duisburg-Essen, Ellernstrasse 29, 45326 Essen, Germany
{bleul, erwin.rathgeb}@exp-math.uni-essen.de

Abstract. Applications based on peer-to-peer (P2P) protocols have become tremendously popular over the last few years, now accounting for a significant share of the total network traffic. To avoid restrictions imposed by network administrators for various reasons, the protocols used have become more sophisticated and employ various techniques to avoid detection and recognition with standard measurement tools. In this paper, we describe a measurement approach based on application level signatures, which reliably detects and measures P2P traffic for these protocols and which is also easily adaptable to the new P2P protocols that now frequently appear. A presentation of the results from a successful field test in a large university network indicates that this approach is not only accurate, but also provides the performance and scalability required for practical applications.

1 Introduction: Peer-to-Peer Networks and Their Characteristics

While communication based on the conventional client-server model, e.g., for accessing web pages, clearly dominated in the past, peer-to-peer overlay networks evolved dramatically with respect to traffic volume and sophistication over the last few years. These overlay networks became enormously popular due to file sharing applications like Napster and Gnutella, which rapidly attracted a large number of users. Furthermore, sharing not only the relatively small MP3 audio files but also huge video files became more and more attractive with the increasing penetration of broadband access for residential users. Therefore, P2P traffic nowadays accounts for a significant part of overall traffic in the internet.

In addition to the increase in traffic volume, the introduction of P2P networks with no semantic differentiation of clients and servers also resulted in a significant change in user behaviour and traffic patterns. In particular, while residential access, e.g. via xDSL, was optimized for the highly asymmetrical client-server traffic, P2P traffic tends to be much more symmetric, thus creating new challenges for network design and dimensioning.

The first P2P protocols such as Napster and Gnutella were relatively simple and inflexible. They used, e.g. fixed TCP ports and well known addresses for

P. Lorenz and P. Dini (Eds.): ICN 2005, LNCS 3421, pp. 606–616, 2005.

accessing and building the overlay network. Due to this behaviour it was fairly simple to detect, recognize and measure the P2P based network traffic and also to restrict it by simple firewall policies.

To avoid the legal implications associated with sharing of copyrighted files as well as to limit the amount of (charged) traffic, administrators in corporate and university networks started to restrict P2P traffic as much as possible. As a reaction to these efforts, the existing protocols were modified and new, more sophisticated protocols appeared. Some investigations have shown that nowadays a significant share of P2P traffic is not using the default ports anymore, most likely to avoid detection (e.g. [1]) and subsequent restriction. Known techniques used to camouflage P2P traffic are (dynamic) port hopping, tunneling and encryption [2, 3].

As shown in fig. 1, which was compiled from netflow [4] data available for the Abilene network [5], the volume of P2P traffic had a sharp drop two years ago and since doesn't show the increase observable for http and unidentified traffic. This development can partially be attributed to the increased awareness of legal and copyright issues, but this is not the only suspected reason. Around the same time, P2P protocols, e.g. KaZaA [6], with more sophisticated camouflage capabilities, such as encryption of the protocol information or camouflage P2P traffic as "normal" web traffic, started to spread. This makes P2P traffic unrecognizable for standard measurement tools.

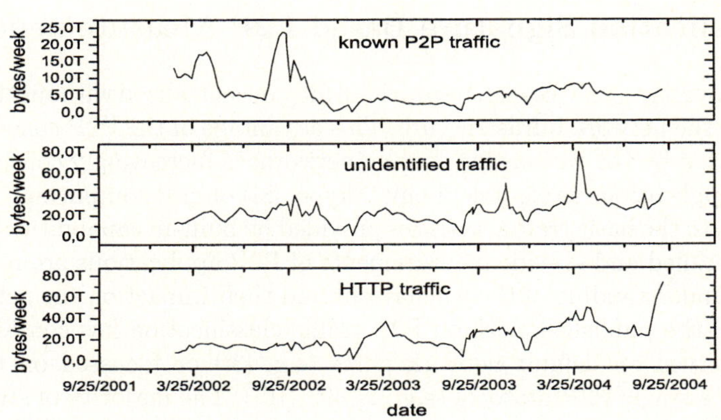

Fig. 1. Long term traffic trends in the Abilene backbone based on netflow measurements

The ability to accurately identify the network traffic associated with different (P2P) applications is crucial for a broad range of network management tasks including application-specific traffic engineering and capacity planning. However, traditional methods of mapping traffic to applications such as default server TCP or UDP network-port based identification are highly inaccurate for camouflaged P2P applications. To again be able to properly attribute the camouflaged traffic to P2P activity, more sophisticated methods to monitor and measure P2P traffic became necessary. Inspired by pattern-matching techniques of Intrusion Detection Systems (IDS) like Snort [7], which uses packet content analysis to identify malicious packets, new approaches have been proposed applying signature detection at the application level [1].

In this paper, we first evaluate two options for realizing this application signature based approach, the one being based on IDS techniques and the other being based on methods used for firewalls. We then describe our approach which is based on the latter concept. We show that this technique is efficient and easily adaptable to new protocols. This is particularly important as the innovation cycle for P2P applications has been very short. As a result, the popularity of specific protocols changes dramatically within a few months. Therefore, it is mandatory to provide flexible multi-protocol detection and measurement capabilities. We present a prototype implementation of this approach which is based on standard PC hardware and a combination of open source software components. The basic detection capabilities provided by these components are extended by adding a comprehensive set of signatures for the currently most popular P2P protocols. We also provide some results from a first field trial where the system was used to analyze the traffic at the central routing node of the university which handles several hundred Mbits of traffic per second. These tests confirm that application level signature analysis is feasible at high speeds and that the system is able to detect camouflaged P2P traffic otherwise not identified.

2   Application Signature Based P2P Measurements

With P2P overlays, services can be provided fully decentralized without the need for support in the network infrastructure. This advantage of the P2P concept clearly demonstrated by the various file sharing networks is increasingly considered also for other applications as e.g. telephony (Skype, [8]) or grid computing. Therefore, in addition to the basic traffic statistics provided by built-in capabilities of modern routers, detailed and specific measurements of P2P applications are required, to allow full understanding of their behaviour and their impact on the networks.

Most of the published work on P2P traffic classification has considered only traffic observed on default network ports (e.g. [3]) or focussed on the traffic analysis of a single P2P protocol (e.g. [2], [9], [10]). The majority of studies were based on the Gnutella protocol, because it was one of the first P2P protocols and also easy to analyze due to its open source character.

2.1   Challenges for Detailed P2P Measurement

A fundamental problem for detailed studies is the fact that most of the newer P2P protocols are proprietary, like KaZaA and Skype, and their control information is encrypted. Therefore, detailed information about the protocol itself can only be found by reverse protocol engineering. Furthermore the data payload of popular P2P protocols is encrypted and the traffic is often camouflaged as http traffic. Most of the current P2P clients use variable ports if a communication on standard ports cannot be established (port hopping).

Another issue is the wide variety and short innovation cycle for P2P protocols and applications, where the popularity – and therefore the impact on the networks – of a specific protocol changes dramatically within a few months and first generation protocols, like e.g. Gnutella, have already become outdated.

In P2P networks, control and signaling traffic is necessary to establish and maintain the overlay structures and to search for the target content or resource. This overhead – which is quite significant compared to client-server based concepts – has to be observed to learn about the topology and dynamic behaviour of the P2P network itself. However, to really understand the impact on the traffic patterns, the control traffic has to be linked to the high volume user traffic (e.g. file download) which is performed using independent point-to-point connections and non-P2P specific protocols, e.g. http. Thus, the correlation between control and user data is only visible at the application layer.

From this discussion it becomes obvious that standard measurements based mainly on layer 3 and 4 information are not sufficient to perform in-depth P2P analysis. This led to the idea of using application layer signatures [1]. This approach allows the set-up of measurement systems providing multi-protocol monitoring capabilities also for camouflaged protocols, and to correlate user and control traffic.

2.2   Options for Application Layer Signature Analysis

Application layer signatures are widely used for anomaly detection in current Intrusion Detection Systems (IDS). The popular freeware IDS "Snort" [7] for Linux systems already provides basic signature definitions for Gnutella, KaZaA and eDonkey in its rule set. Therefore, a concept for a measurement setup based on an extension of Snort has been defined and evaluated. By extending the signature set available already, a sufficiently accurate detection of P2P traffic would be feasible. However, with such a system only the number of packets containing a specific signature could be extracted from the log files, not the actual data volume (in bytes). In addition, tracking of related user data in such a system would be complex and resource intensive. As an IDS, Snort is designed to report specific incidents, but not to dynamically correlate specific sequences of incidents or dynamically track specific packet flows. The limited capability to define "active" rules which can be activated if a signature pattern has been matched are not sufficient for efficient stateful realtime analysis.

Another area where application layer signatures are already used are stateful firewall concepts. Stateful firewalls – other than IDSs – are specifically designed to identify, track and manipulate traffic flows in realtime. Therefore, we decided to base our system on this approach as described below.

2.3   Popular P2P Protocols and Their Signatures

To be able to apply application based signature methods, obviously the availability of appropriate signatures is crucial. We found that for popular protocols, relevant information can be found in the internet or has been published [1, 11] such that there is no need to do intensive re-engineering in most cases. Below we introduce the P2P protocols used in our study and their relevant signatures.

Gnutella: Gnutella was one of the first P2P protocols. It is open source and was used by various clients. It is the most analyzed P2P protocol but its popularity and usage have been diminishing rapidly. Gnutella is an example of a relatively simple P2P protocol without elaborate camouflage capabilities. Gnutella clients use ports 6346 and 6347 for TCP and UDP traffic by default and also http requests for file transfers. File downloads are initiated by a "GET /uri-res/" command. Many TCP packets start with strings "Gnutella" or "GIV", UDP packets with "GND". Gnutella supports connection on dynamic ports (port hopping), but it has been shown [11] that Gnutella clients mostly use standard ports either for source or destination of connections, so that simple port based firewalls can efficiently restrict Gnutella traffic.

DirectConnect: DirectConnect is also an example for a protocol which is relatively simple to identify and restrict. The DirectConnect protocol normally uses dedicated port numbers (411, 412) for TCP and UDP but also offers a passive mode for clients in firewalled environments. However, clients have to register at central nodes called "hubs" before becoming active in the network by using the well known ports. Therefore, usage can be restricted by conventional firewalling techniques preventing clients from registering – although the protocol supports port hopping for subsequent communication. In addition, clients have to provide a significant amount of content to share before allowed to register which also limits the popularity. All signaling packets start with the character "$", which makes DirectConnect flows easily recognizable by analyzing the first few bytes of their data payload. Furthermore the last byte of each packet starting with "$" is a "|" character.

FastTrack: FastTrack has been the most popular P2P network for a long time, used mostly for sharing audio and small video files. FastTrack uses port 1214 for both TCP and UDP traffic by default, but dynamic port adaptation allows switching to other ports, in particular to the default http port 80 (port tunneling). According to [11] FastTrack clients like the most popular KaZaA client use mostly arbitrary ports. Due to the similarity of packet format and block sizes with http traffic and the encryption of packet payload, those flows are very hard to identify and block by layer 4 firewalls. FastTrack downloads start with a TCP packet beginning with a string "GET /.hash=", others with a "GIVE" command. Furthermore the most popular client application KaZaA uses some additional descriptors containing the string "X-Kazaa".

eDonkey/Overnet: The eDonkey/Overnet protocol is particularly popular for sharing large movie files and uses two well known TCP ports (4661, 4662) and the UDP port 4665. While signaling packets are transferred either by TCP or UDP, the data itself is transported via TCP. eDonkey supports port hopping and [11] has shown that approximately 70% of all eDonkey bytes are transferred using arbitrary ports. Due to that characteristic it is impossible to block that kind of P2P traffic using classic firewalling mechanisms.

Packet analysis [1] has shown that TCP packets for signaling as well as for downloads contain a common eDonkey header using a special header marker. The marker value is always set to 0xe3 and the value contained in the 4 subsequent bytes reflects the packet size (excluding TCP and IP header bytes, the marker byte and the 4 byte length field). Control packets of other overnet clients like eMule start with a different byte (0xc5).

The signatures collected for these protocols, e.g. from [1,11], are summarized in table 1. The basic signatures which are implemented by default in the IPP2P module (see section 3.1) are underlined.

Table 1. Characteristic signatures for popular P2P protocols

Protocol	Sinature Elements	Transp. Protocol	Def. ports
Gnutella	"GNUTELLA", "GIV", "GET /uri-res/", "GET /get/",	TCP	
	, "X-Dynami", "X-Query", "X-Ultrap","X-Max"	TCP	
	"X-Quess", "X-Try", "X-Ext", "X-Degree",	TCP	6346, 6347
	"X-Versio", " X-Gnutel"	TCP	
	"GND"	UDP	
FastTrack	"Get/.hash", "GIVE", "X-Kazaa"	TCP	
	0x270000002980, 0x280000002900, 0x29000000,	UDP	1214
	0xc028 ,0xc1 (5 bytes), 0x2a (3 bytes)	UDP	
eDonkey	0xe3, 0xc5	TCP/UDP	4661-4665
DirectConnect	"$Send", "$Search", "$Connect", "$Get",	TCP	
	"$MyNick", "$Direction", "$Hello", "$Quit",	TCP	411, 412
	, "$Lock", "$Key", "$MyInfo"	TCP	
	"$SR", "$Pin"	UDP	

3 Concept and Realization of the Measurement System

3.1 Basic Concept of the Measurement System

The concept of our measurement system is based on the capabilities provided by some software modules for Linux systems, specifically the netfilter framework [12] with its various building blocks and extensions. Netfilter is intended to build stateful firewalls and to manipulate packet flows, e.g. to implement QoS schemes. Specific netfilter extensions support the type of signature analysis required. Furthermore, extensions for stateful firewalling like *conntrack* and *connmark* [12] allow the efficient tracking of identified connections and mark them for advanced analysis. Netfilter also offers detailed information about traffic volume both in number of packets and in number of bytes.

Another module used in our system is IPP2P, a netfilter extension to identify P2P traffic which was originally developed to shape P2P traffic to a given rate [13]. IPP2P analyzes the payload of TCP traffic for signaling patterns of P2P networks and can be used in netfilter rule sets. At the moment version 0.6.1 of IPP2P is available, which supports the detection of eDonkey, DirectConnect, Gnutella, KazAa, AppleJuice, SoulSeek and BitTorrent. However, the functionality provided only allows TCP analysis and provides only a rather limited signature set. Therefore, we extended the signature set as described in section 2.3.

Thus, the netfilter framework and the various freely configurable extensions allow the identification, counting and analysis of the traffic and also shaping, limiting or denying if so desired. With these features, it offers advanced mechanisms which are normally only available in commercial routers.

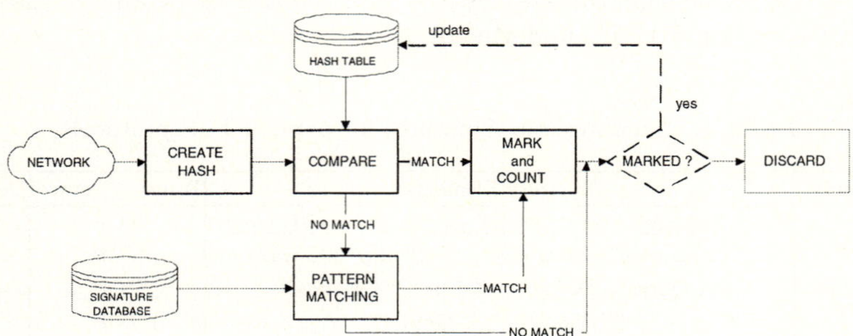

Fig. 2. Logical structure of the measurement setup

A typical configuration for passive measurement is shown in fig. 2. The measurement system acts as a traffic sink for the traffic mirrored on the monitor port of a switch. This setup has the advantage of not interfering with the actual network traffic in the production network. However, active systems inserted in the actual data path able to modify (filtering, shaping, prioritization) traffic flows can also be easily implemented. For all incoming packets, the relevant layer 4 information is hashed and compared to a hash table containing information on already identified P2P connections. If an unknown hash value is found, the packet is fed into the pattern matching process provided by IPP2P netfilter extension. In case a known P2P signature is found, a new mark is created and the packet is marked with it. All marked packets (of the new and the already identified flows) are counted and the data relevant for statistical analysis is collected. New hash values relevant for P2P traffic are used to automatically update the hash table. For all packets not related to P2P traffic, only a volume statistic is performed.

3.2   Implementation of the Measurement System

The hardware used for the prototype measurement system was a standard Pentium 4 PC equipped with a 3.2 GHz CPU and two Gigabit fiber NICs allowing scenarios where the actual traffic is passed through the system. A Linux 2.4.23 kernel was used which was configured as an Ethernet bridge by applying the ebtables-brnf kernel patch [14]. This allowed a flexible configuration of the system as router, transparent bridge or traffic-sink connected to a monitor port of a switch.

To enable traffic analysis, the network interface operated in promiscuous mode to accept all traffic. The IPP2P (version 0.5) module was used to identify the P2P traffic with the extended signature set described in section 2.3. We used iptables-1.2.9 [12] with available pom[1] patches for connection tracking and marking. Simple iptables rules have been used to assign the extended IPP2P submodules for different P2P networks. The `mark` and `restore-mark` commands have been used to provide connection tagging. Furthermore we used a general iptables rule to count the overall traffic in volume (bytes) and packets. Periodic polling of the counter values by using scheduled cronjobs provided a temporal correlation with adjustable time resolution. A bash script produced a formatted output of all counters together with an appropriate time stamp.

4   Field Trial and Results

To validate and evaluate the feasibility of our P2P analyzer we performed a field trial in a real life environment. The primary goal here was not to actually analyze the P2P traffic but to validate that performance and stability of the measurement system are sufficient. Therefore, a traffic volume measurement for the four protocols mentioned in section 2.3 was performed, which enabled us to roughly compare the results against data measured by the university computing centre via other methods.

The measurement PC was connected to the central routing node of the University Duisburg-Essen which also houses the POS 155 Mbit uplink to the G-WiN[2]. The complete external traffic of the Essen campus, the traffic between Essen and Duisburg as well as a part of the internal traffic of the Essen campus was mirrored to the router port connected to the measurement PC. Thus, we didn't interfere with the traffic and possible performance bottlenecks in our setup could not result in traffic degradation.

The measurement ran stably over a period from 6/9/04 to 06/22/04, only being shut down for a day due to reconfiguration. The measurement intervals were set to 10 minutes. Although the traffic peaked close to an average of 400 Mbit/s in some of the intervals, no packet losses could be observed within our system. This indicates that also more specific measurements with a larger set or more complex rules can be supported at these speeds. With respect to the detection accuracy of the system – which had also been tested in some small, local scenarios with specific P2P clients – the results were in the range that had been expected based on measurements performed by the computing centre using state of the art router based methods. The low overall percentage of P2P traffic can be explained by the fact that also high volume internal (backup-) traffic was included in the mirrored traffic and that the university computing centre spends significant effort in restricting P2P traffic as much as possible with conventional firewalling.

Fig. 3 shows traffic measured for the protocols eDonkey/Overnet, Gnutella, DirectConnect and FastTrack. Obviously, eDonkey and FastTrack overlay net-

---

[1] pom: patch-o-matic, patches to the the netfilter sources for module extension.

[2] G-WiN is the national gigabit-backbone of DFN (Deutsches Forschungsnetz) and integral part of the worldwide research and educational networks.

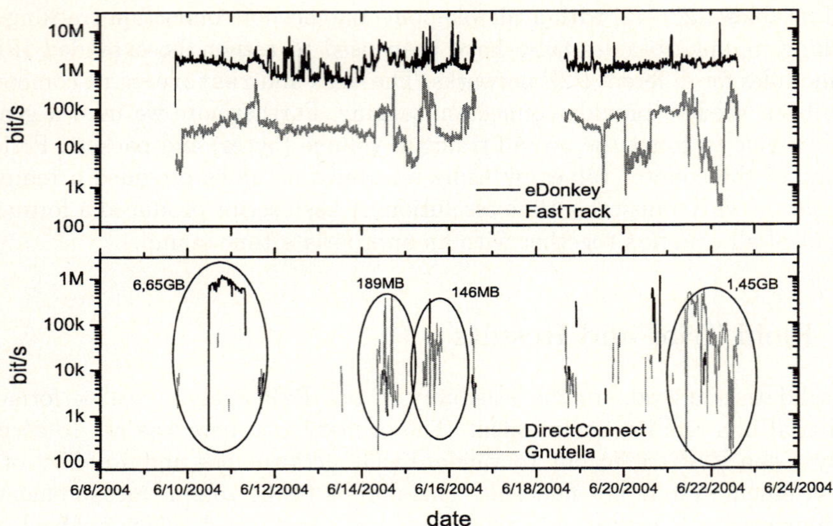

Fig. 3. Measured P2P traffic (logarithmic scale, traffic intensity averages over 10 minute intervals)

Table 2. Statistical summary of the measured data

statistical record	overall traffic	eDonkey	Direct Connect	Gnutella	FastTrack
Mean [Mbit/s]	141.2	1.46	0.06	0.02	0.05
Max [Mbit/s]	389.22	15.88	1.14	0.52	0.65
perc. to overall	100%	1.51%	0.04%	0.02%	0.05%
total measured (GB)	16442.98	170.42	7.06	1.91	5.74

works are the dominating sources of P2P traffic in the university network. They show a continuous activity level without significant periods of inactivity. This result coincides with our predictions, because those protocols are able to camouflage their activity on one side and are used by the currently most popular client applications.

Gnutella and DirectConnect can be tightly controlled and successfully restricted by the mechanisms applied by the computing centre. That explains the low volume and intermittent traffic pattern of these protocols. It can be assumed that the relatively short term activities (reflected by isolated vertical lines in this time scale) generating a few MByte of total traffic are successful attempts from individual users manually avoiding restricted ports. It appears that DirectConnect attempts typically generate more traffic than Gnutella attempts which can be explained by the fact that DirectConnect clients have to provide content to share right from the start. The longer periods of Gnutella and DirectConnect activity (regions marked with circles in the graph) can be interpreted as successful attempts to establish more permanent P2P connectivity. However, the latter interpretations could not be fully validated because the evaluation has been done offline after the field trial.

5   Conclusion and Outlook

In this paper we have presented a flexible and efficient P2P measurement system using application signature analysis which has been implemented with standard hardware and open source software. We demonstrated the feasibility of this approach in a real network environment and showed that the performance is sufficient to accurately measure high volume traffic on high speed links in realtime. The approach has the advantage that it is not specifically targeted towards one specific (P2P) protocol and can easily be extended to new signatures and other protocols. By separating the signatures of one protocol in different groups, more detailed studies with respect to specific protocol mechanisms can be performed.

We are currently planning more detailed and more comprehensive studies including a wider range of P2P protocols and applications, e.g. BitTorrent, which is becoming quite popular. In addition, we plan to perform also more specific measurements related, e.g., to the topology of P2P networks and to the dynamics (node lifetime) of these networks. However, since this involves, e.g., correlation of traffic with specific IP addresses, we are currently also investigating methods how to anonymize the measurements automatically such that legal and privacy concerns are respected.

References

1. S. Sen, O. Spatscheck, D. Wang, Accurate, Scalable In-network Identification of P2P Traffic Using Application Signatures, in Proceedings of the 13th international conference on World Wide Web, New York, 2004
2. R. Schollmeier, A. Dumanois, Peer-to-Peer Traffic Characteristics, in Proceedings of the 9th Eunice Open European Summer School, Budapest, Hungary, 2003
3. S. Sen, J. Wang, Analyzing peer-to-peer traffic across large networks, in Proceedings of ACM SIGCOMM Internet Measurement Workshop, Marseilles, France, 2002
4. Internet2 Netflow, Weekly Reports, data sets 02/18/2002 to 09/13/2004, `http://netflow.internet2.edu/weekly/longit/long.dat`
5. The Abilene Network homepage, 2004, `http://abilene.internet2.edu`
6. The KaZaA homepage, 2004, `http://www.kazaa.com`
7. The Snort Homepage, 2004, `http://www.snort.org`
8. The Skype homepage, 2004, `http://www.skype.com`
9. K.P. Gummadi et al., Measurement, Modeling and Analysis of a Peer-to-Peer File-Sharing Workload, in Proceedings of the 19th ACM Symposium on Operating Systems Principles, 2003
10. K. Tutschku, A Measurement-based Traffic Profile of the eDonkey Filesharing Service, Passive and Active Measurement Workshop, Cambridge, 2004

11. T. Karagiannis. A. Broido, M. Faloutsos, File-sharing in the Internet: A characterization of P2P traffic in the backbone, Tech. Rep., University of California, Riverside, 2003,
    http://www.cs.ucr.edu/~tkarag/papers/tech.pdf
12. The netfilter/iptables homepage, 2004, http://www.netfilter.org
13. The IPP2P homepage, University of Leipzig, 2004,
    http://rnvs.informatik.uni-leipzig.de/ipp2p
14. The ebtables homepage, 2004, http://ebtables.sourceforge.net

Secure Identity and Location Decoupling Using Peer-to-Peer Networks

Stephen Herborn[1], Tim Hsin-Ting Hu[1], Roksana Boreli[2], and Aruna Seneviratne[2]

[1] School of Electrical Engineering and Telecommunications,
University of New South Wales, Sydney, Australia
`{timhu, stephen}@mobqos.ee.unsw.edu.au`
[2,*] National ICT Australia Limited, Bay 15, Locomotive Workshop, Australian Technology, Park, Eveleigh, NSW 1430, Australia
`{roksana.boreli, aruna.seneviratne}@nicta.com.au`

Abstract. The emerging issues of personal and network mobility have created a need for the decoupling of identity from location in Internet addressing. This decoupling requires name resolution systems that can provide scalable resolution of globally unique persistent identifiers of communication endpoints, which may be users, devices, content or services. Recent developments in structured peer-to-peer overlay networks have made possible the scalable resolution of flat names, which opens up new possibilities in the area of naming and name resolution systems. In this paper we propose a scheme to provide authentication and verification in a name resolution system based on structured peer to peer networks such as distributed hash tables (DHTs). We specify how namespace security and global uniqueness may be managed with the use of public key cryptography. We also propose a framework within which independent overlay networks may compose a global namespace.

1   Introduction

Recent years have seen a rapid evolution of technologies that utilize the virtually instantaneous communication and global connectivity enabled by the Internet. Unfortunately, the internet's underlying architecture has not evolved to meet the requirements of new applications and devices. In emerging applications and technologies, a network endpoint can have a succession of addresses over a certain period of time (mobility); may have many addresses simultaneously (multi-homing), and these addresses may belong to different address spaces (heterogeneity),.

To demonstrate the issues of mobility, heterogeneity and multi-homing with an intentionally extreme example, we consider a user with a next generation mobile phone making a call to a friend while walking from his hotel room to a café. The phone has two network interfaces, to GSM and Wi-Fi respectively, the GSM interface is subscribed to a service provider (SP), and the Wi-Fi is capable of connecting to

---

* National ICT Australia is funded through the Australian Government's *Backing Australia's Ability* initiative, in part through the Australian Research Council.

P. Lorenz and P. Dini (Eds.): ICN 2005, LNCS 3421, pp. 617–624, 2005.

local wireless hotspots. As Wi-Fi is cheaper, the user has specified a preference to use local access points if available. The SP's network uses IPv4 and the WLAN uses IPv6 behind a 6-to-4 Internet gateway.

Thus, when the user dials his friend's number as he leaves his hotel room the phone will connect over the first interface (GSM) and commence the call. When he arrives within the coverage area of the café the phone will configure an address on the second interface. As the phone now has two configured address, both of which are usable if desired, we are presented with an instance of *multi-homing*. According to the user's preference, the phone will then elect to reroute the call through the wireless LAN interface. This is an example of *mobility* i.e. the change in the network level address of the phone call endpoint. It is also an example of *heterogeneity* since the address of the endpoint moves from IPv4 address space IPv6 space. It is important to note that the application level endpoint has not changed despite changes in its network address. Upon discovering the phone battery is flat, the user switches to a laptop software phone, which re-routes the call through the laptop wireless interface on the WLAN. This is another example of *mobility* i.e. application mobility between devices.

In the above situation, a number of factors such as network availability, user preferences, and device limitations may drive the data flow destined for a single endpoint to be routed through another network interface or even another device. Applying current solutions to such a scenario would not produce a meaningful result.

There are many unique problems to solve before the above scenario is realizable - one of the most significant is the problem of how to consistently resolve the network endpoint that corresponds to the actual communication endpoint of the data flow. An intuitive way to manage this is by placing an additional overlay network on top of the existing network layers and to allocate persistent, globally unique overlay level names to communication endpoints, as in [3].

Assignment and resolution of appropriate names is essential to ensure that the overlay solution is viable and that it enhances the mobility, heterogeneity and multi-homing abilities of various levels of network endpoints (e.g. users, devices, services and content). The second problem is how to associate the globally unique endpoint names to network addresses that are routable in the underlying networks. In this paper, we concentrate on a global naming and resolution scheme, and the routing and location based parts of the proposed system will be addressed at a later time.

2   Background

Internet name resolution is usually regarded as the direct mapping of a static name to a homogeneous network level address. For this DNS and DDNS were developed, and have served well thus far. However, overlay networks place new requirements on naming resolution that are outside the scope of what DNS was originally designed to handle and to support the development of novel internet based technology and applications, new approaches to naming, and name resolution are needed. Name resolution systems should provide timely and accurate resolution of persistent, general purpose names in a heterogeneous, mobile, and multi-homed network environment.

The structure and capabilities of a name resolution system depend on how the names are constructed. Whether names are semantically loaded or semantic free [4], flat or completely hierarchical or somewhere in between are important issues that depend heavily upon physical and commercial factors. If endpoint names are given specific semantic value, then they are liable to cause contention between owners of other endpoints that wish to use the same name. On the other hand, names with no embedded meaning are hard for humans to deal with and contain no indication of who owns the endpoint. One solution to limit the scope for contention is to build names hierarchically, for example there are many files in the world called *index.html* but only one called *mobqos.ee.unsw.edu.au/index.html*. However hierarchy inhibits the flexibility of the name resolution system by tying names to specific locations. The advantage of a completely flat namespace is that the name should be resolvable regardless of failures at other distant points in the system.

Another important requirement is the insulation of overlay level names from their respective names/addresses on lower layers, which enables implicit physical and logical mobility. As seen from different perspectives in [3] and [8], name/address decoupling also results in pushing ownership of names towards the endpoints.

Given the requirement for decoupling of names from addresses (or identity from location), names should ideally be flat as such names assume no specific network topology. The case for flat names is detailed further in [2] and [5]. The next issue is how the mechanics of the name resolution system that is used to distribute, manage and resolve such names might work. The most unique requirement is imposed by the lack of hierarchical structure in the names, which implies that the name resolution system should be a collection of equally capable peers rather than a strict hierarchy.

Recent developments in structured peer to peer networks, specifically the development of distributed hash tables (DHT) have made the scalable lookup of flat names a more achievable reality. Thus several prior works [3], [4], [5] have mentioned the use of DHTs as a basis for name resolution.

3   Using Structured Peer-to-Peer Networks

Our proposal centres on the use of the consistent hashing of some public key as an additional identifier for globally visible keys. Thus fully resolvable keys corresponding to the identity of communication endpoints, consist of a hash of the endpoint administrator's public key prefixed to some DHT or peer-to-peer level identifier. In this way, we embed a form of 'weak' hierarchy into the system which enables more flexible placement of keys and values throughout an otherwise flat and purely distributed lookup system.

One major problem with this approach is that current DHT implementations all assume global connectivity between the participant nodes at the underlying network level. However, as is apparent from our example scenario, and as mentioned in [1], this may not always be a realistic expectation. Firewalls and network gateways inhibit direct communication between nodes that are otherwise physically connected.

Our solution to this problem is to provide global scope by defining the system to consist of two (or more) layers of resolution, with the top layer being a lightweight overlay residing on the borders between administrative domains. This top layer performs the function of linking the lower level resolution provided by various systems existing on private administrative domains.

Another observation made by [1] is that since most of these systems use randomized identifiers for both keys and nodes in order to provide load balancing and robustness, applications have no way of ensuring that a key is stored within the organisation or specific network domain of the inserting entity.

While some existing proposals provide varying levels of administrative control and name security, their scope does not cover methods to secure the placement of names in a global name resolution system. Thus, it is also clear that if DHTs are to be used for name resolution, there needs to be an integrated way to provide security and administrative autonomy despite an inherently flat structure [8].

In order to secure the keyspace we propose several options that provide different levels of security and entail different levels of complexity. The DHT or peer to peer (overlay) identifier part of the endpoint ID may be obfuscated by encryption with the private key of the administrative entity, or with the administrator' public key. In addition, encryption of the corresponding values provides another type of protection.

Distribution of keys, especially sensitive private keys, is an ongoing research issue not limited to the context of our particular problem. In this paper we assume the existence of a secure out of band method of key distribution. However we hope to further investigate the issue within our scope in future work.

It is beneficial at this point to mention why we do not consider the use of DNS, since it is an established and thoroughly tested technology designed specifically for the purpose of name resolution. The reason for this is that the only capability that DNS has to resolve semantic free flat names is through the use of an 'SRV' type record, but this is not scalable since there is no way of ascertaining the correct DNS server to query for any given name. However, DNS can be used as a building block element for point of attachment discovery in future work.

4   Distributed Global Resolution

The first issue we consider is how to spread a lookup system across different administrative domains. Our solution is necessarily uncomplicated, so as not to place unnecessary restrictions on the applications that will use it. To this end we specify a few simple requirements and aspects of system design with regards to naming, name construction, and overlay architecture.

Naming. Names, or key values, in our system consist of the object administrator's public key hash appended to a DHT identifier or similar peer to peer lookup identifier. Figure 1 shows these two respective elements. The public key part of the name is either embedded into an operating system level service, or provided by an application. In either case the public key will most likely be distributed to users and devices when they join the administrative domain. The DHT identifier part of the name can be generated from any object meta-tag, as long as it is relevant to the lookup system of that particular lower level lookup system and of a set length.

The DHT identifier part may be encrypted before insertion to prevent unauthorized parties from being able to generate the name. This may be useful for applications like distributed private file systems or virtual private networks. (See section 5.)

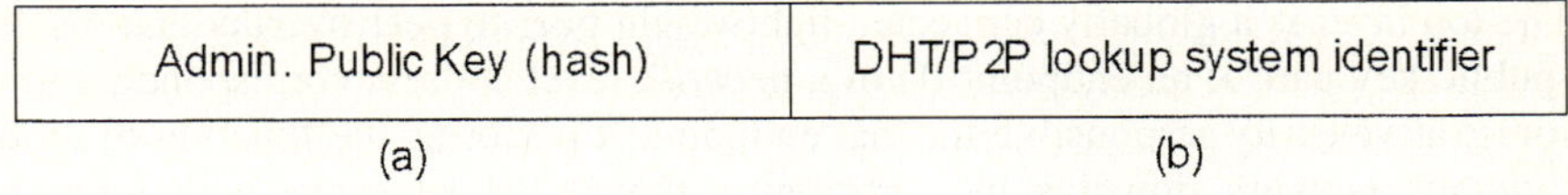

Fig. 1. Communication endpoint names (keys)

It is important to note that in case of name to network address resolution, these names are only used for initial lookup to establish a network level rendezvous point, since the identifiers are quite large and it is not feasible to perform peer to peer lookup and cryptography operations for every data packet. But, given the flexibility of the resolution system, it could be used to resolve just about anything since the manner in which the key value is generated depends entirely on the policy specified by the administrative domain to which it belongs. For example, a particular administrative domain may choose to generate all its keys from fully qualified file names; another may specify that keys are generated from user login details.

Name Construction. Name construction refers to the manner by which names can be generated from communication endpoint attributes or other information, so that lookup can be performed. It can be divided into two related issues, names for local objects and for non-local objects.

A client wishing to perform a lookup for a certain endpoint within the scope of it's own administrative domain takes the object attribute(s) that are specified by the policy of the administrative domain (e.g. a filename) and hashes them according to a common or previously obtained consistent hash function (e.g. SHA-1). It then prefixes the hash string with a hash of the local public key and submits the query.

The process of name construction for non-local endpoints, i.e. endpoints that do not belong to the same administrative scope as the client, is slightly more complicated. This is because the client may not be aware of the administrative domain to which the desired endpoint belongs and thus does not know either the public key or the hash function to apply to object attribute(s) for name construction. It is assumed, however, that the client knows which attribute(s) to use when constructing the name.

As mentioned in several works regarding DHTs and key identity/location decoupling ([1], [3]), the only reasonable way to address this issue is by assuming the existence of some out-of-band means to first determine the overlay level name or at least the correct public key from which the overlay name can then be derived. This could come in the form of a lookup or search service that returns results for human language queries.

Overlay Architecture. In order to provide global scope we define our system to consist of two or more overlay layers. The bottom layer(s) correspond to the lookup system existing within a particular administrative domain. The top level system is a lightweight overlay that bridges the individual lookup systems of participating administrative domains.

The term 'lower-level lookup system' refers to the lookup system existing within a particular administrative domain. This lookup system may or may not be a DHT, in practice all it needs to provide is an interface to some key based lookup API similar to that specified in [7].

The top layer is a globally connected lightweight peer to peer overlay that resolves the public key part of an endpoint ID to a network level address corresponding to the administrative entity responsible for that endpoint. It performs the function of uniting the various network domains by partitioning the global keyspace with respect to public key hash of the various participating administrative domains. For example, any keys under the administration of Organisation A will be prefixed with the public key hash of Organisation A.

5   Securing Keys and Values

As noted in [8] one of the main theoretical drawbacks of systems that separate identity from topological location is the sacrifice of the natural autonomy and authentication available in a strictly hierarchical system. We attempt to mitigate this with simple privacy and authentication mechanisms using public key cryptography.

In our system, full keys consist of a DHT or other lookup system identifier prefixed by a public key hash. The public key hash provides autonomy by ensuring that keys belonging to a certain administrative domain are stored in a known location.

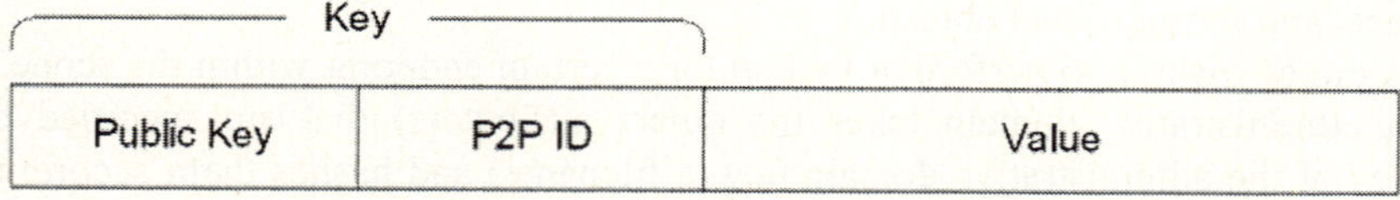

Fig. 2. Unsecured Key/Value pair

Securing Keys. If the DHT identifier part is encrypted with the private key of the administrative entity before insertion into the overlay, then only the administrative entity can generate the key from application level identifiers. Of course, the administrative entity may then pass these identifiers on to other entities that are authorized to have access to the referenced object. This provides a form of weak privacy in the global keyspace that may assist network file system style applications.

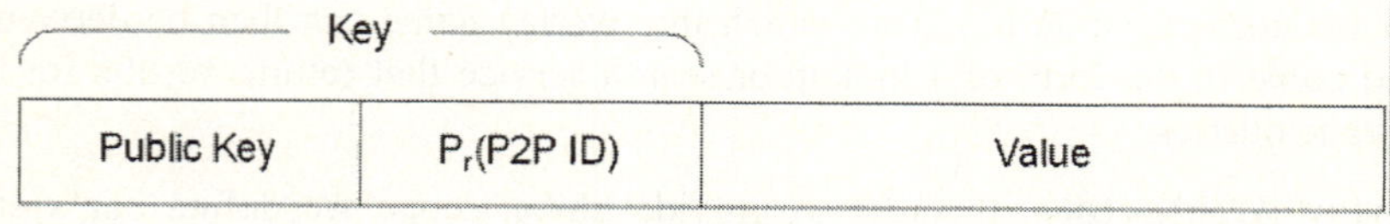

Fig. 3. Key secured with private key

Securing Values. A strong level of security and verifiable authenticity can be obtained by encrypting the values that are referenced by lookup system keys. Using public key cryptography methods, different effects can be generated by encrypting values with either the public or private keys of the administrative entity.

If values are encrypted with the public key of the administrative entity before insertion into the lookup system, then they can only be decrypted by an entity possessing the corresponding private key. This provides read security to the resolution system, meaning that even if keys are globally visible the values that they resolve to are meaningless to unauthorized entities.

If values are encrypted with the private key of the administrative entity before insertion into the lookup system, then they can be decrypted by anyone since the corresponding public key should be freely available. However it can be verified that the value is correct, or at least that it has only been modified by an authorized entity since the correct value can only be generated by an entity possessing the private key.

6   Related and Future Work

The decoupling of endpoint names from lower level addresses is advocated in [3], [5] and [8]. Specifically, [3] and [5] mention the possible use of peer-to-peer networks for name resolution. In [1] a proposal was made to provide keyspace autonomy and administrative control in structured peer-to-peer networks. The proposal suggests partitioning of the keyspace by requiring each participating organization to maintain a private key/value resolution system, and to enlist any globally connected nodes into a large scale key/value resolution ring. However our proposal provides authenticity and security within the overlay rather than just administrative control, and focuses more on using the concept to solve the larger problem of identity/location decoupling.

Host identity protocol [2] decouples identity from location with the use of cryptographic identifiers, but is more focused on the communication flow aspects as opposed to name resolution. HIP, together with SFR [4] also advocate flat naming. The use of public keys as additional identifiers, a similar concept is touched upon in [3] and explored more fully in [28] with somewhat different intent to the use of public key cryptography in our work.

Security for structured peer-to-peer overlay networks is discussed from the perspective of routing in [6].

Our future work will focus on: name resolution mechanics and infrastructure, a top level resolution algorithm, and event triggering (e.g. mid-session endpoint mobility).

7   Conclusions

In this paper, we have specified techniques for the management of persistent names for communication endpoints by securing distributed lookup systems. Our solution provides scalable resolution of flat names and enables the full decoupling of identity from location. We propose the linking of lookup systems belonging to different administrative domains into a global overlay, and mechanisms to provide fine-grained security in this global lookup system.

Our namespace is essentially contention free since the manner in which names are generated is determined by local administrative domain policy. The names are not human readable, which also minimizes contention over specific names. Our names are naturally persistent; the only limiting factor is the duration that a given administrative domain maintains the keys.

The emphasis of our proposal is on securing globally resolvable names while maintaining the flexibility of flat namespaces. We specify means by which names can be secured in a distributed name resolution system based on structured peer to peer overlay networks. A major aspect of our proposal is the use of public key cryptography to encrypt both communication endpoint names and their corresponding values and thus provide a fine-grained authentication mechanism.

Acknowledgements

The authors wish to thank the reviewers, Yuri Ismailov and Zhe-Guang Zhou for their valuable insights.

References

1. A. Mislove and P.Druschel. Providing Administrative Control and Autonomy in Structured Peer-to-Peer Overlays. *IPTPS*, 2004.
2. P. Nikander, J. Ylitalo, and J.Wall, Integrating Security, Mobility, and Multi-Homing in a HIP Way. *NDSS*, 2003
3. H. Balakrishnan, K. Lakshminarayanan, S. Ratnasamy, S. Shenker, I. Stoica, M. Walfish, A Layered Naming Architecture for the Internet. *SIGCOMM*, 2004.
4. M. Walfish, H. Balakrishnan, S. Shenker. Untangling the Web from DNS. In *NSDIS* 2004.
5. B. Ford. Unmanaged Internet Protocol: Taming the edge network management crisis. *HotNets-II*, 2003.
6. M. Castro, P. Drushel, A. Ganesh, A. Rowstron, D. Wallach, Secure routing for structured peer-to-peer overlay networks. *OSDI, 2002*
7. Frank Dabek, Ben Y. Zhao, Peter Druschel, John Kubiatowicz, Ion Stoica. Towards a Common API for Structured Peer-to-Peer Overlays. *IPTPS*, 2003
8. D. Clark, R. Braden, A. Falk, V. Pingali. FARA: Reorganizing the Addressing Architecture. *ACM SIGCOMM*, 2003.
9. D. Mazieres, M. Kaminsky, M. F. Kaashoek, E. Witchel. Separating key management from file system security. *SOSP*,1999
10. H. Balakrishnan, M. F. Kaashoek, D. Karger, R. Morris. Looking up data in P2P systems. *CACM*, Feb 2003.
11. J. Eriksson, M. Faloutsos, S. Krishnamurthy, PeerNet: Pushing peer-to-peer down the stack., *IPTPS*, 2003.
12. A.C. Snoeren, H. Balakrishnan. An end-to-end approach to host mobility. *Mobicom* 2000.
13. A.C. Snoeren, H. Balakrishnan, M.F. Kaashoek, Reconsidering Internet mobility. *HotOS-VIII*, 2001.
14. I. Stoica, R. Morris, D.Liben-Nowell et al. Chord: A scalable peer-to-peer lookup protocol for Internet applications. *IEEE/ACM Transactions on Networking,*, Feb 2003.

Live Streaming on a Peer-to-Peer Overlay: Implementation and Validation

Joaquín Caraballo Moreno[1] and Olivier Fourmaux[2]

[1] Universidad de Las Palmas de Gran Canaria[**]
[2] Laboratoire d'Informatique de Paris 6 (CNRS / Université Pierre et Marie Curie)

Abstract. Peer-to-peer based applications have demonstrated their interest in specific fields like file-sharing or large-scale distributed computing. Few works study the viability of peer-to-peer based continuous media applications. In this context, we get concerned with developing a framework that provides a service of transmission of a single source data flow with the characteristics and requirements of non-interactive audio and video live streaming. To perform experimentation, we have developed an application that uses a simple peer-to-peer protocol that can be easily changed, and can be used as a framework to test different protocols to build and maintain overlay network. We have carried out some tests and measurements in an hybrid approach to evaluate the quality of the application with its protocol and to offer a base to the evaluation of future protocols.

1 Introduction

Nowadays popular radio stations offer the possibility of retrieving their emission over the Internet. However, if the public massively used this service, required resources would be huge. Also, when some program is especially requested, if available resources are not heavily oversized it will not be possible to offer a quality emission to all listeners. With an application like ours, where user nodes also act as relayers, a solution to those difficulties could be provided.

In another scenario, a peer-to-peer application-level multicast approach enables anyone with an ordinary Internet connection to broadcast to potentially any number of listeners. Such a tool could contribute to the use of Internet as a communication media from anyone to many others, and, therefore, assisting to develop the right to free speech.

Radio and video live streaming from a single source to multiple receivers is an area of growing interest that poses some difficulties. With non-interactive streaming, latency constraints are softened but jitter constraints remain. In this work we have concentrated on the issues of efficient data transmission from one sender to a high number of receivers, addressed to non-interactive radio or video live streaming.

[**] This work has been carried out during an internship at LIP6.

P. Lorenz and P. Dini (Eds.): ICN 2005, LNCS 3421, pp. 625–633, 2005.
© Springer-Verlag Berlin Heidelberg 2005

The traditional approach uses a different *unicast* data stream between the sender and every receiver. As lower layers does not use the fact that the data to send are the same for every receiver, the required resources of broadcasting to N receivers are N times those of sending to one of the receivers. Also processing and network resources will have to be proportioned to the number of receivers.

Some standards have been developed providing a *multicast* service, At network layer, some additions to IP have been thought up, and the new IPv6 includes them. Also, at transport layer several protocols have been proposed. However no protocol have been really deployed for general use.

Application-level multicast offers a certain efficiency for one-to-many transmissions keeping lower layer protocols unchanged. Even if remaining out of underlying layers imposes us certain limitations in efficiency, application-level multicast provides us with several possibilities. On the one hand, it can be used to put in practice some multicast techniques and improve them. On the other hand, it permits to develop and use applications that require and can take advantage of a multicast service. Our work uses this kind of approach.

To develop an application-level multicast service, some agents must cooperate relaying received data to others. We have chosen a non-hierarchical approach, where every node has the same category, being a client when requesting the data flow and being a server when retransmitting the data flow to the nodes that require it. That is, we use a *peer-to-peer* network to put into practice the multicast service. All nodes have the same behavior except for the root, which, instead of requesting the data flow to relay it, provides the data flow itself and introduces it at first into the network. Many peer to peer networks have been developed where users transmit and share files, that is, static content; but what we are transmitting is a data flow, whit dynamic nature and time constraints.

The rest of this paper is organized as follows. Section 2 outlines our framework, decribing the service provided by our application, the underlying protocol and the hybrid approach we use for our experimentation. The results are proposed in section 3. Section 4 describes related works and section 5 concludes.

2   Live Streaming Framework

2.1   Service Offered by the Application

The transport service provided by the application is transparent to its users, letting them to be completely unaware of how the flow is being sent. At the sender side, a source connects to a node and sends through it a data flow. After the other nodes, a sink obtains its flow from the node to which it is connected. Figure 1 illustrates this service model.

2.2   A Simple Protocol

We have kept protocol definition as independent as possible from the remainder of the software. Next we describe the one we have developed and included.

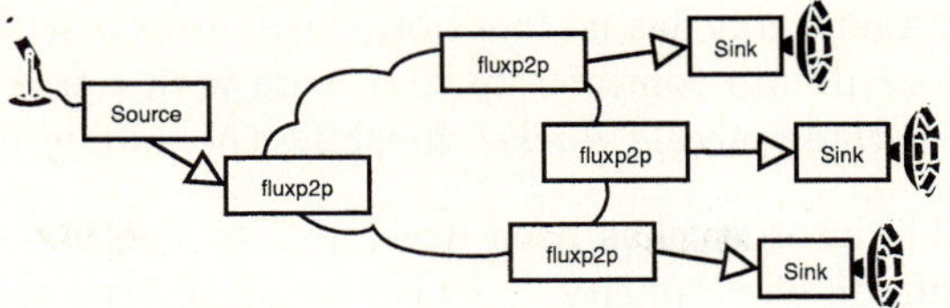

Fig. 1. Offered service

Inter-node communication is made by means of a peer-to-peer protocol. For every broadcasting network, one node will act as root, being the primary source of the data flow, while the others will receive it and possibly retransmit it.

To build the delivery tree five different control packets are used. A *Request Packet* is sent by a node who wants to join the network to the node to which it wants to be connected. If the latter can accept the connection, it answers with an *Accept Packet*, otherwise, a *Forward Packet* can be sent expressing that the node will not accept the connection and informing of other possible candidates. To mesure the quality of the connection with another node, a node can send a *Ping Packet*, that will be answered with a *Pong Packet*, thus, permitting to work out the *round trip time*.

Joining Process. Joining process is illustrated in Fig. 2. When a node wants to join the broadcast network, only the root is in its candidates list (C in the figure). At first, it sends a *Request Packet* to the root node. Usually, the root will not have any free slot, so it will answer with a *Forward Packet*, indicating that it is not able to directly treat the connection and informing how to reach its direct descendants. Consequently, the node will discard the root node and will add to C those received within the *Forward Packet*. Next, it sends *Ping Packets*, that will be probably answered with *Pong Packets*, to measure the quality of the candidates. In accordance with those RTTs, a new *Request Packet* will be sent

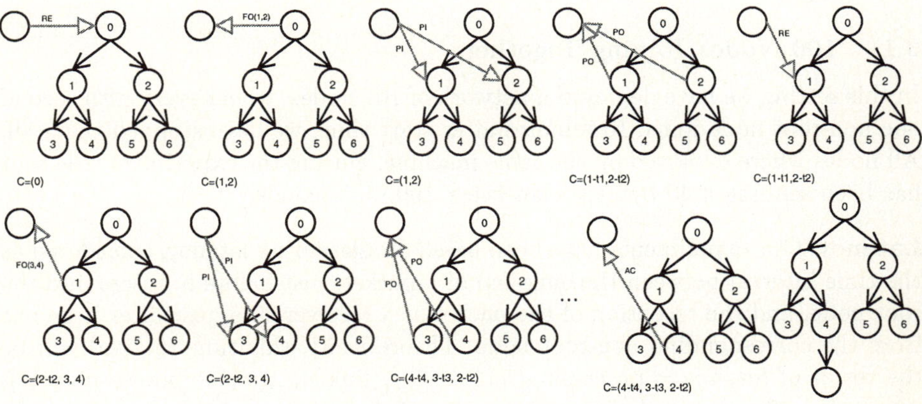

Fig. 2. Joining process

to the best candidate. Again, if it has no free slot, it will answer with a *Forward Packet* and the process will still continue up to a node with a free slot will be contacted and answer with an *Accept Packet*, finishing the joining process.

Scalability Issues. The protocol has been developed to test the software architecture with a maximum of simplicity, but also trying to keep the scalability of the system. In a strict sense, as the proposed joining procedure always starts by contacting the root, if the number of simultaneous joining nodes is very high, the root will became overcharged and therefore the system does not really scale up. However, instead of making every node to start by the same root, we can set, in the non-root nodes configuration, different nodes as root nodes.

2.3   Hybrid Prototype

To evaluate the quality of a protocol or a flow broadcasting application in general, the deployment of a big amount of nodes is necessary. In an ideal situation, we should have a host available for each node. However, as the desired number of nodes for a meaningful test is considerably high, in most of situations such a number of computers will not be available. As a workable solution, we can deploy the set of N nodes on M hosts, where $N >> M$. But if those nodes are located in the same machine or in machines close together, the experienced characteristics of network connections will not be realistic. To deal with this difference we have used a network topology model, composed of a set of nodes and delays between any two of them. For every node in the broadcasting network, we assign to it a node in the topology model, enabling us to obtain a simulated network delay between any pair of nodes. Whenever we receive a packet we simulate its corresponding network delay by waiting before we treat it. This system enables us to really test a broadcast network of a considerably high number of nodes without having to deploy them in a so big number of hosts.

3   Experimental Results

3.1   100 Nodes Joining Together

In this setting we have deployed a network of 101 nodes, where every fifth second one non-root node joins. 10 minutes after start time, we have stopped them all. All nodes where deployed on the same machine. During the experiment, the root has been sending a 40 bytes packet every 100 miliseconds.

Latency. Our experiments have been based on measuring latency, considered as the time interval between the emission of a packet, just before it enters into the root node, and the reception of the packet at a receiver side, just after it go out from the corresponding non-root node. Theoretically, this time interval will be the result of $latency = \text{processing}(n_0) + \sum_{i=1}^{r}(\text{link}(n_{i-1}, n_i) + \text{processing}(n_i))$, where n_0 is the root node, n_r the reception node, n_i a node in the way from the root to the reception node, $\text{processing}(n_i)$ the processing time of the flow in n_i, and $\text{link}(n_i, n_j)$ the network delay between nodes n_i and n_j. In our experiments we have simulated network delays, and, thus, also link times are produced in the host and processing times are incremented because of the cost of this simulation.

In Fig. 3, we can watch how latency is growing as the number of nodes receiving packets is incremented. To study performance, we have taken an arbitrary node, we have taken an arbitrary node, *node 0*, and observed how the latency experienced evolves. In Fig. 4 we can watch this evolution related with the number of nodes receiving data.

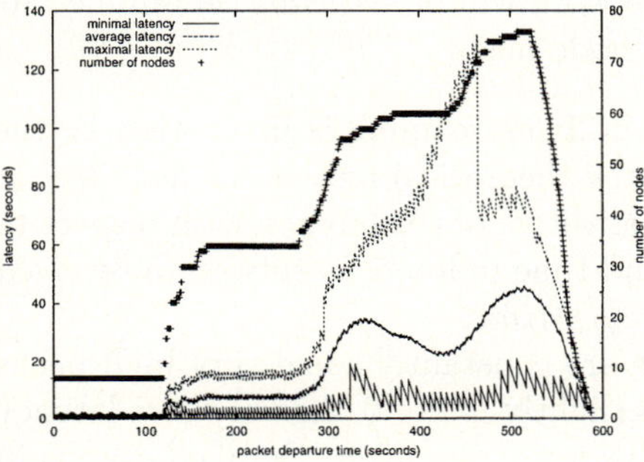

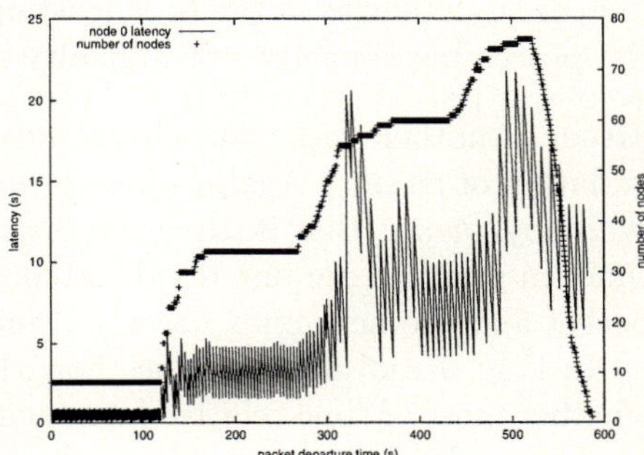

Fig. 3. Minimal, average and maximal latency with number of nodes (*to-100*)

Fig. 4. Node 0 latency with number of nodes (*to-100*)

We can observe that periodically latency is abruptly incremented to quickly recover its usual lower level. Figures 5 and 6 show this behavior in detail. Between 0 and 10 seconds, we can see this fast peeks, that surge from much lower values. At 200 seconds peeks have become greater and involve more blocked packets.

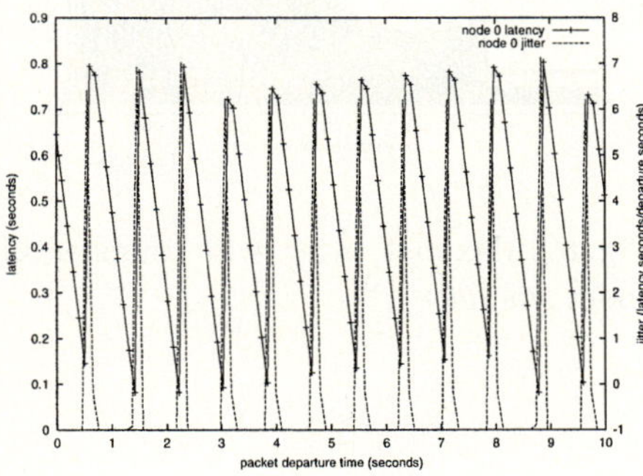

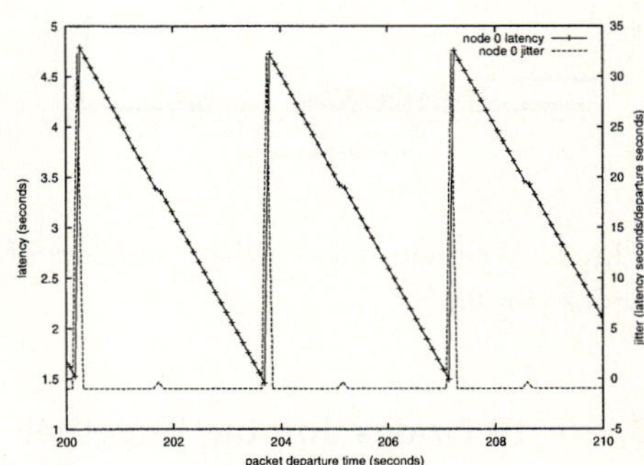

Fig. 5. Node 0 latency and jitter between 0 and 10 seconds (*to-100*)

Fig. 6. Node 0 latency and jitter between 200 and 210 seconds (*to-100*)

We think these peeks are caused by the way in which incoming packets are treated. Nodes are permanently waiting for incoming packets. But, as the system does not notify fast enough their arrival, when a node knows about them there are already several ones waiting to be treated. Those packets are treated, but the first has already waited a lot, the second a bit less, etc., until the last of them which almost has just arrived and so its delay is small. Once it has not other packets to treat, the node come back to wait for new packets and the initiative is passed to the operating system. It will not be notified about new packets until a certain quantity of them have arrived, and the same problem is repeated again.

The size of those jumps start about 0.7 seconds and goes to 5 seconds and further. As this component is much greater than network delays, we deduce that, in general, latencies in the broadcasting network will be basically determined by the processing capacity—or difficulty—of node hosts.

Jitter. Another important characteristic in live streaming is jitter, that is, the variation of latency. Again, we start with the theoretical measurement: $jitter = \partial latency/\partial time$, that is, the jitter is the derivative of the latency with respect to time. In practice, we just divide latency and time differences between a received packet and the next one: $jitter = \triangle latency/\triangle time$.

In Figs. 7 and 8, we can see how there are constantly some very high peeks whether most of time relative increment is slightly under 0. Again, those latency peeks we talked about are causing the behavior of the jitter. Detailed graphs in Figs. 5 and 6 also show the jitter with its originating latency.

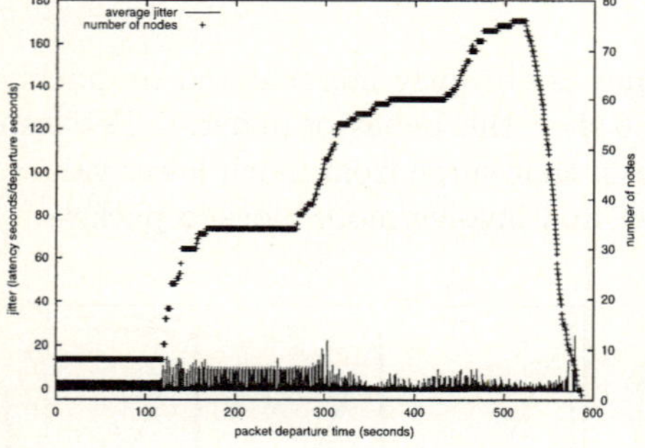

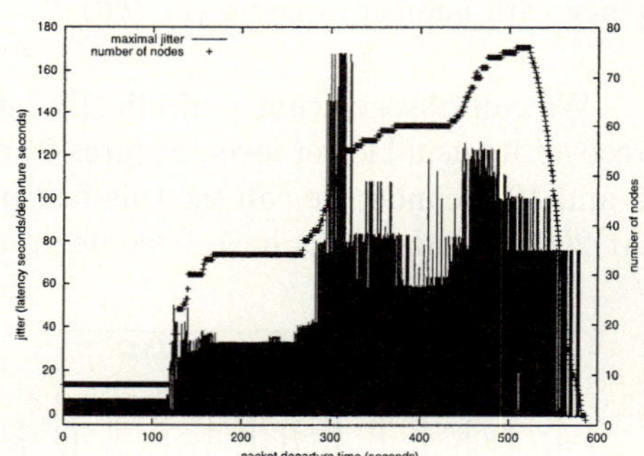

Fig. 7. Average jitter with number of nodes (*to-100*)

Fig. 8. Maximal jitter with number of nodes (*to-100*)

3.2   10 Nodes Joining Together

In this new setting 11 nodes have been deployed. Like before, 40 bytes data packets are sent by the root every 100 milliseconds for 10 minutes and all nodes, including root, are executed on the same machine.

We observe that all the 10 receivers are receiving packets during almost all the experiment and the resulting latency measures are more or less stable. Again the zigzag effect can be observed. We can watch in the detailed graph, in Fig. 9, referred to nodes 4, 3, 7, and 5 for the first 10 seconds, how the same pattern is repeated, keeping a jump size approximately of one second. Besides, we can

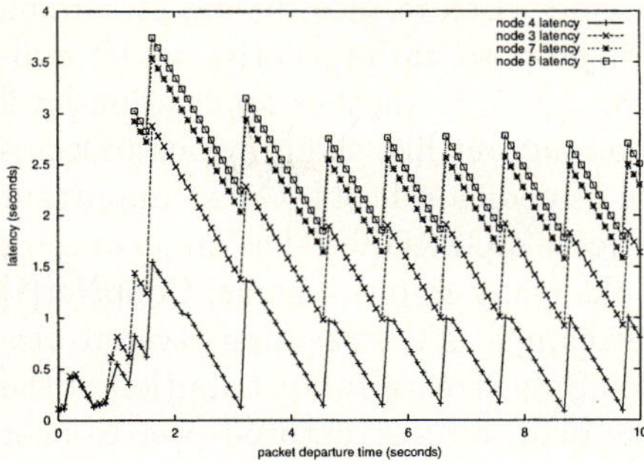

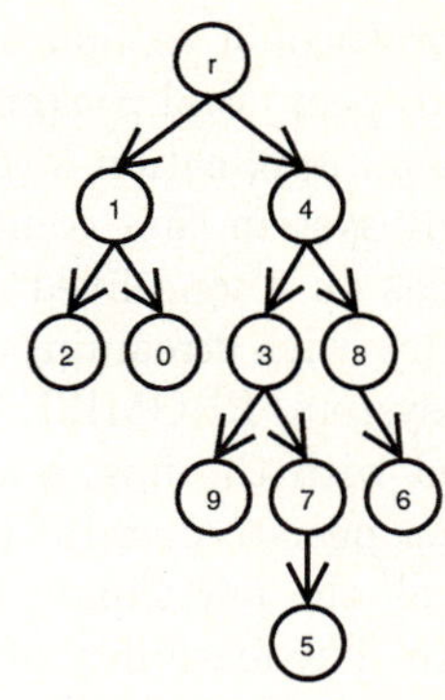

Fig. 9. Nodes 4, 3, 7, and 5 latency between 0 and 10 seconds (*to-10*)

Fig. 10. Resulting broadcast network topology (*to-10*)

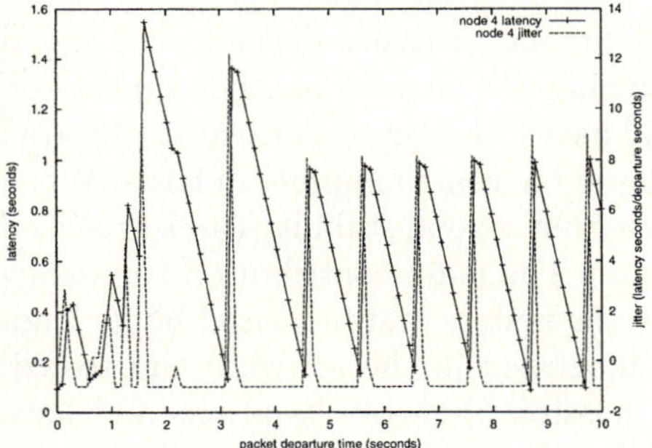

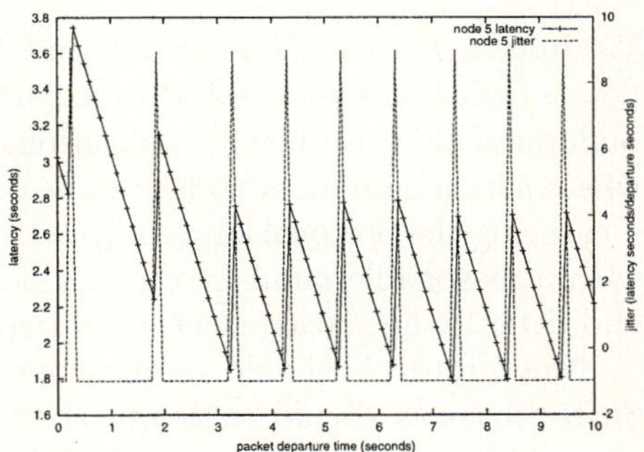

Fig. 11. Node 4 latency and jitter between 0 and 10 seconds (*to-10*)

Fig. 12. Node 5 latency and jitter between 0 and 10 seconds (*to-10*)

also observe that jumps are practically simultaneous in the different nodes, as this effect is produced by the system that they share.

Figure 10 shows the topology created by the joining process. Nodes 4, 3, 7 and 5 form a path from root to a leaf. Looking again at Fig. 9, we observe how the interval between the latency experienced by a node and the latency experienced by its descendant stands stable. For every pair of nodes it stands with a fixed value $l_{i,j}$, where $0.1\,\mathrm{s} < l_{i,j} < 1\,\mathrm{s}$. The most interesting of this jitter is that it will allow us to size the reception buffer quite accurately.

4   Related Work

A first article from Bawa, Deshpande and Garcia-Molina [1] focus on streaming media application with peer-to-peer approach. Their work result in an real application[2] that permits to multicast an audio flow to a number of receivers. It implements a peer-to-peer application-level multicast approach and uses the

Gnutella protocol[3] to find appropriates nodes. Before introducing streaming over peer-to-peer, End System Multicast[4] propose an alternative to IP multicast with an application level overlay. Overcast[5] is another application-level multicasting system that could handle streaming but like the previous proposition, it relies on a long-lived infrastructure of dedicate hosts. Other important works dealing with streaming over peer-to-peer: ZIGZAG[6] relies on a complex clustering system; PROMISE [7] insists on the many-to-one scheme; CoopNet[8] is fully tied with the flow semantic; GnuStream[9] is a streaming layer to use over another peer-to-peer infrastructure; and SplitStream[10] distribution of the flows over all the participant but must take place on a structured peer-to-peer architecture. The feasibility of large-scale peer-to-peer live streaming is analysed with network traffic in [11].

5   Conclusion

We have achieved a framework that will permit to easily develop and test a flow broadcasting protocol. Protocol developing will be freed of other application building details. Resulting applications will have a testing scenario at its disposal where a high number of nodes can be deployed on a small number of hosts. When measuring the protocol efficiency we can see that a good stability can be reached with a reasonable number of nodes per host. The main contribution to latency and jitter is the zigzag effect, produced by the system and increased by its load.

New protocols should be developed, integrated in the software and tested. Such protocols should take into account possible network variations and have a dynamic vision of the network. Our application transparently gives a service of transmission to multiple receivers. Neither sender nor receivers are concerned about how we carry it out. However, sometimes is interesting to allow ends to exercise some kind of regulation. Therefore, we could modify the software architecture and the protocol to dynamically respond to user needs.

This work results of an internship continued the work *Distribution pair à pair de flux audio* carried out at LIP6 by the autors and Alejandro Rodríguez San José. Sources and documentation are available as a GPL Savannah non-gnu project at `http://savannah.nongnu.org/projects/fluxp2p`. See `http://savannah.nongnu.org/cvs/?group=fluxp2p` to retrieve it from CVS.

References

1. Bawa, M., Deshpande, H., Garcia-Molina, H.: Transience of peers and streaming media. HotNets-I (2002)
2. : peercast.org. Website (accessed in 2004) `http://www.peercast.org/`.
3. Clip2: The gnutella protocol specification v0.4 (accessed in 2004) `http://rfc-gnutella.sourceforge.net/developer/stable/index.html`
4. Chu, Y.H., Rao, S.G., Zhang, H.: A case for end system multicast. In: Measurement and Modeling of Computer Systems. (2000)

5. Janotti, J., Gifford, D.K., Johnson, K.L., Kaashoek, M.F., O'Toole, Jr., J.W.: Overcast: Reliable multicasting with an overlay network. In: USENIX OSDI. (2000)
6. Tran, D., Hua, K., Do, T.: Zigzag: An efficient peer-to-peer scheme for media streaming. In: IEEE INFOCOM. (2003)
7. Hefeeda, M., Habib, A., Boyan, B., Xu, D., Bhargava, B.: Promise: peer-to-peer media streaming using collectcast. Technical Report CS-TR 03-016, Purdue University (2003)
8. Padmanabhan, V., Wang, H., Chou, P., Sripanidkulchai, K.: Distributing streaming media content using cooperative networking. ACM/IEEE NOSSDAV (2002)
9. Jiang, X., Dong, Y., Xu, D., Bhargava, B.: Gnustream: A p2p media streaming system prototype. In: International Conference on Multimedia and Expo. (2003)
10. Castro, M., Druschel, P., Kermarrec, A., Nandi, A., Rowstron, A., Singh, A.: Splitstream: High-bandwidth multicast in cooperative environments. ACM Symposium on Operating Systems Principles (2003)
11. Sripanidkulchai, K., Ganjam, A., Maggs, B., Zang, H.: The feasibility of supporting large-scale live streaming applications with dynamic application end-points. ACM SIGCOMM (2004)

Distributed Object Location with Queue Management Provision in Peer-to-Peer Content Management Systems

Vassilios M. Stathopoulos, Nikolaos D. Dragios, and Nikolas M. Mitrou

National Technical University of Athens,
9, Heroon Polytechniou, 15773 Zographou, Athens, Greece
{vstath, ndragios}@telecom.ntua.gr

Abstract. Content Management Systems (CMS) organize and facilitate collaborative content creation following a centralized approach to allow authorized requests from registered users. Distributed Content Management Systems (D-CMS) behave and manage content as a centralized system but they also identify remotely residing content inheriting properties from Peer-to-Peer systems. Herein the Tandem Search Algorithm is proposed as an efficient way to locate remote nodes that host requested data items in a D-CMS. Simulation shows the applicability of our algorithm and supports some architectural enhancements proposed for the network's nodes.

1 Introduction

A Content Management System (CMS) is used to organize and facilitate collaborative content creation [1]. CMS mainly produces content while maintaining control over it. Such a system operates in a centralized way without communicating with remote content resources. Distributed – CMS overcome this restriction. D-CMS not only manages content as CMS do, but also identifies remotely residing content. Operating D-CMS, peer-to-peer (P2P) systems' characteristics should be incorporated. P2P systems are distributed systems without any centralized control or hierarchical organization, where the software running at each node is equivalent in functionality. Real P2P systems follow centralized [2], structured decentralized [3][4][5][6] or fully decentralized and unstructured [7] approaches. CMS nodes with P2P functionality should guaranty fast search within the nodes of the peer network, efficient resource allocation and avoid traffic congestion bottlenecks. CMS nodes should offer a seamless operational environment with the minimum complexity, keep the newly incorporated functionality transparent to the end user and offer a scalable architecture.

In this paper and in section 2, we propose a newly designed algorithm, aiming in successfully locating and withdrawing content in this P2P CMS environment. This is the Tandem Search Algorithm (TSA). Two techniques are proposed in section 3, contributing their offerings at the microscopic and macroscopic level of the network. TSA uses these techniques to accelerate searching and overcome node congestion problems. The performance of the proposed architecture, as well as this of the algorithm and the incorporated techniques, is evaluated through simulation presented in section 4. In Section 5 main conclusions are recapitulated.

P. Lorenz and P. Dini (Eds.): ICN 2005, LNCS 3421, pp. 634–642, 2005.

2   The TSA Algorithm

The proposed algorithm should identify any requested asset by using as less signaling messages as possible which implies that it should be based upon a simplified operational model. Moreover, it should efficiently manage the available resources and protect the network nodes from congestion. For achieving these requirements and create an operable algorithm the following three processes should be defined.

- Building a seamless overlay network that guaranties a successful communication among peer nodes.
- Defining signaling messages for the communication among peer nodes.
- Internal node architectural improvements, based upon a queue management schema.

2.1   The Registration Mechanism of TSA

The initial consideration in a P2P network is to build up the topology of the overlay network, that is, a virtual communication infrastructure upon the real network. Peer nodes identify their neighbors, communicate with each other, following specific rules and build a virtual network according to their performance needs. To achieve these aims an affective *registration mechanism* is incorporated, as part of the algorithm.

Each peer node incorporates server facilities. It accepts registration messages by other peer nodes and manages their registration. Apparently, only one peer node acts as a *server* of the overlay network at a time. A second peer node can only be activated, as a *server*, when the initial one fails. The *server* accepts registration messages from new incoming peer nodes, keeping their characteristics, such as network addresses, in a registration list. Based upon this list and an overlay topology model, it produces a table with the neighbors of each new registered node. The *server* forwards this table to each specific node and the peer node, after accepting the list of its neighbor(s), is ready to communicate with them for requesting an asset. Moreover, the overlay topology also guaranties protection in cases of isolation of a node when no neighbors can be contacted.

2.2   Signaling Messages of the Algorithm and Add-on Techniques

When an asset is requested, signaling messages need to be routed to identify its position. Routing is analyzed through a simplified example illustrated in Fig. 1. An authenticated user a is registered within CMS-node A and requests for an asset named k. This request initiates a session within node A. A recognizes that it does not contain this specific asset into its local repository, issues a request message for this asset and forwards it to its neighbor node B. If B does not contain the asset in its local repository, it similarly issues and forwards a new request message to the next (its neighbor) peer node (i.e. node C). The above-described process is continuously repeated until the requested asset is identified into the repository of a peer node. We refer to this process as the node-to-node (*in tandem*) search process.

According to the overlay network architecture, some of the peer nodes are connected with more than two neighbor nodes, forming local tree topologies. Every local tree has a root (i.e. node B at Fig. 1), which is the upper node in the hierarchy and two or more children sub-trees (i.e. sub-tree B-C, sub-tree B-D at Fig. 1). When a request for an asset reaches any node being a root for some sub-tree, the node forwards the request only to one of its sub-trees (i.e. sub-tree B-C). When the entire sub-tree is searched without locating the asset a *failure* response is sent back to B. B in turn forwards a *new request* to its other sub-trees or a *failure* message to its father A in case B has no more sub-trees. When the asset is found in a sub-tree a *reply* message (message 3 of Fig. 1) is returned to the root B and from there to node A (message 10 of Fig. 1) and no more sub-trees of B are accessed.

All nodes that participate in the identification of the asset are actually participants of the same session. The node that has started the session is called *initiator* while the one that contains the session is called *carrier*. The path that is established by a request from the originator towards the carrier node and passing through all intermediate nodes is called *session path* while the opposite direction, from the carrier towards the originator is called *opposite session path*.

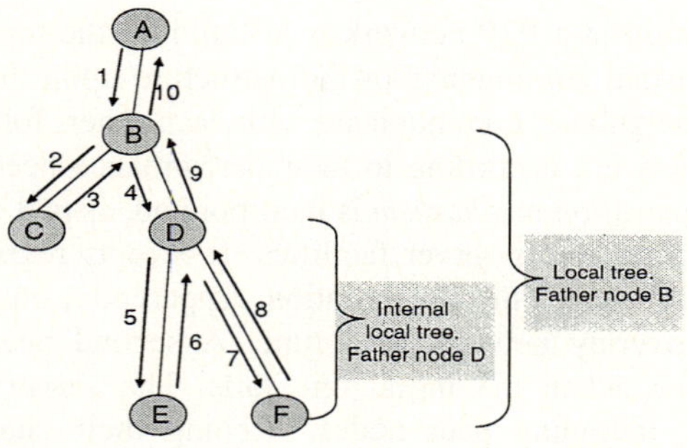

Fig. 1. Identifying the asset through routing process

The messages of the protocol of the algorithm are following analyzed:

IdentifyAssetHolder() – IAH(): This message represents the request message. It is the first message issued during an asset identification process and initiates the session. This message travels through the overlay network nodes where each node forwards it to its neighbor(s).

IdentifyAssetHolderReply() – IAHR(): This is the reply message and it is issued by the carrier towards the originator for completing the active session.

IdentificationFailled() – IF(): This is the failure message. It actually issued if a sub-tree is searched without the asset to be identified. Following the last node of the sub-tree issues this message and sent it backwards.

GetAsset() – GA(): Calling of the specific message is executed if the requested asset has been identified. For example, if node A is informed that the asset is located at remote node Y, then node *A* issues the *GetAsset()* message towards node *Y* for receiving the content of the asset.

A problem that may influence the network performance is the possibility of encountering the same request exchange among the same nodes that are referred to the same session (looping problem). Each request, through its carried information, informs the nodes about its session and so nodes avoid the looping problem.

Another problem related with network performance is the existence of many parallel (needless) sessions that request for the same asset but generated from different initiators. Their respective requests that pass common intermediate nodes may be merged into one "common request" when passing through their first common node, by so improving the network performance. Continuously, if the "common request" reaches a carrier, a "common reply" travels back to the common node. By the time the reply reaches the common node, it informs all the originators. We refer to this mechanism as the "merging" mechanism.

Moreover, the time that a session remains active may be noticeably reduced if special mechanisms are applied. Two mechanisms have been identified, where performance results prove that their application noticeably increases the algorithm's and the network's performance. These are:

- Signaling messages are classified into categories while priorities are allocated towards each of these categories. Hence the messages with the higher priority prioritize their execution within a peer node, against others with lower priorities.
- Sending a *direct reply* from the carrier towards the originator by initially ignoring the "opposite session path". Following, the *carrier* issues a second reply back to the originator by now following the "opposite session path". The second replay message is used for closing the session registrations within the intermediate nodes.

3 Node Architectural Enhancements

If the resource capacity of a D-CMS node is exceeded, its queue starts building, by increasing message's delay within it. In order to overcome the above obstacles and improve network and algorithm performance an alternative technique is proposed regarding the queue management. The central process that manages the algorithm's operations is divided into a number of smaller processes (threads) and consequently a lengthy queue into a number of smaller queues for each process (thread). Hence, some messages can be executed in parallel and others may be prioritized.

The first defined queue, the *Receiver* queue, represents the entry point. It accepts all signaling messages (User Initial Request - *UIR()*, IAH(), IAHR(), IdF(), and GA()) and filters them, according to their type, towards the next appropriate queue. A second queue, named as the "DB access", aggregates IAH() or UIR(). Its thread accesses the local DB, which is a time consuming process and so sensitive to load, for servicing these messages' requests. To efficiently control the loaded "DB access" queue, the "User Request" queue is added in the queue model. This queue accepts only UIR() messages. The rest of the messages (i.e. IAHR(), IdF()) are forwarded towards a fourth queue named as the "Track Session" which is the output gate to any neighbors. Following, if the UIR() and IAH() cannot identify the requested asset, are forwarded towards the "Track Session" queue too. Fig. 2 illustrates this enhanced queue model. Fig. 2 also depicts two thresholds (*Th*) associated with the "User Request" and the "DB access" queues respectively. Thresholds are used as sensors for protecting their queues by extensive buffering. Their actual values are determined at next section through simulation.

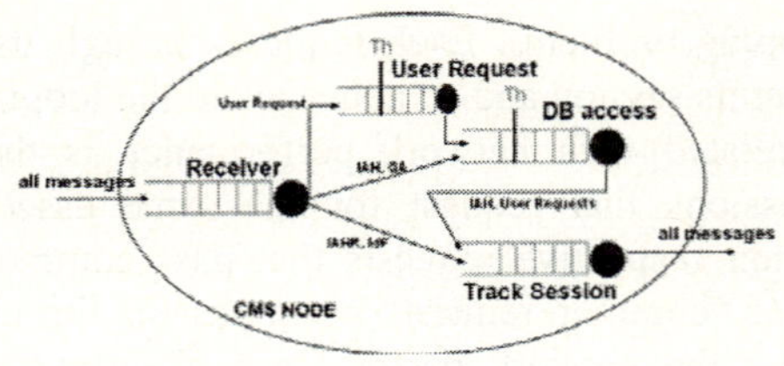

Fig. 2. The internal architecture of a CMS node

When the "User Request" processor that monitors "DB access" queue detects an overflow, which appears if the "DB access" queue exceeds threshold *Th*, it reduces the rate of sending UIR() messages towards the overloaded "DB access" queue. It actually applies a delay window. This delay is determined by the equation:

$$\Delta\tau(i) = \alpha + \beta \cdot i^3$$

This equation is the product of extended number of experiments. Variable i is given non-negative integer values. Hence, for the first violation the processor assigns the value 1 for variable i, holds UIR() for time equal to $\Delta\tau(1)$ and then forwards it towards the "DB access" queue (see Fig. 2). If the processor, at the next checking time, finds the queue violated for a second time, then variable i is assigned the value of 2 and the previous procedure is repeated. The value of variable i linearly increase as long as these violations continue to be repeated. Following, if the queue decreases below Th then $\Delta\tau(i)$ becomes equals α which denotes the value of the delay window, for forwarding UIR() messages, at a minimum secure rate. Several tests were executed by experimenting with various values regarding the constants α, β, the exponent of variable i as well as the threshold Th. The value 3 for the exponent of variable i, values 30 and 100 for constants α and β respectively and the value 20 for Th were chosen as the most effective ones.

Finally, when the "User Request" processor, mainly due to the applied delay, detects that its queue exceeds threshold *Th*, it creates a "Wrapper UIR" with high priority and forwards it directly to the "Track Session" queue. Details of these results are presented in next sections.

4   Simulation

To evaluate the performance of the proposed algorithm we simulate the internal architecture and operations of a number of D-CMS nodes under a ring network topology. The algorithm is enhanced with two additional techniques, that is, the 'Priority' and 'Queue–Control' mechanisms. These two enhanced versions of the algorithm are evaluated and compared with the simple version. Simulations scenarios are executed for different number of CMS nodes up to 200. The results herein refer to the case of 50 nodes. In all executions a high number of assets (more than 5000 assets) are almost equally distributed in the network nodes. A high number of requests (more than 10000 requests) enter the simulated network following the Poisson process while the mean departure rate of requests per second (rps) is denoted with

$\lambda' = 1000\lambda$, where λ is the mean departure rate of assets per simulation unit (1 second equals to 1000 simulation units).

To measure the execution times of fundamental operations, we create an adequate platform from real PCs and a number of experiments are executed. The results are:

- **db_access_t (15 ms):** This is the time that is required for a Java object to access an SQL DB to locate or not the name of a requested asset.
- **interprocess_t (1 ms):** This is the time that is required for a Java invocation to be executed during the communication among ordinary Java objects.
- **search_list_t (1 ms):** This is the time that is required for a Java object to access and operate in a local list of values.
- **marchalling_t (7 ms):** This is the time required for invoking a remote J2EE object from peer node A towards a remote one, such as peer node B.

4.1 Simulating the Network

Simple Sub-scenario. Initially the saturation throughput of the network is identified. Results regarding the session completion time are illustrated in Fig.3. The x-marked curve represents the simple version of the algorithm. The rate is 90 rps (per node).The network successfully controls the generated traffic and executes all sessions within reasonable time delays (less than 14 seconds). The rectangular–marked curve represents the same simple-based algorithm at the rate at 92 rps.

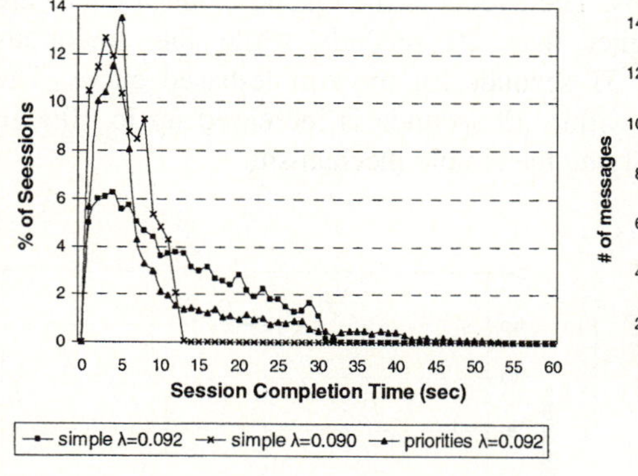

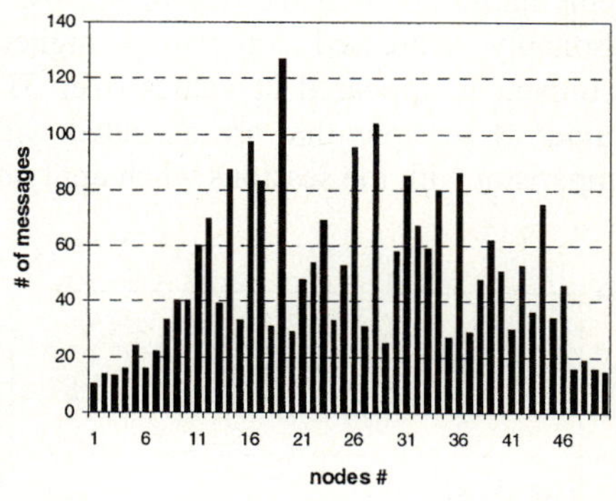

Fig. 3. Simple vs Priorities-based algorithm   **Fig. 4.** DB-access queue contents

The sessions' execution time is dangerously increased and a reasonable percentage of sessions exceed the threshold of 20 seconds. The nodes are not anymore able to control the specific rate, resources are given out, queues start building and the network is getting overloaded. This estimation is proven from the results appeared at Fig.4. This figure illustrates a trace of contents of every "DB access" queue, as the most time consuming process, concerning their maximum values, during the operation at rate 92 rps. All nodes contain high number of messages. Results denote that these values are not instants but maintain their high values for large period of times.

Priority Based Sub-scenario. The priority-based version of TSA attempts to manage resources, messages and active sessions, by classifying messages and giving them priorities. Messages that deactivate a session, such as, IAHR, IdF, GA are given greater servicing priority compared with messages that activate a new session or searching for the requested asset (i.e. IAH()). Hence if a session is active, it is completed at the minimum time duration.

The triangular-marked priority-based curve of Fig.3 illustrates the priority-based version at the rate of 92 rps. Indeed, the sessions' execution time has been decreased, compared with the previous case, while the curve approaches the simple-based curve for the rate of 90 rps. Hence, by using the priority-based algorithm, an increment of the nodes' throughput is achieved (i.e. the rate 92 rps is now acceptable). The number of sessions that are executed within 20 seconds is increased up to 6% in comparison with the sessions when applying the simple mechanism. However, the priority-based algorithm is not able to protect the nodes from overloading, as this has been proven by executing extensive simulation results.

Another measurement is the number of active sessions maintained in parallel within each node. Simulation results for the simple version prove that the number of parallel active sessions was 400 for the less loaded nodes and 700 for the highly loaded ones. Even with the priority-based experiment the values remain almost the same.

Queue-Control (QC) Sub-scenario. This experiment is executed at the rate of 92 rps. Results are illustrated at Fig.5 by the x-marked curve, which is compared with the simple-based curve at the rate of 92 rps. The events at the Queue Control curve are reasonably decreased for values higher than 20 seconds while the analogous decrement is appeared at values over 31 seconds for the simple-based curve. The number of sessions that are executed within 20 seconds is increased up to 10% in comparison with the sessions when applying the simple mechanism.

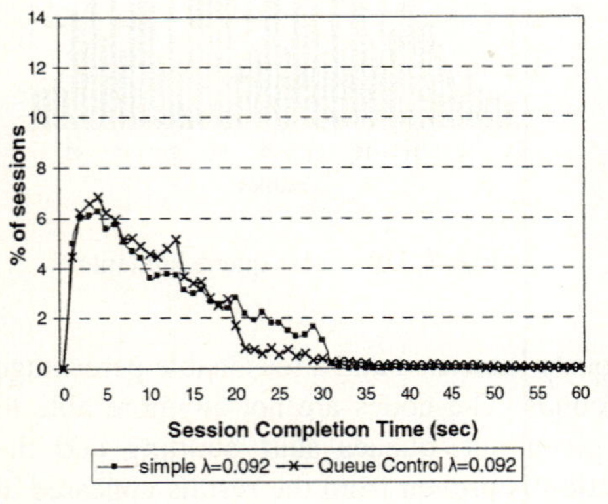

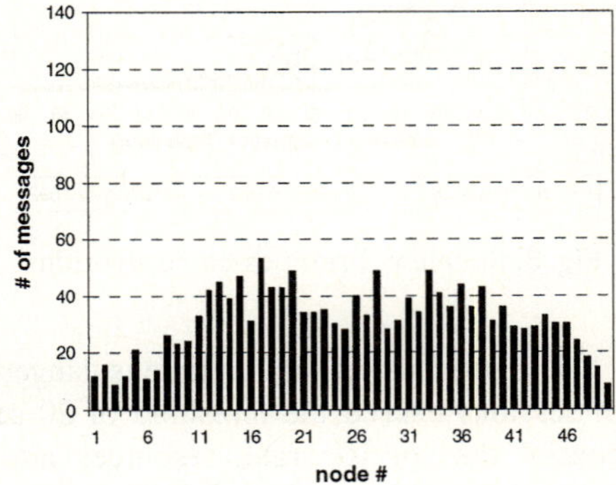

Fig. 5. Queue Control vs Simple version   **Fig. 6.** DB-access queue contents

Fig.6 illustrates the maximum number of messages that are accumulated within the "DB access" queue for every node. By comparing Fig.6 with Fig.4 we observe that the queue has intensively decreased. This proves that the Queue Control algorithm successfully control messages and hereupon managing resources.

Simulation results show that when operating the QC-based experiment the maximum number of active sessions maintained in parallel within each node has been drastically decreased. Indeed, for the less loaded nodes the active sessions are around 200 and for the highly loaded ones around 400, reaching half the number of the Simple-based case. Consequently, the QC mechanism successfully increment throughput over 10% compared with the Simple-based case. Further simulation experiments show that for further increase at the incoming request rate, queues remain controlled and the effective resource management is preserved.

5   Conclusions

In this paper, a search algorithm, named as TSA, is proposed, aiming in efficiently locating assets and also withdrawing it through a P2P environment. Each node is able to efficiently search at remote repositories by so locating a requested assets and following fetches its content.

This algorithm bears additional functionalities, such as, a mechanism for avoiding the circulation of messages of the same session among the same peer nodes and achieving searching with the minimum cost in resource consuming. Moreover, for protecting nodes from congestion and for succeeding an effective queue management policy and efficient resource allocation it bears two additional mechanisms, that is, the Priorities-based (Pr) and Queue Control-based (QC) mechanisms.

The performance of the network that operates with the proposed algorithm and the two mechanisms is evaluated through simulation. Indeed, for the Pr case the number of sessions that are executed within 20 seconds is increased up to 6% in comparison with the sessions when applying the simple mechanism, while an increment of 10% is appeared for the QC one.

Acknowledgments

This work was partially funded by the General Secretary of Research and Technology Union in Greece ΠΒΝ -127 project. The authors wish to express their gratitude to the members of the consortium for fruitful discussions.

References

[1]   www.contentmanager.eu.com/

[2]   www.napster.com

[3]   Rowstron, A. and DruscheL, P. "Pastry: Scalable, distributed object location and routing for large-scale peer-to-peer systems', In Proceedings of IFIP/ACM International Conference on Distributed Systems Platforms (Middleware), Germany, Nov, 2001, 329-350.

[4] Zhao, B.Y., Kubiatowicz, J.D., Joseph, A.D. Tapestry: An infrastructure for fault tolerant wide-area location and routing. Technical Report UCB/CSD-01-1141, UC Berkeley (2001).

[5] Stoica, I., Morris, R., Karger, D., Kaashoek, M. F., Balakrishnan, H. Chord: A scalable peer-to-peer lookup service for Internet applications. In Proceedings of the 2001 ACM SIGCOMM Conference, San Diego, California, USA (2001), 149-160.

[6] Ratnasamy, S., Francis, P., Handley, M., Karp, R., Shenker, S. A scalable content-addressable network. In Proceedings of ACM SIGCOMM, August 2001, 161-172.

[7] gnutella.wego.com

An Approach to Fair Resource Sharing
in Peer-to-Peer Systems

Yongquan Ma and Dongsheng Wang

Department of Computer Science and Technology, Tsinghua University,
Beijing 100084, P. R. China
myq02@mails.tsinghua.edu.cn, wds@tsinghua.edu.cn

Abstract. Peer-to-Peer systems are designed to share resources of each computer for the common good of everyone in the system. As more and more p2p applications have been deployed, an important problem occurs that is the absence of incentive mechanism which encourages participants to contribute their resources to the system. To solve this problem, we present a novel mechanism of fair sharing, which intends to ensure that the resources amount consumed by one node is in proportion to that it contributes to the system. Moreover, this mechanism is able to defend itself against malicious attacks to some extent. Simulation results show that the commutation overhead of this mechanism is small in a very large scale.

1 Introduction

In recent years, a large number of p2p systems have been developed and put into use, such as Chord [1], Pastry [2], CAN [3], Tapestry [4], CFS [5], Gnutella [6] and Ka-Zaa [7].

P2P systems are designed to share various resources, for example, files, information, CPU circles and so on. Each node contributes its resources to the system and gets resources from the system. As a result, the whole system reaches the maximum utility. However, there is an absence of incentive mechanism, which results in such a situation that nodes could get many resources while contribute few.

Most existing p2p systems assume that all nodes will behave as predefined algorithms and protocols without deviation. However, as more and more users join the system, "free riding" problem has appeared. For example, a study result of Gnutella [8] showed that there were 70% of the users who downloaded files from the system but contributed none, and that 50% of the shared files were contributed by only 1% of the users.

In fact, because of the large scale and low requirement of trust of p2p systems [9], nodes in the system belong to different people and organizations, who have different behalf, and there are nearly no effect cooperation and responsibility. As a result, nodes only care for their own interests even at the cost of the common good of the system. Thus most nodes in p2p systems have their own strategies and behalf. In order to solve this problem, we should establish an incentive mechanism, which make nodes willing to share their resources with other nodes in the interests of themselves.

P. Lorenz and P. Dini (Eds.): ICN 2005, LNCS 3421, pp. 643–652, 2005.

This paper presents a contract-based third-party-auditing mechanism, which ensures the resources amount that a node can get from the system is in proportion to that it contributes to the system. In this way, nodes will share their resources in the system and participate in maintenance of this mechanism for their own interests.

To illustrate the power of this mechanism, we focus on the specific resource of storage space and suppose that the information is stored as file. Moreover, this mechanism can be potentially extended to enforce fair sharing of other resources. In this paper, section 2 discusses the basic ideas. Section 3 introduces the system model of our design. Section 4 discusses auditing mechanism in details. Section 5 presents simulation results. Finally, section 6 summarizes related work and section 7 concludes.

2   Basic Ideas

Nodes in p2p systems can be classified into four types [10] [11] [14] according to their behavior mode:

- **Obedient nodes.** Nodes of this type act as predefined algorithms and protocols without deviation.
- **Faulty nodes.** Nodes of this type don't behave normally.
- **Rational nodes.** Nodes of this type participate in the system for their own behalf. When conflict occurs between behalf of their own and that of the whole system, they will give the former priority. In order to fulfill their requirement, they can make use of information collected from the system to establish their own strategies.
- **Adversary nodes.** Nodes of this type intend to do harm to the system and establish their strategies aiming at this objective.

We think that most of nodes in the system are rational nodes, which is the focus of our research. And there are a lot of obedient nodes. Both of them account for majority of the system. However, there are still some adversary nodes, which have some irrational activities in order to undermine the system. This type of nodes involves security problem and we'll consider them to some extent.

In this contract-based third-party-auditing mechanism, how much resources nodes can obtain from the system depends on the amount of resources they contribute to the system. When a node obtains resources from the system or shares its resources with other nodes, related information will be recorded and distributed in the system. At the same time, nodes audit these track recorders to enforce fair sharing. If there are disputes about the results of auditing, they can start arbitration program. Once cheating is found, nodes concerned will be punished. In this way, nodes will share their resources in order to get resources.

3   System Model

We assume that there is a public key infrastructure in the system, allowing nodes to digitally sign a document, which can be verified by other nodes. And it is computationally infeasible for others to forge.

In p2p storage systems, each node wants to store some files into the system and each node is demanded to store files belonging to other nodes. The matter that node A want to store a file F on node B and B agrees to store this file is termed a "**deal**". Information about the deal is kept in a document, called a "**contract**". Node A is named "**debit node**" and node B is named "**credit node**". A contract should at least record information shown in table 1.

The two parties signing the contract should check up the contract and add encrypted code resulting from encrypting items of 4, 5, 6 and 7 by themselves. Every contract has a period of validity, after which the contract will expire and the credit node can delete the file. If the debit node wishes to continue the storage, the both sides need sign a new contract.

Firstly, node A creates a contract and fills in the 1st, 2nd, 4th, 5th, 6th, 7th and 8th items. Then it sends the contract to node B with the file F. B should check up the items written by A, fill in other items and sent the contract back to A. Because the contract encompasses encrypted code of both sides, neither of them can tamper with the contract. Subsequently, both A and B select nodes as many as D in the system and distribute the contract to them. The number D is a well-known constant in the system.

Table 1. Basic content of a contract

No.	Item	Note
1	Contract ID	The id of this contract
2	Debit Node ID	The id of the node requesting for storage
3	Credit Node ID	The id of the node fulfilling the request
4	File Name	The name of the file wanted to be stored
5	File Size	The size of the file wanted to be stored
6	Time of Deal	The time when this contract is signed
7	Period of Validity	The period of validity of this contract
8	Encrypted Code of Debit Node	Code resulting from encrypting items of 4, 5, 6 and 7 by the debit node
9	Encrypted Code of Credit Node	Code resulting from encrypting items of 4, 5, 6 and 7 by the credit node

Besides, all nodes check periodically whether contracts stored on them expire. The expiring ones should be discarded.

Each node in p2p storage systems provides some storage space to the system, the amount of which is called **Advertised Storage Capability** (ASC). ASC is determined by the node itself according to its capability, its storage requirement, and some other factors. At the same time, each node stores some files on other nodes. The maximum of storage space a node can get from the system goes by the name of **Usage Quota** (**UQ**), which depends on ASC. There is a well-known coefficient prescribing the ratio of ASC to UQ, which is generally equal to 1. We call the total amount of storage space in the system a node has used **Current Usage Amount** (**CUA**). All these three parameters can be queried by any other node.

When a node's CUA is no more than its UQ, we say the node is "under quota". A node in the state of under quota is allowed to write files into the system. We say a node is "beyond quota" if the node's CUA is more than its UQ, which is not allowed. Nodes in this state should be punished.

Every node maintains two lists named **debit list** and **credit list**. The former contains the IDs of contracts which the node itself signed as debit node. It also encompasses the size of requested storage space in the contracts. The latter records the IDs of contracts which the node itself signed as credit node. It also keeps the size of storage space provided by the node in the contracts. Both lists are available for any other node to read.

For rational nodes, in order to get more storage space illegally, they tend to inflate credit list or deflate debit list. The means to inflate credit list is to add some spurious contracts into the list or keep expired contracts in the list while delete related files. In order to deflate debit list, nodes skip some contracts which should be add into the list.

Rational nodes will increase ASC or decrease CUA owing to increasing their UQ. It is supposed that increasing ASC will result in more storage request than upper limit of the node. Therefore, nodes will inflate their credit lists when increasing ASC. If nodes want to get more storage space from the system while keeping the UQ changeless, they must deflate their debit lists.

As discussed above, only when a node is in the state of under quota and is still in this state after writing a file to the system, it can advance storage request. However, it's "rational" for rational nodes to store as many files into the system as possible. So it is essential to establish a kind of auditing mechanism to monitor the behavior of nodes, which will be described in details in next section.

4 Auditing

In our design, each node in the system performs auditing periodically, aiming at examining whether nodes keep real records of related deals and perform their obligations actually. The auditing mechanism consists of several parts described as follows.

4.1 Basic Auditing

According to the contract distributing process, each node will store some contracts locally. Nodes should perform periodical audit. Every time a node can audit all the contracts stored on it, or only part of them.

After choosing the contract to be audited, the auditing node asks both sides of the contract whether the contract has expired. If the debit node informs the credit node of canceling the storage request within the period of validity, the latter can delete related file and the contract expires. When the two parties both give affirmative replies, the node stops auditing. If the debit node gives an affirmative reply while the credit node gives a negative reply, the auditor should request the debit node to inform the credit node of expiration of the contract. In the contrary instance, as long as the credit node is able to offer the expiration notification from the debit node, the auditing node will agree with expiration of the contract and think the debit node lying.

When both sides reply that the contract is still in effect, the auditor should examine whether the credit node stores the file concerned by verification program, which will be described below. If the debit node passes in verification program, it also passes in basic auditing. Otherwise, it should get some kind of punishment. If the credit node doesn't agree with the result of auditing, it can start the arbitration program that will be described in section 4.4.

For any deal, if only we are able to confirm two points, which are the existence and validity of the contract and the fulfillment of this contract by both sides, we can accomplish auditing. It is apparent that the two points can be achieved if all nodes in the system carry out auditing periodically.

All contracts and some kinds of important messages contain encrypted code, which can both prove authenticity of the contracts and be the evidence when cheating is found.

4.2 Full Auditing

There are still some unfinished issues for basic auditing, such as the authenticity of ASC, UQ and CUA of nodes. These issues will be accomplished by full auditing.

As discussed in section 3, nodes have two means to cheating, which are inflating credit list and deflating debit list. We can check the authenticity of ASC by auditing credit list and check the authenticity of CUA by auditing debit list.

The auditing node chooses a target node at random, which is debit node or credit node of one of contracts stored on the auditor.

Firstly, the auditor asks for ASC, CUA, credit list and debit list of the target node. Then it starts basic auditing of contracts contained by the two lists by sending messages to some nodes. These nodes hold the contracts in which the target node is either credit node or debit node. The underlying overlay network will provide the way to find nodes concerned. After finishing basic auditing, the result, in addition to the id and size of related files, should be returned to the auditor.

In the third step, the auditor adds up the size of files in debit list and check whether the result is equal to CUA. It's obvious that the target node is cheating if it's not equal. Similarly, the auditor adds up the size of files in credit list and compares the result with ASC. If the result is no less than ASC, the target node must be cheating.

Subsequently the auditor compares the credit list of target node with all the contracts in which the target node is the credit node. In the same way, the node compares the debit list of target node with all contracts whose debit node is the target node. If there is no disaccord between them, the target node passes in the full auditing.

If the target node is found cheating, it should be punished. If a node disagrees with the auditing result, it can start arbitration program described following.

4.3 Verification Program

What is prerequisite to auditing is to verify whether a node really stores a specific file. Verification program accomplishes this task.

When node A wants to verify whether node B really stores file F, A firstly finds a node other than B, who stores the same file F. Then A asks the node for the hash code of some specific bytes of F. The hash function is well-known to all nodes in the system. This message of query should avoid being caught by B, which needs supports from the overlay network. In the same message, A tells the node not to respond to the same query from B.

Subsequently, A asks B for the same hash code. A can think B doesn't store file F when either B return a false hash code or B can't return the hash code.

By the way, because p2p storage system ensures availability and reliability by replication, there always are several duplicates of a file. How to find out a duplicate depends on the protocol of overlay network.

4.4 Arbitration Program

In auditing, if nodes concerned disagree with the result, they are entitled to start arbitration program.

In this program, all the nodes storing the specific contract perform the auditing respectively. The conclusion of most nodes is the final result.

This program can defend the system against malicious attacks to some extent.

5 Experiments

In this section, we present some simulation results of the communication overhead of this mechanism. For our simulation, we assume that no nodes are cheating. The numbers of files all nodes storing is chosen from truncated normal distributions and range between 50 less than and 50 more than the expected value.

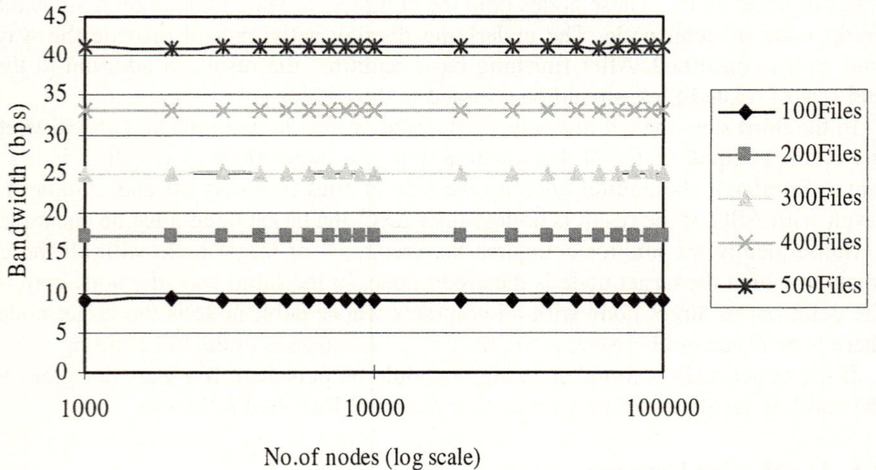

Fig. 1. Overhead with different number of nodes

Note that we only measured the communication cost due to auditing. In particular, we exclude the overhead of maintenance of overlay network and fetching of files. In all simulations, there is an average node life time of 7 days.

Figure 1 shows the average bandwidth requirement as a function of the number of nodes. The bandwidth requirement of each node is almost constant. Therefore the system scales well with the size of overlay network. And we can find that the per-node bandwidth requirement increased with the number of average files stored on each node, but the rate of increase is under linear rate.

As far as we know, one of the primary advantages of p2p system is that every node is both client and server. It means that when the number of nodes increases, both processing capability and workload will increase. More nodes mean not only more contracts need to be audited but also more nodes take part in auditing. Therefore, the rise of nodes doesn't bring more requirement of bandwidth. However, the more files are stored in the system, the more contract need to be audited. Workload rises, but processing capability remains the same. As a result, bandwidth requirement increases.

Figure 2 shows the average bandwidth requirement as a function of interval of basic auditing. The longer the interval between two basic auditing, the less bandwidth is required. But as you can see, the influence of interval of basic auditing is usually very feeble. We believe this is due to the communication overhead of basic auditing is relative small. Accordingly we should think more about the other aspects of the system when we choose basic auditing interval.

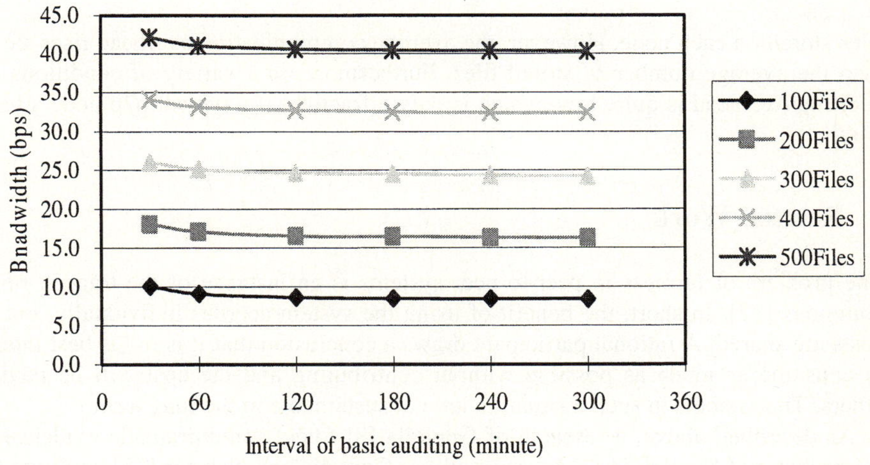

Fig. 2. Overhead with different interval of basic auditing

Figure 3 shows the per-node bandwidth requirement versus interval of full auditing. The communication overhead drops quickly when interval gets longer. In contrast with figure2, it is evident that the impact of full auditing on the overhead is more distinct. As described in section 4.2, a full auditing includes many times basic auditing and a lot of other actions. Therefore, the overhead of full auditing is quite heavy. To select an appropriate interval of full auditing is much more subtle on account of both communication cost and auditing efficiency.

In summary, the system with this auditing mechanism scales well. The primary factor causing variation of per-node bandwidth requirement is the average number of

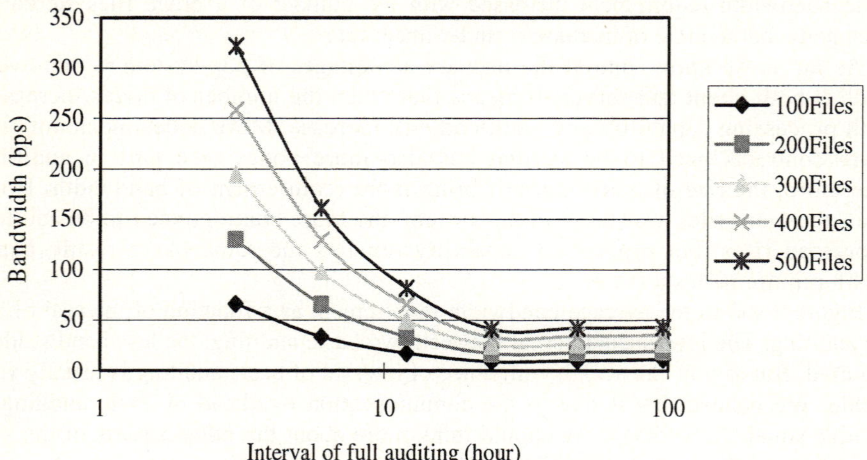

Fig. 3. Overhead with different interval of full auditing

files stored on each node. However, the average communication overhead rises slower than the average number of stored files. Furthermore, in a variety of conditions, the auditing overhead is quite low, which is only a fraction of a typical p2p node's bandwidth.

6   Related Work

The problem of fairness in peer-to-peer systems is an instance of the tragedy of the commons [12]. In short, the benefit of using the system accrues individually, but the costs are shared. A rational participant draws a conclusion that it is in his best interest to consume as much as possible without contributing and the cost will be paid by others. The systems in such a situation are not sustainable in the long term.

As described above, a research of Gnutella [8] found some dramatic evidence for the problem of "free riding". A later study of Gnutella and Napster [13] confirms this trend. This study also found that almost one third of the users of Napster under-reported their available bandwidth in order to avoid being chosen by other nodes to download files from them.

Feigenbaum [14] has outlined the basics of Distributed Algorithmic Mechanism Design (DAMD) and then reviewed previous DAMD results on multicast sharing and interdomain routing. The Internet's decentralized nature, in which distributed computation and autonomous agents prevail, makes DAMD a very natural approach for many Internet problems. This paper discussed the incentive problem of p2p file sharing systems and proposed three open problems of this field.

Shneidman [11] has advocated a model in which peer-to-peer users are expected to be rational and self-interested. And he tries his best to convince the reader that rationality is a real issue in p2p networks. Furthermore, three open problems were described, which were relevant in the peer-to-peer systems but were unsolved in existing ADM/DAMD work.

Ngan [10] has proposed a distributed accounting infrastructure. Each peer maintains a record of every data object it stores, and every data object stored on its behalf elsewhere. Each node periodically chooses another random node to audit. For each file the node claims to hold for some other peer, the auditor retrieves the corresponding peer's record, and compares the two. This framework differs from our design in several ways. First, each peer keeps every concerned record itself. And there is no objective evidence when disagreement occurs. Nor is the solution to dispute. Second, the auditor wants to fetch the related records from other nodes before auditing. In our design, the information of each deal is kept in a contract, which is a kind of objective evidence. When there is dispute about a deal, the corresponding contract is the basis of judgment. Each peer only needs to perform auditing of contracts stored on it, which reduces communication overhead remarkably.

SHARP [15] is a framework for secure distributed resource management in an Internet-scale computing infrastructure. The cornerstone of SHARP is a construct to represent cryptographically protected resource claims—promises or rights to control resources for designated time intervals—together with secure mechanisms to subdivide and delegate claims across a network of resource managers.

In Samsara [16], each peer that requests storage of another must agree to hold a "claim" in return—a placeholder that accounts for available space. After exchange, each partner checks the other to ensure faithfulness. Samsara punishes unresponsive nodes probabilistically.

7 Conclusion

This paper has presented a mechanism which is able to achieve fair resource sharing in p2p systems by means of auditing. Experimental results show that this architecture has small communication overhead and scales well to large number of nodes. It will provide incentives for nodes to contribute more resources to the system in the interests of themselves.

Acknowledgement

The work presented in this paper was supported by National Natural Science Foundation of China (NFCS) under grant number 60273006 and 60433040.

References

[1] Ion Stoica, Robert Morris, David Karger, M.Frans Kaashoek, Hari Balakrishnan. Chord: A Scalable Peer-to-peer Lookup Service for Internet Applications. In Proceedings of ACM SIGCOMM 2001, San Diego, California, USA, 2001.

[2] Antony Rowstron and Peter Druschel. Pastry: Scalable, decentralized object location androuting for large-scale peer-to-peer systems. In proceedings of the 18th IFIP/ACM International Conference on Distributed, 2001.

[3] Sylvia Ratnasamy, Paul Francis, Mark Handley, Richard Karp, Scott Shenker. A ScalableContent-Addressable Network. In proceedings of ACM SIGCOMM 2001, San Diego, California, USA, 2001.

[4] Ben Y. Zhao, John Kubiatowicz and Anthony D. Joseph. Tapestry: An Infrastructure for Fault-tolerant Wide-area Location and Routing. Technical Report No. UCB/CSD-01-1141, University of California Berkeley, 2001.

[5] Frank Dabek, M. Frans Kaashoek, David Karger, Robert Morris, Ion Stoica. Wide-area cooperative storage with CFS. In proceedings of the eighteenth ACM symposium on Operating systems principles 2001.

[6] M Ripeanu. Peer-to-peer Architecture Case Study: Gnutella. In Proceedings of International Conference on P2P Computing, 2001.

[7] Kazaa. http:// www.kazaa.com.

[8] Eytan Adar, Bernardo A.Huberman. Free Riding on Gnutella. First Monday, 5(10), October 2000.

[9] Ian Foster, Adriana Tamnitchi. On Death, Taxes, and the Convergence of Peer-to-Peer and Grid Computing. IPTPS'03, 2003.

[10] Tsuen-Wan "Johnny" Ngan, Dan S. Wallach, and Peter Druschel. Enforcing Fair Sharing of Peer-to-Peer Resources. IPTPS'03, 2003.

[11] Jeffrey Shneidman, David C Parkes. Rationality and Self-Interest in Peer to Peer Networks. IPTPS'03, 2003.

[12] G Hardin. The tragedy of the commons. Science, 162: 1243-1248, 1968.

[13] S Sarou, G P Krishna and S D Gribble. A measurement study of peer-to-peer files sharing systems. In Proceedings of the SPIE Conference on Multimedia Computing and Networking, pages 156-170, January 2002.

[14] Joan Feigenbaum, Scott Shenker. Distributed algorithmic mechanism design: recent results and future directions. In Proceedings of the Dial-M'02, Sep 2002.

[15] Yun Fu, Jeffrey Chase, Brent Chun, Stephen Schwab, and Amin vahdat. SHARP: an architecture for secure resource peering. In Proceedings of SOSP'03, Oct 2003.

[16] Landon P Cox, Brian D Noble. Samsara: honor among thieves in peer-to-peer storage. In Proceedings of SOSP'03, Oct 2003.

Discovery and Routing in the HEN Heterogeneous Peer-to-Peer Network

Tim Schattkowsky

Paderborn University, C-LAB,
D-33102 Paderborn, Germany
`tim@c-lab.de`

Abstract. Network infrastructures are nowadays getting more and more complex as security considerations and technical needs like network address translation are blocking traffic and protocols in IP-based networks. Applications in such networks should transparently overcome these limitations. Examples for such applications range from simple chat clients to collaborative work environments spanning different enterprises with different heterogeneous network infrastructure and different security policies, e.g., different firewalls with different configurations. Overlay networks are a convenient way to overcome this problem. In many cases, diverse barriers like multiple facing firewalls would require significant user knowledge to establish a connection. Self-organizing peer-to-peer networks appear to be a convenient solution, but contemporary systems still have limitations in overcoming connectivity problems in heterogeneous networks. Thus, we introduce a self-organizing peer-to-peer infrastructure that overcomes these issues by transparently interconnecting networks with different protocols and address schemes.

1 Introduction

The Internet led the way to new distance-spanning applications enabling communication and collaboration between distant computers and individuals. However, when trying to interconnect applications and their users in heterogeneous networks, e.g., between protected Intranets, connectivity problems arise. These problems are often caused by different network technologies and security measures.

The same problem also applies to many typical end user network applications. Many interesting new applications like IP telephony clients and instant messengers suffer from the same problems (e.g., when two DSL users with routers try to interact).

Current overlay networks including peer-to-peer networks cannot resolve the problem in many of the outlined scenarios, especially not when routing over multiple intermediate nodes is required. In order to overcome these problems, we introduce HEN as a self-organizing peer-to-peer network infrastructure that transparently overcomes communication barriers when possible. It establishes a peer-to-peer network interconnecting nodes in different physical networks using different transport protocols. It includes support for features like routing, relay, polling, tunnelling to overcome connectivity problems between heterogeneous networks like those caused

P. Lorenz and P. Dini (Eds.): ICN 2005, LNCS 3421, pp. 653–661, 2005.

by firewalls and network address translations. It is designed with the goal to allow the transmission of single data packets from one Node in the network to another while dealing transparently with the involved obstacles like routing and relay.

The remainder of this paper is structured as follows. The next section discusses related works in the field of peer-to-peer based systems. Section 3 introduces the HEN network model. The EGG protocol used for network discovery and routing is described in section 4 before section 5 closes with conclusions and future work.

2 Related Work

Many peer-to-peer systems like Napster and Gnutella seem to explicitly focus on sharing data. These networks usually are bound to an IP protocol and do not deal with heterogeneity in the underlying networks. Centralized networks like Napster overcome some of the connectivity problems by the use of centralized servers or relaying supernodes (e.g, FastTrack), but this does not solve the general problem. Other approaches like FreeNet [1] focus on anonymity at the cost of performance.

Pastry [6] and Tapestry [10] are peer-to-peer architectures that both use dynamically assigned random unique identifiers to identify network nodes. Thus, named nodes are not supported at this level. Routing is performed by forwarding a message to a known node having a node identifier sharing a longer prefix with the target node identifier than the current hop. This mechanism is originally inspired by Paxton et al. (PRR)[4][2] and is to some extent similar to RFC1518 [5]. The join protocol presented in [3] illustrates how several nodes can concurrently join such a network. As it seems that both Pastry and Tapestry target towards homogeneous IP networks without communication barriers, the routing actually introduces a significant overhead by most messages through $O(\log_2 \#nodes)$ nodes where an initial discovery using the routing mechanism could also enable direct communication on the underlying network. This also generally avoids the effects of latency differences in an overlay network that is elaborated in [9].

Tapestry uses the same mechanism to locate named objects. The object identifiers in Tapestry are syntactically equal to node identifiers. For a particular object, the node with the most similar node identifier stores a reference to the node containing the object. Thus, the routing to the node can be used to discover the actual object. The root node is the root of a tree that is dynamically created by back-propagation of the object reference once messages for the object arrive at the root node. The reference will be propagated back the path to the original sender. Each node on that path will later redirect messages for the object to the actual node containing the object. However, the messages are still subject to the original routing mechanism.

Other approaches like Chord [8] introduce centralized lookup services for object discovery that could be used to discover a named node. However, this does not resolve the routing problem and introduces a single point of failure.

In previous work, we have introduced ANTS [7] as a peer-to-peer infrastructure for interconnecting Web Services across heterogeneous networks. However, ANTS had serious limitations in scalability. All communication was SOAP-based and thus required an application server. Furthermore, fixed relay nodes where needed. While this was sufficient for small-scale collaborative workspaces, it is not sufficient as a general-purpose platform for network applications. Thus, we decided to completely separate the transport from the actual services and significantly extend its capabilities to a general purpose self-organzing peer-to-peer architecture for heterogeneous networks, which we called the HEterogenous Network (HEN).

3   The HEN Network Model

The network model underlying HEN is shown in Figure 1. In our model, a computer in the network is a *Node*. Each Node is identified by a fixed location-independent NodeID. This NodeID is used by the Node when connecting to the HEN network.

A real-world network enabling direct bi-directional communication between all the nodes in that network is a *Network* in our model. Thus, Nodes in the same Network must support exactly the same *Protocol*. Both are again identified by a GUID. For the Internet, these protocols are derived from the IP. In our notion, Nodes in the Internet supporting HEN Connections using the TCP Protocol form a Network among many others. It is important to notice that the same nodes may be part of other Networks as well, e.g., just by supporting an additional Protocol like HTTP. We aim at creating an overlay peer-to-peer network that contains Nodes from several different Networks. Nodes within the same Network as a Node and Nodes in Networks directly connectable by the Node are said to be *local Nodes*.

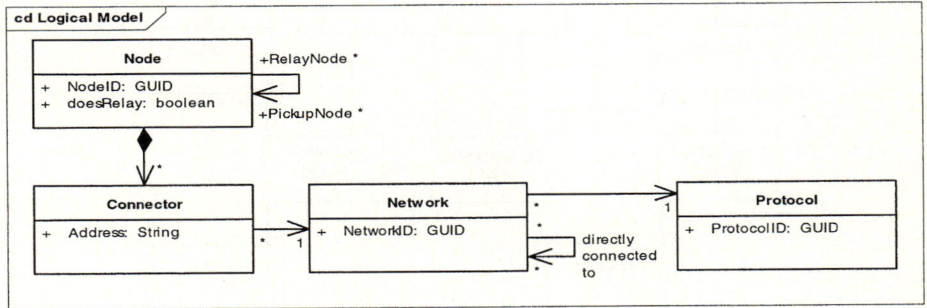

Fig. 1. Basic Network Model

The ability to use certain protocols in certain real-world networks may differ from Node to Node. Thus, each Node is connected to each of these Networks using a Connector exposing its identity in that network. This Connector is a piece of software implementing the Network Protocol (e.g., TCP for the Internet).

A Node residing in two networks is called a *Gateway* between these Networks. These Gateways are essential to routing in the HEN. Furthermore, a network maybe directly connected to another network. This implies that nodes of the originating network can act as nodes from the target network without being reachable in the target network. In practice, this means usually that these nodes are behind a router or firewall. However, this indicates also the need to establish a relay to enable direct communication between the networks. This situation is actually detected by nodes joining the source network and detecting a missing relay connection. These nodes act as a *PickupNode* by establishing a connection to a *RelayNode* node from the target network to enable direct communication between the networks. The node thus becomes a full member of the target Network. However, it is important to notice that it is assumed here that established connections in the underlying network are bi-directional, which is essential to both the relay mechanism and network discovery.

Figure 2 shows an example model showing the communication path from A to C. This example corresponds to Figure 4, which is elaborated later in this paper. It describes how Node A can connect to Node C because it is in a protected sub-network of the Internet, use C and D as intermediate Node to have E forwarding its messages picked up from a direct connection to D to the final destination Node B.

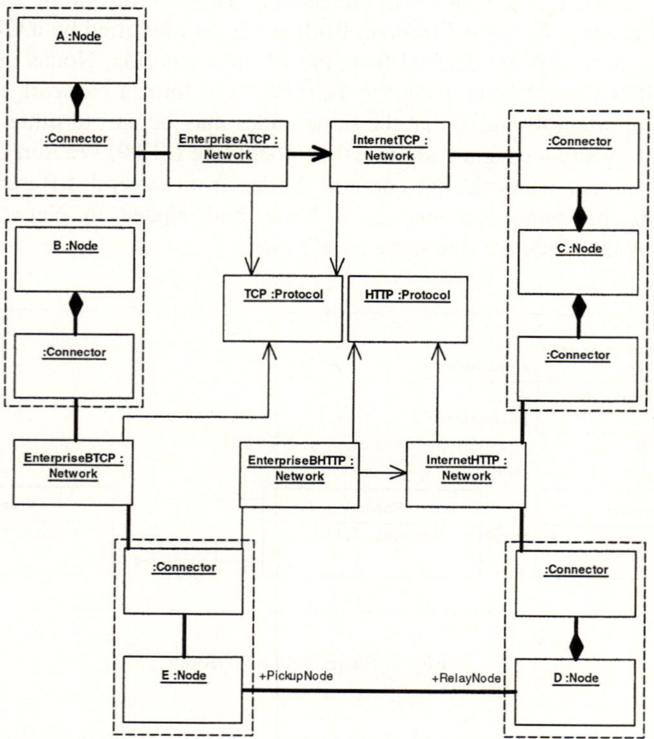

Fig. 2. Communication Model Example - (Path from A to B highlighted)

4 The EGG Network Protocol

The EGG protocol for network discovery and transport is responsible for all data exchange in the network. It is based on these messages:

- `PACKET(NodeID, (byte)*)` sends a packet to a local Node. The packet may be subject to routing there to reach its final destination Node.
- `DROPPED(NodeID, Number)` indicates packets dropped at a distant node.

- `DISCOVER(NodeID, [NetworkID,Address,timestamp] (Network ID)*)` to discover a certain Node. If NetworkID and Address of the Node are believed to be known, these are included. The message contains a history to avoid loops.
- `ROUTE(NodeID, (NodeID,NetworkID,Address,timestamp)*)` returns routing information to a discovering node. All Nodes in the message will share this information to enable packet transport through these Nodes.
- `PING(id,timestamp)` causes the destination node to respond with a
- `PONG(id,timestamp)` message containing the local time. This is also used to coordinate the time between nodes.
- `DO_PING(NodeID)` ask a remote Node to ping the given node results in a
- `DO_PING_RESULT(responsetime)` message containing the response time for the PING message.
- `NEW_GATEWAY(NetworkID,NodeID,Stamp,Address)` indicates the presence of a new local Gateway to the given Network.
- `GET_GATEWAYS()` causes the remote node to respond with a
- `GATEWAYS((NetworkID,NodeID,Stamp)*)` message containing a list of all known local Gateways.

All messages additionally contain the NodeID of the originating Node. These messages are sent directly through the Connectors of a Node without being subject to routing. Messages may be either data packets contained in `PACKET` messages that are subject to routing or protocol messages. Messages are limited to a total size of 64k.

During runtime, each HEN Node holds four tables that are essential for routing packets in the network. The *Networks Table* contains an entry for each Network ever discovered. Such an entry consists of three items, a Network and a Protocol identifier as well as a time stamp. Both identifiers are GUIDs identifying the network and the protocol to be used in the network respectively. While Protocol identifiers are most likely to be taken from the set of pre-defined identifiers including identifiers for TCP, HTTP, HTTPS and SMTP/POP3, the network identifiers are likely to be randomly created except for the Internet, which has a pre-defined identifier.

The *Nodes Table* contains an entry for each Node in every Network ever discovered. Each entry consists of five items. The first two items are an UID for the Node and the Network. Thus, the table may contain different entries for the same Node in different Networks. The third item indicates whether the node is believed to be active or down. The remaining two items are a time stamp for the entry and a string containing the actual protocol-address of the Node in the Network.

The *Gateways Table* stores time-stamped pairs of Network identifiers. Such a pair indicates that the first network contains a direct gateway to the second network enabling data transport from the first network to the second network. It is important to notice that the Gateways Table cannot be derived from the Nodes Table as certain Nodes may refuse to gate traffic for other Nodes although it is technically possible.

The *Forward Table* contains pairs of NodeIDs indicating that packets send to the first Node have to be routed over the second Node.

The time stamps in the tables are 64 bit numbers indicating the coordinated time at which the original information was created. Thus, time stamps from different nodes are comparable. By including a tolerance (i.e., a minute), time differences between the nodes can be largely ignored and items from different nodes become comparable.

4.1 Joining the Network

When a Node starts up, it has to discover the existing network. Initially, it has to connect to the network at all. This is done by using a stored Nodes Table. This Nodes Tables is either initially provided somewhere (for a new node) or has been persistently stored from the when the Node was last connected.

A Node has to determine its current Network if the physical address of a Node has changed since its last connection to the HEN network, the Node connects to the HEN network for the first time or it obviously has trouble to connect any Node in its current Network. The last situation indicates that the node has trouble using its underlying network connection, which may be caused by a failure or configuration change. However, to actually determine the current Network, a Node attempts to contact some of the most recently known Nodes of each Network from its Node Table that uses a protocol compatible the Node's capabilities. The order in which the Networks are tested is determined by a history of Networks the Node has been in. The Node will first attempt to locate itself in these networks, trying the most recent one first. However, later it will attempt to test all known Networks including the Internet.

If the Node detects the ability to connect to a certain network, it is still necessary to test if Nodes from that Network can access the current node or are blocked (e.g., by NAT or a firewall). The Node will send a DO_PING message containing its NodeID, the currently used ProtocolID and the Node address. The remote address will attempt send a PING message to the node to test the connection and finally indicate a successful attempt where a PONG response has been received by responding with a DO_PING_RESULT message with a positive result. This indicates that the current Node is in the remote Node's Network. Otherwise, the current Node is not, but can access that network. Thus, the Node can establish a connection between these Networks. However, first the test for the current network will be completed. If the Node is not located in any of the known networks, it will use a random NetworkID.

4.2 Gateways

When a Node identifies itself as a potential Gateway between Networks because it resides in both Networks, the node should publish itself as a Gateway by sending a NEW_GATEWAY message containing the NetworkID of the remote Network to some other Nodes in the same Network from its Node Table unless this is prohibited by its configuration. Upon reception of such a message, a Node will check its Gateway and Node Tables and update it with the new information. If the Gateway was already known, the message is discarded, otherwise it will be forwarded to other local Nodes.

However, each Node must use the GET_GATEWAYS message first to ask a local Node if there are already enough such Gateways published. Currently, HEN Nodes only publish themselves as a Gateway when less then 16 other Gateways to the same Network are known. This avoids useless traffic in situations where actually are Nodes in one Network could establish a Relay-Connection (e.g., this is the case for the firewall scenario). Anyway, this mechanism still currently holds back the approach as it has scalability issues especially for large networks connected to many very small networks (i.e., the internet connected to a lot of NAT users).

For Sub-Networks, each Node is its own Gateway to the parent Network, but does not publish this information as this is a property of the sub-network. However, each of these nodes needs to open a persistent connection to the parent network to enable bi-directional communication. The relay node will start acting as a proxy for the original Node. Thus, no Gateway announcement is necessary as the Node is now effectively a member of the parent network. This actually avoids the broadcast problem for most real life configurations.

4.3 Discovery and Routing

Routing in the HEN network is based on finding a route to the destination node based on the Nodes and Gateways Tables. These tables must contain a path to the destination Node where each hop is a Network. If no such path exists or the destination node is unknown, the HEN node will start a discovery to find such a path.

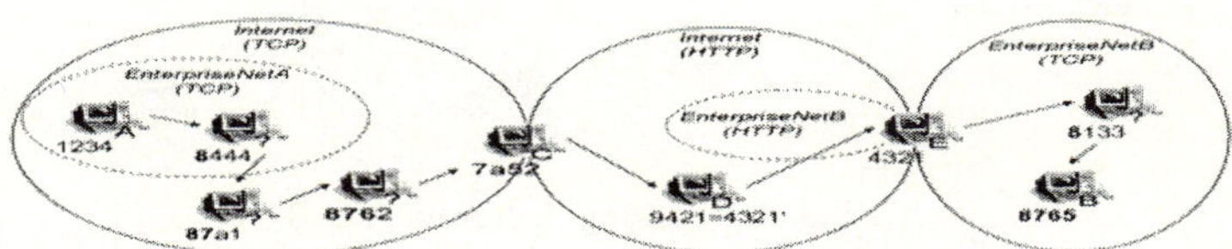

Fig. 3. Discovery of a path (ACDEB) from 1234 to 8765

Discovery of a Node in a Network starts by attempting to locate that node using a variant of the PRR approach. Thus, the discovery is based on sending a DISCOVERY message that traverses the local network by moving from each Node to a Node sharing a longer prefix with the requested Node than the current Node does. If a Node has an entry for the destination Node in its Nodes Table and this entry is newer than the information contained in the DISCOVERY message (if any), it replaces this information. If the message addresses a Node in a Network containing the Node currently processing the message, it will attempt to forward the message to that Node. If this fails, all information in the DISCOVERY message is cleared. If no more Networks are reachable via Gateways, the message is discarded.

Once the message arrives at the desired destination node, the node responds with a ROUTE message containing all networks passed as in the DISCOVERY message. This message travels back the path along the Gateways from the DISCOVERY messages and updates the forward table of each of the Gateways accordingly.

At each Node, when a packet has to be sent its final destination address is examined. A route to this address is computed from the Nodes and Networks Tables. The next hop on this route is a Network because otherwise the message would have been delivered directly to a local Node. This Network ID is used to determine a Gateway to the Network which will be used as the next hop and the message will be transmitted. If this fails, the Gateway will be marked as "down" and another Gateway will be used. This next hop NodeID and NetworkID are cached to be directly available when the next packet for that node arrives.

Figure 3 shows example with simplified NodeIDs where Node A (1234) attempts to discover Node B (8765). Node A is in an Enterprise Network with unlimited access to the Internet (i.e., NAT). All nodes within this network have established relay connection and proxies. These are not shown here, as the proxies are only involved in receiving a message, as shown for Node E which has Internet access using an HTTP proxy while Node B is in a protected enterprise network. This Node uses Node D as a relay and proxy for E, which is a Gateway like Node C.

Discovery starts by moving to Node 8444 and 8721 sharing an increasing part of the destination NodeID. However, no Nodes with a longer prefix exist. Thus, the search moves through a Gateway (8721 itself) into the Internet and ends again at 8762. Here the search moves to Node D as a Getaway into EnterpriseNetB(HTTP) and end continues on E. As E is also in EnterpriseNetB(TCP), it initates the search there which finally leads to B. Node B responds by sending a ROUTE message the same path back resulting in all Nodes along the path to receive the necessary updates

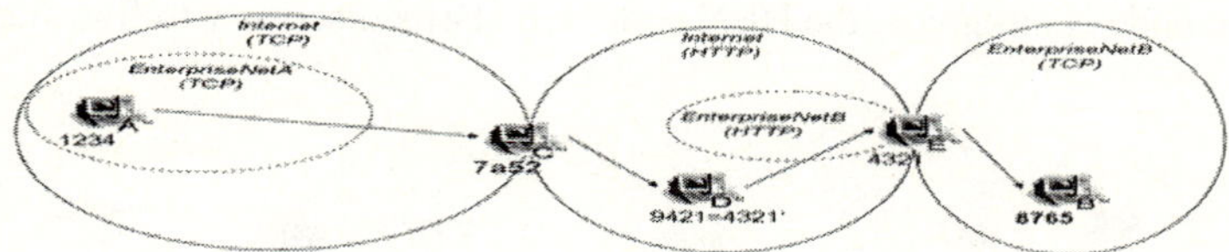

Fig. 4. Resulting Path from A to B

for their tables to directly transport packets to B as needed. The resulting path consists of all Gateways taken and is shown in Figure 4. This situation is modelled in the communication model in Figure 2.

5   Conclusions and Future Work

We have introduced HEN as a self-organizing de-centralized peer-to-peer network spanning heterogeneous networks with communication barriers. HEN demonstrates how the concept of self-organizing networks can be extended to support heterogeneous networks with multiple protocols, relay and routing. HEN has been implemented in JAVA and successfully applied to interconnect Web services in a collaborative engineering environment. Future work will include a layer 3 transport protocol implementation and detailed simulation work on the network behavior.

References

[1] Clarke, I., Sandberg, O., Wiley, B., Hong, T.W.: Freenet: A Distributed Anonymous Information Storage and Retrieval System. In Proc. of the ICSI Workshop on Design Issues in Anonymity and Unobservability, Berkeley, CA, 2000.
[2] Hildrum, K., Kubiatowicz, J. D., Rao, S., Zhao, B. Y.: Distributed object location in a dynamic network. In Proc. SPAA, August 2002.

[3] Liu, H., Lam, S. S.: Consistency-preserving neighbor table optimization for p2p networks. TR-04-01, Dept. of CS, Univ. of Texas at Austin, January 2004.

[4] Plaxton, C. G., Rajaraman, R., Richa, A. W. : Accessing nearby copies of replicated objects in a distributed environment. In Proc. of ACM SPAA. 1997.

[5] Rekhter, Y.: An architecture for IP address allocation with CIDR. RFC 1518, 1993.

[6] Rowstron, A. I. T., Druschel, P.: Pastry: Scalable, distributed object location and routing for large-scale peer-topeer systems. In Proc. Middleware 2001, 2001.

[7] Schattkowsky, T., Loeser, C., Müller, W.: Peer-To-Peer Technology for Interconnecting Web Services in Heterogeneous Networks. In Proc. 18th International Conference on Advanced Information Networking and Applications (AINA 2004), April 2004.

[8] Stoica, I., Morris, R., Karger, D., Kaashoek, M. F., Balakrishnan, H.: Chord: A scalable peer-to-peer lookup service for Internet applications. TR-819, MIT, March 2001.

[9] Zhang, H.. Goel, A., Govindan, R.: Incrementally improving lookup latency in distributed hash table systems. In Proc. of SIGMETRICS, June 2003.

[10] Zhao, B. Y., Huang, L., Stribling, J., Rhea, S. C., Joseph, A. D., Kubiatowicz, J. D.: Tapestry: A resilient global-scale overlay for service deployment. IEEE Journal on Selected Areas in Communications, Vol.22(No.1), January 2004.

Scalable Group Key Management with Partially Trusted Controllers

Himanshu Khurana[1], Rafael Bonilla[1], Adam Slagell[1], Raja Afandi[2],
Hyung-Seok Hahm[1], and Jim Basney[1]

[1] NCSA, University of Illinois, Urbana-Champaign, USA
{hkhurana, bonillla, slagell, hahm, jbasney}@ncsa.uiuc.edu
[2] Computer Science Department, University of Illinois, Urbana-Champaign, USA
afandi@cs.uiuc.edu

Abstract. Scalable group key management solutions are crucial for supporting Internet applications that are based on a group communication model. Many solutions have been proposed and of these the most efficient and scalable ones are based on logical key hierarchies (LKH) with symmetric keys organized in a tree. However, these solutions centralize trust in the group controller and make it an attractive attack target for access to communication keys for all groups supported by the controller. In this paper we propose a novel group key management approach, which uses a partially trusted controller that does not have access to communication keys and yet provides the same level of efficiency and scalability as LKH schemes. For this we develop a new public-key encryption scheme, which is based on El Gamal, and we show that the scheme is as secure as El Gamal.

1   Introduction

Many collaborative applications such as conferencing, command-and-control systems, white-boards, publish-subscribe systems, interactive distance-learning, and shared instruments need reliable and secure communication for large groups comprising as many as thousands of members that may leave or join groups at any time. Recently proposed solutions for reliable multicast communications for both IP multicast (e.g., Adamson *et al.* [1]) and applications-layer multicast (e.g., SpinGlass [2]) can support such large groups. Group key management is the cornerstone of providing secure communication, and the key management problem translates to ensuring that only the current group members have the session key. Therefore, the key management scheme must be efficient in changing the session key on member join and leave events and scalable to support large groups. Solutions proposed by Wong *et al.* [21], Wallner *et al.* [20], and Carroni *et al.* [5] that are based on logical key hierarchies (LKH) of symmetric keys organized in a tree provide both efficiency and scalability by reducing both group rekey operation complexity and key storage requirement to $\mathcal{O}(logN)$ for group size N.

P. Lorenz and P. Dini (Eds.): ICN 2005, LNCS 3421, pp. 662–672, 2005.

Consider the tree-based, group-oriented LKH key management scheme of Wong *et al.* [21]. In this scheme a trusted group controller (GC) creates and distributes symmetric keys to group members, and maintains the user-key relations. For N users GC organizes keys as the nodes of a full and balanced d-ary tree where the root node key serves as the group's session key while the remaining keys serve as *key encrypting keys* that are used to efficiently update the session key on member join and leave events. When a user leaves the group the server only needs to perform approximately $d*log_d(n)$ symmetric encryptions and send a message of size $\mathcal{O}(d * log_d(n))$ to update the session key. This illustrates the efficiency and scalability provided by this approach. However, this scheme requires GC to be fully trusted because its compromise would result in disclosure of the session key and the key encrypting keys managed by GC. Extending that argument, when a group controller manages multiple groups, its compromise leads to compromise of keys for all those groups; e.g., a GC supporting an interactive distance learning application with different groups representing different classrooms, or a GC supporting a command-and-control system where a large number of groups have permissions to access different components of the system. This has two consequences. First, the data protected by the session keys can be deciphered by the adversary. Second, the only way to recover from such a compromise is to re-key all groups managed by GC since the key encrypting keys are known to the adversary.

Both decentralized key distribution solutions (e.g., [15]) and contributory key agreement solutions (e.g., [4], [10], [18], [17]) solve this problem by eliminating the GC but, unfortunately, they do not scale to large groups. In this paper we address this problem by developing a group key management approach that offers the efficiency and scalability of LKH schemes while minimizing trust in GC at the same time. The new approach is based on a (discrete-log) public key encryption scheme that we have developed utilizing concepts of proxy re-encryption [3] [8] [11]. The encryption scheme enables GC to maintain user-key relations but does not allow GC access to session keys directly or indirectly via decryption keys. Consequently, GC's compromise does not provide the adversary with access to session keys or to (session) key decrypting keys. The latter implies that recovery from GC's compromise does not require re-keying of the entire group. We feel that this is a useful contribution that aims to mitigate the consequences of server compromise, which is an inescapable reality indicated by recent statistics on electronic crime [6].

The rest of this paper is organized as follows. In Section 2 we present our group key management scheme and analyze its costs in Section 3. In Section 4 we discuss related work and conclude in Section 5. Due to space limitations the security analyses and proofs are provided in [9].

2   The Proposed Scheme: TASK

In this section we describe our group key management scheme, TASK – a Tree-based scheme that uses Asymmetric Split Keys. TASK comprises a communication group with n members M_1, M_2, ..., M_n, and a partially trusted group controller GC. The group is created by the first member and GC. Members then join (leave) the group with the help of a *sponsor* (an existing group member) and GC. TASK is similar to LKH schemes in that GC manages a d-ary key tree except that TASK uses split asymmetric keys, and GC only manages shares of the split keys (members manage the other part of the shares). Session keys are computed with member shares and, consequently, GC does not have access to them. To ensure security, session keys are updated whenever members join and leave the group.

We now provide some defintions, present the TASK public-key encryption scheme, provide an instance of the key-tree, and show how TASK supports member join and leave events.

2.1   Definitions

1. *El Gamal.* Let $\mathcal{E}_{eg} = (Gen, Enc, Dec)$ be the notation for standard El Gamal encryption [7]. Gen is the key generating function. Hence $Gen(1^k)$ outputs parameters (g, p, q, a, g^a) where g, p and q are group parameters, (p being k bits), a is the private key, and $y = g^a \bmod p$ is the public key. The Enc algorithm is the standard El Gamal encryption algorithm and is defined as $e = (mg^{ar} \bmod p, g^r \bmod p)$, where r is chosen at random from Z_q and m is the message. To denote the action of encrypting message m with public key y, we write $Enc_{PK_y}(m)$. Dec is the standard El Gamal decryption algorithm and requires dividing mg^{ar} (from e) by $(g^r)^a \bmod p$.

2. *TASK Encryption Keys.* All members manage *private keys* that they use for encrypting and decrypting protocol messages. Each member has a unique private key and a set of $\mathcal{O}(log_d n)$ private keys in its path to the root node that will be common to other members. For example, K_1 denotes member M_1's private key, $PK_1 = g^{K_1} \bmod p$ is M_1's *public key*, and K_{1-9} and K_{1-8} are the root private keys as illustrated in Figure 1. GC manages *corresponding private keys* for members and for intermediate and root nodes that it uses for re-encrypting protocol messages; e.g., K_1' is M_1's corresponding private key, and K_{1-9}' and K_{1-8}' are the corresponding root private keys. All private and corresponding private keys are essentially El Gamal keys and every pair of private and corresponding private keys adds up to the same value – $GKEK$, the *Group Key Encrypting Key*. However, no entity knows the value of $GKEK$.

3. *Session Keys.* The session key, also known as the Data Encrypting Key (DEK), is computed by members to be the (one-way) hash of the root private key.

4. *Key Randomization.* TASK private keys are often updated by either adding or subtracting random numbers. To simplify representation of key updates we define two functions. $A_{(r_1, r_2, ..., r_n)}(K)$ denotes the addition of n random numbers $r_1, r_2, ..., r_n$ to key K $(\bmod\ q)$. Similarly, $S_{(r_1, r_2, ..., r_n)}(K)$ denotes the subtraction of n random numbers $r_1, r_2, ..., r_n$ from key K $(\bmod\ q)$.

5. *External PKI Keys.* Some protocol messages are encrypted with external PKI keys (using El Gamal) and all protocol messages are signed and verified with external PKI keys[1] (using the DSA signature algorithm). We differentiate between TASK keys and external PKI keys by placing a bar on top of external PKI keys (e.g., $Enc_{\overline{PK_{GC}}}(m)$ denotes encryption of message m with GC's external PKI public-key).

2.2   TASK Public-Key Encryption Scheme $\mathcal{E}$

We denote the TASK Asymmetric Encryption scheme by $\mathcal{E} = (IGen, UGen, KU, AEnc, ADec, \Gamma)$. Here $IGen$ is a distributed protocol executed by M_1 (the first member of the group) and GC to generate group parameters g, p and q, private and corresponding private keys K_1 and K_1' and public key $PK_1 = g^{K_1}$. (K_1 is simply a random number modulo q chosen by M_1, and K_1' is a random number modulo q chosen by GC). $UGen$ is a distributed protocol executed by joining member M_i, an existing member sponsor M_s, and GC to generate (1) a public-private key pair for M_i, (2) if required, a private key for a (new) intermediate node along the path from M_i's location in the key-tree to the root, and (3) the corresponding private keys for GC. The $UGen$ protocol requires M_i, M_s, and GC to generate random numbers and add/subtract them from K_s and K_s'. It is guaranteed that $K_i + K_i' = K_s + K_s' = GKEK$. KU is a key update protocol initiated by an existing member sponsor and executed by all group members and GC to update $GKEK$. It involves the sponsor choosing a random value r and distributing it to all other group members who add it (mod q) to their private keys, and GC choosing a random value r_{GC} and adding it (mod q) to all corresponding private keys; i.e., $GKEK$ is modified by adding r and r_{GC}. $AEnc$ and $ADec$ are identical to Enc and Dec defined above for El Gamal. Γ is a transformation function which transforms a message encrypted with one TASK public key into a message encrypted with another TASK public key. For example, once $UGen$ has been executed for members M_i and M_j, $\Gamma_{(K_i', K_j')}$ would take as input an encrypted message of the form $(g^{RK_i}S, g^R)$ and output $(g^{RK_j}S, g^R) = ((g^{RK_j'})^{-1}g^{RK_i'}g^{RK_i}S, g^R)$ where S is the message. The fact that $K_j + K_j' = K_i + K_i'$ is crucial in enabling this transformation.

The encryption scheme $\mathcal{E}$ is secure if (1) it retains the same level of security as the standard El Gamal scheme against all adversaries $\mathcal{A}$, (2) the GC cannot distinguish between encryptions of two messages even with access to multiple *corresponding private keys*, and (3) a group member cannot distinguish between encryptions of two messages for which he does not have a decryption key (even if he has other private keys). The proofs are given in [9].

---

[1] Most other group key management schemes (e.g., [10], [17], [21]) assume the presence of an external PKI for signature verification but not for encryption. We assume an external PKI for encryption as well but argue that this imposes little or no additional costs in terms of certificate distribution; e.g., PGP certificates typically contain both encryption and signature verification keys.

Theorem 1. Let $\mathcal{E} = (IGen, UGen, KU, AEnc, ADec, \Gamma)$ be the TASK encryption scheme. $\mathcal{E}$ is CPA (chosen plaintext attack) secure against the GC, group members, and any adversary $\mathcal{A}$.

2.3   Member Join Protocol

The first member of the group, M_1, creates the group along with GC. First, M_1 generates group parameters g, p and q, chooses a private key $K_1 \pmod{q}$, and computes the public key $PK_1 = g^{K_1}$. M_1 then sends the group parameters to GC who chooses the corresponding private key K_1' (also mod q). Both M_1 and GC implicitly agree that $K_1 + K_1' = GKEK \bmod q$, though neither knows its value.

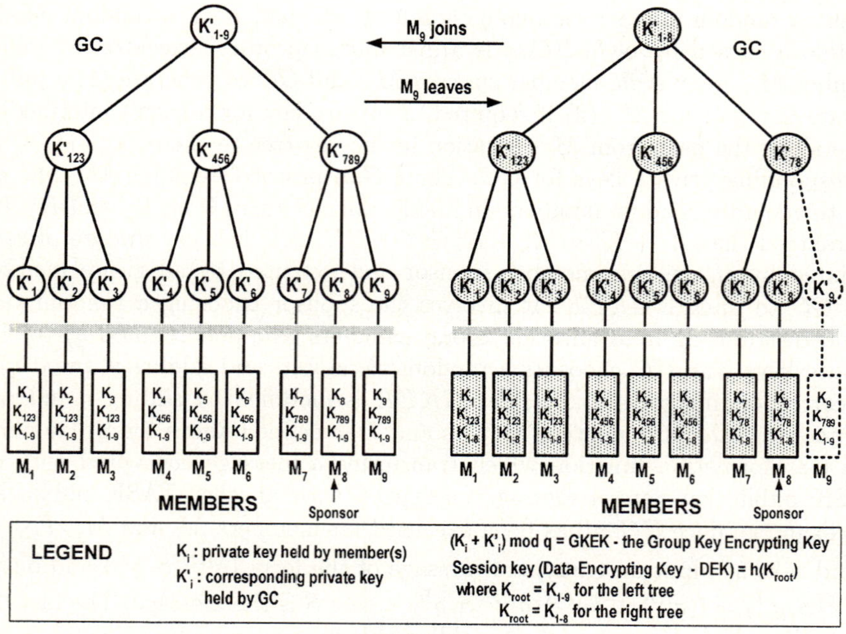

Fig. 1. An Example of a TASK Key Tree with Member Join and Leave Events

All other members join the group with the help of a sponsor (an exisiting group member) and GC. We illustrate the join of member M_9 in an existing group of eight members in Figure 1. To join the group, M_9 sends a JOIN message to GC. GC then determines the insertion node in the tree, which will either be shallowest rightmost node if the join does not increase the height of the tree, or the root node otherwise. GC also determines (and then informs) the sponsor for the join event to be the rightmost leaf node in the subtree rooted at the insertion point; M_8 in this case. These join and sponsor determination messages are separate from any join messages generated by the underlying group communication system, though they may be combined for increased efficiency. The

TASK join protocol is a three-step protocol illustrated below but consists of only two communication rounds: one in which the sponsor M_8 sends unicast messages to M_9 and GC, and a multicast message to members $M_1 \dots M_7$, and the second in which M_9 sends a unicast message to GC. All TASK multicast messages are encrypted with TASK keys while unicast messages are encrypted with either the receiver's external PKI public-key or with TASK keys.

TASK PROTOCOL FOR MEMBER M_9's JOIN EVENT

Step 1:

- M_8 generates random value r and adds it all private keys; i.e., $K_8 \leftarrow A_r(K_8)$, $K_{789} \leftarrow A_r(K_{78})$, and $K_{1-9} \leftarrow A_r(K_{1-8})$. Compute new session key: $DEK \leftarrow h(K_{1-9})$.
- M_8 generates random value r_9 and computes a temporary key for M_9: $TK_9 \leftarrow A_{r_9}(K_8)$.
- $M_8 \longrightarrow M_9$: $Enc_{\overline{PK_9}}(g, p, q, TK_9, K_{789}, K_{1-9})$
- $M_8 \longrightarrow GC$: $Enc_{\overline{PK_{GC}}}(r_9)$.
- $M_8 \longrightarrow M_1 \dots M_7$: $AEnc_{PK_{1-8}}(r)$.

Step 2:

- Members $M_1 \dots M_7$ decrypt r from M_8's message using private key K_{78}, and add $r \pmod q$ to all their private keys; e.g., member M_1 computes $K_1 \leftarrow A_r(K_1)$, $K_{123} \leftarrow A_r(K_{123})$, and $K_{1-9} \leftarrow A_r(K_{1-8})$. Members also compute the new session key: $DEK \leftarrow h(K_{1-9})$.
- GC decrypts random values from M_8 and stores them temporarily.
- M_9 decrypts and stores private keys from M_8, computes new session key from K_{1-9}, generates random value $r_{\bar{9}}$ and computes his private key: $K_9 \leftarrow A_{r_{\bar{9}}}(TK_9)$.
- $M_9 \longrightarrow GC$: $Enc_{\overline{PK_{GC}}}(r_{\bar{9}})$

Step 3:

- GC computes corresponding private key for M_9: $K_9' \leftarrow S_{r_9, r_{\bar{9}}}(K_8')$.
- GC chooses a random value r_{GC} and adds it $\pmod q$ to all corresponding private keys; i.e., $\forall i \; K_i' = A_{r_{GC}}(K_i')$.

To add M_9 to the group, the sponsor M_8 generates private key K_{789}, M_9 generates private key K_9, and GC generates corresponding private keys K_{789}' and K_9'. All key generations are accomplished by adding and subtracting random values from previously held keys, and are therefore computationally inexpensive operations. Furthermore, in order to ensure that M_9 cannot access the previous session key, all private keys in the path from the joining node to the root node are updated by adding a random value r to the previous values. However, since TASK requires all private and corresponding private keys to add up to the same value $(GKEK)$, r is added to all private keys held by the group members. A new session key is then computed from the updated root private key. In Figure 1, the change in key values is denoted by different subscripts and fill patterns for the tree nodes. The three-step join protocol is detailed below.

In the first step, M_8 generates a random value r that it adds to all private keys and a temporary key for M_9. M_8 also computes the new session key from K_{1-9}. M_8 then sends (1) to M_9 the group parameters, the temporary key, and the private keys in the path from M_9's location in the tree to the root (encrypted with M_9's public key), (2) to GC the random values for generating corresponding private keys (encrypted with GC's public key), and (3) to members $M_1...M_7$ the random value r (encrypted with public key PK_{1-8}). In the second step, members $M_1...M_7$ decrypt r with private key K_{1-8}, update their private keys by adding r, and compute the new session key $DEK = h(K_{1-9})$. In addition, M_9 generates private key K_9 and sends to GC a random value for generating corresponding private key K_9'. Note that M_9 has a private key that is not known to the sponsor even though the sponsor's key was used to generate it. In the third step GC computes corresponding private keys K_9' and chooses a random value r_{GC}, which it adds (mod q) to all corresponding private keys. At the end of the join protocol we have added r and r_{GC} to $GKEK$ and all the splits of this key add up to the new value.

The join protocol also presents an easy way for a sponsor to *refresh* the private keys and the session key at any point in time. The sponsor simply chooses a random value r, adds it to all her private keys, and broadcasts it to the entire group by encrypting it with the (old) root private key. On receiving this message, all other group members decrypt r and add it to their private keys. The new session key will be the hash of the new root private key.

2.4   Member Leave Protocol

We illustrate the leave of member M_9 in Figure 1 above. M_9 informs GC of his desire to leave the group and GC then determines (and informs) the sponsor to the right-most leaf node of the subtree rooted at the leaving member's sibling node; M_8 in this case. The leave of member M_9 is enforced by deleting corresponding private key K_9', updating private keys in the path from M_9's node to the root node, and changing the session key. Private keys are changed by adding a random number to previous value and since TASK requires all private and corresponding values to add to $GKEK$, the random number is added to all private keys. The TASK leave protocol is a three-step protocol that requires two communication rounds; one in which the sponsor sends a unicast message to GC, and the second in which GC sends a multicast message to members $M_1...M_7$. The three-step leave protocol is detailed below.

In the first step, M_8 generates a random value r to be added to all private keys and sends r to GC encrypted with his TASK public key PK_8. In the second step, GC deletes the key (and node) K_9', transforms the encrypted random value using a minimal set of corresponding private keys (in this case $\mathcal{O}(log_d n)$ keys) such that all remaining members can decrypt r (but M_9 cannot), and sends the multicast message to the remaining group members. (See Section 2.2 for details on the transformation function.) The transformation is used to distribute r because we cannot use the old root key to encrypt r (since M_9 has that key). In the third step, group members decrypt r and compute the new session key, and GC adds a random value r_{GC} to all corresponding private keys. Note that at the end of the leave protocol we have added r and r_{GC} to $GKEK$ and that all the splits of this key add up to the new value.

TASK PROTOCOL FOR MEMBER M_9's LEAVE EVENT
Step 1:

- M_8 generates random value r and it adds it (mod q) to all its remaining private keys; i.e., $K_8 \leftarrow A_r(K_8)$, $K_{78} \leftarrow A_r(K_{789})$, and $K_{1-8} \leftarrow A_r(K_{1-9})$. Compute the new session key: $DEK \leftarrow h(K_{1-8})$.
- $M_8 \longrightarrow GC$: $X = AEnc_{PK_8}(r)$

Step 2:

- GC deletes member node K_9'.
- $GC \longrightarrow M_1...M_7$: $\Gamma_{(K_8', K_{123}')}(X), \Gamma_{(K_8', K_{456}')}(X), \Gamma_{(K_8', K_7')}(X)$

Step 3:

- Members $M_1...M_7$ decrypt r from the message sent by GC and add it (mod q) to all their private keys. They compute the new session key: $DEK \leftarrow h(K_{1-8})$.
- GC chooses a random value r_{GC} and adds it (mod q) to all corresponding private keys.

3   Analysis

In this section we analyze the communication and computation costs for TASK and discuss its advantages over LKH schemes. In Table 1 we provide costs focusing on the number of communication rounds, total number of messages, serial number of expensive modular exponentiations, and the serial number of signature generations and verifications (we use DSA signatures for message authentication). The serial cost is the greatest cost incurred by any member in any given round, and assumes parallelization within each round. The total number of messages is the sum of all messages sent by the members in any given round (unicast and multicast messages counted separately). In addition, we provide the maximum size of any unicast and multicast message sent by the members as well as the key storage costs for executing TASK protocols. We separate costs incurred by members and by GC. We compare the costs of our scheme with those for the LKH scheme of Wong *et al.* [21]. We ignore tree balancing costs, which are computed to be $\mathcal{O}(log_2n)$ for binary trees by Moyer *et al.* [12] and we expect balancing costs of similar complexity for our scheme.

Comparison. From Table 1 we can see that TASK scales to large groups just like LKH. There are only minor differences in both communication and computation costs except for modular exponentiations, which are an artifact of using asymmetric keys. Even there the costs are constant except when GC needs to re-encrypt messages for a member leave event in which case the costs are $\mathcal{O}(log_d n)$ and scale to large groups. (We have not included the average $2(h + 1)$ cost for symmetric encrytions incurred by GC in LKH [21] because these costs are orders of magnitude smaller than exponentiation costs).

Table 1. TASK and LKH Costs

Schemes	Entity	Events	Rounds	Unicast		Multicast		Exponentiations	Signatures	Verifications	Storage
				# Msgs	Msg Size	# Msgs	Msg Size				
TASK	GC	Join	2	1	O(1)	0	N/A	0	1	3	O(n)
		Leave	2	0	N/A	1	O(dh)	$\lceil dh \rceil$	1	1	
	Member	Join	2	3	O(h)	1	O(1)	7	3	1	O(h)
		Leave	2	1	O(1)	0	N/A	3	1	1	
LKH**	GC	Join	1	2	O(h)	1	O(h)	0	3	1	O(n)
		Leave	1	0	N/A	1	O(dh)	0	1	0	
	Member	Join	1	1	O(1)	0	N/A	0	1	2	O(h)
		Leave	1	0	N/A	0	N/A	0	0	1	

** We provide the costs for a group-oriented scheme presented in Wong *et al.* [20]

LEGEND	n: number of current group members
	d: degree of tree
	h: height of tree

Advantages. While providing the efficiency and scalability of LKH, TASK also provides minimization of trust in GC. If GC is compromised, the adversary will get access to all corresponding private keys but will not be able to get any session keys since the corresponding private keys cannot be used to decrypt any session keys. (Note that GC is a *partially trusted* entity because it manages corresponding private key trees, and yet (1) it can get compromised and (2) it cannot discover any session keys). Furthermore, once GC has been re-instated after recovery, the adversary's advantage will be nullified by updating the private and corresponding private keys without having to re-key the entire group. A member will initiate and execute a refresh operation to update private keys and establish a new session key, and GC will update corresponding private keys by adding a random number to them. However, if GC and one group member are simultaneously compromised, the adversary gets access to the current session key, the member's private keys, GC's corresponding private keys, and to the $GKEK$ for that member's group — by adding the compromised member's private and corresponding private keys. In that case the group will have to be re-keyed, but this will not affect any other groups supported by GC.

4   Related Work

A large number of group key management solutions have been proposed and [13] provides a comprehensive survey of these solutions. Recently, several performance optimization and reliable rekeying techniques have been proposed for LKH schemes. Periodic rekeying techniques [16], [22] and tree management based on usage patterns [23] further improve the scalability and efficiency of LKH schemes while reliable rekeying [24], [22] enables secure group communication in environments where reliable multicast communication may not be available. However, all of these schemes fully trust the GC, and we provide an efficient and scalable group key management scheme that minimizes this trust.

5   Conclusion

In this paper we presented a novel group key management scheme, TASK, which minimizes trust in the group controller and yet retains the efficiency and scalability of LKH schemes. We have shown that TASK can be extended to support application driven group merges and partitions, and will present that work in the near future. In the future, we will also apply reliable re-keyeing and performance optimization techniques proposed for LKH to TASK and analyze the results.

Acknowledgements

This work was funded by the Office of Naval Research under contract numbers N00014-03-1-0765 and N00014-04-1-0562. The views and conclusions contained in this document are those of the authors and should not be interpreted as representing the official policies, either expressed or implied, of the Office of Naval Research or the United States Government.

References

1. B. Adamson, C. Bormann, M. Handley, J. Macker, "NACK-Oriented Reliable Multicast Protocol (NORM)", RMT Working Group INTERNET-DRAFT, draft-ietf-rmt-pi-norm-09, January 2004.
2. K. P. Birman, M. Hayden, O. Ozkasap, Z. Xiao, M. Budiu, and Y. Minsky, "Bimodal Multicast", *ACM Transactions on Computer Systems*, Vol. 17, No. 2, pp 41-88, 1999.
3. M. Blaze, G. Bleumer, and M. Strauss, "Divertible protocols and atomic proxy cryptography", in Eurocrypt'98, LNCS 1403, Springer-Verlag, 1998.
4. M. Burmester and Y. Desmedt, "A Secure and Efficient Conference Key Distribution System", (Extended Abstract), EUROCRYPT 1994: 275-286.
5. G. Caronni, M. Waldvogel, D. Sun, and B. Plattner, "Efficient security for large and dynamic groups," in Proceedings of the 7th Workshop Enabling Technologies, Cupertino, CA: IEEE Comp. Soc. Press, 1998.
6. CERT E-crime Watch Survey. Carnegie Mellon Software Engineering Institute, May 2004. *http://www.cert.org/about/ecrime.html*

7. T. E. Gamal, "A Public Key Cryptosystem and a Signature Scheme Based on the Discrete Logarithm", *IEEE Transactions of Information Theory*, 31(4): 469-472, 1985.

8. A. Ivan and Y. Dodis, "Proxy Cryptography Revisited", in Proceedings of the Network and Distributed System Security Symposium (NDSS), February 2003.

9. H. Khurana *et al.*, "Scalable Group Key Management with Partially Trusted Controllers", (full length manuscript), November 2004, available at *http://www.ncsa.uiuc.edu/people/hkhurana*.

10. Y. Kim, A. Perrig and G. Tsudik, "Simple and Fault-Tolerant Key Agreement for Dynamic Collaborative Groups", in Proceedings of 7th ACM Conference on Computer and Communication Security (CCS), 2000.

11. M. Mambo and E. Okamoto, "Proxy Cryptosystems: Delegation of the Power to Decrypt Ciphertexts", *IEICE Transactions on Fundamentals*, vol. E80-A, No. 1, 1997.

12. M. Moyer, J. Rao, and P. Rohatgi, "Maintaining Balanced Key Trees for Secure Multicast", draft-irtf-smug-key-tree-balance-00.txt, IETF Secure Multicast Group, June 1999.

13. S. Rafaeli and D. Hutchison, "A survey of key management for secure group communication", *ACM Computing Surveys*, Vol.35, No.3, September 2003, pp. 309-329.

14. R. V. Renesse, K. P. Birman, M. Hayden, A. Vaysburd, and D, Karr, "Building adaptive systems using ensemble," *Software–Practice and Experience* Vol28, No. 9, pp, 963-979, August 1998.

15. O. Rodeh, K. Birman, and D. Dolev, "The Architecture and Performance of the Security Protocols in the Ensemble Group Communication System", *Journal of ACM Transactions on Information Systems and Security (TISSEC)*, 2001.

16. S. Setia, S. Koussih, S. Jajodia and E. Harder, "Kronos: A Scalable Group ReKeying Approach for Secure Multicast", in Proceedings of the 2000 IEEE Symposium on Security and Privacy, pp. 215-228, 2000.

17. M. Steiner, G. Tsudik and M. Waidner, "Key Agreement in Dynamic Peer Groups", *IEEE Transactions on Parallel and Distributed Systems*, August 2000.

18. D. Steer, L. Strawczynski, W. Diffie, and M. Wiener, "A secure audio teleconference system", in CRYPTO '88, 1988.

19. M. Waldvogel, G. Caronni, D. Sun, N. Weiler, and B. Plattner, "The VersaKey Framework: Versatile Group Key Management",*IEEE Journal on Selected Areas in Communications*,17(9),pp: 1614–1631, September 1999.

20. D. Wallner, E. Harder, and R. Agee, "Key Management for Multicast: Issues and Architectures", Internet-draft, September 1998.

21. C. K. Wong, M. G. Gouda, S. S. Lam, "Secure group communications using key graphs", *IEEE/ACM Transactions on Networking* 8(1): 16-30, 2000.

22. X. B. Zhang, S. S. Lam, D.-Y. Lee, Y. R. Yang, " Protocol design for scalable and reliable group rekeying", *IEEE/ACM Transactions on Networking*, 11(6): 908-922, 2003.

23. S. Zhu, S. Setia, and S. Jajodia, "Performance Optimizations for Group Key Management Schemes for Secure Multicast", in Proc.of the 23rd IEEE International Conference on Distributed Computing Systems (ICDCS 2003), May 2003.

24. S. Zhu, S. Setia, S. Jajodia, "Adding Reliable and Self-healing Key Distribution to the Subset Difference Group Rekeying Method for Secure Multicast", in Proc. of Networked Group Communication Conference, 2003.

H.323 Client-Independent Security Approach

Lubomir Cvrk, Vaclav Zeman, and Dan Komosny

Brno University of Technology, Dept. of Telecommunications, Purkynova 118,
61200 Brno, Czech Republic, phone +420 5 4114 3020
{cvrk, zeman, komosny}@feec.vutbr.cz
http://www.feec.vutbr.cz

Abstract. The security of videoconferencing or telephony data transmissions is a very important problem. There are several approaches how to protect communication in general. Videoconferencing or telephony over IP networks is a specific case of security assurance. There are many H.323 clients in the world, no matter whether software or hardware black box. Many of them are closed systems, which are used a lot but with no security. It could be expensive and in some cases complicated to update these clients to provide security functions. This article deals with a relatively cheap, secure and general enough solution of security provision in H.323 networks without need to modify the current H.323 software.

1 Introduction

Today, if we need to secure videoconferencing or telephony communication (multimedia communication) transported over IP networks, we can do so in several ways.

The first is to use virtual private networks based on the IPsec protocol [3], which works at the kernel level of an operating system. Using virtual private networks (VPN) requires configuring of both sides of communication. The client must configure the VPN gateway's IP address and the VPN gateway must set access for this client, both of them requiring administrator's action.

The second way is to use SSL/TLS [7] based virtual private networks [8]. It creates VPN tunnels over a single TCP or UDP port and is based on a virtual network device. The virtual network device can be viewed as a simple Point-to-Point or Ethernet device, which instead of receiving packets from a physical media, receives them from user space program, and instead of sending packets via physical media it sends them to the user space program.

The third way how to protect multimedia communication is using the H.323 security standards based on H.235 [2] with the use of SRTP protocol [5]. This requires a complete re-compilation of the source codes of the H.323 client programs.

The fourth way of security application to data transmissions is called Opportunistic encryption, which has been invented by John Gilmore in the FreeS/WAN project [6]. This approach is based on the idea of an automatically secure created VPN tunnel between communicating nodes if both nodes support it, otherwise communication proceeds in the ordinary way. For authentication this approach requires DNSSEC [9] in full employment or TXT field in reverse-DNS records.

P. Lorenz and P. Dini (Eds.): ICN 2005, LNCS 3421, pp. 673–680, 2005.

There are many H.323 client programs and black boxes – for example Microsoft NetMeeting (software) or Tandberg 1000 (hardware black box). Lots of them do not provide any encryption functions so the multimedia communication "as is" can be easily sniffed. All possible methods of securing multimedia communication that have been mentioned above do not completely satisfy the requirements on simple implementation of security on multimedia communication. They provide enough security but there are several problems why they do not seem to be a good solution for securing H.323 communication. Why?

1.1 Security of Multimedia Communication

When we talk about the security of multimedia communication we must define the requirements that the implementation of security onto current H.323 networks must satisfy.

Security must be (1) deployed easily, (2) must not need modification of client software, (3) should be applied with the lowest possible administration requirements, (4) communication must be fast enough so the users do not recognize that the data transmission is secured, (5) if one of H.323 clients does not provide security, other peers must be able to communicate with him, (6) approach must be able to implement very fast (must be simple and general enough), (7) approach must be secure enough.

An approach that fits these rules is our goal.

1.2 Application of Security to H.323 Communication Using Standard Approaches

Let us take a look at the possible security mechanisms with respect to these rules.

First the virtual private network has been mentioned, implemented by IPsec protocols. It can not be used because the installation and configuration of IPsec for VPN (in connection with rule #1) is one of its weaknesses. It requires a well informed administrator to set up correctly all the security rules and conditions of IPsec. Another problem of IPsec is its static configuration (see next paragraph).

The second to be mentioned were the SSL/TLS based VPNs. This also involves a problem, because they are statically configured. The configuration is simpler than that of IPsec, but we suppose that the other side of H.323 communication is not known at the time of configuration of SSL VPN on the user machine. If another peer appears (it means somebody else wants just to communicate securely with me), the security must work automatically without need for any reconfiguration on the client's side. The current implementation requires establishing a VPN tunnel first[1].

Securing H.323 communication based on the H.235 ITU-T standard would look like the most suitable solution at first sight, but it is the most complicated task. It would require recompiling all participating client software. Updating the clients by their producers could be (especially in the case of black boxes) expensive and is completely out of anybody's control.

Opportunistic encryption as implemented in FreeS/WAN requires DNSSEC in full production or it requires access to the reverse-DNS records to add a TXT field for

---

[1] IPsec suffers from this problem too because it would require a new tunnel configuration and establishment.

publishing the public key. Nobody knows how long DNSSEC will take to be fully accessible according to the DNSSEC standards being currently redesigned. The FreeS/WAN project has been stopped, because its contributors did not see the goals to be reached. From the point of view of our goals, opportunistic encryption is the most suitable solution but it is still too complicated for common users that do not know anything about DNS system, for example[2].

As we can see there is no completely suitable and simple enough solution for this problem right now[3].

2   Client-Independent Security (CIS)

The task is to find a suitable approach to apply the security mechanisms on H.323 communication based on rules #1 to #7. The most important rule is not to modify the current client software and black boxes. The solution must work in such a way that the clients (and thus also MCU) do not know anything about the security processes over the data they generate. In connection with this the rule #5 is important – the security system must be auto-configurable so it has to be able to switch encryption off if a peer does not support this concept of security. It must work fine in the case of multipoint conferencing with MCU and in the case of peer-to-peer dialogue.

The approach can be divided into three main sections: security, communication, and configuration. All these sections permeate in the process of client-independent security operation.

2.1   Secured Communication Negotiation

Our design is based on the fact that the opposite end of communication may not support this concept of security. That is why at the beginning of the communication the initiator must ask the responder whether it supports CIS. This will be done by the Simple Tunnelling Negotiation Protocol (see 2.1.1). If it does not support CIS, the encryption of data for a remote peer must not start and must be transported with no modification.

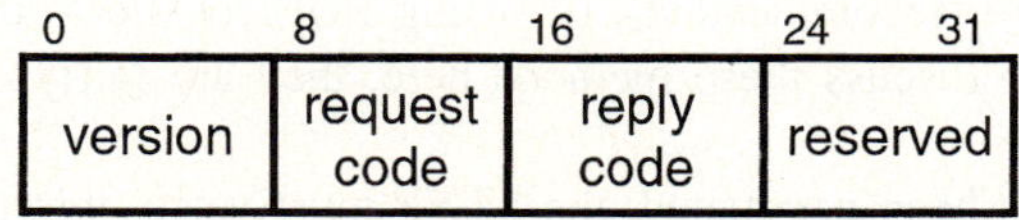

Fig. 1. Simple Tunnelling Negotiation Protocol format

2.1.1   Simple Tunnelling Negotiation Protocol
Simple Tunnelling Negotiation Protocol operates over TCP. Its service listens on port 4077.

The protocol format is shown in fig. 1.

---

[2] Opportunistic encryption is currently implemented only for the Linux operating system.
[3] Except IPv6, but we are living in the "epoch" of IPv4 now.

The field *version* tells the receiver's STNP daemon (negotiation service) the version of STNP protocol of the sender. Right now it is 1.

The field *request code* is set (>0) when the sender needs to request something from the other side. This field can acquire value 1 – which means "Do you support Client independent security?" Value 2 means "Is tunnel established OK?" (see 2.3).

The field *reply code* is set (>0) whenever the sender of the message replies to some previously received request message. The reply code can contain 1 – "Yes, I support. Negotiate encryption." Value 2 means "Tunnel established OK!" Value 255 means "Unrecognized request" (see 2.3).

Opening Connection: STNP messages are sent over TCP port 4077 to the responder. If the responder implements CIS (i.e. also STNP) it must reply with the STNP message, where it sets its version of STNP. The request code must remain as set by the initiator (1) and the reply code is set to 1. Otherwise the connection is closed. If the port is not open or the connection is closed, it means that encryption cannot be set.

Closing Connection: When the sender wants to stop encrypting, he sets the appropriate flag in Simple Tunnelling Encryption Protocol, see 2.2.4.

Packets with other settings must be discarded.

The IP address of the opposite end of communication is given to the negotiation service by the API kernel-to-user space, which communicates with the kernel level packet capturer (see 2.3).

Subsequent Connection Establishment: Let us have two nodes – *A* and *B*. *A* supports CIS and *B* does not. When *A* detects packets for *B* and initiates the STNP session, it is unsuccessful. Communication between them will then be insecure. If *B* later runs CIS, then (in the event of packet directed to *A*) it asks *A* to open the CIS-encrypted communication, which will immediately negotiate and start.

2.1.2 Authentication and Key-Negotiation

Authentication and key-negotiation are the initial part of every secured communication and will be done by the negotiation service based on the ISAKMP protocol [4]. Our approach supports two ways of authentication and key-negotiation (1) using public-key infrastructure (recommended), (2) using PGP certificate database (optional). There is no need to discuss these methods here, they are fairly described and well known.

After the key has been negotiated, the STNP must verify that the tunnel is established OK, (see 2.3).

2.1.3 STNP Service (Daemon)

The STNP daemon must have a user interface available that is to inform a user about the IP addresses that communicate securely and about the addresses where the key-negotiation is running. From this interface the daemon must be able to shut down the CIS. This interface also interacts with the user in the matter of peer's certificate acceptance.

2.2 Encryption

The process of encryption is the main task to be done. Our approach requires the use of stream ciphers operating in the counter mode (CTR [1]) because it does not need two different algorithms for encryption and decryption and its use and implementation are easy [1]. Block cipher modes are not allowed. Because of simplicity, we require just the support of AES-128 in the CTR mode, others are not allowed. The HMAC-SHA-1 hash function must be supported.

The encryption must be done using stream ciphers because the system will encrypt captured IP packets payload and add some overhead data. What is the overhead? First, each packet must be authenticated before it is encrypted. Second, the encryption must count on the independence of each IP packet with the possibility of some of them getting lost. That is why it is necessary to design a very simple protocol solving this problem. In general it is the problem of initialization vector (IV) [1].

2.2.1 Initialization Vector Computations

The encryption (and decryption) of each packet must start with the same IV to generate a repeatable key stream on both sides of the communication. We use the IV generation method introduced in [1] and called Nonce[4]-generated IV.

Our nonce value is based on the message number and sender's IP address. The counter of messages starts at 0, which is the first encrypted message transported between nodes and is iteratively incremented by 1. Its storage space is 48 bits. The nonce computation is performed by concatenating the IP address and the message counter.

The nonce is set by the following formula[5].

$$IP_{reverse} := ip[1] \parallel ip[2] \parallel ip[3] \parallel ip[4]$$
$$IP_{common} := ip[4] \parallel ip[3] \parallel ip[2] \parallel ip[1]$$
$$IP_{mixed} := ip[3] \parallel ip[1]$$
$$Nonce := counter \parallel IP_{reverse} \parallel IP_{mixed} \parallel IP_{common}$$

The nonce is constructed to be 128 bits long. It is encrypted by AES-128 to generate the first segment of the key-stream.

2.2.2 Encryption Process

Key-stream segments are then computed as follows:

$$S_0 = E(k, Nonce);$$
$$S_i = E(k, Nonce + i \bmod 2^{128}); i = 1,..,m$$

S_i is the i-th key-stream segment; E is the AES-128 encryption algorithm.

Each of the key-stream segments is then XORed with a 128bit (or shorter) plaintext or ciphertext block.

---

[4] Nonce means "**n**umber used **once**" [1].

[5] ip[x] means xth byte of IP address memory storage bytes. The left-most byte (nearest to :=) represents the highest bits, e.g. IP_{common} 's ip[4] are the 25th to the 32nd bits; ip[1] are the 8th to the 1st bits.

After the message counter hits 2^{48} it must be reset to 0 and a new key negotiation process must begin because a nonce can not be used more than once with a single key [1].

2.2.3 Message Authentication

Before encryption, each IP packet's payload must be authenticated using the 128 bit HMAC-SHA-1 [10]. Authentication is computed for the *total length, protocol, source IP address, destination IP address* fields from the IP protocol header [12] plus six bytes long *message number* and one byte long *command* fields in the STEP protocol (2.2.4) plus *IP payload*.

2.2.4 Simple Tunnelling Encryption Protocol

This protocol is used for the modification of captured IP packets that have to be encrypted with respect to the authentication of each packet and the possibility of packets being lost. STEP modifies the IP packet according to this scheme:

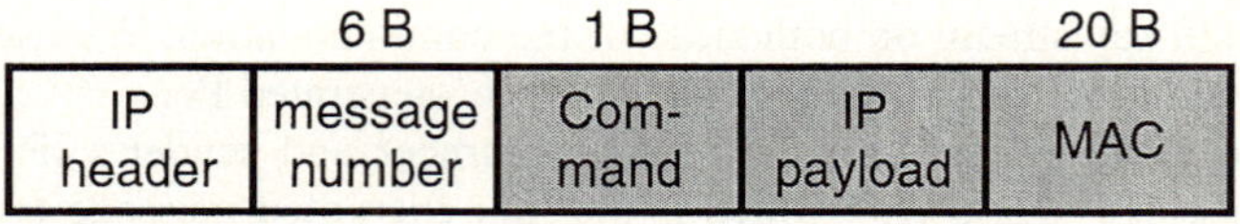

	6 B	1 B		20 B
IP header	message number	Com- mand	IP payload	MAC

Fig. 2. Simple Tunnelling Encryption Protocol in the context of IP protocol. Grey color means later encrypted fields

When the IP packet has been processed, STEP encapsulates the IP payload between its overhead fields. The process computes MAC and adds it after the IP payload. Then it runs the encryption function and encrypts the string made of these three fields (command, IP payload, MAC).

The *message number* contains a 48bit value representing the number of the message being sent to one destination (determined by the destination IP address).

The *command* field controls the communication process. It is used to stop the encryption. Whenever the source node needs to stop the CIS encryption (CIS is shutting down), it sets the *command* value to 1. After the destination decrypts and authenticates that IP packet and recognizes the command to be 1, it stops packet processing from that source IP address. All subsequent packets are expected to be common IP packets so they are passed to the upper layers unmodified.

All IP packets encapsulated by STEP have in the *protocol* field of IP header the value 99 (IANA's "any private encryption scheme" [11]).

2.3 Communication Basis and Packet Processing

The core of CIS approach is the IP packet capture at the operating system kernel level. Windows implementation needs an NDIS intermediate driver that executes the processing. The Linux operating system offers several kernel hooks for packet capturing. The driver itself is operating system dependent so the implementation requires an internal API which eliminates the platform-dependent amount of source code needed for packet processing implementation.

The whole system works in the following steps:

1. The CIS daemon and kernel driver are started and the driver captures the incoming and outgoing IP packets.
2. In the event of the first outgoing packet to destination IP address A, the driver tells the STNP daemon the destination IP address.
3. The STNP daemon checks whether the destination supports CIS. If it does, then it negotiates the keys. While checking and negotiating, the packets with destination address A pass the kernel driver unencrypted. After the STNP session is finished the daemon tells the driver the negotiated key associated with the IP address.
4. Once the driver has a key associated with the IP address it starts the encryption – the tunnel appears to be established.
5. In the event of the first packet with protocol number 99 from the opposite side of the previous key-negotiating communication the driver switches to the decryption/encryption state and encrypts and decrypts all traffic between the two nodes (authentication within).
6. For tunnel verification the STNP daemon on the initiators side sends the STNP request "Is tunnel established OK?" and waits for the reply, which must be "Tunnel established OK!" since both messages will be encrypted and decrypted only if all is OK. If the daemon does not receive the STNP message "Tunnel established OK!" within 10 seconds, it tells the driver to switch off the encryption into IP address A and resets the protocol state.

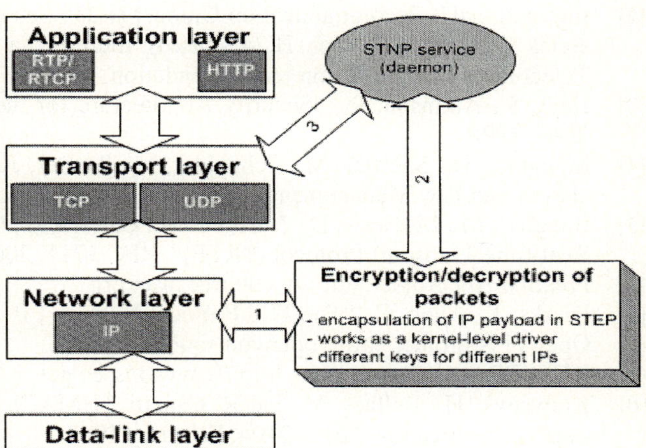

Fig. 3. Schema of Client-independent security. The whole traffic between the nodes implementing this system is secured. The numbers inside the arrows relate to the numbers in the paragraph above

For clarity, Fig. 3 shows the entire system simplified.

3   Conclusion

This work concerns an application of security to H.323 data transmissions. It defines the conditions for a simple application of security and confronts them with available security solutions like IPsec, SSL/TLS, H.235 based security, etc. None of the current security approaches fits completely the conditions and therefore a new approach is introduced.

The approach is based on packet capture and its encryption with authentication. For a correct functionality the approach defines two communication protocols which implement the mechanisms of secured tunnel establishment, data encryption and encryption process control. The security system is application independent, so no H.323 client software recompilation is needed for security deployment. The deployment is very easy and does not need any configuration. It just needs installation on a machine and a public key cryptography certificate. The tunnel establishment and encryption process are controlled automatically.

It works in networks with MCU, in peer-to-peer and works with network address translation (NAT) if the NAT box runs the CIS system. It does not work with multicast yet but it is a topic for further study.

References

[1]   Fergusson, N., Schneier, B., *Practical Cryptography*, Wiley Publishing, Inc., Indianopolis USA, 2003

[2]   International Telecommunication Union, "H.235 Security and encryption for H-series (H.323 and other H.245-based) multimedia terminals", International Telecommunication Union recommendation, 2003

[3]   Kent, S., Atkinson, R., "Security Architecture for the Internet Protocol", RFC 2401, 1998

[4]   Maughan, D., Shertler, M., Schneider, M., Turner, J., "Internet Security Association and Key Management Protocol (ISAKMP)", RFC 2408, 1998

[5]   Baugher, M., McGrew, D., Naslund, M., Carrara, E., Norrman, K., "The Secure Real-time Transport Protocol (SRTP)", RFC 3711, 2004

[6]   FreeS/WAN project, http://www.freeswan.org

[7]   Dierks, T., Allen, C., "The TLS Protocol Version 1.0", RFC 2246, 1999

[8]   OpenVPN project, http://openvpn.sourceforge.net

[9]   DNSSEC project web page, http://www.dnssec.net

[10] Krawczyk, H., Bellare, M. and R. Canetti, "HMAC: Keyed- Hashing for Message Authentication", RFC 2104, February 1997

[11] Internet Assigned Numbers Authority, http://www.iana.org/assignments/protocol-numbers

[12] Information Sciences Institute at University of Southern California, "Internet protocol", RFC 791, 1981

Architecture of Distributed Network Processors: Specifics of Application in Information Security Systems

V.S. Zaborovskii[1], Y.A. Shemanin[2], and A. Rudskoy[1]

[1] St. Petersburg Politechnic University,
[2] Robotics Institute, St. Petersburg,
{vlad, yuri}@neva.ru

1 Introduction

Modern telematic networks or Internet are distributed hierarchical systems consisting of basic components: nodes and communication lines. Telematic network nodes are computers with network interfaces employed for data exchange. A node with several network interfaces is called the router or network processor (NP). Each NP interface is provided by one or several identifiers called addresses. There are several types of addresses: physical or MAC, network or IP, application or Port Number. The set of network addresses form specific spaces with its topology and metric. Topology is the measure of nearness in the network. The metric is defined by communication line. If the number of addresses that connected by the line is more than two, the communication line is termed broadcasting. The number of communication lines determines the distance between nodes. The distance between the nodes without network addresses is undefined. By combining nodes into a telematic network, one can provide information exchange among computer applications, which are executed at the network nodes. Information exchange is based on forwarding and receiving network packets. A packet is a specific logical sequential/recursive structure, which is formed at network nodes to execute information exchange. The sequential part of this structure consists of two, header and payload, fields. The recursiveness of a packet stems from the fact that the payload itself may be another packet with its specific structure and addresses (Fig.1). A packet originating from an application running on a node and destined to node in different network, arrives at an NP and is forwarded by it to the appropriate network on the basis of destination addresses in the packet's header.

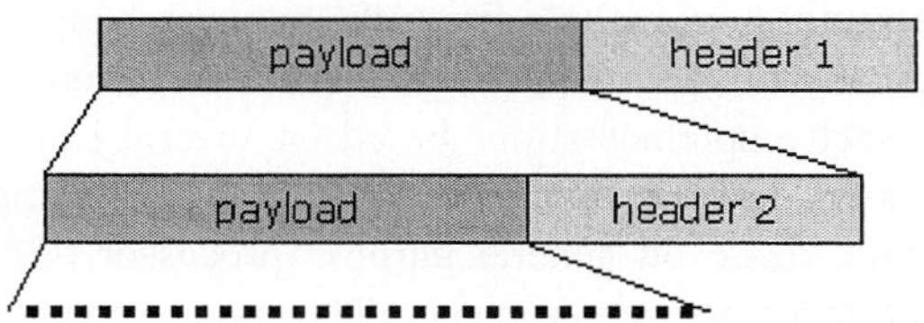

Fig. 1.

The lines in telematic network are used for bits transmission only. No data processing is executed in a communication line itself. The processing culminates in

P. Lorenz and P. Dini (Eds.): ICN 2005, LNCS 3421, pp. 681–688, 2005.

selection of the NP network interface from which the packet will be sent into the network. Should the processing produce a decision not to send the packet into the network, it is assumed that the packet has reached the required network node, or it will be dropped. Thus, the basic functionality of an NP or router is determined by two sequential processing stages of packets after their arrival from a communication line, namely, **store-and-forward**. However, with extensively growing in size and shifting into more and more sophisticated applications, NP become more complex and incorporated new functionality. The hundreds of scientific papers are being published proposing changes to existing NP architecture or introducing new communication mechanisms. In practice only very few modifications to the current Internet are deployed. One reason is that most improvements require that current routers have to be replaced.

We are considering here a new approach to selecting the NP architecture, by which extension of functional demands on the various packet processing stages in particular, those involved in addressing information security issues, is executed by distributing the procedures of their execution among different network devices. One of the key issues in this approach is that the devices logically belong to one NP. The specific feature of this distribution lies in that it does not interfere with the existing address connections or routing policy among the network nodes. This means that new devices that do not change network address space supplement expensive routing equipment that is already in place. Using a special functioning mode called the stealth mode attains the address invariance of the transformation of NP under the extension of their functionality.

2   Trends in Telematics Systems Progress

As the data transfer rate over communication lines increases and the protocol spectrum broadens, we are witnessing a growth in demands on the performance of the NP employed in packet handling at network nodes. The architecture and specific features of operation of such processing engines has become a subject of a large number of studies [1--3]. Rather than drawing on a systematic analysis of the various specific requirements and design alternatives, however, most of these studies invoked the well-known results of application of multi-processor architectures to increasing the speed of data flow processing. The solutions proposed to improve the functionality of the router now include firewalls, network address translators, means for implementing quality-of-service (QoS) guarantees to different packets flows and other mechanisms. Such implementations based on several primary operations with packets: parse, search, resolve, and modify (Fig.2). To implemented all this operations in real-time mode on general purpose processor (GPP) often becoming unfeasible due to performance requirements. This issue motivates solutions where packet-processing functionality of NP is implemented in specific pooled and pipeline hardware. Such a decision has restricted flexibility. Complex nature of packets operations favor software based implementations on GPP. To address these conflicting issues and organizing the stages in packet processing, recently a new **store--process-and-forward** scenario has been proposed.

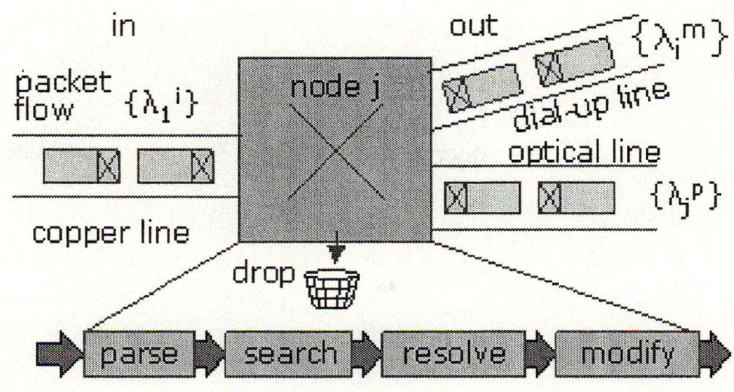

Fig. 2.

In a general case, all solutions may be separated in two classes. Grouped in the first class are the solutions aimed at boosting the pure router performance. The main parameters governing the router operation are the packet destination addresses, and, therefore, the solutions chosen are directed at accelerating data search in the router lookup tables. The second class of solutions involves implementation of various procedures without routing decisions: packet classification, data processing, providing the required QoS, bandwidth allocation, and so on. In principle, this separation of the handling processes permits one to break up the integrated performance of a NP into components that can be distributed among the individual processes.

So, if a packet operation among such components occurs without the use of routing decision, they can be functionally assigned to communication lines. This approach modifies the basic network scenario from **store-and-forward** to **process—store-and-forward**. This scenario offer a solution to providing necessary flexibility in telematic network by keeping basic routing operation, adding new functionality without changing network topology and redistributing computation power between all components of network.

3   Information Security Issues

The principle underlying modern computer-based telecommunications is packet switching (Fig.3). In practice, this principle actually uses the open-system interaction (OSI) model to provide several control levels. At each level special data structures or packets are controlled by specific rules. The corresponding control processes can be broken down into the following stages: (1) collection of a data to be transmitted through the network; (2) configuring a structure to quantitatively determine the volume of the data to be transmitted; (3) attaching to the data a special header specifying the set of parameters to be used in handling the packet in network nodes; (4) formation of a frame meeting the requirements of the communication line hardware; and (5) frame transmission over the communication line connecting two network nodes.

Packets are transmitted over network nodes of several types, more specifically, generation nodes, nodes handling packet headers only, and nodes to process both

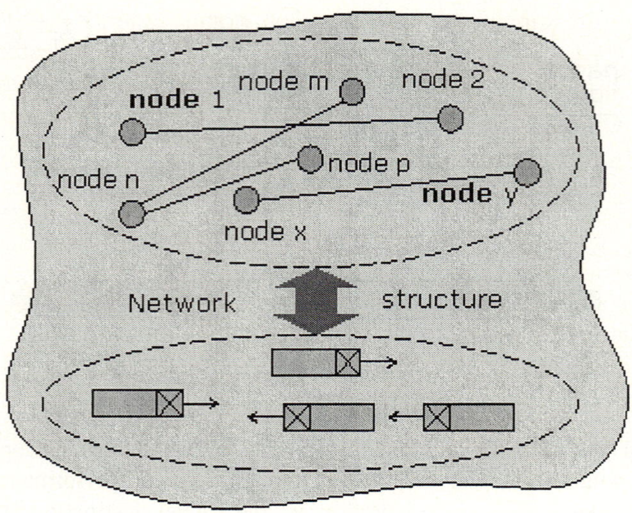

Fig. 3.

headers and data. The routing or selection of the interface where the packet is forwarded after processing is a process of a local character, i.e., it is executed on each network node through which a packet passes. Routing is based on the packet destination node address which is specified in the corresponding header field and on the lookup table relating the network node addresses to the router interface numbers.

The above process is prone to various malicious actions which are capable of interfering with the standard procedure of packet transmission or of substituting packets on the way from their generation to reception.

One can conceive of the following basic protection measures: (1) designing a special packet path through the network nodes which support processing rules denying transmission of packets with preset addresses and header parameters; (2) executing the tunneling mode, in which the packet to be protected is transmitted in the data field of another network packet; and (3) using special packet transmission modes in which the header parameters are protected cryptographic algorithms.

All these protective measures can be implemented by several means, which can be divided into methods of packet filtration, and of cryptographic data processing. The first group of methods protects the network address space by means of special NP called firewall network processors (FNP) [3]. In common configuration FNP does not becomes an end point of packet transmission and has to be installed in the network segments crossed by packet flows. These segments are customarily placed between the protected network and the interface of the router connected to this network. To keep the basic functionality of telematic network we need to have routing policy to be invariant to the place where FNP has been placed. It is possible if network metrics does not changes by FNP due to filtering interfaces have no physical and network addresses. Protective measures of the second type require designing special network gateways supporting the tunneling mode, with packet encoding being optional in this case. If such gateways are provided by routing functions, one of the promising network protocols may in this case be IPSec. This protocol permits different implementation possibilities one of which is based on approach that separate routing and cryptographic tasks between different processes which formed specific processor network connected by communication lines (Fig.4).

While telecommunication industry featuring an excess throughput of physical lines, experiences nevertheless an ever-increasing demand for efficient packet processing methods. These demands have stimulated a broad spectrum of studies dealing with development of special NP for use in the network security systems. Development of such NPs should be carried out taking into account the trends predicting the growth of throughput of communication lines based on optical media and wave division multiplexing technologies. General solutions to the problem of boosting NP performance may be found in network technologies or by mean of spreading out needed power between different nodes. The well known possible means can be judiciously divided into the following groups: development of NPs based on parallel processors with a shared RAM; development of pipeline NPs with RAM resources distributed between different processing phases; hybrid network specific architectures, in which the stages of sequential and parallel processing are matched to the number of independent data flows.

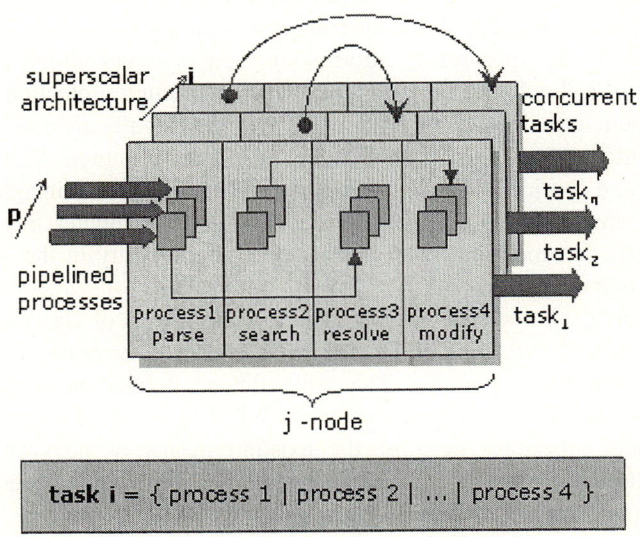

Fig. 4.

The efficiency of such solutions is fully determined by the specific algorithmic features of the problems to be solved and the way the relevant data are supplied. In the case of packet processing in network security systems, the factors of particular significance are: the parallel character of the flow in time space, in which the number of simultaneously processed connections depends on that of nodes with different network addresses, and the sequential character of packet transmission in network address space. Because the transmission of packets is executed in an asynchronous mode, i.e., it is initiated independently by each node, the number of logical connections passing through the routers is a random quantity obeying a fractal distribution function [4]. The packet switching processes having a complex character, the nominal number of parallel processors in the architecture of an NP does not determine fully its performance, so that the optimum number of pipeline processing stages depends on the actual character of the problem awaiting solution and, thus, can vary. All these factors stimulate a search for new approaches to a better organization of network packet processing.

4   Distributed NP Architecture

Development of NPs for security systems can be based on separation of packet processing functions into base and additional operations. Among the base operations is packet routing, and to the additional ones one could assign the other packet operations connected with extension of the NP functionality, for instance, packet filtration. The proposed separation permits one to consider a network node as a part of a special packet processing network. The connection topology of the processing devices should be such that packet transmission among them does not involve the addresses of the nodes included into the routing lookup table.

Application of this approach to information security issues allows the use of the network control technologies based on the "security through protection of protection devices" system principle. This principle places the significance of the two key aspects of information security underlying standard Common Criteria, namely, functionality and confidence, on equal footing. Adhering to this principle implies that the devices employed to protect information in a computer network should incorporate efficient mechanisms to ensure their own security in the stages of both development and operation itself.

To reach this goal, one should undertake at several levels of OSI model measures which would make localization of the protection devices in the network address space by remote monitoring impossible. This concealment of functioning gives rise to a modification of the protection model using the NP without addressing interfaces (NP in stealth mode), because most of the existing means of network attacks and destructive interference are based on remote neutralization of the devices employed to protect information resources in a network.

Development of protective devices in the stealth mode with the use of distributed NPs becomes possible because such devices do not act in most of their operational regimes as sources or destinations of network packets.

Therefore, network interfaces of these devices may have no physical or logical addresses altogether and, hence, transmission of IP packets or MAC frames through them becomes similar in character to their passing through a HUB or cable line segments used in packets exchange. To operate successfully, the NP should work like

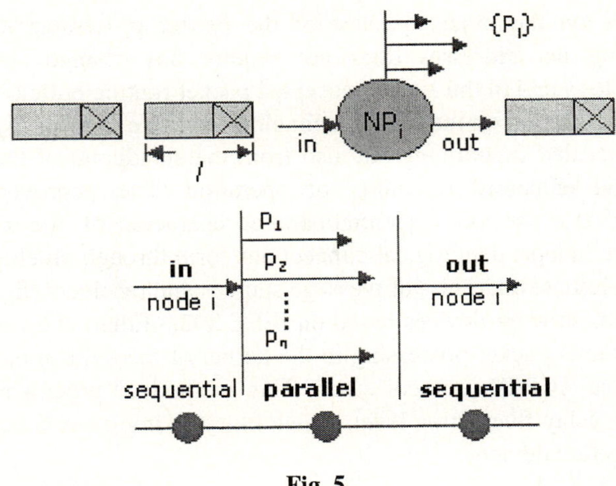

Fig. 5.

a "sophisticated" parallel bundle of network cables or a transparent but secure logical channel between the network nodes (Fig 5).

The next step of decomposition based on sequential-parallel-sequential stages in packet-handling processes. This offers a possibility to cut packet delays in the packet reception and processing mode. Operation sequence in the second mode can be integrating into a specialized pipeline cluster and spread out between its nodes (Fig.6).

In this scheme, the NP is able to either "bridge" or "route" traffic. In the first case, the NP functions as a layer-2 network bridge with IP transparent or "stealth" interfaces. This means that each interface has a MAC address but network or IP address space is the same on both sides of the firewall.

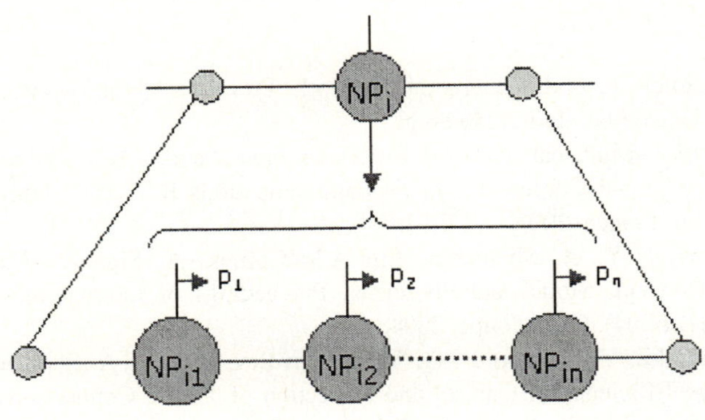

Fig. 6.

This method of concealing the network address of information protection devices, on the one hand, provides conditions necessary for execution of the protection functions, while on the other, because of the packet processing device network interfaces having no addresses does not require any changes in the network connection topology and in the already accepted packet routing policy.

Security devices based on the stealth technologies have a number of assets not only due to their concealed functioning but also from the standpoint of the scalability of performance and enhanced reliability of operation. The improved performance originates from the use of sequential/parallel character of the network traffic employed, where independent logical connections form through pipeline transmission of packets with definite addresses of message sources and receivers (fig.6).

Operation with network devices based on IEEE 802.3 Ethernet technologies in the stealth mode permits packet processing in the kernel of the built-in operating system without using the TCP/IP protocol stack. This method of processing reduces the packet buffering delay fluctuation level, which likewise improves concealment of the location of protection devices.

5   Conclusion

Application of network processors with a distributed architecture broadens substantially the range of use of information protection systems in telematic networks. The concealed character of operation of the protection devices offers a possibility of integrating additional packet processing procedures into the standard switching process while not changing in any way the routing policy. Application of the stealth technology cuts the costs of network upgrading, because its implementation permits redistribution of the required processing power among various network devices. The NP clusterization technologies provide a possibility of scaling up the performance of network nodes and increase overall system reliability.

References

1. Intel Corp. Intel Second Generation Network Processor, http://www.intel.com/design/network/products/npfamily/ixp2400.htm
2. V.S. Zaborovsky «Multiscale Network Processes: Fractal and p-Adic analysis», Proceedings of 10-th International Conference on telecommunications ICT`2003, University of Haute Alsace, Colmar, France, 2003.
3. V.S. Zaborovsky, Y. A. Shemanin, Jim A.McCombs, A. Sigalov «Firewall Network Processors: Concept, Model and Platform», Proceedings of International Conference on Networking (ICN'04), Guadeloupe, 2004.
4. N. O. Vil'chevskii, V. S. Zaborovsky, V. E. Klavdiev, and Yu. A. Shemanin, Methods of Evaluating the Efficiency of Control and Protection of Traffic Connections in High-Speed Computer Networks, Proc. Conf. "Mathematics and the Security of Information Technologies" (MaBIT-03), Lomonosov MSU, October 23--24, 2003.

Active Host Information-Based
Abnormal IP Address Detection

Gaeil Ahn and Kiyoung Kim

Security Gateway Research Team,
Electronics and Telecommunications Research Institute (ETRI),
161 Gajeong-dong, Yuseong-gu, Daejeon, 305-350, Korea
{fogone, kykim}@etri.re.kr

Abstract. In this paper, we propose an abnormal IP address detection scheme, which is capable of detecting IP spoofing and network scan attacks. Our scheme learns active host information such as incoming interface number, whether or not working as Web server, whether or not working as DNS server, and etc., by collecting and verifying flow information on networks. By using active host information learned, we can check if IP address is normal or abnormal. Through simulation, the performance of the proposed scheme is evaluated. The simulation results show that our scheme is able to detect source IP spoofing attacks that forge using the IP address of subnet that attacker belongs to as well as using the external IP address. And also, they show that our scheme is able to detect network scan attacks with low false alarm rate.

1   Introduction

One of the most common forms of network intrusion is network scan and most of network attacks start with fake source IP [1].

Network scan is used to know the configuration of a victim networks [2]. Attacker is interested in identifying active hosts and application services that run on those hosts. For example, worm virus, such as Limda and Slammer, scans network to find victim systems with weak point. To detect network scan attack, the existing several schemes observe whether TCP connection request succeeds or fails [3][4]. Those schemes rely on that only a small fraction of addresses generated randomly by scanner are likely to respond to a connection request. So, if the fail count/rate of the connection requests initiated by a source is very high, then the source is regarded as network scanner. However, those approaches may generate false alarms because the connection requests initiated by normal user may fail by reason of network congestion, destination system failure, and etc.

IP spoofing is used to hide attacker's identity by sending IP packets with forged source address [5]. IP spoofing is commonly found in DoS (Denial of Service) attack [6] that generate a huge volume of traffic for the purpose of assailing a victim system or networks. Current Internet architecture does not specify any method for validating the authenticity of the packet's source. So attacker can forge source address to any he/she desires. Currently, uRPF (unicast Routing Path Forwarding) [7] is commonly

P. Lorenz and P. Dini (Eds.): ICN 2005, LNCS 3421, pp. 689–698, 2005.

used to detect attack packets with fake source IP address. However, uRPF has a limitation in case that attacker's source IP address is spoofed to that of hosts on the same subnet. This is because it uses routing table that has location information not by a host, but by a group of hosts.

In this paper, we propose an active host information-based IP Address Inspection scheme, which is capable of detecting IP spoofing and network scan attacks by checking if IP address is normal or abnormal. Our scheme learns active host information such as incoming interface number, whether or not working as Web server, whether or not working as DNS server, and etc., by collecting and verifying flow information on networks. When a packet arrives, its source and destination IP address is inspected using active host information learned. If the result of IP address inspection indicates that the source IP is abnormal, it is thought of as source IP spoofing attack. If it indicates that the destination IP is abnormal, it is thought of as network scan attack. If such attack count exceeds a threshold, then we regard it as real attack and block it.

The rest of this paper is organized as follows. Section 2 overviews IP spoofing and network scan attacks and shortly describes the other works that tackle them. Our scheme is described in details in section 3 and its performance is evaluated in section 4. Finally, conclusion is given in section 5.

2   IP Spoofing and Network Scan Attacks

The typical examples of attacks using abnormal IP address are network scan and IP spoofing attacks.

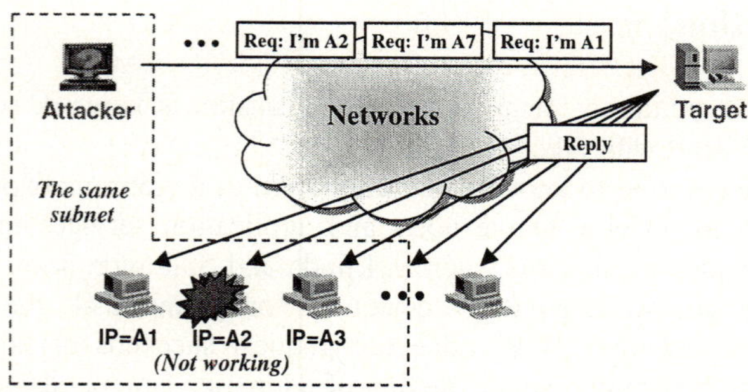

Fig. 1. Example of IP spoofing attack

Fig. 1 shows an example of IP spoofing attacks. In Fig. 1, attacker use A1, A2, A3, etc. as source IP address to hide his/her identity.

To detect IP spoofing attacks, there have been proposed a few approaches: ingress filtering [8], uRPF [7], packet marking [9] and ICMP Traceback [10]. Ingress filtering employs the scheme that boundary router filters all traffic coming from the customer that has a source address of something other than the addresses that have been assigned to the customer. uRPF discards faked packet by accepting only packet from interface if and only if the forwarding table entry for the source IP address matches the ingress interface. Packet marking scheme is one that probabilistically marks packets with partial path information as they arrive at routers in order that the victim may reconstruct the entire path. ICMP Traceback scheme uses a new ICMP message, emitted randomly by routers along the transmission path of packet and sent to the destination. The destination can determine the traffic source and path by using the ICMP messages.

The existing schemes have a difficulty in detecting IP spoofing attacks if attacker generate fake source IP address using address of the subnet that the attacker belongs to. However, our scheme can detect IP spoofing attacks by checking if attacker uses an IP address of host that is not working as a source IP address. In Fig. 1, the host with IP address, A2, is not working. So we can know that A2 is fake IP address.

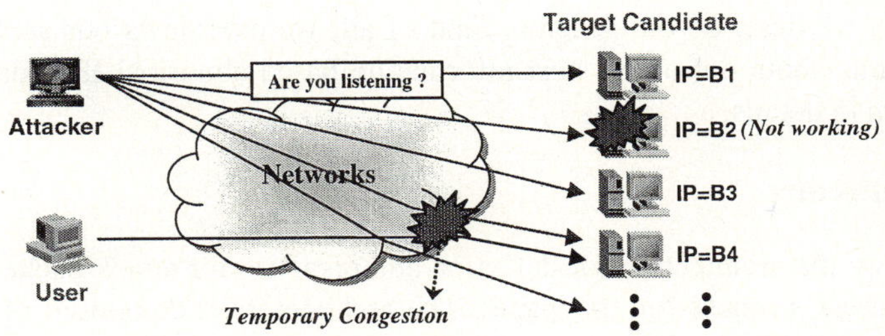

Fig. 2. Example of network scan attack

Fig. 2 shows an example of network scan attack. Network scan attack is used to know the configuration of a victim networks. Attacker is interested in identifying active hosts and application services that run on those hosts. Network scan can be classified by two kinds of attacks according to attacker's behavior: host scan attack and port scan attack. Host scan attack scans a range of network address in order to find out which addresses represent an active host. On the other hand, port scan attack scans a range of TCP or UPD ports on a host in order to identify which network services are active on that host.

There have been proposed many schemes to detect network scan. [1] employs a method to count the number of events of interest that occur within a given time period. For example, if the number of accesses to hosts in the last 5 minutes exceeds 100 then it is regarded as the behavior of IP scan attack. [11] employs data mining techniques to learn the signature of an attack that executes network scan.

Simple form of network scan attacks is to access hosts/posts sequentially in a short period of time. This kind of attacks can be easily detected. But, more sophisticated form of network scan attacks can be used to hide attacker's intentions using a range of measures, such as randomizing the order of their accesses or scanning over a longer period of time. So it's not easy to detect network scan attacks.

To such a sophisticated form of network scan attack, several schemes have been proposed. Those schemes commonly assume that lots of connection requests initiated by scanner will fail because scanner uses randomly generated IP address as destination address [3][4]. However, this approach may generate false alarms because connection requests initiated by normal user may fail by reason of network congestion, destination system failure, and etc. For example, in Fig. 2, when a temporary congestion occurs on networks, both normal user and attacker cannot access the host, B4. In that case, it may results in a problem that normal user falls under suspicion.

In this paper, we propose simple but accurate network scan detection scheme. Our scheme detects network scan attack by checking if attacker uses IP address of host that is not working as the destination address. In Fig. 2, the host with IP address, B2, is not working. So we can know that attacker executes network scan attack.

3   Abnormal IP Address Detection

From now, we describe our scheme. First of all, we overviews our architecture, and then explains about our active host information based abnormal IP address detection mechanism in details.

3.1   Architecture

Fig. 3 shows the architecture for detection and response for attack packet with abnormal IP address, proposed in this paper. The architecture is composed of five components: Host-information-learning, IP-address-inspection, Response-decision, Response-Enforcement, and Aging-check.

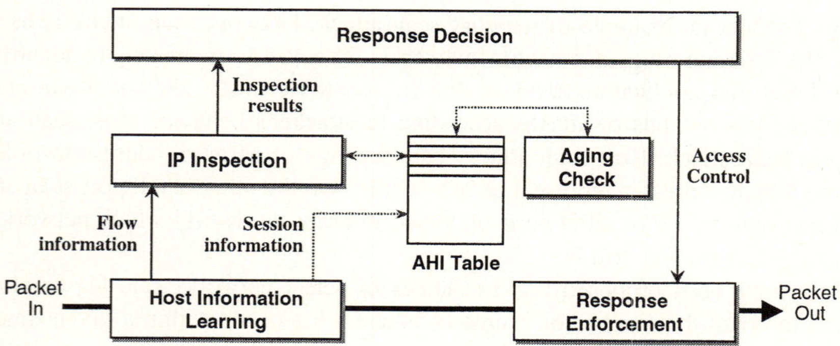

Fig. 3. Architecture for detection and response of attack packet with abnormal IP address

Host-Information-Learning extracts flow information of the incoming packet and checks if a new session is established. And then, it passes the flow information to IP-Inspection component and if a new session is established then it stores the session information in AHI (Active Host Information) table. IP-Address-Inspection component inspects the source and destination IP address of the incoming flow by comparing it with active host information registered in AHI table. The inspection result can be normal IP or abnormal IP. The IP inspection results are analyzed by Response-Decision component. If Response-Decision decides a host to be attacker or victim, then it creates ACL (Access Control List) and applies it to Response-Enforcement component for the purpose of defeating/mitigating attacks. AHI table learned has location information and application service information of active host. To make AHI table information fresh, Aging-Check component deletes old AHI table entries that have not been re-registered by Host-Information-Learning for a given time period.

3.2 Learning of Active Host Information

In this paper, a flow is identified by source IP, source port, destination IP, destination port, protocol, incoming interface, and outgoing interface. Learning of active host information is done whenever a new full or semi-session is established.

In TCP protocol, source and destination hosts execute three-way handshaking (SYN packet generated by source host, ACK packet generated by destination host, and ACK packet generated by source host) to establish a session between both hosts. By checking the three-way handshaking, we can know whether or not a session is established. In case of UDP protocol, source and destination hosts do not execute three-way handshaking. But, we can know if there is a semi-session between two hosts by checking if there is a bi-directional communication between both. For example, a DNS service is started by request of DNS client and finished by reply of DNS server. In case of semi-session (i.e. UDP protocol) information, we trust only its reply part. Its request part is not reflected in AHI table.

The session information is stored in AHI (Active Host Information) table. AHI table has attributes such as IP Address, incoming interface number, and running application services as follows;

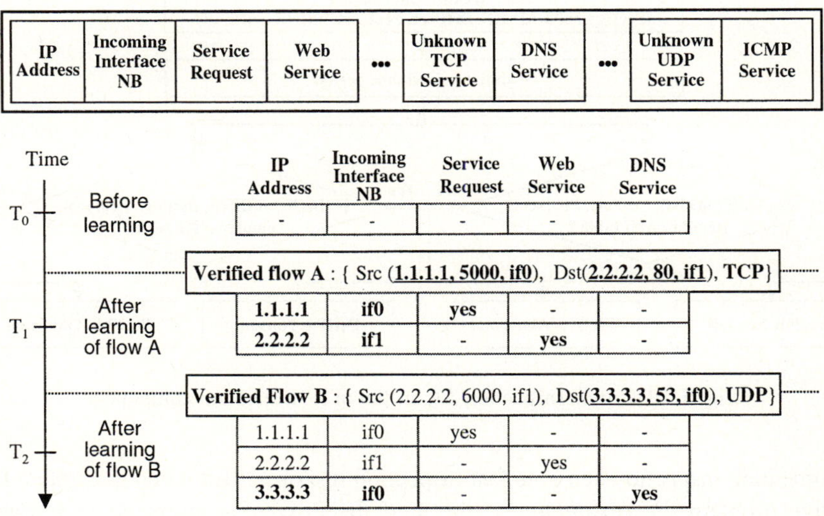

Fig. 4. Example of active host information learning

Fig. 4 shows an example of active host information learning. If flow-A using TCP is verified (i.e. new session is established), we creates two AHI table entries and set

IP address, incoming interface number, and service request fields of one entry to 1.1.1.1, if0, and YES, respectively, and also set IP address, incoming interface number, and Web server fields of the other entry to 2.2.2.2, if1, and YES, respectively. If flow-B using UDP is verified (i.e. new semi-session is established), we creates a AHI table entry and set IP address, incoming interface number, and DNS service fields to 3.3.3.3, if0, and YES, respectively. The source flow of flow-B is not registered in AHI table.

Through active host information learning shown in Fig. 4, we can now know the following information about hosts:

1. Host with IP, 1.1.1.1 is qualified to request a service and located on interface number, 0.
2. Host with IP, 2.2.2.2 runs Web server and is located on interface number, 1.
3. Host with IP, 3.3.3.3 runs DNS server and is located on interface number, 0.

3.3 Attack Detection and Response

Fig. 5 show IP address inspection algorithm, which is executed in IP-Inspection component. The source/destination IP address of a flow is normal if and only if its IP address and interface number is found in AHI table and the application corresponding to its port number is YES. IP inspection can be executed as the following two modes:

– Immediate inspection -- executes the inspection as soon as it receives a flow.
– Delayed inspection -- delays the inspection for a given time period.

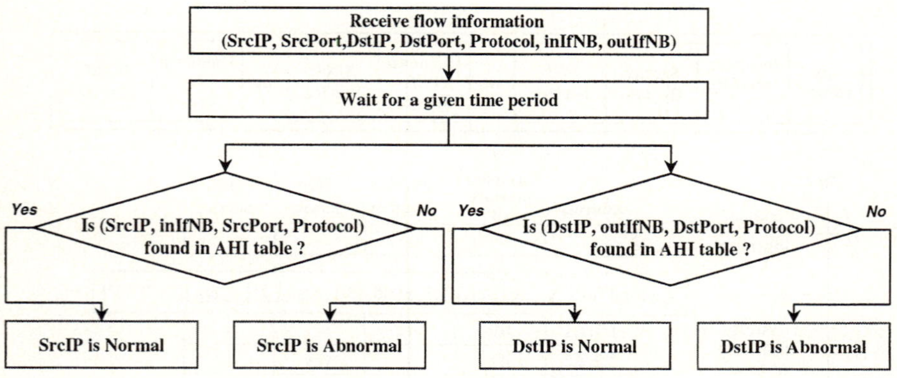

Fig. 5. IP address Inspection Algorithm of IP-Inspection component

Immediate inspection can give fast inspection response. But it can generate a false-positive problem. For example, in Fig. 4, if the flow-A is inspected in immediate inspection mode (e.g., at time T_0), its source and destination IP is all abnormal. But, if it is inspected in delayed inspection mode (e.g., at time T_1), its source and destination IP is all normal. For such reason, we use delayed inspection mode. The delay time can be determined by considering packet RTT (Round Trip Time).

Our response strategy for defeating attacks using abnormal IP address is based on policy that consists of condition and action parts. Our scheme performs the following four kinds of response according to IP address inspection results:

```
(1) Results: Source IP = Normal, Destination IP = Normal
    • Condition: None
    • Action    : These IPs are included in Normal_User_List,
                  which is a list of normal users to protect.

(2) Results: Source IP = Normal, Destination IP = Abnormal
    // The host with the source IP is
    // suspected of network scan attacker.
    • Condition: If the source IP's abnormality count
                 exceeds a THRESHOLD defined by policy
    • Action    : The attacker with the source IP is blocked.

(3) Results: Source IP = Abnormal, Destination IP = Normal
    // The host with the destination IP may
    // be a victim of source IP spoofing attacker.
    • Condition: If the destination IP's victim count
                 exceeds a THRESHOLD defined by policy
    • Action    : Only Normal_User_List is allowed to access
                  the victim host with the destination IP.

(4) Results: Source IP = Abnormal, Destination IP = Abnormal
    // It may be a random attack.
    • Condition: If the random attacks count
                 exceeds a THRESHOLD defined by policy
    • Action    : Only Normal_User_List is allowed to use
                  network service.
```

4 Performance Evaluation

In order to evaluate the performance of our scheme, we extended ns-2 (Network Simulator - version 2) simulator [12].

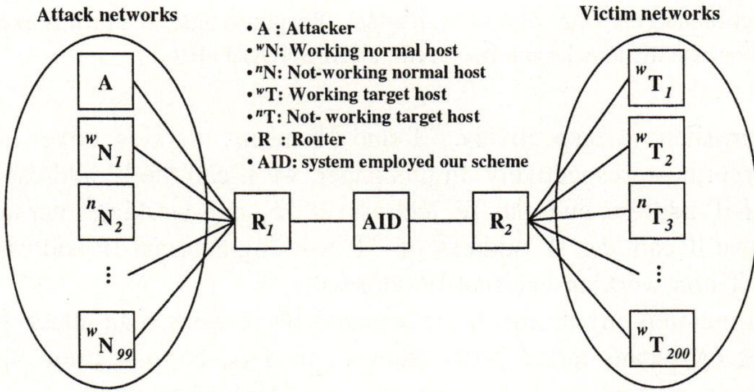

Fig. 6. Simulation networks

Fig. 6 shows simulation networks. In Fig. 6, attack networks have 100 hosts and victim networks 200 hosts. Some of the hosts are not running. Our scheme is executed on AID between R_1 and R_2. In Fig. 6, ^{w}N and ^{n}N means working normal host and non-

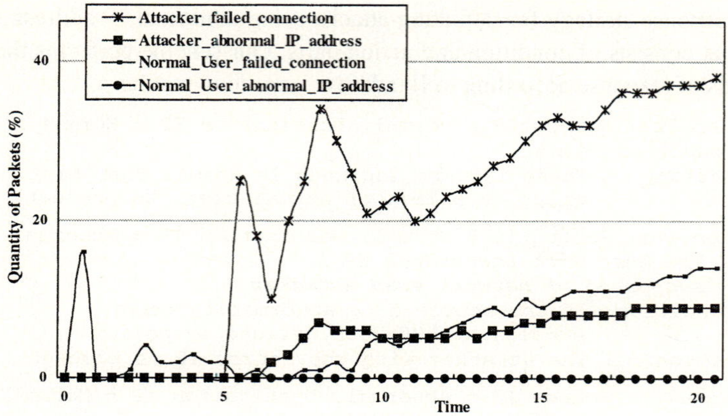

Fig. 7. Connection failure vs. IP address inspection of attacker (network scanner) and normal user under network congestion

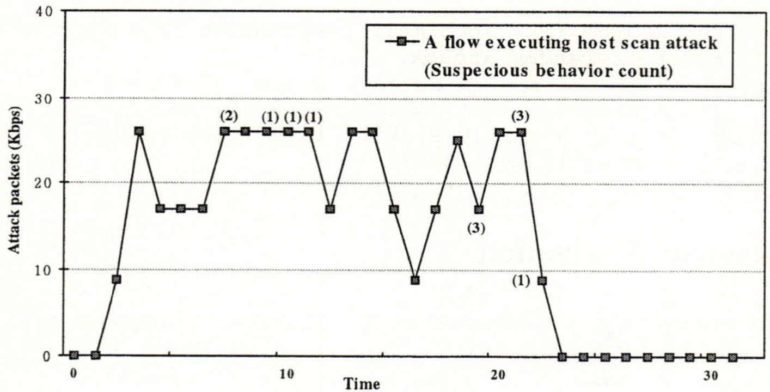

Fig. 8. Detection and response of host scan attack: Response against attack is executed at about 21 seconds because the attacks count exceeds THRESHOLD (10)

working normal host, respectively. ^{w}T and ^{n}T means working target host and non-working target host, respectively. In this paper, we'll call the IP address of ^{w}N working-internal-IP-address and the IP address of ^{n}N non-working-internal-IP-address. Similarly, we'll call the IP address of ^{w}T working-external-IP-address and the IP address of ^{n}T non-working-external-IP-address.

In the simulation environment, the scenario of network scan attack is as follows. An attacker (A) scans target hosts from T_1 to T_{200}. Normal users with working-internal-IP-address make a connection with only target hosts with working-external-IP-address. The 15% of target hosts on victim networks is not working. Temporary network congestion occurs during simulation.

The scenario of IP spoofing attack is as follows. An attacker (A) attacks the target host (T_1) by sending packets with fake source IP address. The attacker (A) uses the IP address of all hosts (i.e., from N_1 to N_{100} and from T_1 to T_{100}) on the simulation networks as its source IP address. Normal users with working-internal-IP-address make a connection with only target hosts with working-external-IP-address. 15% of normal hosts on attack networks are not working.

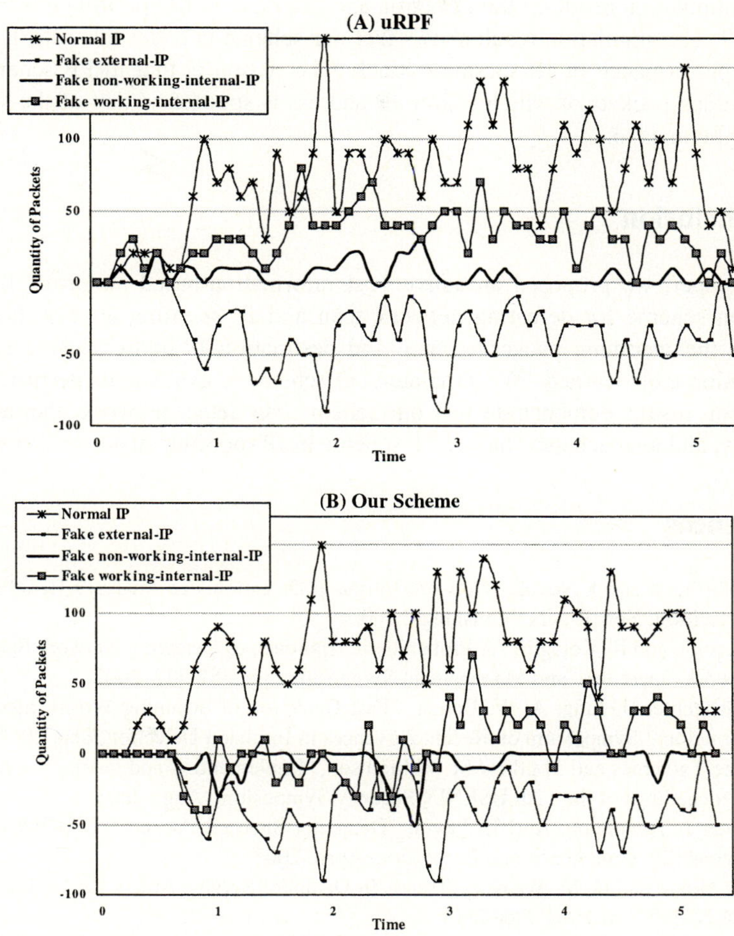

Fig. 9. uRPF scheme vs. our scheme under IP Spoofing attack: In y axis, + means the quantity of packets concluded to use normal IP address by uRPF or our scheme. – means the quantity of packets concluded to use abnormal IP address (i.e. fake IP address) by uRPF or our scheme

Fig. 7 shows connection failure vs. IP address inspection results of attacker (network scanner) and normal user under network congestion. When network congestion occurs, the chance to make a connection with a host is not high. So, normal user as well as attacker experiences connection failure under congestion as shown in Fig. 7. However, our scheme is fully aware that normal user uses normal IP address, but attacker uses abnormal IP address. As shown in Fig. 7, our scheme can detect network scan attacks through the IP address inspection results that about 10 % of attack packets use abnormal IP address (i.e, non-working-external-IP-address).

The simulation result on the detection and response of host scan attack is shown in Fig. 8. In the simulation, we set attack response THRESHOLD to 10. The simulation result shows that our scheme perceives host scan attack at about 8 seconds and finally defeats it at about 21 seconds.

The simulation result on the detection and response of IP Spoofing attack is shown in Fig. 9. The simulation result shows that our scheme is better than uRPF in the detection performance of IP spoofing attack. The reason is that uRPF scheme cannot detect attack packets of which source IP address is spoofed to that of the subnet that attacker belongs to.

5 Conclusion

In this paper, we proposed an active host information-based abnormal IP Address detection scheme for defeating network scan and IP spoofing attacks. Our scheme inspects the incoming packet's source and destination IP address using active host information table learned. We simulated our scheme to evaluate its performance. The simulation results demonstrate that our scheme can detect network scan attack very correctly, and also is better than uRPF scheme in IP spoofing attack detection.

References

1. S. Northcutt and J. Novak, "Network Intrusion Detection: An Analyst's Handbook, " Second Edition, New Riders Publishing, 2000.
2. C. Leckie and R. Kotagiri, "A Probabilistic Approach to Detecting Network Scans, " IEEE Network Operations and Management Symposium, pp. 359-372, 2002.
3. S. Schechter, J. Jung, A. W. Berger, "Fast Detection of Scanning Worm Infections," 7th International Symposium on Recent Advances in Intrusion Detection, Sep. 2004.
4. Jamie Twycross and Matthew M. Williamson, "Implementing and testing a virus throttle," In Proceedings of the 12th USENIX Security Symposium, Aug., 2003.
5. Steven J. Templeton, Karl E. Levitt, "Detecting Spoofed Packets," DARPA Information Survivability Conference and Exposition, April, 2003
6. K. J. Houle and G. M. Weaver. "Trends in Denial of Service Attack Technology," The fall 2001 NANOG meeting, Oct. 2001
7. Cisco, "Unicast Reverse Path Forwarding (uRPF) Enhancements for the ISP-ISP Edge," http://www. cisco.com/.../uRPF_Enhancement.pdf, Feb. 2001.
8. P. Ferguson and D. Senie, "Network Ingress Filtering: Defeating Denial of Service Attacks which employ IP Source Address Spoofing," RFC 2827, May 2000.
9. S. Savage, A. Karlin and T. Anderson, "Network Support for IP Traceback," IEEE/ACM Transactions on Networking, Vol. 9, No. 3, June 2001, pp. 226-237
10. S. Bellovin, M. Leech, and T. Taylor, "ICMP Traceback Messages," Internet draft, Oct. 2001.
11. W. Lee, S. Stolfo and K. Mok, "Mining in a Data-flow Environment: Experience in Network Intrusion Detection," in Proceedings of the 5th ACM International Conference on Knowledge Discovery and Data Mining (KDD'99), August 1999.
12. UCB/LBNL/VINT, "ns Notes and Documentation," http://www.isi.edu/nsnam/ns.

Securing Layer 2 in Local Area Networks

Hayriye Altunbasak[1], Sven Krasser[1], Henry L. Owen[1],
Jochen Grimminger[2], Hans-Peter Huth[2], and Joachim Sokol[2]

[1] Georgia Institute of Technology, Atlanta, GA 30332-0250, USA
{hayriye, sven, owen}@ece.gatech.edu
[2] Siemens AG, CT IC2 Corporate Technology, 81730 Munich, Germany
{jochen.grimminger, hans-peter.huth, joachim.sokol}@mchp.siemens.de

Abstract. Network security problems have been well known and addressed in the application, transport, or network layers. However, the Data Link Layer (Layer 2) security has not been adequately addressed yet. To secure Local or Metropolitan Area Networks, the IEEE 802.1AE Media Access Control (MAC) Security Task Group has proposed the IEEE P802.1AE Standard for Local and Metropolitan Area Networks: MAC Security (MACsec). MACsec introduces a new tag field, Security TAG (SecTAG), in Layer 2 frames. In this paper, we discuss the security concerns in Layer 2 and summarize some of the possible attacks in Layer 2 in Internet Protocol (IP) over Ethernet networks. We also provide an overview of the MACsec. Lastly, we propose to incorporate additional fields into the SecTAG to improve security in local area networks.

1 Introduction

Network security has become a concern with the rapid growth of the Internet. There are several ways to provide security in the application, transport, or network layer of a network. However, the network security is only as strong as the weakest section. Since the Data Link Layer security has not been adequately addressed yet, the weakest section may be the Data Link Layer (Layer 2) [1]. Layer 2 enables interoperability and interconnectivity in networks. However, a compromise in Layer 2, which enables internal attacks, may not be detected by the upper layers. In this paper, we focus on the Layer 2 security issues in IP over Ethernet networks.

1.1 Security Concerns in Layer 2

Layer 2 in IP over Ethernet networks is prone to several attacks. Three most commonly known Layer 2 sniffing attacks are Address Resolution Protocol (ARP) poisoning, Media Access Control (MAC) flooding, and port stealing.

ARP is a network layer protocol used to map an IP address to a physical machine address recognizable in the local network, such as an Ethernet address. When a host machine wishes to find a physical address for an IP address, it broadcasts an ARP request, which includes the IP address, on to the network.

P. Lorenz and P. Dini (Eds.): ICN 2005, LNCS 3421, pp. 699–706, 2005.

The host that owns the IP address sends an ARP reply message with its physical address. Each host machine maintains a table, called ARP cache, used to convert MAC addresses to IP addresses. Since ARP is a stateless protocol, every time a host gets an ARP reply from another host, even though it has not sent an ARP request for that reply, it accepts that ARP entry and updates its ARP cache [2]. The process of updating a target host's ARP cache with a forged entry is referred to as poisoning.

Content-Addressable Memory (CAM) tables store MAC addresses, switch port numbers, and Virtual Local Area Network (VLAN) information at switches. They are fixed sizes. In the MAC flooding attack, the attacker floods the switch with MAC addresses using forged gratuitous ARP packets until the CAM table is full. Then, the switch goes into hub-like mode and starts broadcasting the traffic without a CAM entry.

The port stealing attack uses the ability of the switches to learn to bind MAC addresses to ports. When a switch receives traffic from a port with a MAC source address, it binds the port number and the MAC address. In this attack, the attacker floods the switch with forged gratuitous ARP frames with the target host's MAC address as the source address and the attacker's MAC address as the destination address. Since the target host sends frames as well, there is a race condition. If the attacker is fast enough, frames intended for the target host are sent to the attacker's switch port and not to the target host.

In addition to these attacks, there are Layer 2-based broadcasting, Denial of Service (DoS), MAC cloning, and hijacking attacks. In the broadcasting attack, the attacker sends spoofed ARP replies to the network. These ARP replies set the MAC address of the network router to the broadcast address. This causes all the outbound traffic to be broadcast enabling sniffing. This type of attack also affects the network capacity. In the Layer 2-based DoS attacks, the attacker updates the ARP caches in the network with non-existent MAC addresses. The MAC address of each network interface card in a network is supposed to be globally unique. However, it can easily be changed enabling MAC cloning. The attacker uses a DoS attack to disable the network connection of the victim and then uses the IP and MAC addresses of the victim. In the Layer 2-based hijacking attack, an attacker takes control of a connection between two computers in the network. For instance, the attacker takes control of a telnet session after the victim logs in to a remote computer.

There are several ways to mitigate these types of attacks. One of these actions is to enable port security on switches. Port security ties a physical port on a switch to a MAC address/es. A change in the specified MAC address/es for a port or flooding of a port can be controlled in many different ways through switch administration. The port can be configured to shut down or block the MAC addresses that exceed a specified limit. The recommended best practice is to shut down the port that exceeds the limit [1]. Port security prevents MAC flooding and cloning attacks. However, port security does not prevent ARP spoofing. Port security validates the MAC source address in the frame header, but ARP frames contain an additional MAC source field in the data payload, and clients use this field to populate their caches [3]. Another recommended action is to employ static ARP entries. Static ARP entries are permanent entries in an ARP cache. It prevents most of the attacks. However, this method is impractical. Furthermore, it does not allow the use of some Dynamic Host Configuration Protocol (DHCP) configurations. The third method of defense is to utilize Intrusion Detection Systems (IDSs). These can be configured to listen for high amounts of ARP traffic. However, IDSs are prone to reporting false positives. There are also tools specifically designed to listen for ARP traffic on the networks. It is possible to utilize Reverse ARP to detect MAC cloning as well. In addition, there are methods to detect machines in promiscuous mode on the network.

In local networks, VLANs are employed as a security measure to limit the number of clients susceptible to attacks. VLANs create network boundaries, which ARP traffic cannot cross. Then again, VLANs are not always an option and have their own set of vulnerabilities. VLAN Hopping, Spanning Tree, and Private VLAN attacks are some of the possible attacks in VLANs.

VLAN hopping attacks allow attackers to bypass a Layer 3 device when communicating from one VLAN to another. The attack works by taking advantage of an incorrectly configured trunk port [1]. Trunk ports are generally used between switches to route traffic for multiple VLANs across the same physical link. Since the basic VLAN hopping attack is prevented in the new versions of switches, attackers have developed Double Encapsulated VLAN Hopping attacks [4]. This attack uses the fact that switches perform only one level of decapsulation. To mitigate this type of attack, administrators should disable Auto-trunking, use a dedicated VLAN ID for all trunk ports, disable unused ports and put them in an unused VLAN, and avoid using VLAN 1 (only the defaults are allowed in VLAN 1).

Spanning Tree Protocol (STP) is a link management protocol that provides loop-free topologies in a redundant Layer 2 infrastructure. The STP elects a root bridge to prevent loops in a network. In the STP, messages are sent using Bridge Protocol Data Units (BPDUs). The Standard 802.1D STP takes about 30-50 seconds to deal with a failure or root bridge change. The attacker sends BPDUs to force these changes creating a DoS condition in the network [5]. There are two features on switches used to mitigate this type of attack: BPDU Guard and Root Guard. BPDU Guard disables ports upon detection of a BPDU message on the interface. Root Guard disables interfaces that become the root bridge due to their BPDU advertisement.

Private VLANs (PVLANs) are used to create distinct networks within a VLAN. PVLANs work by limiting which ports within a VLAN can communicate with the other ports in the same VLAN. The attacker sends a frame with a rogue MAC address (the one of the router) but with the IP address of the victim. Switches do not forward the frame to the victim, but the router forwards the packet to the victim. To mitigate this attack, an ingress Access Control List (ACL) can be setup on the router interface or VLAN ACL (VACL) can be used.

Lastly, Dynamic Host Configuration Protocol (DHCP) is used to dynamically allocate IP addresses to computers for a time period. It is possible to attack DHCP servers causing DoS in the network or impersonate a DHCP server. For instance, in the DHCP starvation attack, the attacker requests all of the available DHCP addresses. This results in a DoS attack on the network. The attacker can also use a rogue DHCP server to provide addresses to the clients. The attacker can point the users to a different default gateway with the DHCP responses. Authentication of the DHCP messages is required to prevent this type of attack.

We have presented eleven possible Layer 2 attacks in this section. However, this list is not comprehensive. Other attacks worth mentioning are Multicast Brute-Force Failover Analysis, Random Frame Stress attack, and attacks based on proprietary protocols. Furthermore, most of the network management protocols are insecure causing additional vulnerabilities.

2   MACsec

Recently, the 802.1AE Media Access Control (MAC) Security Task Group has been formed to secure Local or Metropolitan Area Networks. The IEEE P802.1AE Standard (Draft) for Local and Metropolitan Area Networks (LAN/MANs): MAC Security specifies how all or a part of a network can be secured transparently to peer protocol entities that use the MAC Service provided by IEEE 802 LANs to communicate. The draft defines MAC security (MACsec) entities in end stations that provide connectionless user data confidentiality, frame data integrity, and data origin authenticity. However, the standard's scope does not include the key management and the establishment of secure associations [6].

MACsec provides security services on a frame by frame basis without introducing any additional frames. MACsec introduces an additional transit delay due to the increase in the MAC Service Data Unit (MSDU) size. MACsec defines how a MAC Security Entity (SecY) operates with a MAC Security Key Agreement Entity (KaY). Each KaY discovers the KaYs present in other stations attached to the same LAN, mutually authenticates, and authorizes those stations, and creates and maintains the secure relationships between the stations that are used by the SecYs to transmit and receive frames [6]. However, MACsec does not specify how the KaY works.

There is only one Connectivity Association (CA) per LAN service. In [6], the abbreviation LAN is used exclusively to refer to an individual LAN specified by a MAC technology without the inclusion of bridges. Each SecY only participates in a single CA at any one time. Each CA is supported by Secure Channels (SCs). There is one SC for secure transmission of frames from one of the systems to all the others in the CA. All the SCs use the same cipher suite at any one time. Each SC comprises a succession of SAs, each SA representing a single value of the transient session key(s) used for a period by the cipher suite to support the SC. Each SA is identified by the SC Identifier (SCI) concatenated with an Association Number (AN). The Secure Association Identifier (SAI) thus created allows the receiving SecY to identify the SA, and thus the session key(s) to be used to decrypt and authenticate the received frame. When the service guarantees provided include replay protection, the MACsec protocol requires a separate replay protection sequence number counter for each SA, as well.

The SecY provides both secure and insecure services to the users of its Controlled Port and Uncontrolled Port respectively, which are part of the IEEE 802.1X. The SecY operates without integrity, origin, or confidentiality protection if the Null Cipher Suite is selected. The services provided by the SecY when a cipher suite is selected include the MAC Service Data Unit (MSDU) encryption, Integrity Check Value (ICV) calculation to protect the MAC Protocol Data Unit (MPDU), and inclusion of a SC field. The SC presents the address where the encryption is applied. In a multipoint or Provider Bridge network, the MAC source address (SA) and destination address (DA) are not the addresses of the intermediate devices that are encrypting and decrypting, they are the original addresses, the end-to-end addresses. If the SecY is a part of the Bridge stack, its address will not be seen at the end points. In that case, in order to make the knowledge of where the encryption is applied available, the SC is used to provide the address of the Bridge port.

Fig. 1. Ethernet frame format with MACsec

The SecY can include a Security TAG (SecTAG) in the initial octets of the provider MSDU, prior to the user data and ICV. The MACsec protocol specifies the mandatory cipher and integrity suites as Null, Galois/Counter Mode-Advanced Encryption Standard (GCM-AES), and GCM as Message Authentication Code (GMAC). The cipher suites, except the Null Cipher Suite, provide confidentiality or integrity or both confidentiality and integrity. The addition of a SecTAG is required for the cipher suites, except the Null Cipher Suite. KaY associated with the SecY provides and periodically updates the keys for the cipher suite. If confidentiality is required, the user data parameter (MSDU) is encrypted. The destination address, source address, and the SecTAG fields are not encrypted. If data integrity is required, an Integrity Check Value (ICV) is calculated over the destination address, source address, SecTAG, and user data (after encryption, if applicable). A simple frame format for Ethernet using MACsec is presented in Figure 1.

When a Cipher Suite is selected (except the Null Cipher Suite), the SAI decoded from the SecTAG of a valid MPDU, the MAC destination and source addresses, the octets of the SecTAG, the octets of the Secure Data, and the ICV are presented to the Cipher Suite implementation. The Cipher Suite implementation identifies the validation parameters associated with the SA for the received frame using the SAI. Then, (using the validation parameters) it validates the addresses, the SecTAG, and the User Data and decrypts (if encrypted) the User Data. If any of the parameters are invalid or MSDU extraction fails, the received frame is discarded, and no indication is made to the user of the SecY.

The ICV is encoded in the last eight octets of the MACsec PDU. It authenticates the MAC destination and source addresses as well as all the fields of the MPDU (including the SecTAG if present). The ICV is computed over the encrypted or clear text data. There are significant advantages of computing the ICV over the encrypted text.

MACsec provides point to point integrity, but not global integrity [6]. There is no protection against legitimate users in ARP spoofing. It is a case of a legitimate user with bad intentions. MACsec does offer the ability to identify the bad user. The ICV is recomputed every time the frame presented to the MAC layer, so a man-in-the-middle (legitimate user) can make unauthorized changes and it will pass the integrity check. If the security transform is applied only to data, not to control frames, it will not offer protection against disclosure due to ARP spoofing, in which an attacker sends gratuitous ARP messages claiming to have IP addresses of other stations. The attacker can then intercept, read, and alter messages between any two points in a point to point topology. The ICV will be recalculated and the altered data will pass the integrity check. If the message is cryptographically protected above Layer 2, for example with IPsec, the data cannot be read or changed by the attacker. This threat occurs at the intersection between the Layer 2 and Layer 3 layers [6].

3   Layer 2/Layer 3 Impact

Layer 2 switches/bridges are used to provide connectivity in LANs, whereas Layer 3 routers are typically used to provide connectivity between LANs. For instance, a switch forwards Ethernet frames based on the MAC addresses or frame headers. It does not alter the MAC addresses when it forwards a frame. On the other hand, a router extracts the network layer header to match the destination address in the routing table, identifies the new MAC destination address for the packet, and transmits the packet to the port associated with the destination address in the routing table. As a result, when a frame/packet goes through a router, the Layer 2 header of the frame/packet is changed. Hence, the information regarding the original MAC source and destination addresses is lost.

The goal of MACsec is to secure LAN/MANs. However, most of the LAN/MANs are composed of bridges/switches and Layer 3 devices (routers). Routers support the Internet Protocol (IP). When a router receives a data frame destined to itself, it removes all the link layer (Layer 2) headers and adds a new link layer header before transmitting it. MACsec mentions an optional Secure Origin Authenticity (SOA) field. Nevertheless, it is not clear how this field is used at switches/bridges. In addition, MACsec does not consider the case when a Layer 3 device forwards a frame. Since MACsec is end-to-end in Layer 2, in the case that the router is the end point for a transmission in Layer 2, all the information regarding the origin of security (original MAC source address) is removed. This limits the capability of tracking spoofed IP/MAC packets/frames because the IP and MAC address pair in the outgoing packet/frame at the router does not provide any information regarding the original MAC source address used. In addition, it removes the IP source address and the MAC source address binding created in the original frame. When MACsec is used with a SecTAG, it provides security for the data frame and binds the MAC and IP addresses via the ICV. The ICV is calculated over the MAC destination and source addresses, SecTAG, and user data. Thus, it prevents unauthorized modifications of the MAC and IP addresses. The binding between the IP and MAC source addresses is critical to provide security and aid in billing procedures. Note that a DHCP server may assign a different IP address to the same subscriber each time it joins the network even though the MAC address of the subscriber stays the same.

We propose to transmit the original MAC source address in the Ethernet frame when a router transmits the frame in a LAN/MAN. If MACsec is not used to secure the ARP messages, the inclusion of the original MAC source address may be used to prevent the ARP attacks in a LAN/MAN. At a router, before forwarding a frame, the original MAC source address should be included in the frame in addition to the SecTAG. The maximum size of the data field in a frame should be reduced to convey the original 48 bit MAC source address field. In addition, this field may be made optional to conserve bandwidth in a network.

Header	SecTAG w/ LC	Original MAC Source Address	Data	ICV	CRC

Fig. 2. Proposed Ethernet frame format with MACsec

Furthermore, we propose to add a hop/link count field into the SecTAG to track the number of hops/links that a frame travels. This is necessary because switches/bridges do not have MAC addresses. They are invisible in Layer 2/3. Even though the port number is contained in the SecTAG, it is not possible to identify the number of switches/bridges that a frame passes through in a network. The port number in the SecTAG identifies the port number of the last end point bridge/switch. A hop/link count field makes end devices aware of intermediate switching/bridging devices. Moreover, it helps to track the traffic in a network providing the information regarding the number of Layer 2 devices on the path. In conjunction with topology plans, network administrators can also examine whether frames take the expected path and IDSs can utilize this field to recognize spoofed or misguided frames. Note that a hop/link count field denotes the number of hops/links in Layer 2 instead of Layer 3. An example for a similar concept with a different intention would be the IP time to live field at Layer 3. The proposed field should have a fixed size to prevent frame fragmentations later in the network. Each SecY should increment the hop/link count before transmitting a frame. Routers require additional features to transfer this link layer information between the ports, as well. However, it may always not be desirable to reveal the number of Layer 2 devices in a network. In such a case, the TAG Control Information (TCI) field, which comprises bits 3 through 8 of the octet 3 of the SecTAG, may be used to indicate whether the hop/link count field is being utilized. A simple presentation of the proposed Ethernet frame is shown in Figure 2. The proposed additional fields should be protected by the ICV in each frame. The TCI field in the SecTAG can be used to facilitate the optional use of these additional fields. In Figure 2, LC stands for Link Count and is included in the SecTAG.

4   Conclusions

In this paper, we focus on the Data Link Layer (Layer 2) security issues in IP over Ethernet networks. We introduce the security concerns in Layer 2 and summarize some of the possible attacks in this layer. In addition, we provide an overview of the IEEE P802.1AE Standard for Local and Metropolitan Area Networks (LAN/MANs): Media Access Control (MAC) Security. We discuss the Layer 2/Layer 3 impact of MACsec, as well. Finally, as an initial approach to improve security in LANs, we propose to use a hop/link count field in the SecTAG and an original MAC source address field in addition to the SecTAG at a Layer 3 device before transmitting a frame. These proposed additional fields provide some level of visibility for Layer 2 devices and protect the original MAC and IP address binding with additional bandwidth requirements.

Future work should focus on performance of MACsec and Layer 2 security threats with MACsec. Moreover, methods to secure ARP should be investigated. Finally, the IP and MAC address binding problem should be studied and addressed in general.

References

1. Howard, C.: Layer 2 – The Weakest Link: Security Considerations at the Data Link Layer. Available at http://www.cisco.com/en/US/about/ac123/ac114/ac173/ac222/about_cisco_packet_feature09186a0080142deb.html
2. Bashir, M. S.: ARP Cache Poisoning with Ettercap. (August 2003) Available at http://www.giac.org/practical/GSEC/Mohammad_Bashir_GSEC.pdf
3. Plummer, D. C.: Ethernet Address Resolution Protocol: Or converting network protocol addresses to 48.bit Ethernet address for transmission on Ethernet hardware. RFC 826 (November 1982)
4. Rouiller, S. A.: Virtual LAN Security: weaknesses and counter measures. Available at http://www.sans.org/rr/papers/38/1090.pdf
5. Convery, S.: Hacking Layer 2: Fun with Ethernet Switches. (Blackhat, 2002) Available at http://www.blackhat.com/presentations/bh-usa-02/bh-us-02-converyswitches.pdf
6. IEEE P802.1AE/D2.0 Draft Standard for Local and Metropolitan Area Networks: Media Access Control (MAC) Security. Available at http://www.ieee802.org/1/files/private/ae-drafts/d2/802-1ae-d2-01.pdf

A Practical and Secure Communication Protocol in the Bounded Storage Model

E. Savaş[1] and Berk Sunar[2]

[1]Faculty of Engineering and Natural Sciences,
Sabanci University, Istanbul, Turkey TR-34956
[2]Electrical & Computer Engineering,
Worcester Polytechnic Institute, Worcester, Massachusetts 01609
erkays@sabanciuniv.edu, sunar@wpi.edu

Abstract. Proposed by Maurer the bounded storage model has received much academic attention in the recent years. Perhaps the main reason for this attention is that the model facilitates a unique private key encryption scheme called hyper-encryption which provides everlasting unconditional security. So far the work on the bounded storage model has been largely on the theoretical basis. In this paper, we make a first attempt to outline a secure communication protocol based on this model. We describe a protocol which defines means for successfully establishing and carrying out an encryption session and address potential problems such as protocol failures and attacks. Furthermore, we outline a novel method for authenticating and ensuring the integrity of a channel against errors.

Keywords: Bounded storage model, hyper-encryption, information theoretical security, pervasive networks.

1   Introduction

Proposed by Maurer [1] the Bounded Storage Model (BSM) has recently received much attention. In this model each entity has a bounded storage capacity and there is a single source which generates a very fast stream of truly random bits. Whenever Alice and Bob want to communicate they simply tap into the random stream and collect a number of bits from a window of bits according to a short shared secret key. Once a sufficient number of random bits are accumulated on both sides, encryption is performed by simply using one time pad encryption. The ciphertext is sent after the window of bits has passed. What keeps an adversary from searching through the random bits is the limitation on the storage space. The model assumes that the adversary may not store the entire sampling window and therefore has only partial information of the sampled random stream.

A major advance was achieved by Aumann, Ding and Rabin [2, 3] who showed that the BSM provides so-called "everlasting security", which simply means that even if the key is compromised, any messages encrypted using this private key prior to the compromise remain perfectly secure. Furthermore, it was also shown

P. Lorenz and P. Dini (Eds.): ICN 2005, LNCS 3421, pp. 707–717, 2005.

that even if the same key is used repeatedly, the encryption remains provably secure. These two powerful features of this model have attracted sustained academic [1, 2, 3, 4, 5] and popular interest [6, 7] despite the apparent practical difficulties.

As promising as it seems there are serious practical problems preventing the BSM from immediate employment in real life communication: implementation of a publicly accessible high rate random source, authentication of the random stream generator, resilience to broadcast errors and noise, synchronization of communicating parties, implementation of the one-time pad generator. In this paper we make a first attempt to address these problems.

2   The Network Architecture and Related Assumptions

In this section, we briefly describe the network architecture and explain related assumptions. We envision a network consisting of many common nodes that desire to communicate securely using the random stream provided by a broadcast station. The parties and important entities in the network are described as follows:

- *Super-node*: The super-node is capable of broadcasting a truly random sequence at an extremely high bit rate that is easily accessible by all common nodes. A typical example for a super-node is a satellite broadcasting such a sequence to its subscribers.
- *Random Stream*: We refer to the broadcast of the super-node as the random stream. For the protocol in this paper it is sufficient to have a weak random source with sufficient min-entropy as described in [5]. We propose the super-node to transmit the random stream over multiple sub-channels with moderate to high bit rates. Note that an ultra high bit rate is unlikely to be sustained by a single channel and due to their constrained nature. Also at high transmission rates it would be difficult to keep the error rate sufficiently small.
- *Common Nodes*: The common nodes represent the end users in the system that want to communicate securely with each other utilizing the service of the super-node. Compared to the super-node they are more constrained and only able to read at the speed of the sub-channels. We also assume that a common node can read any of the sub-channels at any time. Furthermore, during the generation of the one-time-pad (OTP), we allow our nodes to hop from one sub-channel to another. The hopping pattern is determined by the shared secret between the two nodes. Therefore, an adversary has to store all the data broadcast in all channels in order to mount a successful attack.
- *Adversary*: An adversary is a party with malicious intent such as accessing confidential information, injecting false information into the network, disrupting the protocol etc. In our protocol the adversary is computationally unlimited (unlike in other security protocols) but is limited in storage capacity.

3   The Encryption Protocol

In this section, we briefly outline the steps in the proposed protocol. Basically, there are four steps the two parties have to take:

1. *Setup*: Initially the two communicating parties, i.e. Alice and Bob, have to agree on a shared secret which will be needed to derive the OTP from the random stream. For this, they can use a public or secret key based key agreement protocol or simply pre-distribute the secret keys. In the BSM pre-distributing keys makes perfect sense, since in the hyper-encryption scheme keys can be reused with only linear degradation in the security level. The two parties synchronize their clocks before they start searching for a synchronization pattern in the random stream. For this they can use a publicly-available time service with a good accuracy such as Network Time Protocol (NTP) or GPS Time Transfer Protocol [8].

2. *Communication request*: Alice issues a communication request to Bob. The request includes Alice's identity, the length of the message Alice will send, and the time Alice will start looking for the synchronization pattern. Alice may send the request over a public authenticated channel.

3. *Synchronization*: The nodes search for a fixed synchronization pattern in the random stream in order to make sure they start a sampling window at the same bit position in the random stream. Alice and Bob read a predetermined number of bits from the random stream once they are synchronized. They hash these bits and compare the hashes over an authenticated channel to make sure they are truly synchronized.

4. *Generation of the OTP*: Alice and Bob start sampling and extracting the OTP (see Section 6) from the random stream. The length of the stream they collect is the same as the total length of the messages they intend to exchange. The duration of the sampling is an important parameter in the system (window size) and determined by the storage capacity of the attacker.

4   Synchronization

We start our treatment by defining the following system parameters:

- T (bits/sec): Bit rate of random source.
- w (bits) : Maximum number of bits an adversary can afford to store.
- r (bits) : Key length.
- k (bits) : Length of synchronization pattern.
- m (bits): Maximum length of message.
- s: Number of sub-channels.
- γ (sec): Duration of a session.
- τ (sec) : Maximum delay between any two nodes in the network.
- ϵ (sec) : Bias in the random stream, i.e. the distribution of the random source is ϵ-close to uniform distribution.

- δ (sec) : Maximum allowable difference between any two clocks in the system.
- θ (sec): Synchronization delay.
- e : Error rate in the random stream.

The synchronization of the sender and the receiver nodes is a serious problem that must be properly addressed as the random source rate is very high. Especially, when the random stream is broadcast as an unstructured stream of bits where there are no regular frames, special techniques must be used to allow two users to synchronize before the bit extraction phase may begin. Two communicating nodes must start sampling the broadcast channel exactly at the same time. Even if they start exactly at the same time, any drift between the clocks of two nodes or any discrepancy in the broadcast delays to the two nodes may result in synchronization loss. We propose a synchronization technique that is easily applicable in practical settings. In our analysis we do not consider the discrepancy in the broadcast delays, which can be easily taken into account by adding it to the clock drift parameter.

The synchronization of the nodes is achieved through the use of a fixed binary sequence $p \in \{0,1\}^k$, called *synchronization pattern*. Since the super-node broadcasts a weakly random binary stream, we may conclude that the synchronization pattern p appears in the broadcast channel in a fixed time interval with probability proportional to the randomness in the stream. Two nodes that want to communicate securely start searching for p at an agreed time. The synchronization is established once they observed the pattern p for the first time after this agreed time. We expect the clocks of the two nodes to be synchronized before they start searching for the synchronization pattern. However, we cannot assume perfectly synchronized clocks and therefore there is a probability of the node with a faster clock finding a match while the other node is still waiting. We will identify the magnitude of the synchronization failure probability later in this section.

It is important to provide some insight about how to choose the length of the synchronization pattern and to estimate the time needed for getting two parties synchronized. The two nodes are loosely synchronized. Thus the maximum clock difference between any two nodes cannot be larger than a pre-specified value δ. Our aim is to decrease the probability of the synchronization pattern appearing within the time period of δ seconds after the request is sent. The probability of a fixed pattern of length k appearing in a randomly picked bit string of length n (with $n \geq k$) may be found by considering the $n - k + 1$ sub-windows of length k within the n bit string. The matching probability in a single sub-window is $1/2^k$. The probability of the pattern *not* being matched in any of the sub-windows is *approximated* as $P \approx (1 - \frac{1}{2^k})^{n-k+1}$. In the approximation we assumed that the sub-windows are independent which is not correct. However, the fact that the pattern did not appear in a sub-window eliminates only a few possible values in the adjacent (one bit shifted) sub-window. Hence the probability of the pattern not appearing in a window conditioned on the probability of the pattern not appearing in one of the previous windows is roughly approximated as $(1 - \frac{1}{2^k})$.

Any sub-channel broadcasts $n = \delta T/s$ bits in δ seconds. Therefore, the probability of a *synchronization failure* (i.e. the synchronization pattern appears before δ seconds after t_j elapsed) occurring will be

$$1 - P = 1 - (1 - \frac{1}{2^k})^{\delta T/s - k + 1} . \tag{1}$$

This probability can be made arbitrarily small by choosing larger values for k. In practice, due to the $1/2^k$ term inside the parenthesis a relatively small k will suffice to practically eliminate the failure probability. For large k using

$$(1 - \frac{1}{2^k})^{n-k+1} \approx 1 - \frac{n-k+1}{2^k}$$

we obtain the following approximation $1 - P \approx \frac{\delta T/s - k + 1}{2^k}$. Note that for a nonuniform weak random source the unsuccessful synchronization probability will be even smaller.

We also want to determine the synchronization delay for a uniformly distributed random source. Ideally one would determine the value of n for which the expected value of the number of matches becomes one to determine the synchronization delay. Unfortunately, the dependencies make it difficult to directly calculate the expected value. Instead we determine the value of n for which the matching probability is very high. This gives us a maximum on the synchronization delay.

The probability of a pattern miss occurring in a window of n bits is $\eta = (1 - \frac{1}{2^k})^{n-k+1} \approx 1 - \frac{n-k+1}{2^k}$. We obtain $n = (1 - \eta)2^k + k - 1$. Ignoring $k - 1$ for a uniform random source the average synchronization delay will be

$$\theta \approx \frac{(1 - \eta)2^k \cdot s}{T} . \tag{2}$$

Obviously, we would like to keep k as short as possible to minimize the synchronization delay. On the other hand, we need to make it sufficiently large to reduce the synchronization failure probability. These two constraints will determine the actual length of the synchronization pattern.

5   Coping with Errors

The authenticated delivery of the random stream to the users is crucial since an adversary (or random errors) may corrupt the bits of the random stream and thus the extracted bits at both sides of the communication may differ. Nevertheless, it is possible to discard sampled bits if corrupted since the sampling step of the OTP generation method is flexible. It is sufficient to sample new bits. A corruption may be detected by a simple integrity check mechanism. The super-node and common nodes are assumed to share a secret key that can be used for the integrity check. The super-node computes a message digest of the random stream under the shared key and broadcasts it as a part of the random stream.

The common nodes having the same secret key can easily check the integrity of the random stream they observed. Here it is important that the message digest algorithm is sufficiently fast to match the rate of the random stream. Message authentication codes (MACs) [9, 10] allow for low power, low area, and high speed implementations and therefore are suitable for this application.

Another difficulty that common nodes may face is to locate the message digest in the random stream when the random stream has no framed structure. This difficulty can easily be overcome by employing a technique that is similar to the method used for synchronization. The super-node searches for a pattern in the random stream. When it observes the first occurrence of the pattern it starts computing a running MAC of the random stream. It stops the computation when it observes the next occurrence of the pattern. The resulting digest is broadcast after the pattern. With the next occurrence of the pattern the super-node starts the subsequent message digest. Any common node can perform the integrity check of the random stream by calculating the same message digest between any two occurrences of the pattern in the random stream.

In order to develop a proper integrity ensuring mechanism we start by assuming a uniformly distributed error of rate e. Then the probability of successfully collecting t bits from a window of w bits is $P_{suc} = (1 - e)^t$. Since $(1 - e)$ is less than one, even for small error rates the success probability decreases exponentially with increasing number of sampled bits, i.e. t. This poses a serious problem. To overcome this difficulty we develop a simple technique which checks the integrity of the random stream in sub-windows delimited by a pattern. If the integrity of a sub-window cannot be verified the entire sub-window is dropped and the bits collected from this sub-window are marked as corrupted. When the collection process is over the two parties exchange lists of corrupted bits. In this scheme the integrity of each sub-window is ensured separately. The probability of a sub-window of size n_{sw} being received correctly is $P_{sw} = (1 - e)^{n_{sw}}$. Note that for a j bit pattern we expect the sub-window size to be $n_{sw} = 2^j$. Hence, even for moderate length patterns we obtain a very large sub-window and therefore relatively small success probability. The pattern length should be selected so as to yield a high success probability for the given error rate of the channel. For $e \ll 1$ and large n_{sw} the success probability can be approximated as $P_{sw} \approx 1 - n_{sw}e$.

Since some of the sampled bits will be dropped, collecting t bits almost always yields less than t useful bits. Therefore, we employ an oversampling strategy. Assuming we sample a total of zt bits (where $z > 1$) from x sub-windows with a success probability of P_{sw} we can expect on average to obtain $(P_{sw}x)\frac{zt}{x} = ztP_{sw}$ uncorrupted bits. We want $ztP_{sw} = t$ and therefore $t_s = t/P_{sw} = t/(1 - n_{sw}e)$ samples should be taken to achieve t uncorrupted bits on average.

6   The OTP Generator

In this section we outline the one-time-pad generator module which is the unit all common nodes posses and use to collect bits off the random stream. Since this unit will be used in common nodes, it is crucial that it is implemented in the most efficient manner. We follow the sample then extract strategy proposed by Vadhan [5] since it provides the most promising implementation. For instance, the construction of a *locally computable* extractor makes perfect sense from an implementation point of view, as it minimizes the number of bits that need to be collected off the random stream.

We follow the same strategy to construct an OTP generator. To sample the stream a number of random bits (which may be considered as part of the key) are needed. An efficient sampler may be constructed by using a random walk on an expander graph. One such explicit expander graph construction was given by Gabber and Galil [11]. Their construction builds an expander graph as a bipartite graph (U, V) where the sets U, V contain an identical list of vertices labeled using elements of $Z_n \times Z_n$. Each vertex in U is neighbor (has common edge) with exactly 5 vertices in V. A random walk on the graph is initiated after randomly picking a starting node. The following node is determined by randomly picking an edge to follow from the 5 possible edges. This leads to the next node which is now considering as the new starting node. The walk continues until sufficiently many nodes are visited and their indices recorded. In the Gabber and Galil [11] expander construction the edges are chosen by evaluating a randomly picked function from a set of five fixed functions. The input to each function is the label ($\in Z_n \times Z_n$) of the current node and the output is the label of the chosen neighbor node. To make the selection, a linear number of random bits are needed. For the starting node $2 \log n$ random bits are needed. For each following node $\log 5$ random bits are needed to select a function. To create a set of t indices $2 \log n + (t-1) \log 5$ random bits are needed.

Once t bits are sampled using the indices generated by the random walk, a strong randomness extractor is used to extract the one-time pad from the weak source. For this an extractor $E(t, m, \epsilon) : \{0,1\}^t \times \{0,1\}^r \mapsto \{0,1\}^m$ is needed. Such an extractor extracts m ϵ-close to uniform bits from a t-bit input string of certain min-entropy. In Vadhan's construction [5] the length of the t-bit input string is related to the m-bit output string by $t = m + \log(1/\epsilon)$. Here ϵ determines how close the one-time pad is to a random string and therefore determines the security level and must be chosen to be a very small quantity (e.g. $\epsilon = 2^{-128}$). To construct such an extractor one may utilize 2-wise independent hash functions families, a.k.a. universal hash families [12, 13, 14, 9]. It is shown in [15] that universal hash families provide close to ideal extractor constructions. An efficient construction based toeplitz matrices [16] is especially attractive for our application[1]. The hash family is defined in terms of a matrix vector product over the binary field $GF(2)$. A toeplitz matrix of size $m \times t$ is filled with random bits. This requires only $t + m - 1$ random bits since the first row and the first column define a toeplitz matrix. The input to the hash function are the t bits which are assembled into a column vector and multiplied with the matrix to

---

[1] A variation to this theme based on LFSRs [13] may also be of interest for practical applications.

obtain the output of the hash function in a column vector containing the m output bits. In our setting we first want to create indices $i \in [1, w]$ which will be used to sample bits from a window of size w. In the expander construction the number of vertices is determined as $n = \sqrt{w}$. To create a set of t indices $2 \log n + (t - 1) \log 5$ random bits are needed. For the extractor another set of $t + m - 1 = 2m + \log(1/\epsilon) - 1$ random bits are required. To generate m output bits which are ϵ-close to the uniform distribution it is sufficient to use a total of

$$\log w + 4 \log(1/\epsilon) + 5m \tag{3}$$

random bits.

7   An Example

In this section we come up with actual parameter values in order to see whether realistic implementation parameters can be achieved. The most stringent assumption the BSM makes is that the attacker has limited storage space and therefore may collect only a subset of bits from a transmission channel when the data transmission rate is sufficiently high. We assume an attacker is financially limited by \$100 million. Including the cost for the hardware and engineering cost to extend the storage with additional hardware to read the channel we roughly estimate the cost to be \$10 per Gigabyte which yields a storage bound (and window size) of $w = 10^7$ Gigabytes.

The window size is fixed and is determined by the adversaries' storage capacity. However, by spreading a window over time it is possible to reduce the transmission rate. But this also means that two nodes have to wait longer before they can establish a new session. We choose a transmission rate of $T = 10$ Terabits/sec which we believe is practical considering that this rate will be achieved collectively by many sub-channels, e.g. $s = 100$ sub-channels. Under this assumption a session (or sampling window) lasts $\gamma = w/T = 8000$ seconds or 2 hours 13 minutes. This represents the duration Alice has to wait after synchronization before she can transmit the ciphertext. If this latency is not acceptable it is possible to decrease it by increasing the transmission rate. Another alternative is to precompute the OTP in advance of time and store it until needed. Furthermore, assume that the clock discrepancy between any two nodes is $\delta = 1\mu$s. This is a realistic assumption since clock synchronization methods such as the GPS time transfer can provide sub-microsecond accuracy [8]. For this value we want to avoid synchronization failure in a window of size $n = \delta T/s = 10^5$ bits. For $k \geq 32$ the failure probability is found to be less than 0.01% using equation (1). Picking $k = 32$ and $\eta = 0.01$ and using equation (2) we determine the synchronization delay as $\theta = 43$ ms. For a window size of $w = 10^7$ Gigabytes and $\epsilon = 2^{-128}$ using equation (3) we determine the key length as $r = 1080 + 5m$ where m is the message length. We summarize the results as follows. We also must to take the errors into account. We assume an error rate of $e = 10^{-6}$. First we want to maximize the probability of a successful reception of the synchronization pattern. For $k = 32$ the success probability becomes $P = (1 - e)^k = 0.99996$.

To delimit sub-windows we choose a fixed pattern of length $l = 16$. The size of a sub-window is determined as $n_{sw} = 2^{16} = 65,536$. This gives us a probability of receiving a sub-window correctly $P_{sw} = (1 - e)^{2^l} = 0.936$. The oversampling rate is found as $1/P_{sw} = 1.068$. Hence, it is sufficient to oversample only by 7%.

Storage bound	10^4 Terabytes
Transmission rate	10 Terabits/sec
Session length	2 hours 13 minutes
Key length	$1080 + 5m$ bits
Clock precision	1 μsec
Synch. failure rate	less than 0.01 %
Synch. delay	43 msec

8 Attacks and Failures

In this section we outline some of the shortcomings of the proposed protocol:

1. Overwriting (jamming) of the random stream. The adversary may attack the channel by overwriting the public random source with a predictable and known stream generated using a pseudo random number generator. Since the adversary may now generate the stream from the short key the storage bound is invalidated and the unconditional security property is lost. The prevention of this kind of attack is difficult but it is possible for users to detect jamming attacks by making use of a MAC. For this it is necessary that the users and the super-node share a secret key. The MAC is regularly computed and broadcast to the users as described earlier. For the authentication of the channel the users have to carry the burden of computing a MAC whenever they want to communicate. The MAC computation takes place concurrently with OTP generation. Since the super-node does not know which bits will be sampled by the users, the entire broadcast window needs to be authenticated.
2. Denial of service (DOS) attacks on the nodes. The attacker may overwhelm nodes with transmission and synchronization requests. Since the request is sent over an authenticated channel this will not cause the nodes to generate OTPs. However, the nodes may be overwhelmed from continuously trying to authenticate the source of the request. The protocol is not protected against DOS attacks.
3. Loss of synchronization. Due to clock drift it is possible that synchronization is lost. In this case, the nodes will have to resynchronize and retransmit all information.
4. Transmission Errors: Our protocol handles transmission errors by implementing an integrity check on sub-windows of the random stream. The described method works only for moderate to very low error rates. For instance, for an error rate higher than $e = 10^{-3}$ the sub-window size shrinks below a reasonable size to provide a high success rate. The error rate may be reduced by employing error detection/correction schemes.

9   Conclusion

We made a first attempt in constructing a secure communication protocol based on the bounded storage model which facilitates a unique encryption method that provides unconditional and everlasting security. We described a protocol which defines means for successfully establishing and carrying out an encryption session. In particular, we described novel methods for synchronization, handling transmission errors and OTP generation. We showed that such a communication protocol is indeed feasible by providing realistic values for system parameters.

References

1. Maurer, U.: Conditionally-perfect secrecy and a provably-secure randomized cipher. Journal of Cryptology **1** (1992) 53–66
2. Ding, Y.Z., Rabin, M.O.: Hyper-encryption and everlasting security (extended abstract). In: STACS 2003 – 19th Annual Symposium on Theoretical Aspect of Computer Science. Volume 2285., Springer Verlag (2002) 1–26
3. Aumann, Y., Ding, Y.Z., Rabin, M.O.: Everlasting security in the bounded storage model. IEEE Transactions on Information Theory **6** (2002) 1668–1680
4. Lu, C.J.: Hyper-encryption against space-bounded adversaries from on-line strong extractors. In Yung, M., ed.: Advances in Cryptology — CRYPTO 2002. Volume 2442., Springer-Verlag (2002) 257–271
5. Vadhan, S.: On Constructing Locally Computable Extractors and Cryptosystems in the Bounded Storage Model. In Boneh, D., ed.: Advances in Cryptology — CRYPTO 2003. Volume 2729., Springer-Verlag (2003) 61–77
6. Kolata, G.: The Key Vanishes: Scientist Outlines Unbreakable Code. New York Times (2001)
7. Cromie, W.J.: Code conquers computer snoops: Offers promise of 'everlasting' security for senders. Harvard University Gazette (2001)
8. Observatory, U.S.N.: GPS timing data & information. http://tycho.usno.navy.mil/gps_datafiles.html (2004)
9. Halevi, S., Krawczyk, H.: MMH: Software message authentication in the gbit/second rates. In: 4th Workshop on Fast Software Encryption. Volume 1267 of Lecture Notes in Computer Science., Springer (1997) 172–189
10. Black, J., Halevi, S., Krawczyk, H., Krovetz, T., Rogaway, P.: UMAC: Fast and secure message authentication. In: Advances in Cryptology - CRYPTO '99. Volume 1666 of Lecture Notes in Computer Science., Springer-Verlag (1999) 216–233
11. Gabber, O., Galil, Z.: Explicit constructions of linear-sized superconcentrators. Journal of Computer and System Sciences **3** (1981) 407–420
12. Carter, J.L., Wegman, M.: Universal classes of hash functions. Journal of Computer and System Sciences **18** (1978) 143–154
13. Krawczyk, H.: LFSR-based hashing and authentication. In: Advances in Cryptology - CRYPTO '94. Volume 839 of Lecture Notes in Computer Science., Springer-Verlag (1994) 129–139
14. Rogaway, P.: Bucket hashing and its applications to fast message authentication. In: Advances in Cryptology - CRYPTO '95. Volume 963 of Lecture Notes in Computer Science., New York, Springer-Verlag (1995) 313–328

15. Barak, B., Shaltiel, R., Tomer, E.: True Random Number Generators Secure in a Changing Environment. In Çetin K. Koç, Paar, C., eds.: Workshop on Cryptographic Hardware and Embedded Systems — CHES 2003, Berlin, Germany, Springer-Verlag (2003) 166–180
16. Mansour, Y., Nissan, N., Tiwari, P.: The computational complexity of universal hashing. In: 22nd Annual ACM Symposium on Theory of Computing, ACM Press (1990) 235–243

Measuring Quality of Service Parameters over Heterogeneous IP Networks[*]

A. Pescapé[1], L. Vollero[1], G. Iannello[2], and G. Ventre[1]

[1] Universitá Degli Studi di Napoli, "Federico II"
{pescape, vollero, giorgio}@unina.it
[2] Universitá Campus Bio-Medico (Roma)
g.iannello@unicampus.it

Abstract. In real networks, an experimental measure of Quality of Service parameters is fundamental in the planning process of new services over novel network infrastructures. In this work we provide an empirical performance study of a real heterogeneous wireless network with respect to delay, jitter, throughput and packet loss, in UDP and TCP environments, by using an innovative tool for network performance evaluation that we called D-ITG (*Distributed Internet Traffic Generator*). A comparative analysis between our practical results and an analytical model recently presented in literature is presented.

1   Introduction

In the field of heterogeneous, integrated and mobile IP networks, among the many factors that determine the feasibility of a given network scenario for the given set of application requirements, there are network performance. Network performance is generally affected by different aspects at the physical, data link, network, and transport layers. In a generic real network and in particular in a heterogeneous scenario, it is extremely difficult (i) to define a general framework for empirical performance evaluation and (ii) to determine the causes of the experimented performance. This paper focuses on the area of performance evaluation of heterogeneous wireless networks from the application level point of view. First, we introduce a network performance methodology dividing our experimentation on several traffic classes. Second, we measure TCP and UDP performance in more than network scenario where there is interoperability among different network technologies, different end-user devices, different operating systems and finally different user application with different QoS (*Quality of Service*) traffic requirements. The performance evaluation study has been performed following the indication of IP Performance Metrics (IPPM) IETF Working Group [9]. The network behavior has been studied introducing an innovative synthetic traffic generator that we called D-ITG (*Distributed*

---

[*] This work has been partially supported by the Italian Ministry for Education, University and Research (MIUR) in the framework of the FIRB Project "Middleware for advanced services over large-scale, wired-wireless distributed systems" (WEB-MINDS), by Regione Campania in the framework of "Centro di Competenza Regionale ICT", and finally by the E-NEXT IST European project.

P. Lorenz and P. Dini (Eds.): ICN 2005, LNCS 3421, pp. 718–727, 2005.
© Springer-Verlag Berlin Heidelberg 2005

Internet Traffic Generator) [8] and which provides a set of powerful tools for traffic patterns generation and results analysis. We present our experimental results and at the same time we analyze and compare our results with respect to theoretical assumptions on wireless performance behavior carried out in [1]. Finally, we present a clear definition of which system's elements are responsible of network performances degradation and how the used different protocols impact on the observed network performance. In this work we extend our seminal results presented in [12].

The paper is organized in 6 sections. After this introduction, in Section 2 motivations and the reference framework on which our work is based are presented. The experimental setup where our work has been carried out is presented in Section 3, discussing the main issues related to our heterogeneous scenario and describing the measuring procedure. Section 4 reports the obtained results with respect to throughput, delay, jitter, and packet loss. As far as achieved throughput, in Section 5 we provide a summary of our results and we compare and comment our conclusions in the framework of Bianchi model. Finally, Section 6 provides some concluding remarks and issues for future research.

2   Motivation and Related Work

There are several simulation [10] and analytical [11] studies on wireless channel performance, whereas in this work, we test a real heterogeneous mobile environment and present a performance evaluation from the application point of view for a wide range of parameters. Our scenario is heterogeneous in terms of (i) access network technologies (*WLAN 802.11, wired Ethernet 10/100 Mbps*), (ii) end-users' devices (*PDA, Laptop, PC desktop*) and (iii) end-users' operating systems (*Linux Embedded, Linux, Windows XP/ 2k/CE*).

Other experimental analysis are present in the literature. A performance characterization of ad hoc wireless networks is presented in [2]. The paper examines impact of varying packet size, beaconing interval, and route hop count on communication throughput, end-to-end delay, and packet loss. In [3] a new performance model for the IEEE 802.11 WLAN in ad hoc mode is presented. In [4] there is a study on network performance of commercial IEEE 802.11 compliant WLANs measured at the MAC sublayer in order to characterize their behavior in terms of throughput and response time under different network load conditions. A performance study on wireless LAN in a vehicular mobility scenario is presented in [5]. In [6] the performance of a real campus area network are measured. In [7] the authors present a comprehensive study on TCP and UDP behavior over WLAN taking into account radio hardware, device drivers and network protocols. In [13] a discussion on the problems arising when the TCP/IP protocol suite is used to provide Internet connectivity over existing wireless links is presented. [14] studies the capabilities of an IEEE 802.11 wireless LAN. For the test phases, three wireless laptop computers, a wireless and wired desktop computers and an Access Point (AP) are used.

To the best of our knowledge, our work extends previous works on TCP and UDP performance in many directions. First, we present a complete evaluation, from the application point of view, of heterogeneous wireless networks in terms of a wide range of QoS parameters. Second, measured parameters are obtained for different packet size:

in this way we can determine the optimal packet size for each analyzed network condition. Third, we take into account several factors like Operating Systems, End-Users' Device, and Network Technologies and relationships among them, while previous works point their attention only on the wireless channel performance. Finally, after a measurement phase we place our throughput results in the framework of the model proposed by Bianchi in [1] and we use our results as performance references for development of wireless communication applications over multiservice heterogeneous networks.

3   Testbed Infrastructure, Tools and Experimental Methodology

All tests can be collapsed in a same general scenario, depicted in Fig. 1, where two communication entities, a D-ITG transmitter and a D-ITG receiver, are directly connected through an IP network channel. Indeed, as represented in Fig. 1, the tests differ for the type of used network, its configuration and the type of host that executes the D-ITG platform; by changing these parameters we tested several strictly related *"Service Conditions"*. Others parametric elements, like generated traffic patterns, have not been changed: we used only periodical sources, with fixed Packet Size (PS) and fixed Inter-Departure Times (IDT) between packets since our intention for this study was mainly to focus on the impact of heterogeneity. In the case of an ad-hoc scenario, we have experimented more configurations, allowing the two communicating hosts to roam in three classes of end-to-end mutual distances ($d \leq 5\,m, 5\,m \leq d \leq 10\,m, 10\,m \leq d \leq 15\,m$).

In the following, the measures are organized such that to distinguish three types of traffic conditions: (i) *low* traffic load ($\leq 1.2 Mbps$, far from the saturated channel condition), (ii) *medium* traffic load ($\leq 4.0 Mbps$, close to the saturated channel condition) and (iii) *high* traffic load ($\leq 10 Mbps$, in the saturated channel condition). These three traffic conditions are related to three different real traffic loads where we used different packet size. Indeed, in the first traffic profile we used PS equal to {64, 128, 256, 512, 1024, 1500} bytes and IDT equal to $\frac{1}{100}$ (according to *low* traffic load). In the second traffic profile we used PS equal to {64, 128, 256, 512} bytes and IDT equal to $\frac{1}{1000}$ (according to *medium* traffic load). Finally in the third traffic profile we adopted PS equal to {64, 128} bytes and IDT equal to $\frac{1}{10000}$ (according to *high* traffic load). For each traffic condition, we organized the data in three types of configurations: (i) a classic configuration, with only laptop and workstation devices, (ii) a second configuration, where the transmitting host is always a Palmtop and (iii) a third configuration, where the receiving host is always a Palmtop. In order to characterize our system, we used the following QoS parameters by using the recommendations of IPPM working group [9]: (i) the (source/destination)-bandwidth (UDP and TCP protocols); (ii) the delay (UDP

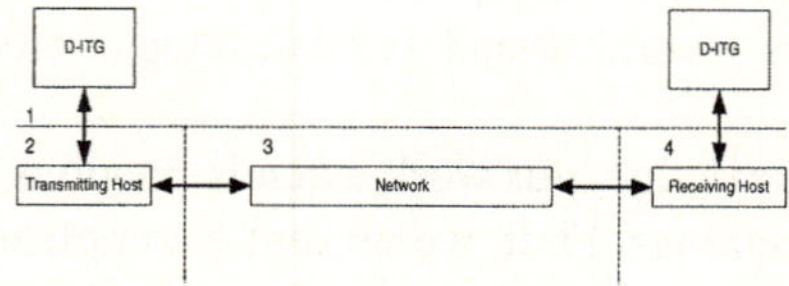

Fig. 1. The Experimental Testbed Infrastructure

only); (iii) the jitter (UDP only) and finally (iv) the packet loss (UDP only). For each measured parameter, several trials have been performed in the same operating conditions. The values reported in table 3 represent a mean value across twenty test repetitions. In order to achieve our target, we built an innovative tool, named D-ITG [8], that we introduced for heterogeneous wireless network performance evaluation.

4   Performance Analysis and Experimentation

In Table 1 the complete reference for the symbols used in Table 3 is reported whereas in Table 2 details on devices used are depicted.

Table 1. Table Legend

Network Scenario	Description
AP-1	Connection between the workstation and the laptop/palmtop through AP
AP-2	Connection between laptop and palmtop/lap-top through AP
AH-i	Connection between laptop and palmtop/lap-top in ad-hoc mode in a range of $5 \cdot i$ meters

Table 2. Technical details on the experimental setup

Device	Description
Laptop1	Mobile Intel PIII 1133 Mhz, Main Memory 128 MB, Cache 256 KB, O.S. Linux Red Hat 9.0 – kernel 2.4.20-18.9
Laptop2	PIII 700 Mhz, Main Memory 128 MB
Workstation1	Intel PII 850 Mhz, Main Memory 128 MB, Cache 256 KB, dual boot Operating Systems: Linux(2.4), Windows XP Professional Service Pack 1
Workstation2	Intel C 400 Mhz, Main Memory 64 MB, Cache 128 KB, O.S. Linux(2.4)
Palmtop	Intel StrongARM 206 Mhz, Main Memory 64 MB, Flash ROM 32 MB, O.S. Linux FAMILIAR – kernel 2.4.18
Access Point	Orinoco Ap1000, 11Mbps (802.11b), Multi Channel support
Wireless LAN cards	WiFi ORINOCO 11Mbps GOLD

Throughput Analysis: Table 3 shows obtained results for different network configurations (Laptop2Laptop (l2l), Laptop2Palmtop (l2p) and Palmtop2Laptop (p2l)) in the case of medium and high traffic load. In the case of *low* traffic load we are far from the saturated wireless channel condition, and test results prove that for each communication scenario sent and received throughput are equals. In the case of *medium* traffic load, we are close to the saturated wireless channel condition. In order to quantify the proximity to the saturated channel condition, in the table it has been brought back also data obtained from the Bianchi theoretical model (BTM column)[1]. Such model gives us a bound to the maximum traffic load that can cross the channel at the MAC layer of the ISO/OSI stack, therefore it supplies a useful bound for the traffic at the upper layer. Using our experimental results, we can also provide a practical validation of the Bianchi theoretical model (see Section 5). In this case of *medium* traffic load, it turns out with more evidence the dependency from the host typology and the used transport protocol. TCP demonstrates of being more sensitive to the losses respect to UDP. Moreover, we

Table 3. Experimental results

IDT	PS	SC	throughput (SS)			throughput (RS)			UDP RTT			Jitter			PL			TCP throughput			BTM
ms	B		l2l	l2p	p2l	l2l	l2p	p2l	l2l	l2p	p2l	l2l	l2p	p2l	l2l	l2p	p2l	l2l	l2p	p2l	kb/s
			kb/s						ms			ms			%			kb/s			
1.0	64	AP-1	512	512	512	511	511	474	276	654	280	1.22	0.07	1.56	0.0	0.2	6.9	512	512	340	890
1.0	64	AP-2	512	512	512	317	147	167	660	321	692	2.40	0.68	3.73	38.3	70.9	66.7	507	507	315	890
1.0	64	AH-1	512	512	512	512	405	512	198	310	398	1.74	0.29	1.23	18.6	51.8	0.0	512	512	495	890
1.0	64	AH-2	512	512	512	512	457	512	196	186	423	1.72	0.24	1.35	18.5	10.4	0.0	512	512	327	890
1.0	64	AH-3	512	512	512	480	435	480	188	1227	335	1.99	0.37	4.24	27.3	14.3	0.0	512	512	358	890
1.0	128	AP-1	1024	1024	1024	1022	1024	1008	302	700	283	1.10	0.08	1.52	0.2	0.2	1.1	1024	1024	688	1547
1.0	128	AP-2	1024	1024	1024	585	290	336	697	343	741	2.51	0.70	3.74	42.5	71.3	66.6	1024	1024	627	1547
1.0	128	AH-1	1024	1024	1024	1024	775	1024	229	348	518	1.80	0.33	1.20	22.1	22.2	0.0	1024	1024	986	1547
1.0	128	AH-2	1024	1024	1024	1005	787	1024	257	228	469	1.83	0.38	1.24	21.3	22.8	0.0	1024	1015	657	1547
1.0	128	AH-3	1024	1024	1024	1021	656	1000	187	1284	352	2.00	1.24	3.33	31.2	37.13	0.0	1001	1007	710	1547
1.0	256	AP-1	2048	2048	2048	2040	2048	1852	340	643	313	0.93	.14	1.70	0.4	0.2	9.0	2048	1958	1415	2611
1.0	256	AP-2	2048	2048	2048	1030	558	667	750	386	908	2.71	0.73	3.64	49.3	72.4	66.9	2048	2048	1895	2611
1.0	256	AH-1	2048	2048	2048	1917	1503	1910	263	370	633	1.6	0.40	1.15	20.7	22.5	6.2	2050	2048	1855	2611
1.0	256	AH-2	2048	2048	2048	1998	1559	1923	268	282	488	1.99	0.57	1.22	31.4	23.5	5.5	2050	2050	1237	2611
1.0	256	AH-3	2048	2048	2048	1989	1468	1900	209	1775	421	1.84	0.5	3.09	0.0	20.0	0.0	2043	2037	1199	2611
1.0	512	AP-1	4096	4096	3147	3750	3755	3128	380	924	414	0.70	0.33	2.05	9.0	7.9	0.0	4100	4096	2595	4503
1.0	512	AP-2	1675	1044	1294	1662	1036	1273	794	502	1248	4.66	7.52	5.20	0.0	74.5	0.0	2591	2532	1928	4603
1.0	512	AH-1	3350	3302	3264	3338	2394	3243	457	621	972	3.00	2.88	1.18	0.0	37.8	0.0	3900	3887	2904	4603
1.0	512	AH-2	3343	3300	3286	3276	2387	3267	446	311	663	3.19	3.11	1.27	0.0	39.0	0.0	3894	3794	2894	4603
1.0	512	AH-3	3220	3200	3120	3134	2106	3100	379	1801	994	3.33	3.73	5.59	0.0	0.0	31.8	3700	3699	2860	4603
0.1	64	AP-1	5120	5120	657	669	667	543	285	491	273	0.93	0.26	1.43	87.6	87.0	2.2	2774	1365	187	890
0.1	64	AP-2	5120	5120	1015	279	147	167	657	316	742	2.59	0.48	3.79	95.0	97.0	17.0	2384	460	924	890
0.1	64	AH-1	5120	5120	643	393	415	525	322	430	500	1.81	0.22	1.31	92.3	91.8	2.3	2387	1413	547	890
0.1	64	AH-2	5120	5120	625	321	450	556	217	196	388	2.02	0.22	1.22	93.1	91.2	1.3	2573	1537	654	890
0.1	64	AH-3	5120	5120	625	329	437	548	207	733	379	1.98	0.24	1.24	93.5	91,3	1.2	2908	1537	424	890
0.1	128	AP-1	10240	10240	10240	1262	1252	1040	298	492	291	0.98	0.28	1.51	87.6	86.9	2.1	3828	2203	372	2611
0.1	128	AP-2	10240	10240	10240	513	287	334	695	338	770	2.68	0.50	3.77	95.0	97.1	16.0	2424	1135	1836	2611
0.1	128	AH-1	10240	10240	10240	633	762	1033	634	397	520	2.14	0.22	1.19	93.8	92.3	1.5	2151	2219	1172	2611
0.1	128	AH-2	10240	10240	10240	562	872	1055	201	191	440	2.45	0.21	1.23	94.5	91.0	1.3	3090	2356	639	2611
0.1	128	AH-3	10240	10240	10240	673	745	1056	233	814	365	2.00	0.24	1.22	03.4	91.7	1.3	3028	2452	990	2611

can observe the greater sensitivity respect to packet dimension of the wireless configurations, especially of those with the Palmtop. A detailed results analysis is reported in Section 5. With respect to previous cases we have analyzed a transmission condition where the packet size is equal to 64 and 128 bytes. Indeed, for whichever packet dimension the channel turns out saturated: longer packets carry to a greater channel busy time for delivered or collided packet, and it only leads to a greater number of losses from the sender side for network interface saturation. It is interesting to notice the behavior of UDP and TCP in the several analyzed configurations: TCP reacts to the saturation condition limiting the demanded transmission bandwidth, while UDP endures a highest packet loss. This behavior is caused from the presence of a flow-control mechanism in the former protocol, and from the ability to the congestion control of TCP to optimize the use of a high loaded channel. Also in this case a deep analysis is reported in Section 5.

Delay Analysis: We measured the Round Trip Time (RTT). As far as RTT thanks to our experimentation we learned that: (i) in the case of *low* traffic condition the configuration with AP presents the lowest performance for high packet size (for PS equal to 1500 bytes we measured RTT $\approx$ 500 ms); (ii) the previous statement is not true in the case of l2p configuration where we experimented the lowest performance in the case of ad-hoc configuration, with a distance d between sender and receiver equal to $10\,m \leq d \leq 15\,m$ (in this case we measured RTT until to 1000 ms). In the case of *medium* and *high* traffic condition we experimented the same behavior described for *low* traffic condition with same trend at a much high values. More precisely, as far as *medium* traffic condition: (i) the configuration AP-2 remains the scenario with lowest performance in the case of l2l and p2l communications; (ii) in the case of l2l communications the RTT for the configuration AP-2 is under the 800 ms ($657\,ms \leq RTT \leq 800\,ms$); (iii) in the case of p2l communications the RTT for the configuration with AP reaches RTT $\approx$ 1200 ms for PS equal to 512 bytes; (iv) in the case of l2p configuration we experimented the lowest performance in the case of ad-hoc configuration, with a distance d between sender and receiver equal to $10\,m \leq d \leq 15\,m$ (in this case we measured an $1200\,ms \leq RTT \leq 1801\,ms$). As far as *high* traffic condition we have the same behavior of the *medium* traffic condition with the following differences in terms of achieved results: (i) in the case of l2l communication lower RTT performance have been reached by using the AP configuration and obtaining RTT $\approx$ 700 ms for PS equal to 128 bytes; (ii) in the case of p2l communication lower RTT performance have been reached by using the AP configuration and obtaining RTT $\approx$ 800 ms for PS equal to 128 bytes; (iii) in the case of l2p communication lower RTT performance have been reached by using the ad-hoc configuration, with a distance d between sender and receiver equal to $10\,m \leq d \leq 15\,m$ and obtaining RTT $\approx$ 800 ms for PS equal to 128 bytes.

Jitter Analysis: In an Internet Telephony architecture, excessive jitter may cause packet loss in the receiver jitter buffers thus affecting the playback of the voice stream. In almost all the analyzed *"Service Conditions"* and for each packet size we experimented the worst case in the configuration with AP. The jitter results confirm that there is a weak sensitivity of the jitter as a function of the used configuration and the used hosts. Digging into details, the experimented jitter values are the following: (i) in the *low* traffic condition the worst case is under the 4 ms (we experimented a jitter equal to 2.5 ms in the l2l configuration); (ii) in the *medium* traffic condition the worst case is under the 8 ms (we experimented a jitter equal to 2.5 ms in the l2l configuration); (iii) in the *high* traffic condition the worst case is under the 4 ms (we experimented a jitter equal to 0.5 ms in the l2p configuration). Highest values of the jitter have been experimented in the case of *medium* traffic load for high values of packet size (512 bytes) and for communications between l2p and p2l: this behavior is due to low capacity of Palmtops.

Packet Loss Analysis: In the case of low traffic load condition, except some singularities in the ad-hoc configuration with $10\,m \leq d \leq 15\,m$, all considered *"Service Conditions"* showed a packet loss under the 0.5% and substantially equal to 0. More precisely, only when at receiver side a Palmtop is present and the packet size is lower than 512 bytes we measured a packet loss diverse from 0 and in all cases lower than 0.5%. At the opposite site, in the cases of medium and high traffic load condition, results show dramatic values for the packet loss. Moreover, in such a case, the configuration with AP presents the lowest performance in terms of packet loss. More precisely, we experimented: (i) UDP packet loss for IDT $= \frac{1}{1000}$ s up to 70%; (ii) UDP packet loss for IDT $= \frac{1}{10000}$ s up to 95%. As far as these last two traffic conditions, we experimented acceptable packet loss values: (i) for a *medium* traffic load only in the case of l2l and l2p configuration, packet size up to 256 bytes and wired2wireless connection; (ii) in the case of *high* traffic load only when the sender was the Palmtop: this behavior is due to low transmission rate of the Palmtop that guarantees the reception of almost all sent packets. Finally, by analyzing packet loss behavior we learned that: (i) the lowest packet loss values are obtained for high packet size; (ii) the worst case is obtained in the case of Palmtop at receiver side; (iii) with the exception of a Palmtop at sender side, at higher data rate the bottleneck are the wireless links (both in ad-hoc and with AP) and not the the end-users' device. The experimented packet loss results are strictly related to throughput behavior (presented in next Section).

5   Summary of Results

TCP over wireless issues have been extensively discussed and several innovative proposals have been presented [13] [15]. We present novel results that take into account a wide range of factors: different devices, different OSs and different network technologies are considered. Over this complex environment TCP performance are extremely difficult to understand. TCP assumption that all losses are due to congestion becomes quite problematic over wireless links. In [16] G.T. Nguyen *et al.* show that (i) WLAN suffers from a frame error rate (FER) of 1.55% when transmitting 1400 byte frames over an 85 ft distance, with clustered losses and that reducing the frame size by 300 bytes halves FER there is an increase of framing overhead; (ii) mobility also increases FER for the WLAN by about 30%; (iii) FER is caused by the frequent invocations of congestion control mechanisms which repeatedly reduce TCP's transmission rate; (iv) if errors were uniformly distributed rather than clustered, throughput would increase. In addition, in [7] G. Xylomenos *et al.* show that in shared medium WLANs, forward TCP traffic (data) contends with reverse traffic (acknowledgments). In the WLAN this can lead to collisions that dramatically increase FER. As fas as maximum throughput, in [13] G. Xylomenos *et al.* show that the maximum throughput over a single wireless link, using either an IEEE 802.11 (2 Mbps) or an IEEE 802.11b (11 Mbps) WLAN is respectively equal to 0.98 Mbps and 4.3 Mbps. Thus, in the case of IEEE 802.11 there is an efficiency equal to 49% whereas in the case of IEEE 802.11b the efficiency is equal to 39.1%. This behavior is due to higher speed links are affected more by losses, since TCP takes longer to reach its peak throughput after each loss. In addition to these already know phenomena we present our innovative results that highlight the dependencies with (i) an high heterogeneity level, (ii) the properties of Palmtop device and (iii) three different traffic classes made by several combinations of IDTs and PSs. Furthermore, we present the TCP performance over wireless link varying the "*application level*" packet size: thanks to this "*modus operandi*" we can simple highlight which is the real TCP behavior over heterogeneous wireless network for different packet size values. Comparing the behavior for the same "*application level*" packet size, our analysis permits to clarify the conditions in which TCP performs better than UDP. In this section we analyze and comment our results with respect to achieved throughput.

Low Traffic Load: As we have anticipated in Section 4, in this case we discuss only the results related to the p2l configuration. As far as the throughput analysis, in the case of UDP protocol, we learn that in the case of *low* traffic load there is substantially the same behavior in all considered configurations. In the case of TCP protocol we observed a similar behavior, with the following difference: in the case of a Palmtop at sender side and in the case of the ad-hoc configuration, with a distance d between sender and receiver equal to $10\ m \le d \le 15\ m$, a light throughput reduction (starting from a packet size equal to 1024 bytes) was experimented. Thus, in this case for the several configurations two aspects are clearly depicted: (a) the communication is reliable and (b) the light degradation of the performance is due to the smaller computational power of the adopted devices (PDAs). Also in this case, we have demonstrate that TCP suffers the losses mainly, having a different behavior with respect to UDP; TCP, indeed, interprets the losses like due to congestion phenomena and reacts consequently, reducing the maximum transmittable rate and emphasizing the phenomenon of bandwidth reduction. Indeed, of particular interest is the case of 1500 bytes packets, where the packet dimension exceeds MTU (*Maximum Transfer Unit*), the maximum allowable dimension of a MAC data unit. The fragmentation produces the duplication of the total number of transmitted packet and it exacerbates the throughput reduction of the wireless channel. Thus due to this behavior we experimented a (little) throughput reduction in the case of *low* traffic load. Finally, in the *low* traffic low and with a packet size close to the MTU, UDP performs better than TCP.

Medium Traffic Load: Probably results obtained in the *medium* traffic load analysis represent one the most important contributes of this paper. Indeed, we learned that in the case of *medium* traffic load there is a throughput behavior strictly coupled with network, device and traffic characteristics. In this case we are close to the Bianchi model hypothesis. Thanks to our results we can demonstrate that: (i) the Bianchi model represents an optimal upper bound; (ii) due to network dynamics present among TCP/IP application and data link layer and due to heterogeneity of considered elements there is a divergence between the theoretical Bianchi results and our real measures. Digging into more details, as far as the throughput analysis, in the case of UDP protocol, from the results we learn that: (i) there is a progressive throughput reduction, at sender side, starting from PS equal to 256 bytes; (ii) both at sender and receiver side the configuration with AP shows lowest performance (indeed, in this case, the generated traffic present a double channel occupation); (iii) in the case of a Palmtop at receiver side there are, in all configuration, lowest performance. This reduction is higher in the case of configuration with AP. For example, in this case with PS equal to 512 bytes there is a difference with the model proposed by Bianchi equal to 3.5 Mbps; (iv) at higher packet size (PS $>$ 512 bytes) all tested configurations are far from the values of the model proposed by Bianchi (except for the wired-to-wireless configuration); (v) in the ad-hoc configuration there is a clear dependence between the achieved throughput and the end-nodes distances; Moreover, in the case of TCP protocol, we learn that: (i) TCP shows better performance than UDP: this behavior is due to TCP capacity of putting more data into a single (TCP) segment. When we transmit UDP, our IP frame will carry only 512 bytes. When we transmit TCP traffic, TCP fits more data into the packet before transmitting (if they are available right away). This can happen until the proximity of MTU: in the *medium* traffic load when we reach the MTU, UDP presents better performance than TCP. Digging into numerical details, at low packet size (PS $<$ 512 bytes) TCP presents, in almost all considered configurations, 1 Mbps more than UDP achieved throughput; (ii) also in this case the configuration with AP shows lowest performance (but in the case of TCP we reach, for the same reason due to fit more data into a single segment, a better throughput with respect to the same configuration).

High Traffic Load: In the case of *high* traffic load, results obtained in this analysis show that the model proposed by Bianchi can not be used as an upper bound in all analyzed configuration. More precisely, in the case of UDP protocol the Bianchi curve represents still an upper bound. At the opposite site, in the case of TCP protocol, we measured real throughput that overcame the values indicated in the Bianchi model. In the case of UDP protocol, both at sender and receiver side the configuration with AP shows lowest performance. The other configurations show substantially the same performance. In the case of TCP protocol, where there is a palmtop at receiver side, the configuration with AP shows lowest performance. Finally, using the TCP protocol we observed that all analyzed ad-hoc configuration show best performance. This behavior is due to the same motivation presented in the previous subsection: in this case we are far from MTU and with a saturated channel. We repeated the experiment with a packet size equal to 1500 bytes and the same IDT and we measured that UDP performs better than TCP. We don't provide this graphics because we have a low achieved throughput (in the case of PS = 1500 bytes, IDT = $\frac{1}{10000}$ we have a data rate equal to 120 Mbps over a 11 Mbps channel).

6   Conclusions

We presented a general framework for empirical performance study of heterogeneous wireless networks introducing a per *"traffic load"* class analysis: we defined three traffic conditions and we divided our experimentation in three stages: *low* traffic load, *medium* traffic load, *high* traffic load. A number of tests conducted on our real testbed yielded important characteristics such as throughput, delay, jitter and packet loss under various network loads in UDP and TCP scenarios. We carried out our results by introducing a novel traffic generator, named D-ITG.

References

1. G. Bianchi, "Performance Analysis of the 802.11 Distributed Coordination Function", *JSAC* Vol. 18 Issue: 3, March '00 pp. 535-547
2. C. K. Toh, M. Delwar, D. Allen, "Evaluating the communication performance of an ad hoc wireless network", *Wireless Comm., IEEE Transactions on* , Vol. 1 Issue: 3 , July '02 pp. 402 -414
3. F. Eshghi, A. K. Elhakeem, "Performance analysis of ad hoc wireless LANs for real-time traffic", *JSAC*, Vol. 21 Issue: 2, Feb. '03 pp. 204 -215
4. B. Bing, "Measured performance of the IEEE 802.11 wireless LAN", *LCN'99*, Oct.'99
5. J. P. Singh, N. Bambos, B. Srinivasan, D. Clawin, "Wireless LAN performance under varied stress conditions in vehicular traffic scenarios", *VTC, 2002.* Vol. 2, Sept. 02 pp. 743 -747
6. A. Messier, J. Robinson, K. Pahlavan, "Performance monitoring of a wireless campus area network", *LCN97*, Nov. 1997 pp. 232 -238
7. G. Xylomenos, G. C. Polyzos, "TCP and UDP performance over a wireless LAN", *INFOCOM '99*. Vol. 2, March 1999 pp. 439 -446
8. D-ITG, Distributed Internet Traffic Generator: http://www.grid.unina.it/software/ITG
9. IP Performance Metrics (IPPP), IETF Working Group http://www.ietf.org/html.charters/ippm-charter.html.
10. M. Carvalho and J.J. Garcia-Luna-Aceves, "Delay Analysis of IEEE 802.11 in Single-Hop Networks" *Proc. 11th IEEE ICNP*, Nov. 03.
11. H. Wu, Y. Peng, K. Long, S. Cheng, J. Ma, "Performance of Reliable Transport Protocol over IEEE 802.11 Wireless LAN: Analysis And Enhancement", *IEEE INFOCOM'2002*.

12. G. Iannello, A. Pescapé, G. Ventre, L. Vollero, "Experimental analysis of heterogeneous wireless networks", WWIC 2004, LNCS Vol. %2957. pp. 153 - 164
13. G. Xylomenos, G. C. Polyzos, P. Mahonen, M. Saaranen, "TCP Performance Issues over Wireless Links", *IEEE Communications Magazine*, Vol. 39 N. 4, 2001, pp. 52-58.
14. T. Demir, C. Komar, C. Ersoy, "Measured Performance of an IEEE 802.11 Wireless LAN"
15. H. Balakrishnan, V. N. Padmanabhan, S. Seshan, R. H. Katz, "A comparison of mechanisms for improving TCP performance over wireless links", *SIGCOMM '96*, Aug. 96, pp. 256-267.
16. G.T. Nguyen, R.H. Katz, B. Noble, M. Satyanarayanan, "A tracebased approach for modeling wireless channel behavior", *Winter Simulation Conference*, Dec. 96, pp. 597-604.

Performance Improvement of Hardware-Based Packet Classification Algorithm*

Yaw-Chung Chen[1], Pi-Chung Wang[2], Chun-Liang Lee[2], and Chia-Tai Chan[2]

[1] Department of Computer Science and Information Engineering,
National Chiao Tung University, HsinChu, 300 Taiwan, R.O.C.
ycchen@csie.nctu.edu.tw
[2] Telecommunication Laboratories, Chunghwa Telecom Co., Ltd.,
7F, No. 9 Lane 74 Hsin-Yi Rd. Sec. 4, Taipei, 106 Taiwan, R.O.C.
{abu, chlilee, ctchan}@cht.com.tw

Abstract. Packet classification is important in fulfilling the requirements of differentiated services in next generation networks. In the previous work, we presented an efficient hardware scheme, *Condensate Bit Vector*, based on bit vectors. The scheme significantly improves the scalability of packet classification. In this work, the characteristics of *Condensate Bit Vector* are further illustrated, and two drawbacks that may negatively affect the performance of *Condensate Bit Vector* are revealed. We show the solution to resolve the weaknesses and introduce the new schemes, *Condensate Ordered Bit Vector* and *Condensate and Aggregate Ordered Bit Vector*. Experiments show that our new algorithms drastically improve the search speed as compared to the original algorithm.

1 Introduction

Packet classification has been extensively employed in the Internet for secure filtering and service differentiation by administrators to reflect policies of network operations and resource allocation. Using the pre-defined policies, the packets can be assigned to various classes. However, packet classification with a potentially large number of policies is difficult and exhibits poor worst-case performance.

In the previous work [1], we presented an efficient hardware scheme, *Condensate Bit Vector*, based on bit vectors. The scheme significantly improves the scalability of packet classification. In this work, the characteristics of *Condensate Bit Vector* are further illustrated, and two drawbacks that may negatively affect the performance of *Condensate Bit Vector* are revealed. In the following, we present how to resolve the weaknesses and introduce the new *Condensate Ordered Bit Vector* (CoBV) and *Condensate and Aggregate Ordered Bit Vector* (CAoBV) schemes. Experiments demonstrate that the new schemes drastically improve the search speed as compared to the original algorithm.

* This work is supported in part by the National Science Council under Grant No. NSC93-2752-E-009-006-PAE.

P. Lorenz and P. Dini (Eds.): ICN 2005, LNCS 3421, pp. 728–736, 2005.

The rest of this paper is organized as follows. Section 2 introduces related works on packet classification and describes the ideas of Lucent BV, Aggregate Bit Vector (ABV) and Condensate Bit Vector (CBV) in details. Section 3 describes the main drawbacks of the CBV algorithm and presents CoBV and CAoBV schemes. Section 4 evaluates the performance of the proposed scheme. Finally, Section 5 concludes the work.

2   Related Works

Recently, researchers got more interested in the packet classification and had proposed several algorithms to solve the issue [2]. The related studies can be categorized into two classes: software-based and hardware-based. Several software-based schemes have been proposed in the literature such as grid of tries/cross-producting, tuple space search, recursive-flow classification, fat inverted segment trees, and hierarchical intelligent cuttings solutions [3–8]. The software-based solutions do not scale well in either time or storage, thus we focus on the hardware-based solutions.

In [9], Lakshman and Stiliadis proposed a scheme called Lucent BV scheme. By constructing k one-dimensional tries, each prefix node in every one-dimensional trie is associated with a bit vector (bv). Each bit position maps to a corresponding policy in the database. The policy database is sorted by an descendent order of priority. By performing **AND** operation on all matched bv, the first matched result is the best matching policy. Since the length of bv increases proportional to the number of policies, the Lucent BV scheme is only suitable for small-size policy databases.

Baboescu *et al.* [10] proposed a bit-vector aggregation method to enhance the BV scheme. By adding *aggregate bit vector* (abv), the number of memory accesses is decreased significantly. While the ABV scheme improves the average speed of the BV scheme, it increases the required storage by appending extra abv for each bv as well.

The CBV scheme [1] further improves the performance of ABV by merging multiple policies into one. Thus, both the speed and storage performance can be improved. The CBV scheme consists of three steps. In the first step, the prefixes extracted from the policies are used to construct the binary trie. For each one-dimensional binary trie, the following procedure is executed to mark the subtrie root according to the pre-defined threshold. It checks whether the number of prefixes under the current node is equal to the threshold by traversing the binary trie with the depth first order. If *yes*, a subtrie root is marked. Otherwise, its left and right child nodes will be traversed recursively. Second, the prefixes in each policy are replaced by the bit-streams corresponding to the nearest ascending subtrie roots. Since new policies might be redundant, the duplicate copies are merged and the indices of the original policies are appended to the new policy. In the third step, the bvs are generated based on the new policies, namely cbv. Because the number of policies in the new database is reduced, the required bits

Table 1. Two-dimensional policy database with 16 rules

Index	Source Prefix	Dest. Prefix	Index	Source Prefix	Dest. Prefix	Index	Source Prefix	Dest. Prefix	Index	Source Prefix	Dest. Prefix
F_0	000*	11*	F_4	111*	00*	F_8	10*	110*	F_{12}	000*	010*
F_1	0*	0000*	F_5	0*	110*	F_9	011*	10*	F_{13}	011*	010*
F_2	1*	1111*	F_6	000*	1111*	F_{10}	000*	0110*	F_{14}	1*	00*
F_3	01*	010*	F_7	10*	1110*	F_{11}	011*	0*	F_{15}	*	00*

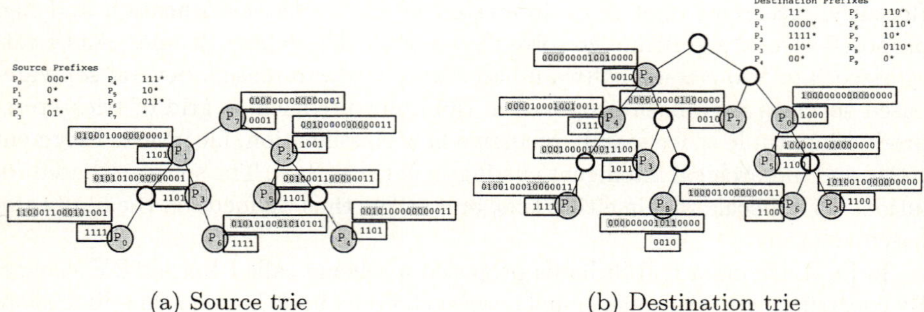

(a) Source trie   (b) Destination trie

Fig. 1. Bit vectors constructed by BV and ABV scheme

in each bit vector is less than that in the Lucent BV. Furthermore, the number of different prefixes in each dimension is reduced to generate fewer vectors.

We use a two-dimensional policy (*source prefix, destination prefix*) database with 16 policies in Table 1 to illustrate the bit-vector construction of the Lucent BV, ABV and CBV schemes. The prefix nodes for the source and destination tries are depicted as the shady nodes in Fig. 1. Each prefix node in both tries is labeled with two bit vectors, *bv* (16 bits) and *abv* (4 bits). The *bv* length is equal to the number of policies in the database and the length of *abv* is altered by the aggregate size (4). For an incoming address pair (00001010, 01001100), Lucent BV scheme uses the source address 00001010 to walk through the source trie and receives the *bv* value "1100011000101001". Similarly, it gets a *bv* value "0001000010011100" in the destination trie. Finally, Lucent BV scheme performs an **AND** operation on these two *bvs* and obtains a result of "0000000000001000" showing that the matched policy is F_{12}. With the ABV scheme, an **AND** operation on the *abvs* "1111" and "1011" yields result "1011". It indicates that the second 4-bit segment will not contain any set bits and could be ignored.

Next, the procedures of CBV scheme is presented. Assuming that the number of clustered prefixes is 2. After performing the proposed algorithm to the trie in Fig. 1, the constructed subtrie roots are shown in Fig. 2. Each dark-gray circle represents the position of the subtrie root. Next, the original prefixes in each policy are replaced by the bit-streams corresponding to the nearest ascending subtrie roots. For example, F_0 (000*, 11*) in Table 1 is changed to (0*, 11*), and F_6 (000*, 1111*) is changed to (0*, 111*). Some new policies are redundant,

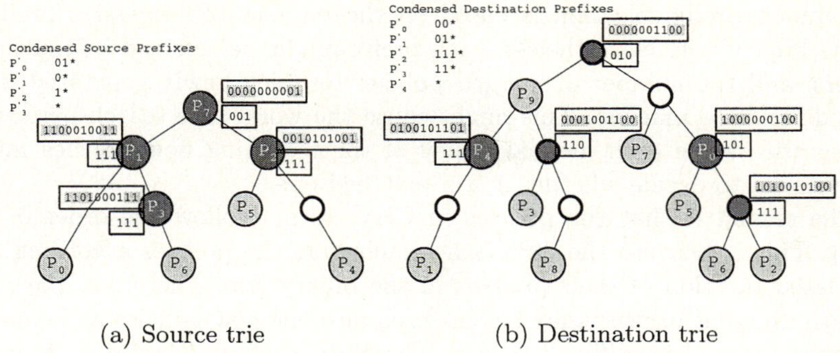

(a) Source trie   (b) Destination trie

Fig. 2. Bit Vector Condensation

Table 2. New Policy Database

Index	First Dimension	Second Dimension	Included Policies	Index	First Dimension	Second Dimension	Included Policies
F_0'	0*	11*	F_0, F_5	F_5'	0*	111*	F_6
F_1'	0*	00*	F_1	F_6'	1*	11*	F_8
F_2'	1*	111*	F_2, F_7	F_7'	01*	*	F_9, F_{11}
F_3'	01*	01*	F_3, F_{13}	F_8'	0*	01*	F_{10}, F_{12}
F_4'	1*	00*	F_4, F_{14}	F_9'	*	00*	F_{15}

e.g. the new policies generated by F_0 and F_5. These policies are merged, and
their indices are appended to the new policy. Table 2 presents the new policy
database. Then, the bv is constructed based on the new policy database. The
16-bit bvs is transformed into new 10-bit bvs, as shown in Fig. 2. In addition, the
number of bvs is reduced from 18 to 9 with the proposed scheme. Also in Fig. 2,
the condensate bit vector with aggregation is illustrated, namely *condensate and
aggregate bit vector* (*cabv*). The aggregate size is defined as 4 in these cases. The
construction of *cabv* follows the same procedure as that of *abv*.

3   Condensate Ordered Bit Vector (CoBV)

As described above, the CBV scheme is simple and efficient. By merging policies according to their prefixes, the number of bit vectors and their lengths can
be significantly reduced, hence the totally required storage is dramatically decreased. In addition, the CBV scheme combining ABV can further improve the
performance of packet classification. Nevertheless, the native CBV scheme has
two drawbacks.

 – The first is the way that the CBV scheme clusters the policies. As described
   in the first step of *cbv* generation, the common prefixes of the original prefixes in the policies are selected from the binary trie. The threshold for the

common prefix selection is based on the number of successive prefixes in the binary trie. Nevertheless, each prefix might be referred by multiple filters, and the number of merged policies for each newly generated policies is difficult to manage. This might cause the worst-case search performance degraded since each merged policy of the matching new policies must be retrieved to decide whether it is a matched one.

– The second is that the policies in CBV do not follow the order of priority. This is because the CBV scheme clusters the policies according to the related position of their prefixes in the binary trie. Therefore, the policies with different priority may be clustered into one new policies and causes the generated policies difficult to sort. Therefore, each matched policies must be retrieved to decide the one with least cost. Consequently, the average search performance is decreased. However, the worst-case search performance is not affected since the whole *cbv* is traversed in the worst situation. The same phenomenon also occurs in ABV, where the policies are reordered in order to improve the performance of aggregation.

Aiming to the two drawbacks described above, we propose two new ideas, *clustering based on the number of filters* and *bit vector ordering*, to improve the CBV scheme. Next, the detailed procedures of these two ideas are described.

The first drawback is resolved by introducing a different threshold for prefix generating. The original CBV scheme generates prefixes according to the number of successive prefixes. However, the number of successive prefixes is less meaningful for packet classification since it only reflects one-dimensional information. To correct this, the number of filters, which refer to the successive prefixes, is used to decide whether the successive prefixes are clustered. The first step of CBV construction procedure is modified as follows. While constructing the binary tries, each prefix node is tagged with the number of referring policies. Next, the prefix generating procedure decides whether the current node is marked as a prefix by summing up the number of referring policies in every successive prefix. If the resulted value is larger than the predefined threshold, the current node is marked as a prefix node. The path from root to the prefix node is then extracted as a new prefix.

To resolve the second drawback, we cluster only the policies with identical priority. The first step of CBV construction procedure is the same. After generating the prefixes according to the predefined threshold, the original prefixes in the policies are replaced by the newly generated prefixes. In the CBV scheme, the policies with identical prefixes are merged. The step is modified by only merging the policies with identical prefixes and priority. The resulted bit vector is called *condensate ordered bit vector* (cobv). Consequently, the search procedure can be terminated by retrieving the first matching policies. Hence the average search performance can be improved. However, the effect of policies clustering is reduced and might increase the required storage. To alleviate the storage expansion caused by *cobv*, a larger threshold for prefix generating is necessary.

By combined the new procedures with bit aggregation, the *condensate and aggregated ordered bit vector* (caobv) is generated. In the next section, we demonstrate that CoBV and CAoBV outperform the existing schemes.

4   Performance Evaluation

In this section, we evaluate the performance of the CoBV and CAoBV schemes and compare it with Lucent BV, ABV and CBV schemes. To test the scalability of our CBV scheme, the synthetic databases are used to evaluate whether CBV scheme could accommodate the future network service configurations. The synthetic databases are generated by randomly selecting the source and destination addresses in these 22 classifiers. The size of our synthetic databases varies from 10K to 100K. Assume that the memory word size and the aggregate size of ABV are both 32 bits, identical to the settings used in [10].

Two performance metrics are measured in our experiments: the storage requirement and the classification speed. The required storage mainly ties to the number of bit vectors and their length. The number of bit vectors is equal to the number of prefix nodes, while their lengths is identical to the number of policies. The speed of each packet classification is measured in terms of the number of memory accesses.

In our experiments, the numbers of clustered filters are set to 32, 64, 128, 256. Since the numbers of clustered filters do not affect the Lucent BV and ABV schemes, their storage requirement and speed will remain constant. In the

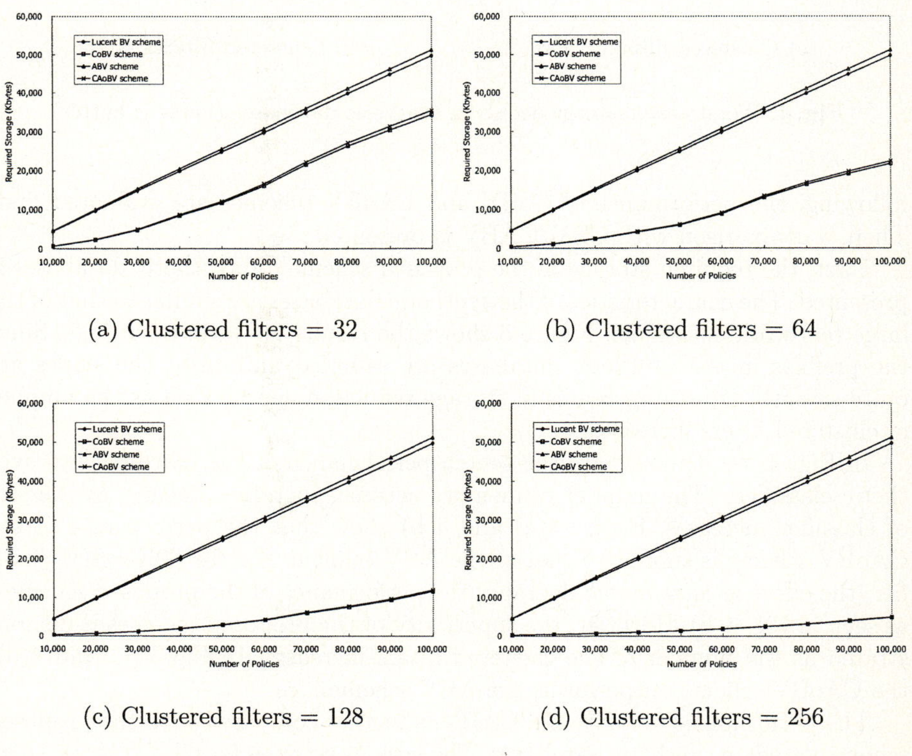

(a) Clustered filters = 32   (b) Clustered filters = 64

(c) Clustered filters = 128   (d) Clustered filters = 256

Fig. 3. Storage requirement in synthetic databases (lower is better)

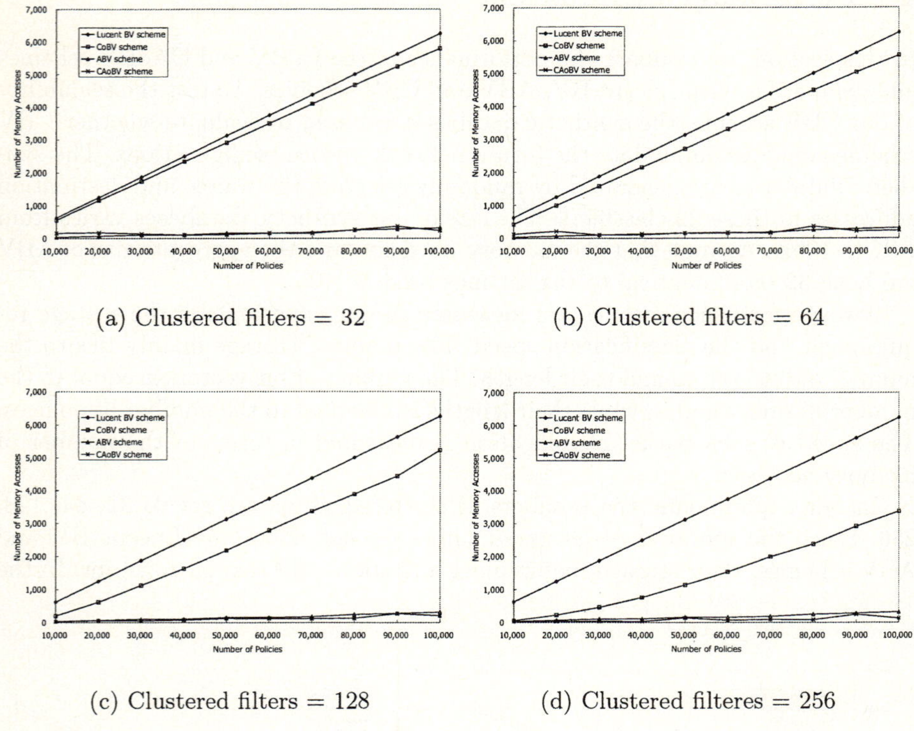

(a) Clustered filters = 32

(b) Clustered filters = 64

(c) Clustered filters = 128

(d) Clustered filteres = 256

Fig. 4. Worst case memory access in synthetic databases (lower is better)

following, the performance of CoBV and CAoBV schemes are evaluated first. Then, a comparison with CBV/CABV is presented.

First, the required storage of the proposed scheme for synthetic databases is presented. The characteristics of the synthetic databases are similar to that of the large real-world classifiers. Figure 3 shows the results for various settings. Since the prefixes in the synthetic databases are sampled uniformly, the slopes are quite smooth. Drastic increases in storage reduction can be seen as the number of clustered filters increases.

In Fig. 4, we demonstrate the search performance of our schemes with synthetic classifiers. The number of memory accesses increases linearly as the size of classifier increases. Figure 4(a) and 4(b) show that the performance of the CAoBV scheme is similar to that of the ABV scheme. For the 80K-entry classifier, the effect of *false match* degrades the performance of the proposed schemes, as seen in Fig. 4(b). However, the superiority of the proposed schemes is demonstrated as the number of the clustered filters increases. In Fig. 4(c) and 4(d), the CAoBV scheme outperforms the ABV scheme.

The performance of CBV and CABV is further compared with the proposed scheme based on synthetic databases. The various settings for CBV/CABV (p=2 or 4) and CoBV/CAoBV (f=128 or 256) are adopted since their properties are

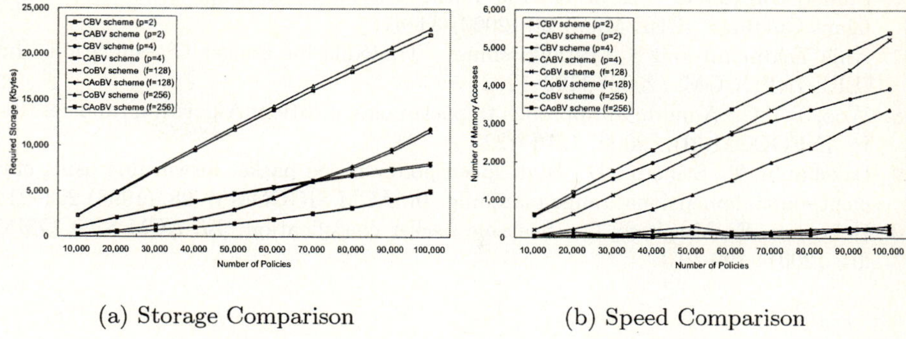

(a) Storage Comparison   (b) Speed Comparison

Fig. 5. Comparisons with CBV and CABV

quite different. As the number of clustered filter is large enough, the required storage of CoBV/CAoBV can be less than that of CBV/CABV. In the mean time, CoBV/CAoBV schemes can also achieve better search performance. Therefore, CoBV/CAoBV schemes can feature better scalability as compared with existing bit-vector algorithms.

5   Conclusions

In this work, we proposed the *condensate ordered bit vector* to improve the classification performance of bit-vector algorithms. The ideas of CoBV include *clustering based on the number of filters* and *bit vector ordering*. While the first improves the worst-case search performance, and the second promotes the average performance. By demonstrating the performance based on real and synthetic policies databases, the CoBV and CAoBV schemes can outperform CBV/CABV schemes and other existing bit-vector algorithms. In conclusion, the flexibility of CoBV/CAoBV schemes can significantly improve the performance of packet classification for large policy databases.

References

1. Chang, H., Chan, C., Wang, P., Lee, C.: A scalable hardware solution for packet classification. In: IEEE ICCS '2004. (2004) 542–546
2. Gupta, P., Mckneown, N.: Algorithms for packet classification. In: IEEE Network Magazine. (2001) 24–32
3. Srinivasan, V., et al.: Fast and scalable level four switching. In: ACM SIGCOMM '98. (1998) 191–202
4. Srinivasan, V., Suri, S., Varghese, G.: Packet classification using tuple space search. In: ACM SIGCOMM '99. (1999) 135–146
5. Gupta, P., McKeown, N.: Packet classification on multiple fields. In: ACM SIGCOMM '99. (1999) 147–160

6. Pankaj Gupta and Nick McKeown: Packet Classification using Hierarchical Intelligent Cuttings. IEEE Micro **20** (2000) 34–41
7. Anja Feldmann and S. Muthukrishnan: Tradeoffs for Packet Classification. In: IEEE INFOCOM. (2000) 1193–1202
8. Woo, T.Y.C.: A modular approach to packet classification: Algorithms and results. In: INFOCOM (3). (2000) 1213–1222
9. Lakshman, T., Stiliadis, D.: High-speed policy-based packet forwarding using efficient multi-demensional range matching. In: ACM SIGCOMM '98. (1998) 203–214
10. Baboescu, F., Varghese, G.: Scalable packet classification. In: ACM SIGCOMM '01. (2001) 199–210

Analyzing Performance Data Exchange in Content Delivery Networks[*]

Davide Rossi and Elisa Turrini

Department of Computer Science, University of Bologna,
Mura Anteo Zamboni, 7 - 40127 Bologna - Italy
`{rossi, turrini}@cs.unibo.it`

Abstract. In a Content Delivery Network a set of geographically distributed surrogate servers, holding mirror copies of resources obtained from an origin server, are used to improve the perceived quality of a web portal. The Request Routing System is the component of a CDN responsible for both redirecting user requests to the best surrogate and collecting, usually from the surrogates themselves, the performance information needed to make redirection decisions. The accuracy of these performance information is very important in order to make right decisions that lead to an improvement of the *Quality of Experience* perceived by the users. In this paper we propose and evaluate algorithms for sending performance information in a CDN. The goal of the proposed algorithms is to guarantee the better performance information accuracy with the lowest possible impact on the network traffic. To the best of our knowledge, our work represents the first contribution in this research direction.

1   Introduction

In a Content Delivery Network (CDN) a set of geographically distributed surrogate servers, holding mirror copies of resources obtained from an origin server, are used to improve the perceived quality of a web portal. Client requests that reach the CDN are automatically rerouted to the most appropriate surrogate. The policy to identify the best surrogate depends on the CDN, and it is usually chosen in order to improve the Quality of Experience (QoE) [1] perceived by the users. The QoE is a quality that, in a Web surfing session, is usually bound to the time lag between the user gesture and the loading of the selected resource.

In a CDN, the Request Routing System (RRS) [2] is responsible for making rerouting decisions. These decisions depend on a set of parameters that characterize the CDN (bandwidth of network links, server load, network load, etc...). These parameters are usually heuristically composed in order to identify the best surrogate. Needless to say, the more precise the heuristic is, the better the choice

---

[*] This work was partially supported by the EU project TAPAS (IST-2001-34069) and the WebMinds FIRB project of the Italian Ministry of Education, University and Research.

P. Lorenz and P. Dini (Eds.): ICN 2005, LNCS 3421, pp. 737–745, 2005.

will be. In order to implement good heuristics, the RRS needs a snapshot of the state of the CND as precise and fresh as possible; the main mechanism to obtain fresh measures about the CDN is collecting the reports that each surrogate server sends to the RRS. These reports inform the RRS about parameters such as the state of the network and the load of the surrogates that can be combined with the position of the client in the network in order to make a good routing decision.

Given the paramount number of potential clients, it is impossible to obtain estimates of the link status between a surrogate and each possible client. This leads to partitioning the clients into a reasonable amount of homogeneous zones. Given the complexity of the Internet, the number of zones cannot be too small; since the surrogates should report to the RRS a set of information that includes performance information whose size depends on the number of zones in which the Internet has been split, this can be quite a large amount of data.

In this paper we compare a set of algorithms that reduce the size of surrogate reports trying to minimize the differences between the data seen by the surrogate and those seen by the RRS. The effectiveness of these algorithms is strongly affected by the characteristics of the measured performance information (that cannot be estimated with simple simulation models). For this reason we used as a test bed a set of data resulting from measurements of real network traffic.

This paper is structured as follows: section 2 introduces the routing mechanisms in a CDN; section 3 shows that a large amount of data has to be exchanged among the components of a CDN; section 4 introduces and compares a set of algorithms to optimize surrogate reports; sections 5 and 6 discuss our future work and conclude the paper.

2   Request Routing in a Content Delivery Network

As explained before, the RRS is responsible for redirecting user requests to a surrogate that in turn will satisfy them. Basically, this redirection can be performed using mechanisms such as the Domain Name System (DNS), HTTP redirection, and URL rewriting (a detailed discussion is available in [2]).

Whatever the redirection mechanism implemented by the RRS is, the rerouting decisions are driven by a redirection policy. For instance with a Round-Robin policy the requests are redirected to each surrogate in a fixed order. Similarly, with a Random policy the surrogate is selected at random. Of course policies as simple as that lead to very sub-optimal choices of client-surrogate coupling.

More complex policies are based on some metrics, like server load and network delay. We can broadly separate these metrics in two groups:

- Network proximity: it indicates how much a surrogate is *close* to a given client. It can be measured with different sub-metrics, such as latency, hops count and packets loss.
- Server Load: it is related to the number of users connections and represents the server's capacity to quickly satisfy new requests. It is measured with

sub-metrics, as CPU load, number of dropped requests and number of connections being served in a given time interval (or average time necessary to serve a request).

In this work we call *performance data* the data expressed in these metrics. In order to improve the user QoE, it is important that the redirection policies are based on values that are a good approximation of the response time perceived by the user. Moreover, it is also important to perform a continuous resource monitoring in order to promptly signal the sudden changes in the network and server conditions (see [3] for a good example on how dynamic active measures can improve a replica selection policy). Policies that, on the contrary, do not need surrogates feedback and use only static metrics have the non negligible limit of not taking into account the variations in the network and server load.

3   A Scale Problem

Content Delivery Networks are typically used for web portals that can attract a large amount of traffic, often from every region of the planet. In order to associate each client with the best surrogate, the RRS has to know the network status and the status of the surrogates. To this end, the surrogates continuously measure performance data and periodically send reports to the RRS. In an ideal setting, these reports should include the surrogate load and the network proximity with respect to all the possible clients. This is obviously impossible. First because gathering a network proximity measure (be it estimated or be it real, obtained using active or passive probing) for each host in the Internet is unfeasible; second because, even if this is possible, it would result in a huge amount of data that has to be included in each report. The solution adopted in all CDNs is to partition the Internet in a set of *clusters* and to associate a unique network proximity value to all the nodes inside the same cluster. For this mechanism to work effectively all the nodes inside a cluster must have homogeneous network proximity; this surely prevents from creating clusters as large as a continent (even if it looks that some commercially deployed CDNs use similar partitioning schemes, no wonder they work in a quite sub-optimal way [4]).

In [5], the Internet is split in clusters trying to group nodes under distinct administrative domains on the basis of information gathered from BGP routing tables: in an experiment related to clients that made connections to a single server they identified 9.853 distinct domains. In [6], the Internet is split in clusters trying to group nodes under similar network proximity: in an experiment related to clients that connected to a single server (with even fewer hits than that of the previous experiment) they identified 17,270 distinct domains. It is then obvious that, even after clustering, the amount of data that a surrogate has to report to the RRS is very high. It is true that the surrogates, be they using passive or active probing, will never have fresh proximity information for each cluster (leading to partial reporting), nevertheless, given the high total amount of clusters, the size of the data to be sent can easily impact adversely on the network traffic.

It is clear that two adverse needs have to be conciliated: sending detailed and fresh performance data to the RRS and minimizing the extra network traffic due to CDN's service messages. In order to do this, we have to characterize the values contained in a typical report. Typically these data are not obtained via direct measurements but are heuristically derived composing parameters such as round-trip time, latency, number of drop packets and others. It turns then out that the performance data sent in a report depends on the heuristics used by the CDN (and even from its policy: improving QoE rather than optimizing load or others). Given that, characterizing these data is a task that could depend on the systems used for their measurement/estimate.

Since we want to be able to apply the results of our studies to the broader number of cases we decided, in the analysis of the algorithms, to use the real data that the heuristics try to approximate: the client's access time to a resource. To this end, we collected a set of access time measures to a 10 kilobytes resource [7] from several clients spread in different zones of the Internet.

4   Report Data Optimization Algorithms

In order to test (and implement) a report data optimization algorithm, we have to outline the structure of the system we are targeting. In this paper we assume a very simple CDN structure composed by a set of geographically distributed surrogates and a single, centralized, RRS component. While this oversimplified structure surely eases the task of testing and comparing the report algorithms, it does not alter significantly the results with respect to a more complex system where (as it usually happens) the RRS is a distributed system itself. The point is that whether a surrogate has to send its reports to a centralized entity or to a distributed one the reporting problem holds very similar, so that our analysis can be applied in both cases. Another assumption we have to make relates to the format of a surrogate report. We assume that a report can hold a server load record and several distance records. The distance records are used to report the network proximity (measured or estimated using some heuristic method) between the surrogate and a given zone.

Since the zones can change dynamically and it is not feasible to design a system in which all of the distributed components see the same Internet partitioning scheme at any given point in time, the zones cannot be indexed but have to be explicitly identified with a set of IP addresses. We assume the most commonly used format: a CIDR record [8]. This allows to create groups with different sizes that contain homogeneous hosts.

The best way to evaluate a report algorithm is analyzing how its behavior affects the effectiveness of the rerouting policy and, in the end, the QoE perceived by the users. Unfortunately, putting at work such an experiment requires an implementation of a CDN which is quite a complex system (see [9]). For our comparison we decided to use the following metrics.

Network traffic generated. It depends both on the numbers of records sent from the surrogate to the RRS and on the size of the records. The traffic generated not only can adversely affect the network conditions but also can overload the machine receiving the performance data. In general, the more performance data are sent, the more the RRS needs to allocate resources managing the performance data received.

Performance data coherence. It indicates how much the performance data stored by the RRS are synchronized with the performance data collected by the surrogates. It can be estimated using different metrics.

In our analysis we decided to estimate the performance data coherence with two measures: the inverse of the signal-to-noise ratio and the out-of-range count. The inverse of the signal-to-noise ratio (1/SNR) is a metric widely adopted to compare network traces and is obtained as: $\sum e^2 / \sum s^2$ where s is the performance data at the surrogate and e is the difference between the performance data at the surrogate and the performance data at the RRS.

The out-of-range count (OOR) is a measure that tries to capture how a datum at the RRS relates to the changes of the same datum at the surrogate. To this end we calculated two sequences, called upper and lower sequence, that demarcate the sequence of changes of each datum at the surrogate. The sequences are obtained by low-pass filtering the sequence of the datum and joining all the points of the sequence that are above the filtered values and all the points of the sequence that are below the filtered values. The out-of-range count measures the number of times the datum at the RRS falls outside the upper-lower sequences.

We also investigated discrepancy functions like the FIT proposed in [10] but given the different nature of the data we did not obtain results more significant than with 1/SNR.

We analyzed and compared several algorithms in order to identify the one that gives the better trade-off between data coherence and the generated network traffic. Since sending performance data as soon as they are changed would generate a huge amount of traffic, the idea behind all algorithms is that of trying to reduce the number of records sent affecting as less as possible the performance data coherence. It is also worth noting that the distribution of the performance data strongly affects the report algorithm performance, that means that the algorithms do not have to find the better trade-off given a generic set of data but have to find the better trade-off when applied to web traffic data. As an example of web traffic data, in order to appreciate visually its characteristics, in Figure 1 we report a chart of one of the traces we collected for our evaluations. It must be noted that the access time is far from constant (as shown in the magnification). This is the kind of data our algorithms are expected to serve better (for a good characterization of web traffic see [11]).

In the algorithms we are going to analyze, we say that a record is *s-changed* if it is "significantly" changed with respect to the previous update; we say a record is *Ready To Te Sent* (RTBS) when, according to a specific algorithm policy, the record should be contained in the report to be sent from the surrogate to the RRS. We also assume the updates of the performance data at the surrogate side happen at fixed time intervals. Our time unit is this update interval.

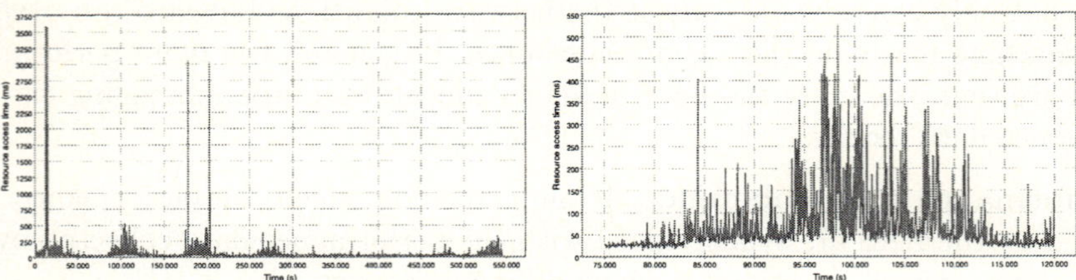

Fig. 1. An access time trace and its magnification

The algorithms are:

Naive. According to this algorithm, a record is marked as RTBS when its value is s-changed with respect to a δ parameter. Every f updates at the surrogate side, a report containing all records marked as RTBS are sent to the RRS. This guarantees that the performance data incoherence at the RRS side cannot last for more than a time corresponding to f updates. However, this approach has the problem that given a report at time t, a record that becomes RTBS at the time $t+1$ has to wait until the time $t+f$ before being sent.

Aging. As shown in Figure 1, the performance data are characterized by the presence of *spikes*, that is temporary instantaneous peaks. The approach implemented by the Aging algorithm is based on the consideration that it might be convenient not to send spikes for two reasons. First: sending to the RRS performance data concerning spikes can be unnecessary expensive, since, in order to guarantee a little bit more of coherence, we have to send two flows of opposite performance data, one after the other in a short period of time. Second, spikes represent temporary and local variations. Temporary because they strongly depend on the instant in which the request has been sent. Local because they may depend on the Web-server scheduling algorithm behavior. Then, since the RRS uses performance data received from the surrogates to redirect the user requests, it does not seem to make much sense to base this decision on spikes. This consideration stands at the basis of the Aging approach. It consists in trying to avoid sending transient changes, sending performance data only when they seem to be steady changed. To this end, we have introduced the *age* parameter. A record becomes RTBS when its value s-changes from when it was sent the last time for at least *age* times consecutively. This guarantees that new performance data are sent only when the change seems to be non-temporary. It also allows to smooth the effect of jitter, that is the values variation around the mean. Specifically, setting the *age* parameter with a value more than one (for instance 4 or 5 resulted as the more appropriate in our tests), most spikes are not sent.

Averaging. The observation at the basis of this algorithm is that not considering spikes at all can be misleading. Indeed, if several spikes are present in a given time interval, this means that the user resource access time can be compromised. The rationale is that sending spikes is a bad idea but ignoring the fact that a link exhibit rapid variations is not good either. The algorithm consists in computing an exponential weight moving average filter:

$$\overline{x}_{t+1} = \alpha\overline{x}_t + (1 - \alpha)x_t$$

where $\overline{x}$ is the average, x is the performance datum and α is a parameter (ranging from 0 to 1) that specifies the impact of fresh data. At every update the average is updated and it is compared with the one previously sent. If the difference is grater than δ then the current average (not the performance datum!) is sent.

Our tests show that Averaging exhibits better performances with a very high values of α (in the range 0.95 - 0.97); this is due to the high jitter we have in all the collected traces.

The table below shows a comparison of the Naive, Aging and Averaging algorithms applied to ten traces (corresponding to ten different zones) relative to 5 days of measures (one each 30 seconds). All the algorithms sent 2700 records.

Table 1. Algorithms comparison: 1/SNR and out-of-range count

Algorithm	1/SNR	OOR count
Naive	6.86	69881
Aging	5.94	38991
Averaging	5.32	53450

Note that this is one of the many possible comparisons. First we decided to compare the algorithms when they send the same amount of records, we could have checked the number of records sent when the algorithms have the same 1/SNR or the same OOR count. Our decision is based on the observation that in a real system network traffic minimization is probably more relevant. We performed, however, different kind of comparison tests not shown here and we collected consistent results. Moreover it should be noted that the same algorithm with different parameters (value of δ, etc...) gives different results. The results we are showing here are based on the best tuned algorithms.

In the evaluation of the algorithms we obtained the results we were expecting: by adding knowledge about the nature of the traffic (as in Aging and in Averaging) we achieved better results in term of 1/SNR; results that are even better with respect to the OOR count, that is reasonable considering the fact that this last measure tries to capture the characteristics of the web traffic more than the 1/SNR does.

While Averaging outperforms Aging in terms of 1/SNR it does worst in term of OOR count. This is not unexpected if we think that Averaging gives a smoothed, somehow delayed, vision of the data.

To visually understand the behavior of Averaging with respect to Aging in Figure 2 the data sent by the two algorithms are compared with the original trace (the same we shown before).

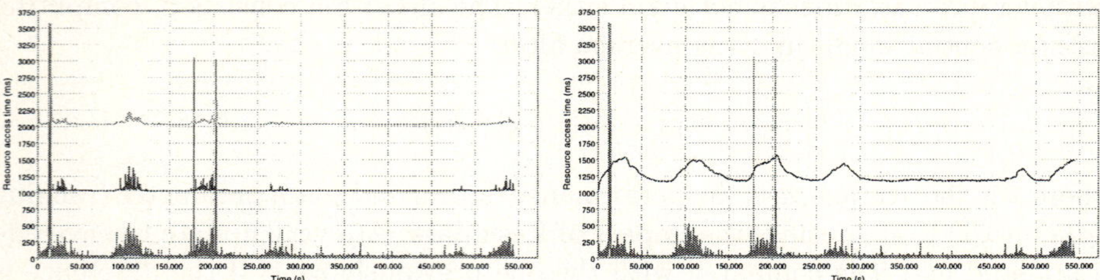

Fig. 2. Left: A comparison between Aging (shifted up 1s) and Averaging (up 2s)

Fig. 3. Right: The penalty function

5 Future Work

A path we are actively pursuing is to find a better metric to compare the algorithms in order to find a measure that better reflect the QoE perceived by the users. The metrics we used so far (1/SNR and out-of-range count) do not take into account how loaded resources are experienced by the users. Consider this very simple example: two surrogates serve the same zone with a constant average access time, but the second exhibits high jittering. Our best algorithm, Averaging, being an average filter, reports to the RRS records that are quite similar for both surrogates. But a user is usually more pleased with a server that exhibits a constant access time: waiting 1 second to access each page is perceived better that having to wait 0.2 seconds 3 times out of four and having to wait 3.4 seconds one time out of four when a link is clicked. We tried to address this problem by introducing a penalty factor based on the variance of access times presented at the surrogate. This way a surrogate with higher variance is reported as slower than it really is to the RRS. Figure 3 shows the penalty factor for the same trace displayed in Figure 1 (note that the range domain for the factor is different than the one of the access time: 1000ms corresponds to a factor of 1). It is easy to see how the factor grows when the jittering is higher. The penality factor is calculated with the formula:

$$p_{t+1} = \beta p_t + (1 - \beta) \frac{1}{(1 + |\overline{x}_t - x_t|)^\gamma}$$

that is similar to Averaging itself but it tries to average the divergences of the data in the sequence. The γ parameter determines how steeply the penalty factor raises with respect to jittering.

6 Conclusions

The more the web becomes the privileged way to get in contact with (potential) customers the more the attention of the content providers focuses on quality aspects that are not just bound to a nice looking web design but deal with a whole set of parameters that influence the QoE of the user. Strong *Quality of Service* constraints for the World Wide Web are unfeasible for its current infrastructure. Nevertheless the improvement of access times (latency, transfer time) related to web content is a relevant issue that becomes even more important with the spreading of high speed connections such as (A)DSL: a delay that can be tolerated by a dial-up user could be just too much for a user with a higher connection speed. Content Delivery Networks are the more advanced systems to improve the quality perceived by the web surfers but, in their actual incarnation, they appear to often use heuristics based on rough estimate of real network parameters. Surely a better knowledge of the network status can lead both to a better usage of the resources that compose a CDN and to a better service delivered the the users. The work we presented in this paper tries to address specific issues related to the increment of the user-perceived performances of a CDN: by improving the knowledge of the network the CDN can make better rerouting choices. We shown how the peculiarities of web traffic can be exploited in order to minimize the amount of data the surrogates have to report to the RRS; we compared different algorithms designed to this end and we tested them with data collected from real web traffic measures.

References

1. CacheFlow Inc.: Shared vs. dedicated content delivery network. White paper, 2001. `http://www.cacheflow.com/files/whitepapers/wp_shared_vs_dedicate.pdf`.
2. Barbir, A., Cain, B., Nair, R., and Spatscheck, O.: Rfc3568: Known content network (cn) request-routing mechanisms. IETF RFC, May 2002.
3. Carter, R.L., and Crovella, M.E.: Server selection using dynamic path characterization in wide-area networks. In *Proc. IEEE INFOCOM '97*, pages 1014–1021, 1997.
4. Johnson, K.L., Carr, J.F., Day, M.S., and Kaashoek, M.F.: The measured performance of content distribution networks. In *Proc. of the 5th International Web Caching and Content Delivery Workshop*, 2000.
5. Andrews, M., Shepherd, B., Srinivasan, A., Winkler, P., and Zane F.: Clustering and server selection using passive monitoring. In *Proc. The 21st Annual Joint Conference of the IEEE Computer and Communications Societies*, 2002.
6. Krishnamurthy, B., and Wang, J.: On network-aware clustering of web clients. In *Proc. ACM SIGCOMM 2000*.
7. Breslau, L., Cao, P., Fan, L., Phillips, G., and Shenker, S.: Web caching and zipf-like distributions: Evidence and implications. In *Proc. IEEE INFOCOM '99*, pages 126–134, 1999.
8. Fuller, V., Li, T., Yu, J., and Varadhan, K.: Rfc3568: Known content network (cn) request-routing mechanisms. IETF RFC, May 2002.
9. Dilley, J., Maggs, B., Parikh, J., Prokop, H., Sitaraman, R., and Weihl, B.: Globally distributed content delivery. *IEEE Internet Computing*, 6(5):50–58, September/October 2002.
10. Mitzenmacher, M., and Tworetzky, B.: New models and methods for file size distributions. In *Allerton 2003*.
11. Crovella, M.E., and Bestavros, A.: Self-similarity in world wide web traffic: evidence and possible causes. *IEEE/ACM Transactions on Networking*, 5(6):835–846, December 1997.

Passive Calibration of Active Measuring Latency

Jianping Yin, Zhiping Cai, Wentao Zhao, and Xianghui Liu

Department of Computer Science and Technology,
National University of Defense Technology, Changsha, 410073 China
jpyin@nudt.edu.cn, {xiaocai, bruce_zhao}@163.net, liuxh@tom.com

Abstract. Network performance obtained from the active probe packets is not equal to the performance experienced by users. To gain more exact result, the characteristics of packets gained by passive measuring are utilized to calibrate the result of active measuring. Taking the number of user data packets arriving between probe packets and the latency alteration of neighborhood probe packets into account, we propose the Pcoam (Passive Calibration of Active Measurement) method. The actual network status could be reflected more exactly, especially when the network is in congestion and packet loss. And the improvement has been validated by simulation.

1 Introduction

Measurements and estimation of performance parameters, such as end-to-end delay, in IP network are becoming increasingly important for today's operators [1], [2], [3], [4], [5]. In general, conventional schemes to measure the network delay are classified into two types, active and passive measurements. Active measurement measures the performance of a network by sending probe packets and monitoring them. In passive measurement, the probe device accessing the network records statistics about the network characteristics of data packets. Unfortunately, both types have drawbacks especially when they are applied to delay measurement [6].

Masaki Aida *et al.* have proposed a new measuring technique, called CoMPACT Monitor [6], [7], [8]. Their scheme requires both active and passive monitoring using easy-to-measure methods. It is based on change-of-measure framework and is an active measurement transformed by using passively monitored data. Their scheme can estimate not only the mixed QoS/performance experienced by users but also the actual QoS/performance for individual users, organizations, and applications. In addition, their scheme is scalable and lightweight.

The CoMPACT scheme supposes that the interval of sending probe packets can be very short and thus ensures that the interval of receiving those probe packets is also short. The error of estimator would increase as the interval of the probe packets arriving increases, especially when the network is in congestion or the loss ratio is high.

We can use the characteristics of user packets gained by passive measuring to calibrate the active measuring for more exact measuring result. We consider not only the number of user data packets, but also the relationship of the adjacent probe

P. Lorenz and P. Dini (Eds.): ICN 2005, LNCS 3421, pp. 746–753, 2005.

packets' delay. We propose the Pcoam method (Passive Calibration of Active Measurement) and our method could reflect the actual network status more exactly in the case of network congestion and packet loss.

The paper proceeds as follows. We analyze the adjacent packets latency alteration and propose the Pcoam method in section 2. In section 3, we show the validity of the Pcoam method by simulation. Finally, conclusions deriving from the effort are presented and future work is discussed in Section 4.

2   Pcoam Method

2.1   Pcoam Method

It is difficult to measure user packets delay directly in that not only should the time clocks of the monitoring devices be synchronized, but also the identification process is hard as the packet volume is huge in a large-scale network.

Although we could not measure user packets delay directly, we can use the active probe packets delay to estimate the users packets delay according to different network status.

Let $V(t)$ be the delay at the time t. When the network is not busy, we can assume the change of delay is little if the interval of sending probe packets Δt is short enough compared to the time variance of $V(t)$. Then

$$\forall s, s' \in [t, t - \Delta t) = V(s) \cong V(s') \tag{1}$$

We can obtain the number of user packets between the neighborhood probe packets through the simplified passive monitoring device as same as the device used in the CoMPACT monitor [6], [7], [8], which is proposed by Masaki Aida *et al.*. The simplified passive monitoring device only monitors the arrival of the probe packets and counts the number of the user data packets.

Suppose there are n user data packets and m probe packets arriving in the measuring period. Let A_i be the delay of active probe packet i, the indicator function ϕ is as:

$$\phi(i, a) = \begin{cases} 1, A_i > a \\ 0, A_i \leq a \end{cases} . \tag{2}$$

Let X be the measuring objective that is the delay of a user's data packets. Supposing ρ_i is the number of user data packets between active probe packet i and active probe packet $i - 1$, the distribution function of user packets delay is obtained by the estimator of active probe packets:

$$\Pr(X > a) = \sum_{i=1}^{m} \phi(i, a) \frac{\rho_i}{n} . \tag{3}$$

The mean delay is as:

$$M(X) = \frac{1}{n} \sum_{i=1}^{m} A_i \rho_i \tag{4}$$

In practice, this method is similar to the CoMPACT monitor in the condition that the network is not busy.

On the other hand, when the network is busy or in congestion, we can assume that the change of link delay is continuous in the period of $[t, t + \Delta t)$, where the interval Δt is short enough compared to the time variance of $V(t)$.

Furthermore, we can assume the delay of packets in the period $[t, t + \Delta t)$ is a linear relationship with the time. Then

$$\forall s \in [t, t + \Delta t) \Rightarrow \frac{V(s) - V(t)}{s - t} \cong \frac{V(t - \Delta t) - V(t)}{\Delta t} \tag{5}$$

Then the weight of active probe packets delay would not merely be the number of user data packets between the former probe packet and this packet, for the difference between the delays of the adjacent probe packets should be taken into account necessarily.

When the delay of adjacent probe packets is both higher than a, we can consider the delay of user packets arriving between those two adjacent probe packets is higher than a. Similarly, when the delay of adjacent probe packets is both lower than a, we can consider the delay of user packets arriving between this two adjacent probe packets is lower than a. If the delay of one probe packet is higher than a and the other one is lower than a, we can assume the user packets delay is distributed uniformly between the adjacent probe packets delay. So the indicator function is as follows:

$$\phi'(i,a) = \begin{cases} 1 & A_i > a, A_{i-1} > a \\ \dfrac{A_i - a}{A_i - A_{i-1}} & A_i > a, A_{i-1} \leq a \\ \dfrac{A_{i-1} - a}{A_{i-1} - A_i} & A_i \leq a, A_{i-1} > a \\ 0 & A_i \leq a, A_{i-1} \leq a \end{cases} \tag{6}$$

Then, the delay distribution is as follows:

$$\Pr(X > a) = \sum_{i=1}^{m} \phi'(i,a) \frac{\rho_i}{n} \tag{7}$$

The mean delay is as:

$$M(X) = \frac{1}{n} \sum_{i=1}^{m} \frac{(A_i - A_{i-1})}{2} \rho_i \tag{8}$$

where $A_0 = 0$.

The Pcoam method would help to do more exact evaluation of network performance even with slightly complex computations comparing to the CoMPACT monitor [7].

2.2 Implementation

We can deduce the status of a network from the probe packets. First we can set up a delay threshold based on experience to decide whether the network is busy or not. As the delay of both adjacent probe packets is beyond the threshold, we can suppose that each router is busy when those probe packets and data packets traverse each hop. In this case we can presume the network is busy. Otherwise we presume the network is not busy.

With establishing the threshold value D, when the adjacent packets delay is higher than D, we should adopt formula (6), whereas the formula (2) should be adopted. Therefore the indicator function is as follows:

$$\phi^{''}(i,a) = \begin{cases} \phi^{'}(i,a) & A_i > D, A_{i-1} > D \\ \phi(i,a) & otherwise \end{cases}. \tag{9}$$

The delay distribution function is:

$$\Pr(X > a) = \sum_{i=1}^{m} \phi^{''}(i,a) \frac{\rho_i}{n}. \tag{10}$$

To decide the status of a network, it is important to determine an appropriate threshold. The threshold depends on network topology, network congestion condition and the interval of sending active probe packets.

3 Simulations

We use the *ns2* [9] network simulator to demonstrate our schemes. We measure the queueing delay at the bottleneck router, which does not include the service time for the packets themselves. We use a few of ON-OFF sources to generate and simulate the network traffic. Fig. 1 shows the network topology for the simulation. Twenty sources are connected to a bottleneck route with 1.5-Mbps links, and two routers are connected with a 10-Mbps link. The queue discipline of these links is FCFS.

Each host on the left in the Fig. 1 is a source and transfers data to the corresponding host on the right. Twenty sources are categorized four different types described in Table 1. Each application traffic type is assigned to five hosts. As the transport protocol for the user packets, both TCP and UDP are evaluated.

Active probe packets are generated every 2 seconds. Simulation time is 1800 second. There are 1931369 data packets generated, including 900 probe packets. As the size of probe packet is fixed at 64 bytes, the extra traffic caused by the active probe packets is only about 0.00256% of the link capacity of 10 Mbps and 0.0047% of the entire network traffic. The influence on user traffic could be negligible.

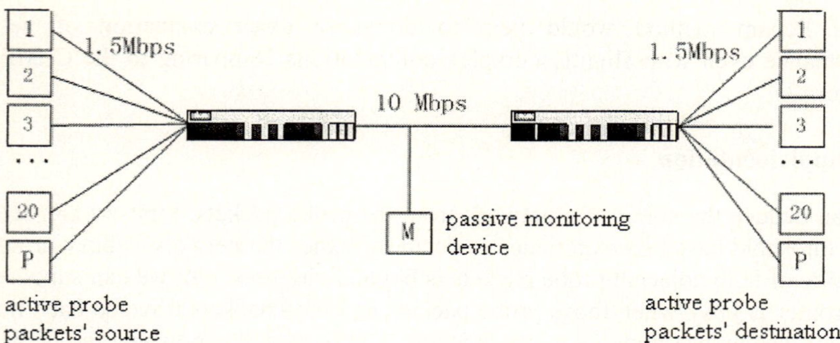

Fig. 1. Simulative single hop network model

Table 1. Simulative single hop network node traffic configure

Node	Protocol	Packet length	Mean ON period	Mean OFF period	ON/OFF length distribution	Shape parameter	Rate at ON period
#1—5	TCP	1.5KB	10 s	5 s	Exponential	--	1Mbps
#6--10	UDP	1.5KB	5 s	10 s	Exponential	--	1Mbps
#11-15	TCP	1.5KB	10 s	10 s	Pareto	1.5	1.5Mbps
#16-20	UDP	1.5KB	5 s	10 s	Pareto	1.5	1.5Mbps

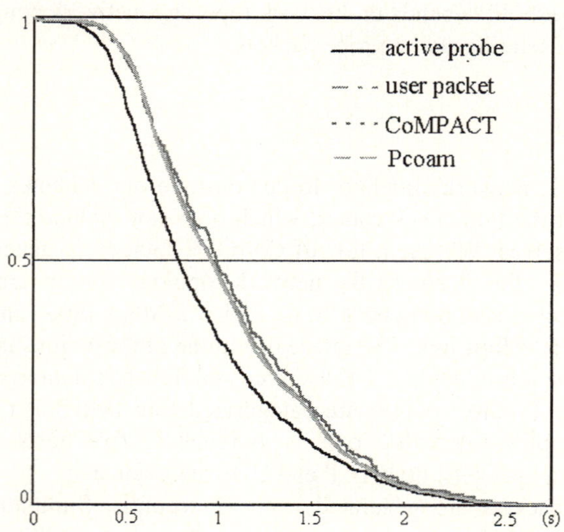

Fig. 2. Delay distribution for connection #6

We chose the connection #6 to do the analysis. This connection use UDP protocol as its transport protocol.

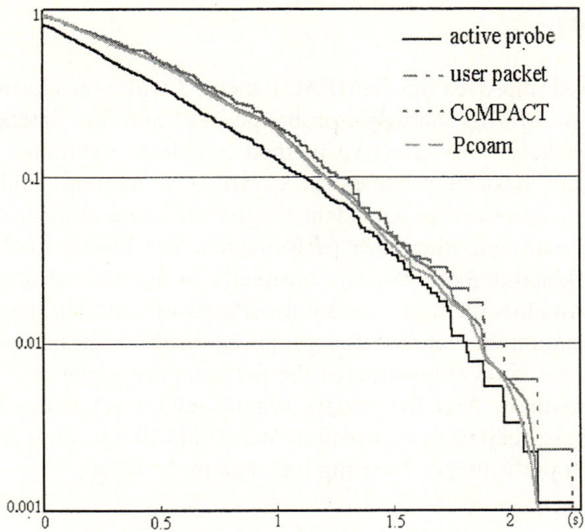

Fig. 3. Delay distribution for connection #6 (logarithm scale)

Fig. 2 and Fig.3 show the queueing delay distributions of the active probe packets and user data packets in connection #6, and the delay distributions obtained by CoMPACT Monitoring and Pcoam method respectively. The logarithm scale is used in Fig.3 for it could clearly analyze the small probability event. We could see that simple active probe packets cannot estimate the delay experienced by users exactly. From two figures, we found that the distribution curve obtained by the Pcoam method is closer to that of user data packets than the curve obtained by the CoMPACT method. So the Pcoam method could reflect the actual network status more exactly.

Table 2 lists the mean delay of four type connections obtained by four methods. We found the simple active probe packets could only get the performance of whole network, but could not reflect the per-flow performance. Whereas the performance experienced by each user could be estimated well with the implementation of Pcoam Method.

Table 2. Mean Delay

Node	Mean delay of user data packets (ms)	Mean delay of active probe packets (ms)	Mean delay by CoMPACT Monitor (ms)	Mean delay by Pcoam Monitor (ms)	Number of user data packets
#1	724.746	910.094	741.219	735.108	46298
#6	1048.390	910.094	1072.290	1038.170	44048
#11	744.670	910.094	761.245	755.633	43458
#16	949.759	910.094	959.408	947.159	176703

4 Conclusions

The Pcoam method improves the CoMPACT method with considering the number of user data packets arriving between probe packets and the latency alteration of adjacent probe packets. This effective method is able to overcome the deficiencies both in active and passive schemes as CoMPACT method could. It has some advantages such as protocol-independent, negligible extra traffic, convenience and being able to estimate individual user performance. The Pcoam method could reflect the actual network status more exactly, especially in the case of network congestion and packet loss. And this method is useful for IP network and Virtual private network.

We used simulation to validate the proposed method. It is shown that Pcoam scheme could give a good estimation of the performance experienced by the user as the CoMPACT method. And the Pcoam scheme could get better result especially when the network is busy or in congestion. We would like to improve our method to deduce more exact performance by using less data in the future.

Acknowledgement

The research reported here has been supported by the National Natural Science Foundation of China(No.60373023). And the authors would like to thank Dr. Masaki Aida for his help and encouragement.

References

1. T. Lindh: A new approach to performance monitoring in IP networks – combining active and passive methods. Proc. of Passive and Active Measurements 2002(2002)
2. G. Almes, S. Kalidindi, and M. Zekauskas: A one-way delay metric for IPPM. RFC2679 (1999)
3. V.Paxon, G.Almes, J.Mahdavi and M.Mathis: Framework for IP performance metric. RFC 2330 (1998).
4. Cooperative association for Internet data analysis(CAIDA): http://www.caida.org/.
5. Breibart Y, Chan C Y, Carofalakis M, Rastogi R, Silberschatz A: Efficiently monitoring bandwidth and latency in IP network. Proc. of IEEE INFOCOM 2000 (2000)
6. M. Aida, N. Miyoshi and K. Ishibashi: A Scalable and Lightweight QoS Monitoring Technique Combining Passive and Active Approaches. IEEE INFOCOM 2003. (2003)
7. M. Aida, K. Ishibashi and T.Kanazawa: CoMPACT-Monitor: Change-of-measure based passive/active monitoring—weighted active sampling scheme to infer QoS--. Proc. of IEEE SAINT 2002 Workshop (2002)
8. K. Ishibashi, T. Kanazawa and M. Aida: Active/Passive combination type performance measurement method using change-of-measure framework. Proc. of IEEE GLOBECOM 2003. (2003)
9. "UCB/LBNL/VINT Network Simulator 2(ns2)": http://www.isi.edu/nsnam/ns.
10. Liu XH, Yin JP, Tang LL, Zhao JM: Analysis of Efficient Monitoring Method for the Network flow. (in Chinese) Journal of Software, 3(2003) 300-304.
11. Liu XH, Yin JP, Lu XC, Zhao JM: A Monitoring Model for Link Bandwidth Usage of Network based on Weak Vertex Cover. (in Chinese) Journal of Software, 4(2004) 545-549

12. Attila Pasztor, Darryl Veitch: On the scope of End-to-End probing methods. IEEE Communications Letters, 11(2002)
13. Susmit H.Patel: Performance Inference Engine (PIE)—Deducing more performance using less data. Proc. of ACM PAM 2000.(2000)
14. K. Ishibashi, M. Aida and S. Kuribayashi: Estimating packet loss-rate by using delay information and combined with change-of-measure framework. Proc. of IEEE GLOBECOM 2003 (2003)
15. V.Paxon. End-to-end Internet packet dynamics. IEEE/ACM Trans. Networking, 7(1999) 277-292
16. N.G.Duffield, C.Lund, M Thorup: Properties and prediction of flow statistics from sampled packet streams. Proc. of ACM SIGCOMM Internet Measurement Workshop 2002 (2002)
17. Kartik Goplalan, Tzi-cker Chiueh, Yow-Jian Lin: Delay Budget Partitioning to Maximize Network Resource Usage Efficiency. Proc. of IEEE INFOCOM 2004(2004).
18. Liu XH, Yin JP, Cai ZP: The Analysis of Algorithm for Efficient Network Flow Monitoring. Proc of 2004 IEEE International Workshop on IP Operations & Management (2004).

Application-Level Multicast Using DINPeer
in P2P Networks

Huaqun Guo[1,2], Lek Heng Ngoh[2], and Wai Choong Wong[1,2]

[1] Department of Electrical & Computer Engineering, National University of Singapore,
2 Engineering Drive 4, Singapore 117584
[2] Institute for Infocomm Research, A*STAR, 21 Heng Mui Keng Terrace, Singapore 119613
{guohq, lhn, lwong}@i2r.a-star.edu.sg

Abstract. In this paper, we propose a DINPeer middleware to overcome limitations in current peer-to-peer (P2P) overlay systems. DINPeer exploits a spiral-ring method to discover an inner ring with relative largest bandwidth to form a DINloop (Data-In-Network loop). DINPeer further integrates DINloop with P2P overlay network. The key features of DINPeer include using DINloop to replace a multicast rendezvous point and turning DINloop into a cache to achieve application data persistency. Simulations show that DINPeer is able to optimize application-level multicast that when the size of the DINloop is capped within a limit, it can achieve a better performance than native IP multicast and P2P overlay multicast systems.

1 Introduction

Recent works on P2P overlay network offer scalability and robustness for the advertisement and discovery of services. Pastry [1], Chord [2], CAN [3] and Tapestry [4] represent typical P2P routing and location schemes. Furthermore, there has been a number of works reported on adding multicast schemes and applications on P2P object platform, e.g., Scribe [5] and CAN-Multicast [6]. Compared to native network-level IP multicast, application-level multicast has a number of advantages. First, most proposals do not require any special support from network routers and can therefore be deployed universally. Second, the deployment of application-level multicast is easier than IP multicast by avoiding issues related to inter-domain multicast. Third, the P2P overlay network is fully decentralized.

However, P2P overlay multicast also suffers some disadvantages. Due to the fact that the underlying physical topology is hidden, using application-level multicast increases the delay to deliver messages compared to IP multicast. A node's neighbors on the overlay network need not be topologically nearby on the underlying IP network and this can lead to inefficient routing.

Recently, there are some works that acknowledged the above limitation of P2P overlay network and inferred network proximity information for topology-aware overlay construction [7]. [8] described and compared three approaches in structured overlay networks: proximity routing, topology-based nodeId assignment, and proximity neighbor selection. In [9], proximity neighbor selection was identified as

P. Lorenz and P. Dini (Eds.): ICN 2005, LNCS 3421, pp. 754–761, 2005.

the most promising technique. In our scheme, we adopt proximity neighbor selection, but use a different method in Section 2.

In addition, the routing in the P2P overlay network does not consider the load on the network. It treats every peer as having the same power and the same bandwidth. Further, in P2P tree-based multicast, the multicast trees are built at the application level and the rendezvous point is the root of the multicast tree. The rendezvous point can potentially be subjected to overloading and single-point of failure.

In this paper, we propose DINPeer to overcome the above limitations of application-level multicast in P2P network. The remainder of this paper is organized as follows: we present the details of DINPeer in Section 2. Section 3 presents our simulation results. Finally we conclude research results in Section 4.

2 Details of DINPeer

DINPeer is a massively scalable middleware that combines the strengths of multicast and P2P systems. DINPeer exploits a spiral-ring method to discover an inner ring with relative largest bandwidth to form a logical DINloop. DINPeer further integrates the DINloop with the P2P overlay network. DINPeer has two key features. First, DINPeer uses the DINloop instead of a rendezvous point as multicast sources. Second, DINPeer uses the DINloop as a cache in the form of continuously circulating data in the DINloop to achieve data persistency. The detail of DINPeer is elaborated in the following sub-sections.

2.1 Generalizing DINloop

We exploit a spiral-ring method to find an inner ring to form a DINloop. First, a big outer ring with all nodes is formed. A new node is always inserted between two nearest connected nodes. Initially, *Node 0* and *Node 1* form a ring (Fig. 1a). Then *Node 2* is added to the ring (Fig. 1b). From *Node 3* onwards, the two nearest nodes will break the link and the new node is added (Fig. 1c and Fig. 1d).

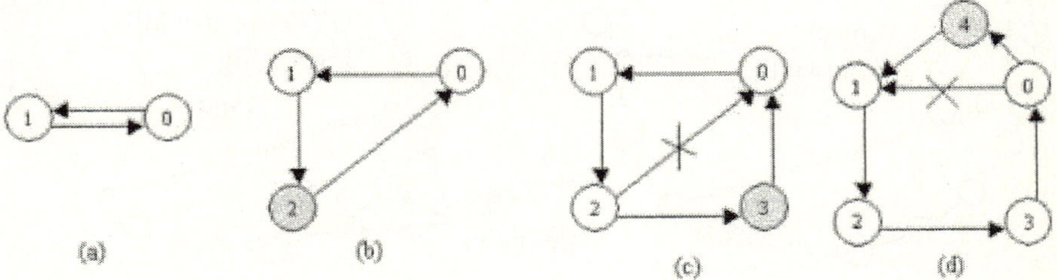

Fig. 1. Automatic formation of ring

The algorithm to obtain two nearest nodes is described here. Assume *Node i* plans to join the ring and knows of a local nearby *Node k* in the ring. *Node i* sends a join message to *Node k*. Then *Node k* and its two adjacent *Node (k-1)* and *Node (k+1)* in the ring will ping *Node i* and get the Round Trip Time (RTT). If *Node k* gets the minimum RTT to *Node i*, *Node k* and one of its two adjacent nodes with lower RTT to

Node i will be determined as two nearest nodes to *Node i*. If *Node* $(k+1)$ gets the minimum RTT to *Node i*, *Node* $(k+2)$ will ping *Node i* and get the RTT. If *Node* $(k+1)$ still gets the minimum RTT to *Node i*, *Node* $(k+1)$ and one of its two adjacent nodes with lower RTT to *Node i* will be determined as two nearest nodes to *Node i*. If *Node* $(k+2)$ gets the minimum RTT to *Node i*, *Node* $(k+3)$ will ping *Node i* and get the RTT. The process continues until two nearest nodes to *Node i* are found.

Second, we use a spiral-ring method to find an inner-most ring with relative largest ring bandwidth (Fig. 2). The inner spiral ring must provide a higher bandwidth than the outer ring. The formation of the inner ring is not limited, as links with lower bandwidth are dropped if enough nodes are available and the ring bandwidth is increased.

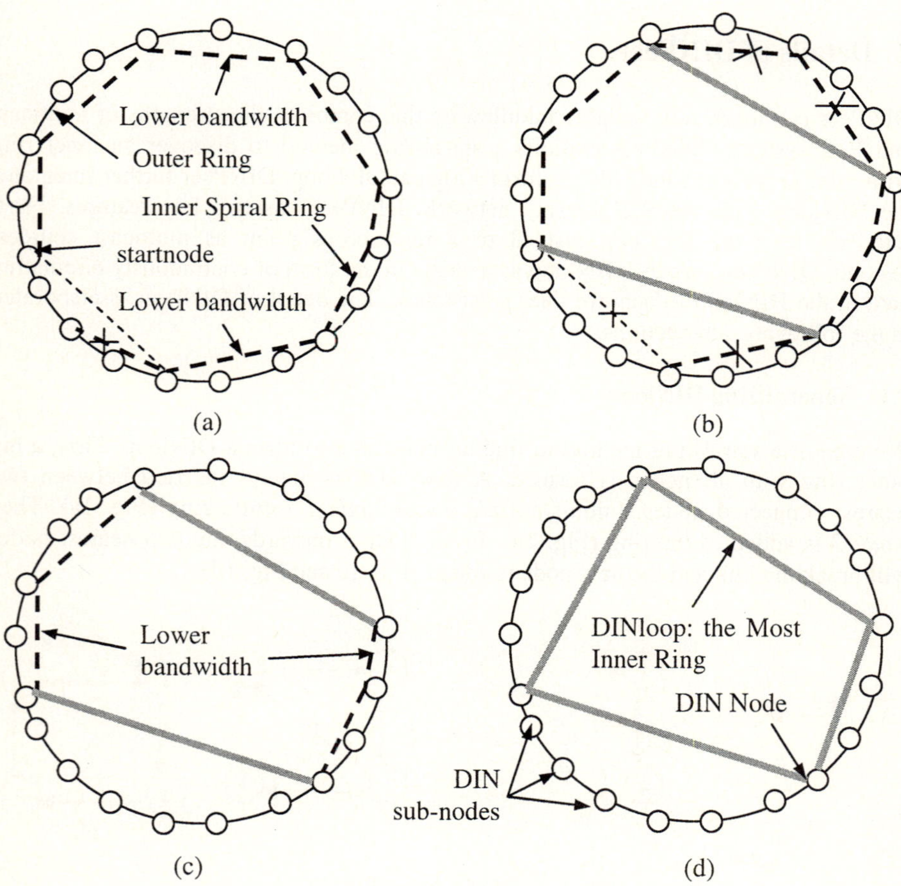

Fig. 2. Spiral-Ring Method

We use f_b as the bandwidth-increasing rate and N as the desirable number of DIN Nodes. Let β be the current inner ring bandwidth. Each recurring process drops inner links where their link bandwidths are less than $\beta*(1+f_b)$ (Fig. 2b) by replacing them with further inner links with higher bandwidth. The process is repeated (Fig. 2c) until the desired number of nodes in the inner ring is achieved or the ring bandwidth can no longer be increased (Fig. 2d).

Now we show how fast the recurrence can converge. We assume the outer ring bandwidth is B_0. We assume B_1 is the ring bandwidth of the inner ring at the 1st iteration of dropping inner links with the lower bandwidth. We assume B_2 is the ring bandwidth of the inner ring at the 2nd iteration of dropping inner links with the lower bandwidth. We assume that after k iterations, B_k is the maximum ring bandwidth of the inner ring. So,

$$B_1 \geq B_0 * (1 + f_b) . \tag{1}$$

$$B_2 \geq B_1 * (1 + f_b) \geq B_0 * (1 + f_b)^2 . \tag{2}$$

$$B_k \geq B_{k-1} * (1 + f_b) \geq B_0 * (1 + f_b)^k . \tag{3}$$

$$\therefore k \leq \log_{(1+f_b)} (B_k \div B_0). \tag{4}$$

Thus, the iteration step k has a small upper bound and therefore the process converges quickly.

In this way, we find the inner ring with largest bandwidth. This inner ring is used as a DINloop and nodes in this inner ring are used as DIN Nodes. The DINloop can therefore be used as a cache called a DIN cache. The other nodes in DINPeer are called DIN sub-nodes (Fig. 2d). Each DIN sub-node finds a nearest DIN Node and becomes a member of a group associated to this DIN Node (Fig. 3).

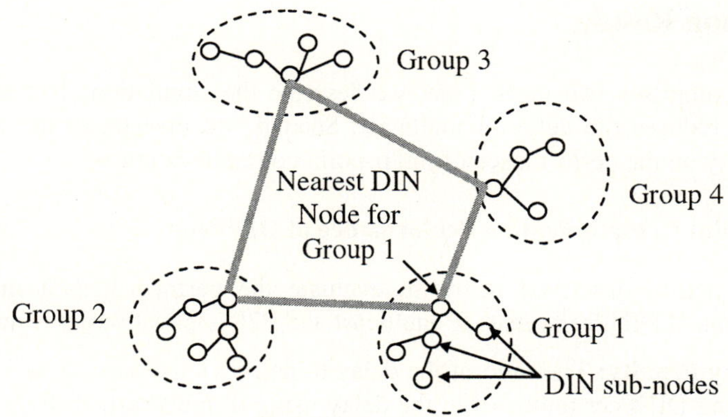

Fig. 3. DIN sub-nodes with associated DIN Nodes

2.2 Integrating with P2P Overlay Network

After the DINloop is formed, the DINloop is integrated with the P2P overlay network [1]. The routing algorithm in DINPeer is similar to the routing algorithm in Pastry [1], but it integrates with DIN Nodes and DIN cache. Given a message, the node uses the routing table that is constructed using nodes in its own group and forwards the message to a node locally. If there is no node that can be routed to, the message is marked to differentiate itself from the beginning message and forwarded to its associated DIN Node. Then the DIN Node checks the DIN cache with this marked message. If the DIN Node has a copy of the object in the DIN cache, the copy of object is returned to the requesting node. If there is no copy in the DIN cache, the DIN Node forwards the marked message to all other DIN Nodes. Then each DIN Node forwards it to the node in its own group respectively. Finally, a copy of the object is returned via the DINloop and the DIN cache is updated. DIN cache is updated using the RLU (Recent Least Use) algorithm.

2.3 Optimizing P2P Application-Level Multicast

In IP multicast, a sender sends data to the rendezvous point and the rendezvous point forwards the data along the multicast tree to all members. In DINPeer, the DINloop with multiple DIN Nodes is used to replace a single rendezvous point. Each DIN Node is the root of its associated multicast tree. Subsequently, every member in its associated group, who wants to join the multicast group, sends a message to the DIN Node, and the registration is recorded at each node along the path in the P2P overlay network. In this way, the multicast tree is formed. When multicasting, a node sends a multicast message to the nearest DIN Node. The nearest DIN Node then forwards the message to its child-nodes, and the neighbor DIN Node along the DINloop. The neighbor DIN Node forwards the message to its associated child-nodes, and its neighbor DIN Node along the DINloop. The process repeats itself until all DIN Nodes receive the message or when the lifetime of the message expires.

3   Evaluation Results

This section comprises two parts. First, we describe the simulations to demonstrate that DINPeer reduces the delay of multicast. Second, we investigate the impact of data persistency on the performance of multi-point communication.

3.1   Multi-point Communication Performance of DINPeer

We use the metrics described below to evaluate the performance of multi-point communications in DINPeer versus IP multicast and P2P application-level multicast.

Relative Delay Penalty: The ratio of the delay to deliver a message to each member of a group using DINPeer multicast to the delay using IP multicast. RMD is the ratio of the maximum delay using DINPeer multicast to the maximum delay using IP multicast. RAD is the ratio of the average delay using DINPeer multicast to the average delay using IP multicast.

The simulations ran on the network topologies, which were generated using the Georgia Tech [10] random graph generator according to the transit-stub model [11].

We used the graph generator to generate different network topologies. The number of nodes in the transit domain was changed and the number of nodes in stub domains was fixed, i.e., 6000. We randomly assigned bandwidths ranging between 1Gbps to

10Gbps to the links in the transit domain, used the range of 100Mbps to 1Gbps for the links from transit domain to stub domains, and used the range of 500kbps to 1Mbps for the links in the stub domains. The topology is similar to one backbone network and multiple access networks. We assumed that all the overlay nodes were members of a single multicast group. Using the spiral-ring method, the different number of DIN Nodes was obtained. For IP multicast, we randomly chose a node as a rendezvous point. We repeated the simulation six times and obtained the average delay. The results of simulation are shown in Fig. 4.

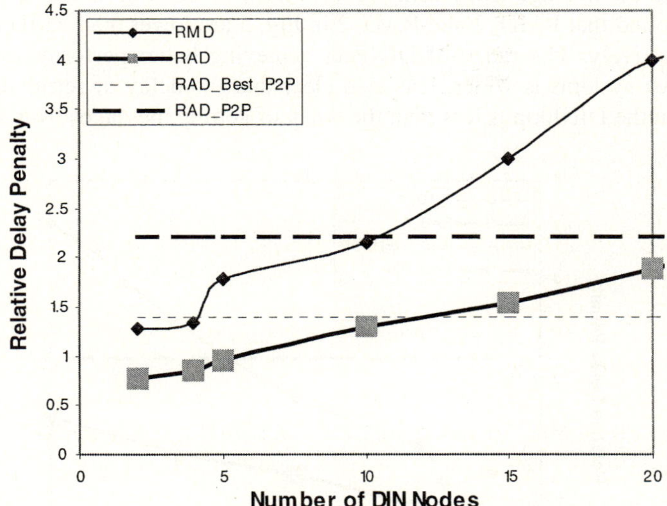

Fig. 4. Delay ratio of message retrieval using DINPeer over IP multicast

From Fig.4, when the number of DIN Nodes is small enough, the average latency in DINPeer overlay network is even lower than that of IP multicast, as RAD is less than 1. On another hand, other topology-aware routing techniques are currently able to achieve an average delay stretch (delay penalty) of 1.4 to 2.2, depending on the Internet topology model [8]. The average delay penalty of 1.4 in other P2P overlay network reported in [8] is shown in the thin dash-line RAD_Best_P2P in Fig. 4. The average delay penalty of 2.2 in other P2P overlay network reported in [8] is shown in the thick dash-line RAD_P2P in Fig. 4. Since the simulation conditions between our scheme and other P2P systems are different, the RAD difference in quantity cannot be calculated. We only show the general difference. DINPeer has the opportunity to achieve better performance than IP multicast while other P2P overlay multicast systems are worse than IP multicast. In Fig. 4, DINPeer, as shown below the thin dash-line, is the best among any other P2P overlay multicast systems. Thus, DINPeer provides the opportunity to achieve better performance than other P2P overlay multicast systems when the number of DIN Nodes is within the certain range.

3.2 Performance of Multi-point Communication with Data Persistency

In this sub-section, simulation was conducted to investigate the impact of data persistency on the performance of multi-point communication. We measured the delay to deliver a message to each member of a group using the DINloop and IP multicast respectively. The same network topologies as in the previous sub-section were used here. The results are shown in Fig. 5. *RMD_2* is the ratio between the maximum delay from the DINloop to receivers and the maximum delay using IP multicast, and *RAD_2* is the ratio between the average delay from the DINloop to receivers and the average delay using IP Multicast. When Fig. 5 was compared with Fig. 4, we found that RMD_2 and RAD_2 in Fig. 5 are lower than RMD and RAD in Fig. 4 respectively. The range of DINPeer achieving better performance than other P2P multicast systems is wider. It is also clear that the delay of retrieving messages directly from the DINloop is less than the delay of getting messages from the sender.

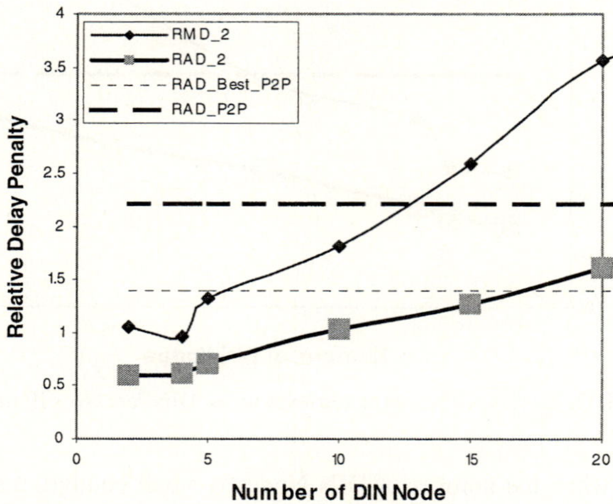

Fig. 5. Delay ratio of message retrieval from DINloop over IP multicast

4 Conclusions

We propose a DINPeer middleware to overcome the existing limitations in current P2P overlay systems. DINPeer exploits the spiral-ring method to discover an inner ring with relative largest bandwidth to form the DINloop. DINPeer further integrates the DINloop with a P2P overlay network. DINPeer has two key features. First, DINPeer uses the DINloop instead of the rendezvous point as multicast sources. Second, DINPeer uses the DINloop as a cache in the form of continuously circulating data in the DINloop to achieve data persistency. Our principle findings are (1) When the number of DIN Nodes is capped within a limit, DINPeer even achieves better performance than native IP multicast and is the best among any other P2P overlay multicast systems reported in the literature. (2) Data persistency further improves the performance of multi-point communication in DINPeer.

References

1. Rowstron, A., Druschel, P.: Pastry: Scalable, distributed object location and routing for large-scale peer-to-peer systems. Proc. of the 18th IFIP/ACM International Conference on Distributed Systems Platforms (Middleware), Heidelberg, Germany (2001) 329-350
2. Stoica, I., Morris, R., Karger, D., Kaashoek, M. F., Balakrishnan, H.: Chord: A scalable peer-to-peer lookup service for Internet applications. ACM SIGCOMM 2001, San Deigo, CA (2001) 149-160
3. Ratnasamy, S., Francis, P., Handley, M., Karp, R., Shenker, S.: A scalable content-addressable network. Proceedings of ACM SIGCOMM 2001 (2001)
4. Zhao, B. Y., Kubiatowicz, J., Joseph, A. D.: Tapestry: an infrastructure for fault-tolerant wide-area location and routing. UCB Technical Report (2001)
5. Castro, M., Druschel, P., Kermarrec, A.-M., Rowstron, A.: SCRIBE: a large-scale and decentralized application-level multicast infrastructure. IEEE Journal on Selected Areas in Communications (JSAC) (2002)
6. Ratnasamy, S., Handley, M., Karp, R., Shenker, S.: Application-level multicast using content-addressable networks. Proc. of NGC 2001 (2001)
7. Ratnasamy, S., Handley, M., Karp, R., Shenker, S.: Topologically-Aware Overlay Construction and Server Selection. Proc. of INFOCOM (2002)
8. Castro, M., Druschel, P., Hu, Y. C., Rowstron, A.: Topology-aware routing in structured peer-to-peer overlay networks. A. Schiper et al. (Eds.), Future Directions in Distributed Computing 2003. LNCS 2584 (2003) 103-107
9. Castro, M., Druschel, P., Hu, Y. C., Rowstron, A.: Topology-aware routing in structured peer-to-peer overlay networks. Technical Report, MSR-TR-2002-82 (2002)
10. Zegura, E., Calvert, K., Bhattacharjee, S.: How to model an internetwork. Proc. of IEEE Infocom (1996)
11. Modeling Topology of Large Internetworks. http://www.cc.gatech.edu/projects/gtitm/

Paradis-Net

A Network Interface for Parallel and Distributed Applications

Guido Malpohl and Florin Isailă

Institute for Program Structures and Data Organization,
University of Karlsruhe, 76128 Karlsruhe, Germany
{Malpohl, Florin}@ipd.uka.de

Abstract. This paper describes Paradis-Net, a typed event-driven message-passing interface for designing distributed systems. Paradis-Net facilitates the development of both peer-to-peer and client-server architectures through a mechanism called "Cooperation". We introduce the programming interface and compare its mechanisms to active messages and remote procedure calls. Finally we demonstrate how the interface can be used to implement communication patterns typical for distributed systems and how peer-to-peer functionality can be mapped onto Paradis-Net.

1   Introduction

There is a growing interest in large-scale distributed systems consisting of a large number of cooperative nodes. Cluster computing, peer-to-peer-systems, grid computing and ad-hoc networks are examples of current active research areas. All these directions have in common the development of complex communication protocols. The architecture of these systems has recently shifted from the traditional client-server paradigm to decentralized cooperative peer-to-peer models and towards hybrid approaches combining both client-server and peer-to-peer paradigms.

Paradis-Net is a typed message-passing interface for distributed applications and operating system services. It is suitable for designing distributed systems for both high speed networks (e.g. Myrinet [1] or Infiniband [2]) or for relatively slow transport mediums (Internet). Paradis-Net offers a simple interface facilitating the implementation of complex communication patterns and abstracting away from particular network hardware.

Paradis-Net emerged from our experience in developing the Clusterfile parallel file system [3, 4] and addresses the problems of communication paradigms such as RPC and active messages, that have been described by the developers of xFS [5]. In a paper describing their experience with xFS [6], the authors identify the mismatch between the service they are providing and the available interfaces as a main source of implementation difficulties. We show in section 2 how our work on Paradis-Net addresses these issues.

P. Lorenz and P. Dini (Eds.): ICN 2005, LNCS 3421, pp. 762–771, 2005.
© Springer-Verlag Berlin Heidelberg 2005

The contributions of this paper are:

- Paradis-Net offers a low-level *transport independent* interface (for both, *user and kernel space*).
- Multi-party protocols are supported through the *Cooperation* mechanism, allowing several nodes to collaborate in order to serve a request.
- A Paradis-Net message handler can delegate the request to a remote handler. This mechanism is similar to *continuation passing*.

In addition to these points there are several notable details about the Paradis-Net interface:

- Paradis-Net and its Cooperation and continuation passing mechanisms match the needs of P2P overlay networks. (see section 4.2)
- The semantics of Cooperations suit the requirements of the RDMA protocol, which consists of memory registration and effective data transfer. Therefore, the low-level RDMA mechanism can be transparently exposed to the applications. (see section 4.3)
- For efficient utilization of SMPs, Paradis-Net has been designed as a multi-threaded library and offers a thread-safe implementation.

We have implemented Paradis-Net on top of TCP/IP sockets in user-level and in kernel-level. A user-level implementation for the Virtual Interface Architecture (VIA [7]) demonstrates that RDMA can be used transparently.

2   Related Work

Active messages (AM [8]) is a low-level message passing communication interface. An AM transaction consists of a pair of request and reply messages. Each request activates a handler associated with the message, that extracts the data from the network and delivers it to the application. The low-level interface of AM allows exposing the features of Network Intelligent Card (NIC) such as zero-copy RDMA to the applications. Paradis-Net can also map low-level NIC intelligence on the *Cooperation* communication abstraction (see section 4.3).

Remote Procedure Calls (RPC [9]) are a common standard for distributed client-server applications, e.g. NFS [10]. Both AM and RPC paradigms are suitable for client-server application due to their point-to-point request-reply nature. However, Wang et al. [6] found them unnatural for the multi-party communication needed by a peer-to-peer system: A point-to-point RPC call has to be followed by a reply from the liable peer. If the request is delegated to an another peer, the reply has to travel back on the same way, as shown in figure 1. On the other hand, Paradis-Net through its continuation passing mechanism, allows a direct reply from the last peer as depicted in figure 2. In general, for n delegations, RPC needs $2n$ messages, Paradis-Net only $n + 1$.

The Parallel Virtual Machine (PVM [11]) and the Message Passing Interface (MPI [12]) are used to specify the communication between a set of processes forming a concurrent program. With many communication and synchronization primitives they target the development of *parallel* applications following the SPMD paradigm and are not well suited for distributed system development.

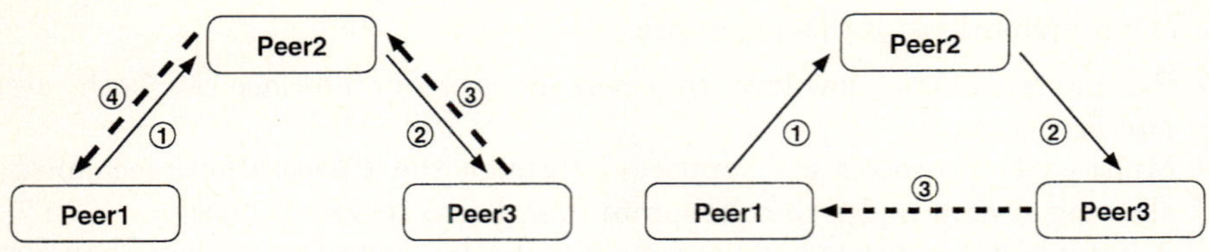

Fig. 1. Delegation through RPC   **Fig. 2.** Delegation in Paradis-Net

Solid arrow: Request – Dashed Arrow: Reply

Table 1. The Paradis-Net API

General	
int <u>initialize</u> (*end_point* ep [])	section 3.1
void <u>finalize</u> ()	section 3.1
peer_id <u>get_peer_id</u>(**char** name[])	section 3.1
Communication	
int <u>send</u>(*peer_id* to, *msg_type* type, *coop_nr* nr, **void** ∗msg, **int** msg_size)	section 3.1
int <u>forward</u>(*peer_id* to, *msg_type* type, **void** ∗msg, **int** msg_size)	section 3.2
Handlers	
void <u>set_handler</u>(*msg_type* msg, **int** opt, *handler_fun* ∗handler)	section 3.2
void <*handler_fun*>(*peer_id* from, *msg_type* type, *coop_nr* nr, **void** ∗msg , **int** msg_size)	section 3.2
Cooperations	
coop_nr <u>start_cooperation</u>(*rcv_desc* ∗rcvec, **int** vec_size)	section 3.3
int <u>end_cooperation</u>(*coop_nr* nr, **int** timeout)	section 3.3

3 The Paradis-Net Architecture

Paradis-Net offers a peer-to-peer communication model in which every communication endpoint can be a server and a client at the same time. The Paradis-Net library is a layer between the application and the native network interface. Applications see a simple, uniform interface that is independent of the actual network technology used, thus easing the development of complex distributed protocols. Paradis-Net can be extended to support different network technologies. For a more detailed description of the API see our technical report on Paradis-Net [13].

3.1 Initialization, Peer IDs and Sending Data

The *initialize* method (see table 1) is needed to initialize the internal data structures and to open local endpoints. The *ep* argument contains the configuration options for the different network interfaces supported by this peer. The inverse operation *finalize* closes down all the endpoints and releases the corresponding data structures.

Every Paradis-Net endpoint has a unique peer name. This name usually consists of protocol and address information used to address other peers and to send messages to them. For example a TCP endpoint name is: *"tcp:<ip-address>:<service-port>"*. For convenience and performance reasons *get_peer_id* (see table 1) returns peer IDs that help refering to remote endpoints through handles rather then full endpoint names.

The Paradis-Net library offers only one explicit communication primitive: *send* (see table 1). This operation sends a typed message to the peer represented by the peer ID (*to*). In case of an error the error number is returned. Upon return of this function the memory area that contains the message can be reused immediately.

3.2   Request Handlers

Paradis-Net does not offer a function to receive data from other peers. Instead, it uses handler functions that are called upon the arrival of a message. A handler function for a certain message type is set using the *set_handler* (see table 1) function. This event-driven mechanism requires an agreement between peers with respect to the message type they use. In contrast to Active Messages, Paradis-Net handlers are not limited in their execution time and can initiate calls to the library, including arbitrary *send* operations.

Handlers facilitate server implementation: Request reception and invocation of the appropriate user-defined handler is being taken care of by the library. A handler usually fulfills the request service and sends the reply back to the client.

The traditional client-server model can be expanded in Paradis-Net with the *forward* (see table 1) function. When called from within a handler, *forward* allows sending the message to a different peer and thereby also delegating the obligation to answer. The handler on the next peer will be invoked with the local *peer_id* that corresponds to the peer which was the original source of the request. Section 4.1 will give an example communication pattern the uses this operation.

3.3   Cooperations

The handler concept on its own is not convenient for the implementation of protocols involving several peers. For this reason Paradis-Net introduces "Cooperations". A Cooperation is a concept that defines a relationship between the outgoing requests and the incoming answers by creating a token that accompanies all involved messages. The function *start_cooperation* (see table 1) registers a Cooperation at the client side. The parameter (*rcvec*) describes the expected reply.

We will exemplify the life-cycle of a typical Cooperation in a client-server scenario: A Cooperation starts at the client by having the function *start_cooperation* registering the Cooperation and returning a token that represents the Cooperation. A Cooperation is registered using a *receive descriptor* containing memory locations to store the replies. Receive descriptors also define criteria to distinguish between different message types and origins. Next, the client will send a request to the server and afterwards call *end_cooperation* (see table 1). This function blocks until the expected result is available. The token accompanies the request message on its way to the server by using the optional parameter of the *send* operation that attaches the token to the message. On the servicing peer, Paradis-Net will invoke the handler that has been assigned to the message type with the Cooperation token as a parameter (see the signature of handler functions in table 1). When the reply is sent, the Cooperation token is again attached to the message and travels back to its origin. On the client site, Paradis-Net identifies the reply as being part of a Cooperation by the type of the message. Although there is still the possibility to invoke a handler function upon the arrival of such a message, the library will first check if the attached cooperation token matches any of the currently active Cooperations on this peer. If this is the case, the service thread will store the message at the memory location that was declared when calling *start_cooperation* and afterwards wake up the thread that is waiting for the cooperation to finish.

4   Applications

In this section we will illustrate flexibility and simplicity of Paradis-Net in parallel and distributed file systems and P2P systems as well as its ability to transparently support remote zero-copy operations (RDMA).

4.1   Parallel and Distributed File Systems

Here are two examples that stem from our own experience developing the parallel file system *Clusterfile* [3] and the observations that were made building the distributed file system *xFS* [6].

Delegation. *xFS* illustrates the delegation pattern with respect to cooperative caching (see figure 2): A client (*Peer1*) reading from a file incurs a read miss in the local cache and sends a block request to the cache manager (*Peer2*) in order to retrieve the cached data from a different peer. The manager consults its map, finds the responsible cache server (*Peer3*) and forwards the request to it. The cache server then responds back to the client with the cached data. The same pattern can be also be employed for routing in a peer-to-peer system (section 4.2). Although this scenario appears to be simple, it is difficult to realize with traditional message passing interfaces. Wang et al. [6] demonstrate that RPCs are unsuitable to implement it because of the strict semantics imposed by the model.

On *Peer1* the method *start* (listing 1) registers a Cooperation, sends the request to a peer (*Peer2*) and waits for an answer.

The *start* function can be used for typical client-server communication as well. If *Peer2* answers the request directly, *start* does not have to be changed at all, since the receive descriptor accepts replies from any peer, as long as the reply carries the Cooperation token issued by the local Paradis-Net library.

```
void start(peer_id to, void *msg, int msg_len) {
  coop_nr coop;
  rcv_desc desc =                      // receive descriptor
    { memory: NULL, size: 0,          // allocate memory for
          reply automatically
      type: REPLY,                     // only accept messages
          with type "REPLY"
      options: RCV_FROM_ANY };        // accept messages from
          any peer

  coop = start_cooperation(&desc, 1);           // register
          cooperation
  send(to, REQUEST, coop, msg, msg_len);        // send
          request
  ...                                            // eventual
          computation
  end_cooperation(coop, 0);                      // wait for
          reply (no timeout)
}
```

Listing 1. Sending the request

When *Peer2* receives the request from *Peer1*, Paradis-Net invokes the handler function *forward_handler*, which has been registered for messages with the type *REQUEST*. The handler first inspects the incoming message to find the peer responsible for answering the request and then forwards the message to it:

```
void forward_handler(peer_id from, msg_type type, coop_nr
        coop, void *msg,
                        int msg_len) {
  peer_id liable_peer = find_liable_peer(msg, msg_len);

  forward(liable_peer, msg_type, msg, msg_len);
}
```

Listing 2. Forwarding the request

This implementation of the *forward_handler* ignores errors that might happen when forwarding the message. In the case of an error, *Peer2* could reply back to *Peer1* with an error message, or try to forward the request to a different peer.

The message from *Peer2* to *Peer3* carries the address of *Peer1* that will allow *Peer3* to identify the requester and to reply to him. On *Peer3* Paradis-Net calls the local handler function (*serve_request*, listing 3) with the peer ID of *Peer1* as first parameter, so that the handler function will not be able to see the mediator that forwarded the request.

```
void serve_request(peer_id from, msg_type type, coop_nr coop,
      void *msg,
                    int msg_len) {
  reply_msg reply;  // this variable will hold the reply
  int reply_len;    // the length of the reply

  fulfill_request(msg, msg_len, &reply, &reply_len);  //
        application specific
  send(from, REPLY, coop, &reply, reply_len);
}
```

Listing 3. Serving the request

Scatter/Gather. Scatter/Gather is a one-to-many communication pattern in which a peer sends requests in parallel to many peers and waits for all individual responses. Figure 3 illustrates the procedure.

Clusterfile [3] uses this pattern to contact several data servers storing stripes of a given file. Although the requests are sent out in a particular order, the order of the replies is arbitrary.

The peers which play the server role in this pattern (*Peer2*, *Peer3*, ...), define a handler function to process the request. This handler will, after assembling an answer, reply back to *Peer1*. The procedure accords with the one of *Peer3* in the delegation example and therefore the implementation is the same: see listing 3.

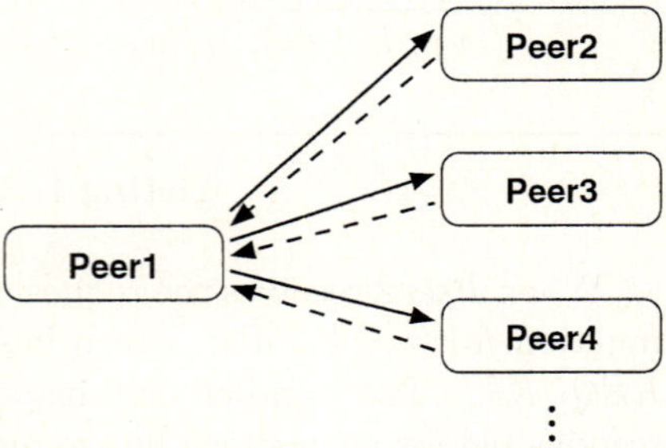

As an extension of the pattern, it is also possible to forward the request to a different peer using the *forward* function, akin to listing 2. This would result in a combination of the Scatter/Gather and the Delegation pattern.

For simplicity reasons we send the same message to every peer and expect reply messages of type *reply_type* (listing 4). After sending the requests, *Peer1* will block in *end_cooperation* until all answers have been received.

Fig. 3. The Scatter/Gather pattern Solid arrow: Request – Dashed Arrow: Reply

4.2   Structured P2P Overlays

Three groups from MIT, Berkeley and Rice University joined their efforts in order to define a common three tier API for structured overlays [14]. The lowest tier 0 is the key-base routing layer, which provides basic communication services. Tier 1 provides higher level abstractions like distributed hash tables, while the applications at tier 2 use these abstractions in order to offer services like file sharing and multicasting. The researchers describe the tier 0 implementation of the four most influential structured overlay systems: CAN [15], Chord [16], Pastry [17] and Tapestry [18]. We will outline how the routing messages operations of tier 0 can be straightforward implemented using Paradis-Net.

An overlay node is identified by a *nodehandle* that encapsulates its network address (e.g. IP), in our case the unique peer name in Paradis-Net (section 3.1). Overlay nodes are assigned uniform random *node_ids* from a large identifier space by hashing their network addresses. The application objects are mapped on the same identifier space by computing a *key*. The objects are placed on the overlay nodes by assigning their keys to *node_ids* (for instance the longest prefix match in Tapestry). In order to efficiently sent a message from one node to the other, each node maintains a routing table. The routing table is used for choosing a next hop whose *node_id* is closer (for instance by using the Hamming distance) to the *node_id* of the destination. The routing strategy is system specific and is not discussed here. Given a node and a message to be routed, we assume the routing table delivers the *node_id* of the next hop.

At tier 0, two sets of API functions are proposed: routing messages and routing state access. The latter set refers to strictly local operations and is therefore not relevant for our discussion. The first set consists of three API functions: *route*, *forward* and *deliver*. The call *route(key K, msg M)* delivers the message M to the node storing the object associated with the key K. The optional argument hint specifies the first hop to be used. The route operation can be implemented using the continuation passing mechanism of Paradis-Net as a chain of handlers assigned to the message type *ROUTE*, running on the peers from the source to the destination. Each handler consults the local routing table in order to chose the next hop and then invokes the local *forward(key K, msg M, nodehandle nextHop)*, as provided by the application. This function may change K, M or *nextHop*, according to the application needs. Upon returning from it, *send* can be used for sending K and M to the node *nextHop*. At the destination, the up-call *deliver(key K, msg M)*, informing the application that a message for object K arrived, will be called from a Paradis-Net handler.

4.3   Transparent RDMA

The capabilities of direct access transport (DAT) standards such as the Virtual Interface Architecture [7], Infiniband [2] and Remote Direct Data Placement (RDDP [19]) reduce memory copy operations and minimize CPU load when transfering data from the network interface to the main memory. The DAT Collaborative defines DAT requirements and standard APIs that include RDMA, memory registration, kernel bypass and asynchronous interfaces. RDMA allows direct memory-to-memory transport of data without CPU involvement. Memory registration facilitates the specification of granting access to the local memory regions used as destinations of RDMA operations. The kernel bypass eliminates unwanted kernel involvement in the communication path.

Paradis-Net cooperations allow the transparent use of RDMA as validated by an implementation of Paradis-Net over VIA. Memory registration is implemented in the *start_cooperation* routine. A prerequisite for the use of RDMA is that the

```
reply_type* start(peer_id* peers, int num_peers, void* msg,
    int msg_len) {
  coop_nr coop;
  int i;
  // We allocate memory for all replies:
  reply_type replies[] = malloc(num_peers * sizeof(reply_type)
      );
  // init_desc allocates receive descriptors to accomodate the
        replies:
  rcv_desc *desc = init_desc(peers, num_peers, replies);

  coop = start_cooperation(desc, num_peers);  // register
        cooperation

  for (i=0; i<num_peers; i++)                 // send requests
    send(peers[i], REQUEST, coop, msg, msg_len);
  ...                                         // eventual
        computation
  end_cooperation(coop, 0);                   // wait for reply
        (no timeout)
  free(desc);                                 // free the
        descriptors
  return replies;
}
```

Listing 4. Sending requests

remote peer is known and identifiable through the *from* flag. When memory is registered, its handles, which are required for remotely accessing the memory, will be attached to the request along with the Cooperation token. The remote peer then has the information needed to directly write the reply into the client's memory without involving the CPU of the other machine.

Similarly, requests can be sent using RDMA and pre-registered buffers. In all cases, the kernel is bypassed with a complete user-level implementation over RDMA.

5 Conclusion

This paper introduced Paradis-Net, a low-level network interface which targets easier implementation of complex multi-party protocols. Paradis-Net emerged from our experience with parallel file systems and was motivated by the need of collaborative communication patterns.

To this end Paradis-Net introduces *Cooperations*, a mechanism that allows the user describing the result of collaborative work between several participating peers. We described the mechanism and its potential use by demonstrating how two common communication patterns used in parallel file systems can be implemented. Aside from distributed system development, we outlined how peer-to-peer functionality can be mapped onto Paradis-Net and how the library can transparently use remote direct memory access (RDMA) when available to increase performance.

References

1. Boden, N.J., Cohen, D., Felderman, R.E., Kulawik, A.E., Seitz, C.L., Seizovic, J.N., Su, W.K.: Myrinet: A gigabit-per-second local area network. IEEE Micro **15** (1995) 29–36
2. InfiniBand Trade Association: InfiniBand Architecture Specification Release 1.1. (2002)
3. Isailă, F., Tichy, W.F.: Clusterfile: A flexible physical layout parallel file system. In: Proceedings of IEEE Cluster Computing Conference, Newport Beach. (2001)
4. Isailă, F., Malpohl, G., Olaru, V., Szeder, G., Tichy, W.F.: Integrating collective I/O and cooperative caching into the Clusterfile parallel file system. In: Proceedings of the ACM International Conference on Supercomputing (ICS). (2004)
5. Anderson, T.E., Dahlin, M., Neefe, J., Patterson, D., Roselli, D., Wang, R.: Serverless network file systems. ACM Transactions on Computer Systems **14** (1996) 41–79
6. Wang, R.Y., Anderson, T.E.: Experience with a distributed file system implementation. Technical Report CSD-98-986, University of California at Berkeley (1998)
7. http://www.viarch.org: VIA: The Virtual Interface Architecture. (1998)
8. von Eicken, T., Culler, D.E., Goldstein, S.C., Schauser, K.E.: Active messages: A mechanism for integrated communication and computation. In: 19th International Symposium on Computer Architecture, Gold Coast, Australia (1992) 256–266
9. Nelson, B.J.: Remote Procedure Call. PhD thesis, Carnegie-Mellon University (1981)
10. Sandberg, R., Goldberg, D., Kleiman, S., Walsh, D., Lyone, B.: Design and implementation of the Sun network file system. In: Proceedings of Usenix 1985 Summer Conference. (1985) 119–130
11. Geist, A., Beguelin, A., Dongarra, J., Jiang, W., Manchek, R., Sunderam, V.: PVM: Parallel Virtual Machine. MIT press (1994)
12. The MPI Forum: MPI: A Message Passing Interface. In: Proceedings of the 1993 ACM/IEEE conference on Supercomputing. (1993) 878–883
13. Malpohl, G., Isailă, F.: The Paradis-Net API. Technical Report 2004/20, Universität Karlsruhe, Fakultät für Informatik, Germany (2004)
14. Dabek, F., Zhao, B., Druschel, P., Stoica, I.: Towards a common api for structured peer-to-peer overlays (2003)
15. Ratnasamy, S., Francis, P., Handley, M., Karp, R., Shenker, S.: A scalable content addressable network. In: Proceedings of ACM SIGCOMM 2001. (2001)
16. Stoica, I., Morris, R., Karger, D., Kaashoek, M., Balakrishnan, H.: Chord: A scalable peer-to-peer lookup service for internet applications. In: Proceedings of the 2001 conference on Applications, technologies, architectures, and protocols for computer communications, ACM Press (2001) 149–160
17. Rowstron, A., Druschel, P.: Pastry: Scalable, distributed object location and routing for large-scale peer-to-peer systems. In: IFIP/ACM International Conference on Distributed Systems Platforms (Middleware). (2001) 329–350
18. Zhao, B., Kubiatowicz, J., Joseph, A.: Tapestry: An infrastructure for fault-tolerant wide-area location and routing. Technical Report UCB/CSD-01-1141, Computer Science Division, U. C. Berkeley (2001)
19. Internet Engineering Task Force: Remote Direct Data Placement Charter. (2002)

Reliable Mobile Ad Hoc P2P Data Sharing

Mee Young Sung[1], Jong Hyuk Lee[1], Jong-Seung Park[1], Seung Sik Choi[1], and Sungtek Kahng[2]

[1] Department of Computer Science & Engineering, University of Incheon
{mysung, hyuki, jong, sschoi}@incheon.ac.kr
[2] Department of Information & Telecommunication Engineering,
University of Incheon
s-kahng@incheon.ac.kr

Abstract. We developed a reliable mobile Peer-to-Peer (P2P) data sharing system and performed some experiments for verifying our routing scheme using our real testbed. Our method for guaranteeing reliable P2P data sharing mainly deals with the problem of disconnection due to the sudden disappearance of one or more nodes involved in the active transmission route. The routing scheme of our system allows for reliable transmission of data via mobile devices. It does this using our reconnection mechanism that finds emergency routes in instances of abrupt disconnection using our scanning table and lookup table. The experiments lead us to conclude that our communication method is effective and assures reliable P2P data sharing over mobile ad hoc networks. In addition, our system allows for finding directly connectable devices at the overlay networks, therefore the application program can distinguish whether the target node is directly accessible or not with the aid of our scanning table.

1   Introduction

Mobile ad hoc networks (MANETs) are envisioned to become key components in 4G (fourth-generation) wireless network architecture. MANETs inherit common characteristics found in most wireless networks, and add characteristics specific to ad hoc networking, such as wireless, ad-hoc-based, autonomous, infrastructureless, multihop routing, and mobility [1]. Peer-to-Peer (P2P) networks act independent from the underlying network infrastructure. The TCP/IP only provides a communication platform protocol and does not supply information about the location of requested content or participants. P2P nodes independently maintain P2P networks and show characteristics of a free infrastructure and decentralized network [2]. P2P as well as mobile ad hoc networks follow the same idea of creating a network without the help of central entities. A combination of both, produces synergies and creates new possibilities, but also poses several problems.

The mobile ad hoc P2P paradigm attempts to create open and collaborative networks of a very diverse functionality and nature. Such functionality extends from the most popular data sharing protocols, like Gnutella, up to P2P instant messengers and chat applications, etc. However, regardless of its various applications, P2P

P. Lorenz and P. Dini (Eds.): ICN 2005, LNCS 3421, pp. 772–780, 2005.

networking always involves unstable transmission due to the mobility of some nodes which were engaged in the transmission. This means that P2P computing needs to assure stable data transmission until the successful termination of services.

A mobile ad hoc P2P data sharing system autonomously connects a dynamically changing set of nodes to allow the discovery and transfer of information among them. Together, a mobile ad hoc P2P data sharing system exhibits common characteristics that come with having rapidly and unpredictably changing users and not set infrastructure. Moreover, mobile devices are constrained by battery energy and computation capabilities, and communicate via low bandwidth wireless links [3]. Therefore, it is important to guarantee the stable and efficient transmission of information even though some service nodes suddenly disappear because of the movement of users or power shortage of the battery.

In this paper, we propose a reconnection mechanism for finding the emergency routes that guarantee reliable P2P data sharing over mobile ad hoc networks. By the term "emergency route" we define a new route which can substitute the broken route of the current service.

The paper is organized as follows: In the following section we briefly present the related work. Subsequently, we describe our ongoing development of mobile ad hoc P2P data sharing system. In Section 4, we discuss our proposed routing scheme for mobile ad hoc P2P data sharing application. In Section 5, we describe our experimental setup and discuss the results obtained in terms of their reliability as well as the throughput (reconnection or transmission time). Finally, we conclude in the last section.

2   Related Work

The primary objective of an ad hoc network routing protocol is the correct and efficient route establishment between a pair of nodes so that messages can be delivered reliably and in a timely manner. In general, ad hoc network routing protocols can be divided into two broad categories: proactive routing protocols and reactive on-demand routing protocols [4]. The most representative proactive protocol is   DSDV (Destination-Sequenced Distance-Vector). The most representative reactive protocol is AODV (Ad-hoc on-Demand Distance Vector).

A P2P network structure can be classified in two ways: a centralized structure and a decentralized structure. A decentralized networking structure can in turn be classified again as structured or unstructured. The unstructured architecture does not manage any linkage information of peers and the locations of files until explicit requests are made. Peers update their networking information only when it is requested. The structured architecture always retains the information about a network whenever a node joins and leaves. An example of a structured type is Chord [5] and an example of an unstructured type is ORION (Optimized Routing Independent Overlay Network) [6]. ORION is a file discovery algorithm based on keyword searching. ORION combines application-layer query processing with the network layer process of route discovery. Chord is based on the DHT (Distributed Hash Table) algorithm which uses the distributed hash values of nodes and keys. Chord solves the problem of storing the lookup data in a decentralized manner. Note that ORION, Chord, and our work can be classified as reactive routing protocols. From a structuring point of view, our work can be classified as both structured and unstructured (The scanning table of our work is composed in structured manner and the lookup table of our work is composed in unstructured manner).

3   Mobile Ad Hoc P2P Data Sharing System

Our mobile ad hoc P2P data sharing system allows users to share any information on files. What distinguishes our system from others is that it provides an automatic configuration of ad hoc networking modes without manual setting or execution of software by users. It also automatically generates the private IPs without any user intervention. Our system starts with the establishment of ESSID (Extended Service Set IDentifier). If one of the peers is started, any other peers with the same ESSID will be started and discover other mobile devices. Once peers are connected to each other, they share information about each other's routing tables and then update their own routing tables. Our system is implemented using the embedded Linux development kits. On the development kit, the embedded Linux kernel 2.4.19 is installed and we used airo.o as a PCMCIA wireless LAN card, libiw.so.26 as our wireless network library, and Qt library ver.3.07 for user interfaces.

Main components of our system are as follows:

- A Discovery Engine: It recognizes BSSID (Basic Service Set IDentifier) or ESSID (Extended Service Set IDentifier), translates the MAC addresses into IPs, and creates the scanning table and the lookup table which will be discussed in the following section.
- A Reconnection Engine: It is responsible for connection management and it provides the reestablishment of alternate routes or queries.
- A Transmission Engine (composed of a Packet Manager, a Control Socket, and a Data Socket): It takes charge of data transmission and it is implemented using Linux sockets. The Packet Manager controls the routing, and the actual file transmission is processed by Data Socket.
- A Management Module: It supports the QoS based data buffering for efficient data transmission.
- Presentation Module: It corresponds to the GUI (Graphical User Interface) part of the system and it is implemented using the Qt library. Any manipulation events entered through the GUI are processed by the Controller in Management module.

4   Routing Scheme

Many routing mechanisms and communication protocols are studied in the field of mobile ad hoc networks [7], [8], [9]. In general, a mobile ad hoc P2P data sharing system consists of a search mechanism to transmit queries (lookup) and search results, and a transfer protocol to download files matching a query. Recent studies on mobile ad hoc P2P computing are mainly focused on the efficient search of nodes that store the desired data item. One of the pioneering works is Chord which aims for a good hit ratio. One other work is ORION, which searches mobile devices that hold target information whilst using the least bandwidth. Both discovery algorithms combine application layer query processing with the network layer process of route discovery.

Our routing algorithm is composed of the following two mechanisms:

- An efficient discovery mechanism: This mechanism discovers the devices first and searches routing information for locating the target item. It is provided by our discovery engine which creates the routing table which is composed of a scanning table and a lookup table. Once the target is found, our reliable transmission algorithm starts to process.
- A reliable transfer mechanism: This takes charge of reliable communication thus guaranteeing the successful completion of requested services. It includes a reconnection engine in cases where a sudden disappearance of service nodes occurs because of the movement of users or a shortage of battery energy.

Our reconnection engine is based on the combined usage of the following two reconnection algorithms:

- A dynamic-alternate algorithm: It tries to find an alternate route in searching the lookup table. The lookup table of our system is similar to the file routing table of ORION. However, our lookup table contains information about the final destination of the target item, while ORION only keeps information about the node who knows the location of the target item. Our dynamic-alternate algorithm allows for finding the alternate route which is directly accessible.
- A dynamic-recruiting algorithm: It first selects one other neighbor node from amongst the nodes on the scanning table (which holds the directly accessible nodes in the active transmission range of a peer) and it retries until it finds a new route for accessing the target item. Whenever a peer is connected to another neighbor node, their lookup tables will be updated by sharing each other's information.

4.1 Automatic Configuration of Ad Hoc Networks

Our system allows for automatic configuration of ad hoc networks. Once a mobile device starts, the mode of each mobile device within an active transmission range has to be changed into the ad hoc mode and each device is set with the same ESSID (Extended Service Set IDentifier). Then the mobile device becomes accessible within the active ad hoc network. However, the application program cannot connect to mobile devices without IP addresses. Therefore, we need to provide a private IP for each mobile device. Our system automatically allocates private IPs to mobile devices in an active transmission range at the moment of initial ad hoc network configuration.

Our algorithm is designed for overlay networks. It assumes that the target information is within two hops and constructs the routing table using a scanning table and a lookup table. The scanning table is constructed when the ad hoc network is initially configured and it contains mobile nodes which are directly connectable. It is possible to find directly connectable devices at the level of link layer using MANET routing algorithms such as AODV. However, the application program still needs to know the IPs for data transmission. Therefore, we build the scanning table containing directly accessible IPs and the corresponding flags at the overlay network level.

We distinguish three values of flags: 0 means a node itself, 1 means the directly connectable node, and 2 means the node out of active area. Whenever a node appears within active transmission range, it can define its own flag. The node with flag 2 is not directly accessible. Only the nodes with flags 0 or 1 remain in the scanning table.

Once node A starts to work, it broadcasts a message to discover its neighbors. If node A receives no answers, it takes an IP and sets its flag to 0. If node B comes in the active area of node A and broadcasts a message for discovering then it receives a response (the content of the scanning table) from node A. In this case, node B sets its IP, which is not the same as node A's IP and its flag to 0, and keeps the reply from node A. If node C enters the active area of node B which is not accessible from node A. Node C receives the content of the scanning table from node B and comes to know about the existence of nodes B and A. Node C sets its IP and its flag to 0 and sets node B's flag to 1 because it is accessible. As node C cannot receive any response from node A, it sets node A's flag to 2.

4.2 Scenarios of Discovery and Transmission

To illustrate the operation of our discovery and transmission algorithm, we should consider the six-node scenario shown in Figure 1. Nodes are connected by an arrow and the rectangles near the nodes show parts of the lookup table. The large circle corresponds to the boundary of an active wireless transmission range. In Figure 1, node A can find 5 nodes (B~F) which are bound with the same ESSID. Once a connection is made between two nodes, they update their lookup tables by referencing the lookup table of the counter part.

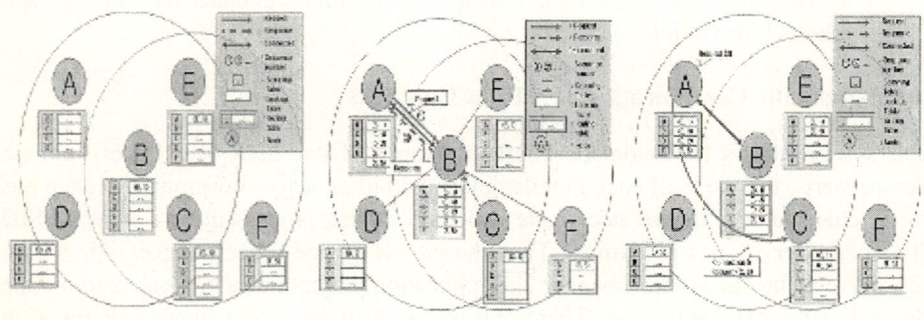

Fig. 1. Six-node Scenario   **Fig. 2.** Scenario 1   **Fig. 3.** Scenario 2

Our routing table is composed of two parts: one is the scanning table and the other is the lookup table. The scanning table contains all of the mobile devices which are found through the scanning broadcasting at the link layer. The scanning broadcasting is implemented using libiw.so.26, which is a C-language library at the MAC layer of IEEE 802.11. The lookup table is created at the application level and it contains pairs of the target item and the node which holds the target item.

There are two types of transmission method. One is to get information in multiple parallel channels from many peers. Another is to get the information from only one peer. To get information in parallel is effective in wired networks. However, it is not desirable to use in wireless networks due to the lack of transmission capability of mobile nodes and the difficulties to allocate a large bandwidth for every channel. Therefore, we limit the number of peers that are connected at the same time to 10 peers and the number of peers transmitting data in parallel to 3. As mentioned before, our discovery and transmission algorithms assume that a connection between two nodes implies sharing and updating of each other's lookup tables.

In Scenario 1 as shown in Figure 2, node A is connected to node B which is not connected to any node. Node A is downloading the target item from node B. In this situation, if node B is removed in the middle of transmission, node A has to try to connect to other nodes in the scanning table because there is no other information concerning the location of the target item in A's lookup table.

Scenario 2 illustrated in Figure 3 corresponds to a situation where node A is connected to node B which is connected to another node C. Node A shares the lookup table of node B and already knows that C contains the target item. If node B disappeared in the middle of transmission, node A could directly connect to node C for downloading the target item.

In Scenario 3 demonstrated in Figure 4, node A is connected to node B which is connected to many other nodes. Node A shares the lookup table of node B which includes information about the node that responded first to the lookup query. Assume that node E responded first in this example, and if node B is removed in the middle of transmission, node A can directly connect to node E for downloading the target item.

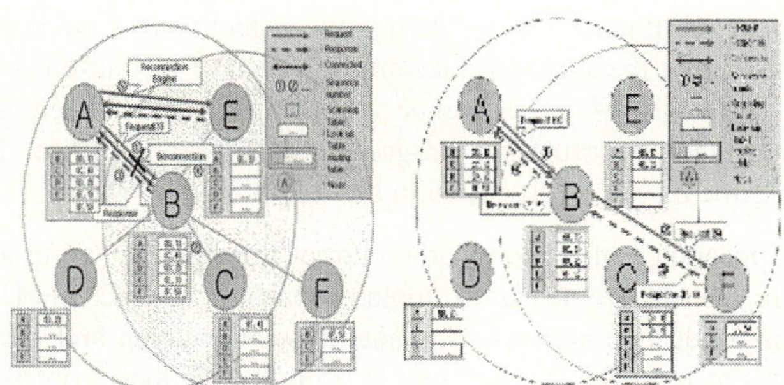

Fig. 4. Scenario 3   **Fig. 5.** Scenarios 4

Scenario 4 presented in Figure 5 is similar to Scenario 3. Node A is connected to node B which is connected to many other nodes. Node A shares the lookup table of node B which includes information about the node that responded first to the lookup query for the target item. Assume that node F responded first. Node F is within the active transmission range of B, however it is out of the transmission range of A. In this situation, if node B disappeared in the middle of transmission, node A has to download the target item from the destination node (node F in this example) via the relay node B.

4.3 What Distinguishes Our Algorithm

Although some features of our routing algorithm are similar to those of ORION, our algorithm distinguishes itself from ORION in the following ways:

- Our method for creating the routing table is different from that of ORION. ORION creates a response routing table (which contains routing information as well as lookup results), instead we separate the routing table into two parts: a scanning table and a lookup table. Processing lookup queries require more bandwidth than creating a simple routing table, therefore constructing the lookup table is more time consuming than constructing a scanning table for a direct connection and the lookup table for the location of items. In addition, our lookup table contains information about the final destination of the target item, while ORION only keeps the information of the relay node who knows the location of target item. Therefore, our dynamic-alternate algorithm allows for discovery and connection to a directly accessible emergency route.
- Our relaying method is also different. Our system provides direct connections to nodes as often as possible and tries to avoid connection to them indirectly. The efficiency of our system comes from the scanning table which contains directly accessible nodes. If the direct connection is not possible, the lookup table is searched for a relayed connection.

5 Experiments

Using the real testbed based on our mobile ad hoc P2P data sharing system, we undertook some experiments for validating the effectiveness of our reconnection mechanism for assuring the successful accomplishment of the current service. We can make diverse scenarios of experiments according to the different numbers of intermediate nodes in the route for on-going transmission service. However, any scenario of experiments can be classified in the following two ways:

- The disappearance of a node in the current transmission route and the system knows the alternate route (for example, let each of A, B, C, and D correspond to a node in an ad hoc network, A wants D's information and A already knows two routes to reach D; the first one is from A to B then from B to D, and the second one is from A to C then from C to D, while communicating through the first route, B is removed and tries to reconnect to D). In this case, our system applies a dynamic-alternate algorithm for reconnection.
- The disappearance of a node in the current transmission route and the system does not know about the alternate route (for example, A wants C's information and A knows B and B knows C, while A communicating with C through B, B is

removed and tries to reconnect to node C). In this case, our system uses a dynamic-recruiting algorithm.

According to four diverse scenarios presented above, we calculated the reconnection time for the target information and the average transfer rate for downloading the target information. For the average transfer rate, we compared the performance of direct transmission (one-to-one, without relays) for mobile ad hoc P2P data sharing of a 1.5 Mbytes mp3 file and that of relayed transmission. We measured the throughput using the following formulas:

$$SearchTime = ST, TransmissionTime = T$$
$$TransmissionCapacity = TC$$
$$Throughput = \frac{TC}{ST - T}$$

The experiment on the reconnection time showed that the time for reconnection increases in proportion to the number of shared files (in average, it takes 1.6 msec more per file). When measuring the performance of transmission, the average direct transfer rate (169 Kbytes/second) is approximately twice that of relayed transmission (83 Kbytes/second) as illustrated in Figure 6 (a) and Figure 6 (b).

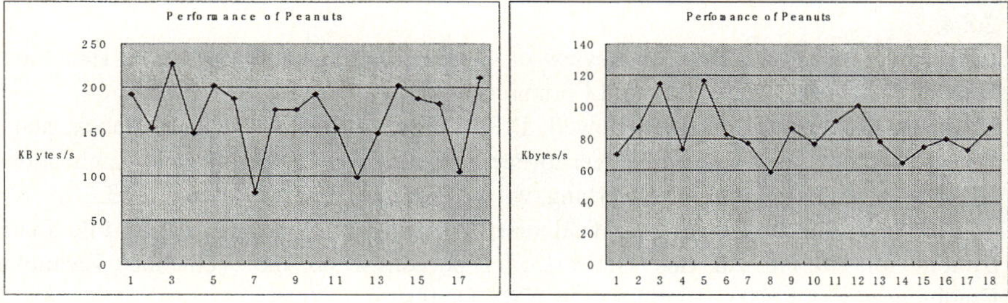

Fig. 6. Average Transfer Rates: (a) Average Transfer rate of direct transmission for downloading a 1.5Mbyte-mp3 file, (b) Average Transfer rate of relayed transmission for downloading a 1.5Mbyte-mp3 file

6 Conclusion

We believe that the most important factor to consider when designing mobile ad hoc applications is to guarantee the completeness of services. Therefore, we developed a mobile P2P data sharing system which assures reliable and stable data transmission even though some service nodes suddenly disappear because of the movement of users or the shortage of power. Our system allows for configuring a MANET without intervention of users by allocating the IPs automatically. The routing scheme of our system selectively applies two reconnection methods; one is the dynamic-alternate algorithm and the other is the dynamic-recruiting algorithm, for finding emergency routes in the case of the abrupt disconnection. Our routing algorithm needs to create a scanning table (which contains the directly connectable nodes) as well as a lookup table (which contains the information about the location of target data) for augmenting reliability in data sharing services. Creating those tables requires some overheads, however it is necessary to eliminate unnecessary transmission relays. Through some experiments, we confirmed that our routing scheme for assuring reliable P2P data sharing over mobile ad hoc networks is effective.

Acknowledgement. This work was supported by the Korea Science and Engineering Foundation (KOSEF) through the Multimedia Research Center at the University of Incheon.

References

1. J.J.-N. Liu and I. Chlamtac, "Mobile Ad hoc Networking with a View of 4G Wireless: Imperatives and Challeges," Mobile Ad Hoc Networking, Chapter 1, S. Basagni, M. Conti, S. Giordano, and I. Stojmenovic, Eds., Wiley, 2004, pp.69-116.
2. R. Schollmeier, I. Gruber, and F. Niethammer, "Protocol for Peer-to-Peer Networking in Mobile Environments," Proceedings on the IEEE International Conference on Computer Communications and Networks (ICCCN) 2003, pp.121-127, 2003.
3. A. Duran and C.C. Shen, "Mobile Ad hoc P2P File Sharing," Proceedings on the IEEE Wireless Communications and Networking Conference (WCNC) 2004, vol.1., pp.114-119, 2004.
4. E.M. Royer and C.-K. Toh, "A Review of Current Routing Protocols for ad-Hoc Mobile Wireless Networks," IEEE Personal Communications, April 1999.
5. I. Stoica, R. Morris, D. Lieben-Nowell, D.R. Karger, M.F. Kaashoek, F. Dabek, and H. Balakrishnan, "Chord: a Scalable Peer-to-Peer Lookup Protocol for Inrternet Applications," IEEE/ACM Transactions on Networking, vol.11, No.1, pp.17-32, February 2003.
6. A. Klemm, C. Lindemann, and O. Waldhorst, "A Special-Purpose Peer-to-Peer File Sharing System for Mobile Ad hoc Networks," Proceedings on the Vehicular Technology Conference (VTC) 2003, vol.4, pp.2758-2763, October 6-9, 2003.
7. G. Anastasi, M. Conti, and E. Gregori, "IEEE 802.11 Ad Ho Networks: Protocols, Performancem, and Open Issues," Mobile Ad Hoc Networking, Chapter 3, S. Basagni, M. Conti, S. Giordano, and I. Stojmenovic, Eds., Wiley, 2004, pp.69-116.
8. J.P. Macker and M. Scott Corson "Mobile Ad Hoc Networks (MANETs): Routing Technology for Dynamic Wireless Networking," Mobile Ad Hoc Networking, Chapter 9, S. Basagni, M. Conti, S. Giordano, and I. Stojmenovic, Eds., Wiley, 2004, pp.255-273.
9. M. Conti, E. Gregori, and G. Turi, "Towards Scalable P2P Computing for Mobile Ad Hoc Networks," Proceedings on the Second IEEE Annual Conference on Pervasive Computing and Communications Workshops (PERCOMW '04), 2004.

The Hybrid Chord Protocol: A Peer-to-Peer Lookup Service for Context-Aware Mobile Applications

Stefan Zöls[1], Rüdiger Schollmeier[1], Wolfgang Kellerer[2], and Anthony Tarlano[2]

[1] Institute of Communication Networks, Technical University of Munich, Germany
[2] Future Networking Lab, DoCoMo Communications Laboratories Europe, Germany
`{stefan.zoels, ruediger.schollmeier}@tum.de`
`{kellerer, tarlano}@docomolab-euro.com`

Abstract. A fundamental problem in Peer-to-Peer (P2P) overlay networks is how to efficiently find a node that shares a requested object. The Chord protocol is a distributed lookup protocol addressing this problem using hash keys to identify the nodes in the network and also the shared objects. However, when a node joins or leaves the Chord ring, object references have to be rearranged in order to maintain the hash key mapping rules. This leads to a heavy traffic load, especially when nodes stay in the Chord ring only for a short time. In mobile scenarios storage capacity, transmission data rate and battery power are limited resources, so the heavy traffic load generated by the shifting of object references can lead to severe problems when using Chord in a mobile scenario. In this paper, we present the Hybrid Chord Protocol (HCP). HCP solves the problem of frequent joins and leaves of nodes. As a further improvement of an efficient search, HCP supports the grouping of shared objects in interest groups. Our concept of using information profiles to describe shared objects allows defining special interest groups (context spaces) and a shared object to be available in multiple context spaces.

1 Introduction

P2P networking refers to a class of systems, applications and architectures that employ distributed resources to perform routing and networking tasks in a decentralized and self organizing way. Within the last few years, P2P traffic has become a major part in IP networks. This lead to an increasing attention of P2P in the research community, and a significant number of new P2P protocols have evolved dealing with the core problem of P2P systems: how to find a node sharing the required object and thereby to generate as little traffic as possible.

A very promising approach to solve this problem is the concept of Distributed Hash Tables (DHT). In contrast to unstructured P2P concepts, DHTs map keys onto nodes allowing direct localization of shared objects and therefore avoid flooding, which reduces effectively the signaling overhead in the peer network.

One realization of DHTs is the Chord protocol [1]. It is a scalable, distributed lookup protocol that uses a hash function to determine a node's m-bit identifier from its IP address. The same hash function is used to produce an identifier for every

P. Lorenz and P. Dini (Eds.): ICN 2005, LNCS 3421, pp. 781–792, 2005.

shared object. The identifier of a shared object is called its key k. The Chord nodes are ordered in an identifier circle modulo 2^m. Every key k is assigned to the first node in the ring whose identifier is equal to or follows k. This node stores references to all shared objects in the network whose key precedes the identifier of the node. Thus flooding of query messages can be avoided, as the query can be routed directly to the responsible node.

When a new node joins the Chord ring, it has to receive all object references according to the keys it is responsible for from its succeeding node. Again when the node leaves the network, all object references have to be transferred to the succeeding node. This can result in a high traffic volume, especially when there are many shared objects (and therefore object references) in the network and when nodes participate in the overlay network only for a short period of time.

In a mobile scenario, a high rate of joining and leaving nodes (= churn rate) is common, e.g. because of high costs for mobile data transfer. Besides high churn rates, also the relatively low transmission data rates and the limited resources of mobile devices demand for a low signaling overhead.

In this paper we aim at a scenario of context aware applications in particular. In a mobile ubiquitous computing environment, a multitude of nodes provide information to be found and accessed by the respective application. However, matching of applications to the relevant context (e.g., provided by sensors) is difficult (if applications and sensors are not 'hard-wired'). Therefore we use the concept of context spaces, introduced in [2], as an application support middleware to allow an application independent access to context information. Context spaces allow sharing of information among applications and context providers by creating and maintaining context spaces as virtual containers of information of particular interest.

In a mobile scenario, such a context spaces middleware could be realized by a multitude of loosely coupled nodes providing information or hosting applications requesting such, which is supported by a P2P infrastructure as described above. We abandon from centralized solutions where all object references are maintained in a highly available centralized server, to avoid the following disadvantages:

- Single point of failure
- Scalability problems caused by a high amount of information that is provided and accessed
- Maintenance work and administration costs for the operator
- Responsibility of the operator for the shared information.

As outlined above, the existing Chord protocol can lead to problems when used in such a mobile scenario. However, we choose Chord as base protocol for our concept to support efficient P2P search in mobile environments. Compared to other DHT-based protocols like CAN [3] or Pastry [4], Chord is rather simple and therefore easy to adapt.

In this work, we propose the Hybrid Chord Protocol (HCP), a modified version of Chord that is able to deal with the problems in mobile, wireless scenarios and to support the concept of context spaces. HCP uses two types of nodes –static nodes and temporary nodes– to reduce significantly the traffic load that is caused by shifting object references.

HCP also addresses the grouping of shared objects in context spaces. We introduce info profiles to label all provided content in the network. Every shared object is described by an info profile which contains several keywords. By means of these keywords the shared object is assigned to multiple context spaces, each containing all available information for a given keyword. This concept allows an easy way of finding required information simply by building the intersection of all context spaces given in the query.

This paper is organized as follows. Section 2 compares HCP to related work. Section 3 gives an overview over the architecture of HCP, while section 4 presents details of the protocol. In section 5, we evaluate the improvements of HCP by comparing the traffic load in Chord and HCP that is generated by shifting object references. Finally, we summarize our contributions and outline items for future work in section 6.

2   Related Work

Currently there exist a high number of DHT based protocols which establish structured P2P networks. The most prominent among them are Chord [1], CAN [3], Pastry [4] and Tapestry [5]. Since their publication, a number of modifications to improve them, e.g. to adapt them to the underlying physical topology [6], to provide anonymity and privacy in the overlay network [7] or to allow complex queries [8] have been proposed.

However, the signaling overhead especially in unstable environments is still a problem, which we try to solve in our approach. Chord, for example, needs at least $O(log^2 n)$ messages to repair the routing tables affected by a single node arrival or departure [9], where n represents the total number of nodes in the Chord network. Furthermore, DHT-based applications often store a large number of object references that have to be transferred to other nodes upon a node arrival or departure. If we take into account the short session lengths measured in FastTrack and Gnutella networks, this may result in high maintenance traffic [10, 11]. The environment can be so dynamic that DHT applications provide little useful service other than constant maintenance operations. Thus we must state that especially in mobile scenarios the current solutions of structured P2P networks are hardly applicable. Another problem of DHT-based networks is that the distribution of keywords describing the shared objects is not uniform. This results in unfair storage consumption between the nodes.

Tang [9] *et al.* developed a new method to minimize the routing distance to one or two hops by distributing all membership and content location changes in form of node profiles to every node in the DHT network. Thus they can reduce routing failures significantly. However, this approach only works in stable environments with low churn rates [9]. The same goal –to decrease routing distances– is targeted by S-Chord [12]. The authors propose a bidirectional ring to improve the node join and leave mechanism. So they can decrease the lookup failure rate and increase the lookup efficiency up to 50%. Hyperchord [13] uses a more aggressive scheme for maintaining routing table entries. It provides in place fixing fingers as an optimization to improve also the efficiency of node joins and leaves. However, both of them do not provide any analysis about unstable environments, and therefore can, from our point of view, not solve the problem of shifting object references caused by high churn rates.

The problem of unfair distribution of object references in a DHT-based network mentioned above is addressed in [14]. The authors propose an architecture called keyword fusion to balance unfair storage consumption among the participating peers on a Chord ring. They can thus achieve a reduction of the storage consumption of the 5% top loaded nodes by 50%, but also can not solve the problem of shifting object references, resulting in a high amount of signaling traffic.

In general, the adaptation of the overlay network to the underlying physical network is certainly a good approach to decrease delays and the overall data rate on the physical layer. Such an approach can avoid zigzag and unnecessary long routes in the physical layer [6]. However, as DHT networks are structured P2P networks, the node IDs would have to be bound to the node's location. Resulting, every change of the ID of a participating node equals to one node leave and one node join, which would cause a significant amount of signaling traffic. Thus such an approach can only be applied in a relatively static environment, where the nodes hardly move, i.e., do not change their physical location frequently.

Koorde [15, 16] is a DHT-based approach which achieves low maintenance overhead by employing de Bruijn graphs [17]. In this work the authors propose an algorithm that can control the maintenance overhead by adjusting the average route length, i.e. the longer the route the smaller is the average maintenance traffic. However, it does not address the problem of high churn rates as they do not differentiate between static and mobile nodes.

The diminished Chord ring approach, proposed in [18], establishes a primal ring and a number of tree-like substructures. These substructures are employed to carry out maintenance and routing tasks that do not require the involvement of all nodes participating in the overlay network. They are mainly based on an adapted finger establishment scheme. In diminished Chord the fingers are not established depending on the distance in the ID space, but on the content and interest shared by each node. Thus the complexity of the whole architecture increases, but the overall maintenance and routing overhead can be reduced. However, especially the overhead caused by high churn rates can still be improved if this approach would differentiate between the nodes according to their resources and their uptime, as proposed in this work.

Assigning a special role to more stable nodes, characterized by longer uptimes, is also the main improvement to Chord suggested in [19]. The authors propose an architecture which increases the routing success probability by replicating shared content on stable nodes, according to the uptime of the nodes. Their basic motivation for this approach is their finding from Gnutella analyses that the longer a node is actively participating in the network, the higher is the probability that a node remains in the overlay network. Thus, data replicated on stable nodes is lost only with a significantly decreased probability compared to the original Chord approach. Furthermore, every node in such a Kademlia network exploits the information routed through itself and can thus reduce the routing and maintenance overhead significantly, as it can reduce the number of stabilize and fix-finger calls. However, Kademlia does not address the problem of high churn rates, which we target in this work, and is therefore from our point of view better suited for stable environments.

To establish the context spaces mentioned in section 1, we employ so-called info profiles, which represent content and context information brought into the network by the participating peers. A similar concept of interlinking objects by means of profiles is also employed in [20] to establish a resource management framework for communication networks. In [20] resource definitions are used to describe shared resources, whereas we employ info profiles to set up and describe context spaces.

3 Architecture of HCP

3.1 Basic Concept

In the conventional Chord protocol all nodes in the network are considered to be equal. This equality between nodes does not exist in real world scenarios. For example in a fixed network, some Chord nodes could have a broadband flat rate connection to the Internet, while other nodes use only slow modem connections. In mobile scenarios, this situation is even more diverse, as the major part of the network is formed by mobile devices like mobile phones or PDAs, which have limited battery and storage capacity, limited transmission data rate and usually remain in the Chord network only for a short time.

HCP addresses these problems by the differentiation between static nodes and temporary nodes:

- Static nodes are highly-available nodes that form a quasi-permanent part of the HCP ring. They usually remain in the network for a longer period of time and have high data rate connections to other static nodes as well as a larger storage capacity. All object references in the network are stored at static nodes.
- Temporary nodes are all nodes that do not form a quasi-permanent part of the HCP network. In most cases, temporary nodes join the network only for a short period of time to find some particular pieces of information. In HCP, temporary nodes do not store object references. When a temporary node joins the network, all object references according to the keys it is responsible for remain with its closest static successor, in order to avoid the shifting of object references. If temporary nodes receive a request for an assigned key, they forward the request to their closest static successor that stores all object references for this key.

By the differentiation between static and temporary nodes, HCP ensures that only static, quasi-permanent nodes store object references. Thus nodes that join the network only for a short time do not cause traffic load due to the shifting of object references, beyond transferring their own object references to static nodes. Furthermore, temporary nodes with limited storage capacity are prevented from storing a lot of object references.

As static nodes store all object references available in the network, a high volume of data has to be transferred when a static node joins or leaves the network. However, this usually does not cause any problems as such joins or leaves are assumed to occur infrequently and the data is transmitted over more stable and usually broadband links between the static nodes. Nodes can become static nodes based on their uptime history and on their hardware and networking requirements.

3.2   Realization of Context Spaces

In HCP, every shared object is described by an info profile. The definition of an info profile is depicted in Fig. 1. It consists of five parts: the name of the shared object, a description (optional), at least one keyword, the IP address of the host, and a timestamp.

<table>
<tr><td colspan="2">Info Profile</td></tr>
<tr><td>+ ObjectName [1]:</td><td>String</td></tr>
<tr><td>+ Description [0…1]:</td><td>String</td></tr>
<tr><td>+ Keywords [1…c]:</td><td>String</td></tr>
<tr><td>+ Host [1]:</td><td>IP address</td></tr>
<tr><td>+ Timestamp [1]:</td><td>long</td></tr>
</table>

Fig. 1. Definition of an info profile

The first string in an info profile represents the name of the shared object, while the second string gives a description about it. Then one or more keywords follow that specify all relevant context spaces for this info profile. We explain the concept of context spaces in detail in the following paragraph. Finally, the info profile contains the IP address of the host offering the object, and a timestamp to ensure that outdated info profiles can be discarded.

HCP groups all available information in context spaces. Every keyword in a shared object's info profile indicates a relevant context space for this object. As an example, Fig. 2 shows an XML version of the info profile describing the file "Beatles – A little help from my friends.mp3".

```
<InfoProfile>
    <ObjectName>Beatles – A little help from my friends.mp3</ObjectName>
    <Description>
        mp3-version of the Beatles' song "A little help from my friends"
    </Description>
    <Keyword>Beatles</Keyword>
    <Keyword>Help</Keyword>
    <Keyword>Friends</Keyword>
    <Host>123.4.5.67</Host>
    <Timestamp>0</Timestamp>
</InfoProfile>
```

Fig. 2. Example for an info profile in XML

By means of the keywords this file is assigned to the context spaces "Beatles", "Help" and "Friends". Thus the sharing host sends this info profile to those static nodes that are responsible for the keywords "Beatles", "Help" and "Friends", i.e. are the first static successors of the hash values of these keywords. Each static node stores a list for every keyword it is responsible for. In this list all info profiles that contain that keyword are collected. These lists establish the context spaces, as they hold all available information for a given keyword. In our example, the static node that is responsible for the keyword "Friends" provides all info profiles with keyword "Friends" in the according context space. Fig. 3 gives an illustration about the organization of info profiles in context spaces.

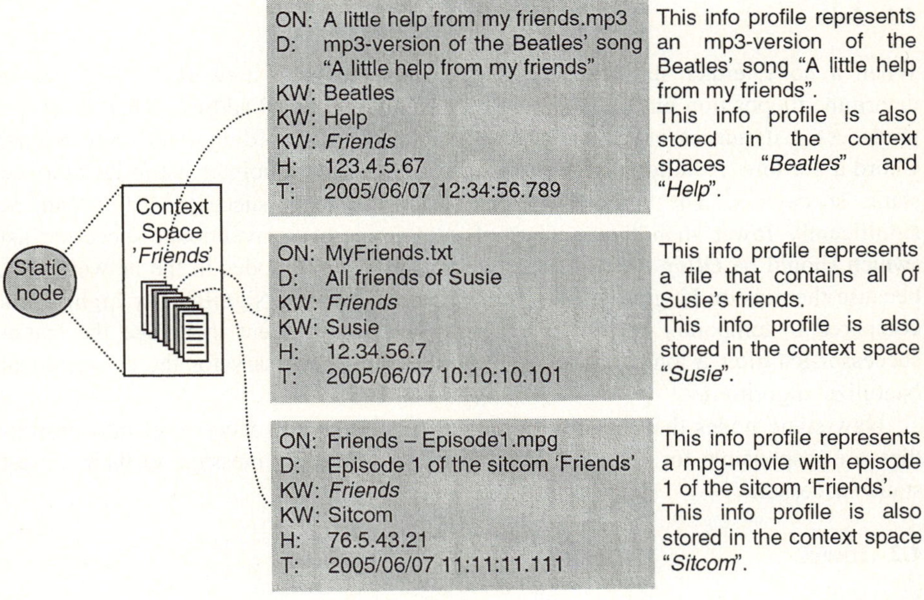

Fig. 3. Example for a context space "Friends". The responsible static node stores all info profiles with keyword "Friends" in this context space

The timestamp in an info profile is used to recognize outdated info profiles, e.g. info profiles from nodes that left the HCP network without deregistering. When a static node receives an info profile for one of its context spaces, it sets the info profile's timestamp to the current time. After a given period of time, the info profile will be removed from the context space if it has not been renewed by the sharing host.

An application that searches the network for a particular piece of information can use the context spaces to efficiently find the requested information. It simply builds the intersection of all context spaces that are relevant for the query. Assume a query with the keywords "Beatles" and "Help". To build the intersection of these two context spaces, an application can either look for all info profiles in the context space

"Beatles" that contain the keyword "Help", or it can look for all info profiles in the context space "Help" that contain the keyword "Beatles". In both cases, it will receive the same results. For this reason, it is sufficient to send the request only to one static node, which is yet another possibility to reduce signaling traffic.

4 Protocol Details

In this section we describe the necessary operations when nodes join or leave the network, what has to be done when nodes want to insert information into the HCP ring, and how queries are routed in HCP.

4.1 Join

When a node, either static or temporary, joins the HCP network, it first has to determine its position on the HCP ring by hashing e.g. its IP address. Then it sets its predecessor, its successor-list and its finger table entries according to the conventional Chord algorithm. Additionally, the joining node must set a pointer-list to its s closest static successors. The number of entries in the static-successor-list s can be significantly lower than the number of entries in the conventional successor-list (which should be $O(log\ n)$, with n as the total number of nodes in the network [3]), because the session length of static nodes is assumed to be significantly higher than compared to temporary nodes. Therefore it is also sufficient to update the static-successor-list after a longer period of time than it is necessary for the conventional 'stabilize' algorithm.

New static nodes that join the HCP ring finally have to receive all info profiles they are responsible for. Therefore they send an according message to their closest static successor.

4.2 Insert

Having joined the HCP ring, a node can insert its shared objects –which are described by info profiles– into the network. For every info profile, the application hashes all keywords and searches for the responsible nodes for this info profile. Then the node contacts the ascertained nodes to ask for their first static successor and sends the info profile to these static nodes. This procedure ensures that only static nodes store info profiles and that temporary nodes have to deal only with low signaling overhead.

4.3 Query

To find a particular piece of information, the user starts a query with one or more appropriate keywords. By hashing the given keywords, the application determines the responsible nodes for the requested information.

According to our example, the Beatles' song "A little help from my friends" from section 3, info profiles containing the keywords "Beatles", "Help" and "Friends", respectively, are stored at the three according static nodes. To find info profiles matching all three keywords, it is sufficient for the application to contact only one of these nodes, e.g. the static node that holds the context space "Beatles", and look there for all info profiles containing also the keywords "Help" and "Friends" (= building the intersection of the context spaces "Beatles", "Help" and "Friends").

Resulting, the querying node sends an according query message to the node that is responsible for the key (= hash value) of "Beatles", and this node –if it is a temporary node– forwards the query message to its closest static successor, which can then reply to the query. In case the responsible node is already a static node, it can immediately send a reply.

4.4 Leave

When a node leaves, it should deregister from the network, i.e. it should inform all static nodes that store info profiles, owned by the leaving node, to delete them. (In addition to that, outdated information can also be recognized by an info profile's timestamp.) Furthermore, a leaving static node has to transfer all its stored info profiles to its closest static successor.

5 Analytical Evaluation of HCP

With our analytical considerations we prove the ability of HCP to reduce significantly the signaling traffic. The traffic analyzed in this section is generated by the shifting of info profiles when nodes join or leave the overlay network. Further maintenance traffic sources are not considered, as they are the same in Chord and in HCP. In the first part of this section we calculate the average traffic load per node in the conventional Chord protocol. Afterwards, we compare the result with the traffic load that is generated in HCP.

Under the assumption that o is the total number of shared objects, and that every info profile that describes a shared object contains c keywords, $o \cdot c$ info profiles have to be stored in the whole network. This results in an average value of

$$I_n = o \cdot c / n \tag{5.1}$$

info profiles per node, where n is the number of nodes participating in the overlay network.

Each node is assumed to leave the network after an average session length T. As info profiles have to be shifted when a node joins the ring and also when it leaves again, the transfer rate per node amounts to

$$\tau_n = 2 / T \tag{5.2}$$

Assuming that b is the average size of an info profile, the average traffic load for a single node λ_n can therefore be calculated by

$$\lambda_n = I_n \cdot b \cdot \tau_n = (o \cdot c / n) \cdot b \cdot (2 / T) = (2 \cdot o \cdot c \cdot b) / (n \cdot T) \tag{5.3}$$

From this formula, we can conclude the advantage of HCP in comparison to the conventional Chord protocol. In HCP, the traffic load that is caused by temporary nodes is zero, as temporary nodes do not store info profiles. Additionally, static nodes –which store all info profiles– usually remain in the network for a long period of time. Due to their high session length T_{Static} the average traffic load per node can be decreased significantly, as shown below.

Assume an HCP ring with n nodes, whereof $x \cdot 100\%$ of the nodes are static and $(1-x) \cdot 100\%$ of the nodes are temporary, with $0 < x < 1$. The static nodes have an average session length which is α times the average session length of a conventional Chord node:

$$T_{Static} = \alpha \cdot T \tag{5.4}$$

The session length of the temporary nodes can be negligible, because temporary nodes do not store info profiles and therefore do not generate traffic load by shifting them.

In summary, every static node stores

$$I_{Static} = (o \cdot c) / (x \cdot n) = (1 / x) \cdot I_n \tag{5.5}$$

info profiles.

These info profiles have to be shifted when a static node joins and leaves the HCP network, so the transfer rate of a static node is given by

$$\tau_{Static} = 2 / T_{Static} = 2 / (\alpha \cdot T) = (1 / \alpha) \cdot \tau_n \tag{5.6}$$

This results in an average traffic load per static node of

$$\lambda_{Static} = I_{Static} \cdot b \cdot \tau_{Static} = (1 / x) \cdot I_n \cdot b \cdot (1 / \alpha) \cdot \tau_n = (1 / (x \cdot \alpha)) \cdot \lambda_n \tag{5.7}$$

As stated above, the traffic load of temporary nodes $\lambda_{Temporary}$ is 0. Resulting, we have an average traffic load for all nodes in the HCP network of

$$\lambda_{HCP} = (x \cdot n \cdot \lambda_{Static} + (1 - x) \cdot n \cdot \lambda_{Temporary}) / n = x \cdot \lambda_{Static} = (1 / \alpha) \cdot \lambda_n \tag{5.8}$$

Thus we can state that the average traffic load per node that is generated by the shifting of info profiles in a HCP network can be decreased by a factor of $1/\alpha$ in comparison to a conventional Chord network, with $\alpha = T_{Static} / T$. This means that the traffic load reduction is direct proportional to the fraction of the length of an average session to the average session length of a static node.

6 Conclusion and Future Work

In this work we proposed a new architecture named HCP that establishes a structured P2P network and shows a good performance especially in unstable, e.g. mobile environments where high churn rates of the participating nodes must be expected.

The architecture is based on the Chord protocol and extends it by two different node classes, namely static nodes and temporary nodes. Therefore we can reduce the overall routing and maintenance traffic of Chord significantly, but additionally keep the advantageous properties of Chord, like the guaranteed availability of shared content. The good performance of HCP, i.e. the significant reduction of traffic load that is generated by the shifting of info profiles, is proven by our analytical evaluation in section 5.

HCP also allows a very efficient implementation of the context spaces concept. Within HCP the context spaces are automatically established and maintained by assigning and transferring info profiles to the according nodes.

Currently we are working on the implementation of Chord and HCP in ns-2 [21]. Thus we want to verify our analytical results and be able to evaluate the performance of HCP in a completely simulated mobile environment. To be able to analyze the performance of HCP in a larger scale, i.e. with a higher number of nodes, we are currently also developing a simulation that scales up to a few million peers.

References

[1] I. Stoica, R. Morris, D. Karger, M. Kaashoek, and H. Balakrishnan., "Chord: A Scalable Peer-to-Peer Lookup Service for Internet Applications" presented at ACM SIG-COMM Conference, 2001.

[2] A. Tarlano and W. Kellerer, "Context Spaces Architectural Framework" presented at SAINT 2004 Workshop on Ubiquitous Services, 2004.

[3] S. Ratnasamy, P. Francis, M. Handley, R. Karp, and S. Shenker, "A Scalable Content-Addressable Network" presented at ACM SIGCOMM Conference, 2001.

[4] A. Rowstron and P. Druschel, "Pastry: Scalable, Distributed Object Location and Routing for Large-Scale Peer-to-Peer Systems" presented at IFIP/ACM International Conference on Distributed Systems Platforms (Middleware), 2001.

[5] B. Y. Zhao, L. Huang, J. Stribling, S. C. Rhea, A. D. Joseph, and J. D. Kubiatowicz, "Tapestry: A Resilient Global-Scape Overlay for Service Deployment" *IEEE Journal on Selected Areas in Communications*, vol. 22, 2004.

[6] L. Zhuang and F. Zhou, "Understanding Chord Performance" Technical Report CS268, 2003.

[7] S. Goel, M. Robson, M. Polte, and E. G. Sirer, "Herbivore: A Scalable and Efficient Protocol for Anonymous Communication" Cornell University Computing and Information Science Technical Report, TR2003-1890, 2003.

[8] M. Harren, J. M. Hellerstein, R. Huebsch, B. T. Loo, S. Shenker, and I. Stoica, "Complex Queries in DHT-based Peer-to-Peer Networks" presented at 1st International Workshop on Peer-to-Peer Systems (IPTPS '02), 2002.

[9] C. Tang, G. Altekar, and S. Dwakadas, "Calot: A Constant-Diameter Low-Traffic Distributed Hash Table" in *Under submission*, 2003.

[10] S. Sen and J. Wang, "Analyzing Peer-to-Peer Traffic Across Large Networks" presented at ACM SIGCOMM Internet Measurement Workshop, 2002.

[11] R. Schollmeier and A. Dumanois, "Peer-to-Peer Traffic Characteristics" presented at EUNICE 2003, 2003.

[12] V. A. Mesaros, B. Carton, and P. V. Roy, "S-Chord: Using Symmetry to Improve Lookup Efficiency in Chord" presented at 2003 International Conference on Parallel and Dis-tributed Processing Techniques and Applications (PDPTA'03), 2003.

[13] K. Lakshminarayanan, A. R. Rao, and S. Surana, "Hyperchord: A Peer-to-Peer data Location Architecture" UC Berkley Technical Report CS-021208, 2001.

[14] L. Liu and K. D. Ryu, "Supporting Efficient Keyword-Based File Search in Peer-to-Peer File Sharing Systems" IBM Research IBM Research Report. RC23145 (W0403-068), 2004.

[15] M. F. Kaashoek and D. R. Karger, "Koorde: A Simple Degree-Optimal Distributed Hash Table" presented at Fifteenth annual ACM-SIAM symposium on Discrete Algorithms, 2004.

[16] D. R. Karger and M. Ruhl, "Simple Efficient Load Balancing Algorithms for Peer-to-Peer Systems" presented at 3rd International Workshop on Peer-to-Peer Systems (IPTPS'04), 2004.

[17] N. D. Bruijn, "A Combinatorial Problem" *Koninklijke Nderlandse Akademie van Wetenschapen*, vol. 49, 1946.

[18] D. R. Karger and M. Ruhl, "Diminished Chord: A Protocol for Heterogeneous Subgroup Formation in Peer-to-Peer Networks" presented at International Workshop on Peer-to-Peer Systems (IPTPS '04), 2004.

[19] P. Maymounkov and D. Mazieres, "Kademlia: A Peer-to-Peer Information System Based on the XOR Metric" presented at International Workshop on Peer-to-Peer Systems (IPTPS'02), 2002.

[20] T. Friese, B. Freisleben, S. Rusitschka, and A. Southall, "A Framework for Resource Management in Peer-to-Peer Networks" presented at NetObjectdays 2002, 2002.

[21] ns-2, "The Network Simulator ns-2 Homepage", http://www.isi.edu/nsnam/ns/

LQPD: An Efficient Long Query Path Driven Replication Strategy in Unstructured P2P Network

Xicheng Lu, Qianbing Zheng, Peidong Zhu, and Wei Peng

School of Computer Science, National University of Defense Technology,
Changsha 410073, P.R.China
nudtzqb@126.com

Abstract. *Random walks* is an excellent search mechanism in unstructured P2P network. However, it generates long delay and makes some resources indiscoverable, especially for uncommon files. File replication strategy can improve the performance of *random walks*. But it has high overhead. An efficient replication strategy *LQPD* is presented. In *LQPD*, file replication operation is triggered only when the length of query path exceeds user-acceptable delay. Experimental results show *LQPD* can get better performance than other strategies with the same number of file replicas in a Gnutella-like overlay network.

1 Introduction

Gnutella-like unstructured P2P file sharing systems cause significant impact on Internet traffic. These systems use flooding-based search policy and TTL-based termination condition. Such search mechanism produces vast redundant messages and wastes enormous bandwidth.

Random walks[1] is an excellent improvement of flooding-based search mechanism. It forwards a query message to a randomly chosen neighbor at each step. And this message is called a "walker". The number of walkers k(typical 16 to 64) is increased to get the user-acceptable delay. The requester peer sends k query messages and each query message takes its own random walk. The termination condition is checking method that the walkers periodically contact the requester peer asking whether the user demand has been satisfied. *Random walks* can achieve significant message reduction compared with the standard flooding scheme[1]. Furthermore, it is identified that *random walks* achieves better results than flooding for searching when the overlay topology is clustered[2].

However, *random walks* needs a long delay to find the uncommon files which don't have many replicas in P2P network for it only visits a small fixed number of peers at each hop. The variations of *random walks*, such as APS[3] and Gia[4], select a neighbor according to heuristic information to accelerate the search process, but they have little effect on finding the uncommon files. In fact, 70% of Gnutella users share no files[5], which sharpens the disadvantage of *random walks* and makes some files indiscoverable by *random walks*.

P. Lorenz and P. Dini (Eds.): ICN 2005, LNCS 3421, pp. 793–799, 2005.

File replication can improve the performance of *random walks*. The *Owner replication* strategy replicates files at the requester peer in Gnutella only when a search is successful. The *Path replication* strategy replicates files at all peers along the path from the requester peer to the provider peer in Freenet[6]. The *Random replication* strategy randomly picks p peers that all walkers visited to replicate files once a search succeeds. p is the number of peers on the path between the requester peer and the provider peer[1]. The *Query-dependence replication* strategy finds square-root allocation which makes the replication of a file be proportional to the square root of its query rate is the optimal allocation and proposes three algorithms to achieve this allocation[7].

Actually, File replication costs extra transfer bandwidth and storage. The more the number of file replicas is, the more the overhead is. However, the above replication strategies don't take the replication overhead into consideration. The queries in Gnutella show a significant amount of temporal locality, which means some queries are submitted many a time in a short period[8]. When the flash crowd of some popular query happens, it will make these replication strategies generate too many replicas on some files and too much replication overhead.

In this paper, we propose an efficient replication strategy—*LQPD(Long Query Path Driven replication)* to improve the performance of *random walks* and make a good tradeoff between replication performance and replication overhead. The rest of this paper is organized as follows. In section 2, we describe *LQPD* in details. In section 3 and 4, we make some experiments and discuss the results. Finally, we end with conclusions in section 5.

2   Long Query Path Driven Replication Strategy

The main idea of *LQPD* is to replicate files moderately and make the performance of *random walks* reach a user-acceptable extent. The performance of *random walks* is mainly indicated by the delay of searching file. If we don't take the underlying network topology into consideration, the delay l can be estimated by the longest query path length of all successful walkers. The query path which a walker w_i traverses from the requester peer v_{i1} is denoted as a peer set $L_i = \{v_{i1}, v_{i2}, ..., v_{im}\}$, and the query path length l_i of the walker w_i is the number of peers in L_i except v_{i1}.

Assuming the user-acceptable delay is l_{max}, the process of *LQPD* is as follow. When the query path length of a walker exceeds l_{max}, the walker begins to log the visited peer information every $l_{max}+1$ hops. When the search is successful, each walker will report its log information to the requester peer. After the requester peer downloads the file successfully, it replicates the file to all peers from the log information of each walker.

Let $D(x)$ denote if the demand of x has been satisfied. A walker can get the value of $D(x)$ by checking method.

$$D(x) = \begin{cases} \text{true,} & \text{if the demand of } x \text{ has been satisfied} \\ \text{false,} & \text{if the demand of } x \text{ has not been satisfied} \end{cases}$$

Let $distance_i$ denote the distance variable which each w_i carries, $distance_i$ equals $l_{max} + 1$ initially. Assuming that s is the requester peer and the total number of peers is N in the P2P network, the pseudo-code of $LQPD$ is shown in Algorithm1.

> **Algorithm 1:** The procedure of $LQPD$.
> **Input:** The requester peer s, the requested file f, a peer set $V = \{v_1, \ldots, v_N\}$, l_{max}.
> **Output:** some file replicas
> LQPD(s,f,V,l_{max})
> /* the operation of each walker w_i */
> **if** (a walker w_i visits a peer v_i)
>   **if** ($D(s) =$**true**)
> w_i reports L_i to s
>   **else**
>   **if** (v_i has not f) **and** ($l_i > l_{max}$)
> $distance_i$ $- -$
>   **if** ($distance_i = 0$)
> w_i logs v_i to L_i
>   **else if** (v_i has f) **and** ($l_i > l_{max}$)
> $distance_i = l_{max} + 1$
>
> /* the operation of s */
> **if** ($D(s) =$**true**)
> s selects all peers from all L_i to replicate f

3 Simulation Methodology

Recent measurements and researches show that Gnutella-like unstructured networks have power-law degree distribution[9, 10, 11], which have similar characteristics to the scale-free model[12]. The scale-free model can be used to simulate Gnutella-like P2P overlay network. We generate a P2P overlay graph OL with 10,004 nodes using Barabais Graph Generator[13]. The graph generator is based on the scale-free model. The average node degree of OL is 3.99 which approximates to the actual value 3.4 of Gnutella[9]. The power-law index k of OL is -2.343 which approximates to the actual value -2.3[11]. So we can use OL to simulate the real Gnutella networks.

Two distributions are used to simulate the P2P system environment.

- file query distribution: it represents how many queries are made for each file. The number of queries for a file indicates its popularity. The file popularity follows a Zipf-like distribution[14]. So the file query distribution also follows Zipf-like distribution. The form of Zipf-like distribution is as follow:

$$f(x) \propto \frac{1}{x^\alpha}$$

Two query sets are generated. The first query set QS_1 comprises 644 queries. The early queries form the flash crowd for some popular files and the rest

queries are for some uncommon files. QS_1 is used to prove that $LQPD$ can avoid generating too many unnecessary file replicas and get a good performance when the flash crowd of some popular query happens. The second query set QS_2 consists 763 queries whose file query distribution follows a Zipf-like distribution with parameter $\alpha = 1$. QS_2 is used to show the performance of $LQPD$ when the flash crowd doesn't happen.
- initial file distribution: it controls the number of each file initially in P2P network. In the simulated environment, 30% of the peers are set to share files and a Zipf-like distribution is used. The reason to choose 30% is based on the report[5].

Two metrics are used to compare the performance of these replication strategies for a query set.

- AH: AH is the average value of l for all successful searches in the query set. It is used to estimate the delay of *random walks* using a replication strategy.
- AM: AM is the average number of messages generated for all successful searches in the query set. It is used to estimate the overhead of *random walks* using a replication strategy.

4   Experimental Results

Two experiments are designed to compare $LQPD$ with owner replication, path replication and random replication strategies in OL. The first experiment evaluates the performance of these replication strategies for QS_1 and the second experiment for QS_2. Let R is the total number limit of file replicas which a replication strategy can generate for a query set and r is the total number of file replicas generated actually. Assuming that l_{max} is 10, each peer can store at most 30 files, the replacement policy is random deletion.

4.1   Performance Evaluation for QS_1

Random walks with no replication, owner replication, path replication, random replication and $LQPD$ run for query set QS_1 in OL.

Fig. 1 shows the change of AH with the increasing of R for these replication strategies. After the flash crowd of some queries happens, $LQPD$ can reduce the delay of *random walks* more rapidly than other replication strategies with the increasing of R. Furthermore, random replication which is proved to be better than owner replication and path replication[1] generates 900 file replicas to get the approximate delay of $LQPD$ with 195 file replicas. The metrics of each replication strategy with $R = 200$ is computed and listed in Table 1. Table 1 shows $LQPD$ reduces the delay of *random walks* from 14.32 to 3.82 with 200 file replicas. However, owner replication, path replication and random replication only reduce the delay of *random walks* to about 9 with the same number of file replicas. $LQPD$ also reduces the overhead of *random walks* with no replication by 73% and the overhead of *random walks* with other replication strategies by about 60%. From Fig. 2(a) and (b), we can find that other replication strategies

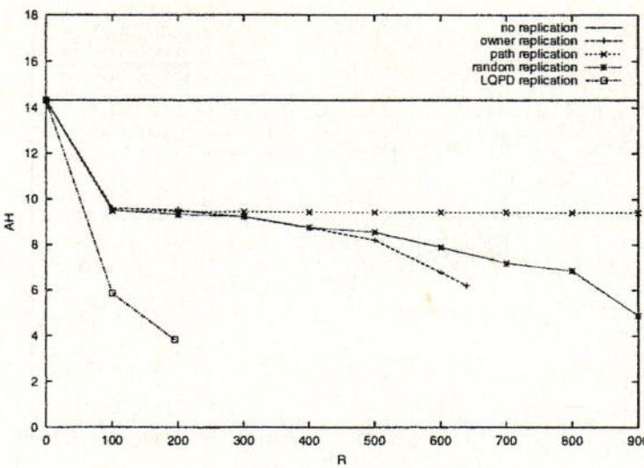

Fig. 1. AH vs. R

Table 1. The performance of four replication strategies with $R = 200$ for QS_1

replication strategy	AH	AM	r
no replication	14.32	316.98	0
owner replication	9.52	216.44	200
path replication	9.45	214.57	200
random replication	9.32	210.81	200
$LQPD$replication	3.82	85.02	195

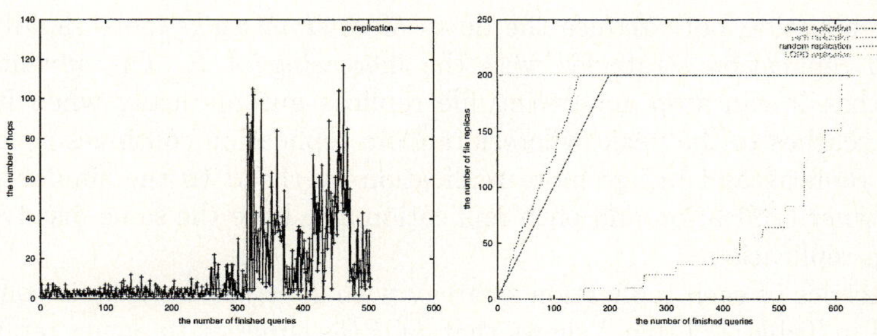

(a) the number of hops vs. the number of finished queries of *random walks*

(b) the number of file replicas vs. the number of finished queries

Fig. 2. The relation between the number of file replicas and popular files which can be find in short delay for different replication strategies with $R = 200$ for QS_1. The number of finished queries increases with the sequential process of each query in QS_1

suffer bad performance due to generating too many unnecessary file replicas on early popular files which can be find in short delay in QS_1.

4.2 Performance Evaluation for QS_2

Random walks with no replication, owner replication, path replication, random replication and $LQPD$run for query set QS_2 in OL. Fig. 3 shows $LQPD$and

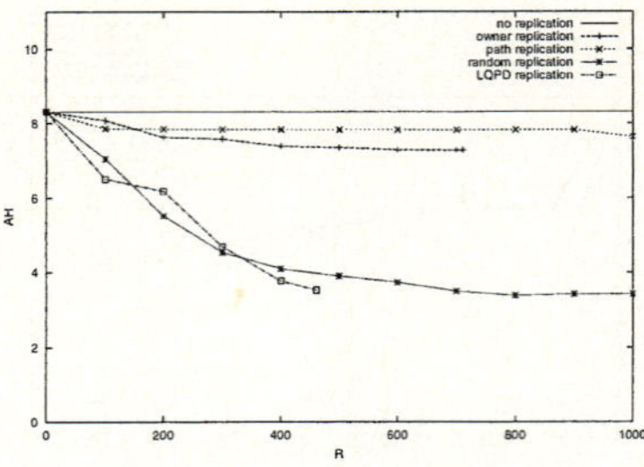

Fig. 3. AH vs. R

Table 2. The performance of four replication strategies with $R = 700$ for QS_2

replication strategy	AH	AM	r
no replication	8.31	199.78	0
owner replication	7.39	174.97	700
path replication	7.83	186.17	700
random replication	3.51	83.70	700
LQPD replication	3.53	85.82	461

random replication both reduce the delay of *random walks* more rapidly than other two replication strategies with the increasing of R. The advantage of *LQPD* is that it can stop generating file replicas automatically when its performance reaches to the peak, whereas random replication continues generating more file replicas and brings more replication overhead to the similar performance. Owner replication and path replication also have the same disadvantage of random replication.

The metrics of each replication strategy with $R = 700$ for QS_2 is computed and listed in Table 2. Table 2 shows that *LQPD* is superior to owner replication and path replication and gets the similar performance to random replication with less file replicas generated actually.

5 Conclusions

We have proposed an efficient replication strategy—*LQPD* to get a good tradeoff between replication overhead and replication performance. Experimental results show *LQPD* can get better performance than other strategies with the same number of file replicas in a Gnutella-like overlay network. When the flash crowd for popular files happens, *LQPD* is superior to other replication strategies greatly. Even when the flash crowd doesn't happens, *LQPD* also can get the better tradeoff between replication performance and replication overhead than other replication strategies.

Acknowledgement

This work is supported in part by NSFC No. 90412011, NSFC No. 90204005 and 863 Program No. 2003AA121510.

References

1. Lv, Q., Cao, P., Cohen, E., Li, K., Shenker, S.: Search and replication in unstructured peer-to-peer networks. In: Proceedings of the annual ACM International Conference on supercomputing. Volume 16th. (2002)
2. Gkantsidis, C., Mihail, M., Saberi, A.: *random walks* in peer-to-peer networks. In: Proceedings of Infocom 2004. (2004)
3. Tsoumakos, D., Roussopoulos, N.: Adaptive probabilistic search for peer-to-peer networks. In: 3rd IEEE Intl Conference on P2P computing. (2003)
4. Chawathe, Y., Ratnasamy, S., Breslau, L., Lanham, N., Shenker, S.: Making gnutella-like p2p systems scalable. In: SIGCOMM. (2003)
5. Adar, E., Huberman, B.A.: Free riding on gnutella. Technical report, Xerox PARC (2000)
6. Clarke, I., Sandberg, O.: Freenet: A distributed anonymous information storage and retrieval system. In: Proceedings of the ICSI Workshop On Design Issues in Anonymity and Unobservability. (2002)
7. Cohen, E., Shenker, S.: Replication strategies in unstructured peer-to-peer networks. In: ACM SIGCOMM. (2002)
8. Markatos, E.: Tracing a large-scale peer to peer system an hour in the life of gnutella. In: Proceedings of IEEE/ACM International Symposium on Cluster Computing and the Grid. (2002)
9. Ripeanu, M., Foster, I., Iamnitchi, A.: Mapping the gnutella network: Properties of large-scale peer-to-peer systems and implications for system design. IEEE Internet Computing Journal **6-1** (2002)
10. Jovanovic, M., Annexstein, F.S., Berman, K.A.: Modeling peer-to-peer network topologies through "small-world" models and power laws. In: Proceedings of the TELFOR. (2001)
11. DSS, C.: Gnutella: To the bandwidth barrier and beyond. Technical report, http://dss.clip2.com (2000)
12. Barabasi, A.L., Albert, R.: Emergence of scaling in random networks. Science **509** (1999)
13. Dreier, D.: Manual of operation: Barabasi graph generator v1.0. Web Pages (2002) http://www.cs.ucr.edu/ ddreier/bman.pdf.
14. Sripanidkulchai, K.: The popularity of gnutella queries and its implications on scalability. In: O'Reilly's www.openp2p.com. (2001)

Content Distribution in Heterogenous Video-on-Demand P2P Networks with ARIMA Forecasts*

Chris Loeser, Gunnar Schomaker, André Brinkmann,
Mario Vodisek, and Michael Heidebuer

Institute of Computer Science, University of Paderborn, Germany
`loeser@uni-paderborn.de`

Abstract. Peer to peer applications have gained high popularity in the past years. In particular, P2P media streaming architectures have attracted much attention, so that audio and video sharing is causing a large fraction of the Internet traffic today.

In this paper we introduce a new peer to peer architecture that focuses on distributed video on demand file sharing and that is based on point-to-point file delivery between the peers. The P2P architecture includes dynamic data distribution and replication schemes that are able to guarantee a fair load balancing among the peers. This load balancing enables the P2P architecture to avoid hot spots inside the distribution network and to ensure a nearly optimal throughput.

A main component of the P2P architecture is an ARIMA based forecasting module, that is able to predict the access probability of individual files. This forecasting module has an impact on the location of documents concerning the characteristic of peers and in addition is used to control the number of replicas of each file. In this paper we present some simulation results indicating not only the feasibility of this architectural approach but also the benefits resulting from dynamic content distribution.

1   Introduction

Peer to Peer networks have gained a lot of attention within the past years. Though the idea is not really new, the potential of P2P networks has grown tremendously with the introduction and availability of high speed internet network accesses. Considering P2P applications, customers typically talk about the sharing of resources like computational power or media files. Each peer offers resources and is also able to occupy resources of other peers simultaneously. Within

---

* Parts of the work are funded by the BMBF within the ITEA-SIRENA project (01ISC09E) and by DFG SFB 376 "Massively Parallel Computation" and by the EU within the 6th Framework Programme under contract 001907 "Dynamically Evolving, Large Scale Information Systems" (DELIS).

P. Lorenz and P. Dini (Eds.): ICN 2005, LNCS 3421, pp. 800–809, 2005.

the P2P area we distinguish between hybrid and pure P2P architectures. Pure P2P techniques are commonly based on distributed indexing and lookup services making use of distributed hash tables (DHTs) like CAN, Chord or Tapestry.

In this paper we propose an architecture for a P2P video on demand (VoD) application. We assume a medium sized set of peers equipped with high-speed synchronous DSL connections where each peer is offering storage capacity. Similar architectures have been proposed in Freenet or Publius where peers are not aware of the content they are storing.

We focus on avoiding (expensive) video on demand streaming servers. Instead, a field of peers acts as a distributed, fast, and fault tolerant video server in e.g. a metropolitan network: Consider households equipped with symmetric DSL, T1 or even T3 lines and set-top boxes. Current SDSL upstream bandwidths reach from 1 to 5 MBit/s.

The major task in a distributed video on demand environment is the seamless and transparent integration of the content delivery process into the peers. The content delivery task has to be evenly distributed among the peers according to their bandwidth and storage capacity. I.e. it has to be avoided that some peers are starving while other peers are getting swamped. To achieve this seamless integration into the peer environments we propose a dynamic, popularity based replica creation within the P2P network. Content has to be distributed according to its future popularity value. This is done by analyzing request time-series and to predict requests for the near future. Within this work we consider both heterogeneous bandwidth with heterogeneous provided HD-capacity and heterogeneous movie sizes. To handle these variables we make use of the Pagoda overlay network: Peers are categorized by their interconnection bandwidth and their stored content.

The outline of this paper is as follows: In the next section we will give an overview about the related work, followed by a detailed description of our architecture which consists of the Seasonal ARIMA forecasting technique and the Pagoda enabled content distribution algorithm. We will conclude with a some simulative results and a short description of our future work.

2   Related Work

P2P networks and world-wide distributed storage systems have become critical components for many application scenarios, including storage administration in global companies, file-sharing, and large-scale database environments. Both include the demand for the distribution of the content and the access to the content but rely on different constraints.

P2P storage projects like CFS, Oceanstore, or PAST distribute content about different locations and ensure a consistent view on this content from each accessing peer. They e.g. enable the concurrent access to a large number of globally spread files from arbitrary locations. To guarantee the availability and accessibility of all files, they are allowed to require the availability of a given fraction of the peers.

P2P file sharing projects, like CAN, Pastry, Chord, or Tapestry, also distribute content about peers. In contrast to P2P storage projects, they have no requirements on the number, availability and kind of these peers and do not support the dynamic change of the content. Therefore they only work on a best effort base and do not make any guarantees on the consistency of the content. The aim of both kinds of projects is to eliminate the need of central instances. Most projects include distributed hash tables for the content indexing [1] [2].

Within the last yeas there have been many VoD applications and architectures for e.g. metropolitan networks. A lot of of them offering centralized streaming instances combined with proxy instances. Due to space limitations we do not consider server/proxy based architectures in this chapter.

Recently there have been several patching approaches as P2Cast [16], P2Vod [17], Pull based approach [18], Splitstream [19], and SpreadIt [5]. These applications create high-level multicast tree. A peer who requested a specific movie caches the video stream (which was received so far) on its disk and offers this segment to the others. This presumes that there are several/many peers requesting the same object.

[6] proposes an architecture of heterogeneous peers supporting *many to one* streaming. Also [7] is based multiple senders serving a single receiver, build as a client server system. In [8] authors recommend to use peers as proxy instances. Content which has been received lately is offered to all other peers.

The generic resource-mapping problem within a distributed, replica containing/ creating storage environment is considered in [9]. Here the current workload of different nodes is taken into account when replicas are created. However the authors just describe the problem and do not give a solution.

ARIMA forecasting techniques are used in multiple fields[11] . They are often used in economics research (e.g. stock exchange forecast), prediction of product sales or airline passengers. Additionally, it is used in core IT area, e.g. for I/O block requests for file systems [10] or workload forecasts for SAP R/3 systems.

Regarding modeling VoD Requests [12] proposes the use of the Zipf/Zeta distribution. They take various items, like access patterns for specific content and the time how long content has been viewed compared with the video clip length, into account for modeling the VoD systems. However just the short term popularity of video content is considered. In previous work Dan et al. combined the Zipf distribution with long term popularity changes [13]. Also [14] assumes a Zipf distribution for caching optimization in VoD servers. In [15] authors say that Poisson and Zipf distribution do not absolutely fit long-term observations. They propose a parameterized decaying e function with a remainder popularity. Authors verified their statement by comparing the function with empirical data. However, they pointed out that their approach is able to model day-to-day changes, but that they need additional information on changes of user behavior throughout the day. Furthermore they mentioned that e.g. children's interests are dominating the afternoon whereas grown up interests are dominating in the evening or at night. Due to authors this could render caches in trivial caching servers as useless. This lack is solved by making use of the S-ARIMA processes combining long- and short-term observations.

3   Architecture

Within this section we describe the main elements of our approach and their marriage. We will show how we have combined a forecasting model with generating replicates, how sharing peers and documents behave within the overlay network to overcome the problems of heterogeneity concerning bandwidth and sharing capacity of peers and the popularity of distributed content.

3.1   The Involvement of an Overly Network

A significant fact of a nowadays peer to peer networks is the heterogeneity of sharing peers. Participating peers and their shared storage capacity may not reflect the accessible bandwidth neither the popularity of their offered contend. Thus to optimize the accessibility of a set of multiple shared documents it is necessary to define an overly network with an deterministic order over all participating peers. This property is significant for exchanging documents if the bandwidth assignment to documents should reflect their popularity and is useful if distributed dictionary strategies based on hashing are deployed to assign responsibilities for depositing lookup information.

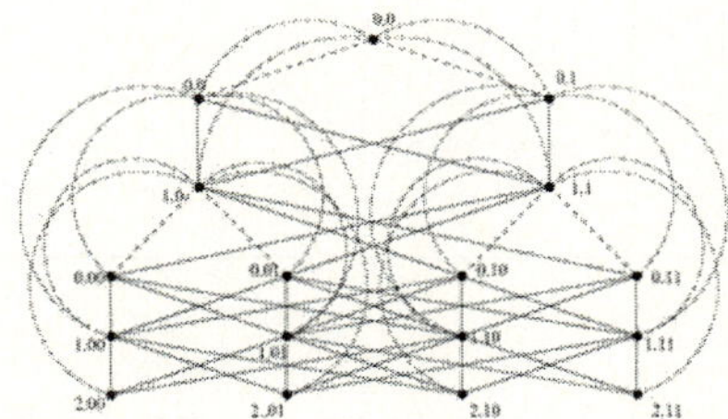

Fig. 1. Exemplary Overlay Network: Pagoda

Eventually we demand a structure where the peers can guess about the capability of their neighborhood to estimate if neighbored peers are capable to exchange and serve content.

Organizing peers by heterogeneous bandwidths in a deterministic structured overlay network with constant degree and still be comparable with randomized unstructured peer to peer systems is already shown in Bahargava et. al. [3]. They have introduced a dynamic overlay network for non uniform peers called *Pagoda*. In our approach we simulate the Pagoda network, but slightly modified concerning the order of peers. Let b_p denote the bandwidth of a specific peer p and $|p|$ denotes the document count of such a peer, than the ratio of $b_p/|p|$ is the order criterion of choice. In addition we believe that the involvement of ARIMA, described below, will cause changes in the arrangement of peers within the Pagoda network. These special peek load cases have only an impact on responsibilities concerning the lookup information and can be compensated by information migration.

3.2   The Use of Timeseries Analysis

The ARIMA (autoregressive integrated moving average) analysis offers the possibility to model stationary and non-stationary processes. It was introduced by Box and Jenkins in [11] and was later extended to deal with seasonal time series (SARIMA).

To optimize the assignment of free resources and their placement, and to obtain a request probability for each shared document x, denoted with $p_{Arima}(x)$ where $\sum_x p_{Arima}(x) = 1$, we have included the ARIMA forecasting model. By means the aim we are chasing for is $b_E(x) = p_{Arima}(x) \cdot \sum_p b_p$ should be the expected fraction of the shared bandwidth gathered by a given document in the whole Network.

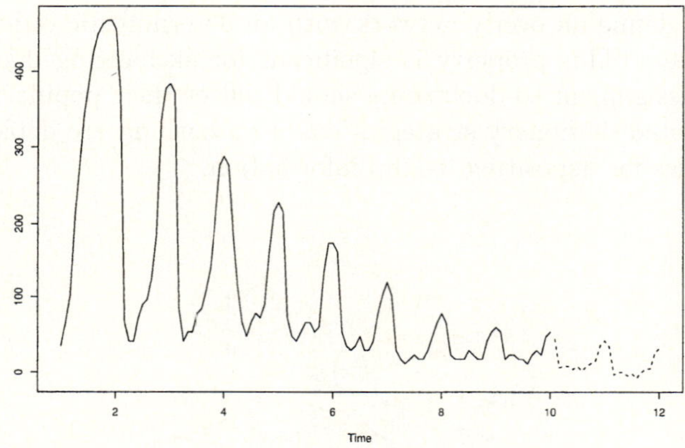

Fig. 2. ARIMA based Popularity forecast instance

Fig. 2 illustrates a typical popularity timeseries for one movie over a few days: The popularity is a combination of a (decaying) long-term observation and changes which occur every day. The graph shows 8 past days and a two-day forecast. In our scenario we presume 24 popularity values for one day.

The classical approach by Box and Jenkins for modeling and forecasting ARIMA time series is performed in three steps [11]:

1. **Model identification** to estimate a model structure by using autocorrelation (ACF) and partial-autocorrelation (PACF) functions to expose dependencies among data. The major task is to transform non-stationary series into stationary series.
2. **Parameter estimation** is a method for fitting the identified model to the observed data. This is done by determining the coefficients of the linear combination. Techniques are max. likelihood estimation, least squares (ELS), Marquardt.
3. **Forecasting** is done to predict future values.

Non-stationary processes contain a trend, seasonality, or both. Here we have 24 h seasonality with a decreasing long-term trend. To fit these processes into an appropriate model [11] proposes **Integrated ARMA** (ARIMA) process. The trend can be eliminated by iteratively differentiate pairs of values. This period-1 differentiation is repeated if the trend should not have been eliminated. In a ARIMA (p,d,q) model the d parameter describes the number of performed differentiations to achieve a stationary process with p AR terms and q MA terms. However it is not possible to eliminate seasonality in time series with iterated period-1 differentiation. The period-S differentiation is responsible for eliminating seasonality. A D-th seasonally differentiation $ARIMA(P, D, Q)_S$ process becomes a stationary ARMA(P,Q) process. The combination of both techniques results in $ARIMA(p, d, q)x(P, D, Q)_S$. This model describes the two components of the time series: The $(P, D, Q)_S$ part describes the seasonal parts whereas the (p, d, q) part is representing the structure within one season.

Due to our presumption of 24 popularity values for 24 hours we get a problem how to consider the request distribution for one hour: As the predicted value describes the sum of all requests which will occur in the next 60 min these requests will probably not be evenly distributed. In the worst case all requests are performed at a single point of time. To ensure the "on demand" character of the application we consider this worst case, i.e. we assume that all requests described by the timeseries prediction are performed at one point in each hour. Doubling the number of popularity values to 48 would increase the granularity, however the basic problem is the same.

3.3 Distribution and Replication Joint with ARIMA

To obtain a fair balanced document assignment concerning the capacity of peers, with the aim to estimate $|p|$, we deployed a distribution technique similar to Share. We will not explain this strategy and refer to the work of Brinkmann et. al. [2]. This allows us to determine $b_p/|p|$ to negotiate the position within Pagoda. In the following section we denote with $C = \sum_{p \in \mathbf{P}} c(p)$ the available storage capacity of all peers, and with $C_V = \sum_{x \in \mathbf{V}} c(x)$ the needed space to serve all documents once, and with C_R the remaining storage left unused. For simplification we assume that the documents are same sized, so we now consider places on peers instead of capacities.

Our simple approach uses the above described initial placement for documents. Thus the ratio $C_\mathbf{V}/C$, is approximately reflected by each peer and its ratio of used and available places. So the probability for x to be assign to specific peer p is given by $c(p)/C$. It is obvious that this strategy can not avoid the problem of magnetism caused by high capacity peers with lower bandwidth. That implies that slow peers with a higher sharing capacity are chronically overloaded by popular files. To overcome this problem we can make use of the overlay network by exchanging content. If there exists a peer in the root directed neighborhood serving less popular file x', x and x' are exchanged. This technique is called *bubble* and is done by each peer in parallel until the system is stable. Finally popular files will bubble up and unpopular files vice versa. It should be noted that these operations will leave the content distribution untouched and the availability for documents increases, because of the chosen Pagoda order.

The second approach includes the use of C_R. If $\lfloor C_R/c(x) \rfloor \geq r \cdot |\mathbf{V}|$ and $r \in \mathbf{N}^*$. So r is the redundancy factor for each document. Now we can distribute in a round robin fashion the originals and than the replicates of the first order, than the second order and so on, up to the k-th order. If the gathered bandwidth $b_r(x) = \sum_{p|x \in p} b_{p(x)}$ for a specific x is not sufficient we can deploy the bubbling technique among the order of replicates for all x until the solution is fine and furthermore we can maximize r. It should be noted that it is reasonable for a specific peer to hold only an original or one replicate.

Within both approaches we still have a competing situation where documents are fighting for better bandwidths. In addition there are many exchange strategies for originals competing replicates or replicates with different orders. To overcome this we modify the assignment of C_R. Peers are offering a fix size of storage capacity, so they expect that it will be used, by means it is fair to use $C_{\mathbf{V}} + C_R$. So if r is determined for each x individually based on $p_{Arima}(x)$, leads to $r_x = \lfloor p_{Arima}(x) \cdot C_R/c(x) \rfloor$. This new assignment has a significant impact on $p_{Arima}(x)$ after the distribution of all documents the expected load is shared over all replicates, thus the number of request a peer has to handle decreases for each x by r_x. Eventually the popularities within the system $p_S(x) = p_{Arima}/r_x$ are nearly equal for all documents. This leads to a different exchange strategy, because bubbling will not work any more. If now $b_{r_x}(x)$ is not sufficient for any specific x, the set of peers holding x choose one of the locations and try to exchange it. Typically the document served on the peer with the minimal $b_p/|p|$ leads to success. If an adequate peer is located among the path to the root node, we simply have to test for each x' on that peer if $|b_E(x') - b_{r'_x}(x')|_{before} - |b_E(x') - b_{r'_x}(x')|_{after}$ is less than $|b_E(x) - b_{r_x}(x)|_{before} - |b_E(x) - b_{r_x}(x)|_{after}$ if this is the case we choose the x' with the smallest deviation and replace it with x. If not we have to choose an other x out of the set.

Too Popular to Serve. One side effect still remains. If we apply the last technique we assume that there are always enough disjoint places left. But if there $\exists x \in V \mid p_{Arima}(x) \cdot |\mathbf{V}| > |\mathbf{P}|$, for such a case we can deploy a simple technique. We just place such documents on each peer by the order of their $p_{Arima}(x)$ and set $C = C - |\mathbf{P}| \cdot c(x)$ until none of such documents is left. It might happen that we dedicate some peers to a fix set of files without any places left. Such peers are blocked for the following phase. After that we can behave as before with a slightly modification for $b_p/|p|$ we have to share the bandwidth heterogeneous concerning to the fix placed documents. This can lead to a worse ratio and has an impact on the position within Pagoda, because $p_{Arima}(x)$ will vary over time as mentioned before.

4   Simulation Results

Within this section we introduce the major result of our simulations: The network consists of 458 nodes with heterogeneous HD capacity and heterogeneous synchronous DSL connections. Furthermore we distributed 800 movies.

As described in the previous section each object gathers bandwidth according to its current popularity. The requests for a specific movie are modelled as follows: The long term popularity is described by a decaying e function: $pop(x) = \exp\left(a - \frac{b}{x} - \frac{x}{b}\right) + c$ where a and b describe the curve and c determines the remaining capacity. This curve is multiplied with a genre-vector (with 24 float values) which describes the daytime dependent access frequencies. The demands look similar as illustrated in 2. For this simulation we presume that the request are coming from the outside of the system. Thus within this simulation the overlay-peers are just serving.

For the simulations we consider three different scenarios, which are described as follows:

1. **No Redundancy:** The content is assigned with a distribution strategy similar to Share [2] concerning the capacity of peers. Each object is represented only once, and there is no object movement.
2. **Max Redundancy:** The content is distributed as before with the slightly modification that each object is represented a constant time more then once and at maximum until no more places are left. The objects do not move, too.
3. **Bubble:** Here we perform the bubble technique which has been described in section 3.3. At the beginning of each hour the ARIMA popularity values for the next 60 minutes are taken into account.

First of all the request latency if of interest, i.e. the number of time steps it takes to download the object from multiple source peers. Documents on peers gather a popularity based bandwidth. Thus we have chosen to illustrate the average of the difference of the expected bandwidth (based on ARIMA forecasts) and the actually used bandwidth.

Fig. 3 illustrates the simulation result. The simulation period covers one week. The two (constant) lines show the results of the first two scenarios. In the second situation the system holds several replicas of each document (about 5 + original). Thus the values of the proportion of the expected bandwidth and the used bandwidth about 5 times better. The results of the third simulation are starting at the same level as in simulation 2. In contrast to the previous two simulations objects move due to their popularity. Thus the value converges after a few hours against a stable value (apart of the variance of the curve).

Remark: When the bubble operation takes place the objects of course would not move upwards the overlay step by step. In practice tickets describing content would be exchanged. Just when fuzzy bubble converges against a stable state real documents would be exchanged.

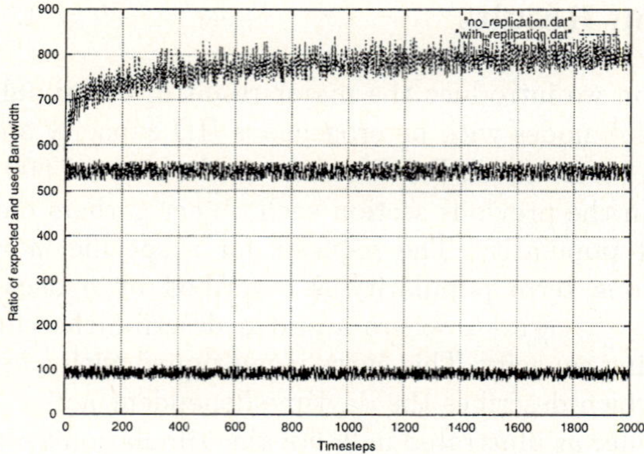

Fig. 3. Content gathered Bandwidth for three simulation scenarios

5   Conclusion and Future Work

Within this paper we have proposed an architecture for distributed P2P Video on Demand application as it might occur in large company intranets or metropolitan networks. We made use of seasonal ARIMA (SARIMA) timeseries analysis to predict future popularity values of individual movies, and additionally we task the Pagoda overlay network as base for our replica distribution. By utilizing a hierarchical overlay network we are able to deal with different bandwidth capacities and heterogeneous HD capacity (i.e. stored documents).

Furthermore we expect the simulative results for the enhanced case, where the document competing is eliminated, in the near future.

If other networks are sufficient enough, for instance a replacement of the embedded *deBruijn networks* by *butterfly networks* to obtain a lower degree and how dynamics created by joining or leaving peers are compensated might be shown in further researches.

References

[1] D. Karger, E. Lehman, F.T. Leighton, M. Levine, D. Lewin, and R. Panigrahy: Consistent Hashing and Random Trees: Distributed Caching Protocols for Relieving Hot Spots on the World Wide Web, pages 654-663, Proceedings of 29th ACM Symposium on Theory of Computing (STOC), May 1997

[2] A. Brinkmann, K. Salzwedel, and C. Scheideler: Compact, Adaptive Placement Schemes for Non-Uniform Distribution Requirements. Proceedings of 14th ACM Symposium on Parallel Algorithms and Architectures (SPAA), pages 53-62, Aug. 2002.

[3] Ankur Bhargava, Kishore Kothapalli, Chris Riley, Christian Scheideler, and Mark Thober: Pagoda: a dynamic overlay network for routing, data management, and multicasting. Proceedings of the sixteenth annual ACM symposium on Parallelism in algorithms and architectures, 2004, ISBN: 1-58113-840-7, pages 170-179, Barcelona, Spain.

[4] Andre; Brinkmann, Kay Salzwedel, and Christian Scheideler: Efficient, distributed data placement strategies for storage area networks. Proceedings of the twelfth annual ACM symposium on Parallel algorithms and architectures, 2000, ISBN 1-58113-185-2, pages 119-128, Bar Harbor, Maine.

[5] H. Deshpande, M. Bawa, and H. Garcia-Molina: Streaming Live Media over a Peer-to-Peer Network, Stanford Database Group Technical Report 2001-30

[6] D. Xu, M. Hefeeda, and S. Hambrusch and B. Bhargava: On Peer-to-Peer Media Streaming, Pursue Computer Science Technical Report, Apr. 2002

[7] T. Nguyen and A. Zakhor: Distributed Video Streaming Over Internet, Proceedings of SPIE/ACM MMCN, 2002

[8] W. Jeon and K. Nahrstedt: Peer-to-peer Multimedia Streaming and Caching Service, ICME 2002, Lausanne, Switzerland, August 2002

[9] J. Chuang: Resource allocation for Stor-Serv: Network storage services with QoS guarantees. Proceedings of NetStore'99 Symposium, 1999

[10] Tran N., and Reed D.A.: ARIMA Time Series Modeling and Forecasting for Adaptive I/O Prefetching, Proc. of the 15th international conference on Supercomputing, 2001, Sorrento, Italy, pages 473-475, ISBN 1-58113-410.

[11] Box, G.E and Jenkins, G.M.: Time Series Analysis Forecasting and Control, 1976, 2nd ed. San Francisco: Holden-day

[12] Dan A. and Sitaram D.: A generalized interval caching policy for mixed interactive and long video workloads, IBM Research, RC20206, 1995

[13] Dan A. and Sitaram D.: Buffer management policy for an on-demand video server, IBM Research, RC 19347, 1993

[14] Jean-Paul Nussbaumer, Baiju V. Patel, Frank Schaffa, and James P. G. Sterbenz: Networking Requirements for Interactive Video on Demand, IEEE Journal of Selected Areas in Communications, volume 13, number 5, pages 779-787, 1995

[15] Griwodz C., Br M, and Wolf L.C.: Long-term movie popularity models in video-on-demand systems: or the life of an on-demand movie. Proc. of the fifth ACM international conference on Multimedia, Seattle, Washington, 1997, ISBN 0-89791-991-2

[16] Yang Guo, Kyoungwon Suh, James F. Kurose, and Donald F. Towsley: P2Cast: peer-to-peer patching scheme for VoD service, Proc. of WWW 2003, pages 301–309

[17] Tai T. Do, Kien A. Hua, Mounir Tantaoui: P2VoD: Providing Fault Tolerant Video-on-Demand Streaming in Peer-to-Peer Environment, Proc. of the IEEE International Conference on Communications, Paris, June 2004.

[18] Anwar Al Hamra, Ernst W. Biersack, and Guillaume Urvoy-Keller: A Pull-Based Approach for a VoD Service in P2P Networks, Proc. of the 7th IEEE International Conference on High Speed Networks and Multimedia Communications, July 2004.

[19] M. Castro, P. Druschel, A. Kermarrec, A. Nandi, A. Rowstron, and A. Singh: SplitStream: High-bandwidth content distribution in a cooperative environment, Proc. of IPTPS'03, Feb. 2003.

Critical Analysis and New Perspective for Securing Voice Networks

Carole Bassil[1], Ahmed Serhrouchni[1], and Nicolas Rouhana[2]

[1] Computer and Networking Department, ENST,
46, rue Barrault, Paris, France
`{carole.bassil, ahmed}@enst.fr`
[2] Cimti, faculty of Engineering, Saint-Joseph University
Mar Roukoz, Mkalles, Lebanon
`nicolas.rouhana@fi.usj.edu.lb`

Abstract. Voice networks evolved from the traditional telephone system, to mobile and wireless networks and now towards a converged voice and data infrastructure. This convergence is based on the spread of the Internet Protocol, where VoIP is developing. Due to IP network characteristics, hackers are able to compromise and take control of different aspects of IP telephony such as signaling information and media packets. Security and privacy become mandatory requirements for this application area. IP telephony requires security services such as confidentiality, integrity, authentication, non-replay and non-repudiation. The available solutions are generic and do not respect voice specificities and constraints. In this paper, we present the security mechanisms as provided in some voice networks, outline major security weaknesses in these different environments and end this paper by a proposition to secure voice over IP packets drawing inspiration from the existing voice security solutions.

Keywords: Security mechanisms, GSM, SRTP, VoIP.

1   Introduction

Plain Old Telephone System (POTS) is the oldest telephone system offering analog voice services. It was commonly considered as a secure network. However, phone tapping may simply involve the installation of a low cost capacitor and the snipping of a wire. Integrated Services Digital Network (ISDN) brings us closer to the goal of a ubiquitous multi-service network, integrating voice, data, and video services in a digital format over a common global network. It was subject to many standards, but no security standards were defined and implemented to protect the network from eavesdropping and accessing critical information carried over its channels. However, security related discussions have been reported by the Integrated OSI, ISDN and Security Program of the Computer Systems Laboratory at the National Institute of Standards and Technology [4]. Digital wireless voice transmission technologies were introduced by mobile networks such as the GSM [3] and Bluetooth [12]. Efforts were made at the IEEE 802.11 [17] working group to implement voice over Wi-Fi by defining new RFCs to assure security and QoS. Wireless technology, by its nature,

P. Lorenz and P. Dini (Eds.): ICN 2005, LNCS 3421, pp. 810–818, 2005.

violates fundamental security principles; it presents security caveats against attacks launched over the radio path. Currently, Internet took the leadership by providing value added services mainly IP telephony services based on H.323 [6] or SIP [14] standards at very low cost. Nevertheless, transferring critical information over communication infrastructure accessible to the public presents security vulnerabilities.

In this paper, we present some technologies that provide voice capabilities and security services. The goal of our investigation is to analyze these different voice platforms with regards to their security applications and understand whether these multiple algorithms and security protocols apply to the voice based on the withheld options and their performance impact on the voice quality. The results of the conducted analysis will lead us to propose a secure voice over IP solution inspired from the points of strengths of the existing deployed security services. Thus, GSM [3] secures the voice communication over the radio link as outlined in section 2. Section 3 introduces the security services as defined in the H.235 accompanying standard of the H.323 [6] VoIP standard and outlines the security constraints of the VoIPsec. The Real-time Transport Protocol [18] provides end-to-end network transport functions suitable for audio applications. The secure RTP profile, SRTP [16], provides security mechanisms to protect the RTP and its control traffic as described in section 4. In Section 5, we introduce a solution to secure voice packets carried over IP by combining the confidentiality, integrity and non-replay mechanisms provided by SRTP along with our proposition to provide authentication and non-repudiation mechanisms. We conclude the paper in Section 6.

2   Security in GSM

Global System for Mobile Communications (GSM) [3] is the most popular mobile telephone network. In order to protect the system, different security features were reinforced mainly authentication and confidentiality. *Authentication [2]* is used to identify the user to the network operator, based on challenge-response encrypted techniques. A specific authentication algorithm (A3) calculates a signature that it sends to the network based on the individual key (K_i) registered on the user SIM card and the challenge sent by the network. The network compares it with the one provided by its database to authenticate the user. *Confidentiality [2]* is provided by the challenge sent by the network, the user's Ki and the specific A8 algorithm to generate a session key (K_c). Each frame crossing the radio link will be encrypted with a different key generated by the A5 encryption algorithm. This algorithm is initialized with the session key (K_c) and the number of frame to be encrypted generating a different key for each frame. *Integrity* is not provided within GSM based on cryptographic algorithm. *Non-repudiation* and *non-replay* are not provided by GSM. However, billing invoices could be used as a proof against repudiation.

GSM Security Weaknesses: In a GSM network, the functions described above are applied only on the radio link between the mobile station and the network. Communications and signaling within the core network and connection with the fixed network are not protected. The implemented comp128 authentication algorithm (A3/A8 algorithms) was broken [19]. The encryption A5 algorithm showed vulnerabilities against attacks [19], [20] conducted on its implementation.

3 Security in VoIP

VoIP can be defined as the ability to make telephone calls over IP-based data networks with a suitable quality of service (QoS) and a much superior cost/benefit. H.235 [7] defines the security requirement in H.323 [6] environment. TLS [5] is used to secure the signaling.

The H.323 standard is made of important building blocks constituting a foundation for audio, video and data communications across IP-based networks. The signaling information is transported reliably over TCP. Voice information is carried by RTP/RTCP packets over UDP. H.235 [7] is part of H.323 building blocks that define security for VoIP. *Authentication* [7] is based on two types of concepts: symmetric encryption-based or shared secret known as "subscription" with three variations: password-based with symmetric encryption; password-based with hashing (also symmetric); and certificate-based with signatures (asymmetric) to authenticate the use itself. As a third option, authentication may be accomplished by TLS [5] or IPSEC [13]. Authentication is brought out during call establishment connection on the signaling channel. The information carried on the signaling channel can secured with TLS. *Confidentiality* can be provided for call control and media channels [7] to protect the data carried on these logical channels. The required cryptography key used for encrypting media channels can be carried on a specific H.235 logical channel. *Integrity* of the exchanged information over all logical channels is provided by using hash functions in conjunction with the deployed authentication mechanisms [8], [10] and [11]. *Non-Replay* was not considered and developed in H.235 recommendation. However, a sequence number and timestamp fields of packet headers could be used to provide protection against replay attacks. *Non-repudiation [9]* could be provided using a digital signature in conjunction with a one-way hash function such as MD5 or SHA1.

Voice over IP Security Weaknesses: There is a lack of security in VoIP implementation by the IP phones vendors and most of firewalls are not VoIP aware. The signaling/control and the media data might be the major target of attacks. Some of the proposed security solution is not suitable for voice such as TLS which secures TCP traffic while voice packets are carried by UDP packets. If we try to secure the VoIP traffic based on the IPsec security framework, two main factors affect voice traffic when IPsec is used [15]: the increased packet size is due to headers added to the original IP packet and the time required to encrypt payload and headers and the construction of new ones is high. Also the authentication as provided with IPsec covers machines only (logical address), users are not identified.

4 Security in SRTP

SRTP (Secure Real-time Transport Protocol) [16] is a security of RTP standard which provides confidentiality, integrity, and authentication and replay protection. SRTP encrypts the payload while leaving the packet header as clear text. SRTP is independent from the underlying layers used by RTP. SRTP is characterized by a high throughput leading to minimize the processing time and low packet expansion. The following is the security services provided by SRTP. *Authentication* is based on a hash function with a key invoked to authenticate the header and payload of the RTP packets. A Message Authentication Code (MAC) is appended to the end of the packet and verified by the receiver by re-computing this MAC using the same process. The default algorithm used to provide authentication and integrity is the HMAC-SHA1 [35], [38]. The source messages authentication is provided in a peer-to-peer communication only. *Confidentiality*: the payload and the authentication tag appended to the RTP packet are encrypted by the transmitter and the receiver using the same session key. One master key is sufficient to insure the confidentiality and integrity of the RTP/RTCP packets. This is performed using a derivation key function which creates session keys derived from the master key. SRTP uses a key-stream approach to encipher the flow of information. Key-stream generation is accomplished by the AES [21] encryption algorithm. *Integrity* is obtained using the hash function provided with the authentication mechanism. *Replay protection* is assured if integrity is enabled. Each SRTP receiver maintains a replay list which indicates the indices of all authenticated received packets using a sliding window technique. After authenticating a packet, this list is updated with the new indices. *Non-repudiation* is not developed within SRTP.

SRTP Security Weaknesses: We need for a separate management key (e.g. IKE, ISAKMP/Oakley, Kerberos or point-to-point mechanisms such as Diffie-Hellman algorithm). We need to modify the protocol in all existing IP phones. There is no user authentication in unicast, multicast and RTP group sessions. Only the source origin of the data is authenticated. The RTP headers are sent in clear text to allow header compression, which leaves certain fields available to attacks.

5   Proposition for Securing Voice over IP

It is well known that the more we add security levels in a system the more the performance is reduced. The target is to reach an acceptable level of security while preserving the necessary performance for the voice service. Therefore a choice has to be made such that the security protocol overhead does not affect the packet processing delay beyond acceptable limits and hence the transmitted voice quality in return. Every security solution described earlier possesses advantages and inconveniences regarding its deployment to secure voice transmission. Our purpose is to analyze the pros and cons, take inspiration from the strong points of each solution and propose a solution to secure voice packets over IP-based networks.

5.1   Security Analysis

POTS/ISDN services provide line number identification (Caller Line ID) through signaling messages. GSM uses a unique user identifier attributed by the network operator and registered on the SIM card to identify and authenticate the user based on challenge-response operations. Data origin authentication is offered by SRTP and IPsec authentication mechanisms while user identity is *authenticated* by means of digital certificates or digital signature as suggested by VoIP standards. However, digital certificates are heavy to implement with voice since it relies on public key cryptography which requires the exponentiation of large numbers, a computationally intensive process which limits their speed. For those reasons applying public key algorithm to secure voice transmission will introduce delays and affects the overall performance. Therefore, a new authentication and identification mechanisms should be proposed which respects the constraints and quality requirement of the voice nature. *Confidentiality* based on different encryption techniques was used to encrypt a point-to-point communication over traditional fixed voice networks. GSM mobile network introduced a new technique over the radio link where each transmitted packet is encrypted with a different key calculated by a specific encryption algorithm to protect the data from attacks. Bluetooth is using a specific stream cipher algorithm to secure the wireless path taking into account devices resources and network characteristics. Amongst the encryption algorithms deployed in the IP based networks, Advanced Encryption Standard (AES) is the most simple, flexible, efficient (computational efficiency and less memory requirements on software and hardware, including smart cards) and secure (key length 128, 192 and 256 bits) symmetric algorithm. *Integrity* is provided by a hash function calculated over the original message or a hashed digital signature performed using public key algorithm. *Replay list* with a sliding window approach based on the sequence numbers and timestamp as defined by SRTP is a good technique to avoid replay attacks. *Non-repudiation functionality* is only provided by public key algorithm while signing the messages with a digital signature along with a timestamp. Since Public key algorithms are very heavy to implement with voice transmission, different approach should be considered to avoid deny of reception or deny of participation in a conference call.

Our proposed solution as highlighted in table 1 will be based on the confidentiality, integrity and replay protection security mechanisms deployed by SRTP. The GSM authentication mechanism will be reconsidered to operate within IP network; non-repudiation will be inspired from that defined within [9].

Table 1. Security mechanisms used in our proposed solution

	GSM	BT	IPsec	VoIP	SRTP
Authentication	+	+	+	+	+
Integrity	-	-	+	+	+
Confidentiality	+	+	+	+	+
Non-Repudiation	-	-	-	+	-
Replay protection	-	-	-	-	+

(+ : service available, - : service not available)

5.2 Proposed Solution

We propose to define the authentication mechanism with an associated non-repudiation feature specific to be deployed within a telephony context. A generic authentication mechanism based on smart cards to authenticate each user will be proposed with a trusted third party (Trusted Authentication Authority) who manages the delivery of such cards. Adding smart cards provides reinforced security and resistance to attacks.

Smart Cards: the usual approach is to store in a permanent memory of the machine the required information such as the subscription identifier in the shape of digital signature for instance. This smart card is a kind of key. If removed from the terminal, the latter cannot be used except for emergency calls, and that is to say it cannot be used for any service which will impact the subscriber's bill. These smart cards can hold besides the user subscription identifier user profiles in terms of abbreviated dialing numbers list with the correspondent alphanumeric index, etc. It can be protected by a password (typically 4-digit) similar to the PINs of credit cards. The main challenge with smart cards is its integration within the actual IP phones.

Trusted Authentication Authority (TAA): Its role is to memorize and attribute for each VoIP subscriber a secret key (digital signature), delivered in the form of smart cards and to manage tokens to be used in a later stage for billing purposes. The Trusted Authentication Authority will handle all user information (name, address, phone number, a signature digitally signed by a public key algorithm) in its User Information Database (UID). A TAA may be associated with a Gatekeeper or implemented as a standalone module at the Internet Service Provider location. Multiple TAAs might be spread over the VoIP network managed by a Global Trusted Authentication Authority (GTAA).

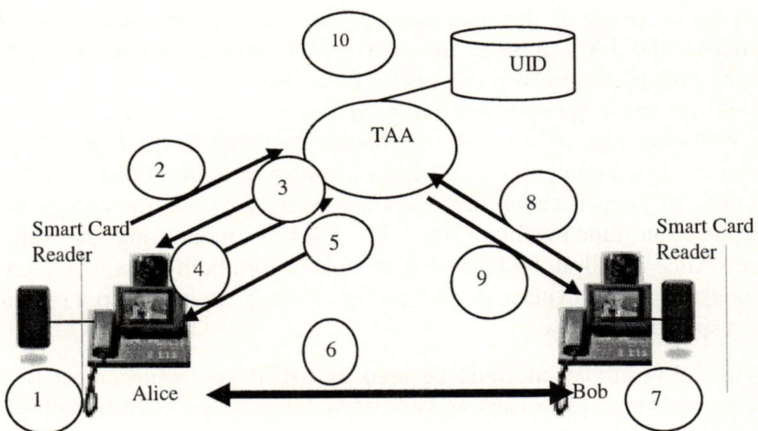

Fig. 1. Authentication mechanism based on smart cards

Authentication Mechanism: Figure 1 shows the proposed security mechanisms as explained below while figure 2 presents the exchanged messages during the authentication and non-repudiation phases.

1. Alice would like to initiate a phone call using an IP phone. She must first insert her authentication card in the smart card reader connected to the IP phone to be able to initiate phone calls through the IP network.
2. Upon inserting the smart card, an authentication request message will be initiated to the TAA asking for access grant.
3. The TAA sends Alice a random number (RAND) (figure 2).
4. Alice's smart card calculates a response based on the random number and her signature using a specific algorithm similar to that used in GSM SIM card. The

resulting value will be hashed with a hash function running on the card, and then sends the resulting digest (HRES) (figure 2) to the TAA to be verified.

5. TAA will calculate the same digest based on the random number sent, the digital signature of the user stored in its database and the hash function. TAA will verify the digest received from the user with the one calculated. After verification of the result, the TAA will grant access to the user by sending a confirmation message along with a Token. The call will be initiated towards Bob's IP phone. Procedures 1 through 5 should be deployed between TAA and Bob's smart card to authenticate Bob's identity. TAA will keep a record of the token delivered to both parties along with some parameters (e.g. starting time of the session, etc.) necessary for issuing billing invoices and for non-repudiation functionality.

Note: steps 1 to 5 may be repeated between Bob and the correspondent TAA in order to authenticate Bob.

6. Upon authenticating both parties, the voice call will be established between the two parties. Integrity and confidentiality features will be deployed to secure the voice communication transported by RTP packet according to the integrity and confidentiality provided by the SRTP standard.

7. A replay list should be managed at the receiver side to protect from replay attacks as stated in the SRTP standard.

8. During the exchange of signaling messages terminating the voice session between participants, the TAA must be informed of the session ending by returning the Token by both parties to keep records for future use.

9. TAA will acknowledge the reception of the Token to both parties.

10. TAA will keep records in a specific database related to the established session between the two parties Alice and Bob, in order to issue billing invoices and to avoid deny of the participant in the session insuring by that the completion of the non-repudiation functionality. Since the TAA is monitoring the call session between Alice and Bob, by retrieving the token from both users, the TAA will be able to issue billing invoices to both parties. Each TAA will keep a record for the call in a specific database.

Note1: The delivered Token could be used by SRTP key management to generate session keys (master keys and slating keys, etc.). Integrating the token with SRTP key management is outside the scope of the current paper.

Note2: Authentication procedure is carried out over signaling messages during the establishment phase of the call. Therefore, the underlying signaling channel should be opened in a secure manner (i.e, using a well known TLS port).

Confidentiality: SRTP uses the strong AES encryption. The encryption transform maps the SRTP packet index and secret key into a pseudo-random key stream segment. Each key stream segment encrypts a single SRTP packet. The process of encryption would be as follows: generating the key stream segment corresponding to the packet, then bitwise exclusive-oring that key stream segment onto the payload of the RTP packet to produce the encryption Portion of the SRTP packets.

Integrity is deployed to protect packet alteration by untruthful parties. This integrity use a one-way hash function calculated over the entire SRTP packets resulting in a digest (authentication tag). This digest will be appended to the end of the packets. RTP headers include sequence numbers which can be used to provide replay protection. Secure replay protection is only possible when integrity protection is present.

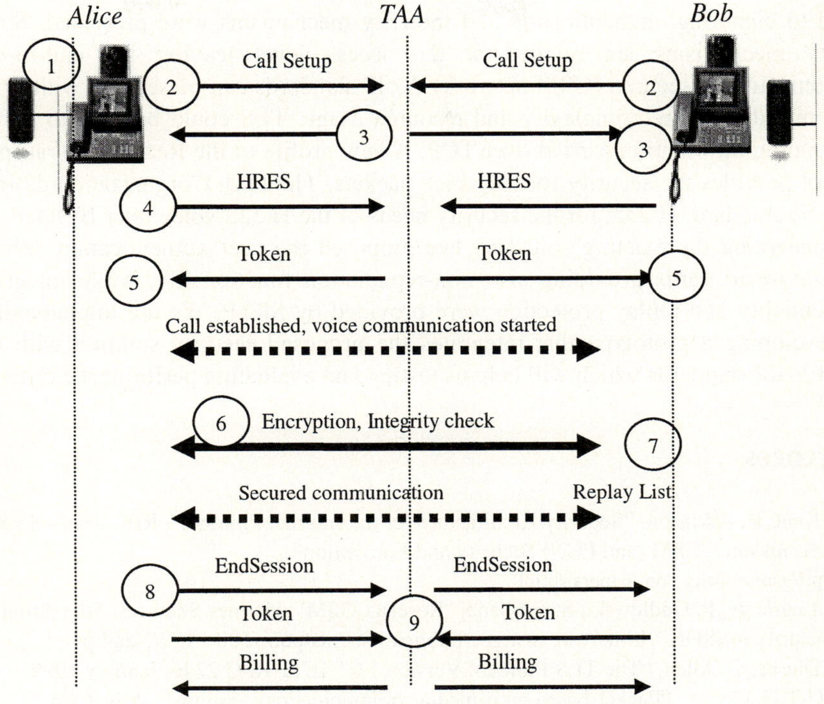

Fig. 2. Message exchange

Non-replay: Replay attacks are avoided by a replay list, maintained by the receiver only containing indices of recently received and authenticated SRTP packets as follows: each SRTP receiver maintains a Replay List, which conceptually contains the indices of all of the packets which have been received and authenticated. In practice, the list can use a "sliding window" approach, so that a fixed amount of storage suffices for replay protection. The receiver checks the index of an incoming packet against the replay list and the window. Only packets with index ahead of the window, or, inside the window but not already received, shall be accepted.

Non-repudiation: Since the TAA is managing the authentication sessions and the delivery of token to parties registered in its database and participating in a particular secured VoIP communication, it will keep records of the footprint (digest) received and the delivered token related to each participant in the voice session for future use to anticipate the denial of any participant in a specific voice call. At the end of the session, each user will send back the token to the appropriate. The TAA will be able to issue billing invoices based on each user's profile. Thus, non-repudiation will be reinforced by the records kept and the billing invoices issued for each recipient.

6 Conclusion

The intent of this paper is to outline the different security techniques implemented in wireless and IP network. Different available solutions within these voice networks related to ciphering, authentication and integrity mechanisms were presented. Some security mechanisms are applied at the access level leaving the end-to-end communication unsecured (GSM network). VoIP standards can be secured with IPsec which introduce more complexity and resource usage. TLS could be used to secure the signaling information carried over TCP. A new profile of the Real-time Transport Protocol provides the security for the voice packets. The ITU-T organization defined a specific standard, H.235, for the security needs of the H.323 voice over IP standard. After analyzing the existing solutions, we proposed the user authentication service based on smart cards providing also non-repudiation functionality, while integrity, confidentiality and replay protection were provided by SRTP. We are implementing and developing a prototype that integrates the proposed security solution with the current VoIP standards which will help us testing and evaluating performance criteria.

References

1. S. Kent, R. Atkinson, "Security Architecture for the Internet Protocol", RFC 2401 - 1998
2. C. Brookson, "GSM (and PCN) Security and Encryption",
    http://gsmsecurity.com/papers.shtml
3. X. Lagrange, P. Godlewski, S. Tabbane, "Réseaux GSM", Hermes Sciences, 5th edition
4. "Security in ISDN", http://csrc.nist.goc/publications/nistpubs/500- 189/isdn4.ps
5. T. Dierks, C. Allen, "The TLS Protocol Version 1.0", IETF RFC 2246, January 1999
6. ITU-T H.323 v4 , "Packet-based multimedia communications systems", July 2003
7. ITU-T H.235 v3, "Security and encryption for H-Series multimedia terminals", August 2003
8. ITU-T Recommendation H.235, "Annex D - Baseline Security Profile"
9. ITU-T Recommendation H.235, "Annex E – Signature Profile"
10. ITU-T Recommendation H.235, "Annex F - Hybrid Security Profile"
11. ITU Recommendation H.235, "Annex I – H.323 Implementation Details"
12. R. Morrow, "Bluetooth: Operation and Use", McGraw-Hill Professional, 1st edition 2002
13. S. Kent, R. Atkinson, "IP Authentication Header", IETF RFC 2402, November 1998
14. M. Handley, H. Schulzrinne, E. Schooler, J. Rosenberg, "SIP: Session Initiation Protocol", IETF RFC 2543 - March 1999
15. R. Barbieri, D. Brushi, E. Rosti, "Voice over IPsec: Analysis and Solutions", 18th Annual Computer Security Applications Conference December 2002, San Diego California
16. M. Baugher, & al. "The Secure Real-time Transport Protocol", RFC 371, March 2004
17. "IEEE 802.11". Available from http://grouper.ieee.org/groups/802/11/main.html
18. H. Schulzrinne, S. Casner, R. Frederick, V. Jacobson, "RTP: A transport Protocol for Real-Time Applications", draft-ietf-avt-rtp-news-12.ps, March 2003
19. L. Pesonen, "GSM Interception", Helsinki University of Technology, November 1999
20. G.Rose, "A precis of the new attacks on GSM encryption", Qualcomm Australia, September 2003.
21. NIST, "Advanced Encryption Standard (AES)", FIPS PUB 1997, http://www.nist.gov/aes

Architecture of a Server-Aided Signature Service (SASS) for Mobile Networks

Liang Cai, Xiaohu Yang, and Chun Chen

College of Computer Science and Technology,
Zhejiang University, Hangzhou, China

Abstract. In recent years there is an explosion in the number of applications for mobile devices. Many of these applications require the ability to issue digital signatures on behalf of their users. Traditionally, digital signatures are based on asymmetric cryptographic techniques which make them computationally expensive. Currently, all mobile devices tend to have limited computational capabilities and equally limited power. Instead of every mobile device performing computationally intensive cryptographic operations, a Server-aided Signature Service (SASS) was designed to offload PKI computation from clients in mobile networks such as GSM and CDMA. The paper details the design of SASS. After thorough performance analysis, we conclude that SASS can significantly improve the signing performance of the mobile clients.

1 Introduction

More and more mobile applications require the ability to issue digital signatures. Traditionally, digital signatures are based on asymmetric cryptographic techniques which make them computationally expensive. Currently, all mobile devices tend to have one feature in common: limited computational capabilities and equally limited power (as most operate on batteries). This makes them ill-suited for complex cryptographic computations such as digital signature generation.

Currently, the RSA cryptosystems is the most widely used PKI cryptosystem for key exchange and digital signatures: SSL commonly uses RSA-based key exchange, most PKI products use RSA certificates, etc. Unfortunately, RSA on a low power mobile device is somewhat problematic. For example, generating a 1024 bit RSA signature on the PalmPilot takes approximately 30 seconds. The problem of generating RSA keys is even worse. Generating a 1024 bit RSA key on the PalmPilot can takes as long as 15 minutes [1].

Instead of every mobile device performing computationally intensive cryptographic operations, we designed a Server-aided Signature Service (SASS) to offload work from clients in mobile networks such as GSM and CDMA.

SASS combines the encryption and key exchange capabilities of Modadugu's protocol [1] and the digital signature generation capability of the S^3 protocol [2]. It can provide full PKI infrastructure service for mobile applications.

P. Lorenz and P. Dini (Eds.): ICN 2005, LNCS 3421, pp. 819–826, 2005.

Use of SASS benefits mobile clients in two ways. First, the SASS have access to cryptographic hardware capable of performing single cryptographic operations faster than they can be performed by the client. Second, offloading cryptographic operations from the client CPU to these remote accelerators can free the client for other operations. Such load reduction can be quite significant, especially given that most cryptographic accelerators on the market today are highly parallel multiprocessors capable of processing many requests at once.

A secure "cell phone-banking" application has been implemented on SASS architecture in CDMA-1X mobile network. After thorough performance analysis, we concluded: (1) SASS can significantly improve the performance of mobile client's cryptographic operation; (2) SASS is a highly scalable service suitable for variant mobile applications and future critical applications which require longer key length.

The rest of the paper is structured as follows. Section 2 gives some background information about the typical message flows between mobile client and server. Section 3 discusses related work. Section 4 presents the overall SASS architecture. Section 5 presents the detailed performance results, and Section 6 concludes.

2   Background

The computationally expensive PKI operations are in contradiction with the incapability of mobile devices. In order to improve the performance of cryptographic operations, large portion of current mobile applications take a simplified version of asymmetric cryptography to achieve secrecy. The typical message flow between mobile clients and application server is as follows (see also Fig. 1):

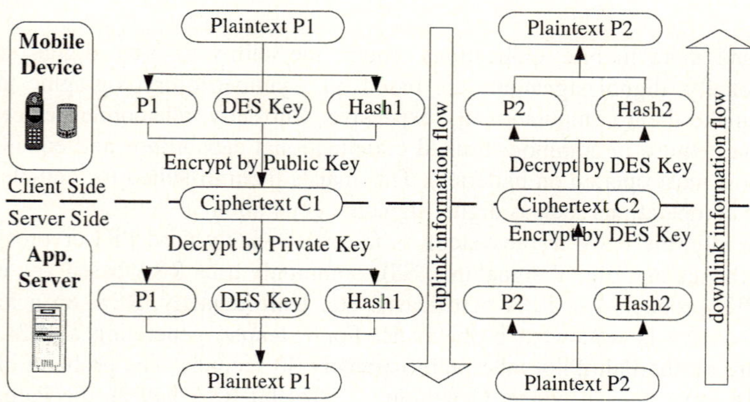

Fig. 1. Typical message flow between mobile client and server

1. Application server generates a RSA key pair. The private key is stored in server, while the public key is distributed to all mobile clients.

2. For uplink communication: (a) Client generates a random session key; (b) The plaintext P1 and the session key are encrypted by server's public key using RSA. The result ciphertext C1 is sent to server; (c) Server decrypts the C1 by its private key using RSA, gets the P1 and the session key, and verifies the integrity.
3. For downlink communication: (a) Server encrypts the plaintext P2 by the session key using DES. The result ciphertext C2 is sent to client. (b) Client decrypts the C2 by the session key using DES, get the plaintext P2, and verify the integrity.

From Fig. 1, we can find that:

1. The uplink and downlink communication between mobile client and application server are unbalanced. The uplink communication is based on asymmetric RSA algorithm, while the downlink communication is based on symmetric DES algorithm.
2. In order to reduce the computational load of the mobile clients, only the server side has the RSA key pair, thus can generate the non-repudiation digital signatures. The client side can only use the symmetric DES algorithm to achieve information privacy, but not non-repudiation.

3   Related Work

The performance problem of signature schemes was originally motivated by the limited computing power of smart cards and smart tokens [3]. Traditional solutions are based on pre-processing or on some asymmetry in the complexity of signature generation and verification (i.e., either sender or recipient must perform complex operations, but not both). On the contrary, ordinary users in SASS are never required to generate the traditional computationally intensive signature.

The RSA key generation protocol in SASS is based on the idea in [1]. Modadugu et al. show how to efficiently generate RSA keys on a low power handheld device with the help of an untrusted server. However the resulting RSA key looks like an RSA key for paranoids [4]. It can be used for encryption and key exchange, but cannot be used for signatures.

The digital signature generation protocol in SASS is based on a weak non-repudiation technique proposed by Asokan et al. in [2]. It is an instantiation of a mediated cryptographic service. Recent work on mediated RSA (mRSA) [5] is another example of mediated cryptography. mRSA provides fast revocation of both signing and decryption capability. However, the computation load on the client end is increased in mRSA, which is something SASS aims to minimize.

Network-attached cryptographic acceleration has been studied by Berson et al [6]. They present a cryptography service which supported secured connection with clients. But they require that client share the private key with the cryptography service. That confines the cryptography service and its clients in the same security perimeter, thus not suitable for mobile networks.

SASS combines the encryption and key exchange capabilities of Modadugu's protocol [1] and the digital signature generation capability of the S^3 protocol [2]. It can provide full PKI infrastructure service for mobile applications.

4   Architecture of SASS Service

SASS is designed to offload work from clients in mobile networks such as GSM and CDMA. As depicted in Fig. 2, SASS service is typically deployed in the mobile service provider's fixed, wired infrastructure. Mobile personal devices can utilize the powerful (both CPU speed and power supply) SASS service to generate keys and digital signatures. The SASS server can even be equipped with a number of hardware cryptographic accelerators. The whole SASS service can be shared among a large number of mobile clients. This sharing allows the cost of the SASS infrastructure service to be amortized over a large number of client machines.

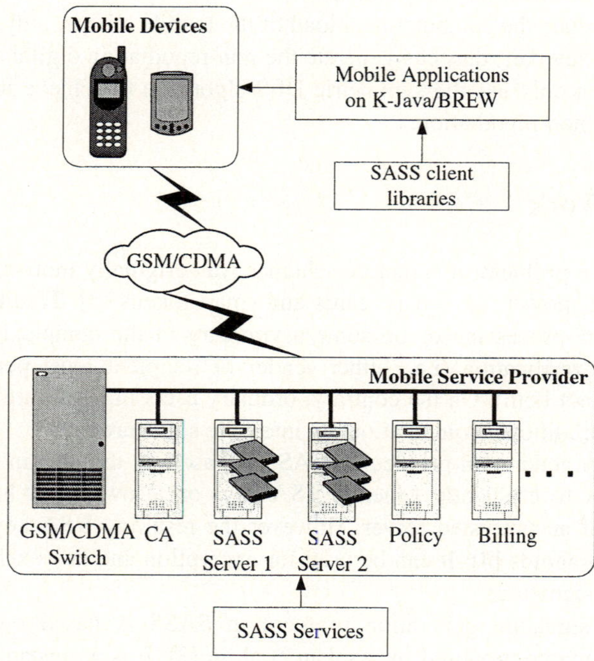

Fig. 2. Architecture of SASS

SASS provides three key interfaces to offload the bulk of cryptographic computation from mobile clients to SASS Service. They are as follows:

1. ISASS_KeyGen( ) help mobile client generate the key used for encryption and key exchange;
2. ISASS_Cert( ) help mobile client initialize the key used for generating the digital signature;
3. ISASS_Sign( ) help mobile client sign a message.

The detailed specification of these interfaces is described in section 4.2.

4.1 Overall Architecture

The overall architecture is made up of three components: CA and its related Admin Utilities, SASS service, and client.

A single CA and its related Admin Utilities (Policy Server for system-wide security policy and Billing Server for accounting) govern several SASS services. Each SASS service, in turn, serves many users. The assignment of clients to SASS services is designated off-line by system security administrator. A client may be served by multiple SASS services.

The CA component is a simple add-on to the existing CA. For each client and SASS service, the CA component issues a certificate. The client component consists of the client libraries that support the SASS operations.

Informally, efficient generation of RSA keys on a low power mobile client with the help of SASS service works as follows. The goal is to generate a modulus of the form $N = p \cdot R$ where p is a 512-bit prime and R is a 4096-bit random number. The primality test will be offloaded onto the servers.

1. The client must hide the modulus p and the exponent p-1. To hide the modulus p the client intends to multiply it by a random number R and send the resulting $N = p \cdot R$ to the servers.
2. The server will perform computations modulo $N = p \cdot R$.
3. The client will run a probabilistic test to verify that p is prime. This is done to ensure that the servers returned correct values.

SASS based signature could be computed as follows [7] (see also Fig. 2):

1. First, a user (Alice) calls the API in client library to start the signing process.
2. The API contacts Alice's SASS service and provides the data to be signed and a one-time ticket.
3. SASS service will ask Policy Server whether Alice is authorized to do the operation. Policy Server makes the decision based on Alice's operation rights, revocation status [5] and billing status. If Alice is authorized, SASS service will compute a half-signature over the data as well as other parameters (including the one-time ticket), and return the results to Alice.
4. Alice verifies SASS's half-signature and produces her own half-signature. Put together, the two respective half-signatures constitute a regular, full SASS signature. This signature is accompanied by SASS service's and Alice's certificates.

Verifying a SASS signature is easy: verifier obtains the signature and verifies the two halves using the accompanying certificates.

Typically, SASS is deployed in the mobile service provider's fixed, wired infrastructure. Mobile devices can utilize the powerful (both CPU speed and power supply) SASS to generate digital signatures. The SASS server can also be equipped with a number of hardware cryptographic accelerators, which as a result may be shared among a large number of mobile clients [6].

The architecture of SASS is especially suitable for cell phone networks such as GSM/CDMA. As shown in Fig. 2, cell phones are only usable when in touch, via a nearby base station, with a fixed infrastructure. Each phone-call needs communication with the infrastructure. This communication channel can be used to carry SASS protocol messages.

4.2 Key Interfaces of SASS Service

The key interfaces related to the specific requirement of offloading the bulk of computation from mobile clients to SASS Service are as follows:

1. ISASS_KeyGen(N, g, s) is called by mobile clients when generating RSA keys. It helps mobile clients offload the expensive exponentiation to SASS Server. To do that, It calculates $X = g^s \bmod N$, and returns X.
2. ISASS_Cert(O, n, PK_O, S) is called when initializing the signing key of the mobile client O. O choose a SASS server S that shall be responsible for generating signatures on O's behalf, generates a random secret key K_O, and constructs the hash chain: $K_o^n = SHA_O(SHA_O(SHA_O(\cdots SHA_O(K_O)\cdots)))$ where SHA_O ( ) is a personalized SHA( ) function used by O. O submits the root public key $PK_O = K_O^n$ to CA. CA return the certificate for O's root public key: $(O, n, PK_O, S)SK_{CA}$.
3. ISASS_Sign(O, MD5(m), i, K_O^i) is called when SASS server S help client O sign a message m. K_O^i is O's current public key. S verifies the received public key based on O's root public key (and O's certificate obtained from CA), i.e., checks that $SHA_O^{n-i}(K_O^i) = PK_O$. S has to ensure that only one signature can be created for a given (O, i, K_O^i). If a message on behalf of O containing K_O^i has not yet been signed, S signs (O, MD5(m), i, K_O^i), records K_O^i as consumed, and sends the signature $(O, MD5(m), i, K_O^i)SK_S$ (which we call the candidate non-repudiation token) back to O.

5 Performance Analysis

We have implemented the SASS architecture for a local mobile service provider to enable a secure "cell phone-banking" application. User of cell phone-banking system can transfer fund between different accounts. To test the performance of SASS, we ran a number of tests with various hardware platforms and different RSA key sizes. SHA algorithm is used as the collision-resistant one-way hash function in SASS implementation, and RSA algorithm with exponent 3 is used as the digital signature scheme for verifiers.

All experiments were conducted over a 100 Mb/s Ethernet LAN in a base station switch and all the cell phones were within the range of this base station. The mobile network provides CDMA-1X service. Banking gateway ran SCO Open Server 5.0.7, and was connected via 64Kb/s DDN. SASS services ran Linux version 2.4 on a Pentium IV 1.2GHz server. The mobile devices used in tests were LG8280 (high-end products) and LG6260 (lower-end products) cell phones.

First, we present in Table 1 traditional RSA timings conducted on the two cell phones via BREW. Table 2 illustrates the SASS based signature timings on the two cell phones with the SASS daemon running on a Pentium IV 1.2GHz server. The timing results of Table 2 include mobile network transmission time as well as SASS daemon processing times. Finally, Table 3 shows the mobile network round-trip communication delay between the user and the SASS daemon, for different key sizes. The size of the signature request is determined by the digest size of the hash function, that is, roughly 164 bytes for 1024-bit RSA key and 656 bytes for a 4096-bit RSA key.

Table 1. Traditional RSA signature timings(s)

Mobile Device	Key Length (bits)		
	1024	2048	4096
Traditional RSA: LG8280	8.2	51.6	350.5
Traditional RSA: LG6260	23.1	165.2	---

Table 2. SASS based signature timings(s)

Mobile Device	Key Length (bits)		
	1024	2048	4096
SASS based: LG8280	4.8	20.2	87.4
SASS based: LG6260	5.0	22.4	99.6

As we can see from Table 2, SASS produces valuable speed-up, as compared with traditional RSA. It is noticed that the mobile clients obtain a factor 2 to 8 speed-up depending on the key size and CPU speed of the cell phone. Besides, the difference in SASS sign time is very small despite large variance in different cell phones' CPU speeds and key lengths. It means that the SASS service is a highly scalable service suitable for different cell phones and future critical applications which require longer key length.

Table 3. Network delay for SASS based signature timings (s)

Mobile Device	Key Length (bits)		
	1024	2048	4096
SASS based: LG8280	1.1	1.4	2.2
SASS based: LG6260	1.5	1.9	2.8

6 Conclusion

Traditionally, digital signatures are based on asymmetric cryptographic techniques which make them computationally expensive. Currently, all mobile devices tend to have limited computational capabilities and equally limited power. This makes them ill-suited for complex cryptographic computations. we designed a SPSS service to offload work from clients in mobile networks. Mobile clients can utilized the powerful SPSS server to generate RSA keys and digital signatures.

SPSS combines the encryption and key exchange capabilities of Modadugu's protocol and the digital signature generation capability of S^3 protocol. It can provide full PKI service for mobile applications.

A secure "cell phone-banking" application has been implemented on SASS architecture in CDMA-1X mobile network. The detailed performance analysis of SASS based signature leads to the conclusion that the mobile clients can obtain a factor 2 to

8 speed-up depending on the key size and CPU speed of the cell phone. Besides, SASS service is a highly scalable service suitable for different cell phones and future critical applications which require longer key length.

References

1. Modadugu, N., Boneh, D., Kim, M.: Generating RSA Keys on a Handheld Using an Untrusted Server. In: Roy, B.K., Okamoto, E. (eds.): Indocrypt 2000. Lecture Notes in Computer Science, Vol. 1977. Springer-Verlag, Berlin Heidelberg New York (2000) 271–282
2. Asokan, N., Tsudik, G., Waidner, M.: Server-Supported Signatures. Journal of Computer Security. 5 (1997) 91–108
3. Schneier, B.: Applied Cryptography: Protocols, Algorithms, and Source Code in C. John Wiley & Sons, Inc. (1996)
4. Shamir, A.: RSA for Paranoids. CryptoBytes 1 (1995) 1–4
5. Boneh, D., Ding, X., Tsudik, G., Wong, C.M.: A Method for Fast Revocation of Public Key Certificates and Security Capabilities. In: Proceedings of the 10th USENIX Security Symposium. USENIX Association, Washington, DC (2001) 297–308
6. Berson, T.A., Dean, R.D., Franklin, M.K., Smetters, D.K., Spreitzer, M.J.: Cryptography as a Network Service. In: Proceedings of Network and Distributed Systems Security Symposium. Internet Society, San Diego (2001)
7. Ding, X., Tsudik, G., Mazzocchi, D.: Experimenting with Server-Aided Signatures. In: Proceedings of Network and Distributed System Security Symposium. Internet Society, San Diego (2002)

Password Authenticated Key Exchange for Resource-Constrained Wireless Communications (Extended Abstract)

Duncan S. Wong[1,*], Agnes H. Chan[2], and Feng Zhu[2]

[1] Department of Computer Science,
City University of Hong Kong, Hong Kong
`duncan@cityu.edu.hk`
[2] College of Computer and Information Science,
Northeastern University,
Boston, MA 02115, U.S.A
`{ahchan, zhufeng}@ccs.neu.edu`

Abstract. With the advancement of wireless technology and the increasing demand for resource-constrained mobile devices, secure and efficient password authenticated key exchange (PAKE) protocols are needed for various kinds of secure communications among low-power wireless devices. In this paper, we introduce an elliptic curve based password-keyed permutation family and use it to construct a PAKE in such a way that it is suitable for efficient implementation on low-power devices. The computation time on each side of our PAKE is estimated to be about 3.4 seconds and can be reduced to 1.5 seconds with precomputation on an embedded device with a low-end 16MHz DragonBall-EZ microprocessor. On its security, we show that the password-keyed permutation family is secure against offline dictionary attack under the assumption that the elliptic curve computational Diffie-Hellman problem is intractable.

Index Terms: Authentication Protocol, Key Exchange, Wireless Communications.

1   Introduction

With the advancement and tremendous development of wireless technology and the increasing demand for low-power mobile devices, secure and efficient communications among low-power wireless devices is becoming important. In this paper, we study the password authenticated key exchange problem and propose a solution to it especially for efficient implementation on low power wireless devices.

---

* The work described in this paper was fully supported by a grant from the Research Grants Council of the Hong Kong Special Administrative Region, China (Project No. 9040904 (RGC Ref. No. CityU 1161/04E)).

P. Lorenz and P. Dini (Eds.): ICN 2005, LNCS 3421, pp. 827–834, 2005.

The *Password Authenticated Key Exchange* (PAKE) problem is about designing a protocol which allows two communicating parties to prove to each other that they know the password (that is, mutual authentication), and to generate a fresh symmetric key securely such that it is known only to these two parties (that is, key exchange). The restriction is that these two parties only have a preshared password possibly some system-wide parameters. There is no cryptographic keys preshared between them and no certified public key preinstalled at any side of the two parties. In addition, the password space is assumed to be so small that an adversary can enumerate all the elements in the space efficiently. Therefore, a PAKE protocol is required to prevent eavesdroppers from getting any information about the password. It is also required to prevent any *active* attacker from guessing *more than* one password[1].

A typical application of PAKE protocols is secure remote access. By using a PAKE protocol, a user only needs a short password to access a remote host securely without worrying his password from being stolen over the communication channel. In addition, it is also very convenient to the user as he does not need to verify or carry any certified public keys, or maintain any preshared cryptographic symmetric keys.

In this paper, we propose an elliptic curve [13] based PAKE protocol for secure communication between two resource-constrained wireless devices. The computation time on each side of our protocol is estimated to be about 3.4 seconds on an embedded device with a 16MHz DragonBall-EZ microprocessor. The protocol also supports precomputation. It reduces the realtime computation time to 1.5 seconds. In our construction, we introduce an elliptic curve based password-keyed permutation family and show under the random oracle model [2] that it is secure against offline dictionary attack if the elliptic curve computational Diffie-Hellman problem [12] is intractable.

1.1 Related Results

Since the first suite of password authenticated key exchange (PAKE) schemes called EKE was proposed by Bellovin and Merritt [3] in 1992, there have been many other schemes proposed [7, 17, 1, 11, 9]. Among those PAKEs which have been specified to run over elliptic curves, SRP [17, 14] requires each communicating party to perform at least three elliptic scalar multiplications and one generation of a random point on an elliptic curve. For SPEKE [7] and PAK-EC [10] on the other side, each party only performs two scalar multiplications (disregarding the validity check of password-entangled public keys [6]) and one random point generation. These schemes use a common random point generation algorithm which is either nondeterministic or having comparable complexity (slightly faster) to one scalar multiplication [6].

---

[1] Note that it is always possible for an active attacker to verify at least one guess of the password in each attack, say by impersonating one party and carrying out the protocol with another honest party [6].

More recently, another PAKE is proposed [9]. Each side of the protocol is expected to carry out 2.34 scalar multiplications when ported to elliptic curve systems. In addition, a MAC (Message Authenticated Code) generating function is required. The MAC function should be chosen such that it is selectively unforgeable against partially chosen message attacks. We will see later in this paper that our protocol is expected to have less number of scalar multiplications and no MAC is required.

Paper organization: The rest of the paper is organized as follows. In Sec. 2, we propose a PAKE which is based on the elliptic curve cryptosystems. It is followed by the performance evaluation in Sec. 3. We conclude the paper in Sec. 4.

2   Our Proposal: PAKE-EC

Our approach has some similarity to that of [16]. We instantiate the concept of password-keyed permutation family, which refers to the special characteristic of a password-entangled public-key generation primitive described in [6].

Let k be a system-wide security parameter which governs the (computational) security of the system. Let $\{0,1\}^*$ denote the set of finite binary strings and $\{0,1\}^n$ denote the set of binary strings of length n. We assume that there is a set of (honest) users, indexed by $i = 1, 2, ...$ For simplicity, we use the indices to represent users in a protocol. Denote a password space PW in which each password shared between two entities, say Alice and Bob, is picked from it according to some probability distribution. Throughout this paper, Alice (Initiator) and Bob (Responder) are indexed by integers A and B respectively. Let $H_1, H_2, H_3, H_4, H_5 : \{0,1\}^* \to \{0,1\}^k$ be five distinct and independent cryptographically strong hash functions. On the security of each hash function, it is viewed as a distinct random oracle [2].

Let $\mathbf{F}$ be a finite field (e.g. an odd prime finite field $\mathrm{GF}(p)$ or a binary finite field $\mathrm{GF}(2^m)$), $\mathscr{C}$ be an elliptic curve defined over $\mathbf{F}$, and G be an element of large prime order q in $\mathscr{C}$. Let $\mathcal{H}$ be a cyclic subgroup of $\mathscr{C}$ generated by the 'base' point G, such that the discrete log problem is intractable. The domain parameters are $(\mathscr{C}, \mathbf{F}, G, q)$ and all of them are publicly known. Since they are parameters independent to any particular entity in a system whilst considered as the system-wide domain parameters by all the entities, these parameters can be 'hard-wired' in each entity during implementation. Currently, $|\mathbf{F}|$ and q in the order of 162 bits and 160 bits long, respectively, would be considered to be reasonably secure [5].

Our protocol, hereafter called PAKE-EC, consists of two major components: a public key encryption algorithm and a permutation family called password-keyed permutation family.

2.1   An ECC-Based Public Key Encryption Function

Let (w, W) be a public key pair where the private key $w \in_R \mathbb{Z}_q$ is a randomly chosen positive integer, and the public key $W = wG$ is computed under the usual scalar multiplication over $\mathscr{C}$. We need one more notation, that is, if Q is a point on $\mathscr{C}$ then Q_x represents the first component of the point Q. The following probabilistic polynomial time algorithm is our ECC-based public key encryption algorithm.

Enc = "On inputs of domain parameters $(\mathscr{C}, \mathbf{F}, G, q)$, a public key W and a message $msg \in \{0,1\}^k$,

1. Verify if $W \in \mathcal{H}$, that is, W is a point on $\mathscr{C}$ defined over $\mathbf{F}$ and $qW = O$ where O is the point at infinity. If any check fails, the algorithm terminates with failure. Otherwise, we proceed to the following steps.
2. Randomly pick $r \in_R \mathbb{Z}_q$, compute $R = rG$, $K = rW$ and then destroy r.
3. Compute $\eta = msg \oplus h(K_x)$ and then destroy K, where $h : \{0,1\}^* \to \{0,1\}^k$ is a cryptographic hash function. As the input of h is an element of $\mathbf{F}$, we herewith assume that there is some appropriate conversion mechanism which converts an element of $\mathbf{F}$ to some binary string implicitly.
4. The ciphertext (i.e. output) y is (R, η)."

When clear, we usually omit the domain parameters and denote the operation of Enc by $(W, msg) \to (R, \eta)$. The decryption algorithm is defined in the obvious way.

2.2 A Password-Keyed Permutation Family

In [4], Gong et al. discovered an 'associativity' problem. It was further exemplified by Jablon in [8] and described as a special characteristic of a password-entangled public-key generation primitive in [6] for avoiding the problem. The problem refers to the ability of an active attacker (through impersonation) to make use of some password-related operations of the underlying PAKE protocol to guess more than one password in each protocol run.

In our protocol, we only have one password-related operation, which is called a password-keyed permutation family and is denoted by

$$\mathcal{P} : \mathcal{H} \times \{0,1\}^k \times \mathcal{H} \times \{0,1\}^k \to \mathcal{H} \times \{0,1\}^k.$$

It takes four inputs, a public key $W \in \mathcal{H}$, a k-bit derivative π of a password and a pair of values $(R, \eta) \in \mathcal{H} \times \{0,1\}^k$ representing a ciphertext with respect to the associating public key encryption function Enc described above. For each $(W, \pi) \in \mathcal{H} \times \{0,1\}^k$, we let $\mathcal{P}_{W,\pi} : \mathcal{H} \times \{0,1\}^k \to \mathcal{H} \times \{0,1\}^k$ be the permutation defined by $\mathcal{P}_{W,\pi}(R, \eta) = \mathcal{P}(W, \pi, R, \eta)$. Therefore $\mathcal{P}_{W,\pi}$ is a permutation *indexed* by (W, π).

The technique we use to prevent the associativity problem is to make sure that $\mathcal{P}_{W,\pi}$ allows the sender to commit himself to only one single password each time. This can be done by devising a password-keyed permutation family to satisfy the following two requirements.

First, we require $\mathcal{P}$ to be *distinct*. That is, for every pair $(\pi_1, \pi_2) \in \{0,1\}^k \times \{0,1\}^k$, $\pi_1 \neq \pi_2$, and for all $W \in \mathcal{H}$ and $(R, \eta) \in \mathcal{H} \times \{0,1\}^k$,

$$\Pr[\mathcal{P}(W, \pi_1, R, \eta) = \mathcal{P}(W, \pi_2, R, \eta)] \leq \epsilon(k)$$

for some negligible function ϵ, where a real-valued function $\epsilon(k)$ is *negligible* if for every $c > 0$, there exists a $k_c > 0$ such that $\epsilon(k) < k^{-c}$ for all $k > k_c$.

Second, we require $\mathcal{P}$ to satisfy the following statement.

Definition 1. *Given a password space PW, the public key encryption function Enc described above in Sec. 2.1 with a public key $W \in \mathcal{H}$, a cryptographic hash function $T : \{0,1\}^* \to \{0,1\}^k$ that behaves like a random oracle [2] and a sufficiently large k, a distinct password-keyed permutation family $\mathcal{P} : \mathcal{H} \times \{0,1\}^k \times \mathcal{H} \times \{0,1\}^k \to \mathcal{H} \times \{0,1\}^k$ is secure (against offline dictionary attack) if for every probabilistic polynomial-time algorithm E^T,*

$$\Pr[\, E^T(1^k, A, B, PW, Enc, W, r_A) \to (R, \eta, r_B, m_1, m_2, pw_1, pw_2) \, : \, r_B \in \{0,1\}^k,$$
$$m_i \in \{0,1\}^k, \; pw_i \in PW, \; (R, \eta) = \mathcal{P}(W, \pi_i, Enc(W, m_i)), i \in \{1,2\}]$$
$$\leq \epsilon(k)$$

for all $r_A \in \{0,1\}^k$, $A, B \in \mathbb{N}$ and for some negligible function ϵ where $\pi_i = T(pw_i, A, B, r_A, r_B)$ for $i = 1, 2$.

It means that an attacker should not be able to compute more than one pair of (pw, m) which transforms to (π, m) using T such that $\mathcal{P}(W, \pi, Enc(W, m))$ produces the same value of (R, η). This limits the number of password guesses that the attacker can make to just one for each (R, η), and for each run of our PAKE protocol.

Below is the password-keyed permutation function we proposed.

PWD_PERMUTE = "On inputs of domain parameters $(\mathscr{C}, \mathbf{F}, G, q)$, a public key $W \in \mathcal{H}$, a password $pw \in PW$, the initiator and the responder identities $A, B \in \mathbb{N}$, some binary strings $r_A, r_B \in \{0,1\}^k$ and a ciphertext $(R, \eta) \in \mathcal{H} \times \{0,1\}^k$,

1. Compute $\pi = T(pw, A, B, r_A, r_B)$ where $T : \{0,1\}^* \to \mathbb{Z}_q$ is a cryptographic hash function which behaves like a random oracle.
2. Output $z = (R + \pi W, \; \eta)$."

This formation can be shown to be secure (Definition 1) under the assumption that the Computational Diffie-Hellman problem [12] is hard.

Theorem 1. *Breaking the password-keyed permutation family $(W, \pi, R, \eta) \to (R + \pi W, \eta)$ is computationally equivalent to solving the EC-CDH (Elliptic Curve Computational Diffie-Hellman Problem).*

The proof is given in the full paper[2].

2.3   The Protocol

We now describe our PAKE protocol using the building blocks above. Let Alice be the initiator indexed by A and Bob be the responder indexed by B. The protocol proceeds as follows and is illustrated in Fig. 1.

---

[2] http://www.cs.cityu.edu.hk/~duncan/

1. Alice randomly picks $r_A \in_R \{0,1\}^k$ and sends (W, r_A) to Bob where $W = wG$ such that $w \in_R \mathbb{Z}_q$ is only known by Alice. It is assumed that A has already generated the public key pair (w, W) as described in Sec. 2.1 beforehand. This public key pair does not need to be certified and will be used each protocol run as long as A is the initiator. For implementation, this can be done when the PAKE software is first installed and initialized on the device of the protocol initiator.

2. Bob randomly picks $r_B, s_B \in_R \{0,1\}^k$, and sets $\pi = T(pw, A, B, r_A, r_B)$. He sends r_B and $z = \mathcal{P}(W, \pi, Enc(W, s_B))$ to Alice.

3. Alice computes π accordingly and obtains s_B from z. She then computes $\alpha = H_1(s_B)$, $c_B = H_3(s_B)$ and $\sigma = H_4(c_A, c_B, A, B)$, randomly picks $c_A \in_R \{0,1\}^k$, and sends $(\alpha \oplus c_A, H_2(\alpha, c_A, A, B))$ to Bob. She destroys s_B, c_A, c_B, α from her memory.

4. Bob computes α and c_B from s_B accordingly. He reveals c_A from the first part of the incoming message and checks if the second part of the incoming message is $H_2(\alpha, c_A, A, B)$. He then computes the session key σ accordingly and sends $H_5(\sigma)$ back to Alice. He destroys s_B, c_A, c_B, α from his memory.

5. Alice finally checks if the incoming message is $H_5(\sigma)$.

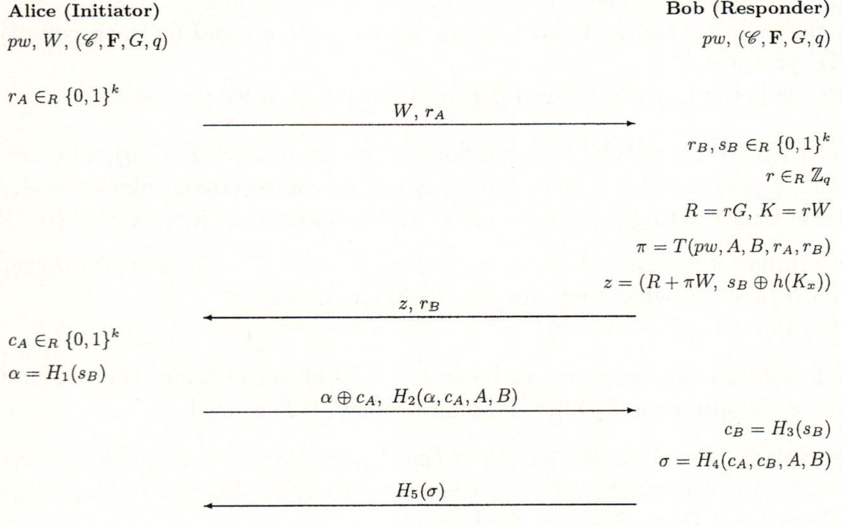

Fig. 1. PAKE-EC

3 Performance

The computational complexity of ECES is symmetric. Other operations of the PAKE-EC, including the password-keyed permutation are also symmetric. Hence the computational complexities on both sides of the protocol are comparable to each other. Hence in the following, we only focus on evaluating the efficiency of the responder, Bob.

In one run of the PAKE-EC, Bob requires to generate three random numbers, perform seven hashes, compute one point addition and three scalar multiplications. Among these operations, the scalar multiplication entails the highest computational complexity while all other operations can be done efficiently and their complexity can be neglected. Therefore, we evaluate the performance of PAKE-EC by essentially counting the number of scalar multiplications required.

Optimizations: The three scalar multiplications performed by each party are rG, rW and πW. Among them, the calculations of rW and πW can be done simultaneously so to reduce the total computation time on each communicating party of the protocol. For example, if add-subtract method [5] is used to compute rW, we need to do point doubling operation $\lg(r)$ times and do point addition operation $\lg(r)/3$ times on average. Without loss of generality, suppose $\lg(r) = \lg(\pi)$, if rW and πW are done simultaneously, then we only need to do doubling operation $\lg(r)$ times and do addition operation $2\lg(r)/3$ times on average. So when the complexity of doubling operation is similar to that of addition operation (for example, when the operations are over affine coordinate representation), the total complexity of computing both rW and πW is only a quarter times more than that of computing only rW. Hence the number of scalar multiplications required to be carried out by each party is only 2.25 on average. According to [15], one scalar multiplication on a Koblitz curve over $GF(2^{163})$ can be done in 1.51 seconds on a 16MHz Handspring Visor. Hence the computation time of each party is expected to be about 3.4 seconds. We obtained similar performance results in our own implementation on a 16MHz Motorola DragonBall-EZ (MC68EZ328) microprocessor.

Precomputation: Similar to some current elliptic curve based PAKE schemes, usually one scalar multiplication can be pre-processed. In PAKE-EC, rG can be precomputed since r is chosen by the responder and G is known beforehand. In addition, if the responder communicates with a fixed initiator for multiple sessions using the PAKE-EC, the public key W of the initiator can also be cached and most of the computations on the responder side before sending the first responder-to-initiator message (the second message in Fig. 1) can be done in advance, except the last step of computing z. Hence the responder only needs to perform one scalar multiplication during the runtime of the protocol and it takes about 1.5 seconds on the 16MHz embedded device described above.

4   Conclusions

In this paper, we propose a PAKE protocol which is suitable for efficient implementation on resource-constrained wireless devices. On an embedded device with a 16MHz DragonBall-EZ microprocessor, the protocol is estimated to take 3.4 seconds of computation time for each party. With appropriate precomputation, the time can be further reduced to 1.5 seconds. By introducing an elliptic curve based password-keyed permutation family, we show under the random oracle model that no active adversary can guess more than one password in each impersonation attack if the elliptic curve computational Diffie-Hellman problem is intractable. That is, it is secure against offline dictionary attack.

References

1. M. Bellare, D. Pointcheval, and P. Rogaway. Authenticated key exchange secure against dictionary attacks. In *Proc. EUROCRYPT 2000*, 2000. LNCS 1807.
2. M. Bellare and P. Rogaway. Random oracles are practical: A paradigm for designing efficient protocols. In *First ACM Conference on Computer and Communications Security*, pages 62–73. ACM, 1993.
3. S. M. Bellovin and M. Merritt. Encrypted key exchange: Password based protocols secure against dictionary attacks. In *Proceedings 1992 IEEE Symposium on Research in Security and Privacy*, pages 72–84. IEEE Computer Society, 1992.
4. L. Gong, M. Lomas, R. Needham, and J. Saltzer. Protecting poorly chosen secrets from guessing attacks. *IEEE J. on Selected Areas in Communications*, 11(5):648–656, 1993.
5. IEEE. *P1363 - 2000: Standard Specifications For Public Key Cryptography*, 2000.
6. IEEE. *P1363.2 / D15: Standard Specifications for Password-based Public Key Cryptographic Techniques*, May 2004.
7. D. Jablon. Strong password-only authenticated key exchange. *Computer Communication Review, ACM*, 26(5):5–26, 1996.
8. D. Jablon. Extended password key exchange protocols immune to dictionary attack. In *Proc. of the WETICE'97 Workshop on Enterprise Security*, Cambridge, MA, USA, Jun 1997.
9. K. Kobara and H. Imai. Pretty-simple password-authenticated key-exchange under standard assumptions. *IEICE Trans.*, E85-A(10):2229–2237, October 2002.
10. P. MacKenzie. More efficient password-authenticated key exchange. In *CT-RSA 2001*, pages 361–377, 2001. LNCS 2020.
11. P. MacKenzie, S. Patel, and R. Swaminathan. Password-authenticated key exchange based on RSA. In *Proc. ASIACRYPT 2000*, pages 599–613, 2000.
12. Ueli M. Maurer and Stefan Wolf. The Diffie-Hellman protocol. *Designs, Codes and Cryptography*, 19:147–171, 2000.
13. A. Menezes. *Elliptic Curve Public Key Cryptosystems*. Kluwer Academic Publishers, 1993.
14. Yongge Wang. EC-SRP. *Submission to IEEE P1363 Study Group, June 2001, updated May 2002*, May 2002.
15. A. Weimerskirch, C. Paar, and S. Chang Shantz. Elliptic curve cryptography on a Palm OS device. In *Information Security and Privacy ACISP 2001*, pages 502–513, 2001. LNCS 2119.
16. D. Wong, A. Chan, and F. Zhu. More efficient password authenticated key exchange based on RSA. In *Progress in Cryptology - INDOCRYPT 2003*, pages 375–387. Springer-Verlag, 2003. LNCS 2904.
17. T. Wu. The secure remote password protocol. In *1998 Internet Society Symposium on Network and Distributed System Security*, pages 97–111, 1998.

An Efficient Anonymous Scheme for Mutual Anonymous Communications[1]

Ray-I Chang[1] and Chih-Chun Chu[2]

[1]Dept of Engineering Science, National Taiwan University, Taipei, Taiwan, ROC
`rayichang@ntu.edu.tw`
[2]Dept of Information Management, National Central University, JungLi, Taiwan, ROC
`s0443006@cc.ncu.edu.tw`

Abstract. Due to the rising of on-line computer and communication applications, users require an anonymous channel to transfer sensitive or confidential data on the networks. How to construct a safe communication environment to guarantee the mutual anonymity during communications becomes an important problem. Recently, researchers have focused on applying the ALR (application-level routing) framework to perform private communications on top of IP networks. The Freenet system is one of the most famous models. Like other ALR solutions, Freenet needs no special device to support mutual anonymous communications. However, it stores and backwards response data over the entire routing path, which causes lots of time and resources wasted. Although the result of our previous research [11] has already improved the performance of Freenet, the location of the source peer still remains unprotected. In this paper, we apply anonymous-shortcut on mutual anonymous communications. The proposed mechanism permits users to choose different security levels on their own demands for not only the request peer but also the source peer. It provides anonymous delivery to support real-time communications [9-10] with acceptable security and controllable delay.

1   Introduction

Recently, the applications of network have become more and more popular. C&C (computer and communication) techniques are increasingly used in various areas such as commerce, military, government and also in our daily life. Due to the rising of on-line e-commerce and e-voting applications, users may require an anonymous channel to transfer sensitive or confidential data on networks. However, network privacy today is still facing a lot of threats. Therefore, it is really important to protect the data packets and to hide their source and destination locations during communications, namely private communications. Due to that applications and users of a system require different levels of security under different circumstances, the development of an efficient and effective private communication protocol is in an urgent demand. Recently, many efforts have been put on the development of ALR (application-level routing) solutions, which perform on top of IP routing and need no special device, to support private communications. Freenet introduced by Clarke et al. [1, 3] is one of

---

[1] *This paper was partially supported by NSC, Taiwan, under grants NSC92-2416-H-002-051- and NSC93-2213-E-002-086-.*

P. Lorenz and P. Dini (Eds.): ICN 2005, LNCS 3421, pp. 835–843, 2005.

the most famous systems for private communications of ALR solutions. Based on Freenet, we propose a new anonymous communication mechanism in this paper.

Freenet is an adaptive peer-to-peer (P2P) network to store and retrieve data files. Usually, P2P systems are taxonomized into two categories. In the hybrid architecture, all peers are managed by a key peer. As shown in Fig. 1(a), the role of a key peer is like a server in the tradition network. However, the key peer does not provide the required data. It provides only the data location to a request peer (RP). A system like CHORD [7] is an example of this architecture.

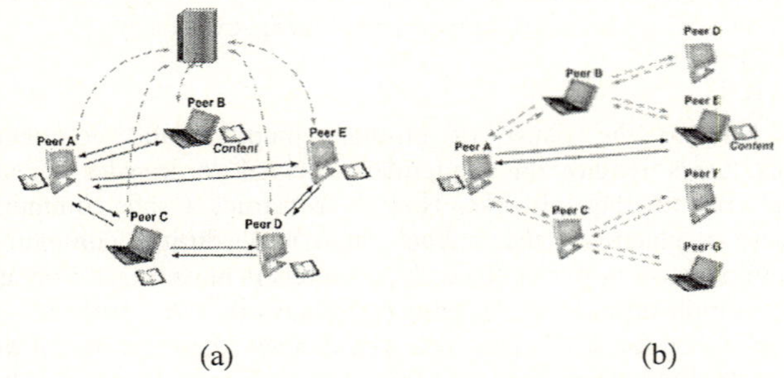

(a)   (b)

Fig. 1. (a) A hybrid P2P network. (b) A purely distributed P2P network system

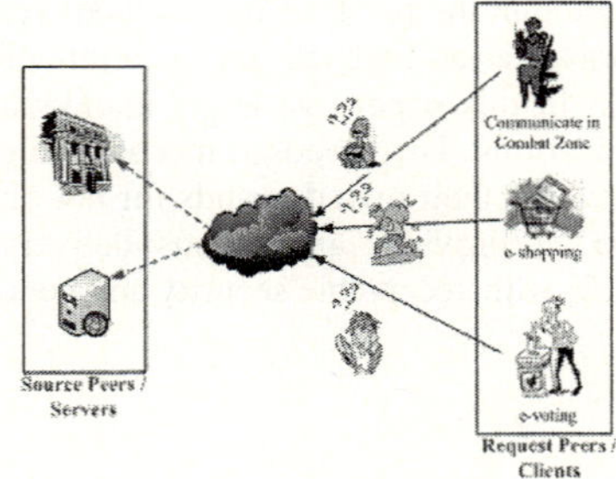

Fig. 2. The server can hide its location by using anonymous communications

Different from CHORD, Freenet is an unstructured and purely distributed system. There is no server peer existed as shown in Fig. 1(b). To retrieve the request content, a RP must search peer by peer. Usually, these systems perform ALR on top of IP routing that allows each node to communicate with the rest nodes of the system. Although the purely distributed system is time consuming and causes heavy traffic load, it provides a high security and fault tolerance system. It is good for applications that require an anonymous channel to transfer sensitive or confidential data. For example, in Fig. 2, the server may need to hide its location in several application areas (e.g., communication in the combat zone, e-shopping, and e-voting). Otherwise, it may suffer from the DoS (denial of service) attack or the risk of disclosing users' private data. For example, if the location of source peer (SP) is exposed, an attacker may obtain the user's personal information through a Trojan horse program.

Freenet is known as one of the most famous P2P networks for distributed information storage and retrieval. Unlike traditional client-server networks, peers in Freenet use an ALR framework to forward data to their next peers. Freenet provides a strong protection for anonymous communications. Unfortunately, with this anonymous communication model, response data should be stored and transmitted backwards over the entire routing path to its client. Lots of time and resource are wasted. Prior researches [1, 4] have shown that Freenet with the storing and backwarding behavior can satisfy the small-world [5-6] hypothesis (where, after a long time, a short path between any two peers would exist in ALR). However, its performance is still arguable especially under a heavy-loaded network

Additionally, in some network applications such as media streaming [9-10], users may need to decide their own security and performance levels. The unwanted short path introduced by the small-world effect may rambunctiously weaken the protection in anonymous communications. How to provide an acceptable performance of data delivery is still a big challenge in Freenet. Our previous paper [11] has tried to improve the performance of Freenet. However, it may expose the location of the SP.

To resolve this drawback, we propose a new anonymous communication mechanism in this paper. Like our previous work [11], the proposed mechanism is also based on the concept of anonymous-shortcut to improve the performance of Freenet. Besides, to improve the security of this mechanism which can protect not only the RP but also the SP, we introduce a pair of parameters, called anonymous depth, to guarantee the security levels of these two individual peers. The experiment results show that our mechanism can provides a safe-enough anonymous communication platform with acceptable performance in network communications. The same concept can be easily applied on other protocols under any kinds of environments. It provides anonymous delivery for application-independent and real-time connections. In 2003, a new version of Freenet with Next Generation Routing (FNGR) was proposed [12]. Although it makes Freenet nodes much smarter to collect extensive statistical information including response times to save routing times, its fundamental concept of remains the same, so that our proposed scheme can be applied directly.

This paper is organized as follows. In Section 2, we present an overview of previous methods. In Section 3, the proposed mechanism for mutual anonymous communication is introduced. Then, a set of simulations is shown in Section 4. Section 5 summarizes the results and gives an outlook on future works.

2 Overview

Freenet is known as one of the most famous P2P network architecture. Freenet is based on ALR that performs on top of IP routing and needs no special device. Usually, three types of keys are used to identify the data files and route their locations in Freenet. The keyword-signed key (KSK) is derived from a descriptive string given by the owner. The signed-subspace key (SSK) is used to identify the owner's personal namespace. The content-hash key (CHK) is derived by directly hashing the contents of the corresponding data file. These keys are obtained by applying the 160-bits SHA-1 hash function currently. In Freenet, each node maintains a routing table with <key, pointer> pairs. The pointer points to the next node that is associated with the key. As there is no global consensus on where data should be stored, a peer on the path forwards the message by checking the most similar key in the routing table. If the response data is found, the same path is used for sending the response data. Fig. 3 shows the forward-sequence of a request message and its backward-sequence of response data in Freenet. Notably, packets were encapsulated peer by peer. Let P_i with address $\&P_i$ be the ith peer in the routing path. Only the peer P_x knows the address of the peer $\&P_{x-1}$.

When a file is ultimately returned for a successful retrieval, the node passes the file to the RP, caches the file in its own data store, as well as creates a new entry in its

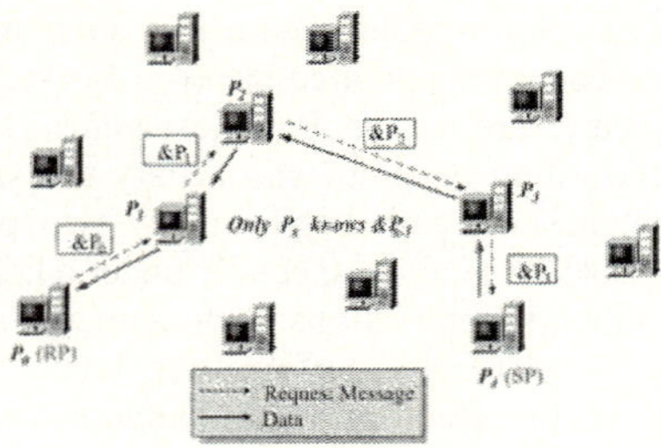

Fig. 3. Typical requesting and backwarding sequences in Freenet

routing table. This mechanism can lead not only network convergence but also forming Freenet into a small-world network. A small-world network follows a scale-free power-law distribution. It permits efficient short paths between arbitrary nodes. However, under heavy load network, Freenet might not satisfy the small-world hypothesis [4]. Even if the system is a small-world network, the unwanted short path may weaken Freenet's protection. It motivates us to design a new mechanism for anonymous communications.

In [11], we had proposed the concept of anonymous-shortcut to improve Freenet's performance and to protect the RP. Like Freenet, we use a table, called *reply_table*, to cache the information of the previous peer for replying the response data. These intermediate peers should replace the packet header and encrypt the packet to protect the address of the RP. They should also cache the original packet header into the *reply_table* table. Thus, the response data can be replied to the correct RP. However, different from Freenet, the RP will generate a positive random number, called *anonymous depth*, to decide the number of consequent peers selected to reply the response data for hiding the location of RP. Therefore, the response data would pass through the anonymous-shortcut instead of the original routing path. For providing anonymous communications, Freenet uses the same path to route the request message and to reply the response data. A simple example of our previous mechanism is shown in Fig. 4.

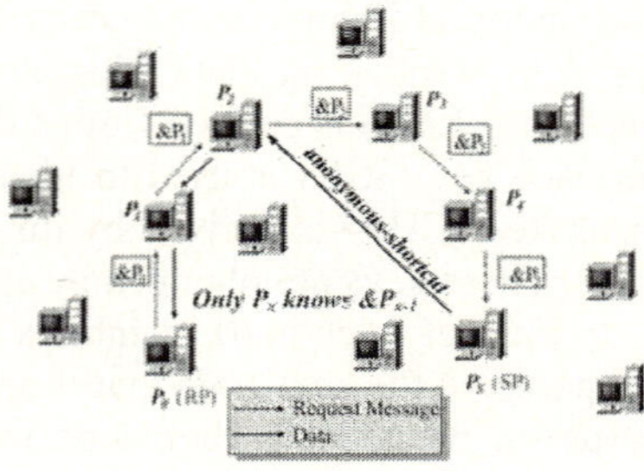

Fig. 4. Requesting and backwarding sequences in our mechanism. [11]

3 Proposed Method

In this paper, the concept of anonymous-shortcut is introduced to protect both RP and SP. There are two types of anonymous depths in our new mechanism. Like our previous work, the RP will generate an anonymous depth number (denotes as R_A_d) to hide its location. On the other hand, to hide the location of SP, the SP will also generate another anonymous depth, denotes as S_A_d, for backwarding. A simple example of our new mechanism is shown in Fig. 5. Both RP and SP are protected in this case. It is not difficult to prove that there is no intermediate peer which knows if its previous peer is the RP (or the SP).

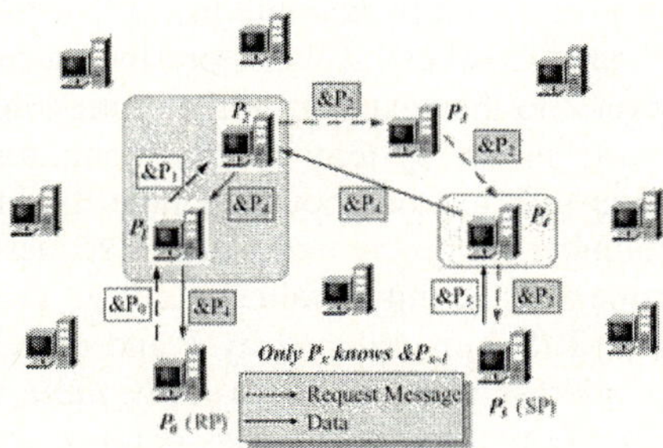

Fig. 5. Requesting and backwarding sequences in our mechanism

In practice, the lower and upper bounds of anonymous depth can be set by users or administrators to fit their requirements in security and efficiency of network transmission. When the lower bound is set to a small value, it can support high efficiency but low security, and vice versa. Note that, even under the small-world model, the proposed mechanism can provide an acceptably secure environment when a large anonymous depth is selected. The detail of algorithm is described as follows.

```
Procedure Query_Process(Pₓ, Z)
Begin
    If (Z.R_Aₐ > 0) {
            Z.R_Aₐ --;
            Store (Z.address, Z.index)
                in reply_table[j] of the current peer Pₓ;
            Z.address = Pₓ; // the current peer
            Z.index = j;
            Pᵧ = the next peer in the routing table;
            Call Query_Process(Pᵧ, Z); }
    Else If (Z.request is stored in the current peer) {
            Z.request = data retrieved;
            Pᵧ = Z.address;
            Call Data_Transfer(Pᵧ, Z); }
        Else {
                Pᵧ = the next peer in the routing table;
            Call Query_Process(Pᵧ, Z); }
End
Procedure Data_Transfer(Pₓ, Z)
Begin
    If (Z.S_Aₐ > 0) {
            Z.S_Aₐ --;
            Z.address = Pₓ; // the current peer
            Z.index = j;
            Pᵧ = the next peer in the routing table;
```

```
    Else {
        Retrieve reply_table[Z.index] in the current peer P_x
            to (P_y, Z.index);
        If (Py == the current peer Px)
            // It is the request peer !
        Else
        Call Data_Transfer(P_y, Z); }
End
```

In the following, we give a simple example to explain the algorithm. While a RP (P_0 in this case) needs data $Z.request$, it will store the current address in *reply_table* and send a request packet Z to the next peer in the routing table. The request packet Z is regarded as three parts in our system: header, communication information, and message. Communication information should include the information of the content identifier, a sequence number (to be the index of this request packet), and the current value of $Z.R\_A_d$. Assume that the initial value of $Z.R\_A_d$ is 3. The value of $Z.address$ is replaced peer-by-peer, $Z.R\_A_d$ is reduced by 1 and each peer on the routing path will store the previous address and index into *reply_table* until Z arrives P_2 (where $Z.R\_A_d$ is reduced to 0). While the other nodes coming after P_2 receive the request packet, they will forward the packet Z directly without any changing. Thus, while the data packet is responded, the SP can forward the request data to the $Z.address = P_2$ peer directly. In other words, the response data can pass through the anonymous-shortcut for improving efficiency. Moreover, in some situations, the location of SP needs to be hidden. Like the RP, the SP can generate a new random number - $Z.S\_A_d$ - to protect himself. Assume that the initial value of $Z.S\_A_d$ is 2. The response data will not be transferred to P_2 directly until it arrives in P_{n-1}.

Note that, $Z.R\_A_d$ /$Z.S\_A_d$ is a positive random number which is decided by the RP/SP. Only the RP/SP knows its initial value, while no intermediate peer knows which peer is the real RP/SP. (A peer is the RP/SP if and only if its $Z.R\_A_d$ /$Z.S\_A_d$ is the same as the initial value. However, only the RP/SP knows the initial value. That means no one but the RP/SP can make the correct decision.) Besides, even if the system is a small-world network and the unwanted short path is introduced, our proposed mechanism can provide sufficient protection for anonymous communications. It is easy to implement and to combine with other protocols.

4 Evaluation

Assume that there are n random requests presented on the network system. The routing length of request i is defined as X_i and the size of the request message is 1 (unit) for each request. The size of the response data is c_i. In Freenet, the path length for replying response data is the same as its routing length. The traffic load introduced by the request i can be defined as

$$X_i + c_i * X_i. \tag{1}$$

It is the total amount of data packets delivered by different network links on the system. Unlike Freenet, our mechanism permits users to apply anonymous-shortcut with the assigned anonymous depth. Assume that the anonymous depth is x_i. The traffic load using our mechanism can be defined as

$$X_i + c_i * x_i. \tag{2}$$

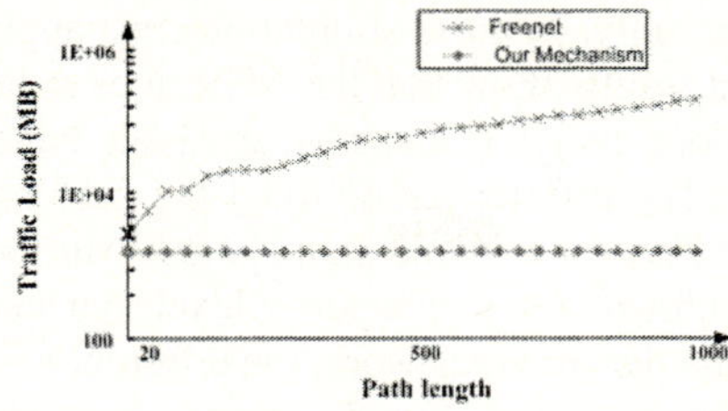

Fig. 6. The traffic loads introduced by our mechanism and by Freenet with different path lengths. (ci = 200000)

Fig. 6 shows the traffic loads introduced by our mechanism and by Freenet with different path lengths.

Note that c_i is usually much larger than that of the request message. For example, c_i is 200000 for a large-sized video stream. While a simple web page is examined, c_i is 100. For not only the large-sized video stream but also the simple web page, we have shown that the proposed mechanism has lower traffic load than Freenet as x_i is usually less than or equal to X_i. Actually, the value of x_i also permits users to choose different security levels on their own demands. There is a tradeoff between the security level and the traffic load.

In large-scale networks, our mechanism is more efficient than Freenet. This implies that our system has higher scalability than Freenet. However, the performance of Freenet may be better than that of our mechanism in a small-scale network with the small-world model since it introduce shortcuts in delivery. (For example, assume that there are 1000 peers and the size of routing table is 200. The routing path in Freenet with a small-world model would be less than 3.) However, according to [8], a good security level (which relates to the path length) should be larger than 50. The unwanted short path may weaken the protection in anonymity. Additionally, as network peers result in high clustering, the protection power of denial-of-service (DoS) attack may be worn down.

In Fig. 7, we evaluate the effect on the Next Generation Routing (NGR) mechanism. In these simulations, we assume that the network size is fixed to 2000, 10000, and 100000 peers. The transmission delay time between any two peers is assigned randomly (The random function follows uniform distribution. And the range of delay time is set from 1 to 10 ms). Each experiment is run 50 times. The RP and

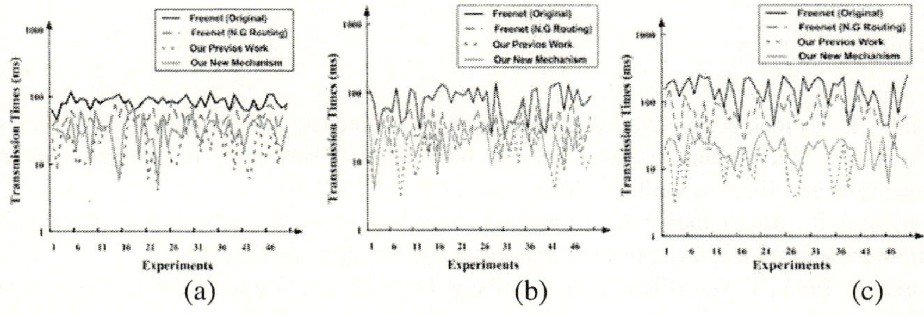

Fig. 7. A comparison between the original Freenet, the FNGR, our previous work, and our new mechanism. The network size is fixed to: (a) 2000 peers, (b) 10000 peers, and (c) 100000 peers

the SP are selected randomly. In these simulations, we applied our mechanisms on the FNGR. The experiment results show that the NGR does reducing the transmission time. The performances of our two mechanisms are both better than either the original Freenet or the FNGR. The differences between the performances of our mechanisms and the two kinds of Freenet become more significant when the scale of network growth. Although, to achieve a higher security level, our new mechanism spends more time in transmission than our previous work, the difference is very mild in statistic.

Although the performance of our previous work may be slightly better than that of our new mechanism in most situations, the protection of the SP is weaker in our previous work. In our previous work, the location of the SP may be exposed by attackers with a higher probability, because the SP will reply data immediately without passing through any consequent peer. For achieving the goal of mutual anonymity, our new mechanism introduces a parameter $Z. S\_A$ to hide the location of SP. Thus, our system provides the same security level as Freenet.

Note that during anonymous communications, any node on the routing path may introduce the fault. The performance of fault tolerance is related to the path length for replying data. In Freenet, the failure rate is related to X_i for any request i. In our mechanism, the failure rate is related to the given value x_i. Because x_i is usually less than or equal to X_i, the failure rate in our mechanism is smaller then that in Freenet. The performance of fault tolerance in our mechanism is better than that in Freenet.

5   Conclusion

We propose a new mechanism in order to provide mutual and real-time anonymous communications. Our mechanism can provide sufficient security and acceptable performance on different purposes. The scalability and the fault tolerance of network in our system are better than those in Freenet. In addition, our mechanism is easy to implement. The presented idea is application independent and can be applied to other network architecture. Our future work is to extend the proposed mechanism to broadcasting anonymous schemes, such as P^5 (Peer-to-Peer Personal Privacy Protocol) [2]. P^5 is a protocol for anonymous communication over Internet. Once a user joins to a group in P^5, its anonymity can be protected by all users in the same group. P^5 provides a strong mechanism to protect users' privacy. However, a terrible traffic load is coming with the broadcasting mechanism. We hope to resolve this problem and construct a more secure network platform with a higher performance.

References

1. Clarke, I., Sandberg, O., Wiley, B., Hong, T. W.: Freenet: A Distributed Anonymous Information Storage and Retrieval System. International Workshop on Design Issues in Anonymity and Unobservability, LNCS, Vol. 2009 (2001) 46-66.
2. Sherwood, R., Bhattacharjee, B., Srinivasav, A.: P5: A Protocol for Scalable Anonymous Communication. IEEE Symposium on Security and Privacy (2002) 1-13.
3. Clarke, I., Hong, T. W., Miller, S. G., Sandberg, O., Wiley, B.: Protecting Free Expression Online with Freenet. IEEE Internet Computing. Vol. 6, No. 1 (2002) 40-49.
4. Zhang, H., Goel, A., Govindan, R.: Using the Small-World Model to Improve Freenet Performance. IEEE INFOCOM (2002) 1228-1237.

5. Watts, D. J., Strogatz, S. H.: Collective dynamics of 'small-world' networks. Nature, Vol. 393, (1998) 440-442.
6. Kleinberg, J.: Navigation in a small-world. Nature, Vol. 406 (2000) 845.
7. Stoica, I., Morris, R., Karger, D., Kaashoek, F., Balalrishnan, H.: Chord: A peer-to-peer loukup service for internet applications. ACM SIGCOMM (2001).
8. Guan, Y., Fu, X., Bettati, R., Zhao, W.: An optimal strategy for anonymous communication protocols. IEEE ICDCS (2002).
9. Chang, R. I., Chen, M. C., Ko, M. T., Ho, J. M.: Schedulable Region for VBR Media Transmission with Optimal Resource Allocation and Utilization. Information Sciences, Vol. 141, Issue 1-2 (2002) 61-79.
10. Chang, R. I., Chen, M. C., Ko, M. T., Ho, J. M.: Online Traffic Smoothing for Delivery of VBR Media Streams. Circuits Systems Signal Processing, Vol. 20, No. 3 (2001) 341-359.
11. Chang, R. I., Chu, C. C., Chiu, Y. L.: An Efficient Anonymous Scheme for Computer and Communication Privacy. IEEE ICCST (2004)
12. Clarke, I.: http://freenetproject.org [Online Avalible] (2003).

GDS Resource Record: Generalization of the Delegation Signer Model

Gilles Guette, Bernard Cousin, and David Fort

IRISA, Campus de Beaulieu, 35042 Rennes CEDEX, France
{gilles.guette, bernard.cousin, david.fort}@irisa.fr

Abstract. Domain Name System Security Extensions (DNSSEC) architecture is based on public-key cryptography. A secure DNS zone has one or more keys to sign its resource records in order to provide two security services: data integrity and authentication. These services allow to protect DNS transactions and permit the detection of attacks on DNS.

The DNSSEC validation process is based on the establishment of a chain of trust between secure zones. To build this chain, a resolver needs a secure entry point: a key of a DNS zone configured in the resolver as trusted. Then, the resolver must find a path from one of its secure entry point toward the DNS name to be validated. But, due to the incremental deployment of DNSSEC, some zones will remain unsecure in the DNS tree. Consequently, numerous trusted keys should be configured in resolvers to be able to build the appropriate chains of trust.

In this paper, we present a model that reduces the number of trusted keys in resolvers and ensures larger secure access to the domain name space. This model has been implemented in BIND.

1 Introduction

Domain Name System (DNS) [1, 2, 3] is a distributed database mostly used to translate computer names into IP addresses. The DNS protocol does not include any security services such as integrity and authentication, which let the protocol vulnerable to several attacks [4, 5, 6]. Therefore the Internet Engineering Task Force (IETF) has developed the DNS security extensions (DNSSEC).

DNSSEC architecture [7, 8, 9, 10] uses public-key cryptography to provide integrity and authentication of the DNS data. Each node of the DNS tree, called a *zone*, owns at least a key pair used to secure the zone information with digital signature.

In order to validate DNS data with DNSSEC, a resolver builds a chain of trust [11] by walking through the DNS tree from a secure entry point [12] (*i.e.*, a *trusted key* statically configured in the resolver, typically a top level zone) to the zone queried. A resolver is able to build a chain of trust if it owns a secure entry point for this query and if there are only secure delegations from the secure entry point to the zone queried.

P. Lorenz and P. Dini (Eds.): ICN 2005, LNCS 3421, pp. 844–851, 2005.

Because of the long[1] and incremental deployement of DNSSEC, resolvers will have to keep many trusted keys. The number of trusted keys needed for a single resolver may be huge and not easy to store. Moreover, these zone keys are periodically renewed and the resolver must update its trusted keys set to keep consistent the keys.

In this paper we propose the creation of a new resource record allowing to reduce the number of trusted keys needed in a resolver. In Section 2, we present the definitions used in this paper and an overview of the DNSSEC processing. In Section 3, we expose the problem of the islands of security model for the resolvers. Section 4 defines the General Delegation Signer Resource Record and its management. Then, in Section 5 we discuss the pros and cons of our method.

2   Validation Process in DNSSEC

2.1   Definitions

In this subsection are explained the notations used in the document.

(1) A DNS domain X is the entire subtree included into the DNS tree beginning at the node X. (2) A DNS zone is a node of the DNS tree. A zone name is the concatenation of the node's labels from its node to the root of the DNS tree. A zone contains all the not delegated DNS names ended by the zone name. For example, the zone `example.com.` contains all the not delegated names `X.example.com.` where X can be composed by several labels. (3) A zone can delegate the responsability of a part of its names. The zone `example.com.` can delegate all the names ended by `test.example.com.` to a new zone. This zone is named the `test.example.com.` zone. (4) RR means Resource Record, the basic data unit in the domain name system. Each RR is associated to a DNS name. Every RR is stored in a zone file and belongs to a zone. (5) Resource records with same name, class and type form a RRset. For example the DNSKEY RRS of a zone form a DNSKEY RRset. (6) DNSKEY ($key1$) is the RR which describes the key named $key1$. (7) RRSIG$(X)_y$ is the RR which is the signature of the RR X generated with the private part of key y. (8) A trusted key is the public part of a zone key, configured in a resolver. (9) A Secure Entry Point is a zone for which the resolver trusts a key.

2.2   DNSSEC Chain of Trust

DNS security extensions define new resource records in order to store keys and signatures needed to provide integrity and authentication.

Each secured zone owns at least one zone key, the public part of this key is stored in a DNSKEY resource record. The private part of this key is kept secret and should be stored in a secure location. The private part of the key generates a digital signature for each resource record in the zone file. These signatures are stored in a RRSIG resource record.

---

[1] Transition duration could be long: for instance IPv6 deployement lasts since 1995.

A resource record is considered valid when the verification of **at least one** of its associated RRSIG RR is complete. Figure 1 shows the signature verification process. In order to verify the signature of a resource record, the resolver cyphers the RRSIG RR with the public key of the zone contained in the DNSKEY RR present in the DNS response message. If the result of this operation, called `Resource Record'` in figure 1 is equal to `Resource Record` present in the DNS response message, the signature is verified. Thus, the resource record is valid.

During the signature verification process, the zone key is needed and must be verified too. This allows to avoid the use of a fake key sent in a message forged by a malicious person. To trust a zone key, DNSSEC uses the DNS-tree model to establish a chain of trust [11] beginning from a secure entry point [12] to the queried zone. To create this chain, a verifiable relation between child zone and parent zone must exist: this is the role of the Delegation Signer resource record (DS RR) [13]. This record, stored in the parent zone, contains information allowing the authentication of one child zone key. Figure 2 shows the validation of a delegation.

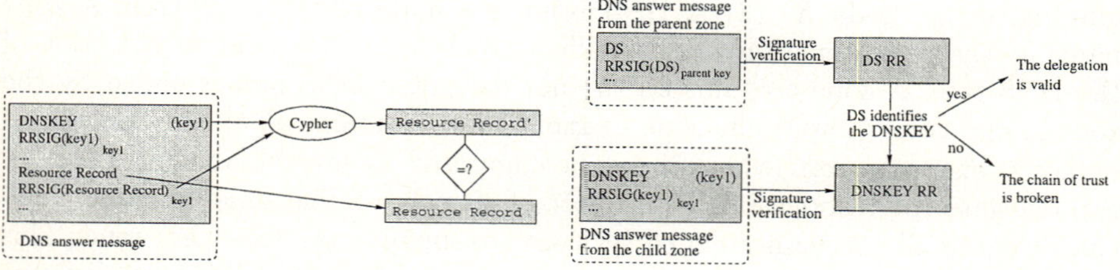

Fig. 1. The signature verification process   **Fig. 2.** The delegation verification process

Once the signatures (the signature of the DS RR provided by the parent zone and the signature of the DNSKEY provided by the child zone) are verified, the resolver checks that information contained in one DS RR identifies one key in the child zone. If one DS RR identifies one DNSKEY RR in the child zone, one link of the chain of trust is built and the name resolution should progress to secure the next link in the DNS tree. If there is no valid DS RR that identifies one valid DNSKEY RR in the child zone, the chain of trust is broken and the name resolution is unsecure.

3   DNSSEC Deployement and Islands of Security Model

Using DNSSEC is not mandatory for zone administrators, they are free to deploy DNSSEC or not. Indeed, some administrators evaluate the overhead to deploy DNSSEC too important compared to the need to secure their zone with DNSSEC. Moreover, deploying DNSSEC software can raise some compatibility problems. Consequently, some parts of the DNS tree may be secure with

DNSSEC while some other parts may be unsecure (DNS only). This leads to an *islands of security* model. An island of security is a subtree of the DNS tree entirely secured with DNSSEC. The incremental deployment of DNSSEC implies some remaining unsecure zones in the DNS tree and hence unsecure delegations. As a resolver may send queries about any DNS name, it should be able to perform secure name resolution about any zone in any existing islands of security. The first naive solution is to configure in each resolver the apex key of all existing islands of security as trusted keys. Two major reasons prove this naive solution can not be implemented: the set of keys needed is hard to manage (the DNS tree is composed by several million of zones). And the zone keys are periodically changed in order to resist to cryptoanalysis attack and should be updated in all resolvers.

The idea discussed in this paper reduces the number of trusted keys needed in a resolver. We define a new resource record: the General Delegation Signer Resource Record (GDS RR) that is a generalization of the Delegation Signer Resource Record (DS RR) [13]. The DS RR makes the link between parent and child zones. With the GDS RR, we generalize this to a link between islands of security. This generalization allows to avoid the *gap of security* created by an unsecure zone and reduce the number of trusted keys needed for a resolver.

4   GDS Resource Record

The GDS resource record is a generalization of the the DS RR and for compatibility reasons we decide to create a new RR and to copy the DS RR format. This allows to keep compatibility with the current validation process implemented in DNS software. Old resolvers that understand only DS RR can continue to validate RRs, even when GDS RRs are present. The management of the GDS RR is slightly different from the DS management. The GDS resource record contains the same four fields than these contained in a DS RR: key tag, algorithm, digest type and the digest of a public key. The key tag is an identifier of the public key. The algorithm field is the algorithm of the key. The digest type identifies the digest algorithm used. The digest field contains the digest of the DNSKEY RR. This format allows concise representation of the keys that the secure descendant will use, thus keeping down the size of the answer for the delegation, reducing the probability of DNS message overflow.

4.1   Management

In the DS management the only involved entities are one zone and its parent zone. Figure 3 shows the current steps to create a DS RR when it is possible. When a zone creates new keys (the zone becomes secured or changes a subset of its keys), this zone notifies its parent zone in order to create DS RRs for these keys. If the parent zone is secure, parent and child zones exchange data needed for the DS RR creation(signed DNSKEY RR). The parent zone checks the received material and then stores the new DS RR in its zone file. A secure link is created between parent and child zones. The delegation signer model is limited to a direct link between a parent zone and one of its child zone.

In the previous algorithm, the requirement for a zone to have a DS RR referencing one of its keys is to have a secure parent. When the parent zone is unsecure it does not store any DS RR for its secure child zones. Consequently, its secure child zones are apex of islands of security. The keys of these zones must be configured as trusted by resolvers that want to perform secure name resolution. The GDS RR solves this problem of additional trusted keys configuration. If a secure zone does not have a secure parent but has a closest secure ancestor (CSA), a GDS RR stored in the CSA creates a secure link between the CSA and the secure zone. GDS RRs stored in the CSA zone file allow to authenticate keys of the secure zone and hence no additional trusted keys configuration is needed in resolvers. The authentication of keys is the same as presented in figure 2 for DS RR. A resolver sends a query about GDS RR identifying the key. It verifies the signatures of the DNSKEY RR and of the GDS RR. Finally, it verifies that the GDS RR identifies the key contained in the DNSKEY RR.

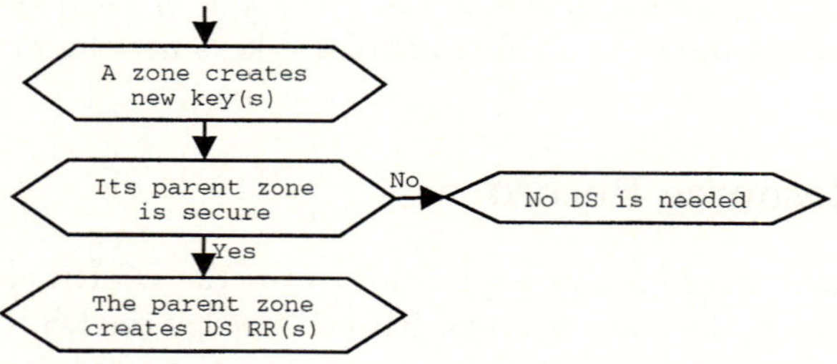

Fig. 3. The current steps of DS creation

A modification of the DS creation algorithm is needed for the creation of the GDS RR when a secure zone creates keys. Two operations on keys can imply the creation of GDS RR: a key rollover in a zone or a unsecure zone becomes secure. Figure 4 shows the steps to follow during the creation of GDS RRs.

The first steps of this algorithm are similar to the current algorithm, because when a DS RR authenticates a key of a given zone, no GDS RR is needed for this key. If the zone that have created new key does not have a secure parent, this secure zone must search its closest secure ancestor. Once, the zone finds its CSA (queries about DNSKEY RR on ancestor zones are sufficient to decide if an ancestor zone is secure or not), the zone exchanges its public keys with its CSA to create GDS RR(s) in the CSA zone file. For the creation of GDS RR in case of a zone key rollover, the previous steps are sufficient (see subsection 4.2).

In case of the creation of a new secure zone, to keep the unicity of secure chain, the CSA must transmit to the new secure zone all the GDS RRs of the zones which are in the subdomain of the new secure zone. When the new secure zone receives these informations from its CSA, the new secure zone can create the secure delegation for zones in its subdomain.

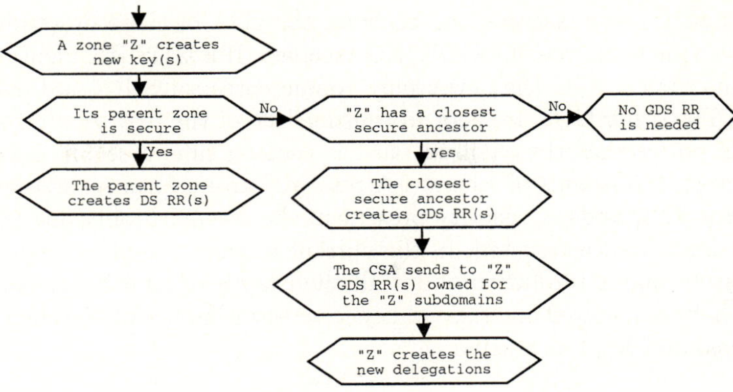

Fig. 4. The steps of GDS creation

The new secure zone has to send queries to its child zones only to verify that RRs are effectively present. No other transactions are needed, because all data needed for the creation of secure delegation are sent by its CSA.

4.2   Proof of Concept

During a key rollover. The current key management in DNSSEC implies that when a zone changes one of its key signing key (KSK), the DS RR identifying this KSK must be updated to keep the chain of trust safe [14]. This management is the same for the GDS RR. When a zone having a CSA, changes one of its KSK the GDS RR identifying this key must be updated. So, the secure zone must contact its CSA to update the GDS RR. The only change to make in the CSA zone is the update of the GDS RR. The CSA does not have to transmit GDS RR, because nothing is changed in the DNS-tree topology.

A zone becomes secure. When an unsecure zone becomes secure GDS RRs must be created, the CSA of the zone should manage this set of GDS RRs. Figure 5 shows the transmission of the GDS RRs between the CSA and the new secure zone. Firstly, the new secure zone finds its CSA, and provides its keys. Then, the CSA creates GDS RRs for the keys of the new secure zone. Finally, the CSA transmits the GDS RR it owns for the descendants of the new secure

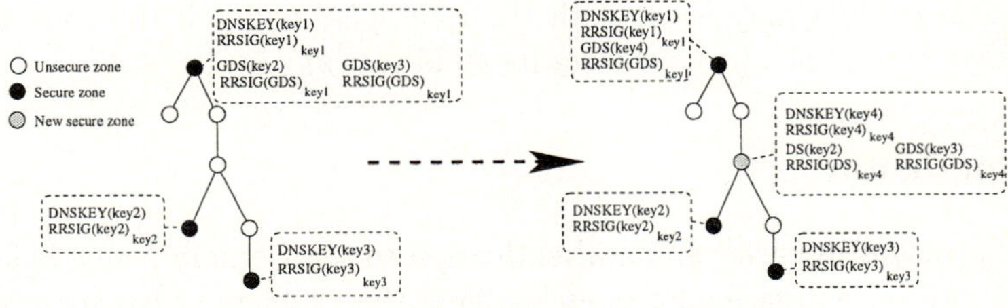

Fig. 5. The GDS RRs update and transmission

zone, because the new secure zone becomes the CSA of these descendant. The new secure zone examines the GDS RRs received. If a GDS RR identifies a key of one of its direct child, the new secure zone creates a DS RR and deletes the GDS RR. The other GDS are kept in the zone file of the new secure zone.

We can notice that the chain of trust is keeping safe, a secure path always exists between the islands of security. A resolver can perform secure name resolution about all islands of security present in the tree with only one key of the root zone configured as trusted. In the current model, a resolver that wants to perform secure name resolution about all the islands of security present in the tree must have configured one trusted key for every islands (for instance in figure 5 *key*1, *key*3 and *key*4 as trusted.

5   Pros and Cons

The proposition described in this paper solves the scalability problem caused by the large number of trusted keys needed in a resolver in order to perform secure name resolutions. With GDS, only the keys of the apex of the upper islands of security have to be configured in the resolver. With this proposition the number of trusted keys in a resolver is easily manageable. Consequently, the incremental deployment of DNSSEC is not anymore a problem to its early use. GDS has been implemented during the IDsA project [15]. This implementation consists on a modification of BIND software [16]. The modified resolver implemented during the IDsA project is able to validate resource records trusting only the root key and following the secure delegations. This resolver validates the secure link existing between islands of security by verifying the GDS RRs.

Moreover, as GDS does not change anything in the existing DNSSEC specification (it does not change the management of existing RRs), it does not raise any compatibility problem. Old DNS softwares just ignore resource records they do not understand. Hence, name resolutions performed with old DNS software succeed even if GDS RRs are present in some zones. The GDS RR provides additional security between the islands of security because it emulates an entire secure tree among secure zones. For instance, experiments show that, when GDS is used, all DNS queries on a secure zone can be validated if one judicious secure entry point has been configured in every resolver despite the presence of numerous unsecure zones.

The main drawback of GDS is the additionnal records stored in the zone files. Possible conflict may occur with the zone local policy if the zone does not want to take care about (except from its child zones).

6   Conclusion

DNSSEC provides integrity and authentication to the Domain Name System. In this paper, we have described a model solving the problem of the large number of trusted keys needed in DNSSEC resolvers to perform secure name resolution. GDS RRs provide secure links between islands of security and reduce the number of trusted keys needed in a resolver. By storing GDS RRs in zone file, the trusted keys management becomes easier for the DNS administrator because of the very small number of keys needed.

Moreover, the GDS model provides a gain of security. Even without a trusted key configured for a given secure zone, a resolver can perform secure name resolution about this zone. The chain of trust is provided automatically by the GDS RRs linking the islands of security.

The GDS RR is implemented and does not raise any compatibility problem with standard DNSSEC. Its use implies a very low overhead, an easier management of the resolver trusted keys set and ensures a larger secure access to the domain name space.

References

1. Mockapetris, P.: Domain Names - Concept and Facilities. RFC 1034 (1987)
2. Mockapetris, P.: Domain Names - Implementation and Specification. RFC 1035 (1987)
3. Albitz, P., Liu, C.: DNS and BIND. fourth edn. O'Reilly & Associates, Inc., Sebastopol, CA. (2002)
4. Bellovin, S.M.: Using the Domain Name System for System Break-Ins. In: Proceedings of the fifth Usenix UNIX Security Symposium, Salt Lake City, UT (1995) 199–208
5. Schuba, C.L.: Addressing Weaknesses in the Domain Name System. Master's Thesis, Purdue University, Department of Computer Sciences (1993)
6. Atkins, D., Austein, R.: Threat Analysis Of The Domain Name System. RFC 3833 (2004)
7. Eastlake, D.: Domain Name System Security Extensions. RFC 2535 (1999)
8. Arends, R., Larson, M., Massey, D., Rose, S.: DNS Security Introduction and Requirements. Draft IETF, work in progress (2004)
9. Arends, R., Austein, R., Larson, M., Massey, D., Rose, S.: Protocol Modifications for the DNS Security Extensions. Draft IETF, work in progress (2004)
10. Arends, R., Austein, R., Larson, M., Massey, D., Rose, S.: Resource Records for the DNS Security Extensions. Draft IETF, work in progress (2004)
11. Gieben, R.: Chain of Trust. Master's Thesis, NLnet Labs (2001)
12. Kolkman, O., Schlyter, J., Lewis, E.: Domain Name System KEY (DNSKEY) Resource Record (RR) Secure Entry Point (SEP) Flag. RFC 3757 (2004)
13. Gundmundsson, O.: Delegation Signer Resource Record. RFC 3658 (2003)
14. Guette, G., Courtay, O.: KRO: A Key RollOver Algorithm for DNSSEC. In: International Conference on Information and Communication (ICICT'03). (2003)
15. IDsA: Infrastructure DNSSEC et ses Applications. http://www.idsa.prd.fr (2004)
16. ISC: Berkeley Internet Naming Daemon. http://www.isc.org (2004)

Secure Initialization Vector Transmission on IP Security

Yoon-Jung Rhee

Department of Computer Science and Statistics,
Cheju National University, Jeju-si, Jeju-do, Korea
`rheeyj@cheju.ac.kr`

Abstract. IPSec is a security protocol suite that provides encryption and authentication for IP messages at the network layer of the Internet. ESP Protocol that is one of the two major protocols of IPSec offers encryption as well as optional authentication and integrity of IP Payloads. IV attacks are a security risk of the CBC encryption mode of block ciphers that can be applied to IPSec. We propose methods for protection against IV attack, which provide IV encapsulation and authentication of ESP payloads.

1   Introduction

Recently, importance and interest about data transmission through network is increasing. Security and reliability for network are weighed according as mass data is transmitted and important data that security is necessary to on-line is passed much through public network. Accordingly, protocol and method that encrypt packet or supply security to protect transferred information safely are used much.

Internet Protocol Security (IPSec) is protocol that provides IP and the higher level (e.g., UDP, TCP) with security. IPSec offers function of encryption, authentication and integrity etc. for message, when packets are exchanged in network layer of the public Internet [2]. There are two main protocols in IPSec; Authentication Header (AH) that offer authentication and integrity for packet, and Encapsulating Security Payload (ESP) that provide encryption as well as authentication [3, 6].

As IPSec supplies Security service in data transmission, an attempt to attack IPSec in middle happens much [4]. Initialization Vector (IV) is initial value used in encrypting data in CBC (Cipher Block Changing) encryption in ESP [5].

ESP's first packet is encrypted through XOR with IV. Receiver decrypts the first ESP packet using IV. IV is included in payload of ESP packet but is transmitted with no-encryption and revealed because it is to be used in decryption by receiver. Therefore, there are lots of dangers of attack during transmission. Decryption in receiver is impossible if IV's value is changed anything, and serious problem of that is changed to information of higher level. Such problem becomes big vulnerable point in IPSec that provide security of data and encryption. Therefore, many studies to defend IV attack that happen altering IV like this are proceeding.

In this paper, present method to prevent IV attack in IPSec. We use Algorithm that apply Electronic Code Book (ECB) mode to original IV that transfer in revealed

P. Lorenz and P. Dini (Eds.): ICN 2005, LNCS 3421, pp. 852–858, 2005.

status, and apply message authentication function for ESP whole payload including IV. After receiver receives packet, it can check integrity for packets through this function.

Composition of this paper is as following. In section 2, we analyze related researches in ESP header, IV attack and existing IV attack prevention. In section 3, we propose new IV attack protection algorithm using ECB encryption against this kind of IV attack and presents performance evaluation of our algorithm and results after comparing other researches. Section 4 shows conclusion.

2   Related Works

In this section, we examine presenting researches related in IV attack.

2.1   Encapsulating Security Payload (ESP)

Two major protocol of IPSec is AH and ESP. AH offers authentication service and integrity service to IP payload, and ESP provides security service as well as optional authentication service [2]. ESP uses symmetric key algorithm (e.g., DES, SKIPJACK, AES, BLOWFISH) for encryption of IP payload. If sender transmits encrypted payloads, receiver decrypts them [3].

IP payload is transferred safely using CBC block cipher. However IV is transferred publicly without any encryption and authentication. At this step, lots of attacks are possible. If some of IV value is changed by attacker in the middle of IP packet transmission, contents of higher-level data are changed according to structure of protocol, as well as receiver does not decrypt the packet.

Therefore receiver does not trust the transferred data. Furthermore it is possible that attacker's wrong data is used. Because IPSec is protocol that offer integrity and confidentiality service, IV attack against IPSec cause more serious problems. To solve these problems, there are various studies about defenses against IV attack [1].

Figure 1 is structure of ESP header. It appears that IV is included inside ESP Header.

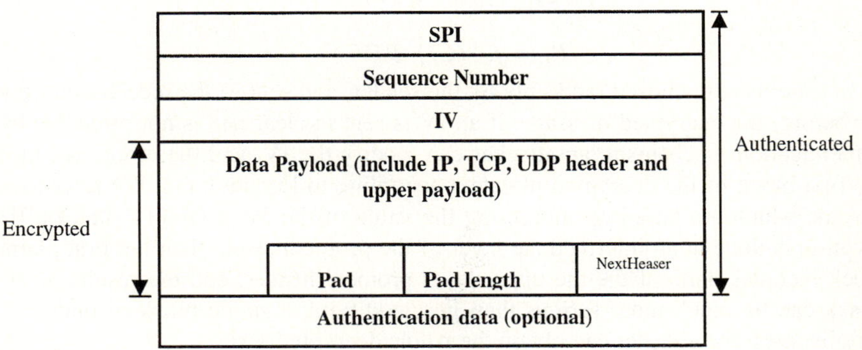

Fig. 1. ESP Header

2.2 IKE (Internet Key Exchange)

IKE is automatic method for Key exchange and various elements used in AH and ESP encapsulation. IKE is consisted of two steps and negotiates various factors such as encryption algorithm, transmission form, key values and lifetime of key etc. for AH or ESP encapsulation between two communication individual before sending packet. In the first step, the Main Mode or the Aggressive Mode perform defining and establishing environments between communication individual, and exchanging key through the Internet Security Association Key Management Protocol (ISAKMP) [9,10].

If the first step completes, the second step, the Quick Mode, is begun for SA negotiation process for data transmit in IPSec. In the step, negotiation for necessary authentication type, algorithm and key etc. is completed. After that data transmission is started.

2.3 Initialization Vector Attack

IV attack is danger on the Cipher Block Chaining (CBC) mode of block ciphers in IPSec. Attacker in the middle of transmission can change IV value. Therefore, it is impossible for receiver to decrypt data, or problems such that original data is changed may occur.

In CBC mode the data stream to be encrypted is divided into as plaintext blocks $P1, P2, ...,$ and each plaintext block is encrypted as

$$C_i = f_k (P_i \oplus C_{i-1})$$

and decrypted as following

$$P_i = f_k^{-1} (C_i) \oplus C_{i-1}$$

where f_K and f'_K denote the block cipher encryption and decryption, and denotes the bit-wise exclusive-or. For the special case of the first block, where there is no ciphertext C_0 to xor, an initialization vector (IV) is used instead:

$$C_1 = f_k (P_1 \oplus IV)$$

$$P_1 = f_k^{-1} (C_1) \oplus IV$$

(1)

IVs are usually chosen randomly by the sender, and sent to the receiver along with (or before) the encrypted message. If an IV is sent in clear and is not protected by an authentication mechanism, an attacker can modify the IV, and therefore, can modify the first block of the decrypted plaintext according to Equation (1). We refer to such attacks, which are based on modifying the value of the IV, as the IV attacks. If encryption is done at an intermediate layer of the protocol stack, then the first plaintext block includes parts of the the upper-layer protocol header, and the results of an IV attack can be much more serious than just modifying a single block of plaintext. In certain cases, the attacker can obtain the whole decrypted text.

IV attacks can be a serious problem for IPsec encryption. IPsec has many possible configurations that enable IV attacks. First, all the IPsec encryption algorithms proposed so far use a block cipher in CBC mode. Moreover, almost all of these algorithms allow the use of clear text unauthenticated IVs. Also in the ESP RFC [5], it is mentioned that the IV may be carried explicitly in the Payload field and usually is not encrypted. Although [5] recommends always using authentication whenever encryption is used, which would prevent the IV attacks, authentication with ESP encryption is neither mandatory nor is it the default mode. Therefore, in practice, IPsec encryption can be very vulnerable to IV attacks. The exact results an attacker can obtain will be dependent on the protocol encapsulated

As several problems above, some of methods to protect IV were suggested. We also present new algorithm to defend IV attack.

2.4   Existing Researches for Prevention of IV Attacks

Researches against IV attack are studied to several directions as follows. First, IV can be protected by sending a value that is hashed by one-way hash function. Receiver can find out if the IV in transmitted packet is modified. Its limit is that we can know only integrity of IV value [1].

The second, it is to change IV value variously. CISCO researches method changing IV value to 4-bit or 8-bit whenever transmitting packet. In this case, initial IV value is set by 8-bit and it is optional to change to 4-bit value. Appointment for the size of IV value should be established between sender and receiver before sending data packet. Its defect is that it can use only two lengths, 4-bit or 8-bit, up to now [2].

The third, modification of the IV can be totally prevented by using a constant agreed-upon IV, such as 0 [1]. It is just alternative way.

3   Proposed Algorithm for Sending Protected IV

In this section, we present new methods that prevent IV attack. We use an algorithm that offers ECB mode encryption of block cipher to IV value and uses message authentication in ESP header value basically. It achieves integrity check of ESP payload and confidentiality of IV value at the same time.

3.1   IV Encryption with ECB Mode Encryption of Block Cipher

The ECB is the simplest encryption mode of block cipher and need not additional IV. In ECB, the data stream to be encrypted is divided into as plaintext blocks $P1, P2, ...,$ and, in contrast to CBC, each plaintext block is encrypted as

$$C_i = f_k(P_i)$$

and decrypted as

$$P_i = f_k^{-1}(C_i)$$

We use ECB mode to encrypt IV value as (2) and decrypted as (3).

$$C_{IV} = f_k(P_{IV}) \tag{2}$$

$$P_{IV} = f_k^{-1}(C_{IV}) \tag{3}$$

The Encrypted IV value is transferred with being inserted in IV field of ESP header. If it is done, the attacker can't see original IV value as IV is encapsulated.

ECB mode also need key agreed between two communication entities when encrypt IV like other block cipher algorithms. If this key for IV ECB encryption should be decided additionally during IKE's Quick Mode, there'll be additional overhead. To avoid the overhead, we use the same key for CBC mod for encryption of ESP payloads.

3.2 Message Authentication for ESP Whole Payload

We provide whole ESP payloads with integrity check using message authentication algorithm such as HMAC-SHA1, HMAC-MD5 etc.

SA with various negotiated elements for secure transmission of IP payloads is decided during IKE's Quick Mode. SA negotiation is achieved If initiator send message with various suggestions and then responder answer to initiator by message that include values for proper algorithms and encryption key. We add the steps to negotiate elements for authentication of ESP payload to SA. After that, The SA is stored in SA database (SADB) that maintains the SAs used for outbound and inbound processing.

The SADB entry contains the active SA entries. Each SA entry is indexed by a triplet consisting of an Security Parameter Index (SPI), a source or destination IP address, and an IPSec protocol. An SADB entry consists of the various fields: Sequence number counter, Sequence counter overflow, Anti-replay window, ESP encryption algorithm, key, IV, and IV mode etc. In proposed method, an algorithm for Message Authentication and ECB mode for IV encryption; to encrypt IV, our algorithm use the same symmetric algorithm as encryption algorithm for ESP payload but different mode from ESP's.

3.3 Proposed ESP Processing

Figure 2 displays new ESP header that IV encryption with ECB and message authentication is added to. Through this, IV's confidentiality and integrity of ESP whole message including SPI, Sequence Number, IV, and Data Payload etc. is achieved.

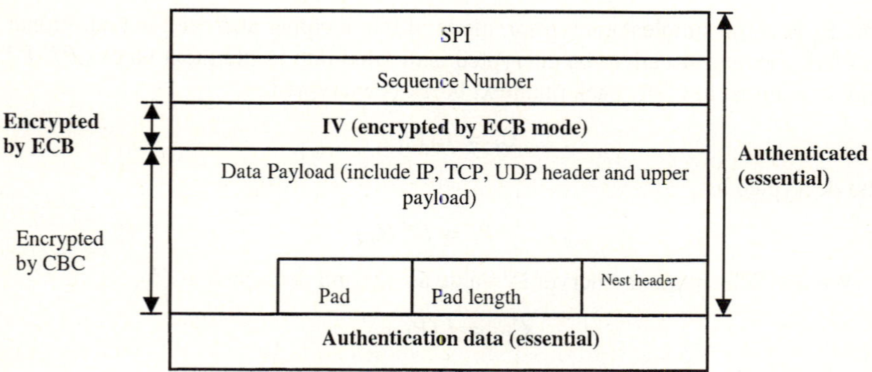

Fig. 2. An IP Packet Protected by New ESP

Outbound processing in new ESP are as following.

1. Proper SA Lookup with SPI in SADB. If no the SA, establish new SA through IKE.
2. Generate or increment Sequence Number.
3. CBC Encryption of the packet.
4. ECB Encryption of IV.
5. Calculate the integrity check value for message authentication.
6. (if necessary) Data fragmentation.
7. Send Packet.

Inbound processing are as following.

1. Receive Packet.
2. SA Lookup with SPI. If no the SA, discard the packet.
3. (if necessary) Data reassemble.
4. Sequence Number and Verification.
5. Message Authentication.
6. ECB decryption of IV.
7. CBC decryption of the packet.

3.4   Performance Evaluation of Proposed ESP Algorithm

In this paper, we suggested solution that prevents attack about IV that is one of problem of existing IPSec. Proposed algorithm proposes two methods, IV encryption and ESP message authentication. In it, we use ECB encryption mode for IV encryption the same key for CBC encryption of ESP payload, which is simple and not necessary additional overhead. In ECB, the same cipher-text blocks are created if the same plain-text blocks are encrypted, then attacker may find out the original texts if he have many kinds of same cipher-text blocks. In this case, However, IV encryption block is just one, the attacker can't guess original IV value. Therefore it provides confidentiality of IV. We also offer message authentication for integrity check to whole ESP payload.

Table 1. Comparison proposed new ESP algorithm with existing studies

	One-way HASH function	IV size changing	IV agreement like O	Proposed Algorithm
Man-in-the-middle attack	Not protect	Not protect	Not protect	Protect
Replay attack	Not protect	Not protect	Not protect	Protect
Integrity check	Possible	Not Possible	Possible	Possible
Confidentiality	Not possible	Not possible	Not possible	Possible

In table 1, we compared our proposed ESP algorithm with existing studies in terms of man-in-the-middle attack, replay attack, integrity check and confidentiality.

4 Conclusion

We showed that IV attacks could be a serious threat for IPSec, if IPSec is not used carefully. In the absence of secure way, an attacker may be able to totally defeat the IPSec encryption. To protect IPSec more securely, we proposed new ESP algorithm that use IV ECB encryption and message authentication in this paper. It is able to satisfy various security requirements such as Integrity check and Confidentiality.

References

1. Chistopher B. Macubbin and Ali Aydin Selcuk, Initialization Vector Attacks on the IPSec Protocol Suite, IEEE Trans. Commun., vol. 17, No. 6. June 2000
2. Naganand Doraswamy and Dan Harkins, IPSec the New Security Standard for the Internet, Intranets, and Virtual Private Networks, Prentice Hall. Networking, vol 4, pp. 885-901, December 1996
3. S. Kent and R. Alkinson, IP Encapsulation Security Payload (ESP), Internet RFC 2406, November 1998.
4. Steven M. Bellovin, "Problem Areas for the IP Security Protocols, 1996.
5. P. Karn, P. Metsger and W. Simpson, The ESP DES-CBC Transform, Internet RFC 1892, August 1995.
6. S. Kent and R. Atkinson, IP Authentication Header, Internet RFC 2402, November 1998.
7. Bruce Shneier, Applied Cryptography Second Edition, John Weliy & Son, 1996.
8. D. Harkins, D. Carrel, The Internet Key Exchange, Internet RFC 2409, November 1998.
9. S. Kent and R. Atkinson, Security Architecture for the Internet Protocol, Internet RFC 2401. November. 1998
10. V. Voydock and S. Kent, Security Mechanisms in High Level Network Protocols, ACM Computing Surveys, 15(2), June 1983.

Multicast Receiver Mobility over Mobile IP Networks Based on Forwarding Router Discovery

Takeshi Takahashi[1], Koichi Asatani[1,2], and Hideyoshi Tominaga[1]

[1] Graduate School of Global Information and Telecommunication Studies,
Waseda University, Tokyo, 169-0051 Japan
`take@tom.comm.waseda.ac.jp`
[2] Graduate School of Electrical and Electronic Engineering,
Kogakuin University, Tokyo, 163-8677 Japan

Abstract. To support mobility over Mobile IP networks, the IETF proposed two schemes, namely bi-directional tunneling (MIP-BT) and remote subscription (MIP-RS). Although MIP-BT enables mobility by establishing bi-directional tunnel from home agent to care-of address, it suffers from redundant routing that causes bandwidth consumption and packet delivery delay. On the other hand, in MIP-RS, mobile node always re-subscribes to its interested multicast group when it enters into a foreign network. Although MIP-RS provides shortest multicast routing path, it suffers from out-of-synch problem. To cope with the problem, we propose new scheme that provides the shortest multicast routing path without out-of-synch problem, and that also minimizes the bandwidth consumption. We evaluate our proposed scheme in the viewpoint of bandwidth consumptions with various parameters and clarify the efficiency of our proposed scheme.

Keywords: Multicast, Mobile IP, Handover, Mobility.

1   Introduction

Recently, multimedia content delivery services such as streaming have been becoming very popular due to the development of the network infrastructure. For these services, multicast is one of the required technologies to efficiently provide one-to-many communication. On the other hand, the Internet Engineering Task Force (IETF) standardized Mobile IP version 6 (Mobile IPv6) [1,2] that supports mobility over IP networks in this June. Hence, plenty of multimedia services will be provided over Mobile IPv6 in the near future. However, the current multicast protocols such as DVMRP [3], MOSPF [4], CBT [5], and PIM [6] are not designed for MNs. Therefore, further protocol is required for efficient multicast support over Mobile IP networks.

To support multicast over Mobile IP networks, the IETF proposed two different schemes, namely bi-directional tunneling (MIP-BT) and remote subscription (MIP-RS). In MIP-BT, MN subscribes multicast groups through its HA. When the MN is visiting foreign network, it establishes bi-directional tunnel with its

P. Lorenz and P. Dini (Eds.): ICN 2005, LNCS 3421, pp. 859–867, 2005.

HA. All the multicast packets will be routed to the HA and will be forwarded to the MN through the bi-directional tunnel. Although this scheme provides mobility for multicast over IP networks, it suffers from redundant routing that causes excess bandwidth consumption. Moreover, this scheme suffers from duplicate packet forwarding. Assuming multiple MNs visiting one same foreign network, all of them will receive the same multicast traffic from their HAs via their tunnels even though they are in the same network. Although the advantage of multicast is to suppress duplicate traffic to the same network, MIP-BT suffers from duplicate traffic forwarding due to each MN's individual tunneling feature.

In MIP-RS, MN subscribes multicast group when it visits foreign network, and it receives multicast packets directly from the closest multicast router without any tunneling. Although this scheme provides shortest multicast routing path, it causes several drawbacks. Since the multicast packet delivery is not synchronized, when MN moves from one multicast router to another, the traffic will be discontinuous, i.e. out-of-synch problem. Moreover, if the visiting network is not the member of multicast group, the AR is required to join and establish the multicast tree.

To cope with these problems, we propose a new scheme to handle handover for multicast receiver so that packet delivery delay and bandwidth consumption will be minimized. Our proposed scheme consists of forwarding router (FwR) discovery and proactive handover. The former enables MN to utilize FwR located on the multicast tree between previous access router (previous AR, PAR) and new AR (NAR) regardless of the PAR's unawareness of the network topology while the latter enhances the handover performance with buffering and packet forwarding at FwR. Here, the FwRs are chosen for each multicast session individually. Moreover, our proposed scheme is compatible with Fast Handovers for Mobile IPv6 (FMIPv6) [7] with proper enhancement. In evaluation, we evaluate our proposed scheme in the viewpoint of bandwidth consumption with various parameters, and clarify the efficiency of our proposed scheme.

2   Related Works

To support multicast over Mobile IP, plenty of researches have been proposed until now. One major approach is the extensions to MIP-BT that suffers from duplicate traffic forwarding problem. To cope with this duplication issue, Mobile Multicast Protocol (MoM) [8] was proposed. MoM chooses designated multicast service provider (DMSP) among HAs that forward the same traffic. By choosing single DMSP, the packet duplication can be avoided. However, MoM lefts some problems such as DMSP handover [8]. Moreover, due to the feature of packet forwarding at HA, MoM still suffers from redundant routing that causes packet delivery delay and excess bandwidth consumption. RBMoM [9, 10] can be also seen as an extensions to MIP-BT. It provides the trade-off between the shortest delivery path and the frequency of the multicast tree reconstruction by controlling the service range of the multicast HA. In this scheme, MIP-RS and MIP-BT are described just as the extreme cases of the scheme. Hence, this scheme inherits the pros and cons of MIP-BT and MIP-RS.

Another major approach is the extensions to MIP-RS that suffers from out-of-synch problem and multicast tree reconstruction problem. To cope with those issues, several schemes [11,12,13,14] has been proposed. In MMROP [11], MN joins multicast group in visiting network as in MIP-RS. However, due to out-of-synch problem, some packets will be lost during the handover process. To compensate the lost packets, the MN's mobility agent forwards the requested packets via tunnel. Different from MIP-BT, the use of tunnel is limited. In case of hierarchical network structure as in HMIP [15], more efficient solution is provided in [12], where the authors propose that, different from MMROP, MAPs [15] perform the packet buffering/forwarding. In SMM [13], authors achieved to keep synchronization regardless of the location by measuring RTT and buffering proper amount of packets in transit routers. In [14], MN establishes tunnel with remote multicast agent if the estimated staying time in visiting network is very short while otherwise it receives multicast traffic from local multicast agent so that the multicast tree reconstruction cost can be suppressed. Although those proposals are effective to solve the out-of-synch problem and multicast tree reconstruction problem of MIP-RS, they still suffer from redundant routing during the handover process by forwarding packets from HA.

In this paper, we propose new scheme that solves out-of-synch problem, and that minimizes redundant routing during the handover process as well.

3   Proposed Scheme

Our proposed scheme introduces FwR that helps handover of MNs by performing packet buffering and packet forwarding. Since it is infeasible to assume that all routers have our protocol specific feature, our proposed scheme is designed so that it can cooperate with routers of non-protocol specific features. Here, we term router with protocol specific features as FwR candidate.

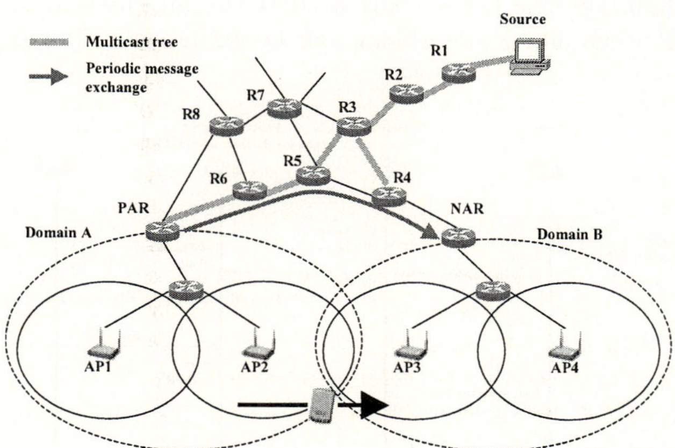

Fig. 1. Concept of our proposed scheme

Figure 1 describes the concept of our proposed scheme. The heavy line in the network describes the multicast tree while the thin line in the network simply describes the link connection between nodes. By inspecting the FwR candidates en route from PAR to NAR, our proposed scheme searches FwR candidate that is already in the part of multicast tree. In Fig. 1, by inspecting routers en route from PAR to NAR whether they are subscribing the interested multicast group or not, PAR will find R4 as the best packet buffering/forwarding point provided R4 is FwR candidate. Since R4 is already in the multicast tree and is closer to NAR compared to HA/PAR, setting R4 as the point of packet buffering/forwarding is more efficient than setting HA or PAR as the point. If R4 is not FwR candidate and R5 is FwR candidate, R5 is chosen as the packet buffering/forwarding point. Although this is not the best point for packet forwarding, it is still sub-optimal point and is more efficient than HA and PAR.

Our proposed scheme consists of the FwR discovery and proactive handover. We elaborate them in the following sections.

3.1   FwR Discovery with Buffering Request

To achieve the concept described in Fig. 1, our proposed scheme utilizes buffering request (bufReq) message. In our proposed scheme, FwR discovery is conducted with buffering request as described in Fig. 2. Buffering request starts when MN realizes the possibility of handover to another network by receiving L2 triggers though the details of the trigger is outside the scope of this paper. Then the MN sends bufReq message to PAR with the access point (AP) identifier of the next AP in visiting network and the subscribing multicast group address. The message may contain some index information such as sequence number of the last received packet so that the FwR can efficiently start buffering though the details are outside the scope of this paper. Upon receiving the message, the PAR looks up its local database to obtain the IP address of NAR from the AP identifier specified in the bufReq message. Note that we assume each AR communicates with its neighboring ARs periodically so that the information of APs belonging to the AR, such as AP identifiers, can be exchanged [7], and so that ARs

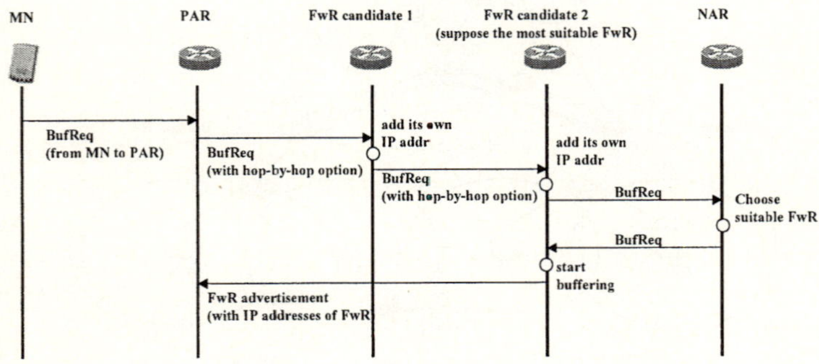

Fig. 2. Buffering request and FwR discovery

can create local database. Then the PAR sends the bufReq message to NAR with hop-by-hop option specified in IPv6 [16]. The hop-by-hop option includes FwR discovery option that can be interpreted by FwR candidates, and all FwR candidates are mandated to handle FwR discovery option. If an FwR candidate receives the message, it checks whether it is subscribing the queried multicast group. It appends its own IP address inside the bufReq message provided it is already subscribing the queried multicast group. Then it forwards the packet to the next hop. Note that the packet was sent to NAR, and FwR candidates are merely transit routers and transacting only the hop-by-hop option. Upon receiving the bufReq message, the NAR analyzes the message and chooses the closest active multicast router as FwR, then sends bufReq message to the chosen FwR. Upon receiving the bufReq message, the chosen FwR sends FwR advertisement message to PAR and starts buffering multicast packets.

In case that multiple MNs belonging to different PARs are preparing handover to one same network, the NAR chooses single FwR for all the PARs. Since the routes from PARs to NAR is not usually identical, it is very possible that the NAR chooses different FwRs if it chooses FwRs individually from the FwR candidates en route from themselves to NAR. When NAR receives bufReq message, it usually choose FwR from the FwR candidates en route from MN to NAR as described in Fig. 2. However, if the NAR has already known FwR for another PAR, it needs to appoint single FwR since newly discovered FwR candidate is most likely different from the already appointed FwR. If the already appointed FwR is better than the newly discovered FwR candidate, the NAR appoints the already appointed FwR as a common FwR for all the PARs. On the other hand, if the newly discovered FwR candidate is better than the already appointed FwR, the NAR appoints the FwR candidate as a common FwR for the PARs. Then, the appointed FwR must inform its address to all its serving PARs.

3.2   Proactive Handover with Forwarding Request

When the FwR discovery procedure with buffering request is completed, the FwR has already started buffering packets sent from the interested multicast group and is waiting for the request for packet forwarding. Figure 3 describes the packet forwarding procedure.

Before disconnecting the link connection, the MN that is about to handover sends forwarding request (fwReq) message to its PAR. The fwReq message contains AP identifier of next AP so that the PAR can retrieve the IP address of NAR from the PAR's own local database. Upon receiving the fwReq message, the PAR forwards the message to the FwR that was chosen in section 3.1. Then the FwR starts forwarding multicast packets preceded by the buffered packets to NAR. Upon receiving the forwarded packets from FwR, the NAR stores those packets in its local buffer until it realizes the existence of the MN under its administrative network. When MN moves into the new network and configured new CoA (NCoA), it sends fwReq message to the NAR, which in return starts forwarding packets sent from FwR. Note that ARs can recognize the differences of the fwReq messages by checking whether a received fwReq message has AP

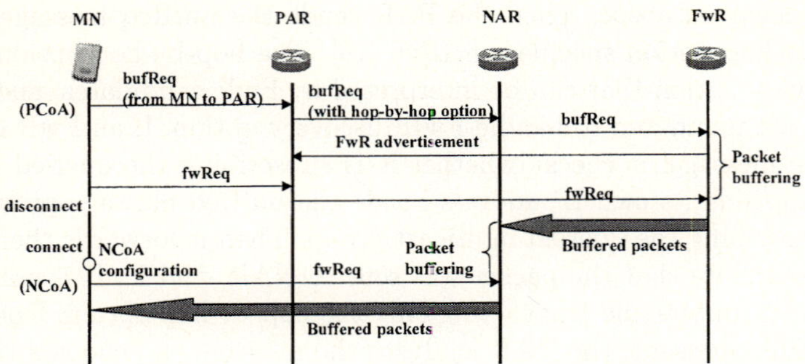

Fig. 3. Packet forwarding with fwReq before handover

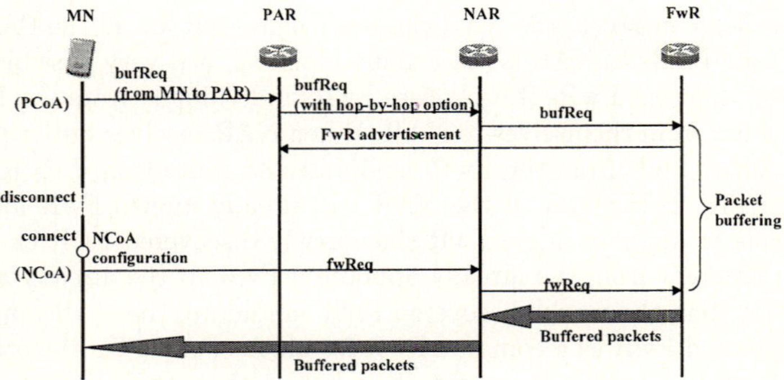

Fig. 4. Packet forwarding without fwReq before handover

identifier inside or not. Simultaneously, the MN starts subscribing to the interested multicast group from its current network, as is specified in MIP-RS.

In Fig. 3, we assumed that the MN can send fwReq message before disconnecting the link connection. However, as described in FMIP [7], MN sometimes cannot notice the timing of the link disconnection. In this case, the MN cannot send fwReq message before the link disconnection. Hence, the protocol procedure will be slightly changing as described in Fig. 4.

Upon entering new network and configuring NCoA, the MN sends fwReq message to NAR, which in return forwards the fwReq message to the FwR that was chosen in section 3.1. Then the FwR starts forwarding packets to NAR, which immediately forwards the packets to MN. Although the behaviors of PAR and NAR are slightly different in case of Fig. 3 and Fig. 4, PAR and NAR are required to handle both scenarios. Note that MN is not required to know the existence of FwR in our proposed scheme and is only required to communicate with ARs.

FwR and NAR stop packet forwarding when timeout expires. They also stop packet forwarding when lost packet compensation has completed and when NAR starts receiving multicast traffic. Then, they immediately terminate the tunneling and stop packet forwarding to MN. If the NAR has been already in the interested multicast group, since it is FwR in this case, it stops packet forwarding after the lost packet compensation.

Consequently, the MN will receive only the traffic from the multicast tree and is able to avoid duplicate traffic delivery. Assuming there are several effective FwRs in the network, our proposed scheme is simple and efficient.

4 Evaluation

In this section, we evaluate our proposed scheme in the viewpoint of bandwidth consumption. The simulation network consists of 5×5 ARs described in Fig. 5 that is derived based on Prim's algorithm and is utilized in [10]. For this evaluation, we prepared time-discrete model. Each AR has its own local area network with single AP for MNs. In this simulation, we compared our proposed scheme with pure MIP-RS and conventional scheme. Here, we call the MIP-RS with temporal packet forwarding from single DMSP [8] as conventional scheme.

Now, we put N nodes randomly in this simulation network as an initial state and let the nodes move randomly during the simulation period. Here we define λ as the time [second] that one MN stays in one network in average. If we denote the probability of handover for each second as ρ, ρ can be calculated in $\rho = 1/\lambda$. The MN can move to the adjacent four networks randomly in each discrete time. Any MN is not allowed to handover more than one network per one discrete time.

By choosing proper value for N, λ and T, we performed simulation with Node 0 as the single multicast source. Figure 6 shows the result when $N = 50, \lambda = 10$, and $T = 2$. X-axis shows the simulation time (between 100 second and 200 second after simulation starts) while Y-axis shows the bandwidth consumption. As can be seen, the proposed scheme has distinct advantage compared to the conventional scheme. Compared to the MIP-RS, proposed scheme shows almost the same performance in the viewpoint of bandwidth consumption without suffering from out-of-synch problem. That is because the distance from NAR to the closest multicast router, i.e. chosen FwR, is usually less than 1, or the NAR itself is already in the multicast tree. Although there are some proposals that let the NAR join the multicast tree before handover occurs, they are not desirable since MN sometimes has several candidate handover networks. Under this circumstance, some ARs are required to join multicast trees even though no

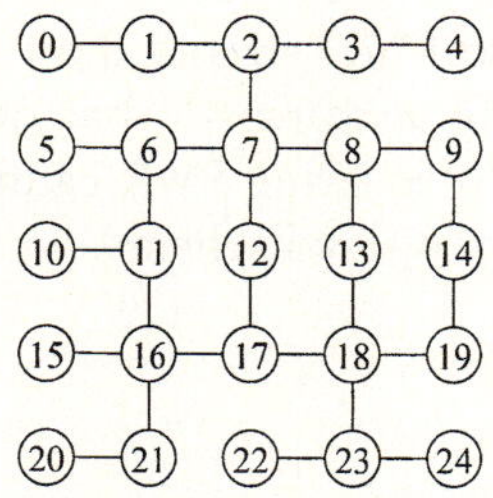

Fig. 5. Simulation topology

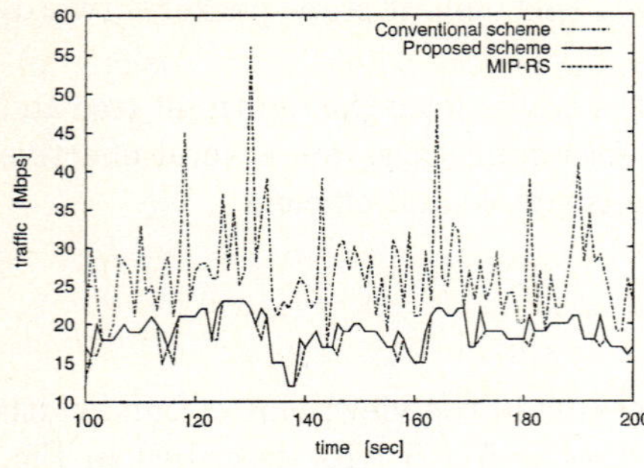
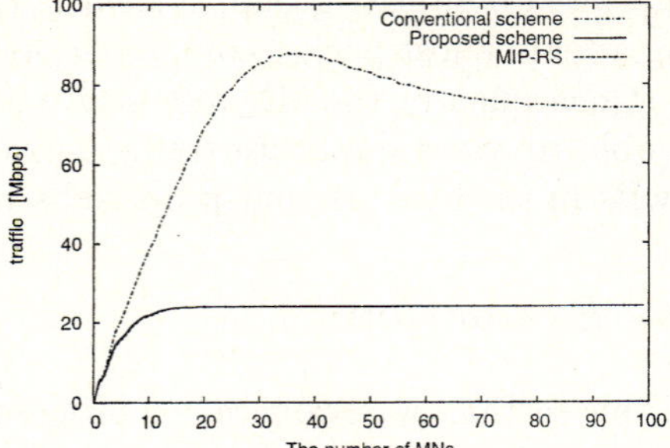

Fig. 6. Bandwidth consumption (N=25, λ=10, T=2)

Fig. 7. Average bandwidth consumption (λ=10, T=2)

new MN is actually coming in. Hence, our proposed scheme reconstructs only the necessary number of the multicast branches, and unnecessary multicast tree reconstructions are suppressed.

Figure 7 shows the average bandwidth consumption when $\lambda = 100$ and $T = 2$, and the value of N changes from 0 to 100. X-axis shows the number of MNs in the simulation network while the Y-axis shows the average bandwidth consumption. As can be seen, our proposed scheme has advantages over conventional scheme all the time, and our proposed scheme barely has extra bandwidth consumption over MIP-RS. Especially, when the number of MN increases, the difference between our proposed scheme and MIP-RS disappeared.

Basically, when the number of MN increases, the benefit of our proposed scheme increases since our proposed scheme is aimed to improve the performance only during the handover process. However, when the number of MN increases to some point, all of the schemes reach to saturation. In case of conventional scheme, since any NAR will have new MN coming from another AR that is just one hop away, the packet forwarding distance will be always set to 1. In case of proposed scheme, since any NAR will have been subscribing to the interested multicast group, the NAR will be the FwR without any exception. Hence, the forwarding distance will be zero, and the amount of bandwidth consumption will be the same value as the one in the case of MIP-RS.

As can be seen, our proposed scheme shows better performance over conventional scheme in the viewpoint of bandwidth consumption. Our proposed scheme can be seen as an extension to FMIPv6 and does not require all the routers to implement our protocol specific features so that our proposed scheme can be a viable solution. By strategically locating FwR candidates in the network, plenty of MNs will benefit from our proposed scheme.

5 Conclusion

To support multicast receiver mobility over Mobile IP networks, plenty of schemes have been proposed until now. However, those schemes suffer from redundant routing during the handover process. To cope with the problems, we introduced FwR that helps handover of MNs by performing packet buffering and packet forwarding. Our proposed scheme solves out-of-synch problem by performing buffering at FwR without suffering from excess bandwidth consumption. In evaluation, we clarified the efficiency of our proposed scheme over conventional schemes in the viewpoint of bandwidth consumption. Our proposed scheme can be seen as an extension to FMIPv6 and does not require all the routers to implement our protocol specific features so that our proposed scheme can be a viable solution. By strategically locating FwR candidates in the network, plenty of MNs will benefit from our proposed scheme.

References

1. D. B. Johnson, C. E. Perkins, and J. Arkko, "Mobility Support in IPv6," *RFC 3775*, June 2004.
2. J. Arkko, V. Devarapalli, and F. Dupont, "Using IPsec to Protect Mobile IPv6 Signaling Between Mobile Nodes and Home Agents," *RFC 3776*, June 2004.
3. D.Waitzman, C.Partridge, and S.Deering, "Distance Vector Multicast Routing Protocol," *RFC 1075*, November 1988.
4. J. Moy, "Multicast Extensions to OSPF," *RFC 1584*, March 1994.
5. A. Ballardie, "Core Based Trees (CBT version 2) Multicast Routing," *RFC 2189*, September 1997.
6. B. Fenner, M. Handley, H. Holbrook, and I. Kouvelas, "Protocol Independent Multicast - Sparse Mode (PIM-SM) : Protocol Specification (Revised)," *Internet Draft*, October 2003.
7. R. Koodli, "Fast Handovers for Mobile IPv6," *Internet Draft (draft-ietf-mipshop-fast-mipv6-02.txt)*, July 2004.
8. T. G. Harrison, C. L. Williamson, W. L. Mackrell, and R. B. Bunt, "Mobile multicast (MoM) protocol : multicast support for mobile hosts," *Proc. ACM/IEEE international conference on Mobile computing and networking*, 1997.
9. C. R. Lin and K.-M. Wang, "Mobile multicast support in IP networks," *Proc. IEEE INFOCOM*, 2000.
10. C. Lin, "Mobile multicast support in IP networks," *Proc. IEEE GLOBECOM*, 2002.
11. J.-R. Lai, W. Liao, M.-Y. Jiang, and C.-A. Ke, "Mobile multicast with routing optimization for recipient mobility," *Proc. IEEE International Conference on Communications*, 2001.
12. H. Omar, T. Saadawi, and M. Lee, "Multicast with reliable delivery support in the regional Mobile-IP environment," *Proc. IEEE Symposium on Computers and Communications*, July 2001.
13. I.-C. Chang and K.-S. Huang, "Synchronized multimedia multicast on mobile IP networks," *Proc. IEEE International Conference on Communications*, May 2003.
14. Y.-J. Suh, D.-H. kwon, and W.-J. Kim, "Multicast routing by mobility prediction for mobile hosts," *Proc. IEEE International Conference on Communications*, May 2003.
15. H. Soliman, C. Castelluccia, K. El-Malki, and L. Bellier, "Hierarchical Mobile IPv6 mobility management (HMIPv6)," *Internet Draft (draft-ietf-mipshop-hmipv6-02.txt)*, June 2004.
16. S. Deering and R. Hinden, "Internet Protocol, Version 6 (IPv6) Specification," *RFC 2460*, December 1998.

Secure Multicast in Micro-Mobility Environments*

Ho-Seok Kang and Young-Chul Shim

Dept. of Computer Engineering, Hongik University, Seoul, Korea
{hskang, shim}@cs.hongik.ac.kr

Abstract. To reduce the control message overhead due to frequent handoffs in mobile IP, micro-mobility protocols have been proposed. In this paper we present an approach for providing secure multicasting services in micro-mobility environments. We first introduce a multicast routing protocol in a micro-mobility environment and then present protocols to add security services to the bare multicasting. The proposed multicast routing protocol builds a shared multicast tree. The added security services include authentication, authorization, confidentiality, and integrity. The security protocols use key hierarchy mechanism in a paging area to efficiently handle frequent group membership changes and handoffs.

1 Introduction

Many applications require secure multicasting services. A secure multicast protocol must ensure that only authorized group members modify a multicast delivery tree and send/receive packets to/from a multicast group. Since cryptography is used to provide such security services, the design of an efficient key management scheme is one of the most important issues in secure multicasting.

Mobile IP is the current standard for supporting macro-mobility and provides a framework for allowing users to roam outside their home networks. Whenever a mobile node moves between neighboring base stations, its movement should be notified to its home agent. If the mobile node is highly mobile, overhead due to this registration becomes excessive. To reduce overhead in mobility management, micro-mobility protocols such as TeleMIP[2] and HAWAII[6] have been proposed. In a micro-mobility environment, a mobile network consisting of a large number of cells is called a domain. There are many domains that are connected to the Internet core. Intra-domain handoffs are handled by micro-mobility protocols and are not visible outside the domain while inter-domain handoffs are processed by macro-mobility protocols that are mobile IP.

In this paper we present an approach for providing secure multicasting services in micro-mobility environments. We first introduce a multicast routing protocol in a micro-mobility environment and then present protocols to add security services to the bare multicasting. The proposed multicast routing protocol builds

---

* This research was supported by University IT Research Center Project, Korea.

P. Lorenz and P. Dini (Eds.): ICN 2005, LNCS 3421, pp. 868–875, 2005.

a shared multicast tree and does not assume any unicast micro-mobility protocols. The added security services include authentication, authorization, confidentiality, and integrity and are based on both symmetric and asymmetric cryptosystems. The security protocols also use key hierarchy mechanism in a paging are to efficiently handle frequent group membership changes and handoffs.

The rest of the paper is organized as follows. Section 2 presents related works. Section 3 explains a multicast routing protocol in a micro-mobility environment. Section 4 presents protocols providing security services to the multicast routing protocol explained in Section 3 and is followed by the conclusion in Section 5.

2   Related Works

IETF proposes two approaches to support mobile multicast: bi-directional tunneling and remote subscription. The former suffers from the tunnel convergence problem. The latter is conceptually simple but its multicast service can be very expensive when a mobile node is highly mobile because of the difficulty in managing the multicast tree. Protocols such as MoM, RBMoM and MMA were proposed to solve these problems[3,9] but they consider only macro-mobility and need to be modified to work with micro-mobility protocols.

In a multicast group senders and receivers can join and leave a multicast group during the existence of the group. But a new member should not be able to read packets exchanged before its joins (backward secrecy) and a leaving member should not be able to read packets exchanged after its leaving (forward secrecy). Thus the message encryption key should be changed whenever there occurs a change in the group membership. There have been many works on scalable approaches for distributing new multicast keys as the group membership changes[5,10]. All the works in this category just present a scalable key distribution method for dynamic groups and provide no other multicast security services. Shields and Garcia-Luna-Aceves proposed a scalable protocol for secure multicast routing[7]. Provided security services include authentication, authorization, confidentiality, and integrity. Recently protocols for secure multicast services in mobile and wireless networks have been proposed[1]. The approach uses hierarchies of multicast keys for distributing multicast keys to a mobile and dynamic group but makes impractical assumption that multicast packets are sent out only by a stationary group manager. Our approach provides all the required security services and makes the most of the micro-mobility architecture to efficiently satisfy backward/forward secrecy in spite of frequent membership changes and handoffs.

3   Multicast Routing Protocol

As in Figure 1, a micro-mobility domain consists of a collection of paging areas that are connected to the domain root router(DRR). A paging area consists of many cells, each of which is managed by a base station(BS). A collection of BSs

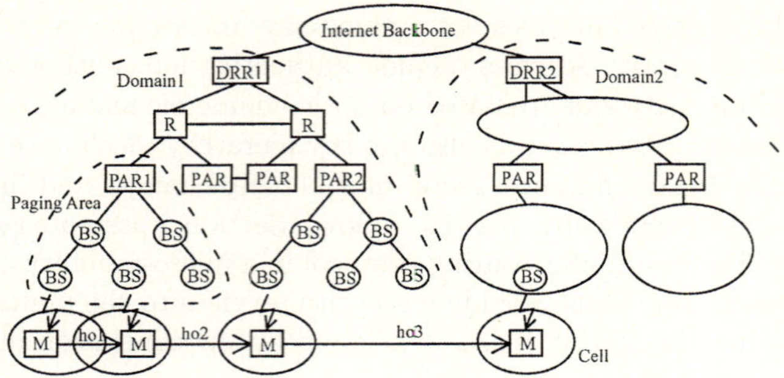

Fig. 1. Micro-Mobility Environment

belonging to a paging area form a tree with the paging area router (PAR) at the root. PARs are connected to the DRR of a micro-mobility domain through connecting routers. DRR, connecting routers, and PARs form an arbitrary topology. The purpose of paging areas is to reduce the control messages required to handle handoffs. If a mobile node is idle and moves from one cell to another in the same paging area, it does not notify its movement and, therefore, saves control messages. Regarding an idle mobile node, a domain knows only the paging area in which the mobile node resides but does not know which cell it belongs to. When a message has to be delivered, the exact location of the mobile node is found using a paging mechanism.

Multicast members build a multicast delivery tree in the form of a bidirectional shared tree around a core. The core is determined by the group initiator that creates the multicast group. If the group initiator can predict domains in which multicast members will reside, it finds the geographical center of DRRs of those domains in the Internet backbone using the algorithm proposed in [8]. If the group initiator cannot predict such domains but can assume that multicast members will be uniformly distributed among domains, it randomly selects one among candidate geographical centers of all DRRs. Candidate centers can be calculated a priori.

A mobile node wishing to join a multicast group as a receiver or sender notifies its BS by sending a join request toward the core. The join message travels until it reaches a multicast tree node (BS or router) or the core. When a mobile node has messages to multicast, it sends them to its BS, which deliver them along the bidirectional multicast tree to all group members.

There are three cases in handoffs: intra-paging area, inter-paging area, and inter-domain. They are depicted as ho1, ho2, and ho3, respectively in Figure 1. We first explain how the intra-paging area handoff works with Figure 2.

```
M notifies new BS of its arrival;
If (M is the first group member in new BS) {
    New BS requests crossover BS(starred BS in the figure) of old/new
        BSs to tunnel multicast packets so that it can relay them to M;
    New BS connects to multicast tree by sending join request;
    After receiving join ack, new BS asks crossover BS to stop tunneling;
```

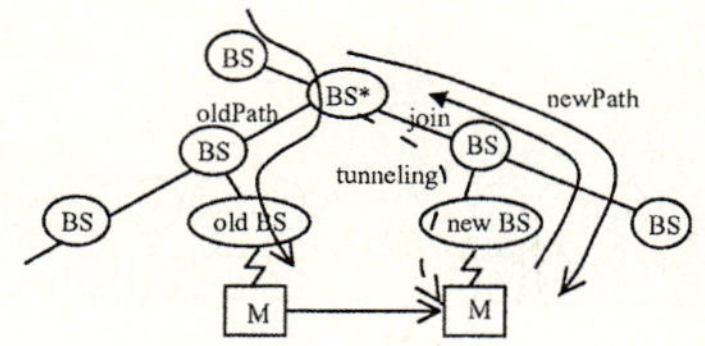

Fig. 2. Intra-Paging Area Handoff

/* Now multicast packets arrive through the new path */ }
Else /* New BS is already connected to the tree. So nothing to do */

If there already exists a group member in the new cell, a mobile node, M, receives packets from the new BS. Otherwise the new BS receives multicast packets from the crossover BS through tunneling and relays them to M. Therefore, the crossover BS is used as a forwarding agent of multicast packets and this happens to be the closest forwarding agent from M. In the meantime, the new BS sends join request toward the core. After it is connected to the tree, it asks the forwarding agent to stop tunneling because multicast packets are arriving through the new path. After detecting M's leaving, the old BS checks if M was the last member in its cell. If so, the old BS waits for some time and then sends leave message toward the core. Inter-paging area handoffs and inter-domain handoffs are handled similarly. The only difference is that the forwarding agent becomes PAR of the old BS: PAR1 in case of ho2 and PAR2 in case of ho3 in Figure 1. So far we explained handoffs for receiver members. Handoffs of sender members are processed similarly except that the direction of the data delivery is reversed.

The proposed algorithm hides node mobility within a domain to the outside and increases the stability of the multicast delivery tree. It uses both remote-subscription and bi-directional tunneling approaches and utilizes the paging area structure to provide seamless packet delivery to group members in spite of frequent handoffs.

4   Secure Multicast Routing Protocol

We will explain protocols that add security services to the multicast routing protocol explained in the previous section. In designing protocols, we assume that BSs and routers can be trusted in delivering control signals and managing key hierarchies but they are not allowed to access the data traffic. We will use the following notations in this section.

- PK-A: the public key of a node A.
- SK-A: the private key of A.
- $CERT_A$: the certificate of A.
- D^K: a message D encrypted with a key K.
- $\{D\}^{SK-A}$: a message D with a signature generated with A's private key.

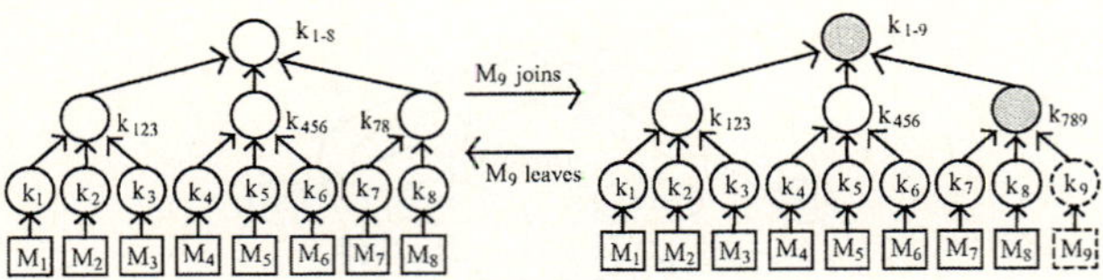

Fig. 3. Micro-Mobility Environment

We assume the existence of following entities.

- AS(Authorization server): It distributes capabilities on request. There are three types of capabilities: Initiator, Sender, and Receiver. The capability for a node N in a multicast group with a multicast address MA has the following format: $CAP_N = \{IP_N, CERT_N, MA, P, TS, L\}^{SK-AS}$. IP_N is the address of the node N and $CERT_N$ is N's certificate. P is the permit allowed for N and can be Initiator, Sender, or Receiver. A sender can send messages to and receive messages from a group while a receiver can only receive messages. TS is the timestamp and L is the lifetime of the capability.
- GI(Group initiator): It is the creator of a multicast group. It starts with an access control list (ACL) for the group. An ACL is an ordered set of a tuple (name, permit).

Before explaining secure multicasting protocols, we first explain keys used in the protocols. Every node N has a public/private key pair. We assume that every node knows the public key of AS, PK-AS. When a multicast group is created, the initiator, I, generates a group key, GK, that is distributed to all the authorized group members when they join the group. To efficiently support both backward/forward secrecy in spite of frequent membership changes and handoffs, PAR and all the group member mobile nodes in its paging area maintain a key hierarchy proposed in [10] as in Figure 3. One key hierarchy is maintained for each multicast group in a paging area. PAR becomes the key server and generates, stores, and modifies all the keys in the key hierarchy. A mobile node stores all the keys on the path from its leaf key to the root key. For example M_5 has k_5, k_{456}, k_{1-8} before M_9 joins. To prevent the joining mobile node from accessing past communications, all keys along the path from the joining point to the root key are changed. For example, if M_9 joins, in addition to creating k_9, PAR changes k_{78} to k_{789} and k_{1-8} to k_{1-9}. To prevent the leaving mobile node from accessing future communications, all keys along the path from the leaving point to the root key are changed. For example, if M_9 leaves, in addition to destroying k_9, PAR changes k_{789} to k_{78}' and k_{1-9} to k_{1-8}'. In a paging area we call the root key of the key hierarchy for a multicast group to be a subgroup key, SGK.

4.1   Group Creation

The group creation is initiated by GI. GI sends ACL to AS. The AS makes a capability list out of this ACL. So upon receiving the following ACL for a multicast group whose multicast address is MA,

- ((I, Group Initiator), (A, Sender), (B, Receiver))

AS generates the capabilities as follows

- GI-CAP$_I$ = {IP$_I$, CERT$_I$, MA, I-P, TS, L}$^{SK-AS}$
- S-CAP$_A$ = {IP$_A$, CERT$_A$, MA, S-P, TS, L}$^{SK-AS}$
- R-CAP$_B$ = {IP$_B$, CERT$_B$, MA, R-P, TS, L}$^{SK-AS}$

and sends them to GI. GI selects the core for the multicast group and stores its address IP$_{CORE}$ and also creates the group key GK.

4.2   Join

A member, M, wishing to join a group contacts GI, authenticates itself, and receives its capability, the core address, and the group key. If M is a sender, it generates a symmetric key called session key, SK, which will be used to encrypt the packets multicast from M. M sends the following message to its BS: {Join, IP$_M$, MA, CAP$_M$}$^{SK-M}$. Upon receiving this message, BS works as follows:

Check the authenticity of the message
    and the capability and store the message;
Notify PAR of its paging area of M's joining and request it to modify &
    distribute the key hierarchy for the subgroup within the paging area;
If (M is the first member of MA in this cell)
    Send Join request toward the core;

4.3   Leave

A member, M, wishing to leave the group sends the following message to its BS: {Leave, IP$_M$, MA, CAP$_M$}$^{SK-M}$. Upon receiving this message, BS works as follows:

Check the authenticity of the message;
Remove information about M from its storage;
Notify PAR of its paging area of M's leaving and request it to modify &
    distribute key hierarchy for the subgroup within the paging area;
If (M is the last member of MA in this cell)
    Send Leave request toward the core;

4.4   Packet Delivery

If a member, M, has a message, D, to be multicast to the group, it first encrypt it with its session key SK and then sign the encrypted message with its private key. Then it double-encrypts SK with GK and its subgroup key, SGK1, and appends the double-encrypted SK as follows: {D^{SK}}$^{SK-M}$, (SK^{GK})^{SGK1}. This packet is first sent to M's BS that checks if M is authorized to send packets. The purpose of this checking is to prevent unauthorized nodes from injecting bogus packets into the multicast tree. The message is multicast as it is inside the paging area. Once the packet arrives at PAR1 and has to travel outside the paging area, PAR1 decrypts SK with SGK1 and sends the following packet outside the paging area: {D^{SK}}$^{SK-M}$, SK^{GK}. Note that BSs and PARs can process and forward packets but cannot read them because they do not have the group key, GK. When the packet arrives at PAR2 of another paging area that has SGK2, PAR2 again encrypts SK with SGK2 and multicast within its paging area the packet modified as follows: $\{D^{SK}\}^{SK-M}$, $(SK^{GK})^{SGK2}$. Members in the new paging area can read this packet because they have both GK and SGK2.

4.5   Handoff

We consider a case where a mobile node M moves from an old cell whose BS and PAR are BS1 and PAR1 to a new cell whose BS and PAR are BS2 and PAR2. In case of an intra-paging area handoff, PAR1 and PAR2 are the same. Otherwise PAR1 is different from PAR2. When M moves into a new cell, it sends the following packet to BS2: $\{$handoff, IP_M, MA, CAP_M, BS1, PAR1$\}^{SK-M}$. After checking the authenticity of the packet, BS2 stores the message and then works as follows:

```
If (PAR1=PAR2) {
      /*intra paging-area handoff so no need to change the key hierarchy*/
      If (M is the first member of the new cell) {
            Request crossover BS of BS1 and BS2 to forward
                multicast packets to M;
            Send Join request toward the core and receive Join-ack;
            Request crossover BS to stop forwarding;}
      Else {/* BS2 can deliver multicast packets to M,
            so there is nothing to do */}
Else /* either inter-paging area or inter-domain handoff */ {
      Request PAR1 to forward packets to M;
      Request PAR2 to change key hierarchy within new paging area;
      If (M is the first member of the new cell) {
            Send Join request toward core and receive Join-ack;}
      Request PAR1 to stop forwarding packets}
```

If the handoff is within the same paging area, the key hierarchy does not need to be changed. So M can receive packets from the new BS if the new BS has already a group member. Otherwise M receives packets from the crossover BS until the new BS joins the group. If M moves into another paging area, it should receive packets from the old paging area because M has keys in the key hierarchy of the old paging area. M can receive packets from the new paging area only after M receives keys in the key hierarchy of the new paging area and the new BS joins the group. After some time, the old BS, BS1, detects that M has moved out of its cell. BS1 notifies its PAR of M's leaving and asks it to change/distribute the key hierarchy in the old paging area. If M was the last group member of BS1, BS1 sends the leave message toward the core.

5   Conclusion

To reduce the control message overhead due to location management, micro-mobility protocols have been proposed. In this paper we presented protocols for efficiently providing secure multicasting services in micro-mobility environments. We first introduced a multicast routing protocol. The proposed multicast routing protocol builds a shared multicast tree and does not assume any unicast micro-mobility protocols. Node mobility within a domain is hidden to the outside and the protocol increases the stability of the multicast tree. It uses both remote-subscription and bi-directional tunneling approaches and utilizes the paging area structure to provide seamless packet delivery to group members in spite of frequent handoffs. We then presented protocols to add security services to the bare multicasting service. The added security services include authentication, authorization, confidentiality, and integrity and use both symmetric and asymmetric cryptosystems. We also used the key hierarchy mechanism in a paging area to efficiently provide forward/backward secrecy in spite of frequent group membership changes and handoffs.

References

1. D. Bruschi and E. Rosti: Secure Multicast in Wireless Networks of Mobile Hosts: Protocols and Issues. Mobile Networks and Applications, 7, 2002.
2. S. Das et al: TeleMIP: Telecommunications-Enhanced Mobile IP Architecture for Fast Intradomain Mobility. IEEE Personal Communications, August 2000.
3. T. Harrison, C. Williamson, W. Mackrell and R. Bunt: Mobile Multicast (MoM) Protocol: Multicast Support for Mobile Hosts. Proc. of ACM MOBICOM, 1997.
4. C. R. Lin and K.-M. Wang: Scalable Multicast Protocol in IP-Based Mobile Networks. Wireless Networks, 8, 2002.
5. S. Mittra: Iolus: A Framework for Scalable Secure Multicasting. Proc. of ACM SIGCOMM Conf., 1997.
6. R. Ramjee et al: HAWAII: A Domain-Based Approach for Supporting Mobility in Wide-Area Wireless Networks. IEEE/ACM Trans. on Networking, 10(3),2002.
7. C. Shields, J.J. Garcia-Luna-Aceves: KHIP - A Scalable Protocol for Secure Multicast Routing. Proc. of ACM SIGCOMM Conf., 1999.
8. Y.-C. Shim and S.-K. Kang,: New Center Location Algorithms for Shared Multicast Trees. Networking 2002, LNCS 2345, 2002.
9. Y.-J. Suh, H.-S. Shin, and D.-H. Kwon: An Efficient Multicast Routing Protocol in Wireless Mobile Networks. Wireless Networks, 7.
10. C. Wong, M. Gouda, and S. Lam: Secure Group Communications using Key Graphs. Proc. of ACM SIGCOMM Conf., 1998.

Scalability and Robustness of Virtual Multicast for Synchronous Multimedia Distribution

Petr Holub[1,2], Eva Hladká[1], and Luděk Matyska[1,2]

[1] Faculty of Informatics
[2] Institute of Computer Science,
Masaryk University, Botanická 68a, 602 00 Brno, Czech Republic
{hopet, ludek}@ics.muni.cz
eva@fi.muni.cz

Abstract. A simple UDP packet reflector for virtual multicast multimedia transfer is extended to form a distributed system of active elements that solves the scalability problem of otherwise centralistic approach. The robustness of such virtual multicast delivery system is also discussed and shown to be better than the native multicast can offer. The maximum latency, important for multimedia transfer and related to the number of hops through the network of active elements, can be kept bounded. Possible support for synchronized multi-stream transfer is also discussed.

1   Introduction

A virtual multicasting environment, based on an active network element called "reflector" [1] has been successfully used for user-empowered synchronous multimedia distribution across wide area networks. While quite robust replacement for native, but not reliable multicast used for videoconferencing and virtual collaborative environment for small groups, its wider deployment is limited by scalability issues. This is especially important when high-bandwidth multimedia formats like Digital Video are used, when processing and/or network capacity of the reflector can easily be saturated.

A simple network of reflectors [2] is a robust solution minimizing additional latency (number of hops within the network), but it still has rather limited scalability. In this paper, we study scalable and robust synchronous multimedia distribution approaches with more efficient application-level distribution schemes. The latency induced by the network is one of the most important parameters, as the primary use is for the real-time collaborative environments. We use the overlay network approach, where active elements operate on an application level orthogonal to the basic network infrastructure. This approach supports stability through components isolation, reducing complex and often unpredictable interactions of components across network layers.

2   Synchronous Multimedia Distribution Networks

Real-time virtual collaboration needs a synchronous multimedia distribution network that operates at high capacity and low latency. Such a network can be

P. Lorenz and P. Dini (Eds.): ICN 2005, LNCS 3421, pp. 876–883, 2005.
© Springer-Verlag Berlin Heidelberg 2005

composed of interconnected service elements—so called *active elements* (AEs). They are a generalization of the user-empowered programmable reflector [1].

The reflector is a programmable network element that replicates and optionally processes incoming data usually in the form of UDP datagrams, using unicast communication only. If the data is sent to all the listening clients, the number of data copies is equal to the number of the clients, and the limiting outbound traffic grows with $n(n-1)$, where n is the number of sending clients. The reflector has been designed and implemented as a user-controlled modular programmable router, which can optionally be linked with special processing modules in run-time. It runs entirely in user-space and thus it works without need for administrative privileges on the host computer.

The AEs add networking capability, i. e. inter-element communication, and also capability to distribute its modules over a tightly coupled cluster. Only the networking capability is important for scalable environments discussed in this paper.

Local service disruption—element outages or link breaks—are common events in large distributed systems like wide area networks and the maximum robustness needs to be naturally incorporated into the design of the synchronous distribution networks. While the maximum robustness is needed for network organization based on out-of-band control messages, in our case based on user empowered peer to peer networks (P2P) approach described in Sections 3.1 and 5, the actual content distribution needs carefully balanced solution between robustness and performance as discussed in Section 4. The content distribution models are based on the idea that even sophisticated, redundant, and computationally demanding approaches can be employed for smaller groups (of users, links, network elements, . . .), as opposed to simpler algorithms necessary for large distributed systems (such as the global Internet). A specialized routing algorithm based on similar ideas has been shown, e. g. as part of the RON approach [3].

3   Active Element with Network Management Capabilities

As already mentioned in Sec. 2, the AE is the extended reflector with the capability to create network of active elements to deploy scalable distribution scenarios. The network management is implemented via two modules dynamically linked to the AE in the run-time: Network Management (NM) and Network Information Service (NIS). The NM takes care of building and managing the network of AEs, joining new content groups and leaving old ones, and reorganizing the network in case of link failure.

The NIS serves multiple purposes. It gathers and publishes information about the specific AE (e. g. available network and processing capacity), about the network of AEs, about properties important for synchronous multimedia distribution (e. g. pairwise one-way delay, RTT, estimated link capacity). Further, it takes care of information on content and available formats distributed by the network. It can also provide information about special capabilities of the specific AE, such as multimedia transcoding capability.

The NM and NIS modules can communicate with the AE administrator using administrative modules of the AE kernel. This provides authentication, authorization, and accounting features built into the AE anyway and it can also use Reflector Administration Protocol (RAP) [4] enriched by commands specific for NM and NIS. The NM communicates with the Session Management module in the AE kernel to modify packet distribution lists according to participation of the AE in selected content/format groups.

3.1   Organization of AE Networks

For the out-of-band control messages, the AE network uses self-organizing principles already successfully implemented in common peer to peer network frameworks [5],[6], namely for AE discovery, available services and content discovery, topology maintenance, and also for control channel management. The P2P approach satisfies requirements on both robustness and user-empowered approach and its lower efficiency has no significant impact as it routes administrative data only.

The AE discovery procedure provides capability to find other AEs to create or join the network. The static discovery relies on a set of predefined IP addresses of other AEs, while the dynamic discovery uses either broadcasting or multicasting capabilities of underlying networks to discover AE neighborhood. Topology maintenance (especially broadcast of link state information), exchange of information from NIS modules, content distribution group joins and keep-alives, client migration requests, and other similar services also use the P2P message passing operations of AEs.

3.2   Re-balancing and Fail-Over Operations

The topology and use pattern of any network changes rather frequently, and these changes must be reflected in the overlay network, too. We consider two basic scenarios: (1) re-balancing is scheduled due to either use pattern change or introduction of new links and/or nodes, i. e. there is no link or AE failure, and (2) a reaction to a sudden failure.

In the first scenario, the infrastructure re-balances to a new topology and then switches to sending data over it. Since it is possible to send data simultaneously over both old and new topology for very short period of time (what might result in short term infrastructure overloading) and either the last reflector on the path or the application itself discards the duplicate data, clients observe seamless migration and are subject to no delay and/or packet loss due to the topology switch. This scenario also applies when a client migrates to other reflector because of insufficient perceived quality of data stream.

On the contrary, a sudden failure in the second scenario is likely to result in packet loss (for unreliable transmission like UDP) or delay (for reliable protocols like TCP), unless the network distribution model has some permanent redundancy built in. While multicast doesn't have such a permanent redundancy property, the client perceives loss/delay until a new route between the source and the client is found. Also in the overlay network of AE without permanent redundancy, the client needs to discover and connect to new AE. This process can be sped up when client uses cached data about other AEs (from the initial discovery or as a result of regular updated of the topology). For some applications, this approach may not be sufficiently fast and permanent redundancy must be applied: the client is continuously connected to at least two AEs and discards the redundant data. When one AE fails, the client immediately tries to restore the degree of redundancy by connecting to another AE. The same redundancy model is employed for data distribution inside the network of AEs, so that re-balancing has no adverse effect on the connected clients.

The probability of failure of a particular link or AE is rather small, despite high frequence of failures in global view of large networks. Thus the two fold redundancy ($k = 2$) might be sufficient for majority of applications, with possibility to increase ($k > 2$) for the most demanding applications.

4   Distribution Models

4.1   Multicast Schemes

In an ideal case, the multicast organization of the data distribution is the most efficient scheme to distribute data to multiple clients. However, it is very difficult for a user to place AEs into the physical network topology in such a way that no data will pass through any physical link twice. The only exception may be when AE network is implemented as a network of active routers, but this goes against the user-empowered approach we support. Thus the multicast paradigm is only an upper-limit on efficiency of the distribution.

There are two basic approaches to build multicast distribution tree: source-based tree also known as shortest path tree (SPT) and shared tree. Regarding the synchronous character of multimedia data distribution, the SPT with reverse path forwarding (RPT) has two major advantages: it minimizes latency compared to shared tree where the data is sent through rendezvous point and it provides shortest paths between the source and the receivers (advantage for large volume of data transmission).

To build SPTs, it is necessary to have underlying unicast routing information. This information can be maintained very efficiently by RON [3]. As an addition to fast convergence in case of network link failure, it is possible to define policy to select the shortest path not based on hop count, but based on path round trip time or even one way propagation delay if available.

Fail-Over Operation. Standard operation when the link failure occurs is to build a new SPT as described above. If even the convergence speed of RON is not acceptable, there is another possible strategy to minimize delay due to SPT reconstruction. It is possible to compute multiple SPTs at the same time, choose single SPT for data distribution and keep the remaining SPTs for fail-over operation. For permanent redundancy scenario, more than one SPT can be

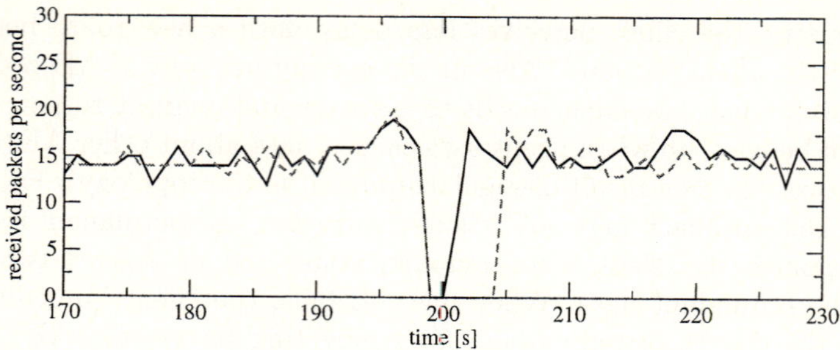

Fig. 1. Recovery time for with backup SPT (solid line) and without it (dashed line) simulated using cnet-based network simulator

used simultaneously and duplicate data will be discarded by client applications. In full graph, there are $n^2 - n$ links between the AEs. For a small number of AEs, alternative SPTs can be computed that don't use one selected link at a time. If that particular link fails, the alternative SPT can be immediately switched on. For larger number of AEs where number of links is too large, it is possible to compute $\lfloor n/2 \rfloor$ possible SPTs with disjunct set of links. When using SPTs or shared trees (ST) with backup based on disjunct sets of links, it is necessary to ensure that not all links from one AE are used in one SPT/ST, since the AE would become isolated in backup SPT/ST. When backup SPT/ST is available, the network recovery is limited just by broadcast speed to announce switching to a new SPT/ST, but when there is no backup, the alternative SPT/ST must be computed first (Fig. 1). During the normal operation, all these SPTs are monitored for their usability and when link fails in the current SPT, the original SPT can be swapped for another working SPT if at least one other usable SPT is available.

4.2   2D Full Mesh

The simplest model with higher redundance, serving also as the worst case estimate in terms of scalability, is a complete graph in which each AE communicates directly with all the remaining AEs. This *2D full-mesh tunneling* model was studied and described in detail in [2].

Let's assume a 2D full mesh of reflectors, each populated with n_r clients. The limiting traffic in this mesh is again the outbound traffic on the AE which scales as $out = n_r^2 m + n_r(m - 2)$.

Fail-Over Operation. When a link or whole AE drops out in the full mesh, the accident only influences data distribution from/to the clients connected to that AE. In case of link failure inside the AE mesh, the client is requested to migrate to an alternative AE. In case that AE itself fails the client initiates migration on its own. Alternative AEs should be selected randomly to distribute load increase more evenly and the load increase will be $\lceil \frac{n_r}{m-1} \rceil$. When even this migration delay is not acceptable, it is possible for a client to be permanently connected to an alternative AE and just switch the communication. For even more demanding applications, the client can use more than one AE for sending in parallel.

Although this model seems to be fairly trivial and not that interesting, it has two basic advantages: first, the model is robust and failure of one node influences only data from/to the clients connected to that AE. Second, it introduces only minimal latency because the data flows over two AEs at most. Next we will examine another model that has the same latency and robustness properties but that scales better.

4.3   3D Layered-Mesh Network

The layered mesh model creates k layers, in which data from a single AE are only distributed. That means each client is connected to one layer for both sending and receiving (sending only if $n_r = 1$; in other cases the client needs to receive data from remaining $n_r - 1$ clients of the AE used for sending) and to all other layers for receiving only. Each layer comprises 2D full mesh of m AEs. For the sake of simplicity, we first assume that $k = m$ and each AE has n_r clients, thus $n_r = \frac{n}{m} = \frac{n}{k}$.

In this scenario, the number of inbound streams is $in = n_r$. Number of outbound streams is $out_{s/r} = n_r^2 + n_r(m - 2)$ if the sending client is connected to this particular AE, and $out_r = n_r^2$ when only receiving clients are connected.

This model is problematic because of increasing the number of AEs used. However it seems to be the last model that doesn't introduce intermediate hops and thus keeps hop-count at minimum.

Intermediate AEs. Let's create q-nary tree used for distributing data from AE with sending clients to $m - 1$ AEs with listening clients. When building q-nary tree with λ intermediate layers $\lambda = \log_q(m - 1) - 1$, the total number of intermediate AEs is $L = \sum_{p=1}^{\lambda} q^p = \frac{m-1-q}{q-1}$.

Flows in this type of network are as follows: $out_{s/r} = n_r(n_r - 1) + qn_r$ for outer AE with sending clients connected, $out_r = n_r^2$ for outer AE with only receiving clients, and $out_i = qn_r$ for inner intermediate AEs. For all types of AEs, the number of inbound flows is n_r.

There are however two disadvantages of this model:

- The number of hops inside the mesh of AEs increases by λ compared to the plain 3D mesh model.
- Compared to the plain 3D model, the number of the intermediate AEs further increases to $m_{tot} = mk + Lk$. For $m = k$, it becomes $m_{tot} = m(m + L)$.

Nevertheless, this model provides the same redundancy while improving scalability compared to the simple 2D mesh.

Fail-Over Operation. Each of the mesh layers monitors its connectivity. When some layer disintegrates and becomes discontinuous, the information is broadcasted throughout the layer and to its clients. The clients that used that layer for sending are requested to migrate to randomly chosen layer from the remaining

$k - 1$ layers and the listening-only clients simply disconnect from this layer. Such behavior increases load on the remaining $k - 1$ layers but as the clients choose the new layer randomly, the load increases in roughly uniform way by $\lceil \frac{n_r}{k-1} \rceil$.

5   Content Organization

The multimedia content can be encoded in many different formats, that suit specific needs or capabilities of the network and the listening clients. In some cases (e. g. MPEG-4 formats) the highest quality format can be decomposed into N different layers (groups) that are sent over network independently. When native multicast is used, the client subscribes for the first $M \in \langle 1; N \rangle$ groups only, thus controlling the quality reduction of received content. With native multicast, there is no easy way to prioritize and synchronize the streams, which may lead to unexpected loss of quality (if data in the first layer are lost, the other layers may render useless).

As AEs support also multimedia transcoding (capable of being *active gateways*), an extended approach can be used. The format decomposition or even transcoding to completely different format may be performed by an AE, providing a flexible on demand service–the transcoding occurs only if really needed by some client. Also, the AEs are capable of synchronizing individual streams—they "understand" the decomposition and may re-synchronize individual streams. In case of severe overload, the higher (less important) stream layers are dropped first (again, AEs know the hierarchy), so the transmission quality is minimally affected.

To formalize our approach, we have designed three layer hierarchy:

- *content groups*—the highest level, an aggregation of several contents; it can be for instance a videoconferencing group (e. g. video and audio streams of individual videoconference participants)
- *content*—intermediate level, a content (a video stream, format independent)
- *format*—the lowest level, format definition.

Each multimedia stream in the network is then characterized by (`content group`, `content`, `format`) triplet which creates one record in the SPT tree. The available formats for each content create an oriented graph where the root is the source format and the child nodes define the formats created from their parents. A client can choose the best suitable format, or different formats for different contents within one content group (e. g. a lecturer's stream with the highest quality).

The information about available content groups, content, and available formats is published via NIS on AEs and is distributed and shared across the network of AEs.

6   Related Work

There are a few known applications for synchronous distribution of multimedia data over IP networks. Probably the most important ones are cascading of

H.323 multi-point connection units (MCUs) and Virtual Room Videoconferencing System (VRVS). The networks of H.323 MCUs are based on a static pre-configured topology and they don't offer user-empowered approach. The VRVS is only provided as a service and the users' traffic is managed by VRVS administrators. Also, although the VRVS team reports some move in favor of more elaborate and dynamic network of reflectors, we believe that creating flexible user-empowered multimedia network is more suited for open systems without centralized administration.

7   Conclusions

In this paper the models for the virtual multicast scalability were introduced with discussion of robustness and fail over capabilities of the proposed solutions. We have implemented a prototype of Active Element suitable for simple networking scenarios for Linux and FreeBSD operating systems and the models have also been verified using network simulator. The AE network organization support is being implemented based on JXTA 2.0 P2P framework. The full application-level multicast data distribution with multicast subgroups as described in Secs. 4.1 and 5 is under development.

Acknowledgment

This research is supported by a research intent "Optical Network of National Research and Its New Applications" (MŠM 6383917201). We would also like to thank to Tomáš Rebok for helping with implementation of network simulations.

References

1. Hladká, E., Holub, P., Denemark, J.: User empowered virtual multicast for multimedia communication. In: Proceedings of ICN 2004. (2004)
2. Hladká, E., Holub, P., Denemark, J.: User empowered programmable network support for collaborative environment. In: ECUMN'04. Volume 3262/2004 of Lecture Notes in Computer Science., Springer-Verlag Heidelberg (2004) 367 – 376
3. Andersen, D., Balakrishnan, H., Kaashoek, F., Morris, R.: Resilient overlay networks. In: 18th ACM Symp. on Operating Systems Princpiles (SOSP), Banff, Canada (2001)
4. Denemark, J., Holub, P., Hladká, E.: RAP – Reflector Administration Protocol. Technical Report 9/2003, CESNET (2003)
5. Rowstron, A., Druschel, P.: Pastry: Scalable, distributed object location and routing for large-scale peer-to-peer systems. In: IFIP/ACM International Conference on Distributed Systems Platforms (Middleware), Heidelberg, Germany (2001) 329–350
6. Yang, B., Garcia-Molina, H.: Designing a super-peer network. In: IEEE International Conference on Data Engineering. (2003) 25

Mobile Multicast Routing Protocol Using Prediction of Dwelling Time of a Mobile Host*

Jae Keun Park, Sung Je Hong, and Jong Kim

Department of Computer Science and Engineering,
Pohang University of Science and Technology(POSTECH),
San 31, Hyoja-dong, Pohang, Kyungbuk, South Korea
{ohora, sjhong, jkim}@postech.ac.kr

Abstract. We propose a mobile multicast routing protocol based on the timer-based mobile multicast (TBMOM). A mobile host that stays in a foreign network receives multicast datagrams from a foreign multicast agent (FMA) through tunneling. When a mobile host hands off, the foreign agent in the foreign network where it moves calculates the expected dwelling time for which the mobile host will stay in the foreign network. The foreign agent decides whether it will be included in the multicast tree using the expected dwelling time during hand-off. The proposed protocol reduces the tunnel length from FMA to the foreign network by predicting the expiration of the timer of a mobile host. Simulation results show that the proposed protocol provides a multicast delivery path closer to the optimal path than TBMOM.

1   Introduction

As the computing power of computers increases and the transmission bandwidth broadens, high-quality multimedia services are available to users. Multicast is a transmission of data to multiple users. it is crucial in providing multimedia services such as Internet television and radio, videoconferencing, and network games. As mobile devices are popular, demand for multimedia services provided through these devices is increasing.

Many multicast routing protocols, such as DVMRP [1], MOSPF   [2, 3], CBT [4], and PIM [5], are proposed mainly for wired networks and thus are not suitable for a mobile environment. Mobile IP [6] has been proposed to provide mobile hosts with continuous communication with its correspondent host but it only supports unicast for mobile hosts. The IETF (Internet Engineering Task Force) has proposed two basic approaches to support multicast for mobile

---

* This research was supported by the MIC(Ministry of Information and Communication), Korea, under the Chung-Ang University HNRC-ITRC(Home Network Research Center) support program supervised by the IITA(Institute of Information Technology Assessment).

P. Lorenz and P. Dini (Eds.): ICN 2005, LNCS 3421, pp. 884–891, 2005.

users using Mobile IP. One is bidirectional tunneling [7], and the other is remote subscription [8]. In bidirectional tunneling, a multicast tree is constructed from the source to the home agents (HA) at which the mobile hosts have registered. A home agent transfers a multicast datagram to mobile hosts that have subscribed to the same multicast group and registered with it using tunneling. In bidirectional tunneling, a foreign agent does not require the tree join when a mobile host hands off and does not need to support multicast. In this approach, the multicast delivery path can be far from optimal since each path is routed via a home agent. When mobile hosts in the same multicast group stay in one foreign network, the foreign agent receives duplicated multicast datagrams that may cause network congestion. Tunneling wastes network resources and produces high processing overhead at the home agent.

In remote subscription, a multicast tree is constructed from the source to foreign agents (FA) in foreign networks (FN) where mobile hosts stay. When a mobile host arrives at a foreign network, the foreign agent joins the tree of the multicast group that the mobile host subscribes to. The foreign agent connected to the multicast tree can provide multicast service to the mobile host. In remote subscription, a foreign agent must support multicast. Remote subscription has high overhead due to tree reconstruction. The number of tree reconstructions may lead to high processing overhead on the network [9].

Mobile Multicast (MOM) protocol [7] prevents duplicated multicast datagrams of bidirectional tunneling. A foreign agent selects one of multiple home agents that subscribe to the same multicast group. The selected home agent tunnels multicast datagrams to the foreign agent and is called the designated multicast service provider (DMSP). MOM protocol wastes network resources because of tunneling.

Timer-based mobile multicast (TBMOM) protocol [10] considers the speed of a mobile host. TBMOM provides the shortest delivery path for low-speed mobile hosts and low tree reconstruction and fast data delivery for high-speed mobile hosts. Each mobile host has a timer which is set to a predetermined value. When the timer of a mobile host expires, the foreign agent in the network where the mobile host stays joins the multicast tree. When the foreign agent receives the first multicast datagram through the multicast tree, it becomes a Foreign Multicast Agent (FMA). FMA delivers multicast datagrams to a foreign agent in the foreign network where the mobile host stays using tunneling. Figure 1 shows the operation of TBMOM. The timer of a mobile host can be expired while remaining in the foreign network. FMA of the mobile host continues to tunnel multicast datagrams to the foreign agent from hand-off to timeout. In

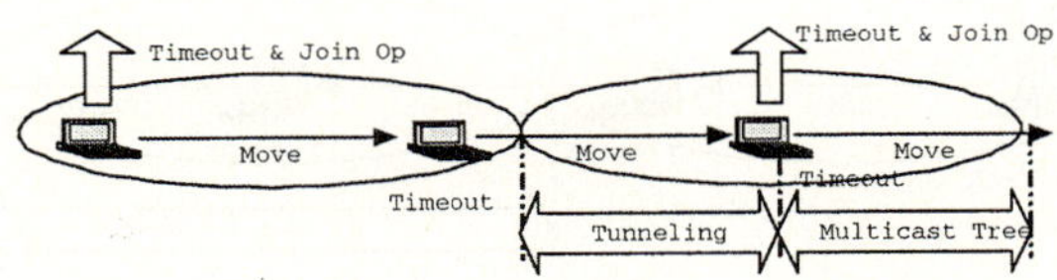

Fig. 1. The operation of TBMOM

this case, the time for which mobile hosts receive multicast datagrams from the multicast tree is reduced while the mobile host remains in the foreign network. This is, tunnel length becomes long, so the waste of network resources increases.

In this paper, we propose a mobile multicast routing protocol based on Timer-based Mobile multicast (TBMOM), reducing the tunnel length from a foreign agent connected to a multicast tree to a mobile host. We can reduce the tunnel length of TBMOM if a foreign agent knows whether the timer of a mobile host expires in its foreign network. In the proposed protocol, a foreign agent predicts the expiration time of a timer in a mobile host during hand-off and decides whether it joins the multicast tree.

The rest of this paper is organized as follows. Proposed protocol is described in Section 2. The performance of the proposed protocol is evaluated and compared with TBMOM in Section 3. Finally, we offer conclusions learned from our research in Section 4.

2   Proposed Protocol

2.1   Overview

To reduce the waste of network resources in mobile environments, multicast routing protocols need to support short tunnel lengths and reduce the number of tree reconstructions due to the mobility of the host. The proposed protocol reduces the tunnel to provide a multicast delivery path close to the optimal path and it is based on TBMOM. When a mobile host hands off, we can reduce the tunnel length of TBMOM if the foreign agent knows whether the timer of a mobile host will expire in its foreign network. When a mobile host hands off, the foreign agent in a new foreign network decides whether it will join the multicast tree by predicting the expiration of the mobile host's timer. If the foreign agent predicts that the timer of a mobile host will expire in the new foreign network soon, it requests to join the multicast tree. Otherwise, the foreign agent receives multicast datagrams from FMA through tunneling. Figure 2 shows the basic operation of the proposed protocol. When the mobile host moves from $(i-1)$-th network to i-th network, the foreign agent in i-th network predicts the expiration of the timer of the mobile host. The foreign agent requests a tree join if expiration of the timer is predicted. Compared with TBMOM in Figure 1, the proposed protocol reduces tunneling time, and provides a delivery path closer to the optimal path than TBMOM.

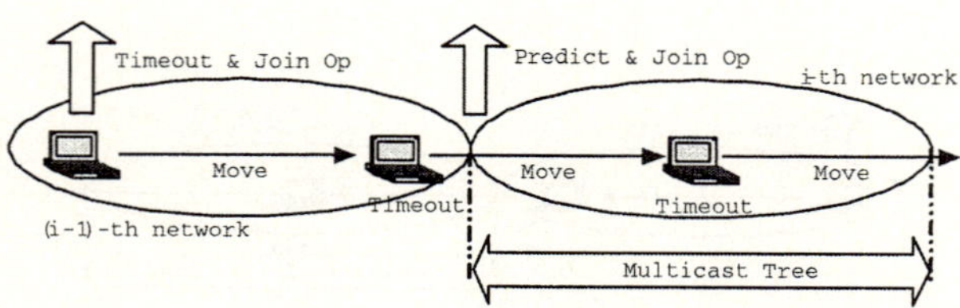

Fig. 2. The operation of the proposed protocol

In the proposed protocol, whenever a mobile host hands off, the foreign agent calculates the dwelling time for which a mobile host will stay at the new foreign network and decides whether to join the multicast or not depending on this time.

2.2   Proposed Protocol Details

The proposed protocol uses the entities of TBMOM, GROUP timer, JOIN timer and FMA. Each mobile host has a JOIN timer which is set to the predetermined value. Each foreign agent has its GROUP timer which is set to the minimum remaining value among JOIN timers of multiple visiting mobile hosts. When the GROUP timer expires, the foreign agent tries to join the multicast tree if it has not connected to the multicast tree, and it makes itself a FMA when receiving the first multicast datagram from the multicast tree. The FMA tunnels multicast datagrams to mobile hosts that have moved into other foreign networks. DMSP hand-off follows the method of TBMOM.

During hand-off, a mobile host sends its information to the foreign agent in a new foreign network and requests the service of the multicast group that it subscribes to. The information provided by the a mobile host includes group identifier, the FMA of the mobile host, the value of JOIN timer, the real dwelling time and the expected dwelling time in the previous foreign network, and a DMSP bit. The real dwelling time and the expected dwelling time are added in the information of TBMOM. The expected dwelling time of a mobile host at a new foreign network is calculated using the expected dwelling time and the real dwelling time of the mobile host at the previous foreign network. We calculate the expected dwelling time (T_{ept}^i) of the mobile host at the i-th network as follows:

$$T_{ept}^i = \{ \begin{matrix} T_{ept}^{i-1} + \alpha(t_{real}^{i-1} - T_{ept}^{i-1}) \; if \; i \geq 2, \\ t_{real}^0 \qquad\qquad\qquad\qquad if \; i = 1. \end{matrix}$$

where t_{real}^{i-1} is the real dwelling time at the $(i-1)$-th network and α is a system parameter between 0 and 1. The difference between the real dwelling time and the expected dwelling time at the $(i-1)$-th network is the compensation value of the expected dwelling time at the $(i-1)$-th network. The mobile host records the real dwelling time and the expected dwelling time at the $(i-1)$-th network to calculate the expected dwelling time at the i-th network.

If the foreign agent in a new foreign network agent is FMA when a mobile host hands off, it starts the multicast service for the visiting mobile host. Otherwise, the foreign agent calculates the expected dwelling time in a new foreign network,

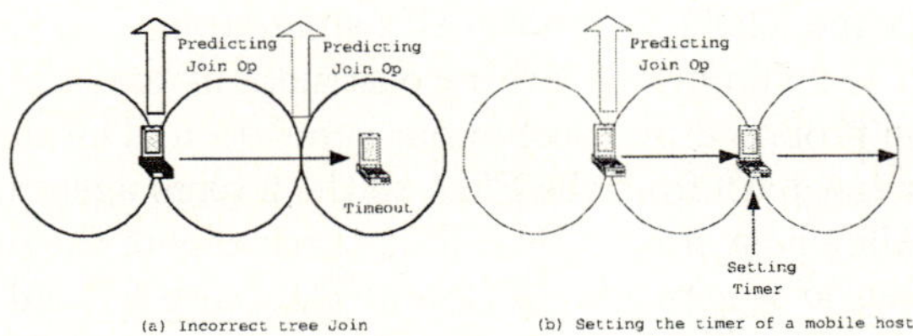

Fig. 3. Handling a false prediction

then it predicts whether the JOIN timer will expire or not while the mobile host stays in the new network, and decides whether to join the multicast tree or not according to the expected dwelling time of the mobile host during hand-off. If the expected dwelling time at the new network exceeds the value of JOIN timer of the mobile host, the foreign agent sends a join request because it predicts that JOIN timer of the mobile host will expire in the new network. Otherwise, the foreign agent does not send a join request.

In the proposed protocol, a false decision of tree join occurs due to the incorrect prediction of the expected dwelling time. The proposed protocol supports the handling of a false of tree join decision. Two cases of incorrect prediction are in the proposed protocol. One case is that the timer expires before the expected time, and the other is that the timer of the mobile host expires after the expected time. In the former case, the foreign agent tries to join the multicast tree because the GROUP timer of the foreign agent expires. In the latter case, a tree join operation happens since the timer of mobile host expires in aother foreign network. Therefore, the number of tree reconstructions increases. Figure 3 (a) shows the incorrect tree join in the proposed protocol. To reduce the number of tree reconstructions, we increase the JOIN timer of the mobile host to prevent expiration of the JOIN timer. The JOIN timer is set to the sum of its remaining value and the value of predetermined timeout interval. Figure 3 (b) shows setting the JOIN timer when the timer of the mobile host expires after the expected time.

3   Performance Evaluation

The performance of the proposed protocol is evaluated by a event-driven simulation with a unit time. The unit time is the minimum time that a mobile host stays in a network. The entire time of each simulation is 10,000 unit times. Topology used in the simulation is a 10×10 mesh network. Each node is one LAN. We assumed that each LAN has one foreign agent that acts as a multicast router. For simplicity, we assume that there is only one multicast group and one source in our simulation. The number of the group members is 20. The multicast source and the initial LANs where mobile hosts stay are selected randomly for each test. The multicast tree is the shortest path tree and the DMSP of a foreign agent is the FMA that has the shortest path. The dwelling time of a mobile host varies from 1 unit time to 42 unit times. The value of α is selected to 0.1 through the simulation. Join Time interval is set to 5 and 15 unit times in the simulation. The random walk model [11] is used in this simulation.

We measured the number of tree reconstructions and the tunnel length of TBMOM and the proposed protocol in our simulation. The tunnel length is the length of the shortest path from the FMA to the foreign agent in the foreign network where a mobile host stays. $[D_{min}\ D_{max}]$ represents the dwelling time that a mobile host stays at a network. D_{min} and D_{max} are defined as the minimum dwelling time and the maximum dwelling time of a mobile host. The dwelling time of a mobile host in a network is selected randomly from D_{min} to D_{max}

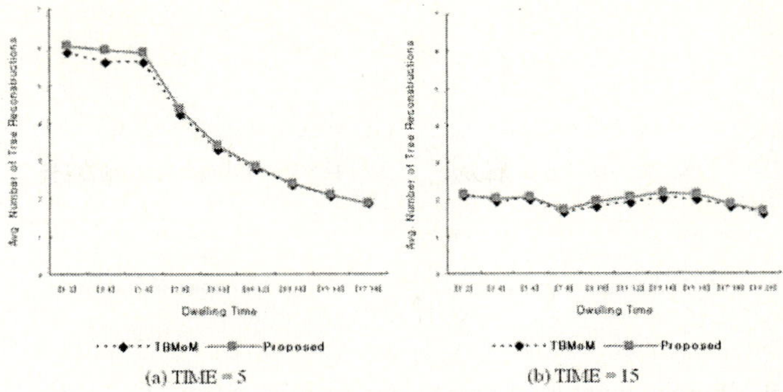

(a) TIME = 5

(b) TIME = 15

Fig. 4. Average number of tree reconstructions when $D_{max} - D_{min} = 1$

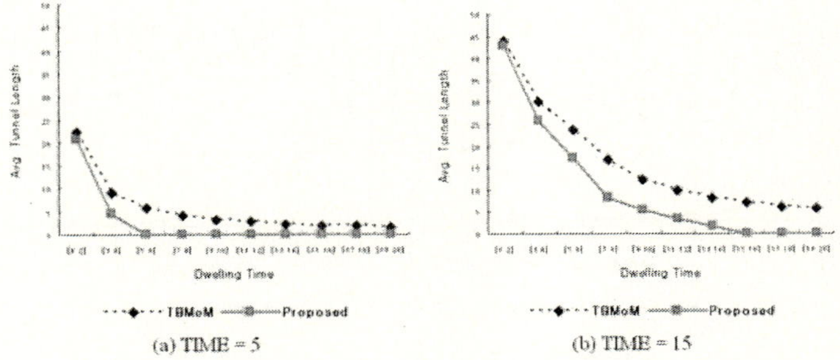

(a) TIME = 5

(b) TIME = 15

Fig. 5. Average tunnel length when $D_{max} - D_{min} = 1$

every time the mobile host hands off. As the difference between D_{max} and D_{min} becomes larger, the dwelling time at a foreign network varies widely.

The first experiment evaluated the tree reconstruction time and the tunnel length of TBMOM and the proposed protocol when the difference value between D_{max} and D_{min} is 1. As the dwelling time of mobile hosts at a foreign network increases, the number of tree reconstructions decreases due to the little movement of mobile host between foreign networks. Figure 4 shows the comparison of TBMOM and the proposed protocol with respect to the number of tree reconstructions when the difference value between D_{max} and D_{min} is 1. The number of tree reconstructions in the proposed protocol is similar with that of TBMOM when the JOIN timer interval is 5 and 15. As the JOIN timer interval grows larger, the average number of tree reconstructions is reduced, but the average tunnel length is longer because the probability that the mobile hosts stay at other foreign networks is high as the JOIN timer interval grows. Figure 5 shows the average tunnel length of TBMOM and the proposed protocol when the difference value between D_{max} and D_{min} is 1. The tunnel length becomes shorter as the movement of mobile hosts decrease. For the number of tree reconstructions, the proposed protocol is similar with TBMOM, but its tunnel length is shorter than that of TBMOM as much by as 5.7 hops and 8.7 hops when each JOIN timer interval is 5 and 15 respectively.

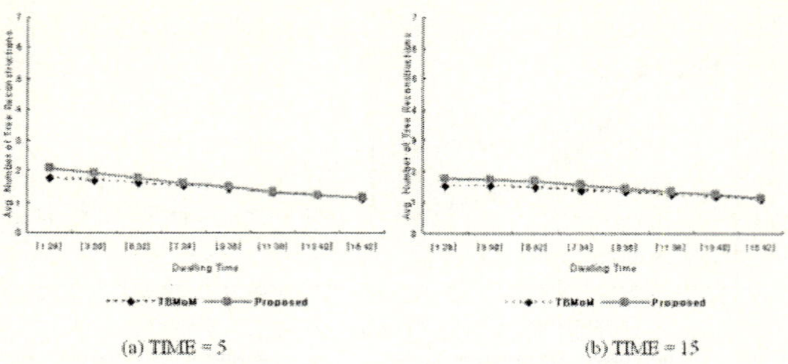

Fig. 6. Average number of tree reconstructions when D_{max} - D_{min} =27

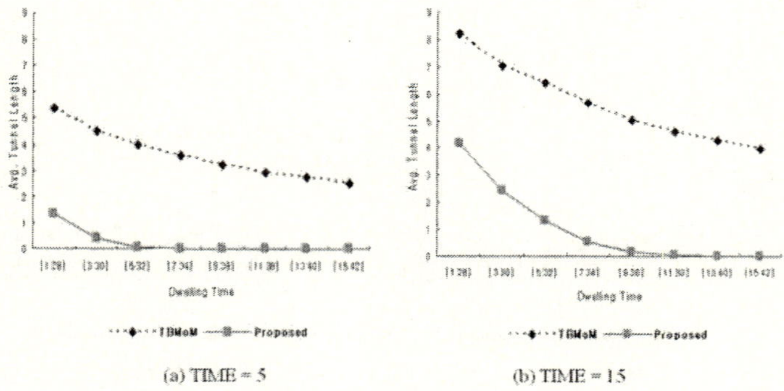

Fig. 7. Average tunnel length when D_{max} - D_{min} = 27

The second experiment compares TBMOM and the proposed protocol when the variation of the dwelling time is large. Figure 6 shows the number of tree reconstructions in TBMOM and the proposed protocol when the difference between D_{max} and D_{min} is 27. The number of tree reconstructions does not change significantly if the dwelling time is greater than the JOIN timer interval. Although the inaccuracy of the prediction of tree join increases as the variation of the dwelling time is larger, the tree reconstruction time of the proposed protocol is little more than that of TBMOM. Figure 7 shows tunnel length of TBMOM and the proposed protocol when the deference between D_{max} and D_{min} is 27. The proposed protocol reduces the influence of inaccurate prediction using the incorrect prediction handling. The average tunnel length of the proposed protocol is shorter than that of TBMOM by as much as 2.3 hops and 5 hops when each JOIN timer interval is 5 and 15 respectively.

4 Conclusion

In this paper, we presented a mobile multicast routing protocol based on TB-MOM. The proposed protocol reduces the tunnel length from the FMA to mobile hosts by predicting the mobile hosts expected dwelling time in the foreign network. When a mobile host hands off, the foreign agent in a new foreign network calculates the expected dwelling time at the new network using the dwelling time for which the mobile host stayed in previous foreign networks. Then the foreign agent decides the multicast tree join by comparing the expected dwelling time and the remaining value of the JOIN timer in the mobile host. A tree join prediction failure increases the number of tree reconstructions. To prevent this increase, we proposed a false prediction handling scheme by setting the JOIN timer. We compared the TBMOM and the proposed protocol through the number of tree reconstructions and the tunnel length. The proposed protocol has shorter tunnel length than TBMOM by as much as 5.7 hops and 8.7 hops when each JOIN timer interval is 5 and 15 and the difference between D_{max} and D_{min} is 1. The proposed protocol provides a multicast delivery path closer to the optimal path than that provided by TBMOM.

References

1. D. Waitzman, C. Partridge and S. Deering, "Distance vector multicast routing protocol," *RFC 1075*, November 1988.
2. J. Moy, "Multicast extensions to OSPF," *RFC 1584*, March 1994.
3. J. Moy, "Multicast Routing Extension for OSPF," *Communications of ACM, vol.37, no.8, pp. 61-66*, Aug. 1994.
4. A. Ballardie, J. Crowcroft and P. Francis, "Core based tree (CBT) : an architecture for scalable inter-domain multicast routing," *In Proceedings of ACM SIGCOMM'93, pp. 85-89*, August 1993.
5. S. Deering, D. Estrin, D. Farinacci and V. Jacobson, "An architecture for wide-area multicast routing," *In Proceeding of ACM SIGCOMM'94, pp. 126-135*, August 1994.
6. C. Perkins, "IP Mobility Support," *RFC 2002, Mobile IP Working Group*, October 1996.
7. T. Harrison, C. Williamson, W. Mackrell and R. Bunt, "Mobile multicast (MOM) protocol: multicast support for mobile hosts," *In Proceeding of ACM MOBICOM 97, pp. 151-160*, September 1997.
8. V. Chikarmane and C. L. Williamson, "Multicast support for mobile host using mobile IP: design issues and proposed architecture," *Mobile Networks and Applications, pp. 365-379*, 1998.
9. C. Jelger and T. Noel, "Multicast for mobile hosts in IP networks: progress and challenges," *IEEE Wireless Communications, vol. 9, issue 5, pp. 58-64*, October 2002.
10. J. Park and Y. J. Suh, "A timer-based mobile multicast routing protocol in mobile network," *Computer Communications, vol. 26, issue 17, pp. 1965-1974*, November 2003.
11. A. Bar-Noy, I. Kessler and M. Sidi, "Mobile users: To update or not to update?," *Proceeding of IEEE INFOCOM 94, pp. 570-576*, Jun 1994.

A Group Management Protocol for Mobile Multicast

Hidetoshi Ueno, Hideharu Suzuki, and Norihiro Ishikawa

NTT DoCoMo, Inc. Network Management Develop Department,
3-5 Hikari-no-oka, Yokosuka, Kanagawa, 239-8536 Japan
`{uenohi, suzukihid, ishikawanor}@nttdocomo.co.jp`

Abstract. IGMP (Internet Group Management Protocol) is a proven technology that is widely used as the group management protocol for multicast. However, IGMP is not always suitable for wireless networks because of its communication costs, especially if clients are moving frequently. IGMP also has some security issues such as denial of service (DoS). Therefore, we propose a new group management protocol called Mobile Multicast Group Management Protocol (MMGP) for mobile multicast. MMGP is designed for use on wireless LANs and mobile networks, and is capable of reducing the communication costs even when clients move frequently. MMGP also considers security and can prevent multicast DoS. We show that MMGP is capable of realizing group management at lower communication costs than IGMP, and is suited as the group management protocol for mobile multicast.

1 Introduction

IP multicast is a technology that can broadcast data to specified clients (data receiver). It is particularly effective for the one-to-many broadcast type of data distribution in which large volumes of data such as video are distributed. IP multicast is being used as a data distribution technology on backbone networks and LANs.

On public wireless LANs or mobile networks, however, since wireless with limited resources is used, communication costs are high compared with wired networks. Accordingly, multicast technology that can broadcast data to multiple receivers by using wireless common channels is considered vital. The 3GPP (3rd Generation Partnership Project), for example, an organization concerned with the standardization of mobile communications, has been working on the standardization of MBMS (Multimedia Broadcast Multicast Services), a broadcast-type data distribution technology for mobile networks [1].

A large number of studies have been made thus far on the application of multicast technology in mobile networks. Most of these studies have concerned multicast routing that can prevent the loss of data when a client receiving it changes the network being accessed due to client movement [2]-[3]. These studies, however, have not looked into applying multicast group management protocols such as IGMP [4] in a mobile environment. IGMP is not always suitable for wireless networks because of its communication costs, especially when clients move frequently. IGMP also has some security issues such as the inability to prevent denial of service (DoS). Our proposal is the new group management protocol called Mobile Multicast Group Management

P. Lorenz and P. Dini (Eds.): ICN 2005, LNCS 3421, pp. 892–903, 2005.

Protocol (MMGP); it resolves the issues that arise when multicast group management protocols such as IGMP are applied to a mobile environment. Note that we do not describe Multicast Listener Discovery (MLD) [5] in this paper. MLD is used for IPv6 instead of IGMP.

In Section 2 of this paper, we identify the issues that occur when IGMP is applied to mobile multicasting and we define the requirements imposed by MMGP; Section 3 describes MMGP in more detail; in section 4, we compare communication costs of MMGP and IGMP and we show how MMGP is more effective in mobile environments. Section 5 discusses related works; and Section 6 presents our conclusions.

2   Issues with IGMP and Requirements of MMGP

Since IGMP [4] was designed mainly for wired LANs, a number of issues occur when it is applied to a mobile environment. In this section, we identify these issues and clarify the requirements imposed by MMGP. First, we discuss the general features that need to be considered in communications utilizing wireless networks.

• Communication Costs for Wireless Networks
Since the frequency resources that can be used for wireless communications are limited, wireless communication costs are generally higher than those of wired communications. It is therefore important to reduce communication costs on wireless networks. If a mobile phone is used as a client, there are limits to its processing capabilities (due to CPU and power supply) which make it necessary to reduce the volume of data that the client is sending and receiving. These reductions will make it possible, for instance, to extend the mobile phone's standby time.

• Mobility of Clients
In mobile communications, it is possible for the power supply to suddenly be cut or for the client to move out of range. If, for instance, the access router (i.e. the first router that clients directory communicate with) manages the status (existence information, etc.) of a client, inconsistencies may occur in the status information. Unnecessary communications to recover from the status inconsistencies may also occur as a result of client movements.

Considering these features, a number of issues in applying IGMP to the mobile environment have been identified. Next, we will address these issues and discuss the requiems for MMGP.

(1) Reducing Communications Overheads
Since IGMP does not use message acknowledgement (ACK), it may not be possible to detect the loss of a message. One way of reducing the message non-arrival rate is by sending the same message more than once (twice is the default setting in Robustness-Variable of IGMP). Since this measure unnecessarily increases the number of communication steps, it is necessary to study a method of group management in MMGP that does not require a message to be sent more than once.

(2) Reducing Client Processing Costs by Eliminating Query Messages
IGMP uses query messages to periodically check for the presence of members. There are two types of IGMP queries – the Group-Specific query and the General query.

The former is directed at group members, while the latter targets all clients. Since both types require processing in most clients, these queries reduce standby time in mobile phones. The IGMP queries may be unsuitable for wireless networks since the original purpose of the group management protocol for the access router is to confirm that there is at least one member. Accordingly, MMGP must minimize the communications associated with queries in order to reduce the communication costs of clients.

(3) Reducing Communication Occurrences Caused by User Movement

In IGMP [4], the access router manages the membership of all members. Accordingly, when a member leaves the group (when an IGMP Leave message is sent), the access router can use the Fast Leave function that deletes status information on the corresponding client without sending the Group-Specific query. In a mobile environment, however, since it is highly possible for a member to move without sending IGMP Leave, the Fast Leave function is not necessarily effective. Furthermore, since Fast Leave does not function, we need a function that can resolve any status inconsistency. MMGP must, therefore, be able to cope with the status inconsistencies in a mobile environment since client movements occur frequently.

(4) Countermeasure to Multicast DoS

In IGMP, since any client can request the start of data reception, it joins any group which raises the problem of multicast DoS leading to the unnecessary construction of multicast distribution paths [6]-[7]. The same sort of problem, unnecessary communication processing, exists even when a client leaves the multicast group. Especially in the mobile environment, since clients are not directly connected via cable or other means and it is difficult to specify which client caused the problem, it is particularly important to counter multicast DoS. With MMGP, as with the multicast DoS countermeasure that was introduced in this paper [6], the first member to join the multicast group and the last member to leave the multicast group must be authenticated and their legitimacy must be confirmed.

3 Overview of MMGP

In this section, we overview MMGP which can satisfy the requirements described in Section 2.

The original purpose of IGMP is to find out whether or not there is at least one client below the access router that wishes to receive certain multicast distribution data. MMGP focuses on this issue and proposes a new method in the access router to intensively managing the group conditions of one client selected from among the members. MMGP uses a special data format, named token, to identify the selected member. The token is given to at least one member selected by the access router, and the member that receives the token (called the token member) is strictly managed by the MMGP router (the access router supporting MMGP) by means such as periodically checking the client's membership status in the group. Since the token is given in principle to one member, the first member to join the multicast group and the last member to leave the multicast group are naturally token members. If the token member leaves the group, the token is reassigned to another member. On the other hand, since it is not important for the MMGP router to manage the status of members who were not given the token (called the non-token member), the MMGP router does not necessarily need to know when non-token members join or leave the group.

The MMGP feature that gives the token to at least one client can also be used to counter multicast DoS. MMGP focuses on the fact that the first member to join the group and the last member to leave the group are always token members, and authenticates clients when the token member joins or leaves the group. This solves the problem of multicast DoS.

3.1 MMGP Details

In this section, we detail MMGP using a concrete example of the communication sequences of MMGP.

1. Join Procedure for the Token Member (Token Member Join)

When the first client joins the multicast group, the join procedure for the token member indicated below is executed (Figure 1). When the client sends the Join Request (1-1), the MMGP router knows that this is the first member to join the corresponding multicast group, and then assigns the token to the client (1-4). By adding the digital certificate to the token, the client can verify that this token was issued by the MMGP router. Next, the client sends an acknowledgement (ACK) to inform the MMGP router that the token has arrived (1-5). If the MMGP router requires a countermeasure for multicast DoS, it may execute the client authentication procedure indicated at (1-2) and (1-3). Client authentication is either challenge-response authentication or authentication using the digital certificate. After the MMGP router receives the join request (1-1) or completes client authentication (1-3), if needed, it builds the multicast tree and starts to relay multicast data.

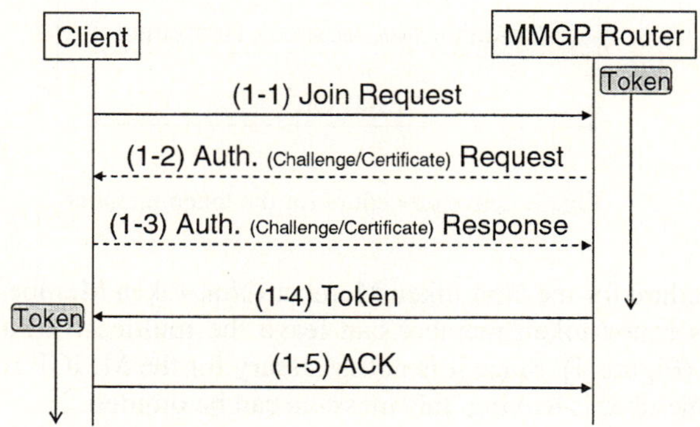

Fig. 1. Join procedure for the token member

2. Join procedure for the non-token member (Non-token Member Join)

If a client joins a multicast group which already has other members, the join procedure for the non-token member is executed (Figure 2). As in the procedure in Figure 1, the client sends the Join Request (2-1). When the MMGP router receives this request, it can judge that this is the second or subsequent member to join (i.e. the non-token member) and processing ends at this time. However, if the client detects that data destined for this multicast group is already being received or can use some other method to detect that other members have already joined the multicast group, it can omit sending the Join Request.

Fig. 2. Join procedure for the non-token member

3. Leave Procedure for the Token Member (Token Member Leave)
When the token member leaves the multicast group, the leave procedure shown in Figure 3 is executed. The client adds the token it received from the MMGP router to the Leave Request message and sends it to the MMGP router. The MMGP router checks the token it has received to verify whether or not it is the one it has issued, and then sends an OK message to indicate it has received the token (3-4). If a countermeasure for multicast DoS is necessary, the MMGP router may execute the client authentication procedure indicated at (3-2) and (3-3).

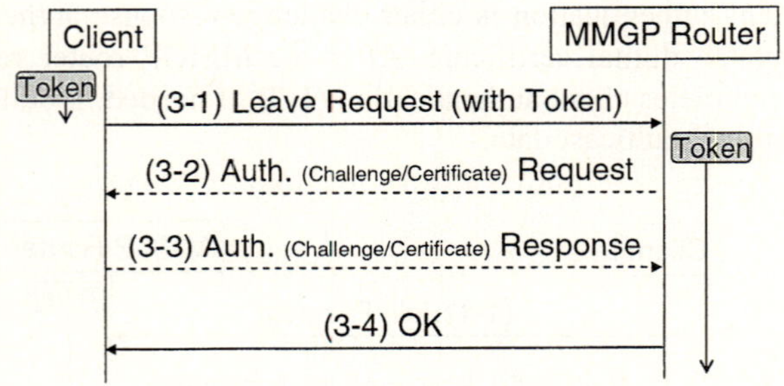

Fig. 3. Leave procedure for the token member

4. Leave Procedure for the Non-token Member (Non-token Member Leave)
A client that is a non-token member can leave the multicast group by sending the Leave Request (Figure 4). Since it is not necessary for the MMGP router to know that the non-token member is leaving, this message can be omitted.

Fig. 4. Leave procedure for the non-token member

5. Token Reassignment

Since MMGP in principle requires that the token be given to one member, if the token member leaves according to the procedure in Figure 3, the token must be reassigned to another member (Figure 5).

If the MMGP router judges that token reassignment is necessary, it sends the Query message to the multicast group to which the member belongs (5-1). All members that received this message send a Join Request message to inform the MMGP router that there are members of the group (5-2). From among the members that sent the Join Request message, the MMGP router selects one to become the token member, and as in Figure 1, the procedure for giving this member the token is then executed (5-5) (5-6). At this time, if a countermeasure for multicast DoS is needed, the MMGP router may execute client authentication processing (5-3) (5-4). If there are no remaining members, however, there will be no members to send the Join Request, in which case the timer for the MMGP router to receive the Join Request will expire and the MMGP router stops relaying the multicast distribution.

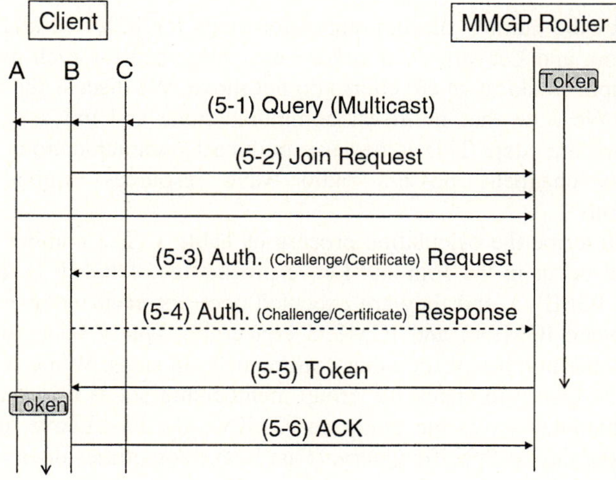

Fig. 5. Token reassignment procedure

6. Token Membership Check (Hello)

The MMGP router periodically checks the membership status of the token member in order to determine the group membership status of the token member (Figure 6). The MMGP router periodically sends a Hello message to the token member (6-1). The client then sends back a Hello ACK message to inform the MMGP router of its membership (6-2). If the timer for the Hello message expires and none is received, the MMGP router can detect that no token member is present, and it then executes the token reassignment procedure shown in Figure 5.

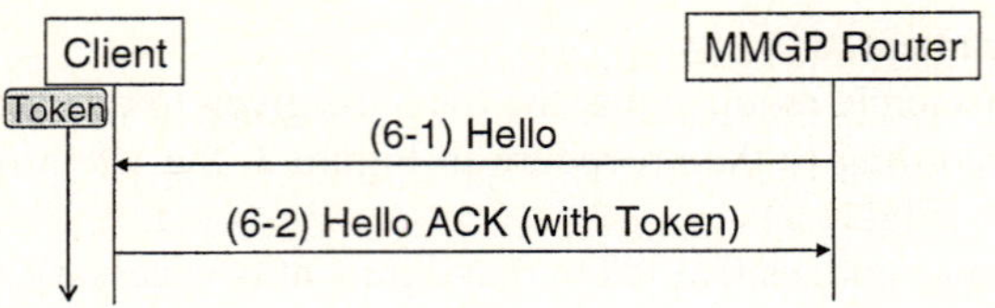

Fig. 6. Token membership check (Hello)

4   Evaluation of MMGP

In this section, we compare the number of communication steps demanded by MMGP and IGMP, and quantitatively analyze the number of communication steps for MMGP. After that, we offer some considerations on MMGP based on the results of our analysis.

4.1   Comparison of the Number of Communication Steps for IGMP and MMGP

Table 1 shows the number of communication steps for IGMPv2 and IGMPv3. Here we assume Join and Leave procedures are executed once for each member. We also assume a simple situation where clients do not move. We discuss the mobility case in Section 4.2. We note that multicast communications and unicast communications equally require one step. This is because multicast communications, by using common wireless channels, possess radio wave resources equivalent to unicast communications.

Next, we describe the calculation process in Table 1. The number of communication steps that occur in the Join and Leave procedures of IGMP is the same in both IGMPv2 and IGMPv3, and they are executed once for each member. However, differences between IGMPv2 and IGMPv3 concerning Query behavior appear in the differences in the number of the other communication steps. While IGMPv2 uses the Group-Specific Query to check the group membership status of remaining members each time a member leaves the group, in IGMPv3, the Fast Leave function obviates the need for the Group-Specific Query. (Fast Leave sometimes does not function due to the movement of the client, but here we simply assume that it does function.) Also, while in IGMPv2, one member sends a Membership Report for the General Query, in IGMPv3, all members of the group send a Membership Report at this time.

Table 1. Number of communication steps in IGMP

	IGMPv2	IGMPv3
Join procedure	nv	nv
Leave procedure	nv	nv
Group-Specific Query procedure	$(2n-1)v$	—
General Query procedure	$2qv$	$q(1+\bar{n})v$
Total	$(4n+2q-1)v$	$\{(2+q/2)n+q\}v$

< Explanation of variables and parameters >
n: Number of members present under one access router
v: Robustness-Variable (default = 2)
q: Number of General Query executions (proportional to IGMP execution period)
$\bar{n}$: Average number members in the group at that time (= n/2)

Table 2 shows the number of communication steps in MMGP. To compare the number of communication steps with IGMP, which has no client authentication function, we assumed that MMGP also had no client authentication function. A member that is the first to join a group performs a total of 5 communication steps when it joins and leaves, since it always becomes the token member. In addition, since the second and subsequent members (n-1 members) are at first non-token members when they join the group, one communication step is executed when they join and again when they leave. Concerning the number of communication steps when the non-token member leaves, however, since there are cases in which token reassignment changes a non-token member into the token member, the number of non-token sequences is reduced by the number of reassignments. If token reassignment occurs, the total number of sequences that are executed when the token member joins or leaves at that time is 5. The number of sequences for instances of the Hello message is 2, including the response to the Hello message.

Table 2. Number of communication sequences in MMGP

		MMGP
First member to join	Join	3
(Token member)	Leave	2
2nd and subsequent member	Join	$n-1$
to join (Non-token member)	Leave	$n-1-r$
Token reassignments		$5r$
Hello		$2h$
Total		$2n+4r+2h+3$

< Explanation of variables and parameters >
h: Number of Hello executions (proportional to MMGP execution period)
r: Number of times token reassignment occurs ($0 \leq r \leq n-1$)

As the calculation results in Table 1 and Table 2 indicate, since the number of communication steps increases in proportion to the number of members n, by determining the value for each parameter, it is possible to obtain the number of communication steps of each protocol corresponding to the number of members n. The number of communication steps, assuming 125-second intervals each for General Query (q) and Hello (h) (125 = default value for General Query) and 30-minute execution time for each protocol, is shown in Figure 7. In MMGP, since the number of communication steps that are executed varies according to the number of times the token member is reassigned, we calculated the best-case scenario in which no token member is reassigned ($r = 0$) and the worst-case scenario in which all members that left are the token member ($r = n\text{-}1$).

We can see from Figure 7 that the required number of communication steps is fewer than IGMP and that they do not depend on the number of token reassignments that occur or the number of members.

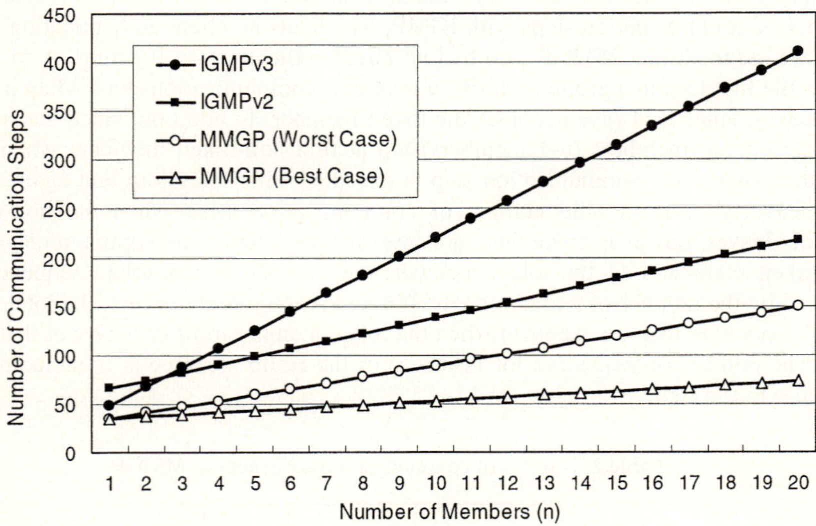

Fig. 7. Comparison of the number of communication steps in IGMP and MMGP

4.2 Considerations on MMGP

In this section, we offer our thoughts on MMGP in each area of study.

- Communication Steps in MMGP

MMGP was originally designed for the purpose of multicast group management and is used to learn whether or not at least one member is present under the access router. MMGP uses the token to distinct a designated member (i.e. the token member) and the other members, and only the communication with the token member is strictly managed. Therefore, MMGP has ACK messages to confirm that communication with the token member has arrived successfully and Hello messages to periodically check the existence of the token member. As a consequence, MMGP can omit the redundant messages that were sent by IGMP to improve its robustness. Consequently, as shown in Figure 7, MMGP makes it possible to reduce the number of communication steps (Requirement 1 in section 2). The reduction is especially important for wireless networks.

- Communication Processing Costs of Clients

In IGMP, all control messages are transacted with multicast addresses used as destinations. In particular, since clients are the target recipients of the Query message, the communication costs become a problem for mobile phones with limited power supplies. In MMGP, membership checks are performed on the token member only by using the Hello message, and since the communication with it is conducted in unicast, the number of communication steps with non-token members is reduced. As a result, a reduction in the communication processing costs for all clients as a whole is possible (Requirement 2 in section 2).

- Mobility of Clients

The number of communication steps of IGMP [4] shall be increased compared with the result of Figure 7 when clients move because the status inconsistency caused by the movement of clients causes failure of the Fast Leave function. It means that the access router needs to use the Group-specific query at all times in order to check the existence of group members. In MMGP, the non-token member can move across sub-networks without sending any messages to the access router because the status of the non-token member is not managed in the access router. This enables a reduction in the status inconsistencies caused by the movement of the non-token member and in the costs of communications that occur as a result (Requirement 3 in Section 2). Therefore, the reduced number of communication steps possible with MMGP is also expected when the clients are mobile.

Regarding the movement of the token member, however, communication costs do arise as a result of movement detection and token reassignment. An important issue for MMGP is, therefore, how to assign the token to a client with the lowest probability of moving. At this point, we looked into a method whereby the token is reassigned to a member that has been in the group the longest time. Since the fact that a member whose has remained in the one sub-network for a long time means that it is less likely for it to move or go out of range, we can expect that the number of token reassignments will decrease. To give an example of how this would actually work, we looked into a method of changing the timing at which a Join Request message was sent after an MMGP Query was received to make it associated with the length of time a client was a member of the group. If a client was a member of the group for a long time, it would send the Join Request quickly, or conversely, if the client was a relatively new member, sending the Joint Request would be delayed, and the MMGP router would reassign the token to the member from which the Join Request was received the most quickly. We are also looking into various other ways of selecting a member to receive a reassigned token, such as detecting the speed at which clients move and reassigning the token to a stationary client.

- Security

Whenever needed, it is possible in MMGP to execute client authentication on the token member that is joining or leaving the group and to take countermeasures against multicast DoS (Requirement 4 in section 2). To prevent third parties from spying on data being distributed, however, client authentication alone is not sufficient and multicast distribution data needs to be encrypted [8]. MMGP is capable of using all types of encryption methods for multicast and using them independently as needed.

- Fairness for Token and the Non-token Members

In MMGP, the communication costs for managing the status of the token member is larger than in the case of the non-token member. If the network being used bases its usage charges on volume, it is possible that unfairness in communication charges will occur among users. This differential in communication costs, however, is considered to occur only once over the short period from the start to finish of group management. If we consider the case in which group management is executed repeatedly over a long period, since all clients have an equal probability of becoming the token member, uniform communication costs are expected. We can also expect that fixed rates for charging usage fees for mobile phones will improve fairness in communication charges.

5   Related Works

There are few works regarding the application of multicast group management protocols to the mobile environment. One paper proposes that data volume can be reduced by combining into one message the IGMP Join message that is sent to a client's destination network if a client moves and the IGMP Leave message that is sent to the client's original network before it leaves [9]. This paper, however, does not touch on security functions and the problem of increased costs of communications caused by Query, and has not considered applying it to the mobile environment.

6   Conclusion

In this paper, we have proposed MMGP for mobile multicast. MMGP has been designed for use mainly on wireless LANs and mobile networks, and it is capable of reducing the communication costs resulting from clients that move frequently. MMGP also considers the aspect of security and can execute countermeasures against multicast DoS. This paper also conducted quantitative analysis on the numbers of communication steps to show that MMGP is superior to IGMP/MLD in the mobile networks. Since no previous studies have adequately addressed the application of multicast group management protocols to the mobile environment, this paper holds an important position with regard to future studies. In our future studies, we plan to perform more detailed assessments of practical installations and simulations, and further clarify the effects of MMGP.

References

1. 3rd Generation Partnership Project, "Multimedia Broadcast Multicast Service (MBMS): Architecture and Functional Description," 3GPP TS 23.246, September 2004.
2. I. Romdhani, M. Kellil, and Hong-Yon Lach, "IP Mobile Multicast: Challenges and Solutions," IEEE Communications Surveys and Tutorials," Vol. 6, No. 1, March 2004.
3. U. Varshney, "Multicast over Wireless Networks," ACM, Vol. 45, No. 12, pp. 31-37, December 2002.
4. B. Cain, S. Deering, I. Kouvelas, B. Fenner and A. Thyagarajan, "Internet Group Management Protocol, Version 3," RFC3376, October 2002.
5. R. Vida and and L. Costa, "Multicast Listener Discovery Version 2 (MLDv2) for IPv6," RFC3810, June 2004.

6. H. Ueno, K. Tanaka, H. Suzuki, N. Ishikawa and O. Takahashi, "Access Controls and Group Key Distribution Protocols for Multicast Communications," 2nd Information Technology Forum (FIT2003), September 2003.

7. H. Ueno, H. Suzuki, F. Miura and N. Ishikawa, "A consideration on countermeasures for DoS attacks of Multicast," 2nd Information Technology Forum (FIT2003), September 2003.

8. T. Hardjono and B. Weis, "The Multicast Group Security Architecture," RFC3740, March 2004.

9. S. Kaur, B. Madan, S. and Ganesan, "Multicast Support for Mobile IP Using a Modified IGMP," IEEE WCNC '99, September 1999.

Propagation Path Analysis for Location Selection of Base-Station in the Microcell Mobile Communications

Sun-kuk Noh[1], Dong-you Choi[2,*], and Chang-kyun Park[3]

[1] Dept. of Radio Communication Engineering, Honam University
[2] Research Institute of Energy Resources Technology,
Chosun University
[3] Dept. of Electronic Engineering, Chosun University

Abstract. In microcell mobile communication using cellular method, we analyze propagation path which can accurately and rapidly interpret mobile communication propagation environment in urban, when subscriber service is done based on the main road in urban. In this paper, to appropriate location of BS in urban street, we suggest a simplified algorithm to interpret a propagation path and a propagation prediction model which can predict receiving field strength of MS, located in a LOS or NLOS within a service area, and simulate. As a result, we present a location condition of BS of microcell CDMA mobile communication systems.

1 Introduction

The cellular method in an early stage was macrocell method whose service radius was between several kilometers and tens of kilometers. But, because of explosive increase of the number of subscribers a microcell method or a picocell method was introduced. And it could enlarge the call processing capacity and improve the transmission quality.[1,2]

Unlike the macrocell method, the service radius of the microcell method is less than 1~2 km. Because microcell method is affected by the surface of the earth and buildings in the cell area, the transmitting power of the base station(BS) should be minimized and an antenna should be installed at the lower elevation than close buildings to reduce the interference by neighboring cells. Therefore, for predicting the optimal power of the BS according to the feature of the surface of the earth and buildings in the cell area, a propagation prediction model is necessary for low cost and high efficiency of the microcell system design.[3-6] For this case, as the representative propagation prediction models, there are two models using the lay-launching method[7-9] and the multiple image method.[10] But, these models are not practical because of complex calculation, impossible execution processing, and much time to predict receiving power.

In reality, when mobile station(MS) is located within line of sight(LOS) in microcell mobile communication propagation environment of urban living space, it's receiv-

* Corresponding Author.

P. Lorenz and P. Dini (Eds.): ICN 2005, LNCS 3421, pp. 904–911, 2005.

ing power can be affected by direct, reflection, or diffraction wave simultaneously. But if MS is located within non-line of sight(NLOS), at the worst, either of reflection and diffraction wave is only received.

In this paper, to appropriate location of BS in urban street, we suggest a simplified algorithm to interpret a propagation path and a propagation prediction model which can predict receiving field strength of MS, located in a LOS or NLOS within a service area, and simulate. As a result, we present a location condition of BS of microcell CDMA mobile communication systems.

2 Suggestion of the Algorithm to Interpret a Propagation Path

Fig. 1 is a model to interpret a path of propagation that have high-lise buildings space between the two lodes. We suggest an algorithm to interpret a propagation path which vary with the changes in road width and the inclined angle of a straight crossing as well as the incident and reflection angle of propagation coming to BS, if MS is located ① at random propagation shadow area, slightly failed to the straight road of LOS ; ② at straight crossing of NLOS which is inclined to certain inclined angle of θ_v against the straight road; ③ or at a random spot, somewhat gone off the straight crossing.

Before the algorithm is suggested until propagation arrives at MS, the following are hypothesized :

First, the difference in height between BS and MS is ignored because it is too small compared with propagation path.

Secondly, as propagation path which is travelled by one reflection is short, the road width of that section is considered regular.

Thirdly, available propagation coming at MS is only one and it is horizontal travelling wave due to the reflection of wall of building.

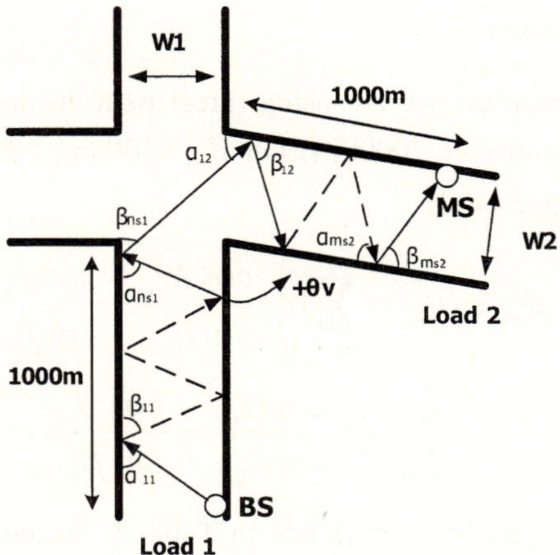

Fig. 1. Propagation prediction model in an urban street

2.1 Algorithm to Interpret the Propagation Path

Transmitting wave from a BS reflects n times by random α_{n1} (incident angle) and β_{n1} (reflection angle) at LOS of W_1 (width of road 1), as shown in Fig. 1, and approaches NLOS of W_2 (width of road 2). While it reflects m times by α_{m2} and β_{m2} again and comes to a MS, r (distance of the total reflection wave path) is the sum of r_{n1} and r_{m2}. The algorithm is as follows.

$$r = r_{n1} + r_{m2} = \sum_{n=1}^{n_s} W_1 \left(\frac{\sin \alpha_{n1} + \sin \beta_{n1}}{\sin \alpha_{n1} \sin \beta_{n1}} \right) + \sum_{m=1}^{m_s} W_2 \left(\frac{\sin \alpha_{m2} + \sin \beta_{m2}}{\sin \alpha_{m2} \sin \beta_{m2}} \right) [m] \tag{1}$$

Where

n_s, m_s : The final reflection number of each area.

r_{n1} : Distance of the total reflection wave path in the LOS

r_{m2} : Distance of the total reflection wave path in the NLOS

On the other hand, the algorithm is as follows to find α_{12}, the first incident angle which approaches NLOS, opposed to $\alpha_{n_s 1}$, the final reflection angle in LOS.

$$\alpha_{12} = 180 - (\beta_{n_s 1} + \theta_v)[°] \tag{2}$$

Where θ_v is sloping angle and considered "+", if rotating counterclockwise, when 6 o'clock is "0".

2.2 Reflection Number

To find reflection number, we are using direct path distance at LOS and NLOS. If total direct path distance is 2000m (LOS $d_{n1} =1000$m, NLOS $d_{m2} =1000$m), direct path distance d is equation(3).

$$d = d_{n1} + d_{m2} = \sum_{n=1}^{n_s} W_1 \left(\frac{\cos \alpha_{n1}}{\sin \alpha_{n1}} + \frac{\cos \beta_{n1}}{\sin \beta_{n1}} \right)$$
$$+ \sum_{m=1}^{m_s} W_2 \left(\frac{\cos \alpha_{m2}}{\sin \alpha_{m2}} + \frac{\cos \beta_{m2}}{\sin \beta_{m2}} \right) \tag{3}$$

Then the result of equation(3) is shown in Table 1. Incident and reflection angle is inversely sloping angle in all road width.

Table 1. Total reflection numbers by θ_v, W, α & β, direct path distance=2000m

Sloping Angle[°]	Incident And reflection angle[°]	Road width [m]			
		10 Reflection number	20 Reflection number	30 Reflection number	40 Reflection number
90	45	198	98	66	48
100	40	166	82	54	40
110	35	140	69	46	34
120	30	114	56	38	28
130	25	92	46	30	22
140	20	72	36	23	17
150	15	52	26	16	12
160	10	34	16	10	8
170	5	16	8	4	4

3 Suggestion of Propagation Prediction Model

3.1 Reflection Coefficient

To find reflection coefficient before prediction model is suggested, all wave-reflection is α, if it is direct reflection and vertical polarization, reflection coefficient $\Gamma(\alpha)$, is found as follows[11].

$$\Gamma(\alpha) = \frac{\sin\alpha - \sqrt{\varepsilon_\gamma - \cos^2\alpha}}{\sin\alpha + \sqrt{\varepsilon_\gamma - \cos^2\alpha}} \tag{4}$$

If $e_c = e_r^{'} - j60\sigma\lambda$ ($e_r^{'} = 15$, $\sigma = 2$) [9], specific inductive capacity of specular surface, $|e_c| = 25$ is substituted ($\lambda = 0.167$ m in case of 1.8GHz) for equation(3), reflection coefficient is found as Table 2.

Table 2. Reflection coefficients by incidence angles and reflection angles

Incident & Reflection angle	Reflection coefficient	Incident & Reflection angle	Reflection coefficient
0°	-1	45°	-0.75
5°	-0.965	50°	-0.7324
10°	-0.9316	55°	-0.7169
15°	-8998	60°	-0.7035
20°	-0.8698	65°	-0.6922
25°	-0.8417	70°	-0.683
30°	-0.8156	75°	-0.6758
35°	-0.7917	80°	-0.6707
40°	-0.7698	85°	-0.6677

3.2 Propagation Prediction Model for LOS

MS is located at propagation shadow area of LOS, L_1 that path loss of MS is found from equation(1) applied. The result is as follows.

$$L_1 = 20\log(\sum_{n=1}^{n_s}\frac{\lambda}{4\pi d_{n1}}\Gamma_{n1})[dB]$$

(5)

If the effective radiation power of BS is P_t, propagation prediction model to interpret $P_{r1}[W]$ receiving power of MS of LOS is found from equation(6) applied. The result is as follows.

$$P_{r1} = P_t[\sum_{n=1}^{n_s}\frac{\lambda\sin\alpha_{n1}\sin\beta_{n1}}{4\pi W_1(\sin\alpha_{n1}+\sin\beta_{n1})}\Gamma_{n1}]^2$$
$$= P_t[\sum_{n=1}^{n_s}\frac{\lambda}{4\pi d_{n1}}\Gamma_{n1}]^2[W]$$

(6)

3.3 Propagation Prediction Model for NLOS

MS is located at propagation shadow area of LOS, L_2 that path loss of MS is found from equation(1) applied. The result is as follows.

$$L_2 = 20\log(\sum_{m=1}^{m_s}\frac{\lambda}{4\pi d_{m2}}\Gamma_{m2})[dB]$$

(7)

If the effective radiation power of BS is P_t, propagation prediction model to interpret $P_{r2}[W]$ receiving power of MS of NLOS is found from equation(7) applied. The result is as follows.

$$P_{r2} = P_t[\sum_{m=1}^{m_s}\frac{\lambda}{4\pi d_{m2}}\Gamma_{m2}]^2[W]$$

(8)

3.4 Propagation Prediction Model for LOS and NLOS

L, total path loss of MS, located at propagation shadow area of NLOS, is the sum of L_1 and L_2 and found from equation(1) applied. The result is as follows.

$$L = L_1 + L_2 = 20\log\sum_{n=1}^{n_s}\sum_{m=1}^{m_s}[\frac{\lambda}{4\pi d}\Gamma_{n1}\Gamma_{m2}]\,[dB]$$

(9)

If effective radiation power of BS is P_t, propagation model to interpret P_r [W, dBm] is found from equation(9) applied. The result is as follows.

$$P_r = P_t \sum_{n=1}^{n_s} \sum_{m=1}^{m_s} [\frac{\lambda}{4\pi d} \Gamma_{n1} \Gamma_{m2}]^2 [W]$$

$$= 20 \log P_t \sum_{n=1}^{n_s} \sum_{m=1}^{m_s} [\frac{10^3 \lambda}{4\pi d} \Gamma_{n1} \Gamma_{m2}][dBm] \tag{10}$$

As a result, if MS is located at propagation shadow area of LOS or NLOS, integration model is found as equation(10) to predict receiving power of MS.

4 Simulation

Microcell PCS mobile communication propagation environment with 1.8GHz of urban living space was selected, as it is relatively similar in using frequency band to IMT-2000 with 2GHz, which is next generation mobile communication method. Then equation(10) was simulated with specification of Table 3 and the result is shown in Table 4.

Table 3. Specification of simulation

Propagation Path Model	Fig. 1
Propagation environment	Urban street microcell
Using frequency	1.8GHz
Effective radiation power	250mW
Transmitting Ant-Receiving Ant (ht-hr)	0m
Number of effective propagation	1
Service distance based on direct path	0~2000m
Section of direct path	1000~2000m(step 200m)
Incident and reflection angle in line of sight	25°
Incident and reflection angle in non-line of sight	Found from equation(2) using variable of θ_v
Sloping of non-line of sight	50°~170°
Road width	10m, 20m, 30m, 40m
Specific inductive capacity of specular surface	25(ε_γ =15, σ=2, λ=0.166)
Reflection coefficient	Table 1

Based on |95dBm| that minimum receiving power of MS possible to service, the shadowing part are not received mobile communication service from Table 4. LOS is received mobile communication service regardless of road width and sloping angle but NLOS, in sloping angle of 130° ~140°, receiving power coming to MS becomes maximum from Table 4. Based on this angle, when sloping angle is increased or decreased, receiving power is decreased.

Table 4. Receiving power by direct path of LOS and NLOS[|dBm|]

Total direct path [m]	Prediction distance [m]	Road width [m]	Sloping angle[°] 50	60	70	80	90	100	110	120	130	140	150	160	170
1,000	Line of sight 500	10	62	62	62	62	62	62	62	62	62	62	62	62	62
		20	53	53	53	53	53	53	53	53	53	53	53	53	53
		30	50	50	50	50	50	50	50	50	50	50	50	50	50
		40	48	48	48	48	48	48	48	48	48	48	48	48	48
	Non-line of sight 500	10	·	·	243	172	130	104	85	73	68	68	73	85	104
		20	·	·	147	111	90	76	67	61	59	59	61	67	76
		30	·	·	116	91	77	67	61	57	55	55	57	61	67
		40	·	·	100	81	70	63	58	55	54	54	55	58	63
1,200	Line of sight 600	10	66	66	66	66	66	66	66	66	66	66	66	66	66
		20	56	56	56	56	56	56	56	56	56	56	56	56	56
		30	53	53	53	53	53	53	53	53	53	53	53	53	53
		40	50	50	50	50	50	50	50	50	50	50	50	50	50
	Non-line of sight 600	10	·	·	281	197	147	115	92	79	73	73	79	92	115
		20	·	·	168	125	99	83	71	65	62	62	65	71	83
		30	·	·	130	101	84	73	65	61	58	58	61	65	73
		40	·	·	112	89	75	67	61	58	56	56	58	61	67
1,400	Line of sight 700	10	71	71	71	71	71	71	71	71	71	71	71	71	71
		20	59	59	59	59	59	59	59	59	59	59	59	59	59
		30	55	55	55	55	55	55	55	55	55	55	55	55	55
		40	53	53	53	53	53	53	53	53	53	53	53	53	53
	Non-line of sight 700	10	·	·	321	223	165	127	101	85	78	78	85	101	127
		20	·	·	189	138	109	90	77	69	65	65	69	77	90
		30	·	·	145	111	91	78	68	63	61	61	63	68	78
		40	·	·	122	96	82	72	65	61	59	59	61	65	72
1,600	Line of sight 800	10	76	76	76	76	76	76	76	76	76	76	76	76	76
		20	62	62	62	62	62	62	62	62	62	62	62	62	62
		30	57	57	57	57	57	57	57	57	57	57	57	57	57
		40	55	55	55	55	55	55	55	55	55	55	55	55	55
	Non-line of sight 800	10	·	·	359	249	182	139	110	92	83	83	92	110	139
		20	·	·	208	153	118	97	81	72	68	68	72	81	97
		30	·	·	158	121	97	82	72	66	63	63	66	72	82
		40	·	·	132	104	86	76	68	63	61	61	63	68	76
1,800	Line of sight 900	10	80	80	80	80	80	80	80	80	80	80	80	80	80
		20	64	64	64	64	64	64	64	64	64	64	64	64	64
		30	59	59	59	59	59	59	59	59	59	59	59	59	59
		40	57	57	57	57	57	57	57	57	57	57	57	57	57
	Non-line of sight 900	10	·	·	399	273	199	151	117	97	87	87	97	117	151
		20	·	·	228	165	127	102	85	76	71	71	76	85	102
		30	·	·	171	128	103	87	75	69	65	65	69	75	87
		40	·	·	144	111	92	79	70	65	63	63	53	70	79
2,000	Line of sight 1,000	10	85	85	85	85	85	85	85	85	85	85	85	85	85
		20	68	68	68	68	68	68	68	68	68	68	68	68	68
		30	62	62	62	62	62	62	62	62	62	62	62	62	62
		40	59	59	59	59	59	59	59	59	59	59	59	59	59
	Non-line of sight 1,000	10	·	·	437	298	216	162	125	103	92	92	103	125	162
		20	·	·	249	179	136	110	91	79	74	74	79	91	110
		30	·	·	185	138	110	91	79	71	68	68	71	79	91
		40	·	·	153	117	96	82	73	67	65	65	67	73	82

5 Conclusion

For microcell CDMA mobile communication in an urban street, suggested prediction model was simulated as shown in Table 4. But, the minimum receiving power of MS was based on -95dBm and the wave coming to the MS is only one.

Although propagation path is irrelative to road width as well as the incident or reflection angle of propagation, because the same incident and reflection angle is inversely proportional to road width, larger road width has less propagation path loss from Table 4 and maximum service-possible direct path from Table 4.

In sloping angle of $130° \sim 140°$ and incident and reflection angle $25°$, receiving power coming to MS becomes maximum from Table 4. Based on this angle, when sloping angle is increased or decreased, receiving power is decreased. And it is maximum service-possible direct path from Table 4. As a result, to appropriate location of BS, we consider a sloping angle of $130° \sim 140°$ and a reflection angle $25°$.

To design microcell CDMA mobile communication systems in urban propagation environment, we confirmed that road situation(sloping angle and road width) within cell coverage are important parameters to select location of BS.

Acknowledgement

This research is supported Chosun University korea, in 2001.

References

1. W.C.Y.Lee, "Mobile Communications Design Fundamentals, Wildy Interscience 1993.
2. W.C.Y.Lee, "Microcell Architecture, IEEE Communications Magazine, Nov. 1991.
3. T.Iwama and M.Mizuno, "Perdiction of propagation characteristics for microcellular land mobile radio, Proc .ISAP, pp.421-424, Sapporo, Japan, 1992.
4. Joseph sarnecki, C.Vinodrai,Alauddin Javed, Patrick O'kelly and Kevin Dick, "Microcell Design Principles, IEEE Communications Magazine, pp.76-82, April, 1993.
5. F.Ikegani, T.Takeuychi, and S.Yoshida, "Theoretical prediction of mean field strength for urban mobile radio, IEEE Trans. Antennas Propagat. Vol.AP-39, pp.299-302, 1991.
6. V.Erceg, S.Ghassemzadh, M.Taylor, D.Li, and D.L.Schilling,"Urban/suburban out-of-sight propagation modeling", IEEE Comm. Mag., pp.56-61, Jun,1992.
7. K.R. Schaubach, N.J.Davis, and T. S. Rappaport,"A ray tracing method for predicting path loss and delay spread in microcelluar environments", in 42nd IEEE Veh. Technol. Conf., Vol. 2, pp.932-935, May. 1992.
8. M. C. Lawton and J. P. McGeehan, "The applications of a deterministic ray launching algorithm for the prediction of radio channel characteristics in small cell environments, IEEE Trans. Veh Tecnol., vol. 43, pp.955-969, Nov. 1994.
9. Scott Y.seidel, T. S. Rappaport, "Site-specific propagation Prediction for Wireless In-Building Personal Communication System Design", IEEE Cellular radio and Communication Vol. 2, pp.223-891, 1994.
10. S. Y. Tan and H. S. Tan, "A microcellular communications propagation model based on the uniform theory of diffraction and multiple image theory," IEEE Trans., Antennas Propagation, Vol.44, pp.1317-1326, Oct. 1996.
11. W. C. Y. Lee, Mobile communication engineering, McGraw-Hill Company, pp.91.

Efficient Radio Resource Management in Integrated WLAN/CDMA Mobile Networks

Fei Yu and Vikram Krishnamurthy

Department of Electrical and Computer Engineering,
the University of British Columbia,
2356 Main Mall, Vancouver, BC, Canada V6T 1Z4
{feiy, vikramk}@ece.ubc.ca

Abstract. The complementary characteristics of wireless local area networks (WLANs) and wideband code division multiple access (CDMA) cellular networks make it attractive to integrate these two technologies. This paper proposes a joint session admission control scheme that maximizes overall network revenue with quality of service (QoS) constraints over both WLANs and CDMA cellular networks. WLANs operate under IEEE 802.11e medium access control (MAC) protocol, which supports QoS for multimedia traffic. A cross-layer optimization approach is used in CDMA networks taking into account both physical layer linear minimum mean square error (LMMSE) receivers and network layer QoS requirements. Numerical examples illustrate that the network revenue earned in the proposed joint admission control scheme is significantly more than that when the individual networks are optimized independently.

1   Introduction

In recent years, wireless local area network (WLAN)-based systems are emerging as a popular means of wireless public access. While WLANs offer relatively high data rates to users with low mobility over smaller areas (hotspots), wideband code division multiple access (CDMA) cellular networks such as Universal Mobile Telecommunications System (UMTS) provide always on, wide-area connectivity with low data rates to users with high mobility. The complementary characteristics of WLANs and CDMA cellular networks make it attractive to integrate these two technologies [1]-[5].

In integrated WLAN/CDMA systems, a mobile user of a laptop/handheld that supports both WLAN and CDMA access capabilities can connect to both networks by roaming agreements [1], [2]. Vertical handoff (also referred as handover in the literature) between WLANs and CDMA networks can be seen as the next evolutionary step from roaming in this integrated environment. Although some work has been done to integrate WLANs and CDMA networks, most of previous work concentrates on architectures and mechanisms to support roaming and vertical handoff, and how to utilize the overall radio resources optimally subject to quality of service (QoS) constraints for multimedia traffic is largely ignored in this coupled environment.

In this paper, we propose an optimal joint session admission control scheme for multimedia traffic in an integrated WLAN/CDMA system with vertical handoff, which

P. Lorenz and P. Dini (Eds.): ICN 2005, LNCS 3421, pp. 912–919, 2005.

maximizes the overall network revenue while satisfying several QoS constraints in both WLANs and CDMA networks. The proposed scheme can optimally control whether or not to admit as well as which network (WLAN or CDMA network) to admit a new session arrival or a vertical handoff session between WLANs and CDMA networks.

We compare our scheme with other WLAN/CDMA integration schemes. It is shown that the proposed scheme results in significant revenue gain over the schemes in which optimization is done independently in individual networks.

The rest of the paper is organized as follows. Section 2 introduces integrated systems and the joint session admission control problem. Section 3 describes the QoS considerations in WLANs and CDMA networks. Section 4 presents our new approach to solve the joint session admission control problem. Some numerical examples are given in Section 5. Finally, we conclude this study in Section 6.

2 Integrated WLAN/CDMA Systems and the Joint Session Admission Control Problem

There are two different ways of designing an integrated WLAN/CDMA network architecture defined as tight coupling and loose coupling inter-working [1], [2]. Fig. 1 shows the architecture for WLAN/CDMA integration, where UMTS is used as a specific example of CDMA networks. In a tightly coupled system, a WLAN is connected to a CDMA core network in the same manner as other CDMA radio access networks. In the loose coupling approach, a WLAN is not connected directly to CDMA network elements. Instead, it is connected to the Internet. In this approach, the WLAN traffic would not go through the CDMA core network. Nevertheless, as peer IP domains, they can share the same subscriber database for functions such as security, billing and customer management.

In order to utilize overall radio resources efficiently, a joint session admission control (JSAC) is very important in the integration of these two wireless access technologies [3], [5]. The problem of JSAC in integrated WLAN/CDMA systems is whether or not to admit and which network (WLAN or CDMA network) to admit a new or handoff

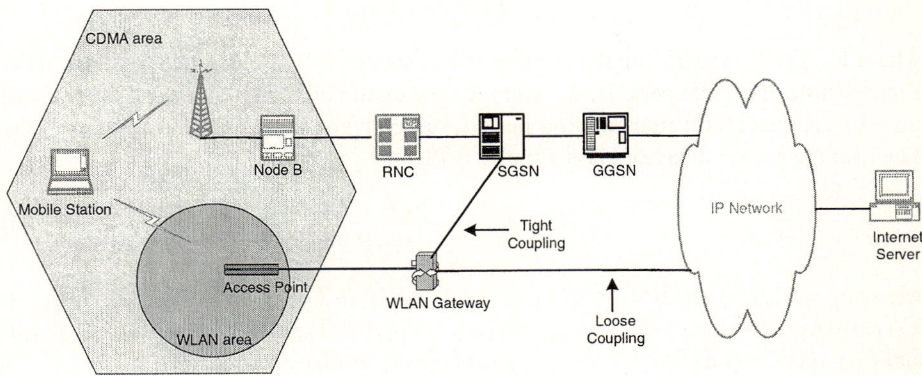

Fig. 1. Integrated WLAN and CDMA cellular networks

session arrival. An optimal JSAC should maximize the long-term network revenue and guarantee QoS constraints for multimedia traffic in both WLANs and CDMA networks.

Since the charging methods in WLANs are usually different from those in CDMA networks, we should consider the different revenue rates in the design of a JSAC. For the QoS constraints in WLANs, throughput and packet delay are important metrics [7], [8]. In CDMA networks, QoS requirements are characterized by SIR. In addition, at the network layer of integrated WLAN/CDMA systems, QoS metrics are blocking probabilities of new or handoff sessions in both networks, which should also be guaranteed.

3   QoS in WLANs and CDMA Cellular Networks

3.1   QoS in WLANs

To support QoS in WLAN, a new MAC protocol, IEEE 802.11e, is proposed [6]. We adopt the derivations of throughput and packet delay for IEEE 802.11e in [8] in this paper. Let ϵ denote the length of a time slot, M denote the average bit rate of the WLAN, T_{SIFS} denote the duration of a short inter-frame space (SIFS), T_{RTS}, T_{CTS}, T_{ACK}, T_{PHY} and T_{MAC} denote the time required to transmit a request-to-send (RTS), a clear-to-send (CTS), an ACK, a physical layer header and an MAC header, respectively. There are J classes of multimedia traffic with distinct QoS requirements in the system. The number of class $j, j = 1, 2, \ldots, J$, sessions in the WLAN is $n_{w,j}$. Assume that a class $j, j = 1, 2, \ldots, J$, packet has a constant probability of collision p_j, all class j packets have the same length S_j, and the propagation delay of all packets is constant π. The average backoff counter of the class j station $\mathbf{E}[BO_j]$ is $(1 - 2p_j)(CW_{j,min} - 1) + p_j CW_{j,min}(1 - (2p_j)^{m_j})/2(1 - 2p_j)$. The probability for a class j station to transmit a packet is $q_j = 2(1 - 2p_j)/(1 - 2p_j)(CW_{j,min} + 1 + AIFSN_j) + p_j CW_{j,min}[1 - (2p_k)^{m_j}]$. The probability of collision can be calculated as $p_j = 1 - (1 - q_j)^{(n_{w,j}-1)} \prod_{1 \leq i \leq J, i \neq j}(1 - q_j)^{n_{w,i}}$. q_j and p_j can be obtained by solving the above equations using numerical techniques. The saturation bandwidth for class j traffic is [8]

$$B_j = \frac{V_j}{T_I + T_C + T_S}, \tag{1}$$

where V_j, T_I, T_C and T_S are the number of bits successfully transmitted for class j, the average time of all idle periods, the average time of all collision periods and the average time of the successful transmission period, respectively, during a transmission cycle. The average packet delay of class j traffic is [8]

$$D_j = b_j + \frac{a_j p_j}{(1 - p_j)^2}, \tag{2}$$

where $b_j = (1/U_j)[\epsilon\mathbf{E}[BO_j] + 4T_{SIFS} + \epsilon AIFSN_j + T_{RTS} + 3\pi + T_{CTS} + T_{PHY} + T_{MAC} + L_j/M + T_{ACK} + (N_j - 1)(2T_{SIFS} + T_{PHY} + T_{MAC} + S_j/M + 2\pi + T_{ACK})]$ and $a_j = (1/U_j)(\epsilon\mathbf{E}[BO_j] + T_{SIFS} + \epsilon AIFSN_j + T_{RTS} + T_{SIFS} + \pi)$.

The throughput constraints and average packet delay constraints will be satisfied if the vector $x_w = (n_{w,1}, n_{w,2}, \ldots, n_{w,J})$ lies within the WLAN admissible set

$$X_W = \left\{ x_w \in \mathbb{Z}_+^J : B_j \geq TB_j, D_j \leq TD_j, j = 1, 2, \ldots, J \right\}, \tag{3}$$

where B_j is defined in (1), D_j is defined in (2), TB_j and TD_j are the target throughput and the target average packet delay, respectively, for class j traffic.

3.2   QoS in CDMA Cellular Networks

Consider a synchronous CDMA system with spreading gain N and K sessions. There are J classes of multimedia traffic in the system. An important physical layer performance measure of class j sessions is the signal-to-interference ratio, SIR_j, which should be kept above the target value ω_j. The signature sequences of all sessions are independent and randomly chosen. Due to multi-path fading, each user appears as L resolvable paths or components at the receiver. The path l of user k is characterized by its estimated average channel gain $\bar{h}_{kl}$ and its estimation error variance ξ_k^2. Linear minimum mean square error (LMMSE) detectors are used at the receiver to recover the transmitted information. In a large system (both N and K are large) with background noise σ^2, the SIR for the LMMSE receiver of a user (say, the first one) can be expressed approximately as [11] $\text{SIR}_1 = (P_1 \sum_{l=1}^{L} |\bar{h}_{1l}|^2 \eta)/(1 + P_1 \xi_1^2 \eta)$, where P_1 is the attenuated transmitted power from user 1, η is the unique fixed point in $(0, \infty)$ that satisfies $\eta = [\sigma^2 + (1/N) \sum_{k=2}^{K} ((L - 1)I(\xi_k^2, \eta) + I(\sum_{l=1}^{L} |\bar{h}_{kl}|^2 + \xi_k^2, \eta))]^{-1}$ and $I(\nu, \eta) = \nu/(1 + \nu\eta)$. In [10], it is shown that a minimum received power solution exists such that all sessions in the system meet their target SIRs if and only if $\omega_j < |\bar{h}_j|^2/\xi_j^2$ and $(1/N) \sum_{j=1}^{J} \sum_{i=1}^{n_{c,j}} R_j^i \Upsilon_j < 1$, where $|\bar{h}_j|^2 = \sum_{l=1}^{L} |\bar{h}_{jl}|^2, j = 1, 2, \ldots, J$, $n_{c,j}$ is the number of class j sessions in the CDMA cell, R_j^i is the number of signature sequences assigned to the ith session of class j to make it transmit at R_j^i times the basic rate (obtained using the highest spreading gain N) and

$$\Upsilon_j = (L - 1)\omega_j \frac{\xi_j^2}{|\bar{h}_j|^2} + \frac{\omega_j \left(1 + \frac{\xi_j^2}{|\bar{h}_j|^2}\right)}{1 + \omega_j}. \tag{4}$$

The SIR constraints will be satisfied if the vector $x_c = (n_{c,1}, n_{c,2}, \ldots, n_{c,J})$ lies within the CDMA admissible set

$$X_C = \left\{ x_c \in \mathbb{Z}_+^J : \frac{1}{N} \sum_{j=1}^{J} \sum_{i=1}^{n_{c,j}} R_j^i \Upsilon_j < 1, j = 1, 2, \ldots, J \right\}. \tag{5}$$

4   Optimal Joint Session Admission Control

In this section, the optimal session admission control (JSAC) problem is formulated as a semi-Markov decision process (SMDP) [12]. For simplicity of the presentation, we consider an integrated WLAN/CDMA system with a single WLAN cell and a single CDMA cell shown in Fig. 1, where the WLAN coverage is within the CDMA cell. We divide the CDMA cell into two areas, CDMA area and WLAN area. Mobile users in the WLAN area can access both the WLAN and the CDMA network, whereas mobile users in the CDMA area can only access the CDMA network. We assume that there are J classes of traffic. Class $j, j = 1, 2, \ldots, J$, new sessions arrive according to a Poisson distribution with the rate of $\lambda_{c,n,j}$ ($\lambda_{w,n,j}$) in the CDMA (WLAN) area. The total new session arrival rate for class j traffic is $\lambda_{n,j} = \lambda_{c,n,j} + \lambda_{w,n,j}$. Class j handoff sessions depart from the CDMA network (WLAN) to the WLAN (CDMA network) according to a Poisson distribution with rate of $\mu_{c,h,j}$ ($\mu_{w,h,j}$). Session duration time for class j traffic is exponentially distributed with the mean $1/\mu_{c,t}$ ($1/\mu_{w,t}$) in the CDMA (WLAN) area. In order to obtain the optimal solution, it is necessary to identify the state space, decision epochs, state dynamics, reward and constraints in the integrated WLAN/CDMA system.

4.1   State Space and Decision Epochs

The state space X of the system comprises of any state vector such that the throughput and average packet delay constraints in the WLAN cell and the SIR constraints in the CDMA cell can be met. Therefore, the state space of the SMDP can be defined as

$$X = \left\{ x = [x_w, x_c] \in \mathbb{Z}_+^{2J} : B_j \geq TB_j, D_j \leq TD_j, \frac{1}{N} \sum_{j=1}^{J} \sum_{i=1}^{n_{c,j}} R_j^i \Upsilon_j < 1, \right\}, \quad (6)$$

where B_j, D_j and Υ_j are defined in (1), (2) and (4), respectively. We choose the decision epochs to be the set of all session arrival and departure instances.

4.2   State Dynamics and Reward Function

The state dynamics of the system can be characterized by the state transition probabilities of the embedded chain and the expected sojourn time $\tau_x(a)$ for each state-action pair. The cumulative event rate is the sum of the rates of all constituent processes and the expected sojourn time is the inverse of the event rate $\tau_x(a) = [\sum_{j=1}^{J}(\lambda_{c,n,j}a_{c,n,j} + \mu_{w,h,j}n_{w,j} + \lambda_{w,n,j}|a_{w,n,j}| + \mu_{c,h,j}n_{c,j}a_{w,h,j}) + \sum_{j=1}^{J}(\mu_{c,t,j}n_{c,j} + \mu_{w,t,j}n_{w,j})]^{-1}$. The state transition probabilities of the embedded Markov chain are

$$p_{xy}(a) = \begin{cases} [\lambda_{c,n,j}a_{c,n,j} + \lambda_{w,n,j}\delta(-a_{w,n,j})]\tau_x(a), & \text{if } y = [x_c + e_j^u, x_w] \\ \lambda_{w,n,j}\delta(a_{w,n,j})\tau_x(a), & \text{if } y = [x_c, x_w + e_j^u] \\ \mu_{w,h,j}n_{w,j}a_{c,h,j}\tau_x(a), & \text{if } y = [x_c + e_j^u, x_w - e_j^u] \\ \mu_{c,h,j}n_{c,j}a_{w,h,j}\tau_x(a), & \text{if } y = [x_c - e_j^u, x_w + e_j^u], \\ \mu_{c,t,j}n_{c,j}\tau_x(a), & \text{if } y = [x_c - e_j^u, x_w] \\ [\mu_{w,t,j} + \mu_{w,h,j}(1 - a_{c,h})]\,n_{w,j}\tau_x(a), & \text{if } y = [x_c, x_w - e_j^u] \\ 0, & \text{otherwise} \end{cases} \quad (7)$$

where $\delta(x) = 0$, if $x \leq 0$ and $\delta(x) = 1$, if $x > 0$.

We define the reward for state-action pair (x, a) as $r(x, a) = \sum_{j=1}^{J}[w_{c,n,j}a_{c,n,j} + w_{c,h,j}a_{c,h,j} + w_{w,n,j}\delta(a_{w,n,j}) + w_{c,n,j}\delta(-a_{w,n,j}) + w_{w,h,j}a_{w,h,j} + w_{c,n,j}(1 - a_{w,h,j})]$, where $w_{c,n,j}, w_{c,h,j}, w_{w,n,j}$ and $w_{w,h,j}$ are the weights associated with $a_{c,n,j}, a_{c,h,j}, a_{w,n,j}$ and $a_{w,h,j}$, respectively.

4.3   Constraints

In the formulated problem, throughput constraints, packet delay constraints in the WLAN and SIR constraints in the CDMA can be guaranteed by restricting the state space in (6). The constraints related to the new session blocking probabilities in the CDMA network can be expressed as $P_{c,n,j}^b \leq \gamma_{c,n,j}$. The new session blocking probability constraints in the CDMA networks can be easily addressed in the linear programming formulation by defining a cost function related to these constraints, $c_{c,n,j}^b(x,a) = 1 - a_{c,n,j}$. Other session blocking probability constraints can be addressed similarly.

4.4   Linear Programming Solution to the SMDP

The optimal policy u^* of the SMDP is obtained by solving the following linear program (LP).

$$\max_{z_{xa} \geq 0, x \in X, a \in A_x} \sum_{x \in X} \sum_{a \in A_x} \sum_{j=1}^{J} r(x,a)\tau_x(a) z_{xa}$$

subject to

$$\sum_{a \in A_y} z_{ya} - \sum_{x \in X} \sum_{a \in A_x} p_{xy}(a) z_{xa} = 0, y \in X,$$

$$\sum_{x \in X} \sum_{a \in A_x} z_{xa}\tau_x(a) = 1,$$

$$\sum_{x \in X} \sum_{a \in A_x} (1 - a_{c,n,j}) z_{xa}\tau_x(a) \leq \gamma_{c,n,j}, j = 1, 2, \ldots, J,$$

$$\sum_{x \in X} \sum_{a \in A_x} (1 - a_{c,h,j}) z_{xa}\tau_x(a) \leq \gamma_{c,h,j}, j = 1, 2, \ldots, J, \tag{8}$$

$$\sum_{x \in X} \sum_{a \in A_x} (1 - |a_{w,n,j}|) z_{xa}\tau_x(a) \leq \gamma_{w,n,j}, j = 1, 2, \ldots, J,$$

$$\sum_{x \in X} \sum_{a \in A_x} (1 - a_{w,h,j}) z_{xa}\tau_x(a) \leq \gamma_{w,h,j}, j = 1, 2, \ldots, J.$$

5   Numerical Results

One class of video traffic is considered in an integrated WLAN/CDMA system with a single WLAN cell and a single CDMA cell. Each video flow is 1.17 Mbps in the WLAN, which is generated by a constant inter-arrival time 10ms with a constant payload size of 1464 bytes. The numerical values for the system parameters are given in Table 1. We compare the average reward earned in the proposed scheme to those in two other WLAN/CDMA integration schemes. In the first scheme, admission control is done independently in individual networks and there is no vertical handoff between the WLAN and the CDMA network. In the second scheme, handoff between these two networks can be supported. Note that there is no joint session admission control and joint radio resource management in both schemes.

The percentage of reward gain is shown in Fig. 2. In this example, 40% of the total new session arrivals in the system occur in the WLAN area. $\mu_{c,t} = \mu_{w,t} = 0.005$, $\mu_{c,h} = 0.004$ and $\mu_{w,h} = 0.0005$. $w_{c,n} = w_{c,h} = 2$. $w_{w,n} = w_{w,h} = 1$. From Fig. 2, we can see that the reward earned in the proposed scheme is always more than those in

Table 1. Parameters used in the numerical examples

Parameter	Notation	Value
average channel bit rate	M	11 Mbps
slot time	ϵ	10 μs
propagation delay	π	1 μs
time required to transmit a PHY header	T_{PHY}	48 μs
time required to transmit an MAC header	T_{MAC}	25 μs
time required to transmit a request-to-send (RTS)	T_{RTS}	15 μs
time required to transmit a clear-to-send (CTS)	T_{CTS}	10 μs
time required to transmit an ACK	T_{ACK}	10 μs
arbitration inter-frame space number (AIFSN) for video traffic	$AIFSN$	1
minimum contention window for video traffic	CW_{min}	16
maximum contention window for video traffic	CW_{max}	32
video packet size	V	1464 bytes
target SIR for video traffic	ω	10 dB
estimated average channel gain for video traffic	$\overline{|h|^2}$	1
channel estimation error variance for video traffic	ξ^2	0.034
number of resolvable paths	L	5
data transmission rate for video traffic	R	240 Kbps

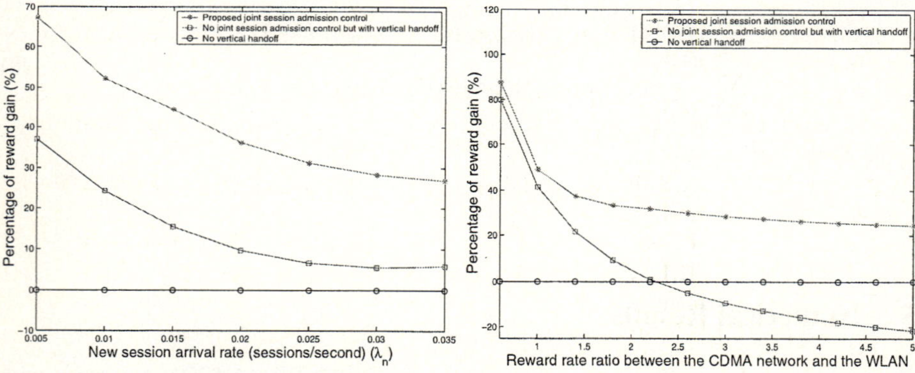

Fig. 2. Percentage of reward gain vs. arrival rate

Fig. 3. Percentage of reward gain vs. reward rate ratio between the CDMA network and the WLAN ($\lambda_n = 0.03$)

two other schemes, and the reward is the least in the scheme in which vertical handoff is not supported. The reward rate ratio between the CDMA network and the WLAN will be different with different network operators. Fig. 3 show the reward gain with $\lambda_n = 0.03$. It is very interesting to observe that the reward in the scheme with vertical handoff but no joint admission control will less than that in the scheme without vertical handoff support when the ratio is larger than 2.2 in Fig. 3. However, the proposed scheme can always have reward gain with a large range of the ratio values.

6   Conclusions

In this paper, we have presented an optimal joint session admission control scheme in integrated WLAN/CDMA networks to utilize the overall radio resources optimally. The proposed scheme can maximize network revenue with QoS constraints in both WLANs and CDMA cellular networks. We illustrated the performance of the proposed scheme by numerical results. It was shown that the proposed optimal joint session admission control scheme can have a very significant gain over the schemes in which admission controls are done independently in individual networks.

References

1. Salkintzis, A. K.:Interworking techniques and architectures for WLAN/3G integration toward 4G mobile data networks. IEEE Wireless Commun., Vol. 11 (2004) 50–61
2. Buddhikot, M., Chandranmenon, G., Han, S., Lee, Y. W., Miller, S., Salgarelli, L.: Integration of 802.11 and third-generation wireless data networks. Proc. IEEE INFOCOM'03 (2003)
3. Luo, L., Mukerjee, R., Dillinger, M., Mohyeldin, E., Schulz, E.: Investigation of radio resource scheduling in WLANs coupled with 3G cellular network. IEEE Commun. Magazine, Vol. 41 (2003) 108–115
4. Zhang, Q., Guo, C., Guo, Z., and Zhu, W.: Efficient mobility management for vertical handoff between WWAN and WLAN. IEEE Commun. Magazine, Vol. 41 (2003) 102–108
5. Zhuang, W., Gan, Y.-S., Loh, K.-J., Chua, K.-C.: Policy-based QoS management architecture in an integrated UMTS and WLAN environment. IEEE Commun. Magazine, Vol.41 (2003) 118–125
6. IEEE: Wireless Medium Access Control (MAC) and Physical Layer (PHY) Specification: Medium Access Control (MAC) Enhancement for Quality of Service (QoS). ANSI/IEEE Std 802.11e, Draft 5.0 (2003)
7. Xiao, Y., Li, H., Choi, S.: Protection and guarantee for voice and video traffic in IEEE 802.11e wireless LANs. Proc. IEEE INFOCOM'04 (2004)
8. Kuo, Y.-L., Lu, C.-H., Wu, E., Chen G.-H.: An admission control strategy for differentiated services in IEEE 802.11. Proc. IEEE Globecom'03 (2003)
9. Singh, S., Krishnamurthy, V., Poor, H. V.: Integrated voice/data call admission control for wireless DS-CDMA systems with fading. IEEE Trans. Signal Proc., Vol. 50 (2002) 1483–1495
10. Comaniciu. C., Poor, H. V.: Jointly optimal power and admission control for delay sensitive traffic in CDMA networks with LMMSE receivers. IEEE Trans. Signal Proc., Vol. 51 (2003) 2031–2042
11. Evans, J., Tse, D. N. C.: Large system performance of linear multiuser receivers in multipath fading channels. IEEE Trans. Inform. Theory, Vol. 46 (2000) 2059–2078
12. Puterman, M.: Markov Decision Processes. John Wiley (1994)

A Study on the Cell Sectorization Using the WBTC and NBTC in CDMA Mobile Communication Systems

Dong-You Choi[1] and Sun-Kuk Noh[2,*]

[1] Research Institute of Energy Resources Technology , Chosun University ,
`dy_choi@chosun.ac.kr`
[2] Dept. of Radio Communication Engineering,
Honam University, Republic of Korea
`nsk7070@honam.ac.kr`

Abstract. CDMA cellular mobile communication service needs to offer good quality of services (QoS) and coverage for new and sophisticated power control to achieve high capacity. In this paper, in order to efficient cell sector and to improve the Ec/Io of the pilot channel in CDMA mobile communication systems, we analyze and compare the wide-beam trisector cell(WBTC) with the narrow-beam trisector cell(NBTC) as a method for cell sectorization. The NBTC method was compared with the WBTC method, using the results of both a theoretical analysis and a simulation, in order to examine its effectiveness and validity. As a result, we confirm that efficient cell sector method to minimize interference from adjacent base stations and to increase cell coverage.

1 Introduction

CDMA cellular mobile communication service needs to offer good quality of services (QoS) and coverage for new and sophisticated power control to achieve high capacity. Various technologies have been developed to improve the speech quality and service of mobile communications.[1-3] On the other hand, en economical approach to enhance spectrum efficiency is to develop a better cellular engineering methodology. This approach is economical in the sense that it minimizes the cost of base station equipment. Cellular engineering includes three major aspects : 1) enhancing frequency planning to reduce interference, 2) selecting a cell architecture to improve the coverage and interference performance, 3) choosing better cell site locations to enhance service coverage.[4]

One way of accomplishing this, cell sectorization and improvement of the Chip Energy/Others Interference(Ec/Io). The cell sectorization techniques are widely in cellular systems to reduce co-channel interference by means of directional antennas and improvement of the Ec/Io of the pilot channel in forward channel, can enhance the speech quality and service of mobile communications and increase the channel capacity, by raising the frequency use efficiency. Furthermore, the costs associated with network construction and operation can be reduced.

---

* Corresponding Author.

P. Lorenz and P. Dini (Eds.): ICN 2005, LNCS 3421, pp. 920–927, 2005.

In this paper, in order to efficient cell sector and to improve the Ec/Io of the pilot channel in CDMA mobile communication systems, we analyze and compare the wide-beam trisector cell(WBTC)[3-6] with the narrow-beam trisector cell(NBTC))[4-7] as a method for cell sectorization. As a result that the comparison of theoretical analysis (Ec/Io, softer hand off, interference ratio) and simulation (Ec/Io, Eb/Nt) of the WBTC and NBTC, we confirm that efficient cell sector method to minimize interference from adjacent base stations and to increase cell coverage.

2 Cell Sectorization Techniques

With the gradual increase in the number of users and base stations used in mobile communications, the problem of adjacent base station interference has become of increasing concern. In order to minimize base station interference and expand base station capacity, cell sectorization has generally been used. This sectorization method has shown itself to be very effective at minimizing interference and expanding base station channel capacity between adjacent base stations when the frequency spectrums being used are different and the traffic channel frequencies used by adjacent base stations are not the same.[8-11] Two important factors influence the effectiveness of sectorization. One is the number of sectors per cell, and the other is the beamwidth of the directional antenna. Base stations in current cellular systems typically have three to six sectors per cell[12-13]. Traditional cell architectures usually have antenna beamwidth of Γ degrees equal to the ratio of 360° over number of sectors per cell S, that is[4]

$$\Gamma = \frac{360}{S} \tag{1}$$

For example, the WBTC in the first generation cellular mobile systems employs three 120° antennas to cover one cell. Instead of using (1), subsequent cellular architectures determine rhe cell contour based on antenna radiation patterns, from which antenna beamwidth is then defined. The NBTC in the second generation cellular mobile systems employs three 60° directional antennas.

2.1 WBTC

The WBTC architecture uses three 100°-120° antennas at each cell site, in which each antenna is designed to cover a rhombus sector, represented in fig.1. Actual antenna radiation pattern is wide beam.[14] A WBTC forms a coverage area with the shape of a hexagon. We observe that the rhombus sector of a WBTC does not match actual coverage contour of the antenna. Side –lobe levels are also significant in adjacent sectors.

In the CDMA system, however, if adjacent base stations use the same frequency, then in the worst case scenario, there will be considerable overlap the main beams between the self-base station and two adjacent base stations at a-area in Fig. 1. (b). Therefore, poor coverage occurs in the corners of the hexagon at the common boundary of the six sectors and that is giving rise to a reduction in speech quality and channel capacity.

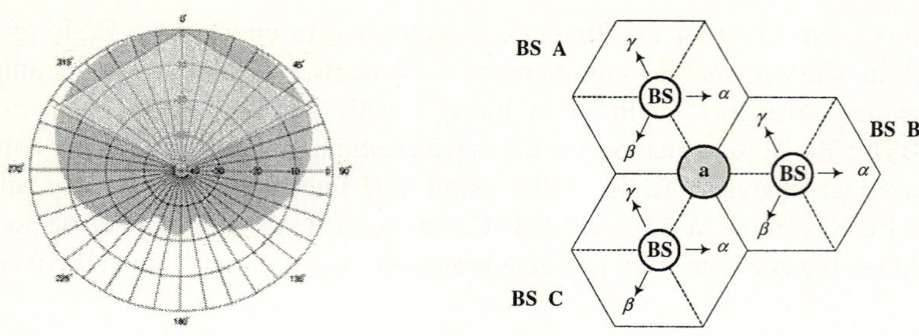

(a) wide beam (the rhombus radiation pattern)   (b) Coverage area of a base station

Fig. 1. WBTC method

2.2 NTBC

The NBTC is covered by a base station with three 60° directional antenna, represented in Fig. 2. Actual antenna radiation pattern is narrow beam.[16] As shown in Fig. 2, the NBTC forms a coverage area with the shape of a hexagon because that the narrow beam radiation pattern match well a hypothetical hexagon sector. With three such antennas, the coverage contour of the NBTC is therefore like a clover leaf. We observe that the ideal hexagon-shaped sector of a NBTC does match actual coverage contour of the antenna. Because of the better match between the cellular contour and the actual cell coverage, the NBTC system performs better than the WBTC system.

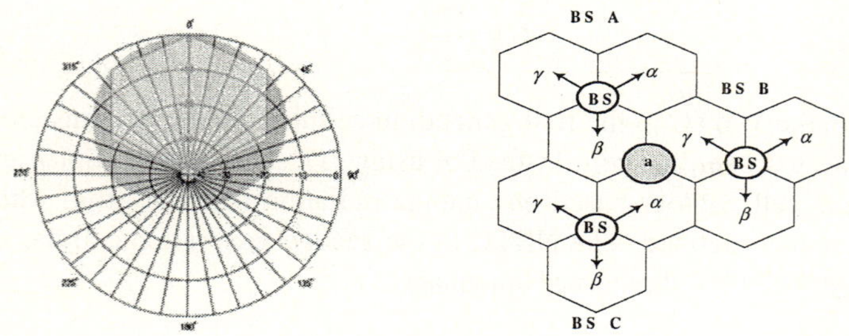

(a) narrow beam(the hexagon radiation pattern)   (b) Coverage area of a base station

Fig. 2. NBTC method

3 Theoretical Analysis of the WBTC and NBTC

In order to theoretical analyze of the WBTC and NBTC, by using an antenna with a narrow beam width instead of a wide beam width at common cell area, the direction of the antenna main beam needs to be placed across in Fig. 3. The strength of the receiving signal at a-area in NBTC will be 1/3 of the transmitted signal of WBTC, if the subscribers are distributed equally by sector, and each sector therefore receives 1/3 of the signal strength.

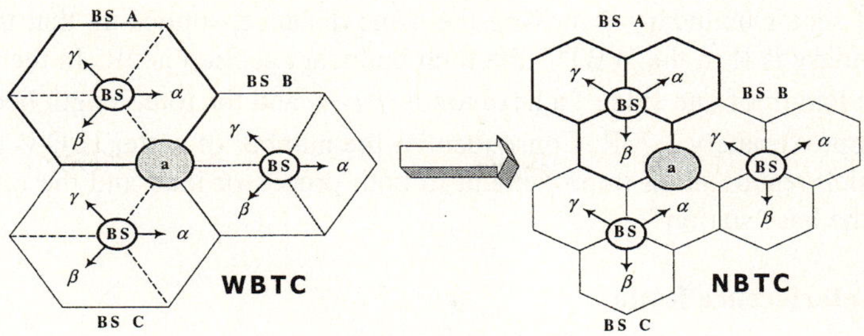

Fig. 3. Comparison of the WBTC and NBTC

3.1 Ec/Io

For the same point in Fig. 3, a-area, let us suppose that the sector power at each base station is the same and that only a pilot signal is transmitted. In the WBTC method, when the terminal station at a-area point picks up β, the sector signal of base station A, the received Ec/Io is given by

$$\frac{E_c}{I_o} = \left(\frac{A_\beta Pilot}{A_\alpha Pilot + A_\beta Pilot + B_\beta Pilot + B_\gamma Pilot + C_\alpha Pilot + C_\gamma Pilot}\right) = 10\log\left(\frac{1}{6}\right) = -7.78 dB \quad (2)$$

On the other hand, in the case of the NBTC method, if the terminal station picks up β, the sector signal of base station A, at the same point a-area, the received Ec/Io is given by

$$\frac{E_c}{I_o} = 10\log\left(\frac{A_\beta Pilot}{A_\beta Pilot + B_\gamma Pilot + C_\alpha Pilot}\right) = 10\log\left(\frac{1}{3}\right) = -4.77 dB \quad (3)$$

It can be seen from equations (2) and (3), that the NBTC method can improve the Ec/Io value by 3.01dB.

3.2 Softer Hand Off

Fig. 4 compares the softer hand off (H/O) of boundary between sector (α, β, γ) of the WBTC to that of the NBTC. Since the number of softer H/O is proportional to the

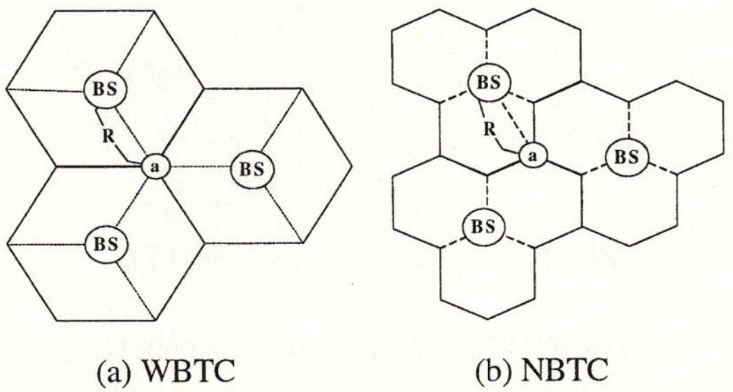

(a) WBTC   (b) NBTC

Fig. 4. Softer H/O comparison of the WBTC and NBTC

distance of sector boundary if moving the same distance, supposing that the distance to the boundary is R in the WBTC, the total boundary section is 3R. In the case of the NBTC, the length of one side of a hexagon is $R/\sqrt{3}$ and the total length becomes $\sqrt{3}R$, which is approximately 1.73R. Consequently, the number of softer H/O is reduced by 42.4%, which results in an improvement in both processor load and the rate of channel use at the base station.

3.3 The Interference Ratio

The capacity of a CDMA mobile communications system varies depending on the capacity of the reverse link. To analyze the reverse capacity, we need to compute the interference ratio(F) of self-cell and other cell, which is given by

$$F = \frac{\eta}{(N-1)S} = \frac{\sum\left(P_{Tmci}D_{ai}^{-\gamma}G_{ai}\right)}{\sum\left(P_{Tm0}R_0^{-\gamma}G_0\right)} = \sum_{i=1}^{N}\left(\frac{P_{Tmci}D_{ai}^{-\gamma}G_{ai}}{P_{Tm0}R_0^{-\gamma}G_0}\right) \tag{4}$$

η : Interference of adjacent cell , N : User number per cell
S : Desired Signal ,   i : active interference
0 : subscript desired signal ,   a : adjacent cell
P_{Tmai} : Transmission power of a mobile station in adjacent cell
P_{Tm0} : Transmission power of a mobile station in self-cell (center cell)
Dai : distance from a mobile station in adjacent cell to base station in self-cell
R_0 : distance from a mobile station to base station in self-cell
Gai : Receive antenna gain of a mobile station in adjacent cell
G_0 : Receive antenna gain of a mobile station in self-cell
γ : path loss exponent = 4.

As shown in Fig 5, if there are 18 interference cells counting from the center(for a total diameter of 15km, a radius of 1.5Km per cell), the computed results of equation (4) were 0.36 for the WBTC and 0.30 for NBTC. This means that the NBTC can reduce other cell interference by 6% compared with the WBTC in the direction of the reverse link, and can give rise to an equivalent improvement in channel capacity.

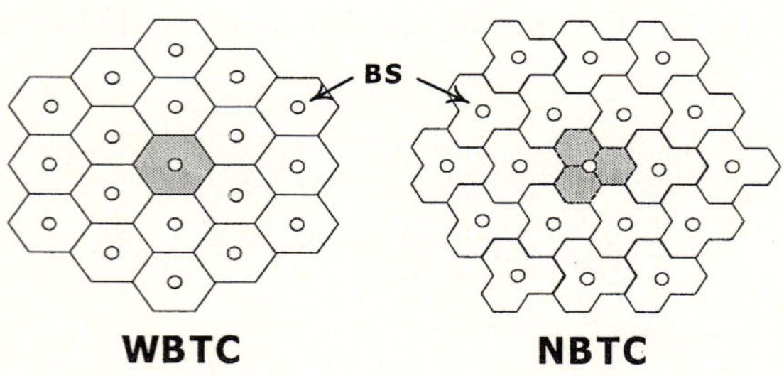

Fig. 5. 19 cells of the WBTC and NBTC

4   Simulation of the WBTC and NBTC

4.1   Simulation

For the simulation, 19 base stations were positioned on even terrain at intervals of 4km. The following values were assumed: base station power: 39 dBm(7.94 W), overhead channel power: 32 dBm(1.58 W), traffic units: 10 Erl per sector. Using the HATTA-OKUMURA[21-22] prediction model, the WBTC and the NBTC were simulated. Based on the value $Rx \geq -94dBm$ Ec/Io=-8dB, and the bit Energy/Total Noise(Eb/Nt)=7dB, which determine the coverage and transmitting quality in a CDMA mobile communication, the simulation results are table 1 and table 2.

Table 1. Comparison of Ec/Io distribution

Cell coverage / Ec/Io	WBTC [%]	NBTC [%]
-4dB over	25.96	41.71
-4 ~ -6dB	40.1	37.62
-6 ~ -8dB	25.89	19.93
-8 ~ -10dB	7.05	0.74
-10dB below	1	0

Table 2. Comparison of Eb/Nt distribution

Cell coverage / Eb/Nt	WBTC [%]	NBTC [%]
11dB over	38.07	52.54
11 ~ 9dB	17.04	16.65
9 ~ 7dB	20.81	17.05
7 ~5dB	24.08	13.76
5dB below	0	0

4.2   Results Analysis

(1) Results of Simulation

In this section, we discuss the improvement obtained in the Ec/Io of the CDMA mobile communication, based on the simulation results, which are as follows.

In the case of the WBTC, a cell coverage was formed, by the interconnection of three rhomboid shapes. In this case, it was determined that at worst, a maximum of 6 interference signals from adjacent base stations were overlapped in the boundary coverage of each sector.

On the other hand, in the case of the NBTC, a cell coverage was formed, by the interconnection of six hexagon shapes. The signal overlap was decreased such that less than 3 interference signals from adjacent base stations were overlapped in the boundary coverage of each sector.

(2) Results of Ec/Io

Table 1 shows that the cell coverage where mobile communication service is possible with Ec/Io $\geq$ -8dB, constituted 99.26% of the NBTC, but only 91.95% of the WBTC. Thus, based on the cell coverage for which an adequate level of service is provided, the NBTC method provides 7.31% more coverage than the WBTC method.

(3) Results of Eb/Nt

Table 2 shows that the cell coverage where mobile communication service is possible with Eb/Nt(Bit energy/Total noise) $\geq$ 7dB, constituted 86.24% of the NBTC, but only 75.92% of the WBTC cell. Thus, in terms of the cell coverage where in an adequate level of service is provided, the NBTC method provides 10.32% more coverage than the WBTC method.

5   Conclusion

In this paper, we analyze and compare the WBTC with the NBTC as a method for cell sectorization, designed to minimize the degradation of call quality and performance in narrowband CDMA mobile communication systems. The effectiveness and validity of the NBTC were verified with a theoretical comparison (Ec/Io, softer hand off, interference ratio) with the WBTC for cell sectorization, and an analysis of the results(Ec/Io, Eb/Nt) of simulation. As a result, we confirm that the NBTC method is an efficient cell sector method to minimize interference from adjacent base stations and to increase cell coverage.

In conclusion, the NBTC for cell sectorization should be considered when constructing the IMT-2000 network and the next generation communication network(NGCN, 4G), since this would provide for more economical network construction, improved speech quality, and the high use efficiency of frequency.

References

1. K.S.Gilhousen, I.M.Jacobs, R.Padovani, A.Viterbi, L.A.Weaver, C.E.Wheatley III,   "On the capacity of a cellular CDMA System", IEEE Transactions on Vehicular Technology, vol. 40. no. 2, pp. 303-312, May 1991.
2. A. Ahmad, "A CDMA Network Architecture Using Optimized Sectoring," IEEE Trans. Veh. Technol., vol.51, pp.404-410, May 2002.
3. Benjamin K. Ng, Elvino S. Sousa, "Performance Enhancement of DS-CDMA System Using Overlapping Sectors With Interference Avoidance," IEEE Trans. Wireless Commum., vol. 2, No. 1, pp.175-185, Jan. 2003.
4. L.C. Wang, K. K. Leung, "A high-capacity wireless network by quad-sector cell and interleaved channel assignment", IEEE Journal on selected areas in comm., vol 18, no. 3, pp.472-480, March. 2000.
5. L.C. Wang, K. C. Chawal, and L. J. Greenstein, "Performance studies of narrow beam tri-sector cellular systems", Int., J. Wireless Information Networks, vol. no.2, pp89-102, 1998.

6. L.C. Wang, K. K. Leung, "Performance enhancement by narrow-beam quad-sector cell and interleaved channel assignment in wireless networks", Global Tel. Conf., pp2719-2724, 1999.
7. L.C. Wang, "A new cellular architecture based on an interleaved cluster concept", IEEE Transactions on Vehicular Technology, vol 48, no. 6, pp.1809-1818, Nov. 1999.
8. Chen, G. K., "Effects of Sectorization on the Spectrum Efficiency of Cellular Radio Systems", IEEE Transactions on Vehicular Technology, vol 41, no. 3, pp. 217-225, Aug. 1992.
9. V.H.MacDonald, "The Cellular Concept", Bell System Technology Journal, vol. 58, pp.15-42, Jan 1992.
10. Bruno Pattan, Robust Modulation Methods and Smart Antennas in Wireless Communication, Prentice Hall PTR, pp. 209-214, 2000.
11. Andrew Miceli, Wireless Technician's Handbook, Artech House, pp. 27-46, 2000.
12. J.Xiang, "A novel two site frequency reuse plan", in Proc. IEEE Vehicular Technology Conf., 1996., pp.441-445.
13. G. K. Chan, "Effecets of sectorization on the spectrum efficiency of cellular radio systems", IEEE Trans. Veh. Technol., vol. 41, pp217-225, Aug. 1992.
14. C. Balanis, Antenna Theory, Harper & Row, Publishers, 1982.
15. Y. Okumura, E. Ohmori, T. Kawano, and K. Fukuda, "Field strength and its variability in VHF and UHF land-mobile radio service," Rev. Elec. Commun. Lab., vol. 16, pp. 825-873, Sept. 1968.
16. M. Hata, "Empirical formula for propagation loss in land mobile radio service," IEEE Trans. Veh. Technol., vol. VT-29, pp317-325, Aug. 1980

DOA-Matrix Decoder for
STBC-MC-CDMA Systems[*]

Yanxing Zeng, Qinye Yin, Le Ding, and Jianguo Zhang

School of Electronics and Information Engineering,
Xi'an Jiaotong University, Xi'an, 710049, P. R. China
qyyin@mail.xjtu.edu.cn

Abstract. The uplink of a space-time block coded multicarrier modulation code division multiple access (MC-CDMA) system equipped with a uniform linear array (ULA) at the base station in macrocell is studied and the base station antenna array is considered largely correlated because of insufficient antenna spacing and lack of scatterings. Under the circumstances, a blind decoder that provides closed-form solutions of both transmitted symbol sequences and directions of arrival (DOAs) for multiple users is developed without uplink space-time vector channel estimation. In particular, the decoder has a desirable property of being automatically paired between the transmitted symbol sequence of an individual user and corresponding DOA, which eliminates performing time-consuming joint diagonalization operations of many matrices [1] and thus saves computation cost. Computer simulations show the proposed receiver outperforms the conventional MMSE receiver based on channel estimation [2,3].

1 Introduction

Space-time block coded MC-CDMA (STBC-MC-CDMA) system is one of the most promising schemes for next-generation wireless communications. To coherently decode the STBC signals, [2,3] proposed a subspace-based channel estimation algorithm and developed a MMSE receiver with the estimated channels. However, it may be difficult or costly to estimate the channel accurately in mobile fading environment. So [4] gave an alternative subspace-based receiver that directly estimates the transmitted symbol sequence without channel information by exploiting the structure of STBC. All of them assumed the channels between all transmit antennas and receive antennas are independent. But in a macrocell environment, the base station is deployed above the surrounding scatters. The received signals at the base station result from the scattering process in the

* This paper was partially supported by the National Sciences Foundation (No.60272071), the Research Fund for Doctoral Program of Higher Education of China (No.20020698024, No.20030698027) and the Science Foundation of Xi'an Jiaotong University.

P. Lorenz and P. Dini (Eds.): ICN 2005, LNCS 3421, pp. 928–935, 2005.
© Springer-Verlag Berlin Heidelberg 2005

vicinity of the mobile station and the multipath components at the base station are thus restricted to a small angular range. Therefore, it is usually assumed the receive antennas at the base station are correlated, meanwhile the transmit antennas at the mobile station are uncorrelated [5]. Under the circumstances, we claim that the aforementioned algorithms are not optimal because they didn't explore the correlations between receive antennas.

In this paper, we apply a full correlated ULA to the base station of STBC-MC-CDMA systems in macrocell and propose a blind receiver algorithm to detect the STBC symbols without uplink channel estimation by use of an ESPRIT-like approach, i.e., direction of arrival Matrix (DOA-Matrix) method [6]. The algorithm separates multiple users with different impinging DOAs in spatial field and obtains a set of signal subspaces, each of them spanned by the transmitted symbol sequences of an individual user. From these signal subspaces, the original symbol sequences of multiple users are estimated by exploiting the structure of STBC. Moreover, the algorithm also obtains the important DOA information. As well-known, in Frequency Division Duplex (FDD) systems, the uplink and the downlink channels are uncorrelated but both links still share many common features, i.e., the number of paths, the path delays and the DOAs, which are not frequency dependent [7]. So, in FDD systems, the DOA estimated from the uplink is necessary to be used by the downlink pre-beamformer. Finally, the algorithm has an interesting property of being automatically paired between the DOA of an individual user and corresponding symbol sequences and hence eliminates performing the time-consuming joint diagonalization operations of many matrices [1]. Computer simulations show the proposed receiver outperforms the conventional MMSE receiver based on channel estimation [2,3].

Notation: $(\cdot)^*$, $(\cdot)^T$, $(\cdot)^H$ denote the complex conjugate, the transpose and the conjugate transpose of matrix.

2 System Model

Fig. 1 displays the baseband model of the STBC-MC-CDMA system equipped with a ULA at the base station, where the subscript k represents the number of user. The antenna separation of the ULA with M antenna elements is half a carrier wavelength. It is assumed that K users uniformly distributed around a macrocell site. All K active users share the same set of subcarriers. As shown in Fig. 1, the input data symbols of each user are coded by Tarokh's rate 1 space-time block encoder [8] from real orthogonal design, with three transmit antennas (specified in Section 4). When MC-CDMA systems over wireless finite impulse response (FIR) channels, a usual approach for combating the resultant inter-block interference (IBI) is via adding cyclic prefix (ACP) to each transmitted data block. Meanwhile, by removing CP (RCP) at the beginning of each received data block, the IBI can be eliminated. For the sake of simplicity, we assume that symbols from different users are synchronized for the uplink.

A frequency domain spreading code is assigned to the k-th user's three transmit antennas, which is defined as a vector $\mathbf{c}^k = [c_1^k, \cdots, c_G^k]^T$. At the n-th symbol

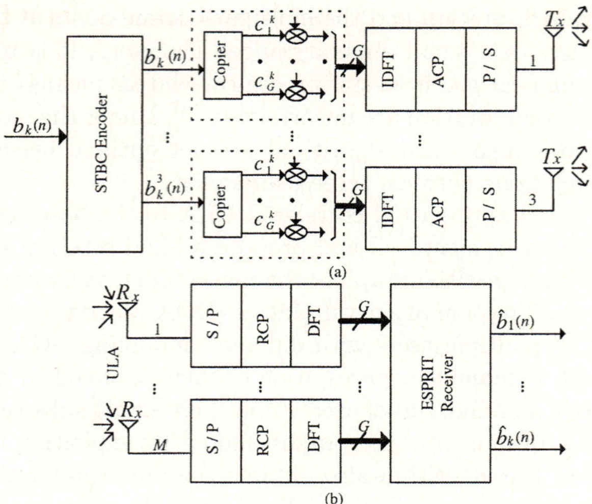

Fig. 1. The baseband model of the STBC-MC-CDMA system equipped with a ULA at the base station (a) user transmitter (b) base station receiver

interval, we define the frequency domain IBI-free uplink received data vector on the reference element 1 of the base station antenna array from the m-th ($m = 1, 2, 3$) transmit antenna of the k-th user as (reference (5) in [3])

$$\mathbf{x}_{1,k}^{m,n} = \mathrm{diag}(\mathbf{c}^k)\boldsymbol{\eta}_{1,k}^{m,n}b_k^m(n) = \mathbf{f}_{1,k}^{m,n}b_k^m(n) \tag{1}$$

where $\boldsymbol{\eta}_{1,k}^{m,n}$ is the frequency domain response [3] of the FIR channel between the m-th transmit antenna of the k-th user and the reference element 1, with dimension of $G \times 1$. $\mathbf{f}_{1,k}^{m,n} = \mathrm{diag}(\mathbf{c}^k)\boldsymbol{\eta}_{1,k}^{m,n}$, $b_k^m(n)$ is the transmitted symbol from the m-th antenna of the k-th user.

On the reference element $r(r = 1, \cdots, M)$ of the base station antenna array, the received data vector consisting of received signals from all K active users over each subcarrier is

$$\mathbf{x}_r^n = \sum_{k=1}^{K}\sum_{m=1}^{3}\mathbf{x}_{r,k}^{m,n} + \mathbf{n}_r^n = \mathbf{F}_r^n\mathbf{b}^n + \mathbf{n}_r^n \tag{2}$$

where $\mathbf{b}^n$ consists of transmitted symbols from all transmit antennas of K users, namely, $\mathbf{b}^n = [b_1^1(n), b_1^2(n), b_1^3(n), \cdots, b_K^3(n)]^T$; $\mathbf{n}_r^n$ is a vector of independent identically distributed complex zero-mean Gaussian noises with variance σ_n^2; $\mathbf{F}_r^n$ is defined as $\mathbf{F}_r^n = [\mathbf{f}_{1,1}^{1,n}e^{j\chi(r-1)\sin\theta_1^n}, \mathbf{f}_{1,1}^{2,n}e^{j\chi(r-1)\sin\theta_1^n}, \mathbf{f}_{1,1}^{3,n}e^{j\chi(r-1)\sin\theta_1^n}, \cdots, \mathbf{f}_{1,K}^{3,n} \cdot e^{j\chi(r-1)\sin\theta_K^n}]$, and $\chi = 2\pi d/\lambda$. Notations λ, d, and θ_k^n represent the wavelength of Radio Frequency (RF) carrier, the inter-element spacing and the incident angle with respect to the array normal, respectively. We assume θ_k^n denotes the DOA of the main cluster from user k's three transmit antennas and there is only one main DOA for each user in the paper.

3   DOA-Matrix Method for Co-channel User Separation

Concatenating uplink received data vectors from element 1 to $M-1$, an extended data vector can be obtained, that is

$$\mathbf{u}^n = \left[\, \mathbf{x}_1^{n^T} \quad \cdots \quad \mathbf{x}_{M-1}^{n^T} \,\right]^T = \mathbf{A}^n \mathbf{b}^n + \mathbf{n}_{head}^n \tag{3}$$

where $\mathbf{u}^n$ is a $(M-1)G \times 1$ vector, $\mathbf{n}_{head}^n$ is given by $\mathbf{n}_{head}^n = [\mathbf{n}_1^{n^T}, \cdots, \mathbf{n}_{M-1}^{n^T}]^T$; $\mathbf{A}^n$ is a $(M-1)G \times 3K$ matrix, and is given by

$$\mathbf{A}^n = \left[\, \mathbf{F}_1^{n^T} \quad \cdots \quad \mathbf{F}_{M-1}^{n^T} \,\right]^T$$
$$= \left[\, \mathbf{a}_1^n \mathbf{f}_{1,1}^{1,n} \quad \mathbf{a}_1^n \mathbf{f}_{1,1}^{2,n} \quad \mathbf{a}_1^n \mathbf{f}_{1,1}^{3,n} \quad \cdots \quad \mathbf{a}_K^n \mathbf{f}_{1,K}^{3,n} \,\right] \tag{4}$$

where $\mathbf{a}_k^n$ is an $(M-1) \times 1$ steering vector for the k-th user, and is defined as $\mathbf{a}_k^n = [1, e^{j\chi\sin\theta_k^n}, \cdots, e^{j\chi(M-2)\sin\theta_k^n}]^T$; $\otimes$ denotes the Kronecker product.

When concatenating the received data vectors from element 2 to M, another extended data vector can be obtained,

$$\mathbf{y}^n = \left[\, \mathbf{x}_2^{n^T} \quad \cdots \quad \mathbf{x}_M^{n^T} \,\right]^T = \mathbf{A}^n \mathbf{\Phi}^n \mathbf{b}^n + \mathbf{n}_{tail}^n \tag{5}$$

where $\mathbf{n}_{tail}^n$ is given by $\mathbf{n}_{tail}^n = [\mathbf{n}_2^{n^T}, \cdots, \mathbf{n}_M^{n^T}]^T$; $\mathbf{\Phi}^n$ is a $3K \times 3K$ diagonal matrix, and is given by

$$\mathbf{\Phi}^n = \mathrm{diag}(e^{j\chi\sin\theta_1^n}, \; e^{j\chi\sin\theta_1^n}, e^{j\chi\sin\theta_1^n}, \cdots, \; e^{j\chi\sin\theta_K^n}) \tag{6}$$

Terms on the main diagonal of the above matrix are associated with users' DOAs. Hence, we call them DOA items.

Under the assumption that the channel is constant during several tens of MC-CDMA symbols, two sets of extended data vectors $\mathbf{u}^n$ and $\mathbf{y}^n$, which are corresponding to successive L MC-CDMA symbols, can be aggregated into two matrices as

$$\mathbf{U}^n = \left[\, \mathbf{u}^n \quad \cdots \quad \mathbf{u}^{n+L-1} \,\right] = \mathbf{A}^n \mathbf{B}^n + \mathbf{N}_{head}^n \tag{7}$$

$$\mathbf{Y}^n = \left[\, \mathbf{y}^n \quad \cdots \quad \mathbf{y}^{n+L-1} \,\right] = \mathbf{A}^n \mathbf{\Phi}^n \mathbf{B}^n + \mathbf{N}_{tail}^n \tag{8}$$

where $\mathbf{U}^n$ and $\mathbf{Y}^n$ are $(M-1)G \times L$ matrices; $\mathbf{B}^n$ is a $3K \times L$ matirx, and is defined as $\mathbf{B}^n = [\mathbf{b}^n, \cdots, \mathbf{b}^{n+L-1}]$; $\mathbf{N}_{head}^n$ and $\mathbf{N}_{tail}^n$ are written as $\mathbf{N}_{head}^n = [\mathbf{n}_{head}^n, \cdots, \mathbf{n}_{head}^{n+L-1}]$ and $\mathbf{N}_{tail}^n = [\mathbf{n}_{tail}^n, \cdots, \mathbf{n}_{tail}^{n+L-1}]$ respectively.

For brevity, we omit the superscript n in the following discussion.

By performing matrix transpose operation, we can obtain $\mathbf{X} = \mathbf{U}^T = \mathbf{B}^T \mathbf{A}^T + \mathbf{N}_{head}^T$ and $\mathbf{Z} = \mathbf{Y}^T = \mathbf{B}^T \mathbf{\Phi} \mathbf{A}^T + \mathbf{N}_{tail}^T$.

Now, the auto-correlation matrix of $\mathbf{X}$ and the cross-correlation matrix between $\mathbf{Z}$ and $\mathbf{X}$ are defined as

$$\mathbf{R}_{XX} = \mathbf{E}[\mathbf{X}\mathbf{X}^H] = \mathbf{B}^T \mathbf{E}[\mathbf{A}^T \mathbf{A}^*]\mathbf{B} + \sigma_n^2 \mathbf{I}$$
$$= \mathbf{B}^T \mathbf{R}_{AA} \mathbf{B} + \sigma_n^2 \mathbf{I} = \mathbf{R}_{XXO} + \sigma_n^2 \mathbf{I} \tag{9}$$

$$\mathbf{R}_{ZX} = \mathbf{E}[\mathbf{Z}\mathbf{X}^H] = \mathbf{B}^T \mathbf{\Phi} \mathbf{E}[\mathbf{A}^T \mathbf{A}^*]\mathbf{B}$$
$$= \mathbf{B}^T \mathbf{\Phi} \mathbf{R}_{AA} \mathbf{B} \tag{10}$$

where $\mathbf{E}[\cdot]$ denotes ensemble average; $\mathbf{R}_{AA}$ is a $3K \times 3K$ matrix, which denotes the auto-correlation matrix of space-time channels; $\mathbf{R}_{XXO}$ is an $L \times L$ matrix; $\mathbf{I}$ is an $L \times L$ identity matrix.

When space-time channels of different users' different transmit antennas are uncorrelated, $\mathbf{R}_{AA}$ is nonsingular, and the rank of $\mathbf{R}_{XXO}$ equals to $3K$. Performing eigen decomposition on $\mathbf{R}_{XXO}$ can obtain $\mathbf{R}_{XXO} = \sum_{l=1}^{3K} \mu_l \mathbf{v}_l \mathbf{v}_l^H$, where μ_l and $\mathbf{v}_l$ are eigenvalues and corresponding eigenvectors of $\mathbf{R}_{XXO}$, respectively.

We define an auxiliary matrix by $\mathbf{R}_{ZX}$ and $\mathbf{R}_{XXO}$ as in [6]

$$\mathbf{R} = \mathbf{R}_{ZX}\mathbf{R}_{XXO}^{+} \tag{11}$$

where $\mathbf{R}_{XXO}^{+}$ is the Penrose-Moore pseudo-inverse of $\mathbf{R}_{XXO}$, and is defined by $\mathbf{R}_{XXO}^{+} = \sum_{l=1}^{3K} \frac{1}{\mu_l}\mathbf{v}_l\mathbf{v}_l^H$.

Theorem 1. *Given* $\mathbf{B}$ *is row full-rank,* $\mathbf{R}_{AA}$ *is nonsingular, then* $\mathbf{R}\mathbf{B}^T = \mathbf{B}^T\mathbf{\Phi}$.

The detailed proof can be found in [9]. Based on Theorem 1, $3K$ eigenvalues can be obtained via eigen decomposition on matrix $\mathbf{R}$. These eigenvalues are DOA items associated with K different users. According to the definition of $\mathbf{\Phi}$ in equation (6), $3K$ eigenvalues in deed include K different values, which are $e^{jx\sin\theta_1^n}, \ldots, e^{jx\sin\theta_K^n}$, and each of them with multiplicity of three. Moreover, each of the three eigenvectors corresponding to the same eigenvalue $e^{jx\sin\theta_k^n}$ is exactly the linear combining of the transmitted symbol sequences associated with three transmit antennas of an individual user. From this interesting observation, we define a matrix consisting of the three eigenvectors corresponding to the same eigenvalue $e^{jx\sin\theta_k^n}$ as $\widetilde{\mathbf{T}}_k$, which can be represented as

$$\widetilde{\mathbf{T}}_k = [\,\mathbf{t}_k^{1,n} \quad \mathbf{t}_k^{2,n} \quad \mathbf{t}_k^{3,n}\,]\mathbf{F} \tag{12}$$

where $\widetilde{\mathbf{T}}_k$ is a column full-rank $L \times 3$ matrix; $\mathbf{F}$ is an unknown full-rank 3×3 matrix; $\mathbf{t}_k^{m,n}$ is the transmitted symbol sequence from the m-th transmit antenna of the k-th user, and is defined as $\mathbf{t}_k^{m,n} = [b_k^m(n), \cdots, b_k^m(n+L-1)]^T$; n denotes the symbol interval.

Since the eigenvalue of $\mathbf{R}$ is associated with its eigenvector, the DOA of an individual user is associated with its signal subspace spanned by transmitted symbol sequences correspondingly. In this sense, we say it can be automatically paired between the signal subspace of a user and corresponding DOA. This property is desirable for it eliminates performing time-comsuming joint diagonalization operations of many matrices [1] and thus saves the cost of computation.

Then, by performing singular value decomposition (SVD) on $\widetilde{\mathbf{T}}_k$, we can obtain

$$\widetilde{\mathbf{T}}_k = [\,\mathbf{U}_{k,s} \quad \mathbf{U}_{k,o}\,]\begin{bmatrix} \mathbf{\Sigma}_k \\ \mathbf{0} \end{bmatrix}\mathbf{V}_{k,s}^H \tag{13}$$

where $\mathbf{U}_{k,s}$ is an $L \times 3$ matrix; $\mathbf{U}_{k,o}$ is an $L \times (L-3)$ matrix; $\mathbf{\Sigma}_k$ is a 3×3 matrix; $\mathbf{0}$ is an $(L-3) \times 3$ zero matrix; $\mathbf{V}_{k,s}^H$ is a 3×3 matrix.

Because $\mathbf{U}_{k,o}^H \perp \text{Range}\{\widetilde{\mathbf{T}}_k\}$, we have

$$\mathbf{U}_{k,o}^H[\,\mathbf{t}_k^{1,n} \quad \mathbf{t}_k^{2,n} \quad \mathbf{t}_k^{3,n}\,] = \mathbf{U}_{k,o}^H\mathbf{T}_k = \mathbf{0} \tag{14}$$

where $\mathbf{0}$ is an $(L-3) \times 3$ zero matrix; $\mathbf{T}_k$ is an $L \times 3$ matrix, and is defined as $\mathbf{T}_k = [\mathbf{t}_k^{1,n}, \mathbf{t}_k^{2,n}, \mathbf{t}_k^{3,n}]$.

4 STBC Decoding

For each user, the input of its encoder is a group of four successive data symbols $\{x_1, x_2, x_3, x_4\}$, and the output of its encoder is a 4×3 coded symbol matrix [8]

$$\mathbf{C} = \begin{bmatrix} x_1 & x_2 & x_3 \\ -x_2 & x_1 & -x_4 \\ -x_3 & x_4 & x_1 \\ -x_4 & -x_3 & x_2 \end{bmatrix} \tag{15}$$

where the m-th $(m = 1, 2, 3)$ column of $\mathbf{C}$ is the transmitted symbol sequence of the m-th transmit antenna in four successive symbol intervals.

Based on the encoder defined in (15), the coded symbol matrix $\mathbf{C}$ and the input symbol vector $\mathbf{c} = [x_1, x_2, x_3, x_4]^T$ have the following relationship:

Theorem 2. Denote an $l \times 4$ matrix $\mathbf{U} = [\mathbf{u}_1, \mathbf{u}_2, \mathbf{u}_3, \mathbf{u}_4]$, where $\mathbf{u}_j$ $(j = 1, \cdots, 4)$ is the j-th column of $\mathbf{U}$. With the columns of $\mathbf{U}$, construct a $3l \times 4$ matrix $\bar{\mathbf{U}}$

$$\bar{\mathbf{U}} = \begin{bmatrix} \mathbf{u}_1 & -\mathbf{u}_2 & -\mathbf{u}_3 & -\mathbf{u}_4 \\ \mathbf{u}_2 & \mathbf{u}_1 & -\mathbf{u}_4 & \mathbf{u}_3 \\ \mathbf{u}_3 & \mathbf{u}_4 & \mathbf{u}_1 & -\mathbf{u}_2 \end{bmatrix} \tag{16}$$

Then, $\bar{\mathbf{U}}\mathbf{c} = \mathrm{vec}(\mathbf{U}\mathbf{C})$, where $\mathrm{vec}(\mathbf{A})$ represents the vectorization of matrix $\mathbf{A}$, that is, stacking the columns of $\mathbf{A}$ one by one as a long vector.

We can prove Theorem 2 by simple substitution. So, the product of a matrix $\mathbf{U}$ and the coded symbol matrix $\mathbf{C}$ can be transformed to the product of a matrix $\bar{\mathbf{U}}$ and the input symbol vector $\mathbf{c}$.

From the previous Section, we know $\mathbf{T}_k$ is an $L \times 3$ coded symbol matrix of the k-th user. Let $L = 4B$, and split $\mathbf{T}_k$ into B sub-matrices, we get

$$\mathbf{T}_k = [\mathbf{T}_{k,1}^T \quad \mathbf{T}_{k,2}^T \quad \cdots \quad \mathbf{T}_{k,B}^T]^T \tag{17}$$

where $\mathbf{T}_{k,i}(i = 1, \cdots, B)$ is a 4×3 coded symbol matrix associated with the i-th group of four successive input symbols $b_k(n + 4i - 4), \cdots, b_k(n + 4i - 1)$. Here we define $\mathbf{x}_{k,i}$ as $\mathbf{x}_{k,i} = [b_k(n + 4i - 4), \cdots, b_k(n + 4i - 1)]$.

$\mathbf{U}_{k,o}^H$ is an $(L - 3) \times L$ matrix, and it is also split into B sub-matrices , that is,

$$\mathbf{U}_{k,o}^H = [\mathbf{U}_{k,1} \quad \mathbf{U}_{k,2} \quad \cdots \quad \mathbf{U}_{k,B}] \tag{18}$$

where $\mathbf{U}_{k,i}$ is an $(L - 3) \times 4$ matrix. Thus, equation (14) can be represented as

$$\mathbf{U}_{k,o}^H \mathbf{T}_k = \mathbf{U}_{k,1}\mathbf{T}_{k,1} + \cdots + \mathbf{U}_{k,B}\mathbf{T}_{k,B} = \mathbf{0} \tag{19}$$

Based on Theorem 2, we have

$$\bar{\mathbf{U}}_{k,i}\mathbf{x}_{k,i}^T = \mathrm{vec}(\mathbf{U}_{k,i}\mathbf{T}_{k,i}) \tag{20}$$

where $\bar{\mathbf{U}}_{k,i}$ is a $3(L - 3) \times 4$ matrix, and is given by

$$\bar{\mathbf{U}}_{k,i} = \begin{bmatrix} \mathbf{u}_{k,i}^1 & -\mathbf{u}_{k,i}^2 & -\mathbf{u}_{k,i}^3 & -\mathbf{u}_{k,i}^4 \\ \mathbf{u}_{k,i}^2 & \mathbf{u}_{k,i}^1 & -\mathbf{u}_{k,i}^4 & \mathbf{u}_{k,i}^3 \\ \mathbf{u}_{k,i}^3 & \mathbf{u}_{k,i}^4 & \mathbf{u}_{k,i}^1 & -\mathbf{u}_{k,i}^2 \end{bmatrix} \tag{21}$$

where $\mathbf{u}_{k,i}^j (j = 1, \cdots, 4)$ is the j-th column of $\mathbf{U}_{k,i}$.

Successively, equation (14) can be transformed to

$$\mathrm{vec}(\mathbf{U}_{k,o}^H \mathbf{T}_k) = \bar{\mathbf{U}}_k \mathbf{x}_k = \mathbf{0} \tag{22}$$

where $\bar{\mathbf{U}}_k$ is a $3(L-3) \times L$ matrix, and is defined as $\bar{\mathbf{U}}_k = [\bar{\mathbf{U}}_{k,1}, \cdots, \bar{\mathbf{U}}_{k,B}]$; $\mathbf{x}_k$ is an $L \times 1$ vector, and is defined as $\mathbf{x}_k = [\mathbf{x}_{k,1}, \cdots, \mathbf{x}_{k,B}]^T$.

Now, by performing SVD on $\bar{\mathbf{U}}_k$, we can easily obtain the estimation of the original input symbol sequence $\hat{\mathbf{x}}$ for the k-th user, which is just the right singular vector associated with the smallest singular value. Obviously, by constructing $\bar{\mathbf{U}}_k$ based on (22) for different user, we can get the input symbol sequences for multiple users.

5   Simulation Results

In this Section, computer simulation results are presented to evaluate the performance of the proposed algorithm. DBPSK modulation mode is used in our simulations. Hadamard codes with length $G = 32$ are assigned to different users. We assume a rich scattering environment near the mobile station and generate the FIR channel coefficients (reference (3) in [3]) as independent identically distributed (i.i.d) complex Gaussian random variables with zero-mean and variance $1/(L_{ch}+1)$. The signal to noise power ratio (SNR) per receive antenna is defined as SNR=$10\log_{10} 1/\sigma_n^2$ in dB. We use samples within 80 MC-CDMA symbols to estimate the auto- and the cross-correlation matrices of the uplink-received data sequence, which are the approximate estimation of auto- and cross-correlation matrices in ensemble-average sense.

The length of FIR channels is fixed to 7 and the number of active users is fixed to 5. The performance of the MMSE receiver with estimated channel information

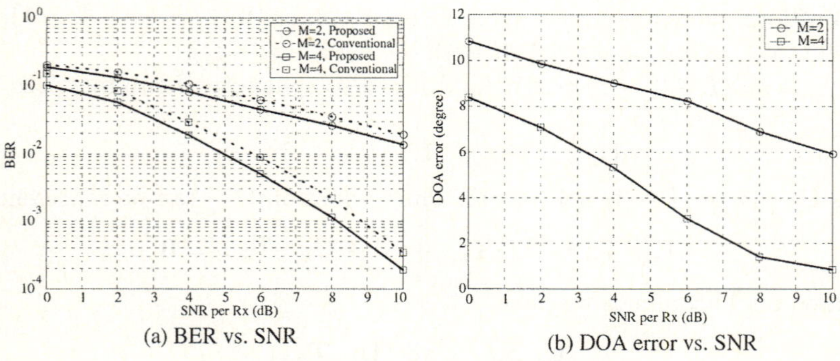

(a) BER vs. SNR   (b) DOA error vs. SNR

Fig. 2. Simulation results

[2,3] is also presented for comparison. Fig.2(a) shows the BER versus the SNR with antenna array of 2 and 4 elements, respectively. Fig.2(b) shows the DOA estimation error versus the SNR with the same parameters. It can be seen that our algorithm outperforms the conventional one. We can also see that both the BER and the DOA estimation error decrease as the number of receive antenna array element increases. Reason behind this is that the antenna array gain of ULA enhances the SNR of received signals when the number of antenna array element increases.

Finally, the computation complexity of the proposed algorithm is $O(L^3)$ flops and that of [2,3] is $O((M-1)^3 G^3)$ flops. Since $L < (M-1)G$ in general, the proposed algorithm has obvious superiority of computation.

References

1. A.J. van der Veen, M.C. Vanderveen, A. Paulraj, "Joint angle and delay estimation using shift-invariance techniques," IEEE Tran. Signal Processing, Feb. 1998, vol. 42, pp. 405-418.
2. W. Sun, H. Li and M. Amin, "MMSE detection for space-time coded MC-CDMA," Proc. IEEE ICC 2003, Anchorage, USA, May 2003, vol. 5, pp.3452-3456.
3. X. Wu, Q. Yin and Y. Zeng, "Downlink channels identification for space-time coded multiple-input multiple-output MC-CDMA systems," Proc. IEEE ICASSP 2003, Hong Kong, April 2003, pp.417-420.
4. A. L. Swindlehurst and G. Leus, "Blind and semi-blind equalization for generalized space-time block codes," IEEE Trans. Signal Processing, vol. 50, Oct. 2002, pp. 2489-2498.
5. L. Dai, S. Sfar and K. Letaief, "Receive antenna selection for MIMO systems in correlated channels," Proc. IEEE ICC 2004, Paris, France, July 2004, vol. 5, pp.2944-2948.
6. Q. Yin, R. Newcomb and L. Zou, "Estimating 2-D angles of arrival via two parallel linear array," Proc. IEEE ICASSP 1989, Glasgow, Scotland, May 1989, pp. 2803-2806.
7. A.J. Paulraj, C.B. Papadias, "Space-time processing for wireless communications," IEEE Signal Processing Mag., Nov. 1997, vol. 14, pp. 49-83.
8. V. Tarokh, H. Jafarkhani and A. R. Calderbank, "Space-time block codes from orthogonal designs," IEEE Trans. Inform. Theory, vol. 45, July 1999, pp. 1456-1467.
9. Y. Zeng and Q. Yin, "Direct Decoder of Uplink Space-Time Block Coded MC-CDMA Systems", IEICE Trans. Com., vol. E88-B, Feb. 2005, pp. 100-112.

Erlang Capacity of Voice/Data CDMA Systems with Service Requirements of Blocking Probability and Delay Constraint

Insoo Koo[1], Jeongrok Yang[2], and Kiseon Kim[1]

[1] University of Ulsan,
Republic of Korea
[2] Korean Intellectual Property Office,
Republic of Korea

Abstract. In this paper, we investigate the Erlang capacity of a CDMA system supporting voice and data services with service requirements of blocking probability and delay. We define the delay confidence defined as the probability that a new data call is accepted within the maximum tolerable delay without being blocked. In this case, the Erlang capacity is confined not only by the required blocking probability of voice call but also by the required delay confidence of data call. For the performance analysis, we develop a two-dimensional Markov model, and we derive the CDF of delay and subsequent relation for the delay requirement. Further we present a numerical procedure to analyze the Erlang capacity under a given blocking probability and delay confidence. As a result, it is necessary to make a balance between the Erlang capacity with respect to the blocking probability of voice call and that with respect to the delay confidence of data call so as to accommodate the more Erlang capacity.

1 Introduction

Future wireless communication systems will provide users with multimedia services such as voice, interactive data, file transfer, internet access and image, comparable to those provided by the wired communication systems. Such multimedia traffic will have different and multiple quality of service (QoS) requirements. In terms of system operation, it is very important task to analyze the capacity of system that can be supportable while multiple service requirements of multimedia traffic being satisfied since the capacity can be used as a measure of system resource. With this reason, many efforts have been taken to analyze the capacity [1, 4, 5].

Multimedia traffic can be roughly classified into "delay-intolerant" and "delay-tolerant" traffic. To achieve higher capacity using the delay-tolerant characteristic, delay-tolerant traffic can be queued until the required resources are available in the system. The blocking probability and the average delay have been typically considered as performance measures for the delay-tolerant traffic [4]. However, more meaningful measurement for delay-tolerant traffic is the delay confidence

P. Lorenz and P. Dini (Eds.): ICN 2005, LNCS 3421, pp. 936–943, 2005.

rather than the average delay where the delay confidence can be defined as the probability that a new data call gets a service within the maximum tolerable delay constraint without being blocked. Noting that the previous works [1, 4, 5] have not considered the delay confidence when evaluating the Erlang capacity, in this paper we adopt the delay confidence as a performance measure of delay-tolerant traffic, and further analyze Erlang capacity of a CDMA system supporting voice and data calls since voice traffic is a typical delay-intolerant traffic while data traffic is a typical delay-tolerant traffic. Here, the Erlang capacity is defined as a set of average offered traffic loads of voice and data calls that can be supported in the system while the required blocking probability of voice call and the required delay confidence of data call are being satisfied, simultaneously. To analyze the Erlang capacity, we develop a two-dimensional Markov model, based on the First-Come-First-Serve (FCFS) service discipline where the queue with finite size is exclusively allocated for delay-tolerant data calls.

The remaining paper is organized as follows. In the next section, we describe system model, and a call admission control (CAC) scheme. In Section 3, we develop a two-dimensional Markov model of voice/data CDMA systems, and further present the blocking probability and the delay confidence. In Section 4, we present a numerical example for analyzing the Erlang capacity under a given blocking probability and delay confidence. Finally, we draw conclusions in Section 5.

2   System Model

In the CDMA systems, although there is no hard limit on the number of concurrent users, there is a practical limit on the number of supportable concurrent users in order to control the interference among users having the same pilot signal; otherwise the system can fall into an outage state where QoS requirements of users cannot be guaranteed. In order to satisfy the QoS requirements for all concurrent users, the capacity of CDMA system supporting voice and data services in the reverse link should be limited, and it is given as like following equation for a single cell case [2].

$$\gamma_v i + \gamma_d j \leq 1 , \qquad i \text{ and } j \geq 0 \tag{1}$$

where

$$\gamma_v = \left(\frac{W}{R_v q_v} + 1\right)^{-1} \text{ and } \gamma_d = \left(\frac{W}{R_d q_d} + 1\right)^{-1} \tag{2}$$

γ_v and γ_d are the amount of system resources that are used by one voice and one data user, respectively. i and j denote the number of supportable users in the voice and data service groups, respectively. W is the allocated frequency bandwidth. q_v and q_d are the required bit energy to interference power spectral density ratio for voice and data calls, respectively, which is necessary to achieve the target bit error rate at the base station. R_v and R_d are the required information data rates of the voice and data calls, respectively. Each user is classified by

QoS requirements such as the required information data rate and the required bit energy to interference spectral density ratio, and all users in same service group have same QoS requirements. Eqn. (1) indicates that the calls with dif-. ferent services take different amount of system resources according to their QoS requirements.

In the aspects of network operation, it is of vital importance to set up a suitable policy for the acceptance of an incoming call in order to guarantee a certain quality of service. In this paper, we adopt a NCAC-type CAC since its simplicity even though the NCAC generally suffers a slight performance degradation over the ICAC [3], and we also use the capacity bound, stipulated by Eqn.1, as a pre-determined CAC threshold. Further, we consider the queue with the finite length of K for delay-tolerant data traffic to exploit a delay-tolerant characteristic, and assume that the service discipline is First-Come-First-Serve (FCFS). Based these assumptions, the call admission control (CAC), for the case $\gamma_d > \gamma_v$ can be summarized as follows.

- If $\gamma_v i + \gamma_d j \leq 1 - \gamma_d$, then both new voice and new data calls are accepted.
- If $1 - \gamma_d < \gamma_v i + \gamma_d j \leq 1 - \gamma_v$, then new voice calls are accepted, and new data calls are queued.
- If $1 - \gamma_v < \gamma_v i + \gamma_d j \leq 1 + (K-1)\gamma_d$, then new voice calls are blocked, and new data calls are queued.
- If $\gamma_v i + \gamma_d j > 1 + (K-1)\gamma_d$, then both new voice and new data calls are blocked.

3   Markov Chain Model of Voice/Data CDMA Systems

In this section, we develop an analytical procedure to investigate the blocking probabilities and delays of voice and data calls. According to the CAC rule based on the number of concurrent users, the set of possible admissible states is given as

$$\Omega_S = \left\{ (i,j) | 0 \leq i \leq \gamma_v^{-1}, \ j \geq 0, \ \gamma_v i + \gamma_d j \leq 1 + \gamma_d K \right\} \tag{3}$$

These admissible states can be further divided into five regions as follows:

$$\Omega_A = \{(i,j) | 0 \leq \gamma_v \cdot i + \gamma_d \cdot j \leq 1 - max(\gamma_v, \gamma_d)\} \tag{4}$$
$$\Omega_B = \{(i,j) | 1 - max(\gamma_v, \gamma_d) < \gamma_v \cdot i + \gamma_d \cdot j \leq 1 - min(\gamma_v, \gamma_d)\}$$
$$\Omega_C = \{(i,j) | 1 - min(\gamma_v, \gamma_d) < \gamma_v \cdot i + \gamma_d \cdot j \leq 1\}$$
$$\Omega_D = \{(i,j) | 1 < \gamma_v \cdot i + \gamma_d \cdot j \leq 1 + \gamma_d \cdot (K-1)\}$$
$$\Omega_E = \{(i,j) | 1 + \gamma_d \cdot (K-1) < \gamma_v \cdot i + \gamma_d \cdot j \leq 1 + \gamma_d \cdot K\}$$

Noting that total rate of flowing into a state (i,j) is equal to that of flowing out, we can get the steady-state balance equation for each state as follows:

$$\textbf{Rate-In} = \textbf{Rate-Out}$$
$$\textbf{Rate-In} = \boldsymbol{a} \cdot P_{i+1,j} + \boldsymbol{b} \cdot P_{i,j+1} + \boldsymbol{c} \cdot P_{i-1,j} + \boldsymbol{d} \cdot P_{i,j-1}$$
$$\textbf{Rate-Out} = (\boldsymbol{i} + \boldsymbol{j} + \boldsymbol{k} + \boldsymbol{l}) \cdot P_{i,j} \tag{5}$$
$$for \ \ all \ \ states$$

Table 1. State transition rates

symbol	definition
a	transition rate from state $(i+1,j)$ to state (i,j)
b	transition rate from state $(i,j+1)$ to state (i,j)
c	transition rate from state $(i-1,j)$ to state (i,j)
d	transition rate from state $(i,j-1)$ to state (i,j)
i	transition rate from state (i,j) to state $(i+1,j)$
j	transition rate from state (i,j) to state $(i,j+1)$
k	transition rate from state (i,j) to state $(i-1,j)$
l	transition rate from state (i,j) to state $(i,j-1)$

where the state transition rates, a, b, c, d, i, j, k and l involved in the Eqn. (5) can be defined in Table 1.

If the total number of all possible states is n_s, the balance equations become $(n_s - 1)$ linearly independent equations. With these $(n_s - 1)$ equations and the normalized equation, $\sum_{(i,j)\in\Omega_S} P_{i,j} = 1$, a set of n_s linearly independent equations for the state diagram can be formed as

$$\mathbf{Q}\pi = \mathbf{P} \tag{6}$$

where $\mathbf{Q}$ is the coefficient matrix of the n_s linear equations, π is the vector of state probabilities, and $\mathbf{P} = [0, \cdots, 0, 1]^T$. The dimensions of $\mathbf{Q}$, π and $\mathbf{P}$ are $n_s \times n_s$, $n_s \times 1$, and $n_s \times 1$, respectively. By solving $\pi = \mathbf{Q}^{-1}\mathbf{P}$, we can obtain the steady-state probabilities of all states [4].

3.1   Blocking Probability and Delay Confidence

Based on the CAC rule, a new voice call will be blocked if the channel resources are not enough to accept the call, and the corresponding blocking probability for voice calls is given by

$$P_{b_v} = \sum_{(i,j)\in\Omega_{(nv,blo)}} P_{i,j} \tag{7}$$

where

$$\Omega_{(nv,blo)} = \{(i,j)|\gamma_v i + \gamma_d j > 1 - \gamma_v\} \tag{8}$$

Similarly, a new data call will be blocked if the queue is full, and the blocking probability for data calls is given by

$$P_{b_d} = \sum_{(i,j)\in\Omega_{(nd,blo)}} P_{i,j} \tag{9}$$

where

$$\Omega_{(nd,blo)} = \{(i,j)\,|\gamma_v i + \gamma_d j > 1 + \gamma_d(K-1)\} \tag{10}$$

Likewise, from the Markov chain model, we can evaluate the delay of data traffics. The delay is defined as the time that a data call waits in a queue until being accepted in the system. To characterize the delay, let's derive the cumulative distribution function (CDF) of delay (τ), $F_d(t)$ defined as $\Pr\{\tau \leq t\}$.

For the convenience of analysis, we separate the CDF of delay into two parts that corresponds to discrete and continuous parts of the random variable τ, respectively such that

$$F_d(t) \equiv \Pr\{\tau \leq t\} = F_d(0) + G(t) \tag{11}$$

where $F_d(0) = \Pr\{\tau \leq 0\}$, and $G(t)$ represents the continuous part of the delay. Since the discrete part $F_d(0)$ represents the case when the delay is zero, it can be calculated as follows :

$$F_d(0) = \Pr\{\tau \leq 0\} = \Pr\{\tau = 0\} \tag{12}$$
$$= \sum_{(i,j)\in\Omega_{(nd,acc)}} P'_{i,j}$$

where $\Omega_{(nd,acc)}$ is the acceptance region of new data calls and it is given as

$$\Omega_{(nd,acc)} = \{(i,j)\,|\,\gamma_v i + \gamma_d j \leq 1 - \gamma_d\} \tag{13}$$

and

$$P'_{i,j} = \frac{P_{i,j}}{1 - P_{b_d}} \tag{14}$$

$P'_{i,j}$ represents the probability that there are i voice and j data calls in the system just before a new data call is admitted. If the state (i,j) belongs to the blocking region of new data calls, $\Omega_{(nd,blo)}$, the call will be blocked.

To investigate the continuous part of delay $G(t)$, let (i',j') denote the number of calls excluding the number of service-completed calls within time τ from (i,j). Consider the case that (i,j) belongs to the queueing region of new data calls just before a new data call is admitted where the queueing region of new data calls is given as

$$\Omega_{(nd,que)} = \{(i,j)\,|\,1 - \gamma_d < \gamma_v i + \gamma_d j \leq 1 + (K-1)\gamma_d\} \tag{15}$$

In order for a new data call to be accepted within the time t according to the FCFS service discipline, (i',j') should fall into the acceptance region of new data calls within the time t. $G(t)$ is the sum of the probabilities of all cases that a state (i,j) in $\Omega_{(nd,que)}$ changes into (i',j') in $\Omega_{(nd,acc)}$ within the time, t, which can be expressed as

$$G(t) = \sum_{(i,j)\in\Omega_{(nd,que)}} \Pr\{(i',j') \in \Omega_{(nd,acc)} \text{ within time } t| \tag{16}$$

the system state is $(i,\,j)$ just before a new data call is admitted$\} \cdot P'_{i,j}$

$$= \sum_{(i,j)\in\Omega_{(nd,que)}} \int_0^t w_{(i,j)}(\tau)d\tau \cdot P'_{i,j}$$

where $w_{(i,j)}(\tau)$ is the delay distribution for the state (i,j), and it represents the probability of a new data call being accepted within time τ, given that the system state is (i,j) just before the call is admitted.

For a delay-tolerant traffic, an important performance measure is related to the delay requirement. Typically, the delay requirement of data calls is that the system should provide the required services to users within the maximum tolerable delay. Here, we introduce the "delay confidence" defined as the probability that new data calls are accepted within the maximum tolerable delay without being blocked, and further we formulate the delay confidence as follows:

$$P_c \equiv (1 - P_{b_d}) \cdot F_d(\tau_{max}) \tag{17}$$

where τ_{max} is the maximum tolerable delay requirement. Here, note that the delay confidence is related to not only the cumulative distribution function (CDF) but also the blocking probability of data calls.

4   Erlang Capacity and Numerical Examples

As a system level performance measure, we utilize Erlang capacity, which is defined as a set of supportable offered traffic loads of voice and data that can be supported while service requirements of blocking probability and delay constraint are satisfied, and further can be formulated as follows:

$$C_{Erlang} \equiv \left\{ (\rho_v, \rho_d) | \, P_{b_v} \leq P_{b_v,req}, \ P_c \geq P_{c_{req}} \right\} \tag{18}$$

where $P_{b_v,req}$ is the required blocking probability for voice traffic and $P_{c_{req}}$ is the required delay confidence for data traffic. For a numerical example, here we consider IS-95B type CDMA systems supporting voice and delay-tolerant data services. Figure 1 (a) shows the voice-limited Erlang capacity and the data-limited Erlang capacity when $K = 0$. The line (i) and line (ii) represent the voice-limited Erlang capacity and the data-limited Erlang capacity, respectively, when $P_{b_v,req} = 1\%$ and $P_{c_{req}} = 99\%$. The Erlang capacity is determined as the overlapped region limited by the line (i) and line (ii) to satisfy both service requirements for voice and data calls at the same time. For the case that there is no queue ($K = 0$), the CDF of delay at the maximum tolerable delay, $F_d(\tau_{max})$ is given 1 such that the delay confidence, P_c becomes $(1 - P_{b_d})$, and the required delay confidence of 99% corresponds to the required blocking probability of 1% of data. The Erlang capacity in Figure 1 (a) corresponds to that analyzed in [5] for the blocking probabilities for voice and data traffics. Figure 1 (a) shows that the Erlang capacity is mainly determined by the data-limited Erlang capacity. The gap between the voice-limited Erlang capacity and the data-limited Erlang capacity comes from the difference in the service requirements for voice and data calls. In this case, the data-limited Erlang capacity is lower than the voice-limited Erlang capacity for the same blocking probability because a data call requires more system resources than a voice call. In reality, the data calls have a distinct characteristic that may allow some delay, however, it was presented

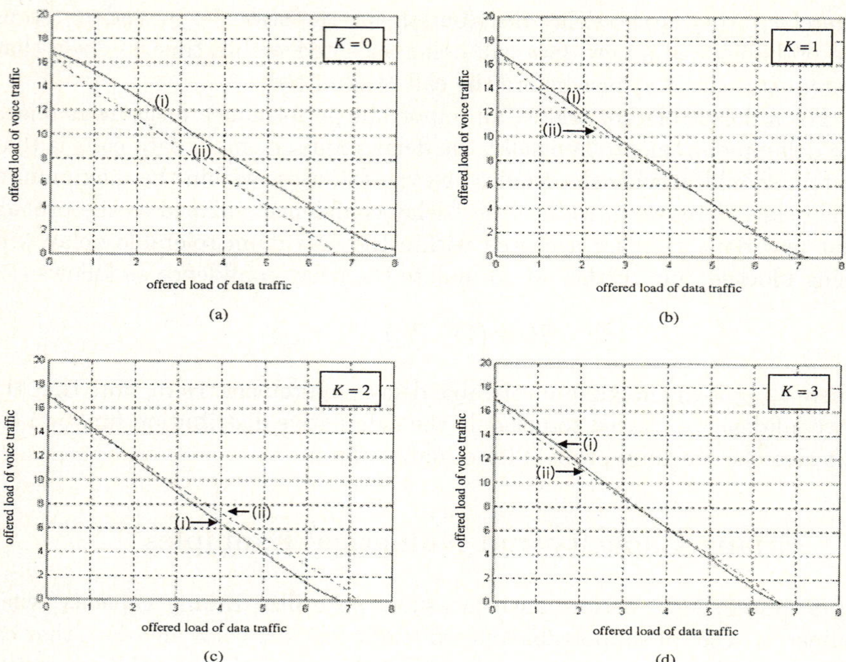

Fig. 1. Effect of the queue size on the Erlang capacity when $P_{b_v,req} = 1\%$, $P_{c_{req}} = 99\%$ and $\tau_{n_{max}} = 0.1$ (a) $K = 0$ (b) $K = 1$ (c) $K = 2$ (d) $K = 3$ where lines (i) and (ii) represents the voice-limited Erlang capacity and the data-limited Erlang capacity, respectively

that the data calls behave like the voice calls which are eventually blocked when there is no instantly available resource.

In order to increase the Erlang capacity, a proper tradeoff is required between the voice-limited Erlang capacity and the data-limited Erlang capacity. One of methods to get the tradeoff is to use queueing for delay-tolerant data calls. Figure 1 shows the effect of the queue size on the Erlang capacity when $P_{b_v,req} = 1\%$, $P_{c_{req}} = 99\%$ and $\tau_{n_{max}} = 0.1$. The solid and dotted line represent the voice-limited Erlang capacity and data-limited Erlang capacity, respectively. We know that the voice-limited Erlang capacity decreases as the queue size increases, which comes from the fact that the call blocking probability of voice increases for a larger queue size. On the other hand, the data-limited Erlang capacity increases until the queue size becomes 2, and after that, it decreases for a larger queue size. It comes from the fact that the call blocking probability of data, and the CDF of delay respectively decreases as the queue size increases. Noting that the delay confidence depends on not only the blocking probability of data calls decreases but also the CDF of delay, we know that the variation of the data-limited Erlang capacity according to the queue size mainly comes from the mutual effects between the improvement in the blocking probability of data calls and the decrease of the CDF of delay. Figure 1 also shows that the Erlang capacity when $K = 3$ is less than that when $K = 0$, which means that the queue size should be properly selected to make a balance between the voice-limited Erlang capacity and the data-limited Erlang capacity as as to accommodate more Erlang capacity. In this case, we observe that the optimum queue size is 1 with respect to maximizing the Erlang capacity.

5   Conclusions

In this paper, we have analyzed the Erlang capacity of a CDMA system supporting voice and delay-tolerant data services with consideration of the delay confidence. As a result, for the case that there is no queue for data calls, it was observed that the Erlang capacity is mainly determined by the data-limited Erlang capacity since one data call requires more system resources than one voice call. For a finite buffer case, the data-limited Erlang capacity increases as the maximum tolerable delay increases or the required delay confidence decreases. By observing the Erlang capacity according to the queue size, we showed that the queue size should be properly selected to make a balance between the voice-limited Erlang capacity and the data-limited Erlang capacity as as to accommodate more Erlang capacity.

Acknowledgement

This work was supported by Korea Science and Engineering Foundation (KOSEF) through the UFON research center at GIST. Insoo Koo particularly was supported by the Basic Research Program of the KOSEF.

References

1. Viterbi, A. M. and A. J. Viterbi, "Erlang Capacity of a Power-Controlled CDMA System," *IEEE Journal on Selected Areas in Communications,* vol. 11, no. 6, pp. 892-900, Aug. 1993.
2. Yang, J. R., Y. Y. Choi, J. H. Ahn and K. Kim, "Capacity Plane of CDMA System for Multimedia Traffic," *Electrononics Letters,* vol. 33, no. 17, pp. 1432-1433, Aug. 1997.
3. Ishikawa, Y. and N. Umeda, "Capacity Design and Performance of Call Admission Cotnrol in Cellular CDMA Systems," *IEEE JSAC,* vol. 15, pp.1627-1635, Oct. 1997.
4. Koo, I., E. Kim and K. Kim, "Erlang Capacity of Voice/Data DS-CDMA Systems with Prioritized Services," *IEICE Transactions on Communications,* vol. E84-B, no. 4, pp. 716-726, Apr. 2001.
5. Cruz-Perez, F. A., D. Lara-Rodriguez, M. Lara. "Performance Analysis of the Fractional Channel Reservation in TDMA and CDMA Integrated Services Networks," *Proc. IEEE Vehicular Technology Conference - Spring,* pp. 1007-1011, 2001.
6. Gross, D. and C. M. Harris, *Fundamentals of Queueing Theory,* John Wiley & Sons, Inc., 1998.

A Simplified Leakage-Resilient Authenticated Key Exchange Protocol with Optimal Memory Size

SeongHan Shin, Kazukuni Kobara, and Hideki Imai

Institute of Industrial Science, The University of Tokyo,
4-6-1 Komaba, Meguro-ku, Tokyo 153-8505, Japan
shinsh@imailab.iis.u-tokyo.ac.jp
{kobara, imai}@iis.u-tokyo.ac.jp
http://imailab-www.iis.u-tokyo.ac.jp/imailab.html

Abstract. At Asiacrypt 2003, Shin et al., have proposed a new authenticated key exchange (AKE) protocol named Leakage-Resilient AKE (for short, LR-AKE) [13]. The authenticity of LR-AKE is based on a user's (relatively short) password *and* his/her stored secrets in both client side and server side. In their LR-AKE protocol, neither PKI (Public Key Infrastructures) nor TRM (Tamper Resistant Modules) is required and leakage of the stored secrets from any side does not reveal any critical information on the password.

As main contributions of this paper, we propose a simplified LR-AKE (LR-AKE) protocol that is more efficient rather than [13]: about 100% decrease in computation costs (excluding some negligible costs) in the initialization phase and about 96% reduction in memory size on client's and servers' devices where the memory size is *optimal*. That makes the LR-AKE protocol applicable for many applications, because password-based AKE protocols have been motivated by the very practical implementations.

1   Introduction

1.1   Background

Along with the rapid advances of technology, internet banking/shopping/aution, electronic commerce/voting, content download and so on have been come true over networks, which have great potential for improving remote access to services. Such implementations over open networks should pay much attention to the fundamental issues of security: authentication and privacy. Both of the security goals can be achieved by establishing a secure channel between two parties (say, a client (a user) and a server) through an authenticated key exchange (AKE) protocol at the end of which the two parties share a common session key. Since AKE protocols are one of the most crucial cryptographic primitives, they have been widely used in SSH (Secure SHell) [7], SSL/TLS (Secure Socket Layer/Transport Layer Security) [4, 8] and many applications.

P. Lorenz and P. Dini (Eds.): ICN 2005, LNCS 3421, pp. 944–952, 2005.

In the literature a variety of AKE protocols have been proposed so far where some (e.g., pages 53-57, Main Document, IEEE P1363 [5]) take fully advantage of PKI (Public Key Infrastructures) and others (e.g., [6]) are based on a human-memorable password to handle authentication. When it comes to the real world where stored secrets (including private keys, symmetric keys and/or any secret values) may not be protected without *perfect* TRM (Tamper-Resistant modules), password-based authentication can be preferable since neither special hardware such as TRM nor security infrastructures is required. However password is a low-entropy secret and chosen from a small dictionary of size so that there exist two attacks on passwords: on-line and off-line dictionary attacks. The on-line dictionary attack is a series of exhaustive search for a secret performed on-line where an adversary is trying to sieve out secret candidates one by one running an authentication protocol with the server (vice versa). In contrast, the off-line dictionary attack is performed off-line massively in parallel by simply guessing secrets and verifying the guessed secret with recorded transcripts of a protocol (with/without leaked secrets). While on-line dictionary attacks are applicable to all of the password-based protocols equally, they can be prevented by letting a server take appropriate intervals between invalid trials. But, we cannot avoid off-line dictionary attacks by such policies, mainly because the attacks are performed off-line and independently of the server. As a result off-line dictionary attacks are critical so that designing a secure password-based AKE protocol is not trivial.

1.2   Previous Works

The problem of off-line attacks has been extensively investigated and well understood in password-only AKE protocols: Password-Authenticated Key Exchange (PAKE). PAKE protocols are designed in the 2-party setting, where a relatively short password is shared between the two parties in advance, to generate a cryptographically-secure session key. While the pre-shared password is short enough and exhaustible with off-line attacks, the generated session key is no longer exhaustible with them even if an adversary completely controls the communications.

Since the first discussion by Bellovin and Merritt [1], many studies on PAKE protocols have appeared in the literature (complete references can be found in [11]). In PAKE protocols, a client is required to keep in mind his/her password whereas the counterpart server should have its verification data that is used to verify the client's knowledge of the password. If stored secrets (or, password verification data) in the server are leaked out, the client's password can be retrieved through off-line dictionary attacks simply by verifying password candidates one by one using the verification data. As a consequence the password-only protocols have faced the following limitation: any AKE protocol, whose authenticity depends only on password (and public information) cannot achieve its immunity against off-line attacks after the leakage of stored secrets (or, verification data) from server side.

At Asiacrypt 2003, Shin et al., have proposed a new class of AKE protocol titled as "Leakage-Resilient Authenticated Key Establishment" (we call LR-

AKE for short) [13], raising a more natural and interesting problem on the password-only AKE protocols: PAKE protocols can't deal with a situation where a client communicates with a lot of different servers, each of which provides an independent service, where he/she remembers only one password. According to [13] the previous password-based AKE protocols (SSH, SSL/TLS and PAKE) turned out to be insecure without TRM. The authenticity of LR-AKE protocol is based on password *and* additional stored secrets in order to cope with the above problem. The protocol provides immunity of client's password to the leakage of stored secrets from a client and servers, respectively, which achieves a much higher level of security without using TRM and PKI. That means, the client need not change his password *even if* stored secrets are leaked out from either the client or servers. Having said this, carrying some insecure devices for storing secrets seems a small price to pay for more strengthened security in the LR-AKE protocol. Refer to [13] for more detailed discussion compared to SSH, SSL/TLS and PAKE protocols.

1.3   Our Contributions

We propose a simplified LR-AKE (LR-AKE) protocol. The LR-AKE protocol is more efficient rather than [13], when considering client's devices with restriction to the computation ability and memory capacity, mainly in aspects of both computation costs and memory size. When registering verification data to multiple different servers, the LR-AKE protocol has about 100% decrease in computation costs in the initialization phase. As for memory size needed for storing secrets on client's and servers' devices, the LR-AKE protocol has a reduction of as much as 96% with the security parameters recommended for use in current practice. Furthermore, the memory size is *optimal*.

Organization. After giving some notations, we propose a simplified LR-AKE (LR-AKE) protocol in the next section. In Section 3, we compare the LR-AKE protocol with [13] in terms of computation costs and memory size. Section 4 is devoted to the security of the LR-AKE protocol. Finally, we give conclusions in Section 5.

2   A Simplified LR-AKE (LR-AKE) Protocol

2.1   Preliminaries

From now on, we briefly give the mathematical background and some notations to be used. The protocol stated in this paper is defined over a finite cyclic group $\mathcal{G} = \langle g \rangle$ whose order is q with the assumption that $\mathcal{G}$ is a prime order subgroup over a finite field $\mathbb{F}_p$. That is, $\mathcal{G} = \{g^i \bmod p : 0 \leq i < q\}$ where p is a large prime number, q is a large prime divisor of $p-1$ and g is an integer such that $1 < g < p-1$, $g^q \equiv 1$ and $g^i \neq 1$ for $0 < i < q$. A generator of $\mathcal{G}$ is any element in $\mathcal{G}$ except 1. In the aftermath, all the subsequent arithmetic operations are performed in modulo p, unless otherwise stated. Let g and h be two generators of $\mathcal{G}$ so that its DLP (Discrete Logarithm Problem), i.e., calculating $a = \log_g h$, should be hard for each entity. Both g and h may be given as system parameters or chosen with an initialization phase between entities.

Let k be the security parameter for p (say, 1024 bits) and let N be a dictionary size (cardinality) of passwords. Let $\{0, 1\}^\star$ denote the set of finite binary strings and $\{0, 1\}^{|N|}$ the set of binary strings of length $|N|$ where $|\cdot|$ indicates its bit-length. Let "$||$" denote the concatenation of bit strings in $\{0, 1\}^\star$. A real-valued function $\epsilon(k)$ of non-negative integers is *negligible* (in k) if for every $c > 0$, there exists $k_0 > 0$ such that $\epsilon(k) \leq 1/k^c$ for all $k > k_0$.

Let us define a secure one-way hash function $\mathcal{H} : \{0, 1\}^\star \rightarrow \{0, 1\}^{|N|}$ and a MAC generation function $\mathsf{MAC}_{km}(m)$ with key km on message m which can be chosen from a family of universal one-way hash functions [10]. Let $\mathcal{C}$ and $\mathcal{S}$ be the identities of client and server, respectively, with representing each ID $\in \{0, 1\}^\star$ as $ID_\mathcal{C}$ and $ID_\mathcal{S}$.

2.2   The LR-AKE Protocol

We consider a scenario where a client is communicating with disparate n servers. Here is a simplified LR-AKE (LR-AKE) protocol using Shamir's $(n + 1, n + 1)$-threshold secret sharing scheme [12], also deployed in the SKI protocol [13]. The rationale is that a client generates n verification data in size of N for server $\mathcal{S}_i$ (for simplicity, we assign the servers consecutive integer $1 \leq i \leq n$) from his password and a pair of client's ID and server's ID. The LR-AKE protocol is *optimal* in terms of memory size needed for storing secrets on client's and servers' devices. More detailed discussion will be followed.

During the initialization phase, a client $\mathcal{C}$ picks a random polynomial $p(x)$ of degree n with coefficients $(\alpha_1, \cdots, \alpha_n)$ also randomly chosen in $\{0, 1\}^{|N|}$:

$$p(x) = \sum_{j=0}^{n} \alpha_j \cdot x^j \bmod N \tag{1}$$

and sets the constant term $\alpha_0 = \mathcal{H}(pw, ID_\mathcal{C}, ID_\mathcal{S})$ where pw is the client's password and $ID_\mathcal{S}$ is a set of $ID_{\mathcal{S}_i}$. After computing the respective share $p(i)$ $(1 \leq i \leq n)$ with the above polynomial, the client registers each of the verification data securely to the corresponding server $\mathcal{S}_i$ $(1 \leq i \leq n)$ as follows:

$$\mathcal{S}_i \leftarrow p(i) \tag{2}$$

where $p(i)$ is a share of Shamir's $(n + 1, n + 1)$-threshold secret sharing scheme. Then the client just stores *a secret polynomial* $p'(x)$ (instead of the polynomial $p(x)$) on devices, such as smart cards or computers that may leak the secret $p'(x)$ eventually, *and* keeps *his password pw in mind*.

$$p'(x) = p(x) - p(0) = \sum_{j=1}^{n} \alpha_j \cdot x^j \bmod N \ . \tag{3}$$

Of course, all the other (intermediate) values should be deleted from the devices.

$$\begin{array}{cc}
\text{Client } \mathcal{C} & \text{Server } \mathcal{S}_i \ (1 \le i \le n)
\end{array}$$

[Initialization]

$$p(x) = \boxed{p'(x)} + \alpha_0 \bmod N \qquad \xrightarrow{\quad p(i) \quad} \qquad \boxed{p(i)}$$

[Protocol Execution]

$$p(x) = \boxed{p'(x)} + \alpha_0 \bmod N$$

$$\begin{array}{ccc}
r_1 \xleftarrow{R} (\mathbb{Z}/q\mathbb{Z})^\star & & r_2 \xleftarrow{R} (\mathbb{Z}/q\mathbb{Z})^\star \\
y_1 \leftarrow g^{r_1} \cdot \underline{h^{-p(i)}} & \xrightarrow{\quad y_1 \quad} & y_2 \leftarrow g^{r_2} \cdot \underline{h^{p(i)}} \\
& \xleftarrow{\quad y_2 \quad} & \\
km_{\mathcal{C}} \leftarrow \left(y_2 \cdot \underline{h^{-p(i)}} \right)^{r_1} & & km_{\mathcal{S}_i} \leftarrow \left(y_1 \cdot \underline{h^{p(i)}} \right)^{r_2} \\
\mathsf{Ver}_1 \leftarrow \mathsf{MAC}_{km_{\mathcal{C}}}(Tag_{\mathcal{C}}\|y_1\|y_2) & \xrightarrow{\quad \mathsf{Ver}_1 \quad} & \mathsf{Ver}_2 \leftarrow \mathsf{MAC}_{km_{\mathcal{S}_i}}(Tag_{\mathcal{S}_i}\|y_1\|y_2) \\
& \xleftarrow{\quad \mathsf{Ver}_2 \quad} &
\end{array}$$

$$\begin{array}{ll}
\text{If } \mathsf{Ver}_2 \ne \mathsf{MAC}_{km_{\mathcal{C}}}(Tag_{\mathcal{S}_i}\|y_1\|y_2), & \text{If } \mathsf{Ver}_1 \ne \mathsf{MAC}_{km_{\mathcal{S}_i}}(Tag_{\mathcal{C}}\|y_1\|y_2), \\
\quad \text{stop the protocol.} & \quad \text{stop the protocol.} \\
\text{Otherwise,} & \text{Otherwise,} \\
\quad sk_{\mathcal{C}} \leftarrow \mathsf{MAC}_{km_{\mathcal{C}}}(Tag_{sk}\|y_1\|y_2). & \quad sk_{\mathcal{S}_i} \leftarrow \mathsf{MAC}_{km_{\mathcal{S}_i}}(Tag_{sk}\|y_1\|y_2).
\end{array}$$

Fig. 1. A simplified LR-AKE (LR-AKE) protocol where the enclosed values in rectangle represent stored secrets of client and server, respectively. The underlined values represent changed parts from the SKI protocol [13]

When establishing an authenticated session key with one of the servers $\mathcal{S}_i$ ($1 \le i \le n$), the client computes his secret value corresponding to the verification data with polynomial $p'(x)$ stored on devices and password pw remembered by himself. In the secrecy amplification phase, client $\mathcal{C}$ chooses a random number $r_1 \leftarrow_R (\mathbb{Z}/q\mathbb{Z})^\star$ and then sends y_1 to server $\mathcal{S}_i$, after calculating $y_1 \leftarrow g^{r_1} \cdot h^{-p(i)}$ using polynomial $p'(x)$ and $\alpha_0 = \mathcal{H}(pw, ID_{\mathcal{C}}, ID_{\mathcal{S}})$. The server $\mathcal{S}_i$ also calculates $y_2 \leftarrow g^{r_2} \cdot h^{p(i)}$ with a random number $r_2 \leftarrow_R (\mathbb{Z}/q\mathbb{Z})^\star$ and its verification data $p(i)$ registered by the client, and then transmits it to the client. On both sides, the client's keying material becomes $km_{\mathcal{C}} \leftarrow (y_2 \cdot h^{-p(i)})^{r_1}$ and the server's one becomes $km_{\mathcal{S}_i} \leftarrow (y_1 \cdot h^{p(i)})^{r_2}$. Only if the client uses the right password pw and the polynomial $p'(x)$ *and* the server $\mathcal{S}_i$ uses the right verification data $p(i)$, both of them can share the same keying material. This phase ends up with only one pass in parallel since both y_1 and y_2 can be calculated and sent independently (where $g^{r_1} \cdot h^{-p'(i)}$ and g^{r_2} are pre-computable). That's why $h^{-p(i)} = h^{-p'(i)} \cdot h^{-\alpha_0}$. Additionally, the implementation cost of this phase is very low because it can be simply obtained from a small modification of widely-used Diffie-Hellman key exchange protocol [3]. The remaining phase of verification and session-key generation can be performed in the same way as the SKI protocol. The resultant protocol is illustrated in Figure 1.

3   Comparison

Since password-based AKE protocols have been motivated by the very practical implementations and widely used even in wireless networks, we have to analyze computation costs and memory size needed for devices (e.g., mobile phones). The main parameters of the LR-AKE protocol are: (1) the number of modular exponentiations in the initialization phase is 0; and (2) the client's memory size is $n|N|$ where n is the number of servers and N is the size of shares (recall that N is a dictionary of size from which passwords are chosen). Since n is usually small (say, $n = O(1)$ and certainly $n \ll N$), we obtain good parameters considering the generality of the LR-AKE protocol. In particular, the parameters are essentially independent of the security parameter k. We summarize comparative results in Table 1 and 2 about how much memory size and computation costs are reduced in the LR-AKE protocol.

Table 1. Comparison between SKI [13] and LR-AKE as for memory size

Protocols	Memory size of		$|p| = 1024,$ $|N| = 36$[3]	$|p| = 1024,$ $|N| = 60$[4]	$|p| = 2048,$ $|N| = 36$	$|p| = 2048,$ $|N| = 60$
	client[1]	server[2]				
SKI [13]	$n|p|$	$|p|$	96.5%	94.1%	98.2%	97%
LR-AKE	$n|N|$	$|N|$	saving[5]	saving[5]	saving[5]	saving[5]

[1]: Memory size needed for storing secrets on client's devices
[2]: Memory size needed for storing verification data on each server's devices
[3]: $|N| = 36$ for alphanumerical passwords with 6 characters
[4]: $|N| = 60$ for alphanumerical passwords with 10 characters
[5]: The percentages are calculated by $\frac{|p|-|N|}{|p|} \times 100$.

As shown in Table 1 in terms of memory size, a client (each server) in the LR-AKE protocol just stores one polynomial $p'(x)$ in size of $n|N|$ (one verification data in size of $|N|$) rather than n secret values in size of $|p|$ (one verification data in size of $|p|$) in the SKI protocol, respectively. Actually, $n|N|$ is the same size as n passwords. However, recall that the client remembers *only one password* in the LR-AKE protocol. For the minimum security parameters recommended for use in current practice: $|p| = 1024$ and $|N| = 36$ (for alphanumerical passwords with 6 characters), the numerical reduction in memory size is about 96%! One can easily see that the longer the prime p, the larger the saving. Furthermore, the memory size in the LR-AKE protocol is *optimal*.

Theorem 1. *The memory size in the* LR-AKE *protocol is* optimal.

Proof. Proving this theorem is trivial according to a well-known fact in the theory of secret sharing schemes [2, 12]. The LR-AKE protocol should guarantee that the password (precisely, α_0) is information-theoretically secure against adversary (see Section 4.1). In order to achieve the information-theoretical security, shares must be of length *at least* as the size of a secret itself. In the LR-AKE protocol, each of the coefficients $(\alpha_1, \cdots, \alpha_n)$ in polynomial $p'(x)$ is the same size of the secret (or, password). That provides the optimal memory size. ∎

Table 2. Comparison between SKI [13] and LR-AKE as for computation costs

| Protocols | Modular exponentiation[*1] | | | Additional operations[*2] | | |
| | Initialization phase | Protocol execution[*3] | | Initialization phase | Protocol execution[*3] | |
		Client $\mathcal{C}$	Server $\mathcal{S}_i$		Client $\mathcal{C}$	Server $\mathcal{S}_i$
SKI [13]	$n+1$	2.34 (2)	2 (1)	$2n$ Mod. q, n Inv.	4 Multi., 3 MAC	2 Multi., 3 MAC
LR-AKE	0	2.34 (2)	2.34 (2)	n Mod. N, n Hash	1 Hash, 1 Mod. N, 2 Multi., 3 MAC	2 Multi., 3 MAC

*1: Modular exponentiation is a major factor to evaluate efficiency of a cryptographic protocol because that is the most power-consuming operation. The number of modular exponentiations where the cost for one simultaneous calculation of two bases is converted into 1.17 due to [9]. The figures in the parentheses are the remaining costs after pre-computation.
*2: As additional operations, Mod., Inv., Multi., MAC and Hash denote modular, modular inversion, modular multiplication, MAC and hash function operations, respectively. But, computation costs of these operations can be negligible.
*3: This is the case that the client establishes a session key with one of the servers.

Table 2 shows that the LR-AKE protocol doesn't require any modular exponentiation in the initialization phase (compared to $n+1$ modular exponentiations in the SKI protocol) when registering n verification data to the corresponding server $\mathcal{S}_i$ ($1 \leq i \leq n$). This represents a 100% reduction of modular exponentiation in the initialization phase. Instead server $\mathcal{S}_i$ is required to compute 2 modular exponentiations if pre-computation is allowed. As a result the LR-AKE protocol is more efficient rather than the SKI one, especially when implementing on client's devices with limited computing power and limited memory capacity.

4   Security

In this section we discuss its security of the LR-AKE protocol.

4.1   Security of Password

The primary goal of adversary after obtaining stored secrets from a client and servers, respectively, is to perform off-line exhaustive search for the client's password that makes possible to impersonate the client to the other servers.

Theorem 2. *The password in the* LR-AKE *protocol of Fig. 1. remains information-theoretically secure against off-line attacks after the leakage of stored secrets from the client C and the servers S_i $(1 \leq i \leq n)$, respectively.*

Proof. First, we prove the security of password against an adversary who obtains stored secret $p'(x)$ of client C and is trying to deduce $\alpha_0 = \sum_{l=1}^{n+1} p(l) \cdot \lambda_l$ for the password where λ_l is a Lagrange coefficient. Be sure that the adversary knows polynomial $p'(x)$, not $p(x) = p'(x) + \alpha_0$.

$$\sum_{l=1}^{n+1} p'(l) \cdot \lambda_l = \alpha_0 - \alpha_0 \sum_{l=1}^{n+1} \lambda_l = 0, \text{ where } \lambda_l = \prod_{m=1, m \neq l}^{n+1} \frac{m}{m-l} \bmod N . \quad (4)$$

Equation (4) means that the polynomial $p'(x)$ doesn't reveal any information on the password pw, simply because the shares $p'(l)$ $(1 \leq l \leq n+1)$ of $(n+1, n+1)$-threshold secret sharing scheme are of a secret 0.

Second, we prove the security of password against an adversary who obtains stored secrets $p(i)$ of all the servers S_i $(1 \leq i \leq n)$ and is trying to deduce α_0 for the password. Although the adversary gathers all of the verification data from servers S_i $(1 \leq i \leq n)$, the number of shares is n. That means the password is information-theoretically secure as a secret value of $(n+1, n+1)$-threshold secret sharing scheme. ∎

4.2   Formal Validation of Security

Due to the lack of space, we omit the security model, definitions and formal security proof (refer to [14]). The intuitive interpretation of Theorem 3 is that if both N and q are large enough and both $\epsilon_{ddh}(k_1, t')$ and $\epsilon_{mac}(k_2, t', q_{se} + 2q_{ex} + q_{re} + 2)$ are negligibly small for appropriate security parameters k_1 and k_2, the advantage for a more powerful adversary (even if stored secret from client is given) can be bounded by a negligibly small value.

Theorem 3. *Let P be the* LR-AKE *protocol of Fig. 1. where passwords are chosen from a dictionary of size N. For any adversary A within a polynomial time t, with less than q_{se} active interactions with the entities (*Send*-queries), q_{ex} passive eavesdropping (*Execute*-queries) and q_{re} queries to the* Reveal *oracle, the advantage of A in attacking the protocol P is upper bounded by*

$$\mathsf{Adv}_{P,A}^{ind_{sk}}(k, t) \leq 2(Q+1) \cdot \epsilon_{ddh}(k_1, t') + \epsilon_{mac}(k_2, t', Q + q_{ex} + q_{re} + 2)$$

$$+ \frac{2(n \cdot q_{se} + 1)}{N} + \frac{2Q}{q} \tag{5}$$

where k, k_1 and k_2 are the security parameters, $Q \leq (q_{se} + q_{ex})$, n is the number of servers and $t' = t + n \cdot t_P$ (t_P is the time required for execution of P by any pair of client and sever).

5   Conclusions

We have proposed a simplified LR-AKE (LR-AKE) protocol that is more efficient rather than [13]: about 100% decrease in computation costs in the initialization phase and about 96% reduction in memory size on client's and servers' devices where the memory size is *optimal*. That makes the LR-AKE protocol applicable for many applications, especially when considering client's devices with restriction to the computation ability and memory capacity.

References

1. S. M. Bellovin and M. Merritt. Encrypted Key Exchange: Password-based Protocols Secure against Dictioinary Attacks. In *Proc. of IEEE Symposium on Security and Privacy*, pages 72-84, 1992.
2. G. R. Blakley. Safeguarding Cryptographic Keys. In *Proc. of National Computer Conference 1979 (AFIPS)*, Vol. 48, pages 313-317, 1979.
3. W. Diffie and M. Hellman. New Directions in Cryptography. In *IEEE Transactions on Information Theory*, Vol. IT-22(6), pages 644-654, 1976.
4. A. Frier, P. Karlton, and P. Kocher. The SSL 3.0 Protocol. Netscape Communications Corp., 1996, http://wp.netscape.com/eng/ssl3/.
5. IEEE P1363. IEEE Standard Specifications for Public Key Cryptography. IEEE, November 12, 1999. http://grouper.ieee.org/groups/1363/P1363/index.html.
6. IEEE P1363.2. Standard Specifications for Password-based Public Key Cryptographic Techniques. Draft version 18, November 15, 2004.
7. IETF (Internet Engineering Task Force). Secure Shell (secsh) Charter. http://www.ietf.org/html.charters/secsh-charter.html.
8. IETF (Internet Engineering Task Force). Transport Layer Security (tls) Charter. http://www.ietf.org/html.charters/tls-charter.html.
9. A. J. Menezes, P. C. Oorschot, and S. A. Vanstone. Simultaneous Multiple Exponentiation. In *Handbook of Applied Cryptography*, pages 617-619. CRC Press, 1997.
10. M. Naor and M. Yung. Universal One-Way Hash Functions and Their Cryptographic Applications. In *Proc. of STOC '89*, pages 33-43, 1989.
11. Phoenix Technologies Inc., Research Papers on Strong Password Authentication. available at http://www.jablon.org/passwordlinks.html.
12. A. Shamir. How to Share a Secret. In *Proc. of Communications of the ACM*, Vol. 22(11), pages 612-613, 1979.
13. S. H. Shin, K. Kobara, and H. Imai. Leakage-Resilient Authenticated Key Establishment Protocols. In *Proc. of ASIACRYPT 2003*, LNCS 2894, pages 155-172. Springer-Verlag, 2003.
14. The full version of this paper will appear.

The Fuzzy Engine for Random Number Generator in Crypto Module

Jinkeun Hong

Division of Information & Communication, Cheonan University,
#115, Anseo-dong, Cheonan-si, Chungnam, 330-704, Korea
jkhong@cheonan.ac.kr

Abstract. Ubiquitous computing raises security issues and a crypto card extend computing to physical spaces. Especially, a crypto module is a small computer in credit card format with no man machine interface. Some crypto module's microprocessors use a random number generator(RNG) and cryptographic processors. Cryptographic procedures always require random numbers. A crypto card requires random numbers for key generation to authenticate the card and for encryption. Random numbers that cannot be predicted or influenced guarantee the security and the ideal solution is a H/W RNG in the crypto module's microcontroller. Critical cryptography applications require the production of an unpredictable and unbiased stream of binary data derived from a fundamental noise mechanism, which is quite difficult to create with a stable random bit stream, as required for statistical randomness, when using a random generator with only a hardware component. Accordingly, this paper proposes a method for stabilizing the input power of a random number generator using fuzzy logic control in crypto module hardware. As such, the proposed scheme is designed to reduce the statistical property of a biased bit stream and optimize the input power to a random number generator engine in a crypto module engine for ubiquitous computing.

1 Introduction

In recent years, ubiquitous computing advocates the construction of massively distributed computing environments that consumer electronics, sensors, GPS(global positioning system) receives. Bluetooth originally thought of as a "serial cable replacement" for small computer peripherals, and 802.11, originally developed as a wireless LAN system for mobile devices(laptop, PDA)[1-3]. In this environment, ubiquitous computing imposes peculiar constraints computational power and energy budget, which make this case significantly different from those contemplated by the canonical doctrine of security in distributed systems. There are many security issues in the ubiquitous environment, including authentication, authorization, accessibility, confidentiality, integrity, and non repudiation. And other issues include convenience, speed, and so on. A H/W random number generator uses a non-deterministic source to produce randomness, and more demanding random number applications, such as

P. Lorenz and P. Dini (Eds.): ICN 2005, LNCS 3421, pp. 953–963, 2005.

cryptography, a crypto module engine, and statistical simulation, benefit from sequences produced by a random number generator, a cryptographic system based on a hardware component [1]. As such, a number generator is a source of unpredictable, irreproducible, and statistically random stream sequences, and a popular method for generating random numbers using a natural phenomenon is the electronic amplification and sampling of a thermal or Gaussian noise signal.

However, since all electronic systems are influenced by a finite bandwidth, 1/ f noise, and other non-random influences, perfect randomness cannot be preserved by any practical system. Thus, when generating random numbers using an electronic circuit, a low-power white noise signal is amplified, then sampled at a constant sampling frequency. Yet, it is quite difficult to create an unbiased and stable random bit stream, as required for statistical randomness, when using a random generator with only a hardware component. The studies reported in [2-4] show that the randomness of a random stream can be enhanced when combining a real random number generator, LFSR number generator, and hash function. However, the randomness of this combined method is still dependent on the security level of the hash function and LFSR number generator.

Therefore, controlling a stable input voltage for a random number generator is an important aspect of the design of a random number generator.

In previous fuzzy studies, Wilson Wang, Fathy Ismail and Farid Golnaraghi(2004) examined a neuro-fuzzy approach to gear system monitoring[6], while Zang and Phillis(2001) proposed the use of fuzzy logic to solve the admission control problem in two simple series paralleled networks[7]. Plus, fuzzy logic has also been applied to admission control in communication networks(Le et al. 2003[9]).

Accordingly, this paper proposes a fuzzy approach to ensuring a stable input power for a random number generator engine. The stability of the input power is a very important factor in the randomness of a random number generator engine. Thus, to consistently guarantee the randomness of an output sequence from a random number generator, the origin must be stabilized, regardless of any change of circumstance elements. Therefore, a random number generator is proposed that applies fuzzy logic control, thereby providing the best input power supply. Additionally we use measure of randomness test to decide fuzzy rule base and its measure is provided the efficiency, which is fast and not weighty due to use test bits of 20,000bits, when it is evaluated the randomness of output stream.

Hereinafter, section 2 reviews the framework of fuzzy logic control. Then, section 3 examines a case study, experimental results and some final conclusions are given in section 4.

2   Framework of Fuzzy Logic Controller (FLC)

Most crypto module's microcomputer chip consists of CPU, ROM, RAM, I/O, EEPROM, etc. The ROM contains the chip operating system and the RAM is the process's working memory.

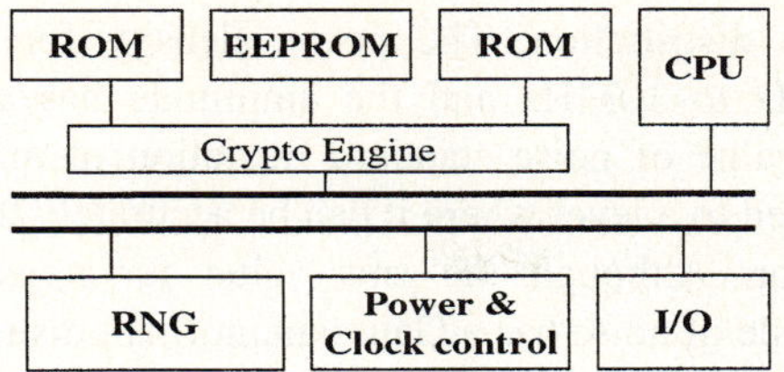

Fig. 1. Microcomputer architecture in crypto module

In the EEPROM memory, data and program can be written to and read from the EEPROM under the control of OS. Within the card, data are passed through a bus under the security logic's control. Crypto module has some form of power and clock control circuitry, BUS, and I/O interface.

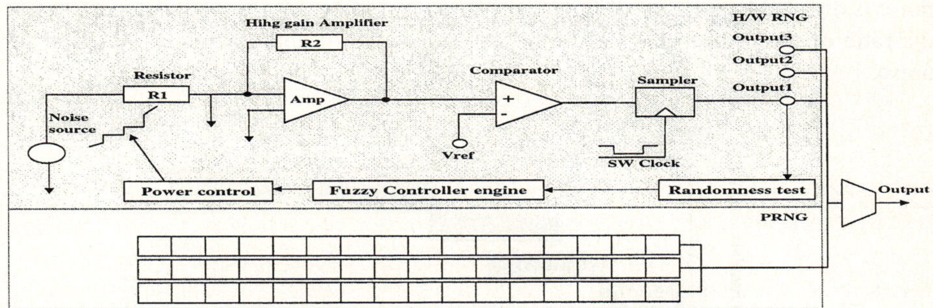

Fig. 2. RNG(H/W RNG & PRNG) chip architecture

A H/W random number generator includes common components for producing random bit-streams, classified as follows: characteristics of the noise source, amplification of the noise source, and sampling for gathering the comparator output[12-13]. The applied noise source uses Gaussian noise, which typically results from the flow of electrons through a highly charged field, such as a semiconductor junction[14-17]. Ultimately, the electron flow is the movement of discrete charges, and the mean flow rate is surrounded by a distribution related to the launch time and momentum of the individual charge carriers entering the charged field. The Gaussian noise generated in a PN junction has the same mathematical form as that of a temperature-limited vacuum diode. The noise seems to be generated by the noise current generator in parallel with the dynamic resistance of the diode. The probability density $f(x)$ of the Gaussian noise voltage distribution function is defined by Eq. (1):

$$f(x) = \frac{1}{\sqrt{2\pi\sigma^2}} e^{-\frac{x^2}{2\sigma^2}} \tag{1}$$

Where σ is the root mean square value of Gaussian noise voltage. However, for designed Gaussian noise random number generator, the noise diode is used the diode with white Gaussian distribution. The power density for noise is constant with frequency from 0.1Hz to 10MHz and the amplitude has a Gaussian distribution. *Vn(rms)* is the rms value of noise standard deviation of distribution function. The noise must be amplified to a level where it can be accurately threshold with no bias by a clocked comparator. Although the rms value for noise is well defined, the instantaneous amplitude of noise has a Gaussian, normal, distribution.

$$V_n(rms) = \sqrt{4kTRB} \tag{2}$$

Where *k* is Boltzmann constant (1.38 x 10E-23 J/deg. K), *T* is absolute temperature (deg. Kelvin), *B* is noise bandwidth (Hz), *R* is resistor (Ohms). In Fig. (3), If *4kT* is 1.66 x 10E-20 and *R* is 1K, *B* is 1Hz, then *Vn(rms)* = √*(4kTRB)* = *4nV/√Hz*. In Fig. (4), the applied voltage is ±15Vdc, and current limiting resistor is 16KΩ. Noise comes from agitation of electrons within a resistance, and it sets a lower limit on the noise present in a circuit. When the frequency range is given, the voltage of noise is decided by a factor of frequency. The crest factor of a waveform is defined as the ratio of the peak to the *rms* value. A crest value of approximately 4 is used for noise.

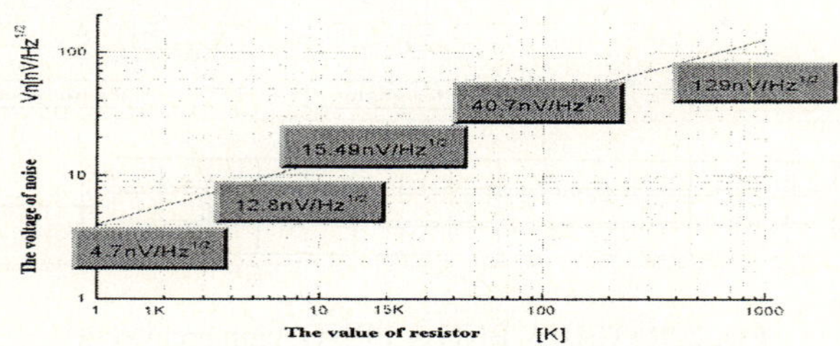

Fig. 3. The plot of Noise voltage vs. Resistor

However, for the proposed real random number generator, the noise diode is a noise diode with a white Gaussian distribution. The noise must be amplified to a level where it can be accurately thresholded with no bias using a clocked comparator.

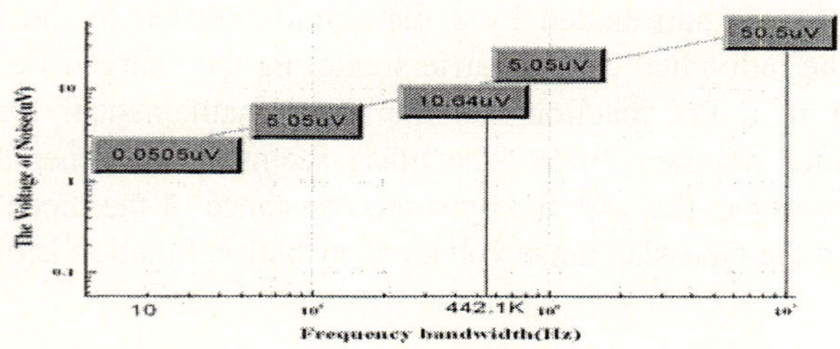

Fig. 4. The plot of Noise voltage vs. Frequency Range

This section provides a short description of the framework of a FLC[6-9] as follows: the input power source(①), generator engine that generates random numbers(②), random test process(③), fuzzification(④), rule base(⑤), inference engine(⑥), and de-fuzzification(⑦).   A fuzzy logic controller consists of three processors; fuzzification, rule base & inference, and de-fuzzification.

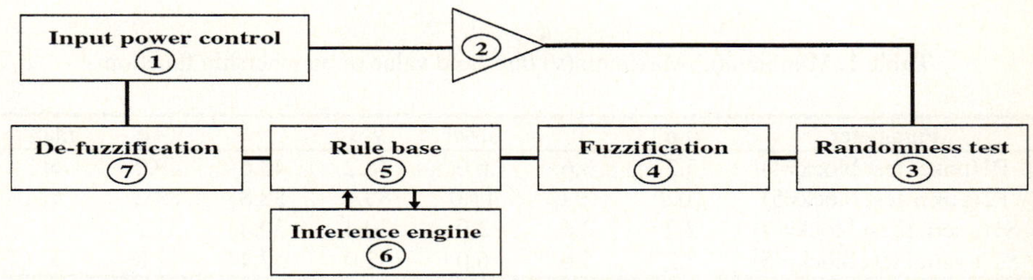

Fig. 5. FLC framework used to generate random numbers

- **Generator Engine That Generates Random Numbers and Randomness Test Block**

The generator engine that generates random numbers includes common components for producing random bit-streams, classified as follows: Gaussian noise process, source amplification process, and sampling process[12-13]. The cryptographic modules that implement a random number generator engine also incorporate the capability to perform statistical tests for randomness. As such, the multiple bit stream of 20,000 consecutive bits of output from the generator is subjected to a poker test, and t-serial test[19-20]. If any of the tests fail, the module then enters an error state. Statistical random number generator test method of FIPS140-1 is faster than conventional test method, which the number of test bits is required the size of 200,000bits.

- **Fuzzification, Rule Base, Inference Engine and De-fuzzification**

Fuzzification is the process of transforming crisp variables into their respective fuzzy variables according to selected membership function variables: poker test value(p), serial test value(s) as given in Fig.6.

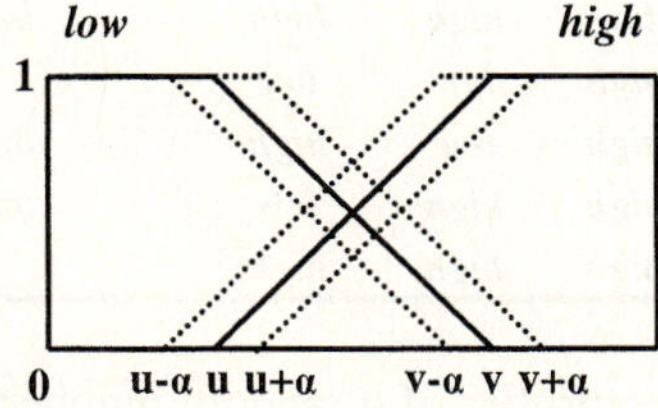

Fig. 6. Membership function of input variables

In run test, a run is defined as a maximal sequence of consecutive bits of either all ones or all zeros, which is part of 20,000 bits output stream. The incidences of runs of all lengths in the output stream should be counted and stored. The test is passed if the number of runs that occur is each within the corresponding interval specified value u and v. where low value of u is "1" and high value of u is "0", and high value of n is "1" and low value of v is "0".

Table 1. Minimum(u)/Maximum(v) threshold value of membership function

Parameter	u-α	u	u+α	v-α	v	v+α	max
P1(poker test block=4)	5.2	5.6	6.0	48.2	48.6	49.0	∞
P2(poker test block=5)	10.2	10.6	11.0	88.4	88.8	89.2	∞
S1(t-serial test block= 4)	2.2	2.6	3.0	30.4	30.8	31.2	∞
S2(t-serial test block= 5)	5.2	5.6	6.0	52.0	52.4	52.8	∞

The rule base of a fuzzy logic controller consists of all the necessary relationships among the input and output feedback variables. In this paper, the rule base is derived using the randomness results of the output bit stream. According to the chosen membership functions in Table2, each variable has linguistic values as follows: zero(*z*), small very low(*svl*), small high(*sh*), small medium(*sm*), small low(*sl*), large very high(*lvl*), large low(*ll*), large very low(*lvl*), and large high(*lh*).

Table 2. Minimum(u)/Maximum(v) threshold value of membership function

Rule base	If p4	& p5	& s4	& s5	then ipcv
Rule1	*low*	*low*	*low*	*low*	*zero*
Rule2	*low*	*low*	*low*	*high*	*small_very_low*
Rule3	*low*	*low*	*high*	*low*	*small_high*
Rule4	*low*	*low*	*high*	*high*	*small_medium*
Rule5	*low*	*high*	*low*	*low*	*small_low*
Rule6	*low*	*high*	*low*	*high*	*small_low*
Rule7	*low*	*high*	*high*	*low*	*small_low*
Rule8	*low*	*high*	*high*	*high*	*small_very_low*
Rule9	*high*	*low*	*low*	*low*	*large_low*
Rule10	*high*	*low*	*low*	*high*	*small_low*
Rule11	*high*	*low*	*high*	*low*	*small_low*
Rule12	*high*	*low*	*high*	*high*	*large_very_low*
Rule13	*high*	*high*	*low*	*low*	*small_low*
Rule14	*high*	*high*	*low*	*high*	*large_very_low*
Rule15	*high*	*high*	*high*	*low*	*large_very_low*
Rule16	*high*	*high*	*high*	*high*	*large_high*

As such, a fuzzy logic controller of a random number generator engine processes crisp input values and produces crisp output values. A Mamdani-type fuzzy system uses fuzzy rules like:

Rule: if x_1 is A and x_2 is B, then y is C,

where A, B, and C are fuzzy sets. Fuzzy sets are usually represented by parameterized membership functions, like triangular functions in Fig.7. Therefore, the fuzzy rule for the membership function suggested in this paper is as follows:

```
Rule1:  if  x₁(p4)=A₁₁(low),  x₂(p5)=A₁₂(low),  x₃(s4)=A₁₃(low),
and  x₄(s5)=  A₁₄(low),  then  y(input  power  control  value)=
C₁(z)
```

where $x_1 \in$ U, $x_2 \in$ V, $x_3 \in$ P, $x_4 \in$ Q, $x_5 \in$ S, and $y \in$ Y. The relationship of the fuzzy sets, a subset of U×V×P×Q×S×Y, is represented as follows: $R_i = (A_{i1} \times A_{i2} \times A_{i3} \times A_{i4} \times A_{i5}) \times C_i$

```
Rule2:if  p4=low,  p5=low,  s4=low,  s5=high,  then  the  input
power  control  value=small_very_low
```

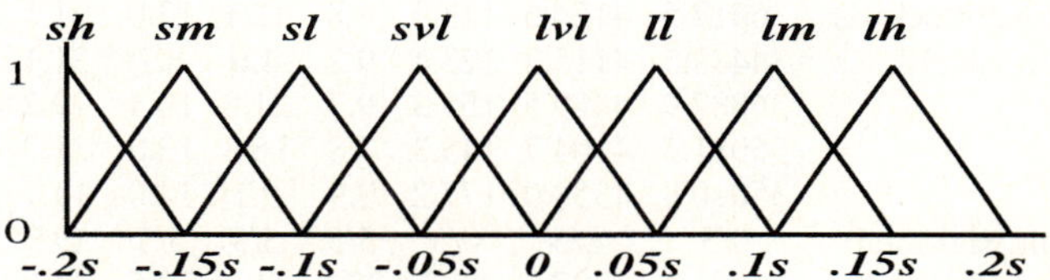

Fig. 7. Membership function of input variable

In the inference stage, the rules are evaluated to determine the values of the fuzzy output variables. s is the scale factor. As such, a conditional condition for a fuzzy condition proposition is analogized if it is given. An assignment function about an n-th fuzz variable is generally represented by an AND operation, as defined by Eq. (3):

$$w_l = \bigcap_{i=1}^{n} w_l(x_l) = \min[w_1(x_1), ..., w_n(x_n)] \tag{3}$$

De-fuzzification is the process of combing the results of the inference process to find crisp outputs, and this study uses the center of average de-fuzzification technique, which is expressed as follows:

$$y_l \approx \frac{\sum_{l=1}^{n} (\bar{y}^l * w_l)}{\sum_{l=1}^{n} w_l} \tag{4}$$

where y_l denotes the decision value, n denotes the total number of implications, $\bar{y}^l$ is the average value of the output variables for the l-th rule, and w_l is the weighting factor for the l-th implication.

3 Experimental Results

A sample random number of 20,000 bits was used in this study. We are applied on variable value which is tuned the result of inference engine. The fuzzy inference variables were converted by a crisp value using a de-fuzzification membership function. When the input power remained within the fuzzy inference border area, the output random number sequence maintained a stable randomness. When 8 levels of input power were given, the randomness of the output random number sequences was as shown in Table 3.

Table 3. Relationship between the result of randomness test and input power

	9.15V	9.3V	9.48V	9.65V	9.83V	10.0V	10.15V	10.3V
Poker test(block=4)	36317.5	4124.5	173.9	7.8	12.1	13.0	21.5	15.7
(X<24.9)	34448.5	4115.3	123.4	9.2	4.6	8.7	31.1	19.5
	36887.9	4429.5	155.3	9.7	21.8	10.4	29.2	26.2
	36081.5	4104.7	145.3	7.8	16.3	12.0	31.2	14.4
	35010.8	4358.0	177.2	9.9	9.1	14.0	14.7	13.7
Result(pass/total)	0/5	0/5	0/5	5/5	5/5	5/5	2/5	4/5
Poker test(block=5)	11989.4	1407.2	82.3	28.6	41.2	30.5	44.0	24.1
(X<44.7)	11705.9	1361.3	62.3	30.1	44.2	23.8	52.5	33.0
	12459.3	1527.6	86.1	41.9	42.0	25.8	28.3	29.1
	11765.7	1412.1	85.6	46.3	38.5	33.8	40.7	25.4
	11610.1	1472.9	91.3	35.6	35.2	29.0	32.6	25.1
Result(pass/total)	0/5	0/5	0/5	4/5	5/5	5/5	4/5	5/5
t-serial test(block=4)	11986.4	1367.5	49.0	4.1	5.0	11.4	4.8	4.8
(X<15.5)	11582.6	1268.7	40.7	5.0	7.5	4.3	15.6	10.2
	12371.1	1448.6	38.7	3.7	9.9	5.4	6.7	8.0
	11889.2	1356.4	49.2	5.1	3.3	6.1	6.1	3.3
	11570.5	1379.7	85.3	7.9	11.5	3.0	8.4	2.2
Result(pass/total)	0/5	0/5	0/5	5/5	5/5	5/5	4/5	5/5
t-serial test(block=5)	37017.1	4200.0	177.6	16.3	23.3	18.9	26.4	22.5
(X<26.3)	35540.3	4184.6	120.6	14.8	23.7	10.7	34.9	22.2
	37572.0	4318.7	138.2	12.4	17.2	29.3	18.7	20.5
	36708.2	4228.6	150.3	17.8	15.5	25.3	16.7	21.7
	35618.4	4145.8	173.2	20.7	17.2	21.3	21.5	20.5
Result(pass/total)	0/5	0/5	0/5	5/5	5/5	4/5	3/5	5/5

The randomness of the output random number sequence reacted sensitively whenever the input power supply was changed. As such, the experimental model highlighted the relationship between the randomness and variations in the input power, where the randomness of the output random number sequences was found to depend on the input power and a threshold value could be used to determine the randomness of the output random number sequence engine. Therefore, modifications in the input power controlled by the proposed FLC were used to stabilize this interdependence between the input power and the randomness of the output random number sequences.

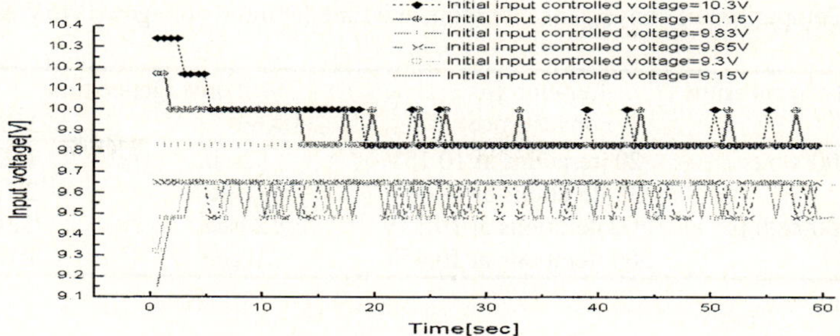

Fig. 8. FLC of input power for each initial control voltage

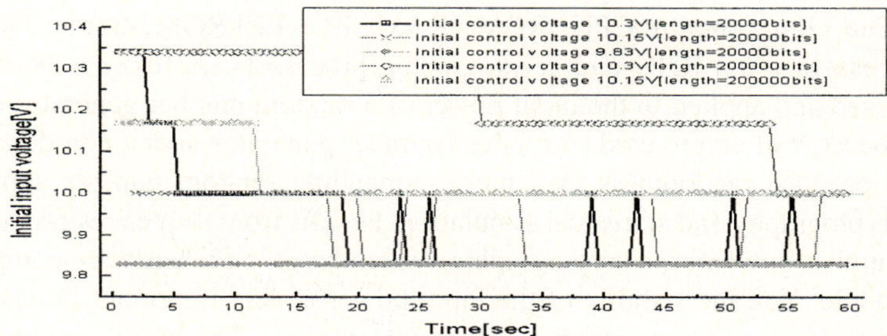

Fig. 9. Convergence time according to the variation of test measure length

In Fig.8-9, the initial input power was set at 9.65V and 9.83V, yet ranged between 9.13V and 10.3V, while the threshold for the input power was set between 9.65V and 10.0V. The adaptability of the FLC was then investigated based on its ability to use the variables to maintain the input power and quantity within the threshold range. As a result, a change in the input power at 10.35V was controlled within the threshold range with a fast slope and converged at 9.83V. When the input power was maintained at 9.83V, this stabilized the output sequence. We present in Table 4 & 5 test measure of 20,000bits other than it use length of 200,000bits, to converge fast and reduce loss of generating bits.

Table 4. Comparison of the number of collected random number bits(at 200Kbps)

(Length of test measure)	Number of collected bits			
	at initial voltage=10.15V		at initial voltage=10.45V	
	Without FLC	With FLC	Without FLC	With FLC
Length= 20000bits	7.2×10^8bits/1h	7.176×10^8/1h	7.2×10^8bits/1h	7.15×10^8bits/1h
Length= 200000bits		5.96×10^8bits/1h		5.78×10^8bits/1h

Table 5. Comparison of convergence iteration and time (at intial voltage=10.15V & 10.45V)

Length of test measure	Iteration for convergence	Convergence time	
		200Kbps	2Mbps
Length= 20000bits	20 iterations at 10.15V	12sec	1.2sec
	35 iterations at 10.45V	21sec	2.1sec
Length= 200000bits	200 iterations at 10.15V	120sec	12sec
	350 iterations at 10.45V	210sec	21sec

4 Conclusion

In ubiquitous computing, a crypto card consists of a chip and an integral operating system. The chip contains CPU, ROM, RAM, I/O, EEPROM. Some crypto card microprocessors use a RNG and cryptographic processors. A fuzzy logic controller was proposed and applied to the input power of a random number generator engine in crypto-processor of crypto card. A random number generator uses a non-deterministic source to produce randomness, and more demanding random number applications, such as cryptography and statistical simulation, benefit from sequences produced by a random number generator, a cryptographic system based on a hardware component in crypto module. Yet, the stability of the input power is very important in ensuring the randomness of a random number generator engine. Therefore, to consistently guarantee the randomness of the output sequences from a random number generator, a method was presented of stabilizing the origin fast, regardless of any changes in the circumstance elements. Tests showed the proposed fuzzy logic controller using the length of 20,000bits to be effective and to be fast in stabilizing the input power of a random number generator engine in crypto module.

References

1. Alireza h., Ingrid V.: High-Throughput Programmable Crypto-coprocessor, IEEE Computer Society(2004).
2. Jalal A. M, Anand R., Roy C., M. D. M: Cerberus: A Context-Aware Security Scheme for Smart Spaces, proc. IEEE PerCom'03(2003).
3. N. O. Attoh-Okine, L. D. Shen: Security Issues of Emerging Smart Cards Fare Collection Application in Mass Transit(1995).
4. WiTness: Interaction of SIM based WiTness security functions and security properties of mobile devices and communication channels, Information society(2003).
5. Thomas S, Ezzat A. Dabbish, Robert H. Sloan: Examing Smart-Card Security under the Threat of Power Analysis Attacks, IEEE Trans. On computers, Vol.51, No.5(2002).
6. Wilson Wang, Fathy Ismail, and Farid Golnaraghi: A Neuro-Fuzzy Approach to Gear System Monitoring, IEEE transactions on fuzzy systems, Vol.12, No.5(2004).
7. R. Zang, Y. A. Phillis: Admission control and scheduling in simple series paralleled networks using fuzzy logic, IEEE transaction on Fuzzy systems, Vol.9, No.2(2001).
8. George J. Klir, Bo Yuan: Fuzzy Sets and Fuzzy Logic Theory and Applications, Prentice-Hall International Inc.(1995).

9. Q. Le, G. M. Knapp: Incorporating Fuzzy Logic Admission Control In Simulation Models, Proc. Winter Simulation Conference, S. Chick, P. J. Sanchez, D. Ferrin, and D. J. Morrice(2003).
10. M. Kimberley: Comparison of two statistical tests for keystream sequences, Electronics Letters, Vol. 23, No. 8(1987).
11. H. M. Gustafson: Statistical Analysis of Symmetric Ciphers, Thesis submitted in accordance with the regulations for Degree of doctor of Philosophy, Queensland University of Technology(1996).
12. C. S. Petrie and J. A. Connelly: A Noise-Based Random Bit Generator IC for Applications in Cryptography, Proc. ISCAS'98, June(1998).
13. M. Delgado-Restituto, F. Medeiro, and A. Rodriguez-Vasquez: Nonlinear switched-current CMOS IC for random signal generation, IEE electronic letters, vol. 29, December(1993).
14. http://www.io.com/~ritter/RES/NOISE.HTM.
15. http://www.clark.net/pub/cme/P1363/ranno.html.
16. http://webnz.com/robert/true_rng.html.
17. Boris Ya, Ryabko and Elena Matchikina: Fast and Efficient Construction of an Unbiased Random Sequence, IEEE Trans. on information theory, vol. 46, no. 3, May(2000).
18. W. Timothy Holman, J. Alvin Connelly, and Ahmad B. Dowlatabadi: An Integrated Analog/Digital Random Noise Source, IEEE Transactions on Circuits and System I: Fundamental Theory and Applications, vol.44, no.6, June(1997).
19. FIPS 140-1: Security Requirements for Cryptographic Modules, Federal Information Processing Standards Publication 140-1, U.S. Department of Commerce/NIST[National Technical Information Service] Springfield, Virginia(1994).
20. http://csrc.ncsl. nist. gov/fips/fips 1401.htm (16 Oct. 1998).
21. Diehard: http://stat.fsu.edu/~geo/diehard.html (16 Oct. 1998).

A Packet Marking Scheme for IP Traceback

Haipeng Qu[1,2], Purui Su[1,2], Dongdai Lin[1], and Dengguo Feng[1]

[1] The State Key Lab of Information Security, Institute of Software,
Chinese Academy of Sciences, Beijing P.O.Box 8718,
100080 Beijing, P.R. China
[2] Graduate School of the Chinese Academy of Sciences, 100039 Beijing, P.R. China
haipeng@acm.org

Abstract. DDoS attack is a big problem to the Internet community due to its high-profile, severe damage, and the difficulty in defending against it. Several countermeasures are proposed for it in the literature, among which, Probabilistic Packet Marking (PPM) first developed by Savage *et al.* is promising. However, the PPM marking schemes have the limitations in two main aspects: high computation overhead and large number of false positives. In this paper, a new packet marking scheme is proposed, which is more practical because of higher precision, and computationally more efficient compared with the PPM scheme proposed by Savage. Furthermore, this scheme can achieve a more higher precision than Advanced Marking Schemes in case of the victim knowing the map of its upstream routers.

1 Introduction

Recently, Internet attacks are increasing greatly [1]. Denial of Service (DoS) attack consumes resources associated with various network elements - e.g., web servers, routers, firewalls and end hosts - which impedes the efficient functioning and provisioning of services in accordance with their intended purposes [2]. There are two types of DoS attacks: the first one takes advantage of the drawbacks of some implementation or algorithmic deficiencies in applications by one or more malformed 'killer' packets; the second takes advantage of the sheer fact that the victim is connected to the Internet by flooding a deluge of packets to the victim. The first type is relatively easy to be addressed by patching up vulnerabilities or filtering out malformed packets. In this work, we focus on the latter, called flood type DoS attack. DoS attack could be more effective if several attackers at different places conspire because the effect is summed up. This is generally called distributed DoS attack (DDoS attack).

DoS attacks are among the hardest security problems to be addressed because they are easy to launch, difficult to defend and difficult to trace [3]. Unlike other types of attacks such as privilege escalation attacks, DoS attacks do not need two-way communication. Therefore the source addresses of DoS attack packets could be spoofed. This feature leaves attackers opportunities to hide their true identities.

P. Lorenz and P. Dini (Eds.): ICN 2005, LNCS 3421, pp. 964–971, 2005.

The need to defend DoS attack and to find the attackers are of growing importance. There are several countermeasures proposed in the literature to deal with it, among which, Probabilistic Packet Marking (PPM) first developed by Savage *et al.* is promising [3]. However, all existing marking schemes are bearing limitations in some aspects.

In this paper, we make two improvements to the technique of Probabilistic Packet Marking. Compared with the basic PPM algorithm, our marking techniques has significantly higher precision (lower false positive rate), simultaneously it has lower computation overhead for the victim to reconstruct the attack paths under highly distributed denial-of-service attacks. We demonstrate the benefits of our approach with an experimental evaluation using simulated packet traces.

The rest of the paper is organized as follows. We present a brief background to this problem and highlight the main challenges of IP marking in Section 2. Section 3 introduces our new Packet Marking Scheme. An analysis and simulation results of the new scheme are provided in Section 4. We conclude the paper in the last section.

2 Background on Packet Marking Scheme

2.1 Overview of Basic PPM

Appending additional data to a packet in flight is a poor choice because it is expensive and may result in fragmentation. Savage *et al.* proposed a marking scheme in which each router marks its address information into packets transiting it with a probability p. In the case of flood-type DoS attack, the victim could get many marked packets. After collecting enough marked information, the victim could reconstruct the attack path through which attack flow traveled.

Identification field of IPv4 header is seldom used because very few packets are fragmented on the fly. Thus, this field could be used to embed some tracing messages. Let's denote two connected nodes on the network as an edge. In Ref.[3], the authors defined edge-ID as XOR of the two IP addresses making up of an edge and "mark" this information into packet identification field probabilistically. After receiving enough packets from attacker(s), the victim could use these edges to reconstruct the full path. For this purpose, the victim needs not only edge-IDs but also corresponding distances. Since distance seldom exceeds 30[4,5], 5 bits are enough to code it. If we rely only on the identification field, we have 11 bits left and that is not enough for 32-bit edge-ID. So edge-ID should be further fragmented into several segments. To ensure reconstructing edge correctly when combining these segments, an error detection code is needed. Ref. [3] uses 32- bit hash of edge-ID as error detector. Therefore, 64 bits should be marked for one edge. The authors further interleave the error detection code with the IP address bit by bit, then separate the 64 bits into 8 blocks. So 5 bits for distance, 3 bits for offset (indicating the place of the block in 64 bits full information), and 8 bits for block are inscribed to identification field, that is 16 bits.

2.2   Limitation of Packet Marking Scheme

PPM mainly considers denial-of-service with a single attacker site, it suffers from two main problems[6]:

(1) High Computation Overhead

Ref. [6] gave an analysis of PPM. In a Reconstruction procedure, the total number of combinations to be checked for all the distances is:

$$|\Gamma| = \sum_{0 \le d \le \max d} (|S_{d-1}| \times \prod_{0 \le f \le 7} |\Psi_{d,f}|) \tag{1}$$

$|S_d|$ denotes the number of distinct routers at distance d. And $\Psi_{d,f}$ denotes the set of unique edge fragments marked with a distance d and fragment ID f.

(2) Large Number of False Positives

For each $z \in \Gamma$, when z is not on the attack paths, the probability of z being a valid IP encoding is $1/2^{32}$, since the hash value is 32 bits. So the expected number of elements in Γ that are valid IP encoding is $|\Gamma|/2^{32}$, denoted as α. Because an IP address is 32 bits, so the expected number of false positives is

$$E[false \quad positives] = (1 - (1 - 1/2^{32})^a) \cdot 2^{32} \tag{2}$$

When $\alpha << 2^{32}$, E[false positives]$\approx \alpha = |\Gamma|/2^{32}$.

Even for a DDoS with 25 attackers, the reconstruction can result in thousands of false positives.

3   A New Packet Marking Scheme

In the proposed new packet marking scheme, we use a similar marking scheme as basic Packet Marking Scheme. We divide the 16-bit IP Identification field into a 5-bit *distance* field, a 3-bit *fID* field, a 1-bit *pID* field and a 7-bit *edge* field. We divide every router's IP address in two 16-bit part, X and Y. We use two independent hash functions, h_1 and h_2 in the encoding of every parts of the routers' IP addresses. h_1 and h_2 both have 16-bit outputs. We also use the first 8 bits of X as a relative checkout code for Y and the first 8 bits of Y as a relative checkout code for X when X or Y is the last node of the marking edge. So, to each part of a router, we can construct four 7×8 matrixes,   as   shown   in   Fig. 1. (Let $X_R = (x_0, x_1, x_2, \cdots x_{15})$ be the first half of the IP address of router R; $Y_R = (y_0, y_1, y_2, \cdots y_{15})$ be the last half of the IP address of router R; $Z_R' = (z_0, z_1, z_2, \cdots z_{15}) = Hash_1(X_R)$; $Z_R'' = (z_{16}, z_{17}, z_{18}, \cdots z_{31}) = Hash_2(X_R)$; $H_R' = (h_0, h_1, h_2, \cdots h_{15}) = Hash_1(Y_R)$; $H_R'' = (h_{16}, h_{17}, h_{18}, \cdots h_{31}) = Hash_2(Y_R)$) Fig. 2   describes the marking procedure of the new Packet Marking Scheme. Fig. 3 describes the reconstruction procedure.

$$U = \begin{pmatrix} u_0 \\ u_1 \\ u_2 \\ u_3 \\ u_4 \\ u_5 \\ u_6 \\ u_7 \end{pmatrix} = \begin{pmatrix} x_0 & x_1 & x_2 & x_3 & x_4 & 0 & 0 \\ x_5 & x_6 & x_7 & x_8 & x_9 & 0 & 0 \\ x_{10} & x_{11} & x_{12} & x_{13} & x_{14} & x_{15} & 0 \\ z_0 & z_1 & z_2 & z_4 & z_5 & z_6 & 0 \\ z_6 & z_7 & z_8 & z_9 & z_{10} & z_{11} & 0 \\ z_{12} & z_{13} & z_{14} & z_{15} & z_{16} & z_{17} & 0 \\ z_{18} & z_{19} & z_{20} & z_{21} & z_{22} & z_{23} & z_{24} \\ z_{25} & z_{26} & z_{27} & z_{28} & z_{29} & z_{30} & z_{31} \end{pmatrix} \quad U' = \begin{pmatrix} u'_0 \\ u'_1 \\ u'_2 \\ u'_3 \\ u'_4 \\ u'_5 \\ u'_6 \\ u'_7 \end{pmatrix} = \begin{pmatrix} x_0 & x_1 & x_2 & x_3 & x_4 & y_0 & y_1 \\ x_5 & x_6 & x_7 & x_8 & x_9 & y_2 & y_3 \\ x_{10} & x_{11} & x_{12} & x_{13} & x_{14} & x_{15} & y_4 \\ z_0 & z_1 & z_2 & z_4 & z_5 & z_6 & y_5 \\ z_6 & z_7 & z_8 & z_9 & z_{10} & z_{11} & y_6 \\ z_{12} & z_{13} & z_{14} & z_{15} & z_{16} & z_{17} & y_7 \\ z_{18} & z_{19} & z_{20} & z_{21} & z_{22} & z_{23} & z_{24} \\ z_{25} & z_{26} & z_{27} & z_{28} & z_{29} & z_{30} & z_{31} \end{pmatrix}$$

$$V = \begin{pmatrix} v_0 \\ v_1 \\ v_2 \\ v_3 \\ v_4 \\ v_5 \\ v_6 \\ v_7 \end{pmatrix} = \begin{pmatrix} y_0 & y_1 & y_2 & y_3 & y_4 & 0 & 0 \\ y_5 & y_6 & y_7 & y_8 & y_9 & 0 & 0 \\ y_{10} & y_{11} & y_{12} & y_{13} & y_{14} & y_{15} & 0 \\ h_0 & h_1 & h_2 & h_4 & h_5 & h_6 & 0 \\ h_6 & h_7 & h_8 & h_9 & h_{10} & h_{11} & 0 \\ h_{12} & h_{13} & h_{14} & h_{15} & h_{16} & h_{17} & 0 \\ h_{18} & h_{19} & h_{20} & h_{21} & h_{22} & h_{23} & h_{24} \\ h_{25} & h_{26} & h_{27} & h_{28} & h_{29} & h_{30} & h_{31} \end{pmatrix} \quad V' = \begin{pmatrix} v'_0 \\ v'_1 \\ v'_2 \\ v'_3 \\ v'_4 \\ v'_5 \\ v'_6 \\ v'_7 \end{pmatrix} = \begin{pmatrix} y_0 & y_1 & y_2 & y_3 & y_4 & x_0 & x_1 \\ y_5 & y_6 & y_7 & y_8 & y_9 & x_2 & x_3 \\ y_{10} & y_{11} & y_{12} & y_{13} & y_{14} & y_{15} & x_4 \\ h_0 & h_1 & h_2 & h_4 & h_5 & h_6 & x_5 \\ h_6 & h_7 & h_8 & h_9 & h_{10} & h_{11} & x_6 \\ h_{12} & h_{13} & h_{14} & h_{15} & h_{16} & h_{17} & x_7 \\ h_{18} & h_{19} & h_{20} & h_{21} & h_{22} & h_{23} & h_{24} \\ h_{25} & h_{26} & h_{27} & h_{28} & h_{29} & h_{30} & h_{31} \end{pmatrix}$$

Fig. 1. The Marking Matrixes: U, U', V and V'

```
[Marking procedure at router R_i]
    for each packet P
        let m be a random number from [0,1]
        if m ≤ q   then
            P.distance ← 0
            let x be a random integer from [0,16)
            if x < 8  then
                P.pID ← 0
                P.edge ← u_x
                P.fID ← x
            else
                P.pID ← 1
                P.edge ← v_{x-8}
                P.fID ← x-8
        else
            if P.distance = 0 then
                y ← P.fID
                if P.pID = 0 then
                    P.edge ← P.edge ⊕ u'_y
                else
                    P.edge ← P.edge ⊕ v'_y
            P.distance = P.distance + 1
```

Fig. 2. Marking procedure of the New Marking Scheme

```
[Path reconstructin procedure at victim v:]
let FragTbl be a table of tuples(edge, pID, fID, distance)
let TmpTbl be a table of tuples(head,tail,end)
let T be a tree with root v
let edges in T be tuples (start, end,distance)
let maxd:=0
let last:=v
for each packet P from attacker
    FragTbl.Insert(P.edge,P.pID,P.fID,P.distanse)
    If P.distance>maxd then
      Maxd:=P.distanse
for d:=0 to maxd
    split last into 2 16-bit parts as last0, last1
    split last into 4 8-bit parts as last0(0), last0(1), last1(0), last1(1)
    for all ordered combinations of fragments from FragTbl
    where pID=0 and distance=d
        construct α, α', α'',β₁, β₂ as Fig.4
      if d≠0 then
          if concatenate(β₁,β₂)=last1(0) then
              α=α ⊕ last0
              α'= α' ⊕ h₁(last0)
              α''= α'' ⊕ h₂(last0)
              if h₁(α) =α' and h₂(α)= α'' then
                  insert ( α,Φ,last) into TmpTbl
    for all ordered combinations of fragments from FragTbl
    where pID=1 and distance=d
        construct α, α', α'',β₁, β₂ as figure 4.
      if d≠0 then
          if concatenate(β₁,β₂)=last0(0) then
              α=α ⊕ last1
              α'= α' ⊕ h₁(last1)
              α|'= α'' ⊕ h₂(last1)
                  if h₁(α) =α' and h₂(α)= α'' then
                      if ∃(x, y, z)((x, y, z)∈TmpTbl∧ y=φ∧ z=last)
                      then    insert edge(x+α,last,d) into T
    last=x+a
    empty Tmplbl
remove any edge (x,y,d) with d != distance from x to v in T
extract path (Ri..Rj ) by enumerating acyclic paths in T
```

Fig. 3. The Reconstruction Procedure

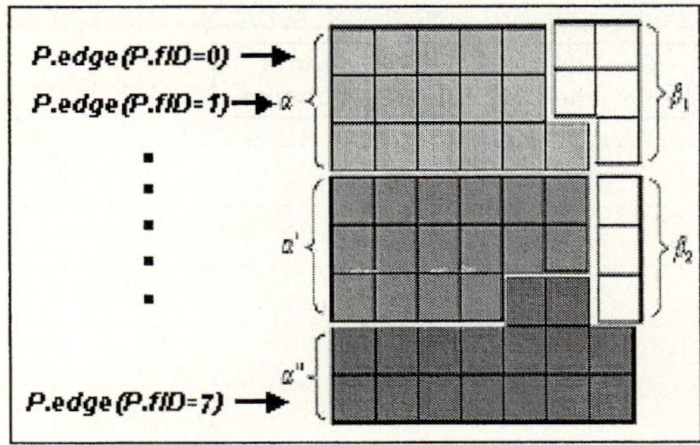

Fig. 4. Construction method for α, α', α'', β₁, β₂

4 Analysis

Assume a DDoS attack where $|S_d|$ denotes the number of routers on the attack paths at distance d from the victim. And S_{d-1} denote the set of routers at distance $d-1$ to the victim in the reconstruction attack path. Assume the number of the attack paths is N, then $|S_d| \mapsto N, |S_{d-1}| \mapsto N$. So, for every element y in S_{d-1}, the expected number of false positive rate is:

$$E \leq \frac{2N^7}{N^7 + 2^{40}} - \left(\frac{N^7}{N^7 + 2^{40}}\right)^2 \tag{3}$$

And the number of computation times of the hash function to reconstruct the attack graph by the victim after receiving all the packets needed is

$$4 \cdot \frac{1}{2^5} |S_{d-1}| \times N_{0,0} \times N_{1,0} \times N_{2,0} \approx \frac{1}{2^3} \cdot N_{d-1} \times N_d^3 \tag{4}$$

Fig. 5 shows the false positive rate of PPM and the new scheme. For example, when the number of the attack paths is 32, the false positive rate of the basic PPM scheme is more than 0.9 and the false positive rate of the new scheme is less than 0.1.
Fig.6 shows the computation times of the hash function to reconstruct the attack graph by the victim after receiving all the packets needed. In the new scheme, the information of the IP Address of the nodes are only in the first 3 fragments of the marking packets(when *fID=0,1 and 2)*, the possible number of α in Fig. 4 are $O(N^3)$. So the computation times of the hash function will be much less than the basic PPM scheme which is $O(N^7)$.

In case that the victim knows its upstream topology (this is just the necessary precondition in Advanced PPM[6]), the reconstruction precision can be improved remarkably. Assume the set of the upstream routers of router r_{d-1} is R. Most false positive edge < r_{d-1}, x> can be ruled out by comparing x with all node in R. So the false positive rate can be about $|R|/2^{32}$ of the false positive rate without the topology information . This false positive rate is also less than that of the Advanced PPM scheme[6].

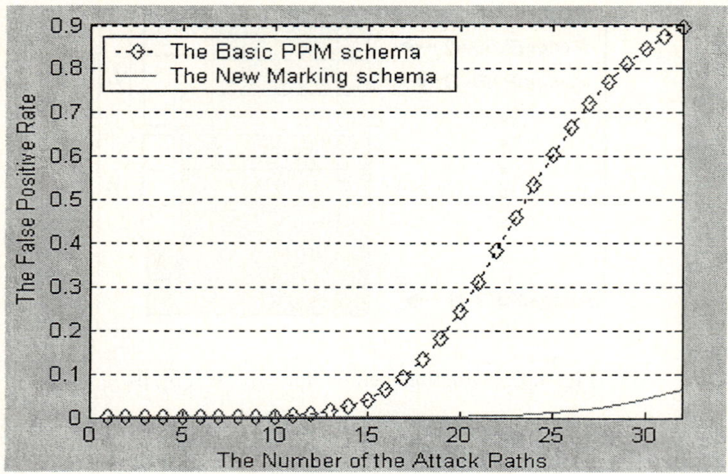

Fig. 5. False Positive Rate of PPM and the New Scheme

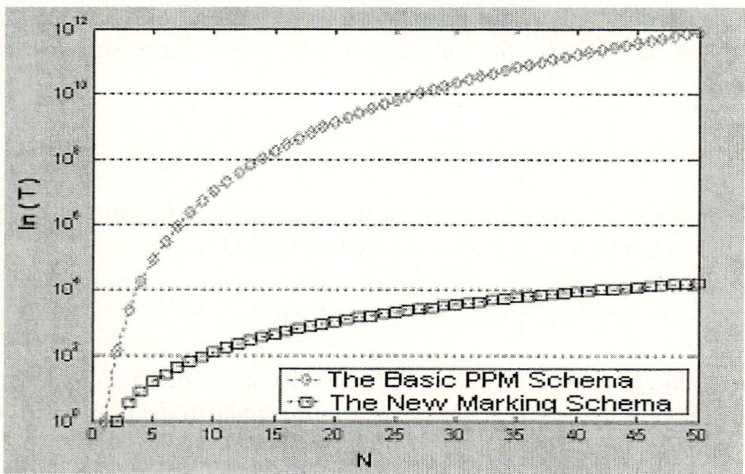

Fig. 6. Computation Overhead for PPM and the New Scheme

5 Conclusions

A brief overview concerning the research in countering flood-type DoS attack is given first. Then, based on the analysis of the typical packet marking schemes, a new packet marking scheme is proposed to overcome the shortcoming of the scheme. Compared with the Basic Probabilistic Packet Marking scheme, the new PPM scheme can reduce the computation overhead from $O(N^7)$ to $O(N^3)$, where N denotes the average path number of distributed attackers.

The proposed new packet marking schemes allows the victim to trace back the approximate origin of spoofed IP packets. It also features as low network and router overhead, a high response time and support incremental deployment. Compared with the basic PPM algorithm, our marking techniques have significantly higher precision (lower false positive rate) and lower computation overhead for reconstructing the attack paths under large scale distributed denial-of-service attacks. At the same time, it also does not need the topology information as in the basic PPM.

References

1. CERT. CERT Statistics 1088-2003, http://www.cert.org/stats/ .
2. Kihong Park and Heejo Lee. A Proactive Approach to Distributed DoS Attack Prevention using Route-Based Packet Filtering. Technical Report, CSD00 -017, Department of Computer Sciences, Purdue University, 2000.
3. Savage S, Wetherall D, Karlin A, Anderson T. Practical network support for IP traceback. In: Proc. of the 2000 ACM SIGCOMM Conf., Stockholm, Sweden, August 2000. 295-306. http://www.acm.org/sigs/sigcomm/sigcomm2000/ conf/paper/sigcomm2000-8-4. ps.gz.
4. Cooperative Association for Internet Data Analysis (CAIDA). The Skitter project. http://www.caida.org/tools/measurement/skitter/.
5. Theilmann W, Rothermel K. Dynamic distance maps of the Internet. In: Proc. of the 2000 IEEE INFOCOM Conf. Tel Aviv, Israel, March 2000.
6. Song DX, Perrig A. Advanced and authenticated marking schemes for IP traceback. In: Proc. of the IEEE INFOCOM'01. http://www.ieee-infocom.org/2001/program.html
7. LI De-Quan, SU Pu-Rui and FENG Deng-Guo. "Notes on Packet Marking for IP Traceback", Journal of Software, Vol.15, No.2, 2004.
8. K. Park and H. Lee. On the effectiveness of probabilistic packet marking for IP traceback under denial of service attack. In: Proceedings of IEEE INFOCOM '01, 2001, pp338-347.
9. Steven Bellovin, Marcus Leech, Tom Taylor. ICMP Traceback Messages. work in progress, Internet Draft, draft-ietf-itrace-02.txt, October 2001.

Securing Admission Control in Ubiquitous Computing Environment

Jong-Phil Yang[1] and Kyung Hyune Rhee[2]

[1] Department of Computer Science, Pukyong National University,
599-1, Daeyeon3-Dong, Nam-Gu, Pusan 608-737, Republic of Korea
bogus@lisia21.net
[2] Division of Electronic, Computer and Telecommunication Engineering,
Pukyong National University,
599-1, Daeyeon3-Dong, Nam-Gu, Pusan 608-737, Republic of Korea
khrhee@pknu.ac.kr

Abstract. In this paper, we introduce a new model which allows an organization to perform secure admission control for foreign users. We adopt a threshold proxy signature scheme to provide precise admission control for the proposed model. Additionally, we propose a new (t, n) threshold proxy signature scheme for realistic implementation of the proposed model.

Keywords: Admission control, Trust management, Ubiquitous Security.

1 Introduction

As computing becomes pervasive, people will live in intelligent spaces. Intelligent spaces provide services and resources with which the user will interact through a laptop, PDA or cell phone, and so on. In ubiquitous computing environments users expect to access resources and services anytime and anywhere, leading to serious security risks and problems with access control as these resources can now be accessed by almost anyone with mobile devices [7]. Therefore, the question arises of how to establish trust relationship in previously unknown devices or resources with such environment. In [4], D. Hutter et al. mentioned that trust management is one of the most important research issues in the security mechanism for ubiquitous computing environment. In this paper, we have an interest in admission control in a small organization as a part of trust management. After reviewing a model to establish trust relationship in ubiquitous environment[7], we propose a new model to enhance admission control over communicating entities in the office scenario. To enhance admission control, we apply a threshold proxy signature scheme to our model and introduce proper rationale for application. Then, we propose a new (t, n) threshold proxy signature scheme which is suitable for our model.

In Section 2, we introduce related works, motivations of our work and system model. Section 3 introduces a new (t, n) threshold proxy signature scheme

P. Lorenz and P. Dini (Eds.): ICN 2005, LNCS 3421, pp. 972–979, 2005.

based on RSA. We introduce protocols to provide an enhanced admission control for our model in Section 4. We focus on system performance and conclude in Section 5.

2   Related Works and System Model

2.1   Related Works

In [7], L. Kagal et al. introduce *Centaurus* system to provide access control to a foreign user in the *Smart Office* scenario. To do so, they use *distributed trust* approach. The notion of distributed trust is similar to Simple Public Key Infrastructure(SPKI) and Pretty Good Privacy(PGP). There is a simple example of the distributed trust approach in the Smart Office as following:

> John is an employee of one of the office's partners, but the service manager is unable to understand his role in the organization, so he is denied access to the service. John approaches one of the managers, Susan, and asks for permission to use the services in the Smart Office. According to the policy, Susan has the right to delegate those rights to anyone she trusts. Susan delegates to John, the right to use the lights, the coffee maker and the printer but no the fax machine, for a short period of time. Susan's laptop sends a short lived signed delegation to John's hand-held device. When John enters the room, the client on his hand-held device sends his identity certificate and the delegation to the service manager. As Susan is trusted and has the ability to delegate, the delegation conforms to the policy and John now has access to the lights, the coffee maker and the print in the room. Once the delegation expires, John is denied access to any services in the room.

We point out two concerning security problems in the above example : *"Is it possible to restrict a fine-grained control for the key which is used to sign a delegation by Susan? Specially, on the valid period of the key"* and *"In case that Susan is malicious, is it possible to prohibit Susan from signing a delegation for a foreign user who conspires with her?"*. Let us consider the first problem. To restrict Susan's signing ability, Susan's key should be restricted to a valid period and usage rules. Therefore, we can use proxy signature schemes to solve this problem. In proxy signature schemes, an original signer can efficiently restrict a valid period and usage rules of proxy signing key through a warrant[2],[8]. For the case of the second problem, if Susan cannot sign a delegation by herself, this problem can be solved. To do so, we apply a threshold signature scheme to our model. By using the threshold signature scheme, a quorum of members in the Smart Office collaboratively sign a delegation for a foreign user[12],[14],[15],[16]. Therefore, we finally decide that threshold proxy signature scheme is the most suitable cryptographic primitive to solve the above mentioned problems.

2.2   System Model

In this paper, we consider an organizational model as the following. In the organization, there is a unique central authority, and it has a key pair for a public key algorithm. The public key is trusted by all entities in the organization. We assume that the organization consists of several departments, and each department has at most n members. Then, in the organization, there are lots of resources such as a fax machine, several lights, a coffee maker and a printer, etc. Someone who owns appropriate rights can access to them. When a foreign user wants to have permission to access to resources in a department of the organization, he approaches one of the members in the department. The member acts as a delegator for the foreign user. Then, the delegator issues an admission credential based on the system policy by collaborating with other members in the department. Then, the foreign user can access to resources in the department under the right in the admission credential. Once the admission credential expires, the foreign user is denied to access to any resources. To design our system, we must consider the following requirements.

- The system must issue an admission credential collaboratively. That is, a member cannot issue an admission credential for a foreign user by oneself.
- The central authority does not involve in a system operation excepting system initialization.
- The foreign user is allowed to access to certain services without creating a new identity for the organization or assigning a temporary role to him.
- The system should be realistic for implementation in the ubiquitous environment.

3   Cryptographic Primitives

Since threshold signature schemes based on discrete logarithm require a distributed key generation(DKG) technique to generate an one-time secret parameter, they have to consume lots of communicational and computational resources [10], [11]. That is, the system must perform n secret-sharings, and each member must perform n parallel verifiable secret sharing(VSS)s to succeed in DKG, where n is the number of members involved in generation of an one-time secret parameter. Therefore, to make our system more efficient, we use a threshold proxy signature scheme based on the *de facto* standard RSA algorithm.

In [9], M.S.Hwang et al. proposed a (t, n) threshold proxy signature scheme based on RSA such that it allows any t or more proxy signers from a designated group of n members to cooperatively sign messages while $t - 1$ or less members cannot generate the legal proxy signature. However, there are two main security concerns in their design. First, an original signer directly shares a proxy signing key among n proxy signers. Second, the combiner must be a trust entity. Therefore, their scheme does not satisfy security requirements for proxy signature schemes such as *strong unforgeability, strong undeniability, prevention of misuse and proxy protected*[2],[8]. So, we design a new (t, n) threshold proxy signature scheme which has an improved security and also suitable for our proposed system.

3.1 (t, n) Threshold Proxy Signature Scheme Based on RSA

In this section, we propose a new (t, n) threshold proxy signature scheme based on RSA. Our scheme is based on generic construction called as "Delegation-by-certificate" which is given in [1].

Setup: Let P_0 be an original signer who has a private key (d_O, N_O) and its corresponding public key (e_O, N_O), and P_1 be a leader of the n proxy signers, $P_1, \ldots, P_n$.

1. The leader P_1 generates a public key (e_p, N_p) and its corresponding private key (d_P, N_P) for the n proxy signers. P_1 sends the public key to both P_0 and P_i, where $2 \leq i \leq n$.
2. P_1 randomly generates a secret polynomial f_1 of degree $t - 1$, of the form $f_1(x) = d_P + a_1 x + \cdots + a_{t-1}x^{t-1} \bmod \phi(N_P)$. P_1 computes P_i's partial signing key, $d_{P_i} = f_1(i)$ and securely sends it to the proxy signer P_i, where $1 \leq i \leq n$.
3. P_1 deletes all secret information about the private key except d_{P_1} from the device.

Proxy key generation: The original signer P_0 performs the following procedures:

1. P_0 generates a new key pair : $N' = p' \cdot q'$, $\delta \cdot \epsilon \equiv 1 \bmod \phi(N')$. We call δ as *temporary secret* and ϵ is public.
2. P_0 generates randomly a secret polynomial f_2 of degree $t - 1$, of the form $f_2(x) = \delta + b_1 x + \cdots + b_{t-1}x^{t-1} \bmod \phi(N')$. P_0 computes P_i's partial temporary secret, $\delta_{P_i} = f_2(i)$ and sends securely it to the proxy signer P_i, where $1 \leq i \leq n$. Then, P_0 deletes all secret information about δ from the device.
3. P_0 computes $K = h(m_w, \epsilon, e_P, e_O)^{d_O} \bmod N_O$ and broadcasts ϵ, N', m_w, K to n proxy signers, where m_w is a warrant.

Proxy-protected signature: Assume that any t or more proxy signers out of n proxy signers want to sign collaboratively a message m on behalf of P_0. Let T denote a set of at least t proxy signers and $h()$ denote an appropriate padding such as PKCS#1 or OAEP padding. Then, the steps of the proxy-protected signature are listed as follows:

1. Each P_i computes partial signature $s_i = h(m, K)^{d_{P_i}} \bmod N_P$ and partial temporary signature $\lambda_i = h(m, K)^{\delta_{P_i}} \bmod N'$, where $1 \leq i \leq n$.
2. One of n proxy signers can combine t partial signatures and t partial temporary signatures, and compute valid proxy signature $S = \prod_{i \in T} s_i^{L_i} \bmod N_P = h(m, K)^{d_P} \bmod N_P$ and temporary signature $\lambda = \prod_{i \in T} \lambda_i^{L_i} \bmod N' = h(m, K)^{\delta} \bmod N'$, where $\ L_i = \prod_{i, j \in T, j \neq i} \frac{-j}{i-j}.$
3. The proxy signature(ρ) consists of $(m, m_w, e_P, N_P, e_O, N_O, \epsilon, N', K, S, \lambda)$.

Verification: To verify the proxy signature(ρ), the verifier checks the following three conditions : $K^{e_O} \bmod N_O \overset{?}{=} h(m_w, \epsilon, e_P, e_O)$, $h(m, K) \overset{?}{=} s^{e_P} \bmod N_P$ and $h(m, K) \overset{?}{=} \lambda^{\epsilon} \bmod N'$. If all are successful, ρ is a valid proxy signature on m.

3.2   Security Considerations

In this paper, we assume that an adversary can compromise at most $t-1$ proxy signers within the lifetime of proxy signing capability of proxy signers. Under this assumption, since the original signer dose not know d_P and proxy signers do not know δ either, our threshold proxy signature scheme guarantees *proxy-protected*. Then, since both the original signer's public key and the proxy signers' public key are used to compute proxy signatures, our scheme guarantees both *strong unforgeability* and *strong undeniability*. Moreover, a warrant m_w is signed by the original signer, our scheme guarantees *prevention of misuse*. In our scheme, K value performs the role of certificate in "Delegation-by-certificate". Although our scheme requires 3 exponential operations for verification, it does not need to perform distributed key generation(DKG) for computing an one-time security parameter in threshold proxy signature schemes based on discrete logarithm[10], [13]. That is, our scheme requires less computational and communicational resource to compute proxy signatures for implementing a real system.

4   Protocol Description

4.1   Initial Setting

For the rest of this paper, we use the following notations and assumptions:

- OTA: Organization Trust Authority. A unique central authority for an organization. It has a key-pair for RSA algorithm: public key (e_O, N_O), private key (d_O, N_O). Also, it has X.509 certificate which is issued by Certificate Authority in PKI. OTA acts as an original signer in the proposed threshold proxy signature scheme.
- Dept^i: We assume that an organization consists of several departments. The i-th department is identified by Dept^i and the number of departments is not limited. Each department has a key-pair for RSA algorithm: In case of Dept^i, public key (e^i, N^i), private key (d^i, N^i). Each department acts as a proxy signer group which consists of n members. Moreover, the valid period and role of the key-pair are not restricted to the proposed threshold proxy signature scheme.
- Mem^i_j: A j-th member who belongs to Dept^i. Note that the number of members in a department is at most n. Each member in the specific department acts as proxy signer. The private key of Dept^i, d^i, is already shared among n members. That is, Mem^i_j has a partial signing key d^i_j for d^i, where $1 \leq j \leq n$.
- FU: A foreign user. He has an X.509 certificate in PKI.
- AC^i_{FU}: FU's Admission Credential for Dept^i. Through AC^i_{FU}, FU has the right to access resources in Dept^i. It is a signature of the proposed threshold proxy signature scheme.

To initialize the system, OTA and each department perform collaboratively the following steps:

1. All leaders of departments send the public keys to OTA.
2. OTA generates a warrant m_w which is used widely in the system. OTA generates RSA key-pairs as many as the number of departments. Let the number of departments be θ. Then, OTA performs only θ times for proxy key generation from step.1 to step.2. Let the temporary secret for Dept^i be δ^i, the corresponding public key be ϵ^i and the modulus be N'^i, where $1 \leq i \leq \theta$. Then, Mem_j^i has the following information.
   - public information: e^i, N^i, ϵ^i, N'^i.
   - private information: δ_j^i as a partial temporary secret, d_j^i as a partial signing key.
3. OTA computes $K = h(m_w, \epsilon^1, \ldots, \epsilon^\theta, e^1, \ldots, e^\theta, e_O)^{d_O} \bmod N_O$ and sends ϵ^i, N'^i, m_w, K to Dept^i respectively, where $1 \leq i \leq \theta$.

The initial setting is performed periodically based on the system policy. Moreover, the next schedule for initial setting defines the valid period of proxy signing capability. Note that setup procedure in the section 3.1 may not be necessary during the next initial setting.

4.2   Admission Credential Issuance Protocol

When a foreign user(FU) enters Dept^i, the following steps are performed:

1. **Participation Request:** FU sends `part-request` to Mem_j^i, who will be a delegator for FU. This request consists of FU's X.509 certificate and additional information which is signed by FU's private key.
2. **Voting:** Upon receipt of `part-request`,
   (a) (Broadcasting) If the delegator has the right to delegate some rights within his/her rights, the delegator first extracts FU's X.509 certificate and verifies the signature. If the verification is successful, the delegator makes `right-info` for FU based on both the system policy and the warrant, and broadcasts it to $n - 1$ members of Dept^i. The `right-info` consists of FU's DN(Distinguished Name) and public key in X.509 certificate and access right information which is generated by the delegator.
   (b) (Partial Voting) If Mem_j^i approves of `right-info`, it replies to the delegator with `partial-voting` for AC_{FU}^i.

$$h(\text{right} - \text{info}, K)^{d_h^i} \bmod N^i, h(\text{right} - \text{info}, K)^{\delta_h^i} \bmod N'^i$$

   (c) (Combining) The delegator combines at least t out of n `partial-votings` into AC_{FU}^i.
3. **AC Issuance:** The delegator sends OTA's X.509 certificate and AC_{FU}^i to FU. Upon receipt of them, FU verifies OTA's X.509 certificate and obtains e_O. Then, FU verifies AC_{FU}^i by the same verification procedure as section 3.1. If the verification is successful, FU can use it to access to resources in Dept^i.

If FU sends AC_{FU}^i to any resources to access to them, the resource must perform the verification of AC_{FU}^i. However, in this case, since K value is already known to all entities in the organization, the resource performs only the verification procedure except $K^{eo} \bmod N_O? = h(m_w, \epsilon, e_P, e_O)$.

4.3   Inter-department Admission Control

When a foreign user who has already owned an admission credential for a department wants to access to resources in the other departments, he can access to them under the restricted bound as following cases:

- In case that each department agrees on a delegation policy and regulation of access rights is unnecessary, the foreign user can access to resources based on the system policy without performing additional steps.
- In case that each department agrees on a delegation policy and regulation of access rights is necessary, the system only performs step.2 and step.3 in admission credential issuance protocol for the other departments.
- In case that each department does not agree on a delegation policy, admission credential issuance protocol must be performed.

5   System Performance and Conclusion

We use JDK 1.3.1 and Cryptix package for simulating the proposed model under Pentium IV processors[3],[6]. In Figure. 1, (a) shows computational costs for issuing an admission credential for a foreign user under the proposed $(4, 7)$ threshold proxy signature scheme according to the various keysizes. Then, (b) shows computational costs for issuing an admission credential under the 1024-bits key size according to the various number of members. As expected, plain RSA shows the best performer in terms of computation time. However, we also see that the time for admission credential issuance exhibits reasonable costs.

Since a foreign user attending in ubiquitous computing does not need to create a new identity for accessing resources in an organization, our model can be properly applied to the ubiquitous environments as a new model for group admission control. Moreover, by means of issuing admission credentials collaboratively, more powerful admission control can be achieved. Also, a central authority can

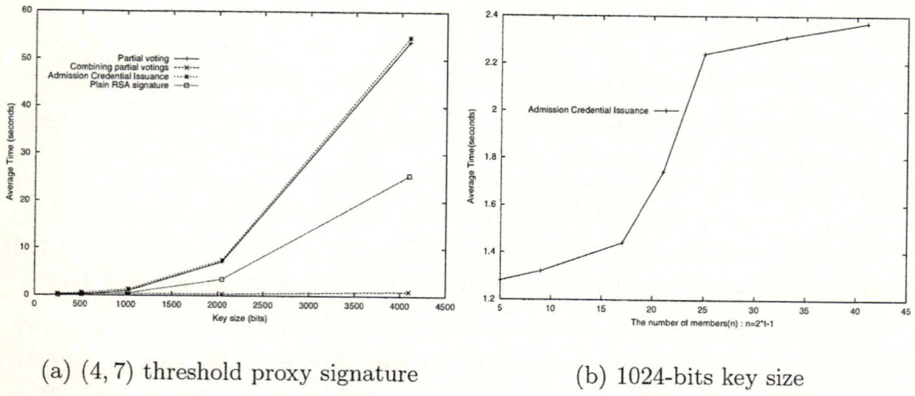

(a) $(4, 7)$ threshold proxy signature   (b) 1024-bits key size

Fig. 1. The computational costs in admission credential issuance

regulate efficiently the function of each members in the organization through the warrant used widely in the system.

In this paper, we considered security requirements for admission control in the Smart Office scenario, and designed a new security model. To guarantee security requirements, we proposed a new threshold proxy signature scheme as cryptographic primitive for our model and design admission control mechanisms. To evaluate the availability of our model, we implemented cryptographic primitive and system parameters, and checked the performance through simulation.

Acknowledgements. This research was supported by the Program for the Training of Graduate Students in Regional Innovation which was conducted by the Ministry of Commerce, Industry and Energy of the Korean Government.

References

1. A. Boldyreva, A. Palacio, and B. Warinschi, "Secure proxy signature schemes for delegation of signing rights". Avaiable at http://eprint.iacr.org/2003/096
2. B. Lee, H. Kim, and K. Kim, "Strong Proxy Signature and its Applications", SCIS2001, vol 2/2, 2001, pp. 603-608
3. Cryptix 3.1.3, http://www.cryptix.org/
4. D. Hutter, W. Stephan, and M. Ullmann, "Security and Privacy in Pervasive Computing State of the Art and Future Directions", Security in Pervasive Computing 2003, LNCS 2802, 2004, pp.285-289
5. Douglas R. Stinson, "Cryptography: Theory and Practice - Second edition", Chapman & Hall/CRC , 2002
6. J. Garms, and D. Somerfield, "Professional Java Security", Wrox Press Ltd. , 2001.
7. L. Kagal, T. Finin and A. Joshi, "Moving from Security to Distributed Trust in Ubiquitous Computing Environments", IEEE Computer, December 2001
8. M. Mambo, K. Usuda, and E. Okamoto, "Proxy signature: Delegation of the power to sign messages", IEICE Trans. Fundamentals, Sep. 1996, Vol.E79-A, No.9, pp.1338-1353
9. M.S. Hwang, Eric J. Lu, and l. Lin, "A Practical (t, n) Threshold Proxy Signature Scheme Based on the RSA Cryptosystem", IEEE Transactions on knowledge and data engineering, VOL.15, NO.6, 2003, pp.1552-1560
10. T. Pedersen, "Non-interactive and information-theoretic secure verifiable secret sharing", CRYPTO'91, 1991, pp.129-140
11. R. Gennaro, S. Jarecki, and H. Krawczyk, "Revisiting the Distributed Key Generation for Discrete-Log Based Cryptosystems", RSA Security' 03, April 2003.
12. R. Gennaro, S. Jarecki, H. Krawczyk, and T. Rabin, "Robust threshold DSS signatures", Eurocrypt'96, LNCS 1070, 1996, pp. 354-371.
13. K.Zhang, "Threshold Proxy Signature Schemes", Proc. Information Security Workshop, LNCS 1396, 1997, pp.282-290
14. S. Xu, and R. Sandhu, "Two Efficient and Provably Secure Schemes for Server-Assisted Threshold Signatures", CT-RSA 2003, LNCS 2612, 2003, pp.355-372
15. T. Rabin, "A simplified Approach to Threshold and Proactive RSA", CRYPTO'98, LNCS 1462, 1998, pp.89-104
16. V. Shoup, "Practical threshold signatures", EUROCRYPT 2000, LNCS 1807, 2000, pp.207-220

Detecting the Deviations of
Privileged Process Execution*

Purui Su[1,2], Dequan Li[1,2], Haipeng Qu[1,2], and Dengguo Feng[1]

[1]State Key Laboratory of Information Security Institute of Software,
Chinese Academy of Sciences, Beijing P.O.Box 8718,
100080 Beijing, P.R. China
[2]Graduate School of the Chinese Academy of Sciences,
100039 Beijing, P.R. China
{supurui, lidequan, hqu, fengdg}@is.iscas.ac.cn

Abstract. Most intruders access system unauthorizedly by exploiting vulnerabilities of privileged processes. Respectively monitoring privileged processes via system call sequences is one of effective methods to detect intrusions. Based on the analysis of popular attacks, we bring forward a new intrusion detection model monitoring the system call sequences, which use locally fuzzy matching to improve the detection accuracy. And the model adopts a novel profile generation method, which could easily generate better profile. The experimental results show that both the accuracy and the efficiency have been improved.

1 Introduction

Anomaly Intrusion Detection identifies intrusions based on deviations from a normal profile. It's difficult for the attackers to cheat the Anomaly Intrusion Detection System. For anomaly intrusion detection, system call is one of the useful data sources in describing the process' behaviors. The system call sequence of the process roughly describes its behavior. In the recent years, many research results have been brought forward, such as *stide*, *N-gram* and so on. [Hofmeyr98] [Warrender99] [Calvin00]

In this paper, we will analyze some of popular attacks, and the difference between normal and abnormal behaviors. Then we will bring forward an intrusion detection model. In the model, we still use the system call sequences with fixed length L to describe the profile. And in the sequences, we introduce the wildcard. A fuzzy match function and a novel method of profile generation will be introduced to decrease the profile size and improve the detection efficiency. By exploiting

* Supported by the National Grand Fundamental Research 973 Program of China under Grant No.G1999035802, the National Foundation of China for Distinguished Young Scholars under Grant No.60025205, the National Natural Science Foundation of China under Grant No.60273027 and the National High-Tech Research and Development Plan of China under Grant No.2003AA142150.

P. Lorenz and P. Dini (Eds.): ICN 2005, LNCS 3421, pp. 980–988, 2005.

the differences between the normal and abnormal behaviors, the model adopts a detection method using locally fuzzy matching to improve detection accuracy. The rest paper is organized as following: In section 2, we will introduce several popular attacks and analyze their effects on process execution. The detecting model will be introduced in Section 3. In Section 4, we will provide some experimental results and evaluate the model. In Section 5, we will discuss related works and the limitations of the model. At last, we will conclude the paper in Section 6.

2   Popular Attacks

For the intruders, taking advantage of vulnerabilities to execute their malicious code is one of the most effective methods to achieve their goals. In the following, we will introduce some methods of exploiting vulnerabilities to execute malicious codes, and analyze their effects on the processes' execution.

Stack-based Buffer Overflow: The stack-based buffer overflow overwrites the stack of a function.If the application has a boundary checking error,the user can inject additional codes into the buffer, and overwrite the return address of the function by the address of the additional codes injected. When the function returns, the process will execute the codes the user injected. [Evan99]

Frame Pointer Overwrite: Frame Pointer Overwrite is caused only by one byte exceeding the buffer space. The overflow could not use this one byte to overwrite the return address of the function, but the saved frame pointer.[Klog99] By this byte, the process could be hijacked to execute the codes injected in. In [Klog99], Klog provides detailed method in exploiting such vulnerabilities.

Runtime Process Infection: Linux provide the function *ptrace()* for debugging the process. [Anonymous02] The intruder could use *ptrace()* to resolve the symbol of the target process, poke codes into process memory, and load an additional lib, which could be used to execute the intruders' malicious operations. In [Anonymous02], one detailed method in exploiting the vulnerability is provided.

Integer Overflow: When a value greater than this maximum value is stored to an integer, it will be overflowed, which is known as integer overflow.[Blexim02] We could exploit them to overwrite the buffer just as the stack-based buffer overflow, which injects malicious codes and changes the process execution flow to execute them. In [Blexim02], some examples of exploiting the vulnerability in real word are provided.

From the attack methods introduced above, we could see that:

1. All of the attacks make the victim execute some additional codes to achieve their malicious aims. Especially for the privileged processes or applications, they will make use of their privileged rights to access the special resources of the system;
2. Because of the additional code injected by intruder, the victim process execution flow has to be changed;

3. Because the intruders need many system calls to access the system resources in their malicious codes, the system call sequence of the process will be changed.
4. Before the malicious codes' execution,there maybe many normal system calls have been triggered by the process. The abnormalities caused by the malicious codes may be concealed by the normal sequences. The abnormalities are negligible from the global view, but obvious from the local view.

3   Detection Model

In this section, we will introduce an intrusion detection model, which monitors the active system call sequences of privileged processes. If it finds obvious deviations in the live system call sequence from the profile, it will mark the process as intrusion. Before discussing the profile generation and the detection method, we will introduce a fuzzy match function, which will be used in both of them.

1. Fuzzy match function
Since perfect matching is extremely rare between strings of any reasonable length, a fuzzy match function is needed. In order to differentiate the similarity of the two sequences, we give a match function, $match(S_1,S_2)$, which does not just return matched or not-matched, but a value scaling the similarity of two strings.

$$match(S_1, S_2) = (\frac{\sum_{i=0}^{L-1} Power(i, S_1, S_2)}{\sum_{i=0}^{L-1} K^i})^{\frac{1}{L-1}} \tag{1}$$

$$Power(i, S_1, S_2) = \begin{cases} 0 & S_1[i] \neq S_2[i] \\ 1 & i = 0 \text{ and } S_1[i] = S_2[i] \\ 1 & i > 0 and S_1[i] \neq S_2[i] \\ & \text{and } Power(i-1, S_1, S_2) = 0 \\ Power(i-1, S_1, S_2) * K & i > 0 and S_1[i] = S_2[i] \\ & \text{and } Power(i-1, S_1, S_2) = 0 \end{cases}$$

The return value of the function is ranging from 0 to 1. The more similar the two sequences, the bigger the return value will be.

In the equation, K ranges from 1 to 2. If $K=1$, the return value of the function will only be affected by the count of identical locations between the two sequences. It means that the more the identical locations between the two sequences, the more similar the two sequences are, just as the method introduced in [Hofmeyr98]. But we think the more continuous the matched locations, the greater the similarity of them, so we set K bigger than 1.

2. Profiling the normal behavior
Profiling the normal behavior is one of the crucial parts of the system. There are many profiling methods[Hofmeyr98] [Yoshinori02] [Lihua02] [Eleazar01] [Wagner01]. In our model, we generate a "rougher" profile based on the system call sequences. In the profile generation, we introduce the wildcard and the

fuzzy match function, by which the size of the profile could be decreased and the detection efficiency could be improved.

First, we will collect system call sequences in secure environments. The more comprehensive the sequences are collected, the better the profile will be. We split the sequences into L-length short sequences by L-length slide window. And the short sequences set will be used as training data to generate the profile. We call the data set as the normal data set, N.After collected enough normal data, we will generate the profile in following steps:

1. Sort the short sequences in the set N by times the short sequence appears in the normal sequences;
2. Get the short sequence S_m from N, which most frequently appears in the normal sequences;
3. Match S_m with all the sequences in the set, and record each location's sum of power, $Sum(i)$. For each location i of the sequence,

$$Sum(i) = \sum_{j=0}^{T} Power(i, S_m, S_j), T = |N| \tag{2}$$

4. Check each location of the sequence S_m:
   For each location i,
      If $Sum(i) \ll \overline{Sum(i)}$
        Set the location as wildcard, represented by -1.
      Endif
   End
   Get the modified sequence, S'_m Add S'_m into the profile set, P;
5. Compare the S'_m with all the sequences in the set N. If the return value of $Match(S'_m, S_i)$ is beyond the limit f_L, we think the S_i has been covered by sequence S'_m and delete the sequence S_i from the set N.
6. If the set N is empty, we terminate the process and the profile set P is the target application's profile we want to get. Otherwise go to step 2).

3. Detecting the intrusions

From the analysis of attack methods, we find there are many continuous abnormalities in the intrusion process.While monitoring a privileged process, we keep a L-size sliding window on the system call sequence. The attacks will cause many continuous strange sequences, and we estimate whether the sequence belongs to the normal profile or not by calculating the "membership" of the sequence to the profile. The sequence's membership to profile P is defined as the maximum of the sequence's similarities with each sequence in the profile.

$$membership(S, P) = max\{match(S, S'_1), match(S, S'_2, ..., match(S, S'_n)\}$$

$$S'_i \in P \text{ and } n = |P|$$

When there are continuous r pieces of short system call sequences whose "memberships" are below the threshold, M_L, we think there is an intrusion. The M_L must not be bigger than the f_L, used to generate the profile. Otherwise, there will be too many false positives. In section 5, we will provide the experimental results and discuss their effects.

4   Experimental Results

In this section, we provide the experimental results. In all the following experiments, the argument K of match function is set as 1.1. And the length of the sequence is 10, which has been discussed in [Hofmeyr98]. And all the data sets used in the experiments are from University of New Mexico.

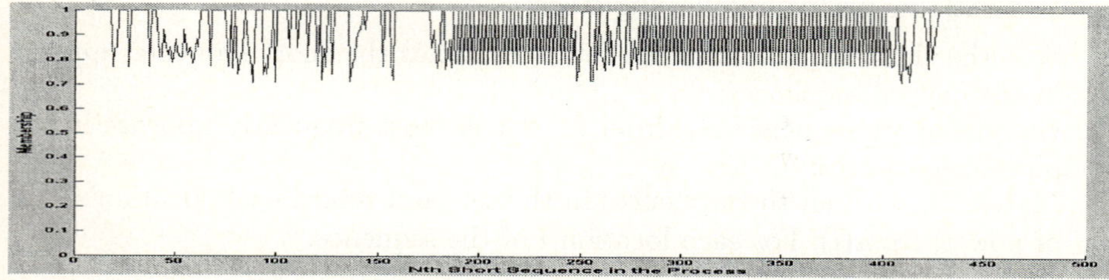

Fig. 1. (a)The membership to the profile in the normal process($xlock$), which is one of the processes used to training the profile, changes within a stable range and at a high level

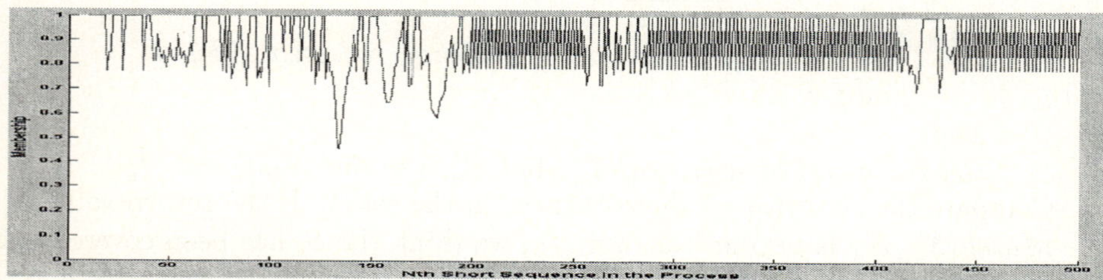

Fig. 2. (b)The membership to the profile in the normal process($xlock$), which is not used to training the profile, changes within a stable range and at a high level and there are rare abnormalities of them

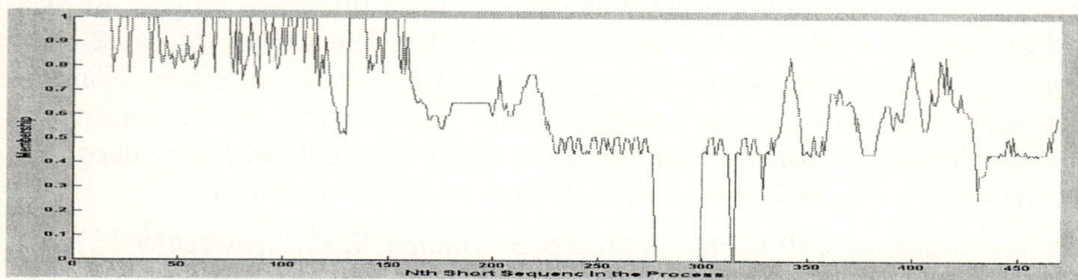

Fig. 3. (c)The membership to the profile in the intrusive process ($xlock$), changes within a much wider range and there are many continuous sequences with low membership

1. Intrusion Behavior's Deviations from Normal

In the Fig.1,2,3, we provide the experimental results of the membership of sequences in normal and abnormal processes. The data sets are from the application *xlock*. And the profile used to calculate the membership is generated with f_L=0.70.

From the graphs, we could find obvious differences between the curves of normal and abnormal processes.The memberships of the normal behavior are centralized in a stable range, most of which in the range, $0.6 \sim 1.0$. But the sequences' membership of the intrusion process spread around a much wider range. The membership is even as small as 0;In the intrusion process, there are many continuous sequences whose membership at a relatively high level, just as graph (c) shows at the beginning of the curve. If we calculate the deviations of the sequences of the process from global view, we may ignore the abnormity coursed by the attacks, which finished only by several system calls.In the intrusion process,there are many continuous sequences, whose memberships at a low level, which coursed the continuous abnormalities from the local view.

2. Profiling the normal behavior

In [Hofmeyr98], Hofmeyr directly splits the normal system call sequences into pieces of L-length sequences by a L-size slide window to describe the normal behaviors. In order to improve the efficiency of the model, we decrease the profile size on the condition that the detection accuracy is acceptable. In

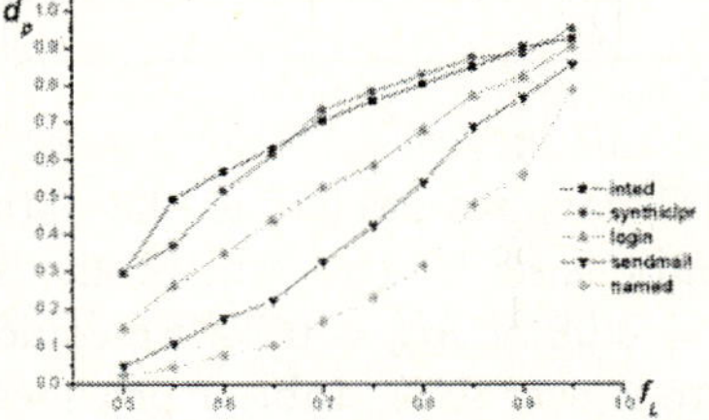

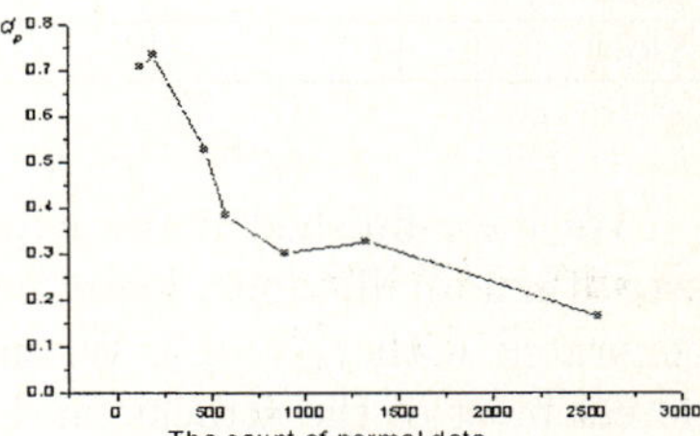

Fig. 4. The d_p increase with f_L increasing, but it decrease with the count of the training date increasing, if we set f_L as a given value ($f_L = 0.7$)

the algorithm of profile generation,the f_L is a key argument in the profile generation. In the experiments, we set f_L from 0.5 to 0.95. Because of the different complexity of applications, the count of the sequences in the profile could not be used to evaluate the method's effect on the profile size. We introduce an argument d_p, the ratio of profile size to the size of normal data set which was directly used as profile in [Hofmeyr98].And from the experimental results, we could find the more the normal data, the more effective the method is. When we set f_L, as 0.7, the profile of the application *named*, is just as 16.5% of its normal data set, which has 2564 sequences, as shown in Fig.4.

3. Detecting intrusions

In the following experiments, the false positive rate is the percentage of decisions in which normal data were flagged as anomalous and the false negative rate is the percentage of decisions, in which the intrusive processes are flagged as normal. After above comparison between different profiles generated with different f_L, we find the profile generated with $f_L = 0.7$ is an acceptable choice.

In the experiments, we test the data sets of the application *xlock*. We find there are few false positive in the test of live date, and the rate of false positive is below 0.05%. When we ensure false positives at a low level, we also should keep the rate of false negatives at an acceptable level.

Table 1. When we set $r = 4$ and use the profile generated with $f_L=0.70$, there are different count of intrusions detected with the different M_L

Application	Intrusion Processes	Detected Processes ($M_L = 0.6$)	Detected Processes ($M_L = 0.65$)	Detected Processes ($M_L = 0.68$)	Detected Processes ($M_L = 0.70$)
Inted	31	1	29	29	30
Login	13	6	9	9	9
Ps	26	13	26	26	26
Named	5	4	5	5	5
ftp	5	5	5	5	5
Sendmail	19	13	14	18	18
Sendmail (CERT)	34	25	31	31	32
Xlock	2	2	2	2	2

We have finished many experiments on UNM data sets,including the data sets of sendmail, xlock, login,inetd and so on. And r is set as 4 and the profile is generated with $f_L = 0.7$. When we set the $M_L = 0.68$ or $M_L = 0.70$, we could detect most of the attacks, and the detection rate is over 98%. Table 1 provides the detailed results.We think it's acceptable for most of the applications.

5   Discussion

Recent research in intrusion detection techniques has shifted the focus from user-based intrusion detection to process-based, which monitors system calls triggered by the process. In the recent years, there are many algorithm used to generate better profile based on system calls, such as Finite State Automata, Hidden Markov Models, Data Mining, and so on.[Warrender99][Sekar01] [Chowalit03] [Wepsi00] [Wagner01].

Based on the model in [Homfeyr98], Wepsi et al. presented a technique to build the profile by variable-length system call sequences.[Wepsi00] It uses the *Teiresias* algorithm to analysis the real time sequences,which is quite time and space consuming. It is not thought as better than Hofmeyr's model. Wagner et al.

proposed the *callgraph model* which is based on the static analysis of the program code.[Wagner01] It is not fit for most of the applications, whose source codes are not available. And then Wagner et al. proposed a complex model - *abstract stack model*, in which the stack forms an abstract version of the program call stack. But it depends on the return address of function, and has not resolve the problem of signals and DLLs, which are loaded at different relative locations.

Compared with the previous methods, the model in this paper could easily generate better profile. The profile generation needs not any other knowledge about the process and attack methods except some normal system call sequences, which could be easily collected by trace when the process in normal behavior.

As all the detection method based on the system call sequence, the model in this paper could not find attacks which do not cause abnormity in system call sequences, such as some DDOS attacks. The limitations have been discussed in [Hofmeyr98] and [Sekar01].

6   Conclusion

In this paper, we have analyzed the difference between the behaviors of intrusion and normal.Based on the analysis, an intrusion model has been brought forward, which adopts locally fuzzy matching to improve the detection accuracy. In the model, a novel profile generation method is introduced. And experimental results show that the detection accuracy of the model based on the profile has been improved. At last, we analyze the limitations of the model.

References

[Evan99]	Evan Thomas:Attack class: Buffer overflows, Hello World!(1999)
[Klog99]	Klog:The Frame Pointer Overwrite, Phrack Magazine,55(1999)
[Anonymous02]	Anonymous:Runtime Process Infection, Phrack Magazine, 59(2002)
[Blexim02]	Blexim:Basic Integer Overflows, Phrack Magazine, 60(2002)
[Hofmeyr98]	Steven A. Hofmeyr, Stephanie Forrest and Anil Somayaji: Intrusion Detection using Sequences of System calls, Journal of Computer Security Vol. 6, 1998
[Yoshinori02]	Yoshinori Okazaki and Izuru Sato: A New Intrusion Detection Method based on Process Profiling, Proceedings of the 2002 Symposium on Applications and the Internet (SAINT'02)
[Lihua02]	Lihua Liao and V.Rao Vemuri: Use of Text Categorization Techniques for Intrusion Detection, in Proceedings of the 11th USENIX Security Symposium, 2002
[Eleazar01]	Eleazar Eskin, Wenke Lee and Salvatore J.Stolfo: Modeling System Calls for Intrusion Detection with Dynamic Window Sizes, Proceedings of DISCEXII. June 2001
[Warrender99]	C.Warrender, S. Forrest, and B. Pearlmutter, Detecting intrusions using system calls: Alternative data models. Proceedings IEEE Symposium on Security and Privacy, pages 133-145,1999

[Sekar01]   R.Sekar, M.Bendre, D. Dhurjati and P. Bollineni: A Fast Automation-Based Method for Detecting Anomalous Program Behaviors, IEEE Symposium on Security and Privacy, 2001

[Chowalit03]   Chowalit Tinnagonsutibout and Pirawat Watanapongse: A Novel Approach to Process-based Intrusion Detection System Using Read-sequence Finite State Automata with Inbound Byte Profiler, ICEP2003, January 2003

[Calvin00]   Calvin Ko: Logic Induction of Valid Behavior Specifications for Intrusion Detection, IEEE Symposium on Security and Privacy, 2000, Berkeley, California

[Wepsi00]   A. Wepsi,M. Dacier and H. Debar: Intrusion Detection Using Variable-Length Audit Trail Patterns,3rd International Workshop on the Recent Advances in Intrusion Detection, LNCS 1907, Springer, pp. 110-129, 2000

[Wagner01]   D. Wagner and D. Dean: Intrusion Detection via Static Analysis, IEEE Symposium on Security and Privacy, Oakland, CA, 2001.

Dynamic Combination of Multiple Host-Based Anomaly Detectors with Broader Detection Coverage and Fewer False Alerts

Zonghua Zhang and Hong Shen

School of Information Science,
Japan Advanced Institute of Science and Technology,
1-1 Tatsunokuchi, Ishikwa, 923-1292, Japan
Tel: 81-761-51-1285, Fax: 81-761-51-1149

Abstract. To achieve broader detection coverage with fewer false alarms, a POMDP-based anomaly detection model combining several sate-of-the-art host-based anomaly detectors is proposed in this paper. An optimal combinatorial manner is expected to be discovered through a policy-gradient reinforcement learning algorithm, based on the independent actions of those detectors, and the behavior of the proposed model can be adjusted through a global reward signal to adapt to various system situations. A primarily experiment with some comparative studies are carried out to validate its performance.

1 Introduction

This paper is motivated by two key observations: first, different anomaly detectors (ADs) have different detection coverage and blind spots; second, diverse operating environments provide different kinds of information to reveal anomalies. We formulate the cooperation between independent elemental detectors as a partially observable Markov decision process (POMDP), and apply a policy-gradient reinforcement learning algorithm to search in an optimal control strategy, with the objective to achieve broader detection coverage with fewer false alerts. Moreover, the combination of different ADs is expected to abstract specific or concrete behaviors sufficiently well to detect families of attacks rather than individual instantiations, thereby allowing for the detection of all the attack variants that attempt to exploit the same system vulnerability.

The modeling of a new POMDP-based multi-agent anomaly detection model is the main contribution of this paper. In addition, we pay some efforts to the analysis of the model's anticipated behavior, based on the insightful understandings and comparative studies of the observation-specific elemental ADs, together with their respective operating environments. Another contribution is that the dynamics of our model allows its adaptability to diverse system situations or various security concerns with admitted criteria, and this can be easily achieved through a global reward function. Also, the distributed architecture of the model facilitate its scalability and dependability. A primarily experiment demonstrates the goals that we expect to achieve with good performance.

P. Lorenz and P. Dini (Eds.): ICN 2005, LNCS 3421, pp. 989–996, 2005.

2   A General Model Formulation

In this paper, we use a host with UNIX OS as the scenario to design our anomaly detection model, therefore the elemental ADs are host-based, and observation-specific. Careful analysis have shown that all the ADs have their own detection capabilities and blind spots, and their detection coverage varies due to the diverse operating environments. A simple comparison between those ADs provides a more insightful understanding and possible complementary operations:

Table 1. A Simple Comparison Among Four Typical Anomaly Detectors

Anomaly Detectors	Observation	Main Property		Detection Cost
		Frequency	Ordering	
MCE [7]	Shell User Commands	$\checkmark$		$O(N * m^2)$
Markov Chains [6]	Audit Events	$\checkmark$	$\checkmark$	$O(N * L)$
STID [3]	Local Ordering of System Calls		$\checkmark$	$O(N * (L - w + 1))$
KNN [4]	Frequency of System Calls	$\checkmark$		$O(N)$

'w' is the sliding window size, 'm' is the number of unique events.

'N' is the size (the number of observation units) of a normal data set that has been constructed, while 'L' is the size of the ongoing trace. What we are concerned with here is the detection cost, rather than the modeling cost.

A general model of POMDP is structurally characterized as follows [1]:

- a finite state space, $S = \{s_1, ..., s_n\}$ of the world;
- a control space $U = \{u_1, ..., u_m\}$ that are available to the policy;
- an observation space $Z = \{z_1, ..., z_q\}$;
- a (possibly stochastic) reward $r(s_i) \in R$ for each state $s_i \in S$.

For each state $s_i \in S$, an observation $z_i \in Z$ is generated independently according to a probability distribution $\nu(s_i)$ over Z, which is denoted as $\nu_{z_i}(s_i)$. For each observation z_i, $\mu(z_i)$ is a distribution over the actions in U to determine a policy, and $\mu_{u_i}(z_i)$ denote the probability under μ of control u_i given observation z_i. To each randomized policy $\mu(\cdot)$ and observation distribution $\nu(\cdot)$, the Markov chain for the state transition s_i and s_j are generated as follows:

$$s_i \in S \xrightarrow{\nu(s_i)} z_i \in Z \xrightarrow{\mu(z_i)} u_i \in U \xrightarrow{p_{ij}(u_i)} s_j \in S \tag{1}$$

To parameterize these chains we parameterize the policies, so that $\mu(\cdot)$ becomes a function $\mu(\theta, z_i)$ of a set of parameters $\theta \in \mathbb{R}^k$ as well as the observation z_i. The markov chain corresponding to θ has state transition matrix $P(\theta) = p_{ij}(\theta)$ given by $p_{ij}(\theta) = E_{z_i \sim \nu(s_i)} E_{u_i \sim \mu(\theta, z_i)} p_{ij}(u_i)$. For the multi-agent environment, the actions set U contains the cross product of all the actions available to each agent, i.e., $U = \{u_1 \times u_2 \times \cdots \times u_n\}$, the overall action distribution therefore is the joint distribution of actions for each agent, $\mu(u_1, u_2, ...u_n | \theta_1, \theta_2, ...\theta_n, y_1, y_2, ...y_n)$. In our model, each AD works in its own

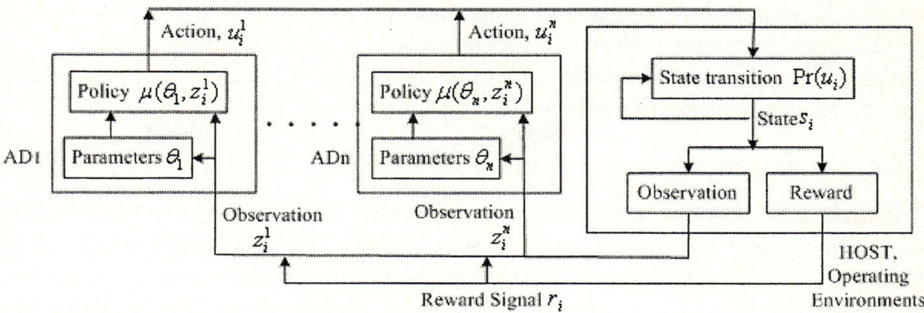

Fig. 1. A General Formulation of POMDP-based Anomaly Detection Model

environment, the true system state can be only indirectly observed through those detectors. Furthermore, for the POMDP-controller, the decision process is a Markov process, because the next state of the system is dependent only upon the current state and the previous decision of the controller. Thus, a multi-agent, partially observable, Markov decision process is formulated here (as figure 1).

3 A Specific Coordination Scheme

To solve the formulated problem, we use a multi-agent variant of the OLPOMDP algorithm [2], a kind of the policy-gradient algorithm, which has been applied to solve a routing problem by Tao et al. [5]. To characterize each AD in term of the model, a formally definition is given as follows:

Definition 1. *All the ADs have no knowledge about the exact system states, in some sense, $|S|$ is infinite; the observation set $Z = \{Normal, Malicous\}$, the action set $U = \{Observe, Alert\}$ according to the specific detection scheme.*

For the sake of simplicity, we assume that each AD has a binary output on per measurement unit ℓ, with each measurement unit's decision rule parameterized by a threshold value λ_ℓ. For any independent AD:

Definition 2. *There is a decision rule $\xi : \Re \to \{0,1\}$ of the form*

$$\xi(\ell) = \begin{cases} 0, & \xi(\ell) \le \lambda_\ell \\ 1, & \xi(\ell) > \lambda_\ell \end{cases} \tag{2}$$

where output "0" denotes the 'Normal' observation and "1" the 'Malicious' observation. Corresponding actions 'Observe' and 'Alert' are taken according to the POMDP controller.

To complete the picture we need to define a performance measure for the detection result, which will be taken as a reward signal to guard the improvement of the general detection performance. In anomaly detection domain, some or all of following cases might happen:

- $\mathcal{N}_n$, legal behavior is detected as normal.
- $\mathcal{N}_a$, legal behavior is detected as anomaly.
- $\mathcal{A}_n$, illegal behavior is detected as normal.
- $\mathcal{A}_a$, illegal behavior is detected as anomaly.

A natural performance metric thus can be defined as:

Definition 3. *Assume that during a particular time period Δt, m activities happened, among those activities, $i \in \mathcal{N}_n$, $j \in \mathcal{N}_a$, $k \in \mathcal{A}_n$, and $l \in \mathcal{A}_a$, if we assign w_1, w_2, w_3, and w_4 to denote their weights respectively, and $\alpha = w_1 \cdot i/m$, $\beta = w_2 \cdot j/m$, $\zeta = w_3 \cdot k/m$, $\delta = w_4 \cdot l/m$, a reward signal can be defined as $r_t = \alpha \cdot \delta - \beta \cdot \zeta$. w_1, w_2, w_3 and w_4 can be defined according to various system situation and security demands.*

Due to the nature of anomaly detection, and the fact that the number of normal activities is much larger than that of anomalies, we usually have $w_1 < w_3 < w_4 < w_2$. In addition, to make the ADs trainable and keep the computations simple, we use a general probabilistic model to describe the behavior of the ADs:

$$Pr(\text{Observe without Alarms}) = Pr(u_t = 0) = \varphi(\varepsilon_t) \tag{3}$$

where $\varphi(x) = e^{-x}$, and $\varphi(x) \in [0, 1)$, while $\varepsilon_t \in (0, \infty)$ is defined as:

$$\varepsilon_t = \frac{\lambda_t^i}{d_t^i(\ell)} \tag{4}$$

where λ_t^i is the threshold of ith AD at time instant t, while $d_t^i(\ell)$ denotes the measurement distance between ongoing observations ℓ and the normal patterns. Our aim is to find a set of parameters for all the detectors that maximizes the expected long-term average reward:

$$\eta(\theta) := \lim_{T \to \infty} \frac{1}{T} E_\theta [\sum_{t=1}^{T} r(o_t)]. \tag{5}$$

where E_θ denotes the expectation over all observation $o_0, o_1, \ldots$ with transitions generated according to $P(\theta)$, and θ is the concatenation of the parameters from all detectors. At each step t, the parameters θ_t of all the ADs are updated by $\theta_{t+1} = \theta_t + \Delta\theta_t$, where the updates $\Delta\theta_t$ lie in the gradient direction $\nabla\eta(\theta)$, and are guided by a reward signal to climb the gradient of the average reward $\eta(\theta)$. Specifically, each AD works in the following manner as the algorithm:

Given:
Coefficient $\rho \in [0, 1)$, step size τ_0, initial system state s_0, concatenation vector θ_0,
begin
1:   **for** discrete time instant $t = 1, 2, \cdots$ **do**
2:     Get ongoing observations and their corresponding measurement stream ℓ.
3:     Generate action u_t^i according to the specific detection scheme and definition 2.

4:     The coordinator broadcasts the reward signal r_t.

5:     Update q_{t+1}^i according to equation: $q_{t+1}^i := \rho \cdot q_{t+1}^i + \frac{\nabla \mu_{u_t}^i(\theta^i, z_t)}{\mu_{u_t}^i(\theta^i, z_t)}$.

6:     Update θ_{t+1}^i according to equation: $\theta_{t+1}^i := \theta_t^i + \tau_t \cdot r_t \cdot q_t^i$.

7: **end for**

8: **end**

where θ_t^i is the parameter vector of the detector i at time t, r_t is the sum of rewards, and q_t^i is a trace of the same dimensionality as θ_t^i with $q_0^i = 0$, $\rho \in [0, 1)$ is a free parameter to control both the bias and the variance of the estimates produced by the algorithm [2], and $\tau_1, \tau_2, \ldots > 0, \sum \tau_t = \infty, \sum \tau_t^2 < \infty$. In addition, we update the action in 5th step according to the following equation:

$$\frac{\frac{\partial}{\partial \lambda_t^i} \mu_{u_t}}{\mu_{u_t}} = \begin{cases} \frac{\varphi'(\varepsilon_t)}{d_t^i(\ell)\varphi(\varepsilon_t)} = \frac{-1}{d_t^i(\ell)} & \text{if } u_t = 0 \\[2ex] \frac{-\varphi'(\varepsilon_t)}{d_t^i(\ell)(1-\varphi(\varepsilon_t))} = \frac{\varphi(\varepsilon_t)}{d_t^i(\ell)(1-\varphi(\varepsilon_t))} & \text{otherwise} \end{cases} \tag{6}$$

4 Performance Validation

4.1 Data Collection and Preprocessing

To the best of our knowledge, there is no true trace data in the open literature that meets our experimental demands. Therefore, we have had to collect and formulate the experimental data according to our particular demands. In addition, to simplify the experiment, all the elemental ADs we use are initialized with the parameters in their original literature. We collected normal behaviors ourselves for 4 weeks using the Solaris 8 operating system (SunOS release 5), combining with several typical host-based attacks (as shown in table 2, 8 cases of *local buffer overflow* and 5 cases of *DoSs*). We usually use text editors, compilers, and some embedded programs on our host. Excluding wrong commands and some noisy data, we got a total of 132,886 records of BSM audit data and 11,240 shell command lines (log all commands using the shell .history file), and these data were

Table 2. The attacks used in our experiments

Attack Type	Attack Description	# of instances
Masquerader	access to programs and data as an imposter by controlling the keyboard.	850 commands
Buffer Overflow	*xlock* heap buffer overflow	2
	eject, to gain a user→root transition.	3
	lp -d, craft input for the argument[-d]	3
DoS	*adb* , to create a kernel panic	2
	UFS file system logging	1
	console device DoS	2

averaged as pure training data and as testing data. Additionally, a small batch of another user's commands history (2127 audit events, 850 command lines) are also added to the testing data set for detecting masquerade intrusion.

4.2   Experimental Results and Discussion

The initial parameters of the ADs are derived from their original version, which is shown in table 3. Thus, the concatenation parameter vector θ_0 is represented as $\theta_0 = (0.80\ 0.50\ 0.75\ 0.70)$.

Table 3. Initial Parameters of Elemental ADs

	MCE	Markov Chain	STID	KNN
Sequence length(L)	30	10	6	variable
Threshold (λ)	0.80	0.50	0.75(LFC=20)	0.70

Testing of False Alerts. Half of 11240 collected shell command lines (i.e., 5620 tokens) were taken as training data, and every unit contains 30 commands (since every login session contains 30 commands or so), therefore a total $\lfloor \frac{5620}{30} \rfloor =$ 187 commands blocks are available. Corresponding system calls (audit events) were extracted as input into their respective AD for creating normal profiles. The testing data, which also includes 187 command blocks are used to evaluate the capability of the model to suppress false alerts. The model reports every command block, while the elemental AD might generate a report sequence during a command block execution. Figure 2 shows the relationship between the average false alert rate (the number of false alerts over the number of command blocks) and the number of command blocks used for testing data. Since the model gives the report with the step of each command block, we compared its performance with that of MCE, which has the same detection measurement unit. The figure shows clearly that the model suppresses false alerts significantly well compared with MCE. For example, both our model and MCE generated the first false alter at the 7th command block (i.e. F.P.=11.43%), but at the 122th command block, MCE has generated 12 alerts, while our model generated 7 with a F.P.=5.74%.

Detection of Common Exploits. To evaluate the masquerade detection, 850 command lines (2127 audit events) from another user were truncated into 28 command blocks, and injected at randomly selected positions, without replacement, into the stream of original 187 command blocks. What we concern here is the detection accuracy of the injected anomalies. As shown in table 4, over total 215 command blocks (187 normal + 28 anomalous), MCE detected 20 out of 28 anomalous command blocks with a false alert rate 11.23% ($\lambda_{MCE} = 0.90$), while our model detected 23 ones with a false alert rate 9.63%($\lambda_{MCE} = 0.93$).

The trained model was also used to detect the injected attacks, and the performance was compared with that of the individual ADs (Table 5). Detection accuracy is defined as the ratio of detected attacks to all the injected attack

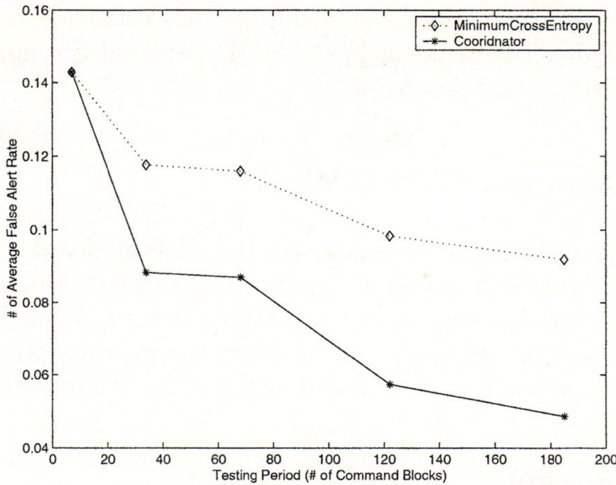

Fig. 2. False Alert Rates on Testing Data

Table 4. Comparison of masquerade detection results between MCE and Our Model

Methods	MCE	Our Model
# of normal command blocks	187	187
# of anomalous Command blocks	28	28
Block size	30	30
Hits%	71.43	82.14
Misses%	28.57	17.86
False Alert Rate%	11.23	9.63

Table 5. Comparison of attacks detection results between Our Model and Individual detectors('B' denotes Buffer overflow, 'D' denotes DoS)

Detectors	Hits(%)	False Alert Rate(%)	Detected Attacks
STID	92.30	10.16	3B+5D
Markov Chain	61.54	12.83	4B+4D
KNN	76.92	3.74*	7B+3D
Coordinator	92.30*	5.34	8B+4D

instances, while false alert rate takes command blocks as the measure unit. The parameters used by the ADs were derived from the coordination strategy rather than by individual training, therefore, we can not rule out the possibility that any individual detector might have better performance after being trained and parameterized carefully. Our detection model had the highest detection accuracy by combing the attack reports from individual detections, while its false alert rate is a little higher than that of KNN. STID detected the most DoS attacks due to its strict string-match scheme.

5   Conclusion and Future Work

A new anomaly detection scheme is cast in a POMDP model in this paper, which coordinates four typical observation-specific ADs through a policy-gradient reinforcement learning algorithm to achieve higher detection accuracy with fewer false alerts. We will extend the model to more complex situations, including a general computer network with several dominated hosts and wireless networks.

Acknowledgement

This research is conducted as a program for the "Fostering Talent in Emergent Research Fields" in Special Coordination Funds for Promoting Science and Technology by Ministry of Education, Culture, Sports, Science and Technology.

References

1. Douglas Aberdeen, *A Survey of Approimate Methods for Solving Partially Observable Markov Decision Processes*, National ICT Australia Report, Australia.
2. Bartlett,P.L., Baxter,J. *Stochastic Optimization of Controlled Partially Observable Markov Decision Processes*, Proceedings of the 39th IEEE Conference on Decision and Control(CDC00).
3. Forrest,S.,Hofmeyr,S.A.,&Longstaff,T.A. *A sense of self for UNIX processes.* In proceedings of 1996 IEEE Symposium on Security and Privacy. Los Alamitos, CA.
4. Yihua Liao, V.Rao Vemuri, Use of K-Nearest Neighbor classifier for intrusion detection, *Computers and Security*, Vol 21, No 5, pp439-448, 2002.
5. Nigel Tao, Jonathan Baxter, Lex Weaver, *A Multi-Agent, Policy-Gradient approach to Network Routing*, 18th International Conference on Machine Learning,ICML 2001.
6. Nong Ye, Xiangyang Li, Qiang Chen, Syed Masum Emran, and Mingming Xu. *Probabilistic Techniques for Intrusion Detection Based on Computer Audit Data.* IEEE Transaction on Systems, Man, and Cybernetics-Part A:Systems and Humans, Vol.31, No.4, July 2001.
7. Dit-Yan Yeung, Yuxin Ding, *Host-based intrusion detection using dynamic and static behavioral models*, Pattern Recognition 36 (2003) 229-243.

Impact of Distributed Denial of Service (DDoS) Attack Due to ARP Storm[1]

Sanjeev Kumar

IEEE Senior Member,
Department of Electrical Engineering, The University of Texas - Pan America,
Edinburg, Texas-78541, USA
Ph: 956-381-2401
sanjeevk@utpa.edu

Abstract. ARP-based Distributed Denial of Service (DDoS) attacks due to ARP-storms can happen in local area networks where many hosts are infected by worms such as code red. In ARP attack, the DDoS agents constantly send a barrage of ARP requests to the gateway, or to another host within the same sub-network, and ties up the resource of attacked gateway or host. In this paper, we measure the impact of ARP-storms on the availability of processing and memory resources of a Window-XP server deploying a high performance Pentium-IV processor. *Index terms*— ARP attack, Computer Network Security, Distributed Denial of Service Attacks.

1   Introduction

A Distributed Denial of Service (DDoS) attack [1] involves multiple DoS agents configured to send attack traffic to a single victim computer (also known as flooding attack). DDoS is a deliberate act that significantly degrades the quality and/or availability of services offered by a computer system by consuming its bandwidth and/or processing time. As a result, the legitimate users are unable to have full access to a web service or services. A Denial of Service attack consumes a victim's system resource such as network bandwidth, CPU time and memory. This may also include data structures such as open file handles, Transmission Control Blocks (TCBs), process slots etc. Because of packet flooding that typically strives to deplete available bandwidth and/or processing resources, the degree of success of a DDoS attack depends on the traffic type, volume of the attack traffic, and the processing power of the victim computer.

According to Computer Emergency Response Team Coordination Center (CERT/CC) [2], there has been an increase in use of Multiple Windows-based DDoS agents. There has been a significant shift from Unix to Windows as an actively used

---

[1] Author is with Networking Research Lab (NRL) at The University of Texas-PanAm, Edinburg, Texas, USA.
Dr. Kumar's research is supported in part by funding from CITeC, FRC, FDC, OBRR5-01, and digital-X grants.

P. Lorenz and P. Dini (Eds.): ICN 2005, LNCS 3421, pp. 997–1002, 2005.

host platform for DDoS agents. Furthermore, there has been an increased targeting of Windows end-users and servers. To raise awareness of such vulnerabilities, the CERT/CC published a tech tip entitled "Home Network Security" in July of 2001 [3].

According to the CERT/CC [2], there is a perception that Windows end-users are generally less security conscious, and less likely to be protected against or prepared to respond to attacks compared to professional industrial systems and network administrators. Furthermore, large populations of Windows end-users of an Internet Service Provider are relatively easy to identify and hence the attackers or intruders are leveraging easily identifiable network blocks to selectively target and exploit Windows end-user servers and computer systems.

In this paper, we consider a Distributed Denial of Service (DDoS) attack that can be caused by a barrage of ARP-requests sent to a victim computer. In order to understand the intensity of the attack and its impact on the end systems, we measure the availability of the processing power and memory resources of the victim computer during an ARP-attack. Since, windows based servers are very commonly deployed, we consider a Window-XP server deploying a 2.66GHz Pentium-IV processor to be used as the victim computer in the attack experiments. Section II presents a background on ARP and how it is used to exploit vulnerability of a computer system, Section III & IV present details on use of ARP requests, and the processing that needs to be done for ARP-request messages. Section V presents the experimental setup; Section VI on performance evaluation discusses results obtained in the experiment for ARP attack and Section VII concludes the paper.

2 Background

The Address Resolution Protocol (ARP) can be used in more than one ways to exploit the vulnerability of a computer system or a network. Some of the security attacks involving ARP can cause Denial of Service (DoS) attack by sending a constant barrage of ARP requests (called ARP storm) to a victim computer or a victim server and tying up its resource. ARP can also be used to create Denial of Service attack by sending a victim computer's outgoing data to a sink by the technique of ARP cache poisoning. Other ARP based attacks can result in unauthorized sniffing of packets, or hijacking of secured Internet sessions. The Denial of service attacks due to ARP storms can also be caused by the worms such as code red due to their rapid scanning activity. The worm initiated ARP storms have been commonly found in networks with high numbers of infected and active computers and servers. In ARP storm, an attacked victim (the gateway or a server) may receive a constant barrage of ARP requests from attacking computers (DDoS agents) in the same sub-network, and this ties up not only its network bandwidth but also degrades availability of its processing power and possibly its memory resources.

In this paper, we investigate the brute force of ARP attack (due to ARP storm) where a constant barrage of ARP requests is directed to a server. We measure server's performance under such attacks in terms of processor exhaustion and occupancy of systems' memory. Since Windows-XP based servers with high performance Pentium-

IV processors are becoming quite affordable and popular with small businesses, we use a Windows-XP server as a victim computer to be stress-tested for the extent of degradation in its availability under the ARP attack.

3 Use of ARP-Request

A gateway or a host on a local area network uses ARP request messages [4] for IP and hardware address bindings. The ARP message contains the IP address of the host for which a hardware address needs to be resolved. All computers on a network receive ARP message and only the matching computer responds by sending a reply that contains the needed hardware address.

4 Processing an ARP-Request Message

ARP is used for a variety of network technologies. The ARP packet format varies depending on the type of network being used. While resolving IP protocol address, the Ethernet hardware uses 28-octet ARP message format [4]. The ARP message format contains fields to hold sender's hardware address and IP address. It also has fields for the target computer's hardware and IP address. When making an ARP request, the sender supplies the target IP address, and leaves the field for the target hardware address empty (which is to be filled by the target computer). In the ARP request, the sender also supplies its own hardware and IP addresses for the target computer to update its ARP cache table for future correspondence with the sender. The main task of the processor in the target computer after receiving the ARP message is to make sure the ARP request message is for it. Thereafter, the processor proceeds to fill in the missing hardware-address in the ARP request format-header, swaps the target and sender hardware & IP address pair, and changes the ARP-request operation to an ARP-reply. Thus the ARP reply carries the IP and hardware addresses of both, the sender and the target computers. ARP replies are directed to just the sender computer and it is not broadcast.

The processing needed for an ARP-request message is fairly simple; nevertheless a barrage of such requests will deliberately exhaust the processing power of the victim computer by forcing it to perform useless computations. The degree of processor exhaustion for a given computer will of course depend on the processor speed and the applied load of such useless ARP-requests. In the following sections, we discuss our experiment to measure the extent of the exhaustion rate of a target computer when inundated with a barrage of such ARP-requests through its fast Ethernet interface.

5 Experimental Setup

In this experiment, an ARP-attack was simulated in a controlled environment of the networking research lab at the UTPA by having different computers send a barrage of ARP-request messages to a victim computer on the same local area network. Windows-XP server was used as the attack target of the simulated ARP-storm. The server deployed a Pentium IV processor with a speed of 2.66GHz.

6 Performance Evaluation

Parameters of performance evaluation considered for this attack experiment were the applied load of the ARP-attack traffic, processor exhaustion during the attack and memory occupied while processing the attack traffic by the target computer. The DDoS attack was simulated as ARP packets coming from multiple different attacking-computers at a maximum aggregate speed of 100 Mbps towards the target server. The attack traffic load (while simulating ARP storm) was incremented at an interval of 10 Mbps from 0% load to 100% load (=100Mbps). In the ARP-storm experiment, the at-tacked target computer continued to receive a barrage of ARP-requests for a period of 20 minutes for a given load, and was obligated to process all of them by creating an ARP-reply. The processor time (Fig.1) during the attack gives an indication of the rate of processor exhaustion for a given applied load of the attack-traffic during the ARP storm.

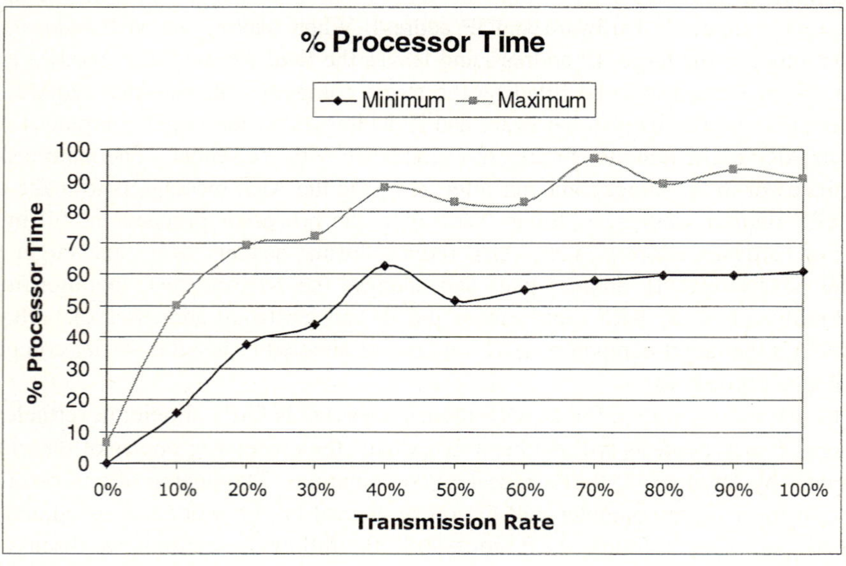

Fig. 1. Processor exhaustion of Windows-XP server deploying a Pentium-IV, 2.66GHz processor as the applied load of the ARP-attack traffic increases

It is observed that as the traffic load of the ARP-requests increases, the processor time also increases rapidly. Fig.1 also shows minimum and maximum processing time for a given load of the ARP-attack traffic. It can be seen that a traffic load of 40% at times could exhaust a Pentium-IV processor to up to 90% of its 2.66GHz processing capacity. The processing of a barrage of ARP-requests could easily consume the useful CPU time and degraded the quality and availability of the web services. Furthermore, it is obvious that if the servers are operated in a Gigabit network deploying higher interfaces such as 1Gbps then it will be easier for such CPU of 2.66GHz to be completely consumed by the Gigabit-flood of ARP-attack traffic, and attacks in such Gigabit environment can completely freeze the system. Complete freeze means that one cannot even move the cursor of the attacked computer, let alone running the security diagnostics. This experiment also shows that a lower capacity (< 2.66GHz) processor can easily be consumed 100% by this type of ARP-storm in commonly available fast Ethernet environment of local area networks.

Fig.2 shows the memory-usage of the victim computer under attack, as the applied traffic load due to ARP-storm increases. The memory occupancy due to such ARP attack seems minimal for a 2.66GHz processor. However, for the processing power less than 2.66GHz, a greater amount of computer's memory resource will be wasted. Slower processing power in the fast Ethernet environment can cause the queue of ARP packets to easily build up waiting for address resolution and computer's response, and hence exhausting also a greater amount of memory resource of the victim computer.

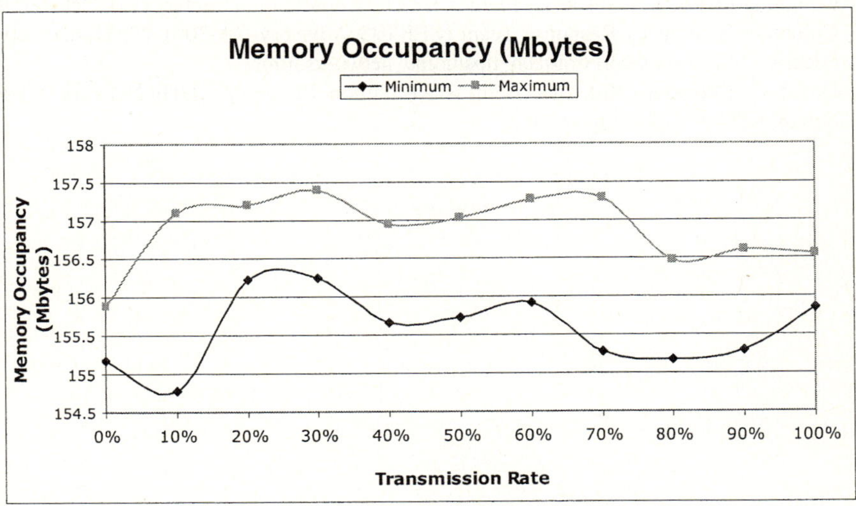

Fig. 2. Occupancy of a Windows-XP server's memory deploying a Pentium-IV, 2.66GHz processor as the applied load of the ARP-attack traffic is increased

7   Conclusion

According to CERT/CC, there has been an increased targeting of Windows end-users and servers for security attacks. Distributed Denial of Service (DoS) attacks due to ARP-storms can be found in local area networks where many hosts are infected by worms such as code red. In this paper, we measure the impact of ARP-storms on the availability of processing and memory resource of a Window-XP server deploying a high performance Pentium-IV processor. It is observed that ARP-storms not only wastes the communication bandwidth but also exhausts a processor's resource rapidly by forcing it to perform computations of replying to a barrage of useless ARP-requests. It is observed that even a 40% (=40Mbps) of ARP-attack traffic in a fast Ethernet environment can easily consume up to 90% of the CPU time of a high-performance Pentium-IV processor of 2.66GHz speed.

Acknowledgement

The author would like to thank Montasir Azad and Uriel Ramirez for their efforts in data collection activities in the Networking Research Lab at UTPA.

References

[1] Neumann, Peter G., Denial-of-Service Attacks, *ACM Communications*, April 2000, vol 43. No. 4.
[2] Kevin J. Houle and George M. Weaver, "Trends in Denial of Service Attack Technology," Computer Emergency Response Team (CERT)® Coordination Center, v1.0, October 2001
[3] Computer Emergency Response Team (CERT)® Advisory CA-2001-20, Home Network Security, http://www.cert.org/tech_tips/home_networks.html
[4] David C. Plummer, "Ethernet Address Resolution Protocol," IETF Network Working Group, RFC-826, November 1982.

Design of Network Management System Employing Secure Multicast SNMP

Deuk-Whee Kwak[1] and JongWon Kim[2]

[1] Backbone Network Research Team, Telecommunication Network Laboratory,
KT, Daejeon, Korea
[2] Networked Media Lab., Department of Information and Communications,
Gwangju Institute of Science and Technology (GIST), Gwangju, Korea
{dwkwak, jongwon}@gist.ac.kr

Abstract. Several operational patterns in common network management system (NMS) exhibit multicast characteristics. However, the latest version of the SNMP (simple network management protocol) standard, SNMPv3, does not support multicast yet. In this paper, we show that the multicast is a very useful mechanism for enhancing the efficiency of network management, and we propose a framework that extends the standard SNMPv3 to support secure multicast. Some modifications on the SNMP engine and the standard MIB-II (the second version of the management information base) are required so that they are interfaced with group membership. We then present a network management architecture where the proposed idea is explained in greater detail and utilized to improve the extensibility and efficiency of network management.

1   Introduction

The purpose of a network management system (NMS) is to collect performance data from the agents, analyze it to determine the current load status, and predict or detect anomalies in the network it manages in order to make the best use of network resources [1]. Most of the time, the network manager sends almost identical commands to the involved agents at the same time (but independently) for collecting the same (or similar) performance data. This operational pattern shows that, if we adopt group communication based on multicast for the NMS, we will derive a number of benefits from it. In addition, we also want the NMS to be secure from most security attacks.

In [2, 3], the authors propose frameworks that combine multicast with the SNMP (simple network management protocol) for the efficient management of a given network. However, the suggested frameworks, which are based on SNMPv1, do not incorporate security considerations such as authentication, integrity and privacy [4, 5]. In this environment, if the messages that are sent by the manager and the agent are not protected from malicious attackers, there will be serious security flaws that expose the network management operation to security attacks such as masquerading and eavesdropping [6, 4].

P. Lorenz and P. Dini (Eds.): ICN 2005, LNCS 3421, pp. 1003–1011, 2005.

As is widely known, the SNMPv3 has been devised to consider such security attacks. It adopts USM (user security model) for authentication and encryption [7]. The USM of SNMPv3 uses a symmetric key system for authentication, integrity and confidentiality services. For authentication and integrity, it uses HMAC-MD5-96 or HMAC-SHA1-96. For confidentiality, it uses DES-CBC (cipher block chaining mode of the US data encryption standard). The USM also provides the key localization concept so that each user can derive many different keys for each server from a single pass phrase.

Thus, in this paper, we make use of USM for the basic multicast security mechanism. However, since the SNMPv3 is designed only for point-to-point communication, it should be modified or extended to accommodate group and key management functions. First, the MIB-II (the second version of the management information base) should be extended, since it does not have definitions for storing group information. Next, we should consider adding group membership and key management modules. The group membership module manages the group key, engineID and membership information. It interacts with the key management module to get the new group key and access the extended MIB directly to update the information. Lastly, the engine part should be modified. To support multicast security, we add modules to the SNMP and modify the dispatcher, one of the engine modules that manipulate incoming and outgoing messages, to make it aware of the new modules that should interact with it. By combining the modified engine and extended MIB with the added modules, we propose a network management framework that provides secure multicast services and is also compatible with the standard SNMPv3.

The remaining sections of this paper are organized as follows: In Section 2, we explain the need for the secure multicast concept in the SNMP environment. In Section 3, we explain the extension of the standard MIB-II and modification of the SNMPv3, and in Section 4, propose a framework for the secure multicast SNMP. We present a network management architecture in Section 5 in which the secure multicast SNMP concept is used for the flexible and efficient management of the network. Finally, we wrap up this paper with Section 6.

2   Why Secure Multicast SNMP?

By adopting secure multicast concept in the security ready SNMPv3 [4], we are expecting following advantages [2, 3]:

- Management efficiency and flexibility: By adopting multicast, we can avoid the overhead of connecting individually to all the managed agents and reduce the overhead of the NMS, and combining it with the sending mode (i.e. unicast or multicast) will bring much flexibility to the NMS organization.
- Enhanced security: In the SNMPv3, user pass-phrase is used as key material for authentication and encryption. However, though the SNMPv3 has key update function, it is hardly updated by the users since no automatic key update mechanism is enforced. This kind of security system is vulnerable to dictionary attack and thus the key should not be used directly or should be

updated periodically. With secure multicast, group key will be renewed regularly or automatically by the GKM whenever there is membership change in the group.

- Fault tolerance and distribution: All the managers in a group can store the management data without an extra backup stage and the full management functions can be distributed among the managers. If the active manager fails, one of the other managers can easily take over the manager's job without interim discontinuance for data management consistency.

By analysis of management efficiency, we show that we can achieve a 50 % improvement in terms of the number of messages, and improve (i.e. reduced) the response time proportionally to the number of managed agents.

3   Extension of SNMPv3 for Secure Multicast SNMP

3.1   Basic Idea and Assumptions

The basic idea and assumptions for supporting secure multicast services on the standard SNMPv3 are as follows: First, we assume that all group members are sharing a pair of engineID and group key. The key is supplied by a general group key management protocol such as GSAKMP (group security association key management protocol) or GDOI (group domain of interpretation) [8]. The group key is only for secure group communication, but not for secure unicast communication. For the later case, normal user keys (pass-phrases), which are assumed to be shared in advance by some secure mechanisms, are used. The group engineID, which is common to all group members, is derived from the group key by truncating it as much as needed according to the adopted crypto algorithm. However, since the concept of key localization is not to enhance the security level but to provide many keys to an user for each agent aceess, the engineID shared by all group members does not cause any additional security problems to the secure multicast SNMP. In fact, we maintain it only for the purpose of making as little modification to the standard SNMPv3 as possible since it only requires a small overhead to the framework.

The second assumption is that the multicast address and port number are determined by the manager and delivered to the agents that are allowed to join the group.

The third assumption is that the IP multicast services are available in the management network; i.e., the multicast services are provided by the network.

Lastly, we assume that the interface (or API: application programming interface) between the SNMP and GKM is so simple that its functions are only for requesting to join a specific group and getting the response, and getting the group key. We want the proposed architecture to be easily adopted to any general GKM. By this assumption, SNMP should manage the membership information by itself using the group key update message from the GKM and the response message from the agent.

Table 1. Group table defined for both manager and agent. Although the field names are the same, the meaning can be different depending on the entity it belongs to

Field Name	Description for Manager	Description for Agent
grpID	The group name to be managed	The group name this agent belongs to
grpAddr	IP address of group to be managed	IP address of the group this agent belongs to
grpPortID	Port number of group to be managed	Port number of the group this agent belongs to
grpKeyMaterial	Group key that received from the GKM	Group key that received from the GKM
grpEngineID	Group engine ID shared with other group members	Group engine ID shared with other group members
destGrpID		Destination group name where the response message should be sent to.
UserID	User name of the agent that belongs to the managed group	User name of a manager where the response message should be sent to
userAddr	IP address of the agent that belongs to the managed group	IP address of a manager where the response message should be sent to
userAuthKey	The authentication key of userID	The authentication key of userID
userPrivKey	The privacy key of userID	The privacy key of userID
status	Current registration status of the agent that belongs to the managed group.	Current registration status to the each group.

3.2   The Extension of Standard SNMPv3

Management Information Base. The MIB that is extended for group membership management is shown in Table 1 [10]. The manager needs the information of each group it manages and agents that belong to the groups, and the agents need the information of the group where they belong to and the destination where they will send the response messages. The manager does not need the destGrpID field, but the agent does. In the agent side, the default value is the value of the grpID. If the field is different from the grpID, it means that the source and the destination groups are different, and the destination group information should be retrieved from the MIB using the destGrpID as a key. If the value is NULL, it means that the response type is unicast, and the userID, userAddr, userAuthKey and userPrivKey are used as the destination manager information. The value of the status should be one of these two: 'ATV' or 'DAV'. The value 'ATV' means that the agent is a member of the group, and 'DAV' means that the agent is not a member of the group anymore, although it was some time ago.

The Modification of SNMP Engine. The standard operation of the SNMP engine is to get a packet from an external entity such as UDP or command application, process the header according to its version and protocol type, perform authentication and encryption functions, and dispatch it to the proper upper application or lower transport protocol [4]. The packet delivery between the MPS (message processing subsystem), one of the engine modules that process the packet header, and SNMP applications or transport protocols is performed by the dispatcher in the standard engine. We can think of the dispatcher as an interface between the engine and the other SNMP functional modules. Thus, we should modify it to support direct interaction with the extended modules. As we mentioned in Section 3.1, we directly updated the group key in the extended MIB and the engineID in the standard MIB through a non-standard manner.

It means that the SNMP needs an extra module to manage the group key and change some existing applications for updating the engineID. However, it has no relevance to the function of the engine. The engine just refers the msgAuthoritativeEngineID, msgAuthenticationParameters, and msgPrivacyParameters values in the message for security check. The values are saved in the MIB by some applications, and the command generator just supplies them to the engine through the message.

In the framework, the reliable-communication management module and group management module communicate directly with the engine. Thus, the dispatcher should have the functions of delivering the commands for the secure group to the reliable-communication management module, and the execution result to the group management module.

4   Proposed Framework and Implementation Issues

In the framework proposed in this paper, we added three modules to the standard functional modules of the SNMPv3: the GM (group management) module, the RM (reliable-communication management) module, and the KM (key management) module. In Fig.1, the three modules are shadowed and their relationships are shown [9].

The function of the GM module is to manage group membership information, including the group key. It interacts with all the other extended modules, the command generator, and the dispatcher. It gets the group key from the KM and generates the engineID from it. If multicast reliability is needed for the return data, it communicates with the RM module. The GM also receives the execution result message of a group control command from the dispatcher to manage the group membership.

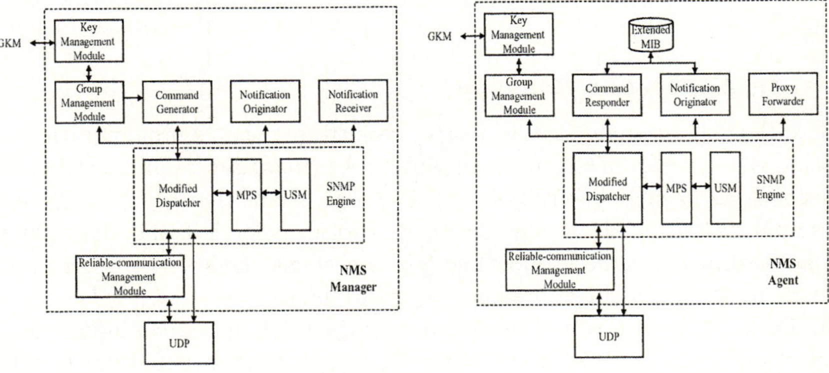

Fig. 1. In the framework, three modules are added to the standard SNMP: group management module, reliable-communication management module, and key management module

The function of the RM is to manage the reliability of multicast communications. It has the responsibility for re-transmission, sequencing, timeout and acknowledgement functions. It also provides the execution result information to the GM.

The function of the KM is to request the group key and get it from the GKM, and instruct the GM to update the group key and membership information. We assume that the interface between this module and the outside GKM is very simple in order to easily adapt most of the general group key management protocols.

The three modules will be the minimum required to support secure multicast in an SNMP environment. In a real NMS, however, other modules such as user interface and traffic analysis will also be needed.

Among the three modules, only two of them, namely GM and KM, are our main concerns for the time being, since the RM does not belong to the essential secure multicast SNMP functionalities. Except for the GM, we did not implement the KM although it is also indispensable for secure group management. Thus, we tested the system under the assumption that the user name used for the group name already exists and that the group keys are always available in some way. Although not yet completed, the module can be implemented by modifying the NET-SNMP [11], which was originally created by CMU (Carnegie Mellon University) and UC Davis (University of California at Davis) and is still being enhanced by many users. We are still modifying and testing it to show that our idea is quite practical.

5   Application to NMS

In Fig. 2, we present a configuration example that shows that the management by the proposed framework provides much flexibility and efficiency to the NMS organization. The example also explains how to apply the framework to a real NMS.

5.1   An Application Example

To describe the grouping status clearly, we introduce two simple notations. The notation $A1->M1$ means message sent by A1 (manager) is unicasted to M1 (a manager or an agent). The notation $A1-\gg M1$ means that the message sent by A1 is multicasted to M1. In the notations, both on the left and the right must be members of the same group. The left is a single member of the group, but the right should be a single member or group name.

In Table 2, we present some sample information of the configuration. The information is impractical in the sense that the real data has a much different form from it. We fabricated it intentionally just to stress the value difference. The M1 has two groups: grp01-M1 and grp02-M1. In the table Manager M1, we show the information of group grp01-M1. In the group, there are two members: EM1 and EM2. The two rows have the same group information, but have different user information for each member. In the table Agent EM1, we can see that it

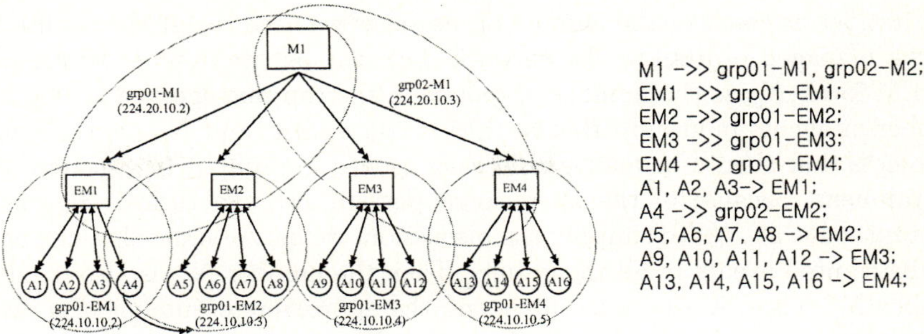

Fig. 2. A group configuration example and its notation

Table 2. The table for group configurations in Fig. 2

Manager M1

Field Name	EM1	EM2
grpID	grp01-M1	grp01-M1
grpAddr	224.20.10.2	224.20.10.2
grpPortID	2010	2010
grpKeyMaterial	keyValueEM01	keyValueEM01
grpEngineID	88000009019	88000009019
UserID	user01-EM1	user01-EM2
userAddr	203.237.52.11	203.237.52.12
userAuthKey	authKeyEM01	authKeyEM02
userPrivKey	privKeyEM01	privKeyEM02
status	ATV	ATV

Manager EM1

Field Name	A1	A2	A3	A4
grpID	grp01-EM1	grp01-EM1	grp01-EM1	grp01-EM1
grpAddr	224.10.10.2	224.10.10.2	224.10.10.2	224.10.10.2
grpPortID	1010	1010	1010	1010
grpKeyMaterial	keyValue01	keyValue01	keyValue01	keyValue01
grpEngineID	88000009029	88000009029	88000009029	88000009029
UserID	user01-A1	user01-A2	user01-A3	user01-A4
userAddr	203.237.53.11	203.237.53.12	203.237.53.13	203.237.53.14
userAuthKey	authKeyAgt01	authKeyAgt02	authKeyAgt03	authKeyAgt04
userPrivKey	privKeyAgt01	privKeyAgt02	privKeyAgt03	privKeyAgt04
status	ATV	ATV	ATV	ATV

Agent EM1

Field Name	M1
grpID	grp01-M1
grpAddr	224.20.10.2
grpPortID	2010
grpKeyMaterial	keyValueEM01
grpEngineID	88000009019
destGrpID	Grp01-EM1
UserID	NULL
userAddr	NULL
userAuthKey	NULL
userPrivKey	NULL
status	NULL

Agent A1

Field Name	EM1
grpID	grp01-EM1
grpAddr	224.10.10.2
grpPortID	1010
grpKeyMaterial	keyValue01
grpEngineID	88000009029
destGrpID	NULL
UserID	user01-A1
userAddr	203.237.52.11
userAuthKey	authKeyAgt01
userPrivKey	privKeyAgt01
status	ATV

Agent A4

Field Name	EM1	EM2
grpID	grp01-EM1	grp01-EM2
grpAddr	224.10.10.2	224.10.20.2
grpPortID	1010	1020
grpKeyMaterial	keyValue01	keyValue02
grpEngineID	88000009029	88000009039
destGrpID	Grp01-EM2	Grp01-EM2
UserID	user01-A4	NULL
userAddr	203.237.52.14	NULL
userAuthKey	authKeyAgt014	NULL
userPrivKey	privKeyAgt014	NULL
status	ATV	NULL

is a member of group grp01-M1. Because the EM should have the dual roles
of manager and agent, it has another table for the manager role. As we can
see in the table Manager EM1, it has four members: A1, A2, A3, and A4. The
rows for each member have the same group information, but have different agent
information. In the last two tables, we show the two information each member
holds. The table Agent A1 has only one row since it is the member of only one
group. Agent A4, however, has two rows for the two groups it belongs to. The first
row in the table shows that the message delivery destination is different from the
source where it receives the command. Agent A4 receives the command in the
group, grp01-EM1, and returns the execution result to the group, grp01-EM2.

5.2 Efficiency Analysis

For analysis purposes, we assume the organization in Fig. 2 but with a little
configuration change. In the changed configuration, agent A4 belonged only to
group grp01-EM1, and it unicasts the response message to manager EM1.

First, let us consider the number of messages we need to get the results from all the managed objects in the network. Let N_{ge} be the number of groups in the EM level, N_{gt} is the number of groups in the top manager level, N_{ai} is the number of agents in group i that an EM is controlling, and N_{oij} is the number of objects that should be managed in each agent j in group i. In a general NMS environment, because of the operational pattern, we can regard N_{oij} as the constant value of N_o. To simplify the discussion, we also assume that the agents are distributed evenly to all the groups. Then the number N_{ai} is also fixed to a number N_a. Thus, in the unicast system, the approximate number of messages in the network is:

$$N_{uni} = 2 \sum_{i=1}^{N_{ge}} \sum_{j=1}^{N_{ai}} N_{oij} = 2 \sum_{i=1}^{N_{ge}} \sum_{j=1}^{N_{ai}} N_o = 2 \sum_{i=1}^{N_{ge}} N_{ai} N_o = 2(N_{ge} N_a N_o). \quad (1)$$

In the proposed framework, the approximate number of messages needed is:

$$N_{mul} = N_o(N_{gt} + N_{ge}) + N_o(N_{ge} N_a) = N_o N_{ge}(1 + N_a) + N_o N_{gt}$$
$$\approx N_{ge} N_a N_o + N_o N_{gt}. \quad (2)$$

The final expression shows that if N_{ge} is sufficiently larger than N_{gt} (as it will be normally), the proposed framework is about 50 % more efficient in terms of number of messages.

We also believe that the proposed framework will reduce the response time. Let us assume that all the agents have the same command processing time, T_p, the command transmission time, T_s, and execution result transmission time, T_r. In the unicast system, the total response time of all objects in the network is:

$$T_{uni} = \sum_{i=1}^{N_{ge}} \sum_{j=1}^{N_{ai}} \sum_{k=1}^{N_{oij}} (T_s + T_p + T_r) + N_{ge} N_o(T_s + T_p + T_r)$$
$$= (1 + N_a) N_{ge} N_o(T_s + T_p + T_r) \quad (3)$$

In the case of a multicast system, the total response time of all objects in the network is:

$$T_{mul} = N_{ge} N_o(T_s + T_p + T_r) + N_{gt} N_o(T_s + T_p + T_r) < 2 N_{ge} N_o(T_s + T_p + T_r) \quad (4)$$

According to the above analysis, under the group configuration of Fig. 2, we expect that the proposed framework will response more quickly by the factor of N_a than the unicast system.

6   Conclusion

In this paper, we propose a framework that is compatible with and provides secure multicast services to the standard SNMP. We present some group configuration examples to show that the concept of secure multicast could be very useful in the real NMS. The current framework we propose here completely depends upon the external GKM for the management of group key. The interface is assumed to be very simple to accommodate most of the general group key management protocols such GSAKMP and GDOI, but it has yet to be defined. Therefore, the next step of our study is to find or define a group key management system and its interface that is very efficient and easily interacts with the proposed framework and is also general enough for all kinds of secure group communication applications.

References

1. J. Philippe and M. Flatin, *WEb-based management of IP networks and systems*, John Wiley and Sons, 2003.
2. E. Al-shaer and Y. Tang, "Toward integrating IP multicasting in Internet network management protocols," June 2000.
3. J. Schoenwaelder, "Using multicast-SNMP to coordinate distributed management agents," in *Proc. 2nd IEEE International Workshop on Systems Management (SMW'96)*, June 1996.
4. W. Stallings, "SNMPv3: A security enhancement for SNMP," *IEEE Communications Survey*, vol. 1 no. 1, 1998.
5. W. Stallings, *SNMP, SNMPv2, SNMPv3, and RMON1 and 2 (3rd Ed.)*, Addison-Wesley, 1999.
6. D. Zeltserman, *A Practical guide to SNMPv3 and network management*, Prentice Hall, 1999
7. U. Blumenthal and B. Wijnen, "User-based security model (USM) for version 3 of the SNMP," *IETF RFC 3414*, Dec. 2002.
8. D. Kwak and J. Kim, "Comparative survey on group key management protocol standards for secure real-time multicast," in *4th RMT Protocol Workshop*, April 2003
9. J. Case, D. Harrington, R. Presuhn, and B. Wijnen, "Message processing and dispatching for the SNMP," *IETF RFC 2572*, April 1999.
10. K. McCloghrie and M. Rose, "Management Information Base for Network Management of TCP/IP-based internets: MIB-II," *IETF RFC 1213*, March 1991.
11. Net-SNMP Homepage, *http://net-snmp.sourceforge.net/*.

Multi-rate Congestion Control over IP Multicast

Yuliang Li, Alistair Munro, and Dritan Kaleshi

Department of Electrical and Electronic Engineering, University of Bristol
{Yuliang.Li, Alistair.Munro, Dritan.Kaleshi}@bristol.ac.uk

Abstract. Multicast is the most efficient way to address shared communication between groups of computers for streaming multimedia content. The implementation of multicast has many challenges: multicast routing algorithms, address allocation, congestion control and multi-session management. In this paper, we give an overview of congestion control, especially for the receiver-based multicast transport layered congestion control protocols. This mainly focuses on the different technical backgrounds analysis and performance evaluation for two important protocols: *Packet-pair receiver-driven cumulative Layered Multicast* (PLM) and *Wave and Equation Based Rate Control* (WEBRC). From our simulation results, PLM can only be used effectively under a *Fair Queuing* (FQ) network environment; WEBRC has good performance results, but its complex definition and scheduling algorithm make it hard to be recommended as a global standard for multimedia streaming without further research. And all of the existing protocols have not addressed the support of wireless environment and multi-session management algorithms. Simulation results in this paper are based on the NS-2 simulator.

1 Introduction

Multicast applications can send one copy of each packet to the terminals that want to receive it. This technique addresses packets to a group of receivers rather than to a single receiver, and it depends on the network to forward the packets to only the receivers that need to receive them. Multicast solves the problem of inefficient use of bandwidth that would arise if unicast distribution was used, and also the insecurity problems of broadcast. It is an efficient way for group communications in IP networks. However deployment of IP multicast on the Internet has not been as rapid as expected. Among the reasons behind this slow deployment is the lack of efficient multicast congestion control schemes [1]. Network congestion is a drastic performance drop in the available bandwidth because of an "unsociable" transport protocol sending more traffic than elements in the network can handle, thus causing the whole network to become unusable. It cannot simply be solved by increasing available bandwidth. According to [2], no matter how much bandwidth is added, the network can still face congestion problems. This is because the network congestion problem is not a static resource shortage problem, but a dynamic resource allocation problem. Increasing additional bandwidth just circumvents the problem, or changes its location, without providing a real solution. Furthermore, it is cost-prohibitive to make networks handle peak loads that only occur for a few hours a day. TCP is a congestion control

P. Lorenz and P. Dini (Eds.): ICN 2005, LNCS 3421, pp. 1012–1022, 2005.

mechanism for unicast, which has been successfully used, in the last decade, and its problems are well documented and researched. If using comparable schemes for multicast, the negative effects would be more dangerous than unicast to the Internet. This is because a single multicast flow can be distributed throughout the Internet via a large global multicast tree [3] and will include network elements of varying capabilities: what is appropriate for one element could be sub-optimal or disastrous for many others. Specific challenges include: adapting to receivers' heterogeneity and QoS requirements, network-resident proactive congestion management, feedback implosion, and round trip time calculation.

2   Classification of Multicast Congestion Control

Sender-Based vs. Receiver-Driven: In the sender-based approach, the sender uses information received from receivers to derive values for metrics of the network congestion status and adjusts accordingly its sending rate. Due to the difficulty of collecting feedback from all the receivers (feedback implosion) and the need to cope better with the heterogeneity of devices in a network (e.g. a single slow receiver may drag down the data rate for the whole group), sender-based congestion control cannot performing very well on a large heterogeneous network.

Receiver-driven congestion control is often used in conjunction with layered encoding. The burden of congestion control is at the receiver's side and the sender has only a passive role. A receiver tunes its reception rate by subscribing to and unsubscribing from layers dynamically according to its own network condition. It is considered a good solution to the heterogeneous network problem.

SR-MCC vs. MR-MCC: With the *Single-Rate Multicast Congestion Control* (SR-MCC) scheme, all receivers in a multicast session receive the data at the same reception rate. The scheme picks one of the slowest receivers as representative. Then, the sender adapts the transmission rate to that of the representative. It uses a (so-called) *non-cumulative* layering mechanism, in which all layers have the same priority, and any subset of the layers can be used for data reconstruction. The SR-MCC scheme has significant limitations in handling large heterogeneous group of receivers. A single sluggish receiver can retard the reception rate of all other receivers that may be in better network conditions.

The *Multi-Rate Multicast Congestion Control* (MR-MCC) scheme uses *cumulative* layering. There is a layer with the highest importance, called a base layer. It contains the most important features of the data for decoding. Additional layers, called enhancement layers, contain data that progressively refine the reconstructed data quality. This information, when combined with the base layer's information, results in improved quality. Separate receivers have their own rights to decide the layers for subscription. The more layers subscribed, the more bandwidth is consumed, and the better quality received. The advantages of this scheme are that it is massively scalable to a high number of receivers because the sender's behaviour is independent of the number of receivers [4]. Secondly the reception rate is not bound to the slowest receiver like SR-MCC; it is more flexible in bandwidth allocation along different network paths. Finally, there is no feedback implosion problem comparing with sender-based congestion control algorithms. All these advantages of MR-MCC make it a very good solution for our complex heterogeneous network and also it is very good at supporting wireless networks with relatively high error-rates, and varying limited bandwidth [5]. All our following study focuses on the receiver-driven MR-MCC.

3   Multi-rate Multicast Congestion Control

There exist three ways for MR-MCC protocols to infer the available bandwidth as well as adjust receivers reception rate to varying network conditions: join-experiment, packet-pair and equation based. According to these technologies, we can roughly divide MR-MCC into three groups (Figure 1).

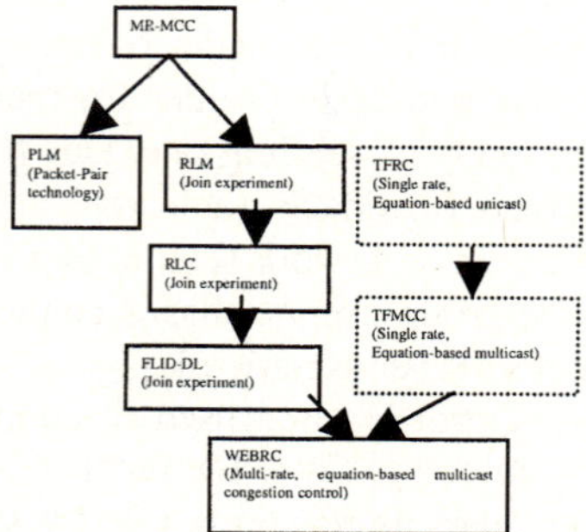

Fig. 1. Classification of MR-MCC

3.1   Join-Experiment vs. Packet-Pair vs. Equation Based

Join-Experiment: The receiver increases its reception rate by periodically subscribing to an additional layer and decreasing its reception rate by unsubscribing from a layer when experiencing packet loss. This mechanism has been used for *Receiver-driven Layered Multicast* (RLM), *Receiver-driven Layered Congestion* control (RLC) and *Fair Layer Increase Decrease* with *Dynamic Layering* (FLID-DL). The join-experiment has relative simple technical theory and can be easily implemented. But it is hard to deal with the coordination problem and the problem of delays in updating the presence of receivers in a group – the IGMP leave latency problem [2].

Packet-Pair **(PP):** With the PP approach, a source periodically sends a pair of its data packets as a burst to infer the bandwidth share of the flow. It uses a one-bit field of the packet header to indicate the first packet of a pair. At the receiver side, the estimation of available bandwidth is calculated from packet size divided by the inter-arrival gap. The first PP that leaves the queue after congestion occurs will be a signal of congestion. The estimated bandwidth is used to adapt the receiving rate. PLM uses the PP approach and set a *Check* period (C) interval to avoid oscillatory rate adaptation. PLM defines a timeout period dealing with the case of packet pair lost during congestion. If no packet is received at the timeout, a layer will be dropped due to the expectation of congestion. If some packets are received but no PP is received, and the loss rate is over a predefined loss threshold, then a layer will be dropped. After this, PLM will do nothing over react to loss, waiting for a predefined blind period before re-estimating the loss rate [6].

Equation Based: The utilization of available bandwidth depends on the receiver's current target reception rate calculation. The receiver continually updates its target reception rate using its measured *Multicast Round Trip Time* (MRTT) and loss rate. WEBRC is the first layered multicast congestion control protocol using equations to calculate the available bandwidth. The available bandwidth equation calculation in WEBRC not only keeps all the advantages from previous MR-MCC protocols, but also merges the bandwidth detection advantage from SR-MCC that has good coordination between senders and receivers and preserves TCP-friendliness.

3.2   RLM, RLC and FLID-DL

RLM: The RLM protocol is the first technique for cumulative layered multi-rate multicast congestion control, proposed by McCanne et al. [10]. It uses a join experiment to adjust receivers' reception rate to the network conditions. The sender splits the data into several layers. A receiver starts receiving by subscribing to the first layer. When the receiver does not experience congestion in the form of packet loss for a certain period of time, it subscribes to the next layer. When a receiver experiences packet loss, it unsubscribes from the highest layer that it is currently receiving. However RLM's mechanism of adding or dropping a single layer based on the detection of packet loss is not TCP-friendly. The regular joining and leaving of layers causes another serious problem of IGMP join/leave latency. Furthermore, a receiver's join experiments can introduce packet losses at other receivers behind the same bottleneck link. It finally results in unfairness among the downstream receivers.

RLC: In order to address the problems of RLM, Vicisano, Crowcroft and Rizzo developed the RLC protocol [7]. The use of bandwidth of each layer in RLC is increased or decreased exponentially. This achieves the same rate adjustment scheme as TCP. In order to prevent inefficient uncoordinated actions of receivers behind a common bottleneck, an implicit coordination is done by using *Synchronisation Points* (SP). A SP corresponds to a special flagged packet in the data stream. A receiver only joins a layer after receiving a SP. In order to decrease the number of failed join experiments, the sender creates a short burst period before a SP. During this burst period the data rate is doubled in each layer. Only if a receiver does not experience any signs of congestion during the burst is it allowed to join the next higher layer. Finally, RLC introduces *Forward Error Correction* (FEC) encoding to support reliable multicast applications. RLC does not take the round-trip time into account when determining the sending rate. This still leads to unfairness towards TCP.

FLID-DL: FLID-DL is a protocol for improving RLC, which is presented in [8]. The protocol uses a Digital Fountain encoding [9], allowing a receiver to recover the original data upon reception of a fixed number of distinct packets, regardless of specific packets losses. FLID-DL introduces the concept of *Dynamic Layering* (DL) to reduce the IGMP leave latencies. With such a dynamic layering, a receiver can reduce its reception rate simply by not joining any additional layer. On the other hand, to maintain or increase its reception rate, the receiver should join more layers. The FLID mechanism is used to tackle the abrupt rate increase in RLC. Instead of having the fixed rate multiplier equal to two, FLID generalizes layering scheme by having the rate multiplier equal to any predefined constant (the recommended rate multiplier is 1.3). FLID-DL has a few improvements over RLC, but it still exhibits not very good fair behavior towards TCP sessions.

3.3 PLM

Legout et al. proposed the PLM congestion control protocol [6]. The difference in this approach, compared with above protocols, is the use of two key mechanisms: *Fair Queueing* (FQ) at routers and receiver-side *Packet-Pair* (PP). Networks with FQ have many characteristics that immensely facilitate the design of congestion control protocols and also improve TCP-friendliness. In the FQ network environment, the bandwidth available to a flow can be determined by using the PP method. PP can sense the bandwidth change in the network before congestion happens, which the join experiment and equation based approach used by other protocols does not. And through use of FQ, PLM becomes very intra-protocol and inter-protocol fair. However, it is still unlikely that the FQ scheduler can be implemented, or configured in the same way, on all the routers over the whole Internet in the future. But for wireless systems it is possible for FQ be implemented at all the *Base Stations* (BS) to support the wired and wireless network. This lets PLM become a possible ideal congestion control protocol for the future hybrid network. Somnuk [3] and some other researchers [1] [5] have already done very good performance analysis on PLM. In this paper, we only focus on the FQ testing.

Simulation Results and Discussion: Two independent TCP sessions share a common bottleneck with a single PLM session. Ideally PLM should share the bottleneck bandwidth equally with TCP sessions as they occur. The topology is shown below (Figure 2). One PLM server and one PLM host connect via two routers (configured to be a bottleneck). Two TCP sessions are connected over the same router pair. The bandwidth on the non-bottleneck links is limited to 100Kbps and 10 Mbps respectively for different scenarios. The bottleneck link is limited to 300Kbps, and the link latency is 5 ms. All routers use FQ and Drop Tail queue principle respectively for each test. The first TCP session (TCP1) starts at time 0. A single infinite duration PLM multicast session starts at time 20. After 60 seconds the second TCP session (TCP2) starts. We run the experiment for 200 seconds.

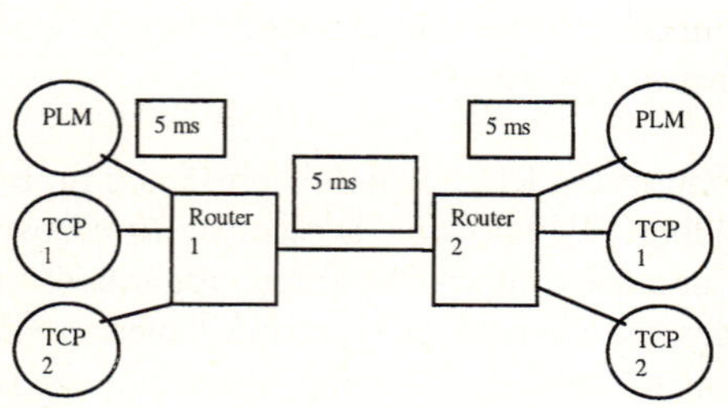

Fig. 2. PLM Simulation Topology

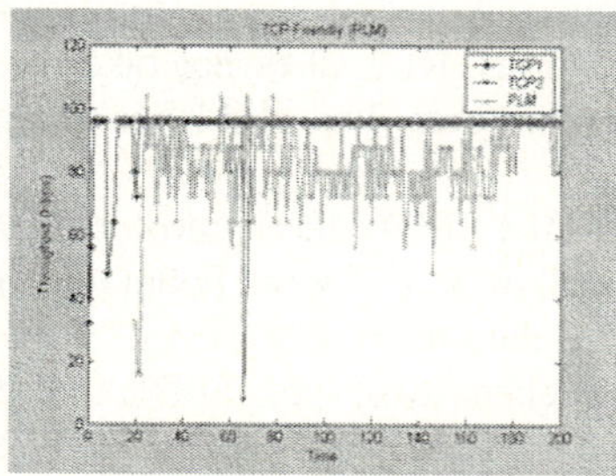

Fig. 3. PLM not under congestion

Scenario 1: We set the bottleneck router bandwidth to 300Kbps; each non-bottleneck link's bandwidth is set to be 100Kbps. This scenario represents the situation when the bottleneck is saturated but not yet congested. Due to no congestion, we can get the same throughput plot (Figure 3) for both FQ and Drop Tail conditions. The two TCP sessions keep their 100Kbps transmission rate steady. For PLM there is some reasonable fluctuation, but in general, the transmission rate of PLM remains about 90Kbps.

Scenario 2: Bottleneck is still set at 300Kbps, but each non-bottleneck link's bandwidth has been changed to 10Mbps. This scenario represents the case of congestion. When congestion happens, under the FQ situation, PLM can coexist in a friendly way with TCP. We can see from Figure 4a, PLM starts at the time 20 sec, it detects the existing TCP1 flow and can share the bandwidth very cooperatively with TCP1 150Kbps respectively. When the new TCP2 arrives, both of the TCP1 and PLM decrease their transmission rate to equally share the bandwidth with TCP2, at about 100Kbps. But when we change the FQ to Drop Tail, we can see from Figure 4b, PLM totally loses its friendliness with TCP and also cause the irregular behaviour of the two TCP sessions.

The experimental results above show PLM can keep a very good TCP-friendliness when queuing management of routers uses FQ. But, under the congestion situation, it cannot maintain TCP-friendliness and shows aggressive behaviour towards TCP when there is no FQ to regulate the fair share at routers as assumed.

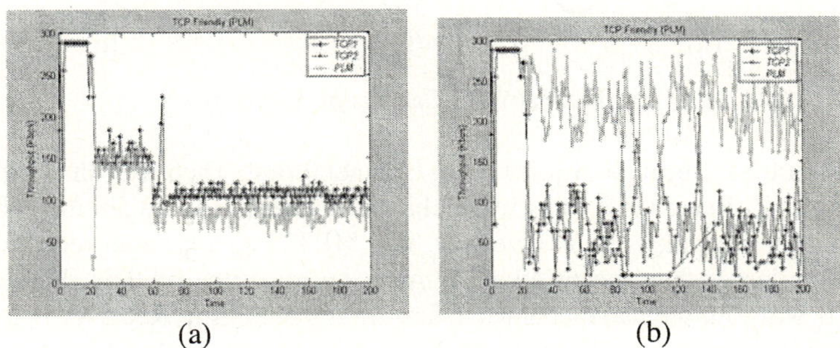

(a)   (b)

Fig. 4. PLM under congestion

3.4 WEBRC

WEBRC is the first multiple rate multicast congestion control protocol to be equation based [11][12][13]. It has two major innovations: MRTT and Wave. MRTT is a multicast analogue of the unicast *Round Trip Time* (RTT). It is defined as the time elapsed between the time sending a join request and the time it receives the first packet of data (the MRTT is equal to the RTT in a session with one receiver). How long it takes to receive the first packet depends on how far up the tree the request needs to go until it reaches a router that is already forwarding packets [14]. This distance depends on which receivers have joined earlier in the same time step. Wave is a process that controls the transmission rate on a channel to be periodic, with an exponentially decreasing form during an active period followed by a quiescent period. A receiver will receive packets from the base channel and some number of consecutive wave channels. The number of wave channels that can be received depends on the receiver's current target reception rate. The receiver at all points in time maintains a target reception rate, and the receiver is allowed to join the next wave channel if joining would increase its reception rate to at most its target reception rate. This target reception rate is continually updated based on a set of measured parameters: *Loss Packet* (LOSSP) and *Average MRTT* (AMRTT). Receivers that have higher target reception rates will have earlier join times and receivers that have lower target reception rates will have lower join times. Thus, the joint time is a reflection of the target reception rate of a receiver before the join. MRTT value varies inversely with target reception rate, as a result of the join impacts the target reception rate after the join.

MRTT Calculation: The example here is based on the scenario given in [14]. MRTT measurements for each wave depend on relative timings of receiver joins to the wave. It can be as large as RTT to the sender (Figure 5a); or it can be as small as RTT to the nearest router (Figure 5b).

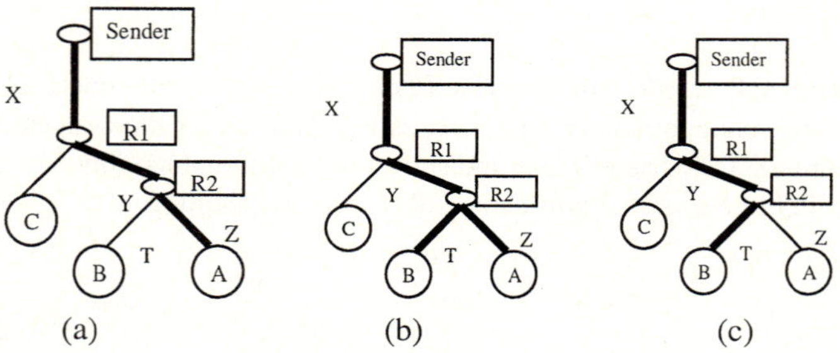

Fig. 5. MRTT Calculation Topology

From Figure 5a, Receiver A joins wave channel 0 first. Then we can get the MRTT (A0) is X+Y+Z. Receiver B joins wave channel 0 later at Figure 5b, the MRTT (B0) is T. For target reception rate is inversely with MRTT, so the target reception rate of receiver B is larger than the receiver A. Also the join time of B is shorter than A. Receiver B will join the wave channel 1 first. In the next turn, Receiver B joins wave channel 1 first (Figure 5c). Then the MRTT (B1) = X+Y+T. Receiver A joins wave channel 1 later, the MRTT (A1) = Z. On a long time period, we can get the average MRTT (AMRTT) under the same bottleneck calculation equation (1):

$$AMRTT = \frac{\{MRTT(A0) + MRTT(B0) + \ldots\ldots + MRTT(An) + MRTT(Bn)\}/n}{2} = \frac{X+Y+Z+T}{2} \qquad (1)$$

Simulation Results and Discussion:
(1) Responsiveness: The constant bit rate (CBR) test is a simple check to see whether WEBRC reacts to available network bottleneck bandwidth smoothly and efficiently. The topology is shown in Figure 6a. In order to make sure the bandwidth can be totally subscribed to (in our examples it has been limited to 1Mbps), we set the available largest WEBRC server and host session rates to 3 Mbps.

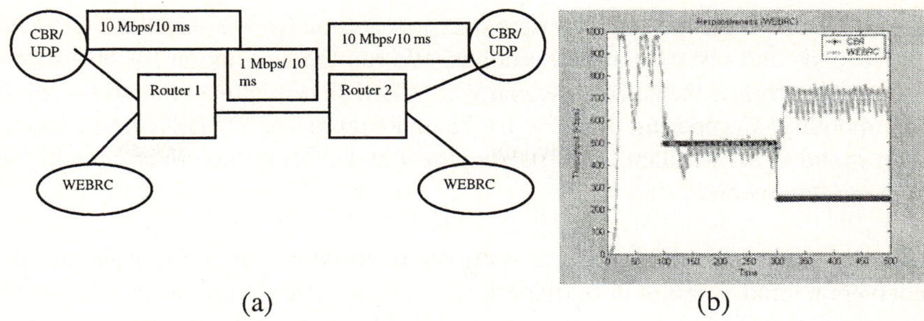

(a)   (b)

Fig. 6. WEBRC Responsiveness Test

Scenario: We use a single WEBRC multicast session across the bottleneck, where the server starts at time 0. After 100 seconds we add a CBR source over the bottleneck link at 500Kbps (half the bottleneck). At 300 seconds we decrease the rate of the CBR stream to 250 Kbps (a quarter of the bottleneck). We run the simulation for 500 seconds. From Figure 6b, we can see that, before 100 sec. WEBRC on its own can make best use of the bandwidth. At time 100 sec, WEBRC detects the 500Kbps CBR stream start. It quickly decreases to 500Kbps to equally share with the CBR stream. When the CBR rate decreases to 250Kbps at 300 sec, WEBRC responds to the additional available bandwidth by increasing its rate to about 750Kbps.

(2) Equal-sharing: In the equal-sharing test, one TCP session shares a common bottleneck with a single WEBRC session. The goal is to see how WEBRC competes with TCP traffic over a common bottleneck. In our simulation experiments, we set the TCP window to 5000 bytes to remove any effect on the data rate generation from the influence of maximum window size (i.e. the window will never fill and lead to TCP back off). We also use the same topology and the same WEBRC session rate setting as in the responsiveness test.

Scenario 1: Both WEBRC and TCP sessions start at time 0. After 100 seconds we stop the WEBRC session, and restart it at 200 seconds. We run the simulation for 500 seconds. From Figure 7a we can see very clearly, at 100 sec, TCP responds to the available bandwidth very quickly and also at 200 sec it decreases its transmission rate in response to the incoming WEBRC session rapidly. After another 50 sec. WEBRC and TCP share the bottleneck not very equally but in a cooperative way. In general, WEBRC takes about 400Kbps, TCP takes about 600Kbps.

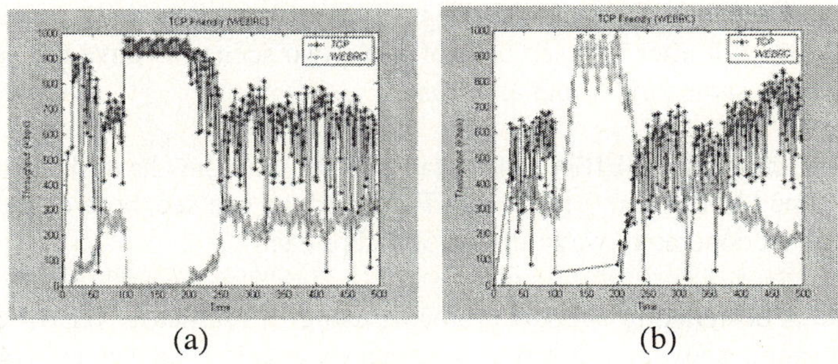

(a)   (b)

Fig. 7. WEBRC Equation-sharing Test

Scenario 2: Both WEBRC and TCP sessions start at time 0. We stop the TCP session at 100 seconds and restart it at 200 seconds. We run the simulation for 500 seconds. The result observed is the same as scenario 1. WEBRC responds to the arrival of TCP traffic smoothly. Comparing with Figure 7a and Figure 7b, WEBRC takes about 20 sec to respond to the availability of bandwidth. But TCP responded very quickly only after a very few seconds.

(3) Coordination: According to [5], it is crucial to coordinate join and leave decisions of receivers behind a common bottleneck: if only some receivers leave a layer while others stay subscribed, no pruning is possible and congestion cannot be reduced. In WEBRC, multiple proximate receivers downstream of a bottleneck link their reception rates and tend to coordinate and equalize due to similar measured MRTT values. As more receivers join the session, the MRTT values of the receivers already in the session tend to decrease. From the topology Figure 8a, we set the bandwidths for all links to 1Mbps. The WEBRC server and host have the same session rate: 500Kbps. The link between the WEBRC server and Router 0 has 0.01 sec RTT. The RTT between two routers is set to 0.2 sec. Two WEBRC hosts have 0.05 sec and 0.1 sec RTT respectively.

Scenario 1: The WEBRC 0 host starts at 0 sec, MRTT = MRTT(0) = 0.01+0.2+0.05 = 0.26. The WEBRC 1 host starts at 200 sec, so we get MRTT(1) = 0.1 and also from the AMRTT calculation equation (1) which we have explained:

$$AMRTT = \frac{0.01 + 0.2 + 0.05 + 0.1}{2} = 0.153$$

Because the target reception rate is inversely related to MRTT, after 200sec the AMRTT was smaller than the MRTT which was calculated before 200 sec, so from the Figure 8b, we can see the throughput has increased after 200 sec, from 300Kbps to about 500Kpbs.

Scenario 2: The WEBRC 1 host starts at 0 sec, MRTT = MRTT(1) = 0.01+0.2+0.1 = 0.31. The WEBRC 0 host starts at 200 sec and we get MRTT(0)=0.05. After 200sec, the average multicast round trip time is the same as scenario 1: AMRTT = 0.153. Figure 8c shows that scenario 2 has the same variation tendency as Scenario 1 after 200 sec. Consistent with the WEBRC definition, we get these interesting results before 200 sec: MRTT(1) of scenario 2 is larger than MRTT(0) from scenario 1, so the throughput of scenario 2 (about 250Kbps) is less than scenario 1 (about 300Kpbs) before 200 sec. But after 200 sec both of these two scenarios have the same MRTT, so they have the same throughput after that.

Scenario 3: Both of the WEBRC hosts start at 0 sec, and they have the same multicast round trip time: AMRTT = 0.153. From Figure 8d we can see, both of them share the bottleneck in a cooperative way (around 500Kbps each).

WEBRC is designed to support protocols using IP multicast. WEBRC gives relatively very good network utilisation, responsiveness, low packet loss rate, smoothness, moderate TCP-friendliness and fast convergence. WEBRC seems like an ideal candidate for the future multi-rate multicast congestion control, but it is still aggressive to TCP. This is because, on congestion, TCP increases the congestion window linearly but in case of WEBRC layers are dropped, which results in an abrupt decrease in the rate. Furthermore the parameter setting and algorithms are quite complex in WEBRC. This means that it is still early to recommend it as a global standard for the traffic scenarios studied here, which are related to the distribution of multimedia content.

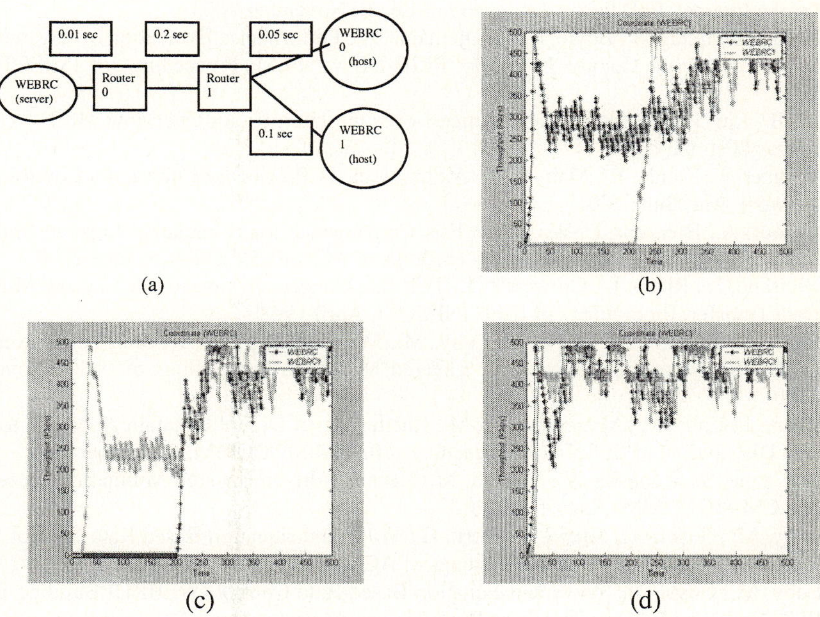

Fig. 8. WEBRC Coordination Test

4 Conclusion and Future Works

In this paper several different schemes proposed for providing multi-rate congestion control for multicast applications have been studied. RLM is not TCP-friendly, PLM can only be used under a FQ environment and WEBRC is too complex to use. None of the existing protocols provide the support for hybrid wired and wireless networks. Also no protocols are ideal for the multi-session management algorithms for multimedia communication.

Multicast congestion control is still a new and active research area. Comprehensive performance evaluation of the current proposed approaches needs to be done to get a deeper understanding of the solution approaches. Further enhancements are needed before any wide-scale deployment on the Internet. Our future intention is to explore the similarities and differences between the approaches, and make our proposal for improved protocols. Such new protocols should aim to combine the advantages from PLM and RLM to support hybrid wired and wireless environments, adding the RTT calculation from WEBRC to address the TCP-friendliness problem. Also it will add multi-session management to better support the future multimedia network.

References

1. Matrawy, A., Lambadaris, I.: A Survey of Congestion Control Schemes for Multicast Video Applications. IEEE Communications Surveys & Tutorials, Fourth Quarter 2003
2. Pauangpronpitag, S.: Design and Performance Evaluation of Multicast Congestion Control for the Internet. PhD thesis, University of Leeds, November 2003
3. Pauangpronpitag, S., Boyle, R. D., Djemame, K.: Performance Evaluation of Layered Multicast Congestion Control Protocols: FLID-DL vs. PLM. Proceedings of ISPECT, July 2003
4. Li, B., Liu, J.: Multirate video Multicast over the Internet: An Overview. IEEE Network. January/February 2003
5. Widmer, J., Denda, R., Mauve, M.: A Survey on TCP-Friendly Congestion Control. IEEE Network, May/June 2001
6. Legout, A., Biersack, E. W.: PLM: Fast Convergence for Cumulative Layered Multicast Layered Multicast Transmission. Proceedings of ACM SIGMETRICS, June 2000
7. Vicisano, L., Rizzo, L., Crowcroft, J.: TCP-like Congestion Control for Layered Multicast Data Transfer. Proceedings of IEEE INFCOM, April 1998
8. Byers, J., Frumin, M., Horn, G., Luby, M., Mitzenmacher, M., Roetter, A., Shaver, W.: FLID-DL: Congestion Control for Layered Multicast. Proceedings of NGC, November 2000
9. Byers, J., Luby, M., Mitzenmacher, M., Roetter, A.: A Digital Fountain Approach to Reliable Distribution of Bulk Data. Proceedings of ACM SIGCOMM, September 1998
10. McCanne, S., Jacobson, V., Vetterli, M.: Receiver-driven Layered Multicast. Proceedings of ACM SIGCOMM, August 1996
11. Luby, M., Goyal, V., Skaria, S., Horn, G.: Wave and Equation Based Rate Control Using Multicast Round Trip Time. Proceedings of ACM SIGCOMM, August 2002
12. Luby, M., Goyal, V.: Wave and Equation Based Rate Control (WEBRC) Building Block. IETF Reliable Multicast Transport Working group Internet-Draft, February 2002
13. Luby, M., Goyal, V.: Wave and Equation Based Rate Control Using Multicast Round Trip Time: Extended report. Technical Report DF2002-27-001, Digital Fountain, July 2002
14. Chaudhuri, K., Maneva, E., Riesenfeld, S.: WEBRC Receiver Coordination. May 2003

A TCP-Friendly Multicast Protocol Suite for Satellite Networks

Giacomo Morabito, Sergio Palazzo, and Antonio Pantò[*]

Dipartimento di Ingegneria Informatica e delle Telecomunicazioni,
University of Catania,
V.le A. Doria 6, 95125 Catania (ITALY)
Tel: 0039-095-7382370 Fax 0039-095-7382397
`{giacomo.morabito, sergio.palazzo,`
`antonio.pantò}@diit.unict.it`

Abstract. In this paper we introduce a novel scheme for multicast communications and describe the implementation choices of a new suite of TCP-friendly transport protocols optimised for reliable and unreliable multicast connections over satellite networks. The suite is composed of two protocol optimised for different applications: the RMT (Reliable Multicast Transport) guarantees a reliable file delivery; the UMT (Unreliable Multicast Transport) is suitable for multimedia streaming. Both RMT and UMT adopt a TCP-Peach like congestion control scheme to optimise the performance in satellite networks. The new protocols have been implemented on top of UDP in a Unix environment and their performance has been evaluated in an emulated satellite network environment.

1 Introduction

An application is said to be *TCP-friendly* if it fairly shares the bandwidth with a concurrent TCP connection. Accordingly we can classify applications as *TCP-friendly* and *non TCP-friendly*. However, in a shared network such as the Internet all traffic flows - reliable and unreliable as well as unicast and multicast - are expected to be *TCP-friendly* [1], i.e., they should comply with the following rules:

- *Rule 1*: Their transmission rate can increase slowly as long as the network is not congested, and
- *Rule 2*: Their transmission rate must decrease immediately when the network is congested.

Next generation IP-routers will penalize traffic flows not compliant with these rules. As a consequence, in the recent past even for real-time and multicast applications there have been several congestion control proposals [2-8, 11, 12, 14, 16]. The transmission rate value is adjusted in such a way that Rules 1 and 2 are satisfied.

---

[*] Corresponding Author.

P. Lorenz and P. Dini (Eds.): ICN 2005, LNCS 3421, pp. 1023–1030, 2005.

The most general architecture for the transport protocol layer can be represented as shown in Figure 1.

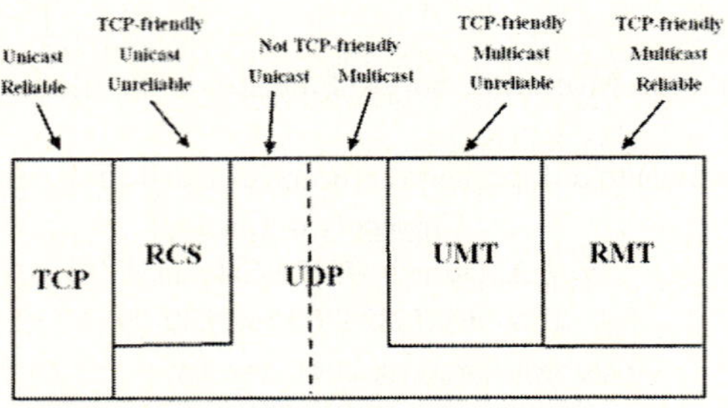

Fig. 1. General Transport Layer Architecture

In order to comply with Rule 2, traffic sources decrease their transmission rate when packet losses are detected because these are the only congestion signal in the current
Internet. If the source decreases its transmission rate when a packet loss occurs due to link errors, then network efficiency decreases drastically.

To solve the above problems, TCP-Peach [9] and RCS [8] have been recently proposed for unicast communications.

In this paper we introduce new solutions for reliable and unreliable multicast transport protocols optimized for satellite networks, namely, the *Reliable Multicast Transport* (RMT) and the *Unreliable Multicast Transport* (UMT).

The rest of this paper is organized as follows. In Section 2 we present the main issues of multicast transport protocols and the related work. We introduce our new solutions, i.e., RMT and UMT in Section 3 and discuss their behavior and performance in Section 4. Finally, in Section 5, we conclude the paper.

2   Multicast Transport Protocols Issues

The requirements for reliable multicast protocols that are to be considered for standardization by the IETF include: congestion control; scalability; security. Moreover, upon request, a multicast transport protocol should provide packet ordering and reliability. In this section we will focus on the reliability and congestion control issues. More specifically, in Section 2.1 we will highlight the major challenges related to reliability, whereas in Section 2.2 we will deal with the problem of congestion control.

2.1   Reliability Issues

The two aspects of reliable multicast that make standardization particularly challenging are explained in [10]: the meaning of reliability varies in the context of different application; multicast protocols may cause congestion disaster if they are widely used and do not provide adequate congestion control.

2.2 Congestion Control Issues

A particular concern for the IETF is the impact of multicast traffic on other traffic in the Internet during congestion. Multicast applications in general have the potential to cause more congestion-related damage to the Internet than unicast applications do, because they have to contend with a potential explosion of complex patterns of control traffic (e.g. ACKs, NACKs, status messages).

Several TCP-like congestion control schemes for multicast application have been studied and proposed, e.g., [11-16]. These solutions are said single-rate solutions.
In this paper we consider single-rate, window-based congestion control schemes. The two most interesting schemes are MTCP, [14], and PGMCC, [11,12]. Recently, TCP-Peachtree has been proposed for satellite networks [16].

The major disadvantages of single rate schemes are that a congested connection to a single receiver causes low transmission rate to all receivers, and multiple, uncorrelated packet losses are difficult to handle. The above features cause dramatic performance problems in satellite scenarios [15].

Objective of RMT and UMT is providing high performance in satellite scenarios while featuring design and implementation simplicity.

3 RMT and UMT Protocols

Both RMT and UMT congestion control algorithms are based on PGMCC [11,12] The section 3.1 describes the PGMCC problems in satellite networks. The new features of the congestion control algorithm implemented by both RMT and UMT will be introduced in Section 3.2; whereas the specific characteristics of RMT and UMT will be presented in Sections 3.3 and 3.4, respectively.

3.1 PGMCC Problems in Satellite Networks

For the multicast reliable transport protocol, we choose to investigate a congestion control scheme based on PGMCC because: it is an end-to-end solution (no changes are required in the intermediate routers); it is scalable; it can be implemented easily.
In PGMCC two issues must be considered very carefully:

- the group's elected acker must necessarily be the slowest receiver, if this is not the case the protocol behaviour in the path between the sender and the truly slowest receiver will be not TCP-friendly;
- the accuracy of the estimation of the resources in the worst path is very important, as it is needed to maximize performance while maintaining a TCP-friendly behaviour.

PGMCC TCP-like congestion control scheme has low performance when satellite system are involved in the communication. In fact:

- Problem 1: The congestion control schemes interpret packet losses as the signal of network congestion. This assumption is not realistic when packet losses occur due to link errors such as it is likely in satellite communications.

- Problem 2: The long propagation delays involved in satellite communications amplify Problem 1 because the time interval required to recover from packet losses is proportional to the round trip time.

3.2 RMT and UMT Congestion Control

A solution to the problem of the correct estimation of the network resources in the path between the sender and the acker is the use of a modified PGMCC.

We introduce a new congestion control scheme for satellite networks that is based on the TCP-Peach [9] approach, and therefore on the use of dummy segments. The sender uses the dummy segments to probe the availability of network resources in the path connecting the sender and the current acker.

The server uses an equation-based scheme for the estimation of the bandwidth available in the path towards the receiver other than the acker. The receivers implement an active feedback algorithm that reduces the amount of transmitted feedback packets, evaluating locally their own throughput.

3.3 RMT Specific Protocol Characteristics

The proposed Reliable Multicast Transport protocol realizes the multicast congestion control scheme illustrated in the previous paragraph.

We applied two different retransmission policies one for the packets lost by the acker and the other for the packets lost by the other receivers:

- The acker implements with the sender all TCP-Peach algorithms, included the retransmission policies, therefore the acker receives all the lost packets until a new acker is selected.
- The other receivers send NACKs when packet losses are detected, and the sender retransmits the NACKed packets.

When the sender sends the last data packet: the current acker sends an ACK packet; all the receivers that have correctly received all the packet exit the multicast transport session.

It is likely at this point that some of the receivers have holes in the packet sequence and therefore send NACKs accordingly. Therefore, sender receives the NACKs and retransmits the requested packets without a congestion control scheme.

3.4 UMT Specific Protocol Characteristics

The UMT protocol implements the same algorithms as the RMT protocol, but the packet retransmission scheme.

Being the UMT suited for multimedia streaming transmission, a new problem arises: if the slowest receiver lead the sender rate under the streaming transmission rate, the other receivers cannot enjoy the service, even if they have adequate bandwidth. The UMT protocol solves this problem introducing an appropriate threshold, and classifying receivers into two lists: a white list, including the receivers that have an evaluated throughput higher than the threshold; a black list, including the receiver that have an evaluated throughput lower than the threshold. Depending on the persistence in the black list, the worst receivers are invited to leave the UMT session, and the IP multicast group.

4   Protocols Behaviour and Performance Evaluation

UMT and RMT have been implemented in Linux environment and assessed in an emulated testbed. In Section 4.1 we describe UMT and RMT implementation choices, in Section 4.2 we illustrate the emulated network characteristics, and, finally, in Section 4.3 and 4.4 we show the performance results in terms of inter- and intra-protocol fairness and delivery time, respectively.

4.1   Implementation Choices

The structural implementation choices of the two multicast protocols, reliable and unreliable, are the same. The two multicast protocols were implemented in a GNU/Linux platform. The IP multicast implementation of the Linux kernel is very efficient and analyzable, being open source. The program was written in C++, using the `pthread` library for the multithreaded structure. The realization of a protocol that at the same time uses a multicast socket, for the data packets, and a unicast socket, for network probing purpose, renders natural the choice of an implementation at user level, using the basic transport functionality of the UDP layer at kernel level.

4.2   Emulated Network

The topology of the testbed has been designed in order to evaluate the performance of the proposed schemes in terms of inter- and intra-protocol fairness, as well as delivery time. In Figure 2 we show the topology of the emulated network.

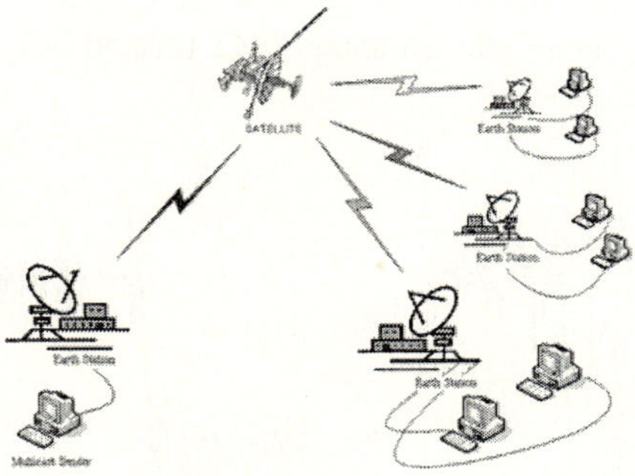

Fig. 2. Emulated Network Topology

In our testbed the packet loss rate, the delay and the bandwidth of every link were emulated through NISTNet [17] running on the end node of each link, given that the NISTNet application operates on the incoming packets of each node.

4.3 Fairness

In the Figure 3 we show the TCP-friendliness of UMT. A UMT instance is running between the multicast sender and two receivers, after 60 sec a TCP connection begins to send packets between the first and second receiver and after 120 sec another TCP connection is established between the two receivers. The Figure 3 shows how the three protocol share the available bandwidth in every phase.

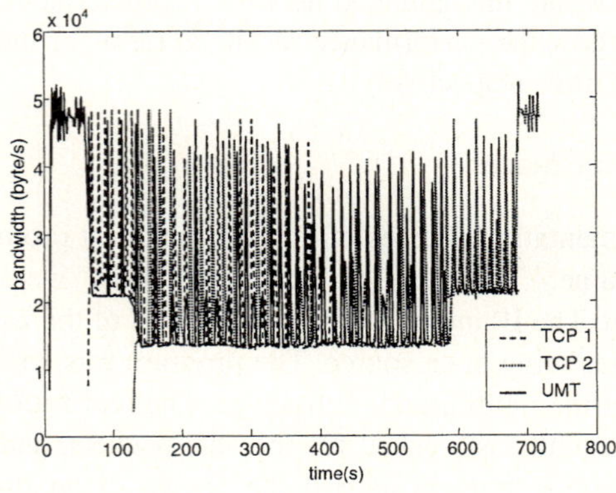

Fig. 3. Inter-protocol Fairness

The Figure 4 shows the UMT intra-protocol fairness. We can observe that:

- when the second instance of the protocol begins to transmit, after a short transition phase, the two instances share the available bandwidth;
- when the first instance ends, the second instance uses the whole bandwidth available.

Similar results have been obtained using RMT instead of UMT.

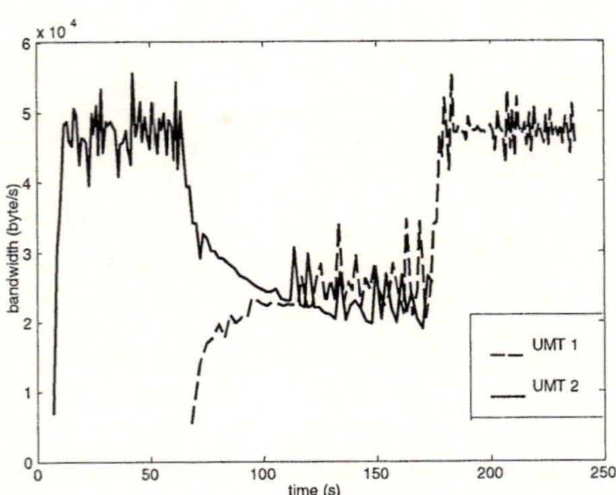

Fig. 4. Intra-protocol Fairness

4.4 Delivery Time

Figure 5 shows the performance comparison between a reliable multicast protocol with PGMCC congestion control scheme and RMT protocol. The figure represents the time required to deliver a 2 Mbyte file versus the packet error probability in a satellite link with a bandwidth of 300 Kbyte/sec:

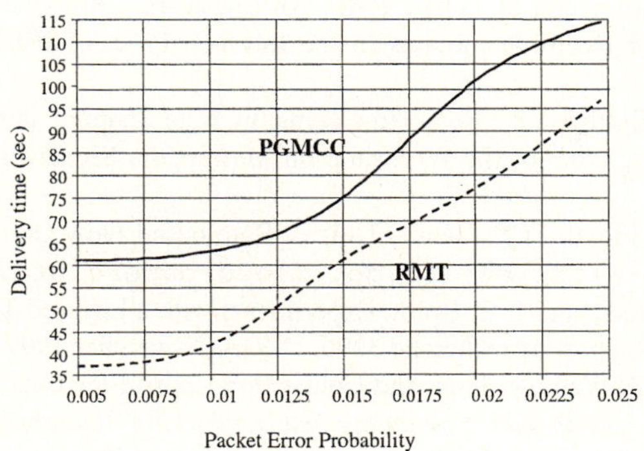

Fig. 5. Delivery Time vs Packet Error Probability

In the left side of the Figure 5, for PEP < 0.01, RMT congestion control achieves significantly better performance than PGMCC. Instead we can notice, for PEP > 0.01, a decrease and then a stabilization of the difference between the two curves. This is due to the increasing portion of the time spent for retransmissions at the end of the session, which is independent of the congestion control scheme applied, as explained in section 3.3.

Lower delivery time means higher throughput, in fact, RMT achieves much higher throughput than PGMCC. This is also the case when UMT is compared to PGMCC.

5 Conclusions

The satellite systems are the natural design choice for multicast communications. Nevertheless, little effort has been devoted so far to the problems of multicast congestion control in satellite communications. In this paper we have introduced two new solutions, for reliable and unreliable multicast congestion control optimised for satellite communications. The new solutions have been implemented in Linux environment and assessed through an extensive experimentation in an emulated testbed.

Packet retransmission at the end of the session does not apply any congestion control algorithm. We are currently developing a solution to render this ending session phase TCP-friendly: when the last data packet is sent and all the receivers that have correctly received the file leave the session, the sender has to initiate another session to recover the holes of the slow receivers; with an iteration of this process we aspect to observe a sequence of transport multicast sessions with a decreasing members number, the data delivery session ends when this number becomes 0.

References

1. Sally Floyd, Kevin R. Fall: Promoting the use of end-to-end congestion control in the Internet. IEEE/ACM Transaction on Networking, Vol. 7, No. 4, pp. 458--472, 1999.
2. R. Rejaie, M. Handley and D. Estrin. RAP: An End-to-End Rate-based Congestion Control Mechanism for Realtime Streams in the Internet. Proc. of IEEE Infocom'99. April 1999.
3. S. Jacobs and A. Eleftheriadis. Real-Time Dynamic Rate Shaping and Control for Internet Video Applications. Proc. of the Workshop on Multimedia Signal Processing 1997. June 1997.
4. J. Mahdavi and S. Floyd. TCP-friendly Unicast Rate-based Flow Control. *Technical Note*. Available at `http://ftp.ee.lbl.gov/floyd/papers.html`. June 1997.
5. T. Turletti, S. F. Prisis, and J.-C. Bolot. Experiments with a Layered Transmission Scheme over the Internet. Rapport de recherche 3296, INRIA. November 1997.
6. S. Cen, C. Pu and J. Walpole. Flow and Congestion Control for Internet Media Streaming Applications. *Proc. Multimedia Computing and Networking*. January 1998.
7. D. Sisalem and H. Schulzrinne. The Loss-Delay Based Adjustment Algorithm: A TCP-Friendly Adaptation Scheme. *Workshop on Network and Operating System Support for Digital Audio and Video*. July 1998.
8. I. F. Akyildiz, O. B. Akan, and G. Morabito, A Rate Control Scheme for Adaptive Real-Time Applications in IP Networks with Lossy Links and Long Round Trip Times. *To appear in IEEE/ACM Transactions on Networking*.
9. I. F. Akyildiz, G. Morabito, and S. Palazzo. TCP-Peach: A New Congestion Control Scheme for Satellite IP Networks. *IEEE/ACM Transactions on Networking*. June 2001.
10. A. Mankin, A. Romanov, S. Bradner, and V. Paxson. IETF Criteria for Evaluating Reliable Multicast Transport and Application Protocol. *RFC 2357*. 1998.
11. L. Rizzo. pgmcc: a TCP-friendly single-rate multicast congestion control scheme. *Proc. Of Sigcomm'2000*. August 2000.
12. L. Rizzo, G. Iannaccone, L. Vicisano, and M. Handley. PGMCC Single Rate Multicast Congestion Control: Protocol Specification. *Internet Draft*: draft-ietf-rmt-bb-pgmcc-00.txt. February 2001.
13. M. Hdley, S. Floyd, and B. Whetten. Strawman Specification for TCP Friendly (Reliable) Multicast Congestion Control. Technical Report – Reliable Multicast Research Group. December 1998.
14. 1I. Rhee, N. Balaguru, and G. Rouskas. MTCP: Scalable TCP-Like Congestion Control for Reliable Multicast. Proc. Of IEEE Infocom'99. March 1999.
15. G. Morabito and S. Palazzo. Modeling and Analysis of TCP-Like Multicast Congestion Control in Hybrid Terrestrial/Satellite IP Networks. *IEEE Journal on Selected Areas of Communications*. February 2004.
16. I. F. Akyildiz and J. Fang. TCP-Peachtree: A Multicast Transport Protocol for Satellite IP Networks. *IEEE Journal of Selected Areas in Communications (JSAC)*, Vol. 22, Issue 2, pp. 388-400, February 2004.
17. `http://snad.ncsl.nist.gov/itg/nistnet/`

An Enhanced Multicast Routing Protocol
for Mobile Hosts in IP Networks

Seung Jei Yang and Sung Han Park

Department of Computer Science and Engineering,
Hanyang University, Ansan, Kyunggi-Do, 425-791, Korea
{sjyang, shpark}@cse.hanyang.ac.kr

Abstract. Conventional multicast routing protocols in IP networks are not suitable for mobile environments since they assume static hosts when building a multicast delivery tree. In this paper, we propose a multicast routing protocol to provide an efficient multicasting service to mobile hosts in IP networks. The main purpose of the proposed protocol is to reduce both the multicast tree reconstruction number and multicast service disruption time. The proposed multicast routing protocol is a hybrid method using the advantages of the bi-directional tunneling and the remote subscription proposed by the IETF Mobile IP working group. The proposed protocol supports the maximum tunneling service satisfying with the maximum tolerable transfer delay time. The simulation results show that the proposed protocol has better performance in the multicast tree reconstruction number and the multicast service disruption time than the previous protocols do.

1   Introduction

The remarkable success of wireless communication networks and the explosive growth of the Internet have led to the Internet user's mobility and a large variety of multimedia services[1]. In particular the provision of various multicast service such as video-conferencing or VOD service to mobile hosts is an important issue of the next generation wireless networks[2,3]. The IETF has proposed two approaches to provide multicast over Mobile IP. There are known as remote subscription[4] and bi-directional tunneling[4]. The remote subscription always resubscribes to its desired multicast groups whenever a mobile host moves to a new foreign network. This protocol is simple since there is no special encapsulation needed and has the advantage of delivering the multicast data on the optimal paths from multicast source to receivers. However a frequent handoff brings about the overhead of reconstructing multicast tree since the number of multicast tree reconstruction is dependent on the frequency of handoff of mobile hosts. In bi-directional tunneling, mobile hosts send and receive all multicast data using the unicast Mobile IP tunneling service from their home agent(HA). This protocol does not reconstruct multicast tree since the multicast routing is not affected by host mobility and so reduces the cost of multicast tree reconstruction. However this protocol has the routing path for multicast delivery that can be far from optimal because the multicast data are forwarded via HA. In addition, this

P. Lorenz and P. Dini (Eds.): ICN 2005, LNCS 3421, pp. 1031–1038, 2005.

protocol has the tunnel convergence problem. Each of the respective HAs creates a separate tunnel to the foreign agent(FA) so that multicast data to their respective mobile hosts can be forwarded. If these mobile hosts are the same members of the multicast group, all of the tunnelings from different HAs to the FA transmit the same multicast data, resulting in packet duplication. To solve the tunnel convergence problem of bi-directional tunneling, the MoM[5] avoids the duplicate data being tunneled to the common FA using the Designated Multicast Service Provider(DMSP), which is in charge of forwarding multicast data to the FA. However, if the number of multicast group members is small and the handoff rates of mobile hosts is increased, this protocol requires frequent DMSP's handoff. This protocol has the problem of long routing path when a mobile host moves to a foreign network far away from its HA. The RBMoM[6] is a hybrid protocol of the remote subscription and the bi-directional tunneling. This protocol intends to trade-off between the tunneling path and the frequency of multicast tree reconstruction and reduces the cost of tunneling by employing the idea of service range and Multicast Home Agent (MHA), as shown in Fig. 1. The MHA can only serve mobile hosts that are roaming around foreign networks and are within its service range. If a mobile host is out of its service range, MHA handoff occurs. However, this protocol does not provide a certain criterion for determining an optimal service range. When the service range is reduced for optimal routing, the number of multicast tree reconstruction is increased since every MHA has the same service range. In addition, when a mobile host also moves to a foreign network which is out of its service range and does not join a multicast group, the problem of multicast packet loss happens during the time required to join the multicast group corresponding to the time of multicast tree reconstruction.

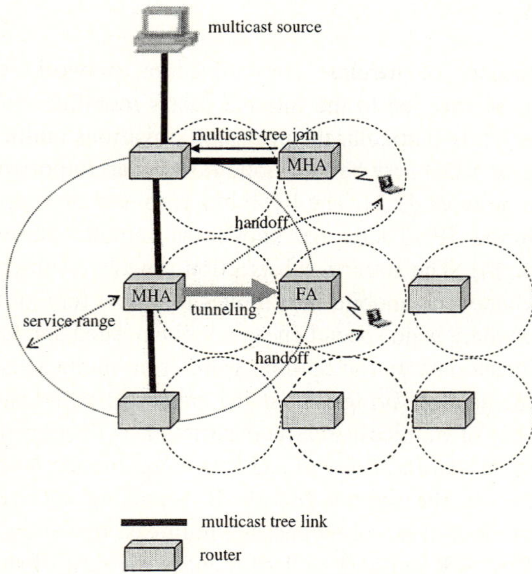

Fig. 1. MHA and service range of RBMoM protocol

To solve these problems, we develop a criterion to determine an optimal service range satisfying with the maximum tolerable transfer delay time and the Advance Join scheme based on the RSVP[7,8] mechanism to reduce the multicast service disruption time. The rest of the paper is organized as follows. The proposed multicast routing protocol is described in Section 2. The performance comparisons of proposed protocol with other protocols are presented in Section 3. Finally the conclusions are included in Section 4.

2 The Proposed Multicast Routing Protocol

Service range is a very important parameter for the performance of the RBMoM since if service range is a big value, the routing path becomes long like the bi-directional tunneling and for a small service range, the multicast tree reconstruction number is increased like the remote subscription because every MHA has the same service range. The RBMoM considers the value of service range simply according to the handoff rates of mobile hosts and the number of multicast group members. However, the above two measures cannot be a criterion to determine an optimal value of service range since they can change dynamically in real circumstances. In this paper, we define a special system parameter called dynamic service range as the maximum tunneling path length that satisfies the maximum tolerable transfer delay time. Each MHA has its own value of dynamic service range. The dynamic service range defined in this paper is determined to satisfy the QoS requirements within the maximum tolerable transfer delay time. Each MHA has the maximum tunneling path length in terms of its dynamic service range and thus the multicast tree reconstruction number is greatly reduced. Another problem of the RBMoM is the multicast service disruption time which occurs when the mobile host moves away out of the present service range and moves to another foreign network. In this case if the foreign network is not a member of the present multicast group, the multicast service disruption corresponding to the amount of time required to join the present multicast group happens. To solve this problem we develop the Advance Join scheme that is based on RSVP mechanism. We define a new terminology called boundary foreign agent(BFA) in this paper. The BFA is a FA which is located at the distance equal to the dynamic service range away from MHA. If a mobile host moves to the BFA, the neighboring FAs join the multicast group in advance to reduce the multicast service disruption time.

2.1 Dynamic Service Range

The proposed multicast routing protocol determines the tunneling path length at MHA according to the parameter of dynamic service range. So the method determining the value of dynamic service range is an important key in this paper. When the multicast tree is set up, each MHA determines the value of dynamic service range. An impact factor for determining the value of dynamic service range is the maximum tolerable transfer delay time. To guarantee the required transfer delay time, the total multicast data transmission delay time must not exceed the required maximum tolerable transfer delay time. Then the value of dynamic service range at each MHA must satisfy Eq. (1).

$$\text{dynamic service range} \leq \frac{T_{TD} - TD_{S\text{-}MHA} - TD_{TUNNEL}}{TD_{LINK}} \tag{1}$$

where, T_{TD} and $TD_{S\text{-}MHA}$ are the maximum tolerable transfer delay time and the multicast data transmission delay time from multicast source to MHA, respectively. TD_{TUNNEL} and TD_{LINK} are the delay time relevant to tunneling service at MHA and packet transmission delay time per link, respectively. We can see that if the value of $TD_{S\text{-}MHA}$ is larger than or equal to the value of T_{TD}, then the value of dynamic service range of MHA is zero. The value of dynamic service range of MHA closer to multicast source is larger than that of MHA far away from multicast source, resulting in having the larger tunneling service range. The proposed protocol supports the maximum tunneling service range satisfying with the maximum tolerable transfer delay time. The dynamic service range of proposed multicast routing protocol is represented in Fig. 2.

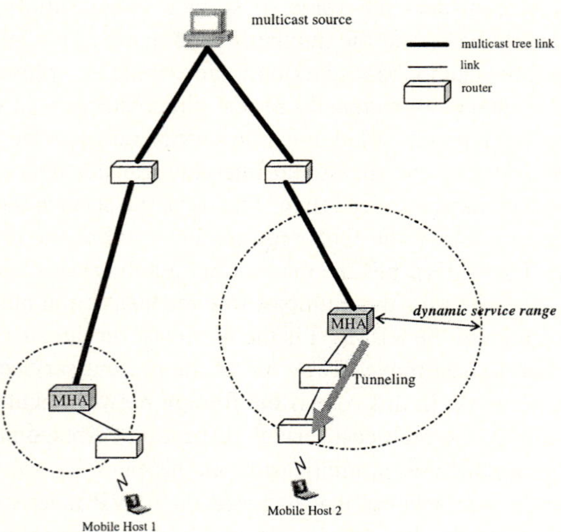

Fig. 2. Dynamic service range of the proposed protocol

2.2 Advance Join Scheme

When a mobile host moves to the foreign network which does not join the multicast group, the multicast service disruption time required to join the multicast group is occurred to the mobile host. If the network size is large and the number of handoffs of mobile hosts is increased, the multicast service disruption time becomes long and the loss rate of a multicast packet is increased[7]. Therefore, the multicast service disruption time becomes critical for the mobile hosts. To solve this problem, we employ the RSVP mechanism. If a mobile host moves to BFA, the neighboring FAs join the multicast group in advance to reduce the multicast service disruption time. For this purpose, in addition to signaling messages defined in the RSVP mechanism, we define new messages below.

- Join_multicast_group : This message is sent by a mobile host to its neighboring FAs to request them to join a multicast group. It contains the multicast address of the group to join.
- Pre_joion : The neighboring FAs which received a Join_multicast_group message send this message to join a multicast group.
- Release_join : This message releases a reserved multicast link.

We assume that a mobile host in a BFA can send the Join_multicast_group message to neighboring FAs. The operation for joining the neighboring FAs to multicast group is shown in Fig. 3. When a mobile host moves to a BFA, it sends a Join_multicast_group message to neighboring FAs to join a multicast group in advance. Neighboring FAs that receive the Join_multicast_group message send a Pre_join message to a multicast source. The neighboring FAs received a Path message from the multicast source and join members of multicast group. When a mobile host moves to one of neighboring FAs and the FA is new MHA of this mobile host, the mobile host can receive the multicast datagrams without multicast service disruption.

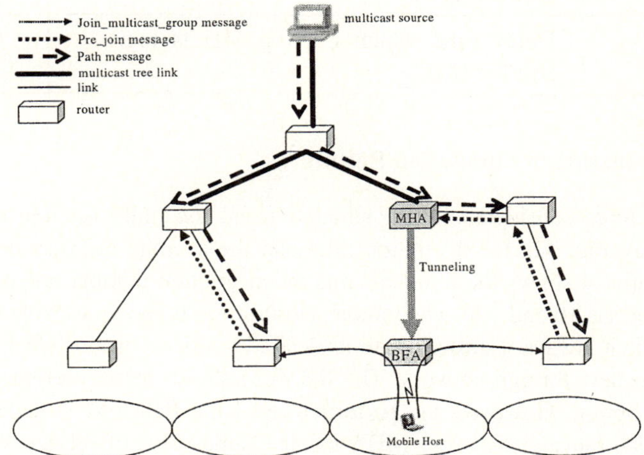

Fig. 3. The operation for joining the neighboring FAs to multicast group

3 Performance Evaluations

3.1 Simulation Model

As a simulation example, a network topology consisting of 100 local networks is used. The initial multicast tree is established for a randomly selected set of network nodes. The initial MHA of a mobile host is set to be its HA. We assume only one multicast group and one source in it. The mobility rate of mobile hosts is from 1 to 15. The value of dynamic service range at each MHA is calculated by Eq. (1). The value of service range of the RBMoM protocol is assumed to be 2. Table 1 summarizes the parameters used in our simulation model where TD_{LINK}, TD_{TUNNEL} and TD_{REG} are those referred to the previous research[9].

Table 1. Simulation Parameters

Parameters	Description	Value
N	Number of LANs	100 (10×10)
M	Number of multicast group	1
G	Number of multicast members	10…100
S	Sources per multicast group	1
H	Average number of handoff of multicast group member	1…15
T_{TD}	Maximum tolerable transfer delay time	77 msec
μ	Average service time	3 min
P_d	Direction probability	1/4
TD_{LINK}	Packet delivery time (packet propagation delay plus routing delay between one hop)	3.5 msec
TD_{TUNNEL}	Protocol processing time to tunnel multicast datagrams	7 msec
TD_{REG}	Delay time which moving MH is registered at MHA	3 msec

3.2 The Evaluation of Simulation Results

The main features considered in our simulation are the multicast tree reconstruction number, the average service disruption time and the average delivery path length per mobile host. Fig. 4 shows the multicast tree reconstruction number when the multicast group member has 10 and 100. The remote subscription is shown to have the worst performance since it always resubscribes to its desired multicast groups whenever a mobile host moves to new foreign network. The RBMoM shows better performance than the remote subscription. However, the performance of the RBMoM is dependent on the value of service range since every MHA has the same value of service range. In comparison, the proposed protocol shows further reduction in the multicast tree reconstruction number and a stable operation regardless of the handoff rate of the mobile hosts.

Fig. 5 shows the average service disruption time when the number of multicast group member is 10 and 100. The remote subscription possesses the worst performance since the service disruption time occurs whenever a mobile host moves to new foreign network. The RBMoM shows that a service disruption time occurs when a mobile host is out of the service range of its MHA and the average service disruption time increase as the handoff rate of mobile hosts increases. The proposed protocol shows further reduction in the multicast service disruption time than the other protocols.

Fig. 6 shows the average delivery path length per a mobile host when the average number of handoff of a mobile host is 5. The remote subscription has the optimal delivery path length since each delivery path is always shortest. The bi-directional tunneling is no limit for the tunneling path and has the worst delivery path length among other protocols. This figure shows that the RBMoM has less delivery path length than the bi-directional tunneling. The proposed protocol is not optimal but has the similar performance to the RBMoM.

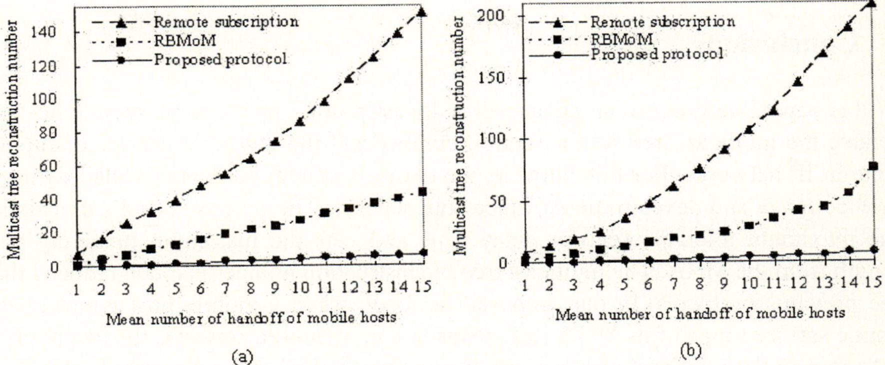

Fig. 4. Multicast tree reconstruction number. (a) multicast group member = 10. (b) multicast group member = 100

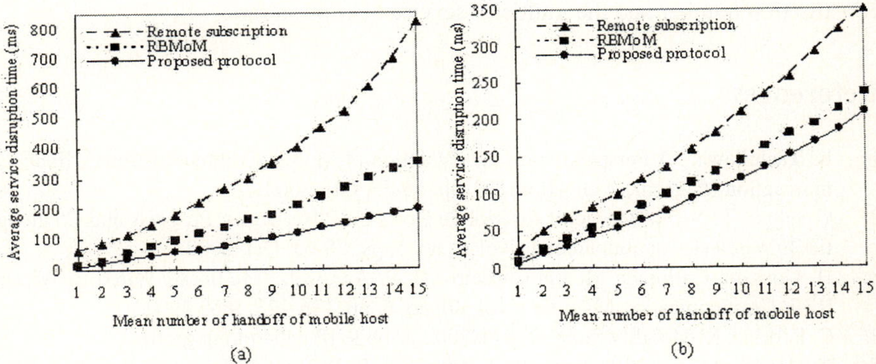

Fig. 5. Average service disruption time. (a) multicast group member = 10. (b) multicast group member = 100

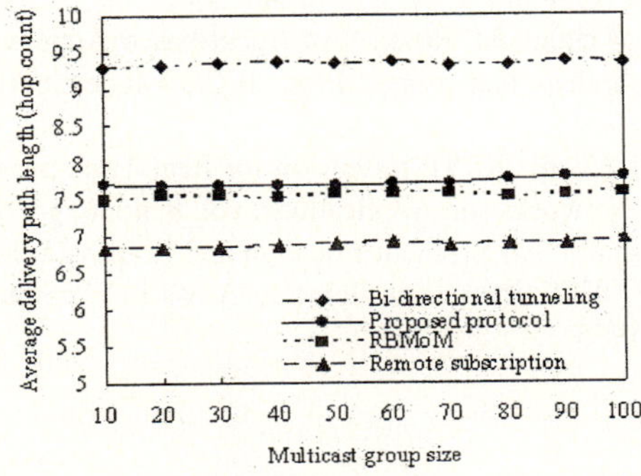

Fig. 6. Average delivery path length. Average number of handoff of a mobile host = 5

With regard to the simulation results of the multicast tree reconstruction number, the multicast service disruption time and the average delivery path length, our protocol seems to be suitable for providing efficient multicasting services to mobile hosts.

4   Conclusions

In this paper, we propose an enhanced multicast routing protocol for mobile hosts to reduce the multicast tree reconstruction number and the multicast service disruption time in IP networks. For this purpose, we define a system parameter called dynamic service range and develop the Advance Join scheme. The proposed protocol satisfies the maximum tolerable transfer delay time and gets the maximum tunneling path length from the MHA. The multicast tree reconstruction number is much reduced than the previous protocols. In our proposed protocol, when a mobile host is out of dynamic service range of its MHA and moves to a new foreign network, the neighboring FAs join to the multicast group in advance using the RSVP mechanism. In this way, the multicast service disruption time appeared during the handoff of a mobile host is much decreased. Simulation results demonstrate that the performances of the proposed protocol are improved over existing solutions in terms of the multicast tree reconstruction number and the multicast service disruption time.

References

1.   K. Tachikawa, "A Perspective on the Evolution of Mobile Communications", IEEE Communication Magazine, vol. 41, no. 10, pp. 66-73, Oct. 2003.
2.   C. Jelger, T. Noel, "Multicast for Mobile Hosts in IP Networks: Progress and Challenges", IEEE Wireless Communications, vol. 9, no. 5, pp. 58-65, Oct. 2002.
3.   H. Gossain, Carlos de Morais Cordeiro, D. P. Agrawal, "Multicast: Wired to Wireless", IEEE Communication Magazine, vol. 40, no. 6, pp. 116-123, June 2002.
4.   C. Perkins, IP Mobility Support, RFC2002, Mobile IP networking group.
5.   G. Harrison, C. L. Williamson, W. L. Mackrell, R. B. Bunt, "Mobile Multicast(MoM) Protocol: Multicast Support for Mobile Hosts", Proc. ACM/IEEE MOBICOM'97, pp. 151-160, Sept. 1997.
6.   C. R. Lin, K. M. Wang, "Scalable Multicast Protocol in IP-Based Mobile Networks", Wireless Networks, vol. 8, no. 1, pp. 27-36, Jan. 2002.
7.   A. Giovanardi, G. Mazzini, M. Rossi, "An Agent-based Approach for Multicast Applications in Mobile Wireless Networks", Proc. IEEE Globecom'00, pp. 1682-1686, Nov. 2000.
8.   N. F. Huang, W. E. Chen, "RSVP Extension for Real-Time Services in Hierarchical Mobile IPv6", Mobile Networks and Applications, vol. 8, no. 6, pp. 625-634, Dec. 2003.
9.   G. Cho, L. F. Marshall, "An Efficient Location and Routing Scheme for Mobile Computing Environments", IEEE Journal on Selected Areas in Communications, vol. 13, no. 5, pp. 868-879, June 1995.

Analysis of Handover Frequencies for Predictive, Reactive and Proxy Schemes and Their Implications on IPv6 and Multicast Mobility

Thomas C. Schmidt[1,2] and Matthias Wählisch[2]

[1] HAW Hamburg, FB Elektrotechnik und Informatik,
Berliner Tor 7, 20099 Hamburg
[2] FHTW Berlin, Hochschulrechenzentrum,
Treskowallee 8, 10318 Berlin
{schmidt, mw}@fhtw-berlin.de

Abstract. Handovers in mobile packet networks commonly produce packet loss, delay and jitter, thereby significantly degrading network performance. Mobile IPv6 handover performance is strongly topology dependent and results in inferior service quality in wide area scenarios. To approach seamless mobility in IPv6 networks predicitive, reactive and proxy schemes have been proposed for improvement. In this article we analyse and compare handover frequencies for the corresponding protocols, as they are an immediate measure on performance quality. Using analytical methods as well as stochastic simulations of walking users within a cell geometry, we calculate the expected number of handovers as functions of mobility and proxy ratios, as well as the mean correctness of predictions. In detail we treat the more delicate case of these rates in mobile multicast communication. It is obtained that hierarchical proxy environments – foremost in regions of high mobility – can significantly reduce the processing of inter–network changes, reliability of handover predictions is found on average at about 50 %.

1   Introduction

Mobility Support in IPv6 Networks [1] has become a proposed standard within these days. Outperforming IPv4, the emerging next generation Internet infrastructure will then be ready for implementation of an *elegant*, transparent solution for offering mobile services to its users.

At first users may be expected to cautiously take advantage of the new mobility capabilities, i.e. by using Home Addresses while away from home or roaming their desktop 'workspaces' between local subnets. Major scenarios in future IPv6 networks, though, move towards the convergence of IP and 3GPP devices, strengthening the vision of ubiquitous computing and real-time communication. The challenge of supporting voice and videoconferencing (VoIP/VCoIP) over Mobile IP remains, as current roaming procedures are too slow to evolve seamlessly, and multicast mobility waits for a convincing design beyond MIPv6 [2].

P. Lorenz and P. Dini (Eds.): ICN 2005, LNCS 3421, pp. 1039–1046, 2005.

Synchronous real-time applications s. a. VoIP and VCoIP place restrictive demands on the quality of IP services: Packet loss, delay and delay variation (jitter) in a constant bit rate scenario need careful simultaneous control. Serverless IPv6 voice or videoconferencing applications need to rely on mobility management for nomadic users and applications [3, 4], as well as multicast support on the Internet layer. Strong efforts have been taken to improve handover operations in a mobile Internet, both in the unicast and the multicast case. Hierarchical mobility management [5], [6] and fast handover operations [7], [8] both lead to accelerated and mainly topology independent schemes. In addition to specific performance issues and infrastructural aspects, these concepts cause a different eagerness to operate handovers.

The occurence of handovers is the major source for degradation in mobile network performance and places additional burdens onto the Internet infrastructure. Reducing their frequencies thus promises to ease roaming and to reduce infrastructural costs. In the present work we quantitatively evaluate handover activities with respect to user mobility and geometry conditions.

This paper is organised as follows. In section 2 we briefly introduce the current proposals for improved unicast and multicast mobility. Section 3 is dedicated to our results on handover frequency analysis, derived from analytical models as well as stochastic simulations. Conclusions and an outlook follow in section 4.

2   Improved Unicast and Multicast Mobility Management

2.1   Hierarchical Mobility and Fast Handovers

Two propositions for improving roaming procedures of Mobile IPv6 are essentially around: A concept for representing Home Agents in a distributed fashion by proxies has been developed within the Hierarchichal Mobile IPv6 (HMIPv6) [5]. While away from home, the MN registeres with a nearby Mobility Anchor Point (MAP) and passes all its traffic through it. The vision of HMIPv6 presents MAPs as part of the regular routing infrastructure. The MN in the concept of HMIPv6 is equipped with a Regional Care-of Address (RCoA) local to the MAP in addition to its link-local address (LCoA). When corresponding to hosts on other links, the RCoA is used as MN's source address, thereby hiding local movements within a MAP domain. HMIPv6 reduces the number of visible handover instances, but - once a MAP domain change occurs - binding update procedures need to be performed with the original HA and the CN.

The complementary approach provides handover delay hiding and is introduced in the Fast Handover for MIPv6 scheme (FMIPv6) [7]. FMIPv6 attempts to anticipate layer 3 handovers and to redirect traffic to the new location, the MN is about to move to. FMIPv6 aims at hiding the entire handover delay to communicating end nodes at the price of placing heavy burdens onto layer 2 intelligence. A severe functional risk arises from a conceptual uncertainty: As the exact moment of layer 2 handover generally cannot be foreseen, and even flickering may occur, a traffic anticipating redirect may lead to data damage largely exceeding a regular MIPv6 handover without any optimization. The significance of this uncertainty has been recently confirmed by empirical studies [9], where even the use of extensive statistical data under fixed geometry condition led to a prediction accuracy of only 72 %.

The two multicast mobility approaches introduced below are built on top of either one of these unicast agent schemes. Minor modifications to HMIPv6 resp. FMIPv6 signaling are requested and both proposals remain neutral with respect to multicast routing protocols in use.

2.2   M-HMIPv6 — Multicast Mobility in a HMIPv6 Environment

"Seamless Multicast Handovers in a Hierarchical Mobile IPv6 Environment (M-HMIPv6)" [6] extends the Hierarchical MIPv6 architecture to support mobile multicast receivers and sources. Mobility Anchor Points (MAPs) as in HMIPv6 act as proxy Home Agents, controlling group membership for multicast listeners and issuing traffic to the network in place of mobile senders.

All multicast traffic between the Mobile Node and its associated MAP is tunneled through the access network, unless MAP or MN decide to turn to a pure remote subscription mode. Handovers within a MAP domain remain invisible in this micro mobility approach. In case of an inter–MAP handover, the previous anchor point will be triggered by a reactive Binding Update and act as a proxy forwarder. A Home Address Destination Option, bare of Binding Cache verification at the Correspondent Node, has been added to streams from a mobile sender. Consequently transparent source addressing is provided to the socket layer. Bi-casting is used to minimize packet loss, while the MN roams from its previous MAP to a new affiliation. A multicast advertisement flag extends the HMIPv6 signaling.

In cases of rapid movement or crossings of multicast unaware domains, the mobile device remains with its previously associated MAP. Given the role of MAPs as Home Agent proxies, the M-HMIPv6 approach may me viewed as a smooth extension of bi-directional tunneling through the Home Agent supported in basic MIPv6.

2.3   M-FMIPv6 — Multicast Mobility in a FMIPv6 Environment

"Fast Multicast Protocol for Mobile IPv6 in the Fast Handover Environments" [8] adds support for mobile multicast receivers to Fast MIPv6. On predicting a handover to a next access router (NAR), the Mobile Node submits its multicast group addresses under subscription with its Fast Binding Update (FBU) to the previous access router (PAR). PAR and NAR thereafter exchange those groups within extended HI/HACK messages. In the ideal case NAR will be enabled to subscribe to all requested groups, even before the MN has disconnected from its previous network. To reduce packet loss during handovers, multicast streams are forwarded by PAR as unicast traffic in the FMIPv6 protocol.

Due to inevitable unreliability in handover predictions — the layer 2 may not (completely) provide prediction information and in general will be unable to foresee the exact moment of handoff — the fast handover protocols depend on fallback strategies. A reactive handover will be performed, if the Mobile Node was unable to submit its Fast Binding Update, regular MIPv6 handover will take place, if the Mobile Node did not succeed in Proxy Router inquiries. Hence the mobile multicast listener has to newly subscribe to its multicast sessions, either through a HA tunnel or at its local link. By means of this fallback procedure fast handover protocols must be recognised as discontinuous extensions of the MIPv6 basic operations.

3   Analysis of Handover Frequencies

3.1   Expected Number of Handovers

As a Mobile Node moves, handovers potentially impose disturbances and put an extra effort onto the routing infrastructure. Thus the expected frequency of network changes can be viewed as a distinctive measure of smoothness for a mobility scheme. The handoff frequency clearly depends on the Mobile Node's motion within cell geometry. Two measures on quantizing mobility have been established in the literature: The cell residence time and the call holding time. Both quantities fluctuate according to the overall scenery and the actual mobility event.

Let us make the common assumption that the cell residence time is exponentially distributed with parameter η and that the call holding time is exponentially distributed, as well, but with parameter α. Then the probability for the occurrence of a handover from MNs residence cell into some neighboring can be calculated analytically to

$$\mathcal{P}_{\mathcal{HO}} = \frac{1}{1+\rho}, \quad \text{where } \rho = \frac{\alpha}{\eta}.$$

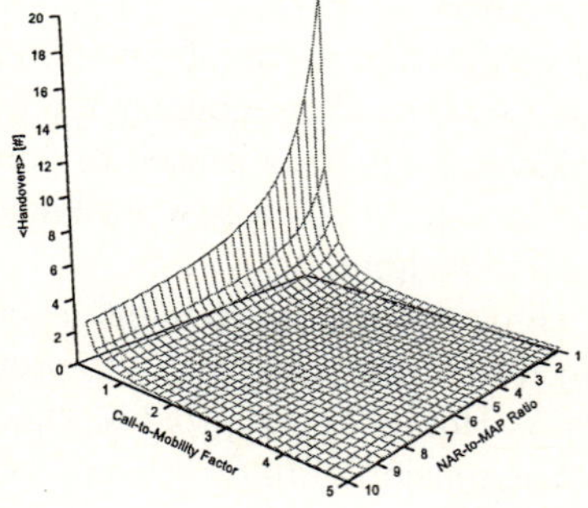

Fig. 1. Expected number of handovers as function of the call–to–mobility factor and the ratio of access routers toMAPs

ρ is known as the *call–to–mobility factor* [10]. It can be observed that the handoff probability increases as ρ decreases. Note that all probability distributions are homogeneous in space, e.g. $\mathcal{P}_{\mathcal{HO}}$ is independent of the current cell or the number of previously occurred handovers. Spatial scaling can be applied, accordingly.

When comparing Fast MIPv6 and Hierarchical MIPv6 approaches, another distinctive quantity becomes relevant: Whereas FMIPv6 operates handovers at Access Routers, HMIPv6 utilizes MAPs, which form a shared infrastructure. In general one MAP is meant to serve k Access Routers, whence the expected number of (inter–MAP) handovers reduces in a HMIPv6 domain.

Let us assume MAP regions to be of approximately circular geometry. Then the expected cell residence time changes to

$$\eta_{MAP}^{-1} = \sqrt{k} \cdot \eta_{AR}^{-1}$$

and the handoff probability transforms into

$$\mathcal{P}_{HO} = \frac{1}{1 + \sqrt{k} \cdot \rho},$$

where k is the ratio of ARs to MAPs.

Now we are ready to calculate the expected number of handovers as a function of the *call–to–mobility* factor ρ and the AR to MAP ratio k

$$\mathcal{E}_{HO} = \sum_{i=1}^{\infty} i \left(\frac{1}{1 + \sqrt{k} \cdot \rho} \right)^i = \frac{1}{k \cdot \rho^2} + \frac{1}{\sqrt{k} \cdot \rho}. \tag{1}$$

It can be observed that in highly mobile regimes ($\rho \ll 1$) $\mathcal{E}_{HO}$ is dominantly a function of the inverse of k, for low mobility ($\rho \gg 1$) of the inverse of $\sqrt{k}$ and attains a singularity at $\rho = 0$. Fig. 1 visualizes this expectation value for common values of ρ and k.

3.2   Stochastic Simulation of Walking Users

To evaluate the distribution of handover prediction types, we perform a stochastic simulation of motion within radio cells. The underlying model combines the following ingredients:

Cell geometry is chosen to be of common honeycomb type, i.e. abutting hexagons completely fill the 2 D plane. The ranges of radio transmission are modeled as (minimal) circles enclosing the combs. Thus, regions of prediction are the overlapping circle edges. A geometry of coherent Hot Spots is assumed here, where cells – without loss of generality – are identified with individually routed subnets.

As *Walking models* a Random Waypoint Model and a Random Direction Model [11] where used, where we consider varying sizes of mobility regions, i.e. squares ranging from cell dimension to infinity. Mobile devices move along (piecewise) straight lines within the preset boundaries, possibly coming to rest at their final destination, even if their current call is ongoing. Note that for finite regions the dynamic of both models is ergodic, whereby our simulated motion is equivalent to a walk continuous in time. Predictions are evaluated along the trajectories, distinguishing their correctness according to the outcome in succeeding steps.

Mean handover occurences for the different geometries are presented in fig. 2. Theoretical results of section 3.1 reproduce nicely for large mobility regions without visible model dependence. Values for small regions are attenuated, as tight geometry borders reduce phase space in 'pushing' trajectories into the inner cells.

Erroneous predictions as the outcome of traversing prediction regions without actually performing the foreseen handover or handovers initiated from outside of prediction areas are shown in fig. 3. Graph (a) displays in parallel the

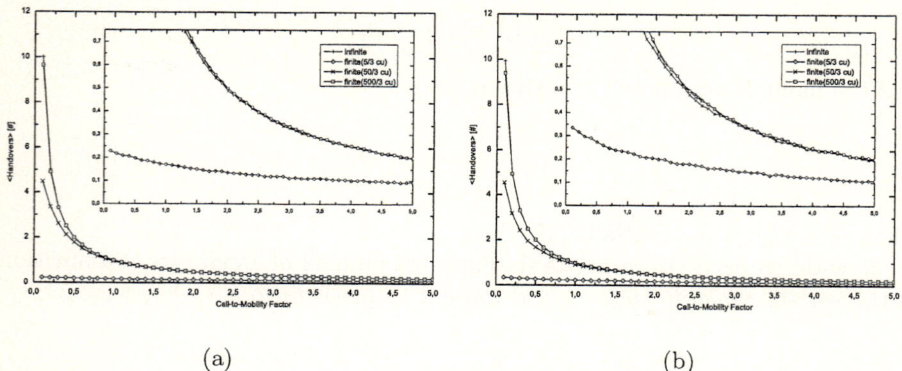

(a)   (b)

Fig. 2. Mean frequencies of handovers as a function of the call-to-mobility factor for different mobility regions (in units of celldiameters cu). (a) Random Waypoint and (b) Random Direction Model

percentage of failures from all predictions in the smallest and infinite mobility region for both walking models. From results variing between 10 and 70 % it can be observed that geometry dependence significantly exceeds the influence of the model. Taken as a rough estimate, the average number of erroneous predictions is about equal to the number of correct ones. Thus the reliability of predictive handover schemes does not exceed 50 %.

Graph (b) compares corresponding results for the random waypoint model with intermediate mobility region (50/3 celldiameter units) and different radii for the circular radio transmission areas, simulating a variation in access point density. The results for systems with optimal radio coverage, i.e. cell radii equal to transmission ranges, show minimal portions of failure predictions. In general a distinctive geometry dependence becomes visible, as well.

To proceed into a more detailed analysis of the sampled predictions, we differentiated the handover events of a simulated trajectory ensemble (model and parameters as in 3 (b)). Fig. 4 visualizes the mean occurrences of correct predictions, false predictions obtained along the path, as the mobile moves contrarily to forecasts derived from radio signal detection, and erroneous predicitons generated by terminating movement or call at a position, where a handover is expected. The latter yields 'on stop' can be identified as almost mobility independent, resulting in a saturated minimal error rate. Incorrect predictions 'on path' as a function of the call–to–mobility factor in contrast scale in correspondence with the correct indicators. It can be concluded from fig. 3 (a) that their exact values clearly depend on geometric and walking type conditions.

3.3   Implications

A common goal in designing HMIPv6 and FMIPv6 has been to approach seamless handovers in mobile IPv6 networks. As was shown in previous work [12], the predictive scheme of FMIPv6 can lead to faster roaming operations, but the

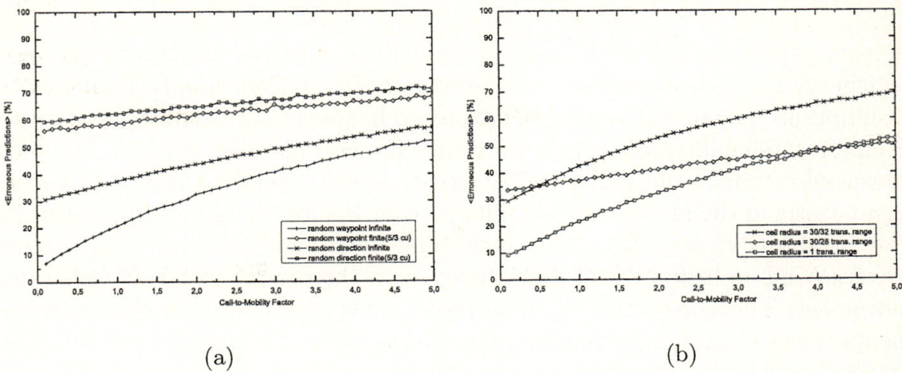

(a)   (b)

Fig. 3. Mean relative yield of erroneous predictions as a function of the call-to-mobility factor for (a) different models and geometries and (b) variing transmission ranges

reactive HMIPv6 procedures admits a comparable timing. In scenarios of significant mobility, i.e. $\rho \leq 1$, this advantage may be easily compensated by reducing the number of attained handovers within an HMIPv6 environment up to an order of magnitude. High prediction error rates, as observed from our simulations, place an additional burden onto the infrastructure, since any handover forecast will lead to a negotiation chain between access routers.

This burden notably increases in the case of multicast communication. A preparatory roaming procedure in M-FMIPv6 will initiate a remote multicast subscription, causing multicast routers to erect new branches for the corresponding distribution trees. In combining the results of section 3.1 and 3.2 we face scenarios, where the same (high) mobile movement leads to three handovers in a M–HMIPv6 domain, but about 40 handover processings under the regime of M–FMIPv6.

Another important aspect must be seen in robustness, i.e. the ability of the Mobile Node to cope with rapid

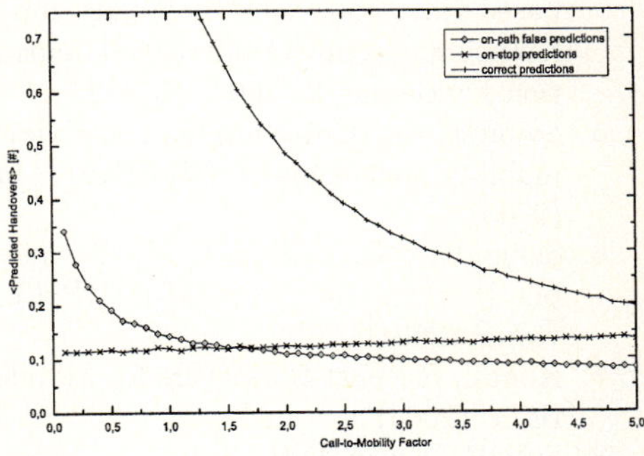

Fig. 4. Detailed view on handovers: Correct predictions, failures along the path and on stop

movement. In the case of a mobile multicast listener leaving its association before a handover completed, an M–HMIPv6 device will remain associated with its previously fully established MAP or Home Agent. On the price of a possible increase of delay, forwarding of multicast data is provided independent of handover frequency. On the contrary M–FMIPv6 forwarding will collapse, as soon as a MN admits a handover frequency incompatible with the signalling completion periods. An M–FMIPv6 device then has to fall back onto MIPv6 by establishing a bi-directional tunnel anew. Meanwhile established services are interrupted.

4   Conclusion and Outlook

In this paper we analyzed common predictive, reactive and proxy mobility schemes w.r.t. their eagerness for handovers. Starting from simple, fundamental assumptions a quantitative study of expected handover occurrences was derived. Prediction reliability was simulated using common mobility models. The 'nervousness' of handovers performed at access routers could be shown to reduce significantly in the presence of Mobility Anchor Points established within the hierarchical MIPv6 approach. This smoothing effect gains additional importance by observing an instability of fast handovers in the case of Mobile Node's rapid movement. The perspective of these results may give rise to further improvements on the smoothing of roaming procedures within the realm of seamlessness, attained at mobile and infrastructure nodes.

References

1. Johnson, D.B., Perkins, C., Arkko, J.: Mobility Support in IPv6. RFC 3775, IETF (2004)
2. Romdhani, I., Kellil, M., Lach, H.Y., Bouabdallah, A., Bettahar, H.: IP Mobile Multicast: Challenges and Solutions. IEEE Comm. Surv. & Tutor. **6** (2004) 18–41
3. Schmidt, T.C., Wählisch, M., Cycon, H.L., Palkow, M.: Global serverless videoconferencing over IP. Future Generation Computer Systems **19** (2003) 219–227
4. Cycon, H.L., Palkow, M., Schmidt, T.C., Wählisch, M., Marpe, D.: A fast waveletbased video codec and its application in an IP version 6-ready serverless videoconferencing system. International Journal of Wavelets, Multiresolution and Information Processing **2** (2004) 165–171
5. Soliman, H., Castelluccia, C., Malki, K., Bellier, L.: Hierarchical Mobile IPv6 mobility management (HMIPv6). Internet Draft – work in progress 02, IETF (2004)
6. Schmidt, T.C., Wählisch, M.: Seamless Multicast Handover in a Hierarchical Mobile IPv6 Environment (M-HMIPv6). Internet Draft – work in progress 02, individual (2004)
7. Koodli, R.: Fast Handovers for Mobile IPv6. Internet Draft – work in progress 02, IETF (2004)
8. Suh, K., Kwon, D.H., Suh, Y.J., Park, Y.: Fast Multicast Protocol for Mobile IPv6 in the fast handovers environments. Internet Draft – work in progress (expired) 00, IETF (2004)
9. Song, L., Kotz, D., Jain, R., He, X.: Evaluating location predictors with extensive Wi-Fi mobility data. In: Proceedings of the 23rd Joint Conference of the IEEE Comp. and Comm. Societies (INFOCOM). Volume 2. (2004) 1414–1424
10. Fang, Y., Chlamtac, I.: Analytical Generalized Results for Handoff Probability in Wireless Networks. IEEE Transactions on Communications **50** (2002) 396–399
11. Bettstetter, C.: Mobility Modeling in Wireless Networks: Categorization, Smooth Movement and Border Effects. ACM Mobile Comp. and Comm. Rev. **5** (2001) 55–67
12. Schmidt, T.C., Wählisch, M.: Topologically Robust Handover Performance for Mobile Multicast Flows. In Lorenz, P., ed.: Proceedings of the International Conference on Netwoking (ICN' 2004). Volume 1., Colmar, University of Haute Alsace (2004) 350–355

Design Architectures for 3G and IEEE 802.11 WLAN Integration

F. Siddiqui[1], S. Zeadally[1], and E. Yaprak[2]

[1] High Speed Networking Laboratory, Department of Computer Science,
Wayne State University, Detroit, MI 48202, USA
{Farhan, Zeadally}@cs.wayne.edu
[2] Division of Engineering Technology,
Wayne State University, Detroit, MI 48202, USA
Yaparak@eng.wayne.edu

Abstract. Wireless LAN access networks show a strong potential in providing a broadband complement to Third Generation cellular systems. 3G networks provide a wider service area, and ubiquitous connectivity with low-speed data rates. WLAN networks offer higher data rate but cover smaller areas. Integrating 3G and WLAN networks can offer subscribers high-speed wireless data services as well as ubiquitous connectivity. The key issue involved in achieving these objectives is the development of integration architectures of WLAN and 3G technologies. The choice of the integration point depends on a number of factors including handoff latency, mobility support, cost-performance benefit, security, authentication, accounting and billing mechanisms. We review 3G-WLAN integration architectures and investigate two such architectures in the case when the UMTS network is connected to a WLAN network at different integration points, namely the SGSN and the GGSN. The evaluation of these integration architectures were conducted through experimental simulation tests using OPNET.

1 Introduction

Mobile communications and wireless networks are developing at a rapid pace. Advanced techniques are emerging in both these disciplines. There exists a strong need for integrating WLANs with 3G networks to develop hybrid mobile data networks capable of ubiquitous data services and very high data rates in strategic locations called "hotspots". 3G wireless systems such as Universal Mobile Telecommunication Systems (UMTS) can provide mobility over a large coverage area, but with relatively low speeds of about 144 Kbits/sec. On the other hand, WLANs provide high speed data services (up to 11 Mbits/sec with 802.11b) over a geographically smaller area. The rest of this paper is organized as follows. Section 2 provides a brief background on 3G and WLAN networks. Section 3 describes the related research and contributions of this work. Section 4 presents a comparison of the various internetworking architectures. In section 5 we compare two integration architectures connecting UMTS

P. Lorenz and P. Dini (Eds.): ICN 2005, LNCS 3421, pp. 1047–1054, 2005.

and 802.11 networks. Section 6 presents our simulation results. Finally in section 7 we present some concluding remarks and future works.

2 Background

802.11b [3] WLAN has been widely deployed in offices, homes and public hotspots such as airports and hotels given its low cost, reasonable bandwidth (11Mbits/s), and ease of deployment. However, a serious disadvantage of 802.11 is the small coverage area (up to 100 meters) [5]. Other 802.11 standards include 802.11a and 802.11g which allow bit rates of up to 54 Mbits/sec.

3G devices can transfer data at up to 384 Kbps. A 3G network (figure 1) consists of three interacting domains- a Core Network (CN), Radio Access Network (RAN) and the User Equipment (UE). 3G operation utilizes two standard suites: UMTS and Code Division Multiple Access (CDMA2000). The main function of the 3G core network is to provide switching, routing for user traffic. The core network is divided into Circuit-switched (CS) and Packet-switched (PS) domains. Circuit switched elements include Mobile Services Switching Center (MSC), Visitor Location Register (VLR), and gateway MSC. These circuit switched entities are common to both the UMTS as well as the CDMA2000 standards. The differences in the CN with respect to the two standards lie in the PS domain.

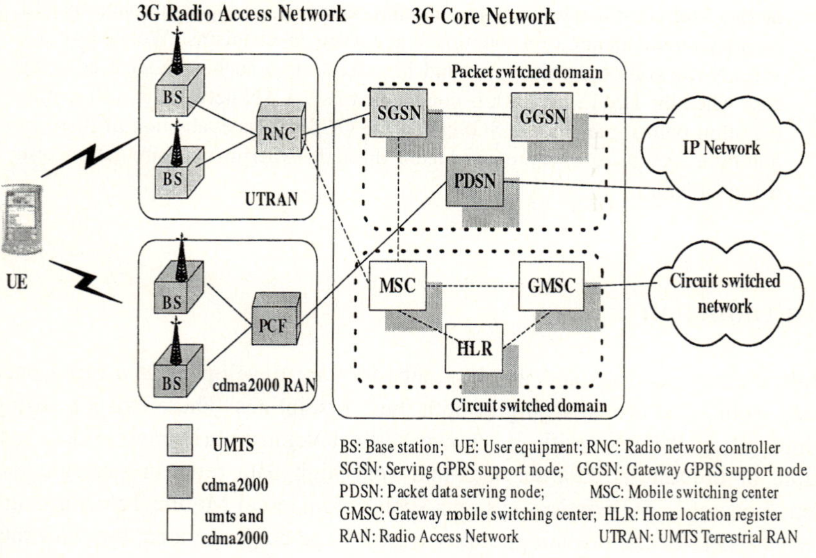

Fig. 1. Components of a 3G Network

GGSN is the gateway to external data networks and provides authentication and IP-address allocation. SGSN provides session management. It also supports inter-system handoff between mobile networks. PDSN incorporates numerous functions within one node such as routing packets to the IP network, assignment of dynamic IP-addresses and maintaining point-to-point protocol (PPP) sessions. The radio access network provides the air interface access method for the UE. A CDMA2000 RAN consists of a base station and 2 logical components- the Packet Control Function (PCF) and the Radio Resources Control (RRC). The primary function of the PCF is to establish, maintain and terminate connections to the PDSN.

3   Related Work and Contributions

A lot of recent works have focused on the design and evaluation of architectures to integrate 3G and WLAN networks. Buddhikot et al. [6] described the implementation of a loosely coupled integrated network that provides roaming between 3G and WLAN networks. Tsao et al. [7] presented another method to support roaming between 3G and WLANs by introducing a new node called the virtual GPRS support node in between the WLAN and the UMTS networks. Tsao et al. [8] evaluated three different internetworking strategies: the mobile IP approach, the gateway approach and the emulator approach with respect to their handoff latencies. Bing et al. [9] discussed mobile IP based vertical handoff management and its performance with respect to signaling cost and handoff latency. All of the above works have focused on evaluating integration architectures in terms of the handoff latency experienced by the users when trying to move across 3G and WLAN networks while accessing a public network such as the Internet. In contrast to the above related efforts, our work mainly focuses on the end-to-end delay experienced when users located in two different networks, namely the IEEE 802.11b and the UMTS communicate with each other directly. We describe two basic integration architectures that are used to connect these networks. Through simulations, the end-to-end packet latencies experienced by users when data is exchanged between these networks through the SGSN and the GGSN nodes are recorded and verified using various types of applications. The results demonstrate the feasibility of these integration architectures.

4   WLAN and 3G Cellular Data Network Integration Architectures – A Review

Table 1 reviews the various 3G-WLAN internetworking strategies and their features. The Mobile IP [8] internetworking architecture allows easy deployment but suffers from long handoff latency and might not be able to support real-time services and applications. The gateway approach [7] permits independent operation of the two networks and provides seamless roaming facility between them. The emulator approach [8] is difficult to deploy since it requires combined ownership of the two networks but does yield low handoff latency. Tight coupling [6] deploys WLAN as an alternative radio access network and offers faster handoffs and high security but requires combined ownership. Loose coupling [6] has low investment costs and permits independent deployment. However it suffers from high handoff delays. The choice of the integration architecture is important since multiple integration points exist with different cost-performance benefits for different scenarios.

Table 1. Comparison of various 3G-WLAN Internetworking Strategies

Internetworking Approach	Deployment	Network ownership	Handoff delay	Mobility scheme
Mobile-IP	Easy	Separate	High	Mobile IP
Gateway	Moderate	Separate	Low	Roaming agreement
Emulator	Very difficult	Combined	Low	UMTS and GPRS mobility
Tight	Moderate	Combined	Low	GPRS mobility
Loose	Difficult	Separate	High	Mobile-IP

5 Simulated Architectures

We evaluated via simulations using OPNET two internetworking architectures to interoperate the 3G (UMTS) and WLAN networks by connecting them at two strategic points- the SGSN node and the GGSN node as shown in figure 2.

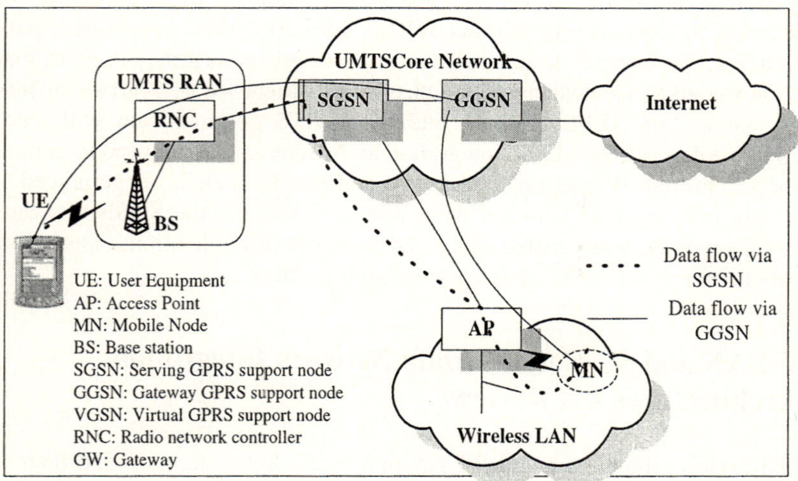

Fig. 2. 3G-WLAN Integration a) SGSN b) GGSN

5.1 UMTS-WLAN Integration at the SGSN Node

When the UMTS and WLAN networks are connected through the SGSN node, the WLAN network does not appear to the UMTS core network as an external packet data network. Instead, it simply appears as another radio access network. The WLAN AP in this case needs to have the capability of processing UMTS messages. Thus, whenever a Mobile Node (MN) in the WLAN network wants to exchange data with the UMTS UE, it first needs to undergo the GMM attach procedure to notify the SGSN of the location on the communicating node and also to establish a packet-switched signaling connection. The WLAN AP is responsible for sending these request messages to the SGSN on behalf of the WLAN MN. The GMM attach procedure is a three-way handshake between the MN, RNC and the SGSN. Upon completion of this procedure, the WLAN MN is authenticated into the UMTS network.

5.2 UMTS-WLAN Integration at the GGSN Node

In this type of integration, whenever a MN in a WLAN network wants to communicate with a UE in the UMTS network, it does so through the GGSN node. The UE in the UMTS network first activates the Packet Data Protocol (PDP) context that it wants to use. This operation makes the UE known to its GGSN and to the external data networks, in this case, the WLAN network. User data is transferred transparently between the UE and the WLAN network with a method known as encapsulation and tunneling. The protocol that takes care of this is the GPRS Tunneling Protocol (GTP). For this kind of internetworking configuration, the WLAN AP is a simple 802.11b access point and does not need to process UMTS messages.

5.3 Simulation Testbed

A network simulation model was constructed using OPNET 10.0.A [10]. OPNET is a discrete event simulator with a sophisticated software package capable of supporting simulation and performance evaluation of communication networks and distributed systems.

Table 2. Descriptions of various applications tested

Application	QoS Class	Measurement (seconds)	Size	Proocol
FTP	Background	File download time	100-1000 Ki-lobytes	TCP
FTP	Background	File upload time	100-1000 Ki-lobytes	TCP
GSM encoded voice	Conversa-tional	End-to-end delay	33 Bytes	UDP
GSM encoded voice	Conversa-tional	Jitter	33 Bytes	UDP
HTTP Web browsing	Interactive	Page response time	3000 Bytes	TCP

The simulation environment we used had a UMTS network connected to a WLAN network. The UMTS network was composed of the RAN and a packet-based CN with SGSN and GGSN nodes The WLAN network is composed of 802.11b wireless MNs configured in Infrastructure Basic Service Set mode. In the GGSN integration case, a simple WLAN access point was used, while in the SGSN integration case, a different access point with additional capability of processing UMTS messages was employed.

The goal of the simulations was to compare the delays involved when user data is exchanged between the UMTS and WLAN networks connected via two methods, namely GGSN and SGSN. Different types of traffic was generated using four different applications including Voice over IP (VoIP) t, FTP, and HTTP (web browsing) as shown in table 2. These applications correspond to the various UMTS QoS classes-Conversational class for real time flows such as VoIP, interactive and background classes for FTP and HTTP respectively. Packet delay, jitter, upload, and download response times were measured. Other parameters associated with each application are summarized in table 2.

6 Simulation Results and Discussion

Simulations performed for both UDP and TCP flows are presented. For the UDP flow (VoIP traffic), end-to-end packet delays and jitter were measured. For TCP flows (FTP, HTTP) the upload/download response times were measured.

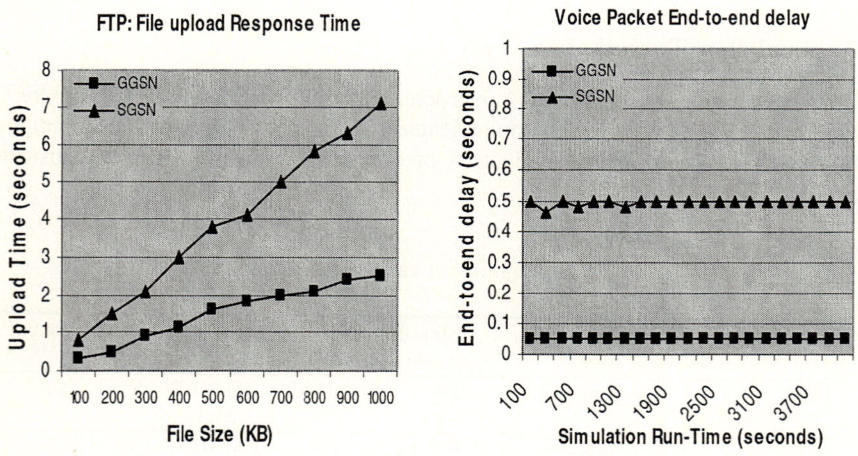

Fig. 3. Application Response Times: a) FTP Upload b) VoIP delay

Figure 3a shows the simulation run-times corresponding to the average file upload times experienced when transferring files of various sizes between the UE and the WLAN MN under two different integration scenarios. In figures 3b and 4a the average delay and jitter for voice is presented. It is observed that both, delay and jitter values are much lower in the GGSN case. Similarly, figure 4b shows the response time to access a web page of size 3000 bytes. As figure 4b illustrates, the page response time is initially high and then decreases as the simulation progresses. We speculate that this reduction in the page response time may be because the web server's cache is initially empty and the first few page requests will cause the page to be fetched from the disk resulting in a high response time. As the more requests are generated with time, the cache is being filled and there is an increasing probability that one or more requests can be satisfied by the cache thereby reducing the overall page response time.

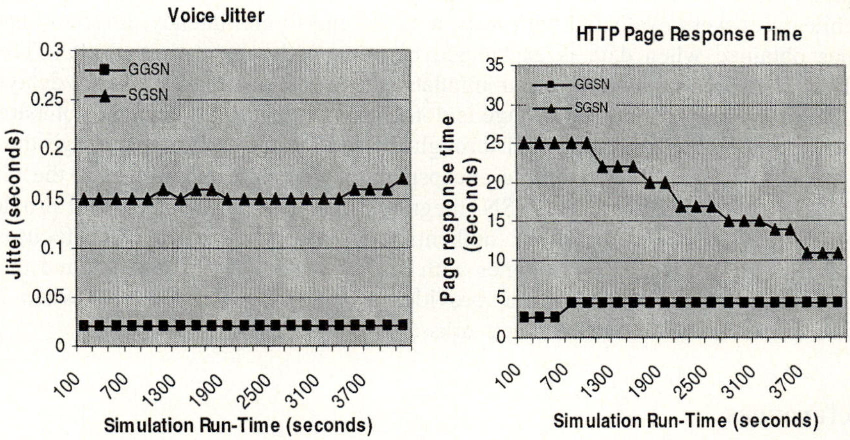

Fig. 4. Application Response Times: a) VoIP Jitter b) HTTP Page Response time

The simulation results reveal that the application response time (delay) is consistently higher in the case where the UMTS and WLAN networks are connected through the SGSN node, as compared to the case where the two networks are connected at the GGSN. This higher response time can be attributed to the additional processing time required at the WLAN access point in the first case. When the two networks are connected at the SGSN node. The WLAN access point performs the functions of a RNC on as well as a WLAN AP. Therefore, it has to perform the additional initialization steps to authenticate the WLAN MN to the UMTS network (GMM Attach procedure and PDP context activations). When integration is done at the GGSN node, the WLAN AP is a simple 802.11b access point and does not require any special capabilities to process UMTS messages. Data packets are transferred between the UE and the WLAN network using encapsulation by the GPRS tunneling protocol. This reduces the packet latency as there is no additional delay due UMTS initialization procedures or packet conversions.

The advantages, however, of using SGSN integration scheme include the reuse of UMTS authentication, authorization, accounting (AAA) mechanisms, usage of common subscriber databases and billing systems, increased security features (since the UMTS security mechanisms are reused), as well as possibility of having continuous sessions as users move across the two networks, since the handoff in this case is very similar to an intra-UMTS handoff as the WLAN AP appears as another RNC to the SGSN node. In the case of GGSN integration, since the WLAN is considered to be an external network, different billing and security mechanisms are needed. Service disruption is also possible during a handoff from one network to another.

7 Conclusions

In this paper we discussed the architecture and performance of a 3G-WLAN integrated system connected at two different points namely the SGSN and GGSN. The architectures were evaluated with respect to the end-to-end latency, jitter and upload times obtained when data is exchanged between nodes located in the UMTS and WLAN networks respectively. Our simulation results show that the overall delays are much lower when the data exchange is done through the GGSN node as compared to when the networks are connected through the SGSN node. Traffic passage through the GGSN is faster due to simple encapsulation procedure employed by the GPRS tunneling protocol. However, SGSN integration has its own advantages of providing strong security, common billing, authentication, etc. Our future work focuses on evaluating these integration schemes with respect to the handoff latency and the development of an architecture that provides seamless session mobility when MNs move across the 3G and WLAN networks.

References

1. Jun-Zhao, S., Sauvola, J., Howie, D., Features in Future: 4G Visions from a Technical Perspective, Vol. 6, IEEE GLOBECOM (November 2001) 3533 – 3537
2. Miah, A., Tan, K., An Overview of 3G Mobile Network Infrastructure, IEEE Student Conference on Research and Development (July 2002) 228-232
3. IEEE 802.11b, Part 11: Wireless LAN Medium Access Control (MAC) and Physical Layer (PHY) Specifications- Higher-Speed Physical Layer Extension in the 2.4 GHz Band (1999)
4. Varshney, U., The Status and Future of 802.11- Based WLANs, Vol. 36, Issue 6, Computer Communications (June 2003) 102- 105
5. Luo, H., Jiang, Z., Kim, B., Shankar, N., Henry, P., Internet Roaming: A WLAN/3G Integration System for Enterprises, AT&T Labs- Research, www.nd.edu/~hluo/publications/SPIE02.pdf
6. Buddhikot, M., Chandranmenon, G., Han, S., Lee, T., Miller, S., Design and Implementation of a WLAN/CDMA2000 Internetworking Architecture, IEEE Communications Magazine (November 2003) 90-100
7. Tsao, S., Lin, C., VSGN: A Gateway Approach to Interconnect UMTS/WLAN Networks, Vol. 1, The 13th IEEE International Symposium on Personal, Indoor and Mobile Radio Communications (2002).275-279
8. Tsao, S., Lin, C., Design and Evaluation of UMTS-WLAN Internetworking Strategies, Vol. 2, Proceedings of IEEE Vehicular Technology Conference (2002) 777-781
9. Hongyang, B., Chen, H., Jiang, L., Performance analysis of vertical handoff in a UMTS-WLAN integrated network, Vol. 1, Proceedings of 14th IEEE conference on Personal, Indoor and Mobile Radio Communications,(September 2003) 187-191
10. http://www.opnet.com

Eliminating the Performance
Anomaly of 802.11b

See-hwan Yoo, Jin-Hee Choi, Jae-Hyun Hwang, and Chuck Yoo

Department of Computer Science and Engineering, Korea University
{shyoo, jhchoi, jhhwang, hxy} @os.korea.ac.kr

Abstract. In this paper, we propose a mechanism to eliminate the performance anomaly of IEEE 802.11b. Performance anomaly happens when nodes that have different transmission rates are in the same wireless cell. All the nodes in the cell might experience the same throughput even though their transmission rates are different because DCF of WLAN provides equal probability of channel access, but it does not guarantee the equal utilization of the wireless channel among the nodes. To reduce such a performance anomaly, we adjust the frame size proportionally depending on the bit rate. Additionally, our scheme eliminates the performance anomaly in multi-hop case. Simulation study shows that our scheme achieves an improvement in the aggregate throughput and the fairness.

1 Introduction

The performance of IEEE 802.11b Medium Access Control (MAC) is a challenging issue. Especially, on the fairness of wireless channel utilization, many MAC layer's fair scheduling algorithms have been proposed [1, 2, 3, 4]. For the long term fairness, IEEE 802.11b defines Distributed Coordination Function (DCF), which gives every node a same probability to access the wireless channel.

In this paper, we address the problems in the Automatic Rate Fallback (ARF) of IEEE 802.11[5]. The basic CSMA/CA is at the root of the performance anomaly[1] of 802.11b MAC. If there are nodes which works at different bit rate in the same cell, the 802.11 cell shows the performance anomaly. The throughput of all hosts that have higher bit rate are degraded, and all the hosts in the wireless cell experience the same throughput regardless of the transmission rate. The main reason is that CSMA/CA mechanism does not provide the same probability to utilize the channel while it guarantees that all the nodes in the same wireless cell have the same probability to access the channel. In terms of the channel utilization, this is quiet unfair because the higher bit rate node defers transmission longer than that of the lower bit rate node.

To eliminate such a performance anomaly, we adopt the Maximum Transfer Unit (MTU) adaptation scheme. By adjusting the MTU size as to the transmission rate, all the nodes can fairly utilize the wireless channel. We show that our scheme achieves higher throughput than that of normal case. Moreover, it avoids the performance anomaly in multi-hop situation.

P. Lorenz and P. Dini (Eds.): ICN 2005, LNCS 3421, pp. 1055–1062, 2005.

This paper is organized as follows. We introduce the motivation of this study in Section 1. In Section 2, the ARF of WLAN and its performance anomaly are introduced. Section 3 presents our analysis which probes that our scheme achieves higher throughput than existing method. In Section 4, we differentiate our work from previous work. Simulation results by the NS-2 simulator are shown in Section 5. Finally, concluding remarks are given in Section 6.

2   Related Work

2.1   ARF of IEEE 802.11b and Its Variants

It is well known that IEEE 802.11 provides multi-rate capability at the physical layer. ARF defines several transmission rate of 802.11 for temporal degradation of wireless channel. When wireless channel is bad, the sender changes the sending rate to lower level.

ARF takes the advantages of positive ACK in explicit link layer. When a sender misses two consecutive ACKs, it drops the sending rate by changing the modulation or channel coding method. In contrast, when timer expires or consecutive 10 ACKs are received successfully, transmission rate is upgraded to the next higher data rate.

While the ARF mechanism provides good estimation of link condition between a fixed pair of nodes, it overlooks the fact that actually not the sender but the receiver need estimation of the channel condition. A direct disadvantage of ARF scheme can be clearly seen when there are multiple nodes communicating with each other in a wireless network. If a node moves to a location with a bad radio characteristics (communication blockage), other nodes communicating with this particular node may experience transmission failures and consequently the transmission rate would be dropped.

Holland [2] proposed the receiver based auto rate scheme (RBAR). In their scheme, the receiver estimates the wireless channel condition and gives the sender feedback information. The sender takes advantage of feedback information and selects the sending rate.

Sandeghi [6] proposed the opportunistic media access scheme (OAR) which extends RBAR. The main idea of OAR is to exploit good channel conditions to transmit as many as possible while retaining the long term fairness provided by 802.11b. OAR achieves fairness of channel utilization by sending a burst of packets for a single RTS-CTS handshake. The number of packets transmitted by OAR in any transmission burst should be limited so as to provide a fair channel access to all nodes. The fair temporal share is determined as the maximum time the channel can be occupied if OAR transmitted a single packet at the base rate. The base rate of a channel is the lowest possible rate with which data can be transmitted. For example, the base rate of 802.11b channel is 2 Mbps. Thus the number of packets sent in every burst is limited to at most 5 packets when the selected transmission rate is 11 Mbps. This guarantees that OAR inherits the same temporal fairness properties of the protocols based on original 802.11.

2.2   Performance Anomaly of 802.11b

Heusse [7] showed the performance anomaly of 802.11b. They analyzed the anomaly theoretically by deriving simple expressions for the useful throughput, validated them by means of simulation, and compared with several performance measurements. In their results, the throughput experienced at each node is same although the data rate is different. The expression for a throughput is as follows:

$$X_s = X_f = \frac{S_d}{(N-1) \cdot T_f + T_s + P_c(N) \times t_{jam} \times N}, \tag{1}$$

Where X_f is throughput at the MAC layer of each of the N-1 fast hosts, X_s is throughput at the MAC layer of slow host. N is number of nodes, and T_f and T_s is transmission time for fast nodes and slow node for a packet, respectively. $P_c(N)$ is probability of collision, S_d is frame size and t_{jam} is the delayed time experienced by collision.

In the above expression, the throughput is not related with the sending rate of a node because all the nodes have the same transmission time and the same frame sizes.

3   Eliminating Performance Anomaly of 802.11b

If we adjust packet size depending on the bit rate of the node, the time to utilize a channel will be fair. MTU size for a node can be expressed as follows:

$$\frac{SMTU}{r} = \frac{LMTU}{R} \tag{2}$$

$$\frac{LMTU}{SMTU} = \frac{R}{r} = K, \ (K > 1) \tag{3}$$

Where LMTU is MTU size of a fast bit rate node, R is the transmission rate of the fast bit rate node at MAC layer, SMTU is MTU size of a slow node, r is the transmission rate of the slow node. The ratio between the high bit rate and the low bit rate is k.

The expression (1) could be modified as follows.

$$X_f = \frac{LMTU}{(N-1) \cdot T_f + T_s + P_c(N) \times t_{jam} \times N}$$

$$= k \cdot \frac{SMTU}{(N-1) \cdot T_f + T_s + P_c(N) \times t_{jam} \times N}, \tag{4}$$

$$X_s = \frac{SMTU}{(N-1) \cdot T_f + T_s + P_c(N) \times t_{jam} \times N}. \tag{5}$$

Therefore, we can say that X_f, X_s is proportional to the bit rate of the node. Furthermore, we show that aggregated throughput in a cell increases. There are N nodes in the same wireless cell, N-1 of them have high bit rate, R, and there is only one node has low bit rate, r. To compare the aggregated throughput of a system, we define two different systems. At first, all the nodes are using the same LMTU frame size (system A). Secondly, lower bit rate node uses small MTU (SMTU), which is proportional to the bit rate (system B). We compare the aggregate throughput of two systems. We have the same assumptions with [7].

$$T_a = \sum_{i=1}^{N-1} X_f(i) + X_s(N) = N \cdot \frac{LMTU}{Time_A}, \tag{6}$$

$$T_b = \sum_{i=1}^{N-1} X_f(i) + X_s(N) = \frac{(N-1) \cdot LMTU + SMTU}{Time_B}, \tag{7}$$

$$T_b \geq T_a, \tag{8}$$

Where $X_f(i)$ is the throughput experienced at high transmission rate host i, $X_s(N)$ is the throughput experienced at low transmission rate host N. $Time_A$, $Time_B$ is the expectation of the time consumed for transmitting a packet. T_a and T_b mean the expectation of aggregated throughput of system A and B respectively.

We have to show that T_b is greater than or equal to T_a. We introduce more symbol conventions with [7].

$$t_{jam} = \frac{2}{N}T_s + (1 - \frac{2}{N})T_f$$

$$t_{jam}' = \frac{2}{N}T_s' + (1 - \frac{2}{N})T_f$$

$$Time_A = (N-1) \cdot T_f + T_s + P_c(N) \cdot N \cdot t_{jam}$$

$$Time_B = (N-1) \cdot T_f + T_s' + P_c(N) \cdot N \cdot t_{jam}'.$$

Applying the above equations to the equation (8), we get following inequality.

$$T_s - T_s' \geq \frac{(N-2) \cdot P_c(N) \cdot (LMTU - SMTU) - N \cdot SMTU}{(1 + P_c(N)) \cdot ((N-1) \cdot LMTU + SMTU)} \cdot T_f. \tag{9}$$

Using equation (2) and (3), we get absolute inequality (10) as follows:

$$(K-1) \cdot \{(K-1)(n-1)+n\} + (K-1) \cdot P_c(n) \cdot \{(K-1)(n-1)+2\} + n \geq 0, \tag{10}$$

Where K is greater than 1, N is natural number greater than 1 and Pc(N) always resides between 0 and 1. So, it proves the inequality expression (8).

4   Discussion

4.1   Comparison with Other Fairness Methods

Many studies proposed the fair scheduling mechanisms in MAC layer. They usually modify the mechanism that controls the contention window. MACAW

[1], Estimation based backoff[3], and Distributed Fair Scheduling[4] consider the fairness problem in 802.11. However, they do not address the problems that occur with multi-rate PHY.

TXOP which is adopted in 802.11e is a good approach to be compared with. TXOP regulates Network Allocation Vector (NAV) to utilize the network fairly for all the nodes in the same wireless cell. By allocating more time to the higher bit rate nodes, QoS for each nodes is discriminated. Thus, TXOP achieves fairness without fragmentation. However, in multi-hop environments, TXOP has problems in cases. When adjacent cells presents different lowest bit rates as in figure 2, several problems occur. At first, the forwarding node is located in the intersection of two cells, the forwarding node's TXOP operation oscillates. The reason of oscillation is that the nodes in the intersection of two cells may communicate to either a cell presents lower bit rate or that presents higher bit rate. In addition, the cell which has lower bit rate nodes have more time utilization. The forwarding node's oscillation makes serious impact on network traffic pattern. Because the nodes in the lower bit rate side may send more packets per channel acquisition than the other side, the forwarding node could not send all the packets that it has received even though the forwarder catches the channel. That is, performance anomaly is diffused from a node to a cell and the forwarder may congested by this inter-cell performance anomaly.

For the same case, on the other hands, our scheme succeeds in keeping fairness of channel utilization among the nodes in both cells. Because all the nodes in the both cells have same time utilization, the forwarder does not congested by the inter-cell performance anomaly.

4.2   Implementation Issue

Our scheme works well with other receiver based feedback mechanisms. We just modified the packet size, so other rate control scheme can be applied transparently. Our scheme can be simply applied by adjusting the MTU size or MSS (Maximum Segment Size) of TCP. That is, by simply modifying a variable, we can achieve fairness of the WLANs while the previous schemes need modification at MAC layer.

5   Simulation Results

We did simulation study through NS-2 simulator[9]. We show that our scheme can easily eliminate the performance anomaly. In the point of view of both the throughput and the fairness sides, our results show higher values. Specifically, the throughput has been increased up to 20% in best case. Moreover, fairness improves up to 70% than normal cases.

All the nodes are in the communication range and only one of them works at 1Mbps and the rest of the nodes work at 11Mbps rate. We use uniform random error model and ricean propagation model. Nodes are placed at randomly-chosen location within 250m ranged area. Simulation time is 150s and all the nodes use the same TCP traffic. We measure the throughput of each TCP sessions, and

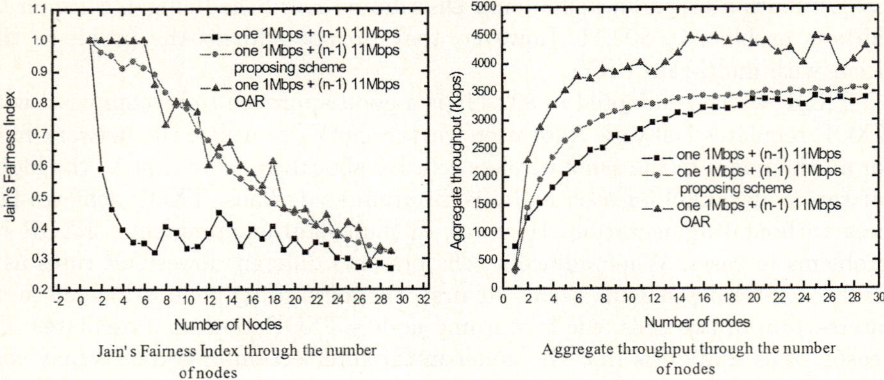

Jain's Fairness Index through the number
of nodes

Aggregate throughput through the number
of nodes

Fig. 1. Fairness Index and Aggregate throughput through the number of nodes

compare the aggregate throughput and throughput fairness. The fairness follows
the Jain's fairness index[8].

Figure 1 shows the improvements in throughput and fairness index in one
hop case. In one hop case, OAR have the highest throughput, but fairness value
oscillates. In the proposed methods, throughput is lower than OAR because
our scheme have more overhead from channel contention. On the other hands,
fairness keeps almost same value with OAR and it shows more stable behavior
than OAR.

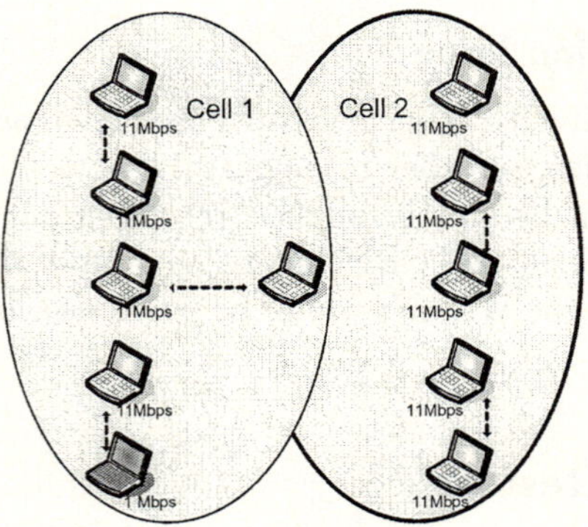

Fig. 2. Topology for Multi-hop case

OAR achieves higher throughput than normal case in one hop case. However,
it presents serious fairness degradation in multi-hop case. To investigate the
anomaly in multi-hop case, we modified the topology of the network as in figure 2.

Through the simulation, we found that the throughput of the cell 2 is seri-
ously degraded. This agrees with our expectation as stated in previous Section.
Moreover, the fairness of cell 2 also degrades. The reason for this degradation
is that the relay node catches the wireless channel longer than neighbor nodes.

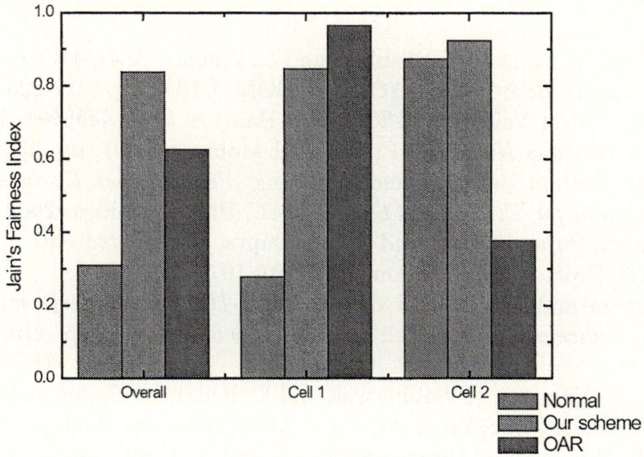

Fig. 3. Comparison of Jain's Fairness Index

While all the nodes in the same cell have the same probability to catch the wireless channel, the relay have higher probability in the same time. We present the throughput fairness index of the both schemes.

Figure 3 shows fairness index of cell 1, cell 2, and overall cases. In normal case, cell 1 has performance anomaly because different rate nodes are transmitting data simultaneously in the cell. On the other hands, OAR has performance anomaly in cell 2. The degradation of OAR is owing to the unfair channel utilization among the nodes. The relay node catches the wireless channel longer than its neighbors. As a result, cell 2 presents the performance anomaly. Our scheme avoids performance anomaly because our scheme keeps utilization same with its neighbor by adjusting the MTU size. Additionally, our scheme has more fine-grained channel access control, so the fairness is kept higher value.

6   Conclusion

In this paper, we present a scheme which eliminates the performance anomaly of 802.11b. By adjusting the packet size to the transmitting bit rate, we can successfully overcome the performance anomaly. Presented solution makes an improvement not only in throughput but also in fairness among the sessions by 20% and 70% in our best cases. Compared with TXOP operation in 802.11e, our solution does not present the fairness problem even in the multi-hop case.

Acknowledgement

This research was supported by the MIC(Ministry of Information and Communication), Korea, under the ITRC(Information Technology Research Center) support program supervised by the IITA(Institute of Information Technology Assessment)

References

1. V. Bharghavan, A. Demers, S. Shenker, and L. Zhang , *MACAW: A Media Access Protocol for Packet Radio,* Proc. ACM SIGCOMM 1994, pp. 212-225
2. Gavin Holland, Nitin Vaidya, and Paramvir Bahl, *A Rate-Adaptive MAC Protocol for Multi-hop Wireless Networks,* Proc. ACM Mobicom 2001, pp. 236-251
3. Zuyuan Fang, Brahim Bensaou, and Yu Wang, *Performance Evlauation of a Fair Backoff Algorithm for IEEE 802.11 DFWMAC,* Proc. Mobicom 2002, pp 48-57
4. Nitin H. Vaidya, Paramir Bahl, and Seema Cupta, *Distributed Fair Scheduling in a Wireless LAN,* Proc. ACM Mobicom 2000, pp. 167-178
5. Ad Kameramam and Leo Monteban, *WaveLan-II: A High-Performance Wireless LAN for the Unlicensed Band,* Bell Labs Technical Journal, pp. 118-133, Summer 1997
6. B. Sandeghi, V. Kanodia, A. Sabharwal, and E. Knightly, *Opportunistic Media Access for Multirate Ad Hoc Networks,* Proc. ACM Mobicom 2002, pp. 24-35
7. M. Heusse, F. Rousseau, G. Berger-Sabbatel, and A. Duda, *Performance Anomaly of 802.11b,* Proc. IEEE INFOCOM 2003, pp. 836-843
8. R. Jain, D.M. Chiu, and W.R. Hawe, *A Quantitative Measure of Fairness and Discrimination for Resource Allocation Shared Computer Systems,* Digital Equipment Corporation technical report TR 301, 1984
9. K. Fall and K. Varadhan, *NS notes and documentation,* the VINT Project, UC Berkeley, LBL USC/ISI, and Xerox PARC, available from http://www-mash.cs.berkeley.edu/ns, November 1997.

Energy Efficiency Analysis of IEEE 802.11 DCF with Variable Packet Length

Bo Gao[1], Yuhang Yang[1], and Huiye Ma[2]

[1] Dept. of Electronic Engineering,
Shanghai Jiao Tong University, Shanghai, China
`gaobo@sjtu.edu.cn`
[2] Dept. of Computer Science and Engineering,
Chinese University of Hong Kong
`hyma@cse.cuhk.edu.hk`

Abstract. The standardized IEEE 802.11 Distributed Coordination Function (DCF) provides a contention-based distributed channel access mechanism for mobile stations to share the wireless medium. However, when stations are mobile or portable units, power consumption becomes a primary issue since terminals are usually battery driven. In this paper, we propose an analytical model that calculates the energy efficiency of the hybrid of the basic and the RTS/CTS access mechanism of the IEEE 802.11 protocol, assuming that the packet lengths are variable. Validation of our new analytical model is carried out by comparison with simulation results using NS-2 simulation package.

1 Introduction

In recent years, Wireless Local Area Networks (WLANs) are becoming popular because they can provide more flexible and convenient connection than wired networks. The IEEE 802.11 standard [1] has achieved worldwide acceptance with WLANs and provides both Medium Access Control (MAC) layer and physical (PHY) layer specification for WLANs. IEEE 802.11 MAC has defined two medium access coordination methods: the compulsory contention-based Distributed Coordination Function (DCF) and optional contention-free based Point Coordination Function (PCF). DCF is an asynchronous data transmission function which better suits delay insensitive data, whereas PCF is used in time-bounded applications. In this paper, we limit our investigation to the DCF.

There is much work on the DCF performance evaluation in the literature since the release of the IEEE 802.11 standard. Simulation studies of the 802.11 performance were presented in [2] and [3]. Bianchi in [4] and [5] used Markov chain models to analyze DCF operation and calculated the saturated throughput of the 802.11 protocol. Yet, almost all existing models assume that the packet length is fixed. The problem of power consumption of IEEE 802.11 interfaces emerged early as a major drawback of the CSMA/CA technique. In [6], in particular, the authors showed through measurements that most power drawn from

P. Lorenz and P. Dini (Eds.): ICN 2005, LNCS 3421, pp. 1063–1070, 2005.

batteries by such devices is consumed by the sensing mechanism (idle mode). From such measurements, the power consumed during receiving/sensing is confirmed to be of the same order of magnitude of the power needed for transmitting. In [7], the authors introduced an analysis method of the average performance for the energy consumption and channel utilization of the p-persistent CSMA access scheme in WLANs, but only the simplified basic mechanism without acknowledgment scheme was considered.

In this paper, by extending the approaches in [7], we introduce a detailed mathematical model to evaluate the energy efficiency of the hybrid access mechanisms of the IEEE 802.11 protocol with the assumption that the packet length is variable and is sampled from a general probability distribution. Using NS-2 [8] simulation results, we validate our mathematical model and show that the proposed model predicts the energy efficiency of the DCF of the IEEE 802.11 standard very accurately. To the best of our knowledge, the literature still lacks a complete analytical model of the energy efficiency of the IEEE 802.11 systems with variable packet length, which represents the main contribution of our paper.

This paper is organized as following. In Section 2, we briefly review both the basic and the RTS/CTS access mechanisms of the DCF of the IEEE 802.11 MAC layer. Section 3 presents our mathematical model and derives the closed form solution for the energy efficiency performance of the IEEE 802.11 protocol. Finally, Section 4 validates theoretical derivations by simulation and makes some conclusions.

2   IEEE 802.11 Distributed Coordination Function

IEEE 802.11 DCF is based on a CSMA/CA access scheme and employs a binary exponential backoff (BEB) technique. The DCF consists of the basic access mode as well as the optional RTS/CTS access mode. These two access modes are described below.

In basic access mode, a station having a packet to transmit must initially "listen" to the channel if another station is transmitting. If no transmission takes place for a distributed interframe space (DIFS) time interval, the transmission may proceed. If the medium is busy, the station has to defer its transmission until the end of the current transmission. It will then wait for an additional DIFS time, and generate a random delay to initiate the backoff timer before transmission. The backoff timer is decreased as long as the medium is sensed as idle and frozen when a transmission is detected on the channel until the end of this transmission and resumes when the medium is sensed as idle again for more than a DIFS interval. The station transmits its packet whenever its backoff timer becomes zero. For efficiency reasons, time is slotted in a basic time unit, denoted by t_{slot}, which is set equal to the time needed at any station to detect the transmission of a packet from any other station. DCF adopts an exponential backoff scheme. The backoff delay is uniformly chosen in the range $[0, W-1]$. The value W is called *contention window*, and depends on the number of transmissions failed for the packet. The initial backoff window size is set to $W = CW_{min}$

where CW_{min} is called minimum contention window. After each unsuccessful transmission, W is doubled, up to a maximum value $CW_{max} = 2^m CW_{min}$. After receiving a packet correctly, an ACK frame is immediately transmitted to the source station by the destination to confirm the correct reception, after a period of time called short interframe space (SIFS), where $t_{SIFS} < t_{DIFS}$. The transmitter reschedules its packet transmission if it does not receive the ACK within a specified ACK_Timeout, or if it detects the transmission of a different packet.

In the second mechanism, which is called RTS/CTS access mechanism, a station that intends to transmit a packet sends a special short RTS frame and the receiving station responds with a CTS frame after SIFS time. The data packet is transmitted after the successful exchange of the RTS and CTS frames. The RTS/CTS mechanism is very effective in terms of system performance, especially when larger packet are considered, as it avoid long collisions and reduces the amount of bandwidth wasted in collisions.

In fact, the common access mechanism is the coexistence of the above two mechanisms, i.e., packets are transmitted by means of the RTS/CTS mechanism if their payload sizes exceed a given threshold L_r, otherwise, the basic access method is used to transmit the packets. We denote this access mechanism as the *hybrid* mechanism in this paper.

3   Mathematical Model

Our analysis assumes that the network consists of M contending stations and each station always has a packet ready for transmission, i.e., they operate in saturation conditions. Let p be the probability that a station transmits in a randomly chosen slot time, and c be the probability that a packet collides. For simplicity, W denotes the minimum contention window size CW_{min}. By the same Markov model used to compute the transmission probability p in [5], we can get the expression for p as following:

$$p = \frac{2(1 - 2c)}{(1 - 2c)(W + 1) + cW(1 - (2c)^m)} \tag{1}$$

where m denotes the maximum backoff stages.

The probability c can be expressed as

$$c = 1 - (1 - p)^{M-1}. \tag{2}$$

Equations (1) and (2) form a non-linear system with two unknown parameters p and c. Note that $p \in (0, 1)$ and $c \in (0, 1)$. This non-linear system can be solved using numerical methods and has a unique solution.

As in [7], the energy efficiency ρ_{energy} can be defined as the ratio between the average energy used by the tagged station in a transmission attempt interval to successfully transmit a message divided by the total energy consumed by the tagged station during a transmission attempt interval. Hence ρ_{energy} can be expressed as

$$\rho_{energy} = \frac{PTX \cdot E[L] \cdot P_{tag\_Succ|N_{tr \geq 1}}}{E\left[Energy_{Idle\_P}\right] + E\left[Energy_{transmission\_attempt}|N_{tr} \geq 1\right]} \tag{3}$$

where $E[L]$ is the average payload length, normalized to the slot time, PTX is the power consumption of the network interface during the transmitting phase, N_{tr} is the number of stations that begin to transmit at the same slot, $P_{tag\_Succ|N_{tr \geq 1}}$ is the probability of a tagged station successful transmission given that there is at least a transmission attempt, $E[Energy_{Idle\_P}]$ is the average energy consumed by the tagged station during the idle period that precedes a transmission attempt, and $E[Energy_{transmission\_attempt}|N_{tr} \geq 1]$ is the average energy consumed during a transmission attempt given that at least one station is transmitting.

$E[Energy_{Idle\_P}]$ is given by

$$E[Energy_{Idle\_P}] = PRX \cdot \frac{(1-p)^M}{1-(1-p)^M} \tag{4}$$

where PRX is the power consumption of the network interface during the listening phase.

$E\left[Energy_{transmission\_attempt}|N_{tr} \geq 1\right]$ is given by

$$E\left[Energy_{transmission\_attempt}|N_{tr} \geq 1\right]$$
$$= E\left[Energy_{tag\_Succ}\right] \cdot P_{tag\_Succ|N_{tr}\geq 1} + E\left[Energy_{not\_tag\_Succ}\right] \cdot P_{not\_tag\_Succ|N_{tr}\geq 1}$$
$$+ E\left[Energy_{tag\_Coll}\right] \cdot P_{tag\_Coll|N_{tr}\geq 1} + E\left[Energy_{not\_tag\_Coll}\right] \cdot P_{not\_tag\_Coll|N_{tr}\geq 1}. \tag{5}$$

Hereafter, we will denote with $E[Energy_{xxx}]$ the average energy consumed by a tagged station with the event xxx occurring (for instance a tagged station success, a tagged station collision, a not tagged station success, etc.). $P_{xxx|N_{tr}\geq 1}$ is the probability of the event xxx occurring given that there is at least a transmission attempt and is given in [7] as following

$$\begin{cases} P_{tag\_Succ|N_{tr}\geq 1} = \frac{p \cdot (1-p)^{M-1}}{1-(1-p)^M}, \\ P_{not\_tag\_Succ|N_{tr}\geq 1} = \frac{(M-1) \cdot p \cdot (1-p)^{M-1}}{1-(1-p)^M}, \\ P_{tag\_Coll|N_{tr}\geq 1} = \frac{p \cdot \left[1-(1-p)^{M-1}\right]}{1-(1-p)^M}, \\ P_{not\_tag\_Coll|N_{tr}\geq 1} = \frac{(1-p)\left[1-(1-p)^{M-1}-(M-1)\cdot p \cdot (1-p)^{M-2}\right]}{1-(1-p)^M}. \end{cases} \tag{6}$$

Next, we compute the values of $E\left[Energy_{tag\_Succ}\right]$, $E\left[Energy_{not\_tag\_Succ}\right]$, $E\left[Energy_{tag\_Coll}\right]$ and $E\left[Energy_{not\_tag\_Coll}\right]$, respectively. In this paper, we always assume that the packet payload size is variable, so these four values are difficult to compute.

In the *hybrid* access scheme, packets are transmitted by means of the RTS/CTS mechanism if their payload size exceeds a given threshold L_r; otherwise, the basic access method is used to transmit the packets. So we have

$$E[Energy_{tag\_Succ}]$$
$$= \sum_{x=1}^{L_r} E[Energy_{tag\_Succ}]^{Bas} \cdot P\{L=x\} + \sum_{x=L_r+1}^{\infty} E[Energy_{tag\_Succ}]^{RTS} \cdot P\{L=x\} \tag{7}$$

where Bas and RTS represent the basic access scheme and the RTS/CTS access scheme, respectively. $P\{L=x\}$ is the probability that a packet payload length L is x.

According to the IEEE 802.11 DCF standard [1], we can get the expressions of $E[Energy_{tag\_Succ}]^{Bas}$ and $E[Energy_{tag\_Succ}]^{RTS}$ as following:

$$\begin{cases} E[Energy_{tag\_Succ}]^{Bas} = PTX \cdot [H+x] + PRX \cdot [ACK + 2\sigma + SIFS + DIFS], \\ E[Energy_{tag\_Succ}]^{RTS} = PTX \cdot [H + x + RTS] \\ \qquad\qquad\qquad + PRX \cdot [CTS + ACK + 4\sigma + 3SIFS + DIFS] \end{cases} \tag{8}$$

By substituting (8) to (7) and after some algebraic manipulations, the expression of (7) becomes

$$\begin{aligned} &E[Energy_{tag\_Succ}] \\ &= PTX \cdot [H + E[L]] + PRX \cdot [ACK + 2\sigma + SIFS + DIFS] \\ &\quad + [PTX \cdot RTS + PRX \cdot (CTS + 2\sigma + 2SIFS)] \cdot (1 - P\{L \le L_r\}) \end{aligned} \tag{9}$$

where $E[L]$ denotes the average packet length and $P\{L \le L_r\}$ is the probability that the packet length is lower than L_r.

In the same way, we can get the expression of $E[Energy_{not\_tag\_Succ}]$

$$\begin{aligned} &E[Energy_{not\_tag\_Succ}] \\ &= \sum_{x=1}^{L_r} E[Energy_{not\_tag\_Succ}]^{Bas} \cdot P\{L=x\} + \sum_{x=L_r+1}^{\infty} E[Energy_{not\_tag\_Succ}]^{RTS} \cdot P\{L=x\} \end{aligned} \tag{10}$$

where Bas and RTS represent the basic access scheme and the RTS/CTS access scheme, respectively. The values of $E[Energy_{not\_tag\_Succ}]^{Bas}$ and $E[Energy_{not\_tag\_Succ}]^{RTS}$ are given by:

$$\begin{cases} E[Energy_{not\_tag\_Succ}]^{Bas} = PRX \cdot [H + x + ACK + 2\sigma + SIFS + DIFS] \\ E[Energy_{not\_tag\_Succ}]^{RTS} = PRX \cdot \begin{bmatrix} H + x + RTS + CTS + ACK + 4\sigma \\ +3SIFS + DIFS \end{bmatrix} \end{cases} \tag{11}$$

where x denotes the packet length.

By substituting (11) to (10) and after some algebraic manipulation, the expression of (10) becomes

$$\begin{aligned} &E[Energy_{not\_tag\_Succ}] \\ &= PRX \cdot [H + E[L] + ACK + 2\sigma + SIFS + DIFS] \\ &\quad + PRX \cdot [RTS + CTS + 2\sigma + 2SIFS] \cdot (1 - P\{L \le L_r\}) \end{aligned} \tag{12}$$

Now, we derive the expression of $E[Energy_{tag\_Coll}]$.

$$\begin{aligned} &E[Energy_{tag\_Coll}] \\ &= \sum_{x=1}^{L_r} P\{L=x\} \cdot E[Energy_{tag\_Coll}]^{Bas} + \sum_{x=L_r+1}^{\infty} P\{L=x\} \cdot E[Energy_{tag\_Coll}]^{RTS} \end{aligned} \tag{13}$$

where $E[Energy_{tag\_Coll}]^{Bas}$ denotes the energy consumed by the colliding tagged station which sends a packet with length less than L_r. $E[Energy_{tag\_Coll}]^{RTS}$ denotes the energy consumed by the colliding tagged station which sends a packet with length greater than L_r.

In the *hybrid* access scheme, there are two kinds of frames involved in one collision: data frame and RTS frame. Since a packet with a payload size longer than L_r is transmitted by the RTS/CTS mechanism, the maximum data frame length involved in a collision must be L_r. Two possible collision scenarios may occur: 1) all other collision frames are RTS frames; 2) there is at least one data frame in all other collision frames. According to [1], the length of an RTS frame is always less than the packet header size H. Thus, in the first scenario, the collision length should be the packet size of the tagged station in $E[Energy_{tag\_Coll}]^{Bas}$ or RTS in $E[Energy_{tag\_Coll}]^{RTS}$. In the second scenario, the collision length is the maximum collision length of all colliding packets. So we have

$$E[Energy_{tag\_Coll}]^{Bas}$$

$$= PTX \cdot [H + x] + PRX \cdot [DIFS + \sigma] + E[Energy_{tag\_Coll}]_{all\_RTS} \qquad (14)$$

$$+ E[Energy_{tag\_Coll}]_{at\_least\_one\_data},$$

where

$$E[Energy_{tag\_Coll}]_{all\_RTS} = 0 \qquad (15)$$

and

$$E[Energy_{tag\_Coll}]_{at\_least\_one\_data}$$
$$= PRX \cdot \sum_{y=1}^{L_r-x} y \cdot \left[\sum_{n=1}^{M-1} [P\{L = x + y | N_{tr} = n\} \cdot P\{N_{tr} = n | N_{tr} \geq 1\}] \right]$$

$$= PRX \cdot \left[\sum_{y=1}^{L_r-x} \left[y \cdot \frac{1}{1-(1-p)^{M-1}} \cdot \left[\begin{array}{l} [p \cdot (P\{L > L_r\} + P\{L \leq y + x\}) + 1 - p]^{M-1} \\ - [p \cdot (P\{L > L_r\} + P\{L < y + x\}) + 1 - p]^{M-1} \end{array} \right] \right] \right] \cdot$$
$$(16)$$

With the same reason,

$$E[Energy_{tag\_Coll}]^{RTS}$$
$$= PTX \cdot RTS + PRX \cdot [DIFS + \sigma] + E[Energy_{tag\_Coll}]_{all\_RTS} \qquad (17)$$
$$+ E[Energy_{tag\_Coll}]_{at\_least\_one\_data},$$

where

$$E[Energy_{tag\_Coll}]_{all\_RTS} = 0 \qquad (18)$$

and

$$E[Energy_{tag\_Coll}]_{at\_least\_one\_data}$$
$$= PRX \cdot \sum_{y=1}^{L_r-x} y \cdot \left[\sum_{n=1}^{M-1} [P\{L = RTS + y | N_{tr} = n\} \cdot P\{N_{tr} = n | N_{tr} \geq 1\}] \right]$$

$$= PRX \cdot \left[\sum_{y=1}^{L_r-x} \left[y \cdot \frac{1}{1-(1-p)^{M-1}} \cdot \left[\begin{array}{l} [p \cdot (P\{L > L_r\} + P\{L \leq RTS + y\}) + 1 - p]^{M-1} \\ - [p \cdot (P\{L > L_r\} + P\{L < RTS + y\}) + 1 - p]^{M-1} \end{array} \right] \right] \right] \cdot$$
$$(19)$$

As to $E[Energy_{not\_tag\_Coll}]$, there are also two possible collision scenarios may occur: 1). All collision frames are RTS frames; 2). There is at least one data frame in all collision frames. So we can get the expression of $E[Energy_{not\_tag\_Coll}]$ as following:

$$E[Energy_{not\_tag\_Coll}]$$

$$= \sum_{n=2}^{M-1} \left[\begin{array}{l} PRX \cdot [RTS + DIFS + \sigma] \\ \cdot P\{L > L_r\}^n \cdot P\{N_{tr} = n | N_{tr} > 1\} \end{array} \right]$$

$$+ \sum_{x=1}^{L_r} \left[\sum_{n=2}^{M-1} \left[\begin{array}{l} \left[\begin{array}{l} PRX \cdot [H + x + DIFS + \sigma] \\ \cdot \left[\sum_{k=0}^{n-1} \left[\binom{n}{k} \cdot P\{L > L_r\}^k \cdot \left[\sum_{k_1=1}^{n-k} \left[\begin{array}{l} \binom{n-k}{k_1} \cdot P\{L = x\}^{k_1} \\ \cdot P\{L < x\}^{n-k-k_1} \end{array} \right] \right] \right] \right] \\ \cdot P\{N_{tr} = n | N_{tr} > 1\} \end{array} \right] \right] \tag{20}$$

After some algebraic manipulation, we can get the closed formula for $E[Energy_{not\_tag\_Coll}]$ as following

$$E[Energy_{not\_tag\_Coll}]$$

$$= \frac{PRX \cdot [RTS + DIFS + \sigma]}{1 - (1-p)^{M-1} - (M-1) \cdot p \cdot (1-p)^{M-2}} \cdot \left[\begin{array}{l} [1 + p \cdot (P\{L > L_r\} - 1)]^{M-1} - (1-p)^{M-1} \\ - (M-1) \cdot p \cdot P\{L > L_r\} \cdot (1-p)^{M-2} \end{array} \right]$$

$$+ \sum_{x=1}^{L_r} \left[\begin{array}{l} \frac{PRX \cdot [H + x + DIFS + \sigma]}{1 - (1-p)^{M-1} - (M-1) \cdot p \cdot (1-p)^{M-2}} \\ \cdot \left[\begin{array}{l} [1 + p \cdot (P\{L > L_r\} + P\{L \le x\} - 1)]^{M-1} \\ - [1 + p \cdot (P\{L > L_r\} + P\{L < x\} - 1)]^{M-1} \\ - (M-1) \cdot p \cdot (1-p)^{M-2} \cdot P\{L = x\} \end{array} \right] \end{array} \right] \tag{21}$$

4   Model Validation and Conclusions

To validate our model, we have compared its results with those obtained with the famous network simulator NS-2 [8]. The values reported in the Fig. 1 for both the model and the simulator have been obtained using the system parameters in Table 1 and are based on the Direct Spread Sequence Spectrum (DSSS)

Table 1. DSSS System Parameters in IEEE 802.11

MAC header	224 bits	Propagation Delay	1 μs
PHY header	192 bits	Slot Time	20 μs
ACK packet	304 bits	SIFS	10 μs
RTS packet	352 bits	DIFS	50 μs
CTS packet	304 bits	Minimum CW	32
Channel Bit Rate	2Mbps	Number of backoff stages	5

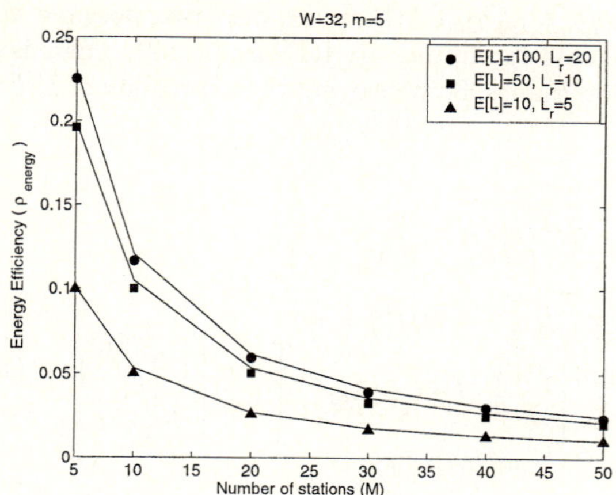

Fig. 1. Energy efficiency—analysis versus simulation

physical layer used in the IEEE 802.11 standard. The network interface power consumption ratio PTX/PRX is set to 1.5. Note that, the probability distribution of the length of packets is chosen as a geometrical distribution with the parameter q because it is reasonable for us to obtain different packet length distributions. Figure 1 validates our analytical model since almost exact matches are observed between analytical results (lines) and simulation outcomes (symbols).

References

1. *IEEE Standard for Wireless Lan Medium Access Control (MAC) and Physical Layer (PHY) Specification*, Aug. 1999.
2. T. S. Ho and K. C. Chen, "Performance analysis of IEEE 802.11 CSMA/CA Medium Access Control Protocol", in Proc. of *IEEE Personal Indoor and Mobile Radio Communication*, Taipei, Taiwan, pp. 407-411, 1996.
3. B. P. Crow, I. Widjaja, J. G. Kim, and P. T. Sakai, "IEEE 802.11 Wireless Local Area Networks", *IEEE Communication Magazine*, 1997.
4. G. Bianchi, "IEEE 802.11-Saturation throughput analysis", *IEEE Communications Letters*, Vol.2, No.12, pp. 318-320, 1998.
5. G. Bianchi, "Performance Analysis of IEEE 802.11 distributed Coordination Function", *IEEE Journal on Selected Areas in Communications*, Vol.18, No.3, pp. 535-547, 2000.
6. L. M. Feeney and M. Nilsson, "Investigating the Energy Consumption of a Wireless Network Interface in an Ad Hoc Networking Environment", in Proc. of *IEEE INFOCOM 2001*, pp. 1548-1557, Ankorage, Alaska, April 22 - 26, 2001.
7. R. Bruno, M. Conti, E. Gregori, "Optimization of Efficiency and Energy Consumption in p-Persistent CSMA-Based Wireless LANs", *IEEE Transactions on Mobile Computing*, vol. 1, no. 1, pp. 10-31, January-March 2002.
8. [Online.] Available: http://www.isi.edu/nsnam/ns/

Scheduling MPEG-4 Video Streams Through the 802.11e Enhanced Distributed Channel Access

Michael Ditze[1], Kay Klobedanz[1], Guido Kämper[1], and Peter Altenbernd[2]

[1] C-LAB, Fürstenallee 11, 33102 Paderborn, Germany
{michael.ditze, kay.klobedanz, guido.kaemper}@c-lab.de
[2] Fachhochschule Darmstadt, Haardtring 100, 64295 Darmstadt, Germany
paltenbernd@fbi.fh-darmstadt.de

Abstract. The upcoming IEEE 802.11e standard improves the Medium Access Control (MAC) of the legacy 802.11 with regard to Quality of Service (QoS) by introducing the Enhanced Distributed Channel Access (EDCA) and the HCF Controlled Channel Access (HCCA). EDCA achieves QoS by providing independent transmit queues and MAC parameters for each traffic class, and hence higher prioritized traffic has a higher probability for transmission. Crucial to the success of such a strategy is a scheduler that assigns the data traffic to the respective transmit queues. This paper develops and accommodates a new dynamic scheduler for EDCA into the MPEG-4 Delivery Framework. Experiments prove that the new scheduling policy timely delivers up to 50% more frames than statical scheduling solutions. To the best of our knowledge this is one of very few scheduling approaches that considers MPEG-4 related traffic priorization in EDCA.

1 Introduction

MPEG-4 applications require QoS support on both, the end-device and the network carriers in order to guarantee the delivery of time-sensitive video data from the source to the sink. As the legacy 802.11 does not provide QoS guarantees due to the random-based medium access scheme in the Data Link Layer, the 802.11e working comitee prepares major amendments to the 802.11 channel access regardless of the physical layer underneath. 802.11e allows to assign prioritized traffic to traffic categories that exhibit different MAC parameters and hence result in different probabilities to gain the medium access. In order to assign data packets to the respective traffic categories, a scheduler is required. Many multimedia applications e.g. video streaming, however, benefit from dynamic scheduling policies that allow the scheduler to adjust to the varying workloads that result from video compression and user interactivity [1]. In cases where a dynamic scheduling policy is deployed that may change priorities at runtime the scheduler becomes all the more the crucial entity that ensures QoS maintenance.

This paper presents a new smart scheduler that dynamically determines priorities for MPEG-4 frames and assigns corresponding data packets to EDCA

P. Lorenz and P. Dini (Eds.): ICN 2005, LNCS 3421, pp. 1071–1079, 2005.

transmit queues accordingly. In contrast to other solutions that are summarized in [2] we do not confuse the importance and the urgency of frame-types and hence schedule lower prioritized frames with a close deadline in the presence of higher prioritized frame in case the latter can still make its deadline. The dynamic priority assignment derives from a modification of the Least Laxity First approach which we already used to suit processor scheduling for MPEG streams on end-devices [2,3], and now adapt for network scheduling. The laxity of a frame hereby denotes the amount of time a data packet can be delayed on its transmission and still arrive before its deadline at the receiving end-device. The scheduler relies on a statistical approach for Admission Control for 802.11e [4] and may provide soft real-time capabilities.

We implemented the scheduling policy on top of the 802.11e MAC into the NS-2 network simulator. Evaluation results prove that using this approach we are able to increase the amount of timely transmitted frames significantly compared to traditional scheduling solutions. Hence, the new scheduling approach schedules up to 50% more frames in overloaded conditions than comparable static solutions. Furthermore, it adapts well to sudden workload changes. To the best of our knowledge this one of very few scheduling approaches for 802.11e [5].

The rest of the paper is organized as follows: Section 2 gives an introduction on 802.11e and MPEG-4. Section 3 describes the new approach followed by the evaluation results in Section 4. Section 5 gives the conclusions.

2   Introduction to 802.11e and MPEG-4

Introduction to 802.11e. The legacy 802.11 standard provides detailed medium access control and operation at the physical layer for Wireless LANs. The fundamental medium access function is referred to as Distributed Coordination function (DCF). DCF operates in contention periods where each station may autonomously access the medium. DCF does neither allow to prioritize traffic nor does it provide timing guarantees and hence QoS to the applications. As a consequence, the 802.11e working group is in the process of developing a new improved medium access scheme for contention periods referred to as Enhanced Distributed Channel Access (EDCA).

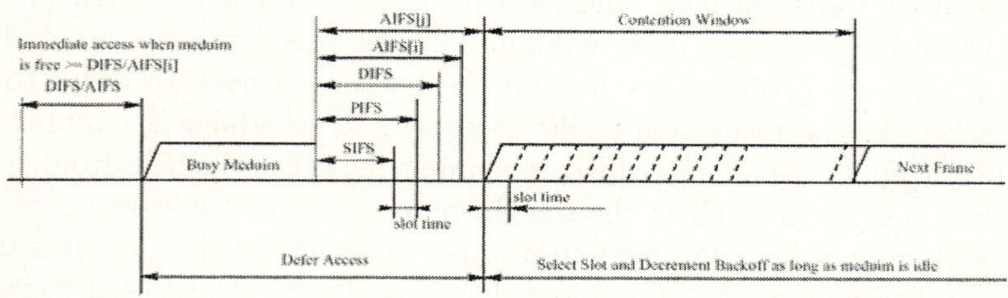

Fig. 1. Medium Access Control through the 802.11e EDCA

In contrast to DCF, EDCA allows to prioritize traffic by introducing four different Access Categories (ACs) to each QoS station (QSTA) [6]. Each AC maintains a separate transmit queue and a dedicated channel access function that features AC-specific parameters. These parameters include different values for minimum and maximum *Contention Windows, Arbitration Interframe Spaces* (AIFS) and a *Transmission Opportunity* (TXOP) duration [7].

AIFS, that is generally larger than DIFS, hereby denotes an individual time for sensing the medium that can be adjusted for each AC. Hence, in order to allow for traffic priorization in 802.11e, higher priority ACs receive shorter AIFSs and lower CWs to increase the probability of a successful channel access (see Fig.1). The channel access itself remains similar to DCF. A TXOP is usually allocated by the QoS Access Point (QAP) and, in contrast to the legacy DCF, hereby grants a station the right to use the medium at a defined point in time for a defined maximum duration. Higher prioritized ACs are granted larger TXOPs which results in a larger throughput per AC.

In case of an internal collision i.e. the backoff timer of at least two ACs simultaneously reaches zero, an internal scheduler grants the access rights to the higher prioritized AC and forces the other station to enter backoff procedure.

Introduction to MPEG-4. In contrast to its predecessors, MPEG-4 [8] allows for the decomposition of video scenes into single audio-visual objects. Each object can be separately encoded and transmitted as a series of frames in one or several Elementary Streams (ES). ESs then pass the Sync Layer before they are transmitted through the Delivery Multimedia Integration Framework that encapsulates them into native transmission protocols, e.g. RTP/IP.

In order to exploit the redundancy in video streams, MPEG-4 defines three particular types of Video Object Planes (VOP) that are temporal instances of an audio-visual object. These VoPs exhibit different compression ratios and are referred to as *I(ntrapicture)*-VOPs, *P(redicted picture)*-VOPs and *B(idirectional predicted picture)*- VOPs.

I-VOPs serve as reference VOPs to P-and B-VOPs whereas P-VOPs are predicted VOPs that collect relevant information encoded in former I-VOPs. They also serve as reference VOPs to B-VOPs. Consequently, I-VOPs and P-VOPs are also referred to as *reference VOPs*. B-VOPs can be either forward or backward predicted and likewise exploit redundant information encoded in previous or subsequent VOPs. However, as the transmission order in MPEG avoids forward precedences, only backward precedences need to be considered for transmission.

A *Group of VOPs* (GOV) is a sequence of VOPs ranging from one I-VOP to the next. As each GOV may be self-contained, it is independent of others which allows for decoding without any knowledge about other groups.

3   A MPEG Scheduler for EDCA

Fig.2 illustrates a simplified architecture that accommodates the MPEG-4 EDCA scheduler into the MPEG-4 Delivery Framework. The scheduler also complies

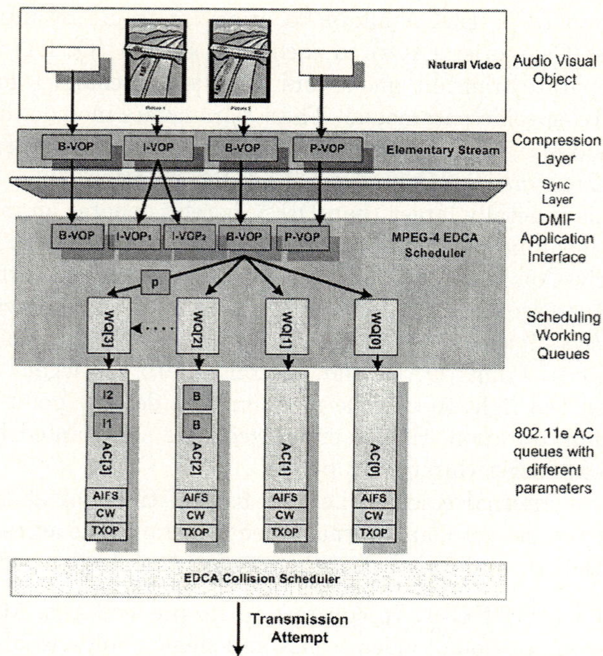

Fig. 2. EDCA Scheduler embedded in the MPEG-4 Delivery Framework

with the latest 802.11e draft [6]. The compression layer encodes VOPs that represent Audio Visual Objects and passes them via the Sync Layer and the DMIF API directly to the new scheduler. The Sync Layer hereby performs time stamp assignment and VOP packet fragmentation. The task of the scheduler is to dynamically determine the priority of each packet according to a scheduling policy that is suited for MPEG-4 and assign the fragmented VOP packets to the EDCA AC transmit queues.

3.1   Priorities and Access Categories Assignment

We use a priority-based scheduling policy that does not confuse the importance of data packets and their urgency. Urgent MPEG-4 packets are those packets that have a close deadline while important packets are required for further frame decoding. As a consequence, I-VOPs usually exhibit the highest importance. B-VOPs on the contrary may temporarily be more urgent as they appear more frequently and have closer deadlines (see Section 2). Corresponding to the EDCA ACs, the scheduler maintains 4 internal working queues WQ that feed the transmit queues of each AC (see Fig.2). The scheduler deploys a Least-Laxity-First (LLF)-extended scheduling policy in order to dynamically assign packets to the queues. Initially, considering the importance, packets are assigned as follows:

Similar to the 802.11e ACs, WQ_3 represents the highest priority and WQ_1 the lowest priority queue. I- and P-VOPs data packets are assigned to top priority WQ_3, whereas B-VOP packets that can be skipped occasionally in favour of

Table 1. Access Categories for MPEG-4 data packets

ACs	Description	Working Queue Order
WQ_3	I- and P-VOP data packets	unique priorities, priority order
WQ_2	B-VOP data packets	unique priorities, priority order
WQ_1	all other data packets	FIFO

other data packets are inserted into the WQ_2 queue. Other remaining packets that do not have a priority are inserted in the WQ_1 queue and hence do not need to undergo Admission Control. The different priorization of VOP packets hereby adheres to the uni- and bidirectional predictive encoding of MPEG.

Further, we prioritize data packets in the working queues in priority order, serving packets with the highest priority first. Priorities are derived as follows:

- WQ_3 gives I-VOP data packets priority over P-VOP data packets. If packets belong to different streams but have the same priority and share the same deadline, they build ordered sets of data packets. Within each such set the lower-sized VOP packets are scheduled first. This is to ensure that large-sized VOPs do not interfere low-sized VOPs with the same priority.
- WQ_2 also builds ordered sets of B-VOP packets in case they belong to different streams and share the same deadline and priority. Similar to WQ_3, lower sized B-VOPs are preferred.

Building ordered sets has a an impact of the throughput performance of MPEG-4 data packets as it allows to use the available bandwidth more efficiently or permits to distinguish between streams with different priorities.

3.2   Scheduling Policy

Data packets in WQ_3 and WQ_2 are concurrently scheduled by an extension to the LLF algorithm. This policy that is often applied in real-time operating systems schedules the most *urgent* data packet, i.e. the data packet with the smallest laxity Y. The laxity hereby denotes the maximum time a data packet can be delayed on its transfer to be transmitted and decoded at the receiver within its deadline. In addition to that our policy

- skips B-VOP packets in WQ_2 if their deadline has already passed or exhibit a negative urgency Y.
- may move single data packets from WQ_2 to WQ_3 in case they can be transferred and decoded without causing the latter to miss their deadline and are kept in order of appearance. Due to the backward precedences in MPEG, WQ_2 VOP packets may only be inserted before those VOP packets that the WQ_2 packets depend on for decoding.
- prefers lower-sized VOPs in case two VOPs in the same AC are equally prioritized and share the same deadline.

3.3   Admission Control and Throughput Analysis

The scheduling approach relies on an Admission Control that founds on a throughput analysis for 802.11e. The throughput analysis estimates the available throughput at saturation conditions [4]. It exploits the statistical collision probability for a flow in an AC in order to compute the transmission probability in a slot and may likewise derive the achievable throughput τ_j for AC_j. We compute τ_j for each AC in order to determine if WQ_2 packets can be moved to WQ_3 (and hence AC_3) without causing WQ_3 packets to miss their deadline.

Once having determined τ_j, we now compute in a second step the expected processing time for each packet in a working queue that we require to decide if urgent packets can be moved to higher prioritized WQs without causing interference. We define the expected processing Time $T_{\rho[i]_j}$ for each data packet i in a working queue j as the expected time required to gain medium access $T_{MAC[i]_j}$ in AC_j, its expected transmission time T_{Θ_i} and the worst-case decoding time C_i of the frame the packet belongs to at the receiving device. $T_{\rho[i]_j}$ calculates as

$$T_{\rho[i]_j} = T_{MAC[i]_j} + T_{\Theta_i} + C_i \tag{1}$$

As we assume a static scenario where devices do not move and the transmission channel does not suffer overlay interference from other transmission protocols operating in the same GHz range, we can assume T_{Θ} to be constant for each packet. Further, C_i can be derived from a Worst-Case Execution Time Analysis for MPEG decoding [9]. The medium access time for each packet in a WQ can thus be computed as the fraction of all packet sizes that precede a packet in the transmit queue and the working queue and the estimated achievable throughput

$$T_{MAC[i]_j} = \frac{\sum_{k=0}^{i-1} s_k | k \epsilon WQ_j \vee k \epsilon AC_j}{\tau_j} \tag{2}$$

where s_i denotes the size of a packet i. Using Eq.1 and Eq.2 we can now simply determine the urgency $Y_{[i]_j}$ of a data packet i in Working Queue j as

$$Y_{[i]_j} = D_i - T_{\rho[i]_j} \tag{3}$$

where D_i denotes the deadline of a packet i and derives from the frame-per-second rate of the video stream. As a consequence, the scheduler may move data packets from lower prioritized WQs to WQs with a higher priority in case the estimated processing time T_ρ of low-prioritized data packets is less than the urgency $Y_{[i]}$ of a higher prioritized packet in another WQ. In case of MPEG-4 VOPs, the scheduler may move entire VOP-packets from WQ_2 to WQ_3, if its estimated processing time $T_{\rho 2}$ is less than the urgency $Y_{[i]_3}$ of corresponding VOP-packet in WQ_3.

4   Testbed and Evaluation

We evaluated the new scheduling policy with the NS-2 network simulator. As a MPEG-4 traffic generator, we chose a model that exploits the Transform Expand Sample Methodology to fill the working queues of the scheduler that is implemented as a class in NS-2.

The traffic generator uses additional separate NS-2 output agents for each VOP-type. This allows to generate graph diagrams in order to analyze the throughput behaviour per VOP-type. The traffic generator then passes the frame packets to the scheduler. The scheduler is built on top of the 802.11e implementation as described in [10]. It routes the VOPs to the EDCA AC queues according to the scheduling policy introduced in the last section.

Each of the AC agents then uses the mandatory *sendmsg* method to transmit the queued packets. Furthermore we extended the senders NS-2 Loss Monitor by a monitoring function and by a separate counter for MPEG-4 VOPs and traffic. Similarly, at the receiver node, a second Loss Monitor records the incoming data and frames. This allows us to calculate the generated video traffic and the actual data received at the client. Thus, we are able to accurately determine the packet loss and likewise the behaviour and results of the scheduling approach.

Experiments. We developed a simple experiment with two wireless stations. One station acts as a video server and generates MPEG-4 video streams that are transmitted over the wireless medium to the receiving media renderer. The generated video exhibits a quality similar to DVD at data-rates up to 5 Mbit/sec. Without loss of generality, we set the physical bandwidth in NS-2 such as the generated video exceeds it by approximately 15% in order to achieve a worst-case situation where abundant frames will de dropped due to the limited bandwidth. The 802.11e parameters are set according to the suggestions by the 802.11e comitee.

We compare the scheduling approach that we introduced in this paper with three other approaches.

- **Mode 1:** The first approach assumes a simple scheduler that assigns each MPEG-4 VOP in order of appearance to the highest-priority working queue on a best-effort basis. This solution corresponds to the proposal of the 802.11e working comitee that suggests to maintain a single AC for video traffic [6].
- **Mode 2:** The second approach assigns VOPs to the Admission Category according to their importance for the GOV. Likewise, it passes I-VOPs to (AC_3), P-VOPs are inserted in (AC_2) and B-VOPs are forwarded to (AC_1).
- **Mode 3:** This approach represents a simplified version of our scheduler and considers the importance of VOP-types. Consequently I- and P-VOPs are inserted into the highest-prioritized working queues whereas less favoured B-VOPs are queued in (AC_2).

Interpretation. The results of the experiment are illustrated below. We only consider those VOPs in the evaluation that can be decompressed by the media

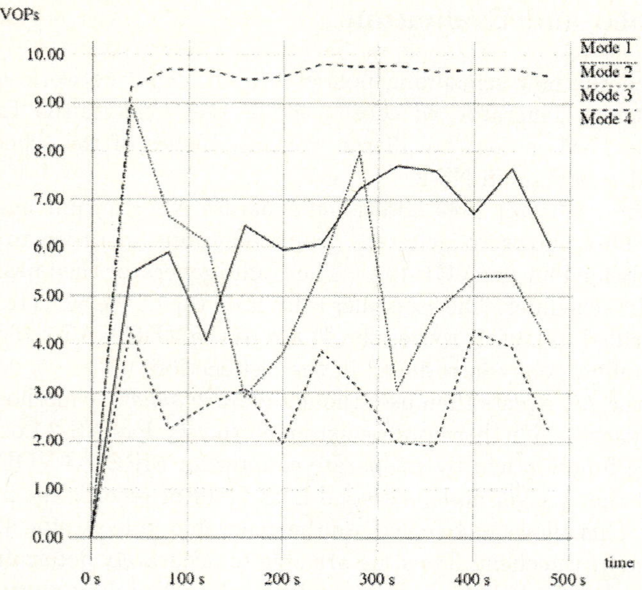

Fig. 3. Timely Delivered VOPs

decoder. Consequently, if a scheduling policy fails to transmit a reference VOP, subsequent VOPs are disregarded even if they are timely delivered. Using the new scheduling approach (Mode 4) we timely (i.e. before the deadline) deliver a considerable amount of 50,65 % more frames compared to the second best solution in Mode 3 (see Fig.3). With regard to the first policy denoted as Mode 1 we even transmit nearly double the amount of frames before the deadline, whereas compared to Mode 2, we still achieve 90% more timely scheduled frames. On the average, the new scheduling policy timely delivers 65% of all generated VOPs in strong overloaded conditions whereas this value decreases dramatically to 43% for Mode 2. Assigning all VOPs to the highest-prioritized Access Category surprisingly proves to be the worst approach in these conditions as only 32% of all VOPs can be timely delivered. The new scheduling approach is insusceptible to sudden workload changes. The amount of delivered VOPs stays constant over the whole duration of the video whereas other solutions suffer from strong jitter which significantly decreases the QoS percepted by the user.

5   Conclusions

This paper presented a new scheduling policy approach for 802.11 EDCA. It is especially designed to dynamically prioritize MPEG traffic and assign it to the EDCA ACs in order to improve the resource utilization and the timely delivery of MPEG VOPs. It relies on a statistical analysis of the available throughput per flow at saturation conditions. Evaluation results prove that this new approach handles MPEG-4 streaming considerably better than traditional solutions that mostly deploy static priority assignment schemes. We are able to schedule 50% more timely frames compared to the second best solution.

Acknowledgements. This work was supported by the EU Integrated Project EuQoS that has been established to bring end to end Quality of Service support for applications.

References

1. Baiceanu, V., Cowan, C.,McNamee, D.,Pu, C.,Walpole, J.: *Multimedia Applications Require Adaptive CPU Scheduling.* In Workshop on Resource Allocation Problems in Multimedia Systems, Washington D.C., December 1996.
2. Ditze, M.: *"A New Method for the Real Time Scheduling and Admission Control of MPEG-2 Streams"* M.Sc. thesis, School of Computer Science, Paderborn University, December 2001.
3. Ditze, M., Altenbernd, P., Loeser, C.: *Improving Resource Utilization for MPEG Decoding in Embedded End-Devices* In Proceedings of the Twenty-Seventh Australasian Computer Science Conference (ACSC2004), Dunedin, New Zealand, January, 2004.
4. Pong, D., Moors, T.: *"Call Admission Control for the IEEE 802.11 Contention Access Mechanism"* In Proceedings of the Globecom 2003, pp. 174-8, December, 2003.
5. Hertrich, D.: *"MPEG4 video transmission in wireless LANs: Basic QoS support on the data link layer of 802.11b"* Minor Thesis, Technical University of Berlin, October, 2002.
6. IEEE Standard for Information Technology - Part 11: Wireless Medium Access Control (MAC) and Physical Layer (PHY) specifications *Amendment7: Medium Access Controll (MAC) Quality of Service (QoS) Enhancements* IEEE 802.11e/D9.0, August, 2004.
7. Mangold, S., Choi, S., May, P., Klein, O., Hiertz, G., Stibor, L.: *"IEEE 802.11e Wireless LAN for Quality of Service"* IEEE Wireless Communications Magazine, Special Issue on Evolution of Wireless LANs and PANs, vol. 10, no. 6, December 2003.
8. International Organisation For Standardisation *"Information Technolgy -Generic Coding Of Audio-Visual Objects Part 2: Visual"* ISO/IEC JTC1/SC29/WG11.
9. Altenbernd, P., Burchard, L., Stappert, F.: *Worst-Case Execution Times Analysis of MPEG-2-Decoding* 12th Euromicro Conference on Real Time Systems, Stockholm, Sweden.
10. Wiethoelter, S., Hoene, C.: *Design and Verification of an IEEE 802.11e EDCF Simulation Model in ns-2.26* Technical Report TKN-03-019, Technische Universitaet Berlin, November 2003.

IEEE 802.11b WLAN Performance with Variable Transmission Rates: In View of High Level Throughput

Namgi Kim[1], Sunwoong Choi[2], and Hyunsoo Yoon[1,*]

[1] Div. of Computer Science, Dept. of EECS, KAIST,
373-1 Kuseong, Yuseong, Daejeon, 305-701, Korea
{ngkim, hyoon}@camars.kaist.ac.kr
[2] School of CSE, SNU,
San 56-1, Sinlim, Gwanak, Seoul, 151-742, Korea
schoi@popeye.snu.ac.kr

Abstract. Wireless networks have been rapidly integrated with the wired Internet and have been widely deployed. In particular, IEEE 802.11b WLAN is the most widespread wireless network today. The IEEE 802.11b WLAN supports multiple transmission rates and the rate is chosen in an adaptive manner by an auto rate control algorithm. This auto rate control algorithm deeply affects the total system performance of the IEEE 802.11b WLAN. In this paper, we examine the WLAN performance with regard to the auto rate control algorithm, especially the ARF scheme, which is the most popular auto rate control algorithm in 802.11b based WLAN products. The experimental results indicate that the ARF scheme works well in the face of signal noise due to node location. However, the ARF scheme severely degrades system performance when multiple nodes contend to obtain the wireless channel and the packet is lost due to signal collision.

1 Introduction

Recently, WLAN (Wireless LAN) has achieved tremendous growth and has become the prevailing technology for wireless access for mobile devices. WLAN has been rapidly integrated with the wired Internet and has been deployed in offices, universities, and even public areas. With this trend, many WLAN technologies based on different physical frequency bands, modulations, and channel coding schemes have been proposed and implemented. In particular, the IEEE 802.11b [1] is cited as the most popular WLAN technology today. The IEEE 802.11a [2] and IEEE 802.11g [3] WLAN have better performance than 802.11b. However, IEEE 802.11b based WLAN products will not disappear for some time, given that they have already been widely deployed throughout the world.

Contrary to the wired network, wireless channel conditions dynamically change over time and space. To cope with variable channel conditions, the IEEE 802.11b WLAN

* This work was supported by the Korea Science and Engineering Foundation (KOSEF) through the advanced Information Technology Research Center (AITrc) and University IT Research Center Project.

P. Lorenz and P. Dini (Eds.): ICN 2005, LNCS 3421, pp. 1080–1087, 2005.

specification provides multiple transmission rates that can maximize the system throughput in the face of radio signal diversity. The transmission rates should be chosen adaptively depending on the conditions of the wireless channel. In general, the transmission rates are automatically changed based on the historical results of previous transmissions and this auto rate control algorithm deeply affects the WLAN performance. Therefore, in this paper, we measure the WLAN performance with variable transmission rates in IEEE 802.11b based products and show the effects of the auto rate control algorithm.

2   IEEE 802.11b WLAN

The IEEE 802.11 WLAN standard [4] defines a single MAC (Medium Access Control) and a few different PHY (Physical Layer) specifications. The MAC specification deals with the framing operation, interaction with a wired network backbone, and interoperation with different physical layers. On the other hand, the PHY specifications deal with radio characteristics, modulations, error correcting codes, physical layer convergence, and other signaling related issues. The IEEE 802.11b is a kind of IEEE 802.11 WLAN standard with a particular PHY specification.

In the IEEE 802.11 WLAN standard, the MAC protocol provides two different access methods for fair access of a shared wireless medium: PCF (Point Coordination Function) and DCF (Distributed Coordination Function). PCF provides contention-free access and DCF provides contention-based access. PCF has been proposed to guarantee a time-bounded service via arbitration by a point coordinator in the access point. However, it is an optional access method and is rarely implemented in currently available 802.11 devices. In the IEEE 802.11 WLAN standard, the DCF is the mandatory and prevailing access method in the market today. Therefore, we also consider the DCF method of the IEEE 802.11 MAC protocol in this paper.

In the IEEE 802.11 WLAN standards, there are various PHY specifications that provide different transmission rates by employing different frequency bands, modulations, and channel coding schemes. The 802.11b PHY [1] provides four different transmission rates ranging from 1Mbps to 11Mbps at the 2.4 GHz ISM (Industrial, Scientific and Medical) frequency band. Another PHY specification, the 802.11a PHY [2], provides high-speed transmission with eight different rates ranging from 6Mbps to 54Mbps in the 5 GHz U-NIII (Unlicensed National Information Infrastructure) frequency band. It boosts up transmission rates by adapting the OFDM (Orthogonal Frequency Division Multiplexing) technique in the physical layer. Lastly, 802.11g PHY [3] has also been developed to adapt the OFDM in the 2.4 GHz ISM band. The 802.11g PHY theoretically has the same transmission rates as the 802.11a PHY. However, it covers a larger range than 802.11a because of operation in the lower frequency band.

Currently, products based on 802.11a PHY or 802.11g PHY are emerging in the consumer market. In particular, products based on 802.11g PHY have the potential to make inroads into the WLAN market because 802.11g PHY is backward compatible with 802.11b PHY. However, the most popular products in the current market are still based on 802.11b PHY. This may remain the status for some time since many 802.11b WLAN products have already been widely deployed. Therefore, we also use the 802.11b WLAN product for our experiments.

2.1 Auto Rate Control Algorithm with IEEE 802.11 WLAN

As noted earlier, the IEEE 802.11 WLAN standard supports multiple transmission rates. The standard, however, does not specify how to change the rates according to channel conditions. In the current WLAN fields, different auto rate control algorithms have been proposed. However, the ARF (Auto Rate Fallback) scheme [5] is the most popular auto rate control algorithm in IEEE 802.11b based WLAN products today. In the ARF scheme, the transmission rate is downgraded to the next lower rate when the transmission continually fails and as a result the ACK (acknowledgement) from the receiver is consecutively missed. The transmission rate is upgraded back to the next higher rate if the next consecutive transmissions are successful or some amount of time has passed. The ARF scheme is simple and easy to implement. However, it degrades WLAN throughput when many nodes attempt to transmit data in contention. In the next section, we show the results of our experiments and analyze the effects of the ARF scheme in IEEE 802.11b WLAN environments.

3 WLAN Performance with Variable Transmission Rates

To measure WLAN performance, we conducted experiments in real IEEE 802.11b based WLAN environments. For the experiments, we used Enterasys's RoamAbout 802.11b WLAN products [6], which adopt the Agere WLAN chipset [7]. These products also adopt the ARF scheme. In these products, the ARF scheme downshifts the transmission rate after missing two consecutive ACKs and upshifts after receiving five successive ACKs. Like other venders' products, Enterasys's WLAN cards support only a contention-based DCF MAC access method. We did not use the CTS/RTS mechanism in our experiments, as is the case in many products' default setting. As mobile nodes, we used four laptops and fifteen iPAQ PDAs. RedHat Linux (kernel version 2.4.7-10) [8] is installed in laptops and LinuxGPE (kernel version 0.7.2) [9] is installed in PDAs. Experiments were conducted in a building. The building has many offices and a long corridor. For each experiment, the mobile nodes send data to the wired nodes via the WLAN AP (Access Point) or vice versa. The size of a packet is 1KByte and each experiment lasted more than 300 seconds.

3.1 Throughput with Node Locations

First, we measured WLAN performance in different locations with one mobile node. Fig. 1 illustrates the locations of the node and the AP. The mobile node was located sequentially from *PT1* to *PT5*. *PT1* and *PT2* are located in the same room with the AP and the other three locations are out of the room in the corridor. In this experiment, a mobile node sends CBR data by UDP packets to the wired node via the AP. The offered traffic is 6.5Mb/s for each experiment. Since only one mobile node attempts to transmit data to the fixed AP in the air, the signal noise depending on the location of the mobile node becomes the most important factor in terms of signal quality.

Fig. 2 shows the total throughput results corresponding to node locations. The bar indicates the throughput depending on the transmission rates at the each location from *PT1* to *PT5*. The error bar denotes the standard deviation of observed throughput. The results indicate that the high transmission rate shows good throughput and small

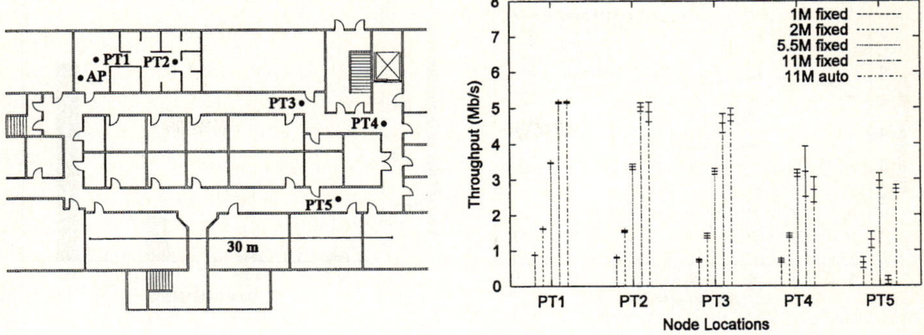

Fig. 1. Locations of mobile nodes and AP   **Fig. 2.** Throughputs with node locations

variation over the near locations. However, as the distance between the location and the AP increases, the high transmission rate suffers from low throughput and large variation because of radio signal noise. The low transmission rate, on the other hand, is relatively robust against signal noise and shows stable performance at all locations. The low transmission rate, however, cannot maximize throughput when the channel condition is good at a near location. As you can see in the results, the 11M auto rate shows relatively good performance in comparison to the fixed transmission rates. The 11M auto rate displays high throughput and low throughput variation over all locations. This is because the ARF scheme works well with packet loss due to signal noise. When only one node attempts to send data, there is no contention and the signal noise depending on location becomes the most important factor in terms of signal quality. Consequently, the 11M auto rate, which adapts the ARF scheme, works well over all locations.

We also conducted the experiments with a downlink traffic pattern. Thus, the wired node sends data to the mobile node. The results of the experiments with downlink traffic are very similar to those obtained from the previous uplink traffic experiments and are thus not presented.

3.2 Throughput with Node Contentions

In this experiment, we measured the effects of node contentions in WLAN environments. Multiple mobile nodes send CBR data through the AP to the wired nodes in a position. The nodes are located in the same place and contend to obtain the wireless channel. The total offered traffic is 6.5Mb/s and the number of nodes is changed for each experiment.

Fig. 3 shows total throughputs corresponding to different numbers of nodes. In the results, we find that the fixed transmission rates are rarely affected by the number of mobile nodes and they show good performance regardless of the number of contending nodes. However, the throughput of the 11M auto rate is severely degraded when the number of nodes is increased. When the number of the mobile nodes is eighteen, the throughput of the 11M auto rate is almost the same as that of the 1M fixed transmission rate.

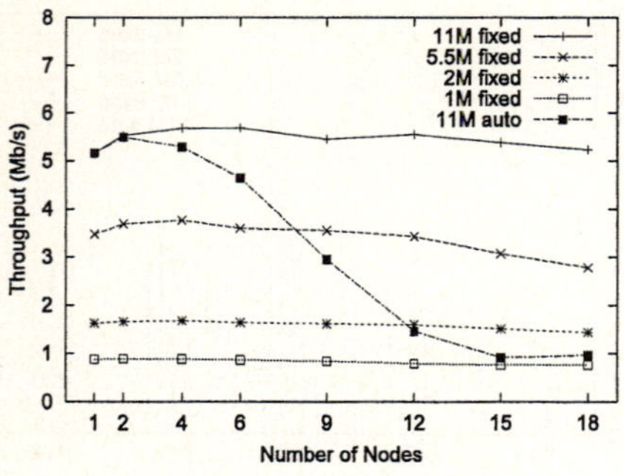

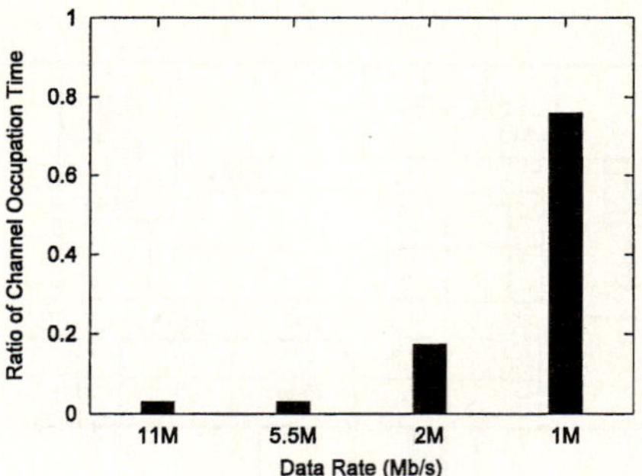

Fig. 3. Throughputs with node contentions   **Fig. 4.** Channel occupancy at 18 nodes

To analyze this phenomenon, we captured the WLAN frames using the Airopeek tool [10]. Through this trace, we examined the channel occupancy according to transmission rates. Fig. 4 shows the ratio of the channel occupation time when the 11M auto transmission rate is employed to eighteen mobile nodes. The results reveal that the 1M rate, the lowest transmission rate, occupies the channel for the greatest period of time and the 5.5M rate, the next lower rate, is second. Consequently, we determined that the throughput of the 11M auto rate was drastically degraded, because the low transmission rates occupied the channel for a lengthy period of time.

According to the results, the ARF scheme does not work correctly in the face of node contentions. When the number of nodes is increased and the nodes contend to obtain a wireless channel in the 802.11 DCF MAC protocol, the packet is lost due to collisions in the air. Packet loss due to collision causes the transmission rate to downshift unnecessarily even though the signal quality of the wireless channel is good. Moreover, this loss obstructs quick upshifting back to higher transmission rates. Therefore, the throughput with the auto rate control algorithm based on the ARF scheme is severely degraded when the node contention is heavy, and consequently packet loss due to signal collision occurs frequently.

3.3 Throughput with TCP

In this experiment, we measured the WLAN performance with TCP in node contentions. To do this, we send TCP data through the AP to the eighteen mobile nodes. TCP is a bidirectional protocol. Thus, the mobile nodes contend to obtain the wireless channel to send back TCP ACK packets. The other experimental setups are the same as those of the previous node contention experiments with UDP packets for CBR traffic.

Fig. 5 shows the TCP throughput corresponding to different transmission rates with eighteen mobile nodes. The throughputs of the fixed transmission rates are similar to the results of the previous CBR experiments. However, the throughput of the 11M auto rate is quite different. The throughput of the 11M auto rate is not severely degraded. It is almost the same as that of the 11M fixed transmission rate, even though the eighteen mobile nodes contend with each other to obtain the wireless channel.

To analyze this effect, we captured the WLAN frames again. Using the raw WLAN trace, we calculated the active nodes attempting to obtain the channel at the same time. In these TCP experiments, the active node is the actual node that is involved in

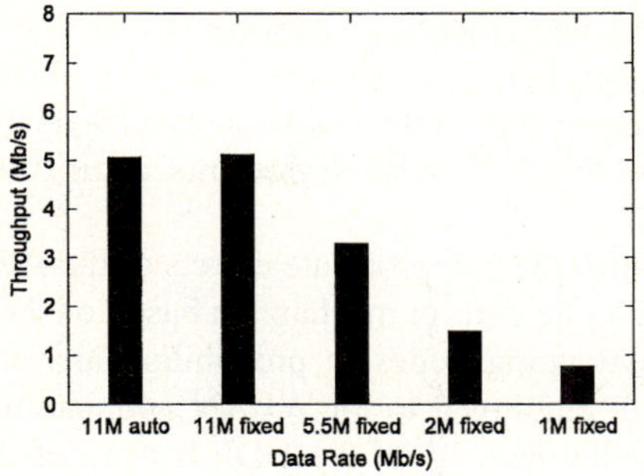
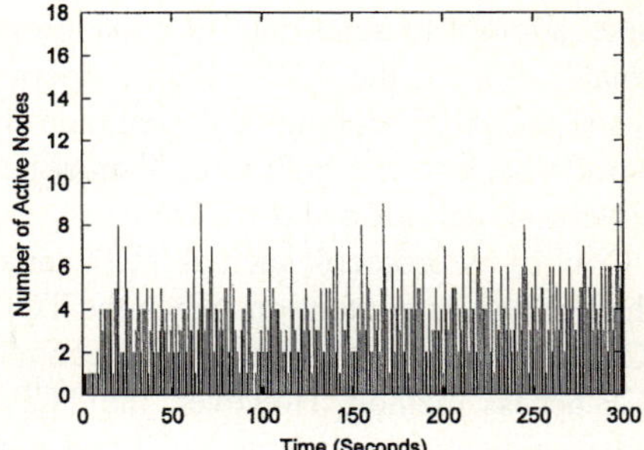

Fig. 5. TCP throughput with 18 nodes

Fig. 6. Active nodes in contentions

contention. We define a node as active if the AP has sent a TCP DATA packet to the mobile node, but the mobile node has yet to send back the TCP ACK packet to the AP. This definition is not strictly correct because the TCP adopts a commutative ACK mechanism. However, we believe this definition is adequate in terms of explaining the TCP performance with the ARF scheme. Fig. 6 shows the number of active nodes per second when the eighteen mobile nodes employ the 11M auto rate. The average number of active nodes is 3.64, which is quite smaller than the number of actual mobile nodes. Accordingly, we infer that the throughput of the 11M auto rate is not degraded because the contention is relieved when using TCP.

TCP reduces the degree of node contentions because it has already adopted rate control mechanism. A TCP ACK packet is generated by successfully transmitted a TCP DATA packet, and a TCP DATA packet is also generated by successfully transmitted a TCP ACK packet. Thus, the mobile node cannot be involved in contentions before receiving a TCP DATA packet from the AP. Also, the AP cannot send the next packet before the mobile node sends back a TCP ACK packet to the AP. In the IEEE 802.11 WLAN, the contention-based DCF access method fairly distributes the wireless channel to all stations. The DCF method does not give greater priority to the AP, which causes to increase the active nodes by consecutive frame transmissions from the AP to every different mobile node. Therefore, when traffic is regulated by TCP, the number of contending nodes is maintained smaller than the actual nodes.

The upstream TCP experiments also showed similar results. This is because TCP regulates contention of nodes attempting to send TCP DATA packets to the wired nodes through the AP reversely.

4   Conclusions and Discussion

The IEEE 802.11b WLAN standard supports multiple transmission rates and the auto rate control algorithm attempts to select the best transmission rate based on the current wireless channel condition. The most popular auto rate control algorithm in the current IEEE 802.11b WLAN products is the ARF scheme. The ARF scheme changes the transmission rates based on the success or failure of previous frame transmissions. This mechanism shows good performance when few mobile nodes exist and channel quality is dominated by signal noise due to location of the node. However, if many nodes attempt to send data in contention and some amount of packets is lost due to signal collision, the ARF scheme operates improperly. Packet loss due to collision causes the ARF scheme to downgrade the transmission rate inaccurately and delays upgrade back to the high rate. This is because the ARF scheme does not distinguish between signal noise and collisions.

Control by protocol, such as TCP, can diminish the inappropriate effects of the ARF scheme in node contention. Since the TCP has a rate control mechanism based on TCP ACK, it decreases the number of actively contending nodes in probabilistically fair DCF access method. However, the TCP is not employed to all WLAN applications. Some applications, such as VoIP and multimedia data, adopt the UDP instead of the TCP. Therefore, the TCP cannot be an optimal solution for all WLAN applications.

Performance abnormality with variable transmission rates in IEEE 802.11b WLAN has been recently noted in some papers [11, 12]. M. Heusse et al. [11] observed performance anomalies with multiple transmission rates. They pointed out that when some mobile nodes use a lower transmission rate than others, the overall WLAN system performance is considerably degraded. However, they only addressed the relationship between the system performance and the mixed transmission rates. They did not determine the effect of an auto rate control algorithm and TCP protocol. In [12], P. Berthou et al. noted that the auto rate control algorithm could degrade the system performance and it should be disabled for multimedia applications. However, they also did not determine why an auto rate control algorithm such as the ARF scheme degrades WLAN system performance. Moreover, they did not perform in-depth study on this phenomenon.

The newly proposed SNR-based algorithm [13] may solve this performance abnormality with the ARF scheme. The SNR-based auto rate control algorithm adaptively changes the transmission rates based on the SNR value of the received packets. It can prevent degradation of the WLAN system performance in node contentions by distinguishing between signal noise and collision. However, the SNR-based algorithm has not been applied in practice. Moreover, unresolved problems such as a means to obtain exact SNR, a practical mapping function between the rate and SNR value, and asymmetry of up and down wireless channels must still be addressed.

We have evaluated the performance of IEEE 802.11b WLAN with variable transmission rates. In particular, we examined the effect of the auto rate control algorithm in real WLAN environments and revealed why the ARF scheme degrades WLAN performance in node contentions. Lastly, we determined the relationship between TCP and the auto rate control algorithm. For future work, we plan to investigate other different auto rate control algorithms with different WLAN standards such as IEEE 802.11a and 802.11g.

References

1. IEEE Std 802.11b-1999, "Part 11: Wireless LAN Medium Access Control (MAC) and Physical Layer (PHY) specifications: High-speed Physical Layer Extension in the 2.4GHz Band," Supplement to ANSI/IEEE Std 802.11, Sep. 1999.
2. IEEE Std 802.11a-1999, "Part 11: Wireless LAN Medium Access Control (MAC) and Physical Layer (PHY) specifications: High-speed Physical Layer Extension in the 5GHz Band," Supplement to ANSI/IEEE Std 802.11, Sep. 1999.

3. IEEE Std 802.11g-2003, "Part 11: Wireless LAN Medium Access Control (MAC) and Physical Layer (PHY) specifications: Further Higher Data Rate Extension in the 2.4GHz Band," Amendment to IEEE 802.11 Std, Jun. 2003.
4. IEEE Std 802.11, "Wireless LAN MEdium Access Control (MAC) and Physical Layer (PHY) specifications," ANSI/IEEE 802.11 Std, Aug. 1999.
5. A. Kamerman and L. Monteban, "WaveLAN-II: A High-Performance Wireless LAN for the Unlicensed Band," Bell Labs Technical Journal, pp. 118-133, Summer 1997.
6. http://www.enterasys.com/
7. http://www.agere.com/
8. http://www.redhat.com/
9. http://www.handhelds.org/
10. http://www.wildpackets.com/
11. M. Heusse, F. Rousseau, G. Berger-Sabbatel, A. Duda, "Performance Anomaly of 802.11b," In proceedings of IEEE INFOCOM 2003, Mar. 2003.
12. P. Berthou, T. Gayraud, O. Alphand, C. Prudhommeaux, M. Diaz, "A Multimedia Architecture for 802.11b Networks," In proceedings of IEEE WCNC 2003, Mar. 2003.
13. J. P. Pavon and S. Choi, "Link Adaptation Strategy for IEEE 802.11 WLAN via Received Signal Strength Measurement," In proceedings of IEEE ICC 2003, May 2003.

Some Principles Incorporating Topology Dependencies for Designing Survivable WDM Optical Networks

Sungwoo Tak

Department of Computer Science and Engineering,
Pusan National University, San-30, Jangjeon-dong,
Geumjeong-gu, Busan, 609-735, Republic of Korea
Tel: +82-51-510-2387
swtak@pnu.edu

Abstract. Many researchers have focused on addressing a few problems of physical and virtual topology dependencies for designing survivable WDM (Wavelength Division Multiplexing) optical networks. If dependencies between the physical topology and the virtual topology are not considered in restoration, it is not possible to achieve optimal restoration. The problem related to physical and virtual topology dependencies is NP-Hard (Nondeterministic Polynomial-Time), therefore, some principles incorporating topology dependencies are required to design a survivable WDM optical network. Consequently, we propose the HRM (Hybrid Restoration Method) principle that takes advantage of the efficient restoration resource utilization of the path-based restoration method and the fast restoration time of the link-based restoration method. The proposed restoration principles incorporating topology dependencies and the HRM principle can be exploited for the design of survivable WDM optical networks.

Keywords: WDM; optical networks; principles; restoration; topology dependencies.

1 Introduction

The possible growth of bandwidths with WDM technology increases the impact of link and node failures. Since WDM optical network resources are expensive, the optimization of resources in designing a survivable WDM optical network is important. Such an optimization results in cost savings. The physical topology in the WDM optical network is physically composed of optical fiber links and photonic nodes. So far, we have analyzed the following physical topology dependencies: 1) optical network architectures, 2) types of physical network topologies: planar and non-planar topologies, 3) link and node connectivity, 4) optimal paths, 5) maximum link disjoint or node disjoint paths, and 6) survivability for multiple link and node failures.

First, in optical network architecture, we need to produce an optimal partial mesh physical topology if the physical topology does not satisfy $(n - 1)$ connectivity given traffic demands with n nodes and a recovery level against link failures and node failures. This problem consists of four sub-problems: (1) the number of optimal node

P. Lorenz and P. Dini (Eds.): ICN 2005, LNCS 3421, pp. 1088–1096, 2005.

degree, (2) the number of optimal links, (3) the number of optimal interconnectivities among regions, and (4) population of optimal nodes and links in a region. Second, a physical topology can be classified into planar and non-planar topologies. Compared to non-planar topologies, planar topologies have a natural advantage for visualization and other NP-Hard problems can be easily solved within a better asymptotic time complexity than the best-known algorithm for general topologies [1]. When a physical topology is nonplanar, the nonplanar topology may need to be transformed into a planar topology in a way as close to planarity as possible. Third, the concept of connectivity is analyzed by mathematical approach. Fourth, it is important to find a set of optimal paths from a global point of view. When finding an optimal path is NP-Hard, its near optimal solution should be considered to find a sequence of paths, which is associated with a sequence of objective function outcomes converging to the near optimal solution. Fifth, Maximum link disjoint paths or node disjoint paths can be used to improve survivability in WDM optical networks. The maximum link disjoint path problem or the node disjoint path problem that is known for its inherent difficulty. Finally, a set of minimum cycles in G_p can be used for survivability of multiple link and node failures [2]. Unfortunately, problems related to physical topology dependencies are NP-hard problems [3-4]. NP-hard problems cannot be solved optimally within polynomial computation times.

The virtual topology in the WDM optical network consists of a graph representing all optical communication paths called lightpaths underlying the physical topology. In order to guarantee high restoration with minimum network resources and costs on a fixed physical topology, virtual topology dependencies should be observed. The virtual topology dependencies for the optimal design of an efficient restoration technique in WDM optical networks are analyzed as follows: First, as a fundamental requirement in WDM optical networks, the distinct wavelength assignment constraint is considered. Second, according to the functionality of optical network nodes, optical connectivity may be static or controllable. A static coupling device on nodes, which consists of passive splitting and combining nodes interconnected by optical fibers, is one of the simplest systems for static optical connectivity. The most controllable optical connectivity contains active nodes, ranging from WSXC (Wavelength Selective Cross Connect) to WIXC (Wavelength Interchanging Cross Connect). The benefit of using controllable optical systems in WDM optical networks is still an open issue and has been questioned by many researchers [5-7]. Third, single-hop and multi-hop virtual WDM optical networks need to be explored. In multi-hop virtual WDM optical networks, there is an operational requirement to provide software that generates these maps for the virtual topology corresponding to a reconfiguration of the physical topology [8-9]. [10] suggests that single-hop virtual WDM optical networks might be better than multiple-hop virtual WDM optical networks for survivable network dimensioning. In developing single-hop virtual WDM optical network architectures, one must remember that restoration techniques must be implemented simply. Namely, restoration techniques should be based on realistic assumptions on the properties of WDM optical components and they are scalable to accommodate a large and expandable user population. Unfortunately, there is no clear winner between single-hop and multi-hop virtual WDM optical networks, so the capabilities and characteristics of both categories of WDM optical networks must be thoroughly examined. Fourth, a process of assigning resources is required to make provision for WDM optical networks. In the absence of any wavelength conversion devices, a lightpath is required to be the same wavelength throughout its path between two end nodes in the network. This requirement is referred to as the wavelength continuity. Given a physical network topology and a number of available wavelengths, the establishment of a given set of available lightpaths is an NP-Hard problem [11]. Finally, wavelengths used for restoration have to be chosen carefully to avoid wavelength conversion as well as wavelength blocking in nodes because wavelength conversion is extremely expensive [12].

2   HRM (Hybrid Restoration Method) Principle

To speed up the restoration time of the path-based restoration method, the link-based restoration method needs to be embodied in the path-based restoration method at the expense of a few additional backup facilities. The HRM (Hybrid Restoration Method) principle proposed in this paper takes advantage of the efficient restoration resource utilization of the path-based restoration method and the fast restoration time of the link-based restoration method. In the HRM, when a link fault or a node fault occurs, an attempt is first made to execute faster restoration based on the link-based restoration method. If this attempt succeeds, then a quick restoration is made. If not, then slower but efficient resource utilization restoration based on the path-based restoration method is executed. To resolve trade-offs between restoration cost and restoration time, the following equation (1) is expressed and analyzed:

$$T_s = \delta \cdot R_{1T} + (1 - \delta) \cdot (R_{1T} + R_{2T}) = R_{1T} + (1 - \delta) \cdot R_{2T} \tag{1}$$

T_S stands for average restoration time. In equation (1), R_1 is a fast but poor resource utilization restoration method. R_2 is a slow but efficient resource utilization restoration method. R_{1T} is the restoration time of R_1. R_{2T} is the restoration time of R_2. δ is a successful restoration execution ratio of R_1. If δ is equal to 0, T_S increases as much as $(R_{1T} + R_{2T})$. If δ is equal to 1, T_S increases as much as R_{1T}. Average restoration time in a WDM optical network is much closer to that of R_1 than that of R_2 when δ becomes higher. Let γ denote the ratio of R_{2T} to R_{1T}. As γ and δ increase higher, T_S decrease lower. Thus, the fast value of R_{1T} as well as the high successful ratio of δ is required to improve T_S in a WDM optical network. All outcomes illustrated in Figures 1 and 6 are analytically measured by the MATLAB tool provided by the Mathworks Incorporation. First, average restoration time T_S is shown in Figure 1. Let γ denote the ratio of R_{2T} to R_{1T}. As γ and δ values increases higher in Figure 1, T_S decrease lower. Thus, fast R_{1T} as well as high δ is required to improve average restoration time T_S in a WDM optical network. The relationship between the average restoration cost and the network resources in the HRM principle is expressed as follows:

$$C_S = (C_{1R} \cdot R_{1W} + C_{2R} \cdot R_{2W}) / (R_{1W} + R_{2W}) \tag{2}$$

In equation (2), C_S is the average network resource cost required to implement the HRM principle. C_{1R} is the average network resource cost of R_1. C_{2R} is the average network resource cost of R_2. R_{1W} is backup wavelengths required by R_1. R_{2W} is backup wavelengths required by R_2. C_S is an approximate optimal cost solution for the HRM principle. Thus, a few additional backup facilities of R_1 are required to achieve an approximate optimal cost solution in the HRM principle, where $C_{1R} \gg C_{2R}$ and $R_{1W} \ll R_{2W}$. Figure 2 illustrates the ratio of C_{2R} to C_S.

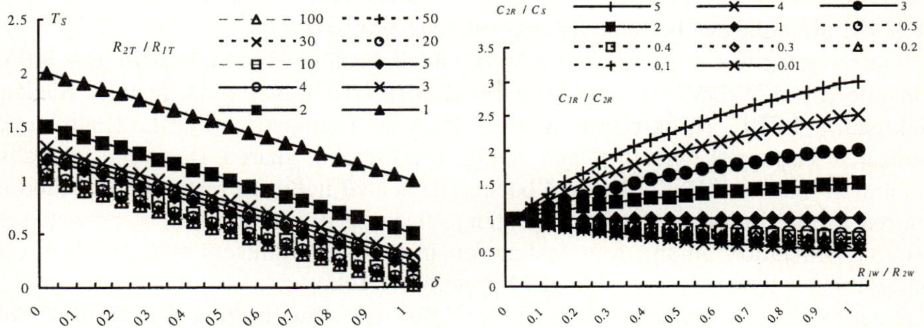

Fig. 1. Average restoration time of HRM   **Fig. 2.** Average restoration cost of HRM

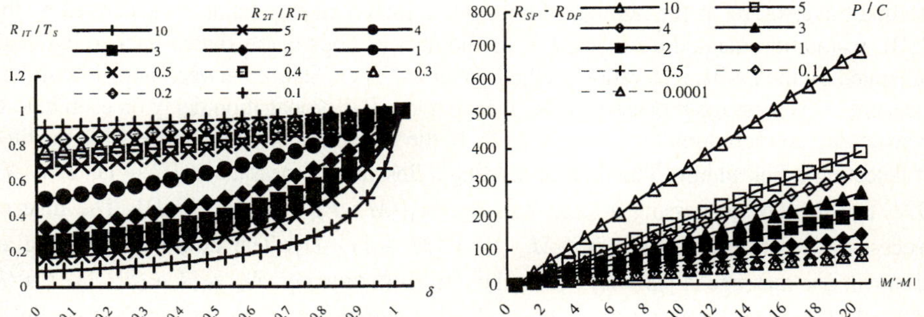

Fig. 3. Restoration access efficiency of HRM   **Fig. 4.** Restoration time difference between R_{SP} and R_{DP}

$$(C_{2R} / C_S) = \{(((C_{1R} / C_{2R}) \cdot \eta) + 1) / (\eta + 1)\}, \text{ where } \eta = (R_{1W} / R_{2W}) \tag{3}$$

When C_{1R} is greater than C_{2R} in Figure 2, the low value of η will reduce the ratio of C_{2R} to C_S. Therefore, fewer backup wavelengths of R_1 is required, compared to those of R_2. When the ratio of C_{1R} to C_{2R} is below 1 in Figure 2, it does not affect the ratio of C_{2R} to C_S. Thus, if C_{2R} is greater than C_{1R}, the HRM principle does not play an important role in achieving cost-efficient and fast restoration at the expense of a few additional backup facilities. In equation (1), the value of δ expected to satisfy the performance requirement is derived as follows:

$$R_{1T} / T_S = 1 / \{ 1 + [(1- \delta) \cdot R_{2T} / R_{1T}]\} \tag{4}$$

Equation (4) represents the restoration access efficiency of the HRM principle. It is a measure of how close average restoration access time T_S is to R_{1T}. As the value δ becomes close to 1 in Figure 3, average restoration access time T_S becomes close to

R_{1T}. When the ratio of R_{2T} to R_{1T} is greater than 2, the HRM principle seems to yield efficient restoration access time. Especially, when the value of δ exists in the range of 0.7 through 1, it generates the good performance of restoration access efficiency of the HRM principle. If the value of δ in equation (1) is 1, a link fault or a node fault is always restored by R_1. As shown in Figure 3, the ideal restoration access efficiency of the HRM principle can be achieved when the value of δ is 1.

Four well-known restoration methods can be achieved to actualize the HRM principle in a WDM optical network: LDBL (Link-based Dedicated Backup Lightpath), LSBL (Link-based Shared Backup Lightpath), PDBL (Path-based Dedicated Backup Lightpath), and PSBL (Path-based Shared Backup Lightpath) restoration methods. The link-based backup dedicated lightpath restoration method is ignored in the design of the HRM principle because it produces the worst network resource utilization among four restoration methods. Parameters and variables are formulated for the analysis of the HRM principle as follows:

R_{DP} stands for the restoration time of the PDBL restoration method. R_{SP} stands for the restoration time of the PSBL restoration method. R_{SL} stands for the restoration time of the LSBL restoration method. N is the number of hops that a primary lightpath traverses. M is the number of hops that a dedicated backup lightpath generated by the PDBL restoration method traverses. M' is the number of hops that a shared backup lightpath generated by the PSBL restoration method traverses. L is the number of hops that a shared backup lightpath generated by the LSBL restoration method traverses. The distance between hops is assumed the same. D is message-processing time at each node. P is propagation delay on each link. C is switching configuration time at each node. In the PDBL restoration method, F is the time of detecting a link failure. The time of sending a link failure message is $\{(N - 1) \cdot P + (N \cdot D)\}$. The time of processing a setup message is $\{(M \cdot P) + (M + 1) \cdot D\}$. The time of processing a confirm message is $\{(M \cdot P) + (M + 1) \cdot D\}$. The time of processing an acknowledge message is $\{(M \cdot P) + (M + 1) \cdot D\}$. Consequently, R_{DP} is $\{F + (N - 1) \cdot P + (N \cdot D) + 2 \cdot (M \cdot P) + 2 \cdot (M + 1) \cdot D\}$. In the PSBL restoration method, the time of detecting a link failure is F. The time of sending a link failure message is $\{(N - 1) \cdot P + (N \cdot D)\}$. The time of processing a setup message is $\{(M' \cdot P) + (M' + 1) \cdot D\}$. The time of switching configuration is $\{(M' + 1) \cdot C\}$. The time of processing a confirm message is $\{(M' \cdot P) + (M' + 1) \cdot D\}$. The time of processing an acknowledge message is $\{(M' \cdot P) + (M' + 1) \cdot D\}$. R_{SP} is $\{F + (N - 1) \cdot P + (N \cdot D) + 2 \cdot (M' \cdot P) + 2 \cdot (M' + 1) \cdot D + (M' + 1) \cdot C\}$.

In the LSBL restoration method, the time of detecting a link failure is F. The time of processing a setup message is $\{L \cdot P + (L + 1) \cdot D\}$. The time of switching configuration is $\{(L + 1) \cdot C\}$. The time of processing a confirm message is $\{L \cdot P + (L + 1) \cdot D\}$. The time of processing an acknowledge message is $\{L \cdot P + (L + 1) \cdot D\}$. R_{SL} is $\{F + 2 \cdot (L \cdot P) + 2 \cdot (L + 1) \cdot D + (L + 1) \cdot C\}$. If $R_{SP} \geq R_{DP}$ in the analysis of restoration time difference between R_{SP} and R_{DP}, then R_{SP} takes $3 \cdot P \cdot (M' - M) + 3 \cdot D \cdot (M' - M) + (M' + 1) \cdot C\}$ longer than R_{DP}. If $M' = M$, R_{SP} takes $\{(M' + 1) \cdot C\}$ longer. If $M' > M$, R_{SP} takes $\{3 \cdot (P \cdot \alpha + 3 \cdot D \cdot \alpha + (M' + 1) \cdot C)$ longer, where $|m'-m| = \alpha$. If $M' < M$, R_{SP} takes $\{(M' + 1) \cdot C - 3 \cdot \alpha \cdot (P + D)\}$ longer. If $R_{DP} \geq R_{SL}$, then $L \geq \{[P - \beta \cdot (N + 3 \cdot M) + C] / (3 \cdot \beta - C)\}$, where $\beta = (P + D)$. Figure 4 shows the restoration time difference between R_{SP} and R_{DP}. In general, R_{SP} takes longer than R_{DP} because the PSBL restoration method requires additional switching configuration time and it requires additional number of hops by the distinct wavelength assignment and the wavelength blocking constraints. If $R_{SP} \geq R_{DP}$, then $\{3 \cdot P \cdot (M' - M) + 3 \cdot D \cdot (M' - M) + (M' + 1) \cdot C\} \geq 0$. If $M' = M$, R_{SP} takes $\{(M' + 1) \cdot C\}$ longer. If $M' > M$, R_{SP}

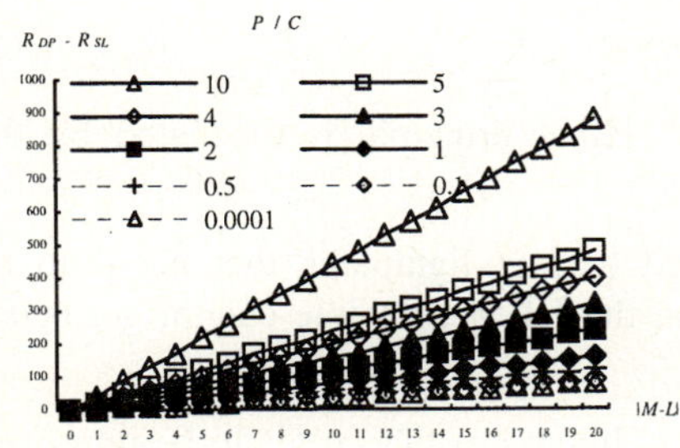

Fig. 5. Restoration time difference between R_{DP} and R_{SL}

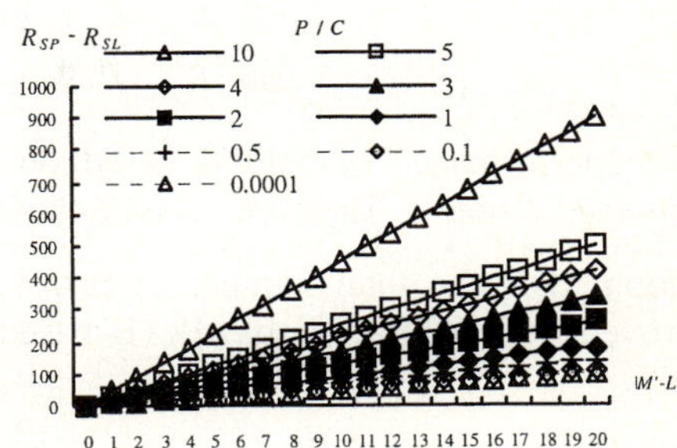

Fig. 6. Restoration time difference between R_{SP} and R_{SL}

takes $\{3 \cdot (P \cdot \alpha + 3 \cdot D \cdot \alpha + (M' + 1) \cdot C)$ longer, where $|m'-m| = \alpha$. If $M' < M$, R_{SP} takes $\{(M' + 1) \cdot C - 3 \cdot \alpha \cdot (P + D)\}$ longer. In Figure 4, as α increases and the ratio of P to C becomes higher, R_{SP} takes much longer than R_{DP}. Figure 5 shows the restoration time difference between R_{DP} and R_{SL}. In general, R_{DP} takes longer than R_{SL} because the number of hops used in the LSBL restoration method is shorter than that used in the PDBL restoration method. As the number of hops that backup lightpaths traverse increases, the propagation delay of backup lightpaths increases and the restoration time of backup lightpaths takes longer. If $R_{DP} \geq R_{SL}$, then $L \geq \{[P - \beta \cdot (N + 3 \cdot M) + C] / (3 \cdot \beta - C)\}$, where $\beta = (P + D)$. Since $L \geq 1$, L becomes $1 \leq L \leq \{[P - \beta \cdot (N + 3 \cdot M) + C] / (3 \cdot \beta - C)\}$. As $|M - L|$ increases and the ratio of P to C becomes higher, R_{DP} takes much longer than R_{SL}. Figure 6 shows the restoration time difference between R_{SP} and R_{SL}. In general, R_{SP} takes longer than R_{SL} because the number of hops used in the LSBL restoration method is shorter than that used in the PSBL restoration method. As the number of hops that backup lightpaths traverse increases, the propagation delay of backup lightpaths increases and the restoration time of backup lightpaths takes longer. If $R_{SP} \geq R_{SL}$, then $L \leq \{[\beta \cdot (N + 3 \cdot M') + (M' \cdot C) - P] / (3 \cdot \beta + C)\}$, where $\beta = (P + D)$. Since $L \geq 1$, L becomes $1 \leq L \leq \{[\beta \cdot (N + 3 \cdot M') + (M' \cdot C) - P] / (3 \cdot \beta + C)\}$. As $|M - L|$ increases and the ratio of P to C becomes higher, R_{DP} takes much longer than R_{SL}.

As a whole, the LSBL restoration method is applicable to R_1 and the PDBL restoration method or the PSBL restoration method is applicable to R_2 for the design of the HRM principle. One of possible heuristic rules required to design the HRM principle is to assign backup lightpaths based on the concept of regions in a physical topology. For example, each region uses the LSBL restoration method for rapid restoration speed and inter-regions use the PDBL (or the PSBL) restoration method for efficient network resource utilization. In the other way, each region uses the

PDBL (or the PSBL) restoration method for efficient network resource utilization and inter-regions use the LSBL restoration method for rapid restoration speed. Thus, region clustering is one of future significant research topics. The MRTB (objective function of Minimizing Restoration Time of Backup lightpaths) can be considered in the HRM principle. MRTB based on the PDBL and MRTB based on the PSBL are formulated in equation (5) and equation (6) respectively.

$$min \, [R_{SL} - R_{DP}], \text{ where } R_{DP} \geq R_{SL} \tag{5}$$
$$min \, [R_{SL} - R_{SP}], \text{ where } R_{SP} \geq R_{SL} \tag{6}$$

The performance of MRTB based on the HRM principle is evaluated by the following Lemma 1, Theorem 1, and Lemma 2.

Lemma 1. If the total number of path-based backup lightpaths that needs to be improved by MRTB is zero, MRTB based on the HRM principle cannot get better results.

Proof. If the total number of path-based backup lightpaths that needs to be improved by MRTB is zero, MRTB based on the HRM principle cannot find any link-based shared backup lightpaths whose restoration time takes shorter than the path-based backup lightpaths. Thus, MRTB based on the HRM principle cannot get better results.

Theorem 1. Let us assume that a full mesh physical topology G_P ($N_P \geq 3$, $|P_{mn}| = 2$, $k(G_P) = N_P - 1$) and a virtual topology G_V that requires one uniform traffic demand among all nodes. W, total number of wavelengths required for the construction of G_V, is 3 when the PDBL restoration method is used.

Proof. N_p denotes the number of nodes in the physical topology. $|P_{mn}|$ denotes the number of optical fiber links between node m and node n. $k(G_P)$ denotes the node connectivity of the physical topology. Each node has uniform traffic demands that will be transmitted to all other nodes. Each traffic stream requires one wavelength. When N_P is 3, G_P becomes G_P' illustrated in Figure 7. Each primary lightpath traverses one physical hop and its backup lightpath traverses two physical hops. One physical link is occupied with at most one primary lightpath and two backup lightpaths. Thus, W required to construct G_V in G_P' is 3. G_P' is the smallest component to construct G_V in the G_P. G_P also takes G_P' pieces. Therefore, W for the construction of G_V in G_P is 3.

Lemma 2. In a full mesh physical topology G_P ($N_P \geq 3$, $|P_{mn}| = 2$, $k(G_P) = N_P - 1$) and G_V (Δ_l, $N_V = N_P$, T_{sd} = one uniform traffic demand among all nodes, $\Gamma_{set} = \{ \Gamma_{MAXDEG}, \Gamma_{IODEG}, \Gamma_{FLOW}, \Gamma_{DCA}, \Gamma_{WA}, \Gamma_{LC}, \Gamma_{NC} \}$, W, S_H), the minimum number of path-based backup lightpaths that needs to be improved by MRTB based on the HRM principle is zero.

Proof. Each node has uniform traffic demands that will be transmitted to all other nodes. Each traffic stream requires one wavelength. By Theorem 1, G_P' ($N_P = 3$, $|P_{mn}| = 2$, $k(G_P') = 2$) is the smallest component in G_P. G_P also takes G_P' pieces vice versa. All path-based dedicated backup lightpaths in G_P' are the same as all link-based

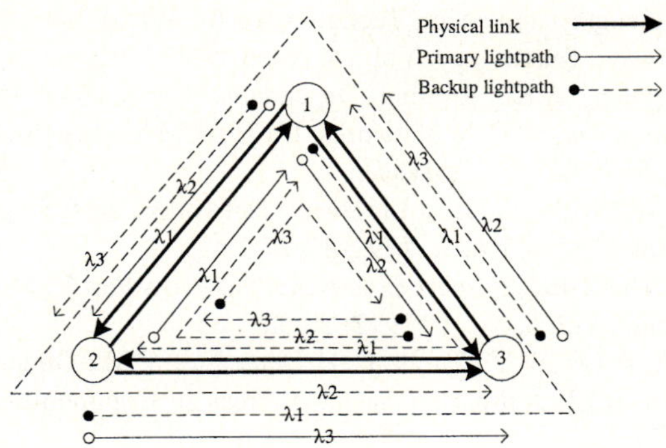

Fig. 7. A physical topology G_P' ($N_P = 3$, $|P_{mn}| = 2$, $k(G_P') = 2$)

shared backup lightpaths as shown in Figure 7. By Lemma 1 and Theorem 1, the minimum number of path-based backup lightpaths that needs to be improved by MRTB based on the HRM principle is zero in G_P and G_V.

3 Conclusion

In this paper, we first analyze some principles incorporating physical topology and virtual topology dependencies for designing survivable WDM optical networks. Additionally, we show the HRM principle that takes advantage of the efficient restoration resource utilization of the path-based restoration method and the fast restoration time of the link-based restoration method. For further study, it is necessary to propose a new efficient restoration technique incorporating principles addressed in this paper. The proposed restoration technique needs to show advantages over existing restoration methods in WDM optical networks.

Acknowledgement

This work has been supported by "Research Center for Future Logistics Information Technology" hosted by the Ministry of Education in Korea and Pusan National University Research Grant (2005).

References

1. Liebers, A.: Planarizing Graphs–A Survey and Annotated Bibliography. Journal of Graph Algorithms and Applications, Vol. 5, No. 1, (2001) 1-74
2. Ellinas, G., Hailemariam, A., Stern, T.E.: Protection Cycles in Mesh WDM Networks. IEEE Journal on Selected Areas in Communications, Vol. 18. No. 10, (2000) 1924-1937
3. Merris, R.: Graph Theory. Wiley-Interscience, 2000.
4. West, D.B.: Introduction to Graph Theory. Prentice Hall, 2000
5. Borella, M.S., Jue, J.P., Banerjee, D., Ramamurthy, D., Mukherjee, B.: Optical Components for WDM Lightwave Networks. Proc. of the IEE, Vol. 85, No. 8, (1997) 1274-1307

6. Yoo, S.J.: Wavelength Conversion Technologies for WDM Network Applications. IEEE Journal of Lightwave Technology, Vol. 14, No. 6, (1996) 955-966
7. Mohan. G., Somani, A.K.: Routing Dependable Connections with Specified Failure Restoration Guarantees in WDM Networks. IEEE/ACM Transactions on Networking, Vol. 9, No. 5, (2001) 553-566
8. Mukherjee, B.: WDM-Based Local Lightwave Networks Part I: Single-hop Systems. IEEE Network Magazine, Vol. 6, No. 3, (1992) 12-27
9. Mukherjee, B.: WDM-Based Local Lightwave Networks Part II: Multihop Systems. IEEE Network Magazine, Vol. 6, No. 4, (1992) 20-32
10. Caenegem, B.V., Parys, W.V., Turck, F.D., Demeester, P.M.: Dimensioning of Survivable WDM Networks. IEEE Journal on Selected Areas in Communications, Vol. 16, No. 7, (1998) 1146-1157
11. Chlamatic, I., Ganz, A., Karmi, G.: Lightpath communications: An approach to high bandwidth optical WAN's. IEEE Transactions on Communications, Vol. 40, No. 7, (1992) 1171-1182
12. Zhou, D., Subramaniam, S.: Survivavility in Optical networks. IEEE Network, Vol. 14, No. 6, (2000) 16-23

Resilient Routing Layers for
Network Disaster Planning

Audun Fosselie Hansen[1,2], Amund Kvalbein[1], Tarik Čičić[1], and Stein Gjessing[1]

[1] Networks and Distributed Systems Group,
Simula Research Laboratory, Oslo, Norway
[2] Telenor R&D, Oslo, Norway
{audunh, amundk, tarikc, steing}@simula.no

Abstract. Most research on network recovery has been centered around
two common assumptions regarding failure characteristics: Failures do
not occur simultaneously and failures do mostly strike links. Even this
may be the characteristics of everyday failures, we argue that disasters
like earthquakes, power outages and terrorist attacks impose other failure
characteristics. In this paper we demonstrate how our method, called
'Resilient Routing Layers', can be used as a tool for recovery from failures
adhering to such disaster characteristics.

Keywords: Network resilience, Recovery in various networks.

1   Introduction

Resilience against physical attacks was one of the primary design goals of the
Internet from the outset [1]. The distributed nature of control information and
routing algorithms allows the Internet to recover from link or node failures by
calculating new valid paths in the remaining network. However, special challenges arise for the Internet when faced with disastrous events like earthquakes,
floods, hurricanes, large-scale accidents, power outages or terrorist attacks.

From a networking point of view, disasters like the ones just mentioned have
two key characteristics. First, a *large number of nodes* can be taken out at the
same time. This makes many traditional protection mechanisms unsuitable, since
they are often designed to protect against single failures, and most methods focus
on link failures only. Second, the failing nodes are geographically near each other,
giving poor connectivity in the disaster area. After the power outage on the US
east coast on September 28 2003, as much as 1% of the Internet was disconnected
for several minutes, and several thousand networks were still unconnected after
eight hours [2].

At the same time, the need to communicate often increases dramatically
in the disaster area. After the 9-11 terrorist attack in New York, there was a
sharp increase in the traffic load in the mobile phone networks, which suffered
from damaged infrastructure and experienced regional congestion [3]. For the
Internet, the overall traffic load did not increase, although the pattern of use

P. Lorenz and P. Dini (Eds.): ICN 2005, LNCS 3421, pp. 1097–1105, 2005.

diverged from the normal. However, with the convergence between the data and the telecommunications infrastructures, we believe that the increased demand experienced in the telephone network, will be increasingly relevant also for the Internet in a disaster situation.

Given the reduced availability and the increased need for communications near and in the disaster area, we believe that a disaster recovery scheme should aim at treating the affected area isolated from the rest of the network. Since the affected area of the network must be considered unreliable, other parts of the network should not be dependent on this area for routing or traffic forwarding. At the same time, the remaining resources in the affected area and between the affected area and the rest of the network, is likely to be scarce, and put under heavy pressure. These communication resources should therefore be available for intra-area traffic and traffic originating or terminating in the area, not transit traffic.

In this paper, we propose using Resilient Routing Layers (RRL) [4] for protecting networks against large-scale disasters. RRL complies well with the goals stated above. Surrounding networks do not rely on the affected area for routing updates, since all forwarding is done using pre-calculated routing information. Traffic that did not pass through the affected area before the failures, is not affected. Our scheme removes transit traffic from the disaster are, while still allowing traffic originating or terminating in the area.

2   Background and Related Work

Planning for communication abnormalities in the event of disasters should be considered in all levels of the communication hierarchy. Research reported in the literature is scarce and what can be found has mainly focused on the application and transport layers.

Because communication during a disaster is mainly a government responsibility, some government funded research has been performed on how network applications and supporting infrastructure should be designed in order to be available in a disaster situation [5][6]. A catastrophe will cause stress in the Internet, and access and admission control methods have been developed that can be beneficial after a disaster [7]. At the transport level special socket communication that is reliable also in the case of catastrophic failures is reported in [8].

The lack of research on network disaster planning may be caused by the fact that the IP protocol and the Internet itself is designed for failures [1]. Many of the common recovery and rerouting techniques developed for more regular errors and failure situation may, however, be used as a starting point when more specific research into network disaster recovery is performed.

One of the most studied topics within the field of recovery research has been efficient algorithms for finding alternative paths between sources and destinations in a network [9] [10]. Network recovery management is, however, difficult if the only view of the network is a large set of unstructured backup paths. The literature provides some alternatives for more structured recovery. Such schemes are based on building a set of subtopologies of the network, serving as a more intuitive abstraction of the alternative paths. These schemes can serve as input to restoration and protection, both global and local. Examples of such approaches are Redundant Trees [11] [12] and Protection Cycles [13] [14].

On the IP level, recovery relies on the IP routing protocols to complete a global rerouting process. In case of normal failure situations this rerouting can take several seconds to complete. Although fast recovery is always welcome, we may tolerate some seconds of disrupted communications during a disaster. However, experiments have demonstrated that IP routing protocols cause very unstable routing for long periods when recovering from failures [15] [16]. During serious outages caused by disasters we expect this instability to be even worse. Another potential obstacle derived from IP rerouting may occur in situations where most network components in an area have failed, but some components have survived and are offering viable routes through that area. This may cause transit traffic to congest the limited amount of resources available in the struck area. This is resources that should be exclusively available for traffic originated and terminated in that area.

The authors have previously described Resilient Routing Layers (RRL) as a method for recovery from node and link failures [4]. RRL can be put in the category of schemes that is based on building subtopologies of the network. RRL differs from other schemes in that it is not bound to a particular kind of subtopologies like trees or cycles. As opposed to most other schemes that focus on link failures and often only single failures, RRL is designed to also isolate many nodes simultaneously [17]. Most schemes presented above have their applicability in connection-oriented networks, while RRL can be applied to both connection-oriented and connectionless networks.

In this paper we will demonstrate how RRL can serve as a tool for pre-planning fast isolation of multiple nodes or even whole areas. RRL only isolates the nodes or areas from carrying transit traffic. Traffic that is originated in or destined for these nodes or areas get exclusively access to all available local resources.

3   Resilient Routing Layers (RRL)

RRL organizes the network topology in subtopologies that we call *routing layers*. In each layer there are some nodes or areas that do not carry transit traffic, and we call these nodes or areas the *safe nodes* of this layer.

Layers are constructed so that all nodes are present in each layer, and there exist a path between all node pairs in each layer. Each node should be safe in at least one layer to guarantee single node fault tolerance. There are numerous ways to construct the layers so that different protection properties are optimized [4] [17] [18].

Figure 1 shows an example of how a network can be covered by two layers. The 8 nodes of figure 1 can also represent 8 areas or subnetworks, as is more

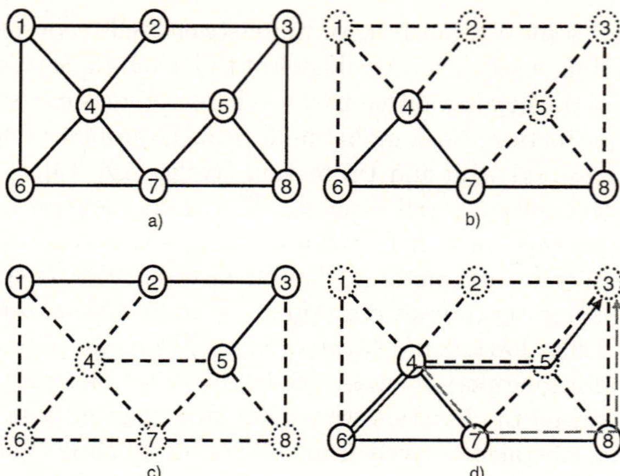

Fig. 1. a): An example network with 8 nodes and 14 links. b): layer 1 (L_1) generated based on a). c): layer 2 (L_2) generated based on a). d): An example on how traffic is routed in a failure situation

thoroughly described later in the article, and depicted in the topology in figure 3. Figure 1a) shows the original full topology. In layer 1 (figure 1b)), nodes 1, 2, 3 and 5 (dashed) are safe. The links connecting these nodes to the rest of the network are also dashed, indicating that the routing on these links must adhere to some specific rules. The rest of the nodes (4, 6, 7 and 8), can be made safe in layer 2 as shown in figure 1c). When a node fails, the traffic affected by the failure will be routed according to the safe layer of the failed node. Traffic not originally routed through the failed node will still be routed according to the full topology.

The following guidelines will arrange that all pairs of nodes can communicate with each other in all layers, and also that safe nodes will not carry any transit traffic, only traffic originating and terminating in the safe nodes:

1) non-dashed links can carry all kinds of traffic originated and terminated anywhere.
2) dashed links can only be used as the first hop or last hop of the communication, meaning that traffic originated in a safe node can use a dashed link as first hop, and that traffic terminated in a safe node can use a dashed link as last hop towards the safe node.

To take advantage of the resilient routing layers a packet network implementation must fulfill certain requirements. Each packet must be marked so that each node on the path will know what layer the packet should be routed on. If n is the maximum number of layers, $log_2(n)$ bits in the packet header should identify the currently valid layer. The node that moves a packet to another layer, marks the packet header with the global identification of the new layer. In the case of failures, only traffic affected by the failed node should be moved to another layer. All packets not affected by the fault will still be routed based on the full topology.

Fig. 1d gives an example of how traffic is switched between layers when node 5 fails. The dotted links may not be used for transit traffic in layer 1, i.e. the safe layer of node 5. Before node 5 fails, all traffic may use the full topology, e.g. traffic from node 6 to node 3 will follow the path 6-4-5-3. When node 5 fails, traffic transiting node 5 must be routed according to layer 1 (using dotted links for first hop and last hop only), while all other traffic (traffic not originally transiting node 5) can still be routed according to the full topology. In the case of local rerouting, traffic is routed from node 6 to 4 according to the full topology. Node 4 detects the failure, and switches traffic to layer 1. The path for traffic between node 6 and node 3 will then be 6-4-7-8-3. If node 6 is notified about the failure (global rerouting) of node 5, the transition to layer 1 could be done by node 6. The path would then be 6-7-8-3.

3.1   RRL Evaluations

Scalability: As demonstrated in [4], RRL seems to scale well, and requires few layers even for very large networks. Table 1 presents a collection of results regarding the number of layers for different real network topologies and some very large synthetic topologies. The real world topologies have been collected from Oliver Heckmann [19] and Rocketfuel [20]. In addition we have generated 100 synthetic topologies using the brite topology generator with 1024 nodes and an average node degree of 4 [21].

Table 1. Percentage of topologies requiring from two till six layers

Topology type	%2	%3	%4	%5	%6
Rocketfuel	17	50	33	0	0
Heckmann	33	17	33	0	17
Brite	0	0	58	42	0

Backup path lengths: Since RRL restricts the number of links that can carry transit traffic in the case of a failure, backup routes will be longer compared to having all links available. Fig. 2 shows that the increased backup path length is relatively modest, i.e., about 0.8 hops in average compared to the most optimal backup path. These figures stem from [4] where we have used 100 brite topologies with 32 nodes and an average node degree of 4.

Resisting multiple node failures: As reviewed in Sec. 2 most contributions on resilience have focused on failures of one component only and particularly link failures. The failure characteristics during disastrous events differ from what have been focus for previous research on recovery. Such events have a tendency to strike numerous nodes instead of a single fiber conduit.

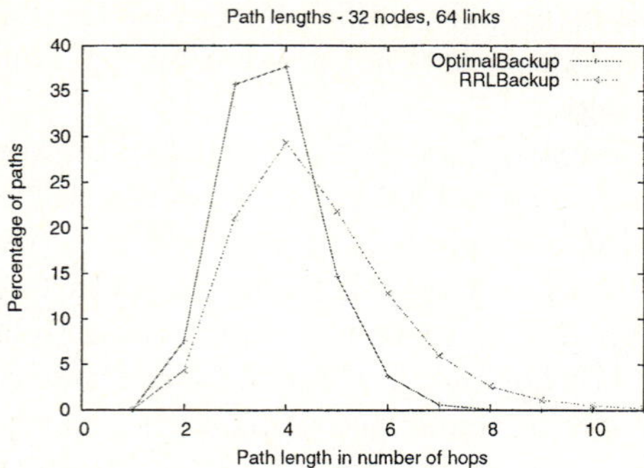

Fig. 2. Distributions of path lengths when introducing one node failure in the original path. The figure shows optimal backup paths and RRL backup paths for 100 brite topologies with 32 nodes and 64 links. Optimal backup path refers to finding a new path in a topology where only the failed node is removed

With respect to this context, RRL offers a flexible way of isolating nodes, including many nodes at the same time. RRL resists simultaneous failures of all nodes that are safe within the same layer. What nodes that should be safe in the same layers can be decided based on knowledge of what nodes have a greater risk of failing simultaneously. Some considerations regarding RRL's flexibility and ability to resist multiple failures can be found in [17].

3.2   Isolating Whole Areas

We have previously argued that disastrous events often strike one or more particular areas. These areas can be states, cities, campuses or buildings. An area that has been struck by a disaster will experience a decreased amount of available communication resources. On the contrary the need for communication to and from that area may increase. Therefore, communications that are transiting this area should find other routes, thus not using precious resources if not absolutely necessary. Also, if the area does not manage to handle any traffic, the traffic only transiting this area must be rerouted.

To accomplish the isolation of whole areas and not only single nodes, we suggest to associate localized nodes as one area. Then we can use RRL on the area-level, meaning that a node represents an area with respect to the RRL overview in Sec. 3.

Fig. 3 shows the hypothetical pan-European cost239 network [22], where the areas correspond to a node in the cost239 network. Each area consists of several intra-area nodes which offer different connections to intra-area nodes in other areas. To arrange that each area is safe at least once, we need two layers.

Upon a disastrous event striking for instance area 6, some intra-area nodes may survive and hence still offer connectivity to other areas, however the capacity has probably decreased. The idea behind RRL is that traffic originally originated and terminated in area 6 will still be routed from or to that area, while traffic

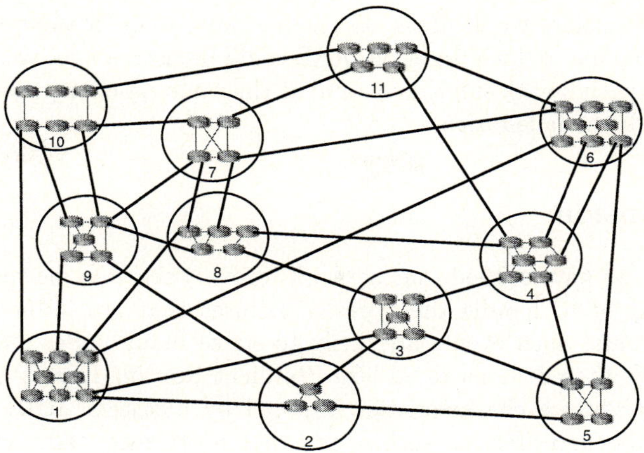

Fig. 3. The original fully connected Cost239 network

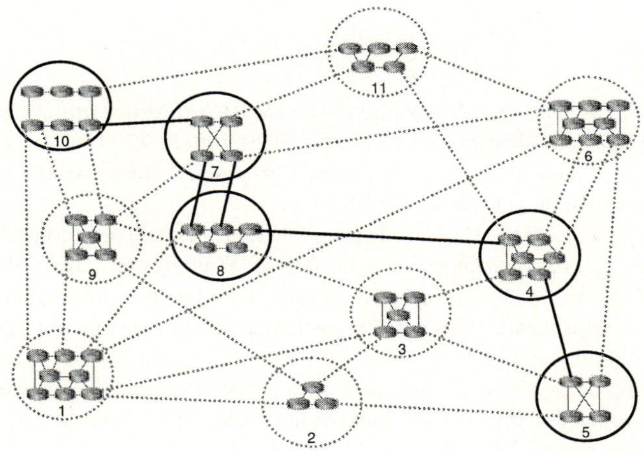

Fig. 4. The layer where area 6 is isolated and safe

originally only transiting area 6 will now be routed around. This is accomplished by letting the affected traffic be routed according to the safe layer of area 6 (Fig. 4). Traffic not originally passing area 6 will still be routed according to the full topology.

Practical support for areas: The most common Internet routing schemes support grouping of nodes and subnetworks into areas. Inter-domain routing, e.g., BGP, is based on modeling Autonomous Systems (ASs) that represent localized nodes and subnetworks [23]. Also Intra-domain routing, e.g., OSPF, support such grouping of localized nodes into areas [24].

Hierarchical RRL: As demonstrated above, RRL can be used to isolate many nodes at the same time, resisting multiple simultaneous failures. In addition, it can be used to isolate whole areas. These two ways of application could be combined to form a hierarchical solution. RRL could be used for recovery of multiple node failures within each single area, and at the same time for preventing transit traffic through the failed area.

4   Conclusion

In this paper we have argued that network recovery schemes, as specified today, are not designed to handle the types of failures that are induced by typical disastrous events. Such events are likely to strike many nodes localized in the same area. We have demonstrated how 'Resilient Routing Layers' (RRL) better adhere to the failure characteristics imposed by disasters. RRL can be used to isolate nodes and areas in such a way that no transit traffic will be routed through a struck node or area, while traffic originated or terminated in such nodes or areas will still be transmitted if surviving connections allows it.

References

1. Clark, D.D.: The design philosophy of the DARPA internet protocols. SIGCOMM, Computer Communications Review **18** (1988) 106–114
2. McGrath, D.: Measuring the 4:11 effect: The power failure and the internet. IEEE Security and Privacy **1** (2003) 16–18
3. Partridge, C., Barford, P.: The Internet Under Crisis Conditions: Learning from September 11. The National Academic press, Washington, D.C. (2003)
4. Hansen, A.F., Cicic, T., Gjessing, S., Lysne, O.: Resilient routing layers: A simple and flexible approach for resilience in packet networks. Technical Report 13, Simula Research Laboratory (2004)
5. Smith, D.R., et al.: Contingence/disaster recovery planning for transmission systems of the defense information system network. IEEE Journal on Selected Areas in Communications **12** (1994) 13–22
6. Luka, G., Fergus, P.: AIN applications to support NS/EP disaster response and recovery. In: Conference Record Military Communications Conference, MILCOM '1995. (1995) 843–847
7. Beard, C.C., Frost, V.S.: Prioritized resource allocation for stressed networks. IEEE/ACM Transactions on Networking **9** (2001) 618–633
8. Haungs, M., Pandy, R., Barr, E.: Handling catastrophic failures in scalable internet applications. In: Proceedings of the 2004 Interantional Symposium on Applications and the Internet (SAINT'04), Tokyo, Japan (2004)
9. Suurballe, J.W.: Disjoint paths in a network. Networks (1974) 125–145
10. Macgregor, M.H., Groover, W.: Optimized k-shortest-paths algorithm for facility restoration. Software-practice and experience **24** (1994) 823–834
11. Medard, M., Finn, S.G., Barry, R.A.: Redundant trees for preplanned recovery in arbitrary vertex-redundant or edge-redundant graphs. IEEE/ACM Transactions on Networking **7** (1999) 641–652
12. Bartos, R., Raman, M.: A heuristic approach to service restoration in MPLS networks. In: Proc. ICC. (2001) 117–121

13. Grover, W.D., Stamatelakis, D.: Cycle-oriented distributed preconfiguration: Ring-like speed with mesh-like capacity for self-planning network restoration. In: Proc. ICC. Volume 1. (1998) 537–543
14. Stamatelakis, D., Grover, W.D.: IP layer restoration and network planning based on virtual protection cycles. IEEE Journal on selected areas in communications **18** (2000)
15. Labovitz, C., et al.: Origins of Internet routing instability. In: Proceedings of IEEE/INFOCOM. (1999)
16. Labovitz, C., Ahuja, A., Bose, A., Jahanian, F.: Delayed Internet routing convergence. IEEE/ACM Transactions on Networking **9** (2001) 293–306
17. Cicic, T., Hansen, A.F., Gjessing, S., Lysne, O.: Applicability of resilient routing layers for k-fault network recovery. In: Proc. ICN'05. (2005)
18. Kvalbein, A., Hansen, A.F., Cicic, T., Gjessing, S., Lysne, O.: Fast recovery from link failures using resilient routing layers. In: submitted to the 10th IEEE Symposium on Computers and Communications (ISCC 2005), La Manga, Spain (2005)
19. See http://dmz02.kom.e-technik.tu-darmstadt.de/~heckmann/.
20. See http://www.cs.washington.edu/research/networking/rocketfuel/.
21. Medina, A., Lakhina, A., Matta, I., Byers, J.: BRITE: An approach to universal topology generation. In: IEEE MASCOTS. (2001) 346–353
22. O'Mahony, M.J.: Results from the COST 239 project. ultra-high capacity optical transmission networks. In: Proceedings of the 22nd European Conference on Optical Communication (ECOC'96), Oslo, Norway (1996) 11–14
23. Rekhter, Y., et al.: A border gateway protocol 4 (BGP-4). IETF, RFC 1771 (1995)
24. Moy, J.: OSPF version 2. In: IETF, RFC 2328 (1998)

Design of a Service Discovery Architecture for Mobility-Supported Wired and Wireless Networks[*]

Hyun-Gon Seo[1] and Ki-Hyung Kim[2,**]

[1] Department of Computer Engineering,
Yeungnam University, Gyungsan, Gyungbuk, Korea
moses@yu.ac.kr
[2] Division of Information and Computer Engineering,
Ajou University, Suwon, Korea
kkim86@korea.com

Abstract. Automatic service discovery will play an essential role in future network scenario and is an important component for MANET and collaboration in ubiquitous computing environment. This paper proposes a service discovery architecture, named as SLPA (Service Location Protocol based on AMAAM) in the mobility-supported wired and wireless environment of which the underlying protocol is AMAAM which is an aggregation-based Mobile IP implementation in MANET. In SLPA, the role of the directory agent is assigned to the mobility agent in AMAAM. The mobility agent periodically beacons an advertisement message which contains both the advertisement of the directory agent in SLP and the advertisement of the mobility agent in Mobile IP. For evaluating the functional correctness of SLPA and the overhead of maintaining a service directory in MANET, we simulate SLPA and analyze the overhead of control overheads for the aggregation. Through the simulation experiments, we investigate the functional correctness of the proposed architecture and analyze the control overheads of the aggregation.

1 Introduction

Discovery of services and other named resources which allows devices to automatically discover network services with their attributes and advertise their own capabilities to the rest of the network, is a major component for such self-configurable networks. Today, there exist several different industrial consortiums or organizations to standardize different service discovery protocols such as *Service Location Protocol (SLP)* of IETF, Sun's *Jini*, Microsoft's *Universal Plug and Play (UPnP)*, IBM's *Salutation*, and Bluetooth's *Service Discovery Protocol (SDP)*.

---

[*] This research was supported by University IT Research Center Project.
[**] Corresponding author: Ki-Hyung Kim (kkim86@korea.com).

P. Lorenz and P. Dini (Eds.): ICN 2005, LNCS 3421, pp. 1106–1113, 2005.

Mobile IP has been proposed for networks with infrastructure by IETF[2]. Mobile IP tries to solve the problem of how a mobile node may roam from its network and still maintain connectivity to the Internet. Mobile IP uses mobility agents to support seamless handoffs, making it possible for a mobile node to roam from subnet to subnet without changing its IP address.

As an another emerging wireless architecture, mobile ad hoc networks (MANET) are infrastructureless networks formed on the fly by devices with wireless communication capabilities, but without the need of infrastructured base stations. MANETs usually have several limitations, such as low communication bandwidth, device power constraints, and device mobility.

Integrating Mobile IP and MANET facilitates the current trend of moving to an all-IP wired and wireless environment[3]. Recently, an architecture of integrating Mobile IP and MANET based on the on-demand routing protocols has been proposed[4]. The architecture, named as AMAAM(Aggregation based Mobility Agent Advertisement Mechanism) uses a periodic beaconing of agent advertisement messages with on-demand routing protocols. A mobility agent floods an agent advertisement message periodically throughout a MANET. Upon receiving an advertisement message, every mobile node in the MANET replies to the mobility agent by a registration request message. To reduce the overhead of the periodic beaconing and the excessive generation of registration request messages from mobile nodes in reply, AMAAM employs an aggregation technique for the registration request messages.

Discovery of services and other named resources is a crucial feature for the usability of MANET. In this dynamic environment, different nodes offering different services may enter and leave the network at any time. Efficient and timely service discovery is a prerequisite for good utilization of shared resources on the network.

In this paper, we propose a service discovery architecture in the mobility-supported wired and wireless environment, named as SLPA(Service Location Protocol based on AMAAM). In SLPA, the role of the Directory Agent(DA) is assigned to the mobility agent(MA) in AMAAM, thereby forming a Mobility & Directory Agent(MDA). The MDA periodically beacons an advertisement message which contains both the advertisement of the directory agent in SLP[1] and the advertisement of the mobility agent in Mobile IP[2] to reduce the overhead of the advertisements of both protocols. The simulation results show that SLPA can effectively reduce the overhead of periodic advertisements.

The rest of the paper is organized as follows. Section 2 presents our motivation and discusses previous works on this topic. Section 3 proposes our protocol. Experimental results are shown in section 4, and section 5 concludes the paper.

2   Preliminaries

In this section, we first describe the underlying architecture of the mobility-supported wired and wireless networks. Then, we show the previous works on the service discovery architectures in MANET.

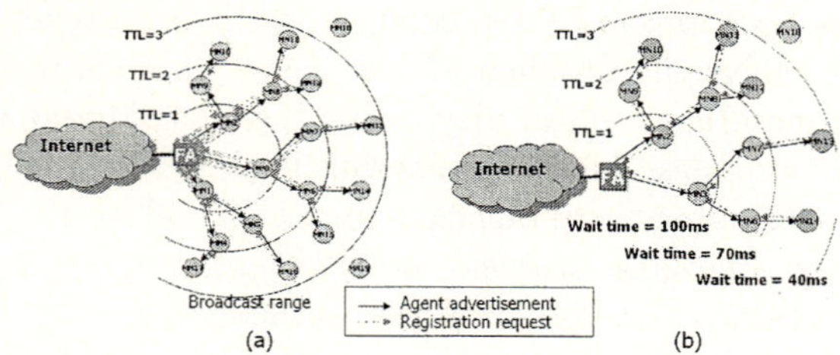

Fig. 1. Agent advertisement and registration request messages delivery in AMAAM. (a) MAAM without aggregation (b) AMAAM

2.1   Integration of Mobile IP and MANET

Mobile IP has been proposed for networks with infrastructure by IETF. In Mobile IP, Home Agent(HA) and Foreign Agent(FA) are the two forms of mobility agents. To be able to receive datagrams while visiting a foreign network, the visiting mobile node has to register its current care-of address with its HA, representing the visiting node within its home network. To do this, the visiting node usually has to register through a FA, located in the foreign network. When a node has registered successfully with the HA, every datagram sent to the node's home address is intercepted by the HA and tunneled to the care-of address. Each FA keeps a visitor list which contains the information about visiting nodes currently registered through it. A HA keeps track of the mapping between each residential node's home address and care-of address in a location dictionary.

Integrating Mobile IP and MANET facilitates the current trend of moving to an all-IP wireless environment. AMAAM (Aggregation-based Mobility Agent Advertisement Mechanism) [4] is an architecture of integrating Mobile IP and MANET based on on-demand routing protocols such as AODV. Basically, AMAAM uses periodic beaconing of agent advertisement messages with on-demand routing protocols. That is, a FA floods agent advertisement messages periodically throughout the MANET. Upon receiving an advertisement, every node in the broadcast range (*i.e.* TTL from the FA is less than or equal to N) of the MANET replies to the FA by a registration request message. Then, the FA receives the registration request messages from all the mobile nodes in the broadcast range. This periodic flooding of agent advertisement messages and the corresponding replies could cause excessive overhead of network traffic and decreases the battery lifetime of nodes. For reducing this overhead, AMAAM aggregates multiple registration request messages into one request message, as shown in Fig. 1.

2.2   Related Works on Service Discovery in MANET

Service discovery protocols enable software components to find each other on a network and to determine if discovered components match their requirements. We briefly describe *SLP* as a representative service discovery protocol in the Internet and as a basis of the proposed service discovery architecture in the mobility-supported network. SLP establishes a framework for service discovery using three types of agents that operate on behalf of network-based softwares; (*i*) a Service Agent(SA) advertises the location and attributes on behalf of services, (*ii*) a Directory Agent(DA) aggregates service information, and (*iii*) a User Agent(UA) performs a service discovery by issuing a 'Service Request' (SrvRqst) on behalf of the client application, specifying the characteristics of the service which the client requires. The UA will receive a Service Reply (SrvRply) specifying the location of all services in the network which satisfy the request.

There have been several research efforts on the service discovery architectures in MANET recently. They can broadly be classified into two ways: centralized and distributed. In the centralized service discovery, there are a couple of DAs which advertise the existence of DAs and collect service registrations/service requests from distributed UAs and SAs in MANET. Since it is a traditional way of the service discovery in fixed infrastructure networks like the Internet, there are many existing well-defined service discovery mechanisms such as SLP, Salutation, and Jini. L. Cheng proposed an approach of implementing the centralized service discovery architecture in MANET by employing ODMRP as the underlying multicast protocol[5].

As an alternative, the distributed service discovery approaches do not rely on particular nodes[6]. An example of this approach is the SLP without DAs, in which a UA multicasts (or broadcasts) a SrvRqst specifying the type of the desired service during a service discovery process. SAs that provide a service satisfying the specified service type respond with a SrvRply including the configuration information of the service. Performing such a service discovery can be an expensive process in MANET since it may cause a large number of generated SrvRqsts to be transmitted all over the network whenever UAs request service discoveries.

3   Service Discovery Architecture in AMAAM

This section presents the proposed service discovery mechanism, named as *SLPA*, which is based on AMAAM. Fig. 2 (a) shows the overall architecture of SLPA. In SLPA, the mobility agent(MA) for Mobile IP also has a role of the directory agent(DA) for service discovery, thereby forming the Mobility & Directory Agent (MDA). Since the mobility agent floods agent advertisement messages periodically in AMAAM, we extend naturally the advertisement to incorporate the directory agent advertisement too. We call the advertisement as MDAAdvert (MDA Advertisement). MDA now floods MDAAdvert messages periodically with the predefined TTL (Time To Live) = N which is the broadcast range (or managed area under the control of the MDA). Upon receiving a MDAAdvert, a SA replies to MDA with the combined message which has both the service registration request for service registration and the mobility registration request for Mobile IP. If a mobile node is just a UA, it just replies with a mobility registration request message without service registration.

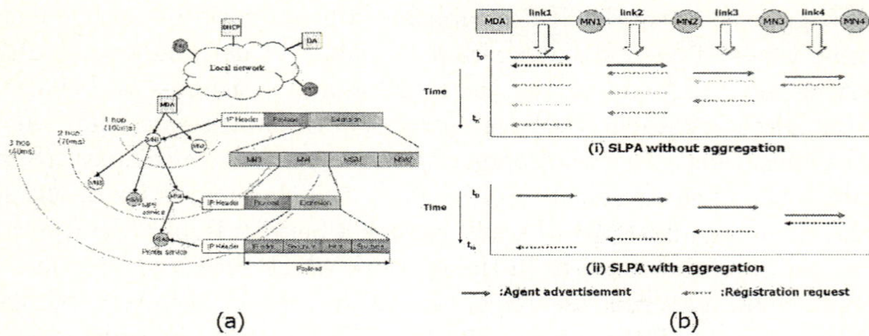

(a)   (b)

Fig. 2. Architecture of SLPA. (a) Overall architecture of SLPA (b) Comparison of the beaconing overheads in SLPA

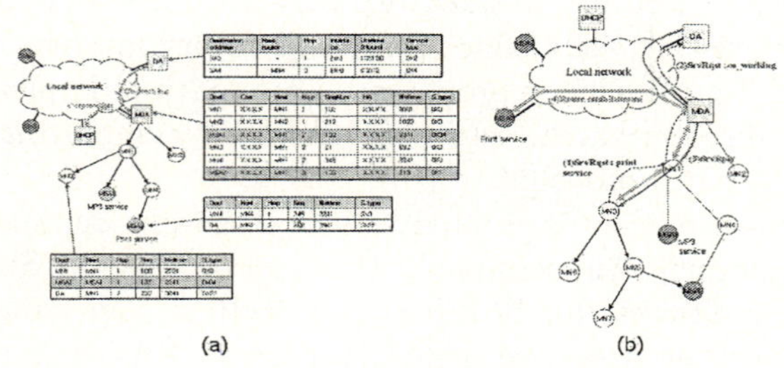

(a)   (b)

Fig. 3. Service Discovery Process of SLPA. (a) Routing table information. (b) Finding and consuming a service across wired and wireless networks

The registration request messages are now aggregated on the intermediate nodes for reducing the number of message communication. Fig. 2 (b) shows the aggregation mechanism of SLPA in detail. When a mobile node receives an agent advertisement message, it forwards the message to the downstream neighbor nodes and waits for the registration request messages from them without immediate reply to the MDA. The waiting time depends on the hop counts from the MDA. Notice that the broadcast range of the agent advertisement is N. Thus, a mobile node whose hop counts distance to the MDA is N immediately replies to the MDA through the upstream nodes without waiting anymore. The waiting time (T_w) for aggregation can be calculated as follows:

$$T_w = (N - H) * T_n * 2 \tag{1}$$

where N is the broadcast range of the agent advertisement, H is the hop counts distance to the MDA, and T_n is the node traversal time.

Fig. 3 (a) shows the routing table information for nodes in wired and wireless networks. The DA is a directory agent for a wired network while MDA is for a wireless network. They cooperate for finding and consuming a service across wired and wireless networks as shown in Fig. 3 (b).

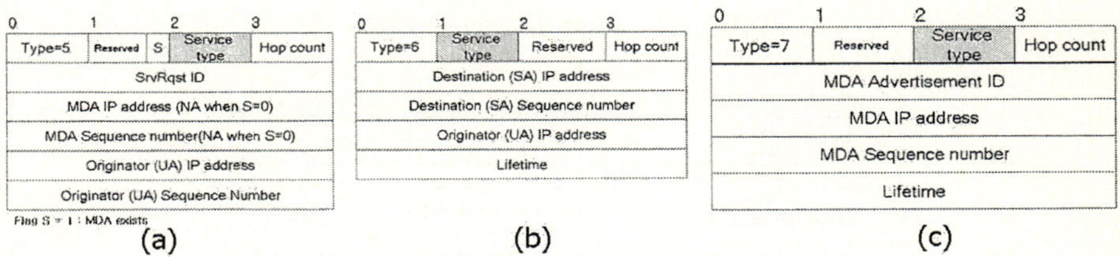

Fig. 4. Message formats for SLPA. (a) SrvRqst format. (b) SrvRply format. (c) MDA's advertisement format

3.1   Implementation on AODV

The proposed architecture can be implemented in AODV-based MANET by defining the following three message types: SrvRqst, SrvRply, and DA's advertisement, as shown in Fig. 4.

To simplify the service type resolution, we predefine supported services and assign them 8 bits service type values. For example, 0x01 is the MP3 streaming service, and 0x02 is the printing service of digital photographs, etc. If a specific service discovery protocol such as SLP is to be implemented, the above three messages can be modified suitably.

4   Experimental Results

In this section, we use a packet-level simulation to explore functional correctness and the performance of the service discovery architecture in mobility-supported wired and wireless networks.

4.1   Simulation Environment

In order to study the performance of SLPA, we simulate the proposed architecture by using ns-2. In the experiments, we generate a variety of MANET fields with different scenarios. Table 1 shows the simulation parameters used for the experiments.

Each node behaves as one of the two agents, *i.e* a UA or a SA. The number of SAs in the network is varied to see the impact of the service availability on the service discovery performance. For instance, if we select 5 nodes (out of 50 nodes) as SAs, the ratio of SAs becomes 10%. The total number of service types during the simulation is 10, and each SA provides two distinct services.

Table 2 shows the simulation performance results while varying the SA ratio which is the ratio of the number of SAs to the total number of mobile nodes. Each simulation result is the average of 5 identical traffic models.

As the number of SAs increases, the service hit ratio is increased, but the delivery ratio of MDA advertisements is decreased. The service hit ratio, the multiplication of the SA discovery ratio and the SA route establishment ratio, is the ratio of the number of service requests to the number services actually

Table 1. Simulation parameters

Parameter	First simulation	Second simulation
Number of nodes	50	50
Max speed (m/sec)	10	10
Simulation time (sec)	120	900
Number of SAs (% is the ratio of SAs to the total num. of nodes)	5(10%), 10(20%), 15(30%), 20(40%), 25(50%), 30(60%), 35(70%)	15(30%)
Area (m²)	1000m x 800m	1000m x 800m
Total number of services (Each SA provides 2 services)	10	10
MDA advertisement messages (sec)	5	5
Service request rate (sec)	0.33 (Each UA requests a service for every 3 sec)	0.33 (Each UA requests a service for every 3 sec)
Energy (Initial_E, RxE, TxE)	100, 1.395, 4.66	100, 0.3, 0.4

Table 2. Performance results while varying the number of SAs

Metrics / Number of SAs	10%	20%	30%	40%	50%	60%
MDA advertisement delivery ratio	62.12	58.87	50.75	54.41	49.07	46.62
SA discovery ratio(r1)	0.352	0.656	0.853	1.088	1.268	1.520
SA route establishment ratio(r2)	0.611	0.545	0.488	0.498	0.494	0.485
Service hit ratio (r1*r2)	0.214	0.356	0.417	0.538	0.627	0.729
Total number of control messages	63761	106855	123762	120094	126203	120482
Service discovery time (msec)	432.29	710.07	988.58	837.41	896.88	822.26
Remaining energy of SAs (J)	51.03	19.54	12.77	9.38	12.192	17.22
Remaining energy of nodes(J)	50.99	20.34	12.50	10.11	11.78	16.35

received. Notice that a MDA just gives the address of the corresponding SA to a UA, not the routing path to the SA. A UA should find a route to a SA by itself after receiving a query reply from a MDA. The service hit ratio is directly proportional to the number of SAs. This is because the probability of whether the service requested by a UA exists in the services offered by SAs is proportional to the number of SAs. Remember that each SA offers two distinct services and the total number of service types is 10.

The control messages of SLPA consist of the following four types: MDA's advertisements, SA's service registration requests, service requests and replies between a MDA and UAs, and route establishments to SAs after finding SAs which provide the requested services. The SA discovery ratio shows the delivery fraction of SA's registration requests to the MDA. The fraction increases as the number of SAs increases. The SA route establishment ratio assumes the underlying routing protocol is AODV. It shows almost same results while varying the number of SAs.

The service discovery time is the time elapsed from the UA's request of a service to a MDA to the UA's reception of the service from the corresponding SA. It shows better performance as the number of SAs decreases. This is because the number of control messages, especially the number of service registration requests, is reduced.

Table 3 shows the simulation results while varying the pause time of mobile nodes to test the impact of mobility. The results show that all metrics do not change significantly for the mobility of nodes.

Table 3. Performance results while varying the pause time of mobile nodes

Metrics \ Pause time	0	100	200	300	400	500	600	700	800	900
MDA advertisement delivery ratio	36.64	36.94	36.79	36.79	36.84	36.81	36.81	36.82	36.82	36.81
SA discovery ratio(r1)	0.622	0.694	0.658	0.658	0.670	0.662	0.663	0.665	0.663	0.664
SA route establishment ratio(r2)	0.557	0.519	0.538	0.538	0.531	0.536	0.535	0.534	0.535	0.535
Service hit ratio(r1•r2)	0.346	0.359	0.353	0.353	0.355	0.353	0.354	0.354	0.354	0.354
Total number of control messages	647180	688216	666888	666888	673197	668858	669518	670545	669661	669929
Search time (msec)	767.65	765.87	766.76	766.76	766.47	766.66	766.63	766.59	766.63	766.61
Remaining energy of SAs	5.002	8.639	6.820	6.820	7.426	7.022	7.090	7.179	7.097	7.122
Remaining energy of mobile nodes	5.474	8.257	6.865	6.865	7.329	7.020	7.072	7.140	7.077	7.096
Number of dropped packets	9191	35754	22473	22473	26900	23948	24440	25096	24495	24677

5   Conclusion

In this paper, we proposed a service discovery architecture, named as SLPA (Service Location Protocol based on AMAAM), in the mobility-supported wired and wireless networks. In SLPA, the role of the directory agent is assigned to the mobility agent in AMAAM. The mobility agent periodically beacons an advertisement message which contains both the advertisement of the directory agent in SLP and the advertisement of the mobility agent in Mobile IP. Through the simulation experiments, we investigated the functional correctness of the proposed architecture and analyzed the control overheads of the aggregation.

References

1. E. Guttman, C. Perkins, J. Veizades, and M. Day, "Service Location Protocol(SLP), Version 2," *IETF RFC2608*, June 1999.
2. C. Perkins, "IP Mobility support for IPv4," RFC3344, August 2002.
3. Y. C. Tseng, C. C. Shen and W. T. Chen, "Integrating Mobile IP with Ad Hoc Networks," IEEE Computer, pp. 48-55, May 2003.
4. H. G. Seo and K. H. Kim, "Mobility Agent Advertisement Mechanism for Supporting Mobile IP in Ad Hoc Networks," Lecture notes in computer science Vol. 3126 (SAPIR2004), August 2004.
5. L. Cheng, "Service Advertisement and Discovery in Mobile Ad hoc Networks," in Conference on Computer Supported Cooperative Work (CSCW2002), 2002.
6. S. Motegi, K. Yoshihara, and H. Horiuchi, "Service discovery for wireless ad hoc networks," The 5th International Symposium on Wireless Personal Multimedia Communications, vol.1 , 27-30 Oct. 2002, pp. 232 - 236

Research on Fuzzy Group Decision Making in Security Risk Assessment

Fang Liu, Kui Dai, Zhiying Wang, and Jun Ma

School of Computer,
National University of Defense Technology,
Changsha, 410073,
China
`liufang_nudt@yahoo.com.cn`

Abstract. Decision making problems in security risk assessment are often associated with multiple criteria and multiple decision makers. In the proposed approach, decision making by multiple decision makers has been considered under uncertain conditions. An optimization model is used to assess criteria weights and then to rank risks. The different preference information from different decision makers are firstly transformed into uniform fuzzy preference relations and aggregated. Then ranking or selection of the alternatives reflects the decision makers' subjective preference based on the objective decision information. It has been found that using fuzzy set theory to represent uncertainties under multiple-participant multi-criteria environment is very promising. The practices indicate that fuzzy group decision making techniques provide concepts and theoretical results that are valuable in formulating and solving problems in security risk assessment.

1 Introduction

An important step in disaster recovery process is to consider the potential impacts of each type of disaster or threat. This is a critical activity for it will determine which scenarios are most likely to occur and which should attract most attention during the emergency planning and preparedness. Fortunately, risk assessment has emerged to simplify these tasks. As an essential element of security risk management, risk assessment provides the means by which systems risks are identified and assessed in order to justify safeguards[1]. Since risks and threats change over time, IT managers should perform a risk assessment every 90 days in order to efficiently protect their information security[2].

This paper presents a multiple-participant multi-criteria risk assessment process. A brief introduction to challenges in security risk assessment is presented in Section 2. The multiple-participant multi-criteria analysis techniques and terminology are presented in section 3. Section 4 discusses the approach used to solve the multi-participant MCDM (Multi-Criteria Decision Making) problem with fuzzy preference information on alternatives in risk assessment. Finally, conclusions deriving from the effort are presented and future work is discussed in Section 5.

P. Lorenz and P. Dini (Eds.): ICN 2005, LNCS 3421, pp. 1114–1121, 2005.

2   Challenges Associated with Security Risk Assessment

Risk has been defined as 'the probability that a threat agent (cause) will exploit a system vulnerability (weakness) and thereby create an effect detrimental to the system.'[1]. Risk assessment approaches are generally estimates the cost of risk and risk reduction techniques based on (1) the likelihood that a threat will occur, (2) the costs of potential losses, and (3) the costs of mitigating actions that could be taken. Reliably assessing information security risks can be more difficult than assessing other types of risks, because the data on the likelihood and costs are often more limited and the risk factors are constantly changing[3]. When reliable data on likelihood and costs are not available, it can be taken by defining risk in more subjective and general terms. In this regard, risk assessments of information security depend more on the expertise, experience, and judgment of those conducting the assessment[3].

2.1   Multi-criteria Decision Problems

Risk assessment often involves multi-criteria decision making to deal with problems having multiple and conflicting objectives. Considering a scenario for deploying the security countermeasures with criteria e.g., to improve customer relations, reduce economic loss and to minimize adverse impacts etc., there exists considerable difficulty in deciding the best mitigation option as the objectives are not comparable on the same scale and the options that are more likely to achieve one objective may be less effective in obtaining the others[4]. Multi-criteria analysis techniques could help decision makers evaluate risks and countermeasures when conflicting criteria must be considered and balanced. In addition, multi-criteria risk assessment provides a systematic and repeatable method for evaluating an information security risk using the best available threat information. The value of a multi-criteria risk assessment is not only in the numbers produced, but also in the insights that security manager's gain during each refinement step of the assessment[5].

2.2   Decision Involving Multiple-Participant

In practice, the process of risk assessment is complicated because assessment data are based on the security manager's subjective estimate of each threat and the anticipated risk reduction from the security technologies. To enable more effective and acceptable decision outcome, it is required that more participation is ensured in the decision making process. Moving from a single decision maker to a multiple decision maker situation, the problem is no longer limited to the selection of the most preferred alternative among the non-dominated solutions by an individual. The analysis must also be extended to account for the conflicts among different decision makers with different objectives. In risk assessment, decision makers' preference information is often used to rank risks. However, their judgments vary in form and depth. Different decision makers may use different ways to express their preference information. Inspired by the study in[6], this paper applies multiple participant MCDM approach to risk assessment which can better reflect decision making situations.

2.3 Uncertainties in Decision Making Problems

Risk analysis decision making is always associated with some degree of uncertainty in a situation where the goals, the constraints and the consequences of the possible actions are not known precisely. The use of fuzzy set theory has become increasingly popular in addressing imprecision, uncertainty and vagueness in group-decision making. In this paper, fuzzy linguistic terms have been used to address the subjective judgment of the decision makers while stating the preference for the alternatives.

To deal with complex risk assessment problems of information security, it is necessary to consider multiple participant with multiple criteria where uncertainties exist at different stages of the decision making process. Appropriate decision making in each of these stages is very important to establish an efficient risk assessment. The objective of this study is to investigate existing methods and to analyze the applicability of the methods to risk assessment.

3 Problem Statements

Since groups instead of individuals make more and more decisions in fast changing world, one of the hottest research topics is the use of fuzzy set theory in solving the multiple person MCDM problems when imprecise preference information is represented in fuzzy terms[6].

3.1 Multi-participant Multi-criteria Problem

It is necessary that the views of the participants be included in a decision making process that will be well accepted to all those involved. The following assumptions and notations[7] are used to represent the multiple participant MCDM problems:

- the decision makers involved are known: let $E = \{e_1, e_2, \cdots, e_K\}$ denote the set of decision makers.

- the alternatives are known: let $S = \{s_1, s_2, \cdots, s_m\}$ denote a discrete set of $m(\geq 2)$ possible alternatives, namely risks in risk assessment.

- the criteria are known: let $C = \{c_1, c_2, \cdots, c_n\}$ denote a set of $n(\geq 2)$ criteria.

- the weights of criteria are unknown: let $w = \{w_1, w_2, \cdots, w_n\}^T$ be the vector of weights, where $\sum_{j=1}^{n} w_j = 1, w_j \geq 0, j = 1, \cdots, n$, and w_j denotes the weight of criteria c_j.

- the decision matrix is known: let $A = \left[a_{ij}\right]_{m \times n}$ denote the decision matrix where $a_{ij} (\geq 0)$ is the consequence with a numerical value for alternative s_i with respect to criterion c_j , $i = 1, \cdots, m, j = 1, \cdots, n$. Since the criteria are generally incommensurate, the decision matrix needs to be normalized so as to transform the various criterion values into comparable values. For the convenience of calculation and extension, the following two functions are used to calculate the membership degrees [8]:

$$b_{ij} = \frac{a_{ij} - a_j^{\min}}{a_j^{\max} - a_j^{\min}}, \qquad i = 1,\cdots,m,\, j = 1,\cdots,n, \quad \text{for benefit criterion,} \quad (1)$$

$$b_{ij} = \frac{a_j^{\max} - a_{ij}}{a_j^{\max} - a_j^{\min}}, \qquad i = 1,\cdots,m,\, j = 1,\cdots,n, \quad \text{for cost criterion,} \quad (2)$$

where $a_j^{\max}$ and $a_j^{\min}$ are given by

$$a_j^{\max} = \max\{a_{1j}, a_{2j}, \cdots, a_{mj}\}, \qquad j = 1,\cdots,n \quad (3)$$
$$a_j^{\min} = \min\{a_{1j}, a_{2j}, \cdots, a_{mj}\}, \qquad j = 1,\cdots,n \quad (4)$$

Then the decision matrix $A = \left[a_{ij}\right]_{m \times n}$ can be transformed into the membership degree matrix:

$$B = \left[b_{ij}\right]_{m \times n} \quad (5)$$

The problem concerned is to rank the risks or select the most severe one, based on the decision matrix A and the preference information on alternatives given by multiple decision makers.

3.2 Formats of Preference Information on Alternatives

Research is in progress to develop the methods to deal with multiple-participant multi-criteria problems under uncertainty. It showed that fuzzy preference over the set of alternatives can be applied to the group decision problem where the decision maker becomes a collective entity and conflicts exist between individual preferences. There are three types of fuzzy preference information on alternatives[7] are given by multiple assessors.

♦ a vector of linguistic terms on $S : L^k = (l_1^k, \cdots, l_m^k)$. $l_i^k = (u_i, \alpha_i, \beta_i)$, represents the linguistic evaluation given by decision maker e_k to alternatives S_i, $i = 1, \cdots, m, e_k \in E$, where $\alpha_i \leq u_i \leq \beta_i$, and u_i is the modal value. α_i and β_i stand for the lower and upper values of the support of the linguistic term respectively. The linguistic terms can embody human's feeling or judgment more accurately when he/she is in complex or fuzzy situations. As l_i^k is defined to be a fuzzy triangular number, following linguistic terms are used by the assessors to express their preference on the alternatives[9]: 'very good' with $(1, 0.8, 1)$, 'good' with $(0.75, 0.6, 0.9)$, 'fair' with $(0.5, 0.3, 0.7)$, 'poor' $(0.25, 0.05, 0.45)$, 'very poor' $(0, 0, 0.2)$.

♦ a fuzzy selected subset of S can be used by a decision maker $e_k (e_k \in E)$ to express his preference on part of the alternatives using fuzzy numbers:

$$\widetilde{S} = \left\{ \left(s_{i1}, l_{i1}^{k}\right), \cdots, \left(s_{in}, l_{in}^{k}\right)\right\}, i_{n} < m, \text{ and } l_{i_{j}}^{k} \text{ is a linguistic term}, i_{j} = 1, \cdots, i_{n}.$$

For example, a decision maker thinks that alternative S_i is 'good', S_j is 'very good', and alternatives S_h and S_l are 'fair'.

♦   fuzzy preference relations on alternatives: the decision maker's preference relation is described by a binary fuzzy relation P on S, where P is a mapping $S \times S \rightarrow [0,1]$ and p_{ij} denotes the preference degree of alternatives S_i over S_j. P is assumed to be reciprocal, by definition. (i) $p_{ij} + p_{ji} = 1$ and (ii) $p_{ii} = -$ (means that the decision maker does not need to give any preference information on alternative S_i), $\forall i, j$.

4   Approach to Multi-participant MCDM for Risk Assessment with Fuzzy Preference

We expand our approach to integrate a fuzzy MCDM model and a structured group decision making process to improve the risk assessment decisions of information security. Based on the social fuzzy preference information, an optimization model is used to assess the criteria weights and then to rank the risk.

4.1   Preference Uniformity

In order to aggregate the individual preferences to reach a group decision, different format preferences must be transformed to unique format preference. As fuzzy preference relation has its merits in aggregation and generality[6], we use it as the basic element for the uniform representation and transform two other formats preferences to fuzzy preference relations as follows[7].

Linguistic Term Vectors to Fuzzy Preference Relations. A linguistic term vector can be used by a decision maker to express his/her preference on the alternatives. Suppose two alternatives S_i and S_j are awarded linguistic terms $l_i^{k} = (u_i, \alpha_i, \beta_i)$ and $l_j^{k} = (u_j, \alpha_j, \beta_j)$ respectively. Following transformation mapping function f and Max-membership defuzzification mapping function g [9][12] can be used to obtain the fuzzy preference relations between S_i and S_j [11].

$$f(l_i^{k}, l_j^{k}) = \frac{l_i^{k} \times l_i^{k}}{l_i^{k} \times l_i^{k} + l_j^{k} \times l_j^{k}} = \left(\frac{u_i^2}{u_i^2 + u_j^2}, \frac{\alpha_i^2}{\alpha_i^2 + \alpha_j^2}, \frac{\beta_i^2}{\beta_i^2 + \beta_j^2} \right), \tag{6}$$

$$p_{ij}^{k} = g\left(f\left(l_i^{k}, l_j^{k}\right)\right) = \frac{u_i^2}{u_i^2 + u_j^2} \tag{7}$$

Fuzzy Selected Subsets to Fuzzy Preference Relations. Sometimes a decision maker $e_k \left(e_k \in E\right)$ can express his/her preference on some alternatives (subset $\widetilde{S}$ of

S) with fuzzy numbers or linguistic terms. For any two alternatives S_i and S_j in S, if they both belong to $\tilde{S}$, where $l_i^k = (u_i, \alpha_i, \beta_i)$ and $l_j^k = (u_j, \alpha_j, \beta_j)$, then the fuzzy preference relations between them can be defined as(similar to (6) and (7)),

$$p_{ij}^k = g\left(f\left(l_i^k, l_j^k\right)\right) = \frac{u_i^2}{u_i^2 + u_j^2}, \qquad 1 \le i \ne j \le m \tag{8}$$

If the two alternatives S_i and S_j do not belong to $\tilde{S}$ either, then

$$p_{ij}^k = 0.5, \qquad 1 \le i \ne j \le m \tag{9}$$

If alternatives S_i belongs to $\tilde{S}$ and S_j does not belong to $\tilde{S}$, then

$$p_{ij}^k = u_i, \qquad 1 \le i \ne j \le m \tag{10}$$

4.2 Preference Aggregation

After the preference information are transformed into uniform fuzzy preference relations, the next step is to aggregate these uniform fuzzy preference relations. We adopt 'simple additive weighting method' to aggregate these fuzzy preference relations[12], i.e. the social fuzzy preference relation between alternatives S_i and S_j is as follows,

$$g_{ij} = \sum_{k=1}^{K} h_k p_{ij}^k, \qquad 1 \le i \ne j \le m \tag{11}$$

where h_k represents the relative importance assigned to the decision maker e_k. If $G = (g_{ij})_{m \times m}$ is not a reciprocal matrix, following operations[12] can be used,

$$g'_{ij} = \frac{g_{ij}}{g_{ij} + g_{ji}}, \qquad 1 \le i \ne j \le m \tag{12}$$

$$g'_{ji} = \frac{g_{ji}}{g_{ij} + g_{ji}}, \qquad 1 \le i \ne j \le m \tag{13}$$

4.3 Preference Approximation

Using the 'simple additive weighting method', the overall value of alternative (risk) S_i can be expressed as[7],

$$d_i = \sum_{j=1}^{n} b_{ij} w_j, \qquad i = 1, \cdots, m \tag{14}$$

Based on all values of d_i, the ranking results of the risks can be obtained. The greater the value d_i is, the more severe the relevant risk S_i will be.

To make the information consistent, the overall values of alternatives can be transformed into fuzzy preference relations. Thus, $\bar{g}_{ik}$ is defined as[13],

$$\overline{g}_{ij} = \frac{d_i}{d_i + d_j} = \frac{\sum\limits_{t=1}^{n} b_{it} w_t}{\sum\limits_{t=1}^{n} (b_{it} + b_{jt}) w_t}, \qquad 1 \le i \ne j \le m \tag{15}$$

where the significance of $\overline{g}_{ij}$ is similar to that of g_{ij}.

The difference between g_{ij} and $\overline{g}_{ij}$ is given[7] by

$$f_{ij}(w) = g_{ij} - \overline{g}_{ij} = g_{ij} - \frac{\sum\limits_{t=1}^{n} b_{it} w_t}{\sum\limits_{t=1}^{n} (b_{it} + b_{jt}) w_t}, \qquad 1 \le i \ne j \le m \tag{16}$$

To reflect the decision maker's fuzzy preference information based on the objective decision matrix, we can minimize $f_{ij}(w)$ by assessing the criteria weights with constrained optimization model[7],

$$Minimize \left| f_{ij}(w) \right| = Minimize \left| \overline{g}_{ij} - g_{ij} \right|, \qquad 1 \le i \ne j \le m \tag{17}$$

According to [13], we have

$$w^* = Q^{-1} e / e^T Q^{-1} e \tag{18}$$

where $e = (1,1,\ldots,1)_{1 \times n}^{T}, Q = (q_{lk})_{n \times n}$. The elements in matrix Q [13] are,

$$q_{lk} = \sum_{i=1}^{m} \sum_{j=1, j \ne i}^{m} \left[g_{ij} (b_{il} + b_{jl}) - b_{il} \right] \left[g_{ij} (b_{ik} + b_{jk}) - b_{ik} \right]$$

$$= \sum_{i=1}^{m} \sum_{j=1, j \ne i}^{m} \left[g_{ij} b_{jl} - g_{ji} b_{il} \right] \left[g_{ij} b_{jk} - g_{ji} b_{ik} \right], \qquad l, k = 1, \ldots, n \tag{19}$$

If equation (14) is substituted with the $w_j^*(j = 1, \cdots, n)$, the overall values of every alternative can be obtained, and the severity of each risk likewise.

5   Conclusion and Future Work

This paper introduces an approach to solve the multi-participant MCDM problem with fuzzy preference information on alternatives in risk assessment. An optimization model is used to assess criteria weights and then to rank risks. The potential future work includes conducting sensitivity analysis to show how sensitive the risk assessment decisions are to the subjective preference information form decision makers and the objective decision information of the alternatives.

Acknowledgements

The research reported here has been supported by the National Natural Science Foundation of China (No.90104025). And the authors would like to acknowledge Dr. Shawn A. Butler for her help and encouragement.

References

[1] Carroll, J. M. (1996). Computer Security, Butterworth-Heinemann.

[2] Cushing, K. (2002). IT Directors Must Review Security Every 90 Days. Computer Weekly: 4

[3] Information Security Risk Assessment, United States General Accounting office(GAO/AIMD-00-33, November (1999))

[4] Fang Liu, Kui Dai, Zhiying Wang. (2004) Improving Security Architecture Development Based on Multiple Criteria Decision Making, AWCC 2004. LNCS 3309. pp. 214-218.

[5] Shawn A. Butler.: Security Attribute Evaluation Method. Carnegie Mellon University, Dissertation of Doctor of Philosophy. May (2003)

[6] Chiclana, F., Herrera, F. and Herrera-Viedma, E. (1998) Integrating Three Representation Models in Fuzzy Multipurpose Decision Making Based on Fuzzy Preference Relations, Fuzzy Sets and Systems, 97, 33-48.

[7] Jian Ma, Quan Zhang, Zhiping Fan, Jiazhi Liang,etc.: An Approach to Multiple Attribute Decision Making based on Preference Information on Alternatives. HICSS (2001)

[8] Feng, S. and Xu, L.D. (1999) Decision Support for Fuzzy Comprehensive Evaluation of Urban Development, Fuzzy Sets and Systems, 105, 1, 1-12.

[9] Cheng, C.H., Yang, K.L. (1999) Evaluating Attack Helicopters by AHP Based on Linguistic Variable Weight, European Journal of Operational Research, 116, 423-435.

[10] Leekwijck, W.V. and Kerre, E.E. (1999) Defuzzification: Criteria and Classification, Fuzzy Sets and Systems, 108, 2, 159-178.

[11] Chen, S.-J. and C.-L. Hwang, Fuzzy Multiple Attribute Decision Making: Methods and Applications, Springer-Verlag, New York, (1992).

[12] Zhou, D.N. (2000) Fuzzy Group Decision Support System Approach to Group Decision Making Under Multiple Criteria. Dissertation of Doctor of Philosophy, City University of Hong Kong, March.

[13] Jian Ma, Zhiping Fan, Quanling Wei: Existence and construction of weight-set for satisfying preference orders of alternatives based on additive multi-attribute value model. IEEE Transactions on Systems, Man, and Cybernetics, Part A 31(1): 66-72 (2001)

A Resilient Multipath Routing Protocol for Wireless Sensor Networks

Ki-Hyung Kim[1], Won-Do Jung[2], Jun-Sung Park[2], Hyun-Gon Seo[2],
Seung-Hwan Jo[3], Chang-Min Shin[3], Seung-Min Park[3], and Heung-Nam Kim[3]

[1,*] Division of Information and Computer Engineering,
Ajou University, Suwon, Korea
`kkim86@korea.com`
[2] Department of Computer Engineering,
Yeungnam University, Gyungsan, Gyungbuk, Korea
[3] Ubiquitous Computing Middleware Research Team,
Embedded S/W Technology Center,
Electronics and Telecommunications Research Institute(ETRI), Korea

Abstract. Wireless sensor networks should be self-configuring, highly
scalable, redundant, and robust in dealing with shifting topologies due
to node failure and environment changes. The energy of a sensor node
is the most important system resource, and one of the best ways of conserving energy is to distribute the routing load across all sensor nodes
as much equally as possible. Considering the load balance for conserving energy, multipath routing protocols have been employed in sensor
networks. In this paper we present a resilient multipath routing protocol based on AODV, named as *RMRP*, for wireless sensor networks.
RMRP is an AODV-based multipath routing protocol in which sensor
nodes setup double routing paths toward sink nodes in an AODV fashion
and select one of them randomly for forwarding a packet. RMRP proposes a resilient route maintenance scheme which repairs broken links
by utilizing localized information around neighbor nodes of the broken
link, thereby reducing the flooding range of control messages. To show
the effectiveness of the proposed route maintenance scheme, we evaluate the performance enhancement of the proposed scheme by ns2-based
simulation.

1 Introduction

Recent technological advances have enabled distributed micro-sensing for large
scale information gathering through a wireless sensor network of tiny, low power
sensor nodes. Sensor networks may be deployed in dynamic, inhospitable environments or remote, inaccessible terrains to sense certain attributes in order to
track activities and events of interest and report them to a control center. Sensor networks should be self-configuring, highly scalable, redundant, and robust

---

* Corresponding author: Ki-Hyung Kim (kkim86@korea.com).

P. Lorenz and P. Dini (Eds.): ICN 2005, LNCS 3421, pp. 1122–1129, 2005.

in dealing with shifting topologies due to node failure and environment changes. Sensor networks are practically stationary, forming a large and dense multihop network, where every node has a limited radio range and serves as a router as well.

Routing algorithms that use fixed paths in traditional wired network are not suitable for sensor networks which have limited resources. Sensor nodes that are located in the fixed path suffer sever energy consumption and exhaust quickly because they provide relaying or routing services to the other nodes. This extreme unfair load-sharing between the sensor nodes on the path and the other nodes incurs the network separating. The simplest routing algorithm in sensor networks could be flooding. However, flooding mechanism consumes too much energy on relaying unnecessary traffic. On-demand routing algorithms developed for MANET, such as AODV[1] could be applied for sensor networks. They, however, might be too heavy to be directly employed because they are developed inherently for dynamic mobile nodes in MANET.

Considering the load balance for conserving the energy of sensor nodes, multipath routing protocols which have the advantage on sharing energy depletion between all sensor nodes have been proposed[2]. However, no research has been conducted for the effects of route maintenance schemes on communication performance and energy efficiency.

In this paper we present a resilient multipath routing protocol based on AODV, named as *RMRP*, for wireless sensor networks. RMRP is an AODV-based multipath routing protocol in which sensor nodes setup double routing paths toward sink nodes in an AODV fashion and select one of them randomly for forwarding a packet. RMRP proposes a resilient route maintenance scheme which repairs broken links by utilizing localized information around neighbor nodes of the broken link, thereby reducing the flooding range of control messages. To show the effectiveness of the proposed route maintenance scheme, we evaluate the performance enhancement of the proposed scheme by ns2-based simulation.

The rest of this paper is organized as follows. Section 2 describes an overview of the multipath routing algorithms in sensor networks. Section 3 presents the proposed multipath routing protocol. Section 4 presents the performance results. Finally, Section 5 concludes the paper.

2   Related Works

Sensor networks are the focus of a growing research effort. Traditional routing schemes have been difficult to adopt, and as a result, many new routing algorithms have been developed.

Directed diffusion is a typical data centric routing protocol for sensor networks. It provides a mechanism for doing a limited flood of a query toward the event, and then setting up reverse gradients to send data back along the best route. Diffusion results in high quality paths, but requires an initial flood of the query for exploration. One of its primary contributions is an architecture that names data and that is intended to support in network processing.

There has been much research on multipath routing algorithms. Classical multipath routing in networks has been explored for two reasons: *load balancing* and *fault tolerance*. For load-balancing, traffic between a source-destination pair is split across multiple paths. In multipath routing schemes[2] especially, sensor nodes have multiple paths to forward their data. Each time data sends back to sink, a sensor node picks up one of its feasible paths based on special constraints such as maximum available energy, minimum delay times, or security. Multipath routing has the advantage on sharing energy depletion between all sensor nodes.

For fault tolerance, multipath routing increases the likelihood of reliable data delivery by sending multiple copies of data along different paths, allowing for resilience to failure of a certain number of paths. Instead of employing expensive reliable transport mechanism, multipath based routing schemes could enhance the communication reliability by duplicating data paths toward sink nodes. Re-InForM is a multipath routing algorithms in sensor networks for reliable information forwarding[3]. Meshed multipath route has been proposed for immediate and successful data delivery toward sink nodes[4]. D. Ganesan et. al. proposed a multipath routing algorithm based on braided multipath routing scheme comparing with disjoint multipath routing in wireless sensor networks[5].

However, no research has been conducted for the effects of route maintenance schemes on communication performance and energy efficiency in sensor networks. Notice that reducing control packets during local repair process and fast local repair are also crucial for energy efficiency.

3   Resilient Multipath Routing Protocol

In this section, we present a resilient multipath routing protocol, named as *RMRP* for sensor network. We describe the architecture and basic operations of RMRP first. Then we present the proposed route maintenance scheme of RMRP.

3.1   Interests Flooding

RMRP is an on-demand multipath routing protocol based on AODV[1]. Based on the basic route discovery mechanism of AODV, RMRP follows the poll-reply model which has been used in sensor networks[2].

A sensing task is disseminated(or flooded) throughout the sensor network as an *interest* for named data from a sink node, which is called the *originator* of interests, as shown in Fig. 1 (a). This flooding of interests sets up the reverse routing paths on all sensor nodes toward a sink node, the originator of interests. RMRP builds double reverse routing paths on each sensor node toward a sink node or distributing routing load all around the network, just as like conventional multipath routing protocols[2]. Maintaining double routing paths on every node toward a sink node could reduce the probability of losing a routing path between a source and a sink.

In AODV, when a node receives the first interest from one of its neighbors, it records the neighbor node (*i.e.* the sender of the interest) as a predecessor toward a sink node. After this recording, the node processes no more RREQs which are from the same neighbor node with the same *rreq_id*; then, it sets up a reverse path through the selected neighbor node toward a sink node. Meanwhile, RMRP records two RREQs from different neighbor nodes in previous hop toward a sink node, thereby having two different reverse route paths toward a sink node as shown in Fig. 1 (a).

When the corresponding sensor nodes which meet the condition of the interest – they are referred to as the targets of the interest and become the source of the events or data packets – receive the interest, they respond with a data or event packet to a sink node.

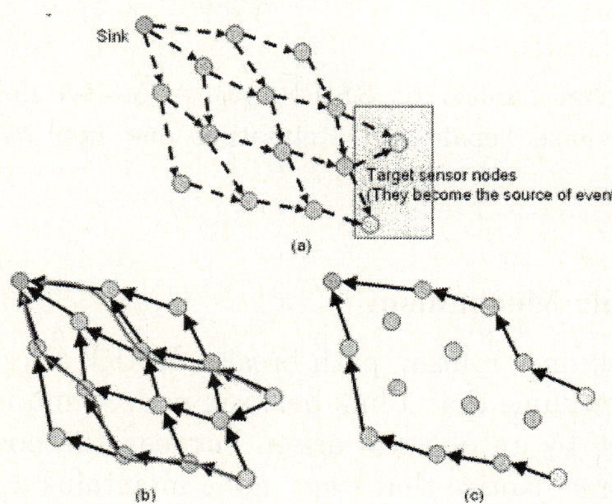

Fig. 1. Route discovery mechanism. (a) Interest flooding in a sensor network. (b) Reverse route paths in RMRP. (c) Reverse route paths in AODV

The routing of a data packet to a sink node is along the corresponding reverse routing paths from a source node to a sink node. Notice that the selection of a specific routing path in RMRP over the double routing paths of each intermediate nodes between a source and a sink is probabilistic, thereby distributing the routing load across all over the network as shown in Fig. 1 (b).

This setup process of reverse routing paths in RMRP is distinguishable from both AODV and Directed diffusion. Unlike AODV, RMRP does not use RREP(route reply) messages for the route discovery phase; every sensor nodes set up double reverse routing paths by receiving multiple RREQs from a sink. In AODV, the setup of a routing path from a source to a sink should be done after an exchange of a RREQ and a RREP between them as shown in Fig. 1 (c). This handshaking of route control packets (RREQs and RREPs) between a source and a sink can be reduced in sensor networks which is assumed to be static in this paper. Since the topology of sensor nodes between sources and a sink does not change frequently, the reverse route paths which are already set up when RREQs are flooded, could be valid in a relatively long time compared to the dynamic MANET.

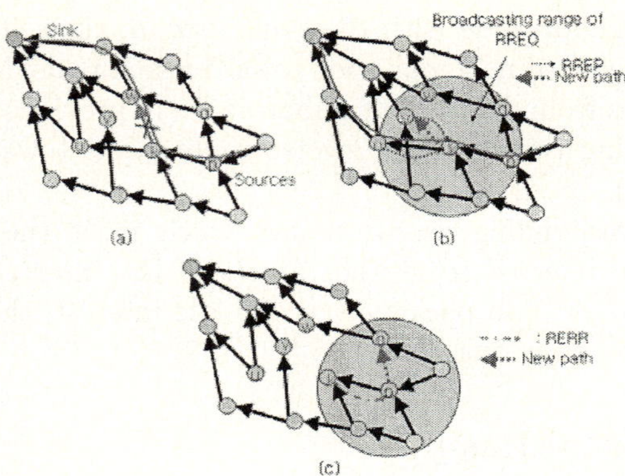

Fig. 2. Local repair mechanism of RMRP. (a) A broken link. (b) Broadcasting RREQ(ttl=1) for a local repair. (c) Retreating one hop back by generating a RERR(ttl=1)

3.2   Local Repair Mechanism

When a wireless link on a routing path breaks, RMRP tries to repair it locally as shown in Fig. 2. Assume that a link between node k and w is broken by some reasons, for instance by an obstacle arisen between the nodes or by a sudden disablement of node w. Notice that every node maintains double routes toward a sink; for instance, node k has two routes via node w and u respectively. Upon a link break between node k and w, node k tries to repair the broken route while utilizing the other route via node u for packet forwarding thereafter.

A local repair process of RMRP starts by the upstream node k's broadcasting of a RREQ with the time-to-live(ttl) value of 1 and the setting of the local-repair flag in RREQ. Upon receiving a RREQ, only nodes whose distance from a sink is smaller than the hop count field in the RREQ (for instance, node v in the Fig. 2 (b)) respond to node k, the upstream node of the link break, with a RREP. As a result of the reply of node v, node k repairs the link break and can maintain double routes again.

During the one-hop broadcasting of a RREQ of node k, there is a chance of no reply from neighbor nodes as shown in Fig. 2 (c), thereby resulting in only one valid route via node u. Now, let's assume that the only valid link between node u and k breaks. Node k tries to repair the link by one-hop broadcasting of a RREQ again, but fails to receive any reply from neighbors to make matters worse. In this case, node k notifies node p of the link break by sending a RERR. Upon receiving a RERR, node p marks the route via node k invalid, and tries to repair the route by repeating the above process, resulting in a repaired link via node q as shown in Fig 2 (c).

The local repair process of RMRP is light weight in terms of the overhead of flooding of control messages such as RREQs compared to AODV. This is because of the basic assumption of RMRP such that the network topology is static and there is little chance of all one-hop neighbors simultaneous losing of routes toward a sink. The intuition of the local repair process of RMRP is that the upstream node of a link break has a good chance of receiving a RREP from one of its immediate neighbors which has a route toward a sink and the distance to the sink is smaller than the upstream node.

Unlike to AODV, RMRP tries to minimize the number of route table entries in each node by considering only sink nodes as destinations. Thus, under the assumption of that sensor nodes do not communicate with each other, the routing table of a sensor node consists of a few sink nodes and its immediate neighbors.

4   Experimental Results

4.1   Simulation Model

To examine the performance of RMRP, ns-2 and AODV-UU are utilized. IEEE 802.11 at 2Mbps is used as the physical, data link and MAC layer protocols. 160 Sensor nodes are uniformly distributed with the inter-node distance of 150m over a 3kmx1.2km area. A sink node (base station) is located at one end of the area, and source nodes are placed at the other end. A CBR source is used at the application layer of the sink node to simulate periodic polling. The size of replied data packet is 40 bytes. The energy dispatching model is that the rates of battery power drain for transmission and reception are set to the same value of 282mW per second. Each node including the sink node has the same initial energy capacity of 300(J).

4.2   Experiments

Most performance comparisons are conducted between the following three protocols: RMRP, MRAODV (Multipath routing with the local repair scheme of AODV), and AODV. MRAODV is a multipath extension of AODV in sensor networks, and AODV is the direct adaptation of AODV in sensor networks. Simulation of MRAODV is done for the purpose of distinguishing the effects of the route maintenance scheme of RMRP from multipath routing. MRAODV has the same route maintenance scheme as AODV except the multipath routing capability.

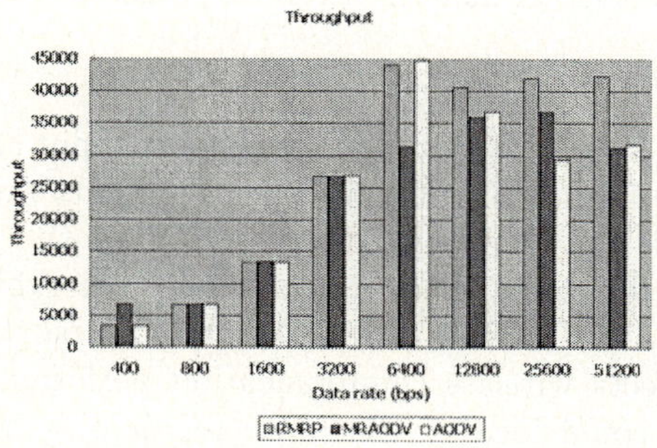

Fig. 3. Throughput vs. data rates

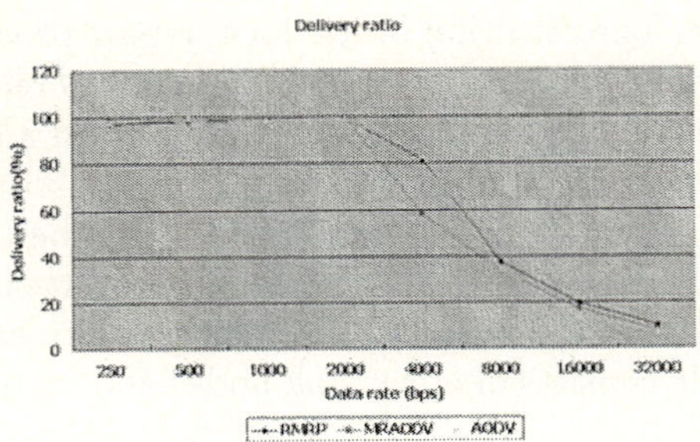

Fig. 4. Delivery ratio vs. data rates

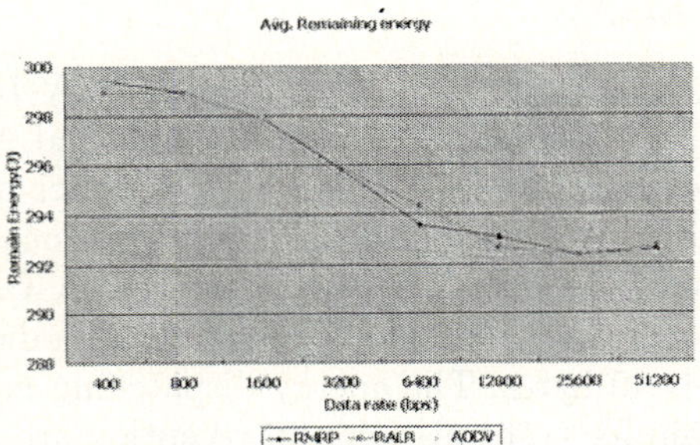

Fig. 5. Average remaining energy vs. data rates

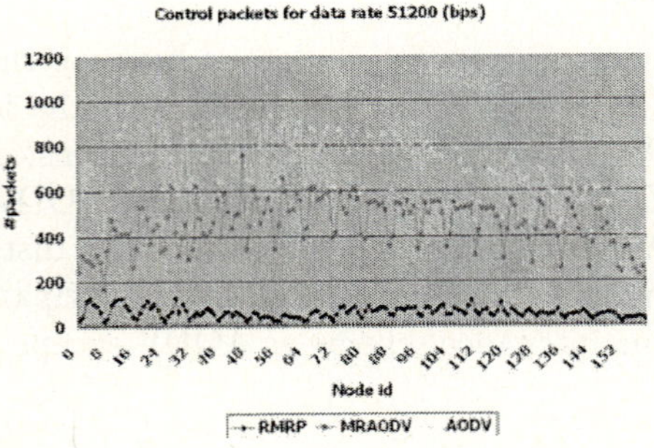

Fig. 6. Average number of control packets vs. data rates

Fig. 3 and Fig. 4 show the comparisons of throughput and data delivery ratio of RMRP, MRAODV, and AODV while increasing data rates. RMRP shows the most resilient result for increasing data rates than the others. As data rate becomes increased, the route path becomes more vulnerable to be broken. Upon a route path break, AODV and MRAODV try to repair it by flooding RREQ messages which affects wireless communication performance over wide range. Meanwhile RMRP could localize the route repair over a limited range of the route break.

Fig. 5 shows the comparison of average remaining energies for the three routing protocols. As the data rate becomes higher, the energy consumption for repairing broken links becomes higher. Also the number of generated and forwarded control packets for this repairing becomes higher as shown in Fig. 6. This means the affected range of local repairs for RMRP is relatively smaller than the others, and shows RMRP could effectively reduce the local repair range.

5   Conclusion

In this paper, we presented a resilient multipath routing protocol based on AODV, named as *RMRP*, for wireless sensor networks. RMRP is an AODV-based multipath routing protocol in which sensor nodes setup double routing paths toward a sink node in an AODV fashion and select one of them randomly for forwarding a packet. Based upon the conventional multipath routing capability, RMRP proposed a resilient route maintenance scheme which repairs broken links by utilizing localized information around neighbor nodes of the broken link, thereby reducing the flooding range of control messages. We evaluated the performance enhancement of the proposed scheme by ns2-based simulation.

Acknowledgements. This research was supported by University IT Research Center Project.

References

1. C. E. Perkins and E. Belding-Royer, "Ad Hoc On-Demand Distance Vector (AODV) Routing," IETF MANET Working Group, RFC3561, July 2003.
2. Xiaoyan Hong, M. Gerla, Hanbiao Wang, and L. Clare, "Load balanced, energy-aware communications for Mars sensor networks," IEEE Aerospace Conference Proceedings, vol. 3, March 2002, pp. 3-1109 - 3-1115.
3. B. Deb, S. Bhatnagar, and B. Nath, "ReInForM: reliable information forwarding using multiple paths in sensor networks," in Proceedings of 28th Annual IEEE International Conference on Local Computer Networks, Oct. 2003, pp. 406 - 415.
4. S. De, Chunming Qiao, and Hongyi Wu, "Meshed multipath routing: an efficient strategy in sensor networks," Wireless Communications and Networking (WCNC 2003), vol.3, March 2003, pp. 1912 - 1917.
5. Deepak Ganesan , Ramesh Govindan , Scott Shenker , and Deborah Estrin, "Highly-resilient, energy-efficient multipath routing in wireless sensor networks," ACM SIGMOBILE Mobile Computing and Communications Review, vol.5 no.4, October 2001.

A Multilaterally Secure, Privacy-Friendly Location-Based Service for Disaster Management and Civil Protection*

Lothar Fritsch and Tobias Scherner

Johann Wolfgang Goethe-University, Chair for M-Commerce & Multilateral Security,
Frankfurt am Main, Germany
Lothar.Fritsch@M-Lehrstuhl.de, Tobias.Scherner@web.de

Abstract. Information technology and modern challenges in civil protection create many questions concerning infrastructure security and reliability. In particular, mobile communications and the spread of mobile phones among citizens offer new opportunities of fine-grained disaster management using location based services (LBS). At the same time, location based services are regarded as very privacy-invading, and are regulated in many countries by law. In our article, we analyze the requirements of a LBS based disaster management scenario and propose a solution on how to build a privacy-friendly, multilaterally secure disaster management infrastructure based on robust, mobile phone infrastructures with high reachability of citizens and the possibility to manage disaster specialists.

1 Introduction

Communication technology can provide new technologies for civil protection and disaster management in the face of natural disasters, accidents or terrorist attacks. Mobile communication infrastructures like GSM[1], UMTS[2] or wireless local area networks are available in many countries around the world. In this article, we sketch a location-based disaster management infrastructure based upon GSM/UMTS infrastructures. Our special focus is the security and privacy properties which we specify according to the principles of Multilateral Security [14]. Disaster planning involves several factors. Regional planning based on geographic information, disaster risk evaluation, disaster forces coordination and information and coordination of citizens in the disaster area are the required tasks. An example can be found in [18]. We focus on the infrastructure responsible for citizen localization, information and management. First, we will describe the diaster management service and its specific technological and security requirements. Then, we introduce the infrastructure we designed, and provide the important details of the design. Finally, we conclude that

---

* This work was supported by the IST PRIME project; however, it represents the view of the authors only.
[1] GSM – Global System for Mobile Commmunication, www.gsm.org, Dec.15,2004.
[2] UMTS – Universal Mobile Telecommunications System, www.3gpp.org, Dec.15,2004.

P. Lorenz and P. Dini (Eds.): ICN 2005, LNCS 3421, pp. 1130–1137, 2005.

multilaterally secure disaster management LBS can be designed and give an outreach on further research.

1.1 Disaster Management Scenario

The high penetration of the population with mobile phones makes GSM/UMTS a high-bandwidth channel for disaster management. Citizens can be informed selectively after being localized by the mobile network to avoid blocking of critical resources. Furthermore, the localization of all mobile phones in a disaster area can be used by the disaster manager to survey the situation and plan the next steps. By pre-registering disaster specialists (e.g. firemen, police forces), they can be identified and notified by the disaster manager. For risk reduction, registered users can monitor their property or family members and receive notification in case of a disaster warning in an area-of-interest. Here, for example neighbours could be asked to secure property before a storm reaches the property. See fig.1 for an illustration of the scenario.

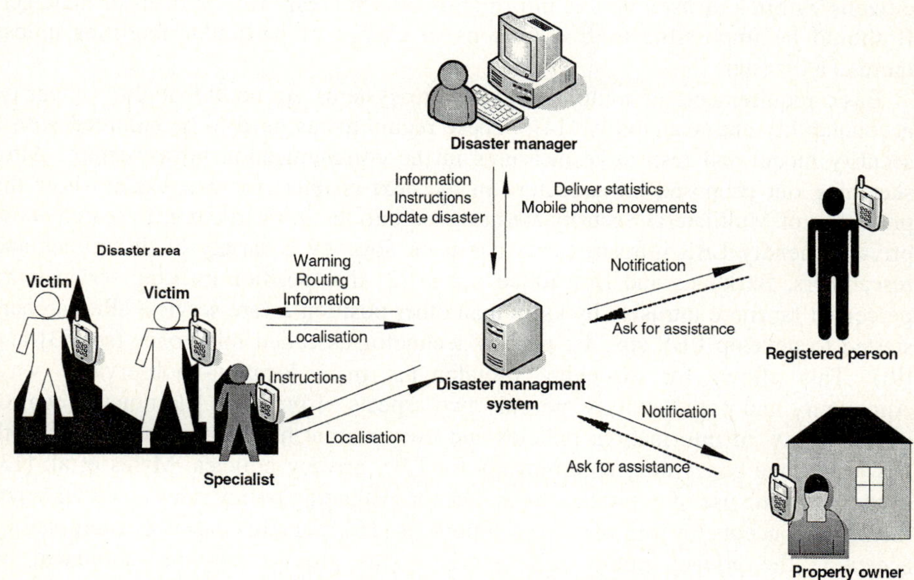

Fig. 1. Disaster Management basic scenario

Summarizing the main system features, the LBS disaster management system is responsible for these functions:

- Geographical administration of disaster areas including weather and traffic data;
- Localization of mobile phones within the disaster area;
- Reporting of statistics of phones, phone movements, population density based on LBS to disaster manager;
- Identification of disaster specialists among the citizens with mobile phones;

- Delivery of messages to mobile phones (either for evacuation, or containing instructions for disaster specialists);
- Notification about citizens threatened by disasters to registered next-of-kin.
- Notification about area threatened by disaster to property owners or persons in charge for areas of interest (e.g. chemical facilities).

The scenario contains several flows of information that is regarded private by some of the participants. Our infrastructure solution provides mechanisms to keep the data private as a measure to ensure high participation rates of the population.

1.2 Security and Privacy Requirements

Privacy is threatened in several places in our scenario. An observer configuration can reveal social networks of registered users. The whereabouts of off-duty disaster specialists can be retrieved with the system. Also, no localization of citizens should happen unless there is a real disaster. Furthermore, the location of citizens within a disaster area is information only relevant for the disaster manager. It should be impossible to find persons in charge of particular facilities unless there is a disaster.

Basic requirements of multilaterally secure systems are confidentiality, integrity, accountability and availability [14]. These requirements need to be enforced with a security model and respective measures in the communication infrastructure. After sketching our proposed infrastructure in the next chapter, we will explain how the principles of Multilateral Security are designed into the infrastructure. Research about privacy-friendly LBS infrastructures has been done by a variety of disciplines and researchers. Barkhuus and Dey found out in [2] that position-tracking services are perceived far more intrusive by users than other position-aware services. Researchers started to develop LBS specific privacy technologies called mix-zones (see [3] and [9]). This allows for switching pseudonyms in a location-unobservable way. Anonymity and pseudonymity are only two aspects of privacy. Additionally control over the flow of information, policies and user consent has to be considered. Work has been done concerning requirements for LBS privacy policies. Myles et al. [13] investigated the use of a middleware server for evaluating policy rules and Snekkenes identifies concepts for formulating such policies [16]. Usually consent is expressed by accepting the privacy policy of a service. This process may be automated by comparing the privacy policy of the service with the privacy preferences of their users. Explicit user consent is a hard requirement in many legal systems, particularly within the European Union (see also [5]).

1.3 Technical Requirements

Disaster management communication technology needs to be reliable and robust against disaster specific risks. Following Held [10], we introduce the technical requirements customized for mobile communication. First of all the establishment of a uniform emergency signal for mobile phones is needed. Through this feature the citizen is able to recognize immediately that it is an emergency notification. This warning function can be divided in [1]:

- Wake up function to direct the citizen's attention to the emergency notification.
- Information function to provide citizens customized information about the emergency and further instruction for action.

Additionally, mobile phones offer the possibility to establish direct communication between rescue forces and victims. Especially the requirements to robustness make mobile technologies attractive for civil protection. The cellular structure of mobile networks and the possibility of battery based service of mobile stations [7] during a power failure grant a high availability of the services even by direct impairment of the network. The use of cell broadcast as a point-to-multipoint technology secures the fast and reliable distribution of notifications to the citizens in the disaster area. The everyday handling by the citizen of mobile phone network and the wide distribution of the technology suggest a high acceptance by the population. The acceptance on the part of the state can be prognosticated by the very low financial commitment needed, the existence of infrastructures already in place, and the potential to co-locate development with efforts like the E911 and E112 emergency call localization systems in the United States and Europe.

2   Multilaterally Secure Disaster Management LBS Architecture

This section explains our approach to implement a disaster management system (DMS) based on mobile networks. As a methodology, we first assessed the security and privacy requirements following the principles of Multilateral Security [14]. The formalized catalog of Common Criteria security properties was used to structure the findings [11]. Then, a system design was modeled in UML, resulting in the scenario we describe in this paper.

2.1   Solution Overview

In our solution we suggest a middleware system with front-end support to control the flow of information and to protect the interests of every party. This system consists of three components shown in figure 2:

- Matcher: Manages locating the citizens in the disaster area. Matches the disaster area with observation rules of the users. Protects the persistent store of observation rules on behalf of the distinct user. Matches profiles of threatened person with their registered contact person.
- Identity management system: Steers information exchange of the disaster management with mobile operator and the citizen in an emergency case. Similarly described by Borking [4].
- Process control: Controls the disaster management system, represents an interface to the disaster manager and is responsible for temporal storage disaster data (observation rules and localisation information).

The warnings of victims and instructions to specialists are to be transmitted via cell broadcast. This bypasses the performance limitations of point-to-point technologies like short message services [8]. Notifications for specialists are encrypted. Unlike the warning mass broadcast, this notification will be sent point-to-point. To be able to use the whole bandwidth of the suggested system it is necessary to register as a user and to deposit observation rules and specialist status including verification of status and claims. Unregistered users are only able to receive warnings and get localized. After a disaster user-specific information will be deleted or anonymously stored. Only the user himself will be able to find out when and how often he was located.

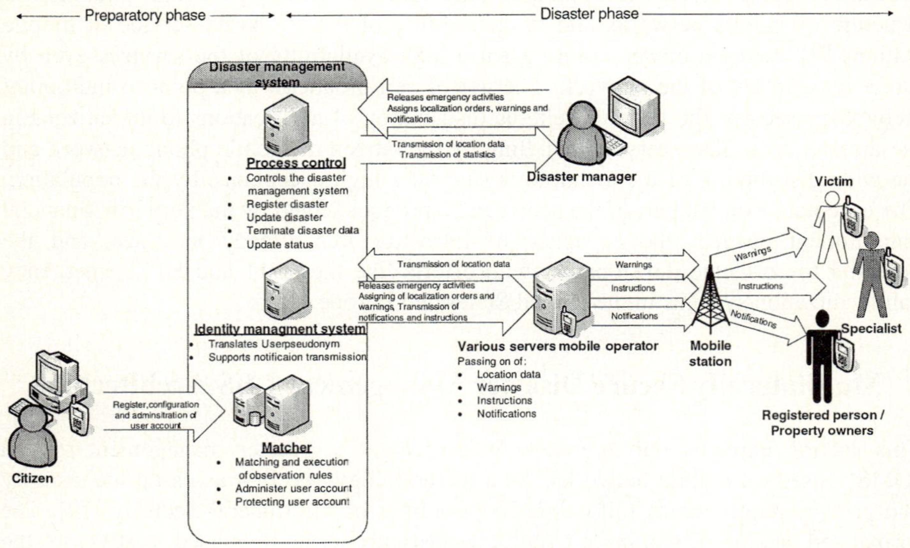

Fig. 2. Disaster Management solution overview

3 Privacy and Security Properties

In the preparatory phase users can enrol themselves in the DMS as a specialist. The disaster manager has the dilemma that he has to rely on this declaration of the status. We solved the problem of trustworthiness of the specialist status within a legitimating process. The DMS hands over a voucher to the user. The emergency service agency/employer confirms the status und sends it digitally singed to the DMS. The confirmation contains a specialist's classification into a qualification-specific group Thereby the disaster manager is able to analyse the distribution of the different kinds of task forces and is able to make further plans accordingly. The employer gets no information about the user's pseudonym and the DMS needs no information about the user's identity. In our approach we provide only the information necessary to each actor. To integrate this property, the identity management system (IDM) uses different pseudonyms for each user on the communication section between mobile operator (MO) and IDM; and between the disaster manager (DM) and IDM. This we call "matching" of pseudonyms. This uncoupling of user and provider secures the user's privacy (e.g. the MO and the DM have no information about the identity of a certain specialist. In the case of using anonymous prepaid SIM-cards neither DM nor MO get knowledge of the user's identity. For the MO the sphere separation ensures business secrets (e.g. current customer numbers). The IDM is illustrated in figure 3.

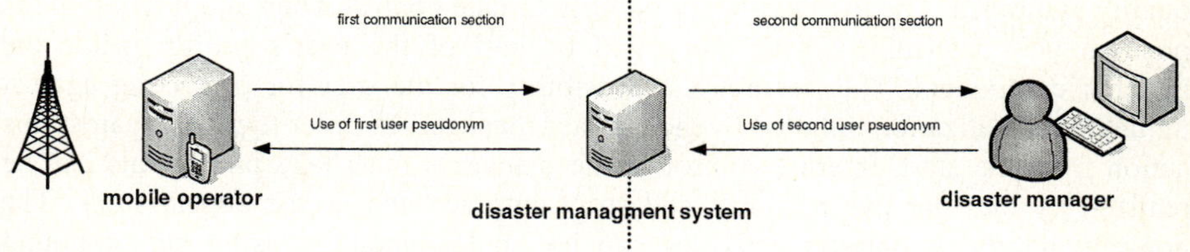

Fig. 3. User's pseudonym

With the occurrence of a disaster the DM sends a digitally signed order to the process control (PC). The management system verifies the authorization on the basis of the DM certificate. After positive evaluation, the DMS creates digitally signed orders for the mobile operators that contain the DM's requests for citizens in the disaster area. The different parties can proof the delivery of the order or data in time by delivery vouchers and by log files. This provides the following advantages: First, we solve the non-repudiation of orders for later legal disputes and examine the authorizing of disaster activities. Secondly this is an instrument for just-in-time quality assurance. It is needed for checking purposes of the accessibility and availability of mobile operators and the disaster manager. This however cannot solve the problem of capacity on the air interface between mobile station and mobile terminal. As mentioned above, cell broadcast (CB) does significantly reduce the traffic over the air. In order to guarantee the receipt however the CB function on the mobile phone must be activated by the provider. This can be managed by a configuration message for SIM application toolkits on the phone cards.

So far there is still potential of abuse of warning notifications with SMS-Messages with a fake sender [12]. In order to avoid uneasiness among victims, warning notifications have to be digitally signed similar to the requests from the DMS to the MO. Thereby Victims can judge autonomously if this is a falsified or authentic notification. This requires technical measures on the mobile terminals. A suitable solution is sketched in [15], where a SIM card based signature system is introduced.

To protect the privacy of users it is necessary to give a notification via CB before the beginning of detection activities, as explicit consent is required in European law before localizing people via telecommunication networks. Registered users expressed their consent while accepting the system's privacy policy. Non-registered users will get a notification, as well, however they have never accepted the privacy policy or given consent. Still, there is a justification to determine their position without consent, which can be well seen in some legal sources, e.g. the directive 2002/58/EC of the European Union [6] and the UN declaration of human rights [17]. Here, the service of saving the person's life can be interpreted as implicit consent.

Special attention of our work lies in the consideration of location data. In order to prevent data collection over several disaster events the DM has only the authorization to read and analyze the victim referred location data which will not be stored permanently. At the end of a disaster, location data inside the disaster management system will be deleted or anonymously stored for education of disaster managers or quality assurance. The user has the possibility to take control when and how often his position was determined. This entry will be part of the user's profile inside the matcher component. This secures the possibility for the user for protecting against unjustified localization initiated by the disaster manager. In order to guard against this action from the start, internal control of the managers must take place. This can be realized by the four eye principle and strict entrance and access supervisions. The non-repudiation of disaster activities can be implemented by using person-bound smartcards with personal certificates. Figure 4 summarizes the security & privacy properties.

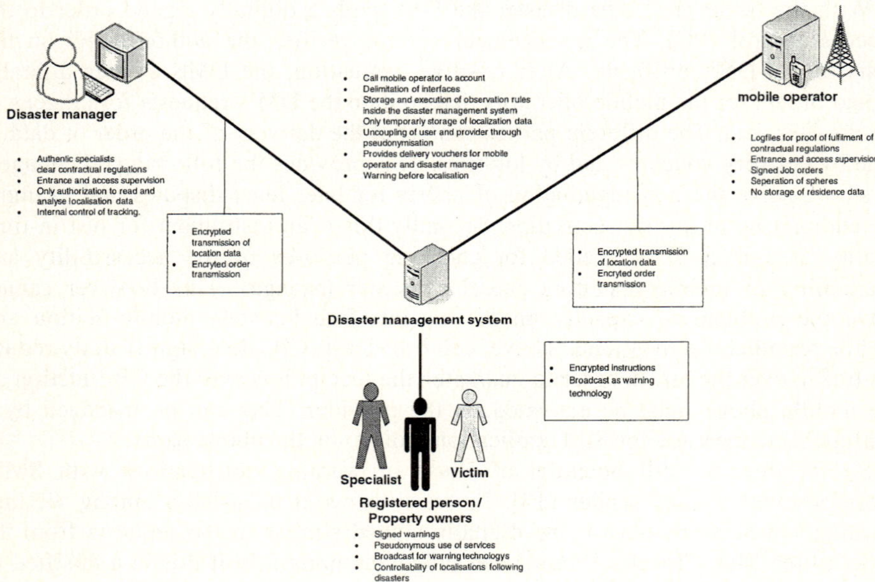

Fig. 4. Security and privacy properties of the disaster management system

4 Summary and Outlook

We show how a multilaterally secure location based service for disaster management can be implemented based on mobile telephone technology. Thus, the advantages of using a wide-spread, wireless and battery-powered personal information infrastructure can be combined with legal and personal requirements of citizens. We are confident that a disaster LBS can be realized with the above properties. Some challenges are left for practical implementation, though. As experiences with the E911 implementation and the European E112 efforts show, standardization efforts, in particular for the interfaces between the actors and their respective quality of service assurance, need to be defined.

References

[1] AGBF-NRW: Planungsgrundlage zur Warnung der Bevölkerung bei Gefahrenlage, 2003, http://www.agbf-nrw.net/ak/zuk/warnung-bei-gefahr.pdf.

[2] Barkhuus, Louise; Dey, Anind: Location Based Services for Mobile Telephony: a study of users' privacy concerns, 2003.

[3] Beresford, Alastair R.; Stajano, Frank: Location Privacy in Pervasive Computing, IEEE Pervasive Computing, 2, 46-55, 2003.

[4] Borking, John: Der Identity Protector, Datenschutz und Datensicherheit (DuD), 20, 654-658, 1996.

[5] Buchta (ed.), Anna: Legel Requirements - part 1 of Deliverable 1.1a of IST PRIME EU project, 2004, http://www.prime-project.eu.org/public/prime_products/deliverables /pub_del_D01.1.a.part1_ec_wp01.1_V2_final.pdf.

[6] European Comission:Directive 2002/58/EC of the European Parliament and of the council,2002,http://europa.eu.int/eur-lex/pri/en/oj/dat/2002/l_201/l_20120020731en00370047.pdf,July 12, 2002.

[7] Gerpott, Torsten J.; Walter, Andreas: Operative und Finanzielle Bewertung von TETRA-, Tetrapol und GSM - BOS Plattformen für das digitale BOS-Mobilfunknetz, 2004, http://www.dialogconsult.com/DCNL/PDF/DCNL020b.PDF.

[8] Grimm, Markus:Sichere transaktionsbasierte Behördenalarmierungssysteme auf GSM-Basis,Lehrstuhl für M-COmmerce und Mehrseitige Sicherheit,Johann Wolfgang Goethe-Universität,Frankfurt am Main,2004.

[9] Gruteser, Marco; Grunwald, Dirk: Anonymous usage of location-based services through spatial and temporal cloaking, First International Conference on Mobile Systems, Applications, and Services (MobiSys'03), 2003.

[10] Held, Volkmar:Technologische Möglichkeiten einer möglichst frühzeitigen Warnung der Bevölkerung, in Schriftenreihe der Schutzkommission beim Bundesminister des Inneren; Vol. 45 (Neue Folge), edited by Bundesminister des Inneren (Bonn, 2001),.

[11] ISO: ISO 15408 The Common Criteria for Information Security Evaluation, 1999,.

[12] Muntermann, Jan; Rossnagel, Heiko; Rannenberg, Kai:Potentiale und Sicherheitsan-forderungen mobiler Finanzinformationsdienste und deren Systeminfrastrukturen, in E-Science und GRID, Ad-hoc-Netze und Medienintegration; Proceedings der 18. DFN-Arbeitstagung über Kommunikationsnetze; Vol. 55, edited by von Knop, Jan; Haverkamp, Wilhelm; Jessen, Eike (Gesellschaft für Informatik (GI), Bonn, 2004), pp.361-376 .

[13] Myles, Ginger; Friday, Adrian; Davies, Nigel: Preserving Privacy in Environments with Location Based Applications, IEEE Pervasive Computing, 2, 56-64, 2003.

[14] Rannenberg, Kai:Multilateral Security - A concept and examples for balanced security, in Proceedings of the 9th ACM New Security Paradims Workshop; ACM Press, Cork, Ireland, 2000), pp.151-162.

[15] Rossnagel, Heiko:Mobile Qualified Electronic Signatures and Certification on Demand, in Proceedings of the 1st European PKI Workshop - Research and Applications; Vol. 3093, Springer, Samos Island, Greece, 2004),.

[16] Snekkenes, Einar: Concepts for personal location privacy policies, 3rd ACM Con-cerence on Electronic Commerce, Tampa, Florida, USA, 2001.

[17] The United Nations:Universal Declaration of Human Rights,1948, http://www.unhchr.ch/udhr/lang/eng.htm,December 10, 1948.

[18] Yoshimura, Fumiaki: Disaster Risk Management Through Hazards Analysis, International Seminar on Disaster Preparedness and Mitigation, New Dehli, 2002.

Survivability-Guaranteed Network Resiliency Methods in DWDM Networks

Jin-Ho Hwang[1], Won Kim[2], Jun-Won Lee[3], and Sung-Un Kim[1],[*]

[1] Pukyong National University, 599-1 Daeyeon 3-Dong Nam-Gu,
Busan, 608-737, Korea
`jhhwang@mail1.pknu.ac.kr`
[2] National Internet Development Agency of Korea, KTF Bldg. 1321-11,
Secho-2 Dong, Secho-Gu, Seoul 137-857, Korea
`wkim@nic.or.kr`
[3] Andong National University, 388 Song-chon Dong,
Andong, Kyoungbuk 760-749, Korea
`leejw@andong.ac.kr`

Abstract. The ability of a network to withstand and recover from failures, survivability, is one of the most important requirements of dense-wavelength division multiplexing (DWDM) networks. And the network resiliency methods through the efficient route computation are critical issues in terms of blocking probability, survivability ratio and service disruption ratio. In this paper, we propose the network resiliency methods used to guarantee survivability by applying shared risk link group (SRLG) and trap avoidance (TA) problem which are important criteria in DWDM networks.

1 Introduction

While coping with the rapid growth of IP and multimedia services, current Internet based on time division multiplexing (TDM) cannot supply sufficient transmission capacity for high bandwidth-needed services. However, the huge potential capacity of one single fiber, which is in Tb/s range, can be exploited by applying DWDM technology which transfers multiple data streams on multiple wavelengths simultaneously. So, DWDM-based optical networks have been a favorable approach for the next generation Internet [1]

In DWDM networks, man-made errors, uncontrollable natural phenomena (e.g. floods and earthquakes), intrinsic vulnerability of optical components and etc. cause equipment, and therefore link or node failures. For example, all the lightpaths that traverse the failed fiber will be disrupted so a fiber cut can lead to tremendous traffic loss. And other network equipments may be affected [2].

A key feature of network survivability is the backup path establishment that keeps physical-diversity (which is also called by physical-disjoint). In DWDM

---
[*] Corresponding Author: `kimsu@pknu.ac.kr`

P. Lorenz and P. Dini (Eds.): ICN 2005, LNCS 3421, pp. 1138–1145, 2005.
© Springer-Verlag Berlin Heidelberg 2005

networks, each link set up in one lightpath may cross one or more optical components, where the fault of optical components may result in the potential failure of the link. A component here essentially presents any part or site involved in the integrity of the links and associated with SRLG defined as resource groups having shared risk in common [3][4].

Moreover, once a primary path is found, one may not be able to have a SRLG-disjoint backup path (even though a pair of SRLG-disjoint paths do exist using a different primary path). This is so-called trap problem, which is rarely present when finding link/node disjoint paths but can occur much more frequently (with a probability of up to 30 percentage in a typical optical network) [5].

The general technique of finding primary and backup paths is generally based on the modified shortest path first (SPF) algorithm, with the constraints-based routing extension [6][7]. But, this approach does not consider SRLG and TA problem, and result in increase of blocking probability. So, in this paper, we propose survivability-guaranteed network resiliency methods by computing paths under SRLG constraint and by considering TA problem. Finally, we simulate the performance of the proposed methods in terms of blocking probability, survivability ratio and service disruption ratio.

The rest of this paper is organized as follows: section 2 analyzes DWDM structure and network survivability requirements in DWDM networks. In section 3, we propose the resiliency methods by applying SRLG constraint and TA problem. Simulations are carried out and analyzed in section 4, and some concluding remarks are made in section 5.

2   Analysis of Network Survivability in DWDM Networks

2.1   DWDM Structure

Architecture of DWDM network is shown in figure 1, in which IP traffics are injected into DWDM ingress nodes for various conventional electric domain based-networks, such as LANs, MANs and ATMs.

Specifically, we can describe three management sections taking into consideration resource types (optical components) and coverage of fault effects [8].

- Optical Channel Section (OCh): Channel management section for one light-path established between ingress node and egress node.
- Optical Multiplexing Section (OMS): Link management section for one link between adjacent nodes. This includes Optical Amplifier Section(OAS) and Fiber Intrusion Section(FIS).
- Node Section: Node management section including demux, optical switch and mux that are divided and managed by sub management sections, i.e. Demultiplexing Section(DS), Switching Section(SS) and Multiplexing Section(MS).

According to the classified sections, this paper classifies faults into two categories: hard faults and soft faults. Hard fault possibilities shown in figure 2 encompass failures in the physical equipment or medium used to transmit the

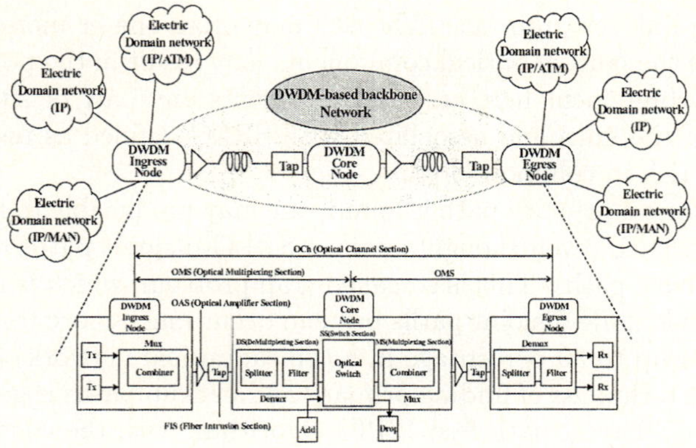

Fig. 1. DWDM Structure

Category	Resource Type	Fault possibility
Channel Fault (OCh)	Receiver (Add port)	Tuned or fixed-tuned filter failure Optical receiver failure Timing recovery circuitry failure
	Transmitter (Drop port)	Tunable or fixed tuned laser failure Monitoring photodiode failure Driver circuitry failure
Link Fault (OMS)	Fiber Fault	Fiber cut Power loss due to damaging fiber
	Optical Amplifier Fault	Fuse or power circuit failure Pump laser failure Input signal monitor failure Amplifier optical path failure
DWDM Node Fault	Optical switch Fault	Optical switches failure
	Wavelength Converter Fault	Saturable absorber failure Loss of signal failure Fuse or power circuit failure Optical path failure
	Mux/Demux Fault	Optical filter failure
	Failure of the power outage of the controller in OXC	
	Loss of signals failure Fuse or power circuit failure Optical path failure	

Category	Soft faults
Noise	Amplified Spontaneous Emission (ASE) Receiver noise Interferometric crosstalk Laser noise Reflections Jitter
Distortion	Chromatic dispersion Filtering effects Polarization Mode Dispersion (PMD) Transmitter driver transfer function Laser diode response Non-linearities (SPM, XPM, FWM) Receiver frequency response and thresholds

Fig. 2. Hard/soft fault possibilities in DWDM networks

data [8][9][10][11]. And these possibilities are directly related to SRLG failure. Namely, each component has the shared risk level, whose a failure in figure 2 may cause all links attached to the component to be broken simultaneously, because all links that go through the same component belong to the same SRLG. So, we compute primary and backup paths not to have the same fault possibility. This is called SRLG-disjointness (also called SRLG-diversity).

On the other side, soft faults described in figure 2 result from slow degradation of the optical signal quality. An optical signal passing through physical components is subject to several perturbations such as noise from random signal fluctuations, pulse-shaped distortion, and crosstalk. All these perturbations affect the signal quality [12][13].

2.2   Network Survivability Requirements in DWDM

Network survivability can be assured by various resilience schemes-protection, restoration, rerouting, etc.-that have very different recovery times and resource

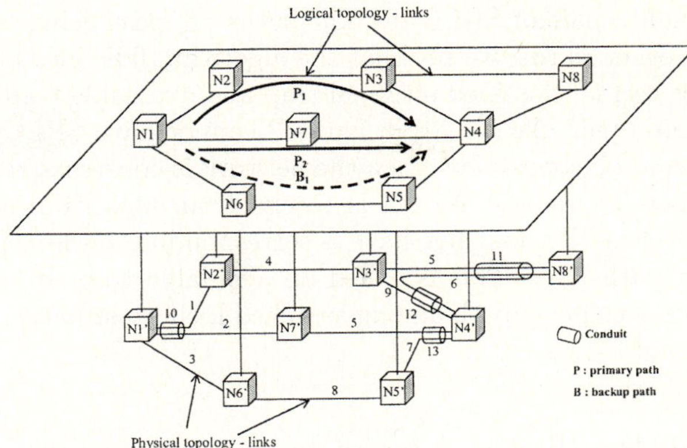

Fig. 3. Basic concept of SRLG and TA

consumptions [14][15]. Since mission-critical data are supposed, we deal with a protection scheme that is the fastest resilience paradigm (with 10-100 ms recovery time) as the backup resources are previously reserved. Because it is necessary to guarantee connectivity even in case of network failures, so the protection plays more and more essential role in backbone networks.

As the key constraint to establish paths, SRLG is being researched intensively. SRLG is defined as a group of links or nodes that share a common risk component, whose any fault in figure 2 can potentially cause the failure of all the links or nodes in the group [4]. For example, all fiber links that go through a common conduit belong to the same SRLG, because the conduit is a shared risk component whose failure, such as a conduit cut, may cause all fibers in the conduit to be broken simultaneously. SRLG is introduced in the generalized multi-protocol label switching (GMPLS) and can be identified by a SRLG identifier, which is typically a 32-bit integer.

Figure 3 illustrates a simple example of the SRLG concept. The upper plane is logical topology controlled by GMPLS and the lower plane is the physical topology in which optical components (i.e., fiber, conduit, EDFA, etc.) are deployed. All links and conduits uniquely have SRLG identifiers. When there is a connection request between node N1 and node N4, N1-N2-N3-N4 {N1'-10-1-N2'-4-N3'-9-6-12-N4'} can be a primary path by a certain algorithm, and there could be two candidates for a backup path, N1-N7-N4 {N1'-10-1-2-N7'-5-7-13-N4'} and N1-N6-N5-N4 {N1'-3-N6'-8-N5'-7-5-13-N4'}. If we only look at the logical topology, both the two backup paths can be allowed under dedicated path protection without SRLG concept. However, if the backup path resolves N1-N7-N4, the primary path and the backup path that go through the same conduit can fail at the same time by one single fault in {10} or in {1}, so the determined backup path is N1-N6-N5-N4. Consequently, in order to make a network survivable against failures, the SRLG concept should be imposed on the selection of backup path.

Also, figure 3 presents TA concept. TA is defined that a routing algorithm fails to find a pair of SRLG-disjoint paths for a source and a destination node pair (even though a pair of SRLG-disjoint paths do exist using a different primary path). In other words, we say that the algorithm falls into trap. From this definition, traps can be classified into real trap and avoidable trap [5]. Real trap means that a node pair like N1-N8 in figure 3 cannot have SRLG-disjoint path pair, so this should be considered when the network is constructed. On the other hand, if a connection request N1-N4 is received, an algorithm can choose P_1 instead of P_2 because P_2 does not have a corresponding backup path while P_1 has the backup path (B_1). This is called an avoidable trap. In this paper, we recursively check a primary path among searched k-shortest paths, thus improve blocking probability.

3   Network Resiliency Methods

In this section, we propose the network resiliency methods considering potential blocking possibilities of future traffic demands. This method chooses a route that does minimize interference for potential future connection requests by avoiding congested links. To achieve this, we fit the previously mentioned "critical links" concept into the context of DWDM networks. These are links with the property that the available wavelengths on the minimum hop routes of one or more node-pairs decreases whenever a lightpath is routed over those links. By reducing the number of wavelengths in a congested link, the number of failed connections by a single failure at a time can be decreased as well.

Figure 4 illustrates the basic concept. For example, our method is to pick route P_2 for connection between (S3, D3) pair that has a minimum affect for other connection requests (S1, D1) as well as (S2, D2) even though the path is longer than P_1 with a congested link L.

To formulate the network resiliency methods, let the directed graph G=(V,E) represent the network with the n-element set of vertices V and the l-element set of directed edges E={e_i|1≤ i ≤ l}⊆ {(u,v)|u,v ∈ V, (u≠v)}. The set of connection requests that are potential source-destination pairs in the future is M={m_i=(s_i,d_i)| 1 ≤ i ≤ m_k)}, where s_i and d_i are the source node and the destination node, respectively. The set of current demands is P={p_i=(a_i, b_i)| 1 ≤ i ≤ p_k)}, where a_i and b_i are the source node and the destination node, respectively. The goal of the route computation is to satisfy the following equation;

$$max \sum_{(s,d)\in M \setminus (a,b)/p} \alpha_{sd} \cdot F_{sd} \tag{1}$$

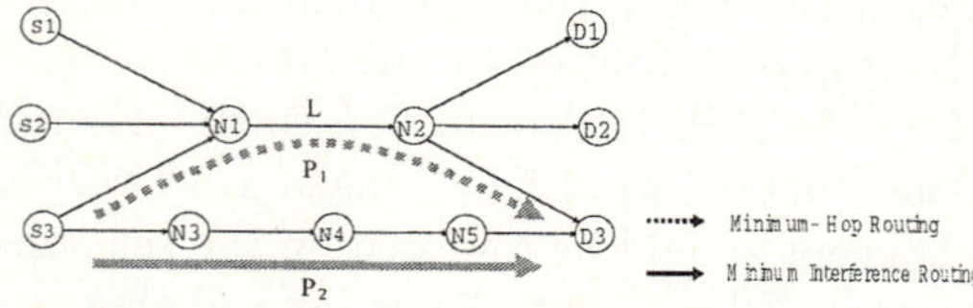

Fig. 4. Basic concept of proposed resiliency methods

where α_{sd} is the weight for each node pair and F_{sd} is the set of available wavelengths of the bottleneck link. This equation represents a maximum available wavelengths problem for each source-destination pair in M except the current demands P. Namely, the current demands along a path do not interfere too much with potential future requests.

The number of available wavelengths on a link is regarded as an important factor to improve network performance in terms of blocking probability. Therefore, we add a new notation $\triangle$ as a threshold value of the available wavelengths on a link to choose the minimum interference path for potential future connection requests with consideration of critical links as well as non-critical links with few wavelengths. Based on notations such as SRLG information and $\triangle$, we determine links with congestion possibility for potential future demands between a (s,d)-pair according to Equation 2, where $\forall$(s,d) $\in$ M \(a,b) and $\forall$l $\in$ L, and call them CL_{sd}^{l}.

$$CL_{sd}^{l} : (srlg_{ab}^{l} \in srlg_{sd}^{p/ab}) \cup (R(l) < \triangle) \qquad (2)$$

where $srlg_{ab}^{l}$ is the set of SRLG ID in which link l has, and $srlg_{sd}^{p/ab}$ is the set of corresponding SRLG IDs in which the primary path has.

The proposed resiliency methods give appropriate weights to each link based on the amount of available wavelengths and SRLG IDs on a link l where $\forall$ l $\in$ L, so that the current request does not have the same risk with which the primary path has. The link weights are estimated by the following procedures. First, let $\partial F_{sd}/\partial R(l)$ indicate the change of available wavelengths on the bottleneck link l for the potential connection request between a (s,d)-pair when the residual wavelengths of link l are changed incrementally. With respect to the residual wavelength of the link, the weight w(l) of a link is set to

$$w(l) = \sum_{(s,d) \in M \setminus (a,b)} \alpha_{sd}(\partial F_{sd}/\partial R(l)), \forall l \in E \qquad (3)$$

where R(l) is the number of currently available wavelengths and $\partial F_{sd}/\partial R(l)$ indicates the change of available wavelengths on the bottleneck link l for the potential connection requests between a (s,d)-pair when the residual wavelengths of link l are changed incrementally. This equation determines the weight of each link for all (s,d)-pair in the set M when setting up a connection between the (a,b)-pair, i.e. (s,d)$\in$M\(a,b)$\in$P.

Equation 3 determines the weight of each link for all (s,d)-pairs in the set M except the current request when setting up a connection between the (a,b)-pair, i.e., (s,d)$\in$M\(a,b), but computing weights for all links is very hard, where $\forall$l$\in$L. To solve this problem, we consider more restricted links than other links, that is, $\partial F_{sd}/\partial R(l)$=1 when [if (s,d):l$\in$CL$_{sd}^{l}$] and $\partial F_{sd}/\partial$ R(l)=0 when [otherwise] if a link belongs to the set of congestion links for a certain (s, d)-pair, i.e., l$\in$CL$_{sd}^{l}$. Therefore, computing the link weights is simplified as shown in Equation 4.

$$w(l) = \sum_{(s,d):l \in CL_{sd}^{l}} \alpha_{sd} \qquad (4)$$

4   Performance Evaluation

For the verification of the proposed network resiliency methods under SRLG information, we use two test networks: (14 nodes, 20 links), (30 nodes, 61 links). And we assume that each link contains two uni-directional fibers, one in each direction and the traffic pattern is dynamic. The connection requests arrive randomly according to the Poisson process, with negative exponentially distributed connection times with unit mean.

We compare the proposed scheme with the modified SPF as shown in figure 5. The proposed method has lower blocking probability than modified SPF in both test networks (improved by about 5-10%).

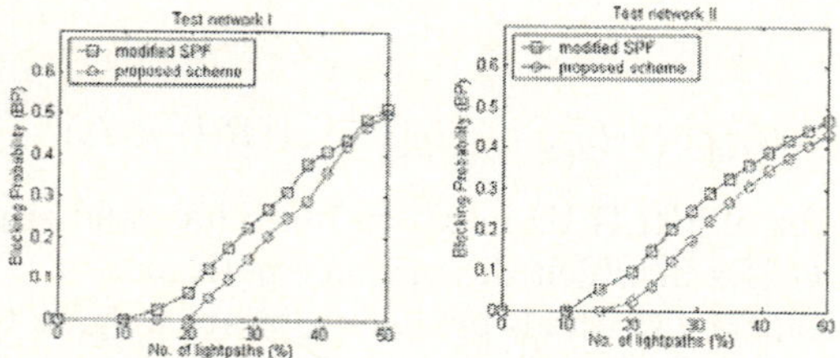

Fig. 5. Blocking probability as a function of percentage of lightpaths requested

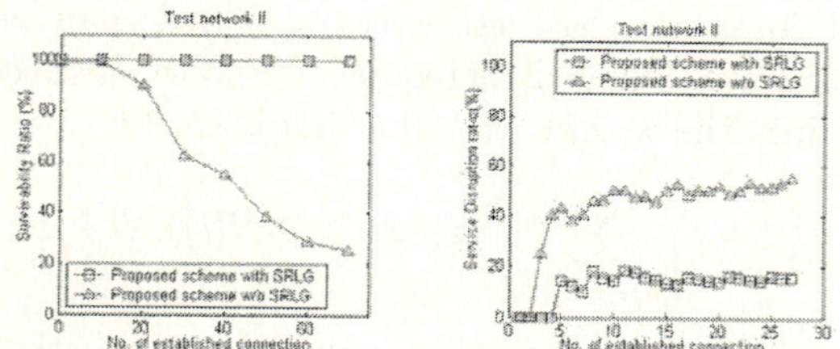

Fig. 6. Survivability ratio and service disruption ratio

In figure 6, the survivability ratio and service disruption ratio show that the proposed scheme is tolerant for single network failure with SRLG constraint than without SRLG constraint.

5   Conclusion

In this paper, we analyzed DWDM structure and fault possibility by separating hard faults and soft faults. Under fault possibility which has the same risk, we proposed the survivability-guaranteed network resiliency methods by deploying SRLG concept and TA problem. The simulation results showed that the blocking probability improves 5-10% than the existing scheme using modified SPF. And survivability ratio and service disruption ratio proved that the proposed method is tolerant in the viewpoint of network survivability in DWDM networks. For the future research, we envisaged that the proposed scheme can be applied to GMPLS used for control protocol in DWDM networks.

Acknowledgment. This work was supported by grant No.(R01-2003-000-10526-0) from Korea Science and Engineering Foundation.

References

1. T. E. Stern and K. Bala: Multiwavelength Optical networks: A layered approach, Addition Wesley Publishers, 1999.
2. Jing Zhang, et al.: A Review of Fault management in WDM Mesh Networks: Basic Concepts and Research Challenges, IEEE, vol.18, no.2, pp.41-48, Mar-Apr 2004.
3. D. Papadimitriou et al.: Inference of Shared Risk Link Groups, draft-many-inference-srlg-02.txt, Internet Draft, Nov. 2001.
4. Sudheer Dharanikota et al.: Inter-domain routing with Shared Risk Groups, OIF2002.20.0, April 2002.
5. Dahai Xu, et al.: Trap avoidance and protection schemes in networks with shared risk link groups, Journal of L.wave Tech., vol.21, no.11, pp.2683-2693, Nov. 2003.
6. Jong-Gyu Hwang, et al.: A RWA Algorithm for Differentiated Services with QoS Guarantees in the Next Gen-eration Internet based on DWDM Networks, Photonic Network Comm., vol.8, no.3, pp.319-334, November 2004.
7. Guido Maier, et al.: Optical Network Survivability: Protection Tech-niques in the WDM Layer, Photonic Network Comm., vol.4, no.3/4, pp. 251-269, July/Dec. 2002.
8. Yun Wang et al.: Dynamic Survivability in WDM Mesh Networks under Dynamic Traffic, Photonic Network Commun., vol.6, no.1, pp.5-24, July 2003.
9. P. Czezowski et al.: Optical Network Failure Recovery Requirements, IETF Internet Draft, draft-czezowski-optical-recovery-reqs-01.txt, Feb. 2003.
10. Chuan-Ching SUE et al.: Fault Tolerant Crossconnect and Wavelength Routing in All-Optical Networks, IEICE TRANS. COMMUN., vol.E83-B, no.10, Oct. 2000.
11. Sung-Un Kim et al.: A Framework for Managing Faults and Attacks in All-Optical Transport Networks, DISCEX 2001, June 2001.
12. C.P. Larsen et al.: Signal Quality Monitoring in Optical Networks, Optical Network Magazine, vol.1, no.4, pp.17-23, Oct. 2000.
13. Stefano Binetti et al.: Impact of Fiber Non-linearity in high Capaity WDM Systems and in Cross-Connected Backbone Networks, Photonic Network Communications, vol.3, no.3, pp.237-243, July 2001.
14. S. Ramamurthy, et al.: Survivable WDM Mesh Networks, Part I - Protection, Proceedings of IEEE INFOCOM'99, pp.744-751, Mar. 1999.
15. O. Crochat, et al.: Design Protection for WDM Optical Networks, IEEE Journal Sel. Areas Comm., vol.16, no.7, pp.1158-1165, Sep. 1998.

Lecture Notes in Computer Science

For information about Vols. 1–3337

please contact your bookseller or Springer

Vol. 3452: F. Baader, A. Voronkov (Eds.), Logic for Programming, Artificial Intelligence, and Reasoning. XI, 562 pages. 2005. (Subseries LNAI).

Vol. 3448: G.R. Raidl, J. Gottlieb (Eds.), Evolutionary Computation in Combinatorial Optimization. XI, 271 pages. 2005.

Vol. 3441: V. Sassone (Ed.), Foundations of Software Science and Computational Structures. XVIII, 521 pages. 2005.

Vol. 3440: N. Halbwachs, L.D. Zuck (Eds.), Tools and Algorithms for the Construction and Analysis of Systems. XVII, 588 pages. 2005.

Vol. 3436: B. Bouyssounouse, J. Sifakis (Eds.), Embedded Systems Design. XV, 492 pages. 2005.

Vol. 3433: S. Bhalla (Ed.), Databases in Networked Information Systems. VII, 319 pages. 2005.

Vol. 3432: M. Beigl, P. Lukowicz (Eds.), Systems Aspects in Organic and Pervasive Computing - ARCS 2005. X, 265 pages. 2005.

Vol. 3427: G. Kotsis, O. Spaniol, Wireless Systems and Mobility in Next Generation Internet. VIII, 249 pages. 2005.

Vol. 3423: J.L. Fiadeiro, P.D. Mosses, F. Orejas (Eds.), Recent Trends in Algebraic Development Techniques. VIII, 271 pages. 2005.

Vol. 3422: R.T. Mittermeir (Ed.), From Computer Literacy to Informatics Fundamentals. X, 203 pages. 2005.

Vol. 3421: P. Lorenz, P. Dini (Eds.), Networking - ICN 2005, Part II. XXXV, 1153 pages. 2005.

Vol. 3420: P. Lorenz, P. Dini (Eds.), Networking - ICN 2005, Part I. XXXV, 933 pages. 2005.

Vol. 3419: B. Faltings, A. Petcu, F. Fages, F. Rossi (Eds.), Constraint Satisfaction and Constraint Logic Programming. X, 217 pages. 2005. (Subseries LNAI).

Vol. 3418: U. Brandes, T. Erlebach (Eds.), Network Analysis. XII, 471 pages. 2005.

Vol. 3416: M. Böhlen, J. Gamper, W. Polasek, M.A. Wimmer (Eds.), E-Government: Towards Electronic Democracy. XIII, 311 pages. 2005. (Subseries LNAI).

Vol. 3415: P. Davidsson, B. Logan, K. Takadama (Eds.), Multi-Agent and Multi-Agent-Based Simulation. X, 265 pages. 2005. (Subseries LNAI).

Vol. 3414: M. Morari, L. Thiele (Eds.), Hybrid Systems: Computation and Control. XII, 684 pages. 2005.

Vol. 3412: X. Franch, D. Port (Eds.), COTS-Based Software Systems. XVI, 312 pages. 2005.

Vol. 3411: S.H. Myaeng, M. Zhou, K.-F. Wong, H.-J. Zhang (Eds.), Information Retrieval Technology. XIII, 337 pages. 2005.

Vol. 3410: C.A. Coello Coello, A. Hernández Aguirre, E. Zitzler (Eds.), Evolutionary Multi-Criterion Optimization. XVI, 912 pages. 2005.

Vol. 3409: N. Guelfi, G. Reggio, A. Romanovsky (Eds.), Scientific Engineering of Distributed Java Applications. X, 127 pages. 2005.

Vol. 3408: D.E. Losada, J.M. Fernández-Luna (Eds.), Advances in Information Retrieval. XVII, 572 pages. 2005.

Vol. 3407: Z. Liu, K. Araki (Eds.), Theoretical Aspects of Computing - ICTAC 2004. XIV, 562 pages. 2005.

Vol. 3406: A. Gelbukh (Ed.), Computational Linguistics and Intelligent Text Processing. XVII, 829 pages. 2005.

Vol. 3404: V. Diekert, B. Durand (Eds.), STACS 2005. XVI, 706 pages. 2005.

Vol. 3403: B. Ganter, R. Godin (Eds.), Formal Concept Analysis. XI, 419 pages. 2005. (Subseries LNAI).

Vol. 3401: Z. Li, L.G. Vulkov, J. Waśniewski (Eds.), Numerical Analysis and Its Applications. XIII, 630 pages. 2005.

Vol. 3398: D.-K. Baik (Ed.), Systems Modeling and Simulation: Theory and Applications. XIV, 733 pages. 2005. (Subseries LNAI).

Vol. 3397: T.G. Kim (Ed.), Artificial Intelligence and Simulation. XV, 711 pages. 2005. (Subseries LNAI).

Vol. 3396: R.M. van Eijk, M.-P. Huget, F. Dignum (Eds.), Agent Communication. X, 261 pages. 2005. (Subseries LNAI).

Vol. 3395: J. Grabowski, B. Nielsen (Eds.), Formal Approaches to Software Testing. X, 225 pages. 2005.

Vol. 3394: D. Kudenko, D. Kazakov, E. Alonso (Eds.), Adaptive Agents and Multi-Agent Systems III. VIII, 313 pages. 2005. (Subseries LNAI).

Vol. 3393: H.-J. Kreowski, U. Montanari, F. Orejas, G. Rozenberg, G. Taentzer (Eds.), Formal Methods in Software and Systems Modeling. XXVII, 413 pages. 2005.

Vol. 3391: C. Kim (Ed.), Information Networking. XVII, 936 pages. 2005.

Vol. 3390: R. Choren, A. Garcia, C. Lucena, A. Romanovsky (Eds.), Software Engineering for Multi-Agent Systems III. XII, 291 pages. 2005.

Vol. 3389: P. Van Roy (Ed.), Multiparadigm Programming in Mozart/OZ. XV, 329 pages. 2005.

Vol. 3388: J. Lagergren (Ed.), Comparative Genomics. VII, 133 pages. 2005. (Subseries LNBI).

Vol. 3387: J. Cardoso, A. Sheth (Eds.), Semantic Web Services and Web Process Composition. VIII, 147 pages. 2005.

Vol. 3386: S. Vaudenay (Ed.), Public Key Cryptography - PKC 2005. IX, 436 pages. 2005.

Vol. 3385: R. Cousot (Ed.), Verification, Model Checking, and Abstract Interpretation. XII, 483 pages. 2005.

Vol. 3383: J. Pach (Ed.), Graph Drawing. XII, 536 pages. 2005.

Vol. 3382: J. Odell, P. Giorgini, J.P. Müller (Eds.), Agent-Oriented Software Engineering V. X, 239 pages. 2005.

Vol. 3381: P. Vojtáš, M. Bieliková, B. Charron-Bost, O. Sýkora (Eds.), SOFSEM 2005: Theory and Practice of Computer Science. XV, 448 pages. 2005.

Vol. 3380: C. Priami, Transactions on Computational Systems Biology I. IX, 111 pages. 2005. (Subseries LNBI).

Vol. 3379: M. Hemmje, C. Niederee, T. Risse (Eds.), From Integrated Publication and Information Systems to Information and Knowledge Environments. XXIV, 321 pages. 2005.

Vol. 3378: J. Kilian (Ed.), Theory of Cryptography. XII, 621 pages. 2005.

Vol. 3377: B. Goethals, A. Siebes (Eds.), Knowledge Discovery in Inductive Databases. VII, 190 pages. 2005.

Vol. 3376: A. Menezes (Ed.), Topics in Cryptology – CT-RSA 2005. X, 385 pages. 2005.

Vol. 3375: M.A. Marsan, G. Bianchi, M. Listanti, M. Meo (Eds.), Quality of Service in Multiservice IP Networks. XIII, 656 pages. 2005.

Vol. 3374: D. Weyns, H.V.D. Parunak, F. Michel (Eds.), Environments for Multi-Agent Systems. X, 279 pages. 2005. (Subseries LNAI).

Vol. 3372: C. Bussler, V. Tannen, I. Fundulaki (Eds.), Semantic Web and Databases. X, 227 pages. 2005.

Vol. 3371: M.W. Barley, N. Kasabov (Eds.), Intelligent Agents and Multi-Agent Systems. X, 329 pages. 2005. (Subseries LNAI).

Vol. 3370: A. Konagaya, K. Satou (Eds.), Grid Computing in Life Science. X, 188 pages. 2005. (Subseries LNBI).

Vol. 3369: V.R. Benjamins, P. Casanovas, J. Breuker, A. Gangemi (Eds.), Law and the Semantic Web. XII, 249 pages. 2005. (Subseries LNAI).

Vol. 3368: L. Paletta, J.K. Tsotsos, E. Rome, G.W. Humphreys (Eds.), Attention and Performance in Computational Vision. VIII, 231 pages. 2005.

Vol. 3367: W.S. Ng, B.C. Ooi, A. Ouksel, C. Sartori (Eds.), Databases, Information Systems, and Peer-to-Peer Computing. X, 231 pages. 2005.

Vol. 3366: I. Rahwan, P. Moraitis, C. Reed (Eds.), Argumentation in Multi-Agent Systems. XII, 263 pages. 2005. (Subseries LNAI).

Vol. 3365: G. Mauri, G. Păun, M.J. Pérez-Jiménez, G. Rozenberg, A. Salomaa (Eds.), Membrane Computing. IX, 415 pages. 2005.

Vol. 3363: T. Eiter, L. Libkin (Eds.), Database Theory - ICDT 2005. XI, 413 pages. 2004.

Vol. 3362: G. Barthe, L. Burdy, M. Huisman, J.-L. Lanet, T. Muntean (Eds.), Construction and Analysis of Safe, Secure, and Interoperable Smart Devices. IX, 257 pages. 2005.

Vol. 3361: S. Bengio, H. Bourlard (Eds.), Machine Learning for Multimodal Interaction. XII, 362 pages. 2005.

Vol. 3360: S. Spaccapietra, E. Bertino, S. Jajodia, R. King, D. McLeod, M.E. Orlowska, L. Strous (Eds.), Journal on Data Semantics II. XI, 223 pages. 2005.

Vol. 3359: G. Grieser, Y. Tanaka (Eds.), Intuitive Human Interfaces for Organizing and Accessing Intellectual Assets. XIV, 257 pages. 2005. (Subseries LNAI).

Vol. 3358: J. Cao, L.T. Yang, M. Guo, F. Lau (Eds.), Parallel and Distributed Processing and Applications. XXIV, 1058 pages. 2004.

Vol. 3357: H. Handschuh, M.A. Hasan (Eds.), Selected Areas in Cryptography. XI, 354 pages. 2004.

Vol. 3356: G. Das, V.P. Gulati (Eds.), Intelligent Information Technology. XII, 428 pages. 2004.

Vol. 3355: R. Murray-Smith, R. Shorten (Eds.), Switching and Learning in Feedback Systems. X, 343 pages. 2005.

Vol. 3354: M. Margenstern (Ed.), Machines, Computations, and Universality. VIII, 329 pages. 2005.

Vol. 3353: J. Hromkovič, M. Nagl, B. Westfechtel (Eds.), Graph-Theoretic Concepts in Computer Science. XI, 404 pages. 2004.

Vol. 3352: C. Blundo, S. Cimato (Eds.), Security in Communication Networks. XI, 381 pages. 2005.

Vol. 3351: G. Persiano, R. Solis-Oba (Eds.), Approximation and Online Algorithms. VIII, 295 pages. 2005.

Vol. 3350: M. Hermenegildo, D. Cabeza (Eds.), Practical Aspects of Declarative Languages. VIII, 269 pages. 2005.

Vol. 3349: B.M. Chapman (Ed.), Shared Memory Parallel Programming with Open MP. X, 149 pages. 2005.

Vol. 3348: A. Canteaut, K. Viswanathan (Eds.), Progress in Cryptology - INDOCRYPT 2004. XIV, 431 pages. 2004.

Vol. 3347: R.K. Ghosh, H. Mohanty (Eds.), Distributed Computing and Internet Technology. XX, 472 pages. 2004.

Vol. 3346: R.H. Bordini, M. Dastani, J. Dix, A.E.F. Seghrouchni (Eds.), Programming Multi-Agent Systems. XIV, 249 pages. 2005. (Subseries LNAI).

Vol. 3345: Y. Cai (Ed.), Ambient Intelligence for Scientific Discovery. XII, 311 pages. 2005. (Subseries LNAI).

Vol. 3344: J. Malenfant, B.M. Østvold (Eds.), Object-Oriented Technology. ECOOP 2004 Workshop Reader. VIII, 215 pages. 2005.

Vol. 3343: C. Freksa, M. Knauff, B. Krieg-Brückner, B. Nebel, T. Barkowsky (Eds.), Spatial Cognition IV. Reasoning, Action, and Interaction. XIII, 519 pages. 2005. (Subseries LNAI).

Vol. 3342: E. Şahin, W.M. Spears (Eds.), Swarm Robotics. IX, 175 pages. 2005.

Vol. 3341: R. Fleischer, G. Trippen (Eds.), Algorithms and Computation. XVII, 935 pages. 2004.

Vol. 3340: C.S. Calude, E. Calude, M.J. Dinneen (Eds.), Developments in Language Theory. XI, 431 pages. 2004.

Vol. 3339: G.I. Webb, X. Yu (Eds.), AI 2004: Advances in Artificial Intelligence. XXII, 1272 pages. 2004. (Subseries LNAI).

Vol. 3338: S.Z. Li, J. Lai, T. Tan, G. Feng, Y. Wang (Eds.), Advances in Biometric Person Authentication. XVIII, 699 pages. 2004.